T0132518

Group Actions in Ergodic Theory, Geometry, and Topology

Group Actions in Ergodic Theory, Geometry, and Topology

Selected Papers

Robert J. Zimmer

Edited by David Fisher
With a Foreword by David Fisher, Alexander Lubotzky,
and Gregory Margulis

The University of Chicago Press || Chicago and London

The University of Chicago Press, Chicago 60637

The University of Chicago Press, Ltd., London

© 2020 by The University of Chicago

Published 2020

Printed in the United States of America

29 28 27 26 25 24 23 22 21 20   1 2 3 4 5

ISBN-13: 978-0-226-56813-3 (cloth)

ISBN-13: 978-0-226-56827-0 (e-book)

DOI: https://doi.org/10.7208/chicago/9780226568270.001.0001

Library of Congress Cataloging-in-Publication Data

Names: Zimmer, Robert J., 1947– author. | Fisher, David (Mathematician), editor,
    writer of foreword. | Lubotzky, Alexander, 1956– writer of foreword. | Margulis, G. A.
    (Gregori Aleksandrovitsch), 1946– writer of foreword.
Title: Group actions in ergodic theory, geometry, and topology : selected papers /
    Robert J. Zimmer ; edited by David Fisher ; with a foreword by David Fisher,
    Alexander Lubotzky, and Gregory Margulis.
Description: Chicago ; London : University of Chicago Press, 2020. | Includes
    bibliographical references. | Contents: Spectra and structure of ergodic actions—
    Amenable actions, equivalence relations, and foliations—Orbit equivalence and
    strong rigidity—Cocycle superrigidity and the program to describe Lie group
    and lattice actions on manifolds—Stabilizers of semisimple Lie group actions :
    invariant random subgroups—Representations and arithmetic properties of actions,
    fundamental groups, and foliations—Geometric structures : automorphisms
    of geometric manifolds and rigid structures ; locally homogeneous manifolds—
    Stationary measures and structure theorems for Lie group actions.
Identifiers: LCCN 2019030986 | ISBN 9780226568133 (cloth) |
    ISBN 9780226568270 (ebook)
Subjects: LCSH: Group actions (Mathematics) | Ergodic theory. | Lie groups.
Classification: LCC QA3 .Z56 2020 | DDC 512/.2—dc23
LC record available at https://lccn.loc.gov/2019030986

♾ This paper meets the requirements of ANSI/NISO Z39.48-1992
(Permanence of Paper).

Contents

1 Spectra and Structure of Ergodic Actions

2 Amenable Actions, Equivalence Relations, and Foliations

3 Orbit Equivalence and Strong Rigidity

4 Cocycle Superrigidity and the Program to Describe Lie Group and Lattice Actions on Manifolds

5 Stabilizers of Semisimple Lie Group Actions: Invariant Random Subgroups

6 Representations and Arithmetic Properties of Actions, Fundamental Groups, and Foliations

7 Geometric Structures: Automorphisms of Geometric Manifolds and Rigid Structures; Locally Homogeneous Manifolds

8 Stationary Measures and Structure Theorems for Lie Group Actions

Foreword

David Fisher, Alexander Lubotzky, and Gregory Margulis

Robert Zimmer has a broad view of mathematics, looking to put mathematical ideas into their best context and most general, usable form and placing a high value on clearly blazing a trail that others can follow. His work and ideas continue to resonate not only in the form of his theorems, but in the form of definitions introduced and conjectures and questions he has asked. We chose in this foreword to illustrate Bob's influence through three well-chosen examples: one important definition, one major theorem, and one collection of conjectures and questions. The reader interested in a more comprehensive view will be best served by studying the papers selected here. Hopefully our examples encourage further study of these papers.

Bob introduced the notion of *amenable actions* in [Zim2]. Inspired by Mackey's notion of a virtual group, he produced a virtual analogue of the already quite ubiquitous amenable groups. In this paper, Bob also made a key connection between this new notion and the Poisson boundaries introduced earlier by Furstenberg [Fur4]. Amenable actions have turned out to be perhaps even more flexible and diverse in their application than the notion of amenable groups that inspired them. The idea quickly found applications in Bob's own work, most notably in his proof of cocycle and orbit equivalence rigidity, our next topic of discussion. The notion of amenable actions has been used repeatedly for proofs of rigidity theorems inspired by Bob's work (see, e.g., [Fur2, BS]), but it has also proven useful for myriad other applications. For example, it was introduced as a tool for computing bounded cohomology by Burger and Monod [BM], an area in which it continues to be used to this day (see, e.g., [BBF+]). The applications in bounded cohomology lead to connections and applications to higher Teichmüller theory and particularly maximal representations of surface groups [BILW]. In an unrelated direction, the notion is used in work by Bowen and Nevo proving ergodic theorems for general actions of general nonamenable groups [BN] and by Caprace and Monod in their study of groups of isometries of nonpositively curved spaces [CM]. This list of areas of application is far from exhaustive as there also seem to be applications to C^* algebras and quantum groups, areas outside our collective expertise.

Zimmer's cocycle superrigidity theorem and his application of it to prove orbit equivalence rigidity theorems were surprising developments that moved rigidity theory, which had till that time focused on discrete subgroups of Lie groups, into substantially new and different realms [Zim3]. This work was clearly motivated by Margulis's superrigidity theorem, by Mostow's earlier work, and by Zimmer's own development of Mackey's theory of virtual groups. Instead of studying homomorphisms of discrete groups to Lie groups, suddenly rigidity theory could study the virtual analogue of orbit equivalence between group actions. Somewhat surprisingly, the most direct impact of this work on future research took some time. Questions Zimmer had raised explicitly about extending his work from Lie groups to lattices were only resolved by Furman almost twenty years later [Fur2, Fur1]. Immediately following Furman's work, more exciting developments began to occur, first in work of Monod and Shalom that combined bounded cohomology with ergodic theory (and amenable actions) on questions of orbit equivalence rigidity [MS]. At around the same time, Popa recast work he had been doing on von Neumann algebras in language of orbit equivalence and cocycle superrigidity, leading to major breakthroughs and explosive growth in the field [Pop]. The state of this field was last comprehensively surveyed by Furman [Fur3] and is long overdue for a new survey, which we will not attempt here.

Our last focus is on the *Zimmer program*, which is the area most frequently associated with Bob's name. Bob wrote a sequence of papers in the early 1980s providing evidence that one might be able to classify actions of higher rank lattices on compact manifolds. He then proceeded to make some conjectures in this direction [Zim4, Zim7, Zim1, Zim6]. A chapter by Fisher in this volume highlights recent dramatic progress in this area and updates his earlier survey [Fis]. Since Fisher's articles survey the main developments in the Zimmer program proper, here we emphasize related research directions inspired by the program. Zimmer's early work on geometry of group actions preserving geometric structures led to a major contribution by Gromov in the 1980s concerning rigid geometric structures and their automorphism groups [Zim5, Gro]. Early work on Zimmer's program led Katok and Spatzier in the 1990s to pioneer the study of differentiable rigidity of actions of higher rank abelian groups [KS2, KS1]. This area has had and continues to have significant interactions with the study of invariant measures for higher rank abelian actions and conjectures of Margulis [Mar]. Zimmer's conjectures and program were also a major impetus for the work of Korevaar and Schoen in the 1990s on harmonic maps into infinite-dimensional nonpositively curved spaces [KS3, KS4].

With the recent remarkable developments concerning Zimmer's conjectures, now is clearly an ideal time to bring out a volume of Zimmer's selected papers. But we hope to have convinced the reader that the papers are not only timely but timeless and that further study of these papers may lead in new, fruitful directions, following the vision of Bob Zimmer.

References

[BS]   U. Bader and Y. Shalom. Factor and normal subgroup theorems for lattices in products of groups. *Invent. Math.*, 163(2):415–454, 2006.

[BN]   L. Bowen and A. Nevo. Pointwise ergodic theorems beyond amenable groups. *Ergodic Theory Dynam. Systems*, 33(3):777–820, 2013.

[BBF+]   M. Bucher, M. Burger, R. Frigerio, A. Iozzi, C. Pagliantini, and M. B. Pozzetti. Isometric embeddings in bounded cohomology. *J. Topol. Anal.*, 6(1):1–25, 2014.

[BM]   M. Burger and N. Monod. Continuous bounded cohomology and applications to rigidity theory. *Geom. Funct. Anal.*, 12(2):219–280, 2002.

[BILW]   M. Burger, A. Iozzi, F. Labourie, and A. Wienhard. Maximal representations of surface groups: Symplectic Anosov structures. *Pure Appl. Math. Q.*, 1(3, Special Issue: In Memory of Armand Borel. Part 2):543–590, 2005.

[CM]   P.-E. Caprace and N. Monod. Isometry groups of non-positively curved spaces: Discrete subgroups. *J. Topol.*, 2(4):701–746, 2009.

[FIS]   D. Fisher. Groups acting on manifolds: Around the Zimmer program. In *Geometry, rigidity, and group actions*, Chicago Lectures in Math., pages 72–157. Univ. Chicago Press, Chicago, IL, 2011.

[FUR1]   A. Furman. Gromov's measure equivalence and rigidity of higher rank lattices. *Ann. of Math. (2)*, 150(3):1059–1081, 1999.

[FUR2]   A. Furman. Orbit equivalence rigidity. *Ann. of Math. (2)*, 150(3):1083–1108, 1999.

[FUR3]   A. Furman. A survey of measured group theory. In *Geometry, rigidity, and group actions*, Chicago Lectures in Math., pages 296–374. Univ. Chicago Press, Chicago, IL, 2011.

[FUR4]   H. Furstenberg. A Poisson formula for semisimple Lie groups. *Ann. of Math. (2)*, 77:335–386, 1963.

[GRO]   M. Gromov. Rigid transformations groups. In *Géométrie différentielle (Paris, 1986)*, volume 33 of *Travaux en Cours*, pages 65–139. Hermann, Paris, 1988.

[KS1]   A. Katok and R. J. Spatzier. Differential rigidity of Anosov actions of higher rank abelian groups and algebraic lattice actions. *Tr. Mat. Inst. Steklova*, 216(Din. Sist. i Smezhnye Vopr.):292–319, 1997.

[KS2]   A. Katok and R. J. Spatzier. First cohomology of Anosov actions of higher rank abelian groups and applications to rigidity. *Inst. Hautes Études Sci. Publ. Math.*, (79):131–156, 1994.

[KS3]   N. J. Korevaar and R. M. Schoen. Sobolev spaces and harmonic maps for metric space targets. *Comm. Anal. Geom.*, 1(3–4):561–659, 1993.

[KS4]   N. J. Korevaar and R. M. Schoen. Global existence theorems for harmonic maps to non-locally compact spaces. *Comm. Anal. Geom.*, 5(2):333–387, 1997.

[MAR]   G. Margulis. Problems and conjectures in rigidity theory. In *Mathematics: Frontiers and perspectives*, pages 161–174. Amer. Math. Soc., Providence, RI, 2000.

[MS]   N. Monod and Y. Shalom. Orbit equivalence rigidity and bounded cohomology. *Ann. of Math. (2)*, 164(3):825–878, 2006.

[POP]   S. Popa. Cocycle and orbit equivalence superrigidity for malleable actions of w-rigid groups. *Invent. Math.*, 170(2):243–295, 2007.

[ZIM1]   R. J. Zimmer. Kazhdan groups acting on compact manifolds. *Invent. Math.*, 75(3):425–436, 1984.

[ZIM2]   R. J. Zimmer. Amenable ergodic group actions and an application to Poisson boundaries of random walks. *J. Functional Analysis*, 27(3):350–372, 1978.

[ZIM3]   R. J. Zimmer. Strong rigidity for ergodic actions of semisimple Lie groups. *Ann. of Math. (2)*, 112(3):511–529, 1980.

[ZIM4]   R. J. Zimmer. Arithmetic groups acting on compact manifolds. *Bull. Amer. Math. Soc. (N.S.)*, 8(1):90–92, 1983.

[ZIM5]   R. J. Zimmer. Semisimple automorphism groups of G-structures. *J. Differential Geom.*, 19(1):117–123, 1984.

[ZIM6]   R. J. Zimmer. Volume preserving actions of lattices in semisimple groups on compact manifolds. *Inst. Hautes Études Sci. Publ. Math.*, (59):5–33, 1984.

[ZIM7]   R. J. Zimmer. Actions of semisimple groups and discrete subgroups. In *Proceedings of the International Congress of Mathematicians, Vol. 1, 2 (Berkeley, Calif., 1986)*, pages 1247–1258. Amer. Math. Soc., Providence, RI, 1987.

INTRODUCTION

Robert J. Zimmer

This volume of selected papers illustrates the main threads of my mathematical work since I was a graduate student at Harvard in the 1970s. Additionally, it provides an opportunity to reflect on the several contexts in which this work took place. In this introduction, I will offer a largely personal view of the historical context and individual influences that led to this work, with no attempt at either completeness or precision.

Central to this work is a set of conjectures I made in the 1980s about actions of Lie groups and their discrete subgroups on compact manifolds. These conjectures absorbed a great deal of my own attention over the years and they were likewise addressed through a variety of approaches in significant work by a number of mathematicians. While the most general conjectures remain open, the recent dramatic achievement of Brown-Fisher-Hurtado (referred to later in this introduction simply as BFH) represents the greatest advance on these conjectures in many years. This result also provides a temporal context for this volume.

As a graduate student at Harvard in the early 1970s, like many enthusiastic graduate students, I wanted to learn everything about mathematics. I particularly wanted to understand how its many components fit together. While mathematics today often reflects porous boundaries between subdisciplines, at that time the mathematics community recognized divisions into algebra, topology, analysis, logic, and applied mathematics to a much greater extent than today. These divisions were not perceived as just a convenient way of designing curriculum, which still persists, but were often taken to reflect genuine differences in the approach to and content of mathematics. Of course there were many individuals whose work provided counterexamples to this general assertion. But unlike today, such work was often viewed as "interdisciplinary," rather than reflective of mathematics as a naturally integrated whole. Russian mathematics at the time, which for reasons of politics in the Soviet Union, language, and available communications vehicles was much more isolated than seems imaginable today, embodied a different perspective, to a greater extent approaching mathematics more holistically. Reading about one topic after another in the library at 2 Divinity Avenue (where the

Harvard mathematics department was housed before moving to the Science Center), the Russian perspective was often on my mind. Needless to say, my desire to understand the "whole" was futile, as it was an ambitious desire far beyond my capacity to fulfill. But the integrative mindset stayed with me in mathematics (and still does in my current role as president of the University of Chicago).

Some of the courses I took as a graduate student were fascinating. I was fortunate to attend Oscar Zariski's last course at Harvard, which was beautiful but unrelenting in its algebraic approach to algebraic geometry, perhaps fittingly so given Zariski's work. I do not remember him even once drawing a curve. A good portion of the mathematics faculty attended the final class. I took several courses from Raoul Bott, a mathematician with a deeply integrative sense of mathematics. Bott's presentations were so elegant and streamlined in combining ideas that, while I could follow everything with some work and found it beautiful, I often had a suspicion (which I later understood to be correct) that there was something I was missing. His enthusiasm was infectious. Another member of the department at the time with an integrative perspective was George Whitelaw Mackey. Mackey had made major contributions to the theory of unitary representations of locally compact groups and his active work in that area was winding down. However, he wrote a number of ambitious expositions, for example describing and attempting to unify the multiple mathematics and physics domains in which induced representations played a central conceptual role. I found it impossible to learn a subject from these writings, but they were wonderful as a motivator for exploring new areas and for providing a richer context for many results I had already studied.

In addition to Mackey's earlier work on unitary representations and his foray into ergodic theory, it was the approach in his expository papers that led me to work with Mackey as my dissertation advisor. While I greatly admired his mathematics, we had a rather distanced relationship as advisor and student. He was happy to let me find my own way, and even though it was not the most comforting approach, I was game for that. In the end, I showed him about two thirds of my dissertation before he even knew what I was working on. Our relationship was friendly though not close. While I was a Dickson Instructor at the University of Chicago, Mackey came to Chicago to deliver a colloquium lecture. Afterward, four of us went to dinner—Mackey, Saunders MacLane, Irving Kaplansky, and I. They had known each other for decades, and I was very much the outlier in terms of age, accomplishment, and experience. At one point Mackey said to the group that I was the perfect graduate student for him—because I wanted to see him as little as he wanted to see me. That was George Mackey—humorous and always honest. As the years passed, we grew closer. In 1985, I came back to Harvard as a visiting faculty member. He wanted to be sure that I had his office, as he was going to be away for that year, a gesture that I greatly appreciated.

As I began thinking seriously about a dissertation, I came to see a particular portion of the mathematical landscape relatively unexplored, which I found perplexing. I had been studying representation theory, and for unitary or finite-dimensional representations one had the following situation at the time. For the group of integers, the theory of finite-dimensional representations was just the theory of an invertible matrix, which was well explored, and the theory of unitary representations was fully described by the spectral theorem. So for this elementary group, one had a clear picture of these representations. Moving beyond abelian groups, Mackey's work provided great insight into the unitary representations for many semidirect products, Kirillov had produced a beautiful and powerful approach for nilpotent Lie groups, and Auslander and Kostant were developing an elaborate approach for solvable groups. In fact, I had the benefit of attending a course that Kostant taught on this work at MIT. For simple Lie groups, one had a beautiful classical theory of finite-dimensional representations of simple Lie algebras on one hand, which essentially dealt with the unitary representations of compact Lie groups at the same time, and a very elaborate if not yet complete understanding of the "unitary dual" in the noncompact case. Even if one did not understand everything fully, the field was extremely active (the difficult and Herculean work of Harish-Chandra as an example) and everyone understood this was a major and mathematically important project. Looking at the area as a whole, this all seemed reasonable—classical results for the integers or reals, and deep results either in hand or being generated for more complicated groups.

On the other hand, I had also been studying ergodic theory and dynamical systems, i.e., the nonlinear actions of the integers (or real line). The nonlinear case was obviously more complicated than that of a single matrix and there was a rich and evolving theory. Aside from the work on ergodic theorems themselves, the theory of Kolmogorov-Sinai entropy was evolving into a very powerful tool, highlighted by the work of Ornstein and his collaborators. The work of Auslander and Moore on flows on homogeneous spaces, Dye's results on orbit equivalence, and the extensive work being done on hyperbolic (or Anosov) transformations are other examples of work generating significant excitement at the time. On the other hand, when one looked at the nonlinear case for more complicated noncompact Lie groups, there was not that much systematic work appearing, in stark contrast to the intense development of representation theory.

For finite groups acting on manifolds, and more generally for compact groups, there was robust development, largely of a topological nature. These cases seemed to absorb much of the energy of those studying actions on measure spaces or manifolds for groups with an interesting algebraic structure. But there was in fact no such systematic theory being developed for simple noncompact Lie groups or their discrete subgroups. An illuminating example is provided by the understanding of the relationship of ergodic theorems and groups. Calderon and others had done work on extending the ergodic theorems to actions of amenable groups. The common wisdom at the time, which lasted for many years, was that amenability was the natural limit of generality for ergodic theorems for group actions. In contrast, the extensive and very successful recent work on ergodic theorems for semisimple groups and their lattice subgroups, for example by Nevo and collaborators, highlights just how undeveloped the thinking about noncompact Lie group actions was at that time.

There were, however, some interesting, important, and suggestive examples. Mackey had generalized the von Neumann–Halmos theorem on actions with discrete spectrum (although the terminology in the context of nonabelian groups was perhaps not ideal). Furstenberg had been exploring various notions of proximality in connection with his work on random processes. The most dramatic development was Mostow's use of the ergodicity of a lattice in hyperbolic space on the boundary in proving his rigidity theorem. While a turning point for the whole field, at first this use of ergodicity looked like something of a singularity. Was this just a brilliant step or would it ultimately fit into a general theory?

Mostow came to Harvard during my last year as a graduate student to talk about his generalization of his rigidity theorem to higher rank. The room (now in the Science Center) was packed, and as Mostow spoke, Bott, who was a compelling figure in every room he was in, was doing what he sometimes did in such circumstances—surveying those in the audience to see whether they were appreciating and understanding what they were hearing. He and I caught each other's eye in this process, and with a gleam in his eye I can still see, he showed once again just how infectious was his enthusiasm for mathematics. Listening to Mostow that day, it was easy to be enthusiastic.

While Mostow was showing how powerful a role ergodic theory could play in understanding subgroups, Mackey was making the converse point based on his work on unitary representations of semidirect products. When his approach to semidirect products worked best, subgroups appeared, but where it did not, ergodic actions appeared. This led Mackey to raise the question of how to gain understanding of ergodic actions by thinking of them in analogy with subgroups. He developed this idea into a formalism of virtual groups and developed some fundamental properties. While the formalism never gained much traction as a widely useful tool, the heuristic in fact proved extremely useful. In some ways, the formalism survives today in the form of measured groupoids and equivalence relations.

This was roughly the state of affairs as I worked on my dissertation and simultaneously began a set of other projects. To say I was determined to contribute to the development of a general approach to the actions of simple Lie groups and their subgroups via ergodic theory would be a retrojection too grandiose and deterministic. But the vacuum was evident to me, it was deeply intriguing, and I was at least determined to make this direction the beginning of my research work.

A recurrent theme in mathematics is the relationship of a linear theory to a nonlinear one, and this plays a strong role in the analysis of (in general nonlinear) measure-class-preserving group actions. For such actions, there is a natural unitary representation on L^2 and it is natural to ask about what this representation says about the action. While it does carry some information for integer actions, the "spectrum" is actually quite a weak invariant for these actions. There is significantly more information of various sorts for actions of simple noncompact groups and their discrete subgroups, and it would be useful to organize and clarify this issue more fully. However, for actions on manifolds there clearly is an entire other "linear" theory, namely on linear spaces that capture the derivative, for example on the tangent bundle, spaces of vector fields, sections of jet bundles, and on Sobolev spaces. These actions are no longer of a unitary nature in general. But the amount of information obviously lying within these spaces, the success of the theory of hyperbolicity and its variants for dynamical systems, and the fact that Harish-Chandra himself moved well beyond the unitary case always suggested viewing the L^2 case as just one small part of the linear story. Another piece of that story is that L^2 itself is a module over L^∞ and hence over G-invariant weak-*-closed subalgebras of L^∞, or equivalently, a module over L^∞ of a measurable-factor G space. This gives another approach to thinking about the linear theory.

I will briefly mention and give some perspective on three directions of my work in those first early years. My dissertation concerned the relationship of L^2 as a module over L^∞ of a factor and the spatial structure of the space over the factor. This provided a generalization to bundles, new even for integer actions, of Mackey's generalization of the von Neumann–Halmos theorem. Beyond the specific results, thinking through this situation led me to a set of techniques and perspectives about actions on measurable bundles that proved valuable in my subsequent work. A second component of my dissertation used these results to prove a general structure theorem for actions. This provided a measurable version of Furstenberg's theorem on the topological structure of distal actions, which he had proven by totally different techniques. Furstenberg himself subsequently reproved the measurable structure theorem in a somewhat different context, as he was focused on number-theoretic applications. It is interesting to compare the impact of the measurable work with the purely topological. For me, this measurable work provided an elementary structure theorem as a step toward a much richer structural theory of actions. Furstenberg and colleagues developed an extensive set of applications to number theory of the measurable work and its generalizations, in the process creating an entire new chapter in the relationship of ergodic theory and number theory. Furstenberg's topological work, on the other hand, stands as a beautiful but somewhat isolated result.

A second direction concerned the question of ergodicity of restricting ergodic simple noncompact group actions to lattices. We established a sharp dichotomy—either the action of the lattice is itself ergodic or the G action is of the form G/C, where C is a compact subgroup (in which case the restriction is obviously not ergodic). One of the most interesting aspects of the proof is that an essential step is establishing the vanishing of the matrix coefficients at infinity for unitary representations without non-0 fixed vectors. At the time, the Howe-Moore result was not yet publicly available and I observed that this vanishing result followed almost immediately by simply applying the Cartan decomposition to a result of Thomas Sherman. When Roger Howe saw a preprint of my paper, he sent me a draft of his beautiful work with Calvin Moore that was in progress. While the Howe-Moore argument is much more elegant and with farther reach, I have on occasion tried to present a more balanced historical record and refer to this vanishing theorem as a result (just about) of Sherman and of Howe-Moore independently. The latter get credit, deservedly so, but Sherman's role has never made it into the general consciousness.

The third direction concerned developing the general notion of an amenable action of a locally compact group, and more generally amenable equivalence relations and foliations. My belief was that this was going to be an important step in any overall program that might be developed. It identified an important and natural class of actions, incorporated the construction of the Poisson boundary, was an invariant of orbit equivalence in the essentially free case, and was exactly the class of actions in the discrete case whose group measure space von Neumann algebra was hyperfinite. The actions of lattices on the boundary of hyperbolic space, so critical for Mostow's work, fell into this category, as did higher rank generalizations. For foliations it allowed a useful analysis of potential geometry along the leaves. In the case of simple Lie groups, we showed every amenable action was induced from an action of an (algebraic) amenable subgroup. The proof of this last result deserves some comment. A central part of the argument involves analyzing the action on the space of measures on a projective space under a linear representation of G. The key result is that the action is quasi-algebraic, i.e., with locally closed orbits and algebraic stabilizers. Furthermore, the Borel density theorem adds to this information by asserting that G-invariant measures are supported on G-fixed points. The proof that amenable actions are induced underscores the importance of the quasi-algebraic property. The same property holds for vectors under a unitary representation by the vanishing of matrix coefficients at infinity. There are other such natural spaces as well, and these prove very useful in the whole theory.

While my dissertation work was completed at Harvard, and some of the roots of the work on amenability had likewise been done there, the bulk of the work on the second and third directions mentioned above were completed while I was teaching at the US Naval Academy in Annapolis. I spent two productive years at USNA between my time at Harvard and my time at Chicago, and I am grateful for the flexibility the Academy showed in allowing me to devote an amount of time on research that was unusual for their faculty. This was particularly impressive to me as it was clear to both the Academy and me that although my and their experience of our relationship was very positive, I was not going to spend my career there. Aside from being very productive mathematically, this was a very informative experience more broadly, as I learned something about the military, the diversity within it, and how to think about training naval officers. For someone who had grown up in Greenwich Village and went to college at Brandeis and graduate school at Harvard, this was a valuable, positive, and personally expanding experience.

By the summer of 1978, I had completed much of this particular work and was now very happily at the University of Chicago. I was also going to Helsinki to attend the International Congress of Mathematicians for the first time. Alain Connes and his colleagues in Norway organized a conference in Oslo to take place before the Helsinki meeting and invited me to attend. Connes was very interested in the work on amenability, which I think accounted for my being invited among a group whose work was focused on operator algebras (and in particular Connes's early ideas about noncommutative geometry that were then in development). As indicated above, I was certainly interested in this area but it was not my primary focus. Alain, as everyone who met him experienced, was a mathematical whirlwind, full of ideas. He talked about Mostow rigidity and orbit equivalence, conjecturing that there was a connection. At a party one evening during the conference, we were sitting together on the stairs drinking beer, and he asked me whether I had studied the work of Margulis. I told him that I had looked at it but not really studied it. Alain urged me to remedy the latter, expressing confidence that there was some connection to what I was doing. (As it happened, Margulis was awarded the Fields Medal in Helsinki the following week. Connes was awarded the Medal four years later.)

Connes's advice was prescient. Upon returning to Chicago, I began to work through Margulis's paper in which he proved his superrigidity theorem. It was a very cryptic paper and not easy. Margulis had rightly achieved great acclaim for his stunning results. But I think it is fair to say that Margulis's approach was so deeply original that, in spite of some expository efforts by Mostow and Tits, his ideas had not been absorbed by the mathematical community. It took me a few months to fully absorb the paper, and when I did I saw an approach to what I wanted to do. This led to what is now known as cocycle superrigidity. As Margulis was concerned with representations of a lattice D in G on a vector space V, one could view his results as describing the structure of the induced G action on a bundle over G/D with fiber V. Cocycle superrigidity addressed precisely this situation in which the base was replaced by any G action with finite invariant measure and the bundle and actions were only measurable.

The first immediate application of cocycle superrigidity was to orbit equivalence rigidity. While the result did not cover the hyperbolic case that Connes and I had discussed, it completely resolved the higher rank cases. After I finished the proof to my satisfaction, I called Alain and told him. He later told me that at the beginning of the call he did not believe me when I said that I had succeeded (this was not many months after we had talked in Oslo) but that after I described the proof on the phone he was convinced that it was plausible.

I had not yet met Margulis because his travel out of the Soviet Union was extremely restricted. In particular, he had not been allowed to go to Helsinki to personally accept the Fields Medal. While restricting travel was standard behavior by the Soviet Union, for Jewish mathematicians in the Soviet Union it was part of a deeply restrictive anti-Semitic attitude pervading the mathematics environment there at the time. In 1979–1980, I spent many months in the Soviet Union on the Academy of Sciences exchange. I met Grisha Margulis for the first time during that trip. Our first meeting was actually (by design) at the end of the platform at a Moscow metro stop. I greatly enjoyed getting to know Grisha on that trip, both personally and through our mathematics conversations. Among mathematicians I have known, Margulis had perhaps the single greatest impact on my mathematical work. The situation with Jewish mathematicians in the Soviet Union continued to be disturbing for years beyond that. One by one, the leading Jewish mathematicians in the Soviet Union found their way out. This was an enormous boon to mathematics in the United States and elsewhere, but quite destructive to mathematics in Russia, which to this day has not recovered its previous eminence. It was very gratifying that many years later, I was presiding as the University of Chicago awarded Grisha an honorary degree.

While orbit equivalence rigidity was the first consequence of cocycle superrigidity, the result clearly had the possibility for much more widespread application. Up to a measurable trivialization, it provided a near-complete picture of the action on the bundle, so in principle it should have consequences for actions on smooth manifolds through applying the result to the tangent bundle or other jet bundles. To achieve these applications, one needed to deal with the question of the regularity of the measurable trivialization. Regularity was not going to be a consequence of a totally general result, as shown by examples for single diffeomorphisms constructed by Katok, and later in a more thorough analysis by Gunesch and Katok. It needed to be a further consequence of the particular group that was acting or some assumption on the action, either by sharpening the proof before the conclusion or in the further analysis of the measurable trivialization. The first result I obtained in this direction was for perturbations of isometric actions, enabling me to show that such perturbations were in fact still isometric. A key part of the argument was utilizing the uniformly slow exponential growth of derivatives (of any order) for perturbations. Understanding conditions related to slow exponential growth continued to play an important role in subsequent work in the area, including in BFH.

This result on perturbations gave confidence to the conjecture I then made in the 1980s that actions of higher rank semisimple groups or their lattices on compact manifolds with an invariant volume density were essentially algebraic in nature, and in particular actions of lattices on (relatively) low-dimensional manifolds were isometric and in a low enough dimension range actually factored through a finite quotient group. The latter result followed, for example, if in addition one assumed a single finite orbit. Interestingly, the latter result also followed if in addition to the invariant volume density one assumed that there was a rigid geometric structure that was also invariant, e.g., a connection. Despite this being an intrinsically interesting case, the proof was relatively straightforward once one had cocycle superrigidity. One significant result toward the general low-dimensional conjecture was showing that the lattice action was measurably isomorphic to an isometric action (although not obviously on a manifold). A number of years later, Dave Witte Morris and I showed how this enabled one to prove the low-dimensional conjectures for $\mathbf{Q}$ points of appropriate algebraic groups, rather than the $\mathbf{Z}$ points (being the lattice case). To move from measurable isometry to true isometry for lattice actions, Katok-Gunesch examples again showed the necessity of argument specific to the groups in question or assumptions on the action. These and many other approaches over the years provided much evidence but no proof of the full conjectures. The proof of the low-dimensional part of the conjecture in general, in most cases, was not resolved until the beautiful work of BFH, more than thirty years after the conjecture was formulated.

This work led me to three other related areas that occupied my attention for a number of years: group actions preserving geometric structures; the existence or not of locally homogeneous structures on manifolds, a classical question in differential geometry; and the nature of the fundamental group of a manifold with a group action. Ultimately, all these are just part of the general question of the relationship between a manifold, its topology and in particular the fundamental group, the geometric structures on it, the groups that act on the manifold, perhaps preserving a geometric structure, and the structure of the group action itself. In other words, it was an attempt to help develop a general theory of these actions, a theory whose nonexistence had struck me from the time I was a graduate student.

One of the first results I obtained about actions preserving a geometric structure applied in particular to the isometry groups of compact Lorentz manifolds. It asserted that the only noncompact simple Lie groups that could act in such a way were locally isomorphic to $SL(2,\mathbf{R})$. The techniques applied much more generally to other geometric structures. I discovered this when I was trying to prove something more complicated, and in fact the proof was simple enough (the Borel density theorem and some features of the adjoint representation being the essential ingredients) that I assumed it was well known. It turned out that it was not known at all. I had two conversations during this period about this result that stand out in my mind. I had occasion to describe the result to Bott, who did not believe it upon first hearing it. Of course once I explained the proof he quickly understood the argument. But I have to admit that surprising Bott on this particular subject matter was very gratifying. I also mentioned the result to Karen Uhlenbeck, asking whether it would be of any interest to physicists. She replied that "physicists think compact Lorentz manifolds are stupid." This was a less gratifying conversation, and I returned to my office in a glum mood and started mulling over the noncompact case. Shortly thereafter, Nadine Kowalsky, who had just started working with me as a doctoral student, came into my office and I told her she was going to prove a great theorem about isometry groups of noncompact Lorentz manifolds (of course without my knowing how she was going to do it). And she did.

I had another conversation with Nadine that was important to me. I was the department chair at the time and one afternoon I saw Nadine examining a notice on the department bulletin board about women in mathematics, mentoring, and role models. I asked Nadine how important this was and how important it was to her. She replied that it was very important. I asked her to come tell me about it, and we spent about an hour discussing the matter. I knew Nadine was brilliant, evidently determined, and intellectually very tough. What I learned from this conversation was that she was also very thoughtful about gender issues in mathematics. It highlighted for me the importance of these issues, a lesson that I never forget. So Nadine taught me more than just about Lorentz manifolds.

Nadine was a wonderful student, who wrote a beautiful dissertation, and went off to spend a year at IAS before taking up a job at Stanford. A few weeks after arriving at Princeton, she collapsed and was diagnosed with leukemia. We arranged for her to come back to Chicago, and she died the following year. Her parents, Walter and Yvonne Kowalsky, named an award for dissertation year doctoral students at Chicago, with the hope that the award would go each year to an outstanding woman student. There have now been nine women at Chicago named Kowalsky Fellows, and because of the generosity of the Kowalskys, this award will continue to be given every year.

One of the people who became interested in this work about actions, fundamental groups, and geometric structures was Misha Gromov. Gromov articulated its role in his thinking about rigid geometric structures, the latter being one of the sequence of areas in which his thinking was transformational. Some of his key work first appeared in a paper of d'Ambra and Gromov and it was evident to me that this work, in addition to its fundamental results, could have a major impact on the program I had undertaken. But I needed to understand his work, which I found quite difficult. I carried his paper on rigid transformation groups with me constantly and worked on it for a year. As with the impact of having understood Margulis's work a number of years before that, finally understanding Gromov's ideas gave me a fundamentally new entry point into the problems I had been working on. In particular, it enabled me to significantly enrich my understanding of the structure of manifolds on which noncompact simple Lie groups act preserving a rigid structure via the analysis of fundamental group and its remarkable representation that Gromov constructed. One of the culminations of this direction appears in the paper "Entropy and arithmetic quotients for simple automorphism groups of geometric manifolds," which is contained in this volume. Later on, I will discuss this paper and how it is suggestive of a general approach to classifying actions on manifolds.

The experiences of reading the watershed papers of Margulis and Gromov had a number of features in common. In both cases, while the results were widely applauded, the deep ideas in the papers had not been absorbed by the mathematics community. Given the difficulty of the papers, it was easy enough to understand why. But the payoff in understanding their ideas and approach, not just the results, was immensely rewarding and productive. I am reminded of the famous remark of Abel attributing his depth of understanding to the fact that he read the masters and not their pupils. Margulis and Gromov were true masters.

One of the other geometric issues I worked on during this period was the classical question of Clifford-Klein forms. Namely, for which homogeneous spaces G/H, where G and H are Lie groups, is there a compact manifold locally modeled on G/H? In work done in part with Francois Labourie and Shahar Mozes, we showed how cocycle superrigidity enabled one to address this for G/H when H had a sufficiently large centralizer in G. This added a new technique to others that had been previously developed. Still others have been developed since. One of the fascinating features of this basic, seemingly elementary question about manifolds is how many different approaches have been developed without any one of them providing a method that subsumes the others. A coherent and unified exposition of the breadth of techniques required and available would be a significant contribution. There are still many fascinating questions remaining.

At the same time this geometric work was taking place, Garrett Stuck and I were investigating a fundamental question about ergodic actions of higher rank Lie groups and their lattices on any measure space with a finite invariant measure. Our main result is that these actions are always essentially free, i.e., almost every stabilizer is trivial, but for the obvious exceptions of the action of G on G/D or the action of a lattice on a finite set. In the vernacular of today, this result identifies the invariant random subgroups of G and D.

There is another elementary question about stabilizers that remains open. If a noncompact simple G acts nontrivially on a compact manifold preserving a volume form, can G have a fixed point? Gromov raised this question to me many years ago when he was visiting Chicago. Obviously lattices can act with a fixed point, and a classic result of Sternberg yields local linearity around a fixed point for a G action. But as far as I know, Gromov's question remains unanswered.

A new chapter in the work on the general structure of actions of noncompact simple Lie groups was my work with Amos Nevo on stationary measures. Furstenberg had developed some important foundations of this area many years before. Amos and I undertook the program of trying to understand how to use stationary measures as a general entry point into the structure of actions without an invariant measure, as the existence of an invariant measure was an assumption while stationary measures always existed. The results we achieved were quite satisfying, showing that with certain additional assumptions the actions had G/Q as a quotient space in which Q was a parabolic, and that there was a relatively invariant measure for the resulting expression of X as a bundle over G/Q. These results were improved recently by Brown, Hertz, and Wang in another beautiful work. Aside from the main structure theorems we proved, there are two other points I want emphasize about this general approach.

First, it seems to me possible that stationary measures are in fact useful in understanding the general action of lattices with a finite invariant measure. The examples of Katok-Lewis, obtained by blowing up a fixed point into a projective space, give examples of actions with finite invariant measure, where the action is a standard example off a singular set, and on the singular set the action is "projective" and hence carries a natural stationary measure. While in the low-dimensional case, we now know by BFH that there are no "singular sets," in the higher dimensional case, which will be nonisometric in general, such may exist. Does identifying the stationary measures help identify singular sets?

Second, we used the structure theorem to characterize some of the "tangential de Rham cohomology" for an action, more precisely for the associated foliation by symmetric spaces. This is the cohomology of measurable bounded forms, smooth along leaves. Aside from the intrinsic appeal of the actual results, I was particularly interested in a variant of this situation, as years before I had worked on the problem of the cohomology of measurable forms smooth along the leaves of a foliation. Part of the interest in this question is that while the cohomology of bounded measurable forms (smooth along the leaves) is an invariant of the action, the cohomology without the boundedness condition is an invariant of orbit equivalence. There are many examples in which the first cohomology (without boundedness assumptions) is nontrivial and relevant to many problems. There are a few other examples in which the second cohomology is nonvanishing. But without some sort of boundedness or integrability property of the forms in question, there is not a good method for showing non-vanishing in certain circumstances in higher degree. (For essentially free amenable group actions, this cohomology is trivial in higher degree.) Is there such a theory for actions of simple noncompact Lie groups, or foliations by symmetric spaces? The results mentioned above are very suggestive, but do not answer this question that has puzzled me for thirty-five years.

Let me return to the paper "Entropy and arithmetic quotients for simple automorphism groups of geometric manifolds" referred to above. This paper addresses how close an action of a higher rank G, preserving a finite measure, is to a standard "arithmetic" example, at least under geometric hypotheses. An action of G is called arithmetic if it is of the form K\H/D, where H is a real algebraic group, D an arithmetic subgroup, and K a (possibly trivial) compact subgroup centralized by G in which G is embedded in H via a homomorphism. Let us call a quotient map X → Y of two G spaces fully entropic if the Kolmogorov-Sinai entropies satisfy h(X,g) = h(Y,g) for every g in G. The main result of the paper in question is that under certain hypotheses, there is a fully entropic measurable arithmetic quotient. While there is also an "engaging" assumption, the main hypothesis is that the action preserves a (real analytic) rigid unimodular geometric structure, for example a connection and a volume density. It is now suggestive to make the following observations about the assumptions and the conclusions.

1. From cocycle superrigidity, there is always a measurable invariant connection, assuming an invariant volume. Is the measurable invariant connection always smooth, at least on an open dense set? (With respect to the latter, recall the blowing-up example of Katok-Lewis.) The recent results of Brown-Fisher-Hurtado show that it is, in the low-dimensional case. Can one handle the more general case? If so, can one bridge the gap between smooth and real analytic? If all of this can be done, can one effectively eliminate the geometric hypotheses on the action?

2. Actions of lattices with zero entropy have a measurable invariant metric (by cocycle superrigidity) and Brown-Fisher-Hurtado implies there is a smooth invariant metric. When do similar techniques enable one to deduce that a fully entropic quotient map is really an "isometric extension"? Will this be an associated bundle of a principal bundle with compact structure group?

3. How much of this can be put together to give a general result that fully describes an action on an open dense set as an associated bundle of a compact principal bundle over an arithmetic quotient?

4. When can the singular set be described as a finite union of projective varieties? Do stationary measures play a role in this analysis?

5. Given the structural results about stationary measures, can any of these speculations be usefully extended beyond the invariant-measure case to the stationary case?

All this is quite speculative at this point (as well as being somewhat imprecise), but the recent Brown-Fisher-Hurtado result makes this tantalizing to think about. The results in "Entropy and arithmetic quotients for simple automorphism groups of geometric manifolds" rely heavily on an analysis that utilizes Gromov's foundational work and earlier work of mine with Lubotzky that proves the existence of a canonical maximal arithmetic quotient under various natural conditions, and ties this to the representations of the fundamental group of M. The latter in turn uses the fundamental work of Marina Ratner on invariant measures on homogeneous spaces.

I began by mentioning context. One major context for almost all of this work has been the mathematics department at the University of Chicago. The department provides an outstanding environment for mathematics research and education, one that I have benefited from since my arrival as a Dickson Instructor in 1977. Aside from wonderful colleagues and large numbers of visitors, the students at the university contribute greatly to this environment. I am especially appreciative of those students I have collaborated with or remained in touch with over the years. Special thanks are of course due to David Fisher, who agreed to provide further context for these collected papers. The work with Amos Nevo, and earlier work with both Alex Lubotzky and Shahar Mozes, brought me to Israel for mathematics with great regularity, both to Hebrew University and to the Technion. It was a great privilege for me, for reasons well beyond the mathematical, to have this opportunity to work in Israel and to forge lasting friendships in the process. As with all mathematicians, I have had the benefit of innumerable conversations with many mathematicians from around the world. I will not attempt to list them, but for those reading this introduction, please know how much I have appreciated our work and our connection.

Chicago, August 2017

1

Spectra and Structure of Ergodic Actions

EXTENSIONS OF ERGODIC GROUP ACTIONS

BY

ROBERT J. ZIMMER

In this paper we shall study extensions in the theory of ergodic actions of a locally compact group. If G is a locally compact group, by an ergodic G-space we mean a Lebesgue space (X, μ) together with a Borel action of G on X, under which μ is invariant and ergodic. If (X, μ) and (Y, ν) are ergodic G-spaces, (X, μ) is called an extension of (Y, ν) and (Y, ν) a factor of (X, μ) if there is a Borel function $p: X \rightarrow Y$, commuting with the G-actions, such that $p_*(\mu) = \nu$. Various properties that one considers for a fixed ergodic G-space have as natural analogues properties of the triple (X, p, Y) in such a way as to reduce to the usual ones in case Y is a point. This is the idea of "relativizing" concepts, which is a popular theme in the study of extensions in topological dynamics. In ergodic theory, relativization is a natural idea from the point of view of Mackey's theory of virtual groups [16]. Although familiarity with virtual groups is not essential for a reading of this paper, this idea does provide motivation for some of the concepts introduced below, and a good framework for understanding our results. We shall therefore briefly review the notion of virtual group and indicate its relevance.

If X is an ergodic G-space, one of two mutually exclusive statements holds:

(i)   There is an orbit whose complement is a null set. In this case, X is called essentially transitive.

(ii)   Every orbit is a null set. X is then called properly ergodic.

In the first case, the action of G on X is essentially equivalent to the action defined by translation on G/H, where H is a closed subgroup of G; furthermore, this action is determined up to equivalence by the conjugacy class of H in G. In the second case, no such simple description of the action is available, but it is often useful to think of the action as being defined by a "virtual subgroup" of G. Many concepts defined for a subgroup H, can be expressed in terms of the action of G on G/H; frequently, this leads to a natural extension of the concept to the case of an arbitrary virtual subgroup, i.e., to the case of an ergodic G-action that is not necessarily essentially transitive. Perhaps the most fundamental notions that can be extended in this way are those of a homomorphism, and the concomitant ideas of kernel and range. These and other related matters are discussed in [16].

From this point of view, the notion of an extension of an ergodic G-space has a simple interpretation. A measure preserving G-map $\phi: X \rightarrow Y$ can be viewed as an embedding of the virtual subgroup defined by X into the virtual subgroup defined by Y. Thus, it is reasonable to hope that many of the concepts that one considers for a given ergodic G-space, i.e., a virtual subgroup of G,

Received October 1, 1975.

can also be defined for extensions, i.e., one virtual subgroup considered as a sub-virtual subgroup of another. We now turn to consideration of one such concept which admits a very fruitful relativization.

For any ergodic G-space X, there is always a naturally defined unitary representation of G on $L^2(X)$, and it is natural to ask what the algebraic structure of the representation implies about the geometric structure of the action. One of the earliest results obtained in this direction, when the group in question is the integers, is the now classical von Neumann–Halmos theory of actions with discrete spectrum [6], [17]. For the integers, a unitary representation is determined by a single unitary operator, and von Neumann and Halmos were able to completely describe those actions for which this operator has discrete (i.e., pure point) spectrum. Their most important results are contained in the uniqueness theorem (asserting that the spectrum is then a complete invariant of the action), the existence theorem (describing what subsets of the circle can appear as the spectrum) and the structure theorem. This last result asserts that every ergodic action of the integers with discrete spectrum is equivalent to a translation on a compact abelian group. As indicated by Mackey in [15], the methods of von Neumann and Halmos enable one to obtain an equally complete theory, for actions with discrete spectrum, of an arbitrary locally compact abelian group. When the group is not abelian, the situation is somewhat more complicated. Using techniques different from those of von Neumann and Halmos, Mackey was able to prove a generalization of the structure theorem for actions of nonabelian groups [15]. He pointed out, however, that the natural analogue of the uniqueness theorem fails to hold. Nevertheless, we shall see that for a suitably restricted class of actions (which includes all actions of abelian groups with discrete spectrum), the uniqueness and existence theorems have natural extensions, even for G nonabelian. These actions are the normal actions, so called because they are the virtual subgroup analogue of normal subgroups. Thus, restricting consideration to normal actions, one has a complete generalization of the von Neumann–Halmos theory.

With this in hand, one can now ask to what extent the whole theory generalizes to the case of extensions. The highly satisfactory answer is that it extends intact, providing a significant new generalization of the von Neumann–Halmos theory even for the group of integers and the real line. If X is an extension of Y, one can define the notion of X having "relatively discrete spectrum" over Y. This can be loosely described as follows. Decompose the measure μ as a direct integral over the fibers of p, with respect to v. This defines a Hilbert space on each fiber, and these Hilbert spaces exhibit $L^2(X)$ as a Borel G-Hilbert bundle over Y. If $L^2(X)$ is the direct sum of finite dimensional G-invariant subbundles over Y, we say that X has relatively discrete spectrum over Y. The structure theorem below, generalizing Mackey's theorem, describes the geometric structure of the extension when this "algebraic" condition is satisfied. It asserts that X can be written as a certain type of skew-product; these are factors of Mackey's "kernel" actions, and are a generalization of Anzai's skew products.

Similarly, the notion of normality can be relativized and it is meaningful to say that X is a normal extension of Y. In virtual group terms, this of course means that X defines a normal subvirtual subgroup of Y. For normal extensions, we prove analogues of the uniqueness and existence theorems. We remark that for a properly ergodic action of the integers, there always exist nonnormal extensions with relatively discrete spectrum. Thus, even for abelian groups the question of normality is highly relevant in these considerations.

In a subsequent paper [23], we use this theory to develop the notion of actions with generalized discrete spectrum. This makes contact with Furstenberg's work in topological dynamics on minimal distal flows, Parry's notion of separating sieves, the theory of affine actions, and quasi-discrete spectrum. It promises other applications as well.

The entire theory sketched above depends in an essential way on the concept of a cocycle of an ergodic G-space. Cocycles have appeared in various considerations in ergodic theory [16], [18], [9], and from the virtual group point of view are the analogue of homomorphisms for subgroups. This "analogy" rests on the fact that for transitive G-spaces, there is a correspondence between cocycles and homomorphisms of the stability groups. This is, in fact, an essential part of Mackey's well-known imprimitivity theorem. (See [20] for an account of the imprimitivity theorem from this point of view.) For properly ergodic ations, the fundamentals of a general theory of cocycles were sketched by Mackey in [16]; some aspects of this theory are worked out in detail by Ramsay in [19]. We have continued the detailed development of certain areas within the theory, particularly the study of cocycles into compact groups. These results are basic to the rest of the paper.

The organization of this paper is as follows. Part I is preparatory, and the material is used throughout this paper as well as [23]. Aside from establishing notation and recalling various results, there are three main features. One is a general existence theorem for factors of a Lebesgue space; this appears in Section 1, and in equivariant form in Section 2. The latter section also discusses the basic connections between extensions and cocycles. This includes the notions of restriction and induction of cocycles, analogous to those for group representations, and a version of the Frobenius reciprocity theorem. Lastly, the general theory of cocycles into compact groups is developed in Section 3. Part II contains the relativized version of the von Neumann–Halmos–Mackey theory. The structure theorem is proved in Section 4, as well as a generalization that subsumes another theorem of Mackey on induced representations (which is in fact a generalization of his own structure theorem.) The virtual subgroup analogue of normality is discussed in Section 5, and is then used to complete the extension theory with the uniqueness-existence theorems in Section 6.

Some of the results of this paper were announced in [22].

The author would like to express his thanks to Professor G. W. Mackey for many helpful discussions and suggestions made during the preparation of this paper.

376   ROBERT J. ZIMMER

I. G-spaces, factors, and cocycles

1. Factors of Lebesgue spaces. We begin by recalling and discussing some facts about Borel spaces. By a Lebesgue space we mean a standard Borel space X, together with a probability measure μ. (See [12] or [19] for detailed definitions of these and other related concepts to follow.) Associated with any Lebesgue space, we have the Boolean σ-algebra $B(X, \mu)$, which consists of the Borel sets of X, any two being identified if their symmetric difference is a null set. If Y is a standard Borel space and $\phi: X \to Y$ is a Borel function, then we have a measure $\phi_*(\mu)$ defined on Y by $\phi_*(\mu)(B) = \mu(\phi^{-1}(B))$, for $B \subset Y$ a Borel set. If (Y, v) is a Lebesgue space, we will call a Borel map $\phi: X \to Y$ a factor map if $\phi_*(\mu) = v$. We will call Y a factor of X or X an extension of Y. Now if $\phi: X \to Y$ is a factor map, we have an induced map $\phi^*: B(Y, v) \to B(X, \mu)$ that is injective. Conversely, it is well known that if $A \subset B(X, \mu)$ is a σ-subalgebra, then there exists a Lebesgue space (Y, v) and a factor map $\phi: X \to Y$ such that $A = \phi^*(B(Y))$ [19, Theorem 2.1].

Since $\phi_*(\mu) = v$, the map $f \to f \circ \phi$ induces an isometric embedding of $L^2(Y)$ as a subspace of $L^2(X)$. It is easy to see that this subspace can be characterized as $\{f \in L^2(X) \mid f$ is measurable with respect to the σ-field of Borel sets in X whose equivalence class in $B(X)$ belongs to $\phi^*(B(Y))\}$. We shall on various occasions need criteria for determining when a given subspace of $L^2(X)$ is of the form $L^2(Y)$ for some factor Y of X. A useful result in this direction is the following theorem.

LEMMA 1.1. *Let A be a collection of subsets of a given set and suppose A is closed under complements. Let B be the set of finite intersections of elements of A, and C be the set of disjoint finite unions of elements of B. Then C is the field generated by A.*

THEOREM 1.2. *Let X be a Lebesgue space and $A \subset L^\infty(X)$ a *-subalgebra (not necessarily closed). Let B be the σ-field of Borel sets in X generated by the functions of A. Then as subspaces of $L^2(X)$,*

$$\bar{A} = L^2(X, B) = \{f \in L^2(X) \mid f \text{ measurable with respect to } B\}.$$

Proof. (i) If $f \in \bar{A}$, then $f = \lim f_n$, $f_n \in A$, where the limit is in $L^2(X)$. Now it follows from the proof of the Riesz-Fisher theorem that there exists a subsequence f_{n_j} such that $f_{n_j} \to f$ pointwise on a conull set. Since each f_{n_j} is measurable with respect to B, so is f. Hence $\bar{A} \subset L^2(X, B)$.

(ii) We now claim that $L^2(X, B) \subset \bar{A}$. Since A is closed under complex conjugation, it is easy to see that B is the σ-field generated by

$$D = \{f^{-1}(M) \mid f \in A, f \in L^\infty(X, \mathbf{R}), M \subset \mathbf{R} \text{ Borel}\}.$$

Let B_0 be the field generated by D. As every element of $L^2(X, B)$ is an L^2-limit of linear combinations of characteristic functions of elements of B, it suffices to see that $\chi_S \in \bar{A}$ for $S \in B$. Since B_0 generates B as a σ-field, it suffices to see

EXTENSIONS OF ERGODIC GROUP ACTIONS   377

that $\chi_S \in \bar{A}$ for $S \in B_0$ [1, p. 21]. By Lemma 1.1, it suffices to see that $\chi_S \in \bar{A}$ whenever S is the finite intersection of sets of D. Suppose $f_i \in A \cap L^{\infty}(X; \mathbf{R})$, and let $R_i = \|f_i\|_{\infty}$, $i = 1, \ldots, k$. Choose $M_i \subset [-R_i, R_i]$ to be a Borel set.

For each positive integer n, suppose $p_{1n}, \ldots, p_{kn}$ are polynomials. Then

$$\prod_{i=1}^{k} p_{in} \circ f_i \in A \quad \text{for all } n,$$

since $f_i \in L^{\infty}(X; \mathbf{R})$ and A is an algebra. Now suppose that g_{in}, $i = 1, \ldots, k$, $n = 1, 2, \ldots$, are bounded Borel functions such that

$$\prod_{i=1}^{k} g_{in} \circ f_i \in \bar{A} \quad \text{and} \quad \lim_{n \to \infty} g_{in} = g_i$$

in bounded pointwise convergence. Then

$$\prod_{i=1}^{k} g_{in} \circ f_i \to \prod_{i=1}^{k} g_i \circ f_i$$

in bounded pointwise convergence, and hence the limit also holds in $L^2(X)$. Thus we also have $\prod_{i=1}^{k} g_i \circ f_i \in \bar{A}$. For each i, the smallest set of functions on $[-R_i, R_i]$ closed under bounded pointwise convergence and containing the polynomials is the set of bounded Borel functions. Hence $\prod_{i=1}^{k} g_i \circ f_i \in \bar{A}$ for all bounded Borel g_i defined on $[-R_i, R_i]$. Letting $g_i = \chi_{M_i}$, we obtain

$$\chi_{\cap_{i=1}^{k} f_i^{-1}(M_i)} = \prod_{i=1}^{k} \chi_{f_i^{-1}(M_i)} = \prod_{i=1}^{k} \chi_{M_i} \circ f_i \in \bar{A}.$$

This completes the proof.

Combining this theorem with the preceding remarks, we have:

COROLLARY 1.3. *Let X be a Lebesgue space and A a *-subalgebra of $L^{\infty}(X)$. Then $\bar{A} = L^2(Y)$ for some factor Y of X.*

We remark that techniques similar to those of the proof above appear in [10, Theorem 2.2].

If $\phi: X \to Y$ is a factor map, the measure μ may be decomposed over the fibers of ϕ. More precisely, for each $y \in Y$, let $F_y = \phi^{-1}(y)$. Then for each y, there exists a measure μ_y on X, that is supported on F_y, such that for each Borel function f on X, $y \mapsto \int f \, d\mu_y$ is Borel on Y, and

$$\int_X f \, d\mu = \int_Y \left(\int f \, d\mu_y \right) d\nu(y).$$

If $\{\mu_y\}$ is such a collection of measures, we write $\mu = \int^{\oplus} \mu_y \, d\nu$. This decomposition of μ is almost unique: If $\mu = \int^{\oplus} \mu_y \, d\nu = \int^{\oplus} \mu'_y \, d\nu$, then $\mu_y = \mu'_y$ almost everywhere. A decomposition of μ yields a decomposition of $L^2(X)$ as a Hilbert bundle over Y:

$$L^2(X) = \int^{\oplus} L^2(F_y, d\mu_y) \, d\nu \qquad \text{(see [19]).}$$

378   ROBERT J. ZIMMER

We now consider a construction which proves to be of much use when studying factors. Suppose $p: (X, \mu) \to (Z, \alpha)$ and $q: (Y, \nu) \to (Z, \alpha)$ are factors. Define

$$X \times_Z Y = \{(x, y) \in X \times Y \mid p(x) = q(y)\}.$$

This is called the fibered product of X and Y over Z, and is a Borel subset of $X \times Y$. There is a natural Borel map $t: X \times_Z Y \to Z$, given by $t(x, y) = p(x) \ (=q(y))$, so $t^{-1}(z) = p^{-1}(z) \times q^{-1}(z)$. Suppose

$$\mu = \int^{\oplus} \mu_z \, d\alpha \qquad \nu = \int^{\oplus} \nu_z \, d\alpha.$$

Then it is easy to check that for $A \subset X \times_Z Y$ Borel,

$$(\mu \times_Z \nu)(A) = \int_Z (\mu_z \times \nu_z)(A) \, d\alpha(z)$$

defines a measure on $X \times_Z Y$, and that $\mu \times_Z \nu = \int^{\oplus} (\mu_z \times \nu_z) \, d\alpha$. In the case that Z is one point, $(X \times_Z Y, \mu \times_Z \nu)$ reduces to the usual Cartesian product, with the product measure.

There is a useful universal characterization of the fibered product. To state this, we first consider the concept of relative independence.

PROPOSITION 1.4.   *Consider a commutative diagram of factor maps of Lebesgue spaces*:

$$
\begin{array}{ccc}
X_0 & \xrightarrow{\psi} & Y \\
\phi \downarrow & & \downarrow q \\
X & \xrightarrow{p} & Z
\end{array}
$$

Let m denote the measure on X_0, and $m = \int^{\oplus} m_z \, d\alpha$. We consider all the L^2-spaces as subspaces of $L^2(X_0)$. Then the following are equivalent:

   (i) $(L^2(X) \ominus L^2(Z)) \perp (L^2(Y) \ominus L^2(Z))$

   (ii) *If $f \in L^2(X)$, $g \in L^2(Y)$, then $E(f \cdot g \mid Z) = E(f \mid Z)E(g \mid Z)$. (Here $E(\cdot \mid Z)$ is a conditional expectation.)*

   (iii) *If $A \subset X$ and $B \subset Y$, then for almost all $z \in Z$, $\phi^{-1}(A)$ and $\psi^{-1}(B)$ are independent sets in (X_0, m_z).*

   Proof. (i) $\Rightarrow$ (ii)   Let $f \in L^2(X), g \in L^2(Y)$. Condition (i) implies $E(g \mid X) = E(g \mid Z)$. Hence $E(f \cdot g \mid X) = fE(g \mid X) = fE(g \mid Z)$. Now take $E(\ \mid Z)$ of this equation; we get $E(f \cdot g \mid Z) = E(f \mid Z)E(g \mid Z)$.

   (ii) $\Rightarrow$ (i)   If $f \in L^2(X) \ominus L^2(Z)$, $g \in L^2(Y) \ominus L^2(Z)$. Then $E(f \mid Z) = E(\bar{g} \mid Z) = 0$. Thus $E(f \cdot \bar{g} \mid Z) = 0$ by (ii) which implies $f \perp g$.

   (ii) $\Rightarrow$ (iii)   This is immediate when one notes that for a set $S \subset X_0$ $E(\chi_S \mid Z)$ is just the function $z \mapsto m_z(S)$.

   (iii) $\Rightarrow$ (ii)   We know $E(f \cdot g \mid Z) = E(f \mid Z)E(g \mid Z)$ when f and g are characteristic functions, and the general result follows by the usual approximation arguments.

EXTENSIONS OF ERGODIC GROUP ACTIONS   379

We remark that when Z is one point, these conditions are equivalent to the σ-fields in X_0 determined by X and Y being independent. Hence, we shall say that X and Y are relatively independent over Z if the conditions of the proposition hold.

We now characterize the fibered product in terms of relative independence.

PROPOSITION 1.5. *Given a commutative diagram as in Proposition 1.4, X and Y are relatively independent over Z if and only if there exists a factor map $h: X_0 \to X \times_Z Y$ such that the following diagram commutes:*

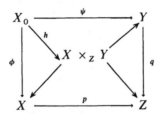

Proof. Define h by $h(x_0) = (\phi(x_0), \psi(x_0))$. If $A \subset X$, $B \subset Y$ are Borel, let $A \times_Z B = (A \times B) \cap (X \times_Z Y)$. To see that h is a factor map, it clearly suffices to show that

$$m(h^{-1}(A \times_Z B)) = (\mu \times_Z v)(A \times_Z B).$$

Now

$$m(h^{-1}(A \times_Z B)) = \int_Z m_z(\phi^{-1}(A) \cap \psi^{-1}(B)) \, dz$$

$$= \int_Z m_z(\phi^{-1}(A)) m_z(\psi^{-1}(B)) \, dz.$$

The uniqueness of decomposition for measures implies $\phi_*(m_z) = \mu_z$ and $\psi_*(m_z) = v_z$ almost everywhere. Thus the integral becomes

$$\int_Z \mu_z(A) v_z(B) \, dz = \int_Z (\mu_z \times v_z)(A \times_Z B) = (\mu \times_Z v)(A \times_Z B).$$

The converse assertion is more or less immediate.

2. G-spaces: Introductory remarks. Let X be a standard Borel space, and G a standard Borel group. We call X a Borel G-space if there is a (right) action of G on X such that the map $X \times G \to X$ is Borel. If X is a Lebesgue space and G is a locally compact group, we shall call X a Lebesgue G-space if it is a Borel G-space and if G preserves the measure. (We shall throughout take "locally compact" to mean locally compact and second countable.) If $X' \subset X$, we will call X' an essential subset if it is Borel, conull, and G-invariant. A factor map $\phi: X \to Y$ between G-spaces will be called a G-map if $\phi(xg) = \phi(x)g$ for all $(x, g) \in X \times G$. We will call Y a factor of X if there exists a factor G-map $X' \to Y$ where $X' \subset X$ is essential. Now G acts on $B(X)$, and if Y is

380   ROBERT J. ZIMMER

a factor of X, $B(Y)$ can be identified with a G-invariant σ-subalgebra of $B(X)$. We now show that every G-invariant subalgebra of $B(X)$ arises in this way.

PROPOSITION 2.1.   *Let (X, μ) be a Lebesgue G-space and $A \subset B(X)$ a G-invariant sub σ-algebra. Then there is a factor Y of X such that $B(Y) = A$.*

   Proof.   This argument is a small modification of the proof of [14, Theorem 2]. A is a Boolean G-space [14] and by [14, Theorem 1], there is a Borel G-space Y and a quasi-invariant measure v such that $B(Y, v) \cong A$ as Boolean G-spaces. Since A has an invariant measure inherited from the measure on $B(X)$, we can assume that v is invariant. Now let $\theta : X \to Y$ be a Borel map such that $\theta^* : B(Y) \to B(X)$ defines the isomorphism $B(Y) \cong A$ [19, Theorem 2.1]. Y is standard, so we can choose a Borel isomorphism $i : Y \to I$ where I is a subset of the unit interval. Let F_G be the universal Borel G-space as defined by Mackey in [14]. Define $\phi : Y \to F_G$ by $\phi(y)(g) = i(yg)$ and $\psi : X \to F_G$ by $\psi(x)(g) = i(\theta(xg))$.

   By the proof of [14, Lemma 2], ϕ and ψ are Borel G-maps, and ϕ is a Borel isomorphism onto an invariant Borel subset of F_G. Since θ^* is a Boolean G-map, it follows that for each g, $\theta(xg) = \theta(x)g$ for almost all x. Thus, by Fubini's theorem,

$$[(\phi \cdot \theta)(x)](g) = i(\theta(x) \cdot g) = i(\theta(xg)) = \psi(x)(g)$$

for almost all (x, g); i.e., $\psi = \phi \circ \theta$ almost everywhere. Thus $X' = \psi^{-1}$ (range ϕ) is conull, Borel and G-invariant. The map $\phi^{-1} \circ \psi : X' \to Y$ is a G-map, and since it agrees with θ almost everywhere, it induces the given Boolean G-isomorphism $\theta^* : B(Y) \to A$.

   When it is convenient, we shall apply (often without explicit mention) various definitions and constructions that we have given for factor maps to factors in general. By this we understand that we have passed to an essential subset for which there is a factor map, and that the notion at hand is independent (at least up to some obvious isomorphism) of the choice of such a set.

   Using the correspondence between factors and σ-subalgebras, it is easy to deduce the following equivariant version of Corollary 1.3.

COROLLARY 2.2.   *Let X be a Lebesgue G-space and A a G-invariant *-subalgebra of $L^\infty(X)$. Then in $L^2(X)$, we have $\bar{A} = L^2(Y)$ for some G-factor Y of X.*

   A Lebesgue G-space X is called ergodic if the action of G on $B(X)$ is irreducible; i.e., there are no elements in $B(X)$ left fixed (by all elements of G) except ϕ and X. Mackey has shown that an equivalent condition is that for any Borel function f on X, $f \cdot g = f$ everywhere (for each $g \in G$) implies that f is constant on a conull set [14, Theorem 3]. It is trivial that a factor of an ergodic G-space is also ergodic.

   If X and Y are transitive G-spaces, then X and Y are essentially isomorphic to G/H and G/K respectively, where H and K are closed subgroups of G. X will be an extension of Y if and only if H is contained in a conjugate of K. The map $X = G/H \to G/K = Y$ is determined by the embedding of H in this conjugate of K. Thus, in terms of Mackey's notion of virtual groups [16], a factor map $X \to Y$ where X and Y are ergodic but not necessarily transitive G-spaces corresponds to an embedding of the virtual subgroup defined by X into the virtual subgroup defined by Y.

We now turn our attention to cocycles of ergodic G-spaces, a concept central to this paper. The reader is referred to [19], [20], [16] as general references for cocycles, and particularly the latter for an explanation of why cocycles are the virtual subgroup analogue of homomorphism (and representation). Most of the remainder of Section 2 is devoted to setting out examples and results for cocycle representations of ergodic G-spaces that have well-known analogues for representations of locally compact groups.

Let S be an ergodic Lebesgue G-space and K a standard Borel group. We call a Borel function $\alpha: S \times G \to K$ a cocycle if for g, $h \in G$, $\alpha(s, gh) = \alpha(s, g)\alpha(sg, h)$ for almost all $s \in S$. A useful extension of this notion arises in the context of Hilbert bundles. Let $\{H_s\}$ be a Hilbert bundle on S, and suppose that for each $(s, g) \in S \times G$, we have a bounded linear map $\alpha(s, g): H_{sg} \to H_s$ such that:

    (i)   For each g, $\alpha(s, g)$ is unitary for almost all s.

    (ii)   For each pair of bounded Borel sections of the bundle $f = \{f_s\}$, $f' = \{f'_s\}$, the function $(s, g) \mapsto \langle \alpha(s, g)f_{sg} \,|\, f'_s \rangle_s$ is Borel.

    (iii)   For each g, $h \in G$, $\alpha(s, gh) = \alpha(s, g)\alpha(sg, h)$ for almost all s.

We then call α a cocycle representation of (S, G) on the Hilbert bundle $\{H_s\}$.

If α, $\beta: S \times G \to K$ are cocycles, we call them cohomologous, or equivalent, if there is a Borel map $\phi: S \to K$ such that for each g,

$$\phi(s)\alpha(s, g)\phi(sg)^{-1} = \beta(s, g) \quad \text{for almost all } s \in S.$$

Similarly, if α is a cocycle representation on the bundle $\{H_s\}$ and β a cocycle representation on the bundle $\{H'_s\}$, we call them equivalent if there exists a Borel field of bounded linear maps $U(s): H_s \to H'_s$ such that:

    (i) $U(s)$ is unitary for almost all s.

    (ii)   For each g, $U(s)\alpha(s, g)U(sg)^{-1} = \beta(s, g)$ for almost all $s \in S$.

Suppose α is a cocycle representation in the product bundle $H = S \times H_0$. From condition (ii) in the definition of cocycle representation, the map $(s, g) \mapsto \alpha(s, g)$ is a Borel map from $S \times G$ into $L(H_0)$, the bounded linear operators on H_0, where the latter is given the weak topology. $L(H_0)$ is standard under the weak Borel structure, and the unitary group $U(H_0)$ is a Borel subset [2]. Hence $\{(s, g) \mid \alpha(s, g) \in U(H_0)\}$ is Borel, and so by changing α on a conull Borel set, we obtain an equivalent cocycle representation β on $S \times H_0$ such that:

    (i) $\beta(s, g)$ is unitary for all $(s, g) \in S \times G$.

    (ii)   For each g, $\alpha(s, g) = \beta(s, g)$ for almost all s.

Thus, up to equivalence, any cocycle representation on the product bundle can be considered as a cocycle into the unitary group $U(H_0)$.

As pointed out by Ramsay [19, p. 264], if $\{H_s\}$ is a Hilbert bundle on S, there exists a decomposition of S into disjoint Borel sets $\{S_\infty, S_0, S_1, \ldots\}$ such that for each n, there exists a Borel field of unitary operators on S_n, $U(s): H_s \to H_n$, where H_n is a fixed Hilbert space of dimension n. Thus

$$\int_{S_n}^{\oplus} H_s \cong S_n \times H_n.$$

If some S_n is conull, we say that $\{H_s\}$ is of uniform multiplicity n. If S is ergodic and $\alpha(s, g)$ is a cocycle representation on the Hilbert bundle $\{H_s\}$, then $\{H_s\}$ is of uniform multiplicity [20, Lemma 9.10]. Thus, every cocycle representation of an ergodic G-space is equivalent to one on a constant field of Hilbert spaces, and hence equivalent to a cocycle into a unitary group.

Example 2.3. We now describe a general method of constructing cocycle representations. Suppose $\phi: X \to Y$ is a G-factor map of ergodic G-spaces. Write $\mu = \int^{\oplus} \mu_y \, dv$. For a fixed $g \in G$, $\mu \cdot g = \mu$, and hence by the uniqueness of decompositions, we have $\mu_y \cdot g = \mu_{yg}$ for almost all $y \in Y$.

LEMMA 2.4.   *The set $A = \{(y, g) \in Y \times G \mid \mu_y \cdot g = \mu_{yg}\}$ is Borel.*

Proof. Let $M(X)$ be the space of measures on X with the usual Borel structure (see e.g., [5] or [11]). Then the map $y \mapsto \mu_y$ is Borel and thus $(y, g) \to yg \to \mu_{yg}$ is Borel. Thus to see that A is Borel, it suffices to see that $(y, g) \to \mu_y \cdot g$ is Borel. But it follows from [11, Theorem 5.2] and [20, Theorem 8.7], that the action of G on $M(X)$ is a Borel action, and hence that $(y, g) \to (\mu_y, g) \to \mu_y \cdot g$ is Borel.

Now

$$L^2(X) = \int_Y^{\oplus} L^2(F_y, \mu_y) \, dv$$

is a Hilbert bundle. If $(y, g) \in A$, then g maps (F_y, μ_y) onto (F_{yg}, μ_{yg}) in a measure preserving way. Let $\alpha(y, g): L^2(F_{yg}) \to L^2(F_y)$ be the induced unitary map. It is straightforward, in light of Lemma 2.4, that α can be extended to a (Borel) cocycle representation (which we also denote by α) of (Y, G) on the Hilbert bundle $L^2(X)$. We call α the natural cocycle representation of the factor map ϕ. We remark that up to equivalence, (in fact up to equality on conull sets) α is independent of the various choices made in its construction.

Example 2.5. The preceding example admits a natural generalization. Suppose $\beta(x, g)$ is a cocycle representation of (X, G) on a Hilbert bundle $\{V_x\}$. We define an associated cocycle representation α of (Y, G) which we call the induced cocycle representation of β. For each y, let

$$H_y = \int_{F_y}^{\oplus} V_x \, d\mu_y(x).$$

Then $\{H_y\}$ is a Hilbert bundle on Y. Furthermore, for $(y, g) \in A$, the map

$$\alpha(y, g): H_{yg} \to H_y, \quad \alpha(y, g) = \int_{x \in F_y}^{\oplus} \beta(x, g),$$

is defined, and for each g, will be unitary for almost all $y \in Y$. Again, α can be extended to a cocycle representation on the Hilbert bundle $\{H_y\}$. In the case where $V_x = \mathbf{C}$ for each $x \in X$, and $\beta(x, g) = 1$ for all (x, g), α is just the cocycle of Example 2.3.

If $X = G/H$ and $Y = G/K$ with $H \subset K$, then a cocycle representation β of $X \times G$ corresponds to a representation Π_β of H [20, Theorem 8.27], and the induced cocycle α will correspond to a representation Π_α of K. One can check that Π_α is the representation induced by Π_β. Thus, in the general case, regarding β as a representation of the virtual subgroup defined by X, we can regard α as the representation of the larger virtual subgroup defined by Y that is induced by β.

In case $Y = \{e\}$, a (Y, G) cocycle representation is simply a unitary representation of the group G. Since $\{e\}$ is a factor of any G-space, the above construction yields, for any cocycle representation α of (X, G), a unitary representation of G, called the representation induced by α, and which we denote by U^α. We recall, for later use, one well-known fact about the relationship between α and U^α.

THEOREM 2.6. *Let α, β be cocycle representations of (X, G) on the Hilbert bundles $\{H_x\}$ and $\{H'_x\}$ respectively. For each $E \subset X$, let $P_E(P'_E)$ be the associated projection operator in $\int^{\oplus} H_x(\int^{\oplus} H'_x)$. Then every intertwining operator T of U^α and U^β with the additional property that $P'_E T = T P_E$ can be written as a bounded Borel field of operators*

$$T = \int^{\oplus} T_x, \quad T_x: H_x \to H'_x$$

such that
(∗) *for each $g \in G$, $T_x \alpha(x, g) = \beta(x, g) T_{xg}$ for almost all $x \in X$.*
Conversely, any bounded Borel field satisfying (∗) defines an intertwining operator T of U^α and U^β satisfying $P'_E T = T P_E$. We will call a field satisfying (∗) an intertwining field for the cocycles α and β.

Proof. Suppose $P'_E T = T P_E$ and $T U^\alpha = U^\beta T$. From the first condition, we have $T = \int^{\oplus} T_x$, $T_x: H_x \to H'_x$, where T is a bounded Borel field. If $f = \int^{\oplus} f_x \in \int^{\oplus} H_x$, then

(∗∗) $\quad (T U^\alpha_g f)_x = T_x \alpha(x, g) f_{xg}$ and $(U^\beta_g T f)_x = \beta(x, g) T_{xg} f_{xg}$.

Choose $f^i \in \int^{\oplus} H_x$ such that $\{f^i_x\}$ is dense in H_x for each $x \in X$. Then for each i, and each $g \in G$, (∗∗) implies

$$T_x \alpha(x, g) f^i_{xg} = \beta(x, g) T_{xg} f^i_{xg}$$

384   ROBERT J. ZIMMER

for all x in a conull set $N_{i,g}$. Then for $x \in \bigcap_i N_{i,g}$, we have

$$T_x \alpha(x, g) = \beta(x, g) T_{xg}.$$

The converse is immediate from (**).

Example 2.7. The notion of induced cocycle is an analogue of the notion of induced representation for groups. We now consider an analogue of restriction of representations.

Suppose $\phi: X \to Y$ is a factor G-map and $\{H_y\}$ a Hilbert bundle on Y. Then the assignment $x \mapsto V_x = H_{\phi(x)}$ is a Hilbert bundle on X, with a fundamental sequence [3] defined as follows. If $\{f_y^i\} \subset \int^\oplus H_y$ is a fundamental sequence for the Hilbert bundle $\{H_y\}$, $g^n \in L^\infty(X)$ is such that $\{g_y^n\}$ is a fundamental sequence for $L^2(X)$ as a Hilbert bundle over Y, and

$$h_{i,n}(x) = g^n(x) f_{\phi(x)}^i \in V_x,$$

then finite linear combinations of $\{h_{i,n}(x)\}$ form a fundamental sequence for the bundle $\{V_x\}$.

Suppose β is a cocycle representation of $Y \times G$ on the Hilbert bundle $\{H_y\}$. Define $\alpha(x, g): V_{xg} \to V_x$ by $\alpha(x, g) = \beta(\phi(x), g)$. Then α is called the restriction of β to (X, G).

If $X = G/H$ and $Y = G/K$ with $H \subset K$, and β is a cocycle corresponding to a representation π of K, then α will be a cocycle corresponding to the restricted representation $\pi \mid H$ of H. Hence, in general, restriction of cocycles can be thought of as the virtual subgroup analogue of restriction of representations.

In further analogy with group representations, we now discuss the algebraic operations of direct sum, tensor product, and conjugation for cocycle representations of ergodic G-spaces. If α and β are cocycle representations of (S, G) on the Hilbert bundles $\{H_s\}$ and $\{H'_s\}$ respectively, then one can form the cocycle $\alpha \oplus \beta$ on the bundle $\{H_s \oplus H'_s\}$, by defining

$$(\alpha \oplus \beta)(s, g) = \alpha(s, g) \oplus \beta(s, g): H_{sg} \oplus H'_{sg} \to H_s \oplus H'_s.$$

Similarly, one can define countable direct sums of cocycles. Given a cocycle representation α, one can ask when it is cohomologous to a representation of the form $\alpha_1 \oplus \alpha_2$. If it is, α will be called reducible, and α_i sub-cocycle representations of α. Otherwise, α is called irreducible. An alternate phrasing of this is made possible by the following easily checked result.

PROPOSITION 2.8. *If $\{V_s\}$ is a sub-Hilbert bundle of $\{H_s\}$, then the following are equivalent:*

(i) $\int^\oplus V_s \, ds$ *is a U^α-invariant subspace.*

(ii) *For each g, $\alpha(s, g)(V_{sg}) = V_s$ for almost all s.*

(iii) α *is cohomologous to $\alpha_1 \oplus \alpha_2$, where α_i are cocycles into $U(H_i)$ and $\int^\oplus V_s$ is unitarily equivalent to $S \times H_1$, $\int^\oplus V_s^\perp$ is unitarily equivalent to $S \times H_2$ (as Hilbert bundles).*

Thus, saying α is irreducible is equivalent to saying that there are no U^α-invariant sub-Hilbert bundles of $\int^\oplus H_s$.

Suppose $\{T_s\}$ is a nontrivial intertwining field for α and β. Let $U(s)$ be the unitary part of the polar decomposition of T_s. Then $U = \int^\oplus U(s)$ intertwines U^α and U^β, and gives a unitary equivalence of

$$U^\alpha \left| \int^\oplus \ker(T_s)^\perp \quad \text{and} \quad U^\beta \right| \int^\oplus \overline{\text{range}(T_s)}.$$

From this it follows that α and β have equivalent subcocycle representations. In particular, if α (or β) is irreducible, and $T_s \neq 0$ on a set of positive measure, then α is a subcocycle representation of β (or vice-versa).

If $\{H_s\}$ and $\{H_s'\}$ are Hilbert bundles, then $\{H_s \otimes H_s'\}$ is also, and one can form the cocycle representation $\alpha \otimes \beta$ on this bundle. Similarly, if α is a cocycle representation on the constant field $S \times H_0$, one can define the conjugate cocycle $\bar\alpha$, as follows: Choose a conjugation $f \mapsto \bar f$ in H_0 [8, p. 15], and for any $A: H_0 \to H_0$, let $\bar A$ be the operator defined by $\langle \bar A f \mid g \rangle = \langle A \bar f \mid \bar g \rangle$, i.e., $\bar A(f) = \overline{A(\bar f)}$. Let $\bar\alpha(s, g) = \overline{\alpha(s, g)}$. It is clear that $\bar\alpha$ is a cocycle and one can check as in [8, p. 16] that the equivalence class of $\bar\alpha$ is independent of the choice of conjugation.

For group representations, a useful relation between these various algebraic concepts is the Frobenius reciprocity theorem. We prove a version of this theorem in the context of induced cocycle representations of ergodic actions. Our theorem is modeled after the group theoretic version given by Mackey [13, Theorem 8.2].

DEFINITION 2.9. If α and β are cocycle representations of $Y \times G$, let $S(\alpha, \beta)$ be the set of intertwining fields $T = \int^\oplus T_y$ such that each T_y is a Hilbert–Schmidt operator. $S(\alpha, \beta)$ is a vector space, and $\dim S(\alpha, \beta) = j(\alpha, \beta)$ is called the strong intertwining number of α and β.

THEOREM 2.10. (Frobenius reciprocity). *Let $\phi: X \to Y$ be a factor G-map, α a cocycle representation of $X \times G$, and β a cocycle representation of $Y \times G$. Let ind (α) and res (β) be the induced and restricted cocycles. Then $j(\text{ind } \alpha, \beta) = j(\alpha, \text{res } \beta)$.*

We begin the proof with several lemmas.

LEMMA 2.11. *Suppose β is a $Y \times G$ cocycle representation on $\{H_y\}$ and I is the 1-dimensional identity cocycle. Then $j(\beta, I)$ is the dimension of the space of G-invariant elements in $\int_Y^\oplus H_y$.*

Proof. If $T = \int^\oplus T_y: \int^\oplus H_y \to \int_Y^\oplus \mathbf{C}$, then each $T_y: H_y \to \mathbf{C}$ and hence there is an element $v = \int^\oplus v_y \in \int^\oplus H_y$ such that for any $f = \int^\oplus f_y$, $(Tf)_y = \langle f_y \mid v_y \rangle$. It is straightforward to see that $T \in S(\beta, I)$ if and only if v is G-invariant. Thus $T \leftrightarrow v$ defines a vector space isomorphism (conjugate linear) between $S(\beta, I)$ and the G-invariant elements.

LEMMA 2.12.   *If α is an $X \times G$ cocycle, then $j(\alpha, I) = j(\text{ind }(\alpha), I)$.*

Proof. If α is a cocycle representation then $U^\alpha \cong U^{\text{ind }(\alpha)}$, so the dimensions of the spaces of G-invariant elements are equal. The result now follows by Lemma 2.11.

LEMMA 2.13.   *Suppose α and β are $S \times G$ cocycles. Then $j(\alpha, \beta) = j(\alpha \otimes \bar{\beta}, I)$.*

Proof. We recall first that for Hilbert spaces H_1 and H_2, there is an isomorphism of $(H_1 \otimes H_2)^*$ with $L_2(H_1; H_2)$ the Hilbert–Schmidt maps from H_1 to H_2, defined by

$$L_2(H_1, H_2) \to (H_1 \otimes H_2)^*, \quad T \to A,$$

where A is given by $A(v \otimes w) = \langle Tv \mid \bar{w} \rangle$.

To prove the lemma, we can suppose that α and β are cocycle representations on the product bundles $S \times H_1$ and $S \times H_2$ respectively. Via the above correspondence, there is a vector space isomorphism between bounded Borel fields of Hilbert–Schmidt operators $T_s: H_1 \to H_2$ and bounded Borel fields $A_s: H_1 \otimes H_2 \to \mathbf{C}$. To prove the lemma it suffices to show that

$$\{T_s\} \in S(\alpha, \beta) \Leftrightarrow \{A_s\} \in S(\alpha \otimes \bar{\beta}, I).$$

Now

$$A_s(\alpha \otimes \bar{\beta})(s, g)(v \otimes w) = A_s(\alpha(s, g)v \otimes \bar{\beta}(s, g)w)$$

$$= A_s(\alpha(s, g)v \otimes \overline{\beta(s, g)\bar{w}})$$

$$= \langle T_s\alpha(s, g)v \mid \beta(s, g)\bar{w} \rangle$$

$$= \langle \beta(s, g)^{-1}T_s\alpha(s, g)v \mid \bar{w} \rangle.$$

But $A_{sg}(v \otimes w) = \langle T_{sg}v \mid \bar{w} \rangle$. Thus

$$A_s(\alpha \otimes \bar{\beta})(s, g) = A_{sg} \quad \text{if and only if} \quad \beta(s, g)^{-1}T_s\alpha(s, g) = T_{sg}.$$

The result now follows.

LEMMA 2.14.   *Let $\phi: X \to Y$ a factor G-map, α an $X \times G$ cocycle representation on the Hilbert bundle $\{H_x\}$ and β a $Y \times G$ cocycle representation on $\{W_y\}$. Let ind (α) and res (β) be the induced and restricted cocycles. Then as $Y \times G$ cocycle representations, ind $(\alpha) \otimes \beta \cong$ ind $(\alpha \otimes$ res $(\beta))$.*

Proof. Let $\gamma_1 =$ ind $(\alpha) \otimes \beta$ and $\gamma_2 =$ ind $(\alpha \otimes$ res $(\beta))$. Let

$$V_y = \int_{x \in \phi^{-1}(y)} H_x \, d\mu_y(x).$$

Then γ_1 is a cocycle on the Hilbert bundle $y \to V_y \otimes W_y$ and

$$\gamma_1(y, g): V_{yg} \otimes W_{yg} \to V_y \otimes W_y$$

is defined by

$$\gamma_1(y, g) = \text{ind}\,(\alpha)(y, g) \otimes \beta(y, g) = \left[\int^{\oplus}_{x \in \phi^{-1}(y)} \alpha(x, g) \right] \otimes \beta(y, g),$$

(for each g, almost all y).

On the other hand, $\alpha \otimes \text{res}\,(\beta)$ is a cocycle representation on the bundle $x \mapsto H_x \otimes W_{\varphi(x)}$ and

$$\gamma_2(y, g): \int^{\oplus}_{x \in \phi^{-1}(yg)} [H_x \otimes W_{yg}]\, d\mu_{yg}(x) \to \int^{\oplus}_{x \in \phi^{-1}(y)} (H_x \otimes W_y)\, d\mu_y(x)$$

is given by (for each g, almost all y),

$$\gamma_2(y, g) = \int^{\oplus}_{x \in \phi^{-1}(y)} [\alpha(x, g) \otimes \beta(y, g)].$$

Now up to ismorphism, tensor products commute with direct integrals. Hence

$$\int^{\oplus}_{\phi^{-1}(y)} (H_x \otimes W_y)\, d\mu_y(x) \cong V_y \otimes W_y,$$

and under this isomorphism, γ_1 corresponds to γ_2.

We are now ready to prove the reciprocity theorem.

Proof of Theorem 2.10. We can suppose α and β are representations on product bundles. By Lemma 2.13, $j(\text{ind}\,(\alpha), \beta) = j(\text{ind}\,(\alpha) \otimes \bar\beta, I)$. By Lemma 2.14, this is $j(\text{ind}\,(\alpha \otimes \text{res}\,(\bar\beta)), I)$ and by Lemma 2.12, equals $j(\alpha \otimes \text{res}\,(\bar\beta), I)$. Lemma 2.13 now implies that this is $j(\alpha, \text{res}\,(\beta))$.

COROLLARY 2.15. *Suppose* $\alpha_i: S \times G \to U(H_i)$ *are equivalent*, $i = 1, 2$; $\dim H_i < \infty$. *Then* $\alpha_1 \otimes \bar\alpha_2$ *contains the identity as a subcocycle representation.*

Proof. This follows immediately from Lemma 2.13.

3. Cocycles into Compact Groups. We now turn to a consideration of some basic facts about cocycles into compact (second countable) groups. Many of the results in this section have natural interpretations in terms of Mackey's definitions of kernel and range for homomorphisms of virtual subgroups [16]. Some of these results have been indicated (without complete proof) by Mackey in [16].

We will find useful a slight weakening of the notion of a G-space.

DEFINITION 3.1. Let S be a Lebesgue space. By a *near-action* of G on S, we mean a Borel map $S \times G \to S$, $(s, g) \mapsto sg$ such that:

(i)   For each $g, h \in G$, $(sg)h = s(gh)$ for almost all s.
(ii) $s \cdot e = s$ for almost all s.
(iii) Each $g \in G$ preserves the measure on S.

Each near-action of G on S determines an essentially unique action of G on S. More precisely, we have the following result:

PROPOSITION 3.2. *If S is a near G-space, then there is a natural induced action of G on $B(S)$. Furthermore, there is a Borel action of G on S, which we denote $s \circ g$ such that:*
   (i) *For each g, $sg = s \circ g$ for almost all s.*
   (ii) *The induced actions on $B(S)$ are equal.*

Proof. The first statement is [19, Lemma 3.1]. The second statement follows from [19, Lemma 3.2] by defining $s \circ g = \psi^{-1}(\psi(s) \cdot g)$ where ψ is as in that lemma.

We shall assume for the remainder of this section that S is an ergodic G-space.

Given a cocycle $\alpha: S \times G \to K$, where K is compact, we define a near-action of G on $S \times K$ by $(s, x)g = (sg, x\alpha(s, g))$. If the cocycle identity holds for $(s, g, h) \in S \times G \times G$, then $(s, x)(gh) = [(s, x)g]h$, so that (i) and (ii) of Definition 3.1 hold, and it is straightforward to check that each $g \in G$ preserves the product measure on $S \times K$. Thus, for a cocycle α, there is, by Proposition 3.2, an essentially unique action of G defined on $S \times K$ agreeing almost everywhere with the near action. When necessary, we shall write $S \times_\alpha K$ to specify the action of G. K also acts on $S \times K$ by $(s, x) \cdot k = (s, k^{-1}x)$. This action commutes with the near-action of G, but not necessarily with the G-action it defines. However, for each $(g, k) \in G \times K$, we have, for almost all $y \in S \times K$, $(y \circ g)k = (yg)k = (yk)g = (yk) \circ g$. Thus we have a naturally induced action of $G \times K$ on $B(S \times K)$.

LEMMA 3.3. *If α and β are cohomologous, then $B(S \times_\alpha K)$ and $B(S \times_\beta K)$ are isomorphic Boolean $G \times K$ spaces.*

Proof. Let $\phi: S \to K$ such that for each $g \in G$,

$$\phi(s)\beta(s, g)\phi(sg)^{-1} = \alpha(s, g) \quad \text{for almost all } s \in S.$$

Define a map $\theta: S \times_\alpha K \to S \times_\beta K$ by $\theta(s, x) = (s, x \cdot \phi(s))$. Then θ is an isomorphism of Lebesgue spaces, and for each $(g, k) \in G \times K$,

$$\theta((s, x) \cdot (g, k)) = \theta(s, x) \cdot (g, k) \quad \text{for almost all } (s, x).$$

Hence $\theta^*: B(S \times_\beta K) \to B(S \times_\alpha K)$ is a Boolean $G \times K$-isomorphism.

COROLLARY 3.4. *If α and β are cohomologous, then $S \times_\alpha K$ and $S \times_\beta K$ are essentially isomorphic G-spaces.*

An important question which we now consider about cocycles into compact groups is to determine when they are equivalent to cocycles into proper closed subgroups. This question is related to the properties of the Boolean $G \times K$-spaces defined above, as well as to Mackey's definition of the kernel and range of α. The reader is again referred to [16] for an explanation of these concepts, and their relation to the $G \times K$-space $S \times_\alpha K$.

Consider the σ-subalgebra B of $B(S \times_\alpha K)$ consisting of elements left fixed by the action of $G \times \{e\}$. Since the actions of $G \times \{e\}$ and $\{e\} \times K$ commute, B is a Boolean K-space. Since the action of $G \times K$ on $B(S \times K)$ is irreducible, the action of K on B must also be irreducible. Hence there is an ergodic K-space Y such that $B(Y) \cong B$ as Boolean K-spaces [14]. Since K is compact, we can choose $Y = K/H_0$ for some closed subgroup $H_0 \subset K$. The following result was indicated without proof by Mackey [16], and motivates the definition of the K-space Y as the "range-closure" of α.

THEOREM 3.5. α is cohomologous to a cocycle into a subgroup $H \subset K$ if and only if K/H is a factor of the K-space Y.

Proof. (i) Suppose $\alpha \sim \beta$ and $\beta(s, g) \in H$ for every (s, g). It follows from Lemma 3.3 that the Boolean K-space B is K-isomorphic to the Boolean K-space B' of elements of $B(S \times_\beta K)$ that are left fixed by $G \times \{e\}$. Now let K/H be the space of left cosets, and $p: S \times K \to K/H$ be $p(s, k) = [k]$. Then p commutes with the K-actions and hence $p^*: B(K/H) \to B(S \times_\beta K)$ is an injective map of Boolean K-spaces. Thus it suffices to see that $p^*(B(K/H)) \subset B'$. Now

$$ p((s, k)g) = p(sg, k\beta(s, g)) = [k\beta(s, g)] = [k] $$

since $\beta(s, g) \in H$. Therefore $p((s, k)g) = p(s, k)$. It follows readily that for $A \in B(K/H)$, $p^*(A) \cdot g \subset p^*(A)$ for each $g \in G$. Since this also holds for g^{-1}, we have $p^*(A) \cdot g = p^*(A)$, i.e., $p^*(B(K/H)) \subset B'$.

(ii) Now suppose that K/H is a factor of $Y = K/H_0$. Then H_0 is contained in a conjugate of H, so it suffices to see that α is cohomologous to a cocycle into H_0. Let $q: B(Y) \to B \subset B(S \times K)$ be an isomorphism of Boolean K-spaces. Then there exists a K-invariant Borel null set $N \subset S \times K$ and a K-map $\lambda: (S \times K) - N \to Y$ such that $\lambda^* = q$. As a K-invariant null set, N must clearly be of the form $A \times K$ where $A \subset S$ is Borel and null. For each $g \in G$, we must have

$$ \lambda((s, k) \cdot g) = \lambda(s, k) \quad \text{for almost all } (s, k) \in [(S - A) \cap (S - A) \cdot g^{-1}] \times K. $$

(This last condition so that the equation is well defined.) This follows since $\lambda^* = q$ and g leaves elements of B fixed. Thus

$$ D = \{(s, g, k) \mid s \in (S - A) \cap (S - A) \cdot g^{-1}; \ \lambda((s, k) \cdot g) = \lambda(s, k)\} $$

is a Borel conull set by Fubini's theorem. Thus, there exists $k_0 \in K$ such that

$$ D_0 = \{(s, g) \mid (s, g, k_0) \in D\} $$

is conull. We phrase this condition in a form we will need: If $(s, g) \in D_0$, then

$$ \lambda(s, k_0) = \lambda(sg, k_0\alpha(s, g)). $$

Since λ is a K-map, we can write this as

$$\lambda(s, e)k_0^{-1} = \lambda(sg, e)\alpha(s, g)^{-1}k_0^{-1}.$$

Equivalently for all $(s, g) \in D_0$,

(*) $\lambda(s, e)\alpha(s, g) = \lambda(sg, e).$

To construct a cocycle equivalent to α with values in H, we first construct a cocycle β such that $\beta(s, g) \in H$ for almost all (s, g). Choose a point $y_0 \in Y$ and let

$$H_0 = \{k \in K \mid y_0 \cdot k = y_0\}.$$

By [13, Lemma 1.1], there exists a Borel map $\theta : Y \to K$ such that $y_0 \cdot \theta(y) = y$ for all $y \in Y$. Define $u : S \to K$ by $u(s) = \theta(\lambda(s, e))$. Now define

$$\beta(s, g) = u(s)\alpha(s, g)u(sg)^{-1}.$$

We claim that for $(s, g) \in D_0$, $\beta(s, g) \in H_0$, i.e., $y_0 \cdot u(s)\alpha(s, g)u(sg)^{-1} = y_0$. We have

$$y_0 \cdot u(s)\alpha(s, g) = y_0 \cdot \theta(\lambda(s, e))\alpha(s, g) = \lambda(s, e)\alpha(s, g).$$

Similarly,

$$y_0 \cdot u(sg) = y_0 \cdot \theta(\lambda(sg, e)) = \lambda(sg, e).$$

Thus for $(s, g) \in D_0$, $y_0 \cdot \beta(y, g) = y_0$ follows from (*). To complete the proof of the theorem, it suffices to prove the following lemma:

LEMMA 3.6.   *Suppose $\beta : S \times G \to K$ is a cocycle, and that $\beta(s, g) \in H$ for almost all (s, g), where $H \subset K$ is a closed subgroup. Then β is cohomologous to a cocycle β_0 such that $\beta_0(s, g) \in H$ for all (s, g).*

Proof. By changing β on a Borel null set, we can obtain a function $\beta' : S \times G \to H$, and using Fubini's theorem, we can see that

$$\beta'(s, g)\beta'(sg, h) = \beta'(s, gh)   \text{ for almost all } (s, g, h) \in S \times G \times G.$$

By [19, Theorem 5.1], we can change β' on a null set to obtain a Borel function $\beta_0 : S \times G \to H$ such that there is a conull set $S_0 \subset S$ with the following property: if $s \in S_0$, $sg \in S_0$, $sgh \in S_0$, then $\beta_0(s, g)\beta_0(sg, h) = \beta_0(s, gh)$. For each $(g, h) \in G \times G$, $S_0 \cap S_0 g^{-1} \cap S_0(gh)^{-1}$ is conull and hence β_0 is a cocycle that differs from β on a null set. We claim that for each $g \in G$, $\beta_0(s, g) = \beta(s, g)$ for almost all $s \in S$. This suffices to prove the lemma. Let

$$G_0 = \{g \in G \mid \beta_0(s, g) = \beta(s, g) \text{ for almost all } s \in S\}.$$

By Fubini's theorem, G_0 is conull. For $g, h \in G_0$, let

$$S_1 = \{s \mid \beta_0(s, g) = \beta(s, g)\}, \quad S_2 = \{s \mid \beta_0(s, h) = \beta(s, h)\},$$

$$S_3 = \{s \mid \beta_0(s, g)\beta_0(sg, h) = \beta_0(s, gh)\}$$

and

$$S_4 = \{s \mid \beta(s, g)\beta(sg, h) = \beta(s, gh)\}.$$

Then if $s \in S_1 \cap S_2 g^{-1} \cap S_3 \cap S_4$, we have

$$\beta_0(s, gh) = \beta_0(s, g)\beta_0(sg, h) = \beta(s, g)\beta(sg, h) = \beta(s, gh).$$

Thus, $gh \in G_0$; therefore G_0 is closed under multiplication and conull, and hence must equal G.

DEFINITION 3.7. If $\alpha: S \times G \to K$ is a cocycle, let K_α be the closed subgroup of K generated by $\{\alpha(s, g)\}$. We will call α a minimal cocycle if there is no cocycle β cohomologous to α such that $K_\beta \subset_{\neq} K_\alpha$.

As a consequence of Theorem 3.5, we have:

COROLLARY 3.8. (i) *Any cocycle (into a compact group) is equivalent to a minimal cocycle.*
(ii) *α is minimal and $K_\alpha = K$ if and only if the action of G on $S \times_\alpha K$ is ergodic.*
(iii) *If α and β are equivalent minimal cocycles, then K_α and K_β are conjugate subgroups of K.*

Proof. From the theorem we see that a subgroup H will be of the form K_α for a minimal α if and only if H is a subgroup conjugate to H_0, so (i) and (iii) are clear. Part (ii) follows since the action of G on $B(S \times_\alpha K)$ is irreducible if and only if $H_0 = K$.

We will need an auxiliary result which asserts that minimality is independent of embedding.

THEOREM 3.9. *Let $\alpha: S \times G \to K$ be a minimal cocycle with $K_\alpha = K$. Suppose there is a compact group K' with $K \subset K'$. Then as a cocycle into K', α is still minimal.*

Proof. Let K'/K be the space of left cosets, and $p: K' \to K'/K$ the projection. Let $t: K'/K \to K'$ be a section of p and define

$$\phi: K' \to K \times K'/K$$

by $\phi(x) = (t(p(x))^{-1}x, p(x))$. Then ϕ is a measure-preserving Borel isomorphism, and the near action of G on $S \times_\alpha K'$ is carried over to the near action of G on $S \times K \times K'/K$ given by $(s, k, [k']) \cdot g = (sg, k\alpha(s, g), [k'])$. Since the action of G on $S \times_\alpha K$ is ergodic, any G-invariant element of the Boolean algebra $B(S \times K \times K'/K)$ must be of the form $S \times K \times A$ where $A \subset K'/K$. Therefore, the σ-algebra of G-invariant elements of $B(S \times K \times K'/K)$ is K'-isomorphic to the Boolean K'-space $B(K'/K)$. It follows from Theorem 3.5 that $\alpha: S \times G \to K'$ is minimal with $K'_\alpha = K$.

Remark. In terms of virtual subgroups, $\alpha: S \times G \to K$ is minimal with $K_\alpha = K$ if and only if "the range of α is dense in K".

We now consider the cocycle representations obtained by composing the representations of a compact group with a minimal cocycle into the group. The result we are aiming for is Theorem 3.14 which asserts that if $\alpha: S \times G \to K$ is minimal, with $K_\alpha = K$, then $\pi \to \pi \circ \alpha$ sets up a one-to-one correspondence between $\hat{K}$ (the dual object of K) and a collection of equivalence classes of irreducible cocycle representations of (S, G). In light of the above remark, this result is highly plausible from the virtual subgroup viewpoint.

PROPOSITION 3.10. *Suppose* $\alpha: S \times G \to K$ *is minimal, with* $K_\alpha = K$. *If* $\rho: K \to H$ *is a surjective homomorphism, then* $\rho \circ \alpha: S \times G \to H$ *is a minimal cocycle with* $H_{\rho \circ \alpha} = H$.

Proof. The map $(1, \rho): S \times K \to S \times H$ is a surjective measure-preserving map. Since $S \times_\alpha K$ is an ergodic G-space and

$$(1, \rho)^*: B(S \times_{\rho \circ \alpha} H) \to B(S \times_\alpha K)$$

is a G-map, $S \times_{\rho \circ \alpha} H$ is ergodic, which implies by Corollary 3.8 that $\rho \circ \alpha$ is minimal, and $H_{\rho \circ \alpha} = H$.

LEMMA 3.11. *Let* $\beta: S \times G \to U(n)$ *be a minimal cocycle. If* β *is reducible, then* $U(n)_\beta$ *is a reducible group of matrices.*

Proof. If β is reducible, then β is equivalent to a cocycle into a subgroup of $U(n)$ of the form $U(p) \times U(n - p)$ for some $1 \leq p < n$. Thus there exists a minimal cocycle α equivalent to β such that $U(n)_\alpha$ is a reducible group. Since α and β are both minimal, Corollary 3.8 implies that $U(n)_\alpha$ and $U(n)_\beta$ are conjugate subgroups of $U(n)$. Since $U(n)_\alpha$ is reducible, so is $U(n)_\beta$.

PROPOSITION 3.12. *Let* $\alpha: S \times G \to K$ *be minimal,* $K_\alpha = K$, *and let* π *be a finite dimensional representation (unitary) of* K. *Then* π *is irreducible if and only if* $\pi \circ \alpha$ *is irreducible.*

Proof. (i) If π is reducible, clearly $\pi \circ \alpha$ is also.
(ii) Let $\beta = \pi \circ \alpha$. By Proposition 3.10 and Theorem 3.9, β is minimal with $U(n)_\beta = \pi(K)$. (Here $n = \dim \pi$). If β is reducible, then Lemma 3.11 implies $\pi(k)$ is reducible.

LEMMA 3.13. *Let* $\alpha: S \times G \to K$ *minimal,* $K_\alpha = K$. *Suppose* π *is a finite dimensional unitary representation of* K *such that* $\pi \circ \alpha$ *contains the one-dimensional identity cocycle representation. Then* π *contains the identity representation.*

Proof. Let $\pi = \sum^\oplus \pi_i$, where π_i are irreducible. Then $\pi \circ \alpha = \sum^\oplus \pi_i \circ \alpha$, and each $\pi_i \circ \alpha$ is an irreducible cocycle by Proposition 3.12. Now consider a G-invariant field of one-dimensional subspaces in $L^2(S, H(\pi))$ under the representation $U^{\pi \circ \alpha}$, say $V = \int^\oplus V_s \, ds$. The projection of V onto $L^2(S, H(\pi_i))$ must be nonzero for some π_i, and this projection will be an intertwining field for $\pi \circ \alpha \mid V$ and $\pi_i \circ \alpha$. Since $\pi \circ \alpha \mid V$ is one dimensional and hence irreducible, and $\pi_i \circ \alpha$ is irreducible, we have $\pi \circ \alpha \mid V$ is equivalent to $\pi_i \circ \alpha$. So π_i is a character χ of K, and $\chi \circ \alpha$ is equivalent to the identity. We claim that this implies $\chi = 1$. Now $\chi \circ \alpha$ equivalent to 1 means there exists a function $u: S \to \mathbf{C}$ such that:

(i) $|u(s)| = 1$
(ii) For each g, $u(s)\chi(\alpha(s, g))u(sg)^{-1} = 1$ for almost all s.

Equation (ii) can be rewritten

$$\frac{\chi(\alpha(s, g))}{u(sg)} = \frac{1}{u(s)}.$$

Now let $f: S \times K \to \mathbf{C}$ be defined by $f(s, k) = \chi(k)/u(s)$. For each $g \in G$, and almost all (s, k), we have

$$f((s, k)g) = f(sg, k\alpha(s, g)) = [\chi(k)\chi(\alpha(s, g))/u(sg)] = \chi(k)/u(s) = f(s, k).$$

Since α is minimal and $K_\alpha = K$, $S \times_\alpha K$ is ergodic, so this implies f is constant on a conull set, which shows $\chi = 1$.

THEOREM 3.14. *Let $\alpha: S \times G \to K$ be a minimal cocycle, $K_\alpha = K$. Let π_1, π_2 be irreducible unitary representations of K. Then $\pi_1 \circ \alpha$ and $\pi_2 \circ \alpha$ are irreducible; they are equivalent if and only if π_1 and π_2 are unitarily equivalent representations.*

Proof. That $\pi_i \circ \alpha$ is irreducible is just Proposition 3.12. If π_1 and π_2 are equivalent, it is clear that $\pi_1 \circ \alpha$ and $\pi_2 \circ \alpha$ are also. If $\pi_1 \circ \alpha$ and $\pi_2 \circ \alpha$ are equivalent, by Corollary 2.15, $(\pi_1 \circ \alpha) \otimes \overline{(\pi_2 \circ \alpha)}$ contains the identity cocycle. But $(\pi_1 \circ \alpha) \otimes \overline{(\pi_2 \circ \alpha)} = (\pi_1 \otimes \bar{\pi}_2) \circ \alpha$. By Lemma 3.13, $\pi_1 \otimes \bar{\pi}_2$ contains the identity representation. Since π_1 and π_2 are irreducible, they must be unitarily equivalent.

II. Extensions with relatively discrete spectrum

4. The Structure Theorem and a generalization. Given a cocycle representation $\alpha(s, g)$ one can try to decompose α into irreducible components. In case α is equivalent to $\sum^{\oplus} \alpha_i$, where α_i are finite dimensional irreducible cocycle representations, we say that α has discrete spectrum. Equivalently, the Hilbert bundle on which α is defined is a direct sum of finite dimensional G-invariant Hilbert subbundles. If $\phi: X \to Y$ is a factor map of ergodic G-spaces, we have a natural $Y \times G$ cocycle representation on the Hilbert bundle $\int^{\oplus} L^2(F_y)$ (Example 2.3). One question that presents itself is what the "spectral" structure of the cocycle α implies about the geometric structure of the triple (X, ϕ, Y). In the case where Y is a point, α just becomes the natural unitary representation of G on $L^2(X)$. The main result of Mackey's paper [15] is a description of the geometric consequences of the representation α of G having discrete spectrum. This generalized a classic result of Halmos and von Neumann in the special case when $G = Z$, the group of integers [6]. The main result of this section is to generalize Mackey's theorem to the case where Y is not necessarily one point.

In virtual group terms, what Mackey does is to describe those virtual subgroups of G for which the representation of G induced by the identity has discrete spectrum. We describe here, for a virtual-subgroup (Y, G), those subvirtual subgroups for which the representation of (Y, G) induced by the identity has discrete spectrum.

Before stating the theorem, we produce a class of examples where the natural $Y \times G$ cocycle will have discrete spectrum.

Example 4.1.   Let K be a compact group and $H \subset K$ a closed subgroup. Suppose $\alpha \colon Y \times G \to K$ is a minimal cocycle, with $K_\alpha = K$ (and Y is ergodic). Define a near-action of G on $Y \times K/H$ by

$$(y, [k]) \cdot g = (yg, [k]\alpha(y, g)).$$

By Proposition 3.2, $X = Y \times K/H$ becomes a G-space, and Y is a factor of X. Since α is minimal, $K_\alpha = K$, and X is a factor of $Y \times_\alpha K$, it follows from Corollary 3.8 that X is ergodic. We claim that the natural (Y, G) cocycle, β, has discrete spectrum. Let σ be the natural representation of K on $L^2(K/H)$ given by right translation. Then $U^\beta = U^{\sigma \circ \alpha}$ since both are just the natural representation of G on $L^2(X)$. Theorem 2.6 shows that β and $\sigma \circ \alpha$ are equivalent. Now $\sigma = \sum^\oplus \sigma_i$ where σ_i are finite dimensional and irreducible, so $\sigma \circ \alpha = \sum^\oplus \sigma_i \circ \alpha$ has discrete spectrum, and hence so does β.

DEFINITION 4.2.   Suppose X and Z are G-extensions of a Lebesgue G-space Y, with $q_1 \colon B(Y) \to B(X)$, $q_2 \colon B(Y) \to B(Z)$ the designated embeddings. We say that X and Z are essentially isomorphic extensions of Y if there exists a Boolean G-isomorphism $p \colon B(X) \to B(Z)$ such that $p \cdot q_1 = q_2$.

Equivalently, there exist essential sets $X' \subset X$, $Z' \subset Z$, a G-isomorphism $f \colon X' \to Z'$, and factor G-maps $\phi_1 \colon X' \to Y$, $\phi_2 \colon Z' \to Y$ such that:

(i) $\phi_2 \circ f = \phi_1$,
(ii) $\phi_i^* = q_i$, $i = 1, 2$.

Remark.   To see the equivalence, choose $X_0 \subset X$ and $Z_0 \subset Z$ essential, and factor G-maps $\lambda_1 \colon X_0 \to Y$, $\lambda_2 \colon Z_0 \to Y$ with $\lambda_i^* = q_i$. This can be done by Proposition 2.1. Now $p \colon B(X_0) \to B(Z_0)$ is an isomorphism, so by [14, Theorem 2], there exist essential $X_1 \subset X_0$, $Z_1 \subset Z_0$ and a Borel isomorphism $f_1 \colon Z_1 \to X_1$ such that $f_1^* = p$. Now $\lambda_2^* = f_1^* \lambda_1^*$, so $\lambda_2 = \lambda_1 f_1$ almost everywhere on Z_1. Since λ_2 and $\lambda_1 f_1$ are both G-maps, they agree on an essential set $Z' \subset Z_1$. Then $X' = f_1(Z')$ is essential, since f_1 is an isomorphism, and we can take $f = (f_1 \mid Z')^{-1}$, $\phi_1 = \lambda_1 \mid X'$, $\phi_2 = \lambda_2 \mid Z'$.

The theorem alluded to above is the following:

THEOREM 4.3 (Structure Theorem).   *If $\phi \colon X \to Y$ is a factor map of ergodic G-spaces and the natural (Y, G) cocycle representation on $\{L^2(F_y)\}$ has discrete spectrum, then there is a compact group K, a closed subgroup $H \subset K$, and a* *minimal cocycle* $\alpha: Y \times G \to K$ *with* $K_\alpha = K$ *such that* X *is essentially isomorphic as an extension of* Y *to the* G-*space* $Y \times K/H$ *of Example* 4.1.

In the case when $Y = \{e\}$, this is exactly Mackey's theorem [15, Theorem 1].

Proof (of Theorem 4.3*).*   We begin the proof by summarizing some facts we will need about the Effros Borel structure. Given a separable Hilbert space H_0, let E be the set of von Neumann algebras on H_0. Effros has shown the following [4, p. 1161]: There exists a standard Borel structure on E such that if S is any standard Borel space, a map $\mathscr{A}: S \to E$ is Borel if and only if there exist countably many Borel functions $A_i: S \to L(H)$ such that for each s, $\{A_i(s)\}$ generate $\mathscr{A}(s)$ as a von Neumann algebra. The unitary group $U(H_0)$ is a standard Borel group with the weak Borel structure [2, Lemma 4] and acts on E by $\mathscr{A} \cdot U = U^{-1} \mathscr{A} U$. In [5, Lemma 2.1], Effros shows that this is a Borel action, i.e., $(\mathscr{A}, U) \mapsto \mathscr{A} \cdot U$ is Borel.

LEMMA 4.4.   *Let* $A = \{\mathscr{A} \in E \mid \mathscr{A}$ *is abelian*$\}$. *Then* A *is a* $U(H_0)$-*invariant Borel set in* E.

Proof.   By [4, Theorem 3], $\mathscr{A} \to \mathscr{A}'$ is Borel and from [4, Corollary 2], $(\mathscr{A}, B) \to \mathscr{A} \cap B$ is Borel. Since $A = \{\mathscr{A} \in E \mid \mathscr{A} \cap \mathscr{A}' = \mathscr{A}\}$, A is Borel. $U(H_0)$-invariance is trivial.

If $\mathscr{A}$ is an abelian von Neumann algebra, let $B(\mathscr{A})$ be the set of projection operators in $\mathscr{A}$. Then it is well known that $\mathscr{A} \to B(\mathscr{A})$ sets up a bijection between abelian von Neumann algebras and Boolean σ-algebras of projections on H_0. Thus we will identify A above with the set $\mathscr{M}$ of such Boolean algebras. Further, for the Borel structure on $\mathscr{M}$ defined by the Effros Borel structure on A, $\mathscr{M}$ is standard, and Borel maps into $\mathscr{M}$ can be identified by the following.

LEMMA 4.5.   *If* $P: Y \to \mathscr{M}$, *with* Y *standard, then* P *is Borel if and only if there exist countably many Borel fields,* $P_i(y)$, *of projections on* H_0, *such that for each* $y \in Y$, $\{P_i(y)\}$ *generates* $P(y)$ *as a Boolean* σ-*algebra.*

Proof.   Since any abelian von Neumann algebra is generated by its projections, the "only if" statement is clear. Conversely, let us suppose we are given bounded Borel fields $\{A_i(y)\}$ of operators on H_0, such that for each y, $\{A_i(y)\}$ generates a von Neumann algebra $D(y)$, where $D: Y \to E$ is Borel. We can assume $A_i(y)$ is self-adjoint for each i and y.

If A is a self-adjoint operator on H_0, the operators $\chi_{[a,b]}(A)$ are projection operators, where $\chi_{[a,b]}$ is the characteristic function of the interval $[a, b]$ of real numbers. It follows from the spectral theorem (see [7, C 40, C 41] for example) that the von Neumann algebra generated by A is equal to the von Neumann algebra generated by $\{\chi_{[a,b]}(A) \mid a, b \text{ rational}\}$. Thus, given a sequence of operators A_i, the von Neumann algebra $W^*(A_1, A_2, \ldots)$ generated by $\{A_1, A_2, \ldots\}$ is equal to the von Neumann algebra generated by

$$Q = \{\chi_{[a,b]}(A_i) \mid a, b \text{ rational, all } i\}.$$

Hence, the Boolean σ-algebra generated by Q is equal to the Boolean σ-algebra associated to $W^*(A_1, A_2, \ldots)$.

In light of these remarks, to complete the proof it suffices to show that if $A(y)$ is a bounded Borel field of self-adjoint operators on H_0, so is $\chi_{[a, b]}(A(y))$. Choose M so that $\|A(y)\| \leq M$ for all y. Then there are polynomials $p_n(x)$ such that $\lim_{n \to \infty} p_n(x) = \chi_{[a, b]}(x)$ in bounded pointwise convergence on $[-2M, 2M]$, and hence

$$\lim_{n \to \infty} p_n(A(y)) = \chi_{[a, b]}(A(y))$$

weakly for each i and y. Since $p_n(A(y))$ is Borel in y, so is $\chi_{[a, b]}(A(y))$.

Now suppose Y is an ergodic G-space and α a cocycle representation on a Hilbert bundle $\{H_y\}$ with discrete spectrum. Thus, there exist G-invariant fields V_i of finite dimensional subspaces such that $\int^\oplus H_y = H = \sum^\oplus V_i$. Using the comments preceding example 2.3, it is not hard to see the following: There exists a conull set $Y_0 \subset Y$, Y_0 Borel, Hilbert spaces H_i with dim $H_i < \infty$, and a Borel field of maps $U(y) : H_y \to H_0 = \sum^\oplus H_i$ such that:

(i) $U(y)$ is unitary for every $y \in Y_0$.
(ii) $U(y)(V_i(y)) = H_i$ for every $y \in Y_0$.

Let us consider further what happens when α is in addition the natural cocycle corresponding to a factor map $\phi : X \to Y$. For each $y \in Y$, we have a Boolean σ-algebra of projections B_y on $L^2(F_y)$, namely the set of multiplication operators corresponding to the elements of $B(F_y)$. Let $\{E_i\}$ be a countable generating sequence of sets for the Borel structure on X and let $P_i(y)$ be multiplication by $\chi_{E_i \cap F_y}$ in $L^2(F_y)$. Then for each i, $\{P_i(y)\}_y$ is a Borel field of operators on $\int^\oplus L^2(F_y)$, and for each y, $\{P_i(y)\}_i$ generates B_y as a Boolean σ-algebra. Now let Y_0 and $\{U(y)\}$ be as above. For $y \in Y_0$, let $A_y = U(y)B_y \cdot U(y)^{-1}$. Then A_y is a Boolean σ-algebra on H_0, so we have a map

$$A : Y_0 \to \mathcal{M}, \quad y \mapsto A_y.$$

If $y \in Y_0$, $\{U(y)P_i(y)U(y)^{-1}\}_i$ generates A_y as a Boolean σ-algebra, and hence A is a Borel function. We can extend A to a Borel function $Y \to \mathcal{M}$, which we continue to denote with the same letter.

If α is the natural cocycle, for each (y, g) such that $y \in Y_0 \cap Y_0 g^{-1}$, let

$$\tilde{\alpha}(y, g) = U(y)\alpha(y, g)U(yg)^{-1}.$$

Then $\tilde{\alpha}$ is Borel and for each g and almost all y, $\tilde{\alpha}(y, g)H_i = H_i$ (see Proposition 2.8). By changing $\tilde{\alpha}$ on a set of measure 0, and extending it to all of $Y \times G$, we can obtain a Borel cocycle $\beta : Y \times G \to U(H_0)$ such that:

(i) $\beta(y, g)(H_i) = H_i$ for every y, g.
(ii) For each g, $\beta(y, g) = U(y)\alpha(y, g)U(yg)^{-1}$ for almost all y.

Now for each g and almost all y, we have $\alpha(y, g)^{-1}B_y\alpha(y, g) = B_{yg}$. Thus, for all g, the following hold for almost all y:

$$\begin{aligned}
\beta(y, g)^{-1}A_y\beta(y, g) &= U(yg)\alpha(y, g)^{-1}U(y)^{-1}A_yU(y)\alpha(y, g)U(yg)^{-1} \\
&= U(yg)\alpha(y, g)^{-1}B_y\alpha(y, g)U(yg)^{-1} \\
&= U(yg)B_{yg}U(yg)^{-1}.
\end{aligned}$$

In other words,

(*) $$\beta(y, g)^{-1}A_y\beta(y, g) = A_{yg}.$$

Now let $K = \{U \in U(H_0) \mid U(H_i) = H_i\}$. Then K is a compact subgroup of $U(H_0)$ (weak topology). Condition (i) on β above just says that $\beta(y, g) \in K$, and (*) says that $A_y \cdot \beta(y, g) = A_{yg}$, (where the expression on the left is the action of $U(H_0)$ on $\mathcal{M}$), the equation holding for almost all y, given any g. By restricting the action of $U(H_0)$, we obtain an action of K on $\mathcal{M}$, and we let $\hat{\mathcal{M}}$ be the space of orbits in $\mathcal{M}$ under K. Since K is compact, $\hat{\mathcal{M}}$ is standard under the quotient Borel structure. Let $p: \mathcal{M} \to \hat{\mathcal{M}}$ be the natural projection and $\lambda: Y \to \hat{\mathcal{M}}$, $\lambda = p \circ A$. Now equation (*) implies that for each g, $p(A_y) = p(A_{yg})$ for almost all y, i.e., $\lambda(y) = \lambda(yg)$ for almost all y. Since $\hat{\mathcal{M}}$ is Borel isomorphic to a subset of $[0, 1]$, and Y is an ergodic G-space, we conclude that λ is constant almost everywhere; i.e., there exists $A_0 \in \mathcal{M}$ such that $p(A_0) = p(A_y)$ for almost all $y \in Y$. Equivalently, if we let $\mathcal{M}_0$ be the orbit of A_0 in $\mathcal{M}$ under K, then $A_y \in \mathcal{M}_0$ for y in a conull Borel set. By changing A on a null set, we obtain a function (Borel) $\tilde{A}: Y \to \mathcal{M}_0$ such that $\tilde{A}_y = A_y$ almost everywhere.

Let $K_0 = \{U \in K \mid A_0 \cdot U = A_0\}$. Then $\mathcal{M}_0$ is Borel isomorphic to K/K_0, and we can choose a Borel section $\theta: \mathcal{M}_0 \to K$. So for each $M \in \mathcal{M}_0$, we have a unitary operator $\theta(M) \in K$ such that $M = A_0 \cdot \theta(M)$, i.e., $M = \theta(M)^{-1} \cdot A_0\theta(M)$. We now use the function $\theta \circ \tilde{A}: Y \to K$ to define a new cocycle, namely

$$\beta_0(y, g) = \theta(\tilde{A}_y)\beta(y, g)\theta(\tilde{A}_{yg})^{-1}.$$

We now claim that for each g, $\beta_0(y, g) \in K_0$ for almost all y. To see $\beta_0(y, g) \in K_0$, it suffices to see $\beta_0(y, g)^{-1}A_0\beta_0(y, g) = A_0$. But the left side is just, for almost all y,

$$\theta(A_{yg})\beta(y, g)^{-1}\theta(A_y)^{-1}A_0\theta(A_y)\beta(y, g)\theta(A_{yg})^{-1}$$

$$\begin{aligned}
&= \theta(A_{yg})\beta(y, g)^{-1}A_y\beta(y, g)\theta(A_{yg})^{-1} \\
&= \theta(A_{yg})A_{yg}\theta(A_{yg})^{-1} \\
&= A_0.
\end{aligned}$$

Thus by changing β_0 on a set of measure 0, we can assume it is a cocycle with values in K_0. As K_0 is compact, β_0 has an equivalent minimal cocycle γ (Corollary 3.8). So there exists a Borel map $S: Y \to K_0$ such that for each g,

$$\gamma(y, g) = S(y)\beta_0(y, g)S(yg)^{-1} \quad \text{for almost all } y.$$

398   ROBERT J. ZIMMER

We now summarize what we have done so far by collecting some relevant facts about γ:

(1) $\gamma: Y \times G \to K_0$ is a minimal cocycle; K_0 is a compact group in $U(H_0)$, and leaves invariant a Boolean σ-algebra of projections A_0 on H_0.

(2) γ is cohomologous to the natural cocycle representation α; more exactly, if $T(y) = S(y)\theta(\tilde{A}_y)U(y)$, then $T(y): H(y) \to H_0$ is a Borel field of operators such that
  (i) $T(y)$ is unitary for almost all y.
  (ii) For each g, $T(y)\alpha(y, g)T(yg)^{-1} = \gamma(y, g)$ for almost all y.
  (3)   For almost all y, $T(y)B_yT(y)^{-1} = A_0$.

We now show the way in which these properties imply the structure theorem. By (1), A_0 is a K_y-space and thus [14, Theorem 1], there exists a K_y-Lebesgue space Z and an isomorphism of Boolean K_y-spaces $p: B(Z) \to A_0$. Form the product space $Y \times Z$; this has a G-action defined by γ, namely $(y, z)g = (yg, z \cdot \gamma(y, g))$. We claim that this action is essentially isomorphic to the action of G on X. Now $B(X) \cong \int^{\oplus} B_y \, dy$ and the G-action is given as follows: If $S \in B(X)$ and $S \cong \int^{\oplus} S_y \, dy$, $S_y \in B_y$, then

$$(S \cdot g)_{yg} = \alpha(y, g)^{-1}S_y\alpha(y, g).$$

Similarly, if $E \in B(Y \times Z)$, and $E = \int^{\oplus} E_y \, dy$, then $(E \cdot g)_{yg} = E_y \cdot g = E_y \cdot \gamma(y, g)$. Since $P: B(Z) \to A_0$ is a K_y-isomorphism, we have

$$P((E \cdot g)_{yg}) = P(E_y) \cdot \gamma(y, g) = \gamma(y, g)^{-1}P(E_y)\gamma(y, g).$$

Define a Boolean isomorphism $\tilde{P}: B(Y \times Z) \to \int_Y^{\oplus} A_0$ by $\tilde{P}(\int^{\oplus} E_y) = \int^{\oplus} P(E_y)$. Then the G-action on $B(Y \times Z)$ is carried over to the G-action on $\int^{\oplus} A_0 \, dy$ given by $F = \int^{\oplus} F_y$, then $(F \cdot g)_{yg} = \gamma(y, g)^{-1}F_y\gamma(y, g)$. We thus have Boolean isomorphisms

$$B(X) \cong \int_Y^{\oplus} B_y \xrightarrow{Q} \int_Y^{\oplus} A_0 \xleftarrow{\tilde{P}} B(Y \times Z)$$

where Q is defined by $Q(\int^{\oplus} S_y) = \int^{\oplus} T(y)S_yT(y)^{-1}$, and is an isomorphism by property (3).

The outside 2 maps defined G-space structures on the inside 2 algebras as shown above. To show that $B(X)$ and $B(Y \times Z)$ are isomorphic, it suffices to show that Q is a G-map. Now

$$Q\left(\left(\int^{\oplus} S_y\right) \cdot g\right)_{yg} = T(yg)(S \cdot g)_{yg}T(yg)^{-1}$$

$$= T(yg)\alpha(y, g)^{-1}S_y\alpha(y, g)T(yg)^{-1}$$

$$= \gamma(y, g)^{-1}T(y)S_yT(y)^{-1}\gamma(y, g) \qquad \text{(by (2)(ii) above)}$$

$$= \left(\left[Q\left(\int^{\oplus} S_y\right)\right] \cdot g\right)_{yg},$$

all these holding for almost all y.

It is clear that under the isomorphism $B(X) \leftrightarrow B(Y \times Z)$, $B(Y)$ is mapped to itself by the identity. Thus, X and $Y \times Z$ are essentially isomorphic extensions of Y. To prove the structure theorem, it remains only to clarify the structure of the K_y-space Z. Since $B(X)$ is an irreducible Boolean G-space, so is $B(Y \times Z)$. If $F \in B(Z)$ is K_y-invariant then clearly $Y \times F$ is G-invariant in $B(Y \times Z)$, and hence F must be θ or Z; i.e., $B(Z)$ is an irreducible K_y-space. Since K_y is compact, we can choose Z to be K_y/H for some closed subgroup H of K_y [15, Lemma 2], and this completes the proof.

If K is a compact abelian group, and $\alpha: S \times G \to K$ is minimal, $K_\alpha = K$, then the natural (Y, G) cocycle of the extension $Y \times_\alpha K$ of Y has discrete spectrum, and each summand of the direct sum of irreducible cocycles is one-dimensional. As a corollary of the structure theorem, we have the following.

COROLLARY 4.6. *Suppose* $\phi: X \to Y$ *is a factor G-map and the natural (Y, G) cocycle has discrete spectrum with each summand one-dimensional. Then there exists a compact abelian group K and a minimal cocycle $\alpha: Y \times G \to K$, $K_\alpha = K$ such that X and $Y \times_\alpha K$ are essentially isomorphic extensions of Y.*

Proof. In the proof of the structure theorem, the compact group

$$\{U \in U(H_0) \mid U(H_i) = H_i\}$$

was constructed. Under the hypothesis of the corollary, each H_i is one-dimensional, and hence this group is abelian. Thus in the proof above, K_y is abelian. If we let $p: K_y \to K_y/H$ the natural homomorphism, then $Y \times_y \cdot K_y/H \cong Y \times_{p \circ y} K_y/H$, and since $p \circ y$ is minimal by Proposition 3.10, the corollary is proven.

The techniques of the proof of Theorem 4.3 actually enable us to prove a stronger result. In [15, Theorem 2], Mackey shows that if S is an ergodic G-space and the representation of G induced by any cocycle representation of $S \times G$ has discrete spectrum, then the representation induced by the identity cocycle also has discrete spectrum (i.e., S can be characterized by the conclusion of his structure theorem). Furthermore, one can give an explicit description of the given $S \times G$ cocycle. We will prove a generalization of this result to the case of extensions. We preface a statement of the theorem with a class of examples.

Example 4.7. Let Y, K, H, α, and X be as in Example 4.1. We construct $X \times G$ cocycles such that the induced $Y \times G$ cocycles have discrete spectrum. Let π be a representation of H on a Hilbert space H_0, and

$$\gamma: K/H \times K \to U(H_0)$$

a cocycle corresponding to this representation [20, Theorem 8.27]. Define a cocycle

$$\beta_\pi: X \times G \to U(H_0)$$

400   ROBERT J. ZIMMER

by $\beta_\pi((y, [k]), g) = \gamma([k], \alpha(y, g))$. One sees that the (Y, G) cocycle ρ induced by β_π,

$$\rho: Y \times G \to U(L^2(K/H; H_0)),$$

is given by $\rho(y, g) = U^\pi(\alpha(y, g))$, where U^π is the representation of K on $L^2(K/H; H_0)$ induced by π. (In case π is the one-dimensional identity representation, $U^\pi = \sigma =$ natural representation of K on $L^2(K/H)$. Then $\rho = \sigma \circ \alpha$ which is the cocycle considered in Example 4.1.) As before, $U^\pi = \sum^\oplus \sigma_i$, so $\rho = \sum^\oplus \sigma_i \circ \alpha$, with $\dim \sigma_i < \infty$. Thus ρ has discrete spectrum.

Loosely, our generalization of the structure theorem says that whenever $\phi: X \to Y$ is a factor map of ergodic G-spaces, and there is a cocycle β on $X \times G$ such that the induced (Y, G) cocycle of β has discrete spectrum, then there exist K, H, α, and π as above so that β is essentially the cocycle β_π constructed above.

THEOREM 4.8.   *Let $\phi: X \to Y$ be a G-factor map. Suppose β is a cocycle representation of (X, G) such that the induced cocycle representation α of (Y, G) has discrete spectrum. Then there exists a compact group K, a closed subgroup $H \subset K$, a unitary representation π of H, a minimal cocycle $\gamma: Y \times G \to K$, $K_\gamma = K$, such that the pair $(U^\beta, B(X))$ is unitarily equivalent to the pair $(U^{\beta_\pi}, B(Y \times_\gamma K/H))$. Thus, under the identification of the G-spaces X and $Y \times_\gamma K/H$, β is cohomologous to the cocycle β_π.*

Before considering the proof, we examine what this says in the case $Y = \{e\}$. Then γ will be a homomorphism $G \to K$ with dense range, and

$$(U^{\beta_\pi}, B(Y \times K/H)) = (U^\pi \circ \gamma, B(K/H)).$$

Thus, Theorem 4.8 asserts a unitary equivalence of $(U^\beta, B(X))$ with $(U^\pi \circ \gamma, B(K/H))$, which is exactly the content of Mackey's Theorem 2 of [15].

Proof (of Theorem 4.8). We construct γ, $T(y)$, and A_0 exactly as in the proof of the structure theorem. Since A_0 is a K_γ-invariant Boolean algebra on H_0, we can write $H_0 \cong L^2(Z; H_1)$ where H_1 is a Hilbert space, Z is a K_γ-space, and $A_0 \cong B(Z)$ as Boolean K_γ-spaces. Since $K_\gamma \subset U(H_0)$, we have (tautologously) a unitary representation σ of K_γ that leaves $B(Z)$ invariant. As in the proof of the structure theorem, one can now check that $T = \int^\oplus T(y)$ will yield a unitary equivalence of the pairs of $(U^\alpha, B(X))$ and $(U^{\sigma \circ \gamma}, B(Y \times Z))$. Reasoning as before, Z must be of the form K_γ/H for a closed subgroup $H \subset K_\gamma$. It then follows from Mackey's imprimitivity theorem that $\sigma = U^\pi$ for some representation π of H. But $U^\pi \circ \gamma = \text{ind}(\beta_\pi)$, and we have an equivalence of

$$(U^{U^\pi \circ \gamma}, B(Y \times K_\gamma/H))$$

with $(U^{\beta_\pi}, B(Y \times K_\gamma/H))$. Hence, the equivalence of $(U^\alpha, B(X))$ with $(U^{\beta_\pi}, B(Y \times K_\gamma/H))$ is established. The last statement of the theorem follows from Theorem 2.6.

5. Normal actions and extensions. In their study of ergodic transformations with discrete spectrum [6], [17], von Neumann and Halmos were able not only to describe the structure of such transformations, but to prove an existence-uniqueness theorem. This theorem says that if two ergodic transformations with discrete spectrum have the same spectrum, they are essentially isomorphic. Furthermore, this (discrete) spectrum is always a countable subgroup of the circle, and conversely, for any countable subgroup S of the circle, there exists an ergodic transformation with S as its spectrum. As Mackey points out in [15], there is an equally complete theorem for actions of a locally compact abelian group (see [21] for a detailed proof). However, as also indicated in [15], the result fails when the hypothesis that the group is abelian is dropped. Our aim in this and the following section is two-fold. First, a natural class of ergodic actions of a group, called normal actions, is defined. We show that this includes all ergodic actions of a locally compact abelian group with discrete spectrum, and that for the class of normal actions, an analogue of the existence-uniqueness theorem is valid. Secondly, and more interestingly, we extend the notion of normality to the relative situation of extensions of arbitrary G-spaces, and prove an existence-uniqueness theorem for normal extensions. Even in the case where $G = Z$, this provides a new direction of generalization of the von Neumann–Halmos theorem.

It will be convenient to establish some notation.

DEFINITION 5.1. If Y is a factor of an ergodic G-space X, and the natural (Y, G) cocycle representation has discrete spectrum, we shall say that X has relatively discrete spectrum over Y. If the cocycle can be written as a direct sum of one-dimensional cocycle representations, we will say that X has relatively elementary spectrum over Y.

Thus Theorem 4.13 and Corollary 4.6 describe the structure of extensions with relatively discrete and relatively elementary spectrum, respectively.

DEFINITION 5.2. Let S be an ergodic Lebesgue G-space and let

$$\beta: G \to U(L^2(S))$$

be the natural representation. Define the cocycle $\alpha. S \times G \to U(L^2(S))$ by $\alpha(s, g) = \beta(g)$. S is called a normal G-space if α is cohomologous to the identity cocycle.

Remark. In terms of virtual groups, α is the "restriction" of β to the virtual subgroup (S, G) (Example 2.7).

PROPOSITION 5.3. *If S is a transitive G-space, i.e., $S = G/H$ for some closed subgroup $H \subset G$, then S is normal if and only if H is a normal subgroup.*

Proof. $\alpha: G/H \times G \to U(L^2(G/H))$ is a cocycle and hence corresponds to a representation of H on $L^2(G/H)$ [20, Theorem 8.27], which in this case is clearly just the restriction of β to H. Furthermore, α is cohomologous to the identity if and only if the representation of H is the identity. It is clear that $\beta \mid H$ is the identity on $L^2(G/H)$ if H is a normal subgroup. Conversely, suppose H acts by the identity on $L^2(G/H)$. Then the Boolean action of H on $B(G/H)$ is the identity and by [14, Theorem 2], almost all points of G/H are left fixed. But the set of points fixed by H is $N/H \subset G/H$ where $N \subset G$ is the normalizer of H. If N/H is conull in G/H, N must be conull in G, and being a subgroup, must equal G. Hence H is normal.

More generally, now suppose that $p: X \to Y$ is a factor G-map of ergodic G-spaces. As usual, let $F_y = p^{-1}(y)$. Then the assignment

$$x \to H_x = L^2(F_{p(x)})$$

defines a Hilbert bundle in a natural way. Let $\beta(y, g): L^2(F_{yg}) \to L^2(F_y)$ be the natural (Y, G) cocycle representation and define a cocycle representation

$$\alpha(x, g): H_{xg} \to H_x$$

by $\alpha(x, g) = \beta(p(x), g)$.

DEFINITION 5.4. X is called a normal extension of Y (or Y a normal factor of X) if the cocycle α is equivalent to the identity.

Remarks. (i)   When $Y = \{e\}$, this construction reduces to that of Definition 5.2.

(ii)   In terms of virtual groups, α is the "restriction" of β to the sub-virtual subgroup (X, G). (See Example 2.7.)

PROPOSITION 5.5.   *If $H \subset K \subset G$ are subgroups of G, and $p: G/H \to G/K$ the natural map, then G/H is a normal extension of G/K if and only if H is a normal subgroup of K.*

Proof. The $G/H \times G$ cocycle representation α is defined by a representation of H on the Hilbert space $H_{[e]} = L^2(F_{[e]})$. But $F_{[e]}$ can be naturally identified with K/H, and the representation of H on $L^2(K/H)$ is the restriction of the natural representation of K. α will be equivalent to the identity if and only if this representation of H is the identity, which is true if and only if H is a normal subgroup of K.

We now consider the question of when an action with discrete spectrum or an extension with relatively discrete spectrum, is normal.

PROPOSITION 5.6.   *Let $\beta: Y \times G \to K$ a minimal cocycle, $K_\beta = K$, $H \subset K$ a closed subgroup, and $X = Y \times_\beta K/H$. Then if H is normal, X is a normal extension of Y.*

Proof. Let $T: K \to K/H$ the natural homomorphism. Replacing K by K/H and β by $T \circ \beta$, we see that we can assume $H = \{e\}$. It is easy to check that the Hilbert bundle $\{H_x\}$ is unitarily equivalent to the Hilbert bundle

$x \mapsto L^2(K)$, and the (X, G) cocycle is given by $\alpha(x, g) = \pi \circ \beta(p(x), g)$, where $p: X \to Y$ is projection and π is the right regular representation of K. For each $x = (y, k)$, let $U(x): H_x \to H_x$ be $U(x) = \pi(k)$. Then

$$U(x)\alpha(x, g)U(xg)^{-1} = \pi(k)\pi(\beta(p(x), g))\pi(k\beta(p(x), g))^{-1} = I,$$

proving the proposition.

The main result of this section is the converse of Proposition 5.6.

THEOREM 5.7.   *Let $\beta: Y \times G \to K$, minimal, with $K_\beta = K$. Let $H \subset K$ be a closed subgroup, and suppose that $X = Y \times_\beta K/H$ is a normal extension of Y. Then H is a normal subgroup of K.*

Proof.   Let π be the natural representation of K on $L^2(K/H)$. Thus

$$\alpha: X \times G \to U(L^2(K/H))$$

is given by $\alpha(x, g) = \pi \circ \beta(p(x), g)$. Suppose now that α is cohomologous to the identity cocycle i. We will denote by α_Y and i_Y the (Y, G) cocycles induced by α and i respectively. To write α_Y and i_Y in a more convenient form, we introduce some notation. Let

$$\tilde{H} = \int_{K/H}^{\oplus} L^2(K/H) = L^2(K/H; L^2(K/H)).$$

We have 2 representations of K on $\tilde{H}$. For $f \in \tilde{H}$, $s \in K/H$, let $(U_k f)(s) = f(s \cdot k)$, $(W_k f)(s) = \pi_k(f(sk))$. Now $\alpha_Y, i_Y: Y \times G \to U(\tilde{H})$, and it is easy to check from the definition of induced cocycle that we can take $\alpha_Y(y, g) = W_{\beta(y, g)}$ and $i_Y(y, g) = U_{\beta(y, g)}$. Since α and i are cohomologous, there exists a Borel function $A: X \to U(L^2(K/H))$ such that for each g, $A(x)\alpha(x, g)A(xg)^{-1} = i(x, g)$, almost everywhere. Define $T: Y \to U(\tilde{H})$ by

$$T(y) = \int_{x \in p^{-1}(y)}^{\oplus} A(x).$$

Then T is a Borel function and for each g,

(*) $T(y)\alpha_Y(y, g)T(yg)^{-1} = i_Y(y, g)$ for almost all $y \in Y$,

Now K acts on $U(\tilde{H})$ by $T \cdot K = U_k^{-1} T W_k$. Under this action, $U(\tilde{H})$ is a standard Borel K-space, and since K is compact, the space of orbits, B, is standard. Let $q: U(\tilde{H}) \to B$ be the natural map. Now by (*), for each g,

$$T(yg) = i_Y(y, g)^{-1}T(y)\alpha_Y(y, g)$$

$$= U_{\beta(y, g)}^{-1}T(y)W_{\beta(y, g)}$$

$$= T(y) \cdot \beta(y, g)   \text{ for almost all } y;$$

i.e., for each g, $q(T(yg)) = q(T(y))$ for almost all y. By the ergodicity of G on Y, and the fact that B is standard, we have that $q \circ T$ is constant on a conull set Y_0. That is, there exists an orbit $R \subset U(\tilde{H})$ such that $T(y) \in R$ for every $y \in Y_0$. Choose an element T_0 of the orbit R. Then there exists a Borel function $\lambda: R \to K$ such that $T \cdot \lambda(T) = T_0$, for all $T \in R$ [13, Lemma 1.1]. Let $\theta: Y_0 \to K$ be $\theta = \lambda \circ T$, and extend θ to a Borel map $Y \to K$. Thus $T(y) \cdot \theta(y) = T_0$ for almost all y. We now use θ to define a new cocycle $\gamma: Y \times G \to K$ equivalent to β, namely $\gamma(y, g) = \theta(y)^{-1}\beta(y, g)\theta(yg)$. We now claim the following, which is basically just unravelling the definitions.

LEMMA. *For each g, $T_0 W_{\gamma(y, g)} T_0^{-1} = U_{\gamma(y, g)}$ for almost all y.*

Proof.

$$\begin{aligned}
T_0 W_{\gamma(y, g)} T_0^{-1} &= (T(y) \cdot \theta(y)) W_{\theta(y)^{-1}\beta(y, g)\theta(yg)} (T(yg) \cdot \theta(yg))^{-1} \\
&= U_{\theta(y)}^{-1} T(y) W_{\beta(y, g)} T(yg)^{-1} U_{\theta(yg)} \\
&= U_{\theta(y)}^{-1} U_{\beta(y, g)} U_{\theta(yg)} \\
&= U_{\gamma(y, g)}.
\end{aligned}$$

Now let $K_0 = \{k \in K \mid T_0 W_k T_0^{-1} = U_k\}$. It is easy to see that K_0 is a closed subgroup of K, and the lemma says that for each g, $\gamma(y, g) \in K_0$ for almost all $y \in Y$. By changing γ on a suitable null set, we get an equivalent cocycle γ_1 such that $\gamma_1(y, g) \in K_0$ for all (y, g). Now γ_1 is equivalent to β, and since β is a minimal cocycle, we must have $K_0 = K$. Thus T_0 is an intertwining operator for U and W. We claim furthermore that T_0 commutes with the natural projection valued measure $P: B(K/H) \to L(\tilde{H})$. To see this, first note that since $T(y) = \int_{x \in F_y}^{\oplus} A(x)$, for each $E \in B(K/H)$, we have $P_E T(y) = T(y) P_E$. In addition, for all k, $U_k^{-1} P_E U_k = P_{Ek}$ and a similar relation holds for W. Combining these identities, we see that $U_k^{-1} T(y) W_k$ commutes with P_E for all E, k, y. Since $T_0 = T(y) \cdot \theta(y)$ for almost all (and hence at least one) y, T_0 commutes with P_E. It follows that the representations of H which induce U and W must be equivalent. But U is the representation induced by the identity representation of H on $L^2(K/H)$, and W is the representation induced by the restriction to H of the natural representation of K on $L^2(K/H)$. For these to be equivalent, H must be normal.

COROLLARY 5.8. *Let X be an ergodic extension of Y with relatively discrete spectrum. Let α be the natural (Y, G) cocycle representation, and let S be the set of equivalence classes of irreducible (Y, G) cocycle representations that are subcocycles of α. Then X is a normal extension of Y if and only if S satisfies the following conditions:*

(i) $\alpha_1, \alpha_2 \in S$ *implies every irreducible component of $\alpha_1 \otimes \alpha_2$ is in S.*

(ii) $\sigma \in S$ *implies $\bar{\sigma} \in S$.*

(iii) $\alpha = \sum_{\sigma \in S} (\dim \sigma)\sigma$.

Proof. Let $X = Y \times_\beta K/H$ where $\beta: Y \times G \to K$ is a minimal cocycle with $K_0 = K$. Let π be the natural representation of K on $L^2(K/H)$. Then α

is equivalent to $\pi \circ \beta$. Let S' be the set of equivalence classes of irreducible representations of K that are subrepresentations of π, and for each $\lambda \in S'$, n_λ the positive integer such that $\pi = \sum_{\lambda \in S'}^{\oplus} n_\lambda \lambda$. Then

$$\alpha = \pi \circ \beta = \sum_{\lambda \in S'}^{\oplus} n_\lambda(\lambda \circ \beta).$$

From this equation and Theorem 3.14, it follows that $S = S' \circ \beta$ and that S satisfies (i), (ii), (iii) above if and only if S' satisfies the analogous properties. By [8, 30. 60] and the Peter-Weyl theorem, this will be true if and only if H is normal, and by Theorem 5.7, this holds if and only if X is a normal extension of Y.

COROLLARY 5.9.   *If X is an ergodic extension of Y with relatively elementary spectrum, then X is a normal extension of Y.*

Proof. By Corollary 4.6, $X = Y \times_\beta K$ where K is compact, abelian, and the result is immediate from Theorem 5.7.

COROLLARY 5.10.   *Any action of a locally compact abelian group with discrete spectrum is normal.*

6. The existence-uniqueness theorems. We now prove analogues of von Neumann's and Halmos' existence-uniqueness theorems for the class of normal actions and extensions. An essential step is the following general result about cocycles into compact groups.

THEOREM 6.1.   *Let $K \subset L$ be compact groups α, $\beta \colon Y \times G \to K$ minimal cocycles with $K_\alpha = K_\beta = K$. Suppose that α and β are equivalent as cocycles into L. Then there is a continuous automorphism σ of K such that $\sigma \circ \alpha$ and β are equivalent as cocycles into K.*

Proof. Let $\phi \colon Y \to L$ be a Borel function such that for each g and almost all y, $\phi(y)\alpha(y, g)\phi(yg)^{-1} = \beta(y, g)$, i.e.,

(*) $\beta(y, g)^{-1}\phi(y)\alpha(y, g) = \phi(yg)$.

Now $K \times K$ acts on L by $l \cdot (k_1, k_2) = k_1^{-1}lk_2$. Since the action is continuous and $K \times K$ is compact, the space of orbits in L under $K \times K$ is a standard Borel space $\hat{L}$. Let $p \colon L \to \hat{L}$ be the natural (Borel) map. Since $\alpha(y, g)$, $\beta(y, g) \in K$, equation (*) implies that for each g, $p(\phi(y)) = p(\phi(yg))$ for almost all y. By the ergodicity of G on Y, $p \circ \phi$ is constant on a conull Borel set Y_0. Choose a point l_0 in the orbit $p(\phi(Y_0))$. Then [13, Lemma 1.1], there exists a Borel map

$$(k_1, k_2) \colon p(l_0) \to K \times K$$

such that for any $l \in p(l_0)$, $l = k_1(l)^{-1}l_0k_2(l)$. Now $k_1 \circ \phi, k_2 \circ \phi \colon Y_0 \to K$ are Borel functions, and can be extended to Borel functions $\phi_1, \phi_2 \colon Y \to K$. Thus, for any $y \in Y_0$, $\phi(y) = \phi_1(y)^{-1} l_0 \phi_2(y)$. From (*), we have for all $g \in G$ and $y \in Y_0 \cap Y_0 g^{-1}$,

$$\phi_1(y)^{-1} l_0 \phi_2(y) \alpha(y, g) \phi_2(yg)^{-1} l_0^{-1} \phi_1(yg) = \beta(y, g).$$

With $\alpha_2(y, g) = \phi_2(y) \alpha(y, g) \phi_2(yg)^{-1}$, this becomes

(**) $l_0 \alpha_2(y, g) l_0^{-1} = \phi_1(y) \beta(y, g) \phi_1(yg)^{-1}.$

Now α is a minimal cocycle with $K_\alpha = K$, and since α_2 is equivalent to α, we also have $K_{\alpha_2} = K$. Consider $K_0 = \{k \in K \mid l_0 k l_0^{-1} \in K\}$. Then K_0 is a closed subgroup of K, and by (**), $\alpha_2(y, g) \in K_0$ for each g and almost all y. Changing α_2 on a suitable null set, we see that α_2 is equivalent to a cocycle with all values in K_0, and since α is minimal, $K_0 = K$. Let σ be the automorphism of K defined by $\sigma(k) = l_0 k l_0^{-1}$. Letting $\phi_0(y) = \sigma(\phi_2(y))$, we have $\phi_0(y) \in K$ for all y, and from (**), for each $g \in G$,

$$\phi_0(y)(\sigma \circ \alpha)(y, g) \phi_0(yg)^{-1} = \phi_1(y) \beta(y, g) \phi_1(yg)^{-1}$$

for almost all y. This completes the proof.

We now apply this result to prove the uniqueness theorem.

THEOREM 6.2 (Uniqueness Theorem).   *Suppose X_1 and X_2 are normal ergodic extensions of Y with relatively discrete spectrum. If the corresponding natural (Y, G) cocycle representations are equivalent, then X_1 and X_2 are essentially isomorphic extensions of Y.*

Remark.   In the case $G = Z$ and $Y = \{e\}$, this is just the von Neumann–Halmos uniqueness theorem.

Proof.   By the Structure Theorem (4.3), and Theorem 5.7, we can write $X_i = Y \times_{\alpha_i} K_i$, where α_i are minimal cocycles with $K_{\alpha_i} = K_i$. Let π_i be the right regular representation of K_i, S_i the dual object of K_i, and form the canonical decompositions

$$L^2(K_1) = \sum_{\sigma \in S_1}^{\oplus} H_\sigma, \quad L^2(K_2) = \sum_{\tau \in S_2}^{\oplus} H_\tau.$$

The hypothesis of the theorem is that $\pi_1 \circ \alpha_1$ and $\pi_2 \circ \alpha_2$ are unitarily equivalent. By Theorem 3.14, for each $\sigma \in S_1$, $\sigma \circ \alpha_1$ will be equivalent to exactly one $\tau \circ \alpha_2$, $\tau \in S_2$. Then it is easy to see that the equivalence of the cocycle representations $\pi_1 \circ \alpha_1$ and $\pi_2 \circ \alpha_2$ can be implemented by a Borel field of unitary operators $U(y): L^2(K_1) \to L^2(K_2)$ such that for σ, τ related as above, $U(y)(H_\sigma) = H_\tau$ for all $y \in Y$. Fix an operator $U: L^2(K_1) \to L^2(K_2)$ such that $U(H_\sigma) = H_\tau$. Let

$$\beta(y, g) = U \pi_1 \alpha_1(y, g) U^{-1},$$

and let

$$K = \{T \in U(L^2(K_2)) \mid T(H_\tau) = H_\tau \quad \text{for all } \tau \in S_2\}.$$

Then K is a compact group and β is a cocycle with values in K. Since $U(y)U^{-1} \in K$ for all y, it follows that β is equivalent as a cocycle into K, to $\pi_2 \circ \alpha_2$. Since α_1 is minimal with $K_{\alpha_1} = K_1$, it follows from Proposition 3.10 and Theorem 3.9 that β is also minimal and $K_\beta = U\pi_1(K_1)U^{-1}$. As $\pi_2 \circ \alpha_2$ is also minimal with $K_{\pi_2 \circ \alpha_2} = \pi_2(K_2)$, Corollary 3.8 implies that there exists $W \in K$ such that

$$W U \pi_1(K_1) U^{-1} W^{-1} = \pi_2(K_2).$$

Let $\phi: K_1 \to K_2$ be defined by $\phi(k) = \pi_2^{-1}(W U \pi_1(k) U^{-1} W^{-1})$. It is clear that ϕ is an isomorphism (since π_i are). Let $\alpha_0: Y \times G \to K_2$ be the cocycle $\alpha_0 = \phi \circ \alpha_1$. Clearly $Y \times_{\alpha_0} K_2$ and $Y \times_{\alpha_1} K_1$ are essentially isomorphic extensions of Y. Thus, to prove the theorem it suffices to show that $Y \times_{\alpha_0} K_2$ and $Y \times_{\alpha_2} K_2$ are essentially isomorphic extensions. Now for all g,

$$W U U(y)^{-1}(\pi_2 \circ \alpha_2)(y, g)U(yg)U^{-1}W^{-1} = (\pi_2 \circ \alpha_0)(y, g)$$

for almost all y. As remarked above, $U \cdot U(y)^{-1} \in K$ for all y, and hence $\pi_2 \circ \alpha_0$ and $\pi_2 \circ \alpha_2$ are equivalent as cocycles into K. Since both of these cocycles take values in $\pi_2(K_2) \subset K$, by Theorem 6.1 there exists a continuous automorphism σ of $\pi_2(K_2)$ such that $\sigma \circ \pi_2 \circ \alpha_0$ and $\pi_2 \circ \alpha_2$ are cohomologous as cocycles into $\pi_2(K_2)$. Since π_2 is an isomorphism, there exists a continuous automorphism γ of K_2 such that $\gamma \circ \alpha_0$ and α_2 are cohomologous. Thus $Y \times_{\alpha_2} K_2$ and $Y \times_{\gamma \circ \alpha_0} K_2$ are essentially isomorphic extensions of Y, and the latter is clearly essentially isomorphic to the extension $Y \times_{\alpha_0} K_2$. By the remarks above, this completes the proof.

COROLLARY 6.3. *For an ergodic G-space Y, let S be the set of equivalence classes of one-dimensional cocycle representations. Then S is a group under the operation of tensor product. If X is an extension of Y with relatively elementary spectrum over Y, let S_X be the subset of S consisting of cocycles appearing in the decomposition of the natural (Y, G) cocycle representation on the Hilbert bundle $L^2(X)$. Then S_X is a countable subgroup of S. If Z is another extension of Y with relatively elementary spectrum, then X and Z are essentially isomorphic extensions of Y if and only if $S_X = S_Z$.*

Proof. By Corollary 4.6, we can take $X = Y \times_\alpha K$, K compact, abelian. The natural (Y, G) cocycle representation, β, on $L^2(X)$ is just $\pi \circ \alpha$, where π is the right regular representation of K. Thus,

$$\beta \cong \sum_{\chi \in K^*}^{\oplus} \chi \circ \alpha,$$

and by Theorem 3.14, S_X is a group isomorphic to $\{\chi \circ \alpha \mid \chi \in K^*\}$. The last assertion follows from Corollary 5.9 and Theorem 6.2.

Remarks. (i) Since S_X is isomorphic to $\{\chi \circ \alpha \mid \chi \in K^*\} \cong K^*$, we can identify the group K in Corollary 4.6 as S_X^*. When $G = Z$, and $Y = \{e\}$, this is a well-known fact in the von Neumann–Halmos theory.

(ii) By Frobenius reciprocity (Theorem 2.10), a one-dimensional cocycle representation γ of (Y, G) will be a subcocycle representation of the natural induced cocycle representation β defined by the extension X, if and only if the restriction of γ to X (Example 2.7) is equivalent to the identity. Thus, S_X can be identified as the group

$$\{\gamma \in S \mid \text{res } (\gamma) \text{ is the class of the identity cocycle of } (X, G)\}.$$

We now proceed to the existence theorem.

THEOREM 6.4.   *Let Y be an ergodic Lebesgue G-space and let $S = \{[\alpha_i]\}_{i \in I}$ be a countable set of equivalence classes of finite dimensional irreducible cocycle representations of (Y, G) where I is some countable index set. Suppose S satisfies:*
  (i) *If $\alpha \in S$ then $\bar{\alpha} \in S$.*
  (ii) *If $\alpha, \beta \in S$ then every irreducible component of $\alpha \otimes \beta$ is in S.*
Then there exists a normal ergodic extension X of Y with relatively discrete spectrum over Y such that the natural (Y, G) cocycle representation on $L^2(X)$ is equivalent to $\sum_{\alpha \in S}^{\oplus} (\dim \alpha)\alpha$.

Proof.   Let J be a sequence of elements from I such that each $i \in I$ appears in J dim (α_i) times. For each $j \in J$, we have

$$\alpha_j : Y \times G \to U(H_j)$$

where H_j is a finite dimensional Hilbert space. Let $H = \sum_J^{\oplus} H_j$ and $\alpha = \sum_{j \in J}^{\oplus} \alpha_j$. Let $K = \{T \in U(H) \mid T(H_j) = H_j \text{ for all } j\}$ with the weak operator topology. Then K is a compact group and $\alpha : Y \times G \to K$. Let $\beta : Y \times G \to K$ be an equivalent minimal cocycle (Corollary 3.8), and for each j, $\pi_j : K_\beta \to U(H_j)$ the map obtained by restricting elements of K_β to H_j. It is easy to see that $\pi_j \circ \beta$ is equivalent to α_j, and hence is irreducible. It follows that for each j, π_j is an irreducible representation of K_β. Let $S' = \{[\pi_j]\}_{j \in J} \subset \hat{K}_\beta$ (the dual object of K_β). It is straightforward to see that S' satisfies properties (i) and (ii) as well as S. Since the representation $\sum_{j \in J}^{\oplus} \pi_j$ is faithful, we have $S' = \hat{K}_\beta$ by [8, Theorem 27.39]. It follows from Theorem 3.14, the construction of J, and the Peter-Weyl theorem, that $\sum_{j \in J}^{\oplus} \pi_j$ is unitarily equivalent to the right regular representation, π, of K_β. Let $X = Y \times_\beta K_\beta$. Then the natural (Y, G) cocycle representation is $\pi \circ \beta$, which is equivalent to

$$\sum_{j \in J}^{\oplus} \pi_j \circ \beta \cong \sum_{j \in J}^{\oplus} \alpha_j.$$

By the choice of J, this is equivalent to the theorem.

COROLLARY 6.5.   *Let Y be an ergodic G-space and S the group described in Corollary 6.3. Let $S' \subset S$ be a countable subgroup. Then there exists an ergodic extension X of Y with relatively elementary spectrum over Y such that $S_X = S'$.*

EXTENSIONS OF ERGODIC GROUP ACTIONS   409

BIBLIOGRAPHY

1. S. BERBERIAN, *Measure and integration*, MacMillan, New York, 1963.
2. J. DIXMIER, *Dual et quasi-dual d'une algebre de Banach involutive*, Trans. Amer. Math. Soc., vol. 104 (1962), pp. 278–283.
3. ————, *Les Algebres d'Operateurs dans l'Espace Hilbertian*, Gauthier-Villars, Paris, 1969.
4. E. EFFROS, *The Borel space of von Neumann algebras on a separable Hilbert space*, Pacific J. Math., vol. 15 (1964), pp. 1153–1164.
5. ————, *Global structure in von Neumann algebras*, Trans. Amer. Math. Soc., vol. 121 (1966), pp. 434–454.
6. P. R. HALMOS AND J. VON NEUMANN, *Operator methods in classical mechanics, II*, Ann. of Math., vol. 43 (1942), pp. 332–350.
7. E. HEWITT AND K. ROSS, *Abstract Harmonic Analysis, I*, Springer-Verlag, Berlin, 1963.
8. ————, *Abstract Harmonic Analysis, II*, Springer-Verlag, Berlin, 1970.
9. A. A. KIRILLOV, *Dynamical systems, factors, and group representations*, Russian Math Surveys, vol. 22 (1967), pp. 63–75.
10. U. KRENGEL, *Weakly wandering vectors and weakly independent partitions*, Trans. Amer. Math. Soc., vol. 164 (1972), pp. 199–226.
11. K. LANGE, *Borel sets of probability measures*, Pacific J. Math., vol. 48 (1973), pp. 141–161.
12. G. W. MACKEY, *Borel structures in groups and their duals,* Trans. Amer. Math. Soc., vol. 85 (1957), pp. 134–165.
13. ————, *Induced representations of locally compact groups, I*, Ann. of Math., vol. 55 (1952), pp. 101–139.
14. ————, *Point realizations of transformation groups*, Illinois J. Math., vol. 6 (1962), pp. 327–335.
15. ————, *Ergodic transformation groups with a pure point spectrum*, Illinois J. Math., vol. 8 (1964), pp. 593–600.
16. ————, *Ergodic theory and virtual groups*, Math. Ann., vol. 166 (1966), pp. 187–207.
17. J. VON NEUMANN, *Zur Operatorenmethode in der Klassichen Mechanik*, Ann. of Math., vol. 33 (1932), pp. 587–642.
18. W. PARRY, *A note on cocycles in ergodic theory*, Comp. Math., vol. 28 (1974), pp. 343–350.
19. A. RAMSAY, *Virtual groups and group actions*, Advances in Math., vol. 6 (1971), pp. 253–322.
20. V. S. VARADARAJAN, *Geometry of quantum theory*, vol. II, Van Nostrand, Princeton, 1970.
21. T. WIETING, *Ergodic affine Lebesgue G-spaces*, doctoral dissertation, Harvard University, 1973.
22. R. ZIMMER, *Extensions of ergodic actions and generalized discrete spectrum*, Bull. Amer. Math. Soc., vol. 81 (1975), pp. 633–636.
23. ————, *Ergodic actions with generalized discrete spectrum*, Illinois J. Math., vol. 20 (1975), to appear in no. 4.

U.S. NAVAL ACADEMY
ANNAPOLIS, MARYLAND

ERGODIC ACTIONS WITH
GENERALIZED DISCRETE SPECTRUM

BY

ROBERT J. ZIMMER

The investigation of extensions in the theory of ergodic actions of locally compact groups was undertaken by the author in [26]. In particular, we considered the notion of extensions with relatively discrete spectrum, and saw how the classical von Neumann-Halmos theory of transformations with discrete spectrum could be generalized to the case of extensions. In this paper, which is a sequel to [26], we study those actions which can be built up from a point by taking extensions with relatively discrete spectrum and inverse limits. We shall say that such actions have generalized discrete spectrum.

A similar construction is well known in topological dynamics. In [4], Furstenberg introduced the notion of an isometric extension of a continuous transformation group, and called an action quasi-isometric if it could be built up from a point by taking isometric extensions and inverse limits. The main result of [4] is the striking theorem that among the minimal transformation groups, the quasi-isometric ones are precisely those that are distal. Thus, one obtains a description of the structure of an arbitrary minimal distal transformation group, and using this, one can answer a variety of questions about such groups.

The structure of extensions with relatively discrete spectrum was described in Theorem 4.3 of [26]. Examination of the conclusion of this theorem shows that extensions with relatively discrete spectrum are a reasonable measure-theoretic analogue of Furstenberg's isometric extensions. Thus, we can consider actions with generalized discrete spectrum as a measure-theoretic analogue of the quasi-isometric transformation groups. Parry has described, at least for actions of the integers, a measure-theoretic analogue of the topological notion of distality [20]. It is not difficult to generalize Parry's definition to arbitrary group actions, and now the question arises as to whether one can prove a measure-theoretic analogue of Furstenberg's theorem. We prove such a theorem below. It asserts that among the nonatomic ergodic actions, those with a separating sieve (as Parry called his actions) are precisely those with generalized discrete spectrum. Using this theorem, one sees immediately, for example, that any minimal distal action preserving a probability measure has generalized discrete spectrum.

Though there are formal similarities between the proof of our theorem and Furstenberg's proof, the proofs are basically quite different. Our proof depends upon, among other things, generalizing the concepts of weak mixing and the

Received October 1, 1975.

Cartesian product action to extensions; this leads to the notions of relative weak mixing and the fibered product. In addition, we make use of a general existence theorem for factors proved in [26]. These notions are all of independent interest and prove useful in other circumstances. Furstenberg's proof rests heavily on topological notions that are not available in the measure-theoretic context.

Given any specific class of actions, one would, of course, like to know which members of this class have generalized discrete spectrum. One class of actions that has attracted considerable attention is the set of affine actions on compact abelian groups. An algebraic criterion for affine transformations to have some nontrivial discrete spectral part was established by Hahn [6], and was extended (along with much of the other theory of affine transformations) to affine actions of a general locally compact abelian group by Wieting [24]. We extend the Hahn-Wieting analysis to establish an algebraic criterion for an affine extension to have some nontrivial relative discrete spectrum. This has two interesting consequences. First, we are able to give an algebraic criterion for an affine action to have generalized discrete spectrum. Second, in the case of a transformation, it enables us to clarify the relationship between generalized discrete spectrum and quasi-discrete spectrum.

Transformations with quasi-discrete spectrum were introduced by von Neumann and Halmos, and first studied systematically by Abramov [1]. The inductive definition of quasi-discrete spectrum can be shown to be a special case of the definition of generalized discrete spectrum. The nilflows considered by Auslander, Green, and Hahn [2] show, however, that not every transformation with generalized discrete spectrum has quasi-discrete spectrum. However, using the algebraic criterion of the preceding paragraph, we show that every totally ergodic affine transformation with generalized discrete spectrum actually has quasi-discrete spectrum. Combined with a result of Abramov, this yields a new characterization of (totally ergodic) transformations with quasi-discrete spectrum, as precisely those that are equivalent to affine transformations with generalized discrete spectrum.

The results of this paper depend heavily on the framework established in [26]. For ease of reference, we have begun this paper with Section 7, all references to Sections 1–6 being to those in [26]. For any unexplained notation the reader is also referred to [26]. The organization of the paper is as follows. Section 7 discusses relative weak mixing and applications of the general existence theorem for factors appearing in Section 2. Section 8 contains the central result, namely the measure-theoretic analogue of the Furstenberg structure theorem. The proof depends heavily on the results of Section 7. Section 9 considers some examples and general properties of actions with generalized discrete spectrum. The connections between affine actions, quasi-discrete spectrum, and generalized discrete spectrum mentioned above are proved in Section 10. This section concludes with a new proof, based on the results of Section 6, of the Abramov-Wieting existence theorem for actions with quasi-discrete spectrum [1], [24].

Some of the main results of this paper were announced in [25].

The author wishes to thank Professor G. W. Mackey for many helpful conversations and suggestions during the preparation of this paper.

7. Relative weak mixing

In preceding sections, we have examined the properties of extensions with relatively discrete spectrum. We now turn to questions involving the appearance or nonappearance of such extensions in various situations.

THEOREM 7.1.   *Suppose X is a Lebesgue G-space. Let $H \subset L^2(X)$ be the closed subspace generated by the finite-dimensional G-invariant subspaces of $L^2(X)$. Then there exists a factor G-space Y of X such that $H = L^2(Y)$ (and hence Y has discrete spectrum).*

Proof.   It is well known that for $f \in L^2(X), f \in H$ if and only if $\{U_g f \mid g \in G\}$ is precompact in $L^2(X)$, where U_g is the natural representation of G on $L^2(X)$. Let $B = \{A \subset X \mid A \text{ is measurable, and } \chi_A \in H\}$.

LEMMA 7.2.   *B is an invariant σ-field of subsets of X.*

Proof.   If $A \in B$, $\chi_{X-A} = 1 - \chi_A \in H$, so $X - A \in B$. If $A_i \in B$, and A_i are mutually disjoint, then

$$\chi_{\cup_1^n A_i} = \sum_1^n \chi_{A_i} \in H,$$

and as $\mu(\bigcup_1^\infty A_i) \leq 1$,

$$\chi_{\cup_1^\infty A_i} = \lim_{n \to \infty} \sum_1^n \chi_{A_i},$$

and so $\bigcup_1^\infty A_i \in B$ since H is closed. B is clearly G-invariant and hence it suffices to see that B is closed under finite intersections. So suppose $A, D \in B$ and $g, h \in G$. Then

$(A \cap D) \cdot g \, \Delta \, (A \cap D) \cdot h$

$$= [(X - Ag \cap Dg) \cap Ah \cap Dh] \cup [Ag \cap Dg \cap (X - Ah \cap Dh)]$$

$$= [(X - Ag) \cap Ah \cap Dh] \cup [(X - Dg) \cap Ah \cap Dh]$$

$$\cup [Ag \cap Dg \cap (X - Ah)] \cup [Ag \cap Dg \cap (X - Dh)]$$

$$\subset (Ag \, \Delta \, Ah) \cup (Dg \, \Delta \, Dh).$$

Thus

$$\|\chi_{(A \cap D)g} - \chi_{(A \cap D)h}\|^2 \leq \|\chi_{Ag} - \chi_{Ah}\|^2 + \|\chi_{Dg} - \chi_{Dh}\|^2.$$

As a set in a metric space is precompact if and only if every sequence has a Cauchy subsequence, to see $A \cap D \in B$, it suffices to show that if $g_i \in G$, then $\chi_{(A \cap D)g_i}$ has a Cauchy subsequence. Since $\{\chi_{Ag}\}$ and $\{\chi_{Dg}\}$ are precompact, there exists a subsequence g_{i_n} such that $\chi_{Ag_{i_n}}$ and $\chi_{Dg_{i_n}}$ are Cauchy, and it follows easily from the inequality above that $\chi_{(A \cap D)g_{i_n}}$ is also.

We now return to the proof of the theorem. It suffices to show that $f \in L^2(X)$ is measurable with respect to B if and only if $f = H$. If f is measurable with respect to B, then f is the limit in $L^2(X)$ of finite linear combinations of characteristic functions of sets in B, and hence $f \in H$. To see the converse, first note that $f \in H$ if and only if $\bar{f} \in H$, and hence it suffices to show the converse for real valued functions. Let $t \in \mathbf{R}$ and $A = \{x \in X \mid f(x) \leq t\}$. It suffices to show that $A \in B$, i.e., the orbit of χ_A is precompact. Suppose not. Then there exists $\varepsilon > 0$ and a sequence $g_i \in G$ such that

$$\|\chi_{Ag_i} - \chi_{Ag_j}\| \geq \varepsilon,$$

i.e.,

$$\mu(Ag_i \, \Delta \, Ag_j) \geq \varepsilon^2.$$

Since $\mu(A) < 1$, it is easy to see that there exists some $\delta > 0$ and a set $D \subset X - A$ such that (i) $\mu(D) \geq 1 - \mu(A) - \varepsilon^2/4$, i.e., $\mu((X - A) - D) \leq \varepsilon^2/4$ and (ii) $f(x) \geq t + \delta$ for every $x \in D$. Now since $\mu(Ag_i \, \Delta \, Ag_j) \geq \varepsilon^2$, we can assume

$$\mu(Ag_i \cap (X - Ag_j)) \geq \varepsilon^2/2.$$

We have $Dg_j \subset X - Ag_j$ and $\mu((X - Ag_j) - Dg_j) \leq \varepsilon^2/4$ by the G-invariance of μ, and hence $\mu(Ag_i \cap Dg_j) \geq \varepsilon^2/4$. For $x \in Ag_i \cap Dg_j$, we have

$$(U_{g_i^{-1}}f)(x) = f(xg_i^{-1}) \leq t,$$

since $xg_i^{-1} \in A$.

Similarly, $(U_{g_j^{-1}}f)(x) \geq t + \delta$ by (ii) above. Thus

$$\int_{Ag_i \cap Dg_j} (U_{g_i^{-1}}f - U_{g_j^{-1}}f)^2 \geq \delta^2(\varepsilon^2/4),$$

and hence

$$\|U_{g_i^{-1}}f - U_{g_j^{-1}}f\| \geq \delta\varepsilon/2.$$

Since this holds for every i, j, the orbit of f is not precompact, which contradicts the assumption that $f \in H$.

This theorem was established in the case $G = Z$ by Krengel [15, Theorem 22].

The above theorem required no assumptions of ergodicity. When the action is ergodic, the following theorem provides a significant generalization; the proof gives a new proof of Theorem 7.1 in the ergodic case.

THEOREM 7.3. *Let $\phi: X \to Y$ be a G-factor map. Let $H \subset L^2(X)$ be the closed subspace generated by the G-invariant fields of finite dimensional spaces, where $L^2(X)$ is considered as a Hilbert bundle over Y. (We suppose X is ergodic.) Then there exists an essential set $X' \subset X$, a G-space Z, and a sequence of factor G-maps $X' \to Z \to Y$ whose composition is ϕ, such that $L^2(Z) = H$ (and hence, Z has relatively discrete spectrum over Y).*

We begin the proof with the following:

LEMMA 7.4.   *Let X be an ergodic G-space. Suppose $H \subset G$ is a countable dense subgroup, and $f: X \to \mathbf{R}$ is a Borel function such that for each $h \in H$, $f(xh) = f(x)$ almost everywhere. Then f is constant on a conull set.*

Proof. If not, there exist α, $\beta \in \mathbf{R}$ such that $A = \{x \mid \alpha \le f(x) \le \beta\}$ has positive measure less than 1. But $f(xh) = f(x)$ almost everywhere implies for each h, $\chi_A(x) = \chi_A(xh)$ almost everywhere. Thus, $U_h(\chi_A) = \chi_A$ in $L^2(X)$, and since U is continuous and $H \subset G$ dense, χ_A is G-invariant in $L^2(X)$. This implies A is null or conull, which is a contradiction.

LEMMA 7.5.   *Let $V = \int^{\oplus} H_y$ be a G-invariant field of subspaces of the Hilbert bundle $L^2(X) = \int^{\oplus} L^2(F_y)$, such that dim $H_y = n < \infty$ for almost all $y \in Y$. Suppose $f_i \in L^2(X)$ such that (i) $\|(f_i)_y\| \le 1$ for every y (here, $(f_i)_y = f_i \mid F_y$); (ii) for almost all $y \in Y$, $\{(f_i)_y\}_{i=1,\ldots,n}$ is an orthonormal basis of H_y. Then each $f_i \in L^\infty(X)$.*

Proof. Let α be the natural cocycle representation; we can assume $\|\alpha(y, g)\| \le 1$ for all (y, g). Let

$$a_{ij}(y, g) = \langle \alpha(y, g)(f_j)_{yg} \mid (f_i)_y \rangle_y.$$

Then $a_{ij}(y, g)$ is Borel, $|a_{ij}(yg)| \le 1$ by (i) above, and for each g, $a_{ij}(y, g)$ is a unitary matrix for almost all y. Choose H to be a countable dense subgroup of G, and define

$$\theta_j(x) = \sup_{g \in H} \left| \sum_{i=1}^{n} a_{ij}(\phi(x), g) f_i(x) \right|.$$

This exists since $|a_{ij}(y, g)| \le 1$ and is Borel since H is countable. Now let $h \in H$. Then

$$\theta_j(xh) = \sup_{g \in H} \left| \sum_{i=1}^{n} a_{ij}(\phi(x)h, g) f_i(xh) \right|.$$

But for almost all x,

$$f_i(xh) = (f_i)_{\phi(x)h}(xh) = [\alpha(\phi(x), h)(f_i)_{\phi(x)h}](x)$$
$$= \left[\sum_k a_{ki}(\phi(x), h)(f_k)_{\phi(x)} \right](x)$$
$$= \sum_k a_{ki}(\phi(x), h) f_k(x).$$

So for almost all x,

$$\theta_j(xh) = \sup_{g \in H} \left| \sum_k \sum_i a_{ki}(\phi(x), h) a_{ij}(\phi(x)h, g) f_k(x) \right|$$
$$= \sup_{g \in H} \left| \sum_{k=1}^{n} a_{kj}(\phi(x), hg) f_k(x) \right|$$
$$= \theta_j(x).$$

Thus, by Lemma 7.4, θ_j is constant on a conull set. For almost all x,

$$\theta_j(x) \geq \left| \sum_i a_{ij}(\phi(x), e)f_i(x) \right| = |f_j(x)|.$$

Thus for each j, $|f_j(x)|$ is bounded on a conull set, i.e., $f_j \in L^\infty(X)$.

Proof of Theorem 7.3. Let $A = \{f \in L^\infty(X) \mid f$ is contained in a G-invariant field of finite dimensional subspaces$\}$. It is straightforward to check that A is a subspace of $L^\infty(X)$, closed under complex conjugation, and multiplication by elements of $L^\infty(Y)$. We claim it is also closed under multiplication. Let f, $h \in A$, $f \in V = \int^\oplus H_1(y)$, $h \in W = \int^\oplus H_2(y)$, V, W G-invariant, dim $H_i(y) < \infty$. Let $f_i(x) \in L^2(X)$ such that $\{(f_i)_y\}_i$ is an orthonormal basis of $H_1(y)$ almost everywhere, and $h_j(x) \in L^2(X)$ such that $\{(h_j)_y\}_j$ is an orthonormal basis of $H_2(y)$ almost everywhere. Since $f \in V$, $h \in W$, there exist Borel functions $a_i(y)$, $b_i(y)$ such that $f(x) = \sum a_i(\phi(x))f_i(x)$ and $h(x) = \sum b_j(\phi(x))h_j(x)$ almost everywhere. For almost all y, $\|f_y\|^2 = \sum_i |a_i(y)|^2$, so

$$|a_i(y)|^2 \leq \|f_y\|^2 \leq \|f\|_\infty^2.$$

Thus, each $a_i \in L^\infty(Y)$, and similarly, $b_j \in L^\infty(Y)$. Now $f \cdot h = \sum_{i,j} (a_i \circ \phi) \cdot (b_j \circ \phi)f_ih_j$, so by the remarks above, to see that $fh \in A$, it suffices to see that $f_ih_j \in A$. We have $f_ih_j \in L^\infty(X)$ by Lemma 7.5, and hence (changing functions on a null set if necessary), for each $y \in Y$, $(f_i)_y(h_j)_y \in L^2(F_y)$. Let $Z(y)$ be the subspace of $L^2(F_y)$ spanned by $\{(f_i)_y(h_j)_y\}_{i,j}$. To see that $f_ih_j \in A$, it suffices to see that $Z = \int^\oplus Z(y)$ is G-invariant, and for this, to see that for a given i, j, g, $U_g(f_ih_j) \in Z$. Let

$$(U_gf_i)(x) = \sum_k a_{ki}(\phi(x))f_k(x) \quad \text{almost everywhere,}$$

and

$$(U_gh_j)(x) = \sum_p b_{pj}(\phi(x))h_p(x) \quad \text{almost everywhere,}$$

where a_{ki}, b_{pj} are Borel functions on Y. Since $U_gf_i \in L^\infty(X)$, we see as above that $a_{ki}(y) \in L^\infty(Y)$ and similarly that $b_{pj} \in L^\infty(Y)$. Thus,

$$U_g(f_ih_j)(x) - (U_gf_i)(U_gh_j)(x) = \sum_{k,p} a_{ki}(\phi(x))b_{pj}(\phi(x))f_k(x)h_p(x) \in Z.$$

Using the same technique in a somewhat simpler setting, one can prove the following companion to Theorem 7.3.

THEOREM 7.6. *Let $\phi: X \to Y$ be a factor map of ergodic G-spaces. Let $H^1 \subset L^2(X)$ be the closed subspace generated by the G-invariant fields of one dimensional subspaces. Then there exists an essential set $X' \subset X$, a G-space Z', and a sequence of factor maps $X' \to Z' \to Y$ whose composition is ϕ, such that $H^1 = L^2(Z')$. (Hence Z' has relatively elementary spectrum.) If Z is as in Theorem 7.3, then $B(Z') \subset B(Z) \subset B(X)$, so Z' is also a factor of Z.*

Theorem 7.3 says that if Y is a factor of X, then the subspace of $L^2(X)$ which corresponds to the discrete part of the spectrum of the natural (Y, G) cocycle is given by $L^2(Z)$, where Z is a space "between" X and Y. We now turn to the question of when the natural cocycle representation has (nontrivial) finite dimensional subrepresentations. If $Y = \{e\}$, this will be the case if and only if the space $X \times X$ is not ergodic [19, Proposition 1]. When $Y \neq \{e\}$, we shall see below that the relevant consideration is the ergodicity of the fibered product $X \times_Y X$. It will be convenient to consider a somewhat more general question, namely the ergodicity of the fibered product $X \times_Y Z$, where Y is a factor of both X and Z. [We note that $X \times_Y Z$ is a G-invariant subset of $X \times Z$ under the product action, and it is straightforward to check that the measure on $X \times_Y Z$ defined in Section 1 is G-invariant.]

LEMMA 7.7.   *Let $\phi\colon X \to Y$ be a factor G-map of ergodic G-spaces. Then the natural cocycle representation α contains the identity one-dimensional cocycle exactly once.*

Proof. $L^2(Y) \subset L^2(X)$ is a G-invariant field of one-dimensional spaces and α restricted to $L^2(Y)$ is the identity. Now suppose $V = \int^\oplus V(y)$ is a one-dimensional G-invariant field, and that restricting $\alpha(y, g)$ to V gives us a cocycle equivalent to 1. Then there exist maps $U(y)\colon V(y) \to \mathbf{C}$ such that $U(y)$ is unitary almost everywhere and for each g, $U(y)\alpha(y, g)U(yg)^{-1} = 1$ almost everywhere. Let $f = \int^\oplus f_y, f \in L^2(X)$, where $f_y = U(y)^{-1}(1)$ almost everywhere. Then

$$(\alpha(y, g)f_{yg})(x) = (\alpha(y, g)U(yg)^{-1}(1))(x) = (U(y)^{-1}(1))(x) = f_y(x)$$

almost everywhere. But by definition of α, $(\alpha(y, g)f_{yg})(x) = f(xg)$ almost everywhere. By the ergodicity of X, f is essentially constant so $f_y \in \mathbf{C}$ for almost all y. Since f_y is a basis of $V(y)$ almost everywhere, $V = L^2(Y)$.

THEOREM 7.8.   *Let (X, μ), (Z, v), (Y, m) be ergodic Lebesgue G-spaces and $\phi\colon X \to Y$, $\psi\colon Z \to Y$ G-factor maps. Then $X \times_Y Z$ is an ergodic G-space if and only if the natural cocycle representations (which we denote α_X and α_Z) do not have a common finite dimensional subcocycle representation other than the identity.*

Proof. (i)   We suppose that α_X and α_Z have a finite dimensional cocycle in common. Then it is easy to see that there exist Borel functions $f_i(x)$, $h_i(z)$, $a_{ij}(y, g)$ such that:

(a)   For almost all y, $(f_1)_y, (f_2)_y, \ldots, (f_n)_y, 1$ are mutually orthogonal; $(h_1)_y, \ldots, (h_n)_y, 1$ are mutually orthogonal.
(b)   For each g, $(a_{ij}(y, g))$ is a unitary matrix for almost all y.
(c)   For each g and almost all x,

$$f_j(xg) = \sum_i a_{ij}(\phi(x), g)f_i(x) \quad \text{and} \quad h_j(zg) = \sum_i a_{ij}(\psi(z), g)h_i(z).$$

Now define $\theta(x, z) = \sum_{i=1}^{n} f_i(x)\overline{h_i(z)}$. Then $\theta \in L^\infty(X \times_Y Z)$ by Lemma 7.5 and it follows from (a) that $\theta \perp 1$ in $L^2(X \times_Y Z)$. Further, for each g and almost all $(x, z) \in X \times_Y Z$, we have

$$\theta(xg, zg) = \sum_j f_j(xg)\overline{h_j(zg)}$$

$$= \sum_j \left(\sum_i a_{ij}(\phi(x), g)f_i(x) \right)\left(\overline{\sum_k a_{kj}(\psi(z), g)h_k(z)} \right)$$

$$= \sum_{i, k} \left(\sum_j a_{ij}(\phi(x), g)\overline{a_{ij}(\psi(z), g)}f_i(x)\overline{h_k(z)} \right).$$

Since $\phi(x) = \psi(z)$, we obtain, using (b),

$$\theta(xg, zg) = \sum_{i, k} \delta_{ik}f_i(x)\overline{h_k(z)} = \theta(x, z).$$

Thus θ is nonconstant and essentially G-invariant, which shows $X \times_Y Z$ is not ergodic.

(ii) We now show the converse. If $X \times_Y Z$ is not ergodic, choose $\theta \in L^\infty(X \times_Y Z)$ to be nonconstant and G-invariant. For each $y \in Y$, define

$$T_y : L^2(\phi^{-1}(y)) \to L^2(\psi^{-1}(y))$$

by

$$(T_y\lambda)(z) = \int_{\phi^{-1}(y)} \theta(x, z)\lambda(x)\, d\mu_y(x).$$

Then $\{T_y\}$ is a Borel field of compact linear operators, and $T = \int^\oplus T_y$ is a bounded linear operator, $T : L^2(X) \to L^2(Z)$. Letting U and W be the natural representations of G on $L^2(X)$ and $L^2(Z)$ respectively, we claim that T is an intertwining operator for U and W. It suffices to see that for each $g \in G$, $T_y\alpha_X(y, g) = \alpha_Z(y, g)T_{yg}$ for almost all y. If $\lambda \in L^2(\phi^{-1}(yg))$, then

$$(T_y \circ \alpha_X(y, g)\lambda)(z) = \int_{\phi^{-1}(y)} \theta(x, z)(\alpha_X(y, g)\lambda)(x)\, d\mu_y(x)$$

$$= \int_{\phi^{-1}(y)} \theta(x, z)\lambda(xg)\, d\mu_y(x).$$

On the other hand,

$$(\alpha_Z(y, g)T_{yg}\lambda)(z) = (T_{yg}\lambda)(zg)$$

$$= \int_{\phi^{-1}(yg)} \theta(w, zg)\lambda(w)\, d\mu_{yg}(w)$$

$$= \int_{\phi^{-1}(y)} \theta(xg, zg)\lambda(xg)\, d\mu_y(x).$$

Since θ is G-invariant, this becomes $\int_{\phi^{-1}(y)} \theta(x, z)\lambda(xg) \, d\mu_y(x)$ and comparing with the equation above, we see that T is an intertwining operator. Now let $A = T^*T = \int^{\oplus} T_y^* T_y$. Then

$$AU_g = T^*TU_g = T^*W_gT = U_gT^*T = U_gA.$$

Thus A is a self-intertwining operator for U, and $A = \int^{\oplus} A_y$, where $A_y = T_y^* T_y$ is compact and self-adjoint. Let $\lambda_1(y) = \sup \{c \mid c$ is an eigenvalue of $A_y\}$. Then $\lambda_1(y)$ is Borel and hence, if $V_1(y) = \{v \in L^2(\phi^{-1}(y)) \mid A_y(v) = \lambda_1(y)v\}$, then $\{V_1(y)\}$ is a subbundle of $L^2(X)$ over Y. Since for each g,

$$\alpha_X(y, g)^{-1} A_y \alpha_X(y, g) = A_{yg}$$

for almost all y, we have $\lambda_1(yg) = \lambda_1(y)$ (for each g and almost all y). By ergodicity of Y, this implies λ_1 is constant on a conull set. Hence $V_1 = \int^{\oplus} V_1(y)$ is G-invariant. Now let

$$\lambda_2(y) = \inf \{c \mid c \text{ is an eigenvalue of } A_y \mid V_1(y)^{\perp}\},$$

and

$$V_2(y) = \{v \in V_1(y)^{\perp} \mid A_y(v) = \lambda_2(y)v\}.$$

As above, λ_2 is essentially constant and $V_2 = \int^{\oplus} V_2(y)$ is a G-invariant subbundle. Continuing inductively, using the spectral theorem for compact self-adjoint operators, we can obtain (after suitable relabelling) the following decomposition: there exist real numbers $\lambda_0 = 0$, $\lambda_i \neq 0$, $i = 1, \ldots$, and G-invariant subbundles of $L^2(X)$, $V_i = \int^{\oplus} V_i(y)$, $i = 0, \ldots$, such that:

(i) $L^2(X) = \sum_{i=0}^{\infty} V_i$ and
(ii) $V_i(y) = \{v \in L^2(\phi^{-1}(y)) \mid A_y(v) = \lambda_i v\}$

for almost all y. Thus $V_0 = \ker (A)$. Since $A = T^*T$, we also have $V_0 = \ker (T)$. If $i > 0$, V_i is a finite-dimensional G-invariant field. So $T(V_i)$ will be a finite dimensional G-invariant subfield of $L^2(Z)$, and the hypothesis of the theorem together with Lemma 7.7 show that $T(V_i) \subset L^2(Y)$. Since this holds for each i, we have $T(L^2(X)) \subset L^2(Y)$, i.e., for almost all y, $T_y(L^2(\phi^{-1}(y)) = \mathbf{C} \subset L^2(\psi^{-1}(y))$. It follows that for almost all y, $\theta(x, z)$ is essentially independent of z. But then the G-invariance of θ and the ergodicity of X imply θ is essentially constant. This is a contradiction and completes the proof.

DEFINITION 7.9. If $X \to Y$ is a factor G-map of ergodic G-spaces, call X relatively weakly mixing over Y if $X \times_Y X$ is ergodic.

When $Y = \{e\}$, this is just the usual notion of weak mixing.
Theorem 7.8 has the following corollaries.

COROLLARY 7.10. *X is relatively weakly mixing over Y if and only if the natural Y × G cocycle representation contains no finite dimensional subcocycle representations other than the identity.*

COROLLARY 7.11. *X is relatively weakly mixing over Y if and only if $X \times_Y Z$ is ergodic for every ergodic extension Z of Y.*

COROLLARY 7.11. *If X is relatively weakly mixing over Y, so is $X \times_Y X$.*

Proof. If Z is an ergodic extension of Y, then $(X \times_Y X) \times_Y Z = X \times_Y (X \times_Y Z)$. Two applications of Corollary 7.11 imply $(X \times_Y X) \times_Y Z$ is ergodic, and it follows from the same corollary that $X \times_Y X$ is relatively weakly mixing over Y.

8. Generalized discrete spectrum and separating sieves

If X is an ergodic extension of Y, we have considered the notion of X having relatively discrete spectrum over Y. We now consider a more general class of extensions, which we shall call extensions with generalized discrete spectrum over Y. Loosely, these will be extensions built up from Y by the operations of taking extensions with relatively discrete spectrum, and taking inverse limits. Formally, this is done in the same way as Furstenberg's notion of quasi-isometric extension of a continuous flow is built up by isometric extensions and limits [4; Definition 2.4]. Thus, some formal aspects of what follows will be similar to those in [4]. Later, we shall discuss the relationship between the content of [4] and the content of the results of this section.

We begin with some remarks on inverse limits of G-spaces. Let η be a countable ordinal. Suppose for each ordinal $\gamma < \eta$ we have a Lebesgue G-space X_γ and for each pair of ordinals $\sigma < \gamma < \eta$ a factor G-map $\phi_{\gamma\sigma}: X_\gamma \to X_\sigma$ such that for any triple $\beta < \sigma < \eta$, the diagram

commutes. Now suppose X is a G-space, $X' \subset X$ essential, and for each γ we have a factor map $p_\gamma: X' \to X_\gamma$ such that for any σ, γ the following diagram commutes:

$$X' \to X_\gamma$$
$$\searrow \downarrow$$
$$X_\sigma$$

Then we call $\{X, p_\gamma, X_\gamma, \phi_{\gamma\sigma}\}$ an ordered system of factors of X. We say that $X = \text{inj lim } X_\gamma$ if $L^2(X) = \overline{\bigcup_{\gamma < \eta} L^2(X_\gamma)}$ or equivalently, $B(X)$ is the σ-algebra generated by $\bigcup B(X_\gamma)$. We also point out that X can be characterized in terms of $\{X_\gamma\}$ by a universal property.

PROPOSITION 8.1. *If Y is a Lebesgue G-space and there exist factor maps $q_\gamma: Y \to X_\gamma$ such that all diagrams*

commute, then there exists an essential set $Y' \subset Y$, and a factor G-map $Y' \to X'$ such that

commutes. Any two such factor maps agree on a conull set. If $\overline{X}$ also (in addition to X) has this property, then X and $\overline{X}$ are essentially isomorphic.

Proof. We have maps $B(X_\gamma) \to B(Y)$. Under the metric $d(A, B) = \mu(A \bigtriangleup B)$, these spaces are complete metric spaces and the maps are isometric. Since the maps are compatible, we have an isometry $\bigcup_\gamma B(X_\gamma) \to B(Y)$ and as $\bigcup B(X_\gamma)$ is dense in $B(X)$, this extends to an isometry $B(X) \to B(Y)$. Since G acts on $B(X)$ and $B(Y)$ by isometries, the map $B(X) \to B(Y)$ is a G-map. Thus, by Proposition 2.1, there is an essential set $Y_0 \subset Y$ and a factor G-map $\theta: Y_0 \to X'$ inducing the Boolean G-map $B(X) \to B(Y)$. For each γ, $p_\gamma \circ \theta = q_\gamma$ on an essential $Y_\alpha \subset Y_0$. Since η is a countable ordinal, $Y' = \bigcap Y_\alpha$ is essential, and $\theta \mid Y': Y' \to X'$ is the required map. The remaining assertions are straightforward.

PROPOSITION 8.2. *If $X = \mathrm{inj}\ \lim X_\gamma$ and each X_γ is ergodic, so is X.*

Proof. If X is not ergodic, there exists $f \in L^2(X)$, $f \perp \mathbf{C}$, $f \neq 0$ such that $U_g f = f$ for every $g \in G$. Now $L^2(X) \ominus \mathbf{C} = \overline{\bigcup L^2(X_\gamma) \ominus \mathbf{C}}$. So if P_γ is orthogonal projection onto $L^2(X_\gamma) \ominus \mathbf{C}$, then $P_\gamma f \neq 0$ for some γ. But since $L^2(X_\gamma) \ominus \mathbf{C}$ is G-invariant, P_γ commutes with all U_g and this implies $P_\gamma f$ is also G-invariant, contradicting the ergodicity of X_γ.

PROPOSITION 8.3. *If $X = \mathrm{inj}\ \lim X_\gamma$, then there exists a conull set $Z \subset X$ such that $x, y \in Z$ implies there exists γ such that $p_\gamma(x) \neq p_\gamma(y)$.*

Proof. We first give an alternative description of $\mathrm{inj}\ \lim X_\gamma$. Namely, let

$$W = \left\{ (x_\gamma) \in \prod_\gamma X_\gamma \mid \phi_{\gamma\sigma}(x_\gamma) = x_\sigma \quad \text{for all } \gamma, \sigma \right\}.$$

Then W is a standard Borel space, and by the Kolmogorov consistency theorem [21; Theorem 5.1], admits a probability measure for which $W = \mathrm{inj}\ \lim X_\gamma$, where $W \to X_\gamma$ is just projection on X_γ. Since W clearly satisfies the requirements from the Proposition, the result now follows from Proposition 8.1.

We now introduce extensions with generalized discrete spectrum.

DEFINITION 8.4. Let X, Y ergodic Lebesgue G-spaces, and X an extension of Y. We say that X has generalized discrete spectrum over Y if there exists a countable ordinal η, and an ordered system of factors $(X_\gamma, \gamma < \eta)$ of X such that, calling $X = X_\eta$,

(i) $X_0 = Y$,

(ii)   For each $\gamma < \eta$, $X_{\gamma+1}$ has relatively discrete spectrum over X_γ (and is a nontrivial extension of X_γ),

(iii)   If $\gamma \leq \eta$ is a limit ordinal, then $X_\gamma = $ inj lim X_σ $\sigma < \gamma$.

If the factors X_γ can be chosen so that $X_{\gamma+1}$ has relatively elementary spectrum over X_γ, we shall say that X has simple generalized discrete spectrum over Y. If $Y = \{e\}$, we shall omit the phrase "over $\{e\}$."

In light of the structure theorem (4.3) and Corollary 4.6, one has a description of the structure of any action with generalized or simple generalized discrete spectrum. The question now arises as to what conditions on a G-space will imply that it has generalized discrete spectrum, or more generally, generalized discrete spectrum over a given factor. We will show that there is a very satisfactory answer to this question. The following definition generalizes a notion due to Parry [20]. It was originally introduced by him as a measure-theoretic analogue of a distal transformation.

DEFINITION 8.5.   Let $\phi: X \to Y$ a factor G-map of ergodic G-spaces, and let $S_1 \supset S_2 \supset \cdots$ be a sequence of Borel sets in X such that $\mu(S_n) > 0$, $\mu(S_n) \to 0$. Then $\{S_n\}$ is called a separating sieve over Y if for every countable set $N \subset G$, there exists a conull set $A \subset X$ such that $x, y \in A$, $\phi(x) = \phi(y)$, and for each n, $xg_n, yg_n \in S_n$ for some $g_n \in N$, implies $x = y$. $\{S_n\}$ will be called a separating sieve if it is a separating sieve over $\{e\}$.

An immediate but important property of separating sieves is the following.

PROPOSITION 8.6.   *Suppose $X \to Y \to Z$ are factor G-maps, and that $\{S_n\}$ is a separating sieve for X over Z. Then it is also a separating sieve for X over Y.*

We now state the main result of this section.

THEOREM 8.7.   *If X is an ergodic extension of Y, then X has generalized discrete spectrum over Y if and only if X is either atomic or has a separating sieve over Y.*

Before proving this theorem, we make some remarks on the relationship of this theorem to Furstenberg's work in topological dynamics. There are numerous analogies between topological dynamics and ergodic theory. (See [5], for example.) An extension of a Lebesgue G-space Y of the form $Y \times_\alpha K/H$ can be considered a measure-theoretic analogue of Furstenberg's topological notion of isometric extension (see [4] for this and other related concepts mentioned below), and in light of the structure theorem (Theorem 4.3), an action with generalized discrete spectrum is analogous to a quasi-isometric flow. The main theorem of [4] asserts that among the minimal flows, the quasi-isometric ones are exactly the distal flows. As Theorem 8.7 asserts, when $Y = \{e\}$, that the actions with generalized discrete spectrum are the actions with a separating sieve (trivial cases aside), we can view this theorem as the measure-theoretic analogue of Furstenberg's structure theorem. Despite some formal similarities, the proofs are basically quite different. The difficult part of Furstenberg's proof makes heavy use of the Ellis semigroup and its properties for minimal distal flows, which is not available in the measure-theoretic situation. Our proof makes use of the results of Section 7.

We begin the proof of Theorem 8.7 with some lemmas.

LEMMA 8.8. *If Y is nonatomic and has relatively discrete spectrum over an atomic factor Z, then Y has a separating sieve.*

Proof. Since Z is atomic and ergodic, it is essentially transitive, and hence we can assume $Z = G/G_0$ for some closed subgroup $G_0 \subset G$. Any $G/G_0 \times G$ cocycle is cohomologous to a strict one [23, Lemma 8.26], and for a strict cocycle α into a compact group K, $(z, [k])g = (zg, [k]\alpha(z, g))$ defines not only a near action of G on $Z \times K/H$, but an action. Thus by the structure theorem, discarding invariant null sets, we can assume $Y = Z \times_\alpha K/H$ for a strict co-cycle α, and that the factor map $Y \to Z$ is given by projection of $Z \times K/H$ onto Z. Since Y is nonatomic, so is K/H, and we can choose a decreasing sequence of open neighborhoods U_i of $[e]$ in K/H such that $\bigcap_i U_i = \{[e]\}$ and $\mu(U_i) \to 0$. Choose an atom $z_0 \in Z$; we claim $\{z_0\} \times U_i$ is a separating sieve for $Z \times K/H$. If

$$(z_1, [k_1])g_n, \ (z_2, [k_2])g_n \in \{z_0\} \times U_n \quad \text{for some } g_n \in G,$$

then $z_1 g_n = z_2 g_n = z_0$ implies $z_1 = z_2$. Further,

$$[k_1]\alpha(z_1, g_n) \in U_n, \qquad [k_2]\alpha(z_1, g_n) = [k_2]\alpha(z_2, g_n) \in U_n$$

implies $[k_1] = [k_2]$, by the existence of a K-invariant metric on K/H.

LEMMA 8.9. *Suppose $\phi: X \to Y$, $\theta: Y \to Z$ are factor G-maps of ergodic G-spaces such that* (i) *X has relatively discrete spectrum over Y, and* (ii) *Y has a separating sieve $\{S_n\}$ over Z. Then X has a separating sieve over Z.*

Proof. By the structure theorem, we can sssume $X = Y \times_\alpha K/H$ and that $\phi(y, [k]) = p(y, [k])$ almost everywhere (here $p(y, [k]) = y$). Choose a decreasing sequence of open neighborhoods U_i of $[e]$ in K/H such that $\bigcap_i U_i = \{[e]\}$, and let $\bar{S}_n = S_n \times U_n$. Then $\bar{S}_n$ is decreasing and $\mu(\bar{S}_n) \to 0$, $\mu(\bar{S}_n) > 0$. We claim that $\bar{S}_n$ is a separating sieve for X over Z. Let N be a countable subset of G, and $A \subset Y$ as in the definition of a separating sieve for Y over Z (see Definition 8.5). For each $g \in N$, let

$$A_g = \{(y, [k]) \mid (y, [k])g = (yg, [k]\alpha(y, g))\},$$

and

$$\bar{A} = \{(y, [k]) \mid \phi(y, [k]) = y\}.$$

Then A_g and $\bar{A}$ are conull, and hence so is

$$A' = \bigcap_{g \in N} A_g \cap \bar{A} \cap p^{-1}(A).$$

568   ROBERT J. ZIMMER

Now suppose $(y_1, [k_1])$, $(y_2, [k_2]) \in A'$ with $\theta \circ \phi(y_1, [k_1]) = \theta \circ \phi(y_2, [k_2])$. Since $A' \subset \bar{A}$, $\theta(y_1) = \theta(y_2)$. If we also have $(y_i, [k_i])g_n \in \bar{S}_n$, $i = 1, 2$ where $g_n \in N$, then $y_i \cdot g_n \in S_n$ since $A' \subset \bigcap A_g$. As $A' \subset p^{-1}(A)$, $y_1, y_2 \in A$, and since S_n is a separating sieve for Y over Z, it follows that $y_1 = y_2$. Furthermore, we have $[k_1]\alpha(y_1, g_n) \in U_n$, and $[k_2]\alpha(y_1, g_n) = [k_2]\alpha(y_2, g_n) \in U_n$. Since K/H admits a K-invariant metric, $[k_1] = [k_2]$. This completes the proof.

LEMMA 8.10.   *Suppose X, Y, X_1, X_2, ... are ergodic G-spaces, and that there exist factor maps $p_n: X \to X_n$ and $\theta_n: X_n \to Y$ such that*

commutes for each n, p. Suppose further that there exists a conull set $Z \subset X$ such that $x, y \in Z$, $x \neq y$ implies there exists n_0 such that $p_{n_0}(x) \neq p_{n_0}(y)$. Then if each X_n has a separating sieve over Y, so does X.

Proof. We recall that if A, B are sets of positive measure in an ergodic G-space, then there exists $g \in G$ such that $Ag \cap B$ has positive measure.

Let $\{S_n^i\}_n$ be a separating sieve for X_i over Y. Because X is ergodic, we can choose $g_n^i \in G$ such that if

$$A_n = \bigcap_{i=1}^{n} p_i^{-1}(S_n^i) \cdot g_n^i,$$

then $\mu(A_n) > 0$. Now let $S_1 = A_1$ and define S_n inductively as follows: choose $h_n \in G$ such that $\mu(S_{n-1} \cap A_n \cdot h_n) > 0$, and let $S_n = S_{n-1} \cap A_n h_n$. Then S_n is a decreasing sequence of sets of positive measure, and $\mu(S_n) \to 0$, since

$$S_n \subset A_n h_n \subset p_1^{-1}(S_n^1 g_n^1 h_n),$$

and $\mu_1(S_n^1) \to 0$ (where μ_1 is the measure on X_1). We now claim that $\{S_n\}$ is a separating sieve for X over Y. Let $N \subset G$ be countable, and for each i, let $N^i = \bigcup_{n=i}^{\infty} N h_n^{-1}(g_n^i)^{-1}$. Let $B^i \subset X_i$ be the corresponding null set for the separating sieve over Y, $\{S_n^i\}$ (i.e., given N^i). Now let $B \subset X$ be $B = \bigcap_i p_i^{-1}(B^i) \cap Z$. Then B is conull. Suppose $x, y \in B$, with $p(x) = p(y)$ (where $p = \theta_n p_n$, which is independent of n), and that $xg_n, yg_n \in S_n$ for some $g_n \in N$. When $i \leq n$, let $h_n^i = g_n h_n^{-1}(g_n^i)^{-1}$, and $h_n^i = h_n^i$ when $i > n$. So $h_n^i \in N^i$ for each i, n. Now for each i, and $n \geq i$,

$$xh_n^i = xg_n h_n^{-1}(g_n^i)^{-1} \in S_n h_n^{-1}(g_n^i)^{-1} \subset A_n(g_n^i)^{-1} \subset p_i^{-1}(S_n^i),$$

i.e., $p_i(x)h_n^i \in S_n^i$ when $n \geq i$; from this it follows that $p_i(x)h_n^i \in S_n^i$ for all (i, n). Similarly, $p_i(y)h_n^i \in S_n^i$ for all (i, n). But since $x, y \in B$, $p_i(x), p_i(y) \in B^i$, we have $p_i(x) = p_i(y)$, and this holds for all i. Hence, since $x, y \in Z$, $x = y$, and this completes the proof.

We are now ready to prove half of Theorem 8.7.

Proof of Theorem 8.7 *(Part One)*.   We suppose that X has generalized discrete spectrum over Y, and that X is not atomic. We claim it has a separating sieve over Y. We consider two cases.

Case 1. Y is not atomic. Then consider the set $S = \{\gamma \le \eta \mid X_\gamma \text{ has a}$ separating sieve over $Y\}$. Since Y is not atomic, $0 \in S$. If $\gamma \in S$, and $\gamma < \eta$, then $\gamma + 1 \in S$ by Lemma 6.9. If γ is a limit ordinal, then $X_\gamma = \text{inj lim } X_\sigma$, $\sigma < \gamma$. If each $\sigma \in S$, it follows by Lemma 8.10 and Proposition 8.3 that $\gamma \in S$. (Recall that η is a countable ordinal.) Thus $\eta \in S$ by transfinite induction.

Case 2. Y is atomic. Let $T = \{\gamma \le \eta \mid X_\gamma \text{ is atomic}\}$. Let $\eta_0 = \sup T$. We consider two subcases:

(a) $\eta_0 \in T$. Then $\eta_0 < \eta$, $X_{\eta_0 + 1}$ is not atomic, and it follows from Lemma 8.8 that $X_{\eta_0 + 1}$ has a separating sieve. Following the argument of Case 1, we conclude that X does also.

(b) $\eta_0 \notin T$. Then η_0 is a limit ordinal and $X_{\eta_0} = \text{inj lim } X_\gamma$, $\gamma \in T$. Since each X_γ is atomic, it has discrete spectrum and hence so does X_{η_0}. Thus X_{η_0} has a separating sieve by Lemma 8.8 (take $Z = \{e\}$), and again one can use the argument of Case 1 to complete the proof.

Given the results of Section 7, we shall see that the essence of what remains to prove the converse assertion of Theorem 8.7 is the following lemma. This lemma generalizes a result of Parry [20, Theorem 3] by adapting his argument to the case at hand.

LEMMA 8.11.   *If $\phi: X \to Y$ is a (nontrivial) factor map of ergodic G-spaces, and X has a separating sieve over Y, then X is not relatively weakly mixing over Y.*

Proof.   Let H be a countable dense subgroup of G. Let $\{S_n\}$ be a separating sieve for X over Y, and let $A \subset X$ be a conull set for the sequence H as in Definition 8.5. It is easy to check that

$$\{(x, y) \mid x, y \in A, \phi(x) = \phi(y), \text{ and } xg_n, yg_n \in S_n \text{ for some sequence } g_n \in H\}$$

$$= \bigcap_{n=1}^{\infty} \left(\bigcup_{g \in H} (S_n \times_Y S_n)g^{-1} \right) \cap A \times_Y A.$$

Saying that $\{S_n\}$ is a separating sieve over Y means, given the choice of A, that this set is contained in the diagonal $D \subset X \times_Y X$. Since ϕ is not an essential isomorphism, $(\mu \times_Y \mu)(D) \ne 1$. If $(\mu \times_Y \mu)(D) > 0$, then D is a nonnull nonconull invariant set, so $X \times_Y X$ is not ergodic. In the case when $(\mu \times_Y \mu)(D) = 0$, we must have

$$\lim_{n \to \infty} (\mu \times_Y \mu) \left(\bigcup_{g \in H} (S_n \times_Y S_n)g^{-1} \right) = 0,$$

since $A \times_Y A$ is conull. Since $\mu(S_n) > 0$, we have $(\mu \times_Y \mu)(S_n \times_Y S_n) > 0$, and thus for some n, we must have

$$0 < (\mu \times_Y \mu)\left(\bigcup_{g \in H} (S_n \times_Y S_n)g^{-1} \right) < 1.$$

But this is an H-invariant Borel set. Since the natural representation of G on $L^2(X \times_Y X)$ is continuous and H is dense, $\bigcup_{g \in H} (S_n \times S_n)g^{-1}$ must be essentially G-invariant, which shows that $X \times_Y X$ is not ergodic.

Proof of Theorem 8.7 (Part Two). If X is atomic, it has discrete spectrum, so we suppose that X has a separating sieve over Y. We consider the collection $\mathscr{C}$ of factor spaces Z of X with generalized discrete spectrum over Y, together with an ordered system of factors of Z, $\{p_\gamma, Z_\gamma, \sigma_{\gamma\sigma}\}$, satisfying Definition 8.4. We identify systems which are isomorphic modulo invariant null sets. Let η_Z be the ordinal such that $Z = Z_{\eta_Z}$. We define an ordering on the set $\mathscr{C}$ as follows. Given $\{Z, p_\gamma, Z_\gamma, \phi_{\gamma\sigma}\}$ and $\{Z', p'_\gamma, Z'_\gamma, \phi'_{\gamma\sigma}\}$, define $Z < Z'$ if $\eta_Z \leq \eta_Z$ and for all γ, $\sigma \leq \eta_Z$, $Z_\gamma = Z'_\gamma$, $p_\gamma = \phi_{\eta_Z\gamma}$, $\phi_{\gamma\sigma} = \phi'_{\gamma\sigma}$ modulo G-invariant null sets. We claim any totally ordered subset $T \subset \mathscr{C}$ has an upper bound. Let $S = \{\eta_Z \mid Z \in T\}$. If S has a maximal element, clearly T does also. If not, let $\eta = \sup S$. Since for each $\sigma < \gamma < \eta$ we have closed subspaces $L^2(Z_\sigma) \subsetneq L^2(Z_\gamma)$ of $L^2(X)$, and $L^2(X)$ is separable, η must be a countable ordinal. Let $Z_\eta = \text{inj lim } Z$, $Z \in T$. It is clear that Z_η is in $\mathscr{C}$ and is an upper bound for T. By Zorn's lemma, there exists a maximal element $Z \in \mathscr{C}$. We claim $Z = X$ (modulo invariant null sets). Suppose not. Then by Proposition 8.6, X has a separating sieve over Z. It follows from Lemma 8.11 that X is not relatively weakly mixing over Z. By Corollary 7.10 the natural (Z, G) cocycle on $L^2(X)$ has nontrivial finite dimensional subcocycle representations, and by Theorem 7.3, there exists a factor space Z' of X such that Z' has relatively discrete spectrum over Z. But then we clearly have $Z' \in \mathscr{C}$ and $Z \nsubseteq Z'$, contradicting the maximality of Z. Thus $Z = X$, and X has generalized discrete spectrum over Y.

We remark that the above proof shows the following:

COROLLARY 8.12.   *Suppose Y is a factor G-space of an ergodic space X, and that for any G-space Z for which there is a sequence of factor G-maps $X' \to Z \to Y$ ($X' \subset X$ essential, and the composition the original factor map) we have X not relatively weakly mixing over Z. Then X has generalized discrete spectrum over Y.*

9. Examples and further properties

An ergodic extension with a relative separating sieve can be viewed as a measure-theoretic analogue of the topological notion of distal extension. We recall the definition of the latter. Let X and Y be compact metric spaces on which G acts continuously, and $\phi : X \to Y$ a continuous surjective G-map.

GENERALIZED DISCRETE SPECTRUM   571

X is called a distal extension of Y if $x, y \in X$, $\phi(x) = \phi(y)$, and $d(xg_n, yg_n) \to 0$ for some sequence $g_n \in G$ implies $x = y$. The following is immediate.

PROPOSITION 9.1. *Suppose X is a distal extension of Y, that X is minimal [4], and that X has nonatomic G-invariant ergodic probability measure μ. Then X has a separating sieve over Y, and hence generalized discrete spectrum over Y.*

Proof. X minimal implies every open set has positive measure; if U_i is a decreasing sequence of open sets whose intersection is a point, it is trivial to check that $\{U_i\}$ is a separating sieve over Y. The remaining statement is just Theorem 8.7.

COROLLARY 9.2. *A minimal distal action preserving an ergodic probability measure has generalized discrete spectrum.*

We consider a specific example to illustrate this corollary. For assertions not proven below, see [2].

Example 9.3. Let N be the nilpotent Lie group consisting of matrices of the form

$$M = \begin{pmatrix} 1 & x & z \\ 0 & 1 & y \\ 0 & 0 & 1 \end{pmatrix}$$

where $x, y, z \in \mathbf{R}$. For notational convenience, we denote M by $[x, y, z]$. Let $D \subset N$ be the discrete subgroup consisting of matrices M such that x, y, z are integers. Then N/D is compact and has an N-invariant probability measure. The commutator subgroup is $[N, N] = \{M \in N \mid x = y = 0\}$, and the quotient $N/D[N, N]$ is a torus. It is easy to see that the functions $g_n(M) = \exp(2\pi inx)$ and $h_n(M) = \exp(2\pi iny)$ factor to functions on $N/D[N, N]$ when n is an integer, and $\{h_n g_j\}_{(n, j) \in Z^2}$ is an orthonormal basis of $L^2(N/D[N, N])$. Now let $A \in N$ be a matrix of the form $[a, b, 0]$ where $a, b, 1$ are rationally independent. Then $A(t) = [ta, tb, \frac{1}{2}abt^2]$ is the 1-parameter subgroup in N with $A(1) = A$ and $[M]t = [MA(t)]$ defines an action of $\mathbf{R}$ on N/D which is measure preserving. By [2, Theorem IV 3, Theorem V 4.2, and Corollary V 4.5] this action is ergodic, minimal and distal. It is immediate that for each $(j, n) \in Z^2$, $g_j h_n$ is an eigenfunction of the flow, and by [2, Theorem V 4.2], constant multiples of these are the only eigenfunctions. Thus $L^2(N/D[N, N]) \subset L^2(N/D)$ is the closed subspace generated by the finite dimensional $\mathbf{R}$-invariant subspaces. Now let $f_n(M) = \exp(2\pi in(z - y[x]))$ where $[x]$ is the largest integer $\le x$. Then the closed subspace of $L^2(N/D)$ generated by $\{g_j h_n f_k\}_{j, n, k \in z}$ is all of $L^2(N/D)$. Now

$$[x, y, z]A(t) = [ta + x, tb + y, \tfrac{1}{2}abt^2 + tbx + z].$$

So

$$(g_j h_n f_k)([x, y, z]A(t)) = \theta(t, x, y) \exp(2\pi ijx) \exp(2\pi iny) \exp(2\pi ik(z - y[x]))$$

572   ROBERT J. ZIMMER

where

$$\theta(t, x, y) = \exp (2\pi i(jta + ntb + kabt^2/2 + ktbx + ky[x]$$
$$- k(tb + y)[ta + x])).$$

Thus

$$(g_j h_n f_k)([x, y, z]A(t)) = \theta(t, x)(g_j h_n f_k)([x, y, z]).$$

But for each t, $(x, y) \mapsto \theta(t, x, y)$ is in $L^\infty(N/D[N, N])$. Thus for each $(j, n, k) \in Z^3$, $g_j h_n f_k$ is contained in a 1-dimensional field of subspaces over $N/D[N, N]$ that is $\mathbf{R}$-invariant. Since these functions generate $L^2(N/D)$, N/D has relatively elementary spectrum over $N/D[N, N]$, and N/D has simple generalized discrete spectrum. In this case, the ordinal $\eta = 2$.

Example 9.4. We remark that for continuous G-actions, the condition of having a separating sieve is more general than being distal. This follows from an example of Kolmogorov of a continuous, ergodic, measure-preserving flow on the torus with discrete spectrum, but no continuous eigenfunctions [see 27]. This flow thus has a separating sieve, but we claim it is not distal. Since every open set has positive measure, and the flow is ergodic, it is also regionally transitive [2, p. 57]. If it were distal, it would be pointwise almost periodic [3, Theorem 1], and hence minimal [2, p. 57]. But minimal distal flows have continuous eigenfunctions [4].

Example 9.5. Another class of actions with generalized discrete spectrum are those with quasi-discrete spectrum. In the case where $G = Z$, these were first studied systematically by Abramov [1]. Subsequently, Wieting has considered these actions when G is an arbitrary locally compact abelian group. We review Wieting's definition. Let G be a locally compact abelian group and X a Lebesgue G-space. We suppose that X is totally ergodic, i.e., that $\{\chi \in G^* \mid \chi$ is a subrepresentation of U_g on $L^2(X)\}$ is torsion free. Let $E_0 = S^1$ (=circle) and define E_n, $n \geq 1$, inductively by

$$E_n = \{f \in L^\infty(X) \mid |f(x)| = 1 \quad \text{and} \quad U_g f/f \in E_{n-1} \quad \forall g \in G\}.$$

If $E = \bigcup_{n=0}^\infty E_n$ generates $L^2(X)$, then X is said to have quasi-discrete spectrum. We show that this implies that X has generalized discrete spectrum. It is clear that E_n is an increasing sequence of G-invariant multiplicative subgroups of the group of functions of absolute value 1 on X. Generalizing a result of Abramov [1, 7°], Wieting showed [24, Theorem P] that if f, $g \in E$ are not constant multiples of one another, they are perpendicular. Now the finite linear combinations of elements of E_n form a G-invariant *-subalgebra of $L^\infty(X)$. By Corollary 2.2, there exists a sequence of factors X_n of X

$$\to X_n \to X_{n-1} \to \cdots \to X_1 \to X_0 = \{e\}$$

such that E_n is an orthonormal basis, together with constant multiples, of $L^2(X_n)$, and

$$L^2(X) = \overline{\bigcup L^2(X_n)}.$$

If $f \in E_n$, then $(U_g f)/f \in E_{n-1}$ and thus $(U_g f)/f$ is a function on X_{n-1} for each $g \in G$. Therefore, each $f \in E_n$ is contained in a 1-dimensional G-invariant field over X_{n-1}, and since E_n generates $L^2(X_n)$, X_n has relatively elementary spectrum over X_{n-1}. Thus, X has simple generalized discrete spectrum, and the ordinal $\eta_X \leq \omega$, the first infinite ordinal.

We remark that even when $G = Z$ and the ordinals are finite, not every G-space with simple generalized discrete spectrum has quasi-discrete spectrum. If we restrict the **R**-action of Example 9.3 to the integers, the resulting Z-action is still ergodic (this follows from [2, Theorem V 4.2]) and has generalized discrete spectrum. However, it cannot have quasi-discrete spectrum since it embeds in an **R**-action [8, Theorem 4.1]. We shall examine in Section 10 the question of how one distinguishes the transformations with quasi-discrete spectrum within the class of transformations with generalized discrete spectrum.

When $G = Z$ or **R**, any transitive action (preserving a probability measure) has discrete spectrum. For more general groups, this statement is, of course, no longer true. The following proposition describes when a transitive action has generalized discrete spectrum.

PROPOSITION 9.6. *Let $H \subset G$ be a closed subgroup such that G/H has finite invariant measure. Then the action of G on G/H has generalized discrete spectrum if and only if there exists a countable ordinal η, and a collection of closed subgroups of G, $H_\gamma \supset H$, $\gamma \leq \eta$, such that:*

(i)   *$H_0 = G$, $H_\eta = H$; if $\sigma \lneqq \gamma$, then $H_\gamma \subsetneqq H_\sigma$.*
(ii)   *The action of H_γ on $H_\gamma/H_{\gamma+1}$ has discrete spectrum.*
(iii) *If γ is a limit ordinal, $H_\gamma = \bigcap_{\sigma < \gamma} H_\sigma$.*

Proof. As every factor of a transitive action is transitive, and is determined by a (conjugacy class of a) closed subgroup, the proof is readily reduced to demonstrating the following statement: If $H \subset K \subset G$ (so $G/H \to G/K$), then G/H has relatively discrete spectrum over G/K if and only if the action of K on K/H has discrete spectrum. Now $(G/K, G)$ cocycles correspond to representations of K [23, Theorem 8.27], and the natural G/K cocycle representation on the Hilbert bundle $L^2(G/H)$ will correspond to the representation of K on the fiber over $[e]$ in $L^2(G/H)$, i.e., to the natural representation of K on $L^2(K/H)$. Under this correspondence, the $(G/K, G)$ cocycle has discrete spectrum if and only if the representation of K does also.

Example 9.7. An example of a transitive action with generalized discrete spectrum is the action of a connected, simply-connected nilpotent Lie group on a nilmanifold. If N is such a group, and $D \subset N$ a uniform, discrete subgroup, the proof of [2, Theorem IV.3] shows that N acts distally on N/D. ([2, Theorem IV.3] states that a one-parameter subgroup of N acts distally on N/D, but an examination of the proof shows that the assumption that the elements of N considered lie in a 1-parameter subgroup was never used.) By Corollary 9.2, this action has generalized discrete spectrum.

We remark that the structure of N/D given by Definition 8.4 (or Proposition 9.6) gives a corresponding decomposition of $L^2(N/D)$ into mutually orthogonal G-invariant subspaces. A thorough study of the decomposition of $L^2(N/D)$ has been made by Moore [18], Richardson [22], and Howe [13]. It would be interesting to see how the decomposition above fits into their scheme.

Another question that arises is to describe which subgroups of a given group, say in particular, which lattice subgroups, define homogeneous spaces with generalized discrete spectrum. One might then try to obtain an understanding of the decomposition of L^2 of the homogeneous space, based upon the L^2-decomposition defined via Definition 8.4.

If G is an abelian group, and X is a transitive G-space, every irreducible (X, G) cocycle representation is one-dimensional. This is because such cocycles correspond to the representations of the stability group of the action. It is thus perhaps somewhat surprising to find that if X is not transitive, there may exist irreducible cocycle representations of dimension greater than one, even if G is abelian. In [17] (see also [14]), Mackey gives an example of an ergodic G-space X, with G abelian, and a minimal cocycle $\alpha \colon X \times G \to K$, where $K = K_\alpha$, and K is compact but not abelian. Thus, by Proposition 3.12, there exist irreducible (X, G) cocycle representations that have dimension greater than one. Equivalently, there exist extensions of X with relatively discrete but not relatively elementary spectrum over X. In virtual group terms, a virtual subgroup of an abelian group can have nonabelian "homomorphic images."

In topological dynamics, there is another example of nonabelian phenomena arising from an abelian situation. If G acts continuously on a compact metric space X, let $\phi(g)$ denote the homeomorphism of X corresponding to $g \in G$. Let $E(G, X)$ be the closure of $\phi(G)$ in X^X under the topology of pointwise convergence. $E(G, X)$ can be shown to be a semigroup (under composition) [4, p. 484] and is called the Ellis semigroup of the action. Now even if G (and hence $\phi(G)$) is abelian, $E(G, X)$ may not be.

We now point out in the consideration of distal actions, the occurrence of these types of nonabelian phenomena are related.

PROPOSITION 9.8. *Suppose G is a locally compact abelian group, and X is a compact metric space, minimal and distal under a continuous G-action, and supporting an ergodic probability measure. If $E(G, X)$ is abelian, then X has simple generalized discrete spectrum.*

Proof. This follows from [28, Theorem 1.2], once one notices that restricted to each fiber, Image (P_λ) is one-dimensional. (Notation as in [28].) One can see this from the definition of P_λ ([28, p. 18]), and the fact that $I_y \subset E(G, X)$ is abelian.

We now turn to consideration of the properties of factors of actions with generalized discrete spectrum.

LEMMA 9.10. *Suppose* $X \to Y \to Z$ *are factor G-maps and that X has generalized discrete spectrum over Z. Then Y is not relatively weakly mixing over Z.*

Proof. Let $\{X_\gamma \mid \gamma \leq \eta\}$ be the factors of X showing that X has generalized discrete spectrum over Z. Consider

$$S = \{\gamma \leq \eta \mid (L^2(X_\gamma) \ominus L^2(Z)) \perp (L^2(Y) \ominus L^2(Z))\}.$$

Assuming that Y is a nontrivial extension of Z, $\eta \notin S$. Let σ be the first ordinal not in S. It follows from property (iii) of Definition 8.4 that σ is not a limit ordinal. Hence, $\sigma - 1$ exists, and is a maximal element of S. Let us denote $X_{\sigma-1}$ by W. By Proposition 1.5, we have a factor map $X \to W \times_Z Y$ such that the following diagram commutes:

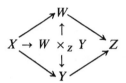

Now projection of $L^2(W \times_Z Y)$ into $L^2(X_\sigma)$ is a G-map commuting with multiplication by $L^\infty(W)$. Furthermore, since $\sigma \notin S$, the image of $L^2(W \times_Z Y)$ in $L^2(X_\sigma)$ is not contained in $L^2(W)$. Since X_σ has relatively discrete spectrum over W, it follows that $W \times_Z Y$ must have some discrete spectrum over W; i.e., $(W \times_Z Y) \times_W (W \times_Z Y)$ is not ergodic. But it is easy to see that this space is isomorphic to $W \times_Z Y \times_Z Y$, which is thus not ergodic. If Y is relatively weakly mixing over Z, then so is $Y \times_Z Y$ (Corollary 7.12), and then $W \times_Z Y \times_Z Y$ would be ergodic (Corollary 7.11). Thus Y is not relatively weakly mixing over Z.

THEOREM 9.11. *If* $X \to Y \to Z$ *are factor G-maps, and X has generalized discrete spectrum over Z, so does Y.*

Proof. By Corollary 8.12, it suffices to show that if $Y \to Y_0 \to Z$ are factor maps, then Y is not relatively weakly mixing over Y_0. But if X has generalized discrete spectrum over Z, it also has generalized discrete spectrum over Y_0 by Theorem 8.7 and Proposition 8.6. The result now follows by Lemma 9.10.

COROLLARY 9.12. *A factor of an action with generalized discrete spectrum also has generalized discrete spectrum.*

We remark that the analogous result holds for distal actions [4, Theorem 3.3], and for transformations with quasi-discrete spectrum [8, Corollary 2.8].

COROLLARY 9.13. *The fibered product of two extensions of a G-space Y does not have generalized discrete spectrum over Y if one of the factors is relatively weakly mixing over* Y. *In particular, the product of two G-spaces will not have generalized discrete spectrum if one of them has continuous spectrum.*

PROPOSITION 9.14. *If T is an invertible transformation such that the associated Z-action has generalized discrete spectrum, then T has entropy 0. More generally, a Z^n action with generalized discrete spectrum has 0 joint entropy* [24].

Proof. By Theorem 8.7, if the space is not atomic, the action has a separating sieve, and an argument of Parry [20] (see also [24, Theorem N]) shows that the entropy is zero.

10. Applications to affine actions and quasi-discrete spectrum

We now consider how the above theory applies to a special class of actions, namely affine actions on compact abelian groups. Affine transformations have been studied by various authors [6], [11], and much of this theory has been generalized to affine actions of arbitrarily locally compact abelian groups by Wieting [24]. Theorem 10.7 below gives an algebraic criterion for an affine action to have generalized discrete spectrum. When $G = Z$, we go on to show in Theorem 10.10 that every totally ergodic affine transformation on a compact connected abelian group with generalized discrete spectrum actually has quasi-discrete spectrum. In light of Abramov's results [1], this enables us to distinguish the class of totally ergodic transformations with quasi-discrete spectrum as the class (up to isomorphism) of totally ergodic affine transformations on compact, connected, abelian groups with generalized discrete spectrum.

We recall the central notions of the theory of affine actions [6], [24]. Let G be a locally compact abelian group and X a compact abelian group. A homeomorphism $\phi: X \to X$ is called affine if it is of the form $\phi(x) = x_0 A(x)$, where $A: X \to X$ is an automorphism, and $x_0 \in X$. If G acts continuously on X, by affine homeomorphisms, we will call X an affine G-space. Then, for each $x \in X$, $g \in G$, we can write $x \cdot g = x_0(g) \cdot A(g)(x)$, where $x_0(g) \in X$ and $A(g) \in \operatorname{Aut}(X)$. The map $A: G \to \operatorname{Aut}(X)$ is a continuous homomorphism, and $x_0: G \to X$ is a continuous crossed homomorphism with respect to A; i.e.,

$$x_0(gh) = x_0(g) \cdot [A(g)(x_0(h))].$$

Conversely, given A and x_0, satisfying the above, they define an affine action. We will thus identify affine actions with pairs (x_0, A). If X and Y are affine G-spaces, we shall call X an affine extension of Y if there exists a surjective G-homomorphism $\phi: X \to Y$.

PROPOSITION 10.1. *Suppose the affine actions of G on X and Y are given by (x_0, A), (y_0, B) respectively. Then a surjective homomorphism $\phi: X \to Y$ is a G-map if and only if $\phi(x_0(g)) = y_0(g)$ and $\phi \circ A(g) = B(g) \circ \phi$.*

If $A: X \to X$ is an automorphism, let $A^*: X^* \to X^*$ be the induced automorphism of dual groups.

GENERALIZED DISCRETE SPECTRUM 577

PROPOSITION 10.2. *Suppose (x_0, A) is an affine action of G on X, and suppose $D \subset X^*$ is an $A(G)^*$-invariant subgroup. Let $\phi: X \to Y = D^*$ be the map induced by inclusion. Let $\tilde{A}: G \to \text{Aut}(Y)$ be the map $\tilde{A}(g) = (A(g)^* \mid D)^*$ and $y_0(g) = \phi(x_0(g))$. Then $(y_0, \tilde{A})$ is an affine action of G on Y, and ϕ is then a G-map.*

A criterion for determining when a given ergodic affine G-space is weakly mixing was established for Z-actions by Hahn [6, Corollary 3], and subsequently extended to arbitrary abelian group actions by Wieting [24, Theorem H]. An extension of this analysis will enable us to determine when an ergodic affine extension is relatively weakly mixing, and more generally, when it has relatively generalized discrete spectrum. We begin with some preparatory lemmas.

If $\phi: X \to Y$ is a surjective homomorphism, let $K = \text{ker } \phi$. Let μ_K, μ_X, μ_Y be the Haar measures. Choose a Borel section $\theta: Y \to X$ for ϕ. The following lemma is then straightforward.

LEMMA 10.3. *$y \mapsto \mu_y = \mu_K \cdot \theta(y)$ is a decomposition of μ_X with respect to μ_Y over the fibers of ϕ.*

We have an induced map $\phi^*: Y^* \to X^*$ that is injective, and we shall identify Y^* with its image in X^*. Then the inclusion $K \to X$ induces an isomorphism $X^*/Y^* \to K^*$.

LEMMA 10.4. *If $f, g \in X^*$ and $f \not\equiv g$ in X^*/Y^*, then $f_y \perp g_y$ for each $y \in Y$ (where $f_y = f \mid \phi^{-1}(y)$).*

Proof. Let $f_0 = f \mid K$. Since $f \not\equiv g$ in X^*/Y^*, $f_0 \neq g_0$ which implies $f_0 \perp g_0$. Hence $f_0 \cdot \theta(y)^{-1} \perp g_0 \cdot \theta(y)^{-1}$ in $L^2(\phi^{-1}(y), \mu_y)$, for each $y \in Y$. But for $x \in \phi^{-1}(y)$,

$$(f_0 \cdot \theta(y)^{-1})(x) = f(x\theta(y)^{-1}) = f_y(x)f(\theta(y)^{-1}).$$

Similarly,

$$(g_0 \cdot \theta(y)^{-1})(x) = g_y(x)g(\theta(y)^{-1}).$$

It follows immediately that $f_y \perp g_y$.

With ϕ as above, $\psi: X \times X \to Y$ defined by $\psi(x, z) = \phi(x)\phi(z)^{-1}$ is a surjective homomorphism, and $\text{ker } \psi = X \times_Y X$. Thus, $X \times_Y X$ is a compact abelian group. We give a realization of its dual.

LEMMA 10.5. *Let $s: X^*/Y^* \to X^*$ be a section (not necessarily homomorphic) of the natural projection $p: X^* \to X^*/Y^*$, with $s([1]) = 1$. Then the map*

$$T: Y^* \times s(X^*/Y^*) \times s(X^*/Y^*) \to (X \times_Y X)^*$$

defined by $T(h, f, g) = h(f \times_Y g)$ is a bijection.

Proof. We note first that the range of T is clearly contained in $(X \times_Y X)^*$. From Lemma 10.4, it follows that the T-images of distinct elements are orthogonal, and hence that T is injective. We now claim T is surjective. Any character of $X \times_Y X$ is of the form $\lambda \times \beta \mid X \times_Y X$ where $\lambda, \beta \in X^*$ [10, 24.12]. We can write $\lambda = h_1 s(p(\lambda))$ and $\beta = h_2 s(p(\beta))$ where $h_i \in Y^*$. Then we have

$$\lambda \times \beta \mid X \times_Y X = h_1 h_2(s(p(\lambda)) \times_Y s(p(\beta)))$$

showing surjectivity.

Now suppose that the homomorphism $\phi: X \to Y$ is an extension of ergodic affine G-spaces, where G is locally compact and abelian. The following result is a partial generalization of [6, Corollary 3] and [24; Theorem H].

THEOREM 10.6. *X is relatively weakly mixing over Y if and only if every nonidentity element in X^*/Y^* has an infinite orbit under $A(G)^*$ (where the action on X is given by (x_0, A)).*

Proof. (i) Suppose $f \in X^*/Y^*$, $f \not\equiv 1$ and that f has a finite orbit, say $f = f_1, f_2, \ldots, f_n$ under $A(G)^*$. Then the closed subspace of $L^2(X)$ generated by

$$\left\{ \sum_{i=1}^n h_i f_i \mid h_i \in L^\infty(Y) \right\}$$

is a G-invariant finite dimensional subbundle of $L^2(X)$ over Y, that is not equal to $L^2(Y)$. Thus X is not relatively weakly mixing by Corollary 7.10.

(ii) Conversely, suppose every nonidentity element in X^*/Y^* has infinite orbit. We claim $X \times_Y X$ is ergodic. Let the affine action of G on Y be given by (y_0, B). The action of G on $X \times_Y X$ is also affine, say (z_0, C). If every nontrivial orbit in $(X \times_Y X)^*$ under $C(G)^*$ is infinite, $X \times_Y X$ is ergodic by [24; Theorem C]. (See also [6, Theorem 1] when $G = Z$.) So suppose $\lambda \in (X \times_Y X)^*$ has a finite orbit. By Lemma 10.5, $\lambda = h \cdot (f \times_Y k)$, for $h \in Y^*, f, k \in s(X^*/Y^*)$. Now for $g \in G$,

$$(*) \qquad C(g)^*(\lambda) = C(g)^*(h \cdot (f \times_Y k)) = B(g)^* h(A(g)^* f \times_Y A(g)^* k).$$

We claim that λ having a finite orbit implies that f and k have finite orbits in X^*/Y^*. To see this, suppose $C(g)^* \lambda = \lambda$. Now $A(g)^* f = \alpha f_0$ and $A(g)^* k = \beta k_0$ where $\alpha, \beta \in Y^*$ and $f_0, k_0 \in s(X^*/Y^*)$. Equation $(*)$ implies

$$C(g)^*(\lambda) = \alpha\beta B(g)^* h(f_0 \times_Y k_0),$$

and since $C(g)^* \lambda = \lambda$, we have

$$\alpha\beta B(g)^* h(f_0 \times_Y k_0) = h(f \times_Y k).$$

By Lemma 10.5, $f = f_0$, $k = k_0$. Thus, $C(g)^* \lambda = \lambda$ implies $A(g)^* f \equiv f$ and $A(g)^* k \equiv k$ in X^*/Y^*, showing that f and k have finite orbits in X^*/Y^*. By the hypothesis of the theorem, $f \equiv k \equiv 1$ in X^*/Y^*, and since $s(1) = 1$, $f = k = 1$. This is turn, via equation $(*)$ implies that h has a finite orbit under $B(G)^*$. By [24, Theorem C] (see also [6, Theorem 4]), the ergodicity of Y implies that there is $g \in G$ such that $h(y_0(g)) \neq 1$. Since $\lambda = h$, it readily follows that $\lambda(z_0(g)) \neq 1$, and by [24, Theorem C] ([6, Theorem 4]), that $X \times_Y X$ is ergodic.

Theorem 10.6 provides an algebraic criterion for determining when an affine extension is relatively weakly mixing. We now establish an algebraic criterion for determining when an affine extension has relatively generalized discrete spectrum.

THEOREM 10.7.   *Suppose $\phi: X \to Y$ is an affine extension. Then the following are equivalent.*

(a)   *X has generalized discrete spectrum over Y.*

(b)   *There exists a countable ordinal ξ, a collection of compact abelian groups X_η, $\eta \leq \xi$, and for each $\eta < \sigma \leq \xi$, a surjective, noninjective homomorphism $\phi_{\sigma\eta}: X_\sigma \to X_\eta$ such that:*

(i)   *For $\rho < \eta < \sigma$, $\phi_{\sigma\rho} = \phi_{\eta\rho}\phi_{\sigma\eta}$.*
(ii)   *$X_0 = Y$, $X_\xi = X$ and $\phi_{\xi 0} = \phi$.*
(iii) *Each X_η is an affine G-space and $\phi_{\sigma\eta}$ is a G-map.*
(iv)   *Every element of $X_{\eta+1}^*/X_\eta^*$ has a finite orbit under $A(G)^*$. (Here dual groups are identified with their images under the induced embeddings.)*
(v)   *If η is a limit ordinal, $X_\eta = \text{inj} \lim_{\sigma < \eta} X_\sigma$.*

(c)   *There exists a countable ordinal ξ and a collection of $A(G)^*$ invariant subgroups D_η of X^*, $\eta \leq \xi$ such that:*

(i)   *$D_0 = Y^*$, $D_\xi = X^*$.*
(ii)   *$\sigma < \eta$ implies $D_\sigma \subsetneqq D_\eta$.*
(iii) *Every element of $D_{\eta+1}/D_\eta$ has finite orbit under $A(G)^*$.*
(iv)   *If η is a limit ordinal, $D_\eta = \bigcup_{\sigma < \eta} D_\sigma$.*

Proof. (b) $\Rightarrow$ (a)   It suffices to see that $X_{\eta+1}$ has relatively discrete spectrum over X_η. This follows from condition (b)(iv), as in the proof of (i) of Theorem 10.6.

(b) $\Rightarrow$ (c)   Let $D_\eta = \phi_{\xi\eta}^*(X_\eta^*)$.

(c) $\Rightarrow$ (b)   Let $X_\eta = D_\eta^*$, and for $\sigma < \eta$, let $\phi_{\eta\sigma}: X_\eta \to X_\sigma$ be the map induced by the inclusion $D_\sigma \to D_\eta$. X_η is an affine G-space by Proposition 10.2, and the remaining assertions follow easily.

(a) $\Rightarrow$ (c)   Let $\{D_\eta \mid \eta \leq \xi\}$ be a maximal collection of subgroups satisfying the conditions of (c), with the possible exception of the condition $D_\xi = X^*$. This exists by Zorn's lemma, as in the proof of Theorem 8.7. Let $Z = D_\xi^*$. Then by Proposition 10.2, Z is an affine G-space, and there are G-homomorphisms

$$X \xrightarrow{\phi_1} Z \xrightarrow{\phi_2} Y$$

such that $\phi_2\phi_1 = \phi$. We claim $Z = X$. Suppose not. Now X has relatively generalized discrete spectrum over Y, and hence also over Z by Theorem 8.7 and Proposition 8.6. In particular X is not relatively weakly mixing over Z.

By Theorem 10.6, there exists $f \in X^*$, $f \notin Z^*$, such that the orbit of f in X^*/Z^* under $A(G)^*$ is finite. Let $D_{\xi+1} = \{h \in X^* \mid \text{orbit of } h \text{ under } A(G)^* \text{ in } X^*/Z^*$ is finite$\}$. Then $D_{\xi+1}$ is an $A(G)^*$-invariant subgroup of X^*, and $D_{\xi+1} \neq D_\xi$. By the definition of $D_{\xi+1}$, we see that $\{D_\eta \mid \eta \leq \xi\}$ is not maximal, which is a contradiction. Thus $Z = X$.

We now use Theorem 10.6 to prove that when $G = Z$, any totally ergodic affine transformation with generalized discrete spectrum actually has quasi-discrete spectrum. We begin with a lemma that is a small modification of Abramov's uniqueness theorem for transformations with quasi-discrete spectrum [1]. (See also [7; Theorem 3]).

LEMMA 10.8. *Let K and X be compact metric spaces, each with a probability measure, positive on open sets. Let S and T be totally ergodic measure-preserving homeomorphisms of K and X respectively. Suppose that S and T have quasi-discrete spectrum, with quasi-eigenfunction groups E_S and E_T such that:*

(i)   *$E_S \subset C(K)$, $E_T \subset C(X)$.*
(ii) *The linear spans $[E_S]$ and $[E_T]$ are uniformly dense in $C(K)$ and $C(X)$ respectively.*
(iii) *The system of quasi-eigenvalues of S and T are equivalent [1], [7].*

Then there exists a homeomorphism $\phi: X \to K$ such that $S\phi = \phi T$.

Proof. Under the assumption that the systems of quasi-eigenvalues are equivalent, Abramov constructs [1; proof of uniqueness theorem] (using somewhat different notation) a unitary map $V: L^2(K) \to L^2(X)$, such that:

(i) $V \mid E_S$ is a group isomorphism $E_S \to E_T$.
(ii) $T^*V = VS^*$, where T^*, S^* are the induced maps in L^2.

It follows as in [9; p. 47] or [24; proof of Theorem A], that $V(L^\infty(K)) = L^\infty(X)$ and that V is an isometry of these Banach spaces. In particular, $V: [E_S] \to [E_T]$ is an involutive, multiplicative isometry (since open sets have positive measure) of dense *-subalgebras of $C(K)$ and $C(X)$, and hence is an involutive multiplicative isometry $C(K) \to C(X)$. It follows that there is a homeomorphism $\phi: X \to K$ such that $\phi^* = V: C(K) \to C(X)$. Since $T^*\phi^* = \phi^*S^*$, we also have $\phi T = S\phi$.

COROLLARY 10.9. *Let T be a totally ergodic affine transformation of a compact abelian group X, with quasi-discrete spectrum, and whose group of quasi-eigenfunctions consists of the constant multiples of elements of X^*. Then T is totally minimal [7].*

Proof. By [7; Theorem 4 and Corollary to Theorem 7], there exist S and K such that all the hypotheses of Lemma 10.8 are satisfied and such that S is totally minimal. The conclusion of Lemma 10.8 implies that T is totally minimal.

THEOREM 10.10.   *A totally ergodic transformation has quasi-discrete spectrum if and only if it is isomorphic to a totally ergodic affine transformation on a compact, connected abelian group with generalized discrete spectrum.*

Proof. (i)   Any totally ergodic transformation with quasi-discrete spectrum is isomorphic to an affine transformation on a compact, connected abelian group by a theorem of Abramov [1], and it has generalized discrete spectrum by Example 9.5.

(ii)   Conversely, let $T = (a, A)$ be a totally ergodic affine transformation of a compact connected abelian group X, and assume that T has generalized discrete spectrum. Consider the set $\mathscr{A}$ of all quotient groups Y of X on which T induces an affine transformation T_Y on Y, such that T_Y is minimal. We can order $\mathscr{A}$ by setting $Y > Z$ if Z is in turn a quotient of Y. If $\{Y_\alpha\}$ is a totally ordered collection in $\mathscr{A}$, we claim inj lim Y_α is again in $\mathscr{A}$. T will induce a transformation on inj lim Y_α by Proposition 10.2, since the dual (inj lim $Y_\alpha)^* = \bigcup_\alpha Y_\alpha^*$ is A^*-invariant. Furthermore, $T_{(\text{inj lim } Y_\alpha)}$ is minimal by the remark of Furstenberg [5; p. 28] that the inverse limit of minimal transformations is minimal. By Zorn's lemma, $\mathscr{A}$ has a maximal element Y. Now T_Y is a minimal affine transformation on a compact connected abelian group Y. Since Y is connected, T_Y is actually totally minimal [7; p. 310]. It follows from the theorem of [12] that for each n, T_Y^n has quasi-discrete spectrum, and by [11; Theorem 3] that the quasi-eigenfunctions of T_Y^n are exactly the multiples of elements of Y^*.

To prove the theorem, it suffices to show $X = Y$. Suppose not. Since X has generalized discrete spectrum, it has generalized discrete spectrum over Y, and hence is not relatively weakly mixing over Y. By Theorem 10.6, there exists $f \in X^*, f \notin Y^*$, such that f has a finite orbit under A^* in X^*/Y^*. Hence, there is an integer n such that $(A^*)^n(f) \equiv f$ in X^*/Y^*. Let B be the subgroup of X^* generated by $Y^*, f, A^*f, \ldots, (A^*)^{n-1}f$. Then B is invariant under A^*, and for each element $h \in B$, $(A^n)^*h \equiv h$ in X^*/Y^*. Let $Z = B^*$. By Proposition 10.2, we have an induced affine transformation T_Z on Z. Furthermore, for $h \in B$, $(T^n)^*h = c\lambda h$, where $c \in S$ and $\lambda \in Y^*$ (here S is the unit circle). Since T_Y^n has quasi-discrete spectrum with quasi-eigenfunctions $S \cdot Y^*$, this implies that T_Z^n also has quasi-discrete spectrum. Since each element of B is a quasi-eigenfunction, and T_Z^n is totally ergodic (being a factor of the totally ergodic transformation T^n), the quasi-eigenfunctions of Z are exactly the elements of $S \cdot B$ [1, 1.7]. By Corollary 10.9, T_Z^n is totally minimal, which implies that T_Z is totally minimal. Since $Z \neq Y$, this contradicts the maximality of Y. Hence $X = Y$. Therefore, T is minimal on X, and by the theorem of [12], T has quasi-discrete spectrum.

We conclude this section with another application, in a somewhat different direction, to quasi-discrete spectrum. Namely, we show how the existence theorem (6.4) can be applied to give a new proof of Abramov's existence theorem for transformations with quasi-discrete spectrum [1, paragraph 3]. This theorem has been generalized by Wieting [24, Theorem S] to actions of a locally compact abelian group, and it is in this context that we shall work. We begin by recalling in more detail (see Example 9.5) Wieting's definition of quasi-discrete spectrum [24, 3.1]. Let G be locally compact and abelian, and X a Lebesgue G-space. We assume the action is totally ergodic; i.e., one of the following equivalent [24, p. 83] statements holds:

  (i) The point spectrum is torsion free.
  (ii) If $H \subset G$ is a subgroup such that G/H is finite, then H acts ergodically on X.

Let $F = F(X) = \{f\colon X \to \mathbf{C} \mid f \text{ Borel}, |f(x)| = 1\}$, with functions identified if they agree almost everywhere. F is an abelian group under pointwise multiplication, and G acts naturally on F by $(f \cdot g)(x) = f(xg)$, for $f \in F$, $g \in G$. If $\gamma\colon G \to F$, γ is called a crossed-homomorphism if $\gamma(g_1 g_2) = \gamma(g_1)(\gamma(g_2)g_1)$. Let

$$\Gamma = \Gamma(X) = \{\gamma\colon G \to F(X) \mid \gamma \text{ is a crossed homomorphism}\}.$$

Then Γ is an abelian group under pointwise multiplication, and $h \in G$ acts on Γ by $(\gamma \cdot h)(g) = \gamma(g) \cdot h$, where the right side is the action of h on F. Define the map $Q\colon F \to \Gamma$ by $Q(f)(g) = (f \cdot g)/f$. It is easy to check that Q is a G-homomorphism. The kth order quasi-eigenfunctions are defined inductively as follows:

$$E_0 = \{f \in F \mid f \text{ is constant}\}, \quad E_k = \{f \in F \mid Q(f)(g) \in E_{k-1} \text{ for each } g\}.$$

A totally ergodic G-space X is said to have quasi-discrete spectrum if $E = \bigcup E_k$ generates $L^2(X)$ as a closed subspace. The order of X is the first integer k such that $E_k = E_{k+1}$ if this exists, and is ∞ otherwise.

$B_k = Q(E_k)$ is called the group of kth order quasi-eigenvalues. $B = \bigcup B_k$ is a G-invariant subgroup of $\Gamma(X)$ and the existence theorem is meant to answer the following question: Given an abelian group A on which G acts by homomorphisms, and an increasing sequence of G-invariant subgroups $A_0 \subset A_1 \subset \cdots \subset A$ such that $A = \bigcup A_n$, when does there exist a totally ergodic G-space with quasi-discrete spectrum such that A_k is (up to compatible isomorphisms) the group of kth order quasi-eigenvalues?

Before answering this, we need one more concept. Because G acts ergodically on X, ker $(Q) = E_0$, which is divisible. Thus, there exists a homomorphic section of Q, i.e., a homomorphism $\phi\colon B \to E$ such that $Q(\phi(\gamma)) = \gamma$ for all $\gamma \in B$. Now $Q(\phi(b \cdot g)) = Q(\phi(b) \cdot g)$ for all b, g, so there is a constant $c(b, g)$ such that

$$c(b, g) = \frac{\phi(b) \cdot g}{\phi(bg)}.$$

It is easy to check that c is a cocycle, and that it is multiplicative, i.e.,

$$c(b_1 b_2, g) = c(b_1, g)c(b_2, g).$$

We call c the cocycle defined by the section ϕ.

Wieting's generalization of Abramov's theorem is:

THEOREM 10.11. (Wieting, Abramov).   *Let A be a torsion free abelian group on which G acts by homomorphisms. For each n, let*

$$A_n = \{a \in A \mid N(g_1) \cdot \ldots \cdot N(g_n)(a) = 1, \text{ for all } g_1, \ldots, g_n \in G\},$$

where $N(g): A \to A$ is defined by $N(g)(a) = (a \cdot g)/a$. Suppose further that there exists a multiplicative cocycle $c: A \times G \to U(1)$ ($= circle$) such that the corresponding map $c_0: A_1 \to G^*$ ($= dual$ of G) defined by $c_0(a)(g) = c(a, g)$ is injective. Then there exists a totally ergodic G-space X with quasi-discrete spectrum, and a G-isomorphism $\psi: A \to B$ such that $\psi(A_n) = B_n$. Moreover, the cocycle

$$d: B \times G \to U(1), \quad d(b, g) = c(\psi^{-1}(b), g)$$

is the cocycle defined by a section.

Proof.   We claim it suffices to construct a sequence of G-spaces $\{X_n\}$ so that the following conditions are satisfied.
   (1)   There exists a factor G-map $p_n: X_n \to X_{n-1}$ ($n \geq 2$). We note that this induces maps $F(X_{n-1}) \to F(X_n)$ and $\Gamma(X_{n-1}) \to \Gamma(X_n)$, both of which we denote by p_n^*.
   (2) X_n has quasi-discrete spectrum of order n. Let B^n be the group of all quasi-eigenvalues on X_n, E^n the eigenfunctions. We further suppose that $p_n^*: B^{n-1} \to B_{n-1}^n$ is an isomorphism, where B_{n-1}^n is the group of $(n - 1)$st order quasi-eigenvalues on X_n.
   (3)   There exist G-isomorphisms $\psi_n: A_n \to B^n$ such that

$$
\begin{array}{ccc}
A_{n-1} & \xrightarrow{\psi_{n-1}} & B^{n-1} \\
\downarrow & & \downarrow{\scriptstyle p_n^*} \\
A_n & \xrightarrow{\psi_n} & B^n
\end{array}
$$

commutes.

   (4)   There exists a homomorphism $\phi_n: B^n \to E^n$ such that:

   (a) $Q \circ \phi_n$ is the identity.
   (b)

$$
\begin{array}{ccc}
B^{n-1} & \xrightarrow{\phi_{n-1}} & E^{n-1} \\
{\scriptstyle p_n^*}\downarrow & & \downarrow{\scriptstyle p_n^*} \\
B^n & \xrightarrow{\phi_n} & E^n
\end{array}
$$

commutes.
   (c)   If $d_n: B^n \times G \to U(1)$ is defined by $d_n(b, g) = c(\psi_n^{-1}(b), g)$, then d_n is the cocycle defined by the section ϕ_n.
In the case where the order of A is $n < \infty$ then X_n is the space, ϕ_n, ψ_n the maps,

required in the theorem. (By the order of A, we mean the first integer n such that $A_n = A_{n+1}$.) If order $(A) = \infty$, let $X = \text{inj lim } X_n$, and $q_n \colon X \to X_n$ the associated factor map. It follows easily that X is totally ergodic, has quasi-discrete spectrum, and using [24, Theorem P], it is clear that the group of nth order quasi-eigenvalues on X is $q_n^*(B^n)$. The compatibility conditions (3) and (4b) allow one to construct a suitable isomorphism and section. Thus, it remains to show that such a sequence of spaces X_n exists. We proceed inductively and begin by constructing X_1. From the hypothesis of the theorem, $c_0(A_1)$ is a countable subgroup of G^*, and by Corollary 6.5, there exists a G-space X_1 with discrete spectrum, and this spectrum is $c_0(A_1)$. We can naturally identify $c_0(A_1)$ with B^1, and let $\psi_1 \colon A_1 \to B^1$ the corresponding isomorphism. It is easy to check that for any section $\phi_1 \colon B^1 \to E^1$. The cocycle defined by ϕ_1 is just $c(\psi_1^{-1}(b), g)$. We now assume $X_1, \ldots, X_n$ have been constructed satisfying the above conditions. We let

$$\Gamma_n = \{\gamma \in \Gamma(X_n) \mid \gamma(g) \in E^n \quad \text{for each } g \in G\}.$$

Step 1. We begin by defining a homomorphism $\psi \colon A_{n+1} \to \Gamma_n$. We first note that if $a \in A_{n+1}$, $((a \cdot g)/a) \in A_n$ for each $g \in G$. For $a \in A_{n+1}$, define $\psi(a) \in \Gamma_n$ by

$$(*)\qquad\qquad \psi(a)(g) = c(a, g)\phi_n\left(\psi_n\left(\frac{a \cdot g}{a}\right)\right).$$

It is immediate that ψ is a homomorphism. We now derive some other properties of ψ that we will need.

(i) If $a \in A_n$, $\psi(a) = \psi_n(a)$.

Proof. By the inductive assumption (4c), for $a \in A_n$, we have

$$c(a, g) = \frac{\phi_n(\psi_n(a)) \cdot g}{\phi_n(\psi_n(a) \cdot g)}.$$

Thus,

$$\psi(a)(g) = \frac{\phi_n(\psi_n(a)) \cdot g}{\phi_n(\psi_n(a))} = Q(\phi_n\psi_n(a))(g) = \psi_n(a)(g)$$

so $\psi(a) = \psi_n(a)$.

(ii) ψ is a G-map, i.e., $\psi(a \cdot h) = \psi(a) \cdot h$ for $a \in A_{n+1}$, $h \in G$.

Proof. It suffices to see that

$$\frac{\psi(a \cdot h)}{\psi(a)} = \frac{\psi(a) \cdot h}{\psi(a)},$$

and since $((a \cdot h)/a) \in A_n$, it suffices to see by (i) that

$$\psi_n\left(\frac{a \cdot h}{a}\right)(g) = \frac{\psi(a) \cdot h}{\psi(a)}(g) \quad \text{for all } g \in G.$$

Now

$$\frac{\psi(a) \cdot h}{\psi(a)}(g) = \frac{\psi(a)(g) \cdot h}{\psi(a)(g)}$$

$$= Q(\psi(a)(g))(h)$$

$$= Q\left(\phi_n \psi_n\left(\frac{a \cdot g}{a}\right)\right)(h)$$

$$= \psi_n\left(\frac{a \cdot g}{a}\right)(h).$$

Thus it suffices to show that

(**) $$\psi_n\left(\frac{a \cdot h}{a}\right)(g) = \psi_n\left(\frac{a \cdot g}{a}\right)(h).$$

The left side is

$$\frac{\phi_n\left(\psi_n\left(\frac{a \cdot h}{a}\right)\right) \cdot g}{\phi_n\left(\psi_n\left(\frac{a \cdot h}{a}\right)\right)} = \frac{c\left(\frac{a \cdot h}{a}, g\right) \phi_n\left(\psi_n\left(\frac{a \cdot h}{a}\right) \cdot g\right)}{\phi_n \psi_n\left(\frac{a \cdot h}{a}\right)}$$

$$= c\left(\frac{a \cdot h}{a}, g\right) \phi_n \psi_n\left(\frac{(a \cdot hg)(a)}{(a \cdot h)(a \cdot g)}\right).$$

A similar expression can be derived for the right side of equation (**), and thus, it suffices to see that

$$c\left(\frac{a \cdot h}{a}, g\right) = c\left(\frac{a \cdot g}{a}, h\right).$$

Since c is multiplicative, this means $c(a, h)c(ah, g) = c(a, g)c(ag, h)$. But this follows from the cocycle identity and the commutativity of G.

(iii) ψ is injective.

Proof. If $\psi(a) = 1$, then

$$\phi_n\left(\psi_n\left(\frac{a \cdot g}{a}\right)\right)$$

is constant for each g. This implies

$$\psi_n\left(\frac{a \cdot g}{a}\right) = 1.$$

Since ψ_n is injective, $(a \cdot g)/a = 1$, which implies $a \in A_1$. So $\psi(a) = \psi_n(a) = 1$, and thus $a = 1$.

(iv) Let $\Gamma_{n-1} = \{\gamma \in \Gamma(X_n) \mid \gamma(g) \in E^n_{n-1}$ (= group of quasi-eigenfunctions of X_n of order $n - 1$)}. Then $\psi(A_{n+1}) \cap \Gamma_{n-1} = B^n$.

Proof. For each $g \in G$, let $M(g): \Gamma \to \Gamma$ be defined by $M(g)(\gamma) = (\gamma \cdot g)/\gamma$. It is easy to check that for $\gamma \in \Gamma_{n-1}$ and $g_1, \ldots, g_n \in G$, $M(g_1), \ldots, M(g_n)(\gamma) = 1$. Since ψ is an injective G-map, it is clear that if $\psi(a) \in \Gamma_{n-1}$, then $a \in A_n$. The result follows.

Step 2. We now construct the space X_{n+1} and verify the inductive conditions. Let S be the set of equivalence classes of one-dimensional cocycle representations of (X_n, G). For each $\gamma \in \Gamma(X_n)$, we have an associated element $[\alpha_\gamma] \in S$, defined by $\alpha_\gamma(x, g) = \gamma(g)(x)$. Furthermore, this map $\alpha: \Gamma(X_n) \to S$ is a homomorphism. By Corollary 6.5, there exists an ergodic extension $X_{n+1}, p_{n+1}: X_{n+1} \to X_n$ with relatively elementary spectrum such that the natural (X_n, G) cocycle representation on $L^2(X_{n+1})$ is equivalent to $\sum^{\oplus}_{\beta \in \alpha(\psi(A_{n+1}))} \beta$.

(i) We now claim $p^*_{n+1}(\psi(A_{n+1})) \subset Q(F(X_{n+1}))$. Let $\gamma \in \psi(A_{n+1})$, and choose any $f \in L^2(X_{n+1})$ such that $|f(x)| = 1$, and f is in the subspace of $L^2(X)$ corresponding to α_γ. Defining $\beta: X_n \times G \to \mathbf{C}$ by

$$\beta(p_{n+1}(x), g) = \frac{f \cdot g}{f}(x),$$

we see that β is a cocycle cohomologous to α_γ. Thus, there exists a Borel function $\theta: X_n \to U(1)$ such that $\beta(x, g) = \theta(x)\alpha_\gamma(x, g)\theta(xg)^{-1}$ (all g, almost all $x \in X_n$). Then a simple calculation shows

$$Q((\theta \circ p_{n+1}) \cdot f) = p^*_{n+1}(\gamma).$$

(ii) Let $D = Q^{-1}(p^*_{n+1}(\psi(A_{n+1})))$. It is clear that D is a G-invariant subgroup of $F(X_{n+1})$. Thus, the finite linear combinations of elements of D form a G-invariant *-subalgebra of $L^\infty(X_{n+1})$. By Corollary 2.2, there exists a factor G-space Z of X_{n+1} such that the closed subspace spanned by D in $L^2(X_{n+1})$ is $L^2(Z)$. Clearly, $p^*_{n+1}(E^n) \subset D$, so we have a sequence of factor maps $X_{n+1} \to Z \to X_n$ (modulo invariant null sets) whose composition is p_{n+1}. For each $\gamma \subset \psi(A_{n+1})$, the function $(\theta \circ p_{n+1})(f)$ constructed above is in both D and the one-dimensional field (over X_n) corresponding to $[\alpha_\gamma]$. From this it follows that this field must be contained in $L^2(Z)$. Since the union of these fields spans $L^2(X_{n+1})$, we have $L^2(Z) = L^2(X_{n+1})$. We thus know that D spans $L^2(X_{n+1})$. Since $\psi(A_{n+1}) \subset \Gamma_n$, it follows that X_{n+1} has quasi-discrete spectrum. Because $\psi(A_{n+1}) \cap \Gamma_{n-1} = B^n$ (see (iv) above), it follows that elements of

$$Q^{-1}(p^*_{n+1}(\psi(A_{n+1} - A_n)))$$

are $(n + 1)$st order but not nth order quasi-eigenvalues. By [24, Theorem P], D contains all quasi-eigenfunctions. Defining $\psi_{n+1} = p^*_{n+1} \circ \psi$, it is easy to see that the inductive assumptions (1), (2), (3) hold. It remains only to construct a section $\phi_{n+1}: B^{n+1} \to E^{n+1}$ satisfying (4).

$Q: E^{n+1} \to B^{n+1}$ has a divisible kernel, and $p_{n+1}^{*}\phi_n(p_{n+1}^{*})^{-1}: B_n^{n+1} \to E^{n+1}$ is a homomorphic section of Q defined on the subgroup $B_n^{n+1} \subset B^{n+1}$. By [10, A.8], it follows that this extends to a homomorphic section $\phi_{n+1}: B^{n+1} \to E^{n+1}$. Thus, (a) and (b) of (4) are satisfied. To verify (c), apply p_{n+1}^{*} to equation $(*)$. We obtain

$$\psi_{n+1}(a)(g) = c(a, g)\phi_{n+1}\psi_{n+1}\left(\frac{a \cdot g}{a}\right).$$

Since ϕ_{n+1} is a section of Q, we can write this as

$$\frac{\phi_{n+1}(\psi_{n+1}(a)) \cdot g}{\phi_{n+1}\psi_{n+1}(a)} = c(a, g)\frac{\phi_{n+1}\psi_{n+1}(ag)}{\phi_{n+1}\psi_{n+1}(a)}.$$

Hence

$$c(a, g) = \frac{\phi_{n+1}(\psi_{n+1}(a)) \cdot g}{\phi_{n+1}(\psi_{n+1}(a) \cdot g)}.$$

This completes the proof.

BIBLIOGRAPHY

1. L. M. ABRAMOV, *Metric automorphisms with quasi-discrete spectrum*, Amer. Math. Soc. Transl., (2), vol. 39 (1962), pp. 37–56.
2. L. AUSLANDER, L. GREEN AND F. HAHN, *Flows on homogeneous spaces*, Annals of Mathematics Studies, no. 53, Princeton, 1963.
3. R. ELLIS, *Distal transformation groups*, Pacific J. Math., vol. 8 (1958), pp. 401–405.
4. H. FURSTENBERG, *The structure of distal flows*, Amer. J. Math., vol. 85 (1963), pp. 477–515.
5. ———, *Disjointness in ergodic theory, minimal sets, and a problem in Diophantine approximation*, Math. Systems Theory, vol. 1 (1967), pp. 1–49.
6. F. HAHN, *On affine transformations of compact abelian groups*, Amer. J. Math., vol. 85 (1963), pp. 428–446.
7. F. HAHN AND W. PARRY, *Minimal dynamical systems with quasi-discrete spectra*, J. London Math. Soc., vol. 40 (1965), pp. 309–323.
8. ———, *Some characteristic properties of dynamical systems with quasi-discrete spectra*, Math Systems Theory, vol. 2 (1968), pp. 179–190.
9. P. HALMOS, *Lectures on ergodic theory*, Chelsea, New York, 1956.
10. E. HEWITT AND K. ROSS, *Abstract harmonic analysis, I*, Springer-Verlag, Berlin, 1963.
11. H. HOARE AND W. PARRY, *Affine transformations with quasi-discrete spectra*, J. London Math. Soc., vol. 41 (1966), pp. 88–96.
12. ———, *Affine transformations with quasi-discrete spectra, II*, J. London Math. Soc., vol. 41 (1966), pp. 529–530.
13. R. HOWE, *On Frobenius reciprocity for unipotent algebraic groups over Q*, Amer. J. Math., vol. 93 (1971), pp. 163–172.
14. A. A. KIRILLOV, *Dynamical systems, factors, and group representations*, Russian Math. Surveys, vol. 22 (1967), pp. 63–75.
15. U. KRENGEL, *Weakly wandering vectors and weakly independent partitions*, Trans. Amer. Math. Soc., vol. 164 (1972), pp. 199–226.
16. G. W. MACKEY, *Borel structures in groups and their duals*, Trans. Amer. Math. Soc., vol. 85 (1957), pp. 134–165.
17. ———, "Virtual groups," in *Topological dynamics*, J. Auslander and W. Gottschalk, eds., Benjamin, New York, 1968.
18. C. C. MOORE, *Decomposition of unitary representations defined by discrete subgroups of nilpotent groups*, Ann. of Math., vol. 82 (1965), pp. 146–182.

588   ROBERT J. ZIMMER

19. ———, *Ergodicity of flows on homogeneous spaces*, Amer. J. Math., vol. 88 (1966), pp. 154–178.
20. W. PARRY, "Zero entropy of distal and related transformations," in *Topological dynamics*, J. Auslander and W. Gottschalk, eds., Benjamin, New York, 1968.
21. K. R. PARTHASARATHY, *Probability measures on metric spaces*, Academic Press, New York, 1967.
22. L. RICHARDSON, *Decomposition of the L^2-space of a general compact nilmanifold*, Amer. J. Math., vol. 93 (1971), pp. 173–190.
23. V. S. VARADARAJAN, *Geometry of quantum theory, vol. II*, Van Nostrand, Princeton, 1970.
24. T. WIETING, *Ergodic affine Lebesgue G-spaces*, doctoral dissertation, Harvard University, 1973.
25. R. ZIMMER, *Extensions of ergodic actions and generalized discrete spectrum*, Bull. Amer. Math. Soc., vol. 81 (1975), pp. 633–636.
26. ———, *Extensions of ergodic group actions*, Illinois J. Math., vol. 20 (1976), pp. 373–409.
27. W. PARRY, *A note on cocycles in ergodic theory*, Composito Math., vol. 28 (1974), pp. 343–350.
28. A. W. KNAPP, *Distal functions on groups*, Trans. Amer. Math. Soc., vol. 128 (1967), pp. 1–40.

UNITED STATES NAVAL ACADEMY
ANNAPOLIS, MARYLAND

Annals of Mathematics, **106** (1977), 573–588

Orbit spaces of unitary representations, ergodic theory, and simple Lie groups[1]

By Robert J. Zimmer

1. Introduction

An important problem in ergodic theory is to determine when the restriction to a subgroup of an ergodic action of a group is still ergodic. Classical results concerning the ergodicity of rotations on compact groups, and ergodicity of the geodesic and horocycle flows are special cases of this problem. The more recent work of L. Auslander, L. Green, and F. Hahn on nilflows [1] and solvflows [2], and of C.C. Moore on actions of subgroups on homogeneous spaces of connected semisimple Lie groups [9] are again special cases, in which the theories of unitary representations and group structure are brought to bear on the problem. In all of these cases, one is largely considering the restriction to a subgroup of an action that is actually transitive and not only ergodic. One of the focuses of this paper will be to consider restrictions to subgroups of properly ergodic actions. Our techniques will depend on the theory of unitary representations of groups and unitary cocycle representations of ergodic actions, or equivalently, ergodic systems of imprimitivity.

The basic idea in using group representations for questions of ergodicity is the following. If G acts ergodically on (S, μ), where μ is a finite G-invariant measure, then there is a naturally induced unitary representation of G on $L^2(S)$ which contains the one dimensional identity representation exactly once. The restriction of the action to a subgroup $H \subset G$ will be ergodic if the restriction of the induced representation to H also contains the one dimensional identity representation only once. Thus, in principle, the question is actually one of examining the restriction of a certain unitary representation to a subgroup. For this approach to be applicable, it is important, of course, that S have a finite invariant measure. In the case in which one is dealing with transitive actions, there is a useful duality principle formulated independently by G.W. Mackey and C.C. Moore, which sometimes allows

0003–486X/77/0106–0003 $00.80

© 1977 by Princeton University Press

For copying information, see inside back cover.

[1] Research supported by the Naval Academy Research Council.

574   ROBERT J. ZIMMER

one to pass from the case of finite invariant measure to a more general situation. This principle asserts that if H_1, H_2 are closed subgroups of G, then H_1 is ergodic on G/H_2 if and only if H_2 is ergodic on G/H_1 [9, Proposition 6]. Thus, for example, if G/H_2 has a finite invariant measure, the question of ergodicity of H_2 on G/H_1 for arbitrary H_1 is equivalent to the question of ergodicity of H_1 on G/H_2, the latter action now preserving a finite measure and hopefully amenable to analysis via the theory of unitary representations. Using this approach, Moore shows for example, that $SL(n, Z)$ acts ergodically on R^n [9, corollary to Theorem 6].

In considering the problem of restricting an arbitrary (not necessarily transitive) ergodic action to a subgroup, we will see that a similar duality principle is applicable. Namely, if $H \subset G$ is a closed subgroup, and S an ergodic G-space (with a quasi-invariant measure, but not necessarily a finite invariant one), then the restriction to H of the G action on S will be ergodic if and only if the product G-action on $S \times G/H$ is ergodic (Theorem 4.2 below). We can regard $S \times G/H$ as a G-space extension of S, and if G/H has a finite invariant measure, the ergodicity of this extension can be studied by means of certain cocycle representations of the action $S \times G$. This becomes, in fact, a special case of the more general problem of studying the ergodicity of a general skew product with an arbitrary ergodic action of a locally compact group as the base, and G/H as the fiber. This type of problem is of considerable interest in its own right, and has been examined by the author for G a compact group in [15] and G a nilpotent Lie group in [16]. In this paper, aside from certain applications of the results of [16], we shall mostly be concerned with the case in which G is a connected non-compact simple Lie group and $H = \Gamma$, a lattice in G. As an example of our results, we quote the following.

THEOREM 5.3. *Let G be a connected non-compact simple Lie group with finite center, and $\Gamma \subset G$ be a lattice subgroup. Then any properly ergodic G-action is still ergodic when restricted to Γ.*

THEOREM 5.4. *Let G_i, $i = 1, \cdots, n$, be connected non-compact simple Lie groups with finite center, $G = \prod G_i$, and $\Gamma \subset G$ an irreducible lattice. Then any properly ergodic G-action is still ergodic when restricted to Γ.*

THEOREM 4.5. *Let N be a connected simply connected nilpotent Lie group which acts ergodically on S. If $[N, N]$ also acts ergodically on S, then Γ acts ergodically on S for any uniform discrete subgroup $\Gamma \subset N$.*

In applying cocycle representation techniques to study ergodicity of extensions in [16], important use was made of facts concerning spaces of orbits of finite dimensional subspaces in a Hilbert space under the action of a certain unitary representation. We will require similar results in a different context here, and hence have begun this paper with a discussion of some general connections between ergodic theory and the spaces of orbits of finite dimensional subspaces under unitary representations. Our application to simple Lie groups of these connections depends heavily on the work of C.C. Moore [9] and T. Sherman [12].

The organization of this paper is as follows. Sections 2 and 3 deal with orbits of finite dimensional subspaces and the connection with ergodic theory. Section 4 presents the duality principle and the application to nilpotent Lie groups. Restricting semi-simple Lie group actions to lattices is dealt with in Section 5, and Section 6 considers the question of ergodicity of general skew products with a semi-simple Lie group modulo a lattice as fiber. Section 7 is concerned with normal ergodic actions (that is, homomorphisms with dense range) of simple groups.

The author wishes to thank the referee for several helpful suggestions, and in particular, for bringing to his attention the relevant paper of R. Howe [20]. We also wish to thank R. Howe for providing us with a copy of that manuscript.

2. Orbit spaces of unitary representations

In this and the following section we discuss some general relationships between orbit spaces of unitary representations and ergodic theory. We shall recall the framework established for this material in [16], and refer the reader there for further details. Let H be a separable Hilbert space and $U(H)$ the unitary group of H with the strong operator topology. Let $B \subset H$ be the closed unit ball of elements whose norm does not exceed one, so that B is a bounded complete separable metric space with the metric d induced by the norm. Let $\mathcal{C}$ be the set of closed subsets of B with the Hausdorff metric ρ,

$$\rho(E, F) = \max \{\sup d(x, F), \sup d(y, E) \,|\, x \in E, y \in F\} \,.$$

For each integer $k \geq 1$, define $\mathcal{F}_k \subset \mathcal{C}$,

$$\mathcal{F}_k = \{V = V' \cap B \,|\, V' \subset H \text{ a non-zero subspace with } \dim V' \leq k\} \,.$$

Then $(\mathcal{F}_k, \rho)$ is a complete separable metric space for each k [16, Corollary 2.3] and the natural action $U(H) \times \mathcal{F}_k \to \mathcal{F}_k$, $(U, V) \to U(V)$, is jointly continuous [16, Proposition 2.5].

Suppose now that G is a second countable locally compact group, and $\pi: G \to U(H)$ is a (strongly continuous) unitary representation. Let B_1 be the

576   ROBERT J. ZIMMER

elements in B of norm one. Then there is an induced right action of G on $\mathcal{F}_k$, $V \cdot g = \pi(g)^{-1}(V)$, a similarly defined action on B_1, and one can inquire as to when the space of orbits under these actions, say $\hat{\mathcal{F}}_k$ and $\hat{B}_1$, will have a reasonably regular structure. This question arose in [16] in consideration of certain ergodicity questions, and we shall see further applications below.

Definition 2.1. Call π a k-smooth representation if $\hat{\mathcal{F}}_k$ is a standard Borel space under the quotient Borel structure, smooth if $\hat{B}_1$ is standard, and totally smooth if it is smooth and k-smooth for all $k \geq 1$.

We recall that the quotient Borel structure being standard is equivalent to a large number of other regularity conditions, and in particular, is equivalent to each orbit being homeomorphic to G/G_0, where G_0 is the stability group of a point in the orbit [4], [6].

PROPOSITION 2.2. (i) *If π is k-smooth, then π is p-smooth for $p < k$.*

(ii) *π smooth implies π is 1-smooth.*

(iii) *If π is 1-smooth and the stability groups in G of points in $\mathcal{F}_1$ are compact, then π is k-smooth for all k.*

Proof. (i) is clear.

(ii) Let K be the unit circle. Then B_1 is a $G \times K$ space, and $\hat{\mathcal{F}}_1$ is naturally isomorphic to the space of orbits in B_1 under $G \times K$. But this is just the space of orbits in $\hat{B}_1$ under K. Since K is compact, $\hat{B}_1$ standard implies $\hat{\mathcal{F}}_1$ is standard.

(iii) Let $V \in \mathcal{F}_k$ and $G_V \subset G$ the stability group at V. It suffices to show that the orbit of V is homeomorphic to G/G_V [4]; that is, if $g_n, g \in G$, then $\rho(V \cdot g_n, V \cdot g) \to 0$ if and only if $[g_n] \to [g]$ in G/G_V. By continuity of the action, it follows that $[g_n] \to [g]$ implies $\rho(V \cdot g_n, V \cdot g) \to 0$, so we need only show the converse assertion. By the usual argument, it suffices to show that a subsequence of g_n converges to g. Let $x_1, \cdots, x_k$ be an orthonormal basis of V. Since $\rho(V \cdot g_n, V \cdot g) \to 0$, it is straightforward to see that there is a subsequence, which we continue to denote by g_n, such that for all i, $x_i g_n \to y_i$, where $y_i \in V \cdot g$. We clearly have that $\{y_i\}$ is an orthonormal basis of $V \cdot g$. Let $T: V \cdot g \to V \cdot g$ be the unitary operator such that $T(x_i g) = y_i$. Let w be an eigenvector of T of norm one with eigenvalue c. Let $z = wg^{-1}$, and choose a_i so that $z = \sum a_i x_i$. Then

$$zg_n = \sum a_i x_i g_n \longrightarrow \sum a_i y_i = \sum a_i T(x_i g) = T(zg) = T(w) = cw = czg .$$

Thus, $[z]g_n \to [z]g$ in $\mathcal{F}_1$, and it follows from the fact that $\hat{\mathcal{F}}_1$ is standard that $[g_n] \to [g]$ in $G/G_{[z]}$, where $G_{[z]}$ is the stability group of $[z]$ in $\mathcal{F}_1$. Since $G_{[z]}$ is compact, by passing to a subsequence we can assume $g_n \to h \in G$. Then

$V \cdot g = V \cdot h$, and so $[g_n] = [h]$ in G/G_V, and we have $[g_n] \to [g]$ in G/G_V.

If $g_n \in G$, we will say $g_n \to \infty$ if $\{g_n\}$ has no convergent subsequences.

PROPOSITION 2.3. *Suppose π is a representation of G and $g_n \to \infty$ implies $\pi(g_n) \to 0$ weakly. Then π is totally smooth.*

Proof. It suffices to show that the orbits in $\mathcal{F}_k$ (and B_1) are closed. Suppose $V \in \mathcal{F}_k$ and $\pi(g_n)V \to W$. Let $x \in V$, $x \neq 0$. Then by passing to a subsequence, we can assume $\pi(g_n)x \to w \in W$. If $\pi(g_n) \to 0$ weakly, this clearly implies $w = 0$, contradicting the fact that $x \neq 0$.

COROLLARY 2.4. *Let $G = R^n$ and suppose that π is a representation with Lebesgue spectrum, that is, equivalent to a multiple of the regular representation. Then $\pi(t) \to 0$ weakly as $t \to \infty$, and so π is totally smooth.*

Let $\mathcal{E}_k = \mathcal{F}_k - \mathcal{F}_{k-1}$. Then $\mathcal{E}_k$ is a $U(H)$-invariant Borel subset of $\hat{\mathcal{F}}_k$. If π is k-smooth, it follows that $\hat{\mathcal{E}}_k$ is also standard.

3. Orbit spaces of representations and ergodic theory

The connection between the orbit spaces of a unitary representation π and ergodic theory with which we shall be dealing concerns the "restriction" of π to a cocycle of the various ergodic actions of G. We begin by reviewing some basic definitions concerning cocycles. The reader is referred to [15] for details, discussion, and further references. As above, we assume G to be a locally compact second countable group. Let S be a standard Borel space with a Borel G-action $S \times G \to S$, and suppose that μ is a probability measure on S, quasi-invariant and ergodic under G. We then call (S, μ) (or sometimes just S) an ergodic G-space. If S is essentially isomorphic as a G-space to G/G_0, where $G_0 \subset G$ is a closed subgroup, S is called essentially transitive. Otherwise, S is called properly ergodic. If M is a standard Borel group, a Borel function $\alpha: S \times G \to M$ is called a cocycle if for all $g, h \in G$, $\alpha(s, gh) = \alpha(s, g)\alpha(sg, h)$ a.e. Two cocycles α and β are called equivalent or cohomologous if there is a Borel function $\varphi: S \to M$ such that for all g, $\varphi(s)\alpha(s, g)\varphi(sg)^{-1} = \beta(s, g)$ a.e. If $M = U(H)$, the unitary group of a separable Hilbert space H with the strong operator topology, α is called a unitary cocycle or a cocycle representation. The set of equivalence classes of unitary cocycles behaves similarly in many respects to the equivalence classes of unitary representations of a group. In particular, one can form direct sums, tensor products, and can speak of subcocycle representations of unitary cocycles. Equivalence classes of cocycles are also in one-one correspondence with equivalence classes of systems of imprimitivity based on S. If $S = G/G_0$, then Mackey's imprimitivity theorem implies that the study of unitary cocycles on $S \times G$ is essentially equivalent to the study of unitary representations of G_0.

If π is a representation of G and S is an ergodic G-space, we define res(π), the restriction of π to $S \times G$, by $\text{res}(\pi)(s, g) = \pi(g)$ [15, Section 2]. If $S = G/G_0$, then res(π) is just a cocycle corresponding to the ordinary restriction of π to the subgroup G_0, and hence our terminology. Just as it is often of interest to examine the decomposition of a unitary representation when restricted to a subgroup, for a variety of problems in ergodic theory it is of interest to examine the restriction of π to an ergodic action. For example, the question of ergodicity of a product action is sometimes equivalent to a statement concerning such a restriction, as in the following result.

PROPOSITION 3.1.   *Let (X, μ) and (Y, ν) be ergodic G-spaces where ν is a G-invariant probability measure. Let π be the natural representation of G on $L^2(Y, \nu) \ominus C$, that is $(\pi(g)f)(y) = f(yg)$. Then $X \times Y$ is an ergodic G-space if and only if res(π) to $X \times G$ does not contain the identity as a subcocycle.*

Proof.   Suppose $X \times Y$ is ergodic and that res(π) does contain the identity. Then there is a one dimensional Borel field of subspaces $x \rightarrow W(x) \subset L^2(Y) \ominus C$ with $\pi(g)W(xg) = W(x)$ a.e., and unitary maps $U(x): W(x) \rightarrow C$ such that for all g, $U(x)\text{res}(\pi)(x, g)U(xg)^{-1} = 1$ a.e., that is, $U(x)\pi(g) = U(xg)$. Let $h(x, y) = (U(x)^{-1}1)(y)$. Then $h(x, y)$ is measurable, and for each g,

$$h(xg, yg) = (U(xg)^{-1}1)(yg) = \pi(g)^{-1}(U(x)^{-1}1)(yg) = (U(x)^{-1}1)(y) = h(x, y) \text{ a.e.}$$

It follows that h is constant a.e., which contradicts the fact that $(U(y)^{-1}1) \perp C$.

Now suppose that $X \times Y$ is not ergodic. Let $h(x, y) \in L^\infty(X \times Y)$ be G-invariant and not essentially constant. Then $f(x, y) = h(x, y) - \int h(x, y)dy$ is G-invariant and not essentially constant, and $f_x(y) = f(x, y)$ is in $L^2(Y) \ominus C$ for all x. (We note that $f \neq 0$ by the ergodicity of X.) Let $W(x) = Cf_x \subset L^2(Y) \ominus C$, and define $U(x): W(x) \rightarrow C$ by $U(x)f_x = 1$. It is straightforward that for each g, $\pi(g)W(xg) = W(x)$ a.e., and $U(x)\pi(g)U(xg)^{-1} = 1$, a.e. This shows that the identity is a subcocycle of res(π).

The basic connection between restrictions of representations to ergodic actions and the action of G on the various spaces $\mathcal{F}_k$ is the following.

THEOREM 3.2.   *Let $\pi: G \rightarrow U(H)$ be a unitary representation of G and (S, μ) an ergodic G-space. Then res(π) to (S, μ) contains the identity cocycle (resp., a cocycle of dimension between 1 and k; a cocycle of dimension k) if and only if there is a measure ν on B_1 (resp. $\mathcal{F}_k$; $\mathcal{E}_k$), quasi-invariant and ergodic under G such that (B_1, ν) (resp., $(\mathcal{F}_k, \nu)$; $(\mathcal{E}_k, \nu)$) is a factor G-space of (S, μ)* [15].

Proof. If res(π) contains a p-dimensional subcocycle representation, then there is a map $V: S \to \mathcal{E}_p$ such that for each g, $\pi(g) V(sg) = V(s)$ a.e. Let $\nu = V_*(\mu)$, a measure on $\mathcal{E}_p$. Then the almost everywhere identity above implies that ν is quasi-invariant and that V induces a G-map of Boolean G-spaces [7] $V^*: B(\mathcal{E}_p, \nu) \to B(S, \mu)$. It follows immediately that $(\mathcal{E}_p, \nu)$ is ergodic and a factor of (S, μ).

Conversely, suppose ν is a quasi-invariant ergodic measure on $\mathcal{E}_p$ and that (possibly replacing S by a conull G-invariant Borel subset) $V: S \to \mathcal{E}_p$ is a G-map with $V_*(\mu) = \nu$. Then $s \to V(s)$ satisfies $\pi(g) V(sg) = V(s)$, and $V(s)$ thus determines a p-dimensional subcocyle of res(π). The proofs for $\mathcal{F}_k$ and B_1 are essentially the same.

Theorem 3.2 makes clear the relevance of the condition of k-smoothness.

COROLLARY 3.3. *If π is k-smooth (resp., smooth), and res(π) to an ergodic G-space (S, μ) contains a k-dimensional subcocycle (resp., the identity cocycle), then S has a factor of the form G/G_0, where $G_0 \subset G$ is a closed subgroup such that π/G_0 contains a finite dimensional representation (resp. the identity representation).*

Proof. If π is k-smooth, then every ergodic measure in $\mathcal{F}_k$ is supported in an orbit [4].

Under certain conditions, having a factor of the form G/G_0 is enough to ensure that S itself is actually essentially transitive.

PROPOSITION 3.4. *If $G_0 \subset G$ is a compact subgroup, then any ergodic extension (S, μ) of the G-space G/G_0 is essentially transitive. Thus, $S \cong G/G_1$ where G_1 is a compact subgroup of G_0.*

Proof. This follows from [19, Theorem 1, p. 75] and the fact that every ergodic action of a compact group is essentially transitive. We remark that an alternate proof can be given using [15, Theorem 4.3].

COROLLARY 3.5. *If π is a k-smooth (resp., smooth) representation and the stability groups in $\mathcal{F}_k$ (resp., B_1) are compact, then for any properly ergodic G-space S, res(π) to $S \times G$ does not contain a finite dimensional (resp., identity) subcocycle.*

The following provides a converse to Corollary 3.5.

PROPOSITION 3.6. *If π is not k-smooth (resp., smooth), then there is a properly ergodic G-space S such that res(π) to $S \times G$ contains a cocycle of dimension $\leq k$ (resp., identity cocycle).*

Proof. By [4, Theorem 2.6] there is a properly G-ergodic probability measure ν on $\mathcal{F}_k$, although not necessarily quasi-invariant. However, because of the local compactness of G, we can take $\mu = \int (\nu \cdot g) d\theta(g)$ where θ is a probability measure equivalent to Haar measure. Then μ will be quasi-invariant and properly ergodic. The result then follows from Theorem 3.2.

COROLLARY 3.7. *If π is a representation of R or Z with absolutely continuous spectrum, then the restriction of π to any ergodic action except translation on R or Z itself does not contain a finite dimensional subcocycle representation.*

Proof. If π has absolutely continuous spectrum, then $\pi(t) \rightarrow 0$ weakly as $t \rightarrow \infty$. It follows that π is totally smooth (Proposition 2.3) and that the stability groups in $\mathcal{F}_k$ are the identity. From Corollary 3.3, if $\mathrm{res}(\pi)$ to $S \times G$ has a finite dimensional subcocycle, S must be an extension of translation on R or Z. But translation on a locally compact group has no non-trivial ergodic extensions (use Proposition 3.4 for example) and the result follows.

COROLLARY 3.8. *If R or Z acts ergodically on X and Y, Y has a finite invariant measure and absolutely continuous spectrum, and X is not essentially isomorphic to translation on R or Z, then $X \times Y$ is ergodic.*

Proof. This follows from Corollary 3.7 and Proposition 3.1.

4. The duality principle

In this section we turn to the question of ergodicity of the restriction of an ergodic action to a subgroup. We begin with some preliminary material.

Suppose X is a compact metric space. Let $H(X)$ be the group of homeomorphisms of X with the topology of uniform convergence. Let $E = C(X)$, the continuous complex-valued functions on X, and let $\mathrm{Iso}(E)$ be the group of isometric isomorphisms of the separable Banach space E. For each $\varphi \in H(X)$, there is a naturally induced element $\varphi^* \in \mathrm{Iso}(E)$. The map $\varphi \rightarrow \varphi^*$ is an injective homomorphism, and we shall identify $H(X)$ with its image subgroup of $\mathrm{Iso}(E)$. With the strong operator topology, $\mathrm{Iso}(E)$ is a second countable metrizable group whose Borel structure is standard [18, Lemma 1.1]. It is not difficult to check that the embedding of $H(X)$ into $\mathrm{Iso}(E)$ is actually a homeomorphism onto its image. Furthermore, this image is exactly

$$\{ T \in \mathrm{Iso}(E) \,|\, T(fg) = T(f)T(g), \, T(\bar{f}) = \overline{T(f)} \} \,,$$

and hence $H(X)$ is a closed subgroup. From these observations we conclude the following.

PROPOSITION 4.1. *$H(X)$ is a second countable metrizable group whose Borel structure is standard.*

We now turn to the generalization of the Mackey-Moore duality principle described in the introduction. Let G act ergodically on S, and let $H \subset G$ be a closed subgroup.

THEOREM 4.2. *H acts ergodically on S if and only if the product action of G on $G/H \times S$ is ergodic.*

Proof. We can suppose that S is a compact metric space and that G acts continuously on S [13, Theorem 8.7]. For each g, let $\pi(g) \in H(S)$ be defined by $\pi(g)(s) = sg^{-1}$, so that $\pi\colon G \to H(S)$ is a continuous homomorphism. Let $Y = G/H$. Then $\alpha(y, g) = \pi(g)$ defines a strict $Y \times G$ cocycle and hence is equivalent to a strict cocycle $\beta\colon Y \times G \to H(S)$ such that $\beta(Y \times G) = \pi(H) = \beta([e] \times H)$ [13, proof of Theorem 8.27]. The [product action $Y \times S$ is then isomorphic to the skew product action $Y \times_\beta S$, $(y, s)g = (yg, \beta(y, g)^{-1}(s))$. Suppose $Y \times S$ is ergodic. If $A \subset S$ is H-invariant, then $Y \times A$ will be invariant in $Y \times_\beta S$ since $\beta(y, g) \in \pi(H)$. Thus $Y \times A$ is null or conull, which implies that A is null or conull. Hence, S is ergodic under H.

Conversely, suppose H is ergodic on S, and let $B \subset Y \times_\beta S$ be G-invariant. Let $B_y = \{s \mid (y, s) \in B\}$. Then for each y and g, $\beta(y, g)B_{yg} = B_y$ by invariance. In particular, $\beta([e], h)B_{[e]} = B_{[e]}$. Since $\beta([e], H) = \pi(H)$, $B_{[e]}$ must be null or conull, and since $\beta(Y \times G) = \pi(H)$, we have $B_y = B_{[e]}$ for all y. Thus B is null or conull, which completes the proof.

We remark that the duality principle for two closed subgroups follows readily from this theorem, by taking $S = G/H_0$. The ergodicity of H on G/H_0 and H_0 on G/H are both equivalent by the theorem to the ergodicity of G on $G/H \times G/H_0$.

Combining the duality principle with Corollary 3.5, we obtain the following result.

COROLLARY 4.3. *Suppose $\Gamma \subset G$ is a closed subgroup and that G/Γ has finite invariant measure. Let π be the representation of G induced by the identity representation of Γ, restricted to $L^2(G/\Gamma) \ominus \mathbf{C}$. Suppose further that π is smooth and that the stability groups of the action of G on B_1 are compact. Then the restriction of any properly ergodic action of G to Γ is still ergodic.*

Proof. Let S be an ergodic G-space. It suffices to see that $G/\Gamma \times S$ is ergodic, and from Proposition 3.1, it suffices to see that $\mathrm{res}(\pi)$ to $S \times G$ does not contain the identity cocycle. This follows from Corollary 3.5.

Theorem 4.2 can be combined with the results of [16] to give the following theorem concerning ergodic actions of nilpotent Lie groups.

THEOREM 4.4. *Let N be a connected, simply connected, nilpotent Lie group, and $\Gamma \subset N$ a uniform discrete subgroup. Suppose S is an ergodic N-space. Then Γ acts ergodically on S if and only if $\Gamma[N, N]$ acts ergodically on S.*

Proof. The main theorem of [16] implies that $S \times N/\Gamma$ is ergodic if and only if $S \times N/\Gamma[N, N]$ is ergodic. (Actually, in [16] the assumption is made that S has an invariant probability measure, but a small modification of the proof in [16] shows that this assumption is not necessary.) The result now follows immediately from Theorem 4.2.

As a corollary we have:

THEOREM 4.5. *Suppose N is a connected, simply connected, nilpotent Lie group, and S is an ergodic N-space. If $[N, N]$ acts ergodically on S, then Γ acts ergodically on S for every uniform discrete subgroup $\Gamma \subset N$.*

We remark that Theorem 4.2 can be viewed in virtual group terms as the duality principle for two subgroups of G, where one of them is virtual.

5. Applications to simple Lie groups

In this section we wish to apply the results of the previous sections to study the restriction of simple Lie group actions to closed subgroups whose quotient has a finite invariant measure. Let G be a connected non-compact simple Lie group with finite center. The following result is due to C.C. Moore.

THEOREM 5.1 [9, Theorem 1]. *If π is any representation of G that does not contain the identity, then the stability groups of points in $\mathcal{F}_k$, $k \geq 1$, and B_1 are compact.*

In light of this result, in order to apply Corollary 4.3 for example, we need only examine the smoothness properties of a representation π. To do this, we first recall some well known structural results. Let $\mathcal{G}$ be the Lie algebra of G and $\mathcal{G} = \mathcal{K} + \mathcal{P}$ a Cartan decomposition. Let $\mathcal{C} \subset \mathcal{P}$ be a maximal abelian subspace of $\mathcal{P}$. Let K be the connected subgroup of G whose Lie algebra is $\mathcal{K}$, and $A = \exp(\mathcal{C})$. Then we have that K is compact, A is a vector group, and $G = KAK$.

THEOREM 5.2. *If π is any unitary representation of G not containing the identity, then $\pi(g) \to 0$ weakly as $g \to \infty$. Thus, π is totally smooth.*

Proof. T. Sherman [12] and C.C. Moore [10, Theorem 8] have shown that the restriction of π to A is equivalent to a multiple of the regular representation of A, and hence by Corollary 2.4, $\pi(a) \to 0$ weakly as $a \to \infty$, for $a \in A$. If $g_n \in G$, and $g_n \to \infty$, then we can write $g_n = x_n a_n y_n$, where x_n, $y_n \in K$ and $a_n \in A$. By the usual argument, it suffices to show that a subsequence of $\pi(g_n)$ converges weakly to 0. Since K is compact, we can assume $x_n \to x$, $y_n \to y$, and $a_n \to \infty$. An elementary argument using the Cauchy-Schwarz inequality will then show that $\pi(g_n) = \pi(x_n)\pi(a_n)\pi(y_n) \to 0$ weakly. The last assertion of the theorem then follows from Proposition 2.3.

As pointed out to us by the referee, the fact that $\pi(g) \to 0$ weakly as $g \to \infty$ has also been observed by R. Howe, who in fact obtains more general results. (In this regard, see Theorem 5.6 below.)

From Theorem 5.2, we obtain the following result.

THEOREM 5.3. *Let G be a connected non-compact simple Lie group with finite center. Let $\Gamma \subset G$ be a closed subgroup such that G/Γ has a finite invariant measure. If S is any properly ergodic G-space, then Γ acts ergodically on S.*

Proof. The conditions of Corollary 4.3 are satisfied.

In certain situations, this result can be extended to the case in which G is semi-simple. We recall that if G is a connected semi-simple Lie group without compact factors, a lattice $\Gamma \subset G$ is called irreducible if for every non-trivial connected normal subgroup $H \subset G$, Γ projects to a dense subgroup in G/H.

THEOREM 5.4. *Suppose G_i, $i = 1, \cdots, n$, are connected simple non-compact Lie groups with finite center, $G = \prod G_i$, and $\Gamma \subset G$ is an irreducible lattice. If S is any properly ergodic G-space, then Γ is ergodic on S.*

Proof. We note first that [9, Theorem 5] implies that the stability groups in $B_1(\subset L^2(G/\Gamma) \ominus \mathbf{C})$ are compact. Thus, it suffices to see that if π is the natural representation of G on $L^2(G/\Gamma) \ominus \mathbf{C}$, then $\pi(g) \to 0$ weakly as $g \to \infty$. By a standard argument, it suffices to show this for almost all irreducible components in a direct integral decomposition of π into irreducibles. Now every irreducible representation U of G can be written as an (external) tensor product $U = U_1 \otimes \cdots \otimes U_n$, where U_i is an irreducible representation of G_i. For each i, Γ is ergodic on G/G_i by the irreducibility of Γ which implies that G_i is ergodic on G/Γ. Thus for almost all U in the decomposition of π, $U|G_i$ does not contain the identity. It follows that U_i does not contain the identity, and hence by Theorem 5.2, $U_i(g_i) \to 0$ weakly as $g_i \to \infty$ $(g_i \in G_i)$. Since this is true for each i, it follows easily that almost all U have the required property.

Example 5.5. If Γ is not assumed to be irreducible, the conclusion of

Theorem 5.4 is no longer true. For example, take $G = G_1 \times G_2$, $\Gamma = \Gamma_1 \times \Gamma_2$. Suppose S_1 is a properly ergodic G_1-space. Then, generalizing the construction of a flow built under a constant, one can construct a naturally associated "induced G-space" S [8], [21] which will be properly ergodic and have G/G_1 as a factor G-space. Since Γ is clearly not ergodic on G/G_1, it cannot be ergodic on S either.

In the work alluded to above, Howe [20] shows that Theorem 5.2 holds as well for simple non-compact connected algebraic groups over local fields of characteristic 0. Using Howe's result and the techniques above, one can then deduce the following.

THEOREM 5.6. *Let G be a simple non-compact connected algebraic group over a local field of characteristic 0. Let $\Gamma \subset G$ be a lattice. Then Γ is ergodic on any properly ergodic G-space.*

Finally, we remark that using the technique in Theorem 5.4, one may formulate similar theorems for products of simple groups over various local fields.

6. Ergodicity of skew products

We now turn to consideration of the ergodicity of extensions of an ergodic action where the fiber is a homogeneous space of a semi-simple Lie group with finite invariant measure. Thus, let G be a locally compact group and (S, μ) an ergodic G-space with quasi-invariant measure. Let H be another locally compact group and $\alpha: S \times G \to H$ a cocycle. If (Y, ν) is any ergodic H-space, then one can form the skew product action of G on $S \times Y$ defined by $(s, y)g = (sg, y\alpha(s, g))$, and this preserves the measure class of $\mu \times \nu$. It is natural to attempt to find conditions under which the skew product will or will not be ergodic. If $H = G$ and $\alpha(s, g) = g$, then one is considering ergodicity of a product action, which we have considered in certain cases above. The case in which H is compact and α arbitrary is considered in [15]. For a detailed analysis in some special cases, the reader is referred to the paper of Veech [14]. For H a connected nilpotent Lie group and $Y = H/\Gamma$ where $\Gamma \subset H$ is a discrete uniform subgroup, the problem of ergodicity of the skew product is studied in [16]. Here, it is our intention to study the case in which H is a simple Lie group and $Y = H/\Gamma$ where H/Γ has finite invariant measure. We shall see that this problem can be reduced to the problem of ergodicity of a certain product action, and we will then be able to apply our previous results. We begin by stating a general result, which we preface by recalling the notion of the Poincaré flow [5] or range of a cocycle [8].

Let $\alpha: S \times G \to H$ be a cocycle, where H is a locally compact (second countable) group. Then $G \times H$ acts on $S \times H$ by $(s, k) \cdot (g, h) = (sg, h^{-1}k\alpha(s, g))$. Let $B(S \times H)$ be the Boolean σ-algebra of Borel sets modulo null sets, so that $B(S \times H)$ is an irreducible (that is, ergodic) Boolean $G \times H$-space [7]. Let B_0 be the Boolean subalgebra of G-invariant elements so that B_0 is an irreducible Boolean H-space. By Mackey's point realization theorem [7], there is an essentially unique ergodic H-space X such that $B(X) \cong B_0$ as Boolean H-spaces. Then X is called the **range**, or Poincaré flow, of α.

THEOREM 6.1. *Let* $\alpha: S \times G \to H$, *and* X *be the range of* α. *Suppose* Y *is an ergodic* H-space. *Then the skew product action of* G *on* $S \times Y$ *determined by* α *is ergodic if and only if the product action of* H *on* $X \times Y$ *is ergodic.*

In terms of Mackey's theory of virtual groups, this result can be interpreted as saying that if one has an ergodic action of a group and a homomorphism of a virtual group into this group, then by composition one obtains an ergodic action of the virtual group if and only if the restriction of the ergodic group action to the closure of the range of the homomorphism is ergodic. As this result is well known if the virtual group is a group, the theorem becomes quite plausible. Theorem 3.3 of [18] is of the same general nature, and the essence of the proof of that theorem can be adapted to provide a proof of Theorem 6.1 in our present (technically less complicated) setting. We leave the details to the reader.

Using this result, we can now derive a criterion for the ergodicity of skew products determined by cocyles into simple Lie groups.

THEOREM 6.2. *Let* (S, μ) *be an ergodic* H-space *where* H *is a locally compact group. Suppose* G *is a connected non-compact simple Lie group with finite center. Let* $\Gamma \subset G$, $\Gamma \neq G$, *be a closed subgroup such that* G/Γ *has finite invariant measure. Let* $\alpha: S \times H \to G$ *be a cocycle. Then the skew product* H-action *on* $S \times G/\Gamma$ *fails to be ergodic if and only if* α *is equivalent to a cocycle into a compact subgroup of* G.

Proof. Let the G-space X be the range of α. By Theorem 6.1, ergodicity of $S \times G/\Gamma$ is equivalent to ergodicity of the product $X \times G/\Gamma$. If we let π be the representation of G induced by the identity representation of Γ restricted to the orthogonal complement of the constants, this is equivalent to $\mathrm{res}(\pi)$ to $X \times G$ not containing the identity (Proposition 3.1). If $\mathrm{res}(\pi)$ does contain the identity, then by the results of Sections 3 and 5, X must be isomorphic to G/G_0 where G_0 is compact. But then α must be equivalent to a cocycle into G_0 [15, proof of Theorem 3.5]. Conversely, if α is equivalent to a cocycle into a compact subgroup $G_0 \subset G$, then G/G_0 is a factor G-space of X [15]. It follows from [9, Theorem 7] that $G/G_0 \times G/\Gamma$ is not ergodic, and so $X \times G/\Gamma$ cannot be ergodic either.

In the case in which the skew product $S \times G/\Gamma$ is ergodic, it is natural to ask whether there is any non-trivial relatively discrete spectrum of $S \times G/\Gamma$ as an extension of S [15]. We again begin with a general result.

LEMMA 6.3. *Let S be an ergodic G-space, $\alpha: S \times G \rightarrow H$, and let the ergodic H-space X be the range of α. Let π be a unitary representation of H. If $\pi \circ \alpha$ has a finite dimensional subcocycle, then $\mathrm{res}(\pi)$ to $X \times H$ also has a finite dimensional subcocycle.*

Proof. Let $H(\pi)$ be the Hilbert space of π, and $\mathcal{F}_k$ be as in Section 2. If $\pi \circ \alpha$ has a finite dimensional subcocycle, then there is a Borel map $V: S \rightarrow \mathcal{F}_p$ such that for all g, $\pi(\alpha(s, g))V(sg) = V(s)$ a.e. Define $\theta: S \times H \rightarrow \mathcal{F}_p$ by $\theta(s, h) = \pi(h)V(s)$. One readily checks that θ is essentially G-invariant, and so there is a Borel map $\tilde{\theta}: X \rightarrow \mathcal{F}_p$ such that $\tilde{\theta} \circ q = \theta$ a.e., where $q: S \times H \rightarrow X$ induces the isomorphism $B(X) \rightarrow B_0 \subset B(S \times H)$. To prove the theorem, it suffices to show that for all $k \in H$, $\pi(k)\tilde{\theta}(xk) = \tilde{\theta}(x)$ for almost all x, and to do this, we need only check this equation for $x = q(s, h)$, for almost all (s, h). But

$$\pi(k)\tilde{\theta}(q(s, h)k) = \pi(k)\tilde{\theta}(q(s, k^{-1}h)) = \pi(k)\theta(s, k^{-1}h) = \pi(h)V(s) = \theta(s, h) \text{ a.e.},$$

completing the proof.

THEOREM 6.4. *Consider the situation in Theorem 6.2, and suppose that $S \times G/\Gamma$ is ergodic. Then $S \times G/\Gamma$ has no non-trivial relatively discrete spectrum as an extension of S.*

Proof. The conclusion is equivalent to the statement that $\pi \circ \alpha$ has no finite dimensional subcocycles, where π is the restriction to the orthogonal complement of the constants of the representation of G induced by the identity representation of Γ. But if $\pi \circ \alpha$ does have finite dimensional subcocycles, Lemma 6.3 implies that $\mathrm{res}(\pi)$ to $X \times G$ does also, where X is the range of α. But then the results of Sections 3 and 5 imply that X must be of the form G/G_0 where G_0 is compact. Thus α is equivalent to a cocycle into G_0 [15, Theorem 3.5], contradicting the ergodicity of $S \times G/\Gamma$ (Theorem 6.2).

As in Section 5, one can formulate Theorems 6.2 and 6.4 where G is a product of simple Lie groups of the type we have been considering and Γ is an irreducible lattice in G.

7. Normal actions of simple groups

The author introduced the notion of a normal ergodic action of a locally compact group in [15] and [17]. (The definition is reproduced below.) If the action is transitive, normality is equivalent to the stability groups being normal. Hence, a simple group has no normal transitive actions other than the trivial ones, namely the action on a point, and translation on itself. For a connected simple Lie group with finite center, the only normal transitive actions will be those of the form G/Z_0, where $Z_0 \subset Z$, the center of G. It is a natural problem to determine when a simple group has no non-trivial normal ergodic actions. This can be considered as a strong form of the notion of simplicity.

In light of the main result of [17], this problem can be phrased in purely topological terms. Namely, normal ergodic actions are just those defined by a dense homomorphic mapping into another locally compact group. Thus while simplicity is the assertion that every surjective homomorphism is trivial, we are asking about the existence of homomorphisms with dense range. From this it becomes clear that simple groups may have normal ergodic actions.

Example 7.1. Let $G = \mathrm{SL}(n, Q)/\mathrm{Center}$. Then there is a dense embedding of G in $\mathrm{SL}(n, R)/\mathrm{Center}$, and hence a non-trivial normal ergodic action.

Example 7.2. One can also have non-trivial normal ergodic actions of simple groups which preserve a probability measure, or equivalently, dense embeddings of simple groups into compact groups. For example, the commutator of the projective orthogonal group over Q is simple in sufficiently large dimensions [3], and clearly this densely embeds in a compact group.

We now point out that for the simple Lie groups we have been considering this type of phenomenon does not occur. We recall the definition of normality. Let G be a locally compact group and S an ergodic G-space. Let $\pi(g)$ be the natural representation of G on $L^2(S)$. Then S is called a normal G-space if $\mathrm{res}(\pi)$ to $S \times G$ is equivalent to the identity. It is straightforward to check that if $S = H$, a locally compact group, and the action of G on H is defined by a homomorphism with dense range, then S is a normal G-space. That the converse is true is the main result of [17].

THEOREM 7.3. *Let G be a connected non-compact simple Lie group with finite center Z. Let S be a normal ergodic G-space. Then S is either a point or $S \cong G/Z_0$, where $Z_0 \subset Z$.*

Proof. If $\mathrm{res}(\pi)$ is equivalent to the identity, and S is not a point, then

588   ROBERT J. ZIMMER

by the results of Sections 3 and 5, S must be essentially transitive, and the conclusion follows.

UNITED STATES NAVAL ACADEMY, ANNAPOLIS, MARYLAND
Current address: University of Chicago, Illinois

REFERENCES

[1] L. AUSLANDER, L. GREEN, and F. HAHN, *Flows on Homogeneous Spaces*, Ann. of Math. Studies, no. 53, Princeton Univ. Press, Princeton, N.J., 1963.

[2] L. AUSLANDER, An exposition of the structure of solvmanifolds, Bull. A.M.S. **79** (1973), 227–285.

[3] J. DIEUDONNÉ, On the orthogonal groups over the rational field, Ann. of Math. **54** (1951), 85–93.

[4] E. G. EFFROS, Transformation groups and C^*-algebras, Ann. of Math. **81** (1965), 38–55.

[5] J. FELDMAN and C.C. MOORE, Ergodic equivalence relations, cohomology, and von Neumann algebras, to appear.

[6] J. GLIMM, Locally compact transformation groups, Trans. A.M.S. **101** (1961), 124–138.

[7] G. W. MACKEY, Point realizations of transformation groups, Illinois J. Math. **6** (1962), 327–335.

[8] ———, Ergodic theory and virtual groups, Math. Ann. **166** (1966), 187–207.

[9] C.C. MOORE, Ergodicity of flows on homogeneous spaces, Amer. J. Math. **88** (1966), 154–178.

[10] ———, Restrictions of unitary representations to subgroups and ergodic theory, in Group representations in mathematics and physics, *Lecture Notes in Physics*, no. 6, Springer-Verlag, New York, 1970.

[11] A. RAMSAY, Virtual groups and group actions, Advances in Math. **6** (1971), 253–322.

[12] T. SHERMAN, A weight theory for unitary representations, Canadian J. Math. **18** (1966), 159–168.

[13] V.S. VARADARAJAN, *Geometry of Quantum Theory*, Volume II, van Nostrand, Princeton, N.J., 1970.

[14] W.A. VEECH, Finite group extensions of irrational rotations, Israel J. Math. **21** (1975), 240–259.

[15] R. J. ZIMMER, Extensions of ergodic group actions, Illinois J. Math. **20** (1976), 373–409.

[16] ———, Compact nilmanifold extensions of ergodic actions, Trans. A.M.S. **223** (1976), 397–406.

[17] ———, Normal ergodic actions, to appear, J. Funct. Anal.

[18] ———, Amenable ergodic group actions and an application to Poisson boundaries of random walks, to appear, J. Funct. Anal.

[19] L. AUSLANDER and C.C. MOORE, Unitary representations of solvable Lie groups, Memoirs of A.M.S. no. **62** (1966).

[20] R. HOWE, On the asymptotic behavior of matrix coefficients, unpublished.

[21] G. W. MACKEY, Ergodic theory and its significance for statistical mechanics and probability theory, Advances in Math. **12** (1974), 178–268.

(Received January 19, 1977)
(Revised May 15, 1977)

2

Amenable Actions, Equivalence Relations, and Foliations

JOURNAL OF FUNCTIONAL ANALYSIS **27**, 350–372 (1978)

Amenable Ergodic Group Actions and an Application to Poisson Boundaries of Random Walks*

ROBERT J. ZIMMER[†]

Department of Mathematics, U.S. Naval Academy, Annapolis, Maryland 21402

Communicated by the Editors

Received September 7, 1976

We introduce and study the class of amenable ergodic group actions which occupy a position in ergodic theory parallel to that of amenable groups in group theory. We apply this notion to questions about skew products, the range (i.e., Poincaré flow) of a cocycle, and to Poisson boundaries.

INTRODUCTION

In this paper we introduce a new notion of amenability for ergodic group actions and use it to give a partial solution to a problem in ergodic theory concerning skew products. We also show how the notion can be used to generalize a result about Poisson boundaries of random walks on groups. Amenable ergodic actions occupy a position in ergodic theory parallel to that of amenable groups in group theory, and one therefore expects the notion to be applicable in diverse situations. Greenleaf [7] has also introduced a notion of amenability which, although related, is quite different from ours. (See Section 4.) We hope the parallel between our results and results from group theory will justify our terminology. We also remark that from Mackey's virtual subgroup point of view [11] what we will be considering are amenable virtual subgroups of locally compact groups.

There are a variety of different equivalent conditions defining amenability for groups. The condition on which we shall concentrate is the fixed point property. Thus, a group G is amenable if every continuous affine action of G on a compact convex set has a fixed point. An affine action is, of course, just a homomorphism into the group of affine automorphisms. There is a strong parallel between homomorphisms in group theory and cocycles in ergodic theory. In fact, for transitive actions the study of cocycles essentially reduces to the study of homomorphisms of the stability group. Thus, we are led to

---

* Research supported by the Naval Academy Research Council.

[†] Present address: Department of Mathematics, University of Chicago, Chicago, Ill. 60637.

consider cocycles into groups of affine automorphisms and some technicalities aside, we will call a G-space S amenable if every cocycle into a group of affine homeomorphisms has a fixed element. Here, a fixed element is no longer a single point, but rather a Borel function from S into the compact convex set. With this definition of amenability the class of amenable ergodic actions behaves quite similarly to the class of amenable groups. In particular, many of the usual combinatorial properties still hold, and the parallel becomes more pronounced if one phrases everything in terms of virtual subgroups. We remark that in the transitive case, the action will be amenable if and only if the stability groups are amenable.

A well-known and useful construction in ergodic theory is the skew product construction. If S is an ergodic G-space and $\alpha\colon S \times G \to H$ is a cocycle into a locally compact group, then one can define a G-action on $S \times H$ by $(s, h)g = (sg, h\alpha(s, g))$. If $G = Z$, the group of integers, the action is determined by an invertible transformation T. A cocycle is determined by the function $f(s) = \alpha(s, 1)$, and the skew product takes the possibly more familiar form $\tilde{T}(s, h) = (Ts, hf(s))$. It is a natural problem, raised for example by Mackey [11], to start with a given ergodic G-space S and ask for which groups H can one find a cocycle so that the skew product action on $S \times H$ is also ergodic. Even for actions of the integers, the answer is unknown. If S is a Z-space with a finite invariant measure, then it is known that for any of the following types of groups, such a cocycle will exist: compact [15, 18]; countable discrete abelian [15]; connected nilpotent Lie [19]; discrete finitely generated nilpotent (use [19, Theorem 3.6], induction, and the fact that a subgroup of a finitely generated nilpotent group is finitely generated); and any finite product of any of these types [19]. All of these groups are amenable, and it follows from this paper that amenability is actually a necessary condition. More generally, we show that any ergodic action (even without invariant measure) of an amenable group is amenable, and that if $\alpha\colon S \times G \to H$ with S amenable and $S \times H$ ergodic, then H must be amenable.

A further generalization of this result is related to a generalization of the flow built under a function construction. If $\alpha\colon S \times G \to H$ is a cocycle, one can form an H-space X called the range of α [11] (or the Poincaré flow [5]) which reduces to the flow built under $f(s) = \alpha(s, 1)$ if $G = Z$, $H = \mathbb{R}$, and f is positive. As every $\mathbb{R}$-flow arises in this way, it is natural to ask which G-actions are the range of a cocycle on a Z-space. It follows from [19] that if there is a cocycle $\alpha\colon S \times Z \to H$ with an ergodic skew product, then every H-action is the range of a cocycle on some Z-space. Here, we establish the result that the range of any cocycle of an amenable action must also be an amenable action. In particular, this excludes the possibility that an action of a non-amenable group with finite invariant measure is the range of a cocycle on a Z-space. We remark that this also provides a large collection of amenable actions of non-amenable groups.

If μ is a probability measure on G, μ defines a random walk on G, and Furstenberg has associated a space to (G, μ), called the Poisson boundary, which enables one, for example, to obtain integral representation theorems for the harmonic functions of the random walk. For μ étalée, we present a different construction of the Poisson boundary which shows that it appears as the range of a naturally defined cocycle of a (semi-group) action of the non-negative integers. A modification of the proof of the result described in the preceding paragraph then shows that the Poisson boundary with a naturally defined invariant measure class is an amenable ergodic G-space. It follows in particular that if G is transitive on a Poisson space of an étalée measure (a situation which occurs in many important cases) the stability groups must be amenable.

The organization of this paper is as follows. Section 1 presents definitions, some preliminary technical material, and deals with amenability in the transitive case. In Section 2 we show that actions of amenable groups and extensions of amenable actions are also amenable. Section 3 deals with the range of a cocycle and skew products. Section 4 concerns amenable pairs, a generalization of Eymard's notion of the conditional fixed point property [4]. We also point out here a connection with Greenleaf's definition of amenability. Section 5 is devoted to the Poisson boundary.

The author wishes to thank J. Feldman and C. C. Moore for discussion on the results of this paper.

1. Definitions and Preliminaries

We begin by describing the concept of amenability and proving some useful auxilliary results. We shall throughout take G to be a second countable locally compact group, S a standard Borel space, and μ a probability measure on S. We suppose that there is a right Borel action of G on S and that μ is quasi-invariant and ergodic under G. Suppose that M is a Borel group (i.e., a group with a Borel structure compatible with the group operations.) A Borel function $\alpha: S \times G \to M$ is called a cocycle if for all $g, h \in G$, $\alpha(s, gh) = \alpha(s, g)\, \alpha(sg, h)$ for almost all s. The study of cocycles is playing an increasingly significant role in many aspects of ergodic theory. (See [5], [16] and the references in these papers as examples.) If S is a point, a cocycle is nothing but a Borel homomorphism $G \to M$ and many properties of homomorphisms have natural analogues as properties of cocycles for general S. Mackey has formalized this analogy in his notion of virtual groups [11], in which an ergodic G-space is considered as a virtual subgroup of G and the cocycles on $S \times G$ as the homomorphisms of this virtual group. (Although it is not necessary for a reading of this paper, the author feels that the virtual subgroup viewpoint is very suggestive.) Amenable groups can be characterized by the properties of a certain class of homomorphisms, namely the existence of fixed points in affine actions on compact convex sets. We now generalize this notion to group actions.

We first establish some preliminary facts. Let E be a separable Banach space and $\mathrm{Iso}(E)$ the group of isometric isomorphisms of E.

LEMMA 1.1.   *With the strong operator topology* $\mathrm{Iso}(E)$ *is a separable metrizable group and the induced Borel structure is standard.*

Proof.   That $\mathrm{Iso}(E)$ is a separable metric group can be seen exactly as in the first paragraph of [14, Lemma 8.34]. Next, note that with the strong operator topology $L_1(E)$, the unit ball in the space of bounded linear maps on E, is metrizable by a complete separable metric. To see that the Borel structure on $\mathrm{Iso}(E)$ is standard, it suffices to show that $\mathrm{Iso}(E)$ is a Borel subset of $L_1(E)$. Let $\{f_i\}$ be a countable dense subset of E_1, the unit ball in E. Then an isometry T is in $\mathrm{Iso}(E)$ if and only if each $f_i \in \mathrm{range}(T)$. Now $f \in \mathrm{range}(T)$ if and only if for all n there exists j such that $\| Tf_j - f \| \leqslant 1/n$. Thus

$$\mathrm{Iso}(E) = \bigcap_i \bigcap_n \left(\bigcup_j \left\{ T \mid \| Tf_j - f_i \| \leqslant \frac{1}{n} \right\} \right) \cap \{ T \mid T \text{ is an isometry} \}.$$

Since the set of isometries is closed in $L_1(E)$, it follows that $\mathrm{Iso}(E)$ is Borel.

COROLLARY 1.2.   *The Borel structure on* $\mathrm{Iso}(E)$ *is the smallest such that all maps* $T \to Tf$ *are Borel,* $f \in E$.

Let E^* be the dual space of E and E_1^* the unit ball in the dual. Then E_1^* is a compact convex set when endowed with the $\sigma(E^*, E)$ topology, and since E is separable, E_1^* is also metrizable. We denote by $\langle\,,\,\rangle$ the dual pairing of E^* and E. For $T \in L_1(E)$, we have an adjoint map $T^* \in L_1(E^*)$.

LEMMA 1.3.   *The map* $\mathrm{Iso}(E) \times E_1^* \to E_1^*$, *defined by* $(T, \lambda) \to T^*(\lambda)$ *is continuous.*

Proof.   Let $T_n \to T$ and $\lambda_n \to \lambda$. Then for $f \in E$,

$$|\langle T_n^*(\lambda_n) - T^*(\lambda), f \rangle| = |\langle \lambda_n, T_n f \rangle - \langle \lambda, Tf \rangle|$$
$$\leqslant |\langle \lambda_n, T_n f - Tf \rangle| + |\langle \lambda_n - \lambda, Tf \rangle|$$
$$\leqslant \| T_n f - Tf \| + |\langle \lambda_n - \lambda, Tf \rangle|.$$

As $n \to \infty$, both terms $\to 0$.

Letting $H(E_1^*)$ be the group of homeomorphisms of E_1^* with the topology of uniform convergence, Lemma 1.3 implies that the induced map $\mathrm{Iso}(E) \to H(E_1^*)$ is continuous and hence Borel.

Now suppose that S is a standard Borel space and for each $s \in S$ we have a non-empty compact convex subset $A_s \subset E^*$. Then $\{A_s\}$ will be called a Borel field of compact convex sets if $\{(s, \lambda) \mid \lambda \in A_s\}$ is a Borel subset of $S \times E_1^*$.

We are now ready to define an amenable ergodic action. With all notation as above, suppose $\alpha: S \times G \to \mathrm{Iso}(E)$ is a Borel cocycle. Then there is an induced adjoint (Borel) cocycle $\alpha^*: S \times G \to H(E_1{}^*)$ defined by $\alpha^*(s, g) = (\alpha(s, g)^{-1})^*$. A Borel field of compact convex sets $\{A_s\}$ is called α-invariant if for each g, $\alpha^*(s, g)A_{sg} = A_s$ for almost all s.

DEFINITION 1.4. If S is an ergodic G-space, S is called amenable if for every separable Banach space E, cocycle $\alpha: S \times G \to \mathrm{Iso}(E)$, and α-invariant Borel field $\{A_s\}$, there is a Borel function $\varphi: S \to E_1{}^*$ such that $\varphi(s) \in A_s$ a.e. and for each g, $\alpha^*(s, g)\,\varphi(sg) = \varphi(s)$ a.e. We will then call φ an α-invariant section in $\{A_s\}$.

If S is a point, α becomes a Borel (and hence continuous [14, Lemma 8.28]) homomorphism $G \to \mathrm{Iso}(E)$, and amenability of the G-space S means that every compact convex G-invariant subset of $E_1{}^*$ contains a fixed point. Aside from the restriction we have made regarding the separability of E, this is the condition of G being amenable. We now deal with the separability hypothesis.

PROPOSITION 1.5. *Let G be locally compact and second countable. Suppose that for every separable Banach space E, continuous homomorphism $G \to \mathrm{Iso}(E)$, and G-invariant compact convex set $A \subset E_1{}^*$, there is a point in A left fixed by G. Then G is amenable.*

Proof. It suffices to show that the hypothesis of the proposition is true for arbitrary (not just separable) Banach spaces. Since G is separable, we can find a collection of separable G-invariant closed subspaces $E_\sigma \subset E$, $\sigma \in I$, where I is some index set, such that $\overline{\bigcup E_\sigma} = E$. Let $\varphi_\sigma : E^* \to E_\sigma{}^*$ be the restriction map and $A_\sigma = \varphi_\sigma(A)$. Thus A_σ is a compact convex G-invariant set, and since E_σ is separable, the set of fixed points $F_\sigma \subset A_\sigma$ is non-empty and closed. For each $\sigma_1, \ldots, \sigma_n \in I$, the subspace $E_{\sigma_1} + \cdots + E_{\sigma_n} \subset E$ will also be separable, and hence there is a fixed point in $\varphi_{\sigma_1 \cdots \sigma_n}(A)$ where $\varphi_{\sigma_1 \cdots \sigma_n} : E^* \to (F_{\sigma_1} + \cdots + F_{\sigma_n})^*$. Denoting this set of fixed points by $F_{\sigma_1 \cdots \sigma_n}$, it is clear that

$$(\varphi_{\sigma_1 \cdots \sigma_n}^{-1}(F_{\sigma_1 \cdots \sigma_n}) \cap A) \subset \left(\bigcap_1^n \varphi_{\sigma_i}^{-1}(F_{\sigma_i}) \cap A \right),$$

and hence this latter intersection is non-empty. By the finite intersection property, it follows that $\bigcap_I \varphi_\sigma^{-1}(F_\sigma) \cap A$ is nonempty. If λ is a point in this intersection, λ is invariant when restricted to each E_σ, and since $\overline{\bigcup E_\sigma} = E$, it follows that λ is G-invariant.

COROLLARY 1.6. *The trivial G-space $\{e\}$ is amenable if and only if G is an amenable group.*

The ergodic actions with the simplest orbit structure are the essentially transitive ones. We now examine what amenability means in this case, generalizing Corollary 1.6. We preface this with a technical lemma which will be of general use.

LEMMA 1.7.   *If $\{A_s\}$ is a Borel field of compact convex sets, then there is a countable collection of Borel functions $a_i : S \to E_1{}^*$ such that for all s in a conull Borel set, $A_s = \overline{\{a_i(s) \mid i = 1, 2, ...\}}$. Conversely, given Borel functions a_i, then $s \to \overline{\{a_i(s) \mid i = 1, 2, ...\}}$ defines a Borel field of compact convex sets if each of these sets is convex.*

Proof.   (i)   If $\{A_s\}$ is Borel, let $\tilde{A} = \{(s, x) \mid x \in A_s\}$, so that $\tilde{A} \subset S \times E_1{}^*$ is Borel. Fix a metric on $E_1{}^*$ and for each n choose a $1/n$-dense subset of $E_1{}^*$, say $x_1{}^n, ..., x_{k(n)}{}^n$. Let $B_j{}^n = S \times B(x_j{}^n; 1/n) \subset S \times E_1{}^*$, so that $B_j{}^n$ is Borel. Letting $p: S \times E_1{}^* \to S$ be projection, $p(B_j{}^n \cap \tilde{A})$ will be analytic. By the von Neumann selection theorem, there is a Borel set $S_j{}^n \subset p(B_j{}^n \cap \tilde{A})$ with the same measure as $p(B_j{}^n \cap \tilde{A})$, and a Borel section of the projection $S_j{}^n \to B_j{}^n \cap \tilde{A}$. Note that $\bigcup_j S_j{}^n$ is a conull Borel set for each n. Combining these sections with just a small amount of finesse, we see that we can obtain, for each n, a finite set of Borel functions $a_j{}^n(s)$ such that $\{a_j{}^n(s)\}_j$ are $2/n$-dense in A_s for s in a conull Borel set. As n is arbitrary, the first assertion of the lemma follows easily.

(ii)   We note that $x \in \overline{\{a_i(s)\}}$ if and only if $x \in \bigcap_n \bigcup_j B(a_j(s), 1/n)$. Thus

$$\{(s, x) \mid x \in A_s\} = \bigcap_n \bigcup_j \{(s, x) \mid x \in B(a_j(s), 1/n)\}$$

and so it suffices to see that each of the latter sets are Borel. But this follows from the fact that the map $S \times E_1{}^* \to \mathbb{R}$, $(s, x) \to (s, a_i(s), x) \to d(a_i(s), x)$ is Borel.

COROLLARY 1.8.   *If α, $\beta: S \times G \to \mathrm{Iso}(E)$ are equivalent, then every α-invariant Borel field $\{A_s\}$ has an α-invariant section if and only if every β-invariant field has a β-invariant section.*

Proof.   Let $T: S \to \mathrm{Iso}(E)$ be such that $T(s)\,\alpha(s, g)\,T(sg)^{-1} = \beta(s, g)$ for each g and almost all s. Suppose every β-invariant field has a β-invariant section, and let $\{A_s\}$ be an α-invariant field. Then it follows from Lemma 1.7 that $s \to T^*(s)^{-1} A_s$ agrees a.e. with a Borel field of compact convex sets, and this field will be β-invariant. If φ is a β-invariant section in $T^*(s)^{-1} A_s$ (a.e.), then $T^*(s)\,\varphi(s)$ will be an α-invariant section in A_s (a.e.).

THEOREM 1.9.   *Let $H \subset G$ be a closed subgroup. Then G/H is an amenable G space if and only if H is amenable.*

Proof. (i) Suppose H is amenable. Let E be a separable Banach space and $\alpha: G/H \times G \to \text{Iso}(E)$ a cocycle. By [14, 8.24–8.28 and particularly the proof of Lemma 8.24], α is equivalent to a strict cocycle β with the following property: if $\lambda: G \to \text{Iso}(E)$ is defined by $\lambda(g) = \beta([e], g)$, then $\beta(G/H \times G) = \lambda(H)$. We remark that β can be expressed in terms of λ by $\beta([k], g) = \beta([e]k, g) = \beta([e], k)^{-1} \beta([e], kg) = \lambda(k)^{-1} \lambda(kg)$. We also note that if $h \in H$ and $g \in G$, then $\lambda(hg) = \lambda(h)\lambda(g)$ from the cocycle identity. In particular $\lambda \mid H$ is a Borel homomorphism. We let $\lambda^*(g) = \beta^*([e], g)$.

Now suppose that $\{A_{[k]}\}$ is a β-invariant Borel field of compact convex sets in E_1^*, so that $\beta^*([k], g)A_{[k]g} = A_{[k]}$ for each g and almost all k. Then for each g, we have $\lambda^*(kg)A_{[kg]} = \lambda^*(k)A_{[k]}$ a.e., and hence, using Fubini's theorem, $\lambda^*(k)A_{[k]} = A$ a.e. where A is a fixed compact convex set. If $h \in H$, we have $\lambda^*(hk)A_{[hk]} = A$ a.e., that is $\lambda^*(h)\lambda^*(k)A_{[k]} = A$ a.e. This clearly implies $\lambda^*(h)A = A$ for all h. Since H is amenable, there is a fixed point $x \in A$, and since $\beta(s, g) \in \lambda(H)$, it follows that $\varphi(s) = x$ is a β-invariant section. To show that G/H is an amenable G space, it suffices to show that $x \in A_{[k]}$ for almost all k. But this follows since $A_{[k]} = \lambda^*(k)^{-1}A$ a.e.

(ii) Conversely, suppose G/H is an amenable G-space. Let $\lambda: H \to \text{Iso}(E)$. Then there is a strict cocycle $\alpha: G/H \times G \to \text{Iso}(E)$ such that $\alpha(G/H \times G) = \lambda(H)$ and $\alpha([e], h) = \lambda(h)$ for $h \in H$. Let $A \subset E_1^*$ be a compact convex set and $A_s = A$ for all s. Choose an α-invariant section $\varphi(s) \in A$. Then for each g and almost all k, $\alpha^*([e], kg) \varphi([k]g) = \alpha^*([e], k) \varphi([k])$, so that for some $x \in A$, $\alpha^*([e], k) \varphi([k]) = x$ a.e. If $h \in H$, $\alpha^*([e], hk) \varphi([hk]) = x$ for almost all k, that is $\lambda^*(h) \alpha^*([e], k) \varphi([k]) = x$. It follows that $\lambda^*(h)x = x$ and hence by Proposition 1.5 that H is amenable.

The reader conversant with Mackey's virtual subgroup viewpoint will recognize that Theorem 1.9 shows that Definition 1.4 is a reasonable definition of an amenable virtual subgroup. This will be reinforced in the succeeding sections, where we exhibit many similarities between the notions of amenability for ergodic actions and for groups.

2. Actions of Amenable Groups and Extensions of Amenable Actions

In this and following sections, we examine how the property of amenability behaves under various natural operations in ergodic theory, and in the process provide a wealth of examples of amenable ergodic actions.

THEOREM 2.1. *Let G be an amenable group and (S, μ) and ergodic G-space. Then S is an amenable G-space.*

We remark that in light of Theorem 1.9, this result is a generalization of the fact that a closed subgroup of an amenable group is amenable. In Mackey's language, a virtual subgroup of an amenable group is amenable. The converse of Theorem 2.1 is not true as we shall see below. A condition under which the converse assertion will hold is given in Section 4. We begin the proof by collecting some properties of vector-valued functions.

Let E be a separable Banach space and let

$$L^1(S, E) = \left\{ f: S \to E \mid f \text{ measurable and } \int \|f(s)\| \, d\mu < \infty \right\}.$$

We identify functions which agree on a conull set. We recall that $f: S \to E$ is measurable if and only if $\theta \circ f: S \to \mathbb{C}$ is measurable for all $\theta \in E^*$ [2, p. 149]. Then $L^1(S, E)$ is a separable Banach space [3, p. 587]. Recall that $\lambda: S \to E^*$ is called weakly measurable if $s \to \langle \lambda(s), x \rangle$ is measurable for all $x \in E$. If λ is weakly measurable, then $s \to \|\lambda(s)\|$ is measurable [3, 8.15.3]. We let $L^\infty(S, E^*) = \{\lambda: S \to E^* \mid \lambda \text{ weakly measurable and } \|\lambda(s)\| \in L^\infty(S)\}$. Then $L^\infty(S, E^*)$ is a Banach space under the essential sup norm [3, p. 578]. If $f \in L^1(S, E)$ and $\lambda \in L^\infty(S, E^*)$, we define $\langle \lambda, f \rangle = \int_S \langle \lambda(s), f(s) \rangle \, d\mu(s)$. For each λ this defines an element $\tilde{\lambda}$ of $L^1(S, E)^*$ and $\lambda \to \tilde{\lambda}$ is an isometric isomorphism of $L^\infty(S, E^*)$ with $(L^1(S, E))^*$ [3, 8.18.2]. Thus the closed unit ball in $L^\infty(S, E^*)$ is a compact metrizable space with the $\sigma(L^\infty(S, E^*), L^1(S, E))$ topology. We denote this ball by $L_1^\infty(S, E^*)$.

Now suppose $\{A_s\}$ is a Borel field of compact convex subsets of E_1^*. Let $B = \{\lambda \in L^\infty(S, E^*) \mid \lambda(s) \in A_s \text{ a.e.}\}$

PROPOSITION 2.2.   *B is a closed convex subset of $L_1^\infty(S, E^*)$.*

Proof.   Since E is separable, it is easy to see that there is a countable set of hyperplanes in E^* that strictly separates points in E_1^* from compact convex sets in E_1^*. For example, we may take as hyperplanes the zeroes of the functions $f_{nq}(\lambda) = \langle \lambda, x_n \rangle - q$ where $\{x_n\}$ is a countable dense set in E and q is rational. Now suppose $\lambda_j \in B$, $\lambda_j \to \lambda \in L_1^\infty(S, E^*)$, and $\lambda(s) \notin A_s$ on a set of positive measure. For each n and q, let

$$S_{nq} = \{s \in S \mid f_{nq}(A_s) < 0 \text{ and } f_{nq}(\lambda(s)) > 0\}.$$

Then $\bigcup S_{nq}$ has positive measure. If $\tilde{A} = \{(s, x) \in S \times E_1^* \mid x \in A_s\}$ and p_1, p_2 are the projections of $S \times E_1^*$ on S and E_1^* respectively, then

$$S - S_{nq} = p_1((f_{nq} \circ p_2)^{-1}([0, \infty)) \cap \tilde{A}) \cup (f_{nq} \circ \lambda)^{-1}((-\infty, 0]).$$

Since $\tilde{A}$ is Borel, the first set in the union is analytic, and it follows that S_{nq} is measurable for all n, q. It follows that S_{nq} has positive measure for some n, q. Define $f \in L^1(S, E)$ by

$$f(s) = \begin{cases} x_n & \text{if } s \in S_{nq} \\ 0 & \text{if } s \notin S_{nq}. \end{cases}$$

Then

$$\langle \lambda_j, f \rangle = \int_{S_{nq}} \langle \lambda_j(s), x_n \rangle \, d\mu \leqslant q\mu(S_{nq})$$

since $\lambda_j \in B$. On the other hand,

$$\langle \lambda, f \rangle = \int_{S_{nq}} \langle \lambda(s), x_n \rangle \, d\mu > q\mu(S_{nq}).$$

Thus we cannot have $\langle \lambda_j, f \rangle \to \langle \lambda, f \rangle$ and the proposition follows.

We now turn to the proof of Theorem 2.1.

Proof of Theorem 2.1. Suppose $\alpha: S \times G \to \mathrm{Iso}(E)$ is a cocycle and $\{A_s\}$ is a α-invariant Borel field. Let $r(s, g)$ be the Radon–Nikodym cocycle of the action of G on S, i.e., a positive Borel function such that for each g, $d\mu(sg) = r(s, g) \, d\mu(s)$. For $g \in G$, define $T(g)$ acting on $L^1(S, E)$ by $(T(g)f)(s) = r(s, g) \, \alpha(s, g) \, f(sg)$. Then

$$\| T(g)f \| = \int r(s, g) \, \| f(sg) \| \, d\mu(s)$$

$$= \int r(sg^{-1}, g) \, \| f(s) \| \, r(s, g^{-1}) \, d\mu(s) = \| f \|,$$

so that $T(g)$ is a representation of G by isometric isomorphisms on $L^1(S, E)$. To see that T is continuous, it suffices to show that $f \in L^1(S, E)$ and $\lambda \in L^\infty(S, E^*)$ implies

$$g \to \langle \lambda, T(g)f \rangle = \int \langle \lambda(s), r(s, g) \, \alpha(s, g) \, f(sg) \rangle \, d\mu(s)$$

is measurable, which it is by Fubini's theorem. We have an induced adjoint action $T^*(g) = (T(g)^{-1})^*$ on $L^1(S, E)^* = L^\infty(S, E^*)$, and one readily verifies that this is defined by $(T^*(g)\lambda)(s) = \alpha^*(s, g) \, \lambda(sg)$. Since $\{A_s\}$ is α-invariant, it follows that the set B of Proposition 2.2 is a non-empty compact convex G-invariant set, and amenability of G implies the existence of $\lambda \in B$ such that $T^*(g)\lambda = \lambda$ for all g. But then it is clear that λ is an α-invariant section in $\{A_s\}$.

COROLLARY 2.3. *If G is a countable discrete group acting freely on (S, μ) and the ergodic equivalence relation defined by the action is hyperfinite* [5], *then S is an amenable G-space.*

Proof. If the action is hyperfinite, the equivalence relation is given by an action of the integers. Cocycles of the G action will then correspond to cocycles of the Z-action: if $\alpha: S \times G \to \mathrm{Iso}(E)$, let $\beta: S \times Z \to \mathrm{Iso}(E)$ be $\beta(s, n) = \alpha(s, g(s, n))$ where $g(s, n) \in G$ is such that $s \cdot g(s, n) = s \cdot n$. Then an α-invariant section, which exists by Theorem 2.1, will also be β-invariant.

This corollary serves to "explain" an example of A. Connes showing that a free hyperfinite action need not be the action of an amenable group. It is the action that must be amenable. We remark that the fact that for a finite type free hyperfinite action the group must be amenable [5, Proposition I.4.5] then follows from Proposition 4.4 below.

We recall that if (X, μ) and (Y, ν) are ergodic G-spaces, X is called an extension of Y if there is a Borel map $p: X \to Y$ such that $p_* \mu$ is in the same measure class as ν, and the induced map of Boolean algebras $p^*: B(Y, \nu) \to B(X, \mu)$ is a G-map. The following result generalizes Theorem 2.1 and reduces to the latter when Y is a point.

THEOREM 2.4. *If X is an extension of Y and Y is an amenable G-space, then X is an amenable G-space.*

Proof. We can assume $\nu = p_* \mu$ and by the the ergodicity assumption, we have, modulo null sets, $X = Y \times I$, where I is the unit interval, $p: X \to Y$ is projection, and $\mu = \nu \times m$ where m is a probability measure on I. Using [10] and [16, Proposition 2.1], we can reduce to the following situation. (X, μ) and (Y, ν) are ergodic G-spaces, $X \subset Y \times I$ is conull and Borel, $\mu = \nu \times m$, and the projection $p: X \to Y$ is a G-map. Denote the projection $X \to I$ by p_2. Let E be a separable Banach space, $\alpha: X \times G \to \mathrm{Iso}(E)$, and $\{A_x\}$ an α-invariant field in E_1^*. Let $F = L^1(I, m, E)$, so that F is also a separable Banach space, with dual $L^\infty(I, m, E^*)$. Define $\beta: Y \times G \to L_1(F)$, the unit ball in the bounded linear operators of F by

$$(\beta(y, g)f)(t) = r(t, y, g)\, \alpha(t, y, g)\, f(p_2((t, y)g)),$$

where r is the Radon–Nikodym cocycle of $X \times G$. As in the proof of Theorem 2.1 we see that β is Borel. Let $\theta(y, g): I \to I$ be $\theta(y, g)(t) = p_2((t, y)g)$, so that $\theta(y, g)$ is defined m-almost everywhere for almost all y. Then it is straightforward to check that $\theta(y, g)^{-1} = \theta(yg, g^{-1})$. Now $d(\nu \times m)((t, y)g) = r(t, y, g)\, d(\nu \times m)(t, y)$, so that for each g and almost all y, we have $dm(\theta(y, g)t) = r(t, y, g)\, dm(t)$. We now claim that for each g, $\beta(y, g) \in \mathrm{Iso}(F)$ for almost all y. We have

$$\|\beta(y, g)f\| = \int_I r(t, y, g)\, \|f(\theta(y, g)t)\|\, dm(t),$$

and replacing t by $\theta(y, g)^{-1}t$, we obtain

$$\int_I r(\theta(yg, g^{-1})t, y, g)\, \|f(t)\|\, r(t, yg, g^{-1})\, dm(t)$$

which for each g and almost all y, $= \int_I \|f(t)\|\, dm(t)$ by the cocycle identity for r. Furthermore, for each g and almost all y, $\beta(y, g)$ is invertible and hence is in $\mathrm{Iso}(F)$. Changing β on a suitable conull set, we can then assume that $\beta: Y \times G \to$

Iso(F) is Borel, and one readily checks that β is a cocycle. For $y \in Y$, let $B_y \subset F_1^*$ be defined by $B_y = \{\varphi: I \to E^* \mid \varphi(t) \in A_{(t,y)}$ for almost all $t\}$. By Proposition 2.2., B_y is a compact convex set. Suppose we knew that $\{B_y\}$ is a Borel field. (We shall show this momentarily.) The α-invariance of $\{A_x\}$ is readily seen to imply the β-invariance of B_y. Amenability of Y then implies the existence of a β-invariant Borel section $\lambda: Y \to F_1^*$ such that $\lambda(y) \in B_y$ a.e. Let $\tilde{\lambda}(t, y) = \lambda(y)(t)$. It is straightforward to check that $\tilde{\lambda}$ is an α-invariant measurable section $X \to E_1^*$ proving the amenability of X. It remains only to show that $\{B_y\}$ is Borel.

Let $K = \{f: I \to I \mid f = \sum q_i \chi_{A_i}\}$, where q_i are rational and $\{A_i\}$ is a finite partition of I into intervals with rational endpoints. Thus K is countable. The fact that B_y is Borel follows from Lemma 1.7 and the following lemma, whose routine proof we omit.

LEMMA 2.5. *If* $a_i : I \to E_1^*$ *are Borel and* $A_t = \overline{\{a_i(t)\}}_i$ *is convex, then functions of the form* $\sum_{i=1}^{n} f_i(t)\, a_i(t)$, *where* $f_i \in K$ *and* $\sum f_i(t) = 1$, *are dense in* $B = \{\lambda: I \to E_1^* \mid \lambda(t) \in A_t$ a.e.$\}$. *We recall that* E_1^* *and* $B \subset L^1(I, E)^*$ *are given the weak-$*$-topology.*

For a certain type of extension, a converse to Theorem 2.4 is available. If $\alpha: S \times G \to K$ is a cocycle, we can form the skew product action of G on $S \times_\alpha K/H$, where $H \subset K$ is a closed subgroup. (See [16] or the beginning of Section 3 below.)

PROPOSITION 2.6. *If* K *is compact and* $S \times_\alpha K/H$ *is ergodic and an amenable* G-space, then S *is an amenable* G-space.

Proof. Let $\gamma: S \times G \to \text{Iso}(E)$ be a cocycle, $\{A_s\}$ a γ-invariant Borel field in E_1^*. Define $\beta(s, x, g) = \gamma(s, g)$, so that $\beta: S \times_\alpha K/H \times G \to \text{Iso}(E)$ is a cocycle. Then $A_{(s,x)} = A_s$ defines a β-invariant field and amenability of $S \times_\alpha K/H$ implies the existence of a β-invariant section $\varphi: S \times K/H \to E_1^*$ in $\{A_{(s,x)}\}$. Let $\psi(s) = \int \varphi(s, x)\, dx$. Then ψ is Borel, and since for almost all s, $\varphi(s, x) \in A_s$ for almost all x, it follows that $\psi(s) \in A_s$ a.e. Finally, for almost all s,

$$\gamma^*(s, g)\, \psi(sg) = \beta^*(s, x, g) \int \varphi(sg, x)\, dx$$

$$= \beta^*(s, x, g) \int \varphi(sg, x\beta(s, g))\, dx$$

$$= \int \varphi(s, x)\, dx = \psi(s).$$

3. AMENABILITY AND THE RANGE OF A COCYCLE

In this section we present the applications, mentioned in the introduction, of amenability to the question of ergodicity of skew products and more generally to the range of a cocycle. If S is an ergodic G-space, and $\alpha: S \times G \to H$ is a cocycle, there is an induced action of G on $S \times H$ defined by $(s, h)g = (sg, h\alpha(s, g))$. (We note that if α is not a strict cocycle then this only defines a near action; however there is an action which agrees with it a.e. for each g [16, Proposition 3.2]). We often denote $S \times H$ by $S \times_\alpha H$ when it is endowed with this action. For a given $S \times G$, it is natural to ask for which groups H is there a cocycle α for which $S \times_\alpha H$ is ergodic. (See [11] for example.) Some positive results are found in [15], [18], [19].

THEOREM 3.1.   *Suppose S is an amenable G-space, $\alpha: S \times G \to H$ is a cocycle, and $S \times_\alpha H$ is ergodic. Then H is an amenable group.*

Proof.   Let E be a separable Banach space, $\pi: H \to \text{Iso}(E)$ a continuous representation, and $A \subset E_1^*$ compact, convex, and H-invariant. It suffices to see that A contains a fixed point (Proposition 1.5). Let $\beta = \pi \circ \alpha$, so that $\beta: S \times G \to \text{Iso}(E)$ is a Borel cocycle. By amenability of S, there is a β-invariant Borel section $\varphi: S \to A$, that is for each g, $\beta^*(s, g) \varphi(sg) = \varphi(s)$ a.e. Define $\theta: S \times H \to A$ by $\theta(s, h) = \pi^*(h) \varphi(s)$. (Recall that $\pi^*(h) = (\pi(h)^{-1})^*$.) Then θ is Borel and for each g, h and almost all s,

$$\theta(sg, h\alpha(s, g)) = \pi^*(h) \pi^*(\alpha(s, g)) \varphi(sg) = \pi^*(h) \varphi(s) = \theta(s, h).$$

By the ergodicity of $S \times_\alpha H$, $\theta(s, h)$ is essentially constant, so on a conull set we have $\pi^*(h) \varphi(s) = \pi^*(k) \varphi(t)$. In particular, for at least one h, $\pi^*(h) \varphi(s) = \pi^*(h) \varphi(t)$ for s, t in a conull set, so $\varphi(s) = \varphi(t) = a$. But then $\pi^*(h)a = \pi^*(k)a$ for h, k in a conull set. Thus, the set of points in H that leaves a invariant is a conull subgroup of H, and hence must be H itself. This completes the proof.

Theorem 3.1 implies a well-known result about random walks, and can in fact be viewed as a generalization of this fact. If ν is a probability measure on G, ν defines, for each $g \in G$, a G-valued stochastic process, X_n, the random walk starting at g at time 0, with independent increments all having distribution ν [6]. We will call the random walk recurrent if for each $A \subset G$ with positive Haar measure $P(X_n \in A$ for infinitely many $n \mid X_0 = g) = 1$ for (Haar) almost all g.

COROLLARY 3.2.   *If G has a recurrent random walk, then G is amenable.*

Proof.   The Haar measure dg is an invariant measure for the random walk and one can form the space of (2-sided) sample sequences of this process with "initial distribution" dg. The recurrence assumption implies that the shift on the sample sequences is ergodic [9]. However, [18] shows that this space is isomorphic to a special type of skew product. Namely, if we let $S = \prod_{-\infty}^{\infty} (G, \nu)$ with the shift transformation and $\alpha: S \times Z \to G$ the cocycle with $\alpha(s, 1)$ equal to evaluation of the zero coordinate, then the space of sample sequences is isomorphic to $S \times_\alpha G$. Ergodicity then implies that G must be amenable.

The reader is referred to Section 5 for further applications to random walks.

We now wish to examine the situation in which $S \times_\alpha H$ is not ergodic. There is an action of H on $S \times_\alpha H$ defined by $(s, h_1)h_2 = (s, h_2^{-1}h_1)$. This commutes with the near action of G, so that the induced Boolean actions of G and H on $B(S \times_\alpha H)$ commute. Let $B_0 \subset B(S \times_\alpha H)$ be the σ-algebra of G-invariant sets modulo 0. Then H acts irreducibly on B_0, and by Mackey's point realization theorem [10], there is an essentially unique ergodic H-space X such that B_0 and $B(X)$ are isomorphic as Boolean H-spaces. The H-space X is called the range of the cocycle α. (See [11] for a discussion.) When $S \times_\alpha H$ is an ergodic G-space, X reduces to a point. The following is a generalization of Theorem 3.1.

THEOREM 3.3.   *If S is an amenable G-space and $\alpha: S \times G \to H$ is a cocycle, then the range of α is an amenable H-space.*

We remark that for $G = Z$, the group of integers, any function $f: S \to H$ defines a cocycle with $\alpha(s, 1) = f(s)$. Hence, Theorem 3.3 provides us with an abundance of amenable actions of non-amenable groups. The proof of Theorem 3.3 is similar in spirit to that of Theorem 3.1 but the technical difficulties are significantly greater. To facilitate the discussion, we shall preface the proof by collecting some notions and facts from [13]. In [13] Ramsay works in a more general framework than we shall require, and we quote his results specified to our situation.

If $S_0 \subset S$ is a conull Borel subset of the ergodic G-space S, let $S_0*G = \{(s, g) \mid s \in S_0, sg \in S_0\}$. Then $S_0 * G$ is called the inessential contraction of $S \times G$ based on S_0. We note that for each g, $\{s \mid (s, g) \in S_0 * G\}$ is conull in S. If $\alpha: S \times G \to M$ is a cocycle with values in the standard Borel group M, α is called strict on $S_0 * G$ if $(s, g), (sg, h) \in S_0 * G$ implies $\alpha(s, gh) = \alpha(s, g) \alpha(sg, h)$.

LEMMA 3.4.   *If $\alpha: S \times G \to M$ is a cocycle, there is a cocycle $\beta: S \times G \to M$ such that for each g, $\beta(s, g) = \alpha(s, g)$ a.e., and β is strict on some inessential contraction $S_0 * G$.*

Proof.   By [13, Theorem 5.1] there is a cocycle β such that β is strict on an inessential contraction and $\beta(s, g) = \alpha(s, g)$ a.e. Arguing as in the proof of [16, Lemma 3.6], we see that for each g, the equality holds for almost all s.

If $S_0 * G$ is an inessential contraction of S and $X_0 * H$ is an inessential contraction of the H-space X, by a strict homomorphism from $S_0 * G$ to to $X_0 * H$ we mean a pair of Borel maps (q, β) where $\beta: S_0 * G \to H$ is a strict cocycle, $q: S_0 \to X_0$, $q(s) \beta(s, g) = q(sg)$ for all $(s, g) \in S_0 * G$, and $(q(s), \beta(s, g)) \in X_0 * H$ for all $(s, g) \in S_0 * G$. We further require that $E \subset X_0$ negligible implies $q^{-1}(E)$ is null. (We recall that E is negligible if the saturation $E \cdot H$ is null. For countable H, E negligible and E null are equivalent, but this of course is far from true for general H.) We will write (q, β) for the map $S_0 \times G \to X_0 * H$ defined by $(q, \beta)(s, g) = (q(s), \beta(s, g))$.

LEMMA 3.5. *Suppose* $\alpha: S \times G \to H$ *is a cocycle and let* X *be the range of* α, *so that* X *is an* H-*space. Then there is a conull set* $S_0 \subset S$, *a measure class preserving* H-*equivariant function* $p: S_0 \times H \to X$, *and a cocycle* $\tilde{\alpha}: S \times G \to H$ *such that for all* g, $\tilde{\alpha}(s, g) = \alpha(s, g)$ *a.e., and* $(\tilde{p}, \tilde{\alpha}): S_0 * G \to X \times H$ *is a strict homomorphism where* $\tilde{p}(s) = p(s, e)$.

Proof. This follows from the proof of [13, Theorem 7.8]. (The statement of [13, Theorem 7.8] is actually somewhat weaker than Lemma 3.5, but the proof in fact shows the stronger statement. The point is that the "similarity" condition in the statement of [13, Theorem 7.8] is actually shown by proving equality a.e.)

If $(\tilde{p}, \tilde{\alpha}): S_0 * G \to X \times H$ is a strict homomorphism and $\theta: S_0 \to H$ is Borel, then $(q, \beta): S_0 * G \to X \times H$ will also be a strict homomorphism where $q(s) = \tilde{p}(s)\,\theta(s)^{-1}$ and $\beta(s, g) = \theta(s)\,\tilde{\alpha}(s, g)\,\theta(sg)^{-1}$.

LEMMA 3.6. *Let* α *be a cocycle and choose* S_0, p, *and* $\tilde{\alpha}$ *as in Lemma* 3.5. *Suppose* $X_0 \subset X$ *is conull Borel. Then there is a conull Borel set* $S_1 \subset S_0$ *and a Borel function* $\theta: S_1 \to H$ *such that* $(q, \beta)(S_1 * G) \subset X_0 * H$.

Proof. This is just [13, Lemma 6.6]. However, as we shall need to examine the proof with some care at a later point, we present Ramsay's proof in our notation and framework. We let μ be the measure on S and ν the measure on X. Let $Z = \{(s, h) \in X \times H \mid x \in X_0 \cdot h\}$. Then Z is Borel and since $\tilde{p}(S_0)$ is not negligible, the projection of Z onto X is $\nu + \tilde{p}_*(\mu)$ conull. By the von Neumann selection theorem, we can choose a Borel function $\lambda: X \to H$ such that $(x, \lambda(x)) \in Z$ for $(\nu + \tilde{p}_*(\mu))$-almost all x. Let $\theta(s) = \lambda(\tilde{p}(s))$. Then there is a conull Borel set $S_1 \subset S_0$ such that $\tilde{p}(s)\,\theta(s)^{-1} \in X_0$ for $s \in S_1$. Furthermore, if $s, sg \in S_1$,

$$\tilde{p}(s)\,\theta(s)^{-1}(\theta(s)\,\tilde{\alpha}(s, g)\,\theta(sg)^{-1}) = \tilde{p}(sg)\,\theta(sg)^{-1} \in X_0$$

and so $(q, \beta)(S_1 * G) \subset X_0 * H$.

One further lemma we shall use is [13, Lemma 5.2].

LEMMA 3.7. *If* $F \subset S \times G$ *contains a conull Borel set, and* $(s, g_1), (sg_1, g_2) \in F$ *implies* $(s, g_1 g_2) \in F$, *then* F *contains an inessential contraction of* $S \times G$.

We are now ready to turn to the proof of Theorem 3.3.

Proof of Theorem 3.3. Let E be a separable Banach space, $\gamma: X \times H \to \mathrm{Iso}(E)$ a Borel cocycle and $\{A_x\}$ a γ-invariant Borel field of compact convex sets in E_1^*. By Lemma 3.4, we can assume γ is strict on some inessential contraction $X_1 * H$. For each h we have $\gamma^*(x, h)A_{xh} = A_x$ a.e. We claim that this identity holds on an inessential contraction of X. By Lemma 1.7, we can choose Borel functions $a_i : X \to E_1^*$ such that $\overline{\{a_i(x)\}} = A_x$ for $x \in X_2 \subset X_1$, a conull Borel set. Then if x, $xh \in X_2$, we have $\gamma^*(x, h)A_{xh} = A_x$ if and only if $\gamma^*(x, h)$ $a_i(xh) \in A_x$ for all i and $\gamma^*(x, h)^{-1} a_i(x) \in A_{xh}$ for all i. For each i, the map $X_2 * H \to X \times E_1^* \times H$, $(x, h) \to (x, \gamma^*(x, h) a_i(xh), h)$ is Borel and injective so the image is Borel. Thus, the projection onto $X \times H$ of this image intersected with $\{(x, a, h) \in X \times E_1^* \times H \mid a \in A_x\}$ is a conull analytic subset of $X \times H$. But this set is precisely $\{(x, h) \in X_2 * H \mid \gamma^*(x, h) a_i(xh) \in A_x\}$. Hence, for each i there is a conull Borel set in $X \times H$ so that $\gamma^*(x, h) a_i(xh) \in A_x$ on this set. A similar argument shows that on a conull Borel set $\gamma^*(x, h)^{-1} a_i(x) \in A_{xh}$, and taking the intersection over all i, we see that $F = \{(x, h) \in X_2 * H \mid \gamma^*(x, h)A_{xh} = A_x\}$ contains a conull Borel set. Now if $(x, h) \in F$ and $(xh, k) \in F$, then

$$\gamma^*(x, hk)A_{xhk} = \gamma^*(x, h)\, \gamma^*(xh, k)A_{(xh)k} = \gamma^*(x, h)A_{xh} = A_x .$$

(Recall γ is strict on $X_2 * H$.) Hence by Lemma 3.7, there is an inessential contraction $X_3 * H \subset F$. For x in a conull Borel subset of X_3, $xh \in X_3$ for almost all h. Call this set X_4. Then again, for x in a conull Borel subset of X_4, $xk \in X_4$ for almost all $k \in H$. Call this set X_0. Then if $x \in X_0$, we have x, xh, $xhk \in X_3$ for almost all $(h, k) \in H \times H$.

Recall that we have the cocycle $\alpha \colon S \times G \to H$. Then there is a conull Borel set $S_1 \subset S$ and a strict homomorphism $(q, \beta) \colon S_1 * G \to X_0 * H$ satisfying the conditions of Lemma 3.6. We let p, $\tilde{\alpha}$, and θ be defined as in that lemma. Let $\delta = \gamma \circ (q, \beta)$, so that $\delta \colon S_1 * G \to \mathrm{Iso}(E)$. Then δ is a cocycle which is strict on $S_1 * G$. Let $A_s = A_{q(s)}$. Then $\{A_s\}$ is a Borel field of compact convex sets and

$$\begin{aligned}
\delta^*(s, g)A_{sg} &= \gamma^*(q(s), \beta(s, g))A_{q(sg)} \\
&= \gamma^*(q(s), \beta(s, g))A_{q(s)\beta(s,g)} \\
&= A_{q(s)} .
\end{aligned}$$

That is, $\{A_s\}$ is a δ-invariant field. Hence there is a δ-invariant section $\varphi \colon S \to E_1^*$. Now $K = \{(s, g) \in S_1 * G \mid \delta^*(s, g) \varphi(sg) = \varphi(s)\}$ is Borel conull, and if (s, g_1), $(sg_1, g_2) \in K$, then $(s, g_1 g_2) \in K$. It follows by Lemma 3.7 that K contains an inessential contraction and hence, by relabelling, we may assume $\delta^*(s, g) \varphi(sg) = \varphi(s)$ for all $(s, g) \in S_1 * G$.

Now define $\psi \colon S_1 \times H \to E_1^*$ by $\psi(s, h) = \gamma^*(q(s), h^{-1})^{-1} \varphi(s)$. Suppose $(s, g) \in S_1 * G$. Then for almost all h,

$$\begin{aligned}
\psi(sg, h\beta(s, g)) &= \gamma^*(q(sg), \beta(s, g)^{-1}h^{-1})^{-1} \varphi(sg) \\
&= \gamma^*(q(s), h^{-1})^{-1} \gamma^*(q(sg) \beta(s, g), \beta(s, g)^{-1})^{-1} \varphi(sg) \\
&= \gamma^*(q(s), h^{-1})^{-1} \gamma^*(q(s), \beta(s, g)) \varphi(sg) \\
&= \gamma^*(q(s), h^{-1})^{-1} \varphi(s) = \psi(s, h).
\end{aligned}$$

Now let $\omega(s,\ h)\ =\ \psi(s,\ h\theta(s)^{-1})$. Then for $(s,\ g)\in S_1 * G$, and almost all h,

$$\omega(sg,\ h\tilde{\alpha}(s,\ g)) = \psi(sg,\ h\tilde{\alpha}(s,\ g)\ \theta(sg)^{-1})$$
$$= \psi(sg,\ h\theta(s)^{-1}\ \beta(s,\ g))$$
$$= \psi(s,\ h\theta(s)^{-1}) = \omega(s,\ h).$$

In other words, ω is essentially G-invariant on $S \times_\alpha H$, and hence there is a map $\sigma\colon X \to E_1{}^*$ such that $\sigma(p(s,\ h)) = \omega(s,\ h)$ a.e.

We now show that σ is a γ-invariant section. We claim first that $\gamma^*(x,\ h)$ $\sigma(xh) = \sigma(x)$ a.e. Now for $s \in S_1$, $q(s)h^{-1} = \tilde{p}(s)\ \theta(s)^{-1}h^{-1} = p(s,\ h\theta(s))$. Hence the map $(s,\ h) \to q(s)h^{-1}$ is measure class preserving. Thus, it suffices to show that

$$\gamma^*(q(s)k^{-1},\ h)\ \sigma(q(s)k^{-1}h) = \sigma(q(s)k^{-1})\quad \text{a.e.}$$

This is equivalent to the following almost everywhere equalities:

$$\gamma^*(q(s)k^{-1},\ h)\ \sigma(\tilde{p}(s)\ \theta(s)^{-1}k^{-1}h^{-1}) = \sigma(\tilde{p}(s)\ \theta(s)^{-1}k^{-1})$$
$$\gamma^*(q(s)k^{-1},\ h)\ \sigma(p(s,\ hk\theta(s))) = \sigma(p(s,\ k\theta(s)))$$
$$\gamma^*(q(s)k^{-1},\ h)\ \omega(s,\ hk\theta(s)) = \omega(s,\ k\theta(s))$$
$$\gamma^*(q(s)k^{-1},\ h)\ \psi(s,\ hk) = \psi(s,\ k)$$
$$\gamma^*(q(s)k^{-1},\ h)\ \gamma^*(q(s),\ k^{-1}h^{-1})^{-1}\ \varphi(s) = \gamma^*(q(s),\ k^{-1})^{-1}\ \varphi(s)$$
$$\gamma^*(q(s),\ k^{-1})\ \gamma^*(q(s)k^{-1},\ h)\ \gamma^*(q(s),\ k^{-1}h^{-1})^{-1}\ \varphi(s) = \varphi(s).$$

Since $q(s) \in X_0$ for $s \in S_1$, the cocycle identity will hold for almost all $(s,\ k,\ h)$, and the equation becomes $\varphi(s) = \varphi(s)$. Thus, we have shown $\gamma^*(x,\ h)\ \sigma(xh) = \sigma(x)$ for almost all $(x,\ h)$. Now consider

$$H_0 = \{h\ |\ \gamma^*(x,\ h)\ \sigma(xh) = \sigma(x)\ \text{for almost all } x\}.$$

Then H_0 is conull and one readily checks that it is closed under multiplication. Hence, $H_0 = H$, and we conclude that σ is a γ-invariant section.

To show that X is amenable, it remains only to show that $\sigma(x) \in A_x$ a.e., and for this it suffices to show that $\sigma(p(s,\ h)) \in A_{p(s,h)}$ a.e. But

$$\sigma(p(s,\ h)) = \psi(s,\ h\theta(s)^{-1}) = \gamma^*(q(s),\ \theta(s)h^{-1})^{-1}\ \varphi(s).$$

Now for almost all s, $\varphi(s) \in A_{q(s)}$, and for almost all $(s,\ h)$,

$$\gamma^*(q(s),\ \theta(s)\ h^{-1})^{-1}\ \varphi(s) \in A_{q(s)\theta(s)h^{-1}} = A_{\tilde{p}(s)h^{-1}} = A_{p(s,h)}\ .$$

This completes the proof.

We shall see in Section 4 that this theorem implies that if the range has finite invariant measure and S is amenable, then H must be amenable.

ROBERT J. ZIMMER

4. Amenable Pairs

In [4], Eymard introduced the notion of the conditional fixed point property for a pair (H, G) where H is a closed subgroup of G. This is the condition that if G acts affinely and continuously in a compact convex set, and if there is a fixed point for H, then there is a fixed point (not necessarily the same point) for G. He relates this notion to the concept of amenable action introduced by Greenleaf [7] (i.e., the existence of a G-invariant mean on $UCB(G/H)$), and then generalizes many known results about amenable groups to amenable pairs.

We now introduce the concept of an amenable pair (S, G) where S is an ergodic G-space.

DEFINITION 4.1. If S is an ergodic G-space, we call (S, G) an amenable pair if for every continuous homomorphism $\pi: G \to \text{Iso}(E)$, E a separable Banach space, and G-invariant compact convex set $A \subset E_1^*$, the existence of an α-invariant section $\varphi: S \to A$ (where $\alpha(s, g) = \pi(g)$) implies the existence of a G-fixed point in A.

PROPOSITION 4.2.   If $S = G/H$, then $(G/H, G)$ is an amenable pair if and only if (H, G) has the conditional fixed point property of Eymard described above [4, p. 11].

Proof. (i) Suppose (H, G) has the conditional fixed point property. If $\pi: G \to \text{Iso}(E)$ and φ is an α-invariant section, it suffices to show there is an H-fixed point. For each g and almost all k, $\pi(g) \, \varphi([k]g) = \varphi([k])$, or $\pi(kg) \, \varphi([k]g) = \pi(k) \, \varphi([k])$. It follows that for almost all k, $\pi(k) \, \varphi([k]) = a$, for some $a \in A$. If $h \in H$, we have $\pi(hk) \, \varphi([k]) = a$ a.e., which clearly implies $\pi(h)a = a$ for all h.

(ii) Conversely, suppose $(G/H, G)$ is amenable. Let E be a separable Banach space, $\pi: G \to \text{Iso}(E)$, and $a \in A \subset E_1^*$ an H-fixed point. There is a strict cocycle $\beta: G/H \times G \to \text{Iso}(E)$ such that β is equivalent to α, the equivalence is implemented by elements of $\pi(G)$, and $\beta(G/H \times G) \subset \pi(H)$ [14, 8.23–8.27]. It follows that $\varphi([k]) = a$ is a β-invariant section, and hence there exists an α-invariant section in A. Therefore G has a fixed point in A. The case of E nonseparable follows by an argument similar to Proposition 1.5.

PROPOSITION 4.3. (i)   *If G is amenable, (S, G) is an amenable pair for every G-space S.*

(ii)   *If S is an amenable G-space and (S, G) is an amenable pair, then G is amenable.*

When S is a transitive G-space, amenability of (S, G) is equivalent to the existence of a G-invariant mean on $L^\infty(S)$ [4, p. 28] which is in turn equivalent to the existence of a G-invariant mean on $CB(S)$, the continuous bounded functions on S [7, Theorem 3.3]. When S is not transitive, the latter equivalence is no longer implied.

PROPOSITION 4.4.   *If there is a G-invariant mean on $L^\infty(S)$ (in particular if S has a finite G-invariant measure), then (S, G) is an amenable pair.*

Proof. Let $\pi: G \to \mathrm{Iso}(E)$ be a continuous representation, $\alpha(s, g) = \pi(g)$, and $A \subset E_1^*$ a G-invariant compact convex set. Suppose $\varphi: S \to A$ is an α-invariant section. Then one can form, as in [4, p. 8], $\int_S \varphi(s) \, dm(s) = a \in A$, where m is a G-invariant mean on $L^\infty(S)$. Since $\pi(g)$ acts affinely on E_1^*, we have by [4, p. 9], $\pi(g)a = \int_S \pi(g) \, \varphi(s) \, dm(s) = \int \pi(g) \, \varphi(sg) \, dm$ by G-invariance, $= \int \varphi(s) \, dm = a$.

COROLLARY 4.5.   *If $\alpha: S \times G \to H$ is a cocycle and S is an amenable G-space, then the existence of an H-invariant mean on $L^\infty(X)$, where X is the range of α, implies that H is amenable.*

Proof.   Theorem 3.3 and Proposition 4.4.

A partial converse to Proposition 4.4 is provided by the following.

PROPOSITION 4.6.   *Suppose S is a compact metric space, G acts continuously on S and μ is quasi-invariant and ergodic. If (S, G) is an amenable pair, then there is a G-invariant probability measure on S.*

Proof. Let $E = C(S)$, $A \subset E_1^*$ the set of probability measures and $\pi(g)$ translation by g in E. For each $s \in S$, let $\varphi(s)$ be the Dirac measure at s. Then φ is Borel and clearly $\pi^*(g) \, \varphi(sg) = \varphi(s)$. Hence φ is an α-invariant section, and so there is a G-fixed point in A.

EXAMPLE 4.7.   Suppose $\theta: G \to H$ is a continuous homomorphism with dense range. Then H is an amenable group if and only if (H, G) is an amenable pair, where H is the G-space defined by θ.

Proof. If H is amenable, there is a G-invariant mean on $L^\infty(H)$, so (H, G) is amenable by Proposition 4.4. Conversely, suppose (H, G) is an amenable pair; let $\pi: H \to \mathrm{Iso}(E)$ and $A \subset E_1^*$ compact, convex, and H-invariant. Since $\theta(G)$ is dense in H, it suffices to show there is a point in A fixed by G. Let $\alpha(h, g) = \pi(\theta(g))$. Choose $a \in A$ and let $\varphi(h) = \pi^*(h)^{-1}a$. Then $\alpha(h, g) \, \varphi(h\theta(g)) = \varphi(h)$, so that φ is an α-invariant section. By the amenability of (H, G), there is a point left fixed by G, completing the proof.

COROLLARY 4.8.   *If $\theta: G \to H$ has dense range and H is an amenable group and an amenable G-space, then G is amenable.*

We remark that dense embeddings of groups are the "normal virtual sub-groups" [17], and Corollary 4.8 can be viewed as the virtual group analogue of the statement that if an amenable normal subgroup has an amenable quotient, then the entire group is amenable. This last statement is of course also a consequence of Corollary 4.8.

We remark that one can also define amenable pairs of G spaces (X, Y), where X is an extension of Y, by a conditional invariant section property. We shall not pursue this here however.

5. AN APPLICATION TO POISSON BOUNDARIES

For any random walk on a group G, Furstenberg has shown how to embed G in a larger space $G \cup P$ in such a way that almost every path of the random walk in G converges to a unique point of P, and every bounded harmonic function on G has an essentially unique representation as a certain type of integral over P. P is called the Poisson boundary and can be used as well for a study of the harmonic functions on homogeneous spaces of G. In a special case, the theory reduces to the classical theory of the Poisson integral representation for harmonic functions in the unit disk. In this section, we present a different method for constructing the Poisson boundary, at least when the law of the walk is étalée, showing it can be obtained by a slight modification of the range construction for a cocycle. A small modification of the proof of Theorem 3.3 will then enable us to show that the Poisson boundary, with a certain naturally defined measure class, is an ergodic amenable G-space.

Let μ be any probability measure on G. Let $\Omega = \prod_{-\infty}^{\infty} G$, $\Omega_0 = \prod_0^{\infty} G$, and $r: \Omega \to \Omega_0$ the projection. Give Ω and Ω_0 the product measures $\eta = \prod_{-\infty}^{\infty} \mu$ and $\eta_0 = \prod_0^{\infty} \mu$. Let $T: \Omega \to \Omega$ be the shift, $(T\omega)(i) = \omega(i+1)$, and T_0 the shift on Ω_0 defined by the same formula. Define $\alpha: \Omega_0 \to G$ by $\alpha(\omega) = \omega(0)$, which we also consider as a function defined on Ω. Then, as we indicated in Corollary 3.2 (see [18] for details), if we let $X = \Omega \times_\alpha G$ with the product measure $\eta \times dg$, then the skew product transformation $\tilde{T}$ is isomorphic to the shift on the space of sample sequences of the (2-sided) random walk. If we let A be the range of α, then Borel functions on A will correspond to invariant functions on the sample sequence space of the 2-sided walk. Harmonic functions on G are in correspondence with the invariant functions on the 1-sided random walk, and so we proceed to construct the range of the 1-sided shift. Thus, if we let $B_0 \subset B(\Omega_0 \times G)$ be the Boolean subalgebra of $\tilde{T}_0$-invariant elements (where $\tilde{T}_0(\omega, g) = (T_0\omega, g\alpha(\omega))$), then B_0 becomes a Boolean G-space, and hence is isomorphic to $B(P, m)$ where (P, m) is an ergodic G-space [10]. By discarding a T_0-invariant Borel null set in Ω_0, we will have a G-map $p: \Omega_0^1 \times G \to P$ which preserves measure class and induces the Boolean isomorphism $p^*(P) \cong B_0$. Now $p \circ \tilde{T}_0$ and p are both G maps from $\Omega_0^1 \times G \to P$ which induce the same map of Boolean algebras. Hence, they agree on a conull G-invariant Borel set in $\Omega_0^1 \times G$, which we will continue to denote $\Omega_0^1 \times G$. Thus, we now have that p is $\tilde{T}_0$-invariant and G-equivariant. Let ν be the measure on P defined by $\nu = (p \circ i)_* (\eta_0)$ where $i : \Omega_0^1 \to \Omega_0^1 \times G$ is given by $i(\omega) = (\omega, e)$. We note that the measure class of m can be recovered from ν by $m = \int_G (\nu \cdot g) \, d\xi(g)$ where ξ is any probability measure in the class of Haar measure. We recall that a measure μ on G is called étalée if the convolution power μ^n has a non-singular component with respect to Haar measure, for some n.

THEOREM 5.1. (P, ν) is a boundary of (G, μ) [6, Definition 3.4] and if μ is étalée, (P, ν) is the Poisson boundary [6, Theorem 3.1].

Thus, this theorem says that the Poisson boundary is the "range" of a naturally defined cocycle of a semi-group action (of the non-negative integers).

Proof.   First, we construct a P-valued μ-process in the sense of [6, p. 16]. Define $z_k : \Omega_0^1 \to P$ by $z_k(\omega) = p(T_0^k\omega, e)$, where $\omega = (x_0, x_1, \ldots)$. It is clear that z_k is a function of x_j, $j \geqslant k$, and since T_0^k preserves measure on Ω_0, z_k all have the same distribution. By invariance of p, and the fact that it is a G-map, we have

$$z_{k+1} x_k^{-1} = p(T_0^{k+1}\omega, x_k) = p(T_0^k\omega, e) = z_k,$$

verifying that z_k is a μ-process. (The difference between our relation $z_{k+1} x_k^{-1} = z_k$ and Furstenberg's condition (d) on [6, p. 16] is only a matter of converting from a right to a left action on P.) Since ν is the common distribution of z_k, it follows that (P, ν) is a boundary [6, Definition 3.4].

For any bounded measurable function φ on P, there corresponds a harmonic function h_φ on G defined by $h_\varphi(g) = \int_P \varphi(gz) \, d\nu(z)$ (where, to conform to Furstenberg's notation, we write the G-action on the left: $gz = zg^{-1}$.) Equivalently, $h_\varphi(g) = \int_{\Omega_0^1} (\varphi \circ p)(\omega, g) \, d\omega$. To see that (P, ν) is the Poisson boundary it suffices to see that every bounded harmonic function can be so represented [6, Theorem 3.1]. But for any bounded harmonic h, there is a bounded function H on the sample sequence space of the 1-sided walk, invariant under the 1-sided shift, for which $h(g) = E_g(H)$ [12, Proposition I.4.2]. Now the proof of [18, Theorem 3] shows that the map

$$\Phi : \Omega_0 \times G \to \Omega_0, \quad (\Phi(\omega, g))(n) = g\omega(0) \cdots \omega(n-1)$$

intertwines $\tilde{T}_0$ and the shift on Ω_0, and that $\Phi(\eta_0 \times \xi) =: P_\xi$ for any probability measure ξ on G, where P_ξ is the probability measure for the random walk with law μ and initial distribution ξ. Thus $H \circ \Phi$ is invariant on $\Omega_0 \times_\alpha G$ and

$$h(g) = E_g(H) = \int_{\Omega_0 \times G} (H \circ \Phi) \, d(\eta_0 \times \delta_g) = \int_{\Omega_0} (H \circ \Phi)(\omega, g) \, d\omega.$$

Since $H \circ \Phi$ is invariant, there is a function $\varphi \colon P \to \mathbb{C}$ such that $H \circ \Phi = \varphi \circ P$ on a conull set. Thus, for almost all g, $h(g) = \int (\varphi \circ p) \, d\omega$. If μ is étalée, every harmonic function is continuous [1, p. 23], and hence two harmonic functions agreeing almost everywhere must be identical.

THEOREM 5.2. *If μ is a measure on a group H, then the boundary (P, m) constructed above is an amenable ergodic H-space.*

Proof. Recall we have $\alpha \colon \Omega \to H$ and A is the range of α considered as a cocycle. We clearly have that P is a factor of A, and after discarding some H-invariant conull Borel sets, we have a commutative diagram of measure class preserving Borel H-maps:

$$
\begin{array}{ccc}
\Omega' \times H & \xrightarrow{\;\mu\;} & A \\
{\scriptstyle r\times \mathrm{id}}\big\downarrow & & \big\downarrow{\scriptstyle t} \\
r(\Omega') \times H & \xrightarrow{\;p_0\;} & P
\end{array}
$$

where Ω', $r(\Omega')$ are Borel, conull, and shift invariant, and p, p_0 are invariant under the shifts. Suppose $\gamma_0 \colon P \times H \to \mathrm{Iso}(E)$ is a cocycle and $\{A_z\}$ is a γ_0-invariant field in $E_1{}^*$. Let γ be the restriction of γ_0 to $A \times H$, i.e., $\gamma(a, h) = \gamma_0(t(a), h)$. In the proof of Theorem 3.3 a γ-invariant section σ was constructed. To prove Theorem 5.2, it suffices to show that in the proof of Theorem 3.3, σ can be chosen so that it actually factors (modulo 0) to a function on P, and in the notation of the proof of Theorem 3.3, it suffices to see that $\omega \colon \Omega' \times H \to E_1{}^*$ factors to a function on $r(\Omega') \times H$. We have

$$
\begin{aligned}
\omega(s, h) &= \gamma^*(\tilde{p}(s)\, \theta(s)^{-1},\ \theta(s)h^{-1})^{-1}\, \varphi(s) \\
&= \gamma_0{}^*(t(\tilde{p}(s))\, \theta(s)^{-1},\ \theta(s)h^{-1})^{-1}\, \varphi(s).
\end{aligned}
$$

Now $t(\tilde{p}(s))$ factors to $r(\Omega')$, so it suffices to see that we can choose $\varphi(s)$ and $\theta(s)$ so that they factor to $r(\Omega')$.

The map $\theta(s)$ is constructed in the proof of Lemma 3.6, where X_0 is the conull Borel set constructed in the proof of Theorem 3.3. Since the cocycle γ is the restriction of γ_0 which is defined on $P \times H$, it is straightforward to check that X_0 can be chosen to be $t^{-1}(P_0)$ for some conull Borel $P_0 \subset P$. Then the map $\lambda \colon A \to H$ in the proof of Lemma 3.6 can be chosen so as to factor to a map on P, and hence $\theta(s)$ factors to a map on $r(\Omega')$.

The map $\varphi \colon \Omega' \to E_1{}^*$ was chosen in the proof of Theorem 3.3 to be an invariant section in $\{A_s\}$ of the cocycle

$$
[\gamma \circ (q, \beta)](s, n) = \gamma_0(t(\tilde{p}(s))\, \theta(s)^{-1},\ \theta(s)\, \tilde{\alpha}(s, n)\, \theta(sn)^{-1}).
$$

Recall that we knew such a section existed by the amenability of the Z-space Ω', which in turn followed from the fact that Z is amenable (Theorem 2.1). In the proof of that theorem, φ was chosen to be a fixed point under the induced action of Z on $L^\infty(\Omega', E^*)$. For $n \geqslant 0$, $(\gamma \circ (q, \beta))(s, n)$ factors to a function on $r(\Omega')$, and hence under the induced Z-action on $L^\infty(\Omega', E^*)$, $L^\infty(r(\Omega'), E^*) \subset L^\infty(\Omega', E^*)$ will be invariant under Z^+, the non-negative integers. Since Z^+ is an amenable semi-group, there will be a Z^+-invariant section in A_s, and this section will then also be Z-invariant. Thus, we can choose φ so that it factors to $r(\Omega')$, completing the proof.

COROLLARY 5.3. *If μ is étaleé, the Poisson boundary with its natural quasi-invariant measure, is an amenable ergodic H-space.*

In many important cases, H is transitive on the Poisson boundary.

COROLLARY 5.4. *If H is transitive on the Poisson boundary of an étalée probability measure, then the stability groups are amenable.*

Proof. Corollary 5.3 and Theorem 1.9.

6. CONCLUDING REMARKS

(a) When $E = C(X)$ for X a compact metric space, the condition of amenability ensures the existence of relatively invariant measures for certain extensions whose fiber is X. More precisely, suppose Y and $Y \times X$ are Borel G-spaces, that the projection $p: Y \times X \to Y$ is a G-map, and that X is compact metric. For each y and g, the action defines a map $\{y\} \times X \to \{yg\} \times X$ which can be identified with a map $\alpha(y, g): X \to X$. Suppose further that these maps are homeomorphisms. Suppose that ν is a quasi-invariant measure on Y such that (Y, ν) is an amenable ergodic G-space. Then there is a probability measure μ on $Y \times X$ that is relatively invariant over ν. That is, one of the two following equivalent conditions holds: (i) if $\mu = \int^{\oplus} \mu_y \, d\nu$ is a decomposition over the fibers of p, then for each g and almost all y, $\alpha(y, g)_* \mu_y = \mu_{yg}$; and (ii) if $r: Y \times G \to \mathbb{R}$ is the Radon–Nikodym cocycle of Y, then $\tilde{r}$ defined by $\tilde{r}(y, x, g) = r(y, g)$ is the Radon–Nikodym cocycle of $Y \times X$.

To see this, by amenability it suffices to show that the induced cocycle

$$\beta: Y \times G \to \mathrm{Iso}(C(X)), \quad (\beta(y, g)f)(x) = f(\alpha(y, g)x)$$

is Borel. But for this it suffices to see that if $\lambda \in M(X)$, then $(y, g) \to \int f(\alpha(y, g)x) \, d\lambda(x)$ is Borel. Since $(y, g, x) \to \alpha(y, g)x$ is Borel, this follows.

(b) One can also define the notion of amenability for a countable ergodic equivalence relation R on a space S using the cohomology of such an object [5] just as we have used the cohomology of $S \times G$. If $R = R_G$ for some amenable action of a countable discrete group G, then R will be amenable. Amenability

372   ROBERT J. ZIMMER

of R implies amenability of S_G if the action of G is free, but this is not true if the freeness condition is dropped.

(c)   There are many other known properties of groups related to amenability and to amenable pairs [4], [8], and one can try to extend these notions and results to our framework. In some cases this is easy, in others it seems more difficult. It would be interesting to determine just how much of the theory does extend.

REFERENCES

1. R. AZENCOTT, "Espaces de Poisson des Groupes Localement Compacts," Lecture Notes in Mathematics, No. 148, Springer–Verlag, New York, 1970.
2. N. DUNFORD AND J. T. SCHWARTZ, "Linear Operators I," Interscience, New York, 1958.
3. R. E. EDWARDS, "Functional Analysis," Holt, Rinehart, & Winston, New York, 1965.
4. P. EYMARD, Moyennes Invariantes et Representations Unitaires, Lecture Notes in Mathematics No. 300, Springer–Verlag, New York, 1972.
5. J. FELDMAN AND C. C. MOORE, Ergodic equivalence relations, cohomology, and von Neumann algebras, I, II, to appear.
6. H. FURSTENBERG, Random walks and discrete subgroups of Lie groups, in "Advances in Probability," Vol. I, pp. 1–63, Marcel Dekker, New York, 1971.
7. F. P. GREENLEAF, Amenable actions of locally compact groups, *J. Functional Analysis* 4 (1969), 295–315.
8. F. P. GREENLEAF, "Invariant Means on Topological Groups," Van Nostrand, New York, 1969.
9. T. E. HARRIS AND H. ROBBINS, Ergodic theory of Markov chains admitting an infinite invariant measure, *Proc. Nat. Acad. Sci. U.S.A.* 39 (1953), 860–864.
10. G. W. MACKEY, Point realizations of transformation groups, *Illinois J. Math.* 6 (1962), 327–335.
11. G. W. MACKEY, Ergodic theory and virtual groups, *Math. Ann.* 166 (1966), 187–207.
12. S. OREY, "Limit Theorems For Markov Chain Transition Probabilities," Van Nostrand, New York, 1971.
13. A. RAMSAY, Virtual groups and group actions, *Advances in Math.* 6 (1971), 253–322.
14. V. S. VARADARAJAN, "Geometry of Quantum Theory," Vol. II, Van Nostrand, Princeton, N.J., 1970.
15. J. J. WESTMAN, Virtual group homomorphisms with dense range, *Illinois J. Math.* 20 (1976), 41–47.
16. R. J. ZIMMER, Extensions of ergodic group actions, *Illinois J. Math.* 20 (1976), 373–409.
17. R. J. ZIMMER, Normal ergodic actions, *J. Functional Analysis* 25 (1977), 286–305.
18. R. J. ZIMMER, Random walks on compact groups and the existence of cocycles, *Israel J. Math.* 26 (1977), 84–90.
19. R. J. ZIMMER, Cocycles and the structure of ergodic group actions, *Israel J. Math.* 26 (1977), 214–220.

Ann. scient. Éc. Norm. Sup.,
4ᵉ série, t. 11, 1978, p. 407 à 408.

INDUCED AND AMENABLE ERGODIC ACTIONS
OF LIE GROUPS

By Robert J. ZIMMER

SUMMARY. — As with unitary representations, one can induce an ergodic action of a closed subgroup of a locally compact group G to obtain an ergodic action of G. We show that every amenable ergodic action of a real algebraic group or a connected semi-simple Lie group with finite center is induced from an action of an amenable subgroup (which is not true for amenable actions of general locally compact groups). The proof depends on the result, of independent interest, that the orbit of any probability measure on real projective space under the action of the general linear group is locally closed in the weak-★-topology. Combined with recent results on the group-measure space construction of von Neumann algebras, this enables us to deduce that any free ergodic action of a real algebraic group or connected semi-simple Lie group with finite center determines a hyperfinite von Neumann algebra via this construction if and only if it is induced from a free ergodic action of an amenable subgroup. Another implication of this result is that a cocycle of an ergodic amenable group action with values in a real algebraic group or connected semi-simple Lie group with finite center is cohomologous to a cocycle taking values in an amenable subgroup.

I. — Introduction

One of the most important methods of constructing unitary representations of groups is that of inducing: to each unitary representation of a closed subgroup of a locally compact group G, there is a naturally associated "induced" unitary representation of G. As pointed out by G. W. Mackey ([12], [13]), one can define induced ergodic actions in an analogous manner. Thus, if G is a locally compact second countable group and H is a closed subgroup, then to each ergodic action of H there is a naturally associated "induced" ergodic action of G. As with unitary representations, the construction is of a very concrete and explicit nature, and one can thus hope to answer many questions about the induced action by examining the action of the smaller and hopefully simpler group H. For a given ergodic action, it is therefore of considerable interest to know when it can be expressed as an action induced from some (perhaps given) subgroup. This is, of course, parallel to a basic theme in the theory of unitary representations. This

408   R. J. ZIMMER

point of view turns out to be quite relevant to the study of a broad class of ergodic actions recently introduced by the author in [20], namely amenable ergodic actions.

Amenable ergodic actions play a role in ergodic theory parallel in many respects to the role played by amenable groups in group theory and arise naturally in a variety of situations. For example, with each ergodic group action there is a naturally associated von Neumann algebra first introduced by Murray and von Neumann and subsequently generalized by a variety of authors, by the group-measure space construction. For free ergodic actions, this von Neumann algebra will be approximately finite dimensional (i. e., hyperfinite) if and only if the action is amenable. This was shown by the author for actions of countable discrete groups in [22] and for actions of general locally compact groups by J. Feldman, P. Hahn, and C. C. Moore in [6], Th. 8.10, using a reduction to the countable case. Other results concerning amenable ergodic actions with applications to problems in ergodic theory and its relation to probability and von Neumann algebras can be found in the author's papers [19]-[23], and the paper of P. Hahn [25].

It follows from the results of [20] that every ergodic action of a group that is induced from an action of an amenable subgroup is an amenable action. Not surprisingly, the converse assertion is false in general as we show by example in section 6 below. However, one of the main points of this paper is to prove the converse for ergodic actions of a suitable class of groups, thus for many purposes reducing the study of amenable actions of such groups to the study of the ergodic actions of amenable subgroups. Specifically, we have:

THEOREMS 5.7, 5.10. — *If* G *is a real algebraic group or a connected semi-simple Lie group with finite center, then every amenable ergodic action of* G *is induced from an ergodic action of an amenable subgroup.*

This theorem has immediate applications to von Neumann algebras and to the cohomology theory of ergodic actions. We also present theorems concerning the structure of amenable ergodic actions of more general Lie groups.

The outline of this paper is a follows. In section 2 we discuss the inducing process for ergodic actions and some of its general properties. In section 3 we recall the definition of amenability for ergodic actions and make some further observations of a general nature concerning these actions. Section 4 is devoted to an examination of the orbits of probability measures on a real projective space under the action of the general linear group. Specifically, we show that every orbit is locally closed in the weak-$\star$-topology. This result is an important step in proving Theorems 5.7 and 5.10, and seems to be of independent interest as well. The proof of this result in turn depends upon a technique of H. Furstenberg for examining the asymptotic behavior of measures on projective space under the general linear group action. Section 5 contains the remainder of the proof of the main theorems, applications of the theorem, and theorems concerning the amenable ergodic actions of more general groups. In section 6 we present an example of an amenable ergodic action that is not induced from an action of an amenable subgroup. Specifically, the action of a lattice subgroup of SL (2, C) on the projective space of a 2-dimensional complex vector space has this property.

II. — Induced ergodic actions

In this section we develop the material we shall need concerning induced ergodic actions. A good deal of this material is implicit in the discussions of Mackey ([12], [13]), but we shall here formulate the notion so as to emphasize the similarity with inducing for unitary representations.

Let G be a locally compact group. Throughout this paper, all locally compact groups will be assumed to be second countable. By a Borel G-space we mean a standard Borel space S together with a jointly Borel action $S \times G \to S$ of G on S. If μ is a σ-finite measure on S, quasi-invariant under the action of G, then μ is called ergodic if $A \subset S$ is a measurable set with $\mu(A g \Delta A) = 0$ for all $g \in G$ implies A is null or conull. We shall then call (S, μ), or sometimes just S, an ergodic G-space. If (S, μ) and (T, ν) are ergodic G-spaces, they are called equivalent, or isomorphic, if there is a G-isomorphism $B(T, \nu) \to B(S, \mu)$ of the associated Boolean σ-algebras of Borel sets modulo null sets. Equivalently [11], there is a conull G-invariant Borel set $S_0 \subset S$ and a measure class preserving G-map $(S_0, \mu) \to (T, \nu)$. In particular, changing the measure μ to a measure in the same measure class (i. e., same null sets) does not change the equivalence class of the action. We may thus assume the measures at hand to be probability measures if we wish. An ergodic G-space is called essentially transitive if there is a conull orbit, or equivalently, if it is isomorphic to the action on a homogeneous space G/H. (We write G/H to be cosets of the form Hg, and $H \backslash G$ to be cosets of the form gH). An ergodic G-space is called properly ergodic if every orbit is a null set. By ergodicity every ergodic action is either essentially transitive or properly ergodic.

Suppose now that $H \subset G$ is a closed subgroup and that (S, μ) is an ergodic H-space. We wish to construct in a natural fashion an associated ergodic G-space. We present two different constructions of this action, both originally described by Mackey. These are in fact analogues of two ways of constructing induced representations: the first as translations on a space of functions on G that transform according to the H-representation (Mackey's original description of induced representations [10]); the second as functions on the quotient G/H where the representation is then defined via a cocycle that corresponds to the representation of H (See [17] for a discussion of induced representations from the latter point of view.)

The first construction of the induced ergodic action is as follows. Let (S, μ) be an ergodic H-space. Then H acts on $S \times G$ by $(s, g) h = (sh, gh)$ and the product of μ with (a probability measure in the class of) Haar measure is quasi-invariant. Let (X, ν) be the space of H-orbits with the quotient Borel structure and quotient measure. As we shall see in a moment, X is a standard Borel space. There is also a G-action on $S \times G$ given by $(s, g) . g_0 = (s, g_0^{-1} g)$ and this commutes with the H-action. There is thus an action of G induced on the orbit space X which clearly leaves ν quasi-invariant. Any G-invariant set in X corresponds to a set $S_0 \times G$ where $S_0 \subset S$ is H-invariant, and so the action of G on X is clearly ergodic.

DEFINITION 2.1. − The space (X, ν) constructed above is called the ergodic G-space induced from the ergodic H-space (S, μ).

To see that X is actually a standard Borel space, it suffices to show that there is a Borel subset of $S \times G$ that meets each H-orbit exactly once. Let $\theta : H \backslash G \to G$ be a Borel section of the natural projection and let $W = \theta (H \backslash G)$, which is a Borel set. Then it is straightforward that $S \times W$ meets each orbit exactly once.

The second construction we present depends on the notion of a cocycle, which the above definition does not require. However, this second approach is also geometrically appealing and often of considerable technical use. Suppose Y is an ergodic G-space and M is a standard Borel group. A Borel function $\alpha : Y \times G \to M$ is called a cocycle if for all $g_1, g_2 \in G$, $\alpha (y, g_1 g_2) = \alpha (y, g_1) \alpha (yg_1, g_2)$ for almost all $y \in Y$. The cocycle is called strict if this identity holds for all (y, g_1, g_2). Two cocycles $\alpha, \beta : Y \times G \to M$ are called equivalent or cohomologous if there is a Borel function $\varphi : Y \to M$ such that for each g, $\varphi (y) \alpha (y, g) \varphi (yg)^{-1} = \beta (y, g)$ for almost all y. If α and β are strict cocycles, they are called strictly equivalent if there exists φ so that this last identity holds for all (y, g). If G is transitive on Y, so that we can write $Y = G/G_0$ for some closed subgroup $G_0 \subset G$, then the strict equivalence classes of strict cocycles $G/G_0 \times G \to M$ correspond to the conjugacy classes of homomorphisms $G_0 \to M$. This correspondence is defined by taking a strict cocycle α and observing that the restriction to $\{ [e] \} \times G_0$ is a homomorphism. Full details of the correspondence can be found in [17]. If $H \subset G$ is a closed subgroup, one has the identity homomorphism $H \to H$ and this will correspond to (an equivalence class of) a strict cocycle $\alpha : G/H \times G \to H$. This cocycle can be defined explicitly as follows: Choose a Borel section $\theta : G/H \to G$ of the natural projection with $\theta ([e]) = e$, and define $\alpha (x, g) = \theta (x) g \theta (xg)^{-1}$.

If $\alpha : Y \times G \to M$ is a strict cocycle and Z is a Borel M-space, then one can define an action of G on $Y \times Z$ by $(y, z).g = (yg, z \alpha (y, g))$. It is exactly the cocycle identity of α that implies that this in fact defines an action. We shall sometimes denote this G-space by $Y \times_\alpha Z$. If α and β are strictly equivalent strict cocycles, then the corresponding actions are easily seen to be equivalent. Applying this procedure to the cocycle defined at the end of the preceding paragraph, we obtain our second description of an induced ergodic action. More precisely, suppose S is an ergodic H-space, where H is a closed subgroup of G. Let $\alpha : G/H \times G \to H$ be a strict cocycle corresponding to the identity homomorphism. Form the G-action $G/H \times_\alpha S$, i. e., $([g_1], s) g = ([g_1] g, s \alpha ([g_1], g))$, which preserves the product measure class. One can easily check that this action is ergodic, but this also follows from the following.

PROPOSITION 2.2. − $G/H \times_\alpha S$ *is equivalent to the ergodic action of G induced from the ergodic H-space S.*

Proof. − Let $\theta : G/H \to G$ be a Borel section of the natural projection and define $\Phi : G/H \times S \to X$ by $\Phi (y, s) = p (s, \theta (y)^{-1})$ where $p : S \times G \to (S \times G)/H = X$ is the natural map. One can readily check that Φ is a measure class preserving Borel isomorphism. To see that Φ is a G-map, it suffices to see that

$$\Phi(yg, s \theta (y) g \theta (yg)^{-1}) = p(s, g^{-1} \theta(y)^{-1}),$$

i. e. that $(s \theta (y) g \theta (yg)^{-1}, \theta (yg)^{-1})$ and $(s, g^{-1} \theta (y)^{-1})$ are in the same H-orbit. But acting upon the latter by $\theta (y) g \theta (yg)^{-1} \in H$ we obtain the former.

There is of course a great similarity between the construction of $G/H \times_\alpha S$ and the construction via cocycles of the unitary representations of G induced from S (*See* [17]). We also remark that the first construction we have given of induced ergodic actions is a special case of Mackey's "range-closure" (or Poincaré flow [5]) construction for arbitrary cocycles into locally compact groups, which generalizes the flow built under a function [12]. The inducing process is exactly the Poincaré flow construction applied to the cocycle $\alpha : S \times H \to G$ defined by $\alpha (s, h) = h$. We shall on occasion make further mention of the range-closure construction and some of its properties, and we refer the reader to [12] and [15] as general references for this material.

Example 2.3. − (*a*) If S is the H-space H/K where K is a closed subgroup, then the induced G-space is G/K. One can see this immediately from Proposition 2.2, since G will clearly be transitive on $G/H \times_\alpha H/K$ and K is a stability group. In particular, the action induced from the trivial action of H (on a point) is just the action of G on G/H, and the action induced from translation of H on H is translation of G on G. We note that this is analogous to facts in representation theory concerning the induced representation of a trivial or regular representation.

(*b*) If $G = \mathbf{R}$ and $H = Z$, then for a Z-space S the induced **R**-action is just the flow built under the constant function 1 ([1], [12]). If $H = Zc$ for some fixed $c \in \mathbf{R}$, then the induced action is just the flow built under the constant function c.

We now present two useful facts concerning induced actions that are direct parallels of results in the theory of unitary representations, namely "inducing in stages" and a parallel of the imprimitivity theorem.

PROPOSITION 2.4. − *Suppose* $K \subset H \subset G$ *are closed subgroups of* G, *and that* S *is ae ergodic K-space. Let* T *be the ergodic H-space induced from* S. *Then the G-spaces obtained by inducing the H-action on* T *to* G *and inducing the K-action on* S *to* G *aru isomorphic.*

Proof. − Let $\alpha : G/H \times G \to H$ and $\beta : H/K \times H \to K$ be strict cocycles corresponding to the identity homomorphism. Then the action of G induced from T is $G/H \times_\alpha T = G/H \times_\alpha (H/K \times_\beta S)$. The G-action on $G/H \times H/K \times S$ is given by $(x, y, s) g = (xg, y \alpha (x, g), s \beta (y, \alpha (x, g)))$. We can identify $G/H \times_\alpha H/K$ with G/K in such a way that $([e], [e])$ corresponds to $[e]$. Then we can consider

$$\gamma : \quad (G/H \times_\alpha H/K) \times G \to K$$

defined by $\gamma ((x, y), g) = \beta (y, \alpha (x, g))$ to be a strict cocycle $G/K \times G \to K$, and this will correspond to the identity homomorphism $K \to K$. It follows that $G/H \times_\alpha T$ is equivalent to $G/K \times_\gamma S$, with proves the proposition.

To describe the analogue of the imprimitivity theorem, we must first recall the notions of extensions and factors of ergodic actions. If (X, μ) and (Y, ν) are ergodic G-spaces, then X is called an extension of Y, and Y a factor of X, if there is a conull G-invariant Borel set $X_0 \subset X$ and a measure-class preserving G-map $X_0 \to Y$. Equivalently

412   R. J. ZIMMER

([18], Prop. 2.1), there is a G-embedding of Boolean σ-algebras $B(Y, \nu) \to B(X, \mu)$. The following analogue of the imprimitivity theorem provides a criterion for determining when a given action is induced.

THEOREM 2.5. — *If* X *is an ergodic* G-*space and* $H \subset G$ *is a closed subgroup, then* X *is induced from an ergodic action of* H *if and only if* G/H *is a factor of* X.

Proof. — We provide an indication of the proof, leaving some measure theoretic details to the reader. One of the implications in the theorem is taken care of by Proposition 2.2, so we assume G/H is a factor of X. Passing to a G-invariant conull Borel set if necessary, we have a measure class preserving G-map $\varphi : X \to G/H$. Let μ be the given probability measure on X and ν the probability measure on G/H. By the transitivity of G on G/H, we can assume $X = G/H \times I$, $\mu = \nu \times m$, where (I, m) is the unit interval with some probability measure. For each g and almost all x, the map $\{ x \} \times I \to \{ xg \} \times I$ defined by the G-action will be measure class preserving and letting $\alpha(x, g)$ be the induced transformation on the Boolean algebra, $\alpha(x, g) : B(\{ xg \} \times I, m) \to B(\{ x \} \times I, m)$, one readily verifies that α is a cocycle on $G/H \times G$ with values in $\text{Aut}(B(I, m))$, the group of automorphisms of the Boolean σ-algebra $B(I, m)$. It is not difficult to see that $\text{Aut}(B(I, m))$ is a standard Borel group (in fact, it is a weakly closed subgroup of the unitary group on $L^2(I, m)$) and that α is Borel. It follows from the discussion of cocycles on transitive G-spaces in [17] that α is equivalent to a strict cocycle into $\text{Aut}(B(I, m))$ which is in turn equivalent to a strict cocycle β with

$$\beta(G/H \times G) = \beta([e] \times H).$$

Each cocycle $G/H \times G \to \text{Aut}(B(I, m))$ defines a Boolean action of G on $B(G/H \times I)$ and hence an action of G that is equivalent to the action on X since α and β are cohomologous. But using Proposition 2.2, this action defined by β is equivalent to the action induced from the H-action defined by the Boolean H-action on $B(I, m)$ given by $\beta \,|\, [e] \times H$.

Next we present another useful criterion that an action be induced.

COROLLARY 2.6. — *If* X *is an ergodic* G-*space, let* $\alpha(x, g) = g$ *so that* $\alpha : X \times G \to G$ *is a cocycle. Then* X *is induced from an ergodic* H-*space if and only if* α *is equivalent to a cocycle taking values in* H.

Proof. — Since the range of the cocycle α is the G-space X, the corollary follows from Theorem 2.5 and the fact that G/H is a factor of the range-closure of α if and only if α is equivalent to a cocycle into H ([18], Th. 3.5).

We now present some other results of a general nature concerning induced actions. We suppose throughout the remainder of this section that $H \subset G$ is a closed subgroup.

PROPOSITION 2.7. — *If* S *is an ergodic* H-*space, then:*

(i) S *is properly ergodic if and only if the induced action of* G *is properly ergodic;*

(ii) S *is essentially free (i. e. almost all stability groups are trivial) if and only if the induced action of* G *is essentially free.*

Proof. — Straightforward.

PROPOSITION 2.8. — *Suppose* S *and* T *are ergodic* H-*spaces and that* X *and* Y *are the corresponding induced* G-*actions. If* S *is an extension of* T, *then* X *is an extension of* Y.

Proof. — Let $\alpha : G/H \times G \to H$ correspond to the identity homomorphism $H \to H$. If $\varphi : S \to T$ is a measure class preserving H-map, then $\psi : G/H \times_\alpha S \to G/H \times_\alpha T$ defined by $\psi(x, s) = (x, \varphi(s))$ is a measure class preserving G-map and the result follows by Proposition 2.2.

It follows from Theorem 2.5 that any extension of an ergodic G-space which is induced from an H-action is also induced from an H-action. The following statement is somewhat sharper. However, as we shall make no use of it below, we omit the proof.

PROPOSITION 2.9. — *Suppose* Y *is an ergodic* G-*space induced from the* H-*space* T. *If* X *is an extension of* Y, *then* X *is induced from an extension of* T.

PROPOSITION 2.10. — *Suppose that* X *is an ergodic* G-*space, and let* X_H *be the* H-*space which has* X *as the underlying set and the restriction of the* G-*action to* H *as the* H-*action. Suppose* H *is ergodic on* X_H. *Then the action of* G *induced from the* H-*action on* X_H *is the product* G-*space* $G/H \times X$.

Proof. — Let $\beta : G/H \times G \to G$ be $\beta(y, g) = g$. Then the product G-space $G/H \times X$ is just the action defined by β, i. e., $G/H \times_\beta X$. But β is strictly equivalent to a strict cocycle α with $\alpha(G/H \times G) = H$ corresponding to the identity homomorphism $H \to H$, and so $G/H \times_\beta X \cong G/H \times_\alpha X$. But the latter is just the action induced from X_H.

We conclude this section with a remark on the ergodic equivalence relation of induced actions. We refer the reader to [5], [6], [16] for the notion of an approximately finite (i. e., hyperfinite) ergodic equivalence relation.

PROPOSITION 2.11. — *If* (X, G) *is induced from* (S, H), *then the ergodic equivalence relation on* X *defined by* G *is approximately finite if and only if the ergodic equivalence relation on* S *defined by* H *is approximately finite.*

Proof. — Writing $X = G/H \times_\alpha S$, the result is clear once we observe that for $x_1 = (y_1, s_1)$, $x_2 = (y_2, s_2)$, we have $x_1 \sim x_2$ if and only if $s_1 \sim s_2$.

III. — Amenable actions

In this section we recall the definition and some properties of amenable ergodic actions and present some further results we shall subsequently require. We refer the reader to [20] for a more detailed and motivated account of amenable actions.

Let E be a separable Banach space, E* the dual Banach space, and E_1^* the unit ball in E*, which is a compact convex set with the $\sigma(E^*, E)$ topology. The group of isometric isomorphisms of E, which we denote by Iso (E), is a separable metrizable group in the strong operator topology, and the associated Borel structure is standard ([20], Lemma 1.1]). If S is a Borel space, by a Borel field of compact convex sets in E_1^* we mean an assignment $s \to A_s$, where $A_s \subset E_1^*$ is compact and convex such that $\{(s, A_s)\} \subset S \times E_1^*$ is Borel.

If S is an ergodic G-space and $\alpha : S \times G \to$ Iso (E) is a cocycle, A_s is called α-invariant if for all g, $\alpha^* (s, g) A_{sg} = A_s$ for almost all s, where α^* is the adjoint cocycle $\alpha^* (s, g) = [\alpha (s, g)^{-1}]^*$. The following "fixed point property" then defines amenability. We call S an amenable ergodic G-space if for all such $(E, \alpha, \{ A_s \})$, there is a Borel function $\varphi : S \to E_1^*$ such that $\varphi (s) \in A_s$ a. e. and $\alpha^* (s, g) \varphi (sg) = \varphi (s)$ a. e. for each $g \in G$. Then φ is called an α-invariant section. The reader should find drawing a sketch in $S \times E_1^*$ helpful in understanding the definition. We record the following for later reference.

PROPOSITION 3.1. — (1) *Any ergodic action of an amenable group is amenable* ([20], Th. 2.1);

(2) *if S is an amenable G-space and there is a G-invariant mean on* L^∞ (S), *then G is amenable* ([20], Prop. 4.3, 4.4);

(3) *an extension of an amenable action is amenable* ([20], Th. 2.4);

(4) *a transitive action is amenable if and only if the stability groups are amenable* ([20], Th. 1.9).

The following relates amenability to inducing.

PROPOSITION 3.2. — *If* (X, G) *is induced from* (S, H), *then* X *is an amenable G-space if and only if* S *is an amenable H-space.*

Proof. — Since X is the range of a cocycle $S \times H \to G$, ([20], Th. 3.3) shows that S amenable implies X amenable. Suppose conversely that $X = G/H \times_\alpha S$ is an amenable G-space where $\alpha : G/H \times G \to H$ corresponds to the identity $H \to H$. Suppose $\gamma : S \times H \to$ Iso (E) is a cocycle and A_s is a γ-invariant field. As in the proof of [20], Th. 2.1, we can define a representation $T : H \to$ Iso $(L^1 (S, E))$ by

$$[T (h) f] (s) = r (s, g) \gamma (s, g) f (sg)$$

where r is the Radon-Nikodym cocycle of the action, and

$$B = \{ f \in L^\infty (S, E^*) \mid f (s) \in A_s \text{ a. e.} \}$$

will be a compact convex subset of the unit ball $L_1^\infty (S, E^*)$ that is invariant under the adjoint representation T^*. It suffices to show that there is a fixed point in B under T^*. Let $F = L^1 (S, E)$. With α as above, we can define a strict cocycle $\beta : G/H \times G \to$ Iso (F) by $\beta (y, g) = T (\alpha (y, g))$. By the argument in part (ii) of the proof of ([20], Th. 1.9), it suffices to show that there is a β-invariant section $G/H \to B \subset F_1^*$. There is a natural isomorphism L^∞ $(G/H \times S, E_1^*) \to L^\infty$ $(G/H, L^\infty (S, E^*))$. Suppose $\varphi : G/H \to L^\infty (S, E^*)$ and $\psi : G/H \times S \to E_1^*$ correspond under the isomorphism. Then it is straightforward that φ is a β-invariant section if and only if ψ is a δ-invariant section, where $\delta : G/H \times_\alpha S \to$ Iso (E) is the cocycle $\delta ((y, s), g) = \gamma (s, \alpha (y, g))$. Furthermore $\varphi (y) \in B$ a. e. if and only if $\psi (y, s) \in A_{y,s} = A_s$ a. e. Amenability of $G/H \times_\alpha S$ ensures the existence of such a function ψ which completes the proof.

Suppose S is an ergodic G-space and H ⊂ G is a closed subgroup such that H is ergodic on S. We wish to examine the relation between amenability of the H-action and amenability of the G-action. We begin with the following.

LEMMA 3.3. — S *is an amenable* H-*space if and only if the product action of* G *on* S × G/H *is amenable.*

Proof. — This follows from Propositions 3.2 and 2.10.

PROPOSITION 3.4. — *Suppose* S *is an ergodic* G-*space and* H ⊂ G *is a closed subgroup such that* H *is also ergodic on* S :

(i) *if* S *is an amenable* G-*space, it is also an amenable* H-*space;*

(ii) *if* S *is an amenable* H-*space and* G/H *has a* G-*invariant probability measure, then* S *is an amenable* G-*space.*

Proof. — (i) follows from Lemma 3.3 and Proposition 3.1 (3). On the other hand, if S × G/H is an amenable G-space and G/H has a finite invariant measure, (ii) follows from the proof of [20], Prop. 2.6.

We now demonstrate the existence of "minimal" amenable ergodic actions.

DEFINITIONS 3.5. — An amenable ergodic action is called minimal amenable if every factor action is non-amenable (other than considering the action as a factor of itself by the identity map).

A transitive G-space G/G$_0$ has only transitive factors and so will be minimal amenable if and only if G$_0$ is a maximal amenable subgroup. Every subgroup of G is contained in a maximal amenable subgroup ([9], Th. IV.1) and the next result is a generalization of this fact.

PROPOSITION 3.6. — *Suppose* X *is an amenable ergodic* G-*space. Then* X *has a minimal amenable factor.*

Proof. — We use Zorn's lemma. Suppose X$_\xi$ is a totally ordered set of factors of X with corresponding G-invariant Boolean σ-algebras B (X$_\xi$) ⊂ B (X). Then ⋂ B (X$_\xi$) = B is a Boolean G-space and hence corresponds to an ergodic G-space Y which is a factor of each X$_\xi$. It suffices to show that Y is an amenable G-space. Let γ : Y × G → Iso (E). Let p : X → Y be the factor map and define α (x, g) = γ (p (x), g). Suppose A$_y$ is a γ-invariant field in E$_1^*$, so that A$_x$ = A$_{p(x)}$ is an α-invariant field. There is an induced representation T* of G on L$^\infty$ (X, E*) defined by (T* (g)f) (x) = α* (x, g)f(xg), and Ã = { f ∈ L$^\infty$ (X, E*) | f(x) ∈ A$_x$ a. e. } is a compact convex G-invariant subset. For each ξ, we can identify L$^\infty$ (X$_\xi$, E*) as a subspace of L$^\infty$ (X, E*) and similarly we can so identify L$^\infty$ (Y, E*). For each ξ, let A$_\xi$ = { f ∈ L$^\infty$ (X$_\xi$, E*) ∩ Ã | T* (g) f = f for all g }. Then A$_\xi$ is a decreasing sequence of compact convex sets, and amenability of X$_\xi$ ensures that A$_\xi$ is non-empty. Choose φ ∈ ⋂ A$_\xi$. Then φ is measurable with respect to each B (X$_\xi$) and hence is measurable with respect to B. Hence φ is the required γ-invariant section.

In section 5 we will explicitly identify all minimal amenable ergodic actions of real algebraic groups and connected semi-simple Lie groups with finite center.

IV. — Orbits of measures on projective space

Let $\mathbf{P}_n$ be the real projective space of lines in $\mathbf{R}^{n+1}$. Let M $(\mathbf{P}^n)$ be the space of probability measures on $\mathbf{P}^n$ with the weak-$\star$-topology. The general linear group GL $(n+1, \mathbf{R})$ (we shall suppress the $\mathbf{R}$ hereafter in this section) acts continuously in a natural fashion on $\mathbf{P}^n$ and hence on M $(\mathbf{P}^n)$. The aim of this section is to prove the following theorem.

THEOREM 4.1. – *For every* $\mu \in$ M $(\mathbf{P}^n)$ *the orbit of* μ *under* GL $(n+1)$ *is locally closed.*

The approach we take in proving this theorem is to employ a technique of H. Furstenberg. If $\mu \in$ M $(\mathbf{P}^n)$ and ν is a limit point of the orbit of μ not contained in the orbit, then Furstenberg has shown in [7], Lemma 1.5, that ν is supported on a union of two proper projective subspaces. Since the space of probability measures supported on a union of two proper projective subspaces is closed, one can deduce immediately that the orbit of any μ which is not so supported must be locally closed. The idea of the proof of Theorem 4.1 is to expand upon these remarks to obtain the theorem for an arbitrary μ.

We begin with notation and some basic facts. $\mathbf{P}^n$ is a compact metrizable space and hence the set $\mathscr{C}$ of closed subsets of $\mathbf{P}^n$ is a compact metric space with the Hausdorf metric. If $A \in \mathscr{C}$, let M (A) denote the set of probability measures on $\mathbf{P}^n$ supported in A. Then M $(A) \subset$ M $(\mathbf{P}^n)$ is a closed subset. If $\mathscr{A} \subset \mathscr{C}$, let $M_{\mathscr{A}} = \bigcup_{A \in \mathscr{A}} M(A)$.

LEMMA 4.2. – *If* $A_n \in \mathscr{C}$, $A_n \to A$, *and* $\mu_n \in$ M (A_n) *with* $\mu_n \to \mu \in$ M $(\mathbf{P}^n)$, *then* $\mu \in$ M (A).

Proof. – Let f be a continuous function on $\mathbf{P}^n$ with supp $(f) \cap A = \varnothing$. Since $A_n \to A$, for sufficiently large n, supp $(f) \cap A_n = \varnothing$. Thus $\int f\, d\mu_n = 0$ for sufficiently large n, so $\int f\, d\mu = 0$.

COROLLARY 4.3. – *If* $\mathscr{A} \subset \mathscr{C}$ *is closed, then* $M_{\mathscr{A}}$ *is closed.*

There is a natural map $\mathbf{R}^{n+1} - \{0\} \to \mathbf{P}^n$ and if $V \subset \mathbf{R}^{n+1}$ is a non-trivial subspace we will denote the image of V in $\mathbf{P}^n$ by $[V]$. We call $[V]$ a projective subspace of $\mathbf{P}^n$. Define $\mathscr{A}$ to be the subset of $\mathscr{C}$ consisting of elements $A \in \mathscr{C}$ of the form $A = \bigcup_{i=1}^{k} [V_i]$ where $[V_i]$ are (non-empty) projective subspaces such that $V_i \not\subset V_j$ for $i \neq j$, and $\sum \dim V_i \leq n+1$. Then define $n(A) = k$, $d(A) = \sum \dim V_i$, and $D(A) = \dim \sum V_i$. These numbers are uniquely determined by the set A. We note that $1 \leq n(A), d(A), D(A) \leq n+1$.

The proofs of the following facts are straightforward.

LEMMA 4.4. – (a) *Let* $\mathscr{A}_k = \{A \in \mathscr{A} \mid n(A) \leq k\}$, $\mathscr{A}^d = \{A \in \mathscr{A} \mid d(A) \leq d\}$ *and* $\mathscr{A}(D) = \{A \in \mathscr{A} \mid D(A) \leq D\}$. *Then these sets are all closed subsets of* $\mathscr{C}$.

(b) *Let* $\mathscr{B}_k^d = \{A \in \mathscr{A} \mid n(A) = k \text{ and } d(A) = d\}$. *If* $A_i \in \mathscr{B}_k^d$ *and* $A_i \to A \in \mathscr{C}$, *then* $A \in \mathscr{B}_k^d \cup \mathscr{A}^{d-1}$. *Hence* $\mathscr{B}_k^d \cup \mathscr{A}^{d-1}$ *is closed in* $\mathscr{C}$.

Now let $\mu \in M(\mathbf{P}^n)$. Define $d(\mu) = \min\{d(A) \mid A \in \mathscr{A}$ and $\mu \in M(A)\}$ and $n(\mu) = \max\{n(A) \mid A \in \mathscr{A}, \mu \in M(A),$ and $d(A) = d(\mu)\}$. Fix an element $A \in \mathscr{A}$ with $\mu \in M(A)$ and $d(A) = d(\mu)$, $n(A) = n(\mu)$, and let $D(\mu) = D(A)$. We note that we actually have $d(A) = D(A)$ for $D(A) < d(A)$ would contradict the definition of $d(\mu)$ since we have $\mu \in M([\sum V_i])$. Let $\mathscr{K}(\mu) \subset \mathscr{C}$ be the closed (by Lemma 4.4) set

$$\mathscr{K}(\mu) = \mathscr{B}^{d(\mu)}_{n(\mu)+1} \cup \mathscr{A}^{d(\mu)-1} \cup \mathscr{A}(D(\mu)-1).$$

Let $\mathscr{U}(\mu)$ be the complement in $M(\mathbf{P}^n)$ of $M_{\mathscr{K}(\mu)}$. Then by the choice of $d(\mu)$, $n(\mu)$, $D(\mu)$, $\mathscr{U}(\mu)$ is an open neighborhood of μ in $M(\mathbf{P}^n)$. Furthermore, the orbit of μ under $GL(n+1)$ is also contained in $\mathscr{U}(\mu)$. Therefore the proof of Theorem 4.1 is reduced to the following.

LEMMA 4.5. — *If $g_i \in GL(n+1)$ and $\mu.g_i \to \nu$ where $\nu \in \mathscr{U}(\mu)$, then ν is in the orbit of μ.*

The proof of Lemma 4.5 depends in turn upon the following formulation of Furstenberg's technique. Let $V \subset \mathbf{R}^{n+1}$ be a subspace and suppose $g_i : V \to \mathbf{R}^{n+1}$ are linear maps of determinant 1. Suppose $[V g_i] \to [W]$ where W is a subspace (necessarily of the same dimension as V). Let μ be the probability measure on $[V] \subset \mathbf{P}^n$ and suppose that $\mu.g_n \to \nu$, so that by Lemma 4.2, ν is a measure on $[W]$.

LEMMA 4.6. — *Either $\{g_i\}$ is bounded, or ν is supported on a union $[Y] \cup [Z]$ where $[Y], [Z]$ are projective subspaces with $\dim Y, \dim Z \geq 1$ and $\dim Y + \dim Z = \dim W$.*

Proof. — The proof is essentially just that of Furstenberg ([7], Lemma 1.5), but we include this minor variation for the reader's convenience. If g_i is not bounded, let $h_i = g_i / \|g_i\|$. Then, perhaps by passing to a subsequence, we have $h_i \to h$ for some linear function $h : V \to \mathbf{R}^{n+1}$ with $\|h\| = 1$ and $\det(h) = 0$. Let $N = \ker(h) \subset V$, $Z = \text{range}(h) \subset W$, so $\dim N + \dim Z = \dim V (= \dim W)$. Again by passing to a subsequence we can assume $[N g_i] \to [Y] \subset [W]$. We claim $\nu = \lim \mu.g_i$ is supported on $[Y] \cup [Z]$. Write $\mu = \mu_1 + \mu_2$ where μ_1 is supported on $[N]$ and μ_2 is supported on $[V] - [N]$. Passing to a subsequence, we have $\nu = \lim(\mu_1.g_i) + \lim(\mu_2.g_i)$. Clearly $\lim(\mu_1.g_i)$ is supported on $[Y]$ so it suffices to see that $\nu_2 = \lim(\mu_2.g_i)$ is supported on $[Z]$. Each g_i can be extended to a linear map $\mathbf{R}^{n+1} \to \mathbf{R}^{n+1}$ and then for f continuous on $\mathbf{P}^n$ with $f = 0$ on $[Z]$ we have

$$\int_{\mathbf{P}^n} f \, d\nu_2 = \lim \int_{\mathbf{P}^n} f \, d(\mu_2 g_i)$$

$$= \lim \int_{\mathbf{P}^n} f(xg_i) \, d\mu_2(x) = \lim \int_{[V]-[N]} f(xg_i) \, d\mu_2(x).$$

But $f(xg_i) \to 0$ pointwise for $x \in [V] - [N]$, so $\int f \, d\nu_2 = 0$ by the dominated convergence theorem, completing the proof of Lemma 4.6.

We now return to the proof of Lemma 4.5 and hence Theorem 4.1.

Proof (of Lemma 4.5). – Let $A = \bigcup [V_i]$ be chosen as above. As we remarked previously, since μ is supported on $[\sum V_i]$, we have $\dim \sum V_i = \sum \dim V_i$. In other words, the subspaces $V_1, \ldots, V_{n(A)}$ are independent. By passing to a subsequence, we can assume $[V_j] g_i \to [W_j]$ for each j, where $\dim W_j = \dim V_j$, and since $v \in M (\bigcup [W_j])$ and $v \in \mathcal{U}(\mu)$, $W_1, \ldots, W_{n(A)}$ will also be independent subspaces. For each j let $\mu_j = \mu \mid [V_j]$, $v_j = v \mid [W_j]$ and $h_{ij} = (g_i \mid V_j) [\det (g_i \mid V_j)]^{-1/\dim V_j}$. Thus $h_{ij} : V_j \to \mathbf{R}^n$, $\det h_{ij} = 1$, $h_{ij} \mid [V_j] = g_i \mid [V_j]$, and $\lim_{i \to \infty} \mu_j h_{ij} = v_j$. We claim that for each j, the sequence h_{ij} is bounded as $i \to \infty$. This is clear if $\dim V_j = 1$. If $\dim V_j \geqq 2$ and h_{ij} is not bounded, then Lemma 4.6 implies that μ_j is supported on $[Y_j] \cup [Z_j]$ where Y_j, $Z_j \neq 0$ and $\dim Y_j + \dim Z_j = \dim W_j$. If $Y_j \cap Z_j \neq 0$, this would imply that v is supported on an element of $\mathcal{A}(D(\mu) - 1)$ which contradicts the fact that $v \in \mathcal{U}(\mu)$. On the other hand, if $Y_j \cap Z_j = 0$, then v is supported on an element of $\mathcal{B}^d_{n(\mu)+1}$ which again contradicts $v \in \mathcal{U}(\mu)$. Thus h_{ij} is bounded for each j and it follows that by passing to a subsequence, as $i \to \infty$ h_{ij} converges to an isomorphism $h_j : V_j \to W_j$ such that $\mu_j h_j = v_j$. Finally, let $h : \mathbf{R}^{n+1} \to \mathbf{R}^{n+1}$ be an invertible linear opertor which agrees with h_j on V_j. Then $h \in GL(n+1)$ and $\mu h = v$. This completes the proof.

V. — Actions of algebraic and semi-simple Lie groups

In this section we prove that amenable ergodic actions of real algebraic groups and connected semi-simple Lie groups with finite center are induced from actions of amenable subgroups, as well as other less precise results for more general groups, and applications. We begin with an analysis of the amenable ergodic actions of GL (n). The essential step involving Theorem 4.1 is Lemma 5.3.

If S is any ergodic G-space, where G is a locally compact group, by a finite ergodic extension of S we mean an ergodic extension $p : T \to S$ such that T is isomorphic as an extension of S to a space of the form $(S \times F, \mu \times m)$ where F is a finite set and m a measure on F. (By ergodicity one can see that this is in fact equivalent to only requiring that under a direct integral decomposition of the measure on T over (S, μ) that almost all fiber measures have finite support). Clearly, a finite extension of a finite extension is again finite. If T is any ergodic extension of S and $\alpha(s, g)$ is a cocycle defined on $S \times G$, then the restriction of α to $T \times G$ is the cocycle $\tilde{\alpha}(t, g) = \alpha(p(t), g)$.

LEMMA 5.1. – *Let M be a locally compact group and* $\alpha : S \times G \to M$ *a cocycle. Suppose* $M_0 \subset M$ *is a normal subgroup of finite index. Then S has a finite ergodic extension T such that the restriction of* α *to T is equivalent to a cocycle into* M_0.

Proof. – Let $q : M \to M/M_0$ be the natural projection. Then $q \circ \alpha$ is a cocycle into the finite group M/M_0. It follows that there is a subgroup $F \subset M/M_0$ such that $q \alpha$ is equivalent to a cocycle β with $\beta(S \times G) \subset F$ and $S \times_\beta F = T$ is ergodic ([18], section 3). Suppose $\varphi : S \to M/M_0$ such that $\varphi(s) q(\alpha(s, g)) \varphi(sg)^{-1} = \beta(s, g)$ for each g, almost all s. Let $\theta : M/M_0 \to M$ be a section of q and define $\psi : S \times_\beta F \to M$ by $\psi(s, x) = \theta(x \varphi(s))$. To prove the lemma it suffices to show that for each g,

$\psi(t) \tilde{\alpha}(t, g) \psi(tg)^{-1} \in M_0$ a. e.   But

$$q(\psi(s, x) \alpha(s, g) \psi(sg, x \beta(s, g))^{-1}) = x \varphi(s) q(\alpha(s, g)) [x \beta(s, g) \varphi(sg)]^{-1}$$
$$= x \beta(s, g) \beta(s, g)^{-1} x^{-1} = e$$

for each x, g and almost all s, and so the result follows.

LEMMA 5.2. – *Suppose* $\alpha : S \times G \to M$ *is a cocycle and* $q : M \to N$ *a surjective homo-morphism. If* $q \circ \alpha$ *is equivalent to a cocycle into a closed subgroup* $N' \subset N$, *then* α *is equivalent to a cocycle into* $M' = q^{-1}(N')$.

Proof. – If $\varphi : S \to N$ with $\varphi(s) q(\alpha(s, g)) \varphi(sg)^{-1} \in N'$, then $\theta(\varphi(s)) \alpha(s, g) \theta(\varphi(sg))^{-1}$ is the required cocycle where θ is a section of q.

The basic application of Theorem 4.1 in proving the main theorems is the following.

LEMMA 5.3. – *Suppose* $\alpha : S \times G \to GL(n)$ *and that* S *is an amenable G-space. Then* S *has a finite ergodic extension* T *such that the restriction of* α *to* T *is equivalent to a cocycle into a subgroup* $M \subset GL(n)$ *that either projects to a compact group in* $SL(n)$ *or leaves a proper subspace invariant.*

Proof. – Let $E = C(\mathbf{P}^{n-1})$ be the Banach space of continuous complex-valued functions on $\mathbf{P}^{n-1}$ and let $\pi : GL(n) \to \mathrm{Iso}(E)$ be the representation induced by the action of $GL(n)$ on $\mathbf{P}^{n-1}$.   Let π^* denote the adjoint representation on E^*.   Then $\pi \circ \alpha : S \times G \to \mathrm{Iso}(E)$ is a cocycle and $M(\mathbf{P}^{n-1})$ is compact convex and $(\pi \circ \alpha)$-invariant.   By amenability, there is a Borel function $\varphi : S \to M(\mathbf{P}^{n-1})$ such that for each g and almost all s.

($\star$) $\pi^*(\alpha(s, g)) \varphi(sg) = \varphi(s)$.

Let $\hat{M}(\mathbf{P}^{n-1}) = M(\mathbf{P}^{n-1})/GL(n)$ be the space of orbits in $M(\mathbf{P}^{n-1})$ under the general linear group.   It is a consequence of Theorem 4.1 and [4] that $\hat{M}(\mathbf{P}^{n-1})$ is a standard Borel space with the quotient Borel structure.   Equation ($\star$) implies that for each g and almost all s, $\varphi(s) \equiv \varphi(sg)$ in $\hat{M}(\mathbf{P}^{n-1})$.   Since $\hat{M}(\mathbf{P}^{n-1})$ is standard and G is ergodic on S, it follows that by changing φ on a null Borel set, $\varphi(S)$ will be contained in a single orbit in $M(\mathbf{P}^{n-1})$.   Choose a point μ_0 in this orbit and let $M \subset GL(n)$ be the stabilizer of μ_0, so that the orbit can be identified with $GL(n)/M$.   *Via* a Borel section $GL(n)/M \to GL(n)$ we can find a Borel map $\theta : \mathrm{Orbit}(\mu_0) \to GL(n)$ such that $\pi^*(\theta(\mu)) \mu_0 = \mu$ for all μ in the orbit.   Define a cocycle $\beta \sim \alpha$ by

$$\beta(s, g) = \theta(\varphi(s))^{-1} \alpha(s, g) \theta(\varphi(sg)).$$

We claim that for each g, $\beta(s, g) \in M$ a. e.   It suffices to show that $\pi^*(\beta(s, g)) \mu_0 = \mu_0$. We can write $\varphi(s) = \pi^*(\theta(\varphi(s))) \mu_0$ for each s, and hence equation ($\star$) becomes $\pi^*(\alpha(s, g) \theta(\varphi(sg))) \mu_0 = \pi^*(\theta(\varphi(s))) \mu_0$ from which the required identity follows immediately.   Since for each g, $\beta(s, g) \in M$ a. e., changing β on a null set we can assume β is a cocycle into M.   Since M is the subgroup of $GL(n)$ leaving μ_0 fixed, it follows from the proof of [7], Lemma 1.5, that either M is compact when projected to $SL(n)$ or has

420   R. J. ZIMMER

a normal subgroup M_0 of finite index such that M_0 leaves a proper subspace invariant. In the former case there is nothing further to show and in the latter case the lemma follows from Lemma 5.2.

We now use an inductive argument to obtain the following result.

LEMMA 5.4. – *Suppose* $\alpha : S \times G \to GL\,(n)$ *and that* S *is an amenable* G-*space. Then* S *has a finite ergodic extension* T *such that the restriction of* α *to* T *is equivalent to a cocycle taking values in a subgroup* $M \subset GL\,(n)$ *which can be described as follows. There is a sequence of subspaces* $0 = V_0 \subset V_1 \subset \ldots \subset V_k = \mathbf{R}^n$ *and inner products* B_i *on* V_i / V_{i-1}, $i = 1, \ldots, k$, *such that* $M = \{\, g \in GL\,(n) \mid V_i\,g = V_i \text{ for all } i, \text{ and } g \text{ induces} \}$ *a similarity on* V_i / V_{i-1} *with respect to* B_i; *i. e.*, $B_i \cdot g$ *is a scalar multiple of* $B_i\,\}$.

Proof. – The proof is by induction on n. If $n = 1$, the lemma is clear. For $n > 1$, suppose the theorem is true for all integers strictly less than n. If S has a finite ergodic extension T such that the restriction of α to T is equivalent to a cocycle with values in $\mathbf{R}\,K$, where $K \subset SL\,(n)$ is compact, we are clearly done. If not, then by Lemma 5.3 we can choose a finite ergodic extension T such that α restricted to T is equivalent to a cocycle β into the group M_1 of all invertible transformations leaving a proper subspace V_1 invariant, and so that V_1 is of minimal dimension among all such choices of T and M_1. There is a natural surjective homomorphism $p : M_1 \to GL\,(V_1)$. Since T is an amenable G-space, we can apply Lemmas 5.3 and 5.2 to $p \circ \beta$ and conclude from the minimality property of V_1 that β is equivalent to a cocycle, which we still denote by β, taking values in the group M_0 consisting of all invertible transformations leaving V_1 invariant and inducing a similarity on (V_1, B_1) for some inner product B_1 on V_1. We can now take the surjective homomorphism $q : M_0 \to GL\,(\mathbf{R}^n / V_1)$ and apply the inductive hypothesis to $q \circ \beta : T \times G \to GL\,(\mathbf{R}^n / V_1)$. An application of Lemma 5.2 then completes the proof.

Let us pause to summarize our situation. The group M in the statement of Lemma 5.4 is a compact extension of a solvable group and is therefore amenable. If we take $G = GL\,(n)$ and $\alpha : S \times GL\,(n) \to GL\,(n)$ to be projection on $GL\,(n)$, then Lemma 5.4 and Corollary 2.6 imply that any amenable ergodic $GL\,(n)$ space has a finite ergodic extension that is induced from an ergodic action of an amenable subgroup M. We now wish to dispose of the need for taking a finite extension (perhaps changing M to another amenable subgroup in the process, of course). In order to do this, we need to recall one important fact about algebraic transformation groups which will be of use to us in another situation below as well.

If a locally compact group G acts in a Borel fashion on a standard Borel space X, the action is called smooth if the orbit space is a standard Borel space with the quotient Borel structure. If X is metrizable by a complete metric and the G-action is continuous, this is equivalent to all orbits being locally closed [4]. (We have used this fact in Lemma 5.3.) The result we will need is that algebraic actions are smooth. More precisely, if G is a real algebraic group acting algebraically on a real algebraic variety, then the action is smooth. For a proof, see the remarks in [3], p. 183-184.

We are now in a position to prove the main theorem for ergodic actions of $GL\,(n)$.

THEOREM 5.5. – *Every amenable ergodic action of* GL (n) *is induced from an action of a closed amenable subgroup.*

Proof. – Let $\alpha : S \times GL\,(n) \to GL\,(n)$ be $\alpha\,(s, g) = g$, and construct the extension $T \to S$ and the sequence of subspaces $0 = V_0 \subset V_1 \subset \ldots \subset V^k = \mathbf{R}^n$ as in Lemma 5.4. Let the cardinality of the fibers of $T \to S$ be the integer p (a. e.). If Z is a finite-dimensional real vector space, let B (Z) be the space of inner products on Z with two inner products identified if they differ by a scalar multiple. Thus B (Z) is a subset of the projective space of the linear space of bilinear maps $Z \times Z \to \mathbf{R}$ and the stability group in GL (Z) of an element in B (Z) is just the group of similarities of the inner product.

Let F be the set of all $2\,k$-tuples, $F = \big\{ (W_1, \ldots, W_k, D_1, \ldots, D_k) \,\big|\, W_i \subset \mathbf{R}^n$ is a subspace with $\dim W_i = \dim V_i$, $W_{i-1} \subset W_i$, and $D_i \in B\,(W_i/W_{i-1}) \big\}$. Then F has a natural Borel structure, GL (n) acts naturally on F and it is not difficult to see that this action is transitive. Thus we can identify F with GL $(n)/M$ where M is the subgroup given in Lemma 5.4. Since M is an algebraic group and algebraic actions are smooth, it follows that GL (n) is smooth on the product space F^p. Let S_p be the symmetric group on p letters which acts naturally on F^p and commutes with the GL (n) action. It follows that GL (n) is smooth on the quotient F^p/S_p as well.

Let $\tilde{\alpha}$ be the restriction of α to T and $\beta : T \times GL\,(n) \to M$ the cocycle equivalent to $\tilde{\alpha}$ given by Lemma 5.4. We can write $T = S \times I$ where I is some finite set. Let $\varphi : T \to GL\,(n)$ implement the equivalence of $\tilde{\alpha}$ and β, i. e., for each g,

$$\varphi(s, y)\,\alpha(s, g)\,\varphi((s, y)\,g)^{-1} = \beta(s, y, g)$$

for almost all $(s, y) \in T$. Let $V \in F$ be the element $V = (V_1, \ldots, V_k, B_1, \ldots, B_k)$ given by Lemma 5.4, and $V\,(s, y) \in F$ be $V \cdot \varphi\,(s, y)$. We then can define a map $\tilde{V} : S \to F^p/S_p$ by $\tilde{V}\,(s) = \big\{ V\,(s, y) \,\big|\, y \in I \big\}$, and it is easy to check that $\tilde{V}$ is Borel. Furthermore, since $V . \beta\,(t, g) = V$, we have for each g, y, and almost all s, $V\,(s, y) . \alpha\,(s, g) = V\,(sg, y')$ for some y'. In other words, $\tilde{V}\,(s)\,\alpha\,(s, g) = \tilde{V}\,(sg)$. This says that $\tilde{V}\,(s)$ and $\tilde{V}\,(sg)$ are in the same GL (n) orbit in F^p/S_p, and since the action of GL (n) on F^p/S_p is smooth, ergodicity of GL (n) on S implies that there is a single orbit in F^p/S_p such that $\tilde{V}\,(s)$ is in that orbit for almost all s. Arguing exactly as in the proof of Lemma 5.3, we see that α is equivalent to a cocycle γ taking values in the group $G_0 = \big\{ g \in GL\,(n) \,\big|\, \tilde{V}_0\, g = \tilde{V}_0 \big\}$ where $\tilde{V}_0 \in F^p/S_p$ is some point in the orbit singled out above. We can write $\tilde{V}_0$ as a set $\{ V^1, V^2, \ldots, V^p \}$ where $V^j \in F$. Let $G_1 = \big\{ g \in G_0 \,\big|\, V^j\, g = V^j$ for all $j \big\}$. Then G_1 is a normal subgroup of G_0 with finite index and G_1 is of course a closed subgroup of $\{ g \in GL\,(n) \,\big|\, V^1\, g = V^1 \}$. The latter is a compact extension of a solvable group and hence amenable, and so G_1 and therefore G_0 is amenable. The theorem now follows from Corollary 2.6.

For the analysis of amenable actions of other groups, we will need the following lemma.

LEMMA 5.6. – *Suppose* X *is a standard Borel G-space and that* $H \subset G$ *is a subgroup of finite index. Then* G *acts smoothly on* X *if and only if* H *acts smoothly on* X.

Proof. — We first observe that it suffices to consider the case where H is normal, since $\bigcap_g g H g^{-1}$ will be normal in G and of finite index in both G and H. If H is normal and smooth on X, then G will be smooth since $X/G \cong (X/H)/G/H$ and finite group actions are smooth. Conversely, if G acts smoothly, then to see that H does as well it suffices to remark that if μ is a properly ergodic quasi-invariant measure under H, then $\sum \mu.g_i$ will be properly ergodic and quasi-invariant under G, where $\{g_i\}$ are a set of representatives of G/H.

We now extend Theorem 5.5 to algebraic groups.

THEOREM 5.7. — *Every amenable ergodic action of a real algebraic group is induced from an action of an amenable subgroup.*

Proof. — Let G be a real algebraic group and (S, μ) an amenable ergodic G-space. Let X be the induced ergodic $GL(n)$ action, and represent $X = GL(n)/G \times_\alpha S$ where $\alpha : GL(n)/G \times GL(n) \to G$ corresponds to the identity as in Proposition 2.2. By Proposition 3.2 and Theorem 5.5, there is a closed amenable subgroup $M \subset GL(n)$ such that $GL(n)/M$ is a factor $GL(n)$-space of X. We can clearly assume that M is a maximal closed amenable subgroup. One can readily check that a conull $GL(n)$-invariant Borel subset of X is of the form $GL(n)/G \times_\alpha S_0$ where $S_0 \subset S$ is conull, Borel, and G-invariant. Thus we can assume that we have a $GL(n)$-map

$$GL(n)/G \times_\alpha S \to GL(n)/M.$$

Restricting this map to $[e] \times S$ gives a (not necessarily measure-class preserving) G-map $\varphi : S \to GL(n)/M$. Let $\nu = \varphi^*(\mu)$, so that ν is quasi-invariant and ergodic under G. Suppose for the moment that ν is supported on a G-orbit. Then S has a G-space factor of the form $G/g M g^{-1} \cap G$ for some $g \in GL(n)$, and since $g M g^{-1} \cap G$ is a closed amenable subgroup of G, the result follows from Theorem 2.5. Thus to prove the theorem, it suffices to show that every G-ergodic measure on $GL(n)/M$ is supported on an orbit, i. e. that G is smooth on $GL(n)/M$. This will be the case if and only if M is smooth on $GL(n)/G$. By [9], Th. IV.2, M has a normal subgroup M_1 of finite index which is a real algebraic group. Since G is algebraic and algebraic actions are smooth, M_1 is smooth on $GL(n)/G$ and by Lemma 5.6, M is as well, completing the proof.

To obtain the corresponding result for connected semi-simple Lie groups with finite center we use the following two lemmas.

LEMMA 5.8. — *Suppose $H \subset G$ is a normal subgroup of finite index and that every ergodic amenable G-action is induced from an action of an amenable subgroup. Then the same is true for H.*

Proof. — If S is an amenable ergodic H-space, form the induced G-space $G/H \times_\alpha S$ and suppose that this has a factor G-space G/G_0 where $G_0 \subset G$ is amenable. Arguing as in the proof of Theorem 5.7, it suffices to show that H is smooth on G/G_0. But this is clear since G_0 is obviously smooth on G/H.

LEMMA 5.9. – *Suppose $p : G \to H$ is a surjective homomorphism with a compact kernel K. If every amenable ergodic H-action is induced from the action of an amenable subgroup, the same is true for G.*

Proof. – Let S be an amenable ergodic G-space. Let $\hat{S} = S/K$, i. e., the space of K-orbits in S. Since K is normal, there is a naturally defined ergodic action of G on $\hat{S}$ which is amenable by [20], Prop. 2.6. This action factors to an action of H which will clearly be amenable, and so there is an amenable subgroup $H_0 \subset H$ and a map of H-spaces $\hat{S} \to H/H_0$. Thus there is a map of G-spaces $S \to G/G_0$ where $G_0 = p^{-1}(H_0)$. Since G_0 satisfies the exact sequence $0 \to K \to G_0 \to H_0 \to 0$ with K, H_0 amenable, it follows that G_0 is amenable. The lemma now follows from Theorem 2.5.

THEOREM 5.10. – *Let G be a connected semi-simple Lie group with finite center. Then every amenable ergodic action of G is induced from an ergodic action of an amenable subgroup.*

Proof. – This follows from Theorem 5.7, Lemmas 5.8 and 5.9, and the fact that $G/Z(G) \cong \text{Ad}(G)$ is the connected component of a real algebraic group.

In the proof of Theorem 5.7, the only point at which the condition that G be algebraic was used was in deducing smoothness of the action of G on $GL(n)/M$. Thus for general closed subgroups of $GL(n)$ one obtains the following by the same proof.

THEOREM 5.11. – *Let G be a closed subgroup of $GL(n)$ and S an amenable ergodic action of G. Then there is an amenable subgroup $M \subset GL(n)$ and a probability measure ν on $GL(n)/M$ quasi-invariant and ergodic under G such that $(GL(n)/M, \nu)$ is a factor of S.*

One can also regard the theorems above as identifying the minimal amenable actions of Lie groups.

COROLLARY 5.12. – (i) *If S is a minimal amenable ergodic action of a closed subgroup $G \subset GL(n)$, then S is of the form $(GL(n)/M, \nu)$ where M is maximal amenable in G and ν is quasi-invariant and ergodic under G.*

(ii) *The minimal amenable ergodic actions of a real algebraic group or a connected semi-simple Lie group with finite center are of the form G/M where M is a maximal amenable subgroup of G.*

It would be interesting to obtain some sort of description and classification of those measures ν appearing in the above corollary, in general or in some specific cases. For example, a case of interest would be to take M to be the triangular matrices and G to be a lattice subgroup of $SL(n)$. Another general case of interest would be that of Poisson boundaries. In [20] we showed that the Poisson boundary of an étalée random walk on G is an amenable ergodic G-space. It follows that if G is a closed subgroup of $GL(n)$ the Poisson boundary has a factor of the form described by Corollary 5.12. It would be interesting to determine which M and ν can arise in this situation. Furstenberg, of course, has very precise results if G is a semi-simple Lie group ([8], Th. 13.4).

Every closed amenable subgroup of $GL(n)$ is a compact extension of a solvable normal subgroup ([9], Th. IV.3). From the above theorems, one can in a certain sense reduce

424   R. J. ZIMMER

amenable actions to compact extensions and actions of solvable subgroups. If $X \to Y$ is an extension of ergodic G-spaces, we call X a compact group extension of Y if X is isomorphic as an extension of Y to $Y \times_\alpha K$ where K is a compact group and $\alpha : Y \times G \to K$ is a cocycle. These extensions have been studied in [18] for example.

COROLLARY 5.13. — *If* S *is amenable ergodic G-space where* G *is a real algebraic group or a connected semi-simple Lie group with finite center, then* S *has a compact group extension* T *which is induced from an ergodic action of a solvable subgroup of* G.

Proof. — Let $\alpha : S \times G \to G$ be projection on G. Then by Theorems 5.7, 5.10 and Corollary 2.6, α is equivalent to a cocycle into an amenable subgroup M. Let $M_0 \subset M$ be a normal solvable subgroup such that M/M_0 is compact ([9], Th. IV.3). Using the argument of Lemma 5.1, we see that there is a compact group extension $T \to S$ such that the restriction of α to T is equivalent to a cocycle into M_0. An application of Corollary 2.6 then completes the proof.

Using Theorems 5.7, 5.10, and the results of [22] and [6], Th. 8.10, we obtain the following corollary.

THEOREM 5.14. — *Suppose* G *is a real algebraic group or a connected semi-simple Lie group with finite center, and that* S *is a free ergodic G-space. Then the von Neumann algebra associated to the action by the group-measure space construction is approximately finite dimensional (i. e., hyperfinite) if and only if the action is induced from a (free) ergodic action of an amenable subgroup of* G.

We note that the statement of Theorem 5.14 involves no concept that depends upon cohomology for its definition. The proof, however, depends from almost beginning to end upon cohomological considerations. One can, of course, formulate similar theorems based on other results above.

Another interesting consequence of Theorems 5.7 and 5.10 concerning cohomology is the following.

THEOREM 5.15. — *Suppose* S *is an amenable G-space and* $\alpha : S \times G \to H$ *where* H *is a real algebraic group or a connected semi-simple Lie group with finite center. Then* α *is equivalent to a cocycle into an amenable subgroup of* H.

Proof. — By [18], Th. 3.5, α is equivalent to a cocycle into an amenable subgroup $M \subset H$ if and only if H/M is a factor of the Poincare flow (range-closure) of α. But the Poincaré flow of α is amenable ([20], Th. 3.3) and so the result follows from Theorems 5.7, 5.10, 2.5.

VI. — A counterexample

In this section we present an amenable ergodic action of a discrete group that is not induced from an ergodic action of any amenable subgroup. In fact, we have the following result.

THEOREM 6.1. — *Let* $M \subset SL(2, \mathbf{C})$ *be the subgroup of all upper triangular matrices, and suppose* Γ *is a lattice subgroup of* $SL(2, \mathbf{C})$, *i. e.,* Γ *is discrete and* $SL(2, \mathbf{C})/\Gamma$ *has finite volume.* Then SL (2, **C**)/M *is an amenable ergodic* Γ-*space that is not induced from an action of any amenable subgroup of* Γ.

Proof. — We remark first that Γ is ergodic on G/M by Moore's theorem [14] and since M is amenable, the Γ action is amenable by Proposition 3.4. To begin the remainder of the proof, we need notation for some subgroups of M. Each element of M is of the form

$$\begin{pmatrix} a & b \\ 0 & a^{-1} \end{pmatrix}$$

where a, $b \in \mathbf{C}$, $a \neq 0$. Let $N \subset M$ be the set of elements with $a = 1$, so in fact $N = [M, M]$. Let A be the elements with $b = 0$, so that M is a semi-direct product AN. The group A is isomorphic to $K \times \mathbf{R}$ where K is the unit circle and we shall identify K and **R** as the corresponding subgroups of M. For brevity, we denote SL (2, **C**) by G for the remainder of the proof.

Suppose that the Γ-action on G/M is induced from an action of an amenable subgroup $\Gamma_0 \subset \Gamma$. Then inducing both Γ-actions to G we *see* by Example 2.3, Theorem 2.5, and Proposition 2.8 that we have an extension of ergodic G-spaces $G/M \times G/\Gamma \to G/\Gamma_0$. By [9], Th. IV.3 and [2], Corol., p. 243, Γ_0 has a normal subgroup Γ_1 of finite index which is triangulable. Since $G/\Gamma_0 \cong G/g\,\Gamma_0\,g^{-1}$ as G-spaces for any $g \in G$, we can assume that $\Gamma_1 \subset M$. We can identify G/M with the complex projective space of $\mathbf{C}^2$. There are two M-orbits in G/M (i. e., two double cosets M\G/M), namely $\{[e]\}$ and its complement. We shall call the latter the non-trivial M-orbit, and we note that as an M-space this orbit is equivalent to M/A. In particular, N is also transitive on the non-trivial M-orbit.

We claim that $\left\{ a \in A \,\middle|\, \begin{pmatrix} a & b \\ 0 & a^{-1} \end{pmatrix} \in \Gamma_1 \text{ for some } b \right\}$ is a discrete set in A. Suppose first that Γ_1 is abelian and that $B = \begin{pmatrix} a & b \\ 0 & a^{-1} \end{pmatrix} \in \Gamma_1$ with $a^2 \neq 1$. Suppose $B_n = \begin{pmatrix} c_n & d_n \\ 0 & c_n^{-1} \end{pmatrix} \in \Gamma_1$ with c_n, c_n^{-1} bounded and c_n distinct. From multiplying out the equation $BB_n B^{-1} = B_n$, it follows that $d_n = ab(c_n - c_n^{-1})(a^2 - 1)^{-1}$. Therefore d_n is bounded which implies that B_n has a convergent subsequence, contradicting the discreteness of Γ_1. Thus, we need only consider the case in which Γ_1 is not abelian. Then $[\Gamma_1, \Gamma_1] \neq \{e\}$, so there is some $\begin{pmatrix} 1 & d \\ 0 & 1 \end{pmatrix} \in \Gamma_1$ with $d \neq 0$. Conjugating this matrix by $\begin{pmatrix} a & b \\ 0 & a^{-1} \end{pmatrix}$ we obtain $\begin{pmatrix} 1 & a^2 d \\ 0 & 1 \end{pmatrix}$. Thus if $\begin{pmatrix} a_n & b_n \\ 0 & a_n^{-1} \end{pmatrix} \in \Gamma_1$ with a_n bounded and distinct, then $\begin{pmatrix} 1 & a_n^2 d \\ 0 & 1 \end{pmatrix} \in \Gamma_1$ and since $d \neq 0$, this again contradicts the discreteness of Γ_1. This verifies our assertion.

We can now observe that $\Gamma_1 N \subset M$ is a closed subgroup. If

$$\begin{pmatrix} a_n & b_n \\ 0 & a_n^{-1} \end{pmatrix} \in \Gamma_1 \quad \text{and} \quad \begin{pmatrix} 1 & x_n \\ 0 & 1 \end{pmatrix} \in N,$$

then the product is

$$\begin{pmatrix} a_n & a_n x_n + b_n \\ 0 & a_n^{-1} \end{pmatrix}.$$

If this converges to $\begin{pmatrix} a & b \\ 0 & a^{-1} \end{pmatrix}$ we can assume by the above remarks that $a_n = a$ for all n

and so

$$\begin{pmatrix} a & b \\ 0 & a^{-1} \end{pmatrix} = \begin{pmatrix} a & b_1 \\ 0 & a^{-1} \end{pmatrix} \begin{pmatrix} 1 & (b-b_1)/a \\ 0 & 1 \end{pmatrix} \in \Gamma_1 N$$

and the latter is closed. Since $\Gamma_1 N$ is normal in M, the action of Γ_1 on $M/\Gamma_1 N$ is trivial. We claim that this implies that the action of Γ_1 on $G/\Gamma_1 N$ has infinitely many ergodic components. To see this, observe that we can consider $G/\Gamma_1 N$ as the G-action induced from the M-action on $M/\Gamma_1 N$. Thus we can write the G-space $G/\Gamma_1 N$ as $G/M \times_\alpha M/\Gamma_1 N$ where $\alpha : G/M \times G \to M$ is the strict cocycle $\alpha(x, g) = s(x) g s(xg)^{-1}$ where $s : G/M \to G$ is a Borel section with $s([e]) = e$. Recall that there is a point, say q, in the non-trivial M-orbit in G/M so that the stabilizer in M of q is A. One can readily check that $J = \begin{pmatrix} 0 & i \\ i & 0 \end{pmatrix}$ is in the coset q. We can also choose the section s above so that $s(q) = J$. Since $\Gamma_1 N$ is transitive on the non-trivial M-orbit, $\Gamma_1 N$ will have infinitely many ergodic components on $G/M \times_\alpha M/\Gamma_1 N$ (which will certainly imply that Γ_1 has infinitely many as well) if and only if the action of the stabilizer of q in $\Gamma_1 N$ has infinitely many ergodic components on $\{q\} \times M/\Gamma_1 N$. To see this, one can argue as in [24], Th. 4.2, for example. The stabilizer of q in $\Gamma_1 N$ is $\Gamma_1 N \cap A$ since $\Gamma_1 N \subset M$. Furthermore, the action of $h \in \Gamma_1 N \cap A$ on $\{q\} \times M/\Gamma_1 N$ is translation of the conjugate $s(q) h s(q)^{-1}$ on $M/\Gamma_1 N$. This just follows from the definition of the action of G on $G/M \times_\alpha M/\Gamma_1 N$. But $s(q) = J$ and on A, conjugation by J is just the map $h \to h^{-1}$. Therefore $\Gamma_1 N \cap A$ acts trivially on $\{q\} \times M/\Gamma_1 N$, and so the verification that Γ_1 has infinitely many ergodic components on $G/\Gamma_1 N$ is complete. Since Γ_1 is of finite index in Γ_0, it is easy to *see* that this implies that Γ_0 cannot be ergodic on $G/\Gamma_1 N$. Equivalently ([24], Th. 4.2), the product action of G on $G/\Gamma_1 N \times G/\Gamma_0$ is not ergodic.

We now consider two cases. Suppose first that $\Gamma_1 \not\subset KN$. We have an extension $G/M \times G/\Gamma \to G/\Gamma_0$ and hence a measure-class preserving G-map

$$G/\Gamma_1 N \times G/M \times G/\Gamma \to G/\Gamma_1 N \times G/\Gamma_0.$$

We have shown that $G/\Gamma_1 N \times G/\Gamma_0$ is not ergodic, so to derive a contradiction, it suffices to show that $G/\Gamma_1 N \times G/M \times G/\Gamma$ is ergodic. Recall that the stabilizer in $\Gamma_1 N$ of a point in the non-trivial M-orbit is a conjugate of $\Gamma_1 N \cap A$. We claim that this is not compact. For any $\gamma \in \Gamma_1$ there is $n \in N$ with $\gamma n \in A$. Since $\Gamma_1 \not\subset KN$ we can find such a product $\gamma n \notin K$. But any compact subgroup of A is actually contained in K, so $\Gamma_1 N \cap A$ is not compact. Since the stabilizer in $\Gamma_1 N$ on the conull orbit is not compact, it follows that the product G-action $G/\Gamma_1 N \times G/M$ has a conull

ERGODIC ACTIONS OF LIE GROUPS   427

orbit with a non-compact stabilizer. By Moore's theorem [14], it follows that $G/\Gamma_1 N \times G/M \times G/\Gamma$ is ergodic, providing the desired contradiction.

It remains to consider the case in which $\Gamma_1 \subset KN$. The group KN is transitive on the non-trivial M-orbit M/A, and as a KN-space this is isomorphic to

$$KN/KN \cap A = KN/K.$$

Since K is compact and Γ_1 is discrete, Γ_1 has infinitely many ergodic components on M/A and hence infinitely many ergodic components on G/M. As above, it follows that Γ_0 is not ergodic on G/M and so $G/M \times G/\Gamma_0$ is not ergodic. However, reasoning as above, $G/M \times G/M \times G/\Gamma$ is ergodic, and so the existence of a measure-class preserving G-map $G/M \times G/M \times G/\Gamma \to G/M \times G/\Gamma_0$ is a contradiction. This completes the proof of the theorem.

REFERENCES

[1] W. AMBROSE, *Representation of Ergodic Flows* (*Annals of Math.*, Vol. 42, 1941, pp. 723-739).

[2] A. BOREL, *Linear Algebraic Groups*, Benjamin, New York, 1969.

[3] J. DIXMIER, *Représentations induites holomorphes des groupes resoluble algébriques* (*Bull. Soc. Math. France*) Vol. 94, 1966, pp. 181-206).

[4] E. G. EFFROS, *Transformation groups and C*-Algebras* (*Annals of Math.*, Vol. 81, 1965, pp. 38-55).

[5] J. FELDMAN and C. C. MOORE, *Ergodic Equivalence Relations, Cohomology, and von Neumann Algebras* (*Trans. Amer. Math. Soc.*, Vol. 234, 1977, pp. 289-360).

[6] J. FELDMAN, P. HAHN, and C. C. MOORE, *Orbit Structure and Countable Sections for Actions of Continuous Groups.* (*Advances in Math.*, Vol. 28, 1978, pp. 186-230).

[7] H. FURSTENBERG, *A Poisson Formula for Semi-Simple Lie Groups* (*Annals of Math.*, Vol. 77, 1963, pp. 335-383).

[8] H. FURSTENBERG, *Boundary Theory and Stochastic Processes on Homogeneous Spaces*, in *Harmonic Analysis on Homogeneous Spaces* (*Symposia in Pure Mathematics*, Williamstown, Mass., 1972).

[9] Y. GUIVARCH, *Croissance polynomiale et périodes des fonctions harmoniques* (*Bull. Soc. Math. France*, Vol. 101, 1973, pp. 333-379).

[10] G. W. MACKEY, *Induced Representations of Locally Compact Groups, I* (*Annals of Math.*, Vol. 55, 1952, pp. 101-139).

[11] G. W. MACKEY, *Point Realizations of Transformation Groups*, (*Illinois J. Math.*, Vol. 6, 1962, pp. 327-335).

[12] G. W. MACKEY, *Ergodic Theory and Virtual Groups* (*Math. Ann.*, Vol. 166, 1966, pp. 187-207).

[13] G. W. MACKEY, *Ergodic Theory and its Significance for Statistical Mechanics and Probability Theory* (*Advances in Math.*, Vol. 12, 1974, pp. 178-268).

[14] C. C. MOORE, *Ergodicity of Flows on Homogeneous Spaces* (*Amer. J. Math.*, Vol. 88, 1966, pp. 154-178).

[15] A. RAMSAY, *Virtual Groups and Group Actions* (*Advances in Math.*, Vol. 6, 1971, pp. 253-322).

[16] C. SERIES, *The Rohlin Theorem and Hyperfiniteness for Actions of Continuous Groups* (to appear).

[17] V. S. VARADARAJAN, *Geometry of Quantum Theory*, Vol. II, van Nostrand, Princeton, N. J., 1970.

[18] R. J. ZIMMER, *Extensions of Ergodic Group Actions* (*Illinois J. Math.*, Vol. 20, 1976, pp. 373-409).

[19] R. J. ZIMMER, *Amenable Ergodic Actions, Hyperfinite Factors, and Poincaré Flows* (*Bull. Amer. Math. Soc.*, Vol. 83, 1977, pp. 1078-1080).

[20] R. J. ZIMMER, *Amenable Ergodic Group Actions and an Application to Poisson Boundaries of Random Walks* (*J. Funct. Anal.*, Vol. 27, 1978, pp. 350-372).

[21] R. J. ZIMMER, *On the von Neumann Algebra of an Ergodic Group Action* (*Proc. Amer. Math. Soc.*, Vol. 66, 1977, pp. 289-293).

428   R. J. ZIMMER

[22] R. J. ZIMMER, *Hyperfinite Factors and Amenable Ergodic Actions* (*Invent. Math.*, Vol. 41, 1977'
     pp. 23-31).
[23] R. J. ZIMMER, *Amenable Pairs of Groups and Ergodic Actions and the Associated von Neumann Algebras*
     [*Trans. Amer. Math. Soc.* (to appear)].
[24] R. J. ZIMMER, *Orbit Spaces of Unitary Representations, Ergodic Theory, and Simple Lie Groups* (*Ann. of Math.*, Vol. 106, 1977, pp. 573-588).
[25] P. HAHN, *The σ-Representations of Amenable Groupoids*, preprint.

(Manuscrit reçu le 20 février 1978.)

R. J. ZIMMER,
Department of Mathematics,
University of Chicago,
5734 University Ave,
Chicago Illinois 60637,
U.S.A.

Inventiones math. 41, 23 – 31 (1977)

Inventiones mathematicae
© by Springer-Verlag 1977

Hyperfinite Factors and Amenable Ergodic Actions *

Robert J. Zimmer **

Department of Mathematics, U.S. Naval Academy, Annapolis, Maryland 21402, USA

We prove that a Krieger factor is hyperfinite if and only if the corresponding ergodic equivalence relation is amenable.

1. Introduction

If a countable discrete group acts ergodically on a standard Borel space with a quasi-invariant measure, there is a von Neumann algebra associated to it by the classical construction of Murray and von Neumann. It is a natural question to determine how various properties of the action and the von Neumann algebra reflect each other. In a previous paper [8], the author has shown that if the algebra has property P as defined by J.T. Schwartz, then the action must be amenable in the sense defined by the author in [7]. In this paper we show conversely that if the action is amenable, the algebra has property E, that is, there is a norm one projection from all bounded operators onto the algebra. In the case in which the action is free, the algebra will be a factor, and we deduce immediately from the work of Connes [1] that the factor will be hyperfinite if and only if the (free) action is amenable. If the action is not free, the algebra will not necessarily be a factor, but Krieger [4] has shown how to modify the Murray-von Neumann construction to obtain a factor which reduces to the original algebra if the action is free. Krieger's factor depends not so much on the group action, but rather on the ergodic equivalence relation it defines. As indicated in [7], one can define amenable ergodic equivalence relations in a manner entirely analogous to the definition of amenable ergodic actions, and the results described above carry over to this context. Thus, we prove that Krieger's factor is hyperfinite if and only if the ergodic equivalence relation is amenable. We also present an application of these results to the general study of amenable actions.

The author wishes to express his thanks to Calvin C. Moore who suggested in discussion of [7] and [8] that the main theorem of this paper might be true.

*   Research supported by the Naval Academy Research Council
**   Current Address: Department of Mathematics. University of Chicago, Chicago, Illinois 60637, USA

2. Amenability and the Murray von Neumann Construction

We shall retain the notation of [8]. Thus G is a countable discrete group acting on the right on a standard Borel space S, and μ is a probability measure on S quasi-invariant and ergodic under G. One has the following operators defined on $L^2(S \times G)$, where $r(s, g)$ is the Radon-Nikodym cocycle of the action of G on (S, μ):

For $g \in G$,

$$(\tilde{U}_g f)(s, h) = f(sg, hg) r(s, g)^{\frac{1}{2}},$$
$$(\tilde{V}_g f)(s, h) = f(s, g^{-1}h);$$

and for $f \in L^\infty(S)$,

$$(M_f h)(s, g) = f(s) h(s, g),$$
$$(N_f h)(s, g) = f(sg) h(s, g).$$

We let L be the von Neumann algebra generated by $\{\tilde{V}_g, N_f\}$ and R the von Neumann algebra generated by $\{\tilde{U}_g, M_f\}$. Then R and L are spatially isomorphic and $R' = L$.

We recall that the action of G on S is called amenable if for every separable Banach space E, Borel cocycle $c: S \times G \to \mathrm{Iso}(E)$, and c-invariant Borel field $\{A_s\}$ of compact convex subsets of E_1^*, there is a c-invariant section in $\{A_s\}$, i.e., a Borel map $\varphi: S \to E_1^*$ such that $\varphi(s) \in A_s$ a.e., and for each g, $c^*(s, g) \varphi(sg) = \varphi(s)$ a.e. Here c^* is the adjoint cocycle on E^*. The reader is referred to [7] for a more detailed discussion of this definition.

Theorem 2.1. If S is an amenable G-space, then L (or equivalently, R) has property E; that is, there is a norm one projection from $B(L^2(S \times G))$ onto L.

We preface the proof of this theorem with some remarks we shall need. If $f \in L^2(S \times G)$, let $f^s(g) = f(s, g)$. Then $f^s \in L^2(G)$ for almost all s, and $f \to \int^{\oplus} f^s \, d\mu$ is a unitary isomorphism of $L^2(S \times G)$ with $\int_S^{\oplus} L^2(G) \, d\mu$. Let us denote by $D \subset B(L^2(S \times G))$ the space of bounded operators decomposable with respect to this direct integral decomposition. It is straightforward to check that if $A = \int^{\oplus} A_s \in D$, then $\tilde{U}_g^{-1} A \tilde{U}_g \in D$ and

$$(\tilde{U}_g^{-1} A \tilde{U}_g)_{sg} = U_g^{-1} A_s U_g, \tag{$*$}$$

where U_g is the right regular representation of G on $L^2(G)$. Let $T(L^2(G))$ be the space of trace class operators on $L^2(G)$. Then $T(L^2(G))$ is a separable Banach space with the tracial norm, and $T(L^2(G))^* = B(L^2(G))$. The group G acts on $T(L^2(G))$ by $g \cdot T = U_g T U_g^{-1}$, and this defines a homomorphism $G \to \mathrm{Iso}(T(L^2(G)))$, where the latter group is the group of isometric isomorphisms of $T(L^2(G))$. Define a cocycle $c: S \times G \to \mathrm{Iso}(T(L^2(G)))$ by $c(s, g)(T) = g \cdot T$. It is straightforward to check that the adjoint cocycle $c^*: S \times G \to \mathrm{Homeo}(B(L^2(G)))$ is also defined by

$c^*(s, g) T = U_g T U_g^{-1}$, where now $T \in B(L^2(G))$. We shall apply the following technical lemma in the above situation.

Lemma 2.2. Suppose E is a separable Banach space and λ_i, $i = 1, 2, \ldots$, a countable collection of Borel functions $\lambda_i: S \to E_1^*$ (where the latter has the $\sigma(E^*, E)$ topology.) Let $F \subset L^\infty(S; E^*)$ be the set of finite sums $\sum f_i(s) \lambda_i(s)$ for all finite collections $\{f_i\}$ where $f_i: S \to [0, 1]$ are Borel functions with $\sum f_i(s) \equiv 1$. Let $\bar{F}$ be the closure of F in the $\sigma(L^\infty(S; E^*), L^1(S; E))$ topology. (Recall $L^\infty(S; E^*) = (L^1(S; E))^*$.) Let $C_s \subset E_1^*$ be the closed convex hull of $\{\lambda_i(s)\}$. Then C_s is a Borel field of compact convex subsets in E_1^* [7, Section 1], $\bar{F}$ is a compact convex set, and

$$\bar{F} = \{\lambda \in L^\infty(S; E^*) \mid \lambda(s) \in C_s \text{ a.e.}\}.$$

Proof. That C_s is a Borel field follows from [7, Lemma 1.7], since elements of the form $\sum q_i \lambda_i(s)$, $\{q_i\}$ a finite set of rationals with $\sum q_i = 1$, are dense in C_s. As F is convex and bounded, $\bar{F}$ will be compact and convex. Clearly, $F \subset \{\lambda \mid \lambda(s) \in C_s \text{ a.e.}\}$, and since the latter is closed [7, Proposition 2.2], $\bar{F} \subset \{\lambda \mid \lambda(s) \in C_s \text{ a.e.}\}$.

To see the reverse inclusion, suppose that $\lambda \in L^\infty(S; E^*)$, $\lambda(s) \in C_s$ a.e., and $\lambda \notin \bar{F}$. Since $\bar{F}$ is compact and convex, there is $f \in L^1(S; E)$, and $a, \delta \in \mathbb{R}$, $\delta > 0$, such that $\operatorname{Re}\langle \varphi, f \rangle < a$ for all $\varphi \in \bar{F}$ and $\operatorname{Re}\langle \lambda, f \rangle > a + \delta$. Let K be the set of finite convex combinations of $\{\lambda_i\}$ with rational coefficients. Then K is a countable subset of F; let us denote the elements of K by $\{\varphi_i\}$. For each s, $\{\varphi_i(s)\}$ is dense in C_s. Let $S_i = \{s \in S \mid |\langle \varphi_i(s) - \lambda(s), f(s) \rangle| < \delta/2\}$. Then S_i is Borel and since $\{\varphi_i(s)\}$ is dense in C_s, $\bigcup S_i = S$. Choose n such that $\mu\left(\bigcup_{i=1}^n S_i\right) > 1 - \delta/4\|f\|$. Define T_i, $i = 1, \ldots, n$ by $T_1 = S_1$, $T_{j+1} = \left(\bigcup_{i=1}^j T_i \cup S_{j+1}\right) - \bigcup_{i=1}^j T_i$, so that T_i are disjoint and $T_i \subset S_i$. Then

$$\varphi(s) = \begin{cases} \sum \chi_{T_i}(s)\, \varphi_i(s), & s \in \bigcup T_i \\ \varphi_1(s), & s \in S - \bigcup T_i \end{cases}$$

defines an element of F. But

$$|\langle \varphi, f \rangle - \langle \lambda, f \rangle| \leq \int_{\bigcup T_i} |\langle \sum \chi_{T_i}(s)\, \varphi_i(s) - \lambda(s), f(s) \rangle| + \int_{S - \bigcup T_i} |\langle \varphi_1(s) - \lambda(s), f(s) \rangle|$$

$$< \delta/2 + 2\|f\|\, \mu(S - \bigcup T_i) < \delta.$$

This is a contradiction, proving the lemma.

We apply Lemma 2.2 to obtain the following result.

Lemma 2.3. Suppose S is an amenable G-space and that $A \in D$ (notation as above). Let $C_S(A)$ be the closed (weak operator topology) convex hull in $B(L^2(S \times G))$ of operators of the form $\sum M_{f_i} \tilde{U}_{g_i}^{-1} A \tilde{U}_{g_i}$, where $\{f_i\}$ is a finite set in $L^\infty(S)$, $f_i: S \to [0, 1]$, $\sum f_i = 1$, and $\{g_i\}$ is a finite subset of G. Then there is an element $W \in C_S(A) \subset D$ such that W commutes with all $\tilde{U}_g$, $g \in G$.

Proof. We can write $A = \int^\oplus A(s)$ where $s \to A(s)$ is a Borel map $S \to B(L^2(G))$ and $\|A(s)\| \leq \|A\|$ for all s. Then the Borel maps $A_g(s) = U_g^{-1} A(sg^{-1}) U_g$ are a countable set with $\|A_g(s)\| \leq \|A\|$ for all s. Let C_s be the (weakly) compact convex hull in $B(L^2(G))$ of $\{A_g(s) | g \in G\}$. Then by Lemma 2.2, C_s is a Borel field of compact convex sets. Further,

$$c^*(s, g) A_h(sg) = U_g U_h^{-1} A(sgh^{-1}) U_h U_g^{-1} = A_x(s), \qquad \text{where } x = hg^{-1}.$$

It follows that $\{C_s\}$ is a c-invariant Borel field. By the amenability of S, there is a c-invariant Borel section $s \to W(s)$ such that $W(s) \in C_s$ a.e. The c-invariance of W means that for each g, $c^*(s, g) W(sg) = W(s)$ a.e., that is, $W(sg) = U_g^{-1} W_s U_g$. Hence, if we let $W = \int^{\oplus} W(s)$, this last equation together with $(*)$ implies $W = \tilde{U}_g^{-1} W \tilde{U}_g$. To complete the proof, it suffices to show that $W \in C_S(A)$. Equation $(*)$ implies $(\tilde{U}_g^{-1} A \tilde{U}_g)(s) = A_g(s)$, and using the natural identification of D with $L^\infty(S; B(L^2(G)))$, Lemma 2.2 implies that W is in the closed convex hull of $\{\sum f_i(s) A_{g_i}(s)\}$ with the $\sigma(L^\infty(S; B(L^2(G))), L^1(S; T(L^2(G))))$-topology. But by [5, Corollary 3.2.2], this is the $\sigma(D, D_*)$ topology, and since the set in question is bounded, the closed convex hull in the $\sigma(D, D_*)$ topology is the same as the closed convex hull in the weak operator topology [5, Proposition 1.15.2].

We now turn to the proof of Theorem 2.1.

Proof (of Theorem 2.1). By the Kakutani-Markov theorem, $\{M_f\}$ has property P and hence the commutant has property E [5, Proposition 4.4.15]. Thus there is a norm one projection from $B(L^2(S \times G))$ onto $\{M_f\}' = D$. Hence, it suffices to construct a norm one projection from D onto $L = D \cap \{\tilde{U}_g\}'$. To do this, we modify the proof of [5, Proposition 4.4.15], that property P implies property E for the commutant, using Lemma 2.3 as a replacement for property P.

D is a von Neumann algebra, and we have the isomorphism $B(D) = (D \otimes_{\max} D_*)^*$. Call the $\sigma(B(D), D \otimes_{\max} D_*)$ topology the σ-topology. For each $g \in G$, define $\Phi_g \in B(D)$ by

$$\Phi_g(T) = \tilde{U}_g^{-1} T \tilde{U}_g,$$

and for each $f \in L^\infty(S)$, define $\Phi_f \in B(D)$ by $\Phi_f(T) = M_f T$. Let $F \subset B(D)$ be the set of all finite sums $\sum \Phi_{f_i} \Phi_{g_i}$ where $g_i \in G$ and $f_i \in L^\infty(S)$ with $f_i \geq 0$ and $\sum f_i = 1$. One verifies readily that F is convex, and for $\Phi \in F$, $\|\Phi\| \leq 1$. Thus, $\bar{F}$, the σ-closure of F, is a compact convex set. We also note that if Φ_α is a net in $B(D)$, and $\Phi_\alpha \to \Phi$, then for all $T \in D$, $\Phi_\alpha(T) \to \Phi(T)$ in the $\sigma(D, D_*)$-topology.

For any $A \in D$, let $C_S(A)$ be as in Lemma 2.3, that is $C_S(A)$ is the weakly closed convex hull of $\{\Phi(A) | \Phi \in F\}$. It follows that $\Phi(A) \in C_S(A)$ for all $\Phi \in \bar{F}$. We also note that one can readily check that $T \in C_S(A)$ implies $C_S(T) \subset C_S(A)$. If $\Phi_1, \Phi_2 \in \bar{F}$, define $\Phi_1 \geq \Phi_2$ if $C_S(\Phi_1(T)) \subset C_S(\Phi_2(T))$ for all $T \in D$. If $\{\Phi_\alpha\}$ is a linearly ordered subset, let $F_\beta = \{\Phi_\alpha | \Phi_\alpha \geq \Phi_\beta\}$, so that $\bar{F}_\beta \subset F$ is compact. Choose $\Phi \in \bigcap \bar{F}_\beta$. Since Φ is an accumulation point of F_β, $\Phi(T)$ is an accumulation point of $\{\Phi_\alpha(T) | \Phi_\alpha \geq \Phi_\beta\}$. But for $\Phi_\alpha \geq \Phi_\beta$, $\Phi_\alpha(T) \in C_S(\Phi_\beta(T))$, and so $\Phi(T) \in C_S(\Phi_\beta(T))$. Thus, $C_S(\Phi(T)) \subset C_S(\Phi_\beta(T))$ for all β, so that Φ is an upper bound for $\{\Phi_\alpha\}$. By Zorn's lemma, we can choose a maximal element $\Phi_0 \in \bar{F}$. We claim that $C_S(\Phi_0(T))$ has exactly one element for all $T \in D$. Suppose $C_S(\Phi_0(T_0))$ has two elements for some $T_0 \in D$.

By Lemma 2.3, $C_S(\Phi_0(T_0)) \cap \{\tilde{U}_g\}'$ is non-empty. Choose an element A in this intersection. Then there is a sequence $\Phi^n \in F$ such that $\Phi^n(\Phi_0(T_0)) \to A$. But since $\|\Phi^n \Phi_0\| \leq 1$, $\{\Phi^n \Phi_0\}$ has a σ-accumulation point Φ_1 with $\|\Phi_1\| \leq 1$ and $\Phi_1(T_0) = A$. Since $A \subset \{\tilde{U}_g\}'$, we have $C_S(\Phi_1(T_0)) \subsetneqq C_S(\Phi_0(T_0))$. Furthermore, for any $T \in D$, $\Phi_1(T)$ is in the closure of $\{\Phi(\Phi_0(T)) \mid \Phi \in F\}$, and so $C_S(\Phi_1(T)) \subset C_S(\Phi_0(T))$. Suppose for the moment that $\Phi_1 \in \bar{F}$. Then the above inclusions contradict the maximality of Φ_0. Therefore, $C_S(\Phi_0(T))$ must be a singleton. Thus for all $T \in D$, we have $\Phi_0(T) \in \{\tilde{U}_g\}'$ (since $C_S(\Phi_0(T)) \cap \{\tilde{U}_g\}'$ is non-empty) and if $T \in \{\tilde{U}_g\}'$, then $\Phi_0(T) = T$. Hence Φ_0 is the required norm one projection. Thus, to complete the proof, it suffices to prove the following.

Lemma. $\Phi_1 \in \bar{F}$.

Proof. It suffices to show that $\Phi \Phi_0 \in \bar{F}$ for each $\Phi \in F$. We note first that $\Phi, \Phi_2 \in F$ implies $\Phi \Phi_2 \in F$. Now $\Phi_0 \in \bar{F}$, so $\Phi_\alpha \to \Phi_0$ for some net Φ_α in F. Since $\Phi \Phi_\alpha \in F$, it suffices to see that $\Phi \Phi_\alpha \to \Phi \Phi_0$. Because all these operators have norm ≤ 1, it suffices to see that for each $A \in D$, $\Phi \Phi_\alpha(A) \to \Phi \Phi_0(A)$ in the $\sigma(D, D_*)$ topology, and since everything is bounded, it suffices to see that this convergence holds in the weak operator topology. But $\Phi_\alpha(A) \to \Phi_0(A)$ in the weak operator topology, and the result follows easily from the form of an element of F.

As a result of Theorem 2.1, we have the following.

Theorem 2.4. If a countable discrete group G acts freely and ergodically on S, then the factor associated to $S \times G$ is hyperfinite if and only if S is an amenable G-space.

Proof. This follows from Theorem 2.1, [8], and the result of A. Connes that for factors, property P, property E, and hyperfiniteness are all equivalent [1].

3. Krieger's Factors

In this section, we indicate the modifications necessary in the proof of Theorem 2.4 to obtain the analogous result for Krieger's factors. We begin by briefly discussing amenability for countable ergodic equivalence relations. The reader is referred to [2] as a general reference for ergodic equivalence relations.

Let S be a standard Borel space with a probability measure μ. Suppose $Q \subset S \times S$ is a countable ergodic equivalence relation on S. We recall that a Borel map $m: Q \to M$, where M is a standard Borel group, is called a cocycle if $(x, y), (y, z) \in Q$ implies $m(x, y) m(y, z) = m(x, z)$. If $M = \text{Iso}(E)$ for some separable Banach space E, there is an induced adjoint cocycle $m^*(x, y)$ defined by $m^*(x, y) = (m(x, y)^*)^{-1}$. A Borel field of compact convex sets $s \to A_s \subset E_1^*$ is called m-invariant if $m^*(x, y) A_y = A_x$ a.e., and $\varphi: S \to E_1^*$ is called a m-invariant section if $m^*(x, y) \varphi(y) = \varphi(x)$ a.e.

Definition 3.1. Q is amenable if for every separable Banach space E, cocycle $m: Q \to \text{Iso}(E)$, and m-invariant Borel field $\{A_s\}$, there is a m-invariant section with $\varphi(s) \in A_s$ a.e.

If a countable discrete group G acts ergodically on S, the orbits define an ergodic equivalence relation on S which we denote by Q_G. Each cocycle m on Q_G defines a cocycle c on $S \times G$ by $c(s, g) = m(s, sg)$. If the G action is free, every $S \times G$ cocycle arises in this way. Thus, one easily sees the following.

Proposition 3.2. If G acts freely on S, then S is an amenable G-space if and only if Q_G is amenable.

The converse of this proposition can readily be seen to be false. However, amenability of Q_G can still be easily described in terms of $S \times G$. Let G_s be the stability group of $s \in S$. Call a cocycle c defined on $S \times G$ a reduced cocycle if $c(s, G_s)$ is the identity. It is easy to check that there is a one-one correspondence between reduced cocycles on $S \times G$ and cocycles on Q_G.

Definition 3.3. Call S a weakly amenable G-space if for every separable Banach space E, reduced cocycle $c: S \times G \to \text{Iso}(E)$, and every c-invariant Borel field $\{A_s\}$ of compact convex subsets of E_1^*, there is a c-invariant section in $\{A_s\}$.

Proposition 3.4. S amenable implies S weakly amenable. S is weakly amenable if and only if Q_G is amenable.

We remark that many of the results of [7] hold for amenable ergodic equivalence relations. For example, the proof of [7, Theorem 3.3] shows the following.

Proposition 3.5. Suppose Q is an amenable countable ergodic equivalence relation, and $m: Q \to H$ is a cocycle, where H is a locally compact group. Then the range-closure of m (i.e., the Poincaré flow of m) is an amenable ergodic H-space.

Krieger's factors can be described as follows. Let Q be a countable ergodic equivalence relation on S, and choose an action such that $Q_G = Q$ [2, Theorem 1]. Let $p: S \times G \to Q \subset S \times S$ be defined by $p(s, g) = (s, sg^{-1})$. Then Q has a σ-finite measure π defined by $\pi(A) = \int_S \text{card}(A \cap \{s\} \times S) \, d\mu(s)$, for $A \subset Q$ [2, Theorem 2]. A function f on $S \times G$ factors (modulo null sets) to a function on Q if and only if for almost all (s, g, h), $sg^{-1} = sh^{-1}$ implies $f(s, g) = f(s, h)$. Alternatively, let us denote by G/G_s the set of cosets $\{G_s g\}$ and by $G_s \backslash G$ the set of cosets $\{g G_s\}$. Then f factors to Q if and only if $f_s(g) = f(s, g)$ factors to $G_s \backslash G$ for almost all s. Then we can identify $L^2(Q, d\pi) \cong \int_S^{\oplus} L^2(G_s \backslash G) \, d\mu \cong \{f: S \times G \to \mathbb{C} \, | \, sg^{-1} = sh^{-1}$ implies $f(s, g) = f(s, h)$ a.e., and $\int |\tilde{f}|^2 \, d\pi < \infty$, where $f = \tilde{f} \circ p\}$.

Let us denote this last set of functions by H. The operators $\tilde{U}_g$, $\tilde{V}_g$, M_f, N_f as defined at the beginning of Section 2 act on the measurable functions on $S \times G$, and it is straightforward to check that they all leave H invariant and that $\tilde{U}_g$ and $\tilde{V}_g$ are unitary on H. We may thus consider these as operators on $L^2(Q)$. As above, we let R be the von Neumann algebra on $L^2(Q)$ generated by $\{\tilde{U}_g, M_f\}$ and L the von Neumann algebra generated by $\{\tilde{V}_g, N_f\}$. Once again, we have that R and L are spatially isomorphic, $R' = L$, and now R and L will be factors. This is the Krieger factor of Q. Krieger has shown that this factor is independent of the choice of G. A construction of the factor without reference to the group appears in [3]. We then have the following generalization of Theorem 2.4.

Theorem 3.6. Q is an amenable ergodic equivalence relation if and only if R (or equivalently, L) is hyperfinite.

Proof. The proof of this theorem follows along the same lines as the proof of Theorem 2.4. We shall thus only indicate the points at which some modification is needed.

Let us consider first the proof that Q amenable (that is, S a weakly amenable G-space) implies that R has property E. Equation (∗), which plays an important role in the proof of Theorem 2.1, must be reinterpreted in the current context. For each s, g we have a natural unitary map $U_g: L^2(G_{sg}\backslash G) \to L^2(G_s\backslash G)$ defined by $(U_g f)(h G_s) = f(h G_s g) = f(hg G_{sg})$, and one can check that (∗) remains true with this definition of U_g. We also remark that if $g \in G_s$, then $U_g = I$. The essense of the proof of Theorem 2.1 was to choose an invariant section for the cocycle $c: S \times G \to \mathrm{Iso}(T(L^2(G)))$. While in Theorem 2.1 we had $L^2(S \times G) = \int_S^{\oplus} L^2(G)$, we now have $L^2(Q) = \int_S^{\oplus} L^2(G_s\backslash G)$, so that the Hilbert spaces in the fibers are not all equal. However, by the ergodicity of G on S, $\int^{\oplus} L^2(G_s\backslash G)$ will be unitary equivalent (by a decomposable unitary) to $\int_S^{\oplus} H_0$, where H_0 is a fixed Hilbert space. Fixing such an equivalence, $T(L^2(G_s\backslash G))$ and $B(L^2(G_s\backslash G))$ can all be identified with $T(H_0)$ and $B(H_0)$ respectively. Thus, a decomposable operator T on $\int^{\oplus} L^2(G_s\backslash G)$ will correspond to a bounded Borel map $S \to B(H_0)$. The cocycle $c(s, g): T(L^2(G_{sg}\backslash G)) \to T(L^2(G_s\backslash G))$ defined by $c(s, g) T = U_g T U_g^{-1}$, and its adjoint c^*, can be identified with cocycles into $\mathrm{Iso}(T(H_0))$ and $\mathrm{Iso}(B(H_0))$ respectively. It is important to observe that c is a reduced cocycle, since $g \in G_s$ implies $U_g = I$. One can then carry out the rest of the proof, working in $\int_S^{\oplus} H_0$, just as in the proof of Theorem 2.1.

We now turn to the proof that R having property P implies that Q is amenable. Here, we indicate the necessary modifications in the proof of the theorem of [8]. For $f \in H$, define $f^s(g) = f(sg, g) r(s, g)^{\frac{1}{2}}$. If $h \in G_s$, we can assume $r(s, h) = 1$ [2, Proposition 2.2], and one can then easily check that f^s factors to a map on G/G_s. Then $f \to f^s$ defines a unitary isomorphism of H with $\int_S^{\oplus} L^2(G/G_s)$. (Use [2, Theorem 2], for example.) Under this isomorphism, $\tilde{U}_g$ corresponds to $\int_S^{\oplus} U_g \, d\mu(s)$, where we write U_g for the translation operator on any $L^2(G/G_s)$. We can define $V_g: L^2(G/G_{sg}) \to L^2(G/G_s)$ as the operator induced by left translation by g^{-1}, and we will have $T = \int T_s \in B(\int^{\oplus} L^2(G/G_s))$ commuting with $\tilde{V}_g$ if and only if $V_g^{-1} T_s V_g = V_{sg}$ a.e. One can now construct a map $\sigma: L^\infty(Q) \to L^\infty(S)$ just as the map σ was constructed in [8], and one sees easily that properties (i)–(v) of [8] continue to hold. (Property (iv) must be modified in an obvious way.) Suppose now that $\gamma: S \times G \to \mathrm{Iso}(E)$ is as in [8], with the additional assumption that γ is a reduced cocycle. Then the map $F: S \times G \to E_1^*$ defined in [8] factors to a map on Q, and hence, using σ as in [8], one can construct $a(s)$. The remainder of the proof goes through without change.

4. An Application to Amenable Actions

In [7], amenable ergodic actions were introduced and it was shown that they possessed several properties analogous to those of amenable groups. Here we use Theorem 2.4 to show that an action of a countable discrete group is amenable if and only if there is an analogue of a G-invariant mean. Such a result may be of general use in elucidating the extent to which properties of amenable groups carry over to amenable ergodic actions. Here we present an application of this result to uniformly bounded cocycle representations.

Proposition 4.1. (i) If G is a countable discrete group acting ergodically on (S, m), then S is an amenable G-space if and only if there is a norm one linear map $\sigma: L^\infty(S \times G) \to L^\infty(S)$ such that
   a) $\sigma(1)=1$, $f \geq 0$ implies $\sigma(f) \geq 0$;
   b) If $A \subset S$ is measurable, $\sigma(f \cdot \chi_{p^{-1}(A)})=\sigma(f)\chi_A$, where $p: S \times G \to S$ is projection;
   c) $\sigma(f \cdot g)=\sigma(f) \cdot g$ where $(f \cdot g)(s, h)=f(sg, hg)$ and $(\varphi \cdot g)(s)=\varphi(sg)$.

   (ii) S is weakly amenable (i.e., Q_G is amenable) if and only if there is a norm one linear map $\sigma: L^\infty(Q_G) \to L^\infty(S)$ satisfying (a), (b), (c), where $p: Q \to S$ is projection on the first factor.

Proof. If such a map σ exists, it follows from the proof of [8] that S (or Q_G) is amenable. Conversely, if S (or Q_G) is amenable, by Theorem 2.1 (or Theorem 3.6), the algebra will be the range of a norm one projection, and by [6], it is the range of a (not necessarily normal) conditional expectation. The proof of [8] then shows that such a σ will exist.

It would be of interest to establish Proposition 4.1 for actions of a general locally compact group.

As an example of how Proposition 4.1 may be applied, consider a Borel cocycle $c: S \times G \to GL(H)$, where the latter is the group of invertible linear operators on a Hilbert space H, and suppose that c is uniformly bounded, that is $\|c(s, g)\| \leq M$ for some M. If S is a point, it is well known that amenability of G implies that c (which is now a representation of G) must be equivalent to a unitary representation.

Corollary 4.2. If G is a countable discrete group and S is an amenable ergodic G-space, then any uniformly bounded cocycle is equivalent to a unitary cocycle.

We remark that if H is finite dimensional, this corollary is true without the assumption of amenability (or the assumption on G.) See [9].

Proof. For each $x, y \in H$, define $f_{xy} \in L^\infty(S \times G)$ by

$$f_{xy}(s, g)=\langle c(s, g^{-1})^{-1} x \mid c(s, g^{-1})^{-1} y \rangle,$$

and define $\langle x \mid y \rangle_s = \sigma(f_{xy})(s)$.

(Here it is convenient to think of σ as already being composed with a lifting of $L^\infty(S)$.) Then each $\langle \ \mid \ \rangle_s$ is a conjugate linear form on H. For each x, $|f_{xx}(s, g)| \leq M^2 \|x\|^2$ and so $\langle x \mid x \rangle_s \leq M^2 \|x\|^2$. On the other hand, $|f_{xx}(s, g)| \geq M^{-2} \|x\|^2$, so $\langle x \mid x \rangle_s \geq M^{-2} \|x\|^2$, and $\langle \ \mid \ \rangle_s$ is equivalent to $\langle \ \mid \ \rangle$. We have $s \to (H, \langle \ \mid \ \rangle_s)$ is a measurable field of Hilbert spaces (taking constant functions as a fundamental sequence.) As $\int^\oplus (H, \langle \ \mid \ \rangle_s) \cong \int^\oplus (H, \langle \ \mid \ \rangle)$ by a decomposable unitary, to see that

c is equivalent to a unitary cocycle, it suffices to see that for each $x, y \in H$, $h \in G$, and almost all s,

$$\langle c(s,h)x \mid c(s,h)y \rangle_s = \langle x \mid y \rangle_{sh}.$$

But the left hand side of this equation is

$$\sigma(\langle c(s,g^{-1})^{-1} c(s,h)x \mid c(s,g^{-1})^{-1} c(s,h)y \rangle)(s)$$

$$= \sigma(\langle c(sg^{-1},gh)x \mid c(sg^{-1},gh)y \rangle)(s),$$

while

$$\langle x \mid y \rangle_{sh} = \sigma(\langle c(s,g^{-1})^{-1} x \mid c(s,g^{-1})^{-1} y \rangle)(sh)$$

$$= \sigma(\langle c(sh,h^{-1}g^{-1})^{-1} x \mid c(sh,h^{-1}g^{-1})^{-1} y \rangle)(s)$$

by property (c) in Proposition 4.1,

$$= \sigma(\langle c(sg^{-1},gh)x \mid c(sg^{-1},gh)y \rangle)(s).$$

5. Concluding Remarks

An examination of the proof of Theorem 2.1 shows that the particular cocycle c was not really relevant to the proof. Thus, essentially the same proof yields the following generalization.

Theorem 5.1. Suppose $c: S \times G \to U(H)$ is a unitary cocycle where S is an amenable G-space. Let U_g^c be the induced representation of G on $L^2(S; H)$, and A the von Neumann algebra generated by $\{U_g^c, M_f \mid g \in G, f \in L^\infty(S)\}$. Then A' (and hence A) has property E. The same result is true if Q_G is amenable, provided that c is a reduced cocycle.

This result can be alternatively phrased as saying that a system of imprimitivity based on an ergodic, amenable G-space necessarily generates a von Neumann algebra with property E.

References

1. Connes, A.: Classification of injective factors. Ann. Math., **104**, 73–115 (1976)
2. Feldman, J., Moore, C.C.: Ergodic equivalence relations, cohomology, and von Neumann algebras, I. To appear
3. Feldman, J., Moore, C.C.: Ergodic equivalence relations, cohomology, and von Neumann algebras, II. To appear
4. Krieger, W.: On constructing non-*-isomorphic hyperfinite factors of type III. J. functional Analysis, **6**, 97–109 (1970)
5. Sakai, S.: C*-algebras and W*-algebras. New York: Springer 1971
6. Tomiyama, J.: On the projection of norm one in W*-algebras. Proc. Japan Acad., **33**, 608–612 (1957)
7. Zimmer, R.J.: Amenable ergodic group actions and an application to Poisson boundaries of random walks. J. functional Analysis (to appear)
8. Zimmer, R.J.: On the von Neumann algebra of an ergodic group action. Proc. Amer. math. Soc. (to appear)
9. Zimmer, R.J.: Compactness conditions on cocycles of ergodic transformation groups, J. London math. Soc. (to appear)

Received December 18, 1976

CURVATURE OF LEAVES IN AMENABLE FOLIATIONS*

By Robert J. Zimmer**

---

1. Statement of Results.   If M is a smooth manifold, a basic problem is to examine the relationship between the fundamental group, $\pi_1(M)$, and the possible Riemannian geometries on M. A fundamental result in this direction, due to S. T. Yau [22] and D. Gromoll and J. Wolf [11], asserts that if M is compact with non-positive sectional curvature and $\pi_1(M)$ is solvable, then M is flat, and in particular $\pi_1(M)$ is a finite extension of an abelian group. This result has been generalized from the compact case to the finite volume case (and in fact to certain infinite volume manifolds as well) by S. S. Chen and P. Eberlein [3]. The point of this paper is to prove an analogue of these results for foliations.

THEOREM 1.   *Let $\mathfrak{F}$ be a foliation of a compact Riemannian manifold M, and assume that $\mathfrak{F}$ has a holonomy invariant measure that is finite on compact subsets of transversals and positive on open subsets. Assume that almost every leaf (with respect to this measure) is a complete simply connected manifold of non-positive sectional curvature. Suppose the foliation is amenable. Then every leaf of $\mathfrak{F}$ is flat.*

Amenability is a measure theoretic condition on the transversals which is a feature of the transversal considered only as a measure space together with an equivalence relation defined on it. (Here, of course, the equivalence relation is defined by belonging to the same leaf.) In a sense, amenable equivalence relations on measure spaces are the simplest and best understood. We discuss this notion briefly in section 2 below. In Theorem 1, the assumption of amenability of the foliation (or equivalently, of the transversals) is the replacement for the assumption of solvability of the fundamental group in the theorems quoted above. Thus, a measure theoretic assumption about the transversals of a foliation is giving us the same type of information about the geometry of the leaves of a

---

Manuscript received January 19, 1981.
*Research partially supported by NSF Grant MCS 80-04026.
**Sloan Foundation Fellow.

foliation that an assumption on $\pi_1(M)$ gives us about the geometry of M. A similar phenomenon appears in the rigidity theorem for ergodic foliations by symmetric spaces that we proved in [26], which is an analogue for foliations of the Mostow-Margulis rigidity theorem for locally symmetric spaces of finite volume. For other results of this nature, see [27].

We in fact prove Theorem 1 under significantly weaker hypotheses. Namely, we need not assume that $\mathfrak{F}$ is a foliation of a compact manifold, but only a foliation of a measure space so that $\mathfrak{F}$ has finite total volume. Correspondingly, the Riemannian structure on the leaves need not come from a Riemannian structure on an ambient manifold, but need only be defined on each leaf so as to vary measurably as we move transversally from leaf to leaf. More precisely, we have the following. (See below for precise definitions of the terms involved.)

THEOREM 2.   *Let $\mathfrak{F}$ be a Riemannian measurable foliation with transversally (i.e. holonomy) invariant measure and finite total volume. Assume that almost every leaf is a complete simply connected manifold of non-positive sectional curvature. If $\mathfrak{F}$ is amenable, then almost every leaf is flat.*

In this generality the theorem is in fact a strict generalization of the results quoted in the first paragraph. By applying Theorem 2 to a suitable foliated bundle, we obtain:

THEOREM 3.   *Let M be a complete finite volume Riemannian manifold of non-positive sectional curvature. If $\pi_1(M)$ is amenable, then M is flat.*

2. Preliminaries.   In this section we shall briefly recall some information we will need, supplying references for a more detailed discussion.

We first recall some geometric information concerning complete simply connected Riemannian manifolds of non-positive sectional curvature, hereafter "Hadamard manifolds." See [2], [6], [7] as general references. If H is such a manifold, then H is diffeomorphic to the unit ball $B \subset \mathbf{R}^{\dim H}$. We can identify the unit sphere ∂B with the "points at infinity," i.e. with the set of equivalence classes of geodesic rays in H, two geodesic rays α, β being called equivalent if $d(\alpha(t), \beta(t))$ is bounded for $t \geq 0$. We denote the equivalence class of a geodesic by $\alpha(\infty)$. Similarly, we can speak of $\alpha(-\infty) \in \partial B$. The topology on $\bar{B} = B \cup \partial B$ can be naturally described in terms of H itself, via the "cone topology" [6]. Given a point $x \in \partial B$, there is a unit C^1-vector field V^x on B whose integral curves are precisely those geodesics α with $\alpha(\infty) = x$. At each point $p \in B$, the ortho-complement $(V^x)^\perp$ is the tangent space to a foliation of B, and the leaves of this foliation are called the horospheres at x. These horospheres can alternatively be described as level sets of Busemann functions. Namely, let γ be a geodesic with $\gamma(\infty) = x$. Define

$$f_\gamma : B \to \mathbf{R} \quad \text{by} \quad f_\gamma(p) = \lim_{t \to \infty} [d(\gamma(t), p) - t].$$

Then f_γ is C^2 [12], convex [6], and is called a Busemann function at x. If α, γ are geodesics with $\gamma(\infty) = \alpha(\infty)$, then $f_\gamma - f_\alpha$ is constant. In fact, $\operatorname{grad} f_\gamma = -V^x$. The level sets of f_γ depend only on x, and are exactly the horospheres at x.

Any Hadamard manifold has a de Rham decomposition [5], [15], [21], i.e. H is isometric to $H_0 \times H_1$ where H_0 is isometric to a euclidean space, and H_1 has no euclidean factor. This representation is essentially unique in that any two such representations define identical pairs of foliations of H corresponding to the vertical and horizontal foliations in the product representation. We shall thus speak of the euclidean and totally non-euclidean foliations on H. The tangent spaces to these foliations at $p \in H$ are the subspaces of $T(H)_p$ on which the holonomy group of H at p acts trivially, and its orthocomplement. Since the leaves of the non-euclidean foliation are parallel, the set of $x \in \partial B$ such that $x = \gamma(\infty)$ for a geodesic γ lying in a given leaf of this foliation will be independent of the leaf. We thus have a subset of ∂B consisting of the points at infinity of any leaf of the non-euclidean foliation, say $N \subset \partial B$. Topologically, of course, N is an embedded sphere in the higher dimensional sphere ∂B.

Suppose now that $\mathcal{F}$ is a foliation of a manifold M with a holonomy invariant measure (i.e. a "transversally invariant measure" in the terminology of [27]). We recall that this entails the assignment to each transversal $T \subset M$ of $\mathcal{F}$ a measure μ_T such that the following invariance condition is satisfied. If $A_i \subset T_i$, $i = 1, 2$ are Borel subsets of the transversals T_i, and $\phi : A_1 \to A_2$ is a Borel isomorphism such that for all $x \in A_1$, x and $\phi(x)$ are in the same leaf, then $\phi_*(\mu_{T_1} | A_1) = \mu_{T_2} | A_2$. For example, any foliation coming from a measure preserving locally free action of a connected unimodular Lie group will have this property [19]. See also [14], [16], [27] for further discussion. Suppose we have such a family of measures, and a Riemannian structure on each leaf. Locally, we can define a measure by integrating the measures defined by the volume form on disks in each leaf with respect to the measure on a transversal. (See [16], [17],

1014   ROBERT J. ZIMMER

[27].) By the invariance property, these locally defined measures will piece together to give us a well defined measure on M, and if this measure is finite, we say that $\mathcal{F}$ has finite total volume (given, of course, the transversally invariant measures and the Riemannian structures). This will be the case for example if M is compact and μ_T assigns finite measure to any compact subset. As indicated above, we will in fact consider the case in which the foliation $\mathcal{F}$ is only a measurable foliation of a measure space and each leaf has a Riemannian structure which varies measurably over the ambient space. (We then speak of a Riemannian measurable foliation.) One can speak of transversally invariant measures and finite total volume in this case as well. We refer the reader to [27] for discussion.

We will find it convenient to use a description of the foliation that we employed in [27], and for this we first recall the notion of cocycles for measurable equivalence relations. Suppose (T, ν) is a measure space with an equivalence relation R on it that is a Borel subset of $T \times T$ and such that each equivalence class is countable. The measure ν is called invariant for R if for every Borel isomorphism $\phi:A_1 \to A_2$, $A_i \subset T$ Borel sets, with $\phi(x) \sim x$, we have

$$\phi_*(\nu|A_1) = \nu|A_2 \qquad [9].$$

(Quasi-invariance is defined similarly.) For example if T is a transversal to a foliation with transversally invariant measure, then μ_T is an invariant measure on T with respect to the equivalence relation $R = (T \times T) \cap \mathcal{F}$. If Γ is a countable group acting in a measure preserving way on a measure space (T, ν), then ν will be invariant with respect to the relation R_Γ on T, $R_\Gamma = \{(t, t \cdot \gamma) | t \in T, \gamma \in \Gamma\}$. If G is a group with a Borel structure, a Borel function $\alpha:R \to G$ is called a cocycle if $\alpha(s, t)\alpha(t, u) = \alpha(s, u)$ for almost all s, t, u with $s \sim t \sim u$. The cocycle is called strict if this identity holds for all s, t, u with $s \sim t \sim u$. Suppose now that $\mathcal{F}$ is a Riemannian measurable foliation of a measure space S with transversally invariant measure. Let T be a transversal that intersects almost every leaf. (Here, we mean transversal in the sense of measurable foliations [27]. Thus, such a T will always exist.) Let ν be the invariant measure on T. Let B be the unit ball in $\mathbf{R}^n$, $\bar{B}$ its closure, and $\text{Diff}_b(B)$ the diffeomorphisms of B that extend to homeomorphisms of $\bar{B}$.

PROPOSITION 4. [27, Proposition 3.1].   *Suppose almost every leaf of $\mathcal{F}$ is a Hadamard manifold. Then there exist:*

CURVATURE OF LEAVES   1015

(i) *A measurable assignment* $s \rightarrow \omega_s$ *of a Riemannian metric* ω_s *on*
    B to a point $s \in T$;

(ii) *A measurable map* $\Phi: T \times B \rightarrow S$; *and*

(iii) *A strict cocycle* $\alpha: \mathcal{F} | T \rightarrow \mathrm{Diff}_b(B)$ *(where* $\mathcal{F} | T = \mathcal{F} \cap (T \times T)$*);*

such that

(a) *For each* $s \in T$, $\Phi_s(0) = s$, *and* $\Phi_s: B \rightarrow S$ *given by* $\Phi_s(v) =$
    $\Phi(x, v)$ *is an isometry of the Riemannian manifold* (B, ω_s)
    with L_s, *the leaf through* s; *and*

(b) *For* $s, t \in T$, $s \sim t$, *we have* $\Phi_t = \Phi_s \circ \alpha(s, t)$, *and hence in*
    light of (a), $\alpha(s, t)^* \omega_s = \omega_t$.

We remark that in particular, we can view α *as a cocycle* $\alpha: \mathcal{F} | T \rightarrow$
Homeo(∂B).

Finally, we recall the notion of an amenable foliation or amenable action. An amenable group is one that has a fixed point in every compact convex set on which it acts by affine transformations [10]. If (S, μ) is a G-space where μ is quasi-invariant under the G-action, there is a natural class of compact convex sets on which G acts by affine transformations, the compact convex sets "over S." These can be briefly described as follows. Suppose E is a separable Banach space, Iso (E) its group of isometric isomorphisms, E^* the dual space, and $\alpha: S \times G \rightarrow$ Iso (E) is a cocycle, i.e. for all $g, h \in G$, $\alpha(s, gh) = \alpha(s, g)\alpha(sg, h)$ a.e. Then there is an induced action of G on $L^\infty(S, E^*)$ given by $(g \cdot f)(s) = \alpha^*(s, g)f(sg)$ where α^* is the adjoint cocycle, $\alpha^*(s, g) = (\alpha(s, g)^{-1})^*$. This action is just translation of f combined with a twist by α. Suppose that for each s, there is a compact convex set A_s which is a subset of the unit ball in E^*, and such that for each g, $\alpha^*(s, g)A_{sg} = A_s$ a.e. If A_s varies measurably in $s \in S$, then

$$A = \{f \in L^\infty(S, E^*) \mid f(s) \in A_s \text{ a.e.}\}$$

is a compact convex G-invariant set, which we call a compact convex set over S. The action of G on S is called amenable if there is a fixed point for every such action. Clearly every action of an amenable group is amenable, but non-amenable groups also have amenable actions. A particularly basic example is the action of a discrete group of isometries of hyperbolic space on the boundary sphere (with Lebesgue measure). This notion was introduced in [23]. See also [4], [24], [25], [26], [28] for a systematic development and applications. A fixed point under the above action is simply a function that satisfies $\alpha(s, g)f(sg) = f(s)$ a.e. For a free action

1016   ROBERT J. ZIMMER

of a countable group G, cocycles can be identified with cocycles on the equivalence relation R_G in the sense defined above. Namely, a cocycle on R_G is defined by $(x, y) \to \alpha(x, g)$ where $y = xg$. Thus, the fixed point property reads $\alpha^*(x, y)f(y) = f(x)$ a.e. It is then easy to extend the definition of amenability to an arbitrary equivalence relation with countable equivalence classes and quasi-invariant measure. Namely for every cocycle $\alpha: R \to \text{Iso}(E)$, Borel assignment $s \to A_s$ of compact convex sets such that $\alpha^*(s, t)A_t = A_s$ a.e., there exists a Borel $\phi: S \to E^*$ such that $\phi(s) \in A_s$ a.e., and $\alpha^*(s, t)\phi(t) = \phi(s)$ a.e. [24]. Amenable equivalence relations can be characterized in a variety of other ways (see [4], for example), and are well understood from the point of view of cohomology [9], [13], [18], [23], [25], the associated von Neumann algebra [24], and orbit equivalence [4]. In fact, the results of Connes, Feldman, and Weiss [4] show that amenable equivalence relations are all defined by an integer action, showing that they are the natural limit of the extension of the Dye-Krieger theory of orbit equivalence. However, because of its appearance in many natural actions of non-amenable groups, the amenability condition is also of much use in studying lattices in semisimple groups and orbit equivalence of non-amenable equivalence relations. See [28] for a survey of some of the ideas involved. A foliation with transversally (i.e. holonomy) quasi-invariant measure is called amenable if a transversal T that intersects almost every leaf is amenable with the equivalence relation $\mathcal{F}|T$. (This is independent of the choice of such a transversal.)

3. Proof of the Theorems.   We begin with the following Lemma.

LEMMA 5.   *Let B be a Hadamard manifold and suppose there is a point $x \in \partial B$ such that all horospheres in B at x are totally geodesic. Then B has a non-trivial euclidean de Rham factor.*

Proof.   By [20], [3, Theorem 2.1, Proposition 2.2], it suffices to show that x has an antipode $y \in \partial B$, i.e. a point y such that $\alpha(-\infty) = y$ for any geodesic α with $\alpha(\infty) = x$. So suppose α, β are geodesics with $\alpha(\infty) = \beta(\infty) = x$. We claim $\alpha(-\infty) = \beta(-\infty)$. Let L be a horosphere at x. We can assume, reparametrizing if necessary, that $\alpha(0)$, $\beta(0) \in L$. For any $a \in \mathbf{R}$, we then have that $\alpha(a)$ and $\beta(a)$ are in the same horosphere, say L^1, by [6, Corollary 3.3] for example. Let γ_0 be the geodesic segment from $\alpha(0)$ to $\beta(0)$, and γ the geodesic segment from $\alpha(a)$ to $\beta(a)$. If f is a Busemann function at x, then f has equal values at the endpoints of γ, so $(f \circ \gamma)'(c) = 0$ for some c. This means

$$\langle \operatorname{grad} f(\gamma(c)) | \gamma'(c) \rangle = 0,$$

so that γ is tangent to a horosphere at some time, which by the totally geo-desic assumption on horospheres implies that γ lies totally within a horo-sphere, which of course must be L^1. Similarly, γ_0 lies in L. Consider the closed curve consisting of the four geodesic segments $\alpha(t)$, $\beta(t)$, $(0 \le t \le a)$, γ_0, γ. Each intersection is a right angle. Let σ be the geodesic from $\alpha(0)$ to $\beta(a)$. We then have two geodesic triangles each of which must clearly have the sum of the angles equal to π. Applying [6, Proposition 5.5] (a version of the Gauss-Bonnet theorem) to each of these triangles and using the Lemma of Synge (cf. [6, Proposition 5.9]), we deduce that each of these triangles span a flat surface, and hence, by the pythagorean theorem, $d(\alpha(s), \beta(s))^2 + a^2 = (\text{length } \sigma)^2$ for $s = 0$, a. Thus, $d(\alpha(t), \beta(t))$ is constant for $t \in \mathbf{R}$, so $\alpha(-\infty) = \beta(-\infty)$.

Proof of Theorem 2.   We suppose $\mathfrak{F}$ is a foliation of S, and we use the notation of Proposition 4. It clearly suffices to show that the totally non-euclidean de Rham factor in almost every Hadamard manifold (B, ω_s), $s \in T$, is trivial. Suppose not. Then on some set T_0 of positive measure the totally non-euclidean de Rham factor is non-trivial, and by condition (b) of Proposition 4, this will clearly be true on the saturation $[T_0]$ of T_0 under the equivalence relation $\mathfrak{F}|T$. The amenability of $\mathfrak{F}|T$ easily implies that of $\mathfrak{F}|[T_0]$, and $\Phi([T_0] \times B) \subset S$ will again be a foliation satisfying the hypotheses of the theorem. Thus, we may assume $T = T_0$. For each $s \in T$, let $N_s \subset \partial B$ be the boundary of the leaves of the totally non-euclidean foli-ation of (B, ω_s) as described above. Since $\alpha(s, t):(B, \omega_t) \to (B, \omega_s)$ is an isometry, we have $\alpha(s, t)N_t = N_s$ for $(s, t) \in \mathfrak{F}|T$. Let A_t be the space of probability measures on ∂B that are supported on N_t. Then $\alpha^*(s, t)A_t = A_s$, so by amenability, there exists a Borel function $\mu:T \to M(\partial B)$ $(M(\partial B)$ being the space of probability measures on ∂B), $t \to \mu_t$, such that $\mu_t \in A_t$ a.e., and on a conull set $\alpha^*(s, t)\mu_t = \mu_s$.

Suppose now that B is a Hadamard manifold, $\mu \in M(\partial B)$, and fix $p \in \partial B$. For any $x \in \partial B$, let $f_p(q, x) = f_\gamma(q)$, the Busemann function at x where γ is the geodesic with $\gamma(0) = p$, $\gamma(\infty) = x$. We remark that by [8, Proposition 2.6] $f_p(q, x)$ is a jointly continuous function of p, q, x. Define $F_p:B \to R$ by $F_p(q) = \int f_p(q, x)d\mu(x)$. For each p, x, $q \to f_p(q, x)$ is C^2 and convex, and hence so is F_p. For fixed x and $p, p' \in B$, $f_p(q, x) - f_{p'}(q, x)$ is independent of q, and hence $F_p - F_{p'}$ is constant. Thus, $\operatorname{grad} F_p$ is a vector field on B independent of $p \in B$. Returning to the situation in the previous paragraph, for $s \in T$ let $f_p^s(q, x)$ be the function defined as above for the space (B, ω_s). Then for each $s \in T$, define $F_p^s(q) = \int f_p^s(q, x) d\mu_s(x)$. Let C^s be the set of critical points of F_p^s (independent of p), $G_s(q) = \| \operatorname{grad} F_p^s(q) \|$ (independent of p), and U^s the unit vector field on $(B, \omega_s) - C^s$ given by $U^s = \operatorname{grad} F_p^s / G_s$. For $s, t \in T$, $s \sim t$, $\alpha(s, t)$ is an isometry, so we clearly have $f_{\alpha(s,t)p}^s(\alpha(s, t)q, \alpha(s, t)x) = f_p^t(q, x)$. By the α^*-invariance condition on μ_s, we have $F_{\alpha(s,t)p}^s \circ \alpha(s, t) = F_p^t$. Hence, $\alpha(s, t)C^t = C^s$, and $\alpha(s, t)_* U^t = U^s$. Thus, in the notation of Proposition 4, $\Phi_s(C^s) = \Phi_t(C^t)$, $(\Phi_s)_*(U^s) = (\Phi_t)_*(U^t)$, and hence the local flows defined by the vector fields U^s induce a local flow on (almost all of) $\tilde{S} = S - \Phi(\{s, C^s\})$ that preserves the leaves of $\mathcal{F}$.

We claim that for almost all $s \in T$, the laplacian $\Delta F_p^s = \operatorname{div}(\operatorname{grad} F_p^s) = 0$. (This is clearly independent of p.) If $\Phi(\{s, C^s\})$ has measure one, this clearly holds. If not, the local flow defined by U^s will in fact be defined on all $B - C^s$ for all $t \geq 0$ by [2, Lemma 2.3]. Thus, we have a flow defined for $t \geq 0$ on $\tilde{S}$ which is a space of finite measure. By the calculation of [2, Proposition 2.2], $\operatorname{div} U^s \geq 0$, which implies that the transformations of $\tilde{S}$ for $t \geq 0$ have Radon-Nikodym derivative at least 1. (Recall that these transformations preserve leaves and that the transversal measures are holonomy invariant.) But since the total measure is finite, this Radon-Nikodym derivative must equal 1 almost everywhere, which implies that for almost all $s \in T$, $\operatorname{div} U^s = 0$. Once again from the calculation in [2, Proposition 2.2], we have (*) $\operatorname{div} U^s = G_s^{-1} \cdot \Delta F_p^s + G_s \cdot (U^s(G_s^{-1}))$. Therefore, at a point, either $\Delta F_p^s = 0$ or (since $\Delta F_p^s \geq 0$) $U^s(G_s^{-1}) < 0$. The latter condition means that if ϕ is an integral curve of the vector field U^s, then $G_s \circ \phi$ is strictly increasing at such a point. As above, using the fact that $\alpha(s, t)$ is an isometry, $\{G_s\}$ and $\{U^s\}$ define a function and vector field, respectively, say G and U, on $\tilde{S}$. Consider $\{x \in \tilde{S} \mid U(G)(x) \neq 0\}$. If this set has positive measure, then there is $r > 0$ such that the set of leaves containing a point x for which $G(x) = r$ and $U(G)(x) \neq 0$ has positive measure (with respect to the transversal measure). [This follows from Fubini's theorem. Namely, if M is a measure space, B a manifold, $E \subset M \times B$ a set of positive measure such that $E_m = \{b \in B \mid (m, b) \in E\}$ is open for all m, and $f : E \to \mathbf{R}$ a measurable function continuous on all E_m, and not locally constant on E_m, at least for m in a set of positive measure, consider $\tilde{f} : E \to M \times \mathbf{R}$, $\tilde{f}(m, b) = (m, f(m, b))$, and let $\tilde{E} = \tilde{f}(E)$. Then clearly $\tilde{E}$ is of positive measure in $M \times \mathbf{R}$. So by Fubini, there exists r such that $\{m \mid f(m, b) = r$ for some $b \in B\}$ has positive measure.] But then under the flow defined by U, $\{x \in \tilde{S} \mid G(x) \geq r\}$ will be a set of positive measure that is taken to a set of strictly smaller measure. (Recall that we always have $U(G) \geq 0$ from (*).) But as remarked above, the Radon-Nikodym derivative is always one, so this is impossible. Hence, $\{x \in \tilde{S} \,|\, U(G)(x) \neq 0\}$ is a null set. From equation (*), we deduce that for almost all s, $\Delta F_p^s = 0$ on $B - C^s$. But we clearly also have $\Delta F_p^s = 0$ on $\text{int}(C^s)$, so by continuity, $\Delta F_p^s = 0$ on all of B.

The equation

$$F_p^s(q) = \int f_p^s(q, x)\,d\mu_s(x)$$

implies $\Delta F_p^s(q) = \int \Delta f_p^s(q, x)\,d\mu_s(x)$ (where Δ is with respect to q, of course), so from convexity we deduce that for each q, $\Delta f_p^s(q, x) = 0$ for μ_s—almost all $x \in \partial B$. This implies that for almost all x, this holds for almost all q, and hence for all q by continuity. In other words, we have shown that for almost all $s \in T$, there is at least one $x \in N_s \subset \partial B$ such that $\Delta f_p^s(q, x) = 0$ for all $q \in B$. (Recall that μ_s is supported on N_s.) Since f_p^s is convex, this implies $\Delta_1 f_p^s(q, x) = 0$ for q in a fixed leaf of the totally non-euclidean foliation of (B, ω_s), where Δ_1 is the laplacian on the leaf. In light of Lemma 5, it suffices to show that if f is a Busemann function on a Hadamard manifold B at a boundary point x with $\Delta f = 0$, then the horocycles at x are totally geodesic. But if $\Delta f = 0$, then $(f \circ \alpha)''(t) = 0$ for any geodesic α (see, e.g. [1, p. 128]), so that $(f \circ \alpha)'(t)$ is constant. Hence if α is tangent to a horosphere at one point, $\alpha(0)$, so that $(f \circ \alpha)'(0) = \langle \text{grad } f(\alpha(0)) \,|\, \alpha'(0) \rangle = 0$, then $(f \circ \alpha)'(t) = 0$ for all t, i.e. $f \circ \alpha$ is constant. This simply means that $\alpha(t)$ is contained in a horosphere, and this completes the proof.

Proof of Theorem 1.   Since the transverse invariant measure is positive on open sets, Theorem 1 follows from continuity of curvature and Theorem 2.

Proof of Theorem 3.   We first observe that any countable discrete group has a free measure preserving action on a finite measure space. Namely, for any such group Γ, let $\Omega = \Pi_\Gamma \{0, 1\}$ with the product measure $\nu = \Pi_\Gamma \mu$, where $\mu(0) = \mu(1) = \frac{1}{2}$. Let $\gamma \in \Gamma$ act on Ω by $(\omega \cdot \gamma)(\gamma_0) = \omega(\gamma_0 \gamma)$. Then if Γ is infinite, it is easy to check that $\{\omega \,|\, \omega \cdot \gamma = \omega\}$ is of measure 0. Thus, discarding a null set, Γ will act freely.

If $\Gamma = \pi_1(M)$ is as in the hypothesis of Theorem 3, we can then form a foliated bundle over M. Namely, let $\tilde{M}$ be the universal covering, on which Γ acts by isometries, and consider the space $S = (\Omega \times \tilde{M})/\Gamma$, where

1020   ROBERT J. ZIMMER

Γ acts on $\Omega \times \tilde{M}$ by the diagonal action. Then S is foliated by the images of $\{s\} \times \tilde{M}$, and the image of $\Omega \times \{m\}$, will be a transversal with equivalence relation isomorphic to R_Γ, for Γ acting on Ω. Since Γ is amenable, R_Γ will be amenable, and since $\tilde{M}/\Gamma$ has finite volume and Ω has a finite measure, the foliation on S will be of finite total volume. Thus, we may apply Theorem 2 to deduce that $\tilde{M}$ is flat.

4. Concluding remarks.

(a) Theorems 1 and 2 are clearly false if we delete the assumption of amenability. For example, simply take a foliation by symmetric spaces constructed from a lattice subgroup of the isometry group of the symmetric space and an action of the lattice as in the proof of Theorem 3 above. We can also have amenable foliations of a compact manifold by spaces of constant negative curvature. For example, if $\Gamma \subset PSL(2, \mathbf{R}) = G$ is a torsion free cocompact discrete subgroup, let Γ act on real projective space $\mathbf{P}^1$. View Γ as $\pi_1(M)$ where $\tilde{M}$ is the Poincare disk G/K, $K = PSO(2, \mathbf{R})$. Once again the above construction yields a foliation of $(\mathbf{P}^1 \times \tilde{M})/\Gamma$ (which is of course compact) by leaves which are two dimensional hyperbolic manifolds, almost all of which are simply connected. Since the Γ action on $\mathbf{P}^1$ is amenable [25], the foliation will be also.

(b) Some assumption along the lines of finite total volume must clearly be made in Theorems 1 and 2. However, there are presumably weaker conditions which will allow some infinite volume foliations. (Cf. [3, Theorem 5.11].)

(c) One can unite Theorems 2 and 3 with the following generalization.

THEOREM 6.   *Let $\mathfrak{F}$ be a Riemannian measurable foliation with transversally invariant measure and finite total volume. Assume each leaf is complete, with non-positive sectional curvature, and that the fundamental group of almost every leaf is amenable. If $\mathfrak{F}$ is amenable, then almost every leaf is flat.*

The technique of proof is essentially the same as that of the proof of Theorem 2. Of course, one has a similar theorem in the framework of Theorem 1.

(d) The hypothesis of Theorem 2 enable us to apply the result to partial foliations or foliations with singularities.

UNIVERSITY OF CHICAGO

CURVATURE OF LEAVES   1021

REFERENCES

[1] M. Berger, P. Gauduchon, E. Mazet, Le Spectre d'une Variété Riemannienne, Lecture Notes in Math., no. 194, Springer-Verlag, New York, 1971.

[2] R. L. Bishop, B. O'Neill, Manifolds of negative curvature, *Trans. Amer. Math. Soc.*, **145** (1969), 1–49.

[3] S. S. Chen, P. Eberlein, Isometry groups of simply connected manifolds of nonpositive curvature, *Ill. J. Math.*, **24** (1980), 73–103.

[4] A. Connes, J. Feldman, B. Weiss, Amenable equivalence relations are generated by a single transformation, to appear, Erg. Thy. and Dyn. Syst.

[5] G. de Rham, Sur la réductibilité d'un espace de Riemann, *Comm. Math. Helv.*, **26** (1952), 328–344.

[6] P. Eberlein, B. O'Neill, Visibility manifolds, *Pac. J. Math.*, **46** (1973), 45–109.

[7] _____, Geodesic flows on negatively curved manifolds, I, *Annals of Math.*, **95** (1972), 492–510.

[8] _____, Geodesic flows on negatively curved manifolds, II, *Trans. Amer. Math. Soc.*, **178** (1973), 57–82.

[9] J. Feldman, C. C. Moore, Ergodic equivalence relations, cohomology, and von Neumann algebras, I, *Trans. Amer. Math. Soc.*, **234** (1977), 289–324.

[10] F. P. Greenleaf, Invariant Means on Topological Groups, Van Nostrand, New York, 1969.

[11] D. Gromoll and J. Wolf, Some relations between the metric structure and the algebraic structure of the fundamental group in manifolds of non-positive curvature, *Bull. Amer. Math. Soc.*, **77** (1971), 545–552.

[12] E. Heintze, H. C. Im Hof, Geometry of horospheres, *J. Diff. Geom.*, **12** (1977), 481–491.

[13] M. R. Herman, *Constructions des difféomorphismes ergodiques*, preprint.

[14] M. Hirsch, W. Thurston, Foliated bundles, invariant measures, and flat manifolds, *Annals of Math.*, **101** (1975), 369–390.

[15] S. Kobayashi, K. Nomizu, *Differential Geometry*, Interscience, 1963.

[16] J. Plante, Foliations with measure preserving holonomy, *Annals of Math.*, **102** (1975), 327–361.

[17] D. Ruelle, D. Sullivan, Currents, flows, and diffeomorphisms, *Topology*, **14** (1975), 319–327.

[18] K. Schmidt, Cocycles of Ergodic Transformation groups, MacMillan Co. of India, 1977.

[19] C. Series, Foliations of polynomial growth are hyperfinite, *Israel J. Math.*, **34** (1979), 245–258.

[20] J. Wolf, Homogeneity and bounded isometries in manifolds of negative curvature, *Ill. J. Math.*, **8** (1964), 14–18.

[21] H. Wu, On the de Rham decomposition, *Ill. J. Math.*, **8** (1964), 291–311.

[22] S. T. Yau, On the fundamental group of compact manifolds of non-positive curvature, *Annals of Math.*, **93** (1971), 579–585.

[23] R. J. Zimmer, Amenable ergodic group actions and an application to Poisson boundaries of random walks, *J. Funct. Anal.*, **27** (1978), 350–372.

[24] _____, Hyperfinite factors and amenable ergodic actions, *Invent. Math.*, **41** (1977), 23–31.

1022   ROBERT J. ZIMMER

[25] _____, Induced and amenable ergodic actions of Lie groups, *Ann. Sci. Ec. Norm. Sup.*, **11** (1978), 407–428.

[26] _____, Strong rigidity for ergodic actions of semisimple Lie groups, *Annals of Math.*, **112** (1980), 511–529.

[27] _____, Ergodic theory, semisimple Lie groups, and foliations by manifolds of negative curvature, *Publ. Math. I.H.E.S.*, **55** (1982), 37–62.

[28] _____, Ergodic theory, group representations, and rigidity, C.I.M.E. Summer Session on Harmonic Analysis and Group Representations, to appear, *Bull. Amer. Math. Soc.*, 1982.

JOURNAL OF FUNCTIONAL ANALYSIS 72, 58–64 (1987)

Amenable Actions and Dense Subgroups of Lie Groups*

ROBERT J. ZIMMER

*Department of Mathematics, University of Chicago,
Chicago, Illinois 60637*

Communicated by the Editors

Received July 15, 1985

Let G be a connected Lie group and H and Γ be closed subgroups with H amenable. Then the action of Γ on G/H will be amenable. This is a basic property of such Γ actions, and plays a significant role in the study of the rigidity properties of discrete groups. (See [6] for the notion of an amenable action and its applications to discrete subgroups.) If Γ is not a closed subgroup, it is no longer true that the action on G/H need be amenable. To our knowledge, this was first observed by Sullivan a number of years ago. The point of this paper is to prove a conjecture of Connes and Sullivan that gives a complete description of when this action is amenable. Our main result is

THEOREM 1.1. *Let G be a connected Lie group, H a closed amenable subgroup. Let Γ be a separable locally compact group with an injective homomorphism $\Gamma \hookrightarrow G$. Let $(\Gamma^-)_0$ be the identity component of the closure of Γ in G. Then the Γ action on G/H is amenable if and only if the image of $\Gamma \cap (\Gamma^-)_0$ under projection to the maximal connected semisimple adjoint factor of $(\Gamma^-)_0$ is closed.*

COROLLARY 1.2. (*Conjecture of Connes and Sullivan*). *If Γ is countable, then Γ acts amenably on G/H if and only if $(\Gamma^-)_0$, the identity component of the closure of Γ in G, is solvable.*

For $G = SL(2, \mathbb{R})$, this was proven by Carriere and Ghys [1]. Our methods are completely different. For noninjective homomorphisms, see Section 5.

One can use Theorem 1.1 (together with Lemma 2.2) to obtain an alternate proof of the classical theorem of Auslander that asserts that any Zariski dense discrete subgroup of a real algebraic group G is still discrete when projected to $G/\mathrm{rad}(G)$, where $\mathrm{rad}(G)$ denotes the radical. (See [2] for

* Research partially supported by NSF Grant DMS-8301882.

0022-1236/87 $3.00

Auslander's theorem.) For an application of Theorem 1.1 to geometry, see [8].

We would like to thank Alex Lubotzky for many stimulating discussions regarding this paper. But for his own protests, he would be a co-author of this work. We would also like to thank the authors of [1] for providing a preprint of their work, M. Nori for some enlightening remarks, and D. Witte for pointing out some inaccuracies in an earlier version of this paper.

2. FIRST REDUCTIONS OF THE PROOF

In this section we make some reductions of the proof which follow from basic results about amenability. While we shall refer the reader to [6] for the actual definition of an amenable action, we here recall one consequence of the definition of amenability, which is the main consequence we shall need.

LEMMA 2.1 [6, Proposition 4.3.9]. *Suppose Γ is a group acting amenably on a measure space S, and that Γ also acts continuously on a compact metric space X. Let $M(X)$ be the space of probability measures on X (with the weak$-*-$topology). Then there is a measurable Γ-map $S \to M(X)$.*

For Lemmas 2.2, 2.3, we assume that Γ, G are locally compact separable groups and that we are given an injective (continuous) homomorphism $\Gamma \hookrightarrow G$.

The following result is a straightforward consequence of the results of [6, Chap. 4], [7].

LEMMA 2.2. *If $Q \subset G$ is a closed amenable subgroup, then Γ acts amenably on G if and only if Γ acts amenably on G/Q.*

The next result follows in a routine manner from the definition of amenability. We omit the proof.

LEMMA 2.3. (a) *Γ acts amenably on G if and only if Γ acts amenably on its closure Γ^-.*

(b) *Suppose $G_0 \subset G$ is a closed subgroup with G/G_0 discrete. If Γ is dense in G, then Γ acts amenably on G if and only if $\Gamma \cap G_0$ acts amenably on G_0.*

(c) *Suppose Γ acts on a measure space X and that the kernel of the action is Λ. If Γ acts amenably on X, so does Γ/Λ. Furthermore Γ acts amenably if and only if Γ/Λ acts amenably and Λ is amenable.*

60   ROBERT J. ZIMMER

(d) *If N is a closed normal subgroup of G and $N \subset \Gamma$, then Γ acts amenably on G if and only if Γ/N acts amenably on G/N.*

As a consequence of the above results, we are easily reduced to the semisimple case.

COROLLARY 2.4. *To prove Theorem 1.1 it suffices to show that if G is a connected, semisimple adjoint Lie group, Γ is acting on G, and Γ is dense and proper in G, then the action is not amenable.*

Proof. Assume the hypotheses of Theorem 1.1. Let Q be the identity component of the closure of Γ, and $\Lambda = \Gamma \cap Q$. By Lemmas 2.2, 2.3, Γ acts amenably on G/H if and only if Λ acts amenably on Q. This in turn, by Lemma 2.2 again, is equivalent to Λ acting amenably on $Ad(Q/\mathrm{rad}(Q)) = Q/\mathrm{rad}(Q) Z^*$, where Z^* is the set of elements in Q projecting into the center of $Q/\mathrm{rad}(Q)$. Let Λ' be the projection of Λ into $Ad(Q/\mathrm{rad}(Q))$. Since $\mathrm{rad}(Q) Z^*$ is solvable, so is the kernel of $\Lambda \to \Lambda'$. By Lemma 2.3, Λ acts amenably on $Ad(Q/\mathrm{rad}(Q))$ if and only if Λ' acts amenably on this group. If Λ' is closed, this action is amenable, and thus we are reduced to the situation in the statement of the corollary.

We now show that we may also assume that G is not compact.

LEMMA 2.5. *If G is a compact, connected, semisimple, adjoint Lie group and Γ is dense and proper in G, then Γ does not act amenably on G.*

Proof. Since Γ is proper and dense, Γ must project to a proper subgroup in one of the simple factors of G, say G'. Let Γ' be the image of Γ in G'. If Γ acts amenably on G, then Γ' acts amenably on G' (Lemma 2.2, 2.3.) Since Γ' is dense and G' is simple, Γ' must have a trivial connected component. Thus Γ' is a separable, totally disconnected locally compact group which admits an injective homomorphism into a Lie group. It follows that Γ' is discrete. Since Γ' acts amenably leaving a finite measure invariant, it follows that Γ' is amenable [6, Proposition 4.3.3]. It then follows by a theorem of Tits [3] that Γ' contains a solvable subgroup of finite index. Since G' is connected, it follows that G' is solvable which is impossible.

3. NONEXTENDABLE HOMOMORPHISMS

To deal with the non-compact semisimple case, the following result will be basic.

THEOREM 3.1. *Let G be a connected, non compact, semisimple adjoint*

Lie group, and Γ a (proper) dense subgroup. Suppose that Γ projects nonsurjectively into a noncompact simple factor of G. Then there is a local field k of characteristic 0, a k-simple k-group H, and a homomorphism $\pi: \Gamma \to H_k$ such that:

(1)   *$\pi(\Gamma)$ is Zariski dense in H;*

(2)   *π does not extend to a continuous homomorphism $G \to H_k$;*

(3)   *$\pi(\Gamma)$ does not have compact closure in H_k.*

In the case in which Γ is not a finitely generated group, we shall use the following result of Wang [4] in the proof of Theorem 3.1.

LEMMA 3.2 ([4], see also [5]). *A dense subgroup of a connected semisimple adjoint Lie group contains a finitely generated dense subgroup.*

Proof of Theorem 3.1.   We can assume that G is the identity component in the Hausdorff topology of the real points of a connected semisimple adjoint algebraic $\mathbb{R}$-group. Let $\mathfrak{A}$ be an absolutely simple factor of this algebraic group, so that the image of G in $\mathfrak{A}$, say G', is non compact, and Γ', the image of Γ in G', is a proper subgroup. Thus, Γ' is a dense proper subgroup of G' and a Zariski dense subgroup of $\mathfrak{A}$. For $g \in G$, we let g' be its image in $\mathfrak{A}$. Let F be the field generated by $\{tr(\gamma)| \gamma \in \Gamma'\}$. By Vinberg's lemma [6, Lemma 6.1.7], we can assume that $\mathfrak{A}$ is defined over F and that $\Gamma' \subset \mathfrak{A}(F)$. Assume first that $[F;\mathbb{Q}] < \infty$. If there is a finite prime $\mathfrak{p}$ of F such that the image of Γ in $\mathfrak{A}(F_\mathfrak{p})$ is not bounded, we may take this as the homomorphism π. Suppose that no such finite prime exists. Let L be the adele group $\mathfrak{A}(\mathbb{A}_F)$, and write $L = L_\infty \times L_f$ the decomposition corresponding to the infinite and finite primes. Since $\Gamma' \subset \mathfrak{A}(F)$, the image of Γ' under the diagonal embedding is discrete in L. Let Γ_1' be the image in Γ' of a finitely generated subgroup $\Gamma_1 \subset \Gamma$, with Γ_1 dense in G. By our assumptions, the image of Γ_1' in L_f is precompact, i.e., we have $\Gamma_1' \subset L_\infty \times K$ where $K \subset L$ is a compact group. Thus the image of Γ_1' in L_∞ must also be discrete. Similarly, the image of Γ_1' in L_0 must be discrete, where L_0 is the product over the set S of infinite primes where the image of Γ' is not precompact. We claim that this shows that one of the representations of Γ into $\mathfrak{A}(F_\mathfrak{p})$ for some $\mathfrak{p} \in S$ cannot be extended to a continuous representation of G. Otherwise, one could extend the representation of Γ into L_0 to a continuous representation $G \to L_0$, which is impossible since Γ_1 is dense in G and Γ_1' is discrete in L_0.

It therefore remains to consider the case in which F is an infinite extension of $\mathbb{Q}$. We recall that every finite dimensional continuous representation of G is rational. We thus deduce that if ρ_1 and ρ_2 are finite dimensional continuous representations of G satisfying $tr(\rho_1(\gamma)) = tr(\rho_2(\gamma))$ for every $\gamma \in \Gamma_1$, then $tr(\rho_1(g)) = tr(\rho_2(g))$ for every $g \in G$. Furthermore, in any dimension, there are only finitely many inequivalent representations of G.

If F is transcendental, choose $\beta \in \Gamma$ such that $\operatorname{tr}(\beta')$ is transcendental. As $\operatorname{Aut}(\mathbb{C})$ acts transitively on transcendental elements, we can find an element $\sigma \in \operatorname{Aut}(\mathbb{C})$ which will take $\operatorname{tr}(\beta')$ to an arbitrarily large element. This ensures that the corresponding representation of $\mathfrak{A}(F)$ (which we still denote by σ) will not have precompact image on Γ', (as $\operatorname{tr}(\sigma(\beta')) = \sigma(\operatorname{tr}(\beta'))$ is too large. Since the number of possibilities for $\operatorname{tr}(\sigma(\beta'))$ is infinite, we have that all but a finite number of such σ will yield representations of Γ which do not extend to G.

We may thus assume that F is algebraic (and infinite over $\mathbb{Q}$.) Since Γ is dense in G and G' is noncompact, we can choose $\beta \in \Gamma$ such that $|\operatorname{tr}(\beta')| > n$, where n is the dimension of the given representation of $\mathfrak{A}$. Let k be the field generated by $\{\operatorname{tr}(\gamma) | \gamma \in \Gamma'_1\}$, so that $[k : \mathbb{Q}] < \infty$ (by Vinberg's lemma, e.g.). Let k_1 be the field generated by k and $\operatorname{tr}(\beta')$, and choose $\gamma \in \Gamma$ such that $c = \operatorname{tr}(\gamma') \notin k_1$. (This exists since F is an infinite extension.) Let σ be an embedding of F in $\mathbb{C}$ such that σ fixes k_1 and $\sigma(c) \neq c$. As above, the corresponding representation of Γ defined by σ will not have precompact image as $\operatorname{tr}(\sigma(\beta')) = \sigma(\operatorname{tr}(\beta')) = \operatorname{tr}(\beta') > n$, and will not extend to a continuous representation of G since σ moves $\operatorname{tr}(\gamma')$. This completes the proof.

We remark that the above proof shows that the theorem is true without any noncompactness assumptions on G if we assume Γ to be finitely generated.

4. COMPLETION OF THE PROOF

The remainder of the proof of Theorem 1.1 depends heavily on techniques and arguments involved in studying measurable equivariant maps between homogeneous spaces, and in particular arguments involved in proving Margulis' superrigidity theorems. We shall use freely (but with references) the results in [6].

Proof of Theorem 1.1. By the results in Section 2, it suffices to suppose G is a connected, semisimple, noncompact, adjoint Lie group, the image of Γ in G is a proper dense subgroup, and to show that the the action of Γ on G is not amenable. We first claim that we can assume Γ is countable. To see this, let Γ_0 be the identity component of Γ. Then Γ_0 is a Lie subgroup of G, normalized by Γ. Since Γ is dense in G, Γ_0 is normal in G. By Lemma 2.3, amenability of Γ on G is equivalent to amenability of Γ/Γ_0 on G/Γ_0. But Γ/Γ_0 is the image of a totally disconnected separable group in a Lie group, and hence is countable.

We now proceed to the main argument. Suppose that Γ acts amenably on G. Choose $\pi : \Gamma \to H_k$ as in Theorem 3.1. Let $P_k \subset H_k$ be the k-points of a minimal k-parabolic subgroup. In particular, H_k/P_k is compact. Let $Q \subset G$ be a minimal parabolic (and hence amenable) subgroup. By amenability of the Γ action on G, there is a measurable Γ-map $\varphi: G/Q \to M(H_k/P_k)$ (Lemmas 2.1, 2.2). By [6, Corollary 3.2.17], the action of H_k on $M(H_k/P_k)$ is smooth, in the sense that every orbit is locally closed. Thus, if we let ψ be the composition of φ with the projection onto the orbit space, then $\psi: G/Q \to M(H_k/P_k)/H_k$ is a measurable Γ-invariant map into a countably separated Borel space. Since Γ is dense in G, it acts ergodically on G/Q [6, Lemma 2.2.13], and hence ψ is essentially constant. Thus $\varphi(G/Q)$ lies (a.e.) in a single H_k-orbit, so we can view φ as a measurable Γ-map $\varphi: G/Q \to H_k/S$, where S is the stabilizer of some measure on H_k/P_k. By [6, Corollary 3.2.19], we have two cases. Either:

(a)   There is a measurable Γ-map $\varphi: G/Q \to H_k/L_k$ where $L \subset H$ is a k-subgroup of strictly smaller dimension; or,

(b) S is compact.

In Case (b), [6, Proposition 5.1.9] implies that $\pi(\Gamma)$ is precompact, contradicting (3) of Theorem 3.1. In case (a), we argue as on [6, p. 128]. We deduce that there is a measurable (and hence continuous) homomorphism $h: G \to H_k$ and a point $x \in H_k/L_k$ such that $\varphi([g]) = xh(g)$ (a.e.). In particular, if k is totally disconnected, φ is essentially constant, which implies that $\pi(\Gamma)$ leaves a point in H_k/L_k invariant. This contradicts the assumption that $\pi(\Gamma)$ is Zariski dense in H_k. If $k = \mathbb{R}$ or $\mathbb{C}$, h is rational, and hence φ is (essentially) rational. Then by [6, Lemma 5.1.3], π extends to a continuous homomorphism of G, which again contradicts the conclusion of Theorem 3.1. This completes the proof of Theorem 1.1.

5. Concluding Remarks

Theorem 1.1 can be generalized to the case of a noninjective homomorphism as follows. This more general result follows from Theorem 1.1 and Lemma 2.3(c).

THEOREM 5.1.   *Let G be a connected Lie group, Γ a locally compact separable group, and $\lambda: \Gamma \to G$ a continuous homomorphism. Then the action of Γ on G is amenable if and only if*

(i)   *ker (λ) is amenable; and*

(ii)   *the image of $\Gamma \cap (\Gamma^-)_0$ under projection to the maximal connected semisimple adjoint factor of $(\Gamma^-)_0$ is closed.*

64 ROBERT J. ZIMMER

REFERENCES

1. Y. CARRIERE AND E. GHYS, Relations d'équivalence moyennables sur les groupes de Lie, *C. R. Acad. Sc. Paris, Serie I*, **300** (1985), 677–680.
2. M. RAGHUNATHAN, "Discrete Subgroups of Lie groups," Springer-Verlag, Berlin, 1972.
3. J. TITS, Free subgroups in linear groups, *J. Algebra* **20** (1972), 250–270.
4. H. C. WANG, On a maximality property of discrete subgroups with fundamental domains of finite volume, *Amer. J. Math.* (1967), 124–132.
5. S. P. WANG, On subgroups with property (*P*) and maximal discrete subgroups, *Amer. J. Math.* **97** 404–414.
6. R. J. ZIMMER, "Ergodic Theory and Semisimple Groups," Birkhauser, Boston, 1984.
7. R. J. ZIMMER, Amenable pairs of groups and ergodic actions and the associated von Neumann algebras, *Trans. Amer. Math. Soc.* **243** (1978), 271–286.
8. R. J. ZIMMER, On connection preserving actions of discrete linear groups, *Ergodic Theory Dynamical Systems*, in press.

3

Orbit Equivalence and Strong Rigidity

Annals of Mathematics, **112** (1980), 511-529

Strong rigidity for ergodic actions of semisimple Lie groups[1]

By ROBERT J. ZIMMER

1. Introduction

The aim of this paper is to prove an analogue of the strong rigidity theorems for lattices in semisimple Lie groups of Mostow and Margulis in the context of ergodic actions of semisimple Lie groups and ergodic foliations by symmetric spaces.

If G is a locally compact group and S is an ergodic G-space with finite invariant measure, one can attempt to study the action by means of the equivalence relation on S which it defines. Two actions (of possibly different groups) are called orbit equivalent if they define isomorphic equivalence relations. Of course for two actions of the same group, conjugacy of the actions (or more generally, automorphic conjugacy, i.e., conjugacy modulo an automorphism of the group) implies orbit equivalence. However in general, orbit equivalence is, to a surprising extent, a far weaker notion. This was first illustrated by the striking result of H. Dye [5] that any two properly ergodic (i.e., ergodic but not transitive modulo a null set) integer actions with finite invariant measure are orbit equivalent. This result has been extended over the years by a number of people ([3], [4], [7], [9], [12], [21], [25]), and one now has, for example, the result that all properly ergodic free actions with finite invariant measure of (possibly different) discrete amenable groups are orbit equivalent, and that a similar result is true for the class of continuous amenable groups. Thus for amenable actions with finite invariant measure, the equivalence relation will not in any way distinguish different actions or groups.

The point of this paper is to show that for semisimple Lie groups, very much the opposite phenomenon occurs. An ergodic action of such a group is called irreducible if the restriction to every non-central normal subgroup is still ergodic. (For simple Lie groups, this is of course no more than

---

[1] Research partially supported by NSF Grant MCS 76-06626 and a Sloan Foundation Fellowship.

512   ROBERT J. ZIMMER

ergodicity.) Our main result is the following. It has been independently conjectured by A. Connes.

THEOREM 4.3. *Let G and G' be centerless connected semisimple Lie groups with no compact factors, and suppose that the real rank of G is at least 2. Let S and S' be (essentially) free irreducible ergodic G- and G'-spaces respectively, with finite invariant measure. Then the following are equivalent.*
  i) *The actions are orbit equivalent.*
  ii) $G = G'$ *and the actions are automorphically conjugate.*

If G and G' are allowed to have finite center, then one can deduce that the groups are locally isomorphic and after factoring out the action of the center we obtain automorphically conjugate actions. From Theorem 4.4 we also deduce the following.

THEOREM 4.5. *Let $\Gamma \subset G$ and $\Gamma' \subset G'$ be lattices in connected simple Lie groups with finite center, and suppose the real rank of G is at least 2. If S and S' are (essentially) free ergodic Γ- and Γ'-spaces (respectively) with finite invariant measure and the actions are orbit equivalent, then G and G' are locally isomorphic.*

In particular, we obtain the following result.

COROLLARY 4.6. *Let $SL(n, Z)$ act on the torus $\mathbf{R}^n/Z^n$ by automorphisms. Then for $n, p \geqq 2$, $n \neq p$, the actions are not orbit equivalent.*

Theorem 4.3 can be thought of as a virtual group analogue ([14], [22]) of the standard rigidity theorems. That is, the Mostow-Margulis theorems ([19], [16]) assert that an isomorphism of certain subgroups, namely lattices, extends to an isomorphism of the ambient Lie groups. An ergodic action of the connected Lie group determines a virtual subgroup and isomorphism of the virtual subgroups is orbit equivalence [8], [9]. Thus Theorem 4.3 asserts that isomorphism of certain virtual subgroups extends to an isomorphism of the ambient Lie group.

The strong rigidity theorems for lattices can of course also be formulated in more geometric terms. Thus, with suitable hypotheses [19], they assert that the Riemannian structure on a locally symmetric space of finite volume is determined up to scalar multiples by the fundamental group. Here we also have a geometric interpretation, in terms of foliations. Namely, we have the following theorem. (See Section 5 for details.)

THEOREM 5.1. *Let $\mathcal{F}$ and $\mathcal{F}'$ be Riemannian measurable foliations by symmetric spaces G/K and G'/K', where G and G' are connected semisimple Lie groups with finite center and no compact factors. Suppose the foliations* *come from free irreducible ergodic actions with finite invariant measure of G and G' respectively, and that G/K has rank at least 2. Then if the foliations are transversally equivalent* (i.e., *the ergodic equivalence relations defined on suitable transversals are isomorphic*) *then $\mathcal{F}$ and $\mathcal{F}'$ are isometrically isomorphic as Riemannian measurable foliations modulo normalizing scalar multiples* (*independent of the leaves*).

Thus, roughly speaking, Theorem 5.1 is asserting that for suitable $\mathcal{F}$, the symmetric Riemannian structure on the leaves is completely determined by the ergodic equivalence relation defined on a transversal.

Our approach to proving Theorem 4.3 is to combine techniques used by Margulis in proving arithmeticity of lattices [16] with various techniques and results concerning the cohomology of ergodic group actions and the notion of an amenable ergodic action ([32], [35]) (which is more general than an action of an amenable group). The algebraic group nature of semisimple Lie groups is used in fundamental ways throughout the proof. Both the techniques of Margulis and Mostow entail the construction of a suitable map on the maximal Furstenberg boundary of G. Here, we shall need to construct an analogous map, this time defined on the product of the G-space S with the Furstenberg boundary, and we achieve this via the fact that this product action is amenable. In order to successfully employ amenability here we need two results concerning the action of a semisimple Lie group on the space of probability measures on the Furstenberg boundary. The first is a result of C. C. Moore which asserts that the stabilizer of any measure is an amenable algebraic group. The second result is the following modification of a result we proved in [35] which is of independent interest.

THEOREM 3.1. *Let P be a parabolic subgroup of G, a connected semisimple Lie group with finite center. Then for every probability measure μ on G/P, the orbit of μ under G is locally closed in the weak *-topology.*

The outline of this paper is as follows. In Section 2 we collect some basic definitions and results we shall need throughout the paper. In Section 3 we discuss the orbits of measures on the Furstenberg boundary, and include here an extension of the main theorem of [35] on the structure of amenable actions from algebraic groups to general locally compact connected groups. Section 4 contains the proofs of the main theorems, and Section 5 gives the geometric formulation in terms of Riemannian measurable foliations. We conclude with some questions in Section 6.

The author would like to express his thanks to Alain Connes and Calvin Moore for several discussions on questions related to this work. The main results of this paper were announced in [38].

2. Preliminaries

We begin by recalling some basic definitions and results concerning ergodic actions and their cohomology that we will be using.

Suppose G is a locally compact second countable group and (S, μ) is a standard Borel space with probability measure on which G acts so that the map $S \times G \to S$ defining the action is Borel. The measure μ is called quasi-invariant if each $g \in G$ preserves the measure class of μ, i.e., for all Borel spaces $A \subset S$ and $g \in G$, $\mu(Ag) = 0$ if and only if $\mu(A) = 0$. The action is called ergodic if $\mu(Ag \Delta A) = 0$ for all $g \in G$ implies A is null or conull. We summarize this situation by calling S an ergodic G-space. We shall be particularly interested in the case in which the measure μ is actually invariant, i.e., $\mu(Ag) = \mu(A)$ for all Borel $A \subset S$ and $g \in G$. If S and T are two ergodic G-spaces, they are called isomorphic or conjugate if, possibly after discarding G-invariant null sets, there is a measure class preserving G-map between them that is a Borel isomorphism. Equivalently [13], there is a G-isomorphism of the associated Boolean G-spaces of Borel sets modulo null sets. The actions are called automorphically conjugate if there is an automorphism A of G such that S is conjugate to the G-space T with action $(t, g) \to t \cdot A(g)$. If S is an ergodic G-space and S' an ergodic G'-space, the actions are called orbit equivalent if (again after discarding null sets) there is a measure class preserving Borel isomorphism $S \to S'$ that takes orbits onto orbits. In other words, the equivalence relations defined by the actions are isomorphic. As indicated in the introduction, for actions of amenable groups there is a very weak relationship between equivalence and orbit equivalence, and the point of this paper is to show that for semisimple Lie groups and their lattice subgroups the relationship is much stronger.

If $H \subset G$ is a closed subgroup and S is an ergodic H-space, there is a naturally associated "induced" ergodic G-space X which we denote by $X = \operatorname{ind}_H^G (S)$. This notion is the analogue for ergodic actions of induced representations and is discussed in some detail in [35]. We briefly recall the definition. Let H act on $S \times G$ by the product action $(s, g)h = (sh, gh)$ and let G act on $S \times G$ by $(s, g)g_0 = (s, g_0^{-1}g)$. Then the H and G actions commute and so G will act on the space of H orbits $X = (S \times G)/H$. By projecting a probability measure in the natural product measure class on $S \times G$ we obtain a quasi-invariant ergodic measure on the G-space X. Given a particular ergodic action of a group, it is natural to ask when it is induced from some (perhaps given) subgroup. One general criterion is provided by an analogue of the

STRONG RIGIDITY FOR ERGODIC ACTIONS   515

imprimitivity theorem.

PROPOSITION 2.1 [35, Theorem 2.5]. $X = \operatorname{ind}_H^G (S)$ *for some ergodic G-space S if and only if there is a G-map $X \to G/H$.*

(As usual, the map may be defined only on a conull G-invariant set. We shall generally disregard this technicality.) If G is an algebraic group, this leads to the following.

PROPOSITION 2.2 [37]. *Let X be an ergodic G-space where G is a real algebraic group. Then there is an algebraic subgroup $H \subset G$ such that X is induced from an H-space but not from an action of any smaller algebraic subgroup. The group H is unique up to conjugacy.*

Thus to each ergodic action of G there is a naturally associated (conjugacy class of an) algebraic subgroup. We call H the algebraic hull of the G-space X, so that H is defined up to conjugacy. The justification of this terminology is that if G is transitive on X, so that $X = G/G_0$ for some closed subgroup $G_0 \subset G$, then H is the algebraic hull of G_0. In particular, we call an action Zariski dense if the algebraic hull is G itself. We then have the following version of the Borel density theorem [1].

THEOREM 2.3 [37]. *If G is a semisimple real algebraic group with no compact factors and X is an ergodic G-space with finite invariant measure, then X is Zariski dense in G.*

Classically Borel's theorem is of course the case in which $X = G/\Gamma$ for some lattice subgroup $\Gamma \subset G$.

A basic notion in the study of ergodic action is that of a cocycle. (See [8], [22], [23], [27], [30] as general references for cocycles.) If X is an ergodic G-space and M is a standard Borel group then a Borel function $\alpha: X \times G \to M$ is called a cocycle if for all $g, h \in G$, $\alpha(x, gh) = \alpha(x, g)\alpha(xg, h)$ for almost all x. Geometrically, these functions enable us to define actions of G on spaces of functions on X with values in a space on which M acts or products of X with M-spaces. They can also be viewed as the Borel 1-cocycles of G in the sense of Eilenberg-MacLane with values in the space of Borel functions from X into M, identified modulo null sets. Two cocycles α, β are called equivalent or cohomologous if there is a Borel function $\varphi: X \to M$ such that for each g, $\varphi(x)\alpha(x, g)\varphi(xg)^{-1} = \beta(x, g)$ for almost all x. This is just the notion of equivalence for Eilenberg-MacLane cohomology. One way in which such cocycles arise is from a homomorphism $\pi: G \to M$. We simply define $\alpha(x, g) = \pi(g)$ and call α the restriction of the homomorphism π to $X \times G$. Cocycles arise also from an orbit equivalence. Namely, suppose S_i is an ergodic G_i-space,

516   ROBERT J. ZIMMER

$i = 1, 2$, and that $\theta: S_1 \to S_2$ is an orbit equivalence. Suppose further that the G_2-action is free. Then define $\alpha: S_1 \times G_1 \to G_2$ by the equation $\theta(s) \cdot \alpha(s, g_1) = \theta(sg_1)$. It is of course the freeness which makes this well-defined. If $G_1 = G_2$, then we can describe the condition that orbit equivalent actions be conjugate or automorphically conjugate in terms of the cocycle.

PROPOSITION 2.4. *Suppose α is a cocycle defined by an orbit equivalence of two free G-actions, say on S_1 and S_2. If α is equivalent to the restriction to $S_1 \times G$ of an automorphism of G, then the G-actions are automorphically conjugate. If the automorphism is inner, then the G-actions are conjugate.*

Proof. If $\theta: S_1 \to S_2$ is the orbit equivalence and $\varphi: S_1 \to G$ is such that $\varphi(s)\alpha(s, g)\varphi(sg)^{-1} = A(g)$ for some $A \in \operatorname{Aut}(G)$, define $\lambda: S_1 \to S_2$ by $\lambda(s) = \theta(s)\varphi(s)^{-1}$. Then for each g and almost all s, $\lambda(s) \cdot A(g) = \lambda(sg)$. It is not difficult to check that λ defines a Boolean isomorphism $B(S_2) \to B(S_1)$, and it follows that the actions are automorphically conjugate on the Boolean algebras, and hence on the spaces themselves [13]. If $A(g) = hgh^{-1}$, then $\tilde{\lambda}(s) = \theta(s)\varphi(s)^{-1}h$ defines the required Boolean G-map.

Suppose now that S is an ergodic H-space and $\alpha: S \times H \to G$ is an arbitrary cocycle, where G is also locally compact. One can then associate to this in a natural way an ergodic G-action which reduces to the inducing construction if H is a closed subgroup of G and α is the restriction to $S \times H$ of the inclusion of H in G. Namely, H acts on $S \times G$ by $(s, g) \cdot h = (sh, g\alpha(s, h))$, and we let X be the space of ergodic components of this H-action. As in the inducing construction, G acts on $S \times G$, the G and H actions commute, and hence we have an ergodic action of G on X. This construction was first described by Mackey [14]. (See also [22], [30] for a more detailed description.) We call this action the Mackey range or the Mackey action of α. The following is straightforward.

PROPOSITION 2.5. *If G_i acts freely and ergodically on S_i, $i = 1, 2$, and $\alpha: S_1 \times G_1 \to G_2$ is the cocycle corresponding to an orbit equivalence, then the Mackey action of α is (isomorphic to) the G_2 action on S_2.*

The following is a basic connection between the cohomology class of α and the Mackey range.

PROPOSITION 2.6. *The cocycle $\alpha: S \times G_1 \to G_2$ is equivalent to a cocycle taking values in the closed subgroup $H \subset G_2$ if and only if the Mackey range is induced from an ergodic action of H.*

Proof. See [30, Theorem 3.5] and use Proposition 2.1.

The final topic we wish to discuss in this section is that of amenability

STRONG RIGIDITY FOR ERGODIC ACTIONS   517

of ergodic actions. Amenable groups are just groups with the fixed point property for affine actions on compact convex sets and this property implies a certain fixed point property for cocycles defined on actions of amenable groups. The general notion of an amenable action is based on the observation that this fixed point property for cocycles can hold for certain ergodic actions of non-amenable groups as well. This property arises naturally in a variety of questions in ergodic theory and in particular for free actions is equivalent to the condition that the associated von Neumann algebra under the Murray-von Neumann group measure space construction is approximately finite dimensional ([33], [9]). We shall now describe the definition and some basic properties of these actions we will need, referring the reader to [32], [35] for further details.

Let E be a separable Banach space, Iso (E) the group of isometric isomorphisms of E, and E_1^* the unit ball in the dual with the $\sigma(E^*, E)$ topology. Suppose S is an ergodic G-space, $\alpha: S \times G \to$ Iso (E) a cocycle, and $s \to A_s$ a Borel assignment of a compact convex set $A_s \subset E_1^*$ to $s \in S$. We call A_s an α-invariant field of sets if for all $g \in G$, $\alpha^*(s, g)A_{sg} = A_s$ a.e., where $\alpha^*(s, g) = (\alpha(s, g)^*)^{-1}$ is the adjoint cocycle. The action of G on S is called amenable [32] if for all E, α, A_s as above, there exists a Borel function $\varphi: S \to E_1^*$ such that $\varphi(s) \in A_s$ a.e., and for all g, $\alpha^*(s, g)\varphi(sg) = \varphi(s)$ a.e. In other words, φ is a fixed point for the action of G on $L^\infty(S, E^*)$, defined by the α-twisting of translation: $(\varphi \cdot g)(s) = \alpha^*(s, g)\varphi(sg)$.

PROPOSITION 2.7. a) *A transitive action is amenable if and only if the stabilizers are amenable* [32, *Theorem* 1.9].

b) *If G is amenable every ergodic G-action is amenable* [32, *Theorem* 2.1].

c) *If a G-action with finite invariant measure is amenable, then G is amenable* [32, *Proposition* 4.3].

d) *If $H \subset G$ and $X = \text{ind}_H^G(S)$, then the G-action on X is amenable if and only if the H-action of S is amenable* [35, *Proposition* 3.2].

e) *Every extension of an amenable action is amenable. That is, if there exists a measure class preserving G-map $X \to Y$ and Y is an amenable G-space, then so is X* [32, *Theorem* 2.4].

Property (d) in the above proposition raises the question as to whether every amenable ergodic action is induced by an action of an amenable subgroup. We showed by example in [35] that this is not true, but also proved the following.

THEOREM 2.8 [35]. *If G is a real algebraic group, then every amenable*

518   ROBERT J. ZIMMER

ergodic action is induced from an action of an amenable subgroup.

We shall observe in Section 3 that this is true for any connected locally compact G.

3. Orbits of measures on homogeneous projective varieties

In this section we present some results we shall need concerning the orbits in certain natural G-spaces where G is an algebraic group. It is well known, for example, that if an algebraic group acts algebraically on a variety, then the orbits are locally closed, which is equivalent to a large number of other regularity conditions for the structure of the space of orbits [6]. The real general linear group GL $(n, \mathbf{R})$ acts naturally on the real projective space $\mathbf{P}^{n-1}$ and hence on the space of probability measures on $\mathbf{P}^{n-1}$, $M(\mathbf{P}^{n-1})$, which is a compact metrizable space with the weak $*$-topology. Based on a technique of H. Furstenberg [10], we showed in [35] that every orbit in $M(\mathbf{P}^{n-1})$ is locally closed. We now observe that this result can be used to show the corresponding theorem for measures on generalized flag manifolds. More precisely, we have

THEOREM 3.1. *Let G be a semisimple Zariski connected real algebraic group or a connected semisimple Lie group with finite center, and suppose P is a parabolic subgroup [28]. Then for any probability measure μ on G/P the orbit of μ under the action of G is locally closed in the weak $*$-topology.*

We recall that for $G = $ SL $(n, \mathbf{R})$ or SL $(n, \mathbf{C})$, the spaces G/P are just the various flag varieties. We begin with the following lemma. An action is called smooth if every orbit is locally closed.

LEMMA 3.2. *For $\mu \in M(\mathbf{P}^{n-1})$, let $G_\mu \subset $ GL $(n, \mathbf{R})$ be the stabilizer of μ. Then for each μ, G_μ is a finite extension of a real quasi-algebraic group.*

(We recall that a real quasi-algebraic group is a subgroup of finite index in a real algebraic group [15, Section 8].)

Proof. We argue by induction on n. The lemma is trivial for $n = 1$. For $n \geq 2$, let $V \subset \mathbf{R}^n$ be a subspace of minimal dimension for which $\mu([V]) > 0$ where $[V] \subset \mathbf{P}^{n-1}$ is the corresponding projective subspace. By the argument of Furstenberg [10, Lemma 1.5], if dim $V = n$, then $G_\mu \cap $ SL $(n, \mathbf{R})$ is compact and hence algebraic, and one can then easily see that G_μ is a finite extension of a quasi-algebraic group. (Use [15, Lemma 8.3] for example.) If dim $V < n$, then by minimality of the dimension of V, $\mu = \mu_1 + \cdots + \mu_k$ where μ_i are disjoint measures supported on projective subspaces $[V_i]$ so that G_μ permutes $\{V_i\}$ and $\{\mu_i\}$. It follows that $\bigcap G_{\mu_i}$ is a normal subgroup in G_μ of finite index.

Therefore, it suffices to show that G_{μ_i} is a finite extension of a quasi-algebraic group. Let G_{V_i} be the subgroup of GL $(n, \mathbf{R})$ leaving V_i invariant. Then G_{V_i} is algebraic and $G_{\mu_i} \subset G_{V_i}$. To see the latter inclusion, note that if $A \in G_{\mu_i}$, then $A_*(\mu_i)$ is supported on $[A(V_i) \cap V_i]$ and since V_i is of minimal dimension among subspaces with positive μ-measure, $A(V_i) = V_i$. There is a natural map $\varphi \colon G_{V_i} \to \mathrm{GL}(V_i)$. Letting H_i be the stabilizer in GL (V_i) of μ_i, we have $G_{\mu_i} = \varphi^{-1}(H_i)$. By the inductive hypothesis, H_i is a finite extension of a quasi-algebraic group, and hence so is G_{μ_i}.

Proof of Theorem 3.1. One can readily reduce to the algebraic case. Then there is a rational representation $\pi \colon G \to \mathrm{GL}(n, \mathbf{R})$ such that P is the stabilizer of a point $x \in \mathbf{P}^{n-1}$. Thus, we can identify $M(G/P)$ as a closed subset of $M(\mathbf{P}^{n-1})$. If $\mu \in M(G/P)$, the GL (n)-orbit of μ in $M(\mathbf{P}^{n-1})$ is locally closed by [35, Theorem 4.1] and can be identified with GL $(n)/H$ where H is a finite extension of a quasi-algebraic group by the preceding lemma. The G-orbit of μ in $M(\mathbf{P}^{n-1})$ is then identified with the G-orbit of $[e]$ in GL $(n)/H$ and hence it suffices to show that the action of G on GL $(n)/H$ is smooth. This is equivalent to the smoothness of $H \times G$ on GL (n), which in turn follows from the smoothness of algebraic actions and [35, Lemma 5.6].

We remark that for P a minimal parabolic, Theorem 3.1 is a natural complement to part of a recent result of C. C. Moore that we shall need later. This result asserts that the stabilizer of any probability measure on G/P is an amenable algebraic group [18]. Thus Moore's theorem describes the homogeneous space structure of each orbit, while Theorem 3.1 gives information about the structure of the space of orbits.

Although we shall not need it in the sequel, we digress briefly to observe that Theorem 2.8 can be extended to connected groups. This extension has also been observed by C. C. Moore.

THEOREM 3.3. *Let G be a connected locally compact group. Then every amenable ergodic action of G is induced from an action of an amenable subgroup.*

Proof. Let $B(G) = G/P$ be the maximal Furstenberg boundary of G [11]. Then P contains a subgroup N which is normal in G such that $H = G/N$ is a semisimple centerless connected Lie group without compact factors and P/N is a minimal parabolic subgroup of H. Then it clearly follows from Theorem 3.1 that G is smooth on $M(G/P)$. Furthermore, by [18, Theorem 2.10], the stabilizer of $\mu \in M(G/P)$ is an amenable subgroup of G. Now let S be an amenable ergodic G-space, $\alpha \colon S \times G \to G$ be $\alpha(s, g) = g$ and view α as a cocycle into Iso $(C(G/P))$, where $C(G/P)$ is the space of continuous functions. Amenability implies the existence of a map $\varphi: S \to M(G/P)$ such that $\alpha^*(s, g)\varphi(sg) = \varphi(s)$, which means exactly that φ is a G-map after we switch to a right action of $M(G/P)$. If m is the given measure on S, $\varphi_*(m)$ is an ergodic quasi-invariant measure on $M(G/P)$ and since the G-action is smooth on this space, $\varphi_*(m)$ must be supported on an orbit. If we let G_0 be the stabilizer of a point in this orbit, we can view φ as a map $\varphi: S \to G/G_0$. But as remarked above, G_0 is amenable, so the theorem follows from [35, Theorem 2.5].

4. Proof of the main theorems

If G is a connected semisimple Lie group and S an ergodic G-space with finite invariant measure, we call S irreducible if the restriction to each non-central normal subgroup is still ergodic. If G is connected, simple, with finite center, irreducibility is of course automatic. If $S = G/\Gamma$ where $\Gamma \subset G$ is a lattice, then the action on S is irreducible in the above sense if and only if Γ is an irreducible lattice. If $G = G_1 \times G_2$, $S = S_1 \times S_2$ where S_i are ergodic G_i-spaces, then S is not irreducible. We begin the proof of the main theorems with the proof of the following theorem, from which the others will follow.

THEOREM 4.1. *Let $\tilde{G}$, $\tilde{G}'$ be Zariski connected semisimple real algebraic groups with trivial center, and G, G' the topological components of the identity. Suppose G, G' have no compact factors and that the real rank of $\tilde{G}$ is at least 2. Suppose S is an irreducible ergodic G-space with finite invariant measure and that $\alpha: S \times G \to G'$ is a cocycle whose Mackey range is Zariski dense in $\tilde{G}'$. Then there exists a surjective homomorphism $\beta: G \to G'$ such that the restriction of β to $S \times G$ is cohomologous to α.*

Our approach to the proof of Theorem 4.1 is to adapt Margulis' techniques used in the proof of arithmeticity of lattices [16] to the present situation via the notion of amenable actions.

Proof. Let $\tilde{P}$ and $\tilde{P}'$ be minimal parabolic subgroups of $\tilde{G}$ and $\tilde{G}'$ respectively, and let $P = \tilde{P} \cap G$, $P' = \tilde{P}' \cap G'$. Then $S \times G/P$ is an ergodic G-space, as ergodicity is equivalent to the ergodicity of P on S [34, Theorem 4.2] which in turn follows from results of Moore [17]. Furthermore, this action is amenable since P is amenable and $S \times G/P$ is an extension of the action of G on G/P. (See Proposition 2.7.) Since $\tilde{G}'/\tilde{P}'$ is compact, $M(\tilde{G}'/\tilde{P}')$ is a compact convex space and $\tilde{G}'$ acts by affine transformations on $M(\tilde{G}'/\tilde{P}')$. Thus if we let $\tilde{\alpha}: S \times G/P \times G \to \tilde{G}'$ be $\tilde{\alpha}(s, x, g) = \alpha(s, g)$, the definition of amenability implies the existence of a measurable map $\varphi: S \times G/P \to M(\tilde{G}'/\tilde{P}')$

such that $\tilde{\alpha}(s, x, g)\varphi(sg, xg) = \varphi(s, x)$ almost everywhere, that is,

(*) $\qquad\qquad\qquad\qquad \alpha(s, g)\varphi(sg, xg) = \varphi(s, x) \quad$ a.e.

By Theorem 3.1, the action of $\breve{G}'$ on $M(\breve{G}'/\tilde{P}')$ is smooth, so the orbit space $\hat{M}$ has a countably separated Borel structure [6]. But (*) implies that $\varphi(s, x)$ and $\varphi(sg, xg)$ are in the same $\breve{G}'$-orbit, so by the ergodicity of G on $S \times G/P$, $q \circ \varphi$ is essentially constant, where $q: M(\breve{G}'/\tilde{P}') \to \hat{M}$ is the natural projection. Thus changing φ on a null set, we can assume $\varphi(s, x)$ lies inside a given orbit. We can identify this orbit as a $\breve{G}'$-space with $\breve{G}'/H'$, where H' is the stabilizer of a point, and hence we can consider φ as a map $\varphi: S \times G/P \to \breve{G}'/H'$, after which (*) reads $\varphi(sg, xg) \cdot \alpha(s, g)^{-1} = \varphi(s, x)$, by transferring to a right action. By results of Moore [18], H' is amenable and we can, in fact, assume we have such a map φ where H' is amenable and algebraic by passing to the algebraic hull.

We now show that with small modifications, Margulis' arguments in [16, proof of Theorem 6] demonstrate the following.

LEMMA 4.2. *For almost all s, $\varphi_s: G/P \to \breve{G}'/H'$ defined by $\varphi_s(x) = \varphi(s, x)$ is a rational mapping.*

Proof. We follow with some modifications the notation of Margulis [16, proof of Theorem 6]. (See also [20, p. 146].) Let A be a maximal connected **R**-split abelian subgroup of P, and $t \in A$, $t \neq 1$. Let C be the centralizer of t in G. We have $\varphi: S \times G/P \to \breve{G}'/H'$ and we can also view this as a map $\varphi: S \times G \to \tilde{G}/H'$. For each $(s, g) \in S \times G$, define

$$w_{s,g}: C \longrightarrow \breve{G}'/H' \qquad \text{by} \qquad w_{s,g}(c) = \varphi(s, cg) .$$

The space Σ of measurable maps $C \to \breve{G}'/H'$ identified modulo equality on conull sets has a countable basis when given the topology of convergence in measure on compact sets, and as in [16, p. 42] the space $\hat{\Sigma} = \Sigma/\breve{G}'$ has a countably generated Borel structure, where $\breve{G}'$ acts naturally on Σ by pointwise multiplication. (See Appendix for a proof.) We view w as a map $S \times G \to \Sigma$ and let $\tilde{w}$ be the composition with projection into $\hat{\Sigma}$. Now

$$w_{s,tg}(c) = \varphi(s, ctg) = \varphi(s, tcg)$$
$$= \varphi(s, cg) \quad (\text{since } t \in P)$$
$$= w_{s,g}(c) ,$$

so we can view w as a map $S \times G/T \to \Sigma$ as well, where $T = \{t^n\}$. For $h \in G$ and almost all (s, g) we also have

$$w_{(s,g)\cdot h}(c) = w_{(sh,gh)}(c) = \varphi(sh, cgh)$$
$$= \varphi(s, cg)\alpha(s, h) = w_{(s,g)}(c) \cdot \alpha(s, h) ,$$

so $\tilde{w}_{(s,g)\cdot h} = \tilde{w}_{(e,g)}$. By Moore's ergodicity theorem [17], T is ergodic on S (using irreducibility of S; cf. [34, Theorem 5.4]), so G is ergodic on $S \times G/T$ [34, Theorem 4.2]. Since $\tilde{w}$ is G-invariant on $S \times G/T$, it follows from the fact that $\hat{\Sigma}$ is countably generated that $\tilde{w}$ is constant on a conull set. Thus, given s, g, g_1, there exists $h(s, g_1, g) \in \check{G}'$ such that for each g_1 and almost all (s, g), $w_{(s,g_1g)} = w_{(s,g)} \cdot h(s, g_1, g)$. Now for $g_1, g_2 \in C$,

$$
\begin{aligned}
w_{(s,g)}(c) \cdot h(s, g_1g_2, g) &= w_{(s,g_1g_2g)}(c) \\
&= \varphi(s, cg_1g_2g) \\
&= w_{(s,g_2g)}(cg_1) \\
&= w_{(s,g)}(cg_1) \cdot h(s, g_2, g) \\
&= \varphi(s, cg_1g) \cdot h(s, g_2, g) \\
&= w_{(s,g_1g)}(c) \cdot h(s, g_2, g) \\
&= w_{(s,g)}(c) \cdot h(s, g_1, g)h(s, g_2, g) \, .
\end{aligned}
$$

Let $\tilde{G}'_{(s,g)}$ be the stabilizer of $w_{(s,g)}$ in $\tilde{G}'$ and $N_{s,g}$ the normalizer of $\tilde{G}'_{(s,g)}$ in $\tilde{G}'$. For $g_1 \in C$ it follows immediately from the definition of $w_{(s,g)}$ that $\tilde{G}'_{(s,g)} = \tilde{G}'_{(s,g_1g)}$, from which it also follows readily that $h(s, c, g) \in N_{s,g}$ for $c \in C$. Hence the above calculation shows that $\tau_{s,g} \colon C \to N_{(s,g)}/\tilde{G}'_{(s,g)}$ is a homomorphism for almost all (s, g), where $\tau_{(s,g)}(c)$ is $h(s, c, g)$ projected to $N_{(s,g)}/\tilde{G}'_{(s,g)}$. Arguing exactly as in [16, p. 43], one deduces that for almost all s, $\varphi(s, hg)$ depends rationally on h in a Zariski open subset of G for almost all $g \in G$. From this it follows that φ_s is a rational map for almost all s, defined at all points of G/P.

Returning to the proof of the theorem, we next make some observations concerning actions in the space of rational functions. Let Q be the set of rational maps from G/P into $\tilde{G}'/H'$. There is a natural action of $G \times \tilde{G}'$ on Q given by $(f \cdot (g, g'))(x) = f(xg^{-1}) \cdot g'$. We claim that the actions of $G \times \tilde{G}'$ and $\tilde{G}'$ on Q are smooth. This follows by passing to the complex varieties, by use of [35, Lemma 5.6], and the following observations. Let H and H' be (complex) algebraic groups acting algebraically on varieties V and V' with V projective and V' quasi-projective, in such a way that V and V' are equivariantly embedded in projective spaces $\mathbf{P}^n$ and $\mathbf{P}^{n'}$ on which H and H' act via rational representations. Let Z be the space of rational maps from V to V' with the $H \times H'$ action as described above. For any $f \in Z$, let W_f be the (closed) set of points in V at which f is not defined and Y_f the Zariski closure of the graph of f, so $Y_f \subset \mathbf{P}^n \times \mathbf{P}^{n'}$. The assignment $f \to (W_f, Y_f)$ is injective. Let $\mathcal{S}$ be the set of closed subvarieties $\mathbf{P}^n$ and $\tilde{\mathcal{S}}$ the set of closed subvarieties of $\mathbf{P}^n \times \mathbf{P}^{n'}$. As is well known, $\mathcal{S}$ and $\tilde{\mathcal{S}}$ both have the structure of a countable union of varieties [26, I. 6.5], and hence so does $\mathcal{S} \times \tilde{\mathcal{S}}$. The

$H \times H'$ action is naturally carried over to an algebraic action on $\tilde{s} \times \tilde{\tilde{s}}$ which is therefore smooth. Thus Z can be viewed as an invariant subspace in a smooth action and hence the $H \times H'$ action on Z is smooth. Similarly, the H' action is smooth.

We define $\Phi: S \to Q$ by $(\Phi(s))(x) = \varphi_s(x)$ and $\tilde{\Phi}: S \to Q/\tilde{G}'$ the composition of Φ which the natural projection $Q \to Q/\tilde{G}'$. We remark that $Q/\tilde{G}'$ is a G-space in a natural way and that G is smooth on $Q/\tilde{G}'$ since $G \times \tilde{G}'$ is smooth on Q. The equation $\varphi(sg, xg) = \varphi(s, x) \cdot \alpha(s, g)$ implies that for each g and almost all s,

(**) $$\Phi(sg) \cdot \alpha(s, g)^{-1} = \Phi(s) \cdot g$$

which implies that $\tilde{\Phi}: S \to Q/\tilde{G}'$ is a G-map. Since the G-action on $Q/\tilde{G}'$ is smooth, $\tilde{\Phi}(S)$ must lie in a single G-orbit. Let us denote by G_0 the stabilizer of a point in this orbit. Since we have a G-map $S \to G/G_0$, G/G_0 must have a finite G-invariant measure. If we choose $y \in Q$ with $[y]$ in the distinguished orbit, then it is straightforward to check that G_0 is the image of projection onto the first factor of the set $A = \{(g, \tilde{g}) \in G \times \tilde{G}' | y \cdot \tilde{g} = y \cdot g\}$. But as indicated above, everything can be embedded in an action on a variety and hence A is a Zariski-closed set. In particular, this implies that G_0 has only finitely many components. But this and the existence of a finite invariant measure on G/G_0 implies $G = G_0$. Hence, aside from a null set, $\tilde{\Phi}(S)$ is a single point in $Q/\tilde{G}'$, or in other words, $\Phi(S)$ lies in a single $\tilde{G}'$-orbit in Q.

Let Φ_0 be a point in this orbit. By choosing a Borel section on the orbit to $\tilde{G}'$, we have a Borel map $\theta:$ orbit $(\Phi_0) \to \tilde{G}'$ such that $\Phi \cdot \theta(\Phi) = \Phi_0$ for all $\Phi \in \text{Orbit}(\Phi_0)$. Let $\lambda: S \to \tilde{G}'$ be $\lambda(s) = \theta(\Phi(s))$, so we have $\Phi(s) \cdot \lambda(s) = \Phi_0$. Thus (**) becomes

(***) $$\Phi_0 \cdot \lambda(sg)\alpha(s, g)^{-1}\lambda(s)^{-1} = \Phi_0 \cdot g .$$

Let $\beta(s, g) = \lambda(s)\alpha(s, g)\lambda(sg)^{-1}$ so that β is a cocycle equivalent to α. We now claim that β is in fact the restriction to $S \times G$ of a homomorphism of of $G \to \tilde{G}'$. It follows from (***) that for each g and almost all s, $\Phi_0(G/P)$, and hence its Zariski closure, is invariant under $\beta(s, g)$. Since the Mackey range of β is Zariski dense in $\tilde{G}'$, such a set of $\beta(s, g)$ cannot lie in a proper algebraic subgroup of $\tilde{G}'$. It follows that $\Phi_0(G/P)$ must be Zariski dense in $\tilde{G}'/H'$. For each g and almost all $s, t \in S$, (***) implies $\Phi_0(x) \cdot \beta(t, g)^{-1}\beta(s, g) = \Phi_0(x)$ for all x, or in other words, $\beta(t, g)^{-1}\beta(s, g)$ leaves $\tilde{G}'/H'$ pointwise fixed. However, $\bigcap \{yH'y^{-1} | y \in \tilde{G}'\}$ is a normal amenable algebraic subgroup, which by the assumptions on $\tilde{G}'$ must be trivial. It follows that β is essentially independent of s, and hence corresponds to a homomorphism $\pi: G \to \tilde{G}'$ which of course has its image in G' by connectedness. The Zariski closure of the

524   ROBERT J. ZIMMER

Mackey range of β is equal to the Zariski closure of the image of π, and it follows that $\pi\colon G \to G'$ is surjective.

The only point that remains to be resolved is that we have shown that α, $\beta\colon S \times G \to G'$ are equivalent as cocycles into $\tilde{G}'$ (i.e., the map λ takes values in $\tilde{G}'$). However, the argument of [30, Lemma 3.4] or [31, Theorem 6.1] shows that there is an automorphism σ of G' (in fact, coming from an inner automorphism of $\tilde{G}'$) such that α and $\sigma \circ \beta$ are equivalent as cocycles into G'. Since $\sigma \circ \beta$ is a homomorphism $G \to G'$, the proof of Theorem 4.1 is complete.

We can now deduce the rigidity theorem.

THEOREM 4.3. *Let G and G' be connected semisimple Lie groups with no center and no compact factors, and suppose the real rank of G is at least 2. Let S and S' be (essentially) free ergodic irreducible G- and G'-spaces respectively, with finite invariant measure. If the actions are orbit equivalent, then G and G' are isomorphic, and the actions are automorphically conjugate.*

Proof. If the actions are orbit equivalent, let α be a cocycle corresponding to an orbit equivalence as in Section 2. By Proposition 2.4, it suffices to show that α is equivalent to the restriction of an isomorphism $G \to G'$. By Theorem 4.1, α is equivalent to the restriction of a surjection $\beta\colon G \to G'$. Let N be the kernel of β and suppose $N \neq \{e\}$. Then for $g \in N$, we have $\lambda(s)\alpha(s, g)\lambda(sg)^{-1} = e$ for almost all s, where $\lambda\colon S \to G'$. If α corresponds to the orbit equivalence $\theta\colon S \to S'$, we have $\theta(s) \cdot \alpha(s, g) = \theta(sg)$, so $\theta(s) \cdot \lambda(s)^{-1} = \theta(sg) \cdot \lambda(sg)^{-1}$. By irreducibility of the action, N is ergodic on S and so $\theta(s) \cdot \lambda(s)^{-1}$ is constant on a conull set. This means $\theta(S)$ is contained in a single G'-orbit which is clearly impossible. Thus, β is an isomorphism.

Theorem 4.3 extends to semisimple groups with center as follows.

THEOREM 4.4. *Let G, G' be connected semisimple Lie groups with finite center and no compact factors, and suppose the real rank of G is at least 2. Let S, S' be free irreducible ergodic G, G' spaces respectively, with finite invariant measure. Suppose the actions are orbit equivalent. Then G and G' are locally isomorphic.*

Proof. Since $Z(G)$ and $Z(G')$ are finite, $S/Z(G)$ and $S/Z(G')$ will be standard Borel spaces and the actions of $G/Z(G)$ and $G'/Z(G')$ will be free, ergodic, irreducible, and with finite invariant measure. We can clearly choose a transversal (or complete lacunary section in the sense of [9]) for the action of G on S that projects injectively to a transversal for the $G/Z(G)$ action on

$S/Z(G)$. From the results of [9] we have that the G and $G/Z(G)$ actions are orbit equivalent. As the same will be true for G', the $G/Z(G)$ and $G'/Z(G')$ actions will be orbit equivalent and hence G and G' are locally isomorphic by Theorem 4.3.

In a similar manner we can extend the results to actions of lattice subgroups.

THEOREM 4.5. *Let* Γ_1, Γ_2 *be lattices in* G_1, G_2, *the latter being connected simple non-compact Lie groups with finite center. Suppose G has real rank at least 2. Suppose that for $i = 1, 2$, S_i is a free ergodic Γ_i space with finite invariant measure. If the Γ_i actions are orbit equivalent, then G_1 and G_2 are locally isomorphic.*

Proof. If we let $X_i = \mathrm{ind}_{\Gamma_i}^{G_i}(S_i)$, then X_i is a free ergodic G_i-space. The G_i actions are orbit equivalent, and hence the result follows from Theorem 4.4.

COROLLARY 4.6. *Let* $\mathrm{SL}(n, Z)$ *act by automorphisms on the torus* T^n. *Then for n, $p \geq 2$, $n \neq p$, the $\mathrm{SL}(n, Z)$ and $\mathrm{SL}(p, Z)$ actions are not orbit equivalent.*

Proof. If $x = (x_1, \cdots, x_n) \in \mathbf{R}^n/Z^n$, and $\{x_1, \cdots, x_n, 1\}$ is a rationally independent set, then the stabilizer of x in $\mathrm{SL}(n, Z)$ is trivial, and hence the $\mathrm{SL}(n, Z)$ action is essentially free. The result now follows from Theorem 4.5.

We remark that in Theorem 4.5 it is not necessarily true that Γ_1 and Γ_2 are isomorphic. For example, if Γ_i are both lattices in G, then the Γ_1 action on G/Γ_2 is stably orbit equivalent to the Γ_2 action on G/Γ_1, in the sense of [8], [9], and as pointed out to the author by J. Feldman and C. C. Moore, these actions will be orbit equivalent as long as G/Γ_1 and G/Γ_2 have the same volume.

5. Geometric formulation

Just as the rigidity theorems of Mostow and Margulis can be formulated in terms of the fundamental group of a locally symmetric space of finite volume determining its Riemannian structure, Theorem 4.3 implies a corresponding statement about the Riemannian structure on the leaves of certain foliations. We shall formulate this in terms of measurable foliations [36] which we briefly describe as follows. Let (S, μ) be a standard Borel space with an analytic equivalence relation such that each leaf (i.e., equivalence class) has a C^∞-structure whose associated Borel structure is compatible with that of S. In addition, we assume a suitable local triviality

property [36] that will not concern us here. Two basic examples are measurable foliations coming from smooth foliations (or partial foliations) of a
manifold and the measurable foliation arising from a Lie group action on a
measure space, where each orbit inherits a C^∞-structure from the group.
One may in a natural way speak of a Riemannian measurable foliation as
one in which a smooth Riemannian structure is assigned to each leaf in such
a way that the assignment is measurable over the ambient measure space.
For example, if S is a free ergodic G-space, where G is a connected semisimple Lie group with finite center, and K is a maximal compact subgroup,
then S/K has a natural measurable foliation on it whose leaves are (a.e.)
isometric to the symmetric space G/K. By a transversal to the foliation we
mean a Borel subset T that intersects almost every leaf in a countable discrete (in the topology of the leaf) subset of the leaf. There is a natural
measure class on a transversal compatible with the measure on the ambient
space, and so a transversal T has the structure of a measure space with a
countable equivalence relation on it, the measure class being invariant under
the partial automorphisms of the equivalence relation [8]. We call two meas-
urable foliations $\mathcal{F}$, $\mathcal{F}'$ transversally equivalent if there are transversals T,
T' that are isomorphic in the sense of measure spaces with countable equi-
valence relations [8]. We then have the following geometric form of Theorem 4.3 which says that an a priori very weak purely measure-theoretic
invariant of th emeasurable foliation in fact determines the Riemannian
structure.

THEOREM 5.1. *Let $\mathcal{F}$, $\mathcal{F}'$ be the Riemannian measurable foliations by
symmetric spaces G/K, G'/K' arising from free ergodic irreducible actions
of G, G' with finite invariant measure. Assume the rank of G/K is at least
2. If $\mathcal{F}$ and $\mathcal{F}'$ are transversally equivalent, then up to normalizing scalar
multiples (independent of the leaf), they are isometrically isomorphic as
Riemannian measurable foliations.*

Proof. Using [9], one can show that transversal equivalence of $\mathcal{F}$ and
$\mathcal{F}'$ implies that the original G and G' actions define similar and hence isomorphic principal measure groupoids. But this just means that the actions
are orbit equivalent. One can now apply Theorem 4.3, and from this the
required properties of the associated foliations follow readily.

6. Concluding remarks

We conclude with an observation and a question. For discrete subgroups, the Mostow-Margulis strong rigidity results were preceded by the

local rigidity (i.e., rigidity under deformation) results of Selberg [24], Calabi and Vesentini [2], and Weil [29]. One can of course formulate the notion of a deformation of an ergodic action of a Lie group on a given measure space where one maintains the same associated measurable foliation (by orbits) under the deformation. Under suitable hypotheses, Theorem 4.3 implies that a deformation of an action yields actions which are automorphically conjugate, which is again very different from the situation of amenable groups.

The techniques of Margulis that we have employed require the hypothesis that the real rank of the semisimple group be at least 2. In light of results on rigidity of lattices in the rank one case [19], it is natural to ask whether Theorem 4.3 extends to this case as well, except as usual for PSL $(2, \mathbf{R})$. The same question of course applies to Theorem 5.1 and measurable foliations. Some progress in this direction appears in [39].

7. Appendix: Proof of a lemma of Margulis

In the proof of Lemma 4.2, we made use of the fact that the space $\hat{\Sigma}$ has a standard Borel structure. As indicated there, this was observed by Margulis in [16, p. 42]. However, as Margulis provides only an indication of the proof, and given its basic role in Lemma 4.2, we present here a simple proof.

LEMMA A. (Margulis). *Let G be a real algebraic group and let M be a variety on which G acts algebraically. Let (X, μ) be a standard measure space and let Σ be the space of Borel maps $X \to M$, identified if there is almost everywhere equality, and endow Σ with the topology of convergence in measure on sets of finite measure. Let G act on Σ via pointwise action on M. Then the orbit space Σ/G is a countably separated (and hence standard) Borel space.*

Proof. For $f \in \Sigma$ and $m \in M$, let G_f and G_m denote the stabilizer of f and m respectively in G. Suppose $g_n \in G$ and $f \cdot g_n \to f$ in Σ. To see that Σ/G is countably separated it suffices to show that $[g_n] \to [e]$ in G/G_f [6], (i.e., that $G/G_f \to$ orbit (f) is a homeomorphism). By passing to a subsequence, we can assume $(f \cdot g_n)(x) \to f(x)$ on a conull set. Let H_f be a countable dense subset of G_f. Then on a smaller conull set, say X_0, we have $H_f \subset \bigcap_{x \in X_0} G_{f(x)}$. This implies $G_f \subset \bigcap_{x \in X_0} G_{f(x)}$ and hence $G_f = \bigcap_{x \in X_0} G_{f(x)}$. Since all the groups in question are algebraic, there is a finite set $x_1, \cdots, x_n \in X_0$ such that $G_f = \bigcap_{i=1}^{n} G_{f(x_i)}$. Thus the diagonal action of G induces an injective map $G/G_f \to \prod_i G/G_{f(x_i)}$ onto the orbit of $([e], \cdots, [e])$. The G action on the variety

528   ROBERT J. ZIMMER

$\prod G/G_{f(x_i)}$ is algebraic, therefore has locally closed orbits, and hence the above map is a homeomorphism. Thus, it suffices to see that $[g_n] \to [e]$ in each $G/G_{f(x_i)}$, which is indeed the case since $f(x_i) \cdot g_n \to f(x_i)$ and the action on M is algebraic.

UNIVERSITY OF CHICAGO

REFERENCES

[1] A. BOREL, Density properties for certain subgroups of semisimple groups without compact factors, Ann. of Math. **72** (1960), 179–188.

[2] E. CALABI and E. VESENTINI, On compact locally symmetric Kähler manifolds, Ann. of Math. **71** (1960), 472–507.

[3] A. CONNES, J. FELDMAN, B. WEISS, to appear.

[4] A. CONNES and W. KRIEGER, Measure space automorphisms, the normalizers of their full groups, and approximate finiteness, J. Funct. Anal. **24** (1977), 336–352.

[5] H. A. DYE, On groups of measure preserving transformations, I, Amer. J. Math. **8** (1959), 119–159.

[6] E. G. EFFROS, Transformation groups and C^*-algebras, Ann. of Math. **81** (1965), 38–55.

[7] J. FELDMAN and D. LIND, Hyperfiniteness and the Halmos-Rohlin theorem for non-singular abelian actions, Proc. A. M. S. **55** (1976), 339.

[8] J. FELDMAN and C. C. MOORE, Ergodic equivalence relations, cohomology, and von Neumann algebras, I, Trans. A. M. S. **234** (1977), 289–324.

[9] J. FELDMAN, P. HAHN, and C. C. MOORE, Orbit structure and countable sections for actions of continuous groups, Adv. in Math. **28** (1978), 186–230.

[10] H. FURSTENBERG, A Poisson formula for semisimple Lie groups, Ann. of Math. **77** (1963), 335–383.

[11] S. GLASNER, *Proximal Flows*, Lecture Notes in Math., No. 517, Springer-Verlag, New York, 1976.

[12] W. KRIEGER, On ergodic flows and the isomorphism of factors, Math. Ann. **223** (1976), 19–70.

[13] G. W. MACKEY, Point realizations of transformation groups, Ill. J. Math. **6** (1962), 327–335.

[14] ———, Ergodic theory and virtual groups, Math. Ann. **166** (1966), 187–207.

[15] G. A. MARGULIS, Non-uniform lattices in semisimple algebraic groups, in *Lie Groups and Their Representations*, ed. I. M. Gelfand, Wiley, New York, 1975.

[16] ———, Discrete groups of motions of manifolds of non-positive curvature, A. M. S. Translations, **109** (1977), 33–45.

[17] C. C. MOORE, Ergodicity of flows on homogeneous spaces, Amer. J. Math. **88** (1966), 154–178.

[18] ———, Amenable subgroups of semisimple groups and proximal flows, preprint.

[19] G. D. MOSTOW, *Strong Rigidity of Locally Symmetric Spaces*, Annals of Math. Studies, No. 78, Princeton Univ. Press, Princeton, N. J., 1973.

[20] ———, Discrete subgroups of Lie groups, in *Lie Theories and their Applications*, Queen's Papers in Pure and Applied Math., No. 48, Queen's Univ., Kingston, Ont., 1978.

[21] D. ORNSTEIN and B. WEISS, to appear.

[22] A. RAMSAY, Virtual groups and group actions, Adv. in Math. **6** (1971), 253–322.

[23] K. SCHMIDT, Cocycles on ergodic transformation groups, Macmillan Lectures in Math., No. 1, Macmillan Co. of India, Delhi, 1977.

[24] A. SELBERG, On discontinuous groups in higher dimensional symmetric spaces, Intl. Colloquium on Function Theory, Tata Institute, Bombay, 1960.

STRONG RIGIDITY FOR ERGODIC ACTIONS   529

[25] C. SERIES, The Rohlin theorem and hyperfiniteness for actions of continuous groups, to appear.

[26] I. R. SHAFAREVICH, *Basic Algebraic Geometry*, Springer-Verlag, New York, 1977.

[27] V. S. VARADARAJAN, Geometry of Quantum Theory, Vol. II, van Nostrand, Princeton. N. J., 1970.

[28] G. WARNER, *Harmonic Analysis on Semisimple Lie Groups*, Vol. I, Springer-Verlag, New York, 1972.

[29] A. WEIL, On discrete subgroups of lie groups, I, II, Ann. of Math. **72** (1960), 369-384, and Ibid., **75** (1962), 578-602.

[30] R. J. ZIMMER, Extensions of ergodic group actions, Ill. J. Math. **20** (1976), 373-409.

[31] ———, Compactness conditions on cocycles of ergodic transformation groups, J. London Math. Soc. **15** (1977), 155-163.

[32] ———, Amenable ergodic group actions and an application to Poisson boundaries of random walks, J. Funct. Anal. **27** (1978), 350-372.

[33] ———, Hyperfinite factors and amenable ergodic actions, Invent. Math. **41** (1977), 23-31.

[34] ———, Orbit spaces of unitary representations, ergodic theory, and simple Lie groups, Ann. of Math. **106** (1977), 573-588.

[35] ———, Induced and amenable ergodic actions of Lie groups, Ann. Sci. Ec. Norm. Sup. **11** (1978), 407-428.

[36] ———, Algebraic topology of ergodic Lie group actions and measurable foliations, submitted.

[37] ———, An algebraic group associated to an ergodic diffeomorphism, preprint.

[38] ———, An analogue of the Mostow-Margulis rigidity theorems for ergodic actions of semisimple Lie groups, Bull. A. M. S. **2** (1980), 168-170.

[39] ———, On the Mostow rigidity theorem and measurable foliations by hyperbolic space, preprint.

(Received June 18, 1979)

Ergod. Th. & Dynam. Sys. (1981), **1**, 237–253
Printed in Great Britain

Orbit equivalence and rigidity of ergodic actions of Lie groups

ROBERT J. ZIMMER†

From the Department of Mathematics, University of Chicago, USA

(*Received* 2 *November* 1980)

Abstract. The rigidity theorem for ergodic actions of semi-simple groups and their lattice subgroups provides results concerning orbit equivalence of the actions of these groups with finite invariant measure. The main point of this paper is to extend the rigidity theorem on one hand to actions of general Lie groups with finite invariant measure, and on the other to actions of lattices on homogeneous spaces of the ambient connected group possibly without invariant measure. For example, this enables us to deduce non-orbit equivalence results for the actions of SL $(n, \mathbb{Z})$ on projective space, Euclidean space, and general flag and Grassman varieties.

1. *Introduction*

If G and G' are locally compact separable groups acting ergodically on measure spaces $(S, \mu), (S', \mu')$ respectively, the actions are called orbit equivalent if there exists (possibly after discarding null sets) a measure class preserving Borel bijection that takes G-orbits onto G'-orbits. For actions of amenable groups, orbit equivalence is a very weak notion as one now has the result that any two free ergodic actions of amenable groups with finite invariant measure are orbit equivalent as long as both groups are either discrete or continuous and unimodular [**16**], [**3**]. On the other hand, we recently showed in [**24**] that for free irreducible ergodic actions of semi-simple Lie groups with finite centre and real rank at least 2, orbit equivalence completely determines the group up to local isomorphism and, in the centrefree case, completely determines the action as well up to an automorphism of the group. As explained in [**24**], this result for semi-simple groups is a direct analogue in ergodic theory of the rigidity theorems for lattices in semi-simple Lie groups of G. D. Mostow [**15**] and G. A. Margulis [**10**]–[**12**]. In fact, a basic ingredient of the proof of the rigidity theorem for ergodic actions in [**24**] is the use of a technique of Margulis for showing the rationality of certain measurable maps between algebraic varieties.

The results of [**24**] can be applied to actions of lattice subgroups and, for example, we deduce in [**24**] that the actions of SL $(n, \mathbb{Z})$ on the torus $\mathbb{R}^n/\mathbb{Z}^n$ are mutually non-orbit equivalent as we vary n, for $n \geq 2$. On the other hand, the results of [**24**]

† Address for correspondence: Dr Robert J. Zimmer, Department of Mathematics, University of Chicago, Chicago, Ill. 60637, USA.

R. J. Zimmer

are strongly dependent upon the actions in question having finite invariant measure and thus do not apply to many natural and classical examples, for instance the action of SL $(n, \mathbb{Z})$ on projective space and more general Grassman and flag varieties. The main point of this paper is to extend the rigidity theorem of [24] to enable us to deduce non-orbit equivalence results for this situation, and more generally for the ergodic actions of lattices in semi-simple Lie groups defined by translation on homogeneous spaces of these Lie groups. As an example of our results, we have the following. (See § 4 for a more general formulation.)

THEOREM 4.2. (a) The actions of SL $(n, \mathbb{Z})$ on the projective spaces $\mathbb{P}^{n-1}$ are mutually non-orbit equivalent as n varies, $n \geq 2$.

(b) The actions of SL $(n, \mathbb{Z})$ on $\mathbb{R}^n$ are mutually non-orbit equivalent as n varies, $n \geq 2$.

(c) For a fixed $n \geq 4$, let $G_{n,k}$ be the Grassman variety of k-planes in $\mathbb{R}^n$. Then the actions of SL $(n, \mathbb{Z})$ on $G_{n,k}$ are mutually non-orbit equivalent as k varies, $1 \leq k \leq [n/2]$.

The extension of the rigidity theorem upon which these results depend is a result concerning orbit equivalence of ergodic actions of general (i.e. not necessarily semi-simple) Lie groups with finite invariant measure.

We recall that in any locally compact separable group H there is a unique maximal normal closed amenable subgroup N. In connected groups, for example, this subgroup can be explicitly described as follows. Let K be a compact normal subgroup such that H/K is a Lie group. Let R be the radical of H/K and Z the centre of $(H/K)/R$. Then

$$((H/K)/R)/Z \cong \prod_1^n G_i,$$

where G_i are connected centreless simple Lie groups. Letting

$$\prod_1^n G_i \to \prod_1^k G_i$$

be projection onto the product of the non-compact factors, we obtain a homomorphism $H \to \prod_1^k G_i$. The kernel of this homomorphism is the group N. Our first extension of the rigidity theorem to connected groups is the following.

THEOREM 3.1. Suppose H, H' are connected locally compact second countable groups, N, N' the maximal normal closed amenable subgroups, and that the real rank of every simple component of H/N is at least 2. Suppose S, S' are (essentially) free ergodic H, H'-spaces, respectively, with finite invariant measure and that the actions are orbit equivalent. Then H/N and H'/N' are isomorphic and N is compact if and only if N' is compact.

One can actually weaken the assumption that the real rank of every simple component of H/N is at least 2 to the assumption that the real rank of H/N itself be at least 2 if one makes in addition certain irreducibility assumptions on the ergodic action of H on S, which, for example, will always hold if the restriction of the *H*-action to *N* is still ergodic. (See theorem 3.3 below.) As in [24], the assumption on the ℝ-rank is necessary in order to employ Margulis' techniques for proving the rationality of appropriate measurable maps. For results in a similar direction in the ℝ-rank one case, the reader is referred to [26] where Mostow's results on quasi-conformal mappings [14] are applied. When both the groups and actions are direct products, we obtain the following sharper result.

THEOREM 3.4. *Let A, A' be amenable groups, $G_i, G'_j, i = 1, \ldots, n, j = 1, \ldots, p$, be connected semi-simple centreless Lie groups of ℝ-rank at least 2, and suppose that neither G_i and G_j nor G'_i and G'_j have common factors for $i \neq j$. Let $S_i(S_j, T, T')$ be a free ergodic $G_i(G'_j, A, A')$-space with finite invariant measure and suppose S_i, S'_j are irreducible. Let $X = T \times \prod S_i (Y = T' \times \prod S'_j)$ be the product ergodic $A \times \prod G_i (A' \times \prod G'_j)$-space. Suppose the actions are orbit equivalent and that either*

    (a) *A and A' are compact, or*

    (b) *the ℝ-rank of every simple ergodic component of each G_i is at least 2.*

Then

    (i) *$n = p$ and by reordering the indices we have $G_i \cong G'_i$.*

    (ii) *The G_i actions on S_i and S'_i are conjugate modulo an automorphism of G_i.*

    (iii) *Under assumption (b), A is compact if and only if A' is compact.*

As in [24], these theorems can be applied to yield non-orbit equivalence theorems for ergodic actions of lattices in connected groups. See, for example, § 4 below.

The proofs of these theorems are based on a cohomological result which is a generalization of the cohomological result that is at the basis of the strong rigidity theorem proved in [24]. The proof of the present cohomological result in turn closely follows the proof in [24], with additional arguments needed to deal with the present more general hypotheses. In order to deduce theorem 3.4, we also make use of a vanishing theorem for cohomology of ergodic actions of semi-simple Lie groups proved in [25] by different techniques (those of representation theory).

§ 2 below contains the proof of the cohomological result and § 3, the proofs of the theorems for connected groups stated above. § 4 contains the applications to (not necessarily measure preserving) actions of lattices.

2. *The main cohomological result*

§ 2 of [24] provides a brief summary of some of the notions from ergodic theory that we shall need. Here we shall only recall the notion of a cocycle and its connection to orbit equivalence, referring the reader to [24] and the references there for further background.

Suppose (S, μ) is an ergodic G-space where $\mu(S) = 1$ and G is locally compact and separable. If M is a standard Borel group, a Borel function $\alpha : S \times G \to M$ is called a cocycle if, for all $g, h \in G$,

$$\alpha(s, gh) = \alpha(s, g)\alpha(sg, h) \quad \text{a.e.}$$

Two cocycles $\alpha, \beta : S \times G \to M$ are called equivalent if there is a Borel function $\phi : S \to M$ such that, for each g,

$$\phi(s)\alpha(s, g)\phi(sg)^{-1} = \beta(s, g) \quad \text{a.e.}$$

240   *R. J. Zimmer*

Cocycles arise naturally in a variety of questions, in particular in problems concerning orbit equivalence.

Suppose S, S' are free ergodic G, G' spaces respectively and that the actions are orbit equivalent. (By 'free' here, we mean essentially free, i.e. almost all stabilizers are trivial.) Then, possibly after discarding Borel null sets, there is a measure class preserving Borel isomorphism $\theta : S \to S'$ which takes orbits onto orbits. Define $\alpha : S \times G \to G'$ by

$$\theta(s) \cdot \alpha(s, g) = \theta(sg).$$

This is well defined by the freeness of the action and it is straightforward to check that α is a cocycle. If $\pi : G \to G'$ is a homomorphism, one also obtains a cocycle $\beta : S \times G \to G'$ simply by defining

$$\beta(s, g) = \pi(g).$$

Then β is called the restriction of π to $S \times G$. A basic connection between orbit equivalence and the cocycle α is the following result proved in [24, proposition 2.4]. (For more details of the technical aspects of the proof in [24] see lemma 3.5 below.) We recall that ergodic G-spaces S, S' are called conjugate if, possibly after discarding invariant Borel null sets, there is a measure-class preserving Borel bijection $S \to S'$ which is a G-map, and are called automorphically conjugate if the action on S is conjugate to the G-action on S' defined by $(s, g) \to s \cdot A(g)$, where A is an automorphism of G.

PROPOSITION 2.1 [24, proposition 2.4]. *Suppose S, S' are orbit equivalent G-spaces and $\alpha : S \times G \to G$ is the cocycle corresponding to an orbit equivalence. If α is equivalent to the restriction of an automorphism of G, then the actions are automorphically conjugate. If the automorphism is inner, the actions are conjugate.*

If $\alpha : S \times G \to H$, where H is a locally compact group, there is a naturally associated ergodic H-space X called the Mackey range of α [9], [24]. If α is a cocycle coming from an orbit equivalence, the Mackey range is just the orbit equivalent H-space. If H is a real algebraic group, any ergodic action has an 'algebraic hull' which is a conjugacy class of algebraic subgroups [23], [24]. This conjugacy class of algebraic subgroups is characterized by the property that the ergodic action in question is induced from an action of an algebraic subgroup if and only if the subgroup is contained in a member of the conjugacy class. An action is called Zariski-dense if the algebraic hull is H itself. If H is a Zariski-connected real algebraic group with no compact factors, then the Borel density theorem [1] implies that any ergodic action of H with finite invariant measure is Zariski-dense [23, theorem 1.4]. The main cohomological result we shall need is the following.

THEOREM 2.2. *Let H be a locally compact group, N the maximal normal amenable subgroup of H, $G = H/N$, and suppose that G is a connected semi-simple Lie group. (For example, take H connected.) Suppose further than each of the simple components of G has $\mathbb{R}$-rank at least 2. Let G' be a connected semi-simple centrefree Lie group with no compact factors and suppose $G' \subset \tilde{G}'$, where $\tilde{G}'$ is a Zariski-connected real* *algebraic group with trivial centre and G' of finite index in $\tilde{G}'$. Let $\alpha : S \times H \to G'$ be a cocycle with Zariski-dense range. Then there is a surjective homomorphism $\beta : G \to G'$ such that α is equivalent to the restriction of $\beta \circ p$ to $S \times H$, where $p : H \to G$ is the natural projection.*

This theorem is, of course, a generalization of [**24**, theorem 4.1]. There are two respects in which the hypotheses in [**24**] are weakened. First, we are considering cocycles defined on an H-space, where H is considerably more general than a semi-simple group. Secondly, we are making no irreducibility assumptions on the ergodic action. Thus, for example, in the case in which $H = G = \prod G_i$, we do not assume each G_i acts ergodically. We do assume, however, that the $\mathbb{R}$-rank of each G_i is at least 2. In [**24**], with an irreducibility assumption, we needed to assume only that the $\mathbb{R}$-rank of G was at least 2. One can formulate the theorem so as to subsume both situations but the statement is somewhat cumbersome and so we postpone presenting it until after the proof of theorem 2.2. However, the modifications of the arguments of [**24**, theorem 4.1] which are needed to prove the result in the more general form are all present in the proof of theorem 2.2.

Proof. We shall, in general, follow the proof of [**24**, theorem 4.1], referring the reader there for those parts of the argument that carry through without essential modification. Let $G = \prod_1^k G_i$, where G_i are connected simple centreless Lie groups of $\mathbb{R}$-rank at least 2. Let P_i be a minimal parabolic subgroup of G_i, $P = \prod P_i$, and $\tilde{P}'$ a minimal parabolic subgroup of $\tilde{G}'$. Let $B = p^{-1}(P)$. Then B is a minimal boundary subgroup of H in the sense of Furstenberg [**7**] and is a maximal amenable (non-normal except in degenerate cases) subgroup of H. We have $N \subset B \subset H$ and $H/B = G/P$ as H-spaces. The action of N on G/P is, of course, trivial. □

LEMMA 2.3. *$S \times G/P$ is an ergodic H-space.*

Proof. $S \times G/P = S \times H/B$ and by [**21**, theorem 4.2] H is ergodic on this product if and only if B is ergodic on S. Let (E, ν) be the space of ergodic components of the action of N on S. Then there is an induced action of G on (E, ν) and this is clearly measure preserving and ergodic, the latter since H is ergodic on S. To see that B is ergodic on S it clearly suffices to show that $B/N = P$ is ergodic on E. However, this follows from Moore's ergodicity theorem [**13**]. □

We now remark that the H-space $S \times G/P = S \times H/B$ is an amenable ergodic H-space in the sense of [**20**]. This follows from the amenability of B as in [**24**]. Arguing exactly as in [**24**] we see that there is an amenable algebraic subgroup $J' \subset \tilde{G}'$ (this is denoted by H' in [**24**]) and a measurable map $\phi : S \times G/P \to \tilde{G}'/J'$ such that for each $h \in H$ and almost all $(s, x) \in S \times G/P$,

$$\phi(sh, xh) \cdot \alpha(s, h)^{-1} = \alpha(s, x).$$

We now show that [**24**, lemma 4.2] is still valid.

LEMMA 2.4. *For almost all s, $\phi_s : G/P \to \tilde{G}'/J'$ defined by $\phi_s(x) = \phi(s, x)$ is a rational mapping.*

Proof. Let A_i be a maximal $\mathbb{R}$-split connected Abelian subgroup of P_i. Fix i and suppose $t \in A_i$, $t \neq e$. Let $H_i = p^{-1}(G_i)$ and Y_i be the space of ergodic components of the H_i action on S. Thus, for almost all $y \in Y_i$, we have H_i is ergodic on (S_y, μ_y), where the latter is an ergodic component. Then for almost all $y \in Y_i$ we have, for all $h \in H_i$, that

$$\phi(sh, xh) \cdot \alpha(s, h)^{-1} = \phi(s, x)$$

for almost all $(s, x) \in S_y \times G/P$. We can then view ϕ as a map from $S_y \times G \to \tilde{G}'/J'$. Let $\hat{G}_i = \prod_{j \neq i} G_j$ and $\hat{P}_i = \prod_{j \neq i} P_j$; we shall then write $G = G_i \times \hat{G}_i$ and similarly for P. Since H_i acts trivially on $\hat{G}_i$, we have, for all $h \in H_i$, that for almost all $(s, g, \hat{g}) \in S_y \times G_i \times \hat{G}_i$,

$$\phi(sh, gh, \hat{g}) \cdot \alpha(s, h)^{-1} = \phi(s, g, \hat{g}).$$

Let C be the centralizer of $t \in G_i$. Define for $(s, g) \in S_y \times G_i$ and $\hat{g} \in \hat{G}_i$, $w_{(s,g),\hat{g}} : C \to \tilde{G}'/J'$ by

$$w_{(s,g),\hat{g}}(c) = \phi(s, cg, \hat{g}).$$

Define $T, \Sigma, \hat{\Sigma}, \tilde{w}$ as in [24]. Arguing exactly as in [24], we can view w as a map from $S_y \times G_i/T \times \hat{G}_i \to \Sigma$ and again as in [24] we have that the map $\tilde{w}$ is H_i-invariant. We now claim that H_i is ergodic on $S_y \times G_i/T$. The latter can also be expressed as $S_y \times H_i/p^{-1}(T)$ and so, by [21, theorem 4.2], it suffices to see that $p^{-1}(T)$ is ergodic on S_y. Let E be the space of ergodic components of N acting on S_y. As in lemma 2.3 there is an induced ergodic action of $G_i = H_i/N$ on E preserving a finite measure. To see that $p^{-1}(T)$ is ergodic on S_y, it suffices to see that T is ergodic on E, which follows from Moore's theorem [13]. With the ergodicity of H_i on $S_y \times G_i/T$ established, one can argue exactly as in [24], fixing $y \in Y_i$ and $\hat{g} \in G_i$. Namely, suppose that $G_i \subset \tilde{G}_i$ is a subgroup of finite index, where $\tilde{G}_i$ is a Zariski-connected real algebraic group with trivial centre, and that $P_i = G_i \cap \tilde{P}_i$, where $\tilde{P}_i$ is a minimal parabolic in $\tilde{G}_i$. Let U_i be the intersection with G_i of the unipotent radical of a parabolic in $\tilde{G}_i$ opposite to $\tilde{P}_i$. Then the arguments of [24, lemma 4.2] and [11, p. 43] show that for almost all $y \in Y_i$, $s \in S_y$, $g \in G_i$, and $\hat{g} \in \hat{G}_i$, $\phi(s, ug, \hat{g})$ depends rationally on $u \in U_i$. By the definition of Y_i, this implies that $\phi(s, ug, \hat{g})$ depends rationally on $u \in U_i$ for almost all $(s, g, \hat{g}) \in S \times G_i \times \hat{G}_i$. Letting i vary, we have that for almost all $s \in S$, $g_1, \ldots, g_k \in G$,

$$\phi(s, u_1 g_1, \ldots, u_k g_k) = \phi(s, (u_1, \ldots, u_k) \cdot (g_1, \ldots, g_k))$$

depends rationally on each u_i for almost all u_j, for any $j \neq i$. Applying Margulis' result that a measurable function on $\mathbb{R}^n \times \mathbb{R}^p$ which is rational in x for almost all y and rational in y for almost all x must be rational [11, p. 43], [12], we deduce that for almost all $s \in S$, $g \in G$, $\phi(s, ug)$ is rational in u for $u \in \prod U_i$. Arguing as in [11, p. 43], an application of [2, 4.10] completes the proof of the lemma. $\square$

The remainder of the proof of [24, theorem 4.1] now carries over with only minor modifications, and we make only one further observation. Define Φ_0 and β as in [24]. Then one obtains a homomorphism $\beta : H \to \tilde{G}'$ such that

$$\Phi_0 \cdot \beta(h) = \Phi_0 \cdot p(h).$$

In particular, for $h \in N$, $\beta(h)$ leaves $\Phi_0(G/P)$ pointwise fixed. Arguing as in [24] (where it is shown that β as first defined is independent of s), we obtain that β factors to a homomorphism $G \to \tilde{G}'$. This observation and the arguments of [24] then complete the proof of theorem 2.2. □

We recall that a semi-simple Lie group acts irreducibly on a probability space if the restriction to each non-central normal subgroup is ergodic. Basically the same proof as above, using irreducibility as in [24, theorem 4.1], shows the following variant of theorem 2.2. In certain situations it allows one to pass from the assumption that every simple component of H/N has $\mathbb{R}$-rank at least 2 to the assumption that H/N has $\mathbb{R}$-rank at least 2.

THEOREM 2.5. *Let H be a locally compact separable group, N the maximal normal amenable subgroup and suppose $H/N = G$ is a product $\prod G_i$ of semi-simple Lie groups with $\mathbb{R}$-rank of G_i at least 2. Suppose further that each G_i acts irreducibly on almost all ergodic components of its action on E, where the latter is the space of ergodic components of the N-action. Let G', α be as in theorem 2.2. Then the conclusion of theorem 2.2 is true.*

The hypotheses of theorem 2.5 hold in particular if N is already ergodic on S. They are also satisfied if $H = N \times G$, G acts irreducibly on S, and the $\mathbb{R}$-rank of G is at least 2. The latter includes the situation dealt with in [24].

3. Rigidity theorems

In this section, we prove the theorems for connected groups stated in the introduction.

THEOREM 3.1. *Suppose H, H' are locally compact separable groups, N, N' the maximal normal amenable subgroups, and suppose $G = H/N$ and $G' = H'/N'$ are connected (or equivalently, that G and G' are centrefree semi-simple connected Lie groups). Suppose further that the $\mathbb{R}$-rank of every simple component of G is at least 2. Let S, S' be free ergodic H, H'-spaces, respectively, with finite invariant measure, and suppose that the actions are orbit equivalent. Then G and G' are isomorphic and N is compact if and only if N' is compact.*

We prepare the proof by recalling some results we shall need. If E is a subset of a space X on which a group acts, its saturation $[E]$ is the union of all orbits intersecting E. If $E \subset A \subset X$, we shall call E saturated in A if $[E] \cap A = E$. In particular, E is saturated in X if and only if it is invariant. A Borel subset of X is called a countable section if it intersects every orbit in at most countably many points, and the countable section is called complete if its saturation is conull [6]. Every action of a locally compact separable group on a standard measure space has a complete countable section [6, theorem 2.8]. Furthermore, there exists a measure class on the countable section whose null sets are precisely the negligible sets (contained in the section) [17, theorem 6.17], [6, prop. 3.6, 3.7]. (We recall that a set is negligible if its saturation is null.) This measure class on the section has the further property that it is invariant under the partial automorphisms of the countable equivalence relation defined on it [5]. It is also not difficult to see that one can choose the section to lie inside any predetermined conull set. We also recall that if a group G acts on a standard measure space (S, μ), by an inessential reduction of the action we mean a subset

$$S_0 * G = \{(s, g) \in S \times G \,|\, s, sg \in S_0\},$$

where S_0 is conull. A basic technical fact proven by A. Ramsay [17, lemma 5.2] is the following.

LEMMA 3.2. *If* $U \subset S \times G$ *is conull, Borel, and satisfies* $(s, g), (sg, h) \in U$ *implies* $(s, gh) \in U$, *then* U *contains an inessential reduction.*

We now turn to the proof of theorem 3.1.

Proof of theorem 3.1. Let $\theta : S \to S'$ be an orbit equivalence and $\alpha : S \times H \to H'$ the corresponding cocycle. Let $p' : H' \to G'$ be the natural projection. The Mackey range of α is just the H' action on S' and it is not difficult to verify that the Mackey range of $p' \circ \alpha$ is then just the G' action induced on the space of ergodic components of the N' action on S'. But this G' action has finite invariant measure and hence the Mackey range of $p' \circ \alpha$ is Zariski-dense in G' [23, theorem 1.4]. It follows from theorem 2.2 that $p' \circ \alpha$ is equivalent to the restriction of a surjective homomorphism $\beta : H \to G'$ which is trivial on N. With the hypothesis of theorem 3.1, one can now quickly deduce that G and G' are isomorphic by observing that the existence of a surjective homomorphism $G \to G'$ implies that the $\mathbb{R}$-rank of every simple component of G' must be at least 2 and so one can apply theorem 2.2 to the cocycle coming from θ^{-1} as well. However, we shall present an alternative proof which will simultaneously yield the conclusion about N and N' and generalize immediately to provide a proof of theorem 3.3 below.

Let K be the kernel of β, so $N \subset K \subset H$. Let $\lambda : S \to G'$ be a Borel function such that for each $h \in H$,

$$\lambda(s)p'(\alpha(s, h))\lambda(sh)^{-1} = \beta(h)$$

for almost all $s \in S$. Composing λ with a Borel section of p' we obtain a Borel function $\tilde{\lambda} : S \to H'$ such that, for all $h \in H$,

$$p'(\tilde{\lambda}(s)\alpha(s, h)\tilde{\lambda}(sh)^{-1}) = \beta(h) \quad \text{a.e.}$$

Define $f : S \to S'$ by

$$f(s) = \theta(s) \cdot \tilde{\lambda}(s)^{-1}.$$

By lemma 3.2 and the definition of orbit equivalence we can find an inessential reduction $S_0 * H$ of the H-action on S such that:

(i) For all $(s, h) \in S_0 * H$, $\lambda(s)p'(\alpha(s, h))\lambda(sh)^{-1} = \beta(h)$.

(ii) α is strict on $S_0 * H$, i.e. the cocycle identity holds for all $(s, g), (sg, h) \in S_0 * H$.

(iii) $\theta(sh) = \theta(s) \cdot \alpha(s, h)$.

Now restrict the H-action to K and choose a complete countable section T for the K-action with $T \subset S_0$. We now claim that for $s, t \in S_0$, s and t are in the same K-orbit if and only if $f(s), f(t)$ are in the same N'-orbit. If $s = t \cdot h$ for $h \in K$, then

$$f(s) = \theta(th) \cdot \tilde{\lambda}(th)^{-1} = \theta(t)\alpha(t, h)\alpha(t, h)^{-1}\tilde{\lambda}(t)^{-1}n = f(t) \cdot n,$$

for some $n \in N'$. Conversely, suppose $f(s) = f(t) \cdot n$ for some $n \in N'$. Then

$$\theta(s)\tilde{\lambda}(s)^{-1} = \theta(t)\tilde{\lambda}(t)^{-1}n$$

and hence $\theta(s)$ and $\theta(t)$ are in the same H' orbit. Since θ is an orbit equivalence, this implies s, t are in the same H orbit, so we have $t = s \cdot h$ for some $h \in H$. Thus

$$\theta(s)\tilde{\lambda}(s)^{-1} = \theta(sh)\tilde{\lambda}(sh)^{-1}n$$

and, by the definition of α, we have

$$\alpha(s, h) = \tilde{\lambda}(s)^{-1}n^{-1}\tilde{\lambda}(sh).$$

Thus $\beta(h) = e$, so $h \in K$.

Let $T' = f(T)$. Let ν be a probability measure on T whose null sets are the K-negligible sets and let $\nu' = f_*(\nu)$, a measure on T'. By discarding a ν'-null set of T', we can choose a Borel section $\sigma: T' \to T$ of f. Let $T_0 = \sigma(T')$ and $\nu_0 = \sigma_*(\nu')$. Then one easily checks that T_0 is also a complete countable section and that ν_0-null sets are the K-negligible sets contained in T_0. In other words, replacing (T, ν) by (T_0, ν_0), we may assume $f: (T, \nu) \to (T', \nu')$ and its inverse are measure preserving Borel isomorphisms which preserve the countable equivalence relations defined by the K and N' actions respectively. Now let dn' be a probability measure on N in the same measure class as Haar measure, and define

$$m = \int (\nu' \cdot n') \, dn'.$$

Then m is a measure quasi-invariant under N' and the ν'-null sets coincide with the m-negligible subsets of T' under the N'-action. Since the K-action on (S, μ) and the N'-action on (S', m) have isomorphic (in the sense of countable ergodic equivalence relations [5]) complete countable sections, it follows from the results of [6] (e.g. propositions 3.6, 3.7, corollary 5.8) that the K-action on (S, μ) and the N' action on (S', m) are orbit equivalent. The notion of an amenable ergodic action [20] can be extended without difficulty to the non-ergodic case and, since N' is amenable, the action on (S', m) will be amenable [20, theorem 2.1]. Since the K-action on (S, μ) is free and orbit equivalent to the N'-action, the K-action on (S, μ) is amenable. But this action has finite invariant measure and hence K is amenable. By the definition of N, this implies $K = N$, and hence β is injective on G. This proves the first assertion of the theorem.

To prove the second assertion of the theorem, suppose N' is compact. Then the action of N' on S' is smooth, i.e. S'/N' is a standard Borel space, and we have an induced action of G' on S'/N' which is clearly ergodic. Furthermore, by the results of [6] (theorem 2.8 and those quoted in the previous paragraph), the G' action on S'/N' and the H' action on S' are orbit equivalent. In other words, we may assume $N' = \{e\}$. However, as shown above, the N action on S is similar in the sense of [6] to the N' action on S' (since they have isomorphic complete countable sections). This implies that the N action on S is also type I, i.e. almost every ergodic component of the N action on S is a transitive N-space. But since the action on S has finite invariant measure, and since the action is free, this implies that Haar measure on N is finite. Hence, N is also compact. This completes the proof of the theorem. $\square$

Using theorem 2.5 in place of theorem 2.2, one obtains the following generalization of theorem 3.1. The proof is basically the same.

THEOREM 3.3. *Let H, H' be locally compact separable groups, N, N' the maximal normal amenable subgroups, $G = H/N$, $G' = H'/N'$, and suppose G, G' connected. Let S, S' be free ergodic H, H'-spaces respectively, with finite invariant measure, and suppose the actions are orbit equivalent. Let E be the space of ergodic components of the N-action on S. Suppose further that $G = \prod G_i$ where each G_i is a semi-simple Lie group of $\mathbb{R}$-rank at least 2; and that G_i acts irreducibly on almost all of its ergodic components on E. Then G and G' are isomorphic and N is compact if and only if N' is compact.*

When the groups are product groups and the actions are product actions, we deduce a sharper result.

THEOREM 3.4. *Let A, A' be amenable groups, G_i, G_j', $i = 1, \ldots, n, j = 1, \ldots, p$, be connected semi-simple centrefree Lie groups of $\mathbb{R}$-rank at least 2 and suppose that neither G_i and G_j nor G_i' and G_j' have common factors for $i \neq j$. Let $S_i (S_j', T, T')$ be a free ergodic $G_i(G_j', A, A')$-space with finite invariant measure and suppose S_i, S_j' are irreducible. Let $X = T \times \prod S_i$ $(Y = T' \times \prod S_i')$ be the product ergodic $A \times \prod G_i (A' \times \prod G_j')$-space. Suppose the actions are orbit equivalent and that either*

    (a) *A and A' are compact, or*

    (b) *the $\mathbb{R}$-rank of every simple ergodic component of each G_i is at least 2.*

Then

    (i) *$n = p$ and by reordering the indices we have $G_i = G_i'$.*

    (ii) *The G_i actions on S_i and S_i' are conjugate modulo and automorphism of G_i.*

    (iii) *Under assumption (b), A is compact if and only if A' is compact.*

Proof. We first remark that, since any amenable ergodic action is hyperfinite [**16**], [**3**], we can assume that A is either the real line, the integers, or the identity (the latter in case A is compact, using [**6**] as above). Let $G = \prod G_i$, $G' = \prod G_j'$, $\theta: X \to Y$ an orbit equivalence and $\alpha: X \times A \times G \to A' \times G'$ the corresponding cocycle. Write $\alpha = (\alpha_1, \alpha_2)$, where α_i is the composition of α with the projection on A' and G' respectively. By theorems 2.5 and 3.3, there is a function $\lambda: X \to G'$ such that, for each $(a, g) \in A \times G$ and almost all $x \in X$,

$$\lambda(x)\alpha_2(x, a, g)\lambda(x \cdot (a, g))^{-1} = \beta(g),$$

where $\beta: G \to G'$ is an isomorphism. If A and A' are not the identity, then by hypothesis (b) and [**25**, theorem A], the cocycle $\alpha_1|X \times G$ is trivial. In other words, we can find $\phi: X \to A'$ such that for all $g \in G$,

$$\phi(x)\alpha_1(x, e, g)\phi(x \cdot (e, g))^{-1} = e \quad \text{for almost all } x.$$

From this it follows that

$$\phi(x \cdot (a, g))^{-1} = \phi((x \cdot (a, e)) \cdot (e, g))^{-1} = \alpha_1(x \cdot (a, e), e, g)^{-1}\phi(x \cdot (a, e))^{-1}.$$

Define $f: X \to Y$ by

$$f(x) = \theta(x) \cdot (\phi(x), \lambda(x))^{-1}.$$

Then, for almost all $x \in X$ and $(a, g) \in H$, we have

$$f(x \cdot (a, g)) = \theta(x \cdot (a, g))(\phi(x \cdot (a, g))^{-1}, \lambda(x \cdot (a, g))^{-1})$$
$$= \theta(x) \cdot \alpha(x, a, g)(\alpha_1(x \cdot (a, e), e, g)^{-1}\phi(x \cdot (a, e))^{-1}, \alpha_2(x, a, g)^{-1}\lambda(x)^{-1}\beta(g))$$
$$= \theta(x) \cdot (\alpha_1(x, a, e)\phi(x \cdot (a, e))^{-1}, \lambda(x)^{-1}\beta(g)).$$

Thus,

$$f(x \cdot (a, g)) = f(x) \cdot (\gamma(x, a), \beta(g)),$$

where $\gamma(x, a)$ is the cocycle

$$\phi(x)\alpha_1(x, a, e)\phi(x \cdot (a, e))^{-1}.$$

Let $f = (f_1, f_2)$, where f_1 and f_2 are the compositions of f with the projection of Y onto T' and $S' = \prod S'_i$ respectively. Writing $x \in X = T \times S$ as $x = (t, s)$ we have

$$f_2((t, s) \cdot (a, g)) = f_2(t, s) \cdot \beta(g),$$

and, setting $g = e$, we obtain

$$f_2(ta, s) = f_2(t, s).$$

For each $a \in A$, this equality holds for almost all (t, s). Thus, for all a in a countable dense subgroup of A, and for almost all s, this equality holds for almost all t. But a Borel function which is essentially invariant under a countable dense subgroup is actually essentially invariant under the entire group. (To see this one can, for example, argue as follows. Choose a Borel isomorphism of S' with $[0, 1]$ so that the function is then identified with an element of $L^\infty \subset L^2$. Then use the continuity of the naturally induced representation on L^2.) Thus for s in a conull set, for all $a \in A$, we have

$$f_2(ta, s) = f_2(t, s) \quad \text{for almost all } t.$$

By ergodicity of the A-action on T this implies that f_2 is essentially independent of t, and so we can write

$$f_2(t, s) = F_2(s) \quad \text{a.e.},$$

where $F_2: S \to S'$. Furthermore, we have

$$f_1((t, s)(e, g)) = f_1(t, s),$$

and so a similar argument shows that

$$f_1(t, s) = F_1(t) \quad \text{a.e.},$$

where $F_1: T \to T'$. We thus have

$$f(t, s) = (F_1(t), F_2(s)) \quad \text{a.e.}$$

We now show, via a technical argument, that F_2 is measure class preserving, essentially bijective, and defines a Boolean map $B(S') \to B(S)$ that behaves well with respect to the G and G' actions.

We first remark that, for all g,

$$F_2(sg) = F_2(s) \cdot \beta(g) \quad \text{a.e.}$$

By lemma 3.2 there is a conull Borel set $S_1 \subset S$ such that this equality holds for all $(s, g) \in S_1 * G$. By the definition of f (and of orbit equivalence), there is a conull set in $T \times S$ such that, if $f(t, s)$ and $f(t', s')$ are in the same $A' \times G'$ orbit for $(t, s), (t', s')$ in this conull set, then (t, s) and (t', s') are in the same $A \times G$ orbit. Since $f = (F_1, F_2)$ a.e., the same statement is true if we replace f by (F_1, F_2). By Fubini's theorem, there is a conull Borel set $S_2 \subset S$ such that, if $s, s' \in S_2$ with $F_2(s)$ and $F_2(s')$ in the same G'-orbit, then s, s' are in the same G orbit. Let $S_0 = S_1 \cap S_2$. Then, if $s, s' \in S_0$ with $F_2(s) = F_2(s')$, we have $s' = sg$ for some $g \in G$ and so

$$F_2(s') = F_2(s) \cdot \beta(g).$$

Since the G' action is free, $\beta(g) = e$, and since β is an isomorphism it follows that $s = s'$. Replacing S_0 by a conull Borel subset, we can assume that S_0 is invariant under a countable dense subgroup $G_0 \subset G$.

Recall that the definition of orbit equivalence entails the existence of a conull Borel subset $(T \times S)_0 \subset T \times S$ such that $\theta : (T \times S)_0 \to \theta((T \times S)_0)$ is a measure class preserving Borel isomorphism onto its image which is conull in $T' \times S'$ and which takes orbits in $(T \times S)_0$ onto orbits in $\theta((T \times S)_0)$. In particular, viewing θ as a map $(T \times S)_0 \to T' \times S'$, for an invariant set $D \subset T' \times S'$ we have $\theta^{-1}(D)$ is null in $T \times S$ if and only if D is null in $T' \times S'$. By passing to a conull subset we can clearly also assume that $f = (F_1, F_2)$ on $(T \times S)_0$. If $B \subset T' \times S'$, then by the definition of f for a point $y \in (T \times S)_0$, we have $f(y) \in [B]$ if and only if $\theta(y) \in [B]$, i.e. (taking f to be also defined on $(T \times S)_0$) we have

$$f^{-1}([B]) = \theta^{-1}([B]).$$

Thus, if $B \subset T' \times S'$ is invariant, we have B is null if and only if $(F_1, F_2)^{-1}(B) \cap (T \times S)_0$ is null. Applying this to $B = T' \times A$, where $A \subset S'$ is invariant, we have that A is null if and only if $(T \times F_2^{-1}(A)) \cap (T \times S)_0$ is null, which is clearly true if and only if $F_2^{-1}(A)$ is null. Since A is invariant, $F_2^{-1}(A) \cap S_0$ is saturated in S_0 by the choice of S_0.

We shall now prove the following lemma, which we shall then apply to F_2. We also remark that this lemma spells out in more detail part of the argument of [24, proposition 2.4].

LEMMA 3.5. *Suppose (S_i, μ_i), $i = 1, 2$, are free ergodic G-spaces with finite invariant measure, where G is a locally compact separable group. Suppose there is a conull Borel set $S_0 \subset S_1$, and a Borel function $F : S_0 \to S_2$ satisfying the following conditions:*

  (i) *There is an automorphism β of G such that, for all $(s, g) \in S_0 * G$, $F(sg) = F(s) \cdot \beta(g)$.*

  (ii) *F is injective on S_0.*

  (iii) *If $s, t \in S_0$ with $F(s)$ and $F(t)$ in the same orbit, then s and t are in the same orbit.*

  (iv) *S_0 is invariant under a countable dense subgroup $G_0 \subset G$.*

  (v) *If $A \subset S_2$ is invariant, then A is null if and only if $F^{-1}(A) \cap S_0$ is negligible. Then $F(S_0)$ is conull in S_2 and $F : S_0 \to F(S_0)$ is a measure class preserving Borel isomorphism.*

Proof. The lemma will follow by an application of the local representation of an invariant measure as described by C. Series [**18**, prop. 1.1], [**19**, lemma 1.4]. Let E be a complete countable section of the G-action on S_1, with $E \subset S_0$, such that E is lacunary in the sense of [**6**]. (In the language of [**19**] E is a sufficient regular transversal.) This means there is a neighbourhood U of the identity in G such that the product map $E \times U \to E \cdot U \subset S$ is injective. Let ν be a measure on E whose null sets are the negligible (in S_1) subsets of E. Let λ be Haar measure on G. Then by Series' results, identifying $E \times U$ with $E \cdot U$, the restriction of the measure μ to $E \cdot U$ is in the same measure class as the product measure $\nu \times \lambda$ on $E \times U$. For $y \in E$, let

$$U_y = \{g \in U \mid yg \in S_0\}, \quad \text{and} \quad E_0 = \{y \mid \lambda(U - U_y) = 0\}.$$

By Fubini's theorem, E_0 is a conull Borel subset of E. Let

$$(E', \nu') = (F(E_0), F_*(\nu)).$$

By the hypothesis on F, $F(E_0)$ is a complete countable section for the G-action on S_2, and $F_*(\nu)$-null sets in $F(E_0)$ are the negligible subsets of S_2 contained in $F(E_0)$. We claim that the product map $F(E_0) \times \beta(U) \to S_2$ is injective. Suppose

$$F(y_1) \cdot \beta(g_1) = F(y_2) \cdot \beta(g_2),$$

where $y_i \in E_0$ and $g_i \in U$. The set

$$\beta(g_1)\beta(g_2)^{-1}\beta(U) \cap \beta(U)$$

is clearly open and non-empty (as it contains $\beta(g_1)$). For g in this set, let

$$h(g) = \beta(g_2)\beta(g_1)^{-1}g,$$

so that $h(g)$ is also in $\beta(U)$. We thus have for a set $g \in \beta(U)$ of positive measure,

$$F(y_1) \cdot g = F(y_2) \cdot h(g).$$

Since β is an automorphism of G, $\beta: U \to \beta(U)$ is measure class preserving. Thus we can find $a \in U_{y_1}$ such that $\beta^{-1}h\beta(a) \in U_{y_2}$ and such that $g = \beta(a)$ satisfies the above condition.

But since $y_1, y_2, y_1 \cdot a, y_2 \cdot \beta^{-1}h\beta(a) \in S_0$, we have

$$F(y_1 \cdot a) = F(y_1)\beta(a) = F(y_2)h(\beta(a)) = F(y_2 \cdot \beta^{-1}h\beta(a)).$$

As F is injective on S_0,

$$y_1 \cdot a = y_2 \cdot \beta^{-1}h\beta(a),$$

which implies $y_1 = y_2$ and $\beta(a) = h\beta(a)$. This implies that $\beta(g_1) = \beta(g_2)$, and injectivity of $F(E_0) \times \beta(U) \to S_2$ is thus established. Hence we can identify $F(E_0) \cdot \beta(U)$ with $F(E_0) \times \beta(U)$ and, by the results of Series mentioned above, the measure μ_2 on the former agrees up to measure class with $F_*(\nu) \times \lambda$. This shows that $F: (E_0 \cdot U) \cap S_0 \to F(E_0) \cdot \beta(U)$ is a Borel bijection onto a conull subset of $F(E_0) \cdot \beta(U)$ and is measure class preserving. For any element $g \in G_0$ we have a commutative diagram (figure 1).

The conclusion of lemma 3.5 then follows by an elementary argument. $\square$

Returning to the proof of theorem 3.4, it follows by an application of lemma 3.5 that F_2 defines a Boolean σ-algebra isomorphism $B(S', \mu') \to B(S, \mu)$ which is a

250   *R. J. Zimmer*

$$((E_0 \cdot g) \cdot g^{-1}Ug) \cap S_0 \overset{g^{-1}}{\to} (E_0 \cdot U) \cap S_0$$

$$\downarrow {\scriptstyle F} \qquad\qquad\qquad\qquad \downarrow {\scriptstyle F}$$

$$(F(E_0) \cdot \beta(g)) \cdot \beta(g^{-1}Ug) \overset{\beta(g)}{\leftarrow} F(E_0) \cdot \beta(U)$$

FIGURE 1

Boolean G-map when G acts on S' via $s' \cdot g = s'\beta(g)$. Thus by [8] the actions of G on S and S' are conjugate. The isomorphism β can be written as a product $\beta_1 \times \cdots \times \beta_n$, where $\beta_i : G_i \to G'_1$ is an isomorphism, possibly after reordering G_i. To complete the proof it clearly suffices to prove the following lemma.

LEMMA 3.6. *Suppose G_i are locally compact groups, S_i, S'_i ergodic G_i-spaces, $G = \prod G_i$, $S = \prod S_i$, $S' = \prod S'_i$. Suppose the G actions on S and S' are conjugate modulo an automorphism β of G of the form $\prod \beta_i$, where β_i is an automorphism of G_i. Then the G_i actions on S_i and S'_i are conjugate modulo β_i.*

Proof. Suppose $f : S \to S'$ is such that

$$f(sg) = f(s)\beta(g) \quad \text{a.e.}$$

Let f_i be the coordinate functions of f and suppose $g_j \in G_j$, $j \neq i$. Then

$$f_i(s_1 g_1, \ldots, s_i, \ldots, s_n g_n) = f_i(s_1, \ldots, s_n).$$

By ergodicity, f_i is essentially independent of s_j, $j \neq i$, and so f_i factors to a measure preserving map $S_i \to S'_i$ such that

$$f_i(s_i g_i) = f(s_i)\beta(g_i) \quad \text{a.e.}$$

Since $f = \prod f_i$ and f is essentially injective so is each f_i, and this completes the proof. $\qquad\square$

4. *Applications to actions of lattices*

Suppose H_1, H_2 are closed subgroups of a locally compact group G. Then the orbit space of the action of H_1 on G/H_2 and that of H_2 on G/H_1 can both be naturally identified with the space of double cosets. It is, of course, possible that G/H_2 has a finite invariant measure while G/H_1 does not. We may thus hope to use results about orbit equivalence of actions with finite invariant measure to obtain results about certain actions which have no invariant measure. (Similar ideas have been used many times before. See, for example, [13], [21].)

THEOREM 4.1. *Let G, G' be connected semi-simple Lie groups with finite centre, and $H \subset G$, $H' \subset G'$ almost connected non-compact subgroups. Let $\Gamma \subset G$, $\Gamma' \subset G'$ be irreducible lattices and suppose the Γ-action on G/H and the Γ'-action on G'/H' are essentially free and orbit equivalent. Let N, N' be the maximal normal amenable subgroups of H, H' respectively, and suppose H/N has $\mathbb{R}$-rank at least 2. Then H/N and H'/N' are locally isomorphic.*

Proof. The orbit space of Γ on G/H is the same as that of H on G/Γ and, since Γ is essentially free on G/H, H will be essentially free on G/Γ. By the results of

[6], we have an orbit equivalence $\theta: G/\Gamma \to G'/\Gamma'$ of the H and H'-actions on these spaces. Let $H_0 \subset H, H_0' \subset H'$ be the connected components of the identity. Let $\alpha_1: G/\Gamma \times H \to H/H_0$ be the cocycle defined by restricting projection of H onto H/H_0 and let $\alpha_2: G/\Gamma \times H \to H'/H_0'$ be the cocycle defined by

$$\alpha_2(x, h) = p'\beta(x, h),$$

where $\beta: G/\Gamma \times H \to H'$ is the cocycle defined by the orbit equivalence θ and $p': H' \to H'/H_0'$ is projection. We thus have a cocycle (α_1, α_2) taking values in $H/H_0 \times H'/H_0'$ and we let $F \subset H/H_0 \times H'/H_0'$ be the Mackey range of this cocycle. (Here, as usual, we abuse notation and identify a transitive Mackey range with a stabilizer of this transitive action.) Let λ be a cocycle equivalent to (α_1, α_2) taking values in F. Since H_0, H_0' are ergodic on $G/\Gamma, G'/\Gamma'$ respectively (by Moore's theorem [13]), α_1 and α_2 have Mackey range $H/H_0, H'/H_0'$ respectively, and so we have an H-map of extensions of $G/\Gamma, G/\Gamma \times_\lambda F \to G/\Gamma \times H/H_0$. The H-space $G/\Gamma \times H/H_0$ clearly has the same orbit space as the H_0 action on G/Γ, and it is easy to see that the H-space $G/\Gamma \times_\lambda F$ has the same orbit space as the H_0-action on a finite ergodic extension Y of G/Γ. On the other hand, via θ, we can identify $(\alpha_1, \alpha_2), \lambda$ as cocycles on the H'-space G'/Γ' and $G/\Gamma \times_\lambda F$ is thus easily seen to be orbit equivalent to an H'-space which factors to $G'/\Gamma' \times H'/H_0'$.

In conclusion, we have orbit equivalent H_0, H_0'-spaces which are essentially free and with finite invariant measure. Let N_0 be the maximal normal amenable subgroup of H_0 and suppose $L \subset H_0$ is a normal subgroup with $N_0 \subset L$ and $N_0 \neq L$. Then, by Moore's theorem [13], L is ergodic on G/Γ and it is easy to deduce from this that the space of ergodic components of L on Y is finite. But since $L \subset H_0$ is normal, H_0 acts on this finite set, and the H_0 action is transitive, since H_0 is ergodic on Y. From the connectedness of H_0, the set is a point so L is ergodic on Y. Thus, the H_0 action satisfies the irreducibility condition in theorem 3.3, and an application of theorem 3.3 then completes the proof. □

Remarks. (i) The proof and the results of [6] show that the theorem is true if we only assume that the Γ action on G/H and the Γ' action on G'/H' are stably orbit equivalent in the sense of [6].

(ii) If G and H are algebraic, $H \supset Z(G)$ (the centre of G) and H does not contain a normal non-central subgroup, then $\Gamma/\Gamma \cap Z(G)$ acts essentially freely on G/H. To see this, observe that for $\gamma \in \Gamma, \{x \in G/H | x\gamma = x\}$ is a closed variety, and hence if it has positive measure γ fixes all of G/H which implies $\gamma \in Z(G)$.

As an example of an application of theorem 4.1, we have the following.

THEOREM 4.2. *(a) The actions of* SL $(n, \mathbb{Z})$ *on the projective spaces* $\mathbb{P}^{n-1}$ *are mutually non-orbit equivalent as n varies, $n \geq 2$.*

(b) The actions of SL $(n, \mathbb{Z})$ *on* $\mathbb{R}^n$ *are mutually non-orbit equivalent as n varies, $n \geq 2$.*

(c) For a fixed $n \geq 4$, let $G_{n,k}$ be the Grassman variety of k-planes in $\mathbb{R}^n$. Then the actions of SL $(n, \mathbb{Z})$ *on $G_{n,k}$ are mutually non-orbit equivalent as k varies, $1 \leq k \leq [n/2]$.*

252   R. J. Zimmer

Proof. If we compare the actions of SL $(n, \mathbb{Z})$ on $\mathbb{P}^{n-1}$ and SL $(p, \mathbb{Z})$ on $\mathbb{P}^{p-1}$ for $n \neq p, n, p \geq 3$, then we can apply theorem 4.1 (and remark (ii) above), since the semi-simple part of the stabilizer of a point in $\mathbb{P}^{n-1}$ has $\mathbb{R}$-rank $n-2$. However, for $n = 2$, the stabilizer is amenable and hence the SL $(2, \mathbb{Z})$ action on $\mathbb{P}^1$ will also be amenable. (This follows from the fact that if P is amenable, G/P is an amenable G-space [**20**] and hence an amenable Γ-space [**22**].) Since the SL $(n, \mathbb{Z})$ action on $\mathbb{P}^{n-1}$ is not amenable for $n \geq 3$ [**22**, proposition 3.4], the proof of (a) is complete. The proofs of (b) and (c) are similar. $\quad\square$

We remark that theorem 4.2 remains true if we replace SL $(n, \mathbb{Z})$ by any lattice in SL $(n, \mathbb{R})$. Furthermore, the same argument yields similar results for actions on more general real and complex flag manifolds. We leave the details to the reader.

As in [**24**], theorems 3.1 and 3.3 can be extended to ergodic actions of lattices in connected groups. Thus, if Γ_i are lattices in groups H_i as in theorem 3.1 and the Γ_i have free ergodic orbit equivalent actions with finite invariant measure, then the H_i/N_i are isomorphic. The proof follows by applying theorem 3.1 to the actions of H_i induced from Γ_i [**22**]. For example, we have the following.

THEOREM 4.3. *Let Γ_n be the semi-direct product of* SL $(n, \mathbb{Z})$ *with $\mathbb{Z}^n$, $n \geq 2$, where the former acts on the latter by matrix multiplication. Then for $i \neq j$, Γ_i and Γ_j do not have orbit equivalent finite measure preserving free ergodic actions.*

In [**25**] it is shown that for $n \geq 3$, Γ_n and SL $(n, \mathbb{Z}) \times \mathbb{Z}^n$ do not have orbit equivalent finite measure preserving free weakly mixing actions.

5. *Concluding remarks*

(a) It would be of interest to extend theorem 3.4 to arbitrary (i.e. not product actions) finite measure preserving free ergodic actions of product groups, or perhaps semi-direct product groups. For example, if H_i is a semi-direct product of G_i and normal A_i, $i = 1, 2$, where A_i is amenable and G_i is a suitable semi-simple group, when can one deduce that orbit equivalence of the H_i actions implies equivalence of associated G_i-actions (e.g. the actions of G_i obtained by restricting the H_i actions; actions of G_i induced on the space of A_i-ergodic components, etc.))?

(b) One can also ask about the converse assertion in (a). Namely if the G_i-actions are orbit equivalent (and hence under suitable hypotheses conjugate modulo an automorphism) are the H_i-actions orbit equivalent? When $G_i = \{e\}$ this, of course, follows from hyperfiniteness of amenable actions [**16**], [**3**] and Dye's theorem [**4**].

(c) Finally, one would like to know to what extent the hypotheses that the $\mathbb{R}$-rank be at least 2 are needed. A start in this direction appears in [**26**].

This reasearch was partially supported by NSF grant MCS79-05036, The Sloan Foundation, and the National Academy of Sciences, USA. The author is a Sloan Foundation Fellow.

REFERENCES

[1] A. Borel. Density properties for certain subgroups of semi-simple groups without compact factors. *Annals of Math.* **72** (1960), 179–188.

[2] A. Borel & J. Tits. Groupes reductifs. *Publ. Math. I.H.E.S.* **27** (1965), 55–150.

[3] A. Connes, J. Feldman & B. Weiss. To appear.

[4] H. A. Dye. On groups of measure preserving transformations, I. *Amer. J. Math.* **81** (1959), 119–159.

[5] J. Feldman & C. C. Moore. Ergodic equivalence relations, cohomology, and Von Neumann algebras, I. *Trans. Amer. Math. Soc.* **234** (1977), 289–324.

[6] J. Feldman, P. Hahn & C. C. Moore. Orbit structure and countable sections for actions of continuous groups. *Advances in Math.* **28** (1978), 186–230.

[7] H. Furstenberg. Boundary theory and stochastic processes on homogeneous spaces. In *Harmonic Analysis on Homogeneous Spaces*, pp. 193–229. Symp. in Pure Math. Williamstown, Mass., 1972.

[8] G. W. Mackey. Point realizations of transformation groups. *Ill. J. Math.* **6** (1962), 327–335.

[9] G. W. Mackey. Ergodic theory and virtual groups. *Math. Ann.* **166** (1966), 187–207.

[10] G. A. Margulis. Non-uniform lattices in semisimple algebraic groups. In *Lie Groups and Their Representations* (ed. I. M. Gelfand), pp. 371–553. Wiley: New York, 1975.

[11] G. A. Margulis. Discrete groups of motions of manifolds of non-positive curvature. *Amer. Math. Society Translations* **109** (1977), 33–45.

[12] G. A. Margulis. Arithmeticity of irreducible lattices in semisimple groups of rank greater than 1. Appendix to Russian translation of M. Ragunathan. *Discrete Subgroups of Lie Groups.* Mir: Moscow, 1977. (In Russian.)

[13] C. C. Moore. Ergodicity of flows on homogeneous spaces. *Amer. J. Math.* **88** (1966), 154–178.

[14] G. D. Mostow. Quasi-conformal mappings in n-space and the rigidity of hyperbolic space forms. *Publ. Math. I.H.E.S.* **34** (1967), 53–104.

[15] G. D. Mostow. *Strong Rigidity of Locally Symmetric Spaces.* Annals of Math. Studies, no. 78. Princeton Univ. Press: Princeton, N.J., 1973.

[16] D. Ornstein & B. Weiss. To appear.

[17] A. Ramsay. Virtual groups and group actions. *Advances in Math.* **6** (1971), 253–322.

[18] C. Series. The Rohlin tower theorem and hyperfiniteness for actions of continuous groups. *Israel J. Math.* **30** (1978), 99–112.

[19] C. Series. Foliations of polynomial growth are hyperfinite. *Israel J. Math.* **34** (1979), 245–258.

[20] R. J. Zimmer. Amenable ergodic group actions and an application to Poisson boundaries of random walks. *J. Funct. Anal.* **27** (1978), 350–372.

[21] R. J. Zimmer. Orbit spaces of unitary representations, ergodic theory, and simple Lie groups. *Annals of Math.* **106** (1977), 573–588.

[22] R. J. Zimmer. Induced and amenable ergodic actions of Lie groups. *Ann. Sci. Ec. Norm. Sup.* **11** (1978), 407–428.

[23] R. J. Zimmer. An algebraic group associated to an ergodic diffeomorphism. *Comp. Math.* **43** (1981), 59–69.

[24] R. J. Zimmer. Strong rigidity for ergodic actions of semisimple Lie groups. *Annals of Math.* **112** (1980), 511–529.

[25] R. J. Zimmer. On the cohomology of ergodic actions of semisimple Lie groups and discrete subgroups. *Amer. J. Math.* (in the press).

[26] R. J. Zimmer. On the Mostow rigidity theorem and measurable foliations by hyperbolic space. Preprint.

Annals of Mathematics, 118 (1983), 9–19

Ergodic actions of semisimple groups and product relations[1]

By Robert J. Zimmer[2]

1. Statement of results

Suppose Γ is a discrete group and (S, μ) is a finite measure space on which Γ acts so that μ is invariant and ergodic. If Γ is amenable, it follows from results of [4], [3] that the Γ-action is orbit equivalent to a product action of a product group, say $\Gamma_1 \times \Gamma_2$ acting on $S_1 \times S_2$. In other words, the equivalence relation on S defined by the Γ-orbits is a product of two other relations in a nontrivial way. The point of this paper is to show that equivalence relations defined by actions of semisimple groups or their discrete subgroups, or equivalence relations defined by foliations in which the leaves have negative curvature, do not have this property. More precisely, we show the following.

THEOREM 1.1. *Let Γ be a lattice in a connected simple Lie group with finite center, and let S be an essentially free ergodic Γ-space with finite invariant measure. Suppose for $i = 1, 2$ that Γ_i are discrete groups acting ergodically on measure spaces S_i, and that the Γ-action is orbit equivalent to the action of $\Gamma_1 \times \Gamma_2$ on $S_1 \times S_2$. Then either S_1 or S_2 is finite (modulo null sets).*

In fact the same result is true if we only assume stable orbit equivalence rather than orbit equivalence. We recall that two actions (of discrete groups) or two relations are called stably isomorphic if there are subsets of positive measure in each space such that the relations restricted to these sets are isomorphic [5]. In Theorem 1.1 we can then also deduce that if S_1 is finite, then the Γ-action on S is stably orbit equivalent to the Γ_2-action on S_2. For an irreducible lattice in a connected semisimple Lie group (rather than simple) with no compact factors, similar assertions hold for actions with discrete spectrum.

The proof of Theorem 1.1 proceeds by inducing the Γ-action to an action of the ambient Lie group, and then proving the corresponding statement for actions of semisimple Lie groups.

[1] Research partially supported by NSF grant MCS-8004026
[2] Sloan Foundation Fellow

10

THEOREM 1.2. *Let G be a connected semisimple Lie group with finite center. Let S be an irreducible ergodic essentially free G-space with finite invariant measure. Then the equivalence relation on S defined by the G-action is not stably isomorphic (in the sense of [6]) to a relation which is a product relation, except if one of the relations is transitive.*

A similar result shows that in Theorem 1.1 we may in fact take Γ to be a lattice in a simple algebraic group over a p-adic field.

THEOREM 1.3. *Let Γ be the fundamental group of a finite volume manifold M with negative sectional curvature bounded away from 0 (e.g. M compact with negative curvature). Then the conclusion of Theorem 1.1 is still true.*

In the proof of Theorem 1.3 the equivalence relation defined by the Γ-action is analysed by expressing it as a transversal to a suitable foliation by manifolds of negative curvature. More generally, we have:

THEOREM 1.4. *Let $\mathcal{F}$ be a Riemannian measurable foliation such that the sectional curvature K of the leaves satisfies $K \leq c < 0$. Assume $\mathcal{F}$ has a transversally (i.e. holonomy) invariant measure, finite total volume, and that almost every leaf is simply connected and complete. Finally, assume $\mathcal{F}$ is ergodic. Let T be a transversal with finite measure and R_T the equivalence relation on T. If R_T is stably isomorphic to a product relation $R_1 \times R_2$ for ergodic countable relations R_i on S_i with finite invariant measure, either S_1 or S_2 is finite.*

The proof of Theorem 1.2 depends in an essential way upon the algebraic group nature of semisimple Lie groups. As with other results concerning ergodic actions of these groups (e.g. [15], [18]), the "boundary behavior" of the action is of basic importance. Similarly, the boundary behavior for foliations, as developed in [18], plays a basic role in the proof of Theorem 1.4. We remark that Theorems 1.2 and 1.4 both imply Theorem 1.1 if Γ is a finitely generated free group with at least two generators. We thus have two distinct approaches to Theorem 1.1 for free groups (although in this case the two proofs are not fundamentally different), one depending on techniques dealing with an associated algebraic group and the other depending on an associated geometric object. It would be of interest to obtain a purely combinatorial proof of Theorem 1.1 for such Γ.

2. Proof for the case of group actions

We shall assume all measure spaces to be standard and all groups to be second countable. A countable equivalence relation R with invariant measure will be taken in the sense of Feldman and Moore [5]. If R is a countable equivalence relation on (S, μ) and $A \subset S$, then $R|A$ will denote the equivalence relation $R \cap (A \times A)$. If I is a set, I^2 will denote the transitive relation on I and $d(I)$ the

ERGODIC ACTIONS OF SEMISIMPLE GROUPS   11

diagonal relation. Following [5], countable relations R_i on (S_i, μ_i), $i = 1, 2$, will be called stably isomorphic if there exist $A_i \subset S_i$, $\mu_i(A_i) > 0$, such that $R_i|A_i$ are isomorphic. Equivalently [5], $R_1 \times Z^2 \cong R_2 \times Z^2$ where Z denotes the integers, and these are relations on $S_i \times Z$. If Γ acts in a measure-preserving fashion on (S, μ), R_Γ will denote the equivalence relation defined by the orbits. If G is a locally compact unimodular group which acts in a measure-preserving fashion on a space (S, μ), there exists a transversal for the action, i.e. a Borel subset $T \subset S$ that meets each orbit in a countable set, and such that $R_G \cong (R_G|T) \times I^2$ where I is some standard measure space [6]. We shall freely use results of [5], [6] concerning isomorphism, stable isomorphism, and transversals. A very brief summary of some of the relevant material is contained in [18, Section 2]. We shall also freely use the notion of a cocycle on an ergodic G-space and on an equivalence relation. (See, e.g., [17], [9] as references.) In particular, we shall need the notion of an amenable equivalence relation ([10], [11]). (Again, see [18, Section 2] for a very brief summary of some relevant facts.)

We now turn to the proof of Theorem 1.2. We recall that an ergodic action of a semisimple Lie group G with finite center is called irreducible if the restriction to each non-compact normal subgroup is still ergodic. We remark first that if S is an essentially free irreducible G-space and $G_0 \subset G$ is compact and normal, then S/G_0 is an essentially free irreducible G/G_0-space, and by results of [6], the action of G on S and G/G_0 on S/G_0 are orbit equivalent. Thus, it suffices to consider the case in which G has trivial center and no compact factors. We assume for $i = 1, 2$ that R_i are countable ergodic equivalence relations with invariant measure on finite measure spaces S_i, and that $\theta: S \to S_1 \times S_2 \times I$ is a measure space isomorphism defining an isomorphism of R_G with the relation $R_1 \times R_2 \times I^2$. We assume that neither R_1 nor R_2 is a transitive relation, and to prove the theorem we wish to derive a contradiction. (We can take θ as above with R_i countable relations by the results of [6].)

For (almost all) $x, y \in S_1 \times S_2 \times I$ with $x \sim y$, let $\alpha(x, y) \in G$ be the unique element with $\theta^{-1}(x) \cdot \alpha(x, y) = \theta^{-1}(y)$. Then $\alpha: R_1 \times R_2 \times I^2 \to G$ is a cocycle. Let $P \subset G$ be a minimal parabolic subgroup, and $M(G/P)$ the space of probability measures on G/P, which is a compact metrizable space with the weak-*-topology. $M(G/P)$ is a G-space in a natural way, and we shall make use of two basic results concerning this action. The first is that the G-orbits in $M(G/P)$ are all locally closed ([12], [15]), and the second is the result of C. C. Moore [8] that the stabilizer of any measure has an amenable algebraic hull, and in particular is a proper algebraic subgroup. Choose properly ergodic (i.e. ergodic but not essentially transitive) subrelations $A_i \subset R_i$ which are amenable. (These always exist. See [5].) Then $A_1 \times A_2 \times I^2$ is an amenable ergodic equivalence relation and so there is an $\alpha|A_1 \times A_2 \times I^2$-invariant function $\phi: S_1 \times S_2 \times I \to M(G/P)$; i.e. for x, y in a conull set in $S_1 \times S_2 \times I$ and $x \sim y$ (with respect to the relation $A_1 \times A_2 \times I^2$), we have $\phi(x) \cdot \alpha(x, y) = \phi(y)$. Thus, on a conull set, $\phi(x) \equiv \phi(y)$ in $M(G/P)/G$. Since the G-orbits in $M(G/P)$ are locally closed, $M(G/P)/G$ is a standard Borel space, and by ergodicity of the relation $A_1 \times A_2 \times I^2$, we deduce that the projection of $\phi(x)$ into $M(G/P)/G$ is constant on a conull set. In other words, we can assume $\phi(S_1 \times S_2 \times I)$ lies in a single G-orbit in $M(G/P)$. Thus, by the result of Moore mentioned above, we have an $\alpha | A_1 \times A_2 \times I^2$-invariant map $\phi \colon S_1 \times S_2 \times I \to G/H$ where $H \subset G$ is a proper algebraic subgroup. (Compare this with the argument in [15, Theorem 4.1].) In particular, ϕ is also an $\alpha | d(S_1) \times A_2 \times I^2$-invariant function. We now develop a uniqueness property of such maps that we will need.

We recall that for the groups G in question, $G \cong H_{\mathbf{R}}^0$ where H is a semisimple algebraic group defined over $\mathbf{R}$. (The superscript here and in the sequel indicates the connected component in the ordinary Hausdorff topology.) We shall consistently abuse notation by referring to a subgroup of G as algebraic if it is of the form $L \cap G$ where $L \subset H$ is an algebraic subgroup defined over $\mathbf{R}$.

Suppose that, for each $s \in S_1$, we have an algebraic subgroup $H_s \subset G$ such that the assignment $s \to H_s$ is measurable [9, Section 9]. Let $p_s \colon G \to G/H_s$ be the natural projection. Let $\phi \colon S_1 \times S_2 \times I \to G$ be a Borel map. Let us say that $(\phi, \{H_s\})$ is α-admissible if for almost all $s \in S_1$, $p_s \circ \phi$ is an $\alpha | \{s\} \times A_2 \times I^2$-invariant function. If $(s, t) \in R_1$ and $x, y \in S_2 \times I$ are equivalent under $A_2 \times I^2$, we have from the cocycle relation that

$$\alpha((s, x), (t, x))\alpha((t, x), (t, y)) = \alpha((s, x), (s, y))\alpha((s, y), (t, y)).$$

Suppose $f \colon S_1 \to S_1$ is a measure space automorphism with $s \sim f(s)$ for all s. Given an α-admissible $(\phi, \{H_s\})$, let $f^*(\phi, \{H_s\})$ be the pair $(\psi, \{\tilde{H}_s\})$ defined by $\psi(s, x) = \phi(f(s), x) \cdot \alpha((f(s), x), (s, x))$, and $\tilde{H}_s = H_{f(s)}$. The above consequence of the cocycle relation is readily seen to imply that $f^*(\phi, \{H_s\})$ is α-admissible. Furthermore, it is clear that if $(\phi_i, \{H_s^i\})$, $i = 1, 2$, are α-admissible, and $S_1 = A \cup B$ is a disjoint decomposition into measurable sets, then defining ϕ to be ϕ_1 on $A \times S_2 \times I$ and ϕ_2 on $B \times S_2 \times I$, and defining H_s similarly from H_s^i, we still obtain an α-admissible pair. It follows from these remarks and ergodicity of the relation R_1 that there exists an α-admissible $(\phi, \{H_s\})$ such that $\dim H_s$ is essentially constant over $s \in S_1$, and that for any admissible $(\psi, \{J_s\})$, we have $\dim H_s \leq \dim J_s$ almost everywhere. The following observation is basic to the proof of Theorem 1.2.

LEMMA 2.1. *Suppose $(\phi, \{H_s\}), (\psi, \{J_s\})$ are α-admissible and both satisfy the above minimality of dimension property. Then*

i) *H_s^0 and J_s^0 are conjugate for almost all s.*

ii) *The conjugacy class of H_s^0 is essentially constant in s. In particular, we can choose H_s satisfying the minimality of dimension property with $H_s^0 = H^0$, independent of s.*

iii) *With such a choice, let* p: $G \to G/N(H^0)$, *where* $N(H^0)$ *is the normalizer of* H^0 (*and is therefore algebraic*). *Suppose we also have* $J_s^0 = H^0$ *almost everywhere, so that* H_s, $J_s \subset N(H^0)$. *Then* $p \circ \phi = p \circ \psi$.

Proof. Consider the map (ϕ, ψ): $S_1 \times S_2 \times I \to G \times G$. For each $s \in S_1$, let p_s: $G \to G/H_s$, q_s: $G \to G/J_s$ be the projections. Then $(p_s \circ \phi, q_s \circ \psi)$ yields an $\alpha|\{s\} \times A_2 \times I^2$-invariant map $\{s\} \times S_2 \times I \to G/H_s \times G/J_s$. Since $A_2 \times I$ is an ergodic relation, and G-orbits are locally closed in $G/H_s \times G/J_s$ (using the fact that we have an algebraic action of an algebraic group) so that the space of orbits is a standard Borel space, we deduce that $(\phi, \psi)(\{s\} \times S_2 \times I)$ is (essentially) contained in a single G-orbit in $G/H_s \times G/J_s$. The stabilizer of such an orbit can be chosen of the form $L_s = H_s \cap bJ_s b^{-1}$ for some $b \in G$. It is a technical exercise to see that $s \to L_s$ can be chosen measurably, and hence $((\phi, \psi), \{L_s\})$ is also α-admissible. By the dimension property of $\{H_s\}$, dim $L_s =$ dim H_s almost everywhere, and hence H_s^0 and J_s^0 are conjugate. To see that the conjugacy class of H_s^0 is independent of s, observe first that if f: $S_1 \to S_1$ is a measure-space automorphism with $(s, f(s)) \in R_1$ for all s, $f^*(\phi, \{H_s\})$ will also satisfy the minimality of dimension condition, and so by the above, H_s^0 and $H_{f(s)}^0$ are conjugate. Thus the map $s \to$ (conjugacy class of H_s^0) is essentially invariant under the relation R_1. But by [1, p. 119], the set of conjugacy classes of connected subgroups of an algebraic group is a standard Borel space, and so by ergodicity, the conjugacy class of H_s^0 is essentially constant. The usual correspondence of cosets of a subgroup and that of a conjugate subgroup then enable us to choose an admissible (ϕ, H_s) satisfying the minimality condition with $H_s^0 = H^0$ almost everywhere. Finally, suppose we also have $J_s^0 = H^0$ almost everywhere. As above, the image of the map (ϕ, ψ) will lie in a single G-orbit, with the stabilizer of a point in the orbit of the form $L_s = H_s \cap bJ_s b^{-1}$, for some $b \in G$. Since dim $H_s = $ dim $J_s = $ dim L_s, $bH^0 b^{-1} = H^0$, i.e. $b \in N(H^0)$. Thus the image of the relevant orbit of $G/H_s \times G/J_s$ in $G/N(H^0) \times G/N(H^0)$ contains a point on the diagonal, and hence is contained in the diagonal. Therefore the image of $(p \circ \phi, p \circ \psi)$ is essentially contained in the diagonal, and this proves the final assertion of the lemma.

LEMMA 2.2. *Let* H_s *be as in the conclusion of Lemma 2.1. Then* $H^0 \neq \{e\}$.

Proof. If $H^0 = \{e\}$, each H_s would be finite. Fix $s \in S_1$ so that all required conditions which are known to hold almost everywhere hold on almost all of $\{s\} \times S_2 \times I$. Let λ: $G/H_s \to G$ be a Borel section of p_s. Then $\alpha|\{s\} \times A_2 \times I^2$-invariance of ϕ implies that for almost all $x, y \in S_2 \times I$ with $x \sim y$ under $A_2 \times I^2$, we have $\lambda(\phi(s, x))\alpha((s, x), (s, y))\lambda(\phi(s, y))^{-1} \in H_s$. Denote this element of H_s by $\beta(x, y)$. Let ψ: $\{s\} \times S_2 \times I \to S$ be defined by $\psi(s, x) = \theta^{-1}(s, x) \cdot \lambda(\phi(s, x))^{-1}$. Then it follows that $\psi(s, x) \cdot \beta(x, y) = \psi(s, y)$. In other words, letting $q: S \to S/H_s$ be the natural map, we have $q(\psi(s, x)) = q(\psi(s, y))$. Since H_s is finite, S/H_s is a standard Borel space. By ergodicity of the relation $A_2 \times I^2$, we have that $q(\psi(s, x))$ is essentially constant in x. Thus for almost all x, $\psi(s, x)$ are all in the same H_s-orbit, so that $\theta^{-1}(s, x)$ are all in the same G-orbit. This implies that the relation A_2 (and hence R_2) is essentially transitive, which contradicts our assumption. This proves the lemma.

To complete the proof of Theorem 1.2, we consider separately the cases in which G is simple or not.

Completion of the Proof of Theorem 1.2 for G simple: As in Lemma 2.1, choose $(\phi, \{H_s\})$ α-admissible, satisfying the minimality of dimension condition, and $H_s^0 = H^0$. As above, let $f: S_1 \to S_1$ be an automorphism with $s \sim f(s)$. As observed preceding Lemma 2.1, $f^*(\phi, \{H_s\})$ will also be α-admissible. It follows from the final conclusion of Lemma 2.1 and the definition of $f^*(\phi, \{H_s\})$ that $p(\phi(f(s), x)) \cdot \alpha((f(s), x), (s, x)) = p(\phi(s, x))$. In other words, $p \circ \phi: S_1 \times S_2 \times I \to G/N(H^0)$ is an $\alpha|R_1 \times d(S_2 \times I)$-invariant map. By our construction at the beginning of the proof, $H^0 \neq G$, and by Lemma 2.2, $H^0 \neq \{e\}$, and so $N(H^0) \neq G$. (Recall that we are already in the center free case.) We can now repeat our entire argument working with $\alpha|R_1 \times d(S_2 \times I)$-invariant functions. Lemmas 2.1 and 2.2 hold in this situation as well, and repeating the arguments immediately above using an automorphism $h: S_2 \to S_2$ with $h(t) \sim t$, we deduce the existence of an $\alpha|R_1 \times R_2 \times d(I)$-invariant map $\phi: S_1 \times S_2 \times I \to G/J$ for some nontrivial proper algebraic subgroup $J \subset G$. But it is straightforward to see that the transitivity of the relation I^2 on I enables us to construct such a map which will be α-invariant (for the entire relation $R_1 \times R_2 \times I^2$).

Recall that α was defined by the equation $\theta^{-1}(x) \cdot \alpha(x, y) = \theta^{-1}(y)$ for $x \sim y$. Let $\psi: S \to G/J$ be defined by $\psi = \phi \circ \theta$. Letting $x = \theta(s)$, $y = \theta(sg)$ in the defining equation for α and using α-invariance of ϕ, we have $\psi(sg) = \phi(\theta(sg)) = \phi(\theta(s)) \cdot g = \psi(s) \cdot g$. Thus, there is a G-map $S \to G/J$ where J is a proper algebraic subgroup. Since S has a finite G-invariant measure, so does G/J, and it is well known that this is impossible [2]. This completes the proof for G simple.

Completion of the Proof of Theorem 1.2 for G semisimple: Let $s \to L_s$ be a measurable assignment of algebraic subgroups $L_s \subset G$ to a point $s \in S_1$ and p_s: $G \to G/L_s$ the projection. Suppose there is a measurable map $S_1 \times S_2 \times I \to G$ such that for each $s \in S_1$, $p_s \circ \phi$ is $\alpha|\{s\} \times R_2 \times d(I)$-invariant. Arguing as in Lemma 2.1, we can assume L_s has the minimality of dimension property for such assignments and that $L_s^0 = L^0$ is constant. Choose $t \to M_t$ a similar assignment where $t \in S_2$ and functions are now required to be $\alpha|R_1 \times \{t\} \times d(I)$-invariant.

If either L^0 or M^0 is not normal, the argument in the simple case again shows that there are a proper algebraic subgroup J and a G-map $S \to G/J$, providing the desired contradiction. Thus, we can suppose both L^0, M^0 are normal. They are nontrivial by the argument of Lemma 2.2, and furthermore, they are proper. The latter assertion follows from the fact that the group H^0 constructed as in the simple case will be amenable (see the beginning of the proof), and hence $N(H^0)$ is proper. Thus, L^0 is proper, and a similar argument shows M^0 is proper. Since L^0, M^0 are normal and the center of G is trivial, it follows that $L_s = L^0$, $M_t = M^0$ almost everywhere. For convenience, set $L^0 = L$, $M^0 = M$.

Fix $(s_2, t) \in S_2 \times I$ contained in a suitable conull set, and let $\alpha_1 \colon R_1 \to G$ be given by $\alpha_1(x, y) = \alpha((x, y), (s_2, s_2), (t, t))$. Using the results of [5], [6], we see that the original orbit equivalence θ can be chosen such that α_1 satisfies the following property. For suitable (s_2, t), there is a subset $W \subset S_1$ of positive measure such that for almost all $x \in W$, if $x_n \sim x$ and $x_n \in W$ are distinct, then $\alpha(x, x_n) \to \infty$ in G, i.e. $\alpha(x, x_n)$ eventually leaves any compact set. (To see this, we recall that $S_1 \times S_2$ was chosen to be stably isomorphic to a set T that intersected every leaf in a countable set. In the terminology of [6], T can be chosen to be a complete lacunary section with respect to a compact neighborhood of the identity. It is then clear that if $s \sim s_n$ in T with $s \cdot g_n = s_n$, then $g_n \to \infty$ in G.) In the terminology of [5], $\tilde{\alpha}_1 = \alpha_1 | W$ is a transient cocycle. We can then form the Mackey range of this cocycle ([15], [17]) (called the Poincaré flow in [5]). This will be an essentially free ergodic G-action on a space X_1 which by transience of the cocycle $\tilde{\alpha}_1$ is stably orbit equivalent to the relation $R_1 | W$, and hence to R_1 ([5], [6]).

Now observe that if $G_0 \subset G$ is a closed subgroup, $\tilde{\alpha}_1$ is equivalent to a cocycle taking values in G_0 if and only if there exists an $\tilde{\alpha}_1$-invariant map $W \to G/G_0$. (This follows immediately from the existence of Borel sections $G/G_0 \to G$.) Furthermore, it is easy to check that there exists an $\tilde{\alpha}_1$-invariant map $W \to G/G_0$ if and only if there is an α_1-invariant map $S_1 \to G/G_0$. By the construction of L, it follows that for almost all (s_2, t), such a map will exist for algebraic G_0 if and only if $G_0 \supset L$. With these observations it follows from [12, Theorem 2.5], [20, Theorem 3.5], that the G-action on X_1 is induced from an action of L, say on a space Y_1, but not from an action of any proper algebraic subgroup. In other words, the L action on Y_1 is Zariski dense in L ([14], [15]). (We remark that at this point we do not know that the L action on Y_1 has finite invariant measure. If we did, then of course Zariski density would follow from the Borel density theorem.) Furthermore, since an action is stably orbit equivalent to its induced action [12], the relation R_1 is stably isomorphic to the relation on Y_1 defined by the L-action. A similar argument shows that R_2 is stably isomorphic to the relation defined by a Zariski dense ergodic M-action, say on Y_2.

16   ROBERT J. ZIMMER

Thus, from [6], we deduce that the G-action on S is orbit equivalent to the $L \times M$ action on $Y_1 \times Y_2$.

We next observe that the $L \times M$ action on $Y_1 \times Y_2$ is Zariski dense in $L \times M$. If not, then there is a proper algebraic subgroup $J \subset L \times M$ and an $L \times M$-map $Y_1 \times Y_2 \to (L \times M)/J$. Thus for almost all $y_2 \in Y_2$, we obtain an L-map $Y_1 \times \{y_2\} \to (L \times M)/J$. Since L is ergodic on Y_1, L and J are algebraic subgroups of $L \times M$, and algebraic actions have locally closed orbits, we deduce that the image of this map must be supported on an L-orbit in $(L \times M)/J$. Hence, we have an L-map $Y_1 \to L/L \cap \tilde{J}$ where $\tilde{J}$ is a conjugate of J in $L \times M$. By Zariski density of Y_1 in L, $L \cap \tilde{J} = L$, and since L is normal, $J \supset L$. Similarly $J \supset M$, so $J = L \times M$, verifying Zariski density. We can now apply the rigidity theorem of [15]. (An examination of the proof of [15, Theorem 4.3] shows that only one action need have finite invariant measure, while the other need only be Zariski dense. The point of course is the hypothesis in [15, Theorem 4.1].) We conclude that there is an isomorphism $\pi: G \to L \times M$ and a measure class preserving Borel isomorphism $\lambda: S \to Y_1 \times Y_2$ such that $\lambda(sg) = \lambda(s)\pi(g)$. However, this clearly implies that $\pi^{-1}(L)$ and $\pi^{-1}(M)$ are not ergodic on S, which contradicts irreducibility of the G-action on S. This completes the proof of Theorem 1.2.

Proof of Theorem 1.1. If $\Gamma \subset G$, apply Theorem 1.2 to $\text{ind}_\Gamma^G(S)$, the induced action of G [12].

We remark that Theorem 1.2 can also be applied to ergodic actions of irreducible lattices in semisimple groups (rather than just simple) whose induced action to the ambient Lie group is irreducible. Not every such induced action is irreducible, as the following example shows.

Example 2.3. Let $\Gamma \subset G_1 \times G_2$ be an irreducible lattice. Let S_i be an ergodic G_i-space with finite invariant measure. Let $S = S_1 \times S_2$ be the product $(G_1 \times G_2)$-space. By restriction, S is then an ergodic Γ-space. However $\text{ind}_\Gamma^{G_1 \times G_2}(S)$ is isomorphic as a $(G_1 \times G_2)$-space to $S_1 \times S_2 \times (G_1 \times G_2)/\Gamma$ [12, Prop. 2.10], which is clearly not irreducible. In this example we can even take S_2 to be a point.

However, for a certain natural class of actions the induced action will be irreducible. We recall that a finite measure preserving action of Γ has discrete spectrum if and only if it is isomorphic to an action of Γ on K/H where K is a compact group, $H \subset K$ is a closed subgroup and the action is defined via a homomorphism $\Gamma \to K$ with dense range [7]. For arithmetic lattices, such examples of course arise from a variety of arithmetic constructions.

PROPOSITION 2.4. *Let G be a connected semisimple Lie group with finite center and no compact factors, and let $\Gamma \subset G$ be an irreducible lattice. Suppose S is an ergodic Γ-space with discrete spectrum. Then $X = \mathrm{ind}_{\Gamma}^{G}(S)$ is an irreducible G-space.*

Proof. Let $G_0 \subset G$ be a normal proper subgroup of positive dimension, and $G_1 = G/G_0$. We first recall that G_0 is ergodic on X if and only if G is ergodic on $X \times G_1$ [13, Theorem 4.2]. The latter action has G/Γ as a factor with fiber $S \times G_1$, and hence ergodicity of G_0 on X is equivalent to ergodicity of Γ on $S \times G_1$. (Recall that by irreducibility of Γ, Γ is ergodic on G_1.) Suppose this Γ action is not ergodic. With the argument as in [13, Section 3] (or [17, Theorem 2.10]), there exists a bounded Γ-map $F: G_1 \to L^2(S) \ominus \mathbf{C}$ such that F is non-zero almost everywhere. Writing $S = K/H$, $L^2(S)$ is a direct sum of finite dimensional Γ-invariant subspaces, and by projecting F onto these subspaces we deduce the existence of an almost everywhere non-zero bounded Γ-map $F: G_1 \to V$ where V is a finite dimensional space on which Γ acts in a unitary and irreducible fashion. Thus there is a finite dimensional Γ-invariant subspace $W \subset L^{\infty}(G_1)$, $W \not\subset \mathbf{C}$, possessing a Γ-invariant inner product. But G_1 acts continuously on $L^{\infty}(G_1)$ with the weak-*-topology, and since Γ is dense in G_1, this will also be G_1 invariant. As is well-known, this is impossible.

Theorem 1.1 is also true for actions of lattices in p-adic groups.

THEOREM 2.5. *Let G be a connected algebraic group defined over a local field k of characteristic 0, and assume G is almost simple over k. Let $\Gamma \subset G_k$ be a lattice and S an essentially free Γ-space with finite invariant measure. Then the conclusion of Theorem 1.1 is valid.*

Proof. In the proof of Theorem 1.1, the only result we used about the real field that is not true for general k is that the stabilizers of measures on G_k/P_k are (k-points of) algebraic groups. The point is that compact matrix groups over $\mathbf{R}$ are real algebraic, while this of course fails for general k. However, if H is the stabilizer of a measure on G_k/P_k, either H is compact or $H \subset L_k$ where L is an algebraic k-group of dimension strictly less than that of G [17, proof of Theorem 9.7]. From an examination of the proof of Theorem 1.2 for G simple (and taking connectedness and connected components to be in the Zariski topology), we see that it suffices to observe the following.

LEMMA 2.6. *Suppose that for each $s \in S_1$, we have a topologically closed subgroup $H_s \subset G_k$ such that the assignment $s \to H_s$ is measurable. Let $p_s: G_k \to G_k/H_s$ be the projection. Suppose $\phi: S_1 \times S_2 \times I \to G_k$ is a Borel map such that for almost all $s \in S_1$, $p_s \circ \phi$ is an $\alpha|\{s\} \times B_2 \times d(I)$-invariant function*

18   ROBERT J. ZIMMER

where $B_2 \subset R_2$ *is an ergodic subrelation. Then for almost all s, H_s is not compact.*

Proof. The proof is exactly the same as that of Lemma 2.2.

3. The case of foliations

We now turn to the proof of Theorem 1.4. For the relevant measure theory of foliations and their transversals, we refer the reader to [18]. We shall freely use results and terminology of that paper.

Let T be a transversal to the foliation, with equivalence relation R, and finite invariant measure μ. Suppose R is stably isomorphic to $R_1 \times R_2$ where R_i is a properly ergodic countable relation with finite invariant measure on (S_i, μ_i). When Z represents the integers, this implies that there is a subset $\tilde{T}$ of $T \times Z$ such that $R_1 \times R_2$ is isomorphic to the relation $\tilde{R} = (R \times Z^2)|\tilde{T}$. Let θ: $\tilde{T} \to S_1 \times S_2$ define the isomorphism, and for each $n \in Z$, let $T_n = \{t \in T | (t, n) \in \tilde{T}\}$. Passing to a subset of Z, we can assume $\mu(T_n) > 0$, so that by ergodicity, T_n is also a transversal to the foliation. If the leaves are of dimension N, let α: $R \to \mathrm{Homeo}(S^{N-1})$ be the boundary cocycle defined in [18, Proposition 3.1]. Let $\tilde{\alpha}$: $\tilde{R} \to \mathrm{Homeo}(S^{N-1})$ be $\tilde{\alpha}((s, n), (t, p)) = \alpha(s, t)$ and β: $R_1 \times R_2 \to \mathrm{Homeo}(S^{N-1})$ the cocycle $\beta(x, y) = \alpha(\theta^{-1}(x), \theta^{-1}(y))$. Let $A_1 \subset R_1$ be an amenable ergodic subrelation. Then there exists a $\beta|A_1 \times d(S_2)$-invariant function ν: $S_1 \times S_2 \to M(S^{N-1})$ (the space of probability measures on the sphere) and by [18, Theorem 3.7], for almost all (s_1, s_2), $\nu(s_1, s_2)$ is supported on at most two points of S^{N-1}. (The ergodicity assumption on the relation in [18, Theorem 3.7] is not required. An examination of that proof shows that all we need is recurrence; i.e. for any measurable set B, for almost all $x \in B$, there exist infinitely many distinct $y_n \in B$ with $x \sim y_n$. This is clearly satisfied by the relations for which we need it, i.e. the relations on T_n induced via θ from $(A_1 \times d(S_2))|\theta(T_n)$.) Furthermore, an examination of the proof of [18, Theorem 3.7] shows that in fact these two points of S^{N-1} are unique. In other words, there is a unique $\beta|A_1 \times d(S_2)$-invariant map ν: $S_1 \times S_2 \to (S^{N-1} \times S^{N-1})/S(2)$ where $S(2)$ is the symmetric group on two letters. We now argue exactly as in Section 2, using this uniqueness in place of (iii) in Lemma 2.1. We deduce that there is an $\tilde{\alpha}$-invariant function $\tilde{T} \to (S^{N-1} \times S^{N-1})/S(2)$, and hence an $\tilde{\alpha}$-invariant function $\tilde{T} \to M(S^{N-1})$. However the proof of [19, Theorem 2] shows that this implies that almost all leaves are flat, contradicting our hypothesis, and thus proving Theorem 1.4. (In [19], the only point at which amenability is used is to construct such an invariant function into $M(S^{N-1})$.)

ERGODIC ACTIONS OF SEMISIMPLE GROUPS   19

Finally, to prove Theorem 1.3 from 1.4, simply apply the foliated bundle construction. See, e.g., [18, Example 2.4(c)].

University of Chicago, Chicago, Illinois

References

[1] L. Auslander, C. C. Moore, Unitary representations of solvable Lie groups, Memoirs A.M.S. **62** (1966).
[2] A. Borel, Density properties for certain subgroups of semisimple groups without compact factors, Ann. of Math. **72** (1960), 179–188.
[3] A. Connes, J. Feldman, B. Weiss, Amenable equivalence relations are generated by a single transformation, preprint.
[4] H. Dye, On groups of measure preserving transformations, I, Amer. J. Math. **81** (1959), 119–159.
[5] J. Feldman, C. C. Moore, Ergodic equivalence relations, cohomology, and von Neumann algebras, I, Trans. A.M.S. **234** (1977), 289–324.
[6] J. Feldman, P. Hahn, C. C. Moore, Orbit structure and countable sections for actions of continuous groups, Adv. Math. **28** (1978), 186–230.
[7] G. W. Mackey, Ergodic transformation groups with a pure point spectrum, Ill. J. Math. **8** (1964), 593–600.
[8] C. C. Moore, Amenable subgroups of semisimple groups and proximal flows, Israel J. Math. **34** (1979), 121–138.
[9] A. Ramsay, Virtual groups and group actions, Adv. Math. **6** (1971), 253–322.
[10] R. J. Zimmer, Amenable ergodic group actions and an application to Poisson boundaries of random walks, J. Funct. Anal. **27** (1978), 350–372.
[11] _____, Hyperfinite factors and amenable ergodic actions, Invent. Math. **41** (1977), 23–31.
[12] _____, Induced and amenable ergodic actions of Lie groups, Ann. Sci. Ec. Norm. Sup. **11** (1978), 407–428.
[13] _____, Orbit spaces of unitary representations, ergodic theory, and simple Lie groups, Ann. of Math. **106** (1977), 573–588.
[14] _____, An algebraic group associated to an ergodic diffeomorphism, Comp. Math., 1981.
[15] _____, Strong rigidity for ergodic actions of semisimple Lie groups, Ann. of Math. **112** (1980), 511–529.
[16] _____, On the Mostow rigidity theorem and measurable foliations by hyperbolic space, to appear, Israel J. Math.
[17] _____, Ergodic theory, group representations, and rigidity, Bull. A.M.S. **6** (1982), 383–416.
[18] _____, Ergodic theory, semisimple Lie groups, and foliations by manifolds of negative curvature, Publ. Math. I.H.E.S. **55** (1982), 37–62.
[19] _____, Curvature of leaves in amenable foliations, to appear, Amer. J. Math.
[20] _____, Extensions of ergodic group actions, Ill. J. Math. **20** (1976), 373–409.

(Received August 14, 1981)

4

Cocycle
Superrigidity and
the Program
to Describe
Lie Group and
Lattice Actions
on Manifolds

VOLUME PRESERVING ACTIONS OF LATTICES IN SEMISIMPLE GROUPS ON COMPACT MANIFOLDS[1]

by ROBERT J. ZIMMER [2]

1. Introduction

In this paper we begin a study of volume preserving actions of lattices in higher rank semisimple groups on compact manifolds. More precisely, let G be a connected semisimple Lie group with finite center, such that the **R**-rank of every simple factor of G is at least 2. Let $\Gamma \subset G$ be a lattice subgroup. Assume M is a compact manifold and ω is a smooth volume density on M. The general question we wish to address is to determine how Γ can act on M so as to preserve ω. All presently known actions of this type are essentially of an algebraic nature, and a fundamental general problem is to determine whether or not the known examples are an essentially complete list or whether there are genuinely " geometric " actions of such groups.

The standard arithmetic construction of cocompact lattices shows that in the cocompact case one may have homomorphisms $\Gamma \to K$ where K is a compact Lie group and the image of Γ is dense in K. Thus, Γ acts isometrically (and ergodically) on the homogeneous spaces of K. One way to attempt to exhibit new actions of Γ would be to start with a given action and try to perturb this action. One of our main results shows that if we start with an isometric action and make a sufficiently smooth perturbation, then at least topologically we stay within the class of isometric actions. More precisely:

Theorem **6.1**. — *Let* G *and* Γ *be as above. Let* M *be a compact Riemannian manifold,* $\dim M = n$. *Set* $r = n^2 + n + 1$. *Assume* Γ *acts by isometries of* M. *Let* $\Gamma_0 \subset \Gamma$ *be a finite generating set. Then any volume preserving action of* Γ *on* M *which*

i) *for elements of* Γ_0 *is a sufficiently small* C^r *perturbation of the original action;* *and*

ii) *is ergodic;*

[1] Research partially supported by NSF grant MCS 8004026.
[2] Sloan Foundation Fellow.

ROBERT J. ZIMMER

actually leaves a C^0 Riemannian metric invariant. In particular, there is a Γ-invariant topological distance function and the action is topologically conjugate to an action of Γ on a homogeneous space of a compact Lie group K defined via a dense range homomorphism of Γ into K.

In addition to showing that we obtain no new ergodic actions of Γ by perturbing an isometric action, Theorem 6.1 can be profitably viewed from other vantage points. We recall that if we have an isometric diffeomorphism (i.e. **Z**-action) a volume preserving perturbation of this diffeomorphism is not likely to be isometric. In fact, hyperbolicity rather than isometry is typical of properties of a diffeomorphism that are preserved under a perturbation. A similar remark obviously applies to actions of free groups as well. Thus Theorem 6.1 shows a sharp contrast between the behavior under perturbation of actions of lattices in higher rank semisimple groups and that of free groups. From another point of view, we recall from the work of Weil [19], Mostow [11], Margulis [9], and others that homomorphisms of lattices in higher rank groups into (finite dimensional) Lie groups have strong rigidity properties. Theorem 6.1 can be viewed as a type of (local) rigidity theorem for a class of homomorphisms of Γ into the infinite dimensional group of diffeomorphisms of M.

As all " algebraic " examples of volume preserving actions of Γ on manifolds have dimension restrictions, we put forth the following conjecture.

Conjecture. — Let G, Γ be as above. Let $d(G)$ be the minimal dimension of a non-trivial real representation of the Lie algebra of G, and $n(G)$ the minimal dimension of a simple factor of G. Let M be a compact manifold, $\dim M = n \ (> 0)$.
Assume

i) $n < d(G)$; and
ii) $n(n + 1) < 2n(G)$.

Then every volume preserving action of Γ on M is an action by a finite quotient of Γ. In particular, there are no volume preserving ergodic actions of Γ on M.

With one additional hypothesis, we can verify the final assertion of this conjecture. We recall that on any compact manifold, the space of Riemannian metrics has a natural (metrizable) C^r-topology for each integer $r \geqslant 0$. If ξ is a metric, $\mathcal{O}$ a C^r-neighborhood of ξ in the space of metrics with the same volume density as ξ, and S is a finite set of diffeomorphisms, we say that ξ is $(\mathcal{O}, S)$-invariant if $f^*\xi \in \mathcal{O}$ for each $f \in S$. In other words ξ is " nearly C^r-invariant " under S. In general, this is not a very strong condition. For example, let φ_t be a smooth flow on M, preserving the volume density of a metric ξ. Then for any $\mathcal{O}$, we have that for t sufficiently small, ξ is $(\mathcal{O}, \varphi_t)$-invariant. Since highly mixing diffeomorphisms can arise this way, e.g. all φ_t, $t \neq 0$, may be Bernoulli, " near isometry " has no implications in general for the existence of an invariant metric. With the additional hypothesis that the generators of Γ act nearly isometrically we can prove the final assertion of the conjecture.

VOLUME PRESERVING ACTIONS OF LATTICES IN SEMISIMPLE GROUPS   7

Theorem **6.2.** — *Let* G, Γ *be as above and* M *a compact manifold,* $\dim M = n$ $(n > 0)$. *Assume* $n(n + 1) < 2n(G)$, *where* $n(G)$ *is as in the conjecture. Set* $r = n^2 + n + 1$. *Let* $\Gamma_0 \subset \Gamma$ *be a finite generating set. Then for any smooth Riemannian metric* ξ *on* M, *there is a* C^r *neighborhood* $\mathcal{O}$ *of* ξ *such that there are no volume preserving ergodic actions of* Γ *on* M *for which* ξ *is* $(\mathcal{O}, \Gamma_0)$-*invariant.*

Both Theorem 6.1 and 6.2 are deduced from the following.

Theorem **5.1.** — *Let* G, Γ *be as above, and* $\Gamma_0 \subset \Gamma$ *a finite generating set. Let* M *be a compact n-manifold, and let* $r = n^2 + n + 1$. *Let* ξ *be a smooth Riemannian metric on* M *with volume density* ω. *Then there is a* C^r *neighborhood* $\mathcal{O}$ *of* ξ *such that any* ω-*preserving smooth ergodic action of* Γ *on* M, *with* ξ $(\mathcal{O}, \Gamma_0)$-*invariant, leaves a* C^0-*Riemannian metric invariant.*

Once again, Theorem 5.1 shows a sharp contrast between actions of Γ and actions of **Z** or more general free groups as shown above by the example coming from a flow.

The framework of both the results and conjecture above can be generalized in a number of directions only the most direct of which we shall consider in this paper. (We hope to return to some of these other directions elsewhere.) For example, let k be a totally disconnected local field of characteristic 0, and suppose that G is a connected semisimple algebraic k-group, such that every k-simple factor of G has k-rank $\geqslant 2$. Let $\Gamma \subset G_k$ be a lattice. From results of Margulis [9] (see also the work of Raghunathan [16]), it follows that for any homomorphism $\Gamma \to GL(n, \mathbf{R})$ the image of Γ is precompact. The following version of the previous conjecture would be a generalization of this result where $GL(n, \mathbf{R})$ is replaced by the infinite dimensional group of volume preserving diffeomorphisms of a compact manifold of arbitrary dimension.

Conjecture. — Let $\Gamma \subset G_k$ be as in the preceding paragraph. Then any smooth volume preserving action of Γ on a compact manifold is an action by isometries.

As in the case of real groups, we can establish the final assertion with one additional hypothesis.

Theorem **6.3.** — *Let* $\Gamma \subset G_k$ *as above. Let* M *be a compact n-manifold* $(n > 0)$. *Let* $r = n^2 + n + 1$. *Let* $\Gamma_0 \subset \Gamma$ *be a finite generating set. Then for any smooth Riemannian metric* ξ *on* M *there is a* C^r *neighborhood* $\mathcal{O}$ *of* ξ *such that any volume preserving ergodic action of* Γ *on* M *for which* ξ *is* $(\mathcal{O}, \Gamma_0)$-*invariant leaves a* C^0-*Riemannian metric invariant.*

We now outline the contents of the remainder of this paper, and in doing so indicate some of the tools needed for the proofs of the above theorems. The first major step in the proofs, and the result which leads one to put forward the above conjectures, is the superrigidity theorem for cocycles proved by the author in [26], [27], [28], [29]. This is a generalization of Margulis' superrigidity theorem [9] and its proof draws in part on Margulis' techniques. (See [28], [29] for an exposition of all of this material.)

ROBERT J. ZIMMER

While in [26] we used the superrigidity theorem to construct homomorphisms between Lie groups, here we use it to construct measurable invariant metrics on vector bundles. In particular, the superrigidity theorem when applied to the derivative cocycle on the tangent bundle, will in our situation give us a Γ-invariant assignment of an inner product to each tangent space on the manifold, but this assignment will only be measurable. To deduce the existence of such a measurable invariant metric from superrigidity requires the use of Kazhdan's property, the Furstenberg-Kesten theorem on the products of random matrices [3], [7], and some general structural properties of algebraic groups. This measure theoretic information that the superrigidity theorem gives us, as well as some other results of a similar type, is developed in section 2. Sections 3-5 are devoted to trying to make the measurable invariant metric continuous.

In section 3 we discuss certain constructions on the frame bundle of a manifold and on some associated jet bundles, and apply some of the results of section 2 to this situation.

Section 4 is devoted to a simple but basic estimate which compares a measurable invariant metric to a smooth metric on a vector bundle over a manifold. Here we make use of the " near isometry " condition, ergodicity of the action, and of Kazhdan's property. We deduce that the measurable invariant metric must satisfy L^2 conditions with respect to a smooth metric.

In section 5 we complete the main argument (the proof of Theorem 5.1). We use the results of sections 2 and 4 applied to a cocycle defined on a jet bundle over the frame bundle of the original manifold. " Near C^r-isometry " and Kazhdan's property are used to construct a fixed point in an associated Sobolev space, and using our L^2 estimates, the Sobolev embedding theorem and some supplementary ergodic theoretic arguments we construct an invariant C^0 metric.

Section 6 contains the deductions of Theorems 6.1, 6.2, 6.3 from the arguments of section 5. In section 7 we present the known examples of volume preserving actions on compact manifolds for the arithmetic groups we have been considering and some questions concerning these actions. Some of the results of this paper were announced in [30].

2. Measurable cocycles and superrigidity

In this section we present some fundamental results we will be using concerning measurable cocycles. We will be applying these in subsequent sections to the actions on various natural vector bundles related to a smooth action on a manifold. As a general background source on measurable cocycles and related matters, see [29].

Let Γ be a locally compact second countable group and (S, μ) a standard Borel Γ-space with μ a Γ-quasi-invariant probability measure. We recall that if H is a standard Borel group, a Borel map $\alpha : S \times \Gamma \to H$ is called a cocycle if for all $\gamma_1, \gamma_2 \in \Gamma$, $\alpha(s, \gamma_1 \gamma_2) = \alpha(s, \gamma_1) \alpha(s\gamma_1, \gamma_2)$ for almost all $s \in S$, and that two cocycles $\alpha, \beta : S \times \Gamma \to H$

VOLUME PRESERVING ACTIONS OF LATTICES IN SEMISIMPLE GROUPS   9

are called equivalent if there is a Borel function $\varphi : S \to H$ such that for each $\gamma \in \Gamma$, $\alpha(s, \gamma) = \varphi(s)\beta(s, \gamma)\varphi(s\gamma)^{-1}$ almost everywhere. We shall be most often concerned with the case in which H is a linear group (usually algebraic). In this case an important invariant of the cocycle is the algebraic hull, introduced in [24] (see also [28]). We summarize some relevant information.

Proposition **2.1** [24], [28]. — *Suppose* $H \subset GL(n, \mathbf{C})$ *is an algebraic* **R**-*group and* $\alpha : S \times \Gamma \to H_{\mathbf{R}}$ *is a cocycle. Suppose the* Γ-*action on* S *is ergodic. Then there exists an* **R**-*subgroup* $L \subset H$ *such that* $\alpha \sim \alpha_1$ *where* α_1 *takes values in* $L_{\mathbf{R}}$, *and there is no proper* **R**-*subgroup* $M \subset L$ *such that* $\alpha \sim \alpha_2$ *where* α_2 *takes values in* $M_{\mathbf{R}}$. *The group* $L_{\mathbf{R}}$ *is unique up to conjugacy by an element of* $H_{\mathbf{R}}$ *and is called the algebraic hull of* α.

If G is a connected semisimple Lie group with finite center $Z(G)$, then $G/Z(G) \cong (G_{\mathbf{R}}^*)^0$ where G^* is a connected semisimple algebraic **R**-group with trivial center. If H is an **R**-group and $\pi : G \to H_{\mathbf{R}}$ is a homomorphism with $\pi \mid Z(G)$ trivial, we shall call π **R**-rational if the induced homomorphism on $G/Z(G)$ is the restriction of an **R**-rational homomorphism $G^* \to H$.

The following is the version of the superrigidity theorem for cocycles that we will need. It is adapted from the version in [27]. (See also [29], Theorem 9.3.14.)

Theorem **2.2** [26], [27], [29]. — *Suppose* G *is a connected semisimple Lie group with finite center such that every simple factor has* **R**-*rank* ≥ 2. *Suppose* $\Gamma \subset G_{\mathbf{R}}$ *is a lattice and that* Γ *acts ergodically on* (S, μ) *where* μ *is an invariant probability measure. Suppose that* H *is a connected, semisimple, adjoint algebraic* **R**-*group, that* $H_{\mathbf{R}}$ *has no compact factors, and that* $\alpha : S \times \Gamma \to H_{\mathbf{R}}$ *is a cocycle whose algebraic hull is* $H_{\mathbf{R}}$ *itself. Then there exists an* **R**-*rational homomorphism* $\pi : G \to H_{\mathbf{R}}$ *(trivial on* $Z(G)$*), such that* $\alpha \sim \alpha_\pi$ *where* $\alpha_\pi : S \times \Gamma \to H_{\mathbf{R}}$ *is the cocycle* $\alpha(s, \gamma) = \pi(\gamma)$.

We shall need information concerning the situation in which the algebraic hull of α is not necessarily $H_{\mathbf{R}}$. To this end we recall the following fact about cocycles in the case in which Γ has Kazhdan's property. (See [1], [6], [18], [29] for Kazhdan's property.)

Theorem **2.3** [17], [25]. — *Suppose* Γ *has Kazhdan's property and that* A *is an amenable group. Suppose* Γ *acts ergodically with finite invariant measure on* S. *If* $\alpha : S \times \Gamma \to A$ *is a cocycle, then* α *is equivalent to a cocycle taking values in a compact subgroup of* A.

We observe that equivalence of a cocycle into a linear group to a cocycle into a compact subgroup is germane to the problem of finding invariant metrics. If $\alpha : S \times \Gamma \to H$ where H is a group acting continuously on the left on a separable metrizable space X, by an α-invariant function $\varphi : S \to X$ we mean a measurable function φ such that for each $g \in \Gamma$, $\alpha(s, g)\varphi(sg) = \varphi(s)$ for almost all $s \in S$. (See [29] for a general discussion.) If $H \subset GL(n, \mathbf{R})$ is a subgroup, then H acts on the space of inner products on $\mathbf{R}^n$, which we denote by $\text{Inn}(\mathbf{R}^n)$. Fixing the standard basis in $\mathbf{R}^n$ we can identify $\text{Inn}(\mathbf{R}^n)$ with the positive definite symmetric matrices, and hence we may speak of the determinant of any element in $\text{Inn}(\mathbf{R}^n)$. If the determinant is 1 we say the inner product is unimodular.

Definition **2.4.** — *If* $\alpha : S \times \Gamma \to H \subset GL(n, \mathbf{R})$ *is a cocycle, an α-invariant function* $B : S \to \text{Inn}(\mathbf{R}^n)$ *is called a measurable α-invariant metric.*

We let $SL'(n, \mathbf{R})$ be the subgroup of $GL(n, \mathbf{R})$ consisting of matrices A with $\det(A)^2 = 1$. We then have:

Proposition **2.5.** — *Suppose* $H \subset GL(n, \mathbf{R})$ *and* $\alpha : S \times \Gamma \to H$ *is a cocycle. If* $\alpha \sim \beta$ *where* $\beta(S \times \Gamma) \subset K$, $K \subset H$ *a compact subgroup, then there is a measurable α-invariant metric B. If* $H \subset SL'(n, \mathbf{R})$, *we can assume* $\det B(s) = 1$ *for all s.*

Conversely, if α is a cocycle into $SL'(n, \mathbf{R})$ and there is a measurable α-invariant metric, then α is equivalent to a cocycle taking values in $O(n, \mathbf{R})$.

Proof. — Let $B_0 \in \text{Inn}(\mathbf{R}^n)$ be a K-invariant inner product and let $B(s) = \varphi(s)B_0$ where $\varphi : S \to H$ satisfies $\alpha(s, \gamma) = \varphi(s)\beta(s, \gamma)\varphi(s\gamma)^{-1}$. Then B is the required α-invariant metric. To see the converse, let B_0 be the standard $O(n, \mathbf{R})$-invariant inner product on $\mathbf{R}^n$, and $\text{Inn}_1(\mathbf{R}^n)$ the inner products of determinant 1. We can choose a measurable function $\varphi : \text{Inn}_1(\mathbf{R}^n) \to SL(n, \mathbf{R})$ such that $P = \varphi(P) \cdot B_0$ for all $P \in \text{Inn}_1(\mathbf{R}^n)$. If $s \to B(s)$ is the measurable α-invariant metric, let $\psi : S \to SL(n, \mathbf{R})$ be $\psi(s) = \varphi(\det(B(s))^{-1/n}B(s))$. Then $\beta(s, \gamma) = \psi(s)^{-1}\alpha(s, \gamma)\psi(sg) \in O(n, \mathbf{R})$.

We can now give an application of Theorems 2.2, 2.3 to the existence of measurable α-invariant metrics. We first record an elementary lemma.

Lemma **2.6.** — i) *Suppose Γ acts ergodically on S, $\alpha : S \times \Gamma \to H$ is a cocycle, and $H_0 \subset H$ is a subgroup of finite index. Then there is a finite ergodic extension $p : T \to S$ (i.e. all fibers are of fixed finite cardinality) such that the cocycle $\tilde\alpha : T \times \Gamma \to H$, $\tilde\alpha(t, \gamma) = \alpha(p(t), \gamma)$ is equivalent to a cocycle into H_0.*

ii) *Suppose $\alpha : S \times \Gamma \to H$ is a cocycle, $p : H \to H_1$ is a surjective homomorphism, and $p \circ \alpha$ is equivalent to the trivial cocycle. Then α is equivalent to a cocycle into $\ker p$.*

iii) *Suppose $T \to S$ is a finite extension of ergodic Γ-spaces and $\alpha : S \times \Gamma \to SL'(n, \mathbf{R})$ is a cocycle. Let $\tilde\alpha$ be defined as in (i). Then there is a measurable α-invariant metric if and only if there is a measurable $\tilde\alpha$-invariant metric.*

Proof. — For (i), (ii) see [29]. For (iii), one direction is clear. Conversely, if there is an $\tilde\alpha$-invariant metric, summing the metrics over the fibers produces an α-invariant metric.

We recall the following notation from the introduction.

VOLUME PRESERVING ACTIONS OF LATTICES IN SEMISIMPLE GROUPS   11

Definition **2.7.** — *If* G *is a semisimple Lie group with no compact factors, let* $d(G)$ *be the minimal dimension of a non-trivial representation of the Lie algebra of* G *on a real vector space.*

Theorem **2.8.** — *Let* G, Γ *be as in Theorem 2.2. Suppose that* (S, μ) *is a* Γ*-space such that (almost) every ergodic component of the* Γ *action on* S *has a finite* Γ*-invariant measure. Let* $\alpha : S \times \Gamma \to SL'(n, \mathbf{R})$ *be a cocycle. If* $n < d(G_{\mathbf{R}})$, *then there is a measurable* α*-invariant metric* B *with* $\det B(s) = 1$ *for all* s.

Proof. — Using a standard ergodic decomposition argument it suffices to consider the case in which Γ acts ergodically with an invariant probability measure. Let $H_{\mathbf{R}} \subset SL'(n, \mathbf{R})$ be the algebraic hull of α (Proposition 2.1). If H is not connected, by Lemma 2.6 (i) we can pass to a finite ergodic extension of S and assume that $\tilde{\alpha}$ (defined as in 2.6) takes values in $(H^0)_{\mathbf{R}}$. If the algebraic hull of $\tilde{\alpha}$ is not $(H^0)_{\mathbf{R}}$, take the algebraic hull. Once again, if this is not connected repeat the process, passing to a finite ergodic extension. In this way, replacing S by a finite ergodic extension if necessary, we can assume (using lemma 2.6 (iii)) that the algebraic hull of α, say $H_{\mathbf{R}}$, is such that H is connected. We can write $H_{\mathbf{R}} = L_{\mathbf{R}} \ltimes U_{\mathbf{R}}$ where L is reductive and U is the unipotent radical of H. Let $q : H_{\mathbf{R}} \to L_{\mathbf{R}}/[L_{\mathbf{R}}, L_{\mathbf{R}}]$ be the natural projection. By Theorem 2.3, $q \circ \alpha$ is equivalent to a cocycle into a compact subgroup of $L_{\mathbf{R}}/[L_{\mathbf{R}}, L_{\mathbf{R}}]$. Since $L_{\mathbf{R}}/[L_{\mathbf{R}}, L_{\mathbf{R}}]$ is the algebraic hull of $q \circ \alpha$, it follows that this group is compact. We have $L = [L, L]Z(L)$ where $Z(L)$ is the center of L and $[L, L] \cap Z(L)$ is finite. It follows that $Z(L)_{\mathbf{R}}$ is also compact.

We can write the **R**-group $L/Z(L)$ as a product of connected semisimple **R**-groups $L/Z(L) = L_1 \times L_2$ where $(L_2)_{\mathbf{R}}$ is compact and $(L_1)_{\mathbf{R}}$ is centerfree with no compact factors. Let $q_1 : H_{\mathbf{R}} \to (L_1)_{\mathbf{R}}$ be the projection. Then $q_1 \circ \alpha$ is a cocycle with algebraic hull $(L_1)_{\mathbf{R}}$ and by the superrigidity theorem for cocycles (Theorem 2.2), there is a **R**-rational homomorphism $\pi : G \to (L_1)_{\mathbf{R}}$ such that $q_1 \circ \alpha$ is equivalent to α_π. We can consider π as a rational homomorphism $\pi : \tilde{G} \to L_1$ where $\tilde{G}$ is the algebraic universal covering group of G^*. (Here G^* is as in the discussion preceding Theorem 2.2.) Then π lifts to a homomorphism $\tilde{\pi} : \tilde{G} \to [L, L] \subset GL(n, \mathbf{C})$. By the definition of $d(G)$, our hypotheses imply $\tilde{\pi}$ is trivial. Since the algebraic hull of $\tilde{\pi}(\tilde{G}_{\mathbf{R}})$ is $(L_1)_{\mathbf{R}}$, L_1 is trivial. It follows that $L_{\mathbf{R}}$ is compact and hence that $H_{\mathbf{R}}$ is amenable. Applying Theorem 2.3 once again, we deduce that α is equivalent to a cocycle taking values in a compact subgroup and hence by Proposition 2.5, there is a measurable α-invariant metric.

If we drop the restriction that $n < d(G)$ the conclusion of Theorem 2.8 is obviously no longer true in general. There are however other hypotheses which will lead to the same conclusion with close to the same proof. If $\alpha : S \times \Gamma \to GL(n, \mathbf{R})$ is a cocycle, $\Gamma_0 \subset \Gamma$ is a finite set, and $\varepsilon > 0$ we say that α is (ε, Γ_0)-admissable if $\|\alpha(s, \gamma)\| \leqslant 1 + \varepsilon$, $\|\alpha(s, \gamma)^{-1}\| \leqslant 1 + \varepsilon$ for all $s \in S$ and all $\gamma \in \Gamma_0$.

Theorem **2.9**. — *Let* Γ, S *be as in Theorem* 2.8, *and suppose* $\Gamma_0 \subset \Gamma$ *is a finite generating set. Fix* $n > 0$. *Then there exists* $\varepsilon > 0$ *such that for any* (ε, Γ_0)-*admissible cocycle* $\alpha : S \times \Gamma \to SL'(n, \mathbf{R})$ *there is a measurable* α-*invariant metric.*

To describe the proof, we first recall the theorem of Furstenberg and Kesten concerning products of random matrices. If Λ is a group, S a Λ-space, and $\alpha : S \times \Lambda \to GL(n, \mathbf{R})$ is a cocycle, we call α a tempered cocycle if for each $h \in \Lambda$, $\|\alpha(., h)\| \in L^\infty(S)$. Now suppose $\Lambda = \mathbf{Z}$, the group of integers. Define

$$e(\alpha)(s) = \lim_{n \to \infty} \frac{1}{n} \log^+ \|\alpha(s, n)\|$$

if it exists. (We remark that if $\alpha(S \times \mathbf{Z}) \subset SL'(n, \mathbf{R})$, then $\|\alpha(s, n)\| \geqslant 1$, so $\log^+ \|\alpha(s, n)\| = \log \|\alpha(s, n)\|$.)

Proposition **2.10** (Furstenberg-Kesten). — *Let* (S, μ) *be a* $\mathbf{Z}$-*space with invariant probability measure. Suppose* $\alpha : S \times \mathbf{Z} \to GL(n, \mathbf{R})$ *is a tempered cocycle. Then:*

i) $e(\alpha)$ *exists and is a* $\mathbf{Z}$-*invariant function on* S. *Hence, if* $\mathbf{Z}$ *acts ergodically,* $e(\alpha)$ *is a constant.*

ii) *If* α *and* β *are tempered and* $\alpha \sim \beta$, *then* $e(\alpha) = e(\beta)$ *a.e.*

For a proof, see [2], [3], [7]. The following proposition summarizes some other useful properties of $e(\alpha)$.

Proposition **2.11**

1) *If* $\alpha(s, n) = A^n$ *for some matrix* $A \in SL'(n, \mathbf{R})$, *then*

$$e(\alpha) = \max\{\log |\lambda| \mid \lambda \text{ is an eigenvalue of } A\}.$$

2) *More generally, suppose* $A \in SL'(n, \mathbf{R})$ *and* $\alpha(s, n) \equiv A^n \bmod K$ *where* $K \subset SL'(n, \mathbf{R})$ *is a compact subgroup normalized by* A. *Then*

$$e(\alpha) = \max\{\log |\lambda| \mid \lambda \text{ is an eigenvalue of } A\}.$$

3) *Let* $H \subset GL(n, \mathbf{C})$ *be a connected* $\mathbf{R}$-*group,* $H = L \ltimes U$ *where* L *and* U *are respectively reductive and unipotent* $\mathbf{R}$-*subgroups. Let* $q : H_\mathbf{R} \to L_\mathbf{R}$ *be projection. If* $\alpha : S \times \mathbf{Z} \to H_\mathbf{R}$ *is tempered, then* $e(\alpha) = e(q \circ \alpha)$.

4) *More generally, with* H, L, q *as in* (3), *suppose* $\sigma : S \times \mathbf{Z} \to GL(n, \mathbf{R})$ *is tempered, that* $\sigma \sim \alpha$ *where* $\alpha(S \times \mathbf{Z}) \subset H_\mathbf{R}$, *and that* $q \circ \alpha$ *is tempered. Then* $e(q \circ \alpha) \leqslant e(\sigma)$.

5) *If* $\rho : T \to S$ *is a finite extension, and* α *is a tempered cocycle on* S, *then* $e(\alpha) = e(\widetilde{\alpha})$ *where* $\widetilde{\alpha}(t, n) = \alpha(p(t), n)$.

We now recall one fact about Lie algebras and one fact about lattices in semisimple algebraic groups.

Proposition **2.12.** — a) *Let $\mathfrak{I}$ be a complex semisimple Lie algebra. Then for any n the number of inequivalent representations of $\mathfrak{I}$ on a space of dimension n is finite.*

b) *Let G be a connected semisimple algebraic $\mathbf{R}$-group such that $G_{\mathbf{R}}$ has no compact factors. Let $\Gamma \subset G_{\mathbf{R}}$ be a lattice. Then for any non-trivial $\mathbf{R}$-rational representation $\pi : G \to GL(n, \mathbf{C})$, there exists $\gamma \in \Gamma$ such that $\pi(\gamma)$ has an eigenvalue λ with $|\lambda| > 1$.*

Proof. — a) is standard and follows for example from the Weyl character formula [20, p. 363].

b) follows from [12], [15].

With these preliminaries, we can now prove Theorem 2.9.

Proof of Theorem 2.9. — As in the proof of 2.8 it suffices to consider the case in which Γ acts ergodically with an invariant probability measure. Let $f : G \to G_{\mathbf{R}}^{*}$ be the projection, where G^{*} is as in the discussion preceding Theorem 2.2. Let $\psi : \widetilde{G} \to G^{*}$ be the algebraic universal covering of G^{*}, $\widetilde{\Gamma} = \psi^{-1}(f(\Gamma))$. By 2.12 we can choose a finite set $\widetilde{F} \subset \widetilde{\Gamma}$ and $r > 1$ such that if σ is any non-trivial $\mathbf{R}$-rational representation $\sigma : \widetilde{G} \to GL(n, \mathbf{C})$, there exists $\widetilde{\gamma} \in \widetilde{F}$ such that $\sigma(\widetilde{\gamma})$ has an eigenvalue λ with $\log|\lambda| > \log r$. Let $F = f^{-1}(\psi(\widetilde{F}))$. Since Γ_{0} generates Γ, there is an integer N with $(\Gamma_{0} \cup \Gamma_{0}^{-1})^{N} \supset F$. Let $\varepsilon > 0$ be such that $(1 + \varepsilon)^{N} < r$. For each $\gamma \in \Gamma$ and each tempered cocycle α, let $e_{\alpha, \gamma} = e(\alpha \mid S \times \{\gamma^{n}\})$. Then if α is $(\varepsilon, \Gamma_{0})$-admissible, we have for $\gamma \in F$ that

$$(*) \qquad |e_{\alpha, \gamma}(s)| \leqslant \log r \quad \text{for all } s \in S.$$

Now consider the proof of Theorem 2.8. We can construct the $\mathbf{R}$-rational homomorphism π and the lifted homomorphism $\widetilde{\pi} : \widetilde{G} \to GL(n, \mathbf{C})$ as in that proof. Let $p : H_{\mathbf{R}} \to L_{\mathbf{R}}$ and $p_{1} : L_{\mathbf{R}} \to (L_{1})_{\mathbf{R}}$ be the projection maps where the groups are defined as in the proof of 2.8. Since $p_{1} \circ p \circ \alpha \sim \alpha_{\pi}$, it follows that $p \circ \alpha \sim \beta$ where $\beta : S \times \Gamma \to L_{\mathbf{R}}$ is a cocycle with $p_{1}(\beta(s, \gamma)) = \pi(\gamma)$. Thus, if $\widetilde{\gamma} \in \widetilde{G}$ projects to $f(\gamma) \in f(\Gamma)$, we have $p_{1}(\beta(s, \gamma)) = p_{1}(\widetilde{\pi}(\widetilde{\gamma}))$. In other words, $\beta(s, \gamma) = \widetilde{\pi}(\widetilde{\gamma}) \cdot b(s, \widetilde{\gamma})$ where $b(s, \widetilde{\gamma}) \in \ker(p_{1})$. However, $\ker(p_{1})$ is a compact normal subgroup of $L_{\mathbf{R}}$. It follows from Proposition 2.11 (2) that $e_{\beta, \gamma} = \max\{\log|\lambda| \mid \gamma$ is an eigenvalue of $\widetilde{\pi}(\widetilde{\gamma})\}$. However, $e_{\beta, \gamma} \leqslant e_{\alpha, \gamma}$ by Proposition 2.10 and 2.11, and hence it follows from $(*)$ that for all $\gamma \subset F$, $e_{\beta, \gamma} \leqslant \log r$. But this contradicts the choice of r unless π is trivial. We can then complete the argument as in the proof of Theorem 2.8.

A similar argument shows the following.

Theorem **2.13.** — *Let Γ be as in 2.8, S a Γ-space with an invariant probability measure. Suppose $\alpha : S \times \Gamma \to SL'(n, \mathbf{R})$ is a tempered cocycle. If $e_{\alpha, \gamma} = 0$ for all $\gamma \in \Gamma$, then there is a measurable α-invariant metric.*

An argument similar to that in the proof of Theorem 2.8 yields another result of the same type.

Theorem **2.14**. — *Let* G, Γ, S *as in* 2.8. *Suppose* $H \subset GL(n, \mathbf{C})$ *is an* **R**-*group and* $\alpha : S \times \Gamma \to H_{\mathbf{R}}$ *is a cocycle. Suppose* **R**-rank(H) < **R**-rank(G_i) *for every non-trivial almost* **R**-*simple factor* G_i *of* G. *Then there is a measurable* α-*invariant metric.*

3. On principal bundles and jet bundles

In this section we recall some basic information concerning principal bundles and jet bundles incorporating some of the measure theoretic information of the previous section.

Let H be a Lie group, M a manifold, and $\pi : P \to M$ a principal H-bundle. We shall take H to be acting on the right of P, so that M is the quotient space P/H. The trivial principal H-bundle is just the product bundle $M \times H$ with H acting on the second coordinate by right translation. We recall that any section $\varphi : M \to P$ of π defines a trivialization of P, i.e. an equivalence of P with the trivial H-bundle. If φ is smooth (resp. continuous, measurable), then the equivalence of P with $M \times H$ will be smooth (resp. continuous, measurable) as well. If X is any left H-space, then H acts on the right of $P \times X$ via $(z, x).h = (zh, h^{-1}x)$ and the quotient space $E = (P \times X)/H$ will be a bundle over M with fiber X, the " associated bundle " to P with fiber X. Suppose now that Γ acts (on the left) by automorphisms of the principal H-bundle P. Letting Γ act trivially on X, the Γ action and the H action on $P \times X$ commute, so that Γ acts on the bundle E in such a way that the projection map $E \to M$ is a Γ-map.

For any principal H-bundle $\pi : P \to M$ one can always find a Borel section of π. This defines an associated measurable trivialization $P \cong M \times H$. If Γ acts by automorphisms of P, then under this trivialization we have the Γ-action given by $\gamma.(m, h) = (\gamma.m, \alpha(m, \gamma)^{-1}h)$ where $\alpha(m, \gamma) \in H$. Since the trivialization is Borel, $\alpha : M \times \Gamma \to H$ will be a Borel function, and if we write the action of Γ on M on the right (as in section 2) instead of the left, it follows directly that α is a cocycle. It is easy to check that different trivializations yield equivalent cocycles. In other words, whenever Γ acts by automorphisms of the principal H-bundle $P \to M$, we have a naturally defined equivalence class of measurable cocycles $M \times \Gamma \to H$. Furthermore, if M is compact we can choose the image of the Borel section to lie in a compact subset of P. It follows that for M compact and H linear we can choose the cocycle α to be tempered. (See the definition preceding Proposition 2.10.)

If $E \to M$ is the bundle associated to P by the action of H on X, the measurable trivialization of P defines a measurable trivialization of $E \cong M \times X$. If Γ acts by automorphisms of P then the associated action on $M \times X$ is simply given by $\gamma.(m, x) = (\gamma m, \alpha(m, \gamma)^{-1}x)$ where α is the above cocycle. We record the following trivial observation.

Lemma **3.1.** — *There is a measurable Γ-invariant section of the associated bundle* $E \to M$ *if and only if there is an α-invariant measurable function* $M \to X$ *(in the sense of the discussion following Theorem 2.3).*

If $X = \mathbf{R}^n$ and H acts by a differentiable linear representation on $\mathbf{R}^n$, then the associated bundle $\pi : E \to M$ will be a vector bundle and Γ will act on E by vector bundle automorphisms. As usual a Γ-invariant (Riemannian) metric on E will mean a Γ-invariant assignment $m \to B_m$, where B_m is an inner product on $\pi^{-1}(m)$. This assignment may be smooth (resp., continuous, measurable, etc.), and for a given bundle we may enquire as to the existence of a smooth (resp., continuous, measurable, etc.) Γ-invariant metric. If $E = TM$, the tangent bundle of M, we speak of a smooth (resp. continuous, measurable) metric on M. Via Lemma 3.1, we may translate some of the results in section 2 into the present situation. It will be convenient to introduce some notation.

If V is a real vector space and η is an inner product on V, let $\| \ \|_\eta$ denote the corresponding norm on V.

Definition **3.2.** — *If η, ξ are two inner products, let*

$$M(\eta/\xi) = \max \{ \|v\|_\eta \mid v \in V, \|v\|_\xi = 1 \}$$
$$= \max_{v \neq 0} \{ \|v\|_\eta / \|v\|_\xi \}.$$

If $E \to M$ is a vector bundle over M and η, ξ are two measurable Riemannian metrics on E, then $M(\eta/\xi)(s) = M(\eta(s)/\xi(s))$ is a measurable function of $s \in M$.

If $\varepsilon > 0$ and F is a finite set of vector bundle automorphisms of E, we call a metric η on E (ε, F)-invariant if for each $f \in F$,

$$M(f^*\eta/\eta), \ M(\eta/f^*\eta) < \varepsilon.$$

For Theorems 3.3-3.5 we make the following hypotheses. Let G be a connected semisimple Lie group with finite center such that every simple factor of G has $\mathbf{R}$-rank ≥ 2. Let $\Gamma \subset G$ be a lattice and $\Gamma_0 \subset \Gamma$ a fixed finite generating set. Suppose (N, ω) is a manifold with a smooth volume density ω, and $E \to N$ is a vector bundle.

Theorem **3.3.** — *Suppose* $\mathrm{degree}(E) < d(G)$. *Then any action of Γ by vector bundle automorphisms of E which covers an ω-preserving smooth action of Γ on N, for which the ergodic components (of the action on N) have a finite invariant measure, leaves a measurable Riemannian metric on E invariant. If $E = TM$, the associated volume density of this metric can be taken to be ω.*

Theorem **3.4**. — *Let* η *be a smooth metric on* E. *Then there exists* ε > 0 *such that any action of* Γ *by vector bundle automorphisms of* E *which covers a smooth* ω-*preserving action of* Γ *on* N *such that:*

i) *the ergodic components of* Γ *on* N *have finite invariant measure; and*

ii) η *is* (ε, Γ₀)-*invariant;*

leaves a measurable Riemannian metric on E *invariant. If* E = TM, *the associated volume density can be chosen to be* ω.

We recall that if H ⊂ GL(n, **R**) is a closed subgroup, n = degree(E), then an H-structure on E is a reduction of the structure group of E to H. Equivalently, it can be defined as a smooth section of the bundle with fiber GL(n, **R**)/H associated to the principal GL(n, **R**)-bundle of frames of E. If Γ acts by vector bundle automorphisms of E, one can clearly speak of a Γ-invariant H-structure on E.

Theorem **3.5**. — *Suppose* Γ *acts by vector bundle automorphisms of* E *covering an* ω-*preserving smooth action on* N *for which the ergodic components have finite invariant measure. Let* n = degree(E) *and* H ⊂ GL(n, **C**) *an* **R**-*group with* **R**-rank(H) < **R**-rank(G_i) *for every non-trivial simple factor of* G_i *of* G. *If* Γ *leaves an* H_**R**-*structure invariant, then there is a measurable* Γ-*invariant Riemannian metric on* E.

Theorems 3.3-3.5 follow from Lemma 3.1 and the results of section 2. When applied to the tangent bundle of a manifold on which Γ acts, Theorems 3.3-3.5 show that the superrigidity theorem for cocycles yields, under very natural and general hypotheses, the existence of a measurable Γ-invariant Riemannian metric. The next main problem, to which most of this paper is devoted, is to obtain the existence of a C⁰ Γ-invariant Riemannian metric, which we achieve via the existence of the measurable invariant metric. Before proceeding, however, we make some general remarks about diffeomorphisms with a measurable invariant metric.

Let φ : M → M be any volume preserving diffeomorphism of a compact manifold. Consider the following properties:

i) there is a C⁰ φ-invariant metric on M;

ii) there is a measurable φ-invariant metric on M;

iii) $h(\varphi) = 0$ where $h(\varphi)$ is the entropy of φ.

For a general diffeomorphism, we have (i) ⇒ (ii) ⇒ (iii). (The first assertion is trivial. As for the second, from Pesin's formula [14] for the entropy of a volume preserving diffeomorphism, to see that $h(\varphi) = 0$ it suffices to see that $e(\alpha) = 0$ where α is the cocycle corresponding to the derivative under a bounded measurable trivialization of the tangent bundle. However, if there is measurable α-invariant metric, then α ~ β where β takes values in O(n), n = dim M. Since α and β are tempered and equivalent, $e(\alpha) = e(\beta)$, and clearly $e(\beta) = 0$.) On the other hand, the converses are not true in general. For example, the time 1 diffeomorphism of a classical horocycle flow

VOLUME PRESERVING ACTIONS OF LATTICES IN SEMISIMPLE GROUPS   17

satisfies (iii) but not (ii). Thus, for a single diffeomorphism the existence of a measurable invariant metric lies properly between the two classical notions of isometry and o entropy. For the groups we have been considering we have:

Theorem **3.6**. — *Let* Γ *be as in* 3.3 *and suppose* Γ *acts on a compact manifold by smooth diffeomorphisms preserving a smooth volume density. Then the following are equivalent*:

a) *There is a measurable* Γ-*invariant metric.*

b) $h(\gamma) = o$ *for all* $\gamma \in \Gamma$.

Proof. — From Pesin's formula for the entropy of a volume preserving diffeomorphism [14], we have that under a bounded Borel trivialization of the tangent bundle $h(\gamma) = o$ implies $e_{\alpha, \gamma} = o$ where α is the cocycle corresponding to the trivialization. The result then follows from Theorem 2.13.

Conjecture. — For Γ as in 3.6, conditions (a) (or (b)) imply the existence of a C^0-invariant Riemannian metric.

It also follows from this discussion that the above consequences of the superrigidity theorem for cocycles, Theorems 3.3-3.5, without approaching the question of the existence of a C^0 invariant metric, already have strong implications for smooth Γ-actions. For example:

Corollary **3.7**. — *Suppose* M *is a compact manifold,* $\dim M < d(G)$. *Then for any smooth volume preserving* Γ-*action on* M *we have* $h(\gamma) = o$ *for all* $\gamma \in \Gamma$.

This of course follows from 3.3. Similar results may be deduced from 3.4, 3.5. For other results of this type, see [4], [29].

In our proof of the existence of Γ-invariant C^0 Riemannian metrics on TM where M is compact (under suitable hypotheses of course), we shall be applying the above consequences of superrigidity not only to TM but to certain jet bundles on the frame bundle of M which is of course a non-compact manifold. This accounts for our formulation of Theorems 3.3-3.5 for certain non-compact manifolds.

We now recall some basic facts and establish some notation concerning jet bundles. Our basic reference here is [13, Chapter 4].

Let N be a manifold and $E \to N$ a (smooth) vector bundle. (Vector bundles will always be assumed to be finite dimensional unless explicitly indicated otherwise.) Then for each integer k we have another vector bundle $J^k(N; E)$ over N, the k-jet bundle of E, where the fiber at each point $x \in N$ consists of smooth sections of E, two sections being identified if they agree up to order k at x. For any vector bundle E we let $C^\infty(N; E)$ be the space of smooth sections of E. There is a natural map

$$j^k : C^\infty(N; E) \to C^\infty(N; J^k(N; E)),$$

$j^k(f)$ being called the k-jet extension of $f \in C^\infty(N; E)$. We can naturally identify the bundles E and $J^0(N; E)$.

If V, W are finite dimensional real vector spaces, we let $S^r(V, W)$ be the space of symmetric r-linear maps $V^r \to W$. Similarly, if E, F are vector bundles over N, we let $S^r(E, F)$ denote the vector bundle whose fiber at x is $S^r(E_x, F_x)$. If $N \subset \mathbf{R}^n$ is open and E is a product bundle over N, then we can naturally identify $J^k(N; E) \cong \sum_{r=0}^{k} {}^{\oplus} S^r(TN, E)$, where TN is the tangent bundle of N. However this decomposition cannot be carried over in a natural way to an arbitrary N (even for trivial bundles) via coordinate charts. In general there are natural maps

$$J^k(N; E) \to J^{k-1}(N; E) \quad \text{and} \quad S^k(TN, E) \to J^k(N; E)$$

such that

$$0 \to S^k(TN, E) \to J^k(N; E) \to J^{k-1}(N; E) \to 0$$

is exact. (This is the jet bundle exact sequence.) There is no natural splitting of this sequence. (One exception of course is that the composition map $J^k(N; \mathbf{C}) \to J^0(N; \mathbf{C}) \to 0$ where $\mathbf{C}$ denotes the trivial 1-dimensional bundle, is naturally split by the " constant map ".) However, given a connection ∇_E on E and a connection ∇_{T^*N} on T^*N, then there is a natural splitting of the jet bundle exact sequence and hence a natural (given the connections) identification $J^k(N; E) \cong \sum_{r=0}^{k} S^r(TN, E)$. (See [13, p. 90].) If $f : (N, E) \to (N', E')$ is a vector bundle isomorphism covering a diffeomorphism $N \to N'$ (which we still denote by f), then f induces natural maps $S^r(f) : S^r(TN, E) \to S^r(TN', E')$ and $J^k(f) : J^k(N, E) \to J^k(N', E')$. If f is connection preserving on E and T^*N, then under the above identification, $J^k(f) \sim \sum_{r=0}^{k} {}^{\oplus} S^r(f)$. This of course is no longer true if f fails to preserve the connections.

Suppose now that $E = \mathbf{C}$, the 1-dimensional trivial bundle. Any smooth Riemannian metric ξ on N defines in a canonical way a connection on TN, hence on T^*N, and hence an isomorphism $J^k(N; \mathbf{C}) \cong \sum_{r=0}^{k} S^r(TN, \mathbf{C})$. This isomorphism depends upon the k-th iteration of the covariant derivative defined by ξ [13, p. 90] and hence, in local coordinates upon derivatives of the coordinates of the metric up to order k. The metric ξ also defines in a natural way a metric on each $S^r(TN, \mathbf{C})$ and hence via the above isomorphism a metric ξ_k on the bundle $J^k(N; \mathbf{C})$. Our remarks imply the following.

Proposition 3.8. — i) The map $\xi \to \xi_k$ is natural; i.e. if $f : N \to N'$ is a diffeomorphism with $f^*(\xi') = \xi$, then the induced map $J^k(f) : J^k(N; \mathbf{C}) \to J^k(N', \mathbf{C})$ satisfies $J^k(f)^*(\xi'_k) = \xi_k$.

ii) The map $\xi \to \xi_k$ is continuous where metrics on N have the C^k-topology (defined by C^k-convergence on compact subsets of N) and metrics on $J^k(N; \mathbf{C})$ have the C^0-topology (defined by C^0-convergence on compact subsets of N).

4. Integrability of a Measurable Invariant Metric

In this section we make an estimate giving conditions under which a measurable invariant metric will satisfy L^p conditions with respect to a smooth metric. This will be a consequence of the following general fact about actions of groups having Kazhdan's property.

Theorem **4.1.** — *Let Γ be a (countable) discrete group having Kazhdan's property, and $\Gamma_0 \subset \Gamma$ a finite generating set. Then there exists $K > 1$ (depending on Γ_0) with the following property: If (S, μ) is any ergodic Γ-space with invariant probability measure, and $f : S \to \mathbf{R}$ is a measurable function satisfying $|f(s\gamma)| \leqslant K |f(s)|$ for all $s \in S$, $\gamma \in \Gamma_0$, then $f \in L^1(S)$.*

We recall the definition of Kazhdan's property. If π is a unitary representation of a locally compact group Γ on a Hilbert space $\mathscr{H}$, $F \subset \Gamma$ is compact, and $\varepsilon > 0$, a unit vector $x \in \mathscr{H}$ is called (ε, F)-invariant if $\|\pi(\gamma)x - x\| < \varepsilon$ for all $\gamma \in F$. We say that Γ has Kazhdan's property if there is some (ε, F) such that any π with (ε, F)-invariant unit vectors actually has invariant unit vectors [1], [6], [29]. Any discrete group with Kazhdan's property is finitely generated and F can clearly be taken to be a generating set. Furthermore, if Γ_0, $F \subset \Gamma$ are two finite sets with $\Gamma_0^n \supset F$ for some integer n, then any $(\varepsilon/n, \Gamma_0)$-invariant unit vector will be (ε, F)-invariant. (To see this just observe that if x is $(\varepsilon/n, \Gamma_0)$-invariant and $\gamma_i \in \Gamma_0$, then

$$\|\pi(\gamma_1 \ldots \gamma_n)x - x\| \leqslant \|\pi(\gamma_1)\pi(\gamma_2 \ldots \gamma_n)x - \pi(\gamma_1)x\| + \|\pi(\gamma_1)x - x\|$$

$$\leqslant \|\pi(\gamma_2 \ldots \gamma_n)x - x\| + \varepsilon/n.$$

Repeating the argument we see that x is $(\varepsilon, \Gamma_0^n)$-invariant.) Thus, in the defining condition of Kazhdan's property we may assume F is any predetermined finite generating set.

Proof of Theorem 4.1. — We apply Kazhdan's property to the unitary representation π of Γ on $L^2(S) \ominus \mathbf{C}$ defined by translation. By ergodicity of Γ on S (with respect to an invariant probability measure), there are no invariant unit vectors for π. Hence, given Γ_0 there exists $\varepsilon > 0$ (independent of the ergodic action) such that there are no (ε, Γ_0)-invariant unit vectors. In other words, for every unit vector x, there exists $\gamma_0 \in \Gamma$ such that $\|\pi(\gamma_0)x - x\| \geqslant \varepsilon$. For any measurable set $A \subset S$, let χ_A be the characteristic function, p_A the projection of χ_A onto $L^2(S) \ominus \mathbf{C}$, and $f_A = p_A/\|p_A\|$ when A is neither null nor conull. If A, B $\subset$ S with $\mu(A) = \mu(B)$, and A neither null nor conull, then a straightforward calculation shows that

$$\|f_A - f_B\|^2 = \mu(A \, \Delta \, B)/\mu(A)(1 - \mu(A)).$$

ROBERT J. ZIMMER

Choose a measurable set $A_0 \subset S$ such that $|f|$ is bounded on A_0, say $|f(x)| \leqslant B$ for $x \in A_0$ and with $\mu(A_0) > 1/2$. If $\mu(A_0) = 1$, we are done. If not, there exists $\gamma_0 \in \Gamma_0$ such that

$$\| \pi(\gamma_0) f_{A_0} - f_{A_0} \| \geqslant \varepsilon, \quad \text{i.e. } \| f_{A_0 \gamma_0} - f_{A_0} \| \geqslant \varepsilon.$$

Thus, $\mu(A_0 \gamma_0 \Delta A_0) \geqslant \dfrac{\varepsilon^2}{2} (1 - \mu(A_0))$, so that

$$\mu(A_0 \gamma_0 \cap (S - A_0)) \geqslant \frac{\varepsilon^2}{4} (1 - \mu(A_0)).$$

Thus we can choose $A_1 \subset (A_0 \gamma_0 \cap (S - A_0))$ such that $\mu(A_1) = \dfrac{\varepsilon^2}{4} (1 - \mu(A_0))$. (We remark that we can assume S is non-atomic for otherwise ergodicity implies S is finite.) Suppose $|f(x\gamma)| \leqslant K |f(x)|$ for $x \in S$, $K > 1$. Then we have $|f(x)| \leqslant KB$ for $x \in A_1$.

We now repeat the argument applied to $A_0 \cup A_1$ instead of A_0. We deduce the existence of $A_2 \subset (S - (A_0 \cup A_1))$ with $\mu(A_2) = \dfrac{\varepsilon^2}{4} (1 - \mu(A_0 \cup A_1))$ and $|f(x)| \leqslant K^2 B$ for $x \in A_2$. Continuing inductively, we find a disjoint collection of measurable sets $A_i \subset S$ such that $|f(x)| \leqslant K^n B$ for $x \in A_n$, and letting $a_i = \mu(A_i)$ we have, for $n \geqslant 0$, $a_{n+1} = \dfrac{\varepsilon^2}{4} (1 - \sum_{i=0}^{n} a_i)$. Thus if $n \geqslant 1$,

$$a_{n+1} = \frac{\varepsilon^2}{4} (1 - \sum_{i=0}^{n-1} a_i - a_n)$$

$$= a_n - \frac{\varepsilon^2}{4} a_n$$

$$= a_n \left(1 - \frac{\varepsilon^2}{4}\right).$$

$\left(\text{For } n = 0, \text{ we have } a_1 = \dfrac{\varepsilon^2}{4} (1 - a_0).\right)$ We clearly have $\sum_{i=0}^{\infty} a_i = 1$, and hence $\int |f| \leqslant \sum_{n=0}^{\infty} a_n K^n B$. This will be finite if $\lim \dfrac{a_{n+1} K^{n+1}}{a_n K^n} < 1$, i.e. if $K < \dfrac{1}{1 - \varepsilon^2/4}$. This completes the proof.

We now apply Theorem 4.1 to a measurable invariant metric. For two measurable metrics η, ξ on a vector bundle, recall from Definition 3.2 the measurable function $M(\eta/\xi)$. We remark that if η, ξ, τ are metrics, then $M(\eta/\tau) \leqslant M(\eta/\xi) M(\xi/\tau)$. We also note that if $T : W \to V$ is a vector space isomorphism and η, ξ are inner products on V, then $M(T^*\eta/T^*\xi) = M(\eta/\xi)$.

VOLUME PRESERVING ACTIONS OF LATTICES IN SEMISIMPLE GROUPS   21

Corollary **4.2.** — *Let* Γ *be a discrete group with Kazhdan's property,* $\Gamma_0 \subset \Gamma$ *a finite generating set. Let* $E \to N$ *be a continuous vector bundle over a separable metrizable space* N *and suppose that* Γ *acts by vector bundle automorphisms of* E *so that the action on* N *preserves a probability measure* μ. *Suppose* η *and* ξ *are measurable Riemannian metrics on* E *such that:*

i) η *is* Γ-*invariant; and*

ii) *for* $\gamma \in \Gamma_0$, $M(\gamma^* \xi / \xi)$ *and* $M(\xi / \gamma^* \xi)$ *are uniformly bounded by* $K^{1/p}$ $(p \geq 1)$ *where* K *is as in Theorem* 4.1.

Then $M(\eta/\xi)$, $M(\xi/\eta) \in L^p(N, \mu)$.

Proof. — For $s \in N$ and $\gamma \in \Gamma_0$, we have

$$M(\eta/\xi)(s\gamma) = M(\gamma^* \eta / \gamma^* \xi)(s)$$
$$\leqslant M(\eta/\xi)(s) . M(\xi/\gamma^* \xi)(s)$$
$$\leqslant K^{1/p} M(\eta/\xi)(s).$$

Thus $M(\eta/\xi) \in L^p(N)$ by Theorem 4.1. That $M(\xi/\eta) \in L^p$ is proved similarly.

5. The Main Argument

The point of this section is to prove the following theorem stated in the introduction.

Theorem **5.1.** — *Let* G *be a connected semisimple Lie group with finite center such that the* **R**-*rank of every simple factor of* G *is at least* 2. *Let* $\Gamma \subset G$ *be a lattice and* $\Gamma_0 \subset \Gamma$ *a finite generating set. Let* M *be a compact n-manifold, and let* $r = n^2 + n + 1$. *Let* ξ *be a smooth Riemannian metric on* M *with volume density* ω. *Then there is a* C^r-*neighborhood* $\mathcal{O}$ *of* ξ *such that any* ω-*preserving smooth ergodic action of* Γ *on* M *with* ξ $(\mathcal{O}, \Gamma_0)$-*invariant leaves a* C^0-*Riemannian metric invariant.*

We begin the proof with some observations on the frame bundle. The volume density ω defines a $SL'(n, \mathbf{R})$-structure on M. We let $P \to M$ be the corresponding principal $SL'(n, \mathbf{R})$-bundle, where the fiber $F_m \subset P$ over $m \in M$ is the space of frames in TM_m spanning a parallelepiped of volume 1 with respect to $\omega(m)$. Thus, $SL'(n, \mathbf{R})$ acts on the right of P and we have an action of Γ on the left of P which commutes with the $SL'(n, \mathbf{R})$-action. Fix an inner product on the Lie algebra $\mathfrak{sl}(n, \mathbf{R})$ which is $Ad(O(n, \mathbf{R}))$-invariant. In the standard manner this defines a smooth metric on each fiber $F_m \subset P$ which is right invariant under $O(n, \mathbf{R})$ and in any admissable chart for P, under which we identify $F_m \cong SL'(n, \mathbf{R})$, we have that the metric is left invariant under $SL'(n, \mathbf{R})$. In other words, we obtain a smooth metric τ on $Vert(P)$, the vertical subbundle of the tangent bundle to P, which is invariant under the right action of $O(n, \mathbf{R})$ on P. Now let ξ be a smooth metric on M whose volume density is ω. This defines a reduction of P to the subgroup $O(n, \mathbf{R}) \subset SL'(n, \mathbf{R})$, or equivalently a smooth section $\varphi : M \to P/O(n, \mathbf{R})$ of the map $P/O(n, \mathbf{R}) \to M$. Let $q : P \to P/O(n, \mathbf{R})$ be the natural map. The space $q^{-1}(\varphi(M)) \subset P$ is $O(n, \mathbf{R})$-invariant, and in fact consists of exactly one $O(n, \mathbf{R})$-orbit in each fiber over M. The metric τ on Vert(P) is $O(n, \mathbf{R})$ invariant on $q^{-1}(\varphi(M))$ and since $SL'(n, \mathbf{R})$ acts freely on P, $\tau \mid q^{-1}(\varphi(M))$ extends in a unique manner to a metric on Vert(P) which is invariant under the right action of $SL'(n, \mathbf{R})$ on P. (We note that this metric on Vert(P) is not equal to τ on all P.) Furthermore, the metric ξ on M defines in a canonical way a connection on P, i.e. a $SL'(n, \mathbf{R})$-invariant subbundle $\mathrm{Hor}(P) \subset T(P)$ complementary to Vert(P). The metric ξ lifts in a canonical way to a $SL'(n, \mathbf{R})$ invariant metric on Hor(P). Putting together these metrics on Vert(P) and Hor(P) we have, given a smooth metric ξ on M, a canonically defined smooth metric ξ^* on P. The map $\xi \to \xi^*$ has the following readily verified properties.

Proposition 5.2. — i) *For any $r \geqslant 1$, the map $\xi \to \xi^*$ is continuous where metrics on M have the C^r-topology and metrics on P have the C^{r-1}-topology. (On a non-compact manifold the C^r-topology is given by C^r-convergence on compact sets. The loss of one degree of differentiability stems from the fact that the connection form on P defined by ξ is expressed locally in terms of the Cristoffel symbols which contain first derivatives of the components of ξ.)*

ii) *The map $\xi \to \xi^*$ is natural. I.e., if $f: M_1 \to M_2$ is a diffeomorphism such that $f^*(\xi_2) = \xi_1$, then the induced map $\widetilde{f}: P_1 \to P_2$ on the (special) frame bundles satisfies $\widetilde{f}^*(\xi_2^*) = \xi_1^*$.*

iii) *ξ^* is $SL'(n, \mathbf{R})$-invariant.*

We also remark that the volume density on P induced by ξ^* is independent of ξ as long as the volume density of ξ is ω. When we speak of the measure on P, we shall henceforth mean this measure.

Combining Propositions 5.2 and 3.8 we obtain:

Proposition 5.3. — *Let M be compact. Then the map $\xi \to \xi_k^*$ from metrics on TM to metrics on $J^k(P; \mathbf{C})$ satisfies:*

i) *If $f: M \to M'$ is a diffeomorphism such that $f^*(\xi') = \xi$, then*

$$J^k(\widetilde{f})^*((\xi')_k^*) = \xi_k^*.$$

ii) *If $\xi^j \to \xi$ is a convergent sequence of smooth metrics on M with the C^{k+1}-topology, then $(\xi^j)_k^* \to \xi_k^*$ uniformly on P in the sense that $M((\xi^j)_k^*/\xi_k^*)$ and $M(\xi_k^*/(\xi^j)_k^*)$ converge to 1 uniformly on P.*

Proof. — (i) and the fact that $M((\xi^j)_k^*/\xi_k^*)$ and $M(\xi_k^*/(\xi^j)_k^*)$ converge to 1 uniformly on compact subsets of P follow directly from corresponding assertions in Propositions 5.2 and 3.8. It remains only to see that this convergence is uniform on P. By construction, the metrics $(\xi^j)^*$ and ξ^* on the bundle $TP \to P$ are invariant under the action of $SL'(n, \mathbf{R})$ on P. Hence, by Proposition 3.8 (i), the metrics $(\xi^j)_k^*$ and ξ_k^* are also invariant under $SL'(n, \mathbf{R})$, and therefore the functions $M((\xi^j)_k^*/\xi_k^*)$ and $M(\xi_k^*/(\xi^j)_k^*)$ are $SL'(n, \mathbf{R})$ invariant as well. Since M is compact, there is a compact subset of P whose saturation

VOLUME PRESERVING ACTIONS OF LATTICES IN SEMISIMPLE GROUPS   23

under SL$'(n, \mathbf{R})$ is all of P. (For example, just take the compact set $q^{-1}(\varphi(M))$ constructed preceding Proposition 5.2.) Thus, uniform convergence on P follows from uniform convergence on compact subsets.

We wish to use Proposition 5.3 in conjunction with Theorem 3.4 to deduce the existence of measurable Γ-invariant metrics on $J^k(P; \mathbf{C})$. To this end (as well as others) we now examine the ergodic decomposition of the Γ action on the non-compact infinite volume manifold P.

Let us choose a Borel section $\varphi : M \to P$ of the natural projection $P \to M$. As in the beginning of section 3, this defines a Borel trivialization of P via the map $\Phi_0 : M \times SL'(n, \mathbf{R}) \to P$, $\Phi_0(m, g) = \varphi(m)g$. Under this trivialization the action of Γ on $M \times SL'(n, \mathbf{R})$ is given by $\gamma(m, g) = (\gamma m, \alpha(m, \gamma)^{-1}g)$ where $\alpha : M \times \Gamma \to SL'(n, \mathbf{R})$ is a cocycle if the Γ-action is written on the right. Applying Theorem 3.4 to the Γ-action on the bundle $TM \to M$, we deduce that we can find a C^0-neighborhood $\mathcal{O}$ of ξ (and hence a C^k neighborhood, $k \geq 1$) such that for any ω preserving Γ-action with ξ $(\mathcal{O}, \Gamma_0)$-invariant there exists a measurable Γ-invariant metric. By 3.1 and 2.5, this implies that $\alpha \sim \beta$ where β is a cocycle taking values in $O(n, \mathbf{R})$. Let $K \subset O(n, \mathbf{R})$ be the compact subgroup which is the algebraic hull of β (Proposition 2.1) and $\delta \sim \beta$ with $\delta(M \times \Gamma) \subset K$. (Recall that compact real matrix groups are real points of algebraic $\mathbf{R}$-groups.) From the theory of cocycles into compact groups developed in [21], it follows that if the Γ-action on M is ergodic, then the action of Γ on $M \times K$ given by $\gamma.(m, k) = (\gamma m, \delta(m, \gamma)^{-1}k)$ is ergodic. Furthermore the ergodic components of the action of Γ on $M \times SL'(n, \mathbf{R})$ defined by the cocycle δ will be exactly the sets of the form $M \times Ka$, where $a \in SL'(n, \mathbf{R})$. (In other words, the ergodic components are explicitly in bijective correspondence with the coset space $K \backslash SL'(n, \mathbf{R})$. Since $\alpha \sim \delta$, there is a measurable map $\lambda : M \to SL'(n, \mathbf{R})$ implementing the equivalence, and hence the ergodic components of the Γ action defined by the cocycle α will be sets of the form $\{(m, \lambda(m)Ka) \in M \times SL'(n, \mathbf{R})\}$ for some $a \in SL'(n, \mathbf{R})$. Transferring this back to P via Φ_0, and letting $\psi(m) = \varphi(m).\lambda(m)$, we deduce the following.

Lemma **5.4.** — *There is a measurable section ψ of the map* $P \to M$ *and a compact subgroup* $K \subset O(n, \mathbf{R})$ *such that the ergodic components of the action of Γ on P are exactly the sets of the form* $E = \bigcup_{m \in M} \psi(m)Ka$ *where* $a \in SL'(n, \mathbf{R})$. *Furthermore, the corresponding measure on* E *(decomposing the measure on P) is* $\int_M^\oplus \text{Haar}(\psi(m)Ka)dm$ *where* $\text{Haar}(\psi(m)Ka)$ *is the measure defined on the submanifold* $\psi(m)Ka \subset P$ *as the image of the Haar measure of K under the map* $k \to \psi(m)ka$. *In particular, each ergodic component has finite measure.*

Combining 5.4, 3.4, and 5.3 we deduce:

Corollary **5.5** *(With the hypotheses of Theorem 5.1).* — *Let ξ be a smooth metric on* M *with volume density ω. Then for any $k > 1$ there is a C^{k+1}-neighborhood $\mathcal{O}$ of ξ such that any* *smooth ω-preserving Γ-action on* M *with* ξ (𝒪, Γ₀)-*invariant leaves invariant a measurable Riemannian metric on* $J^k(P; \mathbf{C})$.

We shall now modify the measurable metric on $J^k(P; \mathbf{C})$ given by this corollary so that it satisfies certain integrability conditions.

Lemma **5.6.** — *Let* M, Γ *as in* 5.1, ξ *a smooth metric on* M *with volume density* ω. *Then for any* $k > 1$, *there is a* C^{k+1} *neighborhood* 𝒪 *of* ξ *such that any smooth ω-preserving ergodic action of* Γ *on* M *with* ξ (𝒪, Γ₀)-*invariant leaves invariant a measurable Riemannian metric* η *on the jet bundle* $J^k(P; \mathbf{C})$ *such that*:

1) $M(\eta/\xi_k^*)$, $M(\xi_k^*/\eta) \in L^2_{loc}(P)$ *(i.e. they are square integrable on any compact subset of* P*)*;

2) *for any compact* $A \subset P$, $\int_A M(\eta/\xi_k^*)^2 \leq c(A)$, *where* $c(A)$ *is a constant depending only on* A *(and on the choice of Haar measure on* $SL'(n, \mathbf{R})$*)*;

3) *on the naturally split trivial subbundle* $J^0(P; \mathbf{C}) \subset J^k(P; \mathbf{C})$, *the metric* η *on each fiber* **C** *agrees with the usual metric (i.e.* $z, w) = z\overline{w})$. *(We note that this is also true of the metric* ξ_k^* *on* $J^0(P; \mathbf{C})$ *by construction.)*

Proof. — From Proposition 5.3 and Corollary 5.5 we deduce that for any $\varepsilon > 0$, there is a C^{k+1} neighborhood 𝒪 of ξ such that for any ω-preserving smooth ergodic Γ-action on M for which ξ (𝒪, Γ₀)-invariant we have:

i) the metric ξ_k^* on $J^k(P; \mathbf{C})$ is (ε, Γ₀)-invariant; and

ii) there exists a measurable Γ-invariant metric η_0 on $J^k(P; \mathbf{C})$.

Since $J^0(P; \mathbf{C}) \subset J^k(P; \mathbf{C})$ is a Γ-invariant subbundle with a natural Γ-invariant complement, by taking the standard metric on $J^0(P; \mathbf{C})$ and η_0 on the complement one can assume that conclusion (3) holds for η_0. Since η_0 is Γ-invariant, $\eta_0 \mid E$ will be Γ-invariant for almost every ergodic component E of Γ acting on P. Fix one such ergodic component E. By Lemma 5.4 we can apply Corollary 4.2 to E, and hence if ε is sufficiently small we deduce that $M((\eta_0 \mid E)/(\xi_k^* \mid E))$, $M((\xi_k^* \mid E)/(\eta_0 \mid E)) \in L^2(E)$. Let us write $\xi_k^* = i \oplus \widetilde{\xi}_k^*$ where i is the standard metric on $J^0(P; \mathbf{C})$, $\widetilde{\xi}_k^*$ is the metric on the complement, and similarly write $\eta_0 = i \oplus \widetilde{\eta}_0$. We note that if V, W are finite dimensional vector spaces, ξ an inner product on V and η_1, η_2 inner products on W, then on $V \oplus W$ we have

$$M(\xi \oplus \eta_1/\xi \oplus \eta_2) \leq \max(1, M(\eta_1/\eta_2)) \leq 1 + M(\eta_1/\eta_2).$$

Hence, for any $t > 0$ we have

$$M(i \oplus t\widetilde{\eta}_0/i \oplus \widetilde{\xi}_k^*) \leq 1 + tM(\widetilde{\eta}_0/\widetilde{\xi}_k^*)$$
$$\leq 1 + tM(\eta_0/\xi_k^*),$$

VOLUME PRESERVING ACTIONS OF LATTICES IN SEMISIMPLE GROUPS   25

and we have a similar expression for $M(i \oplus \tilde{\xi}_k^* / i \oplus t\tilde{\eta}_0)$. Therefore, by choosing t sufficiently small and replacing η_0 by $i \oplus t\tilde{\eta}_0$, we can assume that η_0 satisfies:

a) $\int_E M(\eta_0/\xi_k^*)^2 \leqslant 2$;

b) $M((\xi_k^* \mid E)/(\eta_0 \mid E)) \in L^2(E)$.

Let K be as in lemma 5.4 and choose a measurable map $\lambda : K \backslash SL'(n, \mathbf{R}) \to SL'(n, \mathbf{R})$ such that $\lambda(Kg) \in Kg$. As in 5.4, we write $E = \bigcup_m \psi(m)Ka$ for some $a \in SL'(n, \mathbf{R})$. If E' is any other ergodic component of the Γ action on P, then we can write $E' = \bigcup_m \psi(m)Kb$ for some $b \in SL'(n, \mathbf{R})$. Thus $E = E'b^{-1}a$. We define the metric η on $J^k(P; \mathbf{C}) \mid E'$ to be $(\lambda(Kb)^{-1}a)^* \eta_0$. Since λ is measurable, the resulting Riemannian metric η on the bundle $J^k(P; \mathbf{C}) \to P$ will be measurable. Furthermore, since the Γ action on P (and hence on $J^k(P; \mathbf{C})$) commutes with the $SL'(n, \mathbf{R})$ action and $\eta_0 \mid E$ is Γ-invariant, the metric η will be Γ-invariant.

We now claim that $M(\eta/\xi_k^*) \in L^2_{loc}(P)$, and in fact that the more exact assertion (2) in the statement of the lemma is valid. Let $A \subset P$ be compact and consider $\Phi^{-1}(A)$ where $\Phi : M \times SL'(n, \mathbf{R}) \to P$ is the measurable trivialization defined by ψ, i.e. $\Phi(m, g) = \psi(m)g$. Since Φ is measure preserving, it suffices to see

$$\int_{\Phi^{-1}(A)} (M(\eta/\xi_k^*) \circ \Phi)^2 \leqslant c(A).$$

Via the measurable section λ we can write $SL'(n, \mathbf{R})$ as a measurable product, $K \times K \backslash SL'(n, \mathbf{R}) \cong SL'(n, \mathbf{R})$, namely by the map $(k, Kg) \to k\lambda(Kg)$. We may view Φ as a map defined on $M \times K \times K \backslash SL'(n, \mathbf{R})$. If we let ν be Haar measure on $SL'(n, \mathbf{R})$ we can write $\nu = \nu_1 \times \nu_2$ under above product decomposition where ν_1 is Haar measure on K and ν_2 is an invariant measure on $K \backslash SL'(n, \mathbf{R})$.

Since ξ_k^* is right invariant under $SL'(n, \mathbf{R})$, from the definition of η it follows that $\Psi = M(\eta/\xi_k^*) \circ \Phi$ satisfies the following condition: $\Psi(m, k, x)$ is independent of $x \in K \backslash SL'(n, \mathbf{R})$. When convenient, we shall write this as $\Psi(m, k)$. The assertion (a) above that $\int_E M(\eta_0/\xi_k^*)^2 \leqslant 2$ implies that for each $x \in K \backslash SL'(n, \mathbf{R})$, $\int_{M \times K} \Psi(m, k, x)^2 dm d\nu_1(k) \leqslant 2$. Since A is compact, the sets $A_m \in SL'(n, \mathbf{R})$ given by $A_m = \{g \in SL'(n, \mathbf{R}) \mid \Phi(m, g) \in A\}$ for $m \in M$ have a uniformly bounded Haar measure, say $\nu(A_m) \leqslant c(A)$ for all $m \in M$. (We remark that by the bi-invariance of Haar measure, $c(A)$ depends only on A and the choice of Haar measure, not on the choice of the section ψ.) For $m \in M$, $x \in K \backslash SL'(n, \mathbf{R})$, let $A_{m,x} = \{k \in K \mid \Phi(m, k, x) \in A\}$. Then

$$\int_{\Phi^{-1}(A)} \Psi^2 = \int_{M \times K \backslash SL'} \left(\int_{A_{m,x}} \Psi(m, k, x)^2 d\nu_1(k) \right) dm d\nu_2(x)$$
$$\leqslant \int_{M \times K \backslash SL'} \nu_1(A_{m,x}) \left(\int_K \Psi(m, k, x)^2 dk \right) dm dx.$$

Integrating over x and recalling that $\Psi(m, k, x)$ is independent of x, we obtain

$$\int_{\Phi^{-1}(A)} \Psi^2 \leqslant \int_M \nu(A_m) \int_K \Psi(m, k)^2 dk dm \leqslant c(A) \int_{M \times K} \Psi^2 \leqslant 2c(A).$$

This verifies assertion (2). The proof that $M(\xi_k^*/\eta) \in L^2_{loc}(P)$ is similar, and this completes the proof of Lemma 5.6.

We now consider Sobolev spaces defined using the metric η constructed in Lemma 5.6.

Quite generally, suppose $E \to N$ is a smooth vector bundle over a manifold N and ω is a smooth volume density on N. Suppose σ is a measurable Riemannian metric on E. Then for any p, $1 \leqslant p \leqslant \infty$, we define $L^p(E)_\sigma$ to be the set of measurable sections f of E (with the usual identifications modulo null sets) such that $\|f\|_{p,\sigma} = \left(\int_N \|f(s)\|^p_{\sigma(s)} ds \right)^{1/p} < \infty$. $L^\infty(E)_\sigma$ and $\|f\|_{\infty,\sigma}$ are defined similarly. Then $L^p(E)_\sigma$ is a Banach space, and for $p = 2$ is a Hilbert space. We also recall that we denote by $C^\infty(N; E)$, $C_c^\infty(N; E)$ the space of smooth (respectively smooth with compact support) sections of E. In general, of course, $C^\infty(N; E) \cap L^p(E)_\sigma$ may be trivial if the measurable σ is very badly behaved locally. However, if ξ is a smooth metric on E and $M(\sigma/\xi) \in L^p_{loc}(N)$, then for $f \in C_c^\infty(N; E)$,

$$\|f(s)\|^p_{\sigma(s)} \leqslant M(\sigma/\xi)(s)^p \|f(s)\|^p_\xi \leqslant M(\sigma/\xi)(s)^p \|f\|^p_{\infty,\xi},$$

and since $\mathrm{supp}(f)$ is compact $f \in L^p(E)_\sigma$.

Suppose now that η is a measurable Riemannian metric on $J^k(N; E)$, σ a smooth metric on $J^k(N; E)$ with $M(\eta/\sigma) \in L^p_{loc}(N)$. If $f \in C^\infty(N; E)$, we have the k-jet extension $j^k(f) \in C^\infty(N; J^k(N; E))$, and of course if $f \in C_c^\infty(N; E)$, then $j^k(f)$ also has compact support. Let us set

$$C^\infty(N; E)_{p,k,\eta} = \{f \in C^\infty(N; E) \mid j^k(f) \in L^p(J^k(N; E))_\eta\}.$$

By the remarks of the preceding paragraph, we have $C_c^\infty(N; E) \subset C^\infty(N; E)_{p,k,\eta}$. We have an injective linear map

$$C^\infty(N; E)_{p,k,\eta} \to L^p(J^k(N; E))_\eta, \quad f \to j^k(f),$$

and the completion of $C^\infty(N; E)_{p,k,\eta}$ with the induced norm will be denoted by $L^{p,k}_\eta(N; E)$, the (p, k) Sobolev space with respect to η. The linear embedding j^k extends to a map $L^{p,k}_\eta(N; E) \to L^p(J^k(N; E))_\eta$ which we still denote by j^k.

Returning now to the situation of Lemma 5.6, we observe that because the measurable Riemannian metric η on $J^k(P; \mathbf{C})$ is Γ-invariant the Γ-action on sections of $J^k(P; \mathbf{C})$ defines a unitary representation of Γ on $L^2(J^k(P; \mathbf{C}))_\eta$ and hence on $L^{2,k}_\eta(P; \mathbf{C})$. (This of course is not necessarily the case for $L^{2,k}_{\xi_k^*}(P; \mathbf{C})$.) We also remark that condition (2) of Lemma 5.6 implies that for $f \in C_c^\infty(P)$ with $\mathrm{supp}(f) \subset A$, we have

$$(*) \qquad \|j^k(f)\|_{2,\eta} \leqslant c(A) \|j^k(f)\|_{\infty, \xi_k^*}$$

where $c(A)$ depends only on A and not on η.

We construct an explicit element of $L^{2,k}_\eta(P; \mathbf{C})$. Recall that a smooth metric ξ on M defines in a natural manner a smooth section φ of the natural projection $P/O(n, \mathbf{R}) \to M$, as in the discussion preceding Proposition 5.2. As in that discussion we have the compact submanifold $q^{-1}(\varphi(M)) \subset P$ whose saturation under the right

VOLUME PRESERVING ACTIONS OF LATTICES IN SEMISIMPLE GROUPS 27

$SL'(n, \mathbf{R})$ action is all of P. Fix a non-zero function $F \in C_c^\infty(O(n, \mathbf{R}) \backslash SL'(n, \mathbf{R}))$ such that $\int |F|^2 = 1$. Then for $p \in P$ define $F_\xi(p) = F(g)$ where $g \in SL'(n, \mathbf{R})$ satisfies $pg^{-1} \in q^{-1}(\varphi(M))$. If $pg_1^{-1}, pg_2^{-1} \in q^{-1}(\varphi(M))$, then there exists $a \in O(n, \mathbf{R})$ such that $pg_1^{-1} = pg_2^{-1}a$, so that by freeness of $SL'(n, \mathbf{R})$ on P, $g_1 \in O(n, \mathbf{R})g_2$. Thus F_ξ is independent of the choice of g. It is easy to see that F_ξ is smooth and in fact $F_\xi \in C_c^\infty(P)$. The following properties of the assignment $\xi \to F_\xi$ are readily verified.

Proposition 5.7

1) *The map* $\xi \to F_\xi$ *is continuous where metrics on* M *have the* C^k *topology and* $C_c^\infty(P)$ *has the topology of uniform* C^k-*convergence on compact sets.*

2) $\int_P |F_\xi|^2 = 1.$

3) *Fix* ξ *and a compact set* $A \subset P$ *which contains an open neighborhood of* $\operatorname{supp}(F_\xi)$. *Then there is a* C^0 *neighborhood* $\mathcal{O}$ *of* ξ *such that if* $\sigma \in \mathcal{O}$, *then* $\operatorname{supp}(F_\sigma) \subset A$.

4) *The assignment is natural. I.e. if* $h : M \to M'$ *is a diffeomorphism and* ξ *is a metric on* M, *then* $F_\xi \circ \widetilde{h} = F_{h \cdot \xi}$ *where* $\widetilde{h} : P' \to P$ *is the induced map on the principal bundles.*

Now fix $\xi, \varepsilon > 0$, and a compact set A as in condition (3) of Proposition 5.7. By 5.6 and 5.7, we can choose a C^{k+1}-neighborhood $\mathcal{O}$ of ξ such that the conclusions of 5.6 hold, (3) of 5.7 holds, and using (1) of 5.7, such that $\sigma \in \mathcal{O}$ implies

$$\|j^k(F_\sigma) - j^k(F_\xi)\|_{\infty, \xi_k^*} < \frac{\varepsilon}{c(A)},$$

where $c(A)$ is as in 5.6. By equation (∗) above, this last assertion implies

(∗∗) $\qquad \|j^k(F_\sigma) - j^k(F_\xi)\|_{2, \eta} < \varepsilon.$

We also remark that $\|j^k(F_\xi)\|_{2, \eta} \geq 1$ by (2) of Proposition 5.7 and (3) of Lemma 5.6. Hence, if Γ acts with ξ $(\mathcal{O}, \Gamma_0)$ invariant, it follows from (∗∗) and (4) of Proposition 5.7 (by taking $\sigma = \gamma^* \xi$ in 5.7 (4)) that we have

$$\|\gamma \cdot j^k(F_\xi) - j^k(F_\xi)\|_{2, \eta} < \varepsilon \quad \text{for all } \gamma \in \Gamma_0.$$

In other words, the vector $j^k(F_\xi) \in L_\eta^{2,k}(P)$ is (ε, Γ_0)-invariant under the unitary representation of Γ on $L_\eta^{2,k}(P)$ and has norm at least one. Thus, if ε is sufficiently small, Kazhdan's property implies there is a non-zero Γ-invariant function $f \subset L_\eta^{2,k}(P)$.

Let $B \subset P$ be a relatively compact open set. By Lemma 5.6 (1), $\int_B M(\xi_k^*/\eta)^2 < \infty$. Since $f \in L_\eta^{2,k}(P)$ we can find a sequence $f_j \in C^\infty(P; \mathbf{C})_{2,k,\eta}$ such that $f_j \to f$ in $L^2(P)$ and $\{j^k(f_j)\}$ is a Cauchy sequence in $L^2(J^k(P; \mathbf{C}))_\eta$. It follows that by restricting to B, we have $f_i \to f$ in $L^2(B)$ and $\{j^k(f_i)\}$ is Cauchy in $L^2(J^k(B; \mathbf{C}))_\eta$. Therefore

$$\int_B \|j^k(f_i)(p) - j^k(f_r)(p)\|_{\xi_k^*} dp \leq \int_B M(\xi_k^*/\eta) \|j^k(f_i)(p) - j^k(f_r)(p)\|_\eta dp$$

$$\leq \left(\int_B M(\xi_k^*/\eta)^2 \right)^{1/2} \|j^k(f_i) - j^k(f_r)\|_{2, \eta}.$$

27

28   ROBERT J. ZIMMER

From this we deduce that $f_i \to f$ in $L^1(B)$ and $\{j^k(f_i)\}$ is Cauchy in $L^1(J^c(B; \mathbf{C}))_{\xi_k^*}$. Since ξ_k^* is a smooth metric, it follows that in local coordinate neighborhoods f has weak derivatives up to order k and that these derivatives are in L^1. Hence, the classical Sobolev embedding theorem implies that if $k > \dim P$, $f \in C^0(P)$. Thus, we may summarize our results so far in this section by:

Lemma 5.8. — *Let* $n = \dim M$ *and* $k = n^2 + n$. *Then for any smooth metric* ξ *on* M *there is a* C^{k+1} *neighborhood* $\mathcal{O}$ *of* ξ *such that for any volume preserving smooth ergodic action of* Γ *on* M *with* ξ ($\mathcal{O}, \Gamma_0$)*-invariant there is a non-zero* Γ*-invariant function* $f \in L^2(P) \cap C^0(P)$.

We now proceed to show how the existence of such a function f implies the existence of a Γ-invariant C^0-Riemannian metric on M. For convenience of notation we shall denote $SL'(n, \mathbf{R})$ by H for the remainder of this section.

Let $\Phi : M \times H \to P$ be a measurable trivialization of P. With this identification we identify f as a function $f \in L^2(M \times H)$ and, as usual, the Γ action on P with the Γ-action on $M \times H$ given by $\gamma(m, h) = (\gamma m, \alpha(m, \gamma)^{-1}h)$ where $\alpha : M \times \Gamma \to H$ is a cocycle. For each $m \in M$, let $f_m \in L^2(H)$ be $f_m(h) = f(m, h)$. Let π be the left regular representation of H on $L^2(H)$. Then Γ invariance of f is the assertion that $f_{\gamma m} = \pi(\alpha(m, \gamma))f_m$. The orbits of H in $L^2(H)$ are locally closed in the weak topology [22] and hence the quotient space $L^2(H)/H$ is a countably separated Borel space (see [29], e.g.). Letting $\widetilde{f_m}$ be the H-orbit of f_m, we deduce that $\widetilde{f_{\gamma m}} = \widetilde{f_m}$ in $L^2(H)/H$. Since $L^2(H)/H$ is countably separated and Γ is ergodic on M, $\widetilde{f_m}$ is essentially constant. I.e. there is $\lambda \in L^2(H)$ such that $f_m \in \pi(H)\lambda$ for almost all $m \in M$. (Cf. [27], [28, Theorem 2.10].) In particular, $\lambda \in L^2(H) \cap C^0(H)$ and $\lambda \neq 0$.

We can rephrase this in terms of the original bundle $P \to M$ as follows. Let $\mathcal{H} \to M$ be the associated bundle to the left H-action on $L^2(H)$, so that the fiber of $\mathcal{H}$ is $L^2(H)$. We have $\pi(H)\lambda \subset L^2(H)$ is a G-invariant subset and hence we have an associated bundle $\mathcal{H}_\lambda \to M$ with fiber $\pi(H)\lambda$ and a natural inclusion $\mathcal{H}_\lambda \hookrightarrow \mathcal{H}$. We

$$\mathcal{H}_\lambda \searrow \quad \swarrow \mathcal{H}$$
$$M$$

remark that the stabilizer of λ in H, say H_λ, is compact (this is true for the regular representation of any locally compact group) and that $\pi(H)\lambda$ is homeomorphic as an H-space to H/H_λ [29]. From the preceding paragraph we see that f defines a measurable section $m \to f_m$ of the associated bundle $\mathcal{H} \to M$ which satisfies the condition that $f_m \in \mathcal{H}_\lambda$ for almost all $m \in M$.

Lemma 5.9. — *Let* $W = \{m \in M \,|\, f_m \in \mathcal{H}_\lambda\}$. *Then*

a) *if* $m \notin W$, *then* $f_m = 0$;
b) *the map* $m \to f_m$ *is continuous on* W;
c) W *is open and* Γ*-invariant.*

Proof. — Fix $m \in M$ and an open neighborhood U of m over which P is trivial. Then over U we can consider f as a function in $L^2(U \times H) \cap C^0(U \times H)$. To verify (a), suppose $m \notin W$. Since W is conull we can choose $u_n \in W$, $u_n \to m$. Then choose $h_n \in H$ such that $f_{u_n} = \pi(h_n)\lambda$. We consider two cases: (1) $h_n \to \infty$; and (2) h_n has a convergent subsequence. In case (1), given any compact subset $A \subset H$, and $\varepsilon, \delta > 0$, the fact that $\lambda \in L^2(H)$ implies that for n sufficient large, $\nu(\{a \in A \mid (\pi(h_n)\lambda)(a) > \varepsilon\}) < \delta$, where $\nu =$ Haar measure. Since $\pi(h_n)\lambda = f_{u_n}$ and $f_{u_n} \to f_m$ uniformly on compact sets, it follows that $f_m = 0$ on A, so $f_m = 0$. In case (2), we can assume $h_n \to h \in H$. Then $f_{u_n} = \pi(h_n)\lambda \to \pi(h)\lambda$ uniformly on compact subsets of H, and since $f_{u_n} \to f_m$ uniformly on compact sets, we have $f_m = \pi(h)\lambda$. Thus $m \in W$. This verifies (a).

The proof of (b) is similar. To see continuity at $m \in W$, it suffices to assume $u_n \in W$, $u_n \to m$, and to show the existence of a subsequence u_{n_j} such that $f_{u_{n_j}} \to f_m$ in $L^2(H)$. Choose $h_n \in H$ such that $f_{u_n} = \pi(h_n)\lambda$. If $h_n \to \infty$, then as above we deduce that $f_m = 0$, contradicting the assumption that $m \in W$. If $h_{n_j} \to h \in H$, we deduce as above that $f_m = \pi(h)\lambda$. But by continuity of the left regular representation, $\pi(h_{n_j})\lambda \to \pi(h)\lambda$ in $L^2(H)$, so $f_{u_{n_j}} \to f_m$ in $L^2(H)$.

Finally, to see (c), we have that $W = \{m \mid f_m$ is not identically $0\}$ and from this it is clear that W is open. Since f is Γ invariant, W is as well.

Lemma 5.10. — *With* W *as in* 5.9, *there is a* Γ-*invariant* C^0-*Riemannian metric on* W *(with volume density* ω).

Proof. — By Lemma 5.9 and the discussion preceding it, there is a Γ-invariant C^0-section of the bundle $\mathcal{H}_\lambda \to M$ defined on the open set W. However, by our remarks preceding Lemma 5.9, $\mathcal{H}_\lambda$ is C^0-isomorphic as a bundle over M to the associated bundle of the left action of H on H/H_λ. Since H_λ is compact, a conjugate of H_λ is contained in $O(n, \mathbf{R})$. Thus we have a Γ-invariant C^0 reduction of the bundle of frames (of volume 1 with respect to ω) to the orthogonal group, and this implies the existence of a C^0 Γ-invariant Riemannian metric on W.

We now complete the proof of Theorem 5.1. Choose a connected component $W_0 \subset W$. The Γ-invariant C^0 metric on W_0, say η, defines a topological distance function on W_0 in the usual way: $d(x, y) = \inf \left\{ \int \|\varphi'(t)\|_\eta \mid \varphi \text{ is a continuous piecewise} \right.$ differentiable path from x to $y \}$ and the topology defined by this distance function agrees with the original topology on W_0. Γ permutes the set of connected components of W and since the measure on W is finite and invariant, ΓW_0 contains only finitely many connected components. Thus, the subgroup $\Gamma_0 \subset \Gamma$ leaving W_0 invariant is of finite index. We have $\Gamma_0 \subset \text{Iso}(W_0, d)$, the latter being the group of isometries of W_0 with respect to d. Since W_0 is a connected manifold, $\text{Iso}(W_0, d)$ is locally compact [8] and

ROBERT J. ZIMMER

the stabilizers of points in W_0 are compact. Let $\overline{\Gamma}_0$ be the closure of Γ_0 in $\mathrm{Iso}(W_0, d)$. Then $\overline{\Gamma}_0$ is a locally compact isometry group with compact stabilizers, and since Γ_0 leaves the finite measure $\omega \mid W_0$ invariant, so does $\overline{\Gamma}_0$. Since Γ is ergodic on W, Γ_0 is ergodic on W_0, and hence we can find $x \in W_0$ such that $\Gamma_0 x$ is dense in W_0. It follows from [5, proof of IV, 2.2] that $\overline{\Gamma}_0$ is transitive on W_0. The finiteness of the measure on W_0 and compactness of the stabilizers imply that the Haar measure of $\overline{\Gamma}_0$ is finite and hence that $\overline{\Gamma}_0$ is compact. This implies that W_0 is compact. Ergodicity of Γ on W implies $\Gamma W_0 = W$, so that W is also compact. Since W is conull in M, it is dense in M. Thus, $W = M$ and the proof of Theorem 5.1 is complete.

6. Perturbations and Near Isometries

Theorem 5.1 immediately implies the following perturbation theorem stated in the introduction.

Theorem **6.1.** — *Let* G *be connected semisimple Lie group with finite center such that every simple factor of* G *has* **R**-rank $\geqslant 2$. *Let* $\Gamma \subset G$ *be a lattice. Let* M *be a compact Riemannian manifold,* $\dim M = n$. *Set* $r = n^2 + n + 1$. *Assume* Γ *acts by isometries of* M. *Let* $\Gamma_0 \subset \Gamma$ *be a finite generating set. Then any volume preserving action of* Γ *on* M *which*

i) *for elements of* Γ_0 *is a sufficiently small* C^r *perturbation of the original action; and*
ii) *is ergodic;*

actually leaves a C^0-*Riemannian metric invariant. In particular there is a* Γ-*invariant topological distance function and the action is topologically conjugate to an action of* Γ *on a homogeneous space of a compact Lie group* K *defined via a dense range homomorphism of* Γ *into* K.

The only part of this result we have not proven is that the compact group K is actually Lie. However, this follows from [10, p. 244].

Theorem 5.1 also implies the following result stated in the introduction.

Theorem **6.2.** — *Let* G, Γ *be as in 6.1. Let* M *be a compact manifold,* $\dim M = n$ $(n > 0)$. *Assume* $n(n + 1) < 2n(G)$ *where* $n(G)$ *is the minimal dimension of a simple factor of* G. *Set* $r = n^2 + n + 1$. *Let* $\Gamma_0 \subset \Gamma$ *be a finite generating set. Then for any smooth Riemannian metric* ξ *on* M *there is a* C^r *neighborhood* $\mathcal{O}$ *of* ξ *such that there are no volume preserving ergodic actions of* Γ *on* M *for which* ξ *is* $(\mathcal{O}, \Gamma_0)$-*invariant.*

Proof. — By Theorem 5.1 we can choose $\mathcal{O}$ such that for any ergodic volume preserving action with ξ $(\mathcal{O}, \Gamma_0)$-invariant there is a dense range homomorphism of Γ into a compact Lie (by [10, p. 244]) group of isometries K of M. By results of Margulis [9] concerning homomorphisms of Γ into Lie groups (see also the work of Raghunathan [16]), it follows that $\dim K \geq n(G)$. On the other hand, it is well known

VOLUME PRESERVING ACTIONS OF LATTICES IN SEMISIMPLE GROUPS 31

that $\dim \mathrm{K} \leqslant \dfrac{n(n+1)}{2}$ [10, p. 246]. Thus $2n(\mathrm{G}) \leq n(n+1)$, which contradicts our hypotheses.

We now turn to the p-adic case.

Theorem **6.3.** — *Let k be a totally disconnected local field of characteristic* 0. *Let G be a connected semisimple algebraic k-group such that the k-rank of every k-simple factor of G is $\geqslant 2$. Suppose $\Gamma \subset \mathrm{G}_k$ is a lattice. Let M be a compact n-manifold $(n > 0)$, and let $r = n^2 + n + 1$. Let $\Gamma_0 \subset \Gamma$ be a finite generating set. Then for any smooth Riemannian metric ξ on M there is a C^r neighborhood $\mathcal{O}$ of ξ such that any volume preserving ergodic action of Γ on M for which ξ is $(\mathcal{O}, \Gamma_0)$-invariant leaves a C^0-Riemannian metric invariant.*

Proof. — The proof is basically the same as that of Theorem 6.2 (and Theorem 5.1) once we have two properties of Γ:

(i) superrigidity for measurable cocycles defined on ergodic Γ-spaces with finite invariant measure;
(ii) Kazhdan's property.

For (i), see [28]. In this case the superrigidity theorem asserts that for any cocycle $\alpha : \mathrm{S} \times \Gamma \to \mathrm{H_R}$ where H is a connected semisimple adjoint $\mathbf{R}$-group such that the algebraic hull of α is $\mathrm{H_R}$, we have $\mathrm{H_R}$ is compact. Property (ii) is well-known [1], [6].

7. Concluding Remarks

Let G be a connected semisimple Lie group with finite center such that every simple factor of G has $\mathbf{R}$-rank $\geqslant 2$. Let $\Gamma \subset \mathrm{G}$ be a lattice. We have the following examples of smooth volume preserving actions of Γ on compact manifolds.

Example **7.1**

a) Suppose H is a Lie group, $\Lambda \subset \mathrm{H}$ is a cocompact subgroup such that H/Λ has a finite H-invariant volume density, and there is a continuous homomorphism $\mathrm{G} \to \mathrm{H}$. Then Γ acts on H/Λ.

b) Suppose N is a Lie group and that G acts by unimodular automorphisms of N. Suppose $\Lambda \subset \mathrm{N}$ is a cocompact lattice such that $\gamma(\Lambda) = \Lambda$ for all $\gamma \in \Gamma$. Then Γ acts on N/Λ.

c) If K is a compact Lie group acting smoothly on a compact manifold M and there is a dense range homomorphisms $\Gamma \to \mathrm{K}$, then Γ acts on M. This includes of course the actions of Γ that factor through a finite quotient of Γ.

We can then formulate the following question.

32   ROBERT J. ZIMMER

Problem. — Can every volume preserving Γ-action (or at least every such ergodic action) on a compact manifold be obtained from these examples by elementary constructions (e.g. products, finite covers and quotients)?

In particular, in case the action preserves a stronger structure than a volume density, we put forward the following conjecture.

Conjecture. — Let $H \subset SL'(n, \mathbf{C})$ be an $\mathbf{R}$-group and suppose Γ acts on a compact manifold M so as to preserve an $H_\mathbf{R}$-structure on M. If $\mathbf{R}$-rank$(H) < \mathbf{R}$-rank(G_i) for every simple factor G_i of G, then there is a C^0 Γ-invariant Riemannian metric on M. In this regard, see Theorem 2.14.

REFERENCES

[1] C. Delaroche and A. Kirillov, Sur les relations entre l'espace dual d'un groupe et la structure de ses sous-groupes fermés, *Séminaire Bourbaki*, No. **343** (1967/1968).
[2] Y. Derriennic, Sur le théorème ergodique sous-additif, *C. R. Acad. Sci.* **281**, Paris (1975), Série A, 985-988.
[3] H. Furstenberg and H. Kesten, Products of random matrices, *Ann. Math. Stat.* **31** (1960), 457-469.
[4] H. Furstenberg, Rigidity and cocycles for ergodic actions of semisimple groups [after G. A. Margulis and R. Zimmer], *Séminaire Bourbaki*, No. **559** (1979/1980).
[5] S. Helgason, *Differential Geometry and Symmetric Spaces*, New York, Academic Press, 1962.
[6] D. Kazhdan, Connection of the dual space of a group with the structure of its closed subgroups, *Funct. Anal. Appl.* **1** (1967), 63-65.
[7] J. F. C. Kingman, The ergodic theory of subadditive stochastic processes, *J. Royal Stat. Soc.* B30 (1968), 499-510.
[8] S. Kobayashi, *Transformation groups in differential geometry*, New York, Springer, 1972.
[9] G. A. Margulis, Discrete groups of motions of manifolds of non-positive curvature, *A.M.S. Translations* **109** (1977), 33-45.
[10] D. Montgomery and L. Zippin, *Topological Transformation Groups*, New York, Interscience, 1955.
[11] G. D. Mostow, Strong rigidity of locally symmetric spaces, *Annals of Math. Studies*, No. **78** (1973).
[12] G. D. Mostow, Intersection of discrete subgroups with Cartan subgroups, *J. Indian Math. Soc.* **34** (1970), 203-214.
[13] R. Palais, Seminar on the Atiyah-Singer Index Theorem, *Annals of Math. Studies*, No. **57**.
[14] Ya. B. Pesin, Lyapunov characteristic exponents and smooth ergodic theory, *Russian Math. Surveys* **32** (1977), 55-114.
[15] G. Prasad and M. S. Raghunathan, Cartan subgroups and lattices in semisimple groups, *Annals of Math.* **96** (1972), 296-317.
[16] M. S. Raghunathan, On the congruence subgroup problem, *Publ. Math. I.H.E.S.* **46** (1976), 107-161.
[17] K. Schmidt, Amenability, Kazhdan's property T, strong ergodicity, and invariant means for group actions, *Erg. Th. and Dyn. Sys.* **1** (1981), 223-236.
[18] S. P. Wang, On isolated points in the dual spaces of locally compact groups, *Math. Ann.* **218** (1975), 19-34.
[19] A. Weil, On discrete subgroups of Lie groups, I, II, *Annals of Math.* **72** (1960), 369-384, and *ibid.*, **75** (1962), 578-602.
[20] D. Zelobenko, Compact Lie groups and their representations, *Translations of Math. Monographs*, vol. 40, A.M.S., Providence, R.I., 1973.
[21] R. J. Zimmer, Extensions of ergodic group actions, *Ill. J. Math.* **20** (1976), 373-409.
[22] R. J. Zimmer, Amenable pairs of ergodic actions and the associated von Neumann algebras, *Trans. A.M.S.* **243** (1978), 271-286.
[23] R. J. Zimmer, Orbit spaces of unitary representations, ergodic theory, and simple Lie groups, *Annals of Math.* **106** (1977), 573-588.

[24] R. J. Zimmer, An algebraic group associated to an ergodic diffeomorphism, *Comp. Math.* **43** (1981), 59-69.

[25] R. J. Zimmer, On the cohomology of ergodic actions of semisimple Lie groups and discrete subgroups, *Amer. J. Math.* **103** (1981), 937-950.

[26] R. J. Zimmer, Strong rigidity for ergodic actions of semisimple Lie groups, *Annals of Math.* **112** (1980), 511-529.

[27] R. J. Zimmer, Orbit equivalence and rigidity of ergodic actions of Lie groups, *Erg. Th. and Dyn. Sys.* **1** (1981), 237-253.

[28] R. J. Zimmer, Ergodic theory, group representations, and rigidity, *Bull. A.M.S.* **6** (1982), 383-416.

[29] R. J. Zimmer, *Ergodic Theory and Semisimple Groups*, forthcoming.

[30] R. J. Zimmer, Arithmetic groups acting on compact manifolds, *Bull. A.M.S.*, to appear.

Department of Mathematics,
University of Chicago,
5734 University Avenue,
Chicago, Illinois 60637.

Manuscrit reçu le 6 décembre 1982.

Invent. math. 75, 425–436 (1984)

Inventiones
mathematicae
© Springer-Verlag 1984

Kazhdan groups acting on compact manifolds*

R.J. Zimmer**

Department of Mathematics, University of Chicago, Chicago, Il. 60637, USA

1. Statement of results

In a series of recent papers [21–23], we have examined certain rigidity properties of actions of semisimple Lie groups and their discrete subgroups on
compact manifolds. In these papers it was assumed that the semi-simple Lie
group in question satisfied the property that the R-rank of every simple factor
was at least two. This assumption was made in order to be able to apply two
major tools: the superrigidity theorem for measurable cocycles [19] (a generalization of Margulis superrigidity [7, 20]), and Kazhdan's property [2, 5, 20], or
"property T". In this paper we shall see that in certain special situations, we
can actually use Kazhdan's property itself as a replacement for the superrigidity theorem. Thus, we are able to obtain results (under more restrictive
hypotheses than in [21–23]) for actions of an arbitrary Kazhdan group (i.e.
group with Kazhdan's property) on compact manifolds. In particular, we
obtain results for the R-rank 1 Lie groups $Sp(1,n)$ and lattice subgroups in it.
(These are Kazhdan by [6].) Our main geometric results are summarized in
Theorems 1–9.

Theorem 1. *Let H be a connected non-compact simple Lie group with Kazhdan's
property. Let M be a compact Lorentz manifold. Then every action of H on M
by Lorentz transformations is trivial.*

More generally, the same proof shows:

Theorem 2. *Suppose M is a compact manifold with a G-structure (possibly of
higher order) where G is a real algebraic group such that the reductive component of G_0, the connected component of the identity in G, has compact center
and such that every non-compact simple factor of G_0 fails to have Kazhdan's
property. Let H be as in Theorem 1. Then every action of H on M which
preserves the G-structure is trivial.*

---

* Research partially supported by NSF Grant MCS 8004026
** Sloan Foundation Fellow

If G is a simple Lie group with $\mathbf{R}$-rank$(G) \geq 2$, Theorems 1, 2 follow from a much stronger result proved in [22].

We now consider the case in which the Kazhdan group in question is no longer assumed to be connected. In this event, Theorems 1 and 2 are no longer valid. For example, one may have a dense embedding of a discrete Kazhdan group Γ in a compact Lie group K. Since K clearly admits K-invariant Lorentz structures, the action of Γ on K preserves a Lorentz structure. However, the next theorem asserts that up to topological conjugacy, any action of Γ by Lorentz transformations can be derived from the above class of examples if we make the one further assumption of ergodicity.

Theorem 3. *Let Γ be a Kazhdan group, M a compact Lorentz manifold, and suppose Γ acts ergodically on M so as to preserve the Lorentz structure. Then there is a Γ-invariant C^0-Riemannian metric on M. Hence, there is a compact Lie group K, a closed subgroup $K_0 \subset K$, and a homomorphism $\pi: \Gamma \to K$ such that the action of Γ on M is topologically conjugate to the action of Γ on K/K_0 defined by π.*

More generally, we have:

Theorem 4. *Let Γ be a Kazhdan group, M a compact manifold, and suppose M has a G-structure where G is as in Theorem 2. Assume further that the G-structure is of finite type (in the sense of E. Cartan [12]). Suppose Γ acts ergodically on M so as to preserve the G-structure. Then there is a Γ-invariant C^0-Riemannian metric on M, and hence the final assertion of Theorem 3 holds.*

If we delete the hypotheses that Γ-acts ergodically and that the G-structure is of finite type, then under the remaining hypotheses in Theorem 4 we can deduce the existence of a Γ-invariant measurable Riemannian metric on M. This implies in particular that the entropy of all the diffeomorphisms defined by the Γ-action is 0. Hence, letting $G = SL(2, \mathbf{R})$, we have:

Theorem 5. *Suppose Γ is a Kazhdan group and that Γ acts by volume preserving diffeomorphisms on a compact 2-dimensional manifold M. Then there is a measurable Γ-invariant Riemannian metric on M. In particular, $h(\gamma) = 0$ for all $\gamma \in \Gamma$ (where h denotes entropy).*

Theorem 5 raises the problem of describing the actions of Γ on a surface with a measurable invariant metric. In this direction we have the following result.

Theorem 6. *If Γ is a discrete Kazhdan group and Γ acts isometrically on a compact 2-dimensional manifold M, then the action is trivial on a subgroup of finite index. (I.e. the image of Γ in $\mathrm{Iso}(M)$ is finite.)*

This follows directly from the following algebraic result.

Theorem 7. *Suppose $\pi: \Gamma \to G$ is a homomorphism where Γ is a discrete Kazhdan group and G is a compact Lie group locally isomorphic to $SO(3, \mathbf{R})$ or $SO(4, \mathbf{R})$. Then $\pi(\Gamma)$ is finite.*

Theorem 7 is related to the work of Rosenblatt [10], Schmidt [11], Margulis [8], and Sullivan [13] on the uniqueness of invariant means on S^{n-1}. More precisely, in [8, 13] the uniqueness of a rotationally invariant mean on $L^\infty(S^{n-1})$ is established for $n \geq 5$ by constructing a densely embedded countable Kazhdan subgroup of $SO(n, \mathbf{R}), n \geq 5$. The question of the uniqueness of the rotationally invariant mean on $L^\infty(S^2)$ and $L^\infty(S^3)$ remains open. Theorem 7 implies that the only proof which has appeared so far for $n \geq 5$ will not work in the case $n = 3, 4$.

Combining Theorem 6 with Theorem 4, where we take the G-structure in Theorem 4 to be a connection with volume density, we obtain:

Theorem 8. *Let M be a compact 2-manifold with connection and volume density. Let Γ be a countable Kazhdan group. Then there is no volume preserving ergodic action of Γ on M by affine transformations.*

When Γ is a lattice in a simple Lie group of $\mathbf{R}$-rank at least two, Theorem 8 follows from the results of [23].

Theorems 5, 6, 8 raise the question as to whether there are any volume preserving ergodic actions of a Kazhdan group on a surface. The next result concerns this question and is closely related to the results of [21]. If ξ is a Riemannian metric on a compact manifold M, $\mathcal{O}$ is a neighborhood of ξ in the C^r-topology on the space of all smooth metrics on M, and S is a set of diffeomorphisms of M, we say that ξ is $(S, \mathcal{O})$-invariant if $\phi^* \xi \in \mathcal{O}$ for all $\phi \in S$.

Theorem 9. *Let Γ be a discrete Kazhdan group and $\Gamma_0 \subset \Gamma$ a finite generating set. Let M be a compact 2-manifold and ξ a smooth Riemannian metric on M. Then there is a C^7-neighborhood, $\mathcal{O}$, of ξ in the space of metrics on M such that there is no $\mathrm{vol}(\xi)$-preserving ergodic action of Γ on M for which ξ is $(\Gamma_0, \mathcal{O})$-invariant.*

As indicated in the first paragraph, a major ingredient in the proofs of some of these results is a property of actions of Kazhdan groups which serves as a replacement for one aspect of the superrigidity theorem for measurable cocycles. This result also concerns the measure theoretic behavior of cocycles.

Theorem 10. *Let (S, μ) be a standard probability space and suppose Γ is a Kazhdan group which acts (measurably and) ergodically on S so as to preserve μ. Let G be a real algebraic group and $\alpha: S \times \Gamma \to G$ a (measurable) cocycle. Then α is equivalent to a cocycle β such that $\beta(S \times \Gamma) \subset G_1 \subset G$ where G_1 is an algebraic subgroup of G and G_1 is a Kazhdan group.*

This assertion had been conjectured a few years ago by Furstenberg. The proof depends on representation theoretic results of R. Howe and C.C. Moore [3, 9]. When Γ is discrete it is also valid when the hypotheses that Γ be a Kazhdan group and μ be finite and invariant are replaced by the weaker assumptions that μ be quasi-invariant and the Γ action on S is a Kazhdan action in the sense of [17]. When S is a single point. Theorem 10 reduces to a result of S.P. Wang [15]. The proof of Theorem 10 also depends upon the following.

Theorem 11. *Let (S, μ) be a standard probability space and suppose G is a (locally compact) Kazhdan group acting ergodically on S and preserving μ. Let*

$\alpha: S \times G \to H$ be a cocycle where H is locally compact and amenable. Then $\alpha \sim \beta$ where $\beta(S \times G) \subset K$ where $K \subset H$ is a compact subgroup.

This result has previously been proven when G is discrete [11, 17] or when G admits a lattice subgroup [20]. The idea of the proof of Theorem 11 is basically the same as in the known cases.

The content of the remaining sections is as follows: Sect. 2 contains some preliminaries and the proofs of Theorems 10, 11; Sect. 3 contains the proofs of Theorems 6, 7; in Sect. 4, the remaining geometric results are deduced.

2. Proof of Theorems 10, 11

We recall the definition of a Kazhdan group. Let G be a locally compact second countable group, $\mathscr{H}$ a separable Hilbert space, and π a unitary representation of G on $\mathscr{H}$. If $K \subset G$ is a compact set, $\varepsilon > 0$, and $\chi \in \mathscr{H}$ is a unit vector, we say that χ is (ε, K)-invariant if $\|\pi(g)\chi - \chi\| < \varepsilon$ for all $g \in K$. The group G is called a Kazhdan group if every π which has (ε, K)-invariant vectors for all (ε, K) actually has (non-zero) invariant vectors. Alternatively, if π_n is a sequence of unitary representations on $\mathscr{H}_n$, we say $\pi_n \to I$ if for every (ε, K) there are unit vectors $\chi_n \in \mathscr{H}_n$ such that $\|\pi_n(g)\chi_n - \chi_n\| < \varepsilon$ for all $g \in K$. Then G is a Kazhdan group if and only if for every sequence of irreducible unitary representations π_n with $\pi_n \to I$, we have $\pi_n = I$ for n sufficiently large [2, 5]. Among the connected simple Lie groups with finite center, only those locally isomorphic to $SO(1, n)$ and $SU(1, n)$ fail to be Kazhdan. (Cf. [2, 5, 6, 15, 20].) In particular any simple real algebraic group with **R**-rank at least two is Kazhdan. However $Sp(1, n)$ gives a sequence of Kazhdan groups with **R**-rank equal one. A lattice in a locally compact group is Kazhdan if and only if the ambient group is Kazhdan.

Suppose now that (S, μ) is a standard probability space and G is a locally compact second countable group acting measurably on S, preserving μ. If H is a standard Borel group, we recall that a function $\alpha: S \times G \to H$ is called a cocycle if for all $g, h \in G$, $\alpha(s, gh) = \alpha(s, g)\alpha(sg, h)$ a.e. We shall assume familiarity with this notion, which is discussed in [20] for example. (Any such α can be changed on a set of measure 0 so that the cocycle identity holds for all s, g, h [20, Appendix]. Therefore, for most purposes we may as well assume equality everywhere.) Now let $H = U(\mathscr{H})$, the unitary group of the separable Hilbert space $\mathscr{H}$. If $u_n: S \times G \to U(\mathscr{H})$ is a sequence of cocycles, we shall write $u_n \to I$ if for any compact $K \subset G$, and any $\varepsilon > 0$, there are measurable functions $\phi_n: S \to \mathscr{H}$ such that for all n, $\|\phi_n(s)\| = 1$ for almost all s, and for all n and all $g \in K$,

$$\mu\{s | \|u_n(s, g)\phi_n(sg) - \phi_n(s)\| < \varepsilon\} > 1 - \varepsilon.$$

For any cocycle $u: S \times G \to U(\mathscr{H})$, we can form the induced representation σ_u of G on $L^2(S; \mathscr{H})$ given by $(\sigma_u(g)f)(s) = u(s, g)f(sg)$. If $u_n \to I$, it is clear that $\sigma_{u_n} \to I$ as representations of G. If $\alpha: S \times G \to H$ where H is locally compact and π_n are representations of H, then $u_n = \pi_n \circ \alpha$ will be cocycles on $S \times G$.

Lemma 12. *If $\pi_n \to I$, then $u_n \to I$.*

If G is discrete, this is [17, Lemma 2.6]. The proof of Theorem 11 then follows from Lemma 12 just as the discrete case in [17] follows from [17, Lemma 2.6].

Proof of Lemma 12. Let $K \subset G$ be a compact symmetric set and $\varepsilon > 0$. Let ν denote Haar measure on G. If $A \subset H$ is compact and $g \in G$, let

$$S(g, A) = \{s \in S \mid \alpha(s, g) \in A, \alpha(s, g^{-1}) \in A\}.$$

Since α is measurable and H is σ-compact, there is a compact set $A \subset H$ such that

$$(\mu \times \nu)(\{(s, g) \in S \times K \mid \alpha(s, g) \in A, \alpha(s, g^{-1}) \in A\}) \geq (1 - \varepsilon/3)(\nu(K)).$$

By Fubini's theorem, $\nu(K_0) > 0$, where

$$K_0 = \{g \in K \mid \mu(S(g, A)) \geq 1 - \varepsilon/3\}.$$

Clearly K_0 is symmetric. Thus $K_0^2 = K_0 K_0^{-1}$ contains a neighborhood W of the identity [20, Appendix]. By the cocycle identity $\alpha(s, gh) = \alpha(s, g)\alpha(sg, h)$ and the fact that μ is G-invariant, for any $\chi \in K_0^2$ we have $\mu(S(\chi, A^2)) \geq 1 - 2\varepsilon/3$. Now choose $g_i \in G$, $i = 1, \ldots, n$, such that $\bigcup g_i W \supset K$. Then there is a compact set $B \subset H$ such that $\mu(S(g_i, B)) > 1 - \varepsilon/3$ for all i. It follows as above from the cocycle identity that $\mu(S(g, BA^2)) > 1 - \varepsilon$ for all $g \in K$. Since $\pi_n \to I$, for n large we can choose $v_n \in \mathscr{H}_n$ such that $\|v_n\| = 1$ and $\|\pi_n(h)v_n - v_n\| < \varepsilon$ for all $h \in BA^2$. Define $\phi_n \in L^2(S; \mathscr{H}_n)$ by $\phi_n(s) = v_n$ for all $s \in S$. Then for $g \in K$ we have $\mu(\{s \mid \|u_n(s, g)\phi_n(sg) - \phi_n(s)\| < \varepsilon\}) > 1 - \varepsilon$, proving the Lemma.

We now recall some results about unitary representations of algebraic groups, following [3, 9]. If G is a locally compact group, $H \subset G$ a closed subgroup, and π a unitary representation of G, we say that the matrix coefficients of π vanish at ∞ mod H if for all vectors v, w we have $\langle \pi(g)v \mid w \rangle \to 0$ as $g \to \infty$ in G/H ($g \in G$). The projective kernel of π, denoted by P_π, is defined by $P_\pi = \{g \in G \mid \pi(g) \text{ is a scalar multiple of } I\}$.

Lemma 13. a) [3]. *If π is an irreducible representation of a real algebraic group, then the matrix coefficients vanish at ∞ mod P_π.*

b) [9, Theorem 6.1]. *If π is an irreducible representation of a real algebraic group then P_π is of finite index in its algebraic hull.*

We shall also need the following related result.

Lemma 14. [16, Proposition 3]. *If G is a real algebraic group, π an irreducible representation, and $\pi(P_\pi)$ is closed, then the action of G on the Hilbert space $\mathscr{H}$ defined by π is smooth, in the sense that all orbits are locally closed (or equivalently, $\mathscr{H}/\pi(G)$ is a countably separated Borel space.)*

As further preparation for the proof of Theorem 10, we recall some results about cocycles.

Lemma 15. (*Cocycle reduction Lemma;* [20, Lemma 5.2.11]). *Suppose (S, μ) is an ergodic G-space where G is locally compact and that $\alpha: S \times G \to H$ is a cocycle.*

Suppose X is an H-space on which H acts smoothly (i.e. X/H is countably separated) and that there is a Borel function $\phi: S \rightarrow X$ which is α-invariant, i.e. for each g, $\phi(s\,g) = \phi(s)\,\alpha(s,g)$ a.e. Then $\alpha \sim \beta$ where $\beta(S \times G) \subset H_1$ and H_1 is the stabilizer in H of some point of X.

Lemma 16. *Suppose $\alpha: S \times G \rightarrow H$ is a cocycle where G acts ergodically on S and H is a real algebraic group. Suppose $p: T \rightarrow S$ is a finite extension of G-spaces (i.e. the fibers are a.e. of a fixed finite cardinality). Let $\tilde{\alpha}: T \times G \rightarrow H$ be defined by $\tilde{\alpha}(t,g) = \alpha(p(t),g)$. If $\tilde{\alpha}$ is equivalent to a cocycle $\tilde{\beta}$ such that $\tilde{\beta}(T \times G) \subset H_1$ where H_1 is an algebraic subgroup, then α is equivalent to a cocycle β such that $\beta(S \times G) \subset H_2$ where H_2 is algebraic, $H_2 \supset H_1$, and H_2/H_1 is finite.*

Proof. Let $q: H \rightarrow H/H_1$ denote the projection. Let X be the space of finite subsets of H/H_1. Then H acts on X and the action is smooth since H/H_1 is a variety. There is a function $\phi: T \rightarrow H$ such that $\beta(t,g) = \phi(t)\,\alpha(p(t),g)\,\phi(t\,g)^{-1}$. Define $f: S \rightarrow X$ by $f(s) = \{q(\phi(t))\,|\,p(t) = s\}$. It follows that $f(s\,g) = f(s)\,\alpha(s,g)$. By Lemma 15, α is equivalent to a cocycle γ into the stabilizer of a finite subset of H/H_1. Such a group has a conjugate of H_1 as a subgroup of finite index. Thus, conjugating γ by an element of H gives the required cocycle β.

Finally, we recall the notion of the algebraic hull of a cocycle [18, 20].

Lemma 17. *Suppose $\alpha: S \times G \rightarrow H$ is a cocycle where H is a real algebraic group. Then there is an algebraic subgroup $H_1 \subset H$ such that $\alpha \sim \beta$ where $\beta(S \times G) \subset H_1$ and α is not equivalent to a cocycle taking values in a proper algebraic subgroup of H_1. The subgroup H_1 is unique up to conjugacy.*

H_1 is called the algebraic hull of α and if $H_1 = H$ we shall abuse notation and say that α is Zariski dense in H. If α is Zariski dense in H and $h: H \rightarrow L$ is a surjective rational homomorphism, then $h \circ \alpha$ is Zariski dense in L.

We can now prove Theorem 10.

Proof of Theorem 10. We have $\alpha: S \times \Gamma \rightarrow G$ where Γ is Kazhdan, S has a Γ-invariant probability measure, the action is ergodic, and G is real algebraic. We wish to show α is equivalent to a cocycle into an algebraic Kazhdan subgroup of G. We can clearly assume that α is Zariski dense in G. By [20, Proposition 9.2.7], by passing to a finite ergodic extension $T \rightarrow S$ the cocycle $\tilde{\alpha}$ as defined in Lemma 16 will have G^0 as its algebraic hull, where G^0 is the Zariski connected component of the identity in G. If $A \subset B$ is a subgroup of finite index in a locally compact group B, then A is Kazhdan if and only if B is Kazhdan. Thus, the remarks above and Lemma 16 enable us to assume we are in the case in which G is Zariski connected. We now show that G is Kazhdan.

Let π_n be a sequence of irreducible unitary representations of G and assume $\pi_n \rightarrow I$. We claim $\pi_n = I$ for n sufficiently large. By Lemma 12, $\pi_n \circ \alpha \rightarrow I$ and by the paragraph preceding Lemma 12 and the fact that Γ is Kazhdan, for n sufficiently large there is a non-0 Γ-invariant vector for the induced representation $\sigma_{\pi_n \circ \alpha}$. For ease of notation, for a fixed large n, set $\pi_n = \pi$. Then we have a non-0 function $f: S \rightarrow \mathcal{H}$ such that for all $\gamma \in \Gamma$, $\pi(\alpha(s, \gamma))\,f(s\,\gamma) = f(s)$ a.e. By Lemma 13(b), P_π has only finitely many components (in the Hausdorff topology). It follows from Lemma 14 that the G-action on $\mathcal{H}$ given by π is smooth. Since the action of Γ on S is ergodic, Lemma 15 shows that $\alpha \sim \beta$ where $\beta(S \times \Gamma) \subset G_1$ where G_1 is the stabilizer of a non-0 vector in $\mathcal{H}$. Since $\pi(P_\pi)$ is a closed subgroup of the unit circle (by Lemma 13(b)), $P_\pi/\mathrm{Ker}\,\pi$ is compact, and hence $(G_1/\mathrm{Ker}\,\pi)(P_\pi/\mathrm{Ker}\,\pi)$ is closed in $G/\mathrm{Ker}\,\pi$. It follows that $G_1 P_\pi$ is closed in G. Furthermore, by Lemma 13(a), $G_1 P_\pi/P_\pi$ is compact and hence $G_1 \bar{P}_\pi/\bar{P}_\pi$ is a compact subgroup of $G/\bar{P}_\pi$, where $\bar{P}_\pi$ is the Zariski closure of P_π. Since compact subgroups of real algebraic groups are algebraic and the algebraic hull of $p \circ \beta$ is $G/\bar{P}_\pi$, where $p: G \to G/\bar{P}_\pi$ is projection, it follows that $G_1 \bar{P}_\pi/\bar{P}_\pi = G/\bar{P}_\pi$ and hence $G_1 P_\pi = G$. Thus $G_1 P_\pi$ is of finite index in G. In particular, if we let $G_0 \subset G$ be the connected component of the identity in the Hausdorff topology, we have that $\pi(G_0)$ leaves a line invariant. Since G/G_0 is finite, π is finite dimensional.

We can write $G = L \ltimes U$ where U is unipotent and connected and L is reductive and Zariski connected. We have $L = Z(L)[L,L]$ where $Z(L)$ is the center and $Z(L) \cap [L,L]$ is finite. Let $q: G \to G/[L,L] \ltimes U \cong Z(L)/Z(L) \cap [L,L]$ be the natural map. Then $q \circ \alpha$ is a cocycle which is Zariski dense in an abelian group and since compact subgroups of real algebraic groups are algebraic, Theorem 11 implies $Z(L)/Z(L) \cap [L,L]$ is compact. Thus, $Z(L)$ is compact as well.

We also have $G_0 = L_0 \ltimes U$ where L_0 is the connected component of the identity in L in the Hausdorff topology. Let $\lambda: G_0 \to$ circle be the one dimensional representation defined by a $\pi(G_0)$-invariant line in $\mathcal{H}$. Let $W = (\mathrm{Ker}\,\lambda \,|\, U)_0$ (the subscript denoting connected component), so that $W \subset U$ is a connected, and hence algebraic, subgroup. Clearly, either $W = U$ when $\lambda \,|\, U$ is trivial, or $U/W \cong \mathbf{R}$ when $\lambda \,|\, U$ is non-trivial. Consider first the latter case. Since W is clearly normalized by G_0, the representation λ factors to a homomorphism of $L_0 \ltimes U/W$. Then $\mathrm{Ker}(\lambda \,|\, U/W)$ will be isomorphic to a copy of $\mathbf{Z}$ in $U/W \cong \mathbf{R}$. Since L_0 is connected and $\mathrm{Ker}(\lambda \,|\, (U/W))$ is normalized by L_0, L_0 must commute with $\mathrm{ker}\,\lambda \,|\, (U/W)$ and hence with all U/W (since the action of L_0 on U/W is rational). In other words, $L_0 \ltimes U/W \cong L_0 \times \mathbf{R}$. By passing to a suitable finite ergodic extension $T \to S$ the cocycle $\tilde{\alpha}$ defined in Lemma 16 will be equivalent to a cocycle $\beta: T \times \Gamma \to G_0 = L_0 \ltimes U$. Let $\psi: L_0 \ltimes U \to L_0 \ltimes U/W \cong L_0 \times R$ be the natural projection. Then $p_2 \circ \psi \circ \beta: T \times \Gamma \to \mathbf{R}$ is a cocycle where $p_2: L_0 \times \mathbf{R} \to \mathbf{R}$ is projection on the second factor. By Theorem 11, $p_2 \circ \psi \circ \beta$ is trivial, and hence β (and therefore $\tilde{\alpha}$) is equivalent to a cocycle into L_0. By Lemma 16, α is equivalent to a cocycle into a finite extension of L. Any such group is L itself and since α is Zariski dense, we deduce that U is trivial. This clearly contradicts the assumption that $\lambda \,|\, U$ is non-trivial. Hence, we deduce that $\lambda \,|\, U$ is in fact trivial.

In this case λ factors to a one dimensional representation of $L_0 = Z(L_0)[L_0,L_0]$. Hence, λ is actually trivial on $[L_0,L_0] \ltimes U$, the latter being a normal subgroup of G. Thus, $\pi([L_0,L_0] \ltimes U)$ leaves a non-trivial vector fixed. By irreducibility of π, $\pi([L_0,L_0] \ltimes U)$ leaves all vectors fixed, so that π factors to a representation of $G/[L_0,L_0] \ltimes U$ which is compact since $Z(L)$ is compact.

Summarizing, we have seen that for n sufficiently large, all π_n factor to a representation of a fixed (i.e. independent of n) compact quotient of G. Since any compact group is Kazhdan, $\pi_n = I$, completing the proof of Theorem 10.

We will apply Theorem 10 to the situation in which every Kazhdan subgroup of G is compact.

Proposition 18. *Suppose G is a locally compact group such that for every unitary representation of G with no non-trivial invariant vectors the matrix coefficients vanish at infinity. Suppose G is not Kazhdan. Then every closed Kazhdan subgroup of G is compact.*

Proof. Since G is not Kazhdan there is a sequence of unitary representations $\pi_n \to I$ such that π_n has no invariant vectors. If $H \subset G$ is Kazhdan, $\pi_n | H \to I$ and hence for large n there is a $\pi_n(H)$-invariant unit vector. Since the matrix coefficients of π_n vanish at ∞, H is compact.

Corollary 19. *If k is a local field of characteristic 0, then every Kazhdan subgroup of $SL(2, k)$ is compact.*

Proof. The matrix coefficients for unitary representations of these groups with no non-trivial invariant vectors vanish at infinity by [3], and it is well known that these groups are not Kazhdan. (In fact, except for $k = \mathbb{C}$, $SL(2, k)$ admits a lattice which is isomorphic to a free group [1, 4].)

Corollary 20. *Let G be a real algebraic group with connected component G_0. Suppose every simple non-compact factor of G_0 fails to have Kazhdan's property. Then every closed Kazhdan subgroup of G is compact.*

Proof. It suffices to show every closed Kazhdan subgroup $H \subset G_0$ is compact. Let $G_0 = L \ltimes U$ where L is reductive, U is unipotent, and L and U are connected. If the projection of H into L is contained in a compact subgroup, then H is contained in an amenable subgroup of G_0. Thus H is both amenable and Kazhdan, and therefore is compact. Writing $L = Z(L)[L, L]$ where $Z(L) \cap [L, L]$ is finite, it suffices to see that the projections of H into both $L/Z(L)$ and $L/[L, L]$ are contained in a compact subgroup. Since $H/[H, H]$ is compact, the projection into $L/[L, L]$ is compact. The group $L/Z(L)$ is semisimple and hence it suffices to see that the projection onto each simple factor is precompact. By hypothesis the non-compact simple factors are not Kazhdan and hence the result follows from Proposition 18, Lemma 13(a), and the elementary fact that the closure of the image of a Kazhdan group under a continuous homomorphism of locally compact groups is also Kazhdan.

3. Images of Kazhdan groups in $SO(3, \mathbf{R})$

In this section we prove Theorems 6, 7. We begin with the proof of Theorem 7. Thus, let Γ be a countable discrete Kazhdan group. Since Kazhdan's property is invariant under finite extensions and $SO(4, \mathbf{R})$ is locally isomorphic to $SO(3, \mathbf{R}) \times SO(3, \mathbf{R})$ it suffices to prove the theorem for homomorphisms of Γ into any one compact Lie group locally isomorphic to $SO(3, \mathbf{R})$. We thus assume $\pi \colon \Gamma \to SU(2)$.

We first claim that the eigenvalues of $\pi(\gamma)$ are algebraic numbers for all $\lambda \in \Gamma$. If $\pi(\gamma)$ has a transcendental eigenvalue, then there is $\sigma \in \mathrm{Aut}(\mathbb{C})$ such that $(\sigma \circ \pi)(\gamma)$ has an eigenvalue c such that $c > 1$. (Here we are making the customary abuse of notation by letting σ denote both an element of $\mathrm{Aut}(\mathbb{C})$ and the automorphism of $SL(2, \mathbb{C})$ which it induces.) Thus, $\sigma \circ \pi \colon \Gamma \to SL(2, \mathbb{C})$ has an image which does not have compact closure, which is impossible by Corollary 19. Therefore, all $\pi(\gamma)$ have algebraic eigenvalues. We also observe that we may assume $\pi(\Gamma)$ is Zariski dense in $SL(2, \mathbb{C})$. Otherwise, $\pi(\Gamma)$ would be contained in a solvable subgroup of $SL(2, \mathbb{C})$ and being Kazhdan, $\pi(\Gamma)$ would be finite.

By a standard argument (see [14] or [20, Lemma 6.1.7] e.g.), the assertions of the preceding paragraph imply that there is a non-trivial rational representation $\lambda \colon SL(2, \mathbb{C}) \to GL(n, \mathbb{C})$ for some n such that $\lambda(\pi(\Gamma)) \subset GL(n, K)$ where K is the field of algebraic numbers. Since any Kazhdan group is finitely generated, we can assume $\lambda(\pi(\Gamma)) \subset GL(n, k)$ where $[k : \mathbb{Q}] < \infty$. Set $G = \lambda(SL(2, \mathbb{C}))$, so that G is a rank 1, connected, almost simple algebraic group. Since $\lambda(\pi(\Gamma)) \subset G \cap GL(n, k)$ is Zariski dense in G, G is defined over k [20, Prop. 3.1.8]. It clearly suffices to see that $\lambda(\pi(\Gamma))$ is finite and hence we shall abuse notation and denote $\lambda \circ \pi$ simply by π.

If $\tau \colon k \to \mathbb{C}$ is any field embedding, then $\tau \circ \pi \colon \Gamma \to \tau(G)$ is a homomorphism, and since $\tau(G)$ is locally isomorphic to $SL(2, \mathbb{C})$, by Corollary 19 $\tau \circ \pi(\Gamma)$ must lie in a compact subgroup of $\tau(G)$. On the other hand, suppose K is a completion of k with respect to an non-archimedian valuation. Let $i \colon G_k \to G_K$ be the inclusion. Since G has rank 1, if G is K-split then G_K is locally isomorphic to $SL(2, K)$ and hence by Corollary 19 $i \circ \pi(\Gamma)$ lies in a compact subgroup of G_K. On the other hand, if G is not K-split, then G is K-anisotropic and by a result of Bruhat and Tits G_K is compact. Furthermore, since $\pi(\Gamma)$ is finitely generated only finitely many prime ideals will appear in the factorization of denominators of elements of $\pi(\Gamma)$, and hence for almost all valuations of k we have that $i \circ \pi(\Gamma) \subset G_{O_K}$ where O_K is the ring of integers in K. Taken together, these remarks imply that under the canonical embedding α of G_k in the adele group $G_{A(k)}$ we have $\alpha(\pi(\Gamma))$ is contained in a compact subgroup of $G_{A(k)}$. However, it is well known that $\alpha(G_k)$ is discrete in $G_{A(k)}$ and we deduce that $\alpha(\pi(\Gamma))$ is finite. Since α is injective, $\pi(\Gamma)$ is also finite, and this completes the proof of Theorem 7.

Proof of Theorem 6. Let M be a compact 2-manifold with Riemannian metric and isometry group $I(M)$. It is well known that $\dim(I(M)) \leq 3$ and that $I(M)$ is compact. If $I(M)$ is not simple, its connected component must be abelian in which case any homomorphism $\pi \colon \Gamma \to I(M)$ is known to have finite image. On the other hand, if $I(M)$ is simple, $I(M)$ is locally isomorphic to $SO(3, \mathbb{R})$ and the result than follows from Theorem 7.

4. Geometric consequences

In this section we deduce the remaining unproven geometric results stated in Sect. 1. As we indicated in the first section the basic point is to use Theorem

10 as a partial replacement for superrigidity in the arguments of [21–23]. Thus, we shall only indicate here how to make this substitution and refer the reader to [21–23] for the remainder of the proof.

To prove Theorem 2 we first observe that since the reductive component of G_0 has a compact center, the G-structure also defines a volume density on M which is invariant under the automorphisms of the G-structure. The argument of [22] shows that it suffices to construct a measurable H-invariant Riemannian metric on the vector bundle $TM \to M$ defined on each ergodic component S of the H-action on M. This in turn is implied by the assertion that the cocycle $S \times H \to GL(n, \mathbf{R})$ defined by the derivative is equivalent to a cocycle into a compact subgroup of $GL(n, \mathbf{R})$. Since the H-action leaves a G-structure invariant, the cocycle is equivalent to one into G_1, where G_1 is the first order component of G. We can then simply apply Theorem 10, and Theorem 2 follows.

Theorem 1 is of course just the special case of Theorem 2 in which $G = O(1, n)$.

To prove Theorem 4, the argument in [23] shows that it suffices to see that for any r, the cocycle $\alpha: M \times \Gamma \to G^{(r)}$ is equivalent to a cocycle into a compact subgroup, where $G^{(r)}$ is the r-th prolongation of G, and the cocycle is defined by the action on a measurable trivialization of the r-th prolongation of the G-structure. However the reductive component of the connected component of the algebraic group $G^{(r)}$ is the same as that of G, and hence the required assertion follows from Corollary 20.

Theorem 3 is of course an immediate consequence of Theorem 4. Theorem 5 also follows from Theorem 10 since the existence of a measurable invariant metric follows from the assertion that the derivative cocycle $\alpha: S \times \Gamma \to SL(2, \mathbf{R})$ is equivalent to a cocycle into the circle.

Finally, we indicate the proof of Theorem 9. Let M be a compact 2-manifold with volume density and $P \to M$ the bundle of frames of volume 1. Suppose we have a volume preserving action of the Kazhdan group Γ on M and hence an induced action of Γ on P. Let $J^k(P; \mathbf{C}) \to P$ be the vector bundle of k-jets of $\mathbf{C}$-valued functions on P. An examination of the proof of [21, Theorem 5.1] shows that it suffices to prove the following two assertions:

a) there is a measurable Γ-invariant metric on the tangent bundle of M;

b) there is a measurable Γ-invariant metric on each bundle $J^k(P; \mathbf{C}) \to P$.

Assertion (a) follows as above since $\dim M = 2$. To see (b), we first observe that the frame bundle on P, $F(P) \to P$, which has structure group $GL(5, \mathbf{R})$, has a reduction to the subgroup $G \subset GL(5, \mathbf{R})$,

$$G = \left\{ \begin{pmatrix} I & * \\ 0 & A \end{pmatrix} \middle| I = 3 \times 3 \text{ identity}, \ \det(A)^2 = 1 \right\},$$

which is invariant under the diffeomorphisms of P defined by volume preserving diffeomorphisms of M. (Namely, fix any frame of the Lie algebra $sl(2, \mathbf{R})$. At each $p \in P$ this defines a vertical frame, and the reduction of $F(P)$ to G is defined by bundle of frames which at each point in P contain this vertical frame.) It follows that the k-th frame bundle $F^{(k)} P \to P$, has a Γ-invariant reduction to $G^{(k)}$, the k-th prolongation of G. Since the jet bundle $J^{(k)}(P; \mathbb{C}) \to P$ is naturally (with respect to diffeomorphisms of P) associated to the principal bundle $F^{(k)}P \to P$, it follows that to see the existence of a Γ-invariant measurable metric on $J^{(k)}(P; \mathbb{C}) \to P$ it suffices to see that a cocycle $\alpha: P_0 \times \Gamma \to G^{(k)}$ defined by a measurable trivialization of the Γ-invariant reduction of $F^{(k)}P$ to $G^{(k)}$ is actually equivalent to a cocycle into a compact subgroup of $G^{(k)}$, where P_0 is an ergodic component of the action of Γ on P. (Cf. [21, Lemma 5.4, Corollary 5.5.) By assertion (a), these ergodic components have a finite Γ-invariant measure (cf. [21, Lemma 5.4]). Furthermore the reductive component of the connected component of $G^{(k)}$ is equal to that of G, and hence is isomorphic to $SL(2, \mathbf{R})$. Thus, Theorem 10 and Corollary 20 apply once again to show that α is equivalent to a cocycle into a compact subgroup of $G^{(k)}$.

References

1. Borel, A., Harder, G.: Existence of discrete cocompact subgroups of reductive groups over local fields. J. für Math. **298**, 53–64 (1978)
2. Delaroche, C., Kirillov, A.: Sur les relations entre l'espace dual d'un groupe et la structure de ses sous-groupes fermés, Seminaire Bourbaki. No. **343**, 1967/68
3. Howe, R., Moore, C.C.: Asymptotic properties of unitary representations. J. Funct. Anal. **32**, 72–96 (1979)
4. Ihara, Y.: On discrete subgroups of the two by two projective linear group over p-adic fields. J. Math. Soc. Japan **18**, 219–235 (1966)
5. Kazhdan, D.: Connection of the dual space of a group with the structure of its closed subgroups. Funct. Anal. Appl. **1**, 63–65 (1967)
6. Kostant, B.: On the existence and irreducibility of certain series of representations. Bull. Amer. Math. Soc. **75**, 627–642 (1969)
7. Margulis, G.A.: Discrete groups of motions of manifolds of non-positive curvature. Amer. Math. Soc. Translations **109**, 33–45 (1977)
8. Margulis, G.A.: Some remarks on invariant means. Monat. Math. **90**, 233–235 (1980)
9. Moore, C.C.: The Mautner phenomenon for general unitary representations. Pac. J. Math. **86**, 155–169 (1980)
10. Rosenblatt, J.: Uniqueness of invariant means for measure-preserving transformations. Trans. Amer. Math. Soc. **265**, 623–636 (1981)
11. Schmidt, K.: Amenability, Kazhdan's property T, strong ergodicity, and invariant means for ergodic group actions. Ergod. Th. Dyn. Syst. **1**, 223–236 (1981)
12. Singer, I., Sternberg, S.: The infinite groups of Lie and Cartan. J. d'Anal. Math. **15**, 1–114 (1965)
13. Sullivan, D.: For $n \geq 3$ there is only one finitely-additive rotationally invariant measure on the n-sphere defined on all Lebesgue measurable sets. Bull. Amer. Math. Soc. **4**, 121–123 (1981)
14. Vinberg, E.B.: Rings of definition of dense subgroups of semisimple linear groups. Math. USSR IZV. **5**, 45–55 (1971)
15. Wang, S.P.: On the Mautner phenomenon and groups with property T. Amer. J. Math. **104**, 1191–1210 (1982)
16. Zimmer, R.J.: Uniform subgroups and ergodic actions of exponential Lie groups. Pac. J. Math. **78**, 267–272 (1978)
17. Zimmer, R.J.: On the cohomology of ergodic actions of semisimple Lie groups and discrete subgroups. Amer. J. Math. **103**, 937–950 (1981)
18. Zimmer, R.J.: An algebraic group defined by an ergodic diffeomorphism. Comp. Math. **43**, 59–69 (1981)

436   R.J. Zimmer

19. Zimmer, R.J.: Strong rigidity for ergodic actions of semisimple Lie groups. Annals of Math.
    112, 511–529 (1980)
20. Zimmer, R.J.: Ergodic Theory and Semisimple Groups, to appear
21. Zimmer, R.J.: Volume preserving actions of lattices in semisimple groups on compact mani-
    folds. Publ. Math. I.H.E.S., to appear
22. Zimmer, R.J.: Semisimple automorphism groups of G-structures. J. Diff. Geom. (in press)
23. Zimmer, R.J.: Actions of lattices in semisimple groups preserving a G-structure of finite type.
    Preprint

Oblatum 11-VI-1983

Ergod. Th. & Dynam. Sys. (1985), **5**, 301–306
Printed in Great Britain

Actions of lattices in semisimple groups preserving a *G*-structure of finite type

ROBERT J. ZIMMER

(*Received* 5 *September* 1983)

Department of Mathematics, University of Chicago, Chicago, Illinois 60637, USA

Abstract. In this paper we study actions of lattices in semisimple groups preserving a *G*-structure of finite type.

1. *Statement of results*

In this paper we continue the investigation begun in [9] concerning actions of arithmetic groups on compact manifolds. Here we shall focus attention on actions preserving a *G*-structure (possibly of higher order).

Let H be a connected semisimple Lie group with finite centre such that $\mathbb{R}$-rank $(H') \geq 2$ for every simple factor H' of H. Let $\Gamma \subset H$ be a lattice subgroup, i.e. Γ is discrete and H/Γ has a finite H-invariant measure. Let M be a compact manifold, dim $(M) = n > 0$. For any integer $k \geq 1$, we let GL $(n, \mathbb{R})^{(k)}$ be the group of k-jets of diffeomorphisms of $\mathbb{R}^n$ leaving the origin fixed [1]. Suppose $G \subset$ GL $(n, \mathbb{R})^{(k)}$ is an algebraic subgroup, so that one may speak of a *G*-structure on M (which will be a structure of order k). If Q is a Lie group, we will denote its Lie algebra by $L(Q)$. The main point of this paper is to prove the following theorem.

THEOREM 1. *Let* H, Γ, M *be as above. Suppose that* Γ *acts smoothly on* M *and preserves a smooth volume density. Let* $G \subset$ GL $(n, \mathbb{R})^{(k)}$ *be an algebraic subgroup and suppose that the* Γ-*action on* M *preserves a* G-*structure. If*

    (i) Γ *acts ergodically;*
    (ii) G *is of finite type* (*in the sense of E. Cartan* [3]); *and*
    (iii) *every homomorphism* $L(H) \to L(G)$ *is trivial;*
then there is a Γ-*invariant* C^0-*Riemannian metric on* M, *and hence a* Γ-*invariant topological distance function. Thus there is a compact Lie group* K, *a closed subgroup* $K_0 \subset K$, *and a homomorphism* $\pi: \Gamma \to K$ *with dense range such that the* Γ-*action on* M *is topologically conjugate to the* Γ-*action on* K/K_0 *defined by* π.

We note that condition (iii) will be satisfied for example if $\mathbb{R}$-rank $(G/\mathrm{rad}\,(G)) < \mathbb{R}$-rank (H') for every simple factor H' of H, where rad (G) is the radical of G. We recall that a Lorentzian metric on a manifold is an O $(1, n-1)$ structure and an affine connection on M is a second order G-structure with $\mathbb{R}$-rank $(G/\mathrm{rad}\,(G)) = n - 1$. Both of these are of finite type [1]. Thus we obtain the following corollaries.

R. J. Zimmer

COROLLARY 2. *Let H, Γ be as above. Then any smooth ergodic action of Γ preserving a Lorentzian metric on a compact manifold M actually leaves a C^0-Riemannian metric invariant. Thus the action is topologically conjugate to an action on K/K_0 as described in theorem 1.*

COROLLARY 3. *Let H, Γ be as above. Let M be a compact manifold,* dim $M = n$. *Suppose there are an affine connection and a volume density on M such that Γ acts by volume preserving affine transformations of M. If*

(i) *Γ acts ergodically; and*

(ii) *every representation $L(H) \to \mathfrak{sl}(n, \mathbb{R})$ is trivial (e.g. if $\mathbb{R}$-rank $(H') \geq n$ for every simple factor H' of H);*

then there is a Γ-invariant C^0-Riemannian metric, and the remaining conclusions of theorem 1 hold.

For any Γ as above, it follows from the work of Margulis [2] that any compact Lie group K for which there is a dense range homomorphism $Γ \to K$ satisfies

$$\dim K \geq \min \{\dim H' | H' \text{ a simple factor of } H\}.$$

Since any compact group acting transitively on a topological n-manifold is a Lie group of dimension at most $n(n+1)/2$, Margulis' results combined with theorem 1 yield the non-existence of ergodic G-structure preserving actions in a suitable dimension range. One has for example, from corollary 3 and Margulis' result that the natural action of SL $(n, \mathbb{Z})$ on the torus T^n by group automorphisms is an ergodic affine volume preserving action of minimal dimension. More precisely:

COROLLARY 4. *Let $n \geq 3$. If M is a compact manifold with connection and volume density, and $0 < \dim M < n$ then there is no ergodic affine volume preserving action of SL $(n, \mathbb{Z})$ on M.*

Remarks. (i) Theorem 1 was conjectured in [9] without the assumption that the G-structure be of finite type. It remains open for G-structures of infinite type. Similarly the conjecture in [9] did not assume ergodicity and it remains open as to whether or not this hypothesis is necessary.

(ii) The notion of finite type in the sense of E. Cartan is based on the termination after finitely many steps of an inductively defined natural prolongation scheme. Tanaka [4], [5] has introduced another natural prolongation scheme for G-structures which will be finite for certain infinite Cartan type (in fact even non-elliptic) G-structures. Our proof of theorem 1 will be valid for G-structures of finite type in the sense of Tanaka as well.

(iii) Describing the ways in which Γ can act by automorphisms of a G-structure is a natural geometric generalization of the algebraic question of describing the homomorphisms of Γ into G. If G is an algebraic group, then for Γ as above this latter question has to a large extent been resolved by Margulis [2]. (See also [8].) Margulis' results imply for example that if there are only trivial homomorphisms $L(H) \to L(G/\text{rad}(G))$, then any homomorphism of Γ into G has a precompact image.

Theorem 1 can thus be considered as a geometric extension of this algebraic result of Margulis. We also remark that the group of automorphisms, A, of a G-structure of finite type is known to be a Lie group [1]. Furthermore, every simple subalgebra of $L(A/\text{rad } A)$ must then be a Lie subalgebra of $L(G/\text{rad } G)$ [10]. However, the group of connected components of A may well be infinite and hence one cannot simply apply [10] and Margulis' results.

One also has analogous (in fact stronger) results for actions of suitable lattices over p-adic fields. Namely:

THEOREM 5. *Let k be a totally disconnected local field of characteristic 0, H a connected k-group, almost k-simple, with k-rank$(H) \geq 2$. Let $\Gamma \subset H_k$ be a lattice. Then any volume preserving ergodic action of Γ on any compact manifold (of positive dimension) preserving a (possibly higher order) G-structure of finite type leaves a C^0-Riemannian metric invariant.*

In particular, this applies to ergodic actions by Lorentz transformations and to ergodic volume preserving actions by affine transformations. In [9] we conjectured that any volume preserving action of Γ on a compact manifold is isometric, and theorem 5 can be considered as a result in this direction.

2. *Preliminaries to the proofs*

We shall assume that the reader is familiar with G-structures, G-structures of higher order, and G-structures of finite type. (See [1], [3] for example.) We shall also assume familiarity with the discussion of measurable cocycles and measurable invariant metrics in [9, §§ 2, 3]. As in [9], our basic approach will be first to obtain measure theoretic information about the actions in question, and then to convert this to continuous information which will imply our results. Again as in [9] we will obtain the measure theoretic information by an application of the superrigidity theorem for cocycles which we proved in [6] (a generalization of Margulis' superrigidity theorem [2], [8]) and via Kazhdan's property; the conversion to continuous information will be achieved via the Sobolev embedding theorem.

For the following discussion, by an algebraic group we will mean the $\mathbb{R}$-points of an algebraic $\mathbb{R}$-group. If Q is a Lie group, $L(Q)$ will denote its Lie algebra. We let H and Γ be as in the second paragraph of § 1. Suppose G is an algebraic group. Let (S, μ) be a standard measure space and suppose Γ acts measurably on S so that μ is a finite Γ-invariant measure. Let $\alpha : S \times \Gamma \to G$ be a (measurable) cocycle. The proof of [9, theorem 2.8] (based on the superrigidity theorem and employing Kazhdan's property as well) shows the following.

THEOREM 6. *Let H, Γ, G, α be as above. Suppose every homomorphism $L(H) \to L(G)$ is trivial. Then α is (measurably) equivalent to a cocycle into a compact subgroup of G.*

We re-emphasize that the condition on the Lie algebras holds for example if $\mathbb{R}$-rank $(G/\text{rad } (G)) < \mathbb{R}$-rank (H') for every simple factor H' of H. It holds in many other cases as well of course.

We shall need two other results proved in [9].

LEMMA 7. *Let M be a second countable topological space and μ a finite measure on M which is positive on open sets. Let G be a locally compact second countable group and $P \to M$ a (continuous) principal G-bundle. Suppose Γ is a countable group and that Γ acts by principal bundle automorphisms of P covering a μ-preserving ergodic action of Γ on M. Suppose there is a Γ-invariant function $f \in L^2(P) \cap C(P)$. (We remark that the left Haar measure on G defines a measure on each fibre, and hence together with μ defines a measure on P.) Then there is*

(i) *a compact subgroup $K \subset G$; and*

(ii) *an open Γ-invariant conull set $W \subset M$;*

such that there is a continuous Γ-invariant section $\phi: W \to P/K$ of the natural projection $P/K \to M$.

This result is not explicitly stated in [**9**] but it is easily extracted from [**9**, lemma 5.9] and the discussion preceding it. We remark that although the principal bundle in question in [**9**] is a reduction of the frame bundle of a manifold, this is irrelevant for deducing lemma 7.

The conclusion of the proof of [**9**, theorem 5.1] also shows the following.

LEMMA 8. *Let M be a compact manifold on which a group Γ acts ergodically by volume density preserving diffeomorphisms. Let $W \subset M$ be an open, conull Γ-invariant subset on which there is a Γ-invariant C^0-Riemannian metric. Then $W = M$ and the isometry group of M with respect to the induced topological distance function is transitive.*

3. *Proof of the theorems*

We prove theorem 1. Let $G^{(k)}$ be the kth prolongation of G. Then $G^{(k)}$ is an algebraic group which in fact is a semi-direct product $G \ltimes N_k$ where N_k is a connected unipotent group. We let $P^{(k)} \to M$ be the corresponding prolongation of the G-structure so that we have maps

Each map is a principal bundle projection, q defining an N_k-bundle, π the given G-bundle, and $\pi \circ q$ a $G^{(k)}$-bundle. Since G is of finite type, there is some k for which $G^{(l)} = G^{(k)}$ for all $l \geq k$, and we set $G^{(k)} = \tilde{G}$, $P^{(k)} = \tilde{P}$. The automorphism group of the G-structure, Aut (P), acts on $\tilde{P}$, and as a consequence of the choice of k, there is a complete parallelism on the manifold $\tilde{P}$ that is invariant under Aut (P). In particular, there is a Riemannian metric ξ on $\tilde{P}$ that is invariant under Aut (P). As described in [**9**, §3], for any integer $r \geq 0$ the metric ξ on the vector bundle $T\tilde{P} \to \tilde{P}$ canonically determines a metric ξ_r on the vector bundle $J^r(\tilde{P}; \mathbb{R}) \to \tilde{P}$ of r-jets of smooth real valued functions on $\tilde{P}$. Since ξ is Aut (P)-invariant, it follows that ξ_r is as well. We form the space of L^2-sections of this bundle with respect to the metric ξ_r on the fibres, which we denote by $L^2(J^r(\tilde{P}; \mathbb{R}))_{\xi_r}$. The group Aut (P) acts naturally on the sections of this bundle and since ξ_r is Aut (P)-invariant, we obtain a unitary representation of Aut (P) on the Hilbert space $L^2(J'(\tilde{P};\mathbb{R})_{\xi_r}$. The Sobolev space $L^{2,r}_{\xi_r}(\tilde{P})$ is, as usual, defined to be the completion of

$$\{f \in C^\infty(\tilde{P};\mathbb{R})|j'(f) \in L^2(J'(\tilde{P};\mathbb{R}))_{\xi_r}\}$$

with respect to the norm induced by the embedding in $L^2(J'(\tilde{P};\mathbb{R}))_{\xi_r}$. Thus $L^{2,r}_{\xi_r}(\tilde{P})$ is a Hilbert space which we can identify with a subspace of $L^2(J'(\tilde{P};\mathbb{R}))_{\xi_r}$, and this subspace is clearly Aut (P)-invariant. We thus obtain a unitary representation of Aut (P) on the Sobolev space $L^{2,r}_{\xi_r}(\tilde{P})$. We have a natural norm-decreasing linear map $i: L^{2,r}_{\xi_r}(\tilde{P}) \to L^2(\tilde{P})$ with dense image which intertwines the unitary representations of Aut (P) on these spaces. We let $i^*: L^2(\tilde{P}) \to L^{2,r}_{\xi_r}(\tilde{P})$ be the adjoint map. This is again a norm-decreasing linear map which is injective (owing to density of the image of i) and which intertwines the unitary representations of Aut (P). (This seemingly innocent last remark is in fact basic to the proof. If the representation on $L^{2,r}_{\xi_r}(\tilde{P})$ was not unitary, i^* would intertwine the adjoint actions on these Hilbert spaces, and on the Sobolev space this would not agree with the standard natural action. Thus Aut (P)-invariance of ξ_r, which we obtained from a complete parallelism, plays a crucial role.)

We have a homomorphism of Γ into Aut (P) and we now consider the above unitary representations restricted to Γ. As in [9, § 3], if we measurably trivialize the bundle $\tilde{P} \to M$ by writing $\tilde{P} \cong M \times \tilde{G}$, the action of Γ on $\tilde{P}$ will be given by

$$\gamma \cdot (m, g) = (\gamma m, \alpha(m, \gamma)^{-1}g)$$

where $\alpha: S \times \Gamma \to \tilde{G}$ is a cocycle. By theorem 6, α is (measurably) equivalent to a cocycle β such that $\beta(M \times \Gamma) \subset C$, where $C \subset \tilde{G}$ is a compact subgroup. Thus, if $f \in L^2(\tilde{G})$ is a non-0 function such that $f(ag) = f(g)$ for all $a \in C$, $g \in \tilde{G}$, the function $f' \in L^2(M \times \tilde{G})$ given by $f'(m, g) = f(g)$ will be Γ-invariant under the action on $M \times \tilde{G}$ defined by the cocycle β (i.e. $\gamma(m, g) = (\gamma m, \beta(m, \alpha)^{-1}g)$). Since $\alpha \sim \beta$, the Γ-actions these cocycles define are measurably conjugate and hence there is a non-0 Γ-invariant function $F \in L^2(\tilde{P})$. It follows that for any r, $i^*(F) \in L^{2,r}_{\xi_r}(\tilde{P})$ is also a non-0 Γ-invariant function. Since ξ_r is a smooth metric, if we choose r sufficiently large, the Sobolev embedding theorem implies that $i^*(F) \in L^2(\tilde{P}) \cap C^0(\tilde{P})$.

We can now apply lemma 7. Thus, there is an open, conull, Γ-invariant set $W \subset M$, a compact subgroup $K \subset \tilde{G}$, and a continuous Γ-invariant section $\phi: W \to \tilde{P}/K$. However, $\tilde{G} = G \ltimes N$ where N is a connected unipotent group. Thus, we can assume K actually lies in G. Hence we have a natural Γ-map $\tilde{P}/K \to P/K$. Composing ϕ with this map we obtain a continuous Γ-invariant section $\psi: W \to P/K$. Since P is the frame bundle, K is contained in a conjugate of $O(n, \mathbb{R})$, and ψ thus defines a Γ-invariant C^0-Riemannian metric on W. We then apply lemma 8, from which all assertions in theorem 1 follow.

Theorem 5 follows in the same manner using the following version of theorem 6.

THEOREM 9. *Let Γ be as in theorem 5. Suppose S is a standard Borel Γ-space, that μ is a finite Γ-invariant measure on S, and that $\alpha: S \times \Gamma \to G$ is a cocycle where G is a real algebraic group. Then α is equivalent to a cocycle β with $\beta(S \times \Gamma)$ contained in a compact subgroup of G.*

306   R. J. Zimmer

As with theorem 6, theorem 9 follows from superrigidity for cocycles of Γ-actions (see [7], [8]) and Kazhdan's property via the arguments of [9, theorem 2.8]. (Cf. [9, proof of theorem 6.3].)

The author would like to thank M. Markowitz for a number of useful conversations related to this work. This research was partially supported by NSF grant MCS 8004028. The author is a Sloan Foundation Fellow.

REFERENCES

[1] S. Kobayashi. *Transformation Groups in Differential Geometry.* Springer-Verlag: New York, 1972.

[2] G. A. Margulis. Discrete groups of motions of manifolds of non-positive curvature. *Amer. Math. Soc. Translations* **109** (1977), 33–45.

[3] I. M. Singer & S. Sternberg. The infinite groups of Lie and Cartan. *J. d'Anal. Math.* **15** (1965), 1–114.

[4] N. Tanaka. On generalized graded Lie algebras and geometric structures, I. *J. Math. Soc. Japan* **19** (1967), 215–254.

[5] N. Tanaka. On differential systems, graded Lie algebras, and pseudo-groups. *J. Math. Kyoto Univ.* **10** (1970), 1–82.

[6] R. J. Zimmer. Strong rigidity for ergodic actions of semisimple Lie groups. *Annals of Math.* **112** (1980), 511–529.

[7] R. J. Zimmer. Ergodic theory, group representations, and rigidity. *Bull. Amer. Math. Soc.* **6** (1982), 383–416.

[8] R. J. Zimmer. *Ergodic Theory and Semisimple Groups.* Birkhauser: Boston, 1984.

[9] R. J. Zimmer. Volume preserving actions of lattices in semisimple groups on compact manifolds. *Publ. Math. IHES* **59** (1984), 5–33.

[10] R. J. Zimmer. Semisimple automorphism groups of G-structures. *J. Diff. Geom.* **19** (1984), 117–123.

Proceedings of the International Congress of Mathematicians
Berkeley, California, USA, 1986

Actions of Semisimple Groups and Discrete Subgroups

ROBERT J. ZIMMER

1. Introduction. Let $V = \{\text{primes in } \mathbf{Z}\} \cup \{\infty\}$. As usual, $\mathbf{Q}_p$ will denote the p-adic numbers for p a finite prime, and we set $\mathbf{Q}_\infty = \mathbf{R}$. Let $S \subset V$ be a finite subset. For $p \in S$, let G_p be a connected semisimple algebraic group defined over $\mathbf{Q}_p$, $G_p(\mathbf{Q}_p)$ the group of $\mathbf{Q}_p$-points, and $G = \prod_{p \in S} G_p(\mathbf{Q}_p)$. Then G is a locally compact group. If $S = \{\infty\}$, we say that G is real, and in this case G is a semisimple Lie group. We say that G has *higher rank* if for all $p \in S$, the $\mathbf{Q}_p$-rank of every $\mathbf{Q}_p$-simple factor of G_p is at least 2. If G is real and simple, this is of course equivalent to the condition that the rank of the associated symmetric space be at least 2. Let $\Gamma \subset G$ be a lattice subgroup, i.e., Γ is discrete and G/Γ has a finite G-invariant measure. The basic examples of such groups are the S-arithmetic ones, and by results of Margulis [**M1**] (see also [**Z1**]), under the assumption of higher rank, every lattice is S-arithmetic. The finite-dimensional (continuous) representation theory (say over $\mathbf{C}$) of G is the same as that of $G_\infty(\mathbf{R})$, and hence is classical and well-understood. In higher rank, the finite-dimensional representation theory of Γ is now quite well understood as well, due to the work of Margulis [**M1**] on semisimple representations (Margulis's superrigidity theorem) (see also [**Z5**] and the work of Mostow [**Mo**] and Prasad [**P**]), Raghunathan [**R2, R3**] on general representations for the case in which Γ is not cocompact, and to the work of numerous authors (e.g., Weil [**W**], Matsushima-Murakami [**MM**], Raghunathan [**R1**], Borel [**B1, B2**], Borel-Wallach [**BW**], Garland [**G**], Kazhdan [**K**]) on cohomology vanishing theorems for semisimple representations, which (when combined with Margulis's results) yields essentially complete results in the cocompact case. (Much of this work is also valid under considerably weaker hypotheses than higher rank.) The fundamental conclusion of this theory is that all representations are essentially either orthogonal, or extend to G, or are built up from these two cases. The infinite-dimensional (say unitary) representation theory of G is of course now a highly developed and enormously rich subject. The corresponding theory for Γ is largely undeveloped, except for developments related to the discovery of Kazhdan [**K**] that in higher rank the trivial representation of Γ is isolated in the unitary dual.

Research partially supported by National Science Foundation Grant DMS-8301882.

This property is inherited from G, and is an extraordinarily powerful and flexible property of Γ.

In this report, we will discuss the much more recent program of understanding the realizations of G and Γ in another natural class of groups, namely as smooth transformation groups on compact manifolds. This has both finite-dimensional and infinite-dimensional features. We can view this as the nonlinear finite-dimensional theory, or as the study of homomorphisms into the infinite-dimensional group of diffeomorphisms of a compact manifold. The prevailing theme of the work to date on this program is that one sees very strong manifestations of the rigidity phenomena of the finite-dimensional linear theory in the present context. It does not appear out of the question that one could classify all volume preserving ergodic actions of G and Γ on compact manifolds.

(In many cases below, when considering actions of Γ or G, we will assume for simplicity of exposition that G has higher rank, even though in certain cases less restrictive hypotheses are sufficient.)

We now describe some basic examples:

(a) Let H be a connected Lie group and $\Lambda \subset H$ a cocompact subgroup such that H/Λ has a finite H invariant measure. Let $\rho: G \to H$ be a continuous homomorphism. Then G (and Γ) will act naturally on H/Λ. If H is semisimple, then under very mild assumptions, the fundamental theorem of Moore [**Mr1**] (see also [**Z1**]) implies that these actions are ergodic.

(b) Let K be a compact Lie group, $K_0 \subset K$ a closed subgroup, and suppose $\rho: \Gamma \to K$ is a homomorphism. Then Γ acts on K/K_0, preserving a Riemannian metric. The examples with $\rho(\Gamma)$ dense in K are exactly the isometric ergodic actions of Γ.

(c) Let N be a simply connected nilpotent Lie group on which H acts by automorphisms. Assume that $D \subset N$ is a lattice which is invariant under the action of Γ. Then Γ acts by automorphisms of the compact nilmanifold N/D. A basic example arises from arithmetic realizations of Γ. Namely, if $\rho: \Gamma \to \mathrm{GL}(n, \mathbf{Z})$ is a homomorphism, then Γ acts on the torus $\mathbf{R}^n/\mathbf{Z}^n$.

A basic question is to what extend these examples (and easy modifications of them) represent all the volume preserving examples in the case of higher rank, at least if we assume ergodicity. At present there are no known examples of volume preserving actions not derived from these fundamentally linear situations. While a complete classification does not appear within reach at present, the remainder of this report will describe the present understanding of the actions of these groups under various natural hypotheses: dimension restrictions on M, actions preserving geometric structures, deformations, growth conditions, etc.

2. Invariant geometric structures for semisimple groups.
Let M be a compact, connected, n-manifold and $H \subset \mathrm{GL}(n, \mathbf{R})$ an algebraic subgroup. We recall that an H-structure on M is a reduction of the frame bundle of M to H, and hence is a principal H-bundle $P \to M$ contained in the frame bundle. We let $\mathrm{Aut}(P) \subset \mathrm{Diff}(M)$ be the subgroup of diffeomorphisms of M leaving P

invariant. Let G be as above, and assume G is real. In this section we describe results concerning actions of G preserving an H-structure, i.e., homomorphisms $G \to \mathrm{Aut}(P)$. All these results carry over to the case of higher order structures, i.e., to the case in which $H \subset \mathrm{GL}(n, \mathbf{R})^{(k)}$ is an algebraic subgroup, where $\mathrm{GL}(n, \mathbf{R})^{(k)}$ is the group of k-jets at 0 of diffeomorphisms of $\mathbf{R}^n$ fixing the origin.

THEOREM 2.1 [**Z4**, **Z7**]. *Suppose that G has no compact factors. If G acts effectively on a compact connected M preserving an H-structure where $H \subset \mathrm{SL}'(n, \mathbf{R})$ (the latter being the matrices with $|\det| = 1$), then there is an embedding of Lie algebras $\mathfrak{g} \to \mathfrak{h}$. In fact, this embedding is such that the representation $\mathfrak{g} \to \mathfrak{h} \to \mathfrak{sl}\,(n, \mathbf{R})$ contains $\mathrm{ad}_{\mathfrak{g}}$ as a direct summand.*

This result provides very strong obstructions for G to preserve such an H-structure. For example, if $H = \mathrm{O}(1, n-1)$, so that we are considering actions preserving Lorentz metrics, we deduce

COROLLARY 2.2 [**Z4**]. *If a semisimple group G acts on a compact manifold (of any dimension) preserving a Lorentz metric, then G is locally isomorphic to $\mathrm{SL}(2, \mathbf{R}) \times K$ where K is a compact group.*

We remark that somewhat similar techniques can be used to analyze the solvable component of the automorphism group of a Lorentz manifold, although we shall not pursue this here. See [**Z4**].

The hypothesis of compactness and finite invariant measure on M are both necessary in Theorem 2.1. In case G is transitive on M, Theorem 2.1 can be deduced from the Borel density theorem (asserting that the stabilizer for a homogeneous space of G with finite invariant measure is Zariski dense in G). The proof in general actually makes use of Borel's theorem and the ideas surrounding it. Without the assumption of invariant measure (i.e., for H an arbitrary algebraic subgroup of $\mathrm{GL}(n, \mathbf{R})$), we have

THEOREM 2.3 [**Z7**]. *Suppose G is a semisimple group acting effectively on a compact manifold M preserving an H-structure where H is an algebraic group. Then $\mathbf{R}$-rank$(G) \leq \mathbf{R}$-rank(H).*

(We recall that $\mathbf{R}$-rank is the dimension of a maximal $\mathbf{R}$-split torus.)

3. Invariant geometric structures for discrete subgroups. Suppose now that G has higher rank. We can then state the following general conjecture for actions of lattices that preserve an H-structure.

CONJECTURE I. Suppose G has higher rank and that $\Gamma \subset G$ is a lattice. Suppose that M is a compact n-manifold and that $H \subset \mathrm{SL}'(n, \mathbf{R})$ is an algebraic subgroup. (For higher order structures, we assume $H \subset \mathrm{SL}'(n, \mathbf{R}) \cap \mathrm{GL}(n, \mathbf{R})^{(k)}$.) If Γ acts on M so as to preserve an H-structure, then either:
    (i) there is a Γ-invariant Riemannian metric on M; or
    (ii) there is a nontrivial Lie algebra homomorphism $\mathfrak{g}_{\infty}(\mathbf{R}) \to \mathfrak{h}$.

This conjecture can be viewed as a geometric, or nonlinear, version of Margulis's superrigidity theorem. Margulis's theorem implies (under the above hypotheses on G, Γ, and H) that for a homomorphism $\Gamma \to H$, either the image of Γ is precompact in H or there is a nontrivial Lie algebra homomorphism $\mathfrak{g}_\infty(\mathbf{R}) \to \mathfrak{h}$. We also remark with regard to conclusion (i) in the conjecture that the action is then given by a composition $\Gamma \to K \to \mathrm{Diff}(M)$, where K is a compact Lie group, and that Margulis has also described the compact Lie groups admitting a dense image homomorphism from Γ. In particular, for any such K we have that $\dim(K) \geq n(G)$, where $n(G) = \min\{\dim_{\mathbf{C}} G' | G'$ is a simple factor of $G_\infty\}$. (See [Z5].) If we further assume that the Γ action is ergodic, then in (i) we have that the action is on a homogeneous space of K. Before indicating what is known in the direction of the conjecture, we remark that the following conjecture would be an immediate consequence of Conjecture I, and the fact that the dimension of the isometry group of a compact Riemannian n-manifold is at most $n(n+1)/2$.

CONJECTURE II. Let G, Γ, M be as in Conjecture I. Let $d(G)$ be the smallest integer d for which there is a nontrivial Lie algebra representation $\mathfrak{g}_\infty(\mathbf{R}) \to \mathfrak{sl}(d, \mathbf{R})$. Suppose Γ acts on M so as to preserve a volume density. Then (with $\dim(M) = n$):

(a) If $n < d(G)$, then Γ preserves a Riemannian metric.

(b) If $n < d(G)$ and $\{n(n+1)/2\} < n(G)$, then the Γ action is finite, i.e., factors through a finite quotient of Γ.

For certain H-structures, Conjecture I is known to be true. We recall that E. Cartan has defined the notion of a H-structure of finite type. (See [Ko].) A connection is a (second order) structure of finite type, and any H-structure naturally defining a connection is of finite type. For example, any pseudo-Riemannian structure (i.e., $O(p, q)$-structure) is of finite type. The automorphism group of an H-structure of finite type is a Lie group; however the group of connected components may well be infinite.

THEOREM 3.1 [Z5]. *Conjecture I is true for H-structures of finite type (in the sense of E. Cartan). In particular, if $n < d(G)$, and $\{n(n+1)/2\} < n(G)$, then any action of Γ on M^n preserving a volume density and a H-structure of finite type is a finite action.*

As a concrete example we have

COROLLARY 3.2 [Z5]. *Let $G = \mathrm{SL}(n, \mathbf{R})$, $n \geq 3$, and M a compact manifold with $\dim(M) < n$. Then any action of Γ on M preserving a volume density and a connection is a finite action.*

This shows that the action of $\mathrm{SL}(n, \mathbf{Z})$ by automorphisms of the flat torus is a volume and connection preserving action of minimal dimension for lattices in $\mathrm{SL}(n, \mathbf{R})$. For $\dim(M) = n$, we have the following special result.

THEOREM 3.3 [**Z6**]. *If* Γ *is a lattice in* $\mathrm{SL}(n, \mathbf{R})$, $n \geq 3$, M *is a compact Riemannian* n-*manifold, and* Γ *acts on* M *preserving volume and the connection, then* M *is flat and* Γ *is commensurable with* $\mathrm{SL}(n, \mathbf{Z})$.

We now indicate two other situations in which Conjecture I is known, at least with some further hypotheses. We recall that a linear Lie algebra is called elliptic if it contains no matrices of rank 1. For any H-structure of finite type, $\mathfrak{h}$ will be elliptic.

THEOREM 3.4 [**Z5**]. *Conjecture* I *is true if* $\mathfrak{h}$ *is elliptic, provided we also assume that* $\mathrm{Aut}(P)$ *is transitive on* M.

A linear real algebraic group H is called distal if the reductive Levi component is compact.

THEOREM 3.5 [**Z3**]. *Conjecture* I *is true for* H *distal, provided we assume the* Γ *action on* M *is ergodic. Hence, any ergodic action of* Γ *preserving a distal structure is isometric.*

Theorem 3.5 is closely related to classical notions in dynamical systems. Namely, we recall that an action of a group Γ on a metric space M is called distal if $x, y \in M$, $x \neq y$, implies $\inf\{d(gx, gy)|g \in \Gamma\} > 0$. Furstenberg [**F**] has shown that the structure of every such minimal action can be explicitly described in terms of a tower of bundles, and in case M is a topological manifold, M. Rees [**Re**] showed that the tower consists of topological manifolds. If M is smooth, and the tower is smooth, then Γ will preserve a distal H-structure on M. (However an action may well preserve a smooth distal structure but not be distal in the above sense.) This then raises the purely topological question as to whether every distal action of Γ (where Γ is as in Conjecture I) will preserve a topological distance function. There are natural classes of groups for which this is known to be the case [**MZ, A**].

Finally, we remark that Conjecture I is true if assertion (i) is replaced by the weaker assertion that there is a measurable invariant Riemannian metric [**Z5**].

4. Perturbations, deformations, and cohomology. In this section we discuss the rigidity properties of some of the actions described in §2 under perturbations or deformations. Each group Γ under consideration is finitely generated. For each r, $1 \leq r \leq \infty$, the C^r-topology on Γ actions on M will be the topology of C^r-convergence in $\mathrm{Diff}(M)$ on a fixed finite generating set of Γ.

THEOREM 4.1 [**Z5**]. *Suppose that* Γ *acts isometrically on a compact Riemannian* n-*manifold* M. *For any nonnegaitve integer* k, *there is positive integer* $r = r(n, k)$ *(for* $k = \infty$, *let* $r = \infty$), *so that any action of* Γ *on* M *which*

    (i) *is sufficiently close to the original action in the* C^r-*topology,*

    (ii) *leaves a volume density invariant, and*

    (iii) *which is ergodic,*

must also leave a C^k-*Riemannian metric invariant.*

For $k = 0$, $r = n^2 + n + 1$. For $k \geq 1$, $r = n + k + 4 + \dim(\mathrm{GL}(n, \mathbf{R})^{(k+3)})$, where $\mathrm{GL}(n, \mathbf{R})^{(k)}$ is the group of k-jets at 0 of diffeomorphisms of $\mathbf{R}^n$ fixing the origin.

In other words, roughly speaking, a small perturbation of an isometric action will be isometric. This is of course in very sharp contrast to actions of $\mathbf{R}$ or free groups. It seems possible that the size of r stated above can be improved.

If M is a manifold, we let $\mathrm{Vect}(M)$ be the space of smooth vector fields on M. If Γ acts on M, then $\mathrm{Vect}(M)$ is naturally a Γ-module. The Γ action on M is called *infinitesimally rigid* if $H^1(\Gamma, \mathrm{Vect}(M)) = 0$, locally (r, k)-*rigid* if every action sufficiently close in the C^r-topology is conjugate to the original action by a C^k-homeomorphism, and r-*structurally stable* if it is locally $(r, 0)$-rigid.

Question. For G of higher rank, is every ergodic volume preserving Γ-action infinitesimally rigid and locally rigid? In particular, for $n \geq 3$, is the action of $\mathrm{SL}(n, \mathbf{Z})$ on $\mathbf{R}^n / \mathbf{Z}^n$ locally rigid?

For infinitesimal rigidity, we have the following result.

THEOREM 4.2 [Z8]. *Assume G is real, has higher rank, and that Γ is a cocompact lattice in G. Let H be a semisimple Lie group, $\Lambda \subset H$ a cocompact lattice, and suppose $\Gamma \to H$ is a homomorphism. Let $M = H/\Lambda$, so that Γ then acts on M. Then the Γ action on M is infinitesimally rigid in the following cases:*

(i) *the image of Γ is dense in H; or*

(ii) $H = H_1 \times H_2$, Λ *projects densely into both factors, and Γ maps densely into H_1 and trivially into H_2.*

We expect the techniques of proof to extend to eliminate the hypotheses that G is real, and to apply in at least some situations in which Γ is not cocompact. For Γ cocompact, most of the proof of Theorem 4.2 remains valid for all the ergodic examples of Γ actions considered in §1, and it is possible that the proof may extend to cover these cases as well. The arguments of the proof of Theorem 4.2 can be applied to compute the cohomology of Γ with coefficients in the space of smooth sections of other natural bundles. For example, under the hypotheses of Theorem 4.2 we can deduce that $H^1(\Gamma, C^\infty(M)) = 0$.

Problem. If G has higher rank, and Γ acts (perhaps ergodically) preserving a probability measure in the smooth measure class on a compact manifold M, is $H^1(\Gamma, C^\infty(M)) = 0$?

A positive answer to this question, combined with the techniques of proof of Theorems 3.5 and 4.1, would yield significant progress on a resolution of Conjectures I and II. From Kazhdan's property it follows that the map $H^1(\Gamma, C^\infty(M)) \to H^1(\Gamma, L^2(M))$ is 0.

For nonvolume preserving examples, we have the following theorems of Sullivan and Ghys.

THEOREM 4.3 [S]. *Let $G = O(1, n+1)$, $P \subset G$ a minimal parabolic subgroup, so that G/P can be identified with S^n with the conformal action of G. If $\Gamma \subset G$ is a cocompact lattice, then the action of Γ on S^n is 1-structurally stable.*

THEOREM 4.4 [Gh]. *Let $G = \mathrm{PSL}(2, \mathbf{R})$, acting by conformal transformations of S^1. Let $\Gamma \subset G$ be a cocompact lattice. Then any smooth action of Γ on S^1 sufficiently C^2-close to the given conformal action on S^1 is conjugate (via a smooth diffeomorphism) to the action defined by a linear representation $\Gamma \to G$.*

It would of course be interesting to extend these results to a more general algebraic setting.

5. Actions with fixed points. For actions with a fixed point, or more generally with a finite orbit, we can prove a version of Conjecture II.

THEOREM 5.1 [Z5]. *Assume G has higher rank, and that M is a connected n-manifold (not necessarily compact) and that Γ acts on M. Assume $n < d(G)$ and $\{n(n+1)/2\} < n(G)$. If there is at least one finite Γ-orbit in M, then the Γ action is finite.*

We remark, that in contrast to the conclusion of Theorem 5.1, the standard action of $\mathrm{SL}(n, \mathbf{Z})$ on the n-torus has a dense set of points with finite orbits.

D. Stowe [St] has shown for any group for which one has vanishing of the first cohomology group with coefficients in all finite-dimensional real representations that there is a persistence of fixed points under perturbations. Combined with the cohomological information alluded to in the introduction, we obtain the following instance of Stowe's theorem.

THEOREM 5.2. *Assume G has higher rank. Let Γ act on a manifold and assume that p is a fixed point for the action. Then any action sufficiently close in the C^1-topology to the original action has a fixed point near p. Without the assumption of higher rank, the same conclusion is true for actions of G.*

For compact group actions, a basic tool in the study of fixed points is the fact that one can linearize the action near the fixed point. The following result generalizes this to semisimple groups.

THEOREM 5.3 (GUILLEMIN-STERNBERG [GS]). *Suppose G acts on M with a fixed point p. If the action is real analytic, then in a neighborhood of p, the action is analytically equivalent to the representation of G on TM_p.*

6. Orbit structure. In the classical theory of flows, a significant role is played by the study of the phase portrait of the flow, or equivalently, studying the flow up to equivalence after time change. Similar ideas have played an important role in certain recent developments in ergodic theory [Dy, CFW, Ru]. If for $i = 1, 2$, G_i is a locally compact group acting on a Borel subset M_i of a complete separable metrizable space, preserving the null sets of a finite measure μ_i, we say that the actions are measurably orbit equivalent if there is a measure class preserving bijection (modulo null sets) $M_1 \to M_2$ taking G_1 orbits onto G_2 orbits. If G_i are discrete amenable groups acting ergodically (and with no orbit of full measure) and μ_i are invariant probability measures, then a result of Ornstein and Weiss (see also [**CFW**]), generalizing earlier work of Dye [**Dy**], asserts that the actions are measurably orbit equivalent. In contrast, we have the following result in the semisimple case.

THEOREM 6.1 [**Z1**, **Z2**]. *Assume G_i are as in the introduction and that G_1 has higher rank.*

(i) *If G_i acts ergodically and essentially freely on M_i, does not have a conull orbit, preserves a probability measure, and the G_1 action is measurably orbit equivalent to the G_2 action, then G_1 and G_2 are locally isomorphic. Further, the actions of the adjoint groups $G_i/Z(G_i)$ on $M/Z(G_i)$ are actually conjugate, after the identification of these adjoint groups via an isomorphism.*

(ii) *If $\Gamma_i \subset G_i$ are lattices and Γ_1 and Γ_2 have measurably orbit equivalent ergodic essentially free actions with finite invariant measure (and no conull orbit), then G_1 and G_2 are locally isomorphic.*

For certain isometric actions, Witte [**Wi**] has shown that the conclusion in (ii) can be strengthened to assert isomorphism of the lattices and conjugacy of the actions.

Roughly speaking, Theorem 6.1 asserts that the measurable orbit structure of the action determines both the semisimple group and (in case (i)) the action itself. This is diametrically opposed to the situation for amenable groups. It is not known whether or not this result is true for groups of split rank 1, although some information is available for groups with Kazhdan's property [**Z1**]. One can of course ask about orbit equivalence in the context of smooth actions, and ask that the orbit equivalence be smooth, or at least continuous. If G is real and acts on M, and $K \subset G$ is a maximal compact subgroup, then under suitable hypotheses the orbits of G will project to the leaves of a foliation of M/K, and these leaves will naturally carry the structure of an (infinite volume) locally symmetric space. The next result, which is joint work of P. Pansu and the author, is a result in the same spirit as that of Theorem 6.1 but formulated in the context of foliations.

THEOREM 6.2 [**PZ**]. *Let M be a compact manifold and $\mathcal{F}$ a foliation of M with a holonomy invariant transverse measure (which we assume to be positive on open sets of transversals, and finite on compact sets). For $i = 1, 2$, let ω_i be a Riemannian metric on the tangent bundle to $\mathcal{F}$, such that for $i = 1, 2$, each leaf is a locally symmetric space of negative curvature. Then there is a homeomorphism f of M, taking each leaf to itself diffeomorphically, such that $f^*(\omega_1) = \omega_2$.*

When the foliation has just one leaf, this reduces to a version of the Mostow rigidity theorem. It is natural to conjecture that the same result is true for symmetric spaces of higher rank, but this is not yet known. From Theorem 6.1, one can deduce the existence of a measurable bijection $f \colon M \to M$, taking each leaf to itself diffeomorphically, with $f^*(\omega_1) = \omega_2$. It is also natural to ask whether or not we can ensure that f in Theorem 6.2 can be taken to be smooth.

In the case of nonvolume preserving actions, we have the following result for some standard examples.

THEOREM 6.3 [Z2]. *For $i = 1, 2$, suppose G_i is semisimple and real, $H_i \subset G_i$ is an almost connected closed subgroup, and L_i is the maximal semisimple adjoint quotient group of H_i^0 with no compact factors. Assume L_1 has higher rank. If Γ_i is a lattice in G_i, and the action of Γ_1 on G_1/H_1 is measurably orbit equivalent to the action of Γ_2 on G_2/H_2, then L_1 and L_2 are locally isomorphic.*

A basic property that a measure class preserving action of a locally compact group might have is amenability. This is discussed at length in [Z1]. For essentially free actions (i.e., actions for which almost every stabilizer is trivial) amenability is an invariant of measurable orbit equivalence. One method of constructing amenable actions is to induce from an action of an amenable subgroup. I.e., if $A \subset G$ is a closed amenable subgroup, and X is a measurable A-space, then $M = (X \times G)/A$ will be an amenable G-space. For arbitrary G, this does not yield all amenable actions. However, we have

THEOREM 6.4 [Z1]. *If G is real and semisimple, any amenable ergodic action of G is measurably conjugate to an action induced from a maximal amenable algebraic subgroup.*

We remark that such subgroups have been classified by C. C. Moore [Mr2]. We also observe that for essentially free actions, while amenability depends only upon the action up to measurable orbit equivalence the property of being induced from an action of a subgroup from a given class of subgroups does not. Theorem 6.4 has immediate applications to the measurable cohomology theory of amenable actions of arbitrary groups and to amenable foliations. In this latter form this result has recently been applied by Hurder and Katok [HK] to prove a vanishing theorem for secondary characteristic classes of amenable foliations.

7. Restrictions to unipotent subgroups. In studying the linear representations of G, a basic role is played by the restriction of the representations to unipotent subgroups. Here we discuss some features of these restrictions in the case of actions. The following result was first proved by M. Ratner [Ra1] for $G = \mathrm{PSL}(2, \mathbf{R})$, and extended to more general semisimple groups by Witte [Wi].

THEOREM 7.1 [Ra1, Wi]. *For $i = 1, 2$, suppose G_i is a noncompact semisimple adjoint Lie group and that Γ_i is an irreducible lattice in G_i. Let $U_i \subset G_i$ be a one parameter unipotent subgroup. If the $\mathbf{R}$-actions defined by the U_i actions on G_i/Γ_i are measurably conjugate, then there is an isomorphism of G_1 with G_2 taking Γ_1 onto Γ_2.*

Witte also shows how the arguments of the proof lead to the following information about some of the examples considered in §2.

THEOREM 7.2 [**Wi**]. *Suppose that G is real and adjoint.*

(a) *For $i = 1, 2$, let G_i be a connected semisimple adjoint Lie group, and $\varphi_i \colon G \to G_i$ be an injective homomorphism. Let $\Gamma_i \subset G_i$ be an irreducible lattice, and suppose the G actions on G_1/Γ_1 and G_2/Γ_2 are measurably conjugate. Then there is an isomorphism $G_1 \cong G_2$ taking φ_1 to φ_2.*

(b) *Suppose further that G is of higher rank. Fix a nontrivial unipotent element $g \in G$. Then any measure preserving action of G on G_1/Γ_1 for which g acts by an element of G_1 is actually defined by a homomorphism $G \to G_1$.*

Ratner's theorem for PSL(2) was extended in another direction (concerning horocycle foliations in higher dimensional manifolds of negative curvature) by Flaminio [**Fl**]. For $\mathrm{PSL}(2, \mathbf{R})$, an analogous theorem to Theorem 7.1, assuming continuous orbit equivalence, rather than measurable conjugacy, was proved by Marcus [**Ms**]. This has been extended to a much broader context by Benardette [**Be**]. Further work on the restriction of transitive volume preserving actions to unipotent subgroups can be found in the work of Veech [**V**] and the extensive work of Dani [**D1**].

A basic result in the finite-dimensional linear theory is that for any linear representation of G, the image of a unipotent element is a unipotent matrix. The following problem asks for a generalization of this in the geometric context.

Problem. Assume G is real and of higher rank. Let M be a compact manifold, $E \to M$ a vector bundle of rank n on which G acts by vector bundle automorphisms. Suppose there is a probability measure on M, in the smooth measure class, which is invariant under G. Let U be a unipotent subgroup of G. Is there a distal algebraic subgroup $H \subset \mathrm{GL}(n, \mathbf{R})$ and a smooth U-invariant H-structure on E?

In all known examples, this question has an affirmative answer. In case G acts ergodically on M, it is known [**Z5**] that for some distal H there is a measurable U-invariant H-structure. A positive answer to the above question for some G, combined with the techniques of proof of Theorem 3.5, and results of Howe and Harish-Chandra on asymptotic behavior of matrix coefficients for unitary representations would yield a proof of Conjecture II in the ergodic case, although with weaker control on the dimension of M than that proposed in Conjecture II. See [**Z5**] for a complete discussion of this point and some partial results. We remark that in general one cannot find a U-invariant H-structure where H is a unipotent subgroup of $\mathrm{GL}(n, \mathbf{R})$.

8. Quotient actions. If a group Λ acts on a topological (resp., measure) space X, by a quotient action we mean a topological (resp. measure) space Y and a continuous (resp. measure class preserving) surjective Λ-map $X \to Y$. For some standard actions the quotient actions can all be identified. The following result of M. Ratner accomplishes this for horocycle flows of surfaces.

THEOREM 8.1 [**Ra2**]. *Let $G = \mathrm{PSL}(2, \mathbf{R})$ and let $\mathbf{R}$ act on G/Λ by the upper triangular unipotent subgroup, where Λ is a lattice in H. Then every* *measure theoretic quotient of this* $\mathbf{R}$*-action is of the form* G/Λ'*, where* $\Lambda' \supset \Lambda$ *is a larger lattice, and* $\mathbf{R}$ *acts via the same subgroup.*

For the case of actions on varieties, we have the following theorem, first proved in the measure theoretic case by Margulis [M2] (see also [Z1]), later in the topological case by Dani [D2].

THEOREM 8.2 [M2, D2]. *Assume* G *is adjoint and of higher rank and that* $P \subset G$ *is a parabolic subgroup. Then any quotient of the* Γ *action on* G/P *is of the form* G/P'*, where* P' *is a parabolic subgroup containing* P.

9. Concluding remarks. We have said little about the proofs of the results stated above. In general they involve combinations of arguments of ergodic theory, algebraic groups, representation theory, differential geometry, and global analysis. A basic role is played by various ergodic theoretic generalizations of Margulis's superrigidity theorem [Z1], and Kazhdan's property. A number of the results above can be established assuming only Kazhdan's property (see [Z9], e.g.), and hence are applicable to certain split rank one groups. For some further topics we have not discussed, see [Z1, Z5], and the references therein.

REFERENCES

[A] H. Abels, *Which groups act distally?*, Ergodic Theory Dynamical Systems **3** (1983), 167–186.

[Be] D. Benardette, *Topological equivalence of flows on homogeneous spaces and divergence of one-parameter subgroups of Lie groups*, preprint.

[B1] A. Borel, *Stable real cohomology of arithmetic groups*, Ann. Sci. École Norm. Sup. **7** (1974), 235–272.

[B2] ____, *Stable real cohomology of arithmetic groups. II*, Progress in Math. vol. 14, Birkhauser, Boston, 1981, pp. 21–55.

[BW] A. Borel and N. Wallach, *Continuous cohomology, discrete subgroups, and representations of reductive groups*, Ann. of Math. Stud. **94** (1980).

[CFW] A. Connes, J. Feldman, and B. Weiss, *An amenable equivalence relation is generated by a single transformation*, Ergodic Theory Dynamical Systems **1** (1981), 431–450.

[D1] S. Dani, *Invariant measures and minimal sets of horocycle flows*, Invent. Math. **64** (1981), 357–385.

[D2] ____, *Continuous equivariant images of lattice actions on boundaries*, Ann. of Math. **119** (1984), 111–119.

[Dy] H. Dye, *On groups of measure preserving transformations. I*, Amer. J. Math. **81** (1959), 119–159.

[Fl] L. Flaminio, Thesis, Stanford University, California, 1985.

[F] H. Furstenberg, *The structure of distal flows*, Amer. J. Math. **85** (1963), 477–515.

[G] H. Garland, *p-adic curvature and the cohomology of discrete subgroups of p-adic groups*, Ann. of Math **97** (1973), 375–423.

[Gh] E. Ghys, *Actions localement libres du groupe affine* (to appear).

[GS] V. Guillemin and S. Sternberg, *Remarks on a paper of Hermann*.

[HK] S. Hurder and A. Katok, *Ergodic theory and Weil measures of foliations*, preprint.

[K] D. Kazhdan, *Connection of the dual space of a group with the structure of its closed subgroups*, Functional Anal. Appl. **1** (1967), 63–65.

[Ko] S. Kobayashi, *Transformation groups in differential geometry*, Springer, New York, 1972.

[Ms] B. Marcus, *Topological conjugacy of horocycle flows*, Amer. J. Math. **105** (1983), 623–632.

1258   R. J. ZIMMER

[M1] G. A. Margulis, *Discrete subgroups of motions of manifolds of non-positive curvature*, Amer. Math. Soc. Transl. **109** (1977), 33–45.

[M2] ____, *Quotient groups of discrete groups and measure theory*, Functional Anal. Appl. **12** (1978), 295–305.

[MM] Y. Matsushima and S. Murakami, *On vector bundle valued harmonic forms and automorphic forms on symmetric Riemannian manifolds*, Ann. of Math. **78** (1963), 365–416.

[Mr1] C. C. Moore, *Ergodicity of flows on homogeneous spaces*, Amer. J. Math. **88** (1966), 154–178.

[Mr2] ____, *Amenable subgroups of semisimple groups and proximal flows*, Israel J. Math. **34** (1979), 121–138.

[MZ] C. C. Moore and R. J. Zimmer, *Groups admitting ergodic actions with generalized discrete spectrum*, Invent. Math. **51** (1979), 171–188.

[Mo] G. D. Mostow, *Strong rigidity of locally symmetric spaces*, Ann. of Math. Stud. **78** (1973).

[PZ] P. Pansu and R. J. Zimmer (to appear).

[R1] M. Raghunathan, *On the first cohomology group of subgroups of semisimple Lie groups*, Amer. J. Math. **87** (1965), 103–139.

[R2] ____, *On the congruence subgroup problem*, Inst. Hautes Études Sci. Publ. Math. **46** (1976), 107–161.

[R3] ____, *On the congruence subgroup problem*. II, Invent. Math. **85** (1986), 73–118.

[Ra1] M. Ratner, *Rigidity of horocycle flows*, Ann. of Math. **115** (1982), 597–614.

[Ra2] ____, *Factors of horocycle flows*, Ergodic Theory Dynamical Systems **2** (1982), 465–489.

[Re] M. Rees, *On the structure of minimal distal transformation groups with topological manifolds as phase space*, preprint, 1977.

[Ru] D. Rudolph, *Restricted orbit equivalence*, Mem. Amer. Math. Soc. No. 323 (1985).

[St] D. Stowe, *The stationary set of a group action*, Proc. Amer. Math. Soc. **79** (1980), 139–146.

[S] D. Sullivan, *Quasi-conformal homeomorphisms and dynamics*, II. *Structural stability implies hyperbolicity for Kleinian groups*, Preprint, Inst. Hautes Études Sci. Publ. Math., 1982.

[V] W. A. Veech, *Unique ergodicity of horocycle flows*, Amer. J. Math. **99** (1977), 827–859.

[W] A. Weil, *On discrete subgroups of Lie groups*, Ann. of Math. **72** (1960), 369–384.

[Wi] D. Witte, *Rigidity of some translations on homogeneous spaces*, Invent. Math. **81** (1985), 1–27.

[Z1] R. J. Zimmer, *Ergodic theory and semisimple groups*, Birkhauser, Boston, 1984.

[Z2] ____, *Orbit equivalence and rigidity of ergodic actions of Lie groups*, Ergodic Theory Dynamical Systems **1** (1981), 237–253.

[Z3] ____, *Lattices in semisimple groups and distal geometric structures*, Invent. Math. **80** (1985), 123–137.

[Z4] ____, *On the automorphism group of a compact Lorentz manifold and other geometric manifolds*, Invent. Math. **83** (1986), 411–424.

[Z5] ____, *Lattices in semisimple groups and invariant geometric structures on compact manifolds*, Discrete Groups in Geometry and Analysis: Proceedings of a Conference in Honor of G. D. Mostow, Birkhauser (to appear).

[Z6] ____, *On connection preserving actions of discrete linear groups*, Ergodic Theory Dynamical Systems (to appear).

[Z7] ____, *Split rank and semisimple automorphism groups of G-structures*, J. Differential Geom. (to appear).

[Z8] ____, *Infinitesimal rigidity for smooth actions of discrete subgroups of Lie groups*, preprint.

[Z9] ____, *Kazhdan groups acting on compact manifolds*, Invent. Math. **75** (1984), 425–436.

UNIVERSITY OF CHICAGO, CHICAGO, ILLINOIS 60637, USA

J. DIFFERENTIAL GEOMETRY
26 (1987) 169–173

SPLIT RANK AND SEMISIMPLE
AUTOMORPHISM
GROUPS OF G-STRUCTURES

ROBERT J. ZIMMER

1. Introduction

This paper is a continuation of the investigation begun in [1], [3], [4] concerning the semisimple automorphism groups of G-structures on compact manifolds. In those papers we were concerned with semisimple groups that preserve a structure which is algebraic and which defines a volume density, i.e. where the structure group G is an algebraic subgroup of $SL'(n, \mathbb{R})$, the matrices with $|\det| = 1$. (For higher order structures we assumed that G is an algebraic subgroup of $SL'(n, \mathbb{R}) \cap GL(n, \mathbb{R})^{(k)}$, the latter being the group of k-jets at 0 of diffeomorphisms of $\mathbb{R}^n$ fixing the origin.) One of the basic conclusions in the above papers is that for any simple noncompact Lie group H preserving such a G-structure, we must have that H locally embeds in G. (In fact a stronger assertion is proven. See the above papers and Theorem 2 below.) The main goal of the present paper is to consider the situation in which H is no longer assumed to define a volume density. In this case natural examples easily show that one cannot expect a local embedding of H in G. However, our main result asserts that a basic structural invariant of H must be visible in G. More precisely, we prove:

Theorem 1. *Let H be a semisimple Lie group with finite center and suppose that H acts smoothly on a compact manifold M so as to preserve a G-structure on M, where G is a real algebraic group. Then $\mathbb{R}$-rank$(H) \leqslant \mathbb{R}$-rank(G).*

We recall that the $\mathbb{R}$-rank, or split rank, of a real algebraic group is the maximal dimension of an algebraic torus that is diagonalizable over $\mathbb{R}$. For a semisimple Lie group H, $\mathrm{Ad}(H)$ will be the connected component of the identity of a real algebraic group, and the $\mathbb{R}$-rank, or split rank, of H is defined to be the split rank of this real algebraic group. We shall also clear up

Received June 13, 1986. The author's research was partially supported by a National Science Foundation Grant.

a point that was left open in [1], [3] concerning the case in which G defines a volume density. Namely, the general results in [3] were established for noncompact simple groups, not for semisimple groups. In [3], a special argument was given that clarified the semisimple situation for the case of Lorentz structures. Here we observe that a simple argument enables us to extend the results of [3] to the semisimple case in general, at least in the case of finite center.

Theorem 2 (*cf.* [1], [3], [4]). *Let H be a connected semisimple Lie group with finite center and no compact factors, and suppose that H acts on a compact n-manifold preserving a G-structure, where G is algebraic and defines a volume density. Then there is an embedding of Lie algebras $\mathfrak{h} \to \mathfrak{g}$. Furthermore, the representation $\mathfrak{h} \to \mathfrak{g} \to \mathfrak{sl}(n, \mathbb{R})$ contains $\mathrm{ad}_{\mathfrak{h}}$ as a direct summand.*

Part of this work was done while the author was a visitor at Harvard University. We would like to thank the members of that department for their hospitality.

2. Preliminaries

We establish here some preliminary information we shall need for the proofs of Theorems 1 and 2.

Proposition 3. *Let H be a connected semisimple Lie group with finite center, acting smoothly on a connected manifold M, and assume $p \in M$ is a fixed point. Let $\pi \colon H \to \mathrm{GL}(TM_p)$ be the corresponding representation at p. If π is trivial, then H acts trivially on M.*

Proof. Let $K \subset H$ be a maximal compact subgroup. It suffices to see that K acts trivially. For a compact group, any smooth action can be linearized around fixed points, so the set of invariant frames for the tangent bundle is both open and closed.

Proposition 4. *Suppose H is a connected semisimple Lie group with finite center, acting smoothly on a connected manifold M. If the set of fixed points has positive measure, then H acts trivially.*

Proof. If the set of fixed points, F, has positive measure, choose a density point p for F in the sense of Lebesgue. Then any small ball around p intersects F in a set of positive measure. The action of the maximal compact subgroup $K \subset H$ can be linearized around p, which implies that K leaves a set of vectors in TM_p invariant which has positive measure in TM_p. It follows that this linear representation of K is trivial, and the proof of Proposition 3 completes the proof.

If a Lie group H acts smoothly on a manifold M, and $m \in M$, we let H_m be the stabilizer of m in H, and $\mathfrak{h}_m \subset \mathfrak{h}$ the Lie algebra of H_m. If V is a vector space we let $\mathrm{Gr}_d V)$ be the Grassman variety of d-dimensional linear subspaces.

For a Lie group L, we let L^0 be the identity component. If a Lie group L is the identity component of an algebraically connected real algebraic group, by a rational homomorphism of L into a real algebraic group we mean the restriction of (a necessarily unique) rational homomorphism of the ambient algebraic group. The following is standard.

Lemma 5. *Suppose H is a Lie group acting smoothly on a manifold M. Let d be the minimal dimension of an H-orbit in M. Then $M_1 = \{ m \in M \,|\, \dim(Gm) = d \}$ is closed, and the map $m \to \mathfrak{h}_m$ defines a continuous map $\varphi\colon M_1 \to \mathrm{Gr}_q(\mathfrak{h})$, where $q = \dim(H) - d$. Further, φ is an H-map, where H acts on $\mathrm{Gr}_q(\mathfrak{h})$ via $\mathrm{Ad}(H)$.*

We recall briefly the notion of the algebraic hull of a cocycle defined for an ergodic group action (see [4] or [2] for an elaboration). Suppose that H is a locally compact group acting ergodically on a standard measure space (M, μ). Suppose that G is a real algebraic group and that $\alpha\colon H \times M \to G$ is a cocycle, i.e., the following identity is satisfied (for each $h_1, h_2 \in H$, and almost all $m \in M$): $\alpha(h_1 h_2, m) = \alpha(h_1, h_2 m)\alpha(h_2, m)$. We recall that two cocycles α, β are called equivalent if there is a measurable $\varphi\colon M \to G$ such that for each h and almost all m, $\beta(h, m) = \varphi(hm)^{-1}\alpha(h, m)\varphi(m)$.

Lemma 6 ([2], [4], [5]). *There is an algebraic subgroup $L \subset G$ with the following properties:*

(i) *α is equivalent to a cocycle taking all its values in L.*

(ii) *For any proper algebraic subgroup $L' \subset L$, α is not equivalent to a cocycle taking all its values in L'.*

(iii) *Up to conjugacy in G, L is the unique algebraic subgroup satisfying* (i), (ii).

(iv) *If α is equivalent to a cocycle taking all its values in some closed subgroup $L_0 \subset G$, then some conjugate of L_0 is contained in L.*

L is then called the algebraic hull of α, and it is well defined up to conjugacy in G. The following property is easily established.

Lemma 7. *Suppose $p\colon G_1 \to G_2$ is a rational homomorphism of real algebraic groups. If α is a G_1-valued cocycle with algebraic hull L_1, then the algebraic hull of the G_2-valued cocycle $p \circ \alpha$ is the algebraic hull of $p(L_1)$ (in which, we recall, $p(L_1)$ is a subgroup of finite index).*

3.   Proof of Theorem 1

Let M_1 be as in Lemma 5. Since M_1 is a compact H-space, we can choose a minimal H-space $M_0 \subset M_1$, i.e., a closed H-invariant subset in which every orbit is dense. Then, letting φ be as in Lemma 5 as well, we have that $\varphi(M_0) \subset \mathrm{Gr}_q(\mathfrak{h})$ is minimal. However, the action of H on $\mathrm{Gr}_q(\mathfrak{h})$ is algebraic, and hence every orbit is locally closed. It follows that $\varphi(M_0)$ consists of a single compact H-orbit. Fix $x \in M_0$, and let $\mathfrak{h}_x = \mathfrak{h}_0$. Then we can consider φ as an H-map $\varphi: M_0 \to \mathrm{Ad}(H)/N(\mathfrak{h}_0) \subset \mathrm{Gr}_q(\mathfrak{h})$, where $N(\mathfrak{h}_0)$ is the normalizer of $\mathfrak{h}_0$ in $\mathrm{Ad}(H)$. In particular, the algebraic subgroup $N(\mathfrak{h}_0)$ is cocompact in $\mathrm{Ad}(H)$, and therefore we can find a maximal $\mathbb{R}$-split torus T of $\mathrm{Ad}(H)$, with $T \subset N(\mathfrak{h}_0)$.

Let $\mathfrak{n}$ be the Lie algebra of $N(\mathfrak{h}_0)$, so that $\mathfrak{h}_0 \subset \mathfrak{n}$ is an ideal. The adjoint representation yields a rational (and in particular semisimple) representation $T \to \mathrm{GL}(\mathfrak{h}/\mathfrak{h}_0)$. Let $T_0 \subset T$ be the kernel, so that T_0 is an $\mathbb{R}$-split subtorus. Since the representation of T_0 on $\mathfrak{h}$ is semisimple, we can write $\mathfrak{h} = \mathfrak{h}_0 \oplus W$, where $W \subset \mathfrak{h}$ is a subspace and T_0 acts trivially on W. In particular, $\mathfrak{h}_0$ contains all the root spaces of T_0 acting via Ad_H on $\mathfrak{h}$ corresponding to nontrivial roots. The algebra generated by the nontrivial root spaces for T_0 is an ideal, and hence $\mathfrak{h}_0$ contains an ideal of $\mathfrak{h}$, say $\mathfrak{h}_1$, containing all the nontrivial root spaces for T_0. Thus we can write $\mathfrak{h}$ as a sum of ideals, $\mathfrak{h} = \mathfrak{h}_1 \oplus \mathfrak{h}_2$, where T_0 acts trivially on $\mathfrak{h}_2$. Since $\mathfrak{h}_2$ is semisimple, it follows that $\mathfrak{t}_0 \subset \mathfrak{h}_1 \subset \mathfrak{h}_0$. Let H_1 be the connected normal subgroup of H corresponding to $\mathfrak{h}_1$. Then $\mathfrak{h}_1$, and hence H_1, acts trivially on $\mathfrak{h}/\mathfrak{h}_0$, and by Proposition 3, H_1 acts locally faithfully on $T(M)_x/T(Hx)_x$. In particular, T_0 acts rationally and locally faithfully on $T(M)_x/T(Hx)_x$. Let T_1 be a split torus complementary to T_0 in T. We then have that $T = T_0 \times T_1$, and T_1 acts faithfully on $\mathfrak{h}/\mathfrak{h}_0$.

Now let $M_2 \subset M_0$ be a minimal $N(\mathfrak{h}_0)^0$ space. Since $(H_x)^0$ is normal in $N(\mathfrak{h}_0)$, it fixes all points of $N(\mathfrak{h}_0)x$, and hence fixes all points in the closure of this orbit, in particular all points in M_2. Since the dimension of all stabilizers in H of points in M_0 are the same, we have $\mathfrak{h}_m = \mathfrak{h}_0$ for all $m \in M_2$. Thus for $m \in M_2$, we can identify the tangent space to the H-orbit through m with $\mathfrak{h}/\mathfrak{h}_0$. The representation of H_1 on $T(M)_m/T(Hm)_m$ will vary continuously over $m \in M_2$, and since H_1 is semisimple and M_2 is connected, all these representations are equivalent. In particular, the representations of $(T_0)^0$ on these spaces are all rational, and all equivalent.

Choose a probability measure on M_2 which is invariant and ergodic under T^0 [2, Chapter 4]. Let $\alpha: T^0 \times M_2 \to \mathrm{GL}(n, \mathbb{R})$ be a cocycle corresponding to the action of T^0 on the tangent bundle of M over the space M_2 (cf. [4]). Let L be the algebraic hull of this cocycle. Since H, and in particular T^0, leaves a G-structure on M invariant, we have (up to conjugation) $L \subset G$. By our observations above, we can measurably trivialize TM over M_2 in such a way that $TM \cong M \times \mathbb{R}^n$, $\mathbb{R}^n = V_1 \oplus V_2$, $V_1 \cong \mathfrak{h}/\mathfrak{h}_0$, such that for $t \in T_0^0$, we have

$$\alpha(m, t) = \begin{pmatrix} I & 0 \\ 0 & \pi_2(t) \end{pmatrix},$$

where π_2 is a faithful rational representation, and for $t \in T_1^0$, we have

$$\alpha(m, t) = \begin{pmatrix} \pi_1(t) & * \\ 0 & * \end{pmatrix},$$

where $\pi_1(t)$ is $\mathrm{Ad}(t)$ acting on $\mathfrak{h}/\mathfrak{h}_0$, and, as we remarked above, is a faithful rational representation. Let β be the projection of α in $\mathrm{GL}(V_1) \times \mathrm{GL}(V/V_1)$. To prove the theorem, it suffices to see that the split rank of L is at least as large as $\dim(T)$, and by Lemma 7, to prove this it suffices to see that the split rank of the algebraic hull of β is at least $\dim(T)$. Thus, we need only see that if π is a faithful rational representation of T^0, then the algebraic hull of the cocyle $\beta(m, t) = \pi(t)$ $(m \in M_2)$ is locally isomorphic to T. Let T^* be the algebraic hull of the group $\pi(T^0)$; then T^* is a split torus, $\pi(T^0) \subset T^*$ is of finite index, and $\dim(T) = \dim(T^*)$. If β is equivalent to a cocycle into $Q \subset T^*$, then there is a measurable T^0-map $\varphi\colon M_2 \to T^*/Q$. Since there is a finite T^0-invariant measure on M_2, there is one on T^*/Q as well, and if Q is algebraic, it is clear that $\dim(Q) = \dim(T)$. This complete the proof.

4. Proof of Theorem 2

The argument of [1, Lemma 6], using the Borel density theorem, shows that the Lie algebra of the stabilizer of almost every point is an ideal. By Proposition 4, it follows that almost every point has a discrete stabilizer. The proof then follows as in the simple case, as in [3] or [4].

References

[1]   R. J. Zimmer, *Semisimple automorphism groups of G-structures*, J. Differential Geometry **19** (1984) 117–123.
[2]   _____, *Ergodic theory and semisimple groups*, Birkhauser, Boston, 1984.
[3]   _____, *On the automorphism group of a compact Lorentz manifold and other geometric manifolds*, Invent. Math. **83** (1986) 411–426.
[4]   _____, *Ergodic theory and the automorphism group of a G-structure*, Proc. Conf. in honor of G. W. Mackey, Math. Sci. Res. Inst. Publ., Springer, Berlin, to appear.
[5]   _____, *An algebraic group associated to an ergodic diffeomorphsm*, Compositió Math. **43** (1981) 59–69.

UNIVERSITY OF CHICAGO

ILLINOIS JOURNAL OF MATHEMATICS
Volume 34, Number 4, Winter 1990

MANIFOLDS WITH INFINITELY MANY ACTIONS OF AN ARITHMETIC GROUP[1]

BY

RICHARD K. LASHOF AND ROBERT J. ZIMMER

It is well known that if Γ is a lattice in a simple Lie group of higher split rank then in any finite dimension Γ has only finitely many inequivalent linear representations. This is one manifestation of the strong linear rigidity properties that such groups satisfy. When one considers non-linear representations, say smooth actions of Γ on compact manifolds, one still sees a large number of rigidity phenomena [7]. This is particularly true for actions preserving a connection. On the other hand, the point of this note is to establish the following result.

THEOREM 1. *Let G be the Lie group $SL(n, \mathbf{R})$, $n \geq 3$, or $SU(p, q)$, $p, q \geq 2$. Then there is a cocompact discrete subgroup $\Gamma \subset G$ and a smooth compact manifold M such that there are infinitely many actions of Γ on M with the following properties*:

   i) *The actions are mutually non-conjugate in* Diff(M), Homeo(M), *and* Meas(M), *where the latter is the group of measure class preserving automorphisms of M as a measure space*;

   (ii) *Each action leaves a smooth metric on M invariant, is minimal (i.e., every orbit is dense), and ergodic (with respect to the smooth measure class.)*

Theorem 1 is easily deduced from a certain non-rigidity phenomenon for tori in compact semisimple groups. Namely, fix a compact semisimple Lie group C and call closed subgroups H_1 and H_2 equivalent if there is an automorphism α of C such that $\alpha(H_1) = H_2$. We can then ask to what extent the diffeomorphism class of C/H determines the equivalence class of H. (The natural question is under what circumstance the map from equivalence classes of (a class of) closed subgroups to diffeomorphism classes of manifolds is finite-to-one.) Here we show:

THEOREM 2. *Let $C = SU(n) \times SU(n)$, $n \geq 2$. Then there is a family of mutually non-equivalent tori T_k, $k \in \mathbf{Z}^+$, such that C/T_k are all diffeomorphic.*

Received July 7, 1988.

[1]Research partially supported by the National Science Foundation.

We have not determined whether or not a similar phenomenon can occur if C is a simple Lie group of higher rank. It has been shown by A. Borel (private communication) that this cannot happen for one dimensional tori in $SU(4)$.

To prove Theorem 1 from Theorem 2, we first claim that if G is $SL(n, \mathbf{R})$ or $SU(p, q)$ where $p + q = n$, there is a cocompact lattice $\Gamma \subset G$ and a dense embedding of Γ into $SU(n) \times SU(n)$. This follows via the standard arithmetic construction of cocompact lattices. (See, e.g., [2], [3], [5], [6].) For completeness, we indicate the construction for $SU(p, q)$. The case of $SL(n, \mathbf{R})$ is similar, but a bit more complicated. Let $p \in \mathbf{Q}[X]$ be a cubic with 3 real irrational roots a, b, c with $a < 0 < b, c$. Let k be the splitting field. We assume $[k : \mathbf{Q}] = 3$, and σ, τ the non-trivial elements of $\mathrm{Gal}(k/\mathbf{Q})$. Let B be the Hermitian form on $\mathbf{C}^n$ given by

$$B(z, w) = a\left(\Sigma_1^p z_i \overline{w}_i\right) + \Sigma_{p+1}^{p+q} z_i \overline{w}_i$$

Then $SU(B)$ can be identified with the set of real points of an algebraic group $\mathscr{G}$ defined over k, and as a Lie group it is ismorphic to $SU(p, q)$. Let $\mathscr{O} \subset k$ be the algebraic integers in k, and $\Gamma = \mathscr{G}_{\mathscr{O}} \subset SU(B)$. The real points of the twisted groups $\mathscr{G}^\sigma, \mathscr{G}^\tau$ will be identified with $SU(B^\sigma), SU(B^\tau)$ respectively, which are both isomorphic as Lie groups to $SU(n)$ since these forms are now positive definite, due to the fact $b, c > 0$. It follows from [3] (see also [6] for a discussion) that Γ is a lattice in $SU(B)$ and that $(\sigma, \tau): \Gamma \to SU(B^\sigma) \times SU(B^\tau)$ is a dense embedding.

Now choose T_k as in Theorem 2, and let M be the manifold C/T_k (which is independent of k.) For each k, let Γ act on M via the embedding in C, and the action of C on C/T_k. Since Γ is dense in C, if two of these Γ actions are conjugate in $\mathrm{Homeo}(M)$, then the C actions are conjugate in $\mathrm{Homeo}(M)$, which implies conjugacy of the corresponding tori. In fact, the same is true for a measurable conjugacy. Namely, the set of measurable maps (mod null sets) $C/T_1 \to C/T_2$ is a standard Borel space with the topology of convergence in measure. There is a natural Borel C action on this space (namely $(gf)(x) = gf(g^{-1}x)$), and the stabilizers of points in such actions are closed subgroups [6]. It follows that any Γ-map (which is a fixed point in this function space) is also a C-map. This map must then be a.e. equal to a continuous C-map. This establishes (i) of Theorem 1 and (ii) is obvious from the construction.

We now prove Theorem 2. Suppose more generally that $C = K \times L$ where K and L are simple. Let T be a torus in K and $\rho: T \to L$ a homomorphism. Then

$$T_\rho = \{(t, \rho(t) \in K \times L | t \in T\}$$

is a subgroup of C. Since T is abelian $\rho^*(t) = \rho(t)^{-1}$ is also a homomorphism and $E_\rho = G/T_\rho = K \times_T L, T$ acting on L via ρ^*, is an associated principal bundle to $q:K \to K/T$. The idea of the proof is to choose homomorphisms such that these bundles are equivalent. Also observe that automorphisms of C are of the form $\alpha = (\beta, \delta)$, where β and δ are automorphisms of K and L respectively if $K \neq L$, and such an automorphism composed with a permutation of the factors if $K = L$. This makes it easy to tell when two such groups are not equivalent. As an illustration, consider $K = SU(2), L = SU(n), T$ a maximal torus of K. Since $K/T = S^2$ and $\pi_2(BL) = 0$, where BL is the classifying space for L-bundles, every L-bundle over K/T is trivial. But if $\rho:SU(2) \to SU(n)$ is a non-trivial representation, then T_ρ is not equivalent to T (which corresponds to the trivial representation.)

We now take $K = L = SU(n)$ as in the statement of the theorem. Let T be a maximal torus in K, e.g., the set of diagonal matrices with entries $d_j = \exp(i\theta_j)$ satisfying $\Sigma\theta_j = 0$. The Weyl group W of $SU(n)$ is the group of permutations of the factors d_j. Let $\rho_k:T \to L$ be the homomorphism of T onto the maximal torus $T' = T$ of $L, \rho_k(t) = t^k, k = 1, 2, \ldots$. Note that if $w \in W$ and $\rho = \rho_k$ for some k, then $\rho w = w\rho$. Let $\lambda:K/T \to BT$ be the classifying map for $q:K \to K/T$, let $i:BT' \to BL$ be the map induced by the inclusion of T' in L, and let $B_\rho:BT \to BT'$ be the map induced by ρ. Then the classifying map for E_ρ is $f_\rho = iB_\rho\lambda$.

Now in [1] it is shown that $H^*(SU(n))$ and $H^*(SU(n)/T)$ have no torsion and that this implies that $i^*:H^*(BSU(n)) \to H^*(BT)$ is an isomorphism onto $H^*(BT)^W$, the fixed set under the action of the Weyl group. In particular, this implies $\lambda^*:H^*(BT)^W \to H^*(K/T)$ is trivial. But $\rho w = w\rho$ implies $(B\rho)^*:H^*(BT')^W \to H^*(BT)^W$. Hence $f_\rho^* = \lambda^* \cdot (B\rho)^* \cdot i^*$ is trivial. We claim this implies there are only finitely many equivalence classes of bundles E_ρ for $\rho = \rho_k, k = 1,2,\ldots$

First note that $SU(n)/T$ is a finite CW complex whose cohomology has no torsion. We will use the following theorem of F. Peterson [4].

THEOREM. *Let X be a CW complex of dimension $\leq 2n$ such that $H^*(X)$ has no torsion. Then a complex vector bundle over X is trivial iff all its Chern classes are trivial.*

To apply this result to $SU(n)/T$ we first note the next result.

LEMMA. *Let X be a CW complex and $X^{(n)}$ its n-skeleton. If $H^*(X)$ has no torsion, then $H^*(X^{(n)})$ has no torsion.*

Our claim is an immediate consequence of the following:

PROPOSITION. *Let X be a 1-connected finite CW complex such that $H^*(X)$ has no torsion. Then there are only a finite number of equivalence classes of $SU(n)$ bundles X with all Chern classes zero.*

768   RICHARD K. LASHOF AND ROBERT J. ZIMMER

Proof. By Peterson's theorem and the above lemma, if $f:X \to BU(n)$ is such that $f^*H^*(BU(n)) \to H^*(X)$ is zero, then $f|X^{(2n)}$ is homotopically trivial. Let

$$d:BU(n) \to BU(1)$$

be induced by

$$\det:U(n) \to U(1),$$

so that the fibre of d is $BSU(n)$. If $f = j \cdot g, g:X \to BSU$ and $j:BSU(n) \to BU(n)$ induced by the inclusion, so that df is trivial, then the homotopy of $f|X^{(2n)}$ to the trivial map gives a map of $\Sigma(X^{(2n)})$ to $BU(1)$. Since $\Sigma(X^{(2n)})$ is 2-connected, this last is homotopically trivial rel endpoints. Hence $g|X^{(2n)}$ is homotopically trivial. Since the homotopy of $g|X^{(2n)}$ extends to a homotopy of g, we can assume $g|X^{(2n)}$ is trivial. Since $\pi_i(BSU(n))$ is finite for $i > 2n, [X/X^{(2n)}, BSU(n)]$ is finite. Thus up to equivalence there are only finitely many $SU(n)$ bundles over X with all Chern classes zero.

Thus an infinite number of the E_ρ for $\rho = \rho_k, k = 1, 2, \ldots$, are equivalent. On the other hand, if $T_k = T_\rho$ for $\rho = \rho_k$, no automorphism of C sends T_j to T_k if $j \neq k$. This completes the proof of Theorem 2.

REFERENCES

1. A. BOREL, *Sur la cohomologie des espaces fibre principaux et des espaces homogenes des groupes de Lie compacts*, Ann. of Math., vol. 57 (1953), pp. 115–207.
2. A. BOREL and G. HARDER, *Existence of discrete cocompact subgroups of reductive groups over local fields*, J. fur Math. vol. 298 (1978), pp. 53–64.
3. A. BOREL and HARISH-CHANDRA, *Arithmetic subgroups of algebraic groups*, Ann. of Math, vol. 75 (1962), pp. 485–535.
4. F.P. PETERSON, *Some remarks on Chern classes*, Ann. of Math., vol. 69 (1959), pp. 414–420.
5. M. RAGHUNATHAN, *Discrete subgroups of Lie groups*, Springer, New York, 1972.
6. R.J. ZIMMER, *Ergodic theory and semisimple groups*, Birkhauser, Boston, 1984.
7. _____, "Lattices in semisimple groups and invariant geometric structures on compact manifolds" in *Discrete groups in geometry and analysis*, R. Howe ed., Birkhauser, Boston, 1987, pp. 152–210.

UNIVERSITY OF CHICAGO
CHICAGO, ILLINOIS

ISRAEL JOURNAL OF MATHEMATICS 75 (1991), 65–80

SPECTRUM, ENTROPY, AND GEOMETRIC STRUCTURES FOR SMOOTH ACTIONS OF KAZHDAN GROUPS

BY

ROBERT J. ZIMMER*

Department of Mathematics, University of Chicago

Chicago, IL 60637, USA

ABSTRACT

We show that many natural classes of actions of discrete subgroups of semisimple Lie groups have discrete spectrum, i.e., are measurably conjugate to isometric actions.

1. Statement of Main Results

In this paper we describe a fundamental feature of the measure theoretic structure of certain smooth volume preserving actions of discrete subgroups of higher rank semisimple Lie groups on compact manifolds. This includes all actions in "low" dimensional manifolds, actions in which all elements act with zero entropy, and actions preserving a class of geometric structures. These results constitute a measure theoretic version of results conjectured to be true in [7,9] concerning the smooth structure of such actions.

Suppose G is a locally compact second countable group and (X, μ) is a standard measure space with $\mu(X) = 1$ on which G acts so as to preserve μ. A simple (but *a priori* very restricted) class of examples arises by considering a homomorphism $G \to K$ where K is a compact group, and letting X be a compact metric space on which K acts continuously. In this case, there is a K-invariant (and hence G-invariant) topological distance function on X. We say that an action is measurably isometric if it is measurably conjugate to such an action. If the action is measurably isometric and ergodic, then (up to measurable conjugacy) we may take $X = K/K_0$ where $K_0 \subset K$ is a closed subgroup. Thus, the measurably isometric ergodic actions (and via an ergodic decomposition any measurably

---

*Research partially supported by NSF Grant.
Received December 14, 1990

isometric action) can be completely described, up to measurable conjugacy, in purely algebraic terms. If the compact group K is actually profinite, we say that the action is measurably profinite.

Returning to the general case, we let π be the unitary representation of G on $L^2(X, \mu)$ induced from the action. Following standard (but not entirely satisfactory) terminology, we say that the action of G on X has discrete spectrum if $L^2(X)$ is the direct sum of finite dimensional G-invariant subspaces. If the action is measurably isometric, then the Peter-Weyl theorem implies that the action has discrete spectrum, and it is a classical result of Mackey [1] (generalizing the corresponding result for integer actions of Halmos and von Neumann) that the converse is also true. In other words, one can characterize the spatial condition of being measurably isometric by the purely spectral condition of having discrete spectrum. An action is called weakly mixing if there are no finite dimensional G-invariant subspaces in $L^2(X)$ except C. In the case of integer actions this is, of course, equivalent to the non-existence of an eigenvector for $\pi(g)$, $g \in \mathbf{Z}$, which is in turn equivalent to the classical definition of weak mixing via the limits of certain partial sums.

We also recall that for each $g \in G$, there is the (Kolmogorov-Sinai) entropy $h(g) \in [0, \infty]$. We say that the action has zero entropy if $h(g) = 0$ for all $g \in G$. Actions with discrete spectrum have zero entropy, but for the integers there is a vast array of zero-entropy weakly mixing actions. In fact, from a purely measure theoretic viewpoint, by virtue of Ornstein's theory of Bernoulli shifts, the complication in classifying integer actions arises at the zero-entropy level (or more precisely, in studying extensions with zero relative entropy). One of our main results will be the assertion that for smooth actions on compact manifolds of certain discrete subgroups of Lie groups, zero entropy is in fact equivalent to discrete spectrum.

We now give a precise description of the groups we will be considering and of the geometric aspects of our results. Let S be a finite subset of $\{\infty\} \cup \{\text{primes in } \mathbf{Z}\}$ and for each $p \in S$, let G_p be the set of $\mathbf{Q}_p$-points of a connected semisimple algebraic $\mathbf{Q}_p$-group, where $\mathbf{Q}_p$ is the field of p-adic numbers if p is a prime and $\mathbf{Q}_\infty = \mathbf{R}$. We assume the $\mathbf{Q}_p$-rank of every $\mathbf{Q}_p$-simple factor of G_p is at least 2. We let $G = \prod_{p \in S} G_p$, and shall abbreviate this situation by simply saying G is a semisimple group of higher rank. If $S = \{\infty\}$, we say G is real, in which case G is an almost connected semisimple Lie group with finite center. In fact, for real groups we shall also allow any connected semisimple group with finite center such that all simple factors have $\mathbf{R}$-rank at least 2, not just the algebraic ones. (Of course, any such group is algebraic modulo finite groups.)

If $S = \{p\}$, we simply say G is p-adic. Let Γ be a lattice in G, i.e., a discrete subgroup such that G/Γ has a finite G-invariant measure. The basic examples of such lattices are the S-arithmetic groups, and by a theorem of Margulis [2], [8] (which we will not be using) these are all the examples (in higher rank).

THEOREM 1.1: *Let G be a higher rank semisimple group and $\Gamma \subset G$ a lattice. Suppose Γ acts smoothly on a compact manifold preserving a volume density. Then the following are equivalent:*

(a) *The Γ action has zero entropy.*

(b) *The Γ action is measurably isometric (or equivalently, has discrete spectrum).*

(c) *There is a Γ-invariant measurable Riemannian metric on M.*

The implication (a) $\Rightarrow$ (c) is proven in [5]. (Actually [5] only considers the real case. However, these arguments work in general, using the results of [8, Chapter 10].) We have already remarked that (b) $\Rightarrow$ (a). The main argument of this paper is to show (c) $\Rightarrow$ (b). In [7] it is also shown that natural geometric hypotheses on the action imply condition (c). Namely, suppose $\dim M = n$ and $H \subset \mathrm{SL}(n, \mathbf{R})$ is a real algebraic subgroup. If Γ preserves an H-structure on M, then by [7] either there is a non-trivial Lie algebra homomorphism $\mathcal{G}_\infty \to \mathcal{H}$, or the Γ action has an invariant measurable Riemannian metric. Therefore, we deduce:

THEOREM 1.2: *Suppose every Lie algebra homomorphism $\mathcal{G}_\infty \to \mathcal{H}$ is trivial, and that Γ preserves an H-structure on M. Then the Γ-action is measurably isometric.*

We present two special cases of particular interest. Let $d(G) = \min\{d \,|\, \text{there is a non-trivial Lie algebra representation } \mathcal{G}_\infty \to \mathcal{G}\ell(d, \mathbf{C})\}$.

COROLLARY 1.3: *If $n < d(G)$, then any volume preserving action of Γ on a compact n-manifold is measurably isometric.*

If G is p-adic, then $\mathcal{G}_\infty = (0)$. Hence:

COROLLARY 1.4: *If Γ is a lattice in a higher rank semisimple p-adic group, then any smooth volume preserving action on a compact manifold is measurably isometric. In particular, there are no such weakly mixing actions.*

If G is real, simple, has a Q-structure with Q-rank at least 2, and Γ is the arithmetic lattice $G_{\mathbf{Z}}$ (e.g. $G_{\mathbf{Z}} = \mathrm{SL}(n, \mathbf{Z})$ where $n \geq 3$), then every homomorphism of $G_{\mathbf{Z}}$ into a compact Lie group has finite image. Thus, every measurably isometric action is measurably profinite. For example:

COROLLARY 1.5:   *Let* $\Gamma = SL(n, \mathbf{Z})$ *where* $n \geq 3$. *If* M *is a compact manifold of dimension* d *with* $d < n$, *then any smooth, volume preserving* Γ-*action on* M *is measurably profinite (and in particular is not weakly mixing).*

We have conjectured in [7, 9] that the conditions in Theorem 1.1 are in fact equivalent to the existence of a smooth Γ-invariant Riemannian metric, i.e., that the action is isometric, not just measurably isometric.

Our techniques in this paper also yield results for lattices in those **R**-rank 1 groups with Kazhdan's property.

THEOREM 1.6:   *Let* Γ *be a discrete Kazhdan group. Then any smooth volume preserving action of* Γ *on a compact surface is measurably isometric.*

Our proofs depend very strongly on the local analysis that is available by assuming the actions are smooth actions of manifolds. It would be very interesting to know whether or not conditions (a) and (b) in Theorem 1.1, for example, are equivalent outside the class of smooth actions.

As we indicated above, the proof of Theorem 1.1 is reduced to showing that the existence of a measurable invariant Riemannian metric implies that the action is measurably isometric. For an action of an arbitrary group this implication is false. A. Katok has constructed volume preserving diffeomorphisms of compact manifolds which have a measurable invariant Riemannian metric but which are weakly mixing. However, all the results stated above follow from the next theorem (when combined with the results of [5], [6], [7]). (In [6] it is shown that any Kazhdan group acting on a compact surface preserving a volume has an invariant measurable Riemannian metric.)

THEOREM 1.7:   *Let* Γ *be a discrete Kazhdan group acting smoothly on a compact manifold preserving a smooth volume density. If there is a measurable* Γ-*invariant Riemannian metric then the action has discrete spectrum.*

The idea of the proof of Theorem 1.7 is as follows. Let $m \mapsto \omega(m) \in S^2(T^*M)$ be a measurable invariant Riemannian metric. Extend each $\omega(m)$ to a smooth metric ω_m defined in a neighborhood of $m \in M$. Fix $r > 0$, and for each m, let $f_m \in L^2(M)$ be the normalized characteristic function of the ball of radius r centered at m with respect to the metric ω_m. Define $F_r \in L^2(M \times M)$ by $F_r(m, x) = f_m(x)$. Fix a finite generating set $\Gamma_0 \subset \Gamma$ and let $\delta > 0$. By virtue of the invariance of $\omega(m)$, one can show that for r sufficiently small we have $\|\gamma \cdot F_r - F_r\| < \delta$ for $\gamma \in \Gamma_0$. In other words F_r is (Γ_0, δ)-invariant. By Kazhdan's property, if we choose δ sufficiently small this implies that F_r is close in $L^2(M \times M)$ to a Γ-invariant function. One can verify that this is not a constant function which yields the existence of some non-trivial finite dimensional Γ-invariant subspace in $L^2(M)$. One then needs to see that this general construction can be carried out so that these finite dimensional subspaces generate $L^2(M)$, an issue which presents further non-trivial complications.

2. General Properties of Measurable Actions of Kazhdan Groups

In this section we formulate some consequences of Kazhdan's property for a group acting on a measure space. Throughout this section Γ will be a discrete Kazhdan group and $\Gamma_0 \subset \Gamma$ a fixed finite symmetric generating set. If (N, ν) is a probability space and $B \subset N$ has positive measure, we let ν_B be the probability measure $\nu_B(A) = \nu(A \cap B)/\nu(B)$.

PROPOSITION 2.1:   *Suppose $0 < q < 1$. Then there is some $\varepsilon > 0$ such that for any measure preserving Γ-action on a probability space (N, ν) and any set $D \subset N$ with $\nu_D(\gamma D \cap D) \geq 1 - \varepsilon$ for all $\gamma \in \Gamma_0$, there is a Γ-invariant set $Y \subset N$ satisfying $\nu_Y(Y \cap D), \nu_D(Y \cap D) > q$.*

For the proof, we first recall that Kazhdan's property asserts that there is some $\varepsilon > 0$ such that for any unit vector f in a Hilbert space on which Γ acts unitarily via π with no non-trivial invariant vectors, there is some $\gamma \in \Gamma_0$ such that $\|\pi(\gamma)f - f\| \geq \varepsilon$. (See [8].) This immediately admits the following reformulation.

LEMMA 2.2:   *Given any $\delta > 0$, there is some $\varepsilon' > 0$ such that for any unitary representation (π, H) of Γ and any (ε', Γ_0)-invariant unit vector $f \in H$, there is a Γ-invariant unit vector $h \in H$ with $\|h - f\| < \delta$.*

Proof: Let ε be as above. If f is an $(\varepsilon/n, \Gamma_0)$-invariant unit vector, write $f = f_0 + f_1$ where $f_0 \in H_0$, the space of Γ-invariant vectors, and $f_1 \perp H_0$. Then f_1 is $(\varepsilon/n, \Gamma_0)$-invariant, and since $H_0^\perp$ has only trivial Γ-invariant vectors we have $\|\pi(\gamma)f_1 - f_1\| \geq \varepsilon\|f_1\|$ for some $\gamma \in \Gamma_0$. By ε/n-invariance, we deduce

$\varepsilon/n \geq \varepsilon\|f_1\|$, so $\|f_1\| \leq 1/n$. Therefore, $\|f_0\|^2 \geq 1 - 1/n^2$, $\|f - f_0\|^2 \leq 1/n^2$ and letting $h = f_0/\|f_0\|$, it is clear that for n sufficiently large we can take $\varepsilon' = \varepsilon/n$. $\square$

Proof of Proposition 1.2: Given $0 < q < 1$, let $\delta > 0$ satisfy $q < (1-\delta)(1-\delta^{1/2})^2$ (and in particular $q < (1-\delta)$.) Let ε' be as in Lemma 2.2 for this choice of δ and set $\varepsilon = \varepsilon'/2$. If $A \subset N$, let f_A be the normalized (in L^2) characteristic function of A. Thus, $f_A \mid A = \nu(A)^{-1/2}$ and $f_A \mid A^c = 0$. Since Γ preserves ν we have

$$\|f_D - f_{\gamma D}\|^2 = \nu(\gamma D \triangle D)/\nu(D).$$

By the hypotheses on D we deduce that f_D is an (ε', Γ_0)-invariant unit vector. By Lemma 2.2, we can choose a unit vector $h \in L^2(N)$ that is Γ-invariant with $\|h - f_D\| < \delta$. Let

$$Y = \{x \in N \mid h(x) \geq (1 - \delta^{1/2})\nu(D)^{-1/2}\}.$$

Then Y is Γ-invariant and on $D \cap Y^c$ we have

$$|h - f_D|^2 \geq \delta\nu(D)^{-1}.$$

Therefore,

$$\delta^2 \geq \|h - f_D\|^2 \geq \delta\nu(D)^{-1}\nu(D \cap Y^c),$$

and so

$$\frac{\nu(D \cap Y^c)}{\nu(D)} \leq \delta$$

and

$$\frac{\nu(D \cap Y)}{\nu(D)} \geq 1 - \delta > q.$$

On the other hand, we have $\int_Y h^2 \leq 1$, so

$$(1 - \delta^{1/2})^2\nu(D)^{-1}\nu(Y) \leq 1.$$

Therefore

$$\nu(Y)^{-1} \geq (1 - \delta^{1/2})^2\nu(D)^{-1}$$

and so

$$\frac{\nu(D \cap Y)}{\nu(Y)} \geq \frac{\nu(D \cap Y)}{\nu(D)}(1 - \delta^{1/2})^2 \geq (1-\delta)(1-\delta^{1/2})^2 > q.$$

$\square$

We will be applying Proposition 2.1 to the case in which $N = X \times M$ where X and M both have Γ-invariant probability measures and the Γ action on X is ergodic. To this end, we record the following elementary fact.

LEMMA 2.3:   *Fix $0 < q < 1$, and $0 < \eta < 1$. Then there is some q' with $q < q' < 1$ such that for any $f : X \to [0,1]$ with $\int f(x) \geq q'$ we have $\mu(\{x | f(x) \geq q\}) \geq \eta$. (Here we are assuming (X, μ) is any probability space.)*

Proof: We have

$$q\mu\big(f^{-1}([0,q])\big) + \mu\big(f^{-1}([q,1])\big) \geq \int f \geq q'.$$

Thus, for $\alpha = \mu(f^{-1}[q,1])$, we have

$$q(1 - \alpha) + \alpha \geq q',$$

so

$$\alpha \geq \frac{q' - q}{1 - q}.$$

Thus, we need only choose $q' < 1$ such that $(q' - q)/(1 - q) \geq \eta$. $\square$

We now consider two Γ-spaces X and M with invariant probability measures. If $A \subset X \times M$ and $x \in X$, we let $A_x = \{m \in M | (x, m) \in A\}$.

COROLLARY 2.4:   *Suppose (X, μ_0) and (M, μ) as above and that Γ acts ergodically on (X, μ_0). Let $N = (X \times M, \nu)$ where $\nu = \mu_0 \times \mu$. Given any $0 < q < 1$ and any $0 < \eta < 1$, there is some $\varepsilon > 0$ with the following property: For any set $D \subset X \times M$ satisfying*

(a) $\nu(\gamma D \cap D)/\nu(D) \geq 1 - \varepsilon$ for $\gamma \in \Gamma_0$, and

(b) $\mu(D_x)$ is independent of $x \in X$ (and hence equals $\nu(D)$),

there is a Γ-invariant set $Y \subset X \times M$ such that

$$\mu_0\left(\left\{x \Big| \frac{\mu(Y_x \cap D_x)}{\mu(Y_x)} \geq q, \quad \frac{\mu(Y_x \cap D_x)}{\mu(D_x)} \geq q\right\}\right) > \eta.$$

Proof: Given q and η choose q' as in Lemma 2.3. Apply Proposition 2.1 to q' to find ε. Then if $D \subset X \times M$ satisfies (a) and (b), we have a Γ-invariant $Y \subset X \times M$ such that

$$\frac{\nu(Y \cap D)}{\nu(Y)}, \ \frac{\nu(Y \cap D)}{\nu(D)} \geq q'.$$

We observe that $x \to \mu(Y_x)$ is Γ-invariant since Y is Γ-invariant, hence by ergodicity is constant, and that this constant must then be $\nu(Y)$. Thus

$$\int \frac{\mu(Y_x \cap D_x)}{\mu(Y_x)} = \frac{1}{\nu(Y)} \int \mu(Y_x \cap D_x) = \frac{\nu(Y \cap D)}{\nu(Y)}.$$

One has a similar assertion for $\mu(Y_x \cap D_x)/\mu(D_x)$ by hypotheses (b), and the result then follows from the conclusion of Lemma 2.3.   □

3. Construction of Almost Invariant Sets

We now return to the situation in which (M, μ) is a compact n-dimensional manifold with a smooth Γ-invariant probability measure, and for which there is a measurable Γ-invariant Riemannian metric. We shall use a Taylor series estimate and Corollary 2.4 to produce small Γ-invariant measurable sets in $M \times M$ that are not too badly behaved with respect to balls in M.

We choose a point $p \in M$. For simplicity we may assume $p = 0 \in \mathbf{R}^n$ and that M contains a neighborhood of 0 in $\mathbf{R}^n$. By Moser's theorem [3] we may assume μ is the standard Lebesgue measure on this neighborhood of D. We then fix a smooth Riemannian metric on M which is the standard flat metric in a neighborhood of 0. We denote by $B(x, r)$ the ball of radius r centered at x with respect to this metric. We may assume that the smooth measure associated to this metric is μ. We let $J^k(M) \to M$ be the vector bundle of k-jets of $\mathbf{R}$-valued functions on M. The Riemannian metric on M determines a metric on these bundles, and if $\varphi \in \mathrm{Diff}_\mu(M)$ (the diffeomorphisms preserving μ) we denote by $\|\varphi\|_k$ the norm of the linear map φ induced on $C(M; J^k M)$, the Banach space of continuous sections of the bundle $J^k M$. Let $D_0 = \{\varphi \in \mathrm{Diff}_\mu(M) | \varphi(0) = 0\}$, $J^k(D_0)$ the group of k-jets at 0 of elements of D_0 and $p_k : D_0 \to J^k(D_0)$ the canonical projection. Let $x \mapsto h_x$ be a Borel map $M \to \mathrm{Diff}_\mu(M)$ such that

  (i) $h_x(x) = 0$,
  (ii) $\{\|h_x\|_2, \|h_x^{-1}\|_2 | x \in M\}$ is bounded.

The existence of such a map follows in a routine manner from compactness of M. Define $\alpha(\gamma, x) = h_{\gamma x} \circ \gamma \circ h_x^{-1}$, so that $\alpha : \Gamma \times M \to \mathrm{Diff}_\mu(M)$ is a Borel cocycle. Clearly $\alpha(\Gamma \times M) \subset D_0$, and if $\Gamma_0 \subset \Gamma$ is a finite set, then $\{\|\varphi\|_2 | \varphi \in \alpha(\Gamma_0 \times M)\}$ is a bounded set of real numbers.

The cocycle $p_2 \circ \alpha : \Gamma \times M \to J^2(D_0)$ is measurably equivalent to a cocycle into a compact subgroup of $J^2(D_0)$. This follows from the facts that:

(a) The existence of a measurable invariant Riemannian metric on TM is readily seen to be equivalent to the assertion that $p_1 \circ \alpha$ is equivalent to a cocycle taking all values in a compact subgroup K_1 of $J^1(D_0)$;

(b) $J^2(D_0) \cong J^1(D_0) \ltimes V$, where V is a vector group; and

(c) $K_1 \ltimes V$ is amenable, and hence any cocycle for a Kazhdan group taking values in $K_1 \ltimes V$ must be equivalent to one taking values in a compact subgroup. This last assertion is [8, Theorem 9.1.1] (cf. [7], Prop. 4.7).

We are not assuming that the smooth measure μ is ergodic. We fix a Γ-ergodic component (X, μ_0) of (M, μ) for which the cocycle $p_2 \circ \alpha$ is still equivalent to one into a compact subgroup K of $J^2(D_0)$. (This will be true for almost every ergodic component.) We may clearly assume that $K \subset O(n) \subset J^2(D_0)$, where we identify $O(n)$ with the 2-jets of orthogonal linear maps. We can in fact just let $X = M$, and view μ_0 as simply another Γ-invariant probability measure on M for which $p_2 \circ \alpha$ is equivalent to a cocycle into $K \subset J^2(D_0)$. We shall continue to denote M by X when we are considering it to be endowed with the measure μ_0. We note that we have no *a priori* smoothness properties of μ_0.

We now implement the equivalence of $p_2 \circ \alpha$ to a cocycle into K by a function $\overline{\psi}$. In other words, we choose a measurable $\overline{\psi} : X \to J^2(D_0)$, $x \mapsto \overline{\psi}_x$, such that

$$\beta(\gamma, x) = \overline{\psi}_{\gamma x}(p_2 \circ \alpha)(\gamma, x)\overline{\psi}_x^{-1} \in K \subset J^2(D_0).$$

Lift $\overline{\psi}$ to a measurable map $\psi : X \to D_0$. Then define the cocycle $\sigma : \Gamma \times X \to D_0$ by

$$\sigma(\gamma, x) = \psi_{\gamma x}\alpha(\gamma, x)\psi_x^{-1} = (\psi_{\gamma x} \circ h_{\gamma x}) \circ \gamma \circ (\psi_x \circ h_x)^{-1}.$$

Define $\lambda : X \to \mathbf{R}$ by

$$\lambda(x) = \max\{\|\psi_x \circ h_x\|_2, \quad \|(\psi_x \circ h_x)^{-1}\|_2\},$$

so that λ is a measurable function, and $\lambda(x) \geq 1$ (since these diffeomorphisms are all volume preserving).

THEOREM 3.1: *Let λ be as above and fix $1/2 < q < 1$, and $0 < \eta < 1$. Then for any $r > 0$ sufficiently small, there is a Γ-invariant measurable subset $Y^r \subset X \times M$ (depending upon r) and a measurable subset $X^r \subset X$ (depending upon r) such that:*

(1) $\mu_0(X^r) > \eta$.

(2) *For all $x \in X^r$*

$$\frac{\mu(Y_x^r \cap B(x, \lambda(x)r))}{\mu(Y_x)} \geq q.$$

(3) *For all $x \in X^r$,*

$$\frac{\mu(Y_x^r \cap B(x, \lambda(x)r))}{\mu(B(x, \lambda(x)r))} \geq c_0\lambda(x)^{-n}$$

where c_0 is a constant (independent of r and depending only on the original choice of smooth Riemannian metric on M), and $n = \dim M$.

Proof: Given q and η, choose ε as in Corollary 2.4 when the latter is applied to q and $\frac{\eta+1}{2}$. Now choose a set $X_1 \subset X$ such that for $X_0 = \bigcap_{\gamma \in \Gamma_0} \gamma X_1$ we have

$$\mu_0(X_0) > \max\{\frac{\eta+1}{2}, 1 - \frac{\varepsilon}{2}\}$$

and such that λ is bounded on X_1. From the definition of σ it follows that there is a constant $\xi_0 \in \mathbf{R}$ such that $\|\sigma(\gamma, x)\|_2 \leq \xi_0$ for all $x \in X_0$ and all $\gamma \in \Gamma_0$.

For each $\varphi \in D_0$ such that $p_2 \circ \varphi \in O(n) \subset J^2(D_0)$, from the Taylor series expansion around 0 we have for x in a fixed neighborhood of 0 (independent of φ), that $|\varphi(x)| \leq |x| + \|\varphi\|_2 |x|^2$. Thus, for r sufficiently small, we have for all $\varphi \in \sigma(\Gamma_0 \times X_0)$ that

$$\varphi(B(0,r)) \subset B(0, r + \xi_0 r^2).$$

Assuming (as we may) that Lebesgue measure is suitably normalized, we have $\mu(B(0,r)) = r^n$. It follows that we can find r_1 so that for $r \leq r_1$,

$$\frac{\mu(B(0,r))}{\mu(B(0, r + \xi_0 r^2))} = \frac{r^n}{(r + \xi_0 r^2)^n} \geq 1 - \frac{\varepsilon}{4}.$$

Since any $\varphi \in \sigma(\Gamma_0 \times X_0)$ preserves μ, we also have

$$\frac{\mu(\varphi(B(0,r)))}{\mu(B(0, r + \xi_0 r^2))} \geq 1 - \frac{\varepsilon}{4}$$

for any such φ and $r \leq r_1$. Since both $B(0,r)$ and $\varphi(B(0,r))$ are contained in $B(0, r + \xi_0 r^2)$, we deduce

$$\frac{\mu(B(0,r) \cap \varphi(B(0,r)))}{\mu(B(0, r + \xi_0 r^2))} \geq 1 - \frac{\varepsilon}{2}$$

It follows that

$$\frac{\mu(B(0,r) \cap \varphi(B(0,r)))}{\mu(B(0,r))} \geq 1 - \frac{\varepsilon}{2}$$

as well.

We now use this to construct a set $D \subset X \times M$ satisfying the hypotheses of Corollary 2.4. Fix $r \leq r_1$. For $x \in X$, let $D_x = (\psi_x \circ h_x)^{-1} B(0,r) \subset M$. Then $\mu(D_x) = \mu(B(0,r))$ for all x. If $x \in X_0$ and $\gamma \in \Gamma_0$, then

$$\mu(\gamma D_x \cap D_{\gamma x}) \quad = \mu(\gamma(\psi_x \circ h_x)^{-1} B(0,r) \cap (\psi_{\gamma x} \circ h_{\gamma x})^{-1} B(0,r))$$

and since $\psi_{\gamma x} \circ h_{\gamma x}$ preserves μ this

$$= \mu(\sigma(\gamma, x) B(0,r) \cap B(0,r))$$
$$\geq (1 - \frac{\varepsilon}{2}) \mu(B(0,r))$$

by the preceding paragraph. It follows that

$$\mu((\gamma D)_{\gamma x} \cap D_{\gamma x}) \geq (1 - \frac{\varepsilon}{2})\mu(B(0,r))$$

and therefore that

$$\nu(\gamma D \cap D) = \int_X \mu((\gamma D)_x \cap D_x)d\mu_0(x)$$
$$\geq (1 - \frac{\varepsilon}{2})\mu(B(0,r))\mu_0(X_0).$$

Since

$$\nu(D) = \int \mu(D_x)\, d\mu_0(x) = \mu(B(0,x),$$

and $\mu_0(X_0) \geq 1 - \frac{\varepsilon}{2}$, we have

$$\frac{\nu(\gamma D \cap D)}{\nu(D)} \geq (1 - \frac{\varepsilon}{2})(1 - \frac{\varepsilon}{2}) \geq 1 - \varepsilon.$$

Applying Corollary 2.4 (and recalling that ε was chosen to work for the pair $(q, \frac{\eta+1}{2})$), it follows that for any $r \leq r_1$ we have a Γ-invariant set Y^r satisfying the conclusions of Corollary 2.4 for a set of x of μ_0-measure at least $1 + \frac{\eta}{2}$. Since $\mu_0(X_0) > \frac{1+\eta}{2}$ as well, we deduce that there is a set $X^r \subset X$ (in fact $X^r \subset X_0$) with $\mu_0(X^r) \geq \eta$ such that for $x \in X^r$, we have

(i) $\mu(Y_x^r \cap D_x)/\mu(Y_x^r) \geq q$,
(ii) $\mu(Y_x^r \cap D_x)/\mu(D_x) \geq q$.

From the definition of D_x and $\lambda(x)$, we see that $D_x \subset B(x, \lambda(x)r)$ and hence assertion (i) implies conclusion (2) of Theorem 3.1. It remains only to prove (3). By what we have just verified,

$$\frac{\mu(Y_x^r \cap B(x, \lambda(x)r))}{\mu(B(x, \lambda(x)r))} \geq \frac{\mu(Y_x^r \cap D_x)}{\mu(D_x)} \frac{\mu(D_x)}{\mu(B(x, \lambda(x)r))}$$
$$\geq q\frac{\mu(B(0,r))}{\mu(B(x, \lambda(x)r))}.$$

The proof is then completed by the following remark.

LEMMA 3.2: *Let M be a compact Riemannian n-manifold with volume μ. Then there is some constant c_0 such that for all r sufficiently small, we have for all $m_1, m_2 \in M$ and all $\lambda \geq 1$ that*

$$\frac{\mu(B(m_1,r))}{\mu(B(m_2, \lambda r))} \geq c_0\lambda^{-n}.$$

Proof: This is clear in Euclidean space. By a local uniform comparison with a flat metric, it will be true locally. Compactness then ensures it is true on all M.

$\square$

4. Proof of Theorem 1.7

The simple existence of a non-trivial Γ-invariant set Y in $X \times M$ (which is guaranteed by Theorem 3.1) is enough to show that there is some discrete spectrum in $L^2(M)$. We shall in fact prove the entire spectrum is discrete by making use of the additional information in Theorem 3.1.

We let (Z, ρ) be the maximal measurable factor of (M, μ) with discrete spectrum. That is, we have a Γ-space Z with invariant probability measure ρ, and a measure preserving Γ-map $M \to Z$ such that under the corresponding inclusion $L^2(Z) \hookrightarrow L^2(M)$, $L^2(Z)$ is the closure of the sum of all finite-dimensional Γ-invariant subspaces of $L^2(M)$. Such an (essentially unique) Z exists by [4, Thm. 7.1]. Our aim is to show $M = Z$.

For $z \in Z$, let M_z be the fiber over z and if $A \subset M$, let $A_z = A \cap M_z$. Let $\mu = \int^{\oplus} \mu_z d\rho(z)$ be the decomposition of μ over the fibers of $M \to Z$. We can write $Z = \bigcup Z_i, i \in \{\infty\} \cup \{n \in \mathbb{Z} | n \geq 0\}$ such that $\text{card}(M_z) = i$ if $z \in Z_i$. From this it is easy to see that if $M \neq Z$, we can fix a set $A \subset M$ with $0 < \mu(A) < 1$ such that if $Z_0 = \{z \in Z | A_z \neq \emptyset\}$, then $\mu_z(A_z)$ is constant over $z \in Z_0$, and this constant is not 0 or 1. Define $\theta : M \to \mathbb{R}$ by defining $\theta_z = \theta | M_z$ to be $\theta_z = \chi_{A_z} - \mu_z(A_z)$. Therefore, there are constants $c, c' > 0$ and a decomposition $M = A \cup A' \cup A''$ such that $\theta | A = c$, $\theta | A' = -c'$, $\theta | A'' = 0$. Furthermore, $\theta \perp L^2(Z) \subset L^2(M)$. We fix such A, θ, c, c' throughout the remainder of the proof.

If $f \in L^2(X \times M)$ is Γ-invariant, let $T_f : L^2(M) \to L^2(X)$ be the corresponding integral operator. Then $T_f^* T_f$ is a compact self-adjoint operator in $L^2(M)$ whose eigenspaces for non-0 eigenvalues will be finite dimensional Γ-invariant subspaces. Since these are all contained in $L^2(Z)$ and $\theta \perp L^2(Z)$, it follows that we must have $\theta \in \ker(T_f^* T_f) = \ker(T_f)$ for any such f. To prove the theorem, it therefore suffices to construct such an f for which $\theta \notin \ker(T_f)$, thereby contradicting $M \neq Z$. We shall do this by taking f to be the characteristic function of a suitable set Y^r constructed in Theorem 3.1.

We first choose q and η in Theorem 3.1. First fix $1 > \eta > 1 - \mu_0(a)/8$. (By a suitable choice of ergodic component μ_0, we may assume $\mu_0(A) \neq 0$.) Now choose $X_1 \subset X$ such that $\mu_0(X_1) > \eta$ and there is $\Lambda \in \mathbb{R}$ with $\lambda(x) \leq \Lambda$ for all

$x \in X_1$. Then choose q so that

$$\max(\{c, c'\}) \frac{(1-q)}{q} \le c c_0 \Lambda^{-n}/4$$

(where c, c' are as above and c_0 is as in Theorem 3.1). We claim that for this choice of η and q, by taking r sufficiently small $\theta \notin \ker T_f$, for $f = \chi_{Y^r}$.

By the Lebesgue density theorem, for μ-almost all $x \in A$ we have

$$\lim_{r \to 0} \frac{\mu(A \cap B(x, r))}{\mu(B(x, r))} = 1.$$

It follows that for almost all Γ-ergodic components of (M, μ) this same limit exists almost everywhere with respect to the ergodic component. Therefore, we may suppose that the ergodic component (X, μ_0) of (M, μ) was chosen originally so that for μ_0-almost all $x \in A$,

$$\lim_{r \to 0} \frac{\mu(A \cap B(x, r))}{\mu(B(x, r))} = 1.$$

Define, for $\omega < 1$,

$$A_{\omega, r} = \left\{ x \in A \,\middle|\, \frac{\mu(A \cap B(x, s))}{\mu(B(x, s))} > \omega \quad \text{for all } 0 < s < r \right\}.$$

Choose $\omega < 1$ such that $(1 - \omega)(c + c') \le c c_0 \Lambda^{-n}/2$. Then there is some R_ω such that if $r \le R_\omega$, we have $\mu_0(A_{\omega, r}) \ge \mu_0(A)/2$.

Finally, choose $r > 0$ such that

(a) $\Lambda r < R_\omega$; and

(b) the conclusions of Theorem 3.1 hold with the above choices of η and q.

We then have $\mu_0(X^r) > 1 - \mu_0(A)/8$ (by choice of η), $\mu_0(X_1) > 1 - \mu_0(A)/8$ (by choice of X_1), and $\mu_0(A_{\omega, \Lambda r}) \ge \mu_0(A)/2$. It follows that $\mu_0(A_{\omega, \Lambda r} \cap X^r \cap X_1) > 0$. For ease of notation, with r thus chosen, we set $Y^r = Y$. We show $\theta \notin \ker T_f$ for $f = \chi_{Y^r}$. To do this it suffices to show that if $x \in A_{\omega, \Lambda r} \cap X^r \cap X_1$, then $(T_f \theta)(x) \neq 0$.

We have

$$(T_f \theta)(x) = \int f(x, m) \theta(m) \, d\mu(m)$$

$$= \int_{Y_x} \theta \, d\mu$$

$$= I_1 + I_2$$

where

$$I_1 = \int_{Y_x \cap B(x,\lambda(x)r)} \theta \quad \text{and} \quad I_2 = \int_{Y_x - Y_x \cap B(x,\lambda(x)r)} \theta.$$

Recalling that $\theta|A = c$, $\theta|A' = -c'$ and $\theta|A'' = 0$, we have

$$I_1 = c\mu(Y_x \cap B(x,\lambda(x)r) \cap A) - c'\mu(Y_x \cap B(x,\lambda(x)r) \cap A').$$

Since $x \in A_{\omega,\Lambda r}$ and $\lambda(x) < \Lambda$ (since $x \in X_1$), we have

$$\frac{\mu(B(x,\lambda(x)r) \cap A)}{\mu(B(x,\lambda(x)r))} > \omega,$$

and since $x \in X^r$, we have

$$\frac{\mu(Y_x \cap B(x,\lambda(x)r))}{\mu(B(x,\lambda(x)r))} \geq c_0 \lambda(x)^{-n} \geq c_0 \Lambda^{-n}.$$

Therefore

$$c\mu(Y_x \cap B(x,\lambda(x)r) \cap A) \geq c(c_0\Lambda^{-n} - [1-\omega])\mu(B(x,\lambda(x)r)).$$

For the second term in I_1 we observe

$$\frac{\mu(B(x,\lambda(x)r) \cap A')}{\mu(B(x,\lambda(x)r))} \leq 1-\omega.$$

Therefore

$$-c'\mu(Y_x \cap B(x,\lambda(x)r) \cap A') \geq -c'(1-\omega)\mu(B(x,\lambda(x)r)).$$

Hence

$$I_1 \geq [cc_0\Lambda^{-n} - (1-\omega)(c+c')]\mu(B(x,\lambda(x)r)).$$

By the choice of ω, we deduce

$$I_1 \geq \frac{cc_0\Lambda^{-n}}{2}\mu(B(x,\lambda(x)r)).$$

To estimate I_2, observe that from Theorem 3.1 we have

$$\frac{\mu(Y_x \cap B(x,\lambda(x)r))}{\mu(Y_x)} \geq q,$$

so

$$\mu(Y_x - Y_x \cap B(x,\lambda(x)r)) \leq (1-q)\mu(Y_x).$$

Therefore,

$$|I_2| \leq \max(\{c, c'\})(1 - q)\mu(Y_x).$$

We also have

$$\mu(Y_x) \leq \frac{\mu(Y_x \cap B(x, \lambda(x)r))}{q} \leq \frac{\mu(B(x, \lambda(x)r))}{q}.$$

Hence

$$|I_2| \leq \max(\{c, c'\})\frac{1 - q}{q}\mu(B(x, \lambda(x)r)).$$

By the choice of q we then have

$$|I_2| \leq \frac{cc_0\Lambda^{-n}}{4}\mu(B(x, \lambda(x)r)).$$

This shows

$$I_1 + I_2 \geq \frac{cc_0\Lambda^{-n}}{4}\mu(B(x, \lambda(x)r)),$$

and this completes the proof of the theorem. □

5. Further Questions

Are there smooth volume preserving actions of Kazhdan groups on compact manifolds which have zero entropy but do not have discrete spectrum? For lattices in a higher rank of groups, Theorem 1.1 asserts that this is not the case. The additional information one has available in the higher rank case is of course deduced from superrigidity for cocycles.

Does every measure preserving zero entropy action on a compact metric space of a lattice in a higher rank group have discrete spectrum? More generally, is this true for Kazhdan groups?

References

1. G. W. Mackey, *Ergodic transformation groups with a pure point spectrum*, Ill. J. Math. **6** (1962), 327–335.

2. G. A. Margulis, *Discrete Subgroups of Semisimple Lie Groups*, Springer, Berlin (to appear).

3. J. Moser, *On the volume elements on a manifold*, Trans. Amer. Math. Soc. **120** (1965), 286–294.

4. R. J. Zimmer, *Ergodic actions with generalized discrete spectrum*, Ill. J. Math. **20** (1976), 555–588.

5. R. J. Zimmer, *Volume preserving actions of lattices in semisimple groups on compact manifolds*, Publ. Math. I.H.E.S. **59** (1984), 5–33.

6. R. J. Zimmer, *Kazhdan groups acting on compact manifolds*, Invent. Math. **75** (1984), 425–436.

7. R. J. Zimmer, *Lattices in semismple groups and invariant geometric structures on compact manifolds*, in *Discrete Groups in Geometry and Analysis* (R. Howe, ed.), Birkhauser, Boston, 1987, pp.152–210.

8. R. J. Zimmer, *Ergodic Theory and Semisimple Groups*, Birkhauser, Boston, 1984.

9. R. J. Zimmer, *Actions of semisimple groups and discrete groups*, Proc. ICM, Berkeley, 1986, pp.1247–1258.

COCYCLE SUPERRIGIDITY AND RIGIDITY
FOR LATTICE ACTIONS ON TORI

A. KATOK,[1] J. LEWIS,[2] AND R. ZIMMER[3]

§1. Introduction and statement of results.

This note is part of an ongoing program directed at understanding the actions of lattices in semisimple Lie groups on compact manifolds by diffeomorphisms. A brief account of the history and current state of this program will be given in Section 3. Our main result is the following

Theorem 1.1. *Suppose* Γ *is a subgroup of finite index in* $\mathbf{SL}(n, \mathbb{Z})$, $n \geq 3$, $M = \mathbb{T}^n$, *and* $\rho \colon \Gamma \to \mathrm{Diff}(M)$ *is a smooth action such that*

  i) ρ *preserves a non-atomic probability measure* μ
  ii) *there exists an element* $\gamma_0 \in \Gamma$ *such that the diffeomorphism* $\rho(\gamma_0)$ *is Anosov.*

and in addition one of the following three conditions hold:

  a) *the measure* μ *is positive on open sets, i.e.* $\mathrm{supp}\,\mu = \mathbb{T}^n$ *and* ρ *is ergodic with respect to* μ
  b) μ *is absolutely continuous*
  c) *Let* $\rho_* \colon \Gamma \to \mathbf{GL}(n, \mathbb{Z})$ *denote the homomorphism corresponding to the action on* $H_1(M) \simeq \mathbb{Z}^n$. *Then* $\rho_*(\gamma_0)$ *is an irreducible matrix over* $\mathbb{Q}$ *and either* $n \geq 4$ *or if* $n = 3$ *the eigenvalues of* $\rho_*(\gamma_0)$ *are real.*

Then there exists a 1-cocycle $\alpha \colon \Gamma \to \mathbb{Q}^n/\mathbb{Z}^n$ *(where* Γ *acts on* $\mathbb{Q}^n/\mathbb{Z}^n$ *via* ρ_**) and a diffeomorphism* h *of* M *conjugating* ρ *to the affine action given by* ρ_* *and* α, *i.e.,* $\rho(\gamma) = h(\rho_*(\gamma) + \alpha(\gamma))h^{-1}$ *for every* $\gamma \in \Gamma$. *In particular,* ρ *is smoothly conjugate to* ρ_* *on a subgroup of finite index.*

Here and below we slightly abuse notations by using the same symbol for an integer $n \times n$ matrix and the endomorphism of the n-dimensional torus $\mathbb{T}^n$ induced by that matrix.

At the beginning of the next section we will show that cases b) and c) are reduced to a).

---

[1]Department of Mathematics, Pennsylvania State University, University Park, PA 16802. This work was partially supported by NSF grant DMS9017995.
[2]Renaissance Software, Inc., 175 S. San Antonio Road, Los Altos, CA 94022.
[3]Department of Mathematics, University of Chicago, Chicago, IL 60637. Partially supported by an NSF grant.

Typeset by $\mathcal{A}\mathcal{M}\mathcal{S}$-TEX

Observe also that the homomorphisms $\Gamma \to \mathbf{GL}(n, \mathbb{Z})$ can be completely classified using the (finite-dimensional) superrigidity theorem of Margulis (Theorem 5.1.2 in [Z3]). The consequences for the homomorphism ρ_* corresponding to the action on H_1, given that some element acts by an Anosov diffeomorphism, are worked out in §2 of [K-L2]. The precise conclusion is that there exists a matrix $A \in \mathbf{GL}(n, \mathbb{Q})$ and a homomorphism $\iota \colon \Gamma \to \{\pm I\}$ such that either $\rho_*(\gamma) = \iota(\gamma) A \gamma A^{-1}$ for every $\gamma \in \Gamma$ or $\rho_*(\gamma) = \iota(\gamma) A (\gamma^{-1})^t A^{-1}$ for every $\gamma \in \Gamma$.

Recall that if Γ is any finitely-generated discrete group and G is any topological group whatsoever, we denote by $R(\Gamma, G)$ the space of homomorphisms of Γ into G with the compact/open topology. A homomorphism $\rho_0 \in R(\Gamma, G)$ is said to be *locally rigid* if there exists a neighborhood U of ρ_0 in $R(\Gamma, G)$ such that for every $\rho \in U$ there exists $g \in G$ such that $\rho(\gamma) = g \rho_0(\gamma) g^{-1}$ for every $\gamma \in \Gamma$

An easy argument (2.6 below, which first appeared in [Se]) shows that any C^1 perturbation of a volume-preserving action of a Kazhdan group on a compact manifold must preserve an absolutely continuous probability measure. Furthermore, the standard action of any finite-index subgroup Γ of $\mathbf{SL}(n, \mathbb{Z})$ contains Anosov elements and the property of being an Anosov is C^1 open. These remarks allow to apply the case b) of (1.1) to obtain the local C^1 rigidity of standard actions. For $n \geq 4$ this was proven in [K-L1] but the case $n = 3$ is new. The notion of local rigidity in this statement corresponds to the representations of Γ into the group $\mathrm{Diff}\,\mathbb{T}^n$ of C^∞ diffeomophisms of the n-torus provided with the C^1 topology.

Corollary 1.2. *Let $\Gamma = \mathbf{SL}(n, \mathbb{Z})$ or any subgroup of finite index, $n \geq 3$. Then the standard action of Γ on $\mathbb{T}^n$ is locally C^1 rigid.*

Our proof of (1.1) is based on a result due to Zimmer (extending ideas of Margulis), the "superrigidity theorem for cocycles" (Theorems 5.2.5 and 9.4.14 in [Z3]). The cocycle superrigidity theorem yields a homomorphism $\pi \colon \Gamma \to \mathbf{SL}(n, \mathbb{R})$ and a measurable framing, σ, of the tangent bundle TM with respect to which the derivative $D_x\rho(\gamma)$ is given by $\pi(\gamma)$ for every $\gamma \in \Gamma$ and $x \in M$. The dynamical hypothesis (ii) in (1.1) makes it possible to conclude that the framing σ, which is *a priori* only measurable, is in fact continuous on $\mathrm{supp}\,\mu$. which by assumption a) means everywhere. This implies that elements $\gamma \in \Gamma$ with $\pi(\gamma)$ hyperbolic act by Anosov diffeomorphisms, with Lyapunov exponents determined by the eigenvalues of $\pi(\gamma)$. In particular, we are in a position to apply either Theorem 4.12 in [K-L1] or the argument in [H] to conclude that there exists a free abelian subgroup $\mathcal{A}$ of rank $n - 1$ in Γ whose action is smoothly conjugate to the action on homology. Then it is easy to see that ρ_* and π must coïncide, and that the continuous linearizing frame σ is the image of a constant frame (i.e., constant with respect to the standard trivialization $TM \simeq \mathbb{T}^n \times \mathbb{R}^n$) under the conjugating diffeomorphism for the action of the abelian subgroup $\mathcal{A}$. (1.1) follows.

We should probably remark at the outset that the cocycle superrigidity theorem is applicable in much greater generality, in particular, to actions of more general groups and on other compact manifolds. Indeed, the argument we shall present below applies, with very minor modifications, to some additional cases, such as $\Gamma \subset \mathbf{Sp}(n, \mathbb{Z})$ of finite index acting on $M = \mathbb{T}^{2n}$, $n \geq 2$. However, we have deliberately restricted the scope

2

of this note in order to present the essential new ideas in the most straightforward possible setting. The issues which arise in extending (1.1) to more general actions will be addressed elsewhere; here we shall content ourselves with a brief discussion in the final section.

Finally, we would like to acknowledge a number of helpful conversations with Steve Hurder.

§2. Proofs.

In this section we provide the proofs of theorem (1.1) and its corollary (1.2). By an elementary argument (cf. Lemmas 2.6 and 2.14 in [K-L2]), it will suffice to establish (1.1) on any subgroup $\Gamma' \subset \Gamma$ of finite index. In particular, we may assume without loss of generality that the action ρ is orientation-preserving.

First let us show that the cases b) and c) of the theorem follow from the case a).

Any absolutely continuous invariant measure of an Anosov diffeomorphism is given by a smooth positive density and hence is positive on open sets ([L-S], [L-M-M].) Furthermore, any Anosov diffeomorphism is ergodic with respect to such a measure. ([A]. A more recent exposition of the relevant results appears in §III.2 of [Mañ].) Thus the diffeomorphism $\rho(\gamma_0)$ and hence the whole action ρ is ergodic with respect to μ.

For the case c) we will assume that μ is ergodic (by taking ergodic components if necessary and noticing that the measures will still be non-atomic) and will show that $\operatorname{supp} \mu = \mathbb{T}^n$. First let us recall that any Anosov diffeomorphism of $\mathbb{T}^n$ is topologically conjugate to the linear map given by the action on the first homology group [M]. The conjugacy is a homeomorphism homotopic to identity and is uniquely determined by the image of the origin which, of course must be fixed by our diffeomorphism. Thus $\rho(\gamma_0) = h(\rho_*(\gamma_0))h^{-1}$ for some homeomorphism h homotopic to identity. For any hyperbolic linear automorphism of the torus its centralizer in $\operatorname{Homeo}(\mathbb{T}^n)$ coincides with centralizer in the space of affine maps of the torus (cf e.g. [P-Y]). Furthermore, this centralizer contains a finile index subgroup which belongs to $\mathbf{SL}(n, \mathbb{Z})$, and hence a finite-index subgroup C which lies in $\rho_*(\Gamma)$. Since $\rho(\Gamma)$ contains at most one diffeomorphism in each homotopy class we conclude that $\rho(C) = h(\rho_*(C))h^{-1}$. Since any ρ-invariant measure is in particular $\rho(C)$-invariant, we conclude that $\mu = h_*\nu$ for a $\rho_*(C)$-invariant non-atomic measure ν whose support is obviousy an infinite $\rho_*(C)$-invariant closed set. From irreducibility of the matrix $\rho_*(\gamma_0)$ it follows immediately that its eigenvalues are all different (otherwise the characteristic polynomial of the matrix has non-trivial greatest common divisor with its derivative and is hence reducible over the rationals). The rank of the free part of its centralizer is equal to the rank of the group of units in the ring of integers in the number field $\mathbb{Q}(\gamma_0)$. By the Dirichlet Unit Theorem the latter rank is equal to the number of real eigenvalues of the matrix plus the number of pairs of complex-conjugate eigenvalues minus one. Thus condition c) implies that the centralizer of γ_0 contains $\mathbb{Z}^2$ and hence the rank of C is ≥ 2. Now we invoke the result by D.Berend ([B], Theorem 2.1) which implies that any infinite $\rho_*(C)$-invariant closed set is the whole torus, hence $\operatorname{supp} \mu = h_* \operatorname{supp} \nu = \mathbb{T}^n$ and the case a) applies.

Now we proceed to the proof of the case a). Let PM denote the principal bundle

3

of n-frames in the tangent bundle TM to $M = \mathbb{T}^n$. As usual, we identify each $\varphi \in P_x M$ with an isomorphism $\varphi \colon \mathbb{R}^n \to T_x M$. The following proposition summarizes the consequences of the cocycle superrigidity theorem in the context of (1.1). Although the argument is by now standard, we provide a complete proof for the reader's convenience.

Proposition 2.1. *Suppose that $\Gamma \subset \mathbf{SL}(n, \mathbb{Z})$, $n \geq 3$, $M = \mathbb{T}^n$, and $\rho \colon \Gamma \to \mathrm{Diff}(M)$ satisfy the hypotheses of (1.1), and in addition, that ρ is orientation-preserving. Then there exists a measurable section $\sigma \colon M \to PM$ and a homomorphism $\pi \colon \Gamma \to \mathbf{SL}(n, \mathbb{R})$ such that with respect to the framing σ, the derivative $D_x \rho(\gamma)$ is given by $\pm \pi(\gamma)$, i.e., for every $\gamma \in \Gamma$,*

$$D_x \rho(\gamma)\sigma(x) = \pm \sigma(\rho(\gamma)x)\pi(\gamma)$$

for almost every $x \in M$. Moreover, there is a matrix $A \in \mathbf{GL}(n, \mathbb{R})$ of determinant ± 1 which conjugates the representation π to either the identity representation or the involution $\gamma \mapsto (\gamma^{-1})^t$ (inverse transpose).

Proof. Let $\tau \colon M \to PM$ denote the standard framing on M, i.e., τ corresponds to the constant section $\tau(x) = I$ under the natural identifications $TM \simeq M \times \mathbb{R}^n$, $PM \simeq M \times \mathbf{GL}(n, \mathbb{R})$. Then let $\xi \colon \Gamma \times M \to \mathbf{GL}(n, \mathbb{R})$ denote the derivative cocycle for the action ρ with respect to the section τ, so that

$$D_x \rho(\gamma)\tau(x) = \tau(\rho(\gamma)x)\xi(\gamma, x)$$

for every $\gamma \in \Gamma$ and $x \in M$ Since Γ acts ergodically with respect to μ the algebraic hull $\mathbf{H}_\mathbb{R} \subset \mathbf{GL}(n, \mathbb{R})$ of ξ (or more precisely, its conjugacy class) is well-defined (cf. §9.2 in [Z3]). Let $\tilde{\xi} \colon \Gamma \times M \to \mathbf{H}_\mathbb{R}$ denote a (measurable) cocycle equivalent to ξ, and let $\tilde{\tau} \colon M \to \mathbf{GL}(n, \mathbb{R})$ denote the Borel map which gives the equivalence between ξ and $\tilde{\xi}$, so that for every $\gamma \in \Gamma$,

$$\tilde{\xi}(\gamma, x)\tilde{\tau}(x) = \tilde{\tau}(\rho(\gamma)x)\xi(\gamma, x)$$

for almost every $x \in M$.

The existence of an ergodic invariant probability measure, together with our assumption that ρ is orientation-preserving, imply that we can take $\mathbf{H} \subset \mathbf{SL}(n, \mathbb{C})$. This follows from splitting the cocycle into the $\mathbf{SL}(n, \mathbb{Z})$ part and $\mathbb{R}_*$ part (the latter given by the determinant) and noticing that the second component is cohomologous to identity since Γ is a Kazhdan group.

Let $\mathbf{H}^\circ$ denote the (Zariski) connected component of $\mathbf{H}$, and set $S = \mathbf{H}_\mathbb{R}/(\mathbf{H}^\circ)_\mathbb{R} \times M$ with the $\tilde{\xi}$-twisted action $\gamma(\bar{h}, x) = (\tilde{\xi}(\gamma, x)\bar{h}, \rho(\gamma)x)$. Define $\beta \colon \Gamma \times S \to \mathbf{H}_\mathbb{R}$, $\beta(\gamma, \bar{h}, x) = \tilde{\xi}(\gamma, x)$. Then by (9.2.6) in [Z3], the action of Γ on S is ergodic, and the algebraic hull of the cocycle β is $(\mathbf{H}^\circ)_\mathbb{R}$.

Write $\mathbf{H}^\circ = \mathbf{L} \ltimes \mathbf{U}$ where L is reductive, $\mathbf{U}$ is unipotent, and $\mathbf{L}$ and $\mathbf{U}$ are $\mathbb{R}$-groups. Let $\delta \colon \Gamma \times S \to \mathbf{L}_\mathbb{R}$ denote the cocycle obtained by composing β with the projection onto $\mathbf{L}_\mathbb{R}$. Let $p \colon \mathbf{L}_\mathbb{R} \to \mathbf{L}_\mathbb{R}/[\mathbf{L}_\mathbb{R}, \mathbf{L}_\mathbb{R}]$ denote the projection. Then the algebraic hull of $p \circ \delta$ is $\mathbf{L}_\mathbb{R}/[\mathbf{L}_\mathbb{R}, \mathbf{L}_\mathbb{R}]$, which is amenable, so since Γ is a Kazhdan group, $\mathbf{L}_\mathbb{R}/[\mathbf{L}_\mathbb{R}, \mathbf{L}_\mathbb{R}]$ is compact by (9.1.3) in [Z3]. Since $\mathbf{L} = [\mathbf{L}, \mathbf{L}] \cdot Z(\mathbf{L})$, where $Z(\mathbf{L})$ denotes the center, and $[\mathbf{L}, \mathbf{L}] \cap Z(\mathbf{L})$ is finite, it follows that $Z(\mathbf{L})_\mathbb{R}$ is also compact.

4

Write $\mathbf{L}/Z(\mathbf{L})$ as a product of semisimple $\mathbb{R}$-groups, $\mathbf{L}/Z(\mathbf{L}) = \mathbf{L}_1 \times \mathbf{L}_2$, where $(\mathbf{L}_2)_{\mathbb{R}}$ is compact and $(\mathbf{L}_1)_{\mathbb{R}}$ is center-free with no compact factors. Let $q\colon \mathbf{L}_{\mathbb{R}} \to (\mathbf{L}_1)_{\mathbb{R}}$ denote the projection. Then $q \circ \delta \colon \Gamma \times S \to (\mathbf{L}_1)_{\mathbb{R}}$ is a cocycle with algebraic hull $(\mathbf{L}_1)_{\mathbb{R}}$, so by the superrigidity theorem for cocycles (9.4.14 in [Z3]), there is an $\mathbb{R}$-rational homomorphism $\bar{\pi}\colon \mathbf{SL}(n, \mathbb{C}) \to \mathbf{L}_1$ such that $q \circ \delta$ is equivalent to the cocycle $\xi_{\bar{\pi}|\Gamma}\colon (\gamma, x) \to \bar{\pi}(\gamma)$.

Examining the list of representations of $\mathbf{SL}(n, \mathbb{C})$, we see that there are only two possibilities: either $\mathbf{L}_1$ is trivial, or $\mathbf{L}_1 = \mathbf{SL}(n, \mathbb{C})/Z(\mathbf{SL}(n, \mathbb{C}))$. In the first instance, we could conclude that the algebraic hull of the derivative cocycle ξ is compact. But it's easy to see that this contradicts hypothesis (ii), the existence of $\gamma_0 \in \Gamma$ with $\rho(\gamma_0)$ Anosov. Thus $\mathbf{L}_1 = \mathbf{L}/Z(\mathbf{L}) = \mathbf{H}^{\circ}/Z(\mathbf{H}^{\circ}) = \mathbf{H}/Z(\mathbf{H}) = \mathbf{SL}(n, \mathbb{C})/Z(\mathbf{SL}(n, \mathbb{C}))$. Moreover, $\mathbf{H} = \mathbf{SL}(n, \mathbb{C})$ is connected, $S = M$, and there are essentially only two possibilities for the representation $\bar{\pi}\colon \mathbf{SL}(n, \mathbb{C}) \to \mathbf{L}_1$. Namely, $\bar{\pi} = r \circ \pi$, where $\pi\colon \mathbf{SL}(n, \mathbb{C}) \to \mathbf{SL}(n, \mathbb{C})$ is either the identity map or the involution $\gamma \mapsto (\gamma^{-1})^t$ followed by conjugation by some matrix $A \in r^{-1}((\mathbf{L}_1)_{\mathbb{R}})$, and r denotes the projection $\mathbf{SL}(n, \mathbb{C}) \to \mathbf{L}_1 = \mathbf{SL}(n, \mathbb{C})/Z(\mathbf{SL}(n, \mathbb{C}))$.

Note that $r^{-1}((\mathbf{L}_1)_{\mathbb{R}})$ can be described more concretely:

$$r^{-1}((\mathbf{L}_1)_{\mathbb{R}}) = \{g \in \mathbf{SL}(n, \mathbb{C}) \mid Ad(g) \text{ is an } \mathbb{R}\text{-rational automorphism of } \mathfrak{sl}(n, \mathbb{C})\}.$$

In general, the elements of $r^{-1}((\mathbf{L}_1)_{\mathbb{R}})$ have complex entries. However, it's easy to see that every $A \in r^{-1}((\mathbf{L}_1)_{\mathbb{R}})$ is of the form $A = \lambda A'$, where $\lambda \in \mathbb{C}$ is a scalar (in fact, a root of unity) and $A' \in \mathbf{GL}(n, \mathbb{R})$ with determinant ± 1. In particular, we can replace the conjugating matrix A in the preceding paragraph with a matrix in $\mathbf{GL}(n, \mathbb{R})$, and if $\bar{\eta}\colon M \to (\mathbf{L}_1)_{\mathbb{R}}$ denotes the Borel map which gives the equivalence between $q \circ \delta$ and $\xi_{\bar{\pi}|\Gamma}$, $\bar{\eta}$ lifts to a map $\tilde{\eta}\colon M \to \mathbf{GL}(n, \mathbb{R})$. Then $\tilde{\eta}$ reduces $\tilde{\xi}$ to $\pm \pi$, i.e., for every $\gamma \in \Gamma$,

$$\tilde{\xi}(\gamma, x)\tilde{\eta}(x) = \pm \tilde{\eta}(\rho(\gamma)x)\pi(\gamma)$$

for almost every $x \in M$. Then $\sigma\colon M \to \mathbf{GL}(n, \mathbb{R})$, $\sigma(x) = \tilde{\tau}(x)^{-1}\tilde{\eta}(x)$, or more precisely, the corresponding section $\tau\sigma\colon M \to PM$, $x \mapsto \tau(x)\sigma(x)$, is the required framing. $\qquad\qquad\square$

The statement of (2.1) is perhaps a bit awkward; what the argument actually provides is a Borel map $\bar{\sigma}\colon M \to \mathbf{PGL}(n, \mathbb{R})$ such that for every $\gamma \in \Gamma$,

$$\bar{\xi}(\gamma, x)\bar{\sigma}(x) = \bar{\sigma}(\rho(\gamma)x)\bar{\pi}(\gamma)$$

for almost every $x \in M$. (Here $\bar{\pi}\colon \mathbf{SL}(n, \mathbb{C}) \to \mathbf{L}_1$ as above, and $\bar{\xi}\colon M \to \mathbf{PGL}(n, \mathbb{R})$ is the map corresponding to the derivative cocycle $\xi\colon M \to \mathbf{GL}(n, \mathbb{R})$. Note that

$$\mathbf{PGL}(n, \mathbb{R}) = \mathbf{GL}(n, \mathbb{R})/Z(\mathbf{GL}(n, \mathbb{R}))$$

is naturally identified with

$$(\mathbf{L}_1)_{\mathbb{R}} = (\mathbf{SL}(n, \mathbb{C})/Z(\mathbf{SL}(n, \mathbb{C})))_{\mathbb{R}} :$$

5

every matrix in $\mathbf{GL}(n, \mathbb{R})$ is a scalar multiple of a matrix in $\mathbf{SL}(n, \mathbb{C})$, so there is a natural inclusion $\mathbf{PGL}(n, \mathbb{R}) \hookrightarrow (\mathbf{L}_1)_{\mathbb{R}}$, and our observation in the preceding paragraph shows that this map is surjective.)

Observe that $\bar{\sigma}$ is unique. For if $\bar{\sigma}': M \to \mathbf{PGL}(n, \mathbb{R})$ is another map with the same property, then with $\varphi: M \to \mathbf{PGL}(n, \mathbb{R})$; $x \mapsto \bar{\sigma}'(x)^{-1}\bar{\sigma}(x)$, for every $\gamma \in \Gamma$, we have $\varphi(\rho(\gamma)x) = \bar{\pi}(\gamma)\varphi(x)\bar{\pi}(\gamma)^{-1}$ for almost every $x \in M$. Then the measure $\varphi_*\mu$ on $\mathbf{PGL}(n, \mathbb{R})$, is invariant under conjugation by $\bar{\pi}(\Gamma)$. But the only $\bar{\pi}(\Gamma)$ invariant probability measure on $\mathbf{PGL}(n, \mathbb{R})$ is the point mass at the identity. Although one can give an explicit, elementary proof in this special case, and the general result is probably well-known, we know of no convenient reference and therefore provide a proof which works in general.

Lemma 2.2. *Suppose* $\mathbf{G} \subset \mathbf{GL}(n, \mathbb{C})$ *is a semisimple* $\mathbb{R}$*-group such that* $G = \mathbf{G}_{\mathbb{R}}$ *has no compact factors, and* $\Gamma \subset G^\circ$ *(connected component in the Hausdorff topology) is a lattice. Then the only probability measures on* G *which are invariant under conjugation by the elements of* Γ *are those supported on the center of* G.

Proof. Let μ be a Γ-invariant probability measure on G. By standard arguments (cf., for example, §3 of [Z2]), we may assume without loss of generality that μ is ergodic. Since the action of G on itself by conjugation is algebraic, every conjugacy class is locally closed by [B-S] (cf. 3.1.3 in [Z3]). Then by 2.1.11 in [Z3], any Γ-ergodic measure must be supported on a single conjugacy class. Thus there exists $g_0 \in G$ such that μ is supported on the conjugacy class $\{gg_0g^{-1} \mid g \in G\}$, which is isomorphic to the quotient variety $G/Z_G(g_0)$ with G acting by left translation. By a theorem of S. G. Dani (Corollary 2.6 in [D], which generalizes a theorem of C. C. Moore [Mo]) the stabilizer

$$G_\mu = \{g \in G \mid \text{the } g\text{-action on } G/Z_G(g_0) \text{ preserves } \mu\}$$

is (equal to the $\mathbb{R}$-points of) an $\mathbb{R}$-algebraic group, and

$$J_\mu = \{g \in G \mid gx = x \quad \forall \, x \in \operatorname{supp}\mu\}$$

is a normal, co-compact subgroup in G_μ. Then by the Borel density theorem, $\Gamma \subset G_\mu$ implies $G_\mu = G$, and since G has no compact factors, $J_\mu \supset G^\circ$. In other words, each point in $\operatorname{supp}\mu$ is centralized by G°, hence $\operatorname{supp}\mu \subset Z(G)$. $\qquad\square$

The next step is to show that in the presence of an Anosov diffeomorphism, the section $\bar{\sigma}$, which is *a priori* only measurable, must in fact be continuous.

Lemma 2.3. *Under the hypotheses of (1.1), a) and the notation of the preceding paragraph, the section* $\bar{\sigma}: M \to \mathbf{PGL}(n, \mathbb{R})$ *is continuous.*

Proof. Fix $\gamma \in \Gamma$ and decompose $\mathbb{R}^n$ into characteristic subspaces for the action of $\pi(\gamma) \in \mathbf{SL}(n, \mathbb{R})$:

$$\mathbb{R}^n = \oplus W_i, \quad W_i = \{w \neq 0 \in \mathbb{R}^n \mid \lim_{m \to \pm\infty} (1/m)\ln\|\pi(\gamma)^m w\|/\|w\| \to \chi_i\} \cup \{0\};$$

$\chi_i \in \mathbb{R}$ are called the *characteristic exponents* for $\pi(\gamma)$. Define measurable distributions $\mathcal{W}_i$ on M, $\mathcal{W}_i(x) = \tau(x)\bar{\sigma}(x)W_i$. Let $\|\cdot\|$ denote the standard (Riemannian) fiber metrics on TM and PM.

6

Lemma 2.4. *For μ-almost every $x \in M$,*

$$\mathcal{W}_i(x) - \{0\} = \{w \neq 0 \in T_x M \mid \lim_{m \to \pm\infty} (1/m)\ln\|D_x\rho(\gamma)^m w\|/\|w\| \to \chi_i\} \cup \{0\}.$$

Proof. Given $0 < \varepsilon < 1$, there exists a subset $\mathcal{S} \subset M$ such that $\mu(\mathcal{S}) > 1 - \varepsilon$ and both $\|\sigma(x)\|$ and $\|\sigma(x)^{-1}\|$ are uniformly bounded for $x \in \mathcal{S}$. Then for almost every $x \in \mathcal{S}$, $\rho(\gamma)^{\pm m}x \in \mathcal{S}$ for infinitely many $m \in \mathbb{N}^+$. Thus, for $w \in W_i - \{0\}$ and almost every $x \in \mathcal{S}$, there is a subsequence of m's for which $\lim_{m \to \pm\infty} (1/m)\ln\|D_x\rho(\gamma)^m w(x)\|/\|w(x)\| \to \chi_i$, where $w(x) = \sigma(x)w$.

On the other hand, since the derivative cocycle for $\rho(\gamma)$,

$$\mathbb{Z} \times M \to \mathbf{GL}(n,\mathbb{R}); \ (m,x) \mapsto \xi(\gamma^m, x) = \tau(\rho(\gamma^m)x)^{-1}D_x\rho(\gamma^m)\tau(x),$$

is clearly integrable, Oseledec's multiplicative ergodic theorem [O] (cf. also §V.2 in [Mar]) implies that the limits exist for almost every $x \in M$. Thus the assertion holds for almost every $x \in \mathcal{S}$, and (2.4) follows by taking $\varepsilon \to 0$. $\qquad\square$

Recall that $\gamma_0 \in \Gamma$ such that $\rho(\gamma_0)$ is Anosov. (2.4) implies that $\pi(\gamma_0)$ is hyperbolic (i.e., has no eigenvalues on the unit circle) and that with $W^+ = \bigcup_{|\chi_i|>1} W_i$ and $W^- = \bigcup_{|\chi_i|<1} W_i$, where W_i, χ_i denote the characteristic subspaces and exponents for $\pi(\gamma_0)$, $\mathcal{W}^+(x) = \sigma(x)W^+$ ($\mathcal{W}^-(x) = \sigma(x)W^-$) is equal to the unstable (stable) subspace in T_x for $\rho(\gamma_0)$ for almost every $x \in M$. In particular, the *a priori* only measurable distributions $\mathcal{W}^\pm$ are in fact continuous.

We need to recall part of the discussion from §3 of [K-L2]. Let $\mathbf{G}_k(n,\mathbb{R})$ denote the Grassman variety of k-planes in $\mathbb{R}^n$, $1 \leq k \leq n-1$. For each point $p_0 \in \mathbf{G}_k(n,\mathbb{R})$, the map

$$\mathbf{PGL}(n,\mathbb{R}) \to \mathbf{G}_k(n,\mathbb{R}); \quad \bar{g} \mapsto \bar{g}p_0$$

is smooth. The map

$$\mathbf{PGL}(n,\mathbb{R}) \to \mathbf{G}_k(n,\mathbb{R})^\ell = \underbrace{\mathbf{G}_k(n,\mathbb{R}) \times \cdots \times \mathbf{G}_k(n,\mathbb{R})}_{\ell \text{ times}}; \quad \bar{g} \mapsto (\bar{g}p_1, \ldots, \bar{g}p_\ell)$$

is a local imbedding in an open neighborhood of $\bar{g}_0$ (via the inverse function theorem) if and only if the intersection of the infinitesimal stabilizers $\bigcap_{i=1,\ldots,\ell} \mathfrak{pgl}(n,\mathbb{R})_{\bar{g}_0^{-1}p_i}$ is zero. In particular, this condition is satisfied provided that the intersection of the stabilizers $\bigcap_{i=1,\ldots,\ell} \mathbf{PGL}(n,\mathbb{C})_{\bar{g}_0^{-1}p_i}$ is trivial.

With γ_0, $W^\pm$, and $\mathcal{W}^\pm$ as above, set $k = \dim W^-$ and $p_1 = W^- \in \mathbf{G}_k(n,\mathbb{R})$. By the Borel density theorem, the intersection

$$\bigcap_{\gamma \in \Gamma} \mathbf{PGL}(n,\mathbb{C})_{\pi(\gamma)p_1} = \bigcap_{\gamma \in \Gamma} \pi(\gamma)(\mathbf{PGL}(n,\mathbb{C})_{p_1})\pi(\gamma)^{-1}$$

7

is a (proper) normal subgroup of $\mathbf{PGL}(n, \mathbb{C})$, hence is trivial. Thus we can choose matrices $\gamma_1 = I, \gamma_2, \ldots, \gamma_\ell \in \Gamma$ such that with $p_i = \pi(\gamma_i)p_1$, which is the stable subspace for $\pi(\gamma_i\gamma_0\gamma_i^{-1})$, the stabilizer of $p = (p_1, \ldots, p_\ell) \in \mathbf{G}_k(n, \mathbb{R})^\ell$ in $\mathbf{PSL}(n, \mathbb{C})$ is trivial. Set U equal to the orbit of p under $\mathbf{PGL}(n, \mathbb{R})$, $U = \mathbf{PGL}(n, \mathbb{R})p \subset \mathbf{G}_k(n, \mathbb{R})^\ell$. Note that by the preceding paragraph, for every $\bar{g}_0 \in \mathbf{PGL}(n, \mathbb{R})$, the map $\bar{g} \mapsto \bar{g}q$ is a local imbedding on a neighborhood of $\bar{g}_0$ in $\mathbf{PGL}(n, \mathbb{R})$, where $q = (q_1, \ldots, q_\ell) = (\bar{g}_0 p_1, \ldots, \bar{g}_0 p_l) = \bar{g}_0 p \in U$.

For $1 \le i \le \ell$, let $q_i \colon M \to \mathbf{G}_k(n, \mathbb{R})$ denote the map corresponding to the stable distribution for the Anosov diffeomorphism $\rho(\gamma_i\gamma_0\gamma_i^{-1})$ under the identification of TM with $M \times \mathbb{R}^n$ via the standard trivialization. By (2.4), $q_i = \bar{\sigma} \cdot p_i$. (For example, with the above notation, $q_1(x) = \tau(x)^{-1}\mathcal{W}^-(x)$ for almost every $x \in M$.) Define $q \colon M \to \mathbf{G}_k(n, \mathbb{R})^\ell$, $q = (q_1, \ldots, q_\ell)$. Since each of the maps q_i is continuous (in the Hausdorff topology on $\mathbf{G}_k(n, \mathbb{R})$), it will follow that $\bar{\sigma} \colon M \to \mathbf{PGL}(n, \mathbb{R})$ coincides with the continuous map which is uniquely determined by $\bar{\sigma}(x)p = q(x)$, provided that this makes sense, i.e., provided that $q(x) \in U$ for every $x \in M$.

Set $S = q^{-1}(U) \subset M$. Since $q = \bar{\sigma}p$, $q(x) \in U$ for almost every $x \in M$. In particular, S is dense in M. By (2.4), we can fix a representative for the Borel map $\bar{\sigma}$ and delete a subset of measure zero from S to obtain a set $S' \subset S$ such that S' is dense in M and $q(x) = \bar{\sigma}(x)$ and $\bar{\sigma}(x)W^\pm = \tau(x)^{-1}\mathcal{W}^\pm(x)$ for every $x \in S'$.

Now we make use of an elegant trick due to Furstenburg [Fu]. Suppose that $x_0 \in M - S$ and fix a sequence $x_m \in S'$ with $x_m \to x_0$. As above, for $g \in \mathbf{GL}(n, \mathbb{R})$, we write $\bar{g}$ for its image in $\mathbf{PGL}(n, \mathbb{R})$. Since we are free to multiply by non-zero scalars, we can choose $g_m \in \mathbf{GL}(n, \mathbb{R})$ so that $\|g_m\|$ is bounded and $q(x_m) = \bar{g}_m p$. Then passing to a subsequence, we may assume that $g_m \to g_0 \in \mathbb{R}^{n \times n}$, an $n \times n$ matrix. If g_0 were in $\mathbf{GL}(n, \mathbb{R})$, we would have $q(x_0) = \bar{g}_0 p$ with $\bar{g}_0 \in \mathbf{PGL}(n, \mathbb{R})$, contradicting $x_0 \notin S$. Thus $g_0 \notin \mathbf{GL}(n, \mathbb{R})$, and in particular, $K = \ker(g_0)$ and $V = g_0(\mathbb{R}^n)$ are proper, non-zero subspaces in $\mathbb{R}^n$.

For any subspace $W \subset \mathbb{R}^n$, the set of $g \in \mathbf{SL}(n, \mathbb{C})$ for which $\pi(g)W$ and K are in general position is a non-empty Zariski open set, so by the Borel density theorem, there exists $\gamma \in \Gamma$ such that $\pi(\gamma)W^\pm$ and K are in general position. In other words, replacing γ_0 by $\gamma\gamma_0\gamma^{-1}$, we may assume that $W^\pm$ and K are in general position, i.e.,

$$\dim(W^\pm \cap K) = \sup\{\dim W^\pm + \dim K - n,\ 0\}.$$

Now suppose that W is any subspace of $\mathbb{R}^n$ which is in general position with respect to K and that the sequence $g_m W$ converges in the appropriate Grassman variety, say $g_m W \to W_0$. Then either $\dim W + \dim K \le n$, in which case $W \cap K = (0)$ and $W_0 = \lim g_m W = g_0 W \subset V$, or $\dim W + \dim K > n$, in which case $W + K = \mathbb{R}^n$ and $W_0 = \lim g_m W \supset g_0(\mathbb{R}^n) = V$.

In particular, by compactness of the Grassman varieties, we may again pass to a subsequence and so assume that the sequences $g_m W^\pm$ converge, say $g_m W^\pm \to W_0^\pm$, and there are three possibilities:

(1) $W_0^+, W_0^- \subset V$,
(2) $W_0^+, W_0^- \supset V$, or

(3) $W_0^{\pm} \subset V \subset W_0^{\mp}$.

In any case, $W_0^+ \cap W_0^- \neq (0)$. On the other hand, $\bar{g}_m W^{\pm} = \bar{\sigma}(x_m)W^{\pm} = \tau(x_m)^{-1}\mathcal{W}^{\pm}(x_m)$ by construction, hence $\tau(x_0)^{-1}\mathcal{W}^{\pm}(x_0) = W_0^{\pm}$ by continuity. In other words, the isomorphism $\tau(x_0) \colon \mathbb{R}^n \to T_{x_0}M$ identifies $W_0^{\pm}$ with the stable and unstable subspaces for the Anosov diffeomorphism $\rho(\gamma_0)$, which are transversal. This contradiction establishes $S = M$, and completes the proof of (2.3). $\qquad\square$

One immediate consequence of (2.3) is that the diffeomorphism $\rho(\gamma)$ is Anosov whenever the matrix $\pi(\gamma)$ (equivalently, γ itself) is hyperbolic. Thus we are in a position to apply Theorem (4.12) in [K-L1] to conclude that the action of a suitable abelian subgroup $\mathcal{A}$ in Γ is linear. (In fact, since $\bar{\sigma}$ also determines the periodic data for the action of $\mathcal{A}$, we might just as well apply the regularity theorem in [H].) We summarize this portion of the argument as follows:

Lemma 2.5. *There exists a subgroup $\mathcal{A} \subset \Gamma$, free-abelian of rank $n-1$ (more precisely, $\mathcal{A}$ is a co-compact subgroup in a maximal split Cartan subgroup A of $\mathbf{SL}(n,\mathbb{R})$) acting by Anosov diffeomorphisms and a C^{∞} diffeomorphism $h \colon M \to M$, homotopic to the identity, such that $\rho(\gamma) = h\rho_*(\gamma)h^{-1}$ for every $\gamma \in \mathcal{A}$, where $\rho_* \colon \mathcal{A} \to \mathbf{SL}(n,\mathbb{Z})$ denotes the map induced by the action on $H_1(M) \simeq \mathbb{Z}^n$.*

Define a smooth section $\tau' \colon M \to PM$, $\tau'(x) = (D_{h^{-1}x}h)\tau(h^{-1}x)$, and set σ' equal to the corresponding map $M \to \mathbf{GL}(n,\mathbb{R})$, $\sigma'(x) = \tau(x)^{-1}\tau'(x)$. Then

$$\xi(\gamma,x)\sigma'(x) = \sigma'(\rho(\gamma)x)\rho_*(\gamma) \quad \text{for every } \gamma \in \mathcal{A} \text{ and } x \in M,$$

where, as above, ξ denotes the derivative cocycle for the action ρ with respect to the standard framing τ. As usual, we write $\bar{\sigma}' \colon M \to \mathbf{PGL}(n,\mathbb{R})$ for the map induced by σ'.

Define a continuous map $\psi \colon M \to \mathbf{PGL}(n,\mathbb{R})$, $\psi(x) = \bar{\sigma}'(x)^{-1}\bar{\sigma}(x)$, and observe that

$$\bar{\rho}_*(\gamma)\psi(x) = \psi(\rho(\gamma)x)\bar{\pi}(\gamma) \quad \text{for every } \gamma \in \mathcal{A} \text{ and } x \in M.$$

Since ψ is bounded, this relation implies that ψ takes values in the set of intertwining automorphisms for the representations $\bar{\rho}_* \,|\, \mathcal{A}$ and $\bar{\pi} \,|\, \mathcal{A}$, i.e.,

$$\bar{\rho}_*(\gamma)\psi(x) = \psi(x)\bar{\pi}(\gamma) \quad \text{for every } \gamma \in \mathcal{A} \text{ and } x \in M.$$

Thus ψ is constant along $\mathcal{A}$ orbits. But since the restriction $\rho \,|\, \mathcal{A}$ is conjugate to the linear action $\rho_* \,|\, \mathcal{A}$, we know in particular that $\rho \,|\, \mathcal{A}$ is topologically transitive, i.e., some orbit is dense. Thus ψ is constant, and we conclude that $\bar{\sigma}' = \bar{\sigma} \cdot \bar{g}$, where $\bar{g} \in \mathbf{PGL}(n,\mathbb{R})$ intertwines $\bar{\rho}_* \,|\, \mathcal{A}$ and $\bar{\pi} \,|\, \mathcal{A}$.

Unraveling the notation, we see that

$$D_x(h^{-1}\rho(\gamma)h)\tau(x) = \pm\tau(h^{-1}\rho(\gamma)hx)\rho_*(\gamma) \quad \text{for every } \gamma \in \Gamma \text{ and } x \in M.$$

In other words, $h^{-1}\rho(\gamma)h$ is an affine transformation of the torus, with linear part $\pm\rho_*(\gamma)$. Since the actions on H_1 coïncide, the sign is always positive.

9

Now set $\alpha(\gamma) = h^{-1}\rho(\gamma)h(0)$, so that $\rho(\gamma) = h(\rho_*(\gamma) + \alpha(\gamma))h^{-1}$; $\alpha\colon \Gamma \to \mathbb{T}^n$ is a 1-cocycle, $\alpha \in Z^1(\Gamma, \mathbb{T}^n)$, where Γ acts on the abelian group $\mathbb{T}^n$ via ρ_*. All that remains to complete the proof of (1.1) is to show that the cocycle α takes values in $\mathbb{Q}^n/\mathbb{Z}^n$. To be more precise, we need to show that α is equivalent to a cocycle with values in $\mathbb{Q}^n/\mathbb{Z}^n$. (We can vary α within its cohomology class by adjusting our choice of origin, i.e., by pre-composing h with a translation.) This is equivalent to the purely geometric assertion that $\rho(\Gamma)$ has a finite orbit.

Recall that $H^1(\Gamma, \mathbb{R}^n) = 0$ (this established for a general linear representation of Γ in [Mar]). Thus the short exact sequence of coëfficient modules $\mathbb{Z}^n \to \mathbb{R}^n \to \mathbb{T}^n$ gives a long exact sequence

$$0 \to H^1(\Gamma, \mathbb{T}^n) \to H^2(\Gamma, \mathbb{Z}^n) \to H^2(\Gamma, \mathbb{R}^n).$$

Now by the Künneth formula,

$$H^2(\Gamma, \mathbb{R}^n) = H^2(\Gamma, \mathbb{Z}^n \otimes \mathbb{R}) \simeq H^2(\Gamma, \mathbb{Z}^n) \otimes \mathbb{R},$$

so that the kernel of the natural map $H^2(\Gamma, \mathbb{Z}^n) \to H^2(\Gamma, \mathbb{R}^n)$ is precisely the torsion subgroup of $H^2(\Gamma, \mathbb{Z}^n)$.[4] Combining the two observations, we see that every element of $H^1(\Gamma, \mathbb{T}^n)$ is torsion. In other words, there exist $m \in \mathbb{N}^+$ and $x_0 \in \mathbb{T}^n$ such that

$$m \cdot \alpha(\gamma) = \gamma x_0 - x_0 \quad \text{for every } \gamma \in \Gamma.$$

Fix $x_1 \in \mathbb{T}^n$ such that $m \cdot x_1 = x_0$, and set $h'(x) = h(x - x_1)$, $\alpha'(\gamma) = (h')^{-1}\rho(\gamma)h'(0)$. Then $m \cdot \alpha' \equiv 0$, i.e., α' takes values in the m-division points in $\mathbb{T}^n$. This completes the proof of (1.1).

In order obtain (1.2) from (1.1), we need the following lemma, which first appeared in [Se].

Lemma 2.6. *Suppose Γ is a discrete Kazhdan group (i.e., Γ satisfies Kazhdan's "Property T"), M is a compact manifold, and $\rho_0 \in R(\Gamma, \mathrm{Diff}^1(M))$ is an action of Γ on M by C^1 diffeomorphisms. Then if ρ_0 preserves an absolutely continuous probability measure μ on M, there is a neighborhood U of ρ_0 in $R(\Gamma, \mathrm{Diff}^1(M))$ such that each $\rho \in U$ preserves an absolutely continuous measure μ_ρ.*

Proof. Let $\delta \in L^1(M)$ denote the invariant density, so that for every $\gamma \in \Gamma$,

$$\delta(\rho(\gamma)x) = |D_x\rho(\gamma)|^{-1}\delta(x)$$

for almost every $x \in M$. In other words, $\delta^{1/2} \in L^2$ is a non-trivial invariant vector under the unitary representation

$$\pi_0\colon \Gamma \times L^2 \to L^2; \quad (\pi_0(\gamma)\varphi)(x) = |D_{\rho_0(\gamma)^{-1}x}\rho_0(\gamma)|^{-1/2}\varphi(\rho_0(\gamma)^{-1}x).$$

---

[4]The second author is grateful to M. Raghunathan for pointing this out.

One formulation of Property T for discrete groups is as follows. Γ is finitely-generated, and corresponding to any fixed finite generating set $\{\gamma_1, \ldots, \gamma_m\}$ for Γ, there exists $\varepsilon > 0$ with the following property: If $(\pi, \mathcal{H})$ is any unitary representation of Γ and there exists a unit vector $v \in \mathcal{H}$ with $\|\pi(\gamma_i)v - v\| < \varepsilon$ for $1 \leq i \leq m$, then π actually has non-trivial invariant vectors. (A unit vector v satisfying the preceding condition is said to be ε-*invariant* with respect to the γ_i.)

It is obvious from the definitions that there exists a neighborhood $U \subset R(\Gamma, \text{Diff}^1(M))$ such that for each $\rho \in U$, the vector $\delta^{1/2} \in L^2$ is ε-invariant under the unitary representation π corresponding to ρ. Thus π fixes some unit vector $\varphi_\rho \in L^2(M)$, and the probability density $|\varphi_\rho|^2$ is invariant under ρ. $\qquad\square$

Thus (1.1) implies that there is a neighborhood $U \subset R(\Gamma, \text{Diff}(M))$ (recall that we consider the group $\text{Diff}(M)$ of C^∞ diffeomorphisms with the C^1 topology) of the standard action (by orientation-preserving automorphisms of $M = \mathbb{T}^n$) such that every $\rho \in U$ is smoothly conjugate to a (rational) affine action with standard linear part. As we have already observed, $H^1(\Gamma, \mathbb{R}^n) = 0$. Thus, by a theorem due to D. Stowe [St], the set of $\rho \in R(\Gamma, \text{Diff}(M))$ with a fixed point (near the origin) also contains a neighborhood of the standard action. (1.2) follows.

§3. Concluding Remarks.

Zimmer outlined a general program directed at understanding smooth actions of lattices in semisimple Lie groups of $\mathbb{R}$-rank ≥ 2 on compact manifolds in his 1986 address to the International Congress of Mathematicians [Z1]. (He had previously conjectured that the action of $\mathbf{SL}(n, \mathbb{Z})$ on $\mathbb{T}^n$, $n \geq 3$, was locally rigid during the conference on ergodic theory, differential geometery, and Lie groups in May, 1984, at the Mathematical Sciences Research Institute.) Lewis established an infinitesimal rigidity result for subgroups of finite index in $\mathbf{SL}(n, \mathbb{Z})$ acting on $\mathbb{T}^n$ for $n \geq 7$ in [Le]. Hurder obtained rigidity under *continuous deformations* for $n \geq 3$ in [H]. His key contribution was the use of the theorem by Stowe [St]. A further development in that direction which goes beyond actions containig Anosov elements is due to Qian [Q1].

The first local rigidity result is due to Katok-Lewis for the standard $\mathbf{SL}(n, \mathbb{Z})$ action on $\mathbb{T}^n, n \geq 4$ in [K-L1]; subsequently they extended the technique to obtain global results (again, for $n \geq 4$) in [K-L2]. Their method is based on the recovery of an invariant rational structure using a combination of hyperbolicity ant Stowe's theorem.

We recall that the main result (Corollary 2.14) in [K-L2] yields the same conclusion as (1.1), above, but under the alternative hypotheses that ρ has a finite orbit and with some special restrictions on the element γ_0 such that $\rho(\gamma_0)$ is Anosov. (The hyperbolic matrix γ_0 is required to preserve a non-trivial rational product structure on $\mathbb{Q}^n$, which is equivalent to the requirement that $\rho(\gamma_0)$ preserve a non-trivial decomposition of some finite cover of $\mathbb{T}^n$ as a product of compact subtori. It is this condition that leads to the restriction $n \geq 4$.) Note that the two results are parallel, in that the existence of a finite orbit is equivalent to the existence of an invariant atomic measure.

On the other hand, as we indicated at the outset, the cocycle superrigidity theorem is applicable in much greater generality than we have discussed above, and we believe that it can be used to overcome some essential limitations of the techniques in [H],

11

[K-L1], and [K-L2] which are based on existence of an invariant rational structure. In particular, this approach is applicable in the absence of finite orbits. Very briefly, we will try to describe the main obstacles to extending our technique to more general actions (e.g., to obtain local rigidity for other non-isometric algebraic examples) as well as some approaches to overcome those obstacles.

First, there is the general problem of determining the algebraic hull of the derivative cocycle. It is not at all clear how to use the superrigidity theorem to obtain useful geometric information about the action unless this group is semisimple. In case the lattice Γ is cocompact, Zimmer shows in [Z4] that the algebraic hull is reductive with compact center. We might hope that this is generally the case, but this has not, as yet, been established. The theorem provides definitive information analogous to that summarized in (2.1), above, (at least directly) only in very special cases; essentially when the dimension n of the compact manifold M coïncides with the minimum n for which there exist representations $\Gamma \to \mathbf{SL}(n, \mathbb{R})$ whose image has non-compact closure. R. Feres succeeded in extending this approach under somewhat more general hypotheses, cf. [Fe].

Secondly, the requirement that there exist $\gamma_0 \in \Gamma$ such that $\rho(\gamma_0)$ is Anosov only makes sense in special cases, and must be systematically replaced by more general hypotheses. A first successful step in the partially hyperbolic case has been recently made by V.Nitica and A.Török [N-T1], [N-T2] who developed powerful new analytical techniques allowing to classify cocycles with values in the diffeomorphism groups. Note that the examples described in §4 of [K-L2] make it clear that, at least in the most general setting, we cannot hope to eliminate the dynamical hypotheses entirely.

Finally, the argument (based on (2.5)) for passing from a continuous linearizing frame to a smooth conjugacy needs to be generalized. A relatively straightforward extension to the class of "Cartan actions" has been accomplished in the recent work of Qian [Q2]. More generally, rigidity of hyperbolic actions of higher-rank abelian groups [K-S1], [K-S2] should be brought into the play.

REFERENCES

[A]   D. V. Anosov, *Geodesic flows on closed Riemannian manifolds with negative curvature*, Proc. Steklov Inst. Math., No. 90, 1967 (Russian); English transl., A.M.S., Providence, RI, 1969.

[B]   D.Berend, *Multi-invariant sets on tori*, Trans. A.M.S . **280** (1983), 509–532.

[B-S]   A. Borel and J.-P. Serre, *Théorèmes de finitude en cohomologie galoisienne*, Comm. Math. Helv. **39** (1964), 111–164.

[D]   S. G. Dani, *On ergodic qausi-invariant measures of group automorphisms*, Israel Jour. Math. **43** (1982), 62–74.

[Fe]   R. Feres, *Connection-preserving actions of lattices in* $\mathbf{SL}(n, \mathbb{R})$, Israel Jour. Math. (to appear).

[Fu]   H. Furstenberg, *A Poisson formula for semisimple Lie groups*, Ann. Math. **77** (1963), 335–383.

[H]   S. Hurder, *Rigidity for Anosov actions of higher rank lattices*, Ann. Math. **135** (1992), 361–410.

[K-L1] A. Katok and J. Lewis, *Local rigidity for certain groups of toral automorphisms*, Israel Jour. Math. **75** (1991), 203–241.

12

[K-L2] A. Katok and J. Lewis, *Global rigidity for lattice actions on tori and new examples of volume-preserving actions*, Israel Jour. Math. (to appear).

[K-S1] A. Katok and R. Spatzier, *First cohomology of Anosov actions of higher rank abelian groups and applications to rigidity*, Publ. Math. I.H.E.S. **79** (1994), 131–156.

[K-S2] A. Katok and R. Spatzier, *Subelliptic estimates of polynomial differential operators and applications to rigidity of abelian actions*, Math. Res. Letters **1** (1994), 193–202.

[Le] J. Lewis, *Infinitesimal rigidity for the action of* $\mathbf{SL}(n, \mathbb{Z})$ *on* $\mathbb{T}^n$, Trans. A.M.S. **324** (1991), 421–445.

[Li] A. N. Livshitz, *Homology properties of Y-systems*, Mat. Zametki **10** (1971), 555–564 (Russian); English transl., Math. Notes **10** (1971), 758–763.

[L-S] A.N. Livshitz and Ja.G.Sinai, *Invariant measures that are compatible with smoothness for transitive C-systems*, Soviet Math. Dokl. **13** (1972), 1656–1659.

[Ll-M-M] R. de la Llave, J. M. Marco, and R. Moriyon, *Canonical perturbation theory of Anosov systems and regularity results for the Livshitz cohomology equation*, Ann. Math. **123** (1986), 537–611.

[M] A. Manning, *There are no new Anosov diffeomorphisms on tori*, Amer. J. Math. **96** (1974), 422–429.

[Mañ] R. Mañé, *Ergodic theory and differentiable dynamics*, Springer, 1987.

[Mar] G. A. Margulis, *Discrete subgroups of semisimple Lie groups*, Springer, 1991.

[Mo] C. C. Moore, *Amenable subgroups of semisimple groups and proximal flows*, Israel Jour. Math. **34** (1979), 121–138.

[N-T1] V.Nitica and A.Török, *Cohomology of dynamical systems and rigidity of partially hyperbolic actions of higher rank lattices*, preprint.

[N-T2] V.Nitica and A. Török, *Cohomology abelian group actions and rigidity of higher rank lattice actions*, preprint.

[O] V. I. Oseledec, *A multiplicative ergodic theorem*, Trudy Moskov. Mat. Obsc. **19** (1968), 179–210 (Russian); English transl., Translations A.M.S. **19** (1968), 197–231.

[Q1] N.Qian, *Topological deformation rigidity of higher rank lattice actions*, Math. Res. Letters **1** (1994), 485–499.

[P-Y] J. Palis and J.C. Yoccoz, *Centralizers of Anosov diffeomorphisms on tori*, Ann. Sc. Ec. Norm. Sup. **22** (1989), 99–108.

[Q2] N.Qian, *Tangential flatness and global rigidity of higher rank lattice actions*, preprint.

[Se] G. Seydoux, University of Chicago Doctoral Dissertation, (1991).

[St] D. Stowe, *The stationary set of a group action*, Proc. A.M.S. **79** (1980), 139–146.

[Z1] R. Zimmer, *Actions of semisimple groups and discrete subgroups*, Proceedings of the International Congress of Mathematicians, (1986: Berkeley, CA), A.M.S., Providence, RI, 1987, pp. 1247–1258.

[Z2] R. Zimmer, *Ergodic theory and the automorphism group of a G-structure*, Group representations, ergodic theory, operator algebras, and mathematical physics (C. C. Moore, ed.), Springer, New York, 1987, pp. 247–278.

[Z3] R. Zimmer, *Ergodic theory and semisimple groups*, Birkhäuser, Boston, 1984.

[Z4] R. Zimmer, *On the algebraic hull of an automorphism group of a principal bundle*, Comment. Math. Helv. **65** (1990), 375–387.

13

Volume-preserving actions of simple algebraic $\mathbb{Q}$-groups on low-dimensional manifolds

Dave Witte Morris

Department of Mathematics and Computer Science, University of Lethbridge,
Lethbridge, Alberta, T1K 3M4, Canada

Dave.Morris@uleth.ca, http://people.uleth.ca/~dave.morris/

Robert J. Zimmer

Office of the President, University of Chicago,
Chicago, Illinois 60637, USA

president@uchicago.edu, http://president.uchicago.edu/

December 6, 2013

We prove that $\mathrm{SL}(n, \mathbb{Q})$ has no nontrivial, C^∞, volume-preserving action on any compact manifold of dimension strictly less than n. More generally, suppose $\mathbf{G}$ is a connected, isotropic, almost-simple algebraic group over $\mathbb{Q}$, such that the simple factors of every localization of $\mathbf{G}$ have rank ≥ 2. If there does not exist a nontrivial homomorphism from $\mathbf{G}(\mathbb{R})^\circ$ to $\mathrm{GL}(d, \mathbb{C})$, then every C^∞, volume-preserving action of $\mathbf{G}(\mathbb{Q})$ on any compact d-dimensional manifold must factor through a finite group.

Keywords: group action; algebraic group; volume-preserving; manifold.

AMS Subject Classification: 37C85; 20G30, 22F99, 57S99.

1. Introduction

The second author has conjectured that if $\mathbf{G}$ is a simple algebraic $\mathbb{Q}$-group, and $\mathrm{rank}_\mathbb{R}\, \mathbf{G} \geq 2$, then every C^∞, volume-preserving action of the arithmetic group $\mathbf{G}(\mathbb{Z})$ on an compact manifold of small dimension must be *finite*. (This means that the action factors through the action of a finite group. See [2] for a precise statement of the conjecture and a survey of progress on this problem.) In this paper, we show that known results imply the analogue of the conjecture with $\mathbf{G}(\mathbb{Q})$ in the place of $\mathbf{G}(\mathbb{Z})$. For example, we establish:

Theorem 1.1. $\mathrm{SL}(n, \mathbb{Q})$ *has no nontrivial, C^∞, volume-preserving action on any compact manifold of dimension strictly less than n.*

Remark 1.2. $\mathrm{SL}(n, \mathbb{Q})$ contains large finite subgroups whenever n is large (such as an elementary abelian group of order 2^{n-1}). Therefore, topological arguments imply that if M is any compact manifold, then there is some n, such that $\mathrm{SL}(n, \mathbb{Q})$ has no nontrivial, C^0 action on M (see [6, Thm. 2.5]). However, unlike in Theorem 1.1, the

1

2   *Dave Witte Morris and Robert J. Zimmer*

value of n depends on details of the topology of M, not just its dimension, because every finite group acts freely on some compact, connected, 2-dimensional manifold [1, Thm. 7.12].

The nontrivial part of Theorem 1.1 (namely, when $n \geq 3$) is a special case of the following much more general result:

Theorem 1.3. *Assume:*

(a) $\mathbf{G}$ *is an isotropic, almost-simple, linear algebraic group over* $\mathbb{Q}$, *such that, for every place* v *of* $\mathbb{Q}$, *the* $\mathbb{Q}_v$-*rank of every simple factor of* $\mathbf{G}(\mathbb{Q}_v)$ *is at least two,*

(b) $d \in \mathbb{Z}^+$, *such that there are no nontrivial, continuous homomorphisms from* $\mathbf{G}(\mathbb{R})^\circ$ *to* $\mathrm{GL}(d, \mathbb{C})$, *and*

(c) G *is a subgroup of finite index in* $\mathbf{G}(\mathbb{Q})$.

Then every C^∞, *volume-preserving action of* G *on any d-dimensional compact manifold M is finite.*

Remark 1.4 (anisotropic groups). Assume, for simplicity, that $\mathbf{G}$ is connected. Then the assumption that $\mathbf{G}$ is isotropic can be eliminated if we add two hypotheses on the universal cover $\widetilde{\mathbf{G}}$:

  (i) $\widetilde{\mathbf{G}}(\mathbb{Q})$ is projectively simple, and

 (ii) sufficiently large S-arithmetic subgroups of $\widetilde{\mathbf{G}}$ have the Congruence Subgroup Property.

Both of these hypotheses are known to be true unless $\mathbf{G}$ is anisotropic of type A_n, D_4, or E_6. See Remark 2.8 for more details.

Remarks 1.5. (1) To satisfy the requirement that the $\mathbb{Q}_v$-rank of every simple factor of $\mathbf{G}(\mathbb{Q}_v)$ is at least two, it suffices to let $\mathbf{G}$ be an absolutely almost-simple algebraic group over $\mathbb{Q}$, such that $\mathrm{rank}_\mathbb{Q} \mathbf{G} \geq 2$. In particular, we can take $\mathbf{G} = \mathbf{SL}_n$ with $n \geq 3$. This yields Theorem 1.1.

(2) The assumption that the subgroup G has finite index can be replaced with the weaker assumption that it contains the commutator subgroup $[\mathbf{G}^\circ(\mathbb{Q}), \mathbf{G}^\circ(\mathbb{Q})]$.

(3) Our bound on the dimension d of M is probably not sharp. In particular, we conjecture that $\mathrm{SL}(n, \mathbb{Q})$ has no volume-preserving action on any compact manifold of dimension strictly less than $n^2 - 1$. In the general case, it should suffice to assume that $\mathbf{G}(\mathbb{R})^\circ$ has no simple factor of dimension $\leq d$.

2. Proof of Theorem 1.3

Assume the situation of Theorem 1.3. By passing to a subgroup of finite index, we assume that $\mathbf{G}$ is connected.

Notation 2.1. (1) $\widetilde{\mathbf{G}}$ is the universal cover of $\mathbf{G}$. (We may realize $\widetilde{\mathbf{G}}$ as a Zariski-closed subgroup of $\mathbf{SL}_N$, for some N [5, Thm. 8.6, p. 63], so $\widetilde{\mathbf{G}}(R)$ is defined for any integral domain R of characteristic zero.)

(2) $\pi \colon \widetilde{\mathbf{G}} \to \mathbf{G}$ is the natural homomorphism.

(3) $\mathbf{Z}$ is the kernel of π (so $\mathbf{Z}$ is a finite, central $\mathbb{Q}$-subgroup of $\widetilde{\mathbf{G}}$).

(4) If S is any finite set of prime numbers:

  (a) $\mathbb{Z}_S$ is the ring of S-integers. That is, $\mathbb{Z}_S = \mathbb{Z}[1/p_1, \ldots, 1/p_r]$, where $S = \{p_1, \ldots, p_r\}$.

  (b) $\Gamma_S = \widetilde{\mathbf{G}}(\mathbb{Z}_S)$, so Γ_S is an S-arithmetic subgroup of $\widetilde{\mathbf{G}}$.

  (c) $\widehat{\Gamma}_S$ is the profinite completion of Γ_S.

(5) $\mathbb{Z}_p$ is the ring of p-adic integers, for any prime p.

We begin by recalling a few well-known facts about $\widetilde{\mathbf{G}}(\mathbb{Q})$:

Lemma 2.2. *(1) Every proper, normal subgroup of $\widetilde{\mathbf{G}}(\mathbb{Q})$ is contained in the center of $\widetilde{\mathbf{G}}$, and is therefore finite.*

(2) $\pi\big(\widetilde{\mathbf{G}}(\mathbb{Q})\big) \subseteq G$.

(3) $\mathbf{G}(\mathbb{Q})/\pi\big(\widetilde{\mathbf{G}}(\mathbb{Q})\big)$ is an abelian group whose exponent divides $|\mathbf{Z}(\mathbb{C})|$. (In particular, $\pi\big(\widetilde{\mathbf{G}}(\mathbb{Q})\big)$ is a normal subgroup of $\mathbf{G}(\mathbb{Q})$.)

Proof. (1) See [3, Thm. 8.1]. This relies on our assumption that $\mathbf{G}$ is isotropic.

(2) Since G has finite index in $\mathbf{G}(\mathbb{Q})$, it contains a finite-index subgroup of $\pi\big(\widetilde{\mathbf{G}}(\mathbb{Q})\big)$. However, we know from (1) that $\widetilde{\mathbf{G}}(\mathbb{Q})$ has no proper subgroups of finite index. Therefore G must contain all of $\pi\big(\widetilde{\mathbf{G}}(\mathbb{Q})\big)$.

(3) We have the following long exact sequence of Galois cohomology groups [8, (1.11), p. 22]:

$$H^0\big(\mathbb{Q}; \widetilde{\mathbf{G}}\big) \longrightarrow H^0(\mathbb{Q}; \mathbf{G}) \longrightarrow H^1(\mathbb{Q}; \mathbf{Z}).$$

In other words,

$$\widetilde{\mathbf{G}}(\mathbb{Q}) \longrightarrow \mathbf{G}(\mathbb{Q}) \xrightarrow{\delta} H^1\big(\mathrm{Gal}(\overline{\mathbb{Q}}/\mathbb{Q}), \mathbf{Z}(\mathbb{C})\big).$$

Since $\mathbf{Z}$ is central in $\mathbf{G}$, it is easy to see that the connecting map δ is a group homomorphism. Therefore, the desired conclusion follows from the observation that multiplication by $|\mathbf{Z}(\mathbb{C})|$ annihilates the abelian group $H^1\big(*; \mathbf{Z}(\mathbb{C})\big)$.   □

Since $\Gamma_S \subset \widetilde{\mathbf{G}}(\mathbb{Q})$, and G acts on M, Lemma 2.2(2) provides an action of Γ_S on M (for any S). The following theorem about this action requires our assumption that there are no nontrivial, continuous homomorphisms from $\mathbf{G}(\mathbb{R})^\circ$ to $\mathrm{GL}(\dim M, \mathbb{C})$. It also uses our assumption that simple factors of $\mathbf{G}(\mathbb{Q}_v)$ have rank at least two. (This implies that $\mathbf{G}(\mathbb{Q}_v)$ has Kazhdan's property (T).)

Theorem 2.3 ([13, Cor. 1.3]). *If S is any finite set of prime numbers, then there exist*

- *a continuous action of a compact group K_S on a compact metric space X_S, and*

- *a homomorphism* $\varphi_S \colon \Gamma_S \to K_S$,

such that the resulting action of Γ_S *on* X_S *is measurably isomorphic (a.e.) to the action of* Γ_S *on* M.

We may assume that $\varphi_S(\Gamma_S)$ is dense in K_S. This implies:

Lemma 2.4 (cf. [13, Cor. 1.5]). K_S *is profinite.*

Proof. It is an easy consequence of the Peter-Weyl Theorem that every compact group is a projective limit of compact Lie groups [4, Cor. 2.43, p. 51]. However, since $\mathbf{G}(\mathbb{R})$ has no compact factors, the Margulis Superrigidity Theorem [7, Thm. B(iii), pp. 258–259] tells us that any homomorphism from Γ_S into a compact Lie group must have finite image. Since $\varphi_S(\Gamma_S)$ is dense in K_S, this implies that K_S is a projective limit of finite groups, as desired. □

Therefore, we may assume K_S is the profinite completion $\widehat{\Gamma}_S$ of Γ_S. We have the following well-known description of $\widehat{\Gamma}_S$ (because $\mathbf{G}$ is isotropic).

Theorem 2.5 (Congruence Subgroup Property [9,10,11]). *If S is nonempty, then the natural inclusion* $\Gamma_S \hookrightarrow \times_{p \notin S} \widetilde{\mathbf{G}}(\mathbb{Z}_p)$ *extends to an isomorphism* $\widehat{\Gamma}_S \cong \times_{p \notin S} \widetilde{\mathbf{G}}(\mathbb{Z}_p)$.

Fix a prime number $q \neq 2$. The inclusion $\Gamma_{\{2\}} \subset \Gamma_{\{2,q\}}$ provides us with an action of $\Gamma_{\{2\}}$ on $X_{\{2,q\}}$, but this must be isomorphic to the action of $\Gamma_{\{2\}}$ on $X_{\{2\}}$ (since both are isomorphic to the action on M). Therefore, the action of $\widehat{\Gamma}_{\{2\}}$ on $X_{\{2\}}$ must factor through $\widehat{\Gamma}_{\{2,q\}}$ (a.e.). Furthermore, if we use the Congruence Subgroup Property (2.5) to identify $\widehat{\Gamma}_S$ with $\times_{p \notin S} \widetilde{\mathbf{G}}(\mathbb{Z}_p)$, then it is obvious that $\widetilde{\mathbf{G}}(\mathbb{Z}_q)$ is in the kernel of the homomorphism $\widehat{\Gamma}_{\{2\}} \to \widehat{\Gamma}_{\{2,q\}}$. Therefore, $\widetilde{\mathbf{G}}(\mathbb{Z}_q)$ acts trivially on $X_{\{2\}}$ (a.e.).

Since the subgroups $\widetilde{\mathbf{G}}(\mathbb{Z}_q)$ generate a dense subgroup of $\times_{p \neq 2} \widetilde{\mathbf{G}}(\mathbb{Z}_p) \cong \widehat{\Gamma}_{\{2\}}$, we conclude that $\widehat{\Gamma}_{\{2\}}$ acts trivially (a.e.). Therefore, $\Gamma_{\{2\}}$ acts trivially on M (not just a.e., because $\Gamma_{\{2\}}$ acts continuously on M), so the action of $\widetilde{\mathbf{G}}(\mathbb{Q})$ has an infinite kernel. Hence, Lemma 2.2(1) implies that the kernel is all of $\widetilde{\mathbf{G}}(\mathbb{Q})$. This means that $\widetilde{\mathbf{G}}(\mathbb{Q})$ acts trivially on M.

So the action of G factors through $G/\pi\big(\widetilde{\mathbf{G}}(\mathbb{Q})\big)$. From Lemma 2.2(3), we know that this quotient is an abelian group of finite exponent, so the corollary of the following theorem tells us that the action is finite.

Theorem 2.6 ([6, Thm. 2.5]). *If A is any abelian group of prime exponent, then every C^0 action of A on any compact manifold is finite.*

Corollary 2.7. *If A is any abelian group of finite exponent, then every C^0 action of A on any compact manifold is finite.*

Proof. We can assume the exponent of A is a power of a prime p (because A is the direct product of its finitely many Sylow subgroups). We can also assume that

Volume-preserving actions of simple algebraic Q-groups on low-dimensional manifolds   5

the action of A is faithful, so the theorem tells us that A has only finitely many elements of order p. This means the kernel of the homomorphism $x \mapsto x^p$ is finite, so it is easy to prove by induction that A has only finitely many elements of any order p^k. Since A has finite exponent, this implies that A is finite.   □

This completes the proof of Theorem 1.3. We now discuss the generalization described in Remark 1.4.

Remark 2.8. The assumption that $\mathbf{G}$ is isotropic was used in only two places: the projective simplicity of $\widetilde{\mathbf{G}}(\mathbb{Q})$ (Lemma 2.2(1)) and the Congruence Subgroup Property (2.5).

The projective simplicity is known to be true unless $\mathbf{G}$ is anisotropic of type A_n, $^{3,6}D_4$, or E_6 [8, pp. 513–515]. (Projective simplicity obviously fails if there is a nonarchimedean place v, such that $\mathbf{G}(\mathbb{Q}_v)$ has a compact factor [8, pp. 510–511]. However, compact nonarchimedean factors cannot arise unless $\mathbf{G}$ is of type A_n [8, Thm. 6.5, p. 285]. In any case, we have ruled out compact factors by requiring the simple factors of $\mathbf{G}(\mathbb{Q}_v)$ to have rank at least 2.) When there are no compact factors, projective simplicity is also known to be true for inner forms of type ^1A_n [12, p. 180].

For the Congruence Subgroup Property, it suffices to assume that every prime number q is contained in a finite set S of prime numbers, such that the congruence kernel $C(S, \widetilde{\mathbf{G}})$ is central. (This condition is known to be true unless $\mathbf{G}$ is anisotropic of type A_n, D_4, or E_6 [8, Thms. 9.23 and 9.24, pp. 568–569]. In fact, for our purposes, it would suffice to know that $C(S, \widetilde{\mathbf{G}})$ is abelian.) To see that this assumption suffices, note that, for any finite set S of prime numbers, Strong Approximation [8, Thm. 7.12, p. 427] tells us $\widehat{\Gamma}_S / C(S, \widetilde{\mathbf{G}}) \cong \times_{p \notin S} \widetilde{\mathbf{G}}(\mathbb{Z}_p)$. In particular, $\widehat{\Gamma}_\emptyset / C(\emptyset, \widetilde{\mathbf{G}}) \cong \times_p \widetilde{\mathbf{G}}(\mathbb{Z}_p)$, so, for each prime q, we may let $\widehat{\mathbf{G}}(\mathbb{Z}_q)$ be the inverse image of $\widetilde{\mathbf{G}}(\mathbb{Z}_q)$ in $\widehat{\Gamma}_\emptyset$. The homomorphism $\widehat{\Gamma}_\emptyset \to \widehat{\Gamma}_S$ must map $\widehat{\mathbf{G}}(\mathbb{Z}_q)$ into $C(S, \widetilde{\mathbf{G}})$ for all $q \in S$. If $C(S, \widetilde{\mathbf{G}})$ is abelian, this implies that the image of the commutator subgroup $[\widehat{\mathbf{G}}(\mathbb{Z}_q), \widehat{\mathbf{G}}(\mathbb{Z}_q)]$ is trivial. Since this is true for all q, we conclude that $[\Gamma_\emptyset, \Gamma_\emptyset]$ acts trivially on M. This is sufficient to show that $\widetilde{\mathbf{G}}(\mathbb{Q})$ acts trivially.

Acknowledgments. D. W. M. would like to thank the mathematics departments of Indiana University and the University of Chicago for their excellent hospitality while this paper was being written, and would especially like to thank David Fisher for very helpful conversations about this material.

References

1. B. Farb and D. Margalit, *A Primer on Mapping Class Groups*, (Princeton, 2012). MR2850125, ISBN 978-0-691-14794-9.
2. D. Fisher, Groups acting on manifolds: around the Zimmer program, in *Geometry, Rigidity, and Group Actions*, eds. B. Farb and D. Fisher, (Univ. Chicago Press, 2011), pp. 72–157. MR2807830, ISBN 978-0-226-23788-6.
3. P. Gille, Le problème de Kneser-Tits, *Astérisque* **326** (2009) 39–81. MR2605318.

6   *Dave Witte Morris and Robert J. Zimmer*

4. K. H. Hofmann and S. A. Morris, *The Structure of Compact Groups, 2nd ed.*, (de Gruyter, 2006). MR2261490, ISBN 978-3-11-019006-9.

5. J. E. Humphreys, *Linear Algebraic Groups*, (Springer, 1975). MR0396773, ISBN 0-387-90108-6.

6. L. N. Mann and J. C. Su, Actions of elementary p-groups on manifolds, *Trans. Amer. Math. Soc.* **106** (1963) 115–126. MR0143840.

7. G. A. Margulis, *Discrete Subgroups of Semisimple Lie Groups*, (Springer, 1991). MR1090825, ISBN 3-540-12179-X.

8. V. Platonov and A. Rapinchuk, *Algebraic Groups and Number Theory*, (Academic, 1994). MR1278263, ISBN 0-12-558180-7.

9. G. Prasad and M. S. Raghunathan, On the congruence subgroup problem: determination of the "metaplectic kernel", *Invent. Math.* **71** (1983) 21–42. MR0688260.

10. M. S. Raghunathan, On the congruence subgroup problem, *Inst. Hautes Études Sci. Publ. Math.* **46** (1976) 107–161. MR0507030.

11. M. S. Raghunathan, On the congruence subgroup problem II, *Invent. Math.* **85** (1986) 73–117. MR0842049.

12. A. S. Rapinchuk, The congruence subgroup problem, in *Algebra, K-theory, Groups, and Education (New York, 1997)*, eds. T. Y. Lam and A. R. Magid, (Amer. Math. Soc., 1999), pp. 175–188. MR1732047, ISBN 0-8218-1087-1.

13. R. J. Zimmer, Spectrum, entropy, and geometric structures for smooth actions of Kazhdan groups, *Israel J. Math.* **75** (1991) 65–80. MR1147291.

5

Stabilizers of Semisimple Lie Group Actions: Invariant Random Subgroups

Annals of Mathematics, **139** (1994), 723–747

Stabilizers for ergodic actions of higher rank semisimple groups

By Garrett Stuck and Robert J. Zimmer*

ABSTRACT   Let G be a connected semisimple Lie group with finite center and $\mathbb{R}$-rank ≥ 2. Suppose that each simple factor of G either has $\mathbb{R}$-rank ≥ 2 or is locally isomorphic to $Sp(1, n)$ or $F_{4(-20)}$. We prove that any faithful, irreducible, properly ergodic, finite measure-preserving action of G is essentially free. We extend the result to reducible actions and actions of lattices.

Introduction

The principal goal of this paper is to prove the following:

THEOREM 2.1. *Let G be a connected semisimple Lie group with finite center and $\mathbb{R}$-rank ≥ 2. Suppose that each simple factor of G either has $\mathbb{R}$-rank at least two or is locally isomorphic to $Sp(1,n)$, $n \geq 2$, or $F_{4(-20)}$. Then any faithful, irreducible, properly ergodic, finite measure-preserving action of G is essentially free; i.e., the stabilizer of almost every point is trivial.*

Recall that an ergodic action of G is said to be *irreducible* if every non-central normal subgroup acts ergodically. In Theorem 4.3 we relax the hypothesis of irreducibility, and derive some consequences for stabilizers of ergodic actions of lattices in higher rank groups.

Prior to this it was known that for any finite measure-preserving, faithful, ergodic action of a semisimple Lie group G without compact factors, the stabilizers are almost all discrete ([22, Lemma 6], and [9, Proposition 2.1]). This fact is a consequence of the Borel Density Theorem. It is also known (see Example 2.5) that Theorem 2.1 is false for all rank 1 groups.

Theorem 2.1 can be viewed as a generalization of a special case of Margulis's result on normal subgroups of irreducible lattices in higher rank Lie groups ([13, IV.4.10]). To see this, let G be a simple Lie group of $\mathbb{R}$-rank ≥ 2,

---

*The first author was supported in part by an NSF Postdoctoral Research Fellowship. The second author was supported in part by NSF grant no. DMS9107285, 5-27711. The authors would like to thank S. Adams, D. Witte, S. Mozes, J. Lewis, N.A. Shah, J. Feldman, and G.A. Margulis for helpful comments and conversations regarding this paper. The first author also thanks the University of Chicago and the Mathematical Sciences Research Institute for their hospitality during the preparation of this paper.

and let $\Gamma \subset G$ be a lattice. Let $\Lambda \subset \Gamma$ be a normal subgroup of infinite index. Then there is a free, ergodic, measure-preserving action of Γ/Λ on a probability space (X, μ). The induced action of G on $G/\Gamma \times X$ is properly ergodic, and every stabilizer is conjugate to Λ. It follows from Theorem 2.1 that Λ is in the kernel of the G-action, and therefore that Λ is in the center of G. This is exactly the conclusion of Margulis's theorem.

The proof of Theorem 2.1 is strongly influenced by Margulis's proof of his theorem on normal subgroups of lattices. One of the key ingredients is a result of the second author, [23], generalizing a result of Margulis on measurable equivariant quotients of flag varieties. The other key ingredient is a result, proved in Section 1, ruling out the amenability of an equivalence relation generated by a non-transitive, finite, measure-preserving, ergodic action of a group with Kazhdan's property. This accounts for the hypothesis in the statement of Theorem 2.1 that "each simple factor of G either has $\mathbb{R}$-rank at least two or is locally isomorphic to $\mathrm{Sp}(1, n)$, $n \geq 2$, or $F_{4(-20)}$", which ensures that G has Kazhdan's property.

We describe briefly the structure of the paper: In Section 1 we establish notation and terminology and prove the aforementioned result on equivalence relations generated by actions of groups with Kazhdan's property. In Section 2 we prove the main theorem and illustrate it with a few examples. In Section 3 we study the space of closed subgroups of a Lie group G in preparation for Section 4, where we generalize Theorem 2.1 to include reducible actions and also actions of lattices. We also give, in Section 3, examples of ergodic actions with quasi-invariant measure and non-trivial, non-conjugate stabilizers. In Section 5 we give some applications and extensions of the results of the paper. In particular, we observe that many previous results on essentially free actions of higher rank groups can now be improved by using Theorem 2.1.

1. Preliminaries from ergodic theory

Let G be a second countable, topological group. By a G-space we mean a standard σ-finite Borel measure space (X, μ), and a Borel action $G \times X \to X$ such that μ is quasi-invariant. The action of G is said to be $ergodic$ if every measurable G-invariant subset of X is either null or conull, and $properly\ ergodic$ if it is ergodic and every G-orbit has measure 0. An ergodic action is said to be $irreducible$ if every non-central normal subgroup acts ergodically. The action is said to be $faithful$ if for every $g \in G$, the set $\{x \in X \mid gx \neq x\}$ has positive measure. If (X, μ) is an ergodic G-space and (Y, ν) is an ergodic H-space, then the actions are said to be $orbit\ equivalent$ if there are conull sets $X' \subset X$ and $Y' \subset Y$, and a measure-class-preserving Borel isomorphism $\varphi \colon X' \to Y'$ such that for almost every $x \in X$, $\varphi(Gx \cap X') = H\varphi(x) \cap Y'$.

If (X, μ) is a G-space then there is a Borel equivalence relation $R_G \subset X \times X$ defined by $x \sim y$ if and only if $x = gy$ for some $g \in G$. An essential part of the proof of our main theorem involves relating properties of the action of G to properties of the equivalence relation R_G. In the next few paragraphs we set up the formalism for doing this.

Let (X, μ) be a G-space and H a standard Borel group. A Borel map $\alpha \colon G \times X \to H$ is called a *cocycle* if for all $g, h \in G$, $\alpha(gh, x) = \alpha(g, hx)\alpha(h, x)$ for almost every $x \in X$. Two cocycles α, β are *equivalent* (or *cohomologous*) if there is a Borel map $\varphi \colon X \to H$ such that for each $g \in G$, $\beta(g, x) = \varphi(gx)^{-1}\alpha(g, x)\varphi(x)$ for almost every $x \in X$. A cocycle is *trivial* if it is equivalent to a cocycle into the trivial group $\{e\}$. The cocycle α is *orbital* if for almost every $x \in X$, $\alpha(g, x) = e$ for all $g \in G_x$, where G_x is the stabilizer in G of x. If H acts on a Borel space T, a Borel function $f \colon X \to T$ is said to be α-*invariant* if for each $g \in G$, $f(gx) = \alpha(g, x)f(x)$ for almost every $x \in X$. Associated to the cocycle α there is a Borel action of G on $X \times T$ given by $g \cdot (x, t) = (gx, \alpha(g, x)t)$. We write $X \times_\alpha T$ when we are considering this action on the space $X \times T$.

Let E be a separable Banach space and let $\alpha \colon G \times X \to \mathrm{Iso}(E)$ be a cocycle. Let $\alpha^*(g, s) = (\alpha(g, s)^{-1})^*$ be the dual cocycle. Let E_1^* be the unit ball in E^*, endowed with the weak-$*$-topology.

Definition 1.1. An *affine G-space over X* is a Borel subset $A \subset X \times_{\alpha^*} E_1^*$ satisfying:

(1) for each $x \in X$, the set $A_x = A \cap (\{x\} \times E_1^*)$ is a closed, convex, non-empty subset of E_1^*, and

(2) for each $g \in G$, $\alpha^*(g, x) \cdot A_x = A_{gx}$ for almost every $x \in X$.

If the cocycle α is orbital we say that A is an affine *orbital G-space over X*.

Definition 1.2. The action of G on X is said to be *amenable* (resp., *weakly amenable*) if for every affine (resp., affine orbital) G-space A over X, there is an α^*-invariant Borel function $f \colon X \to E_1^*$ such that for each $x \in X$, $f(x) \in A_x$. Such a function is called a G-invariant section of A.

Now suppose $R \subset X \times X$ is a Borel equivalence relation. A *cocycle* over this equivalence relation is a Borel map $\alpha \colon R \to H$ which satisfies (almost everywhere) the cocycle identity $\alpha(x, y) = \alpha(x, z)\alpha(z, y)$ whenever $(x, z) \in R$ and $(z, y) \in R$. Two cocycles α and β are equivalent if there is a Borel map $\varphi \colon X \to H$ such that $\beta(x, y) = \varphi(y)^{-1}\alpha(x, y)\varphi(x)$ for almost all $(x, y) \in R$. One defines, in analogy with cocycles of group actions, the notions of α-invariant functions, affine spaces over the equivalence relation, and amenability of the equivalence relation (see [20]). Note that if (X, μ) is a G-space, then weak amenability of the action is equivalent to amenability of the equivalence relation R_G, and the orbital cocycles for the G-action are exactly the cocycles for the equivalence relation R_G.

It is a standard fact (see [24, 4.3.3]) that an action with finite invariant measure is amenable if and only if the group which is acting is amenable. Our first goal is to prove a similar result (Lemma 1.5) for weakly amenable actions of groups with Kazhdan's property (see [24, Chapter 7], for the definition of Kazhdan's property). First we prove some intermediate results:

LEMMA 1.3. *Let G be a second countable, locally compact group acting ergodically on (X,μ). If the G action on (X,μ) is weakly amenable then the action is orbit equivalent to an ergodic action of either $\mathbb{R}$ or $\mathbb{Z}$.*

Proof. The proof is primarily a litany of known results. The main technical issues are to keep track of different notions of amenability, and of differences between actions with countable and uncountable orbits.

We may assume without loss of generality that the action is faithful. We consider first the case that G is countable. Then R_G is a discrete, ergodic, measured equivalence relation, and by [20, Theorem 3.6], and [4, Prop. 7], R_G is amenable in the sense of [4, Definition 5]. Then, by the main result of [4], there is an ergodic, measure-class-preserving action of $\mathbb{Z}$ on (X, μ) which generates the equivalence relation R_G. This proves the lemma for countable G.

Suppose then that G is uncountable. Let I be the unit interval with Lebesgue measure λ, and let $\mathcal{I} = I \times I$ be the transitive equivalence relation on I. In the following, all statements should be interpreted modulo null sets. By [5, Theorems 2.8 and 5.6], there is a measurable subset $T \subset X$, a measure ν on T and a Borel isomorphism $\varphi \colon (X,\mu) \to (I \times T, \lambda \times \nu)$, such that:

(1) GT is conull in X,

(2) for each $x \in X$, $T \cap Gx$ is countable,

(3) $\varphi^{-1} \mid \{0\} \times T$ is the inclusion of T in X, and

(4) φ induces an isomorphism $\hat{\varphi} \colon R_G \to \mathcal{I} \times R_G \mid T$.

We claim that $R_G \mid T$ is amenable. Let $\alpha \colon R_G \mid T \to \mathrm{Iso}(E)$ be a cocycle, and $A \subset T \times E_1^*$ be an affine $R_G \mid T$-space over T. Define $\tilde{\alpha} \colon R_G \to \mathrm{Iso}(E)$ by $\tilde{\alpha}(x,y) = \alpha(\varphi_2(x), \varphi_2(y))$, where φ_2 denotes the composition of φ with the projection $I \times T \to T$. Let

$$\tilde{A} = (\varphi^{-1} \times \mathrm{Id})(I \times A) \subset X \times E_1^*.$$

Then $\tilde{\alpha}$ is a cocycle and $\tilde{A}$ is an affine R_G-space over X. By amenability of R_G, there is an $\tilde{\alpha}^*$-invariant section $\tilde{f} \colon X \to \tilde{A}$. The cocycle $\tilde{\alpha}$ is constant on $\varphi^{-1}(I \times \{t\})$ for each $t \in T$, as is the set $\tilde{A}_x$, since

$$\tilde{A}_{\varphi^{-1}(s,t)} = \tilde{A} \cap (\{s\} \times \{t\} \times E_1^*) = A_t.$$

It follows from the $\tilde{\alpha}^*$-invariance of $\tilde{f}$ that for almost every $t \in T$, the map $\tilde{f} \mid \varphi^{-1}(I \times \{t\})$ is an essentially constant map to A_t. Thus $\tilde{f}$ factors through φ_2; i.e., there is a measurable map $f \colon T \to A$ satisfying $f \circ \varphi_2 = \tilde{f}$, almost everywhere. Clearly f is an α^*-invariant section of A, so $R_G \mid T$ is amenable.

Now it follows as in the countable case that $R_G \mid T$ is generated by an ergodic action of $\mathbb{Z}$. The equivalence relation $\mathcal{I} \times R_G \mid T$ is clearly generated by the suspension of this action, so R_G is generated by an action of $\mathbb{R}$. It follows that the action of G is orbit equivalent to an action of $\mathbb{R}$. $\qquad \square$

LEMMA 1.4. *Suppose G acts properly ergodically and weakly amenably on (X, μ). Then there is a non-trivial cocycle $\alpha \colon G \times X \to \mathbb{R}$.*

Proof. By the preceding lemma, the G-action is orbit equivalent to an action of M, where M is $\mathbb{R}$ or $\mathbb{Z}$. A close examination of the proof shows that if the action of G is properly ergodic, then the action of M is properly ergodic and essentially free. (Note that for an action of $\mathbb{Z}$, the sets Fix(n), for $n \in \mathbb{Z}$, are invariant, and by ergodicity must be null or conull. Thus any properly ergodic action of $\mathbb{Z}$ is essentially free.) By [19, Corollary 3.4], there is a non-trivial cocycle $\beta \colon M \times X \to \mathbb{R}$. This cocycle is orbital since the action of M is essentially free. Thus β defines a cocycle $\alpha \colon R_G \to \mathbb{R}$, i.e., α is an orbital cocycle $\alpha \colon G \times X \to \mathbb{R}$. The non-triviality of α follows from the non-triviality of β. $\qquad \square$

LEMMA 1.5. *Suppose G has Kazhdan's property. Let X be an ergodic G-space with finite invariant measure. Then the action of G on X is weakly amenable if and only if the action is essentially transitive.*

Proof. Suppose first that the action is essentially transitive; i.e, there is some $x \in X$ such that $Gx \cong G/G_x$ is conull. It follows from [24, 4.2.12–4.2.14], that an orbital cocycle on $G \times G/G_x$ is trivial. Thus the action of G on G/G_x is weakly amenable. (We have not used Kazhdan's property for this part of the proof.)

. Suppose then that the action of G on X is properly ergodic and weakly amenable. By the preceding lemma, there is a non-trivial cocycle $\alpha \colon G \times X \to \mathbb{R}$. On the other hand, Kazhdan's property implies that α is equivalent to a cocycle into a compact subgroup of $\mathbb{R}$ [24, Theorem 9.1.1]. This is a contradiction. $\qquad \square$

Remark 1.6. There are many non-amenable groups which have weakly amenable, finite measure-preserving actions. In particular, any non-amenable group which admits a surjective homomorphism to an infinite amenable group has such an action. The free group on two or more generators is an example of such a group.

728   GARRETT STUCK AND ROBERT J. ZIMMER

Let G be a locally compact, second countable, topological group. We denote by $\mathcal{S}_G$ the set of all closed subgroups of G. We give $\mathcal{S}_G$ the *Fell* topology ([6]), a basis of which is the collection of sets $U(K, A_1, \ldots, A_k)$, where K is a compact subset of G, A_i is open in G, and

$$U(K, A_1, \ldots, A_k) = \{F \in \mathcal{S}_G \mid F \cap K = \emptyset \text{ and } A_i \cap F \neq \emptyset \text{ for } i = 1, \ldots, k\}.$$

With this topology $\mathcal{S}_G$ is a locally compact, separable, metrizable space. The group G acts on $\mathcal{S}_G$ by conjugation and this action is continuous. If G acts on a Borel space X, then the map $f \colon X \to \mathcal{S}_G$ given by $f(x) = G_x$ is a G-equivariant Borel map ([16]).

Definition 1.7. A G-space X is said to be *tame* if the quotient Borel structure on X/G is countably separated, i.e., if there is a sequence $\{A_i\}_{i \in \mathbb{N}}$ of G-invariant Borel sets in X which separates G-orbits.

It is a fundamental result of ergodic theory (see [24, 2.1.11]) that if G acts ergodically with quasi-invariant measure on X, Y is a tame G-space, and $h \colon X \to Y$ is a G-equivariant Borel map, then the image of h is essentially contained in a G-orbit in Y, i.e., there is a point $y \in Y$ such that $h^{-1}(Gy)$ is conull in X.

LEMMA 1.8. *Let G be a locally compact, second countable topological group acting faithfully with finite invariant measure on X. Let H be a subgroup of G and $M = Z_G(H)$ the centralizer in G of H. If H acts ergodically on X then for almost every $x \in X$, $M_x = \{e\}$.*

Proof. Let $f \colon X \to \mathcal{S}_M$ be the map $f(x) = M_x$. Then f is H-invariant since H commutes with M, and since H acts ergodically on X we conclude that f is essentially constant. Thus there is a subgroup $\Gamma \subset M$ such that for almost every $x \in X$, $M_x = \Gamma$. From the faithfulness of the action we conclude that Γ must be trivial.   □

In Section 4 we need to know that certain subspaces of $\mathcal{S}_G$ are tame G-spaces. We take up this question in Section 3.

2. The main theorem

THEOREM 2.1. *Let G be a connected semisimple Lie group with finite center and $\mathbb{R}$-rank ≥ 2. Suppose that each simple factor of G either has $\mathbb{R}$-rank at least two or is locally isomorphic to $\mathrm{Sp}\,(1,n)$, $n \geq 2$, or $F_{4(-20)}$. Then any faithful, irreducible, properly ergodic, finite measure-preserving action of G is essentially free.*

Proof. Let (X, μ) be a G-space with finite invariant measure and suppose that the action of G is faithful, irreducible, and properly ergodic. Let $Z(G)$ be the center of G. By Lemma 1.8 we know that $Z(G)$ acts essentially freely on X, so replacing G by $G/Z(G)$ and X by $X/Z(G)$ we may assume without loss of generality that G has trivial center.

The group G has Kazhdan's property ([11]). By Lemma 1.5, we conclude that the action is not weakly amenable. Then from the definition of weak amenability it follows that there is a separable Banach space E, an orbital cocycle $\alpha\colon G \times X \to \mathrm{Iso}(E)$, and a G-invariant subset $A \subset X \times_{\alpha^*} E_1^*$ satisfying the properties of Definition 1.1, and for which there is no α^*-invariant section of A.

Let P be a minimal parabolic in G, and define a cocycle

$$\tilde{\alpha}\colon G \times (G/P \times X) \to \mathrm{Iso}(E)$$

by the rule $\tilde{\alpha}(g, hP, x) = \alpha(g, x)$. Let $\tilde{A} = G/P \times A$. Then $\tilde{A}$ is an affine orbital G-space over $G/P \times X$ relative to the cocycle $\tilde{\alpha}^*$. The action of G on G/P is amenable ([24, 4.3.2]), so G acts amenably on $G/P \times X$, and there is an $\tilde{\alpha}^*$-invariant section $\tilde{f}$ of $\tilde{A}$, $\tilde{f}\colon G/P \times X \to E_1^*$. Define

$$f\colon G/P \times X \to X \times_{\alpha^*} E_1^*$$

by $f(hP, x) = (x, \tilde{f}(hP, x))$. Then f takes values in A, and for each $g \in G$, for almost every x,

$$f(g(hP, x)) = (gx, \tilde{f}(g(hP, x))) = (gx, \tilde{\alpha}^*(g, hP, x)\tilde{f}(hP, x))$$
$$= (gx, \alpha^*(g, x)\tilde{f}(hP, x)) = g(x, \tilde{f}(hP, x)) = gf(hP, x).$$

Thus f is G-equivariant.

Let ν be a quasi-invariant probability measure on G/P. Then the measure $\sigma = f_*(\nu \times \mu)$ is a quasi-invariant measure on A, and we have the sequence of measure class preserving G-maps

$$G/P \times X \xrightarrow{f} A \xrightarrow{p} X,$$

where the map p is the projection $A \subset X \times_{\alpha^*} E_1^* \to X$. It follows from [23, Theorem 4.1], that $(A, f_*(\nu \times \mu))$ is isomorphic as a G-space to $G/Q \times X$, where Q is a parabolic subgroup of G containing P, and the isomorphism may be chosen so that the maps $G/P \times X \xrightarrow{f} A$ and $A \xrightarrow{p} X$ are identified with the natural ones. Thus there are a conull G-invariant set $Y \subset G/Q \times X$ and an injective G-map $\varphi\colon Y \to A$ such that $p \circ \varphi = q$, where $q\colon G/Q \times X \to X$ is the projection.

Suppose $Q = G$. Then φ is (essentially) a G-map from X to A. Moreover, for almost every $x \in X$, $p \circ \varphi(x) = x$, so $\varphi(x) \in A_x$. In other words, φ is an α^*-invariant section of A, which we have assumed does not exist.

We may assume therefore that Q is a proper parabolic in G. By Fubini's theorem, there is a conull set $X_0' \subset X$ such that for every $x \in X_0'$, the set $Y_x = Y \cap (G/Q \times \{x\})$ is conull in G/Q. The fact that α is orbital implies that there is a conull set $X_0 \subset X_0'$ such that for every $x \in X_0$, $\alpha^*(G_x, x) = e$. From this it follows that G_x acts trivially on $p^{-1}(x)$ for every $x \in X_0$. The equality $p \circ \varphi = q$ and the fact that φ is an injective G-map imply that for every $x \in X_0$, G_x acts trivially on $q^{-1}(x) = Y_x$. The kernel of the action of G on Y_x is

$$H_x = \bigcap_{gQ \in Y_x} gQg^{-1},$$

so for every $x \in X_0$, $G_x \subset H_x$. For every $x \in X_0$, the normalizer of H_x contains a conull subgroup of G, and must therefore be all of G. Thus H_x is normal in G. We have shown: for almost every $x \in X$, the stabilizer G_x is contained in a proper normal subgroup of G.

Let H be a simple normal subgroup of G, and let $M = Z_G(H)$. By Lemma 1.8, M_x is trivial for almost every $x \in X$. It follows that every proper normal subgroup acts essentially freely. But almost every stabilizer is contained in a proper normal subgroup, and there are only finitely many normal subgroups, so almost every stabilizer is trivial.   □

COROLLARY 2.2. *Let G be a simple Lie group of $\mathbb{R}$-rank greater than 1. Let $\Gamma \subset G$ be a lattice. If Γ acts faithfully and properly ergodically on X preserving a finite measure then the action is essentially free.*

Proof. Consider the induced action of G on $G/\Gamma \times X$. This action is properly ergodic and faithful, so by the preceding theorem it is essentially free. But the stabilizers of the G-action are just conjugates of the stabilizers of the Γ-action on X, so the Γ-action is essentially free.   □

Example 2.3. Let $\Gamma = \mathrm{SL}(n, \mathbb{Z})$, and let $G = \mathrm{SL}(n, \mathbb{R})$, $n \geq 3$. The group Γ acts in a natural way on $T^n = R^n/Z^n$. This action preserves the standard volume on T^n, and the action is ergodic. The set of points in T^n with non-trivial, non-central stabilizer is a countable union of proper closed submanifolds. In particular, the set of points with non-trivial, non-central stabilizer is a dense null subset of T^n.

Example 2.4. Let M and H be simple, center-free, non-compact Lie groups, and suppose there is an irreducible lattice $\Lambda \subset G = M \times H$ (this will be the case if and only if all complex simple factors of the complexification of the Lie algebra of G are isomorphic, see [10]). Let Γ be a lattice in H and let Λ act on H/Γ via the projection of Λ into H. If we induce the action of Λ we get the natural action of G on $G/\Lambda \times H/\Gamma$. This action is not irreducible since M does not act ergodically. This shows that Theorem 2.1 cannot be applied automatically to ergodic actions of irreducible lattices. Note however that in the present example the action of Λ on H/Γ is essentially free. In Section 5 we are going to extend Theorem 2.1 to non-irreducible actions, which will enable us to draw conclusions about actions of lattices.

Example 2.5. Let G be a simple Lie group of $\mathbb{R}$-rank 1 with finite center. Let $\Gamma \subset G$ be a torsion-free lattice. Then Γ is a hyperbolic group, and it follows from [8, Theorem 5.5.A], that there is an infinite, infinite index normal subgroup $\Lambda \subset \Gamma$. Since the normalizer of Λ is of cofinite volume in G, it follows from [1, 4.5], that there is a faithful, properly ergodic, finite measure-preserving action of G such that almost every stabilizer is conjugate to Λ. Thus Theorem 2.1 is false for every simple Lie group of $\mathbb{R}$-rank 1.

It would be interesting to know whether Theorem 2.1 is true for irreducible actions of any semisimple group with $\mathbb{R}$-rank ≥ 2. The more restrictive hypothesis in Theorem 2.1 is required in our proof because we need Lemma 1.5, whose proof requires Kazhdan's property. In order to generalize Theorem 2.1 it would suffice to answer the following question in the affirmative:

Question. Let G be a semisimple Lie group with finite center and no compact factors and $\mathbb{R}$-rank ≥ 2. Suppose G acts irreducibly on X with finite invariant measure. Is every cocycle $\alpha \colon G \times X \to \mathbb{R}$ trivial?

3. The space of lattices

Let G be a Lie group. At the end of Section 1 we defined $\mathcal{S}_G$, the space of closed subgroups of G. Let $\mathcal{D}_G \subset \mathcal{S}_G$ be the space of discrete subgroups of G. Let $\mathcal{L}_G \subset \mathcal{D}_G$ be the space of lattices in G. Let $\mathcal{Q}_G \subset \mathcal{D}_G$ be the space of discrete subgroups of G which are lattices in a normal subgroup of G. We consider the trivial group $\{e\}$ to be a lattice in itself, so $\{e\} \in \mathcal{Q}_G$. When there is no danger of confusion we will drop the subscripts and write simply $\mathcal{S}$, $\mathcal{D}$, $\mathcal{L}$, and $\mathcal{Q}$. Note that $\mathcal{D}$, $\mathcal{L}$ and $\mathcal{Q}$ are G-invariant subsets of $\mathcal{S}$. The space $\mathcal{D}$ is open in $\mathcal{S}$ since

$$\mathcal{D} = \bigcup_{i \in \mathbb{N}} U(\overline{V}_i, \emptyset),$$

where $\{V_i\}_{i \in \mathbb{N}}$ is a neighborhood basis of the identity in G consisting of relatively compact open sets.

LEMMA 3.1. *The G-orbits in $\mathcal{L}$ are closed in $\mathcal{D}$.*

Proof. Let Γ be a lattice in G, Λ a discrete subgroup, and suppose there is a sequence $\{g_n\} \subset G$ such that $g_n \Gamma g_n^{-1} \to \Lambda$ in the Fell topology. We will show that $\Lambda = g\Gamma g^{-1}$ for some $g \in G$.

Choose a relatively compact neighborhood V of the identity in G such that

(1) $V^2 \cap \Lambda = \{e\}$, and

(2) no compact subgroup of G is contained in V.

Let K be the closure of V^2. Choose $N > 0$ so that for $n \geq N$, $g_n \Gamma g_n^{-1} \subset U(K \backslash V, V)$. Then for $n \geq N$,

$$(3.1) \qquad (g_n \Gamma g_n^{-1} \cap K) \subset V.$$

We claim that for $n > N$, $g_n \Gamma g_n^{-1} \cap V = \{e\}$. Suppose $\gamma \in g_n \Gamma g_n^{-1} \cap V$ and $\gamma \neq e$. By discreteness, there are only finitely many distinct powers of γ in V. By (2), there is a positive integer k such that $\gamma^k \in V$ and $\gamma^{k+1} \notin V$. But $\gamma^{k+1} \in K \backslash V$, which contradicts (3.1).

We now apply [15, Lemma 1.12], to conclude that $\pi(g_n)$ has a convergent subsequence, where $\pi \colon G \to G/\Gamma$ is the natural map. We replace g_n by this subsequence and assume that $\pi(g_n)$ converges; i.e., there is a sequence $\gamma_n \in \Gamma$ such that $g_n \gamma_n$ converges to an element $g \in G$. It follows that $g_n \Gamma g_n^{-1}$ converges to $g\Gamma g^{-1}$ in the Fell topology, so $\Lambda = g\Gamma g^{-1}$. $\square$

COROLLARY 3.2. *Let G be a Lie group. Then the action of G on $\mathcal{L}$ is tame.*

Proof. Let Γ be a lattice in G. It follows from the previous lemma that $\overline{G \cdot \Gamma} \cap \mathcal{D} = G \cdot \Gamma$. Thus the G-orbits in $\mathcal{L}$ are locally closed in $\mathcal{S}$. It follows from [7, Theorem 1], that $\mathcal{L}$ is a tame G-space. $\square$

COROLLARY 3.3. *Let G be a connected semisimple Lie group. Then the action of G on $\mathcal{Q}$ is tame.*

Proof. Let $G = G_0, G_1, \ldots, G_k = \{e\}$ be the normal subgroups of G. Let $\mathcal{L}_i$ be the space of lattices in G_i. Then $\mathcal{Q}$ is the disjoint union of the $\mathcal{L}_i$, and each $\mathcal{L}_i$ is a Borel set in $\mathcal{Q}$ (since $\mathcal{L}_i$ is closed in $\mathcal{D}_G$). The G-orbits in $\mathcal{L}_i$ coincide with the G_i-orbits, so the corollary follows from the preceding result. $\square$

LEMMA 3.4. *Let G be a connected semisimple Lie group with trivial center. Suppose G acts ergodically with finite invariant measure on (X, μ). Let H be a normal subgroup of G and suppose that for almost every $x \in X$, H_x is a lattice in a normal subgroup of H. Then there is a decomposition $H = H_1 \times H_2$, and a lattice $\Lambda \subset H_2$, such that H_x is conjugate to Λ for almost every $x \in X$. (The extreme cases $H_2 = H$ and $H_2 = \{e\}$ may occur.)*

Proof. By hypothesis, there is a conull set $X_0 \subset X$, which we may take to be G-invariant, such that for all $x \in X_0$, $H_x \in \mathcal{Q}_H$. Define a map $h: X_0 \to \mathcal{Q}_H$ by $h(x) = H_x$. This map is G-equivariant and Borel, so by ergodicity and Corollary 3.3, the image of h is essentially supported on a single H-orbit in $\mathcal{Q}_H$. Thus for almost every $x \in X$, H_x is conjugate to a fixed lattice in some normal subgroup of H. $\square$

LEMMA 3.5. *Let G be a connected semisimple Lie group without compact factors acting faithfully and ergodically with quasi-invariant measure on a space (X,μ). Suppose that for almost every $x \in X$, G_x is a lattice in G. Then the G-action on X is essentially transitive. Conversely, if the measure is finite and invariant, and the action is essentially transitive, then for almost every $x \in X$, G_x is a lattice in G.*

Proof. Suppose X_0 is a conull invariant subset of X whose stabilizers are lattices in G. Let $f: X_0 \to \mathcal{L}$ be the natural map. By Corollary 3.2, the image of f is essentially contained in a single G-orbit. Thus there is a lattice $\Lambda \in G$ such that f is essentially equivalent to a map $F: X_0 \to G/N_G(\Lambda)$. By ergodicity, for almost every $g \in G$, the group $N_g = N_G(g\Lambda g^{-1})/g\Lambda g^{-1}$ acts ergodically on $F^{-1}(gN_G(\Lambda))$. Since N_g is finite this action must be transitive for almost every $g \in G$. Moreover, this action must be free for almost every $g \in G$, since otherwise there is a set of positive measure in X_0 with stabilizers not conjugate to Λ. It follows that X_0 is equivariantly isomorphic to G/Λ.

Conversely, suppose the action of G on X is finite measure-preserving and essentially transitive. Then there is a point $x \in X$ such that Gx is conull in X. The finite measure on X gives a finite G-invariant measure on $Gx \cong G/G_x$. It follows from the Borel Density Theorem [2] and the faithfulness of the action that G_x is a lattice in G. $\square$

Let G be a connected semisimple Lie group with trivial center, no compact factors, and $\mathbb{R}$-rank ≥ 2. Let H be a second countable, locally compact topological group, and let $\mathcal{H}$ be the space of all homomorphisms of irreducible lattices in G into H. Let $p: G \times H \to G$ be the projection. The map $\eta: \mathcal{H} \to \mathcal{S}_{G \times H}$, $\eta(\varphi) \to \operatorname{graph}(\varphi)$, gives an identification of $\mathcal{H}$ with

$$\{\Gamma \in \mathcal{S}_{G \times H} \mid p(\Gamma) \text{ is an irreducible lattice in } G \text{ and } \Gamma \cap H = \{e\}\,\}.$$

Through this identification $\mathcal{H}$ inherits a topology and a continuous action of $G \times H$. The action is described as follows: If Λ is an irreducible lattice in G, $\varphi: \Lambda \to H$ is a homomorphism, and $(g,h) \in G \times H$, then $(g,h)\varphi$ is the homomorphism

$$(g,h)\varphi : g\Lambda g^{-1} \to H, \qquad (g,h)\varphi(g\lambda g^{-1}) = h\varphi(\lambda)h^{-1}.$$

734   GARRETT STUCK AND ROBERT J. ZIMMER

LEMMA 3.6. *Let G be a connected semisimple Lie group with trivial cen-ter, no compact factors, and $\mathbb{R}$-rank ≥ 2. Let K be a local field of character-istic 0, $\mathbb{H}$ an algebraic K-group, and $H = \mathbb{H}(K)$. Then the action of $G \times H$ on $\mathcal{H}$ is tame.*

Proof. We first claim that there are only countably many conjugacy classes of irreducible lattices in G. Indeed, by the Margulis Arithmeticity Theorem ([13, IX.1.11]), if Λ is an irreducible lattice in G then there are a semi-simple algebraic $\mathbb{Q}$-group M, and a surjective homomorphism $h\colon M(\mathbb{R}) \to G$, such that $h(M(\mathbb{Z}))$ is commensurable with Λ. Note that

(1) there are only countably many $\mathbb{Q}$-isomorphism classes of semisimple algebraic $\mathbb{Q}$-groups (because there are only countably many finite-dimensional $\mathbb{Q}$-Lie algebras),

(2) for a given $\mathbb{Q}$-group M, there are only finitely many G-conjugacy classes of surjective homomorphisms from $M(\mathbb{R})$ to G (since the inner auto-morphisms have finite index in the full automorphism group of a semisimple group),

(3) there are only countably many lattices in G commensurable to a given lattice.

To prove (3), fix a lattice $\Lambda \in G$. Then Γ is commensurable to Λ if and only if $\Lambda \cap \Gamma$ has finite index in both Λ and Γ. There are only countably many finite index subgroups of Λ, and for each finite index subgroup $\Lambda' \subset \Lambda$, there are only finitely many lattices containing Λ' ([15, IX.9.8]). This proves (3), and verifies our claim that the set of conjugacy classes of irreducible lattices in G is countable.

Our next claim is that for a fixed irreducible lattice $\Lambda \subset G$, the space $\mathcal{H}_\Lambda \subset \mathcal{H}$ of homomorphisms from Λ to H is a tame H-space. The group Λ is finitely generated ([15, 6.18,13.15]), and we fix a set of generators $\{\lambda_1, \ldots, \lambda_k\}$. There is a natural identification of $\mathcal{H}_\Lambda$ with the K-points of an algebraic sub-variety of $\mathbb{H}^k$ ([15, 6.2]). The Fell topology on $\mathcal{H}_\Lambda$ coincides with the topology of $\mathcal{H}_\Lambda$ considered as a subspace of $\mathbb{H}(K)^k$, where the latter is endowed with the Hausdorff topology. By [3], the action of H on $\mathcal{H}_\Lambda$ is tame.

Let $\{\Lambda_i\}_{i\in\mathbb{N}}$ be representatives of the conjugacy classes of lattices in G. For each $i \in \mathbb{N}$, let $\{U_{i,j}\}_{j\in\mathbb{N}}$ be a sequence of H-invariant Borel sets in $\mathcal{H}_{\Lambda_i}$ which separates H-orbits. The sets $\mathcal{H}_{\Lambda_i}$ are Borel sets in $\mathcal{H}$ (since $\mathcal{H}_{\Lambda_i}$ is the inverse image of a point in $\mathcal{L}_G$ under the projection map $\mathcal{H} \to \mathcal{L}_G$), so their G-saturations are as well, and the sequence $\{GU_{i,j}\}_{i,j\in\mathbb{N}}$ separates orbits in $\mathcal{H}$.   □

Remark 3.7. It is possible to show, using results of Margulis, that there are in fact only countably many $G \times H$-orbits in $\mathcal{H}$.

LEMMA 3.8. *Let G be a connected semisimple Lie group with trivial center, no compact factors, and $\mathbb{R}$-rank ≥ 2. Let Λ be a lattice in G, and suppose that for each normal subgroup $L \subset G$ of $\mathbb{R}$-rank 1, $\Lambda \cap L = \{e\}$. Let Γ be a non-trivial normal subgroup of Λ. Then there is a non-trivial normal subgroup $M \subset G$ such that Γ is a lattice in M.*

Proof. There is a unique direct product decomposition $G = \prod_{i=1}^{k} G_i$ such that:

(1) for each $i = 1, \dots, k$, $G_i \cap \Lambda$ is an irreducible lattice in G_i,
(2) each G_i has $\mathbb{R}$-rank ≥ 2, and
(3) $\prod_{i=1}^{k}(G_i \cap \Lambda)$ is a subgroup of finite index in Λ.

Then $G_i \cap \Gamma$ is a normal subgroup of $G_i \cap \Lambda$, and by the theorem of Margulis on normal subgroups of lattices ([12]) it follows that either $G_i \cap \Gamma$ is central in G_i (and therefore trivial), or $G_i \cap \Gamma$ is of finite index in $G_i \cap \Lambda$. Let $i_1, \dots, i_l$ be the indices for which $G_i \cap \Gamma$ is non-trivial. Then

$$\Gamma_0 = \prod_{j=1}^{l}(G_{i_j} \cap \Gamma)$$

is a lattice in $M = \prod_{j=1}^{l} G_{i_j}$, and Γ_0 has finite index in Γ.

Let $p \colon G \to G/M$ be the projection. Then $p(\Gamma)$ is a finite subgroup of G/M normalized by $p(\Lambda)$. It follows that $p(\Gamma)$ is central, and therefore trivial, so $\Gamma \subset M$. $\square$

The proof of Theorem 2.1 would be an easy consequence of the theorem of Margulis on normal subgroups of lattices if it were true that for a semisimple Lie group G, the space S_G is a tame G-space. We devote the remainder of this section to showing that S_G is *not* a tame G-space.

LEMMA 3.9. *Let $\mathbb{F}_2$ be the free group on the generators $\{a,b\}$. Let S be the space of subgroups of $\mathbb{F}_2$ with the Fell topology. Then the action of $\mathbb{F}_2$ on S is not tame.*

Proof. By [24, 2.1.14], it suffices to show that the orbits of this action are not locally closed. Thus it suffices to find a subgroup $\Lambda \subset \mathbb{F}_2$ and a sequence $\{g_n\}$ such that the image of $\{g_n\}$ in $\mathbb{F}_2/N_{\mathbb{F}_2}(\Lambda)$ goes to infinity, and $g_n \Lambda g_n^{-1}$ converges to Λ in the Fell topology.

We define recursively a sequence of finite sets $S_i \subset \mathbb{F}_2$ by setting $S_0 = \{a\}$, and for $i > 0$,

$$S_i = b^{3^i}(\bigcup_{j=0}^{i-1} S_j)b^{-3^i}.$$

Let $S_\infty = \bigcup_{i=0}^\infty S_i$, and let Λ be the subgroup of $\mathbb{F}_2$ generated by S_∞. Let $\Lambda_n = b^{-3^n} \Lambda b^{3^n}$. We claim that Λ_n converges to Λ. From the definition of the Fell topology, it suffices to show:

(1) For any finite set $A \subset \Lambda$, $A \subset \Lambda_n$ for n sufficiently large.

(2) Given any finite set K in the complement of Λ, $\Lambda_n \cap K = \emptyset$ for n sufficiently large.

The group Λ_n is generated by

$$b^{-3^n} S_\infty b^{3^n} = b^{-3^n} (\bigcup_{i=0}^{n-1} S_i) b^{3^n} \cup (\bigcup_{i=0}^{n-1} S_i) \cup (\bigcup_{i>n} b^{-3^n} S_i b^{3^n}).$$

In particular, $\Lambda_n \cap \Lambda$ contains $\bigcup_{i=0}^{n-1} S_i$. If A is a finite set in Λ, then A is contained in the subgroup of Λ generated by $\bigcup_{i=0}^n S_i$, for some n, so (1) is verified.

It follows easily from the definition of the S_i that:

(i) Each element of $b^{-3^n} (\bigcup_{i=0}^{n-1} S_i) b^{3^n}$ is of the form $b^{-\alpha} a b^\alpha$, where

$$\frac{3^n + 3}{2} = 3^n - (\sum_{i=1}^{n-1} 3^i) \le \alpha \le 3^n.$$

(ii) Each element of $\bigcup_{i=0}^{n-1} S_i$ is of the form $b^\beta a b^{-\beta}$, where

$$0 \le \beta \le \sum_{i=1}^{n-1} 3^i = \frac{3^n - 3}{2}.$$

(iii) Each element of $\bigcup_{i>n} b^{-3^n} S_i b^{3^n}$ is of the form $b^\gamma a b^{-\gamma}$, where

$$\gamma \ge 3^{n+1} - 3^n = 2 \cdot 3^n.$$

Let $x \in \mathbb{F}_2$ be an element with reduced expression $x = b^{m_1} a^{n_1} \cdots b^{m_k} a^{n_k}$, where $n_i \ne 0$ for $i = 1, \ldots, k-1$. We define the *b-magnitude* of x to be $\sup\{|m_i|\}_{i=1,\ldots,k}$. Let K be a finite subset of the complement of Λ and suppose that $x \in \Lambda_n \cap K$. Since $x \notin \Lambda$, its reduced expression in the elements of $b^{-3^n} S_\infty b^{3^n}$ must contain an element of $b^{-3^n} (\bigcup_{i=0}^{n-1} S_i) b^{3^n}$ or an element of $\bigcup_{i>n} b^{-3^n} S_i b^{3^n}$. It follows from (i), (ii), and (iii), that the b-magnitude of x is at least

$$\inf\{\tfrac{3^n+3}{2}, 2 \cdot 3^n, 2 \cdot 3^n - (\tfrac{3^n+3}{2}), \tfrac{3^n+3}{2} + \tfrac{3^n-3}{2}\} = \tfrac{3^n+3}{2}.$$

Since the b-magnitude of the elements of K is bounded, this verifies (2).

To complete the example we must show that b^{-3^n} goes to infinity mod $N_{\mathbb{F}_2}(\Lambda)$, i.e., given any finite set $K \subset \mathbb{F}_2$,

$$b^{-3^n} \Lambda b^{3^n} \not\subseteq \bigcup_{k \in K} k \Lambda k^{-1}.$$

It is easy to check that the element $b^{-3^n} a b^{3^n} \in b^{-3^n} \Lambda b^{3^n}$ is not in $\bigcup_{k \in K} k \Lambda k^{-1}$ for n sufficiently large.   □

LEMMA 3.10. *Let G be a Lie group which contains a closed connected subgroup H locally isomorphic to* $\mathrm{SL}(2, \mathbb{R})$. *Then the action of G on $\mathcal{S}_G$ is not tame.*

Proof. Let H' be the adjoint group of H, and $\mathrm{Ad} \colon H \to H'$ the adjoint homomorphism. The group $H' \cong \mathrm{PSL}(2, \mathbb{R})$ has a natural structure as a real algebraic group, and we define the *Zariski topology* on H to be the pull-back to H of the Zariski topology on H'. Let $\Lambda' \subset H'$ be a discrete subgroup isomorphic to $\mathbb{F}_2$, and let a' and b' be generators for Λ'. Fix $a \in \mathrm{Ad}^{-1}(a')$ and $b \in \mathrm{Ad}^{-1}(b')$, and let Λ be the group generated by a and b. Then Λ is free on the generators a, b, and any non-abelian subgroup of Λ is Zariski dense in H. To see the last assertion, note that a proper algebraic subgroup of $\mathrm{PSL}(2, \mathbb{R})$ is solvable, and a discrete solvable subgroup of $\mathrm{PSL}(2, \mathbb{R})$ is abelian.

Let $\varphi \colon \mathcal{S}_\Lambda \to \mathcal{S}_G$ be the inclusion. Then $\varphi(\Lambda x) \subset G \varphi(x)$ for all $x \in \mathcal{S}_G$. To show that the G-action on $\mathcal{S}_G$ is not tame it suffices (see [24, p. 13]) to show:

(i) For each $x \in \mathcal{S}_G$, $\varphi(\mathcal{S}_\Lambda) \cap Gx$ is countable.

Statement (i) is equivalent to:

(ii) For each subgroup $\Gamma \subset \Lambda$, the set $\{g \Gamma g^{-1} \mid g \Gamma g^{-1} \subset \Lambda,\ g \in G\}$ is countable.

If $g H g^{-1} \cap H$ is a proper subgroup of H then it is solvable, and every solvable subgroup of Λ is abelian. Since $g \Gamma g^{-1} \subset g H g^{-1} \cap H$ and the set of abelian subgroups of Λ is countable, it suffices to show:

(iii) For each non-abelian subgroup $\Gamma \subset \Lambda$, the set $\{g \Gamma g^{-1} \mid g \Gamma g^{-1} \subset \Lambda,\ g \in N_G(H)\}$ is countable.

The group $N_G(H)$ acts by automorphisms of H, and since the subgroup of inner automorphisms of H has finite index in the full automorphism group, it suffices to show:

(iv) For each non-abelian subgroup $\Gamma \subset \Lambda$, the set $\{h \in H \mid h \Gamma h^{-1} \subset \Lambda\}$ is countable.

Let Γ be a non-abelian subgroup of Λ. Let x and y be non-commuting elements of Γ and let F be the free group they generate. Then F is Zariski dense in H, so two inner automorphisms of H which agree on F must agree on all of H. But there are only countably many finitely generated subgroups of Λ, so there can be at most countably many distinct inner automorphisms of H which carry F into Λ. This verifies (iv) and completes the proof of the lemma.   □

COROLLARY 3.11. *Let G be a non-compact semisimple Lie group. Then S_G is not a tame G-space.*

Proof. It is a standard fact from Lie theory that there is a homomorphism of the universal cover of $SL(2, \mathbb{R})$ to G (with closed image). The corollary then follows from the preceding lemma.   □

LEMMA 3.12. *Let G be a locally compact, second countable topological group. If S_G is tame then for every ergodic action of G with quasi-invariant measure, there is a closed subgroup $H \subset G$ such that the stabilizer of almost every point is conjugate to H.*

Proof. Suppose S is a tame G-space, and G acts ergodically with quasi-invariant measure on (X, μ). Then the image of the map $h: X \to S$, $h(x) = G_x$, is essentially contained in a single G-orbit, so almost every stabilizer is conjugate to a fixed subgroup of G.   □

There are many Lie groups G with the property that S_G is tame. For example, if G is either abelian, compact, or finitely generated nilpotent, then S_G is tame. It would be interesting to

(1) Classify those Lie and/or p-adic groups G for which the action of G on S_G is tame.

(2) Classify those groups G which have the property that every finite invariant ergodic measure on S_G is supported on a single orbit.

Note that Theorem 2.1 implies that if G is a simple Lie group with finite center and $\mathbb{R}$-rank ≥ 2, then every finite invariant ergodic measure on S_G is supported on a single orbit, in spite of the fact that G does not act tamely on S_G.

4. Reducible actions

In order to generalize Theorem 2.1 to reducible actions we are going to use an inductive argument which involves passing to the space of ergodic components of a normal subgroup. We begin this section by recalling briefly some properties of the ergodic decomposition of an action and then establishing some preliminary results to be used in the proof of Theorem 4.3.

Let G be a semisimple Lie group with trivial center acting ergodically on a space (X, μ) with finite invariant measure. Let H be a normal subgroup of G. Then there is a G-space (Y, ν) with finite invariant measure and a measure-preserving G-equivariant map $\varphi: X \to Y$ such that:

(1) If $\{\mu_y\}_{y \in Y}$ is a disintegration of μ with respect to ν, then for almost every $y \in Y$, μ_y is a finite measure on $\varphi^{-1}(y)$.

(2) H acts trivially on Y, and for almost every $y \in Y$, the H-action on $(\varphi^{-1}(y), \mu_y)$ is ergodic and measure-preserving.

We say that $\varphi \colon X \to Y$ is a decomposition of X into H-ergodic components, and call Y the space of H-ergodic components of X. Since H acts trivially on Y, the action of G on Y factors to an ergodic action of G/H. The decomposition of a space into H-ergodic components is unique in the sense that if $\varphi' \colon X \to Y'$ is a measurable, measure-preserving H-invariant map, where H acts trivially on Y', and the action of H on almost every fiber of φ' is ergodic, then there is a measurable G-equivariant isomorphism $h \colon Y \to Y'$ (of conull sets) such that $\varphi' = h \circ \varphi$. Moreover, the ergodic decomposition is natural in the following sense: If $f \colon X \to Z$ is a measure-preserving G-equivariant map, and $\psi \colon Z \to W$ is the decomposition of Z into H-ergodic components, then there is a G-equivariant, measurable, measure-class-preserving map $h \colon Y \to W$ such that $h \circ \varphi = \psi \circ f$. By uniqueness and naturality, it follows that if f is itself an ergodic decomposition for a subgroup $L \subset G$, and L and H commute, then h is a decomposition of Y into L-ergodic components.

Suppose now that the action of G on X is faithful. Let

$$H' = \{g \in G \mid gy = y \text{ for a.e. } y \in Y\}.$$

Then H' is a normal subgroup of G which acts trivially on Y, and since the action of G on X is faithful it is easy to see that for almost every $y \in Y$, the action of H' on $\varphi^{-1}(y)$ is faithful. Note that H' acts ergodically on almost every fiber of φ since $H \subset H'$.

Let $M = Z_G(H')$. Then $G = M \times H'$. The ergodicity of the G-action on X implies that M acts ergodically on Y. Note that M acts faithfully on Y since H' is the kernel of the G-action on Y.

Definition 4.1. Let G be a semisimple Lie group with trivial center and suppose that each simple factor of G either has $\mathbb{R}$-rank ≥ 2 or is locally isomorphic to $\mathrm{Sp}(1, n)$, $n \geq 2$, or $F_{4(-20)}$. Let (X, μ) be a G-space with finite invariant measure. We say that the action is *admissible* if it is faithful and ergodic, and for any normal subgroup $L \subset G$ of $\mathbb{R}$-rank 1, the following holds: If $\psi \colon X \to Z$ is a decomposition of X into L-ergodic components, then either

(i) L acts essentially transitively on $\psi^{-1}(z)$ for almost every $z \in Z$, or

(ii) there is a simple normal subgroup $N \subset G$ distinct from L such that N acts trivially on Z and for almost every $z \in Z$, N acts ergodically on $\psi^{-1}(z)$.

Note that if G does not contain any factor of $\mathbb{R}$-rank 1, then any ergodic, faithful, finite measure-preserving action of G is admissible. Note also that (i) is equivalent (by Lemma 3.5) to the statement: "for almost every $x \in X$, L_x is a lattice in L"; and (ii) of the definition is equivalent to the statement: "there

740   GARRETT STUCK AND ROBERT J. ZIMMER

is a simple normal subgroup $N \subset G$ distinct from L such that $\psi \colon X \to Z$ is also a decomposition of X into N-ergodic components."

LEMMA 4.2. *Let G be a connected semisimple Lie group with trivial center and suppose that each simple factor of G either has $\mathbb{R}$-rank ≥ 2 or is locally isomorphic to $\mathrm{Sp}\,(1,n)$, $n \geq 2$, or $F_{4(-20)}$. Let (X,μ) be an admissible G-space with finite invariant measure. Let H be a normal subgroup of G, $\varphi \colon X \to Y$ a decomposition of X into H-ergodic components, H' the kernel of the action of G on Y, and $M = Z_G(H)$. Then*
    (1) *for almost every $y \in Y$, the H'-action on $\varphi^{-1}(y)$ is admissible,*
    (2) *the M-action on Y is admissible, and*
    (3) *if $L \subset G$ is a simple factor of $\mathbb{R}$-rank 1 for which condition (ii) of Definition 4.1 holds, and if $L \subset M$, then the action of L on Y is essentially free.*

Proof. We have already verified that for almost every $y \in Y$, H' acts ergodically and faithfully on $\varphi^{-1}(y)$, and that M acts faithfully and ergodically on Y. Suppose L is a simple subgroup of G of $\mathbb{R}$-rank 1. Let $\psi \colon X \to Z$ be the space of L-ergodic components. Suppose that (i) of Definition 4.1 holds for the L-action on X. Then for almost every $x \in X$, L_x is a lattice in L. If $L \subset H'$ then (i) holds for the L-action on almost every fiber of φ. If $L \subset M$, then the inclusion $L_x \subset L_{\varphi(x)}$ implies that for almost every $y \in Y$, L_y is a lattice in L.

Suppose then that the L-action on X satisfies (ii) of Definition 4.1. Let N be a simple normal subgroup of G distinct from L such that ψ is an ergodic decomposition of N. If $L \subset H'$, then φ factors through ψ; i.e., there is a map $\eta \colon Z \to Y$ such that $\eta \circ \psi = \varphi$. Since N acts trivially on Z it follows that N acts trivially on Y, so $N \subset H'$. It follows that (ii) holds for the action of H' on $\varphi^{-1}(y)$, for almost every $y \in Y$, and this completes the proof of (1).

If $L \in M$ let $f \colon Y \to W$ be the decomposition of Y into L-ergodic components. Then there is an equivariant map $h \colon Z \to W$ such that the following diagram commutes:

$$
\begin{array}{ccc}
X & \xrightarrow{\ \psi\ } & Z \\
\downarrow{\scriptstyle \varphi} & & \downarrow{\scriptstyle h} \\
Y & \xrightarrow{\ f\ } & W.
\end{array}
$$

The fact that L is in M implies that the fibers of ψ are not contained in the fibers of φ, and since N acts ergodically on the fibers of ψ, it follows that N does not act trivially on Y, so $N \subset M$. Moreover, N acts trivially on W (since it acts trivially on Z). Let $g \colon Y \to U$ be a decomposition of Y into N-ergodic components. Since ψ is the N-ergodic decomposition of X, there is a map $i\colon Z \to U$ which makes the following diagram commute:

$$
\begin{array}{ccc}
X & \xrightarrow{\ \psi\ } & Z \\
\downarrow{\scriptstyle\varphi} & & \downarrow{\scriptstyle i} \\
Y & \xrightarrow{\ g\ } & U.
\end{array}
$$

By naturality and uniqueness, U is the space of H-ergodic components of Z. Again by uniqueness, U is G-equivariantly isomorphic to W. It follows that the N-ergodic components of Y coincide with the L-ergodic components of Y. This completes the proof of (2).

To prove (3), we maintain the notation of the preceding paragraph and note that the subgroup $L \times N \subset M$ acts irreducibly on almost every fiber of f. By Theorem 2.1, the action of $L \times N$ on almost every fiber of f is either essentially transitive, or essentially free. In the former case, the stabilizers for the action of $L \times N$ are irreducible lattices in $L \times N$, which intersect L trivially.   □

THEOREM 4.3. *Let G be a connected semisimple Lie group with finite center, and suppose that each simple factor of G either has $\mathbb{R}$-rank at least two or is locally isomorphic to $\mathrm{Sp}\,(1,n)$, $n \geq 2$, or $F_{4(-20)}$. Suppose G acts admissibly on (X,μ) with finite invariant measure. Then there is an almost direct product decomposition $G = HM$ and a lattice $\Lambda \subset M$ such that for almost every $x \in X$, G_x is conjugate to Λ.*

Proof. We first reduce to the case of trivial center. Let $Z(G)$ be the center of G. We showed in the proof of Theorem 2.1 that faithfulness of the action implies that $Z(G)$ acts essentially freely on X. Let $G' = G/Z(G)$, $X' = X/G$, and let $p\colon G \to G'$ and $q\colon X \to X'$ be the natural maps. If Theorem 4.3 holds in the center-free case, either the action of G' is essentially free, or there are a proper, non-trivial normal subgroup $M' \subset G'$ and a lattice $\Lambda' \to M'$ such that G'_x is conjugate to Λ' for almost every $x \in X'$. The essential freeness of the action of $Z(G)$ on X implies that for almost every $x \in X$, $p\colon G_x \to G'_{q(x)}$ is an isomorphism. It is now a straightforward exercise in group theory (which we leave to the reader) to verify that the assertion of the theorem holds for the action of G on X.

We suppose then that G has trivial center. We argue by induction on the number of simple factors in G. Suppose G is simple. If the action is properly ergodic, then the result follows from Theorem 2.1. If the action is essentially transitive, the result follows from Lemma 3.5.

Suppose the theorem is true for semisimple groups with $n - 1$ or fewer simple factors, and suppose G has n simple factors. The proof divides into two cases according to whether the following condition holds:

(∗) For every proper normal subgroup $H \subset G$, $H_x = H \cap G_x$ is trivial for almost every $x \in X$.

Suppose first of all that (∗) does not hold, i.e., for some normal subgroup $H \subset G$, $H_x \neq \{e\}$ for all x in a set of positive measure. Choose an H which is minimal with respect to this property. We are going to show that H_x is a lattice in H for almost every $x \in X$.

Let $\varphi \colon X \to Y$ be a map defining a decomposition of X into H-ergodic components. We claim that for almost every $y \in Y$, the action of H on $\varphi^{-1}(y)$ is admissible. Suppose that $L \subset H$ is a normal subgroup of $\mathbb{R}$-rank 1. Let $\psi \colon X \to Z$ be a decomposition into L-ergodic components. Then there is a map $\eta \colon Z \to Y$ such that $\eta \circ \psi = \varphi$. By hypothesis, either (i) of the definition holds, in which case $L = H$ and there is nothing further to show, or there is a simple normal subgroup $N \subset G$, $N \neq L$, such that ψ is a decomposition of X into N-ergodic components. The group N acts trivially on Z and hence also on Y. Suppose N is not in H. Then the map $f \colon X \to \mathcal{S}_H$, $f(x) = H_x$, is N-invariant, and therefore constant on each N-ergodic component. Thus f is constant on each L-ergodic component, and in particular on each L-orbit. It follows that for almost every $x \in X$, H_x is normalized by L. Thus for almost every $x \in X$, the projection of H_x into L is normal in L, and therefore is either trivial or all of L. The set of points where the former occurs is an invariant Borel set, and is therefore either null or conull. It cannot be null, because for almost every $x \in X$, H_x is discrete ([9, Lemma 6]). On the other hand it cannot be conull, since then for almost every $x \in X$, H_x is contained in a proper normal subgroup of H, contradicting the minimality of H. We conclude that $N \subset H$, so the H action on almost every fiber of φ is admissible.

Now we can apply the inductive hypothesis to conclude that for almost every $y \in Y$, there are a decomposition $H_y = H_y^1 \times H_y^2$ and a lattice $\Lambda_y \subset H_y^2$ such that H_x is conjugate to Λ_y for almost every $x \in \varphi^{-1}(y)$. This gives a measurable map $f \colon X \to \mathcal{Q}_H$. By Corollary 3.3, f is essentially constant; i.e., there are a decomposition $H = H_1 \times H_2$ and a lattice $\Lambda \subset H_2$ such that for almost every $x \in X$, H_x is conjugate to Λ. On account of the minimality of H we must have $H = H_2$; i.e., Λ is a lattice in H and for almost every $x \in X$, H_x is conjugate to Λ. It follows from Lemma 3.5 that for almost every $y \in Y$, H acts essentially transitively on $\varphi^{-1}(y)$.

Let $M = Z_G(H)$. Then M acts ergodically on Y. We claim that M acts faithfully on Y. Indeed, for each $y \in Y$, M_y acts on $\varphi^{-1}(y)$, and the action of M_y commutes with the action of H. On the other hand, any measurable transformation of H/Λ which commutes with the H-action is essentially of the form $h\Lambda \to hu\Lambda$, where $u \in N_G(\Lambda)$. It follows that for almost every $y \in Y$ the action of M_y on $\varphi^{-1}(y)$ is given by a homomorphism $M_y \to N_G(\Lambda)$. In particular, if L is a connected subgroup of M which pointwise fixes almost every point in Y then L acts essentially trivially on all of X. Since any normal subgroup of M is connected it follows from the faithfulness of the G-action that M acts faithfully on Y.

We may now apply the inductive hypothesis (by Lemma 4.2) to the action of M on Y to conclude that there are a decomposition $M = M_1 \times M_2$ and a lattice $\Gamma \subset M_2$ such that for almost every $y \in Y$, M_y is conjugate to Γ. Let $p\colon G \to M$ be the projection. Then for every $x \in X$, $p(G_x) \subset M_{\varphi(x)}$. We claim that for almost every $x \in X$, $p(G_x) = M_{\varphi(x)}$. Fix $y \in Y$ such that M_y is conjugate to Γ and H acts essentially transitively on $\varphi^{-1}(y)$. Fix $x \in \varphi^{-1}(y)$ such that Hx is conull in $\varphi^{-1}(y)$. Then H_x is a lattice in H, and M_y acts on H/H_x; for $m \in M$ the action of m is given by $m \cdot hH_x = hu_m H_x$, for some $u_m \in N_H(H_x)$. In particular, $(u_m^{-1}, m) \cdot x = u_m^{-1} u_m x$, so $(u_m^{-1}, m) \in G_x$. It follows that $M_y \subset p(G_x)$. This verifies the claim.

To finish the proof of the theorem in the case that $(*)$ fails to hold, it suffices to show that for almost every $x \in X$, G_x is a lattice in $H \times M_2$. For then it follows from ergodicity and Corollary 3.3 that for almost every $x \in X$, G_x is conjugate to a fixed lattice $\Delta \subset H \times M_2$.

Let $q\colon G \to H$ be the natural map. The group H_x is a normal subgroup of G_x, and $q(H_x) = H_x$, so $q(G_x) \subset N_H(H_x)$. Fix $x \in X$ with the property that H_x is a lattice in H and $M_{\varphi(x)}$ is a lattice in M_2. Then

$$[q(G_x) : H_x] \le [N_H(H_x) : H_x] < \infty.$$

Let $G_x^0 = q^{-1}(H_x)$ and $M_{\varphi(x)}^0 = p(G_x^0)$. Then $[G_x : G_x^0] < \infty$ and $[M_{\varphi(x)} : M_{\varphi(x)}^0] < \infty$. We have the split exact sequence

$$\{e\} \to H_x \to G_x^0 \to M_{\varphi(x)}^0 \to \{e\}$$

so G_x^0 is isomorphic to $H_x \times M_{\varphi(x)}^0$, which is a lattice in $H \times M_2$. It follows that G_x is a lattice in $H \times M_2$. This completes the proof of the theorem if $(*)$ fails.

Suppose now that $(*)$ holds. If G acts irreducibly on X the theorem follows from Theorem 2.1. We assume therefore that it does not, and let H_0 be a simple normal subgroup of G which does not act ergodically on X. Let $\varphi\colon X \to Y$ be a map defining a decomposition of X into H_0-ergodic components. Let H be the kernel of the G-action on Y and let $M = Z_G(H)$. Let $p\colon G \to M$ be the projection. Then for almost every $x \in X$, $p\colon G_x \to p(G_x)$ is an isomorphism because $\ker p_x = H \cap G_x = \{e\}$ for almost every $x \in X$. Note that $p(G_x) \subset M_{\varphi(x)}$.

By Lemma 4.2, we may apply the inductive hypothesis to conclude that there is a decomposition $M = M_1 \times M_2$ with a lattice $\Lambda \subset M_2$ such that M_y

is conjugate to Λ for almost every $y \in Y$. The fact that $p(G_x) \subset M_2$ implies that $G_x \subset H \times M_2$ for almost every $x \in X$. We conclude from $(*)$ that either $M_2 = \{e\}$, and $G_x = \{e\}$ for almost every $x \in X$, or $M_2 = M$. We suppose the latter holds and argue to obtain a contradiction. Note that in this case, Y is equivariantly isomorphic to M/Λ by Lemma 3.5.

We claim that for almost every $x \in X$, $p(G_x)$ is an irreducible lattice in M. Note first of all that because of $(*)$, if $L \subset M$ is a simple normal subgroup of $\mathbb{R}$-rank 1, then condition (i) of Definition 4.1 does not hold for L. Thus by (3) of Lemma 4.2, L acts essentially freely on Y. Thus Λ satisfies the hypothesis of Lemma 3.8, and because $(*)$ holds, to prove the claim it suffices to show for almost every $x \in X$ that $p(G_x)$ is a normal subgroup of $M_{\varphi(x)}$. Define $h_y \colon \varphi^{-1}(y) \to \mathcal{S}_M$ by $h_y(x) = p(G_x)$. Then h_y is H-invariant, so by ergodicity it is essentially constant on $\varphi^{-1}(y)$ for almost every $y \in Y$. Thus for almost every $y \in Y$, $p(G_x)$ is constant on almost every M_y-orbit in $\varphi^{-1}(y)$. But for each $m \in M_y$, $mp(G_x)m^{-1} = p(mG_xm^{-1}) = p(G_{mx})$, and the claim follows.

Let X_0 be the G-invariant conull set of points $x \in X$ for which $p(G_x)$ is an irreducible lattice in M. Let $\mathcal{H}$ be the space of homomorphisms of irreducible lattices in M to H. Identifying an element of $\mathcal{H}$ with its graph, we get a map $f \colon X_0 \to \mathcal{H}$ defined by $f(x) = G_x$. This is a Borel G-map, and the action of $G = M \times H$ on $\mathcal{H}$ is tame by Lemma 3.6, so f essentially takes values in a single G-orbit of $\mathcal{H}$. Thus there are an irreducible lattice $\Lambda \in M$, and a homomorphism $\varphi \colon \Lambda \to H$, such that f is essentially equivalent to a G-map $F \colon X_0 \to G/N_G(\mathrm{graph}(\varphi))$. We deduce moreover from $(*)$ that φ is injective and its image is not contained in any proper normal subgroup of H.

The finite invariant measure μ on X_0 pushes down to a finite invariant measure on $G/N_G(\mathrm{graph}(\varphi))$. We will force a contradiction by showing that $G/N_G(\mathrm{graph}(\varphi))$ does not admit a finite invariant measure. Let $\Phi = N_G(\mathrm{graph}(\varphi))$. Then $p(\Phi)$ normalizes $p(\Lambda)$, so $p(\Phi)$ is an irreducible lattice in M. Moreover, if $x \in \Phi \cap M$, then for all $\lambda \in \Lambda$, $\varphi(x\lambda x^{-1}) = \varphi(\lambda)$. It follows from the injectivity of φ that x centralizes λ, so $\Phi \cap M = \{e\}$. Thus Φ is itself the graph of a homomorphism of $p(\Phi)$ into H. In particular, Φ is discrete and isomorphic to $p(\Phi)$. Since $G/N_G(\mathrm{graph}(\varphi))$ has finite invariant measure, we conclude that Φ is a lattice in G. It is an easy consequence of super-rigidity that Φ cannot be both an irreducible lattice in M and a lattice in $M \times H$. We have obtained a contradiction, which implies that G_x is trivial for almost every x. $\square$

COROLLARY 4.4. *Let G be a connected semisimple Lie group with finite center, and suppose that each simple factor of G has $\mathbb{R}$-rank at least two. Let Λ*

be an irreducible lattice in G. *Suppose* Λ *acts faithfully and properly ergodically with finite invariant measure on* (X,μ). *Then the action is essentially free.*

COROLLARY 4.5. *Let* G *be a connected semisimple Lie group with finite center, and suppose that each simple factor of* G *has* $\mathbb{R}$*-rank at least two. Let* Λ *be a lattice in* G. *Suppose* Λ *acts faithfully and properly ergodically with finite invariant measure on* (X,μ). *Then there is a subgroup* $\Gamma \subset \Lambda$ *such that* Γ *is a lattice in a normal subgroup of* G, *and for almost every* $x \in X$, Λ_x *is conjugate to* Γ.

Remark 4.6. The hypothesis of admissibility in Theorem 4.3 is not the weakest condition under which the theorem remains true, but the proof becomes much more complicated when one substitutes weaker conditions.

5. Complements

5.1. *Volume-preserving actions with codimension one orbits.* It is natural to ask whether the conclusions of Theorem 2.1 can be strengthened in the presence of some additional regularity for the group action. In Example 2.3 we showed that even for smooth, essentially free actions, the set of points with non-trivial stabilizer can be rather large. It would be interesting to know whether Theorem 2.1 can be strengthened to show, say, that for a smooth, properly ergodic, volume-preserving action of a higher rank group on a compact manifold, the set of points with non-trivial stabilizer is of first category. Although we have nothing to say about this question in general, we do have the following special result:

THEOREM 5.1. *Let* G *be a connected semisimple Lie group with finite center and suppose that each simple factor of* G *is either of* $\mathbb{R}$*-rank* ≥ 2 *or is locally isomorphic to* $\mathrm{Sp}(1,n)$ *or* $F_{4(-20)}$. *Suppose* G *acts by* C^2 *diffeomorphisms of a compact manifold* M *preserving a finite Borel measure. If all the orbits have codimension one, then there are a cocompact lattice* $\Gamma \subset G$, *and a finite cover* $\hat{M}$ *of* M *such that* $\hat{M}$ *is* G*-equivariantly diffeomorphic to* $G/\Gamma \times S^1$.

Proof. By [14, Lemma 9.1], the G-invariant Borel measure on M yields a transverse invariant measure for the foliation defined by orbits of the action. By [14, Theorem 6.3], any leaf in the support of a transverse invariant measure has polynomial growth. But [17, Theorem B] implies that any leaf of polynomial growth has finite volume and is therefore compact (because the injectivity radius of a leaf is bounded below). Thus there are a cocompact lattice

$\Gamma \subset G$ and a point $m \in M$ such that Gm is diffeomorphic to G/Γ. Now by a straightforward argument using the Reeb-Thurston stability theorem ([18]) and the fact that $H^1(\Gamma, \mathbb{R}) = 0$, one obtains the theorem. $\square$

Theorem 5.1 was in fact the starting point for this paper. Although the present proof does not involve the main results of this paper, our original more complicated proof pointed us to one of the observations crucial for the proof of Theorem 2.1.

5.2. *Orbit equivalence.* There are a number of results in the ergodic theory of semisimple groups of higher rank which are predicated on the essential freeness of the action. In light of Theorem 2.1, the hypothesis of essential freeness can now be removed from many of these results. In particular, we have the following improvement of (a special case of) a result in [21].

THEOREM 5.2. *Suppose G and G' are connected semisimple Lie groups with finite center and no compact factors. Suppose S (resp. S') is a faithful, ergodic, irreducible, G (resp. G') space with finite invariant measure, and assume that the actions are orbit equivalent. Assume that $\mathbb{R}$-rank$(G) \geq 2$ and that each simple factor of G and G' either has $\mathbb{R}$-rank ≥ 2 or is locally isomorphic to $\mathrm{Sp}\,(1,n)$, $n \geq 2$, or $F_{4(-20)}$. Then*

(1) *G and G' are locally isomorphic.*

(2) *In the center-free case, $G \cong G'$, and if G and G' are identified via this isomorphism, the actions of G on S and S' are isomorphic.*

For other examples of results which can be improved using Theorem 2.1 and Theorem 4.3 we refer the reader to [24] and the references therein.

5.3. *Extension to p-adic and S-arithmetic groups.* Let V be the set {primes in $\mathbb{Z}$} $\cup$ {∞}, and let $S \subset V$ be a finite subset. Let $\mathbb{Q}_\infty = \mathbb{R}$. Suppose that for each $p \in S$, G_p is a connected semisimple $\mathbb{Q}_p$ group such that $G_p(\mathbb{Q}_p)$ has no compact factors. Let $G = \prod_{p \in S} G_p(\mathbb{Q}_p)$. Suppose that G has Kazhdan's property and that $\sum_{p \in S} \mathbb{Q}_p$-rank$(G_p) \geq 2$. It is likely that the proof of Theorem 2.1 applies practically verbatim to show that any faithful, irreducible, and properly ergodic action of G with finite invariant measure is essentially free. There is also probably a version of Theorem 4.3 for such groups, and therefore a version for S-arithmetic groups. The one reservation we make is that Theorem 4.1 of [23] is proved only for real Lie groups, but it should be possible to extend that result to p-adic groups using the same techniques as in the proof for real groups (cf. [13, IV.2.11]).

UNIVERSITY OF MARYLAND, COLLEGE PARK, MD
UNIVERSITY OF CHICAGO, CHICAGO, IL

STABILIZERS FOR ERGODIC ACTIONS   747

REFERENCES

[1] S. ADAMS and G. STUCK, Splitting of non-negatively curved leaves in minimal sets of foliations, Duke Math. J. **71**(1993), 71–92.

[2] A. BOREL, Density properties for certain subgroups of semisimple groups without compact factors, Ann. of Math. **72**(1960), 179–188.

[3] A. BOREL and J-P. SERRE, Théorème de finitude en cohomologie galoisienne, Comment. Math. Helv. **39**(1964), 111–164.

[4] A. CONNES, J. FELDMAN and B. WEISS, An amenable equivalence relation is generated by a single transformation, Erg. Theory & Dyn. Sys. **1**(1981), 431–450.

[5] J. FELDMAN and P. HAHN and C.C. MOORE, Orbit structure and countable sections for actions of continuous groups, Adv. Math. **28**(1978), 186–230.

[6] J.M.G. FELL, A Hausdorff topology for the closed subsets of a locally compact non-Hausdorff space, Proc. A.M.S. **13**(1962), 472–476.

[7] J. GLIMM, Locally compact transformation groups, Trans. A.M.S. **101**(1961), 124–138.

[8] M. GROMOV, Hyperbolic groups, in *Essays in Group Theory*, S.M. Gersten, editor, M.S.R.I. Publ., vol. 8, Springer, 1987, 75–263.

[9] A. IOZZI, Equivariant maps and purely atomic spectrum, J. Funct. Anal., to appear.

[10] F.E.A. JOHNSON, On the existence of irreducible discrete subgroups in isotypic Lie groups of classical type, Proc. London Math. Soc. **56**(1988), 51–77.

[11] D. KAZHDAN, Connection of the dual space of a group with the structure of its closed subgroups, Funct. Anal. Appl. **1**(1967), 63–65.

[12] G.A. MARGULIS, Quotient groups of discrete subgroups and measure theory, Funct. Anal. Appl. **12**(1978), 295–305.

[13] _____, *Discrete Subgroups of Semisimple Lie Groups*, Springer, 1991, New York.

[14] J. PLANTE, Foliations with measure preserving holonomy, Ann. of Math. **102**(1975), 327–361.

[15] M.S. RAGHUNATHAN, *Discrete Subgroups of Lie Groups*, Springer, 1972, New York.

[16] A. RAMSAY, Virtual groups and group actions, Adv. Math. **6**(1971), 253–322.

[17] G. STUCK, Growth of homogeneous spaces, density of discrete subgroups and Kazhdan's property (T), Invent. Math. **109**(1992), 505–517.

[18] W. THURSTON, A generalization of the Reeb stability theorem, Topology **13**(1974), 347–352.

[19] R.J. ZIMMER, Cocycles and the structure of ergodic group actions, Israel J. Math. **26**(1977), 214–220.

[20] _____, Hyperfinite factors and amenable ergodic actions, Invent. Math. **41**(1977), 23–31.

[21] _____, Strong rigidity for ergodic actions of semisimple Lie groups, Ann. of Math. **112** (1980), 511–529.

[22] _____, Semisimple automorphism groups of G-structures, J. Diff. Geom. **19**(1984), 117–123.

[23] _____, Ergodic theory, semisimple Lie groups, and foliations by manifolds of negative curvature, I.H.E.S. Publ. Math. **55**(1982), 37–62.

[24] _____, *Ergodic Theory and Semisimple Groups*, Birkhäuser, 1984, Boston.

(Received August 17, 1992)

6

Representations and Arithmetic Properties of Actions, Fundamental Groups, and Foliations

JOURNAL OF THE
AMERICAN MATHEMATICAL SOCIETY
Volume 1, Number 1, January 1988

ARITHMETICITY OF HOLONOMY GROUPS
OF LIE FOLIATIONS

ROBERT J. ZIMMER

1. INTRODUCTION

The point of this paper is to prove arithmeticity of a class of finitely generated groups that arise naturally in a purely geometric context. Namely, we prove arithmeticity of the holonomy group of a Lie foliation of a compact manifold, assuming the existence of a metric for which the leaves are symmetric spaces of nonpositive curvature and rank at least 2. This provides a partial answer to the question raised a number of years ago by A. Haefliger [13] of determining the possible holonomy groups of Lie foliations.

To indicate the nature of the questions we are considering, let us first recall that if N is a Riemannian manifold, a fundamental question is to understand the relationship of the topology of N and the geometry of N (or for many purposes, the geometry of $\tilde{N}$). There has been a wide variety of work on this type of question from a variety of viewpoints. In particular, we can ask how the geometry of N (or $\tilde{N}$) controls the nature of $\pi_1(N)$. One can ask a more general question, namely to understand to what extent the geometry of $\tilde{N}$ controls the actions of $\pi_1(N)$ on a manifold T. A standard construction associates to every such action a foliation of a manifold M which has all leaves locally isometric to $\tilde{N}$, in which T appears naturally as a submanifold of M transverse to the foliation, and the $\pi_1(N)$-action on T controls the way in which the leaves of this foliation are "tied together". We can thus reinterpret our last question, and broaden its context as well, by formulating the following question: Determine the extent to which the geometry of a simply connected Riemannian manifold X controls the way in which the leaves of a foliation are tied together if we assume the leaves are all locally isometric to X. In general, it is of course difficult to understand actions of a group on a manifold, and correspondingly difficult to understand foliations with leaves locally isometric to a given X. It is natural to begin an investigation of such actions with the case in which the action on T preserves some geometric structure, e.g., a Riemannian metric. In terms of the associated foliation, this becomes a special case of the

---

Received by the editors October 26, 1986.

1980 *Mathematics Subject Classification* (1985 *Revision*). Primary 22E40, 28D15, 53C12, 57R30.

Research partially supported by NSF Grant DMS-8301882.

notion of a "Riemannian foliation" (described in more detail below) in which
the leaves are tied together in a way that preserves a metric. In this situation,
generalizing the case of a foliation coming from an action of a fundamental
group, there is a discrete group, called the holonomy group, with an embedding
in a connected Lie group, that controls the way in which the leaves are tied
together. The nature of the results we prove is that under certain assumptions
on the geometry of X, this "tying together", i.e., the holonomy group together
with an embedding in a connected Lie group, is of a very specific arithmetic
nature.

   To be more precise, we first recall the notion of a Lie foliation. Let $\mathscr{F}$ be a
foliation of a compact connected manifold M. If $W \subset M$ is a sufficiently small
open set, the foliation on W is given by the fibers of a submersion $p: W \to U$.
If $p_i: W_i \to U_i$ are two such submersions, we obtain a diffeomorphism (or
homeomorphism if the foliation is only assumed to be continuous)

$$f_{ij}: p_i(W_i \cap W_j) \to p_j(W_i \cap W_j).$$

If X is a manifold and G is an effective transformation group of X, we say
that $\mathscr{F}$ is a (G, X)-foliation if we can choose a covering $\{W_i\}$ of M such
that $U_i \subset X$ is open and each f_{ij} is the restriction of an element of G. This
notion, and related notions, have been studied in [2, 7, 13, 21, 24], for example.
In particular, if X is a Riemannian manifold and G is a group of isometries,
$\mathscr{F}$ is called a Riemannian foliation, and if $X = G$, a connected Lie group,
then $\mathscr{F}$ is called a Lie foliation, or more precisely, a G-foliation. These cases
have been studied in [4, 8, 9, 12, 17, 22], and elsewhere. By passing to a
transverse orthogonal frame bundle, the study of Riemannian foliations can in
most respects be reduced to the study of Lie foliations (see, e.g., [17]). It is clear
that if G and G' are locally isomorphic, a G-foliation is also a G'-foliation.
Thus, one can speak more properly of a $\mathfrak{g}$-foliation, where $\mathfrak{g}$ is a Lie algebra.

   For any G-foliation, there is a natural homomorphism $h: \pi_1(M) \to G$,
the holonomy homomorphism, and a locally trivial fibration $D: \widetilde{M} \to G$ (the
developing map) which commutes with the action of $\pi_1(M)$ (where $\pi_1(M)$
acts on G via translation by $h(\gamma)$, $\gamma \in \pi_1(M)$). Here $\widetilde{M}$ is the covering
of M corresponding to the normal subgroup $\ker(h)$. Furthermore, the leaves
of $\mathscr{F}$ are exactly the images of the fibers of D under the covering projection
$\widetilde{M} \to M$. The group $h(\pi_1(M)) = \Gamma \subset G$ is called the holonomy group of the
G-foliation. Various general results on Lie foliations allow reduction to the case
of foliations with a dense leaf [17]. In this case, the holonomy group Γ will
be dense in G. For the construction of h and D, see [8, 24]. If the fibers of
D are connected, then the leaves of $\mathscr{F}$ are in natural 1-1 correspondence with
the Γ-orbits in G. We then say that $\mathscr{F}$ has good development over G. If G
is simply connected, this will always be the case, but it may or may not be true
for other groups locally isomorphic to G.

   Suppose now that G is a connected semisimple Lie group with finite center
and that $\Gamma \subset G$ is a dense subgroup. Then Γ is called arithmetic if there is a

semisimple algebraic **Q**-group H and a smooth surjection $p: H^0_{\mathbf{R}} \to \mathrm{Ad}_G(G)$ such that $p(H^0_{\mathbf{R}} \cap H_{\mathbf{Z}})$ and $\mathrm{Ad}_G(\Gamma)$ are commensurable (cf. the definition of arithmeticity for discrete subgroups [28]). In light of Margulis' arithmeticity theorem [15, 28], this is equivalent to the condition that there is a semisimple Lie group G' and an irreducible lattice $\Lambda \subset \mathrm{Ad}_G(G) \times G'$ such that $\mathrm{Ad}_G(\Gamma)$ is commensurable with the projection of Λ into $\mathrm{Ad}_G(G)$. While this latter formulation is perhaps more natural in a geometric context (although it carries less information without Margulis' theorem), we wish to emphasize the arithmetic formulation here. This is because we shall show that a suitable holonomy group is arithmetic not by first showing it is a lattice and then applying Margulis' theorem, but rather by showing via other methods that it is naturally a subgroup of a group of the form $H_{\mathbf{Z}}$ where H is a **Q**-group and then using further arguments to show that it must be of finite index in this group. Thus, in our proof, the fact that the group is a lattice will follow after we establish its arithmetic nature, rather than vice versa by an application of Margulis' theorem.

We recall that any symmetric space Y of nonpositive curvature has a (essentially unique) de Rham decomposition $Y = \Pi Y_i$, where each Y_i is an irreducible symmetric space of nonpositive curvature. We let

$$d(Y) = \min\{\dim \mathrm{Isom}(Y_i)\}$$

where $\mathrm{Isom}(Y_i)$ is the group of isometries of Y_i (and is a noncompact simple Lie group if Y_i is not Euclidean).

We can now formulate a precise version of our main results.

Theorem A. *Let* $\mathfrak{g}$ *be a (real, finite dimensional) Lie algebra and* $\mathscr{F}$ *a* $\mathfrak{g}$-*foliation of a compact connected manifold* M. *Suppose there is a dense, simply connected leaf. Assume there is a Riemannian metric on* M *such that each leaf is a locally symmetric space of nonpositive curvature such that all irreducible factors in the de Rham decomposition of the covering symmetric space* Y *have rank at least* 2. *Then:*

(1) $\mathfrak{g}$ *is semisimple.*

(2) *There is a (connected) Lie group* G *with finite center (and Lie algebra* $\mathfrak{g}$) *such that* $\mathscr{F}$ *has good development over* G.

(3) *The holonomy group* $\Gamma \subset G$ *is a dense arithmetic subgroup.*

(4) $\mathrm{Ad}_G(\Gamma)$ *is commensurable with the projection into* $\mathrm{Ad}_G(G)$ *of a lattice in* $\mathrm{Ad}_G(G) \times \mathrm{Isom}(Y)$.

(5) $\mathrm{codim}(\mathscr{F}) = \dim \mathfrak{g} \geq d(Y)$.

As indicated above, the hypothesis that $\mathscr{F}$ has a dense leaf is not a serious restriction. It is possible that Theorem A remains true without the assumption that there is such a leaf which is simply connected. In this direction, our arguments show:

Theorem B. *Assume all hypotheses of Theorem* A, *except the hypothesis that there is a simply connected leaf. (We still assume there is a dense leaf.) Then* $\mathrm{codim} \mathscr{F} \geq d(Y)/2$.

Theorem B implies of course the nonexistence in low codimension of Lie foliations with a dense leaf and leaves locally isometric to Y. Via the reduction of Riemannian foliations to Lie foliations alluded to above (see [17]), we obtain the following consequence.

Corollary C. *Let $\mathscr{F}$ be a Riemannian foliation of a compact manifold M with codimension d. Assume there is a metric on M such that the leaves of $\mathscr{F}$ are locally isometric to a symmetric space Y all of whose irreducible factors have rank at least 2. If $d(d+1) < d(Y)$, then all leaves are compact. Moreover, there is a finite covering $\widetilde{M}$ of M such that the lifted foliation $\widetilde{\mathscr{F}}$ on $\widetilde{M}$ is the foliation defined by a fiber bundle projection.*

A priori, the holonomy group of a Lie foliation is a finitely generated subgroup of a Lie group. Haefliger [13] raised the general problem of understanding which such groups may arise this way. The general nature of Theorem A is of course that it is a theorem relating the geometry of the leaves to the holonomy embedding $\Gamma \hookrightarrow G$. There are other such results. Namely, in [4] it is shown that for a Lie algebra $\mathfrak{g}$, a $\mathfrak{g}$-foliation has leaves of polynomial growth if and only if $\mathfrak{g}$ is nilpotent, and that the leaves are "Følner" (i.e. satisfy a certain isoperimetric inequality) if and only if $\mathfrak{g}$ is solvable. In [8], $(O(1,n), H^n)$-foliations (H^n = hyperbolic n space) are studied in case $\dim \mathscr{F} = 1$, i.e. the leaves are locally isometric to the line.

As we remarked above, our proof of Theorem A does not employ Margulis' arithmeticity theorem, but the proofs do have certain features in common. Namely, the basic step in the proof of Margulis' theorem is his "superrigidity" theorem. In [26] (see also [27, 28]) we generalized Margulis' superrigidity theorem to obtain "superrigidity for cocycles", a result which gives detailed measure-theoretic information about actions of semisimple groups on vector (or principal) bundles. This theorem will in turn play a basic role in the proof of the present arithmeticity theorem (i.e. Theorem A).

Although the hypotheses of Theorem A are of a geometric and topological nature, the more difficult and central part of the argument in the proof involves the relationship of algebraic groups and ergodic theory. In fact, we shall basically deduce Theorem A from the following result in ergodic theory. We refer the reader to [10, 11] and §3 below for the notion of measurable stable orbit equivalence.

Theorem D. *Let G be a connected Lie group and $\Gamma \subset G$ a dense finitely generated subgroup. Let H be a connected semisimple adjoint Lie group, each of whose simple factors has $\mathbf{R}$-rank at least 2. Suppose the action of Γ on G (by translations) is measurably stably orbit equivalent to an essentially free action of H on a (standard) measure space preserving a finite measure. Then:*

(1) *G is semisimple.*
(2) *$\Gamma \cap Z(G)$ is of finite index in $Z(G)$.*
(3) *$\mathrm{Ad}_G(\Gamma)$ is a dense arithmetic subgroup of $\mathrm{Ad}_G(G)$.*

(4) $\mathrm{Ad}_G(\Gamma)$ *is commensurable with the projection into* $\mathrm{Ad}_G(G)$ *of a lattice in* $\mathrm{Ad}_G(G) \times H$.

We remark that the properties of the measurable equivalence relation defined by the action of a dense subgroup on a Lie group also play a role in the results of [4] described above on $\mathfrak{g}$-foliations with $\mathfrak{g}$ solvable. Namely, in [31] we showed that if Γ is a dense subgroup of G, where G is a connected Lie group, then the Γ action on G is amenable if and only if G is solvable. This was applied in [4] in the proof that for a $\mathfrak{g}$-foliation, the leaves are Following if and only if $\mathfrak{g}$ is solvable.

§§2–10 of this paper are devoted to the proof of the results stated above. In §11, we show how one can obtain results similar to Corollary C for foliations in which the leaves are locally isometric to certain rank one symmetric spaces of negative curvature, namely those whose isometry group is a Kazhdan group. This is based on a result proved in §11, concerning stable orbit equivalence of the action defined by a dense subgroup of a Lie group with a finite measure preserving action of a Kazhdan group.

I would like to thank Y. Carriere and E. Ghys for communicating their papers [4, 5] which stimulated my interest in some of these matters.

2. Preliminaries on foliations

We recall from the introduction the following definition.

Definition 2.1. Let $\mathfrak{g}$ be a (real, finite dimensional) Lie algebra and G a connected group with Lie algebra $\mathfrak{g}$. Let $\mathscr{F}$ be a $\mathfrak{g}$-foliation of a compact, connected manifold M, $h\colon \pi_1(M) \to G$ the holonomy homomorphism, and $D\colon \widetilde{M} \to G$ the developing map, where $\widetilde{M}$ is the covering of M corresponding to $\ker(h)$. We say that $\mathscr{F}$ has good development over G if the fibers of D are connected.

We then have the following easily verified assertions.

Proposition 2.2. (a) *If* $G_1 \to G_2$ *is a covering and* $\mathscr{F}$ *has good development over* G_2, *it has good development over* G_1.
   (b) *If* G *is simply connected,* $\mathscr{F}$ *has good development over* G.

We let $M/\mathscr{F}$ denote the (in general non-Hausdorff) space of leaves.

Proposition 2.3. *If* $\mathscr{F}$ *has good development over* G, *with* $h(\pi_1(M)) = \Gamma$, *then the* $\pi_1(M)$-map $D\colon \widetilde{M} \to G$ *induces a bijection of* $M/\mathscr{F}$ *with* G/Γ.
   (*We remark that in general,* Γ *is not closed in* G.)

It will be convenient to express this bijection in an alternate manner.

Proposition 2.4. *If* $\mathscr{F}$ *has good development over* G, *then there is a Borel isomorphism* θ *of* G *with a Borel subset* $T \subset M$ *such that:*

   (a) T *intersects every leaf at least once, and in at most a countable set.*

(b) *If* $x, y \in T$, *then* x *and* y *are in the same leaf if and only if* $\theta^{-1}x$ *and* $\theta^{-1}y$ *are in the same* Γ-*orbit in* G.

Proof. We can choose a Borel section of the Γ-map $D \colon \widetilde{M} \to G$, and with some care the image of this section will map injectively under projection to M. Let T be this image in M.

We shall need the following observation about foliations by locally symmetric spaces.

Proposition 2.5. *Let* $\mathscr{F}$ *be a foliation of a compact manifold* M. *Assume each leaf is locally isometric to a symmetric space* Y *of nonpositive curvature. Let* H *be the connected component of the identity of the isometry group of* Y. *Fix* $y_0 \in Y$ *and let* K *be the stabilizer of* y_0 *in* H, *so that* K *is a compact Lie group. Then there is a principal* K-*bundle* $\rho \colon M^* \to M$ *with a smooth* H-*action on* M^* *such that the* H-*orbits in* M^* *are precisely the inverse images under* ρ *of the leaves of* $\mathscr{F}$. *Further, for* $x \in M^*$, *the stabilizer* H_x *is isomorphic to the fundamental group of the leaf of* $\mathscr{F}$ *through* $\rho(x)$. *If* $\mathscr{F}$ *is a Riemannian or Lie foliation (or more generally if there is a transverse invariant measure of smooth class) then there is a finite* H-*invariant measure on* M *of smooth class.*

Proof. Let M^* be the set of smooth maps $q \colon Y \to M$ such that q is a Riemannian covering of a leaf of $\mathscr{F}$. If $h \in H$, clearly $q \circ h \in M^*$, so that H acts on M^*. Fix a point $y_0 \in Y$ and define $\rho \colon M^* \to M$ by $\rho(q) = q(y_0)$. Then all assertions of the proposition may be verified in a routine manner.

3. Preliminaries on Ergodic Theory

In this section we review a number of ergodic theoretic notions that we will need for the proofs of the main theorems. There will be a number of statements in which equality should be interpreted as equality almost everywhere. We may occasionally be careful in this direction, but for the most part this distinction causes only routine problems and we will usually ignore them. In subsequent sections there will be a significant number of known results we shall use freely, providing references but not in general discussing them as background material. The reader is referred to [28] for a general discussion of the ideas involved.

We begin by recalling the notion of orbit equivalence and stable orbit equivalence. We refer the reader to [10, 11, 20, 27, 28] and the references therein for more detail. Suppose G and H are locally compact groups acting on (standard) measure spaces (X, μ) and (Y, ν) respectively. The measures μ and ν are assumed to be σ-finite and quasi-invariant. The actions are called orbit equivalent if (possibly after discarding null sets) there is a measurable and measure class preserving bijection $\theta \colon X \to Y$ such that θ takes G-orbits onto H-orbits; more precisely, for (a.e.) $x \in X$, $\theta(xG) = \theta(x)H$. A somewhat weaker notion, whose measure theoretic properties were examined in detail in

[10, 11, 20] (although in a slightly different context in [20]) is the notion of stable orbit equivalence.

Proposition 3.1 [11]. *The following are equivalent.*

(i) *The action of $G \times S^1$ on $X \times S^1$ is orbit equivalent to the action of $H \times S^1$ on $Y \times S^1$ (where S^1 is the circle).*

(ii) *After discarding invariant null sets, there is a measurable map $\theta: X \to Y$ such that θ induces a null set preserving bijection $X/G \to Y/H$. (We remark that we do not assume θ itself to be either injective or surjective.)*

Definition 3.2. If the conclusions of 3.1 are satisfied, the G-action on X and the H-action on Y are called stably orbit equivalent.

We say that an action has "continuous orbits" if (almost) every orbit is uncountable.

Proposition 3.3 [11]. (a) *If two actions are stably orbit equivalent, one is ergodic if and only if the other is ergodic.*

(b) *Actions with continuous orbits are stably orbit equivalent if and only if they are orbit equivalent.*

(c) *If the G-action on X has continuous orbits and H is a countable group acting on Y, then the actions are stably orbit equivalent if and only if (after discarding invariant null sets) there is a Borel isomorphism θ of Y with a Borel subset $T \subset X$ such that θ induces a null set preserving bijection $Y/H \to X/G$.*

Example 3.4. Suppose $\mathscr{F}$ is a foliation of a compact manifold M and that the leaves are locally homogeneous spaces with a common universal cover $\tilde{L}$. Let H be the connected component of the identity of the isometry group of $\tilde{L}$. Suppose $\mathscr{F}$ is a $\mathfrak{g}$-foliation with good development over G. Let $\Gamma \subset G$ be the holonomy group. Then the action of H on M^* (where M^* is as in Proposition 2.5) is stably orbit equivalent to the Γ-action on G.

Proof. This follows from 2.4, its proof, 2.5, and 3.3.

A basic property of stable orbit equivalence is that it induces isomorphism on cohomology. Before stating this in the form we need, we recall some facts about cocycles (see [28] as a general reference).

We recall that if G acts on (the right of) X and L is a locally compact group, a Borel function $\alpha: X \times G \to L$ is called a cocycle if for all $g, h \in G$, $\alpha(x, gh) = \alpha(x, g)\alpha(xg, h)$ (a.e. x). Two such cocycles α, β are called equivalent or cohomologous (and we write $\alpha \sim \beta$) if there is a measurable $\phi: X \to L$ such that for each $g \in G$, $\phi(x)\alpha(x, g)\phi(xg)^{-1} = \beta(x, g)$ (a.e.). We denote by $Z^1((X, G); L)$ the space of cocycles and by $H^1((X, G); L)$ the set of cohomology classes of cocycles. If $L_1 \subset L$ is a subgroup, it is often of interest to know when α is cohomologous to a cocycle taking all values in L_1. If Y is an L-space, a function $\phi: X \to Y$ is called α-invariant if $\phi(x)\alpha(x, g) = \phi(xg)$. The following is basically formal.

Proposition 3.5. *The following are equivalent:*

(i) α *is equivalent to a cocycle taking all values in* L_1;

(ii) *there is a measurable* α-*invariant* $\phi: X \to L/L_1$.

Using this, one can define the important notion of the algebraic hull of a cocycle taking values in an algebraic group. (See [28, 30] for a proof and discussion of the following result.)

Proposition 3.6. *Let* k *be a local field of characteristic* $0, L$ *an algebraic* k-*group, and* L_k *the group of* k-*points. Suppose* G *is a locally compact group acting ergodically on* X, *and that* $\alpha: X \times G \to L_k$ *is a cocycle. Then there is an algebraic* k-*group* $H \subset L$ *with the following properties.*

(i) α *is equivalent to a cocycle taking all its values in* H_k.

(ii) *For any algebraic* k-*group* H' *with* $H'_k \subset H_k$ *a proper inclusion,* α *is not equivalent to a cocycle taking all its values in* H'_k.

(iii) *Up to conjugacy in* L_k, H_k *is the unique such group satisfying* (i), (ii).

(iv) *If* α *is equivalent to a cocycle taking all its values in some algebraic group* $H'_k \subset L_k$, *then some conjugate of* H_k *in* L_k *is contained in* H'_k.

Definition 3.7 [28]. The group H_k (or more precisely its conjugacy class in L_k) is called the algebraic hull of α. The algebraic hull is an invariant of the cohomology class defined by α. We say that α is Zariski dense in L_k if the algebraic hull is L_k itself.

Example 3.8 [28]. Suppose G is a semisimple connected real algebraic group with no compact factors, and X is a G-space on which G acts ergodically with finite invariant measure. Let $\alpha(x, g) = g$, so that $\alpha: X \times G \to G$ is a cocycle. If H is a closed subgroup, then α is equivalent to a cocycle taking all values in H if and only if there is a G-map $X \to G/H$. If H is algebraic, the Borel density theorem [3, 28] implies $H = G$. Thus, α is Zariski dense in G.

In general, an ergodic action of a real algebraic group G is called Zariski dense in G if the cocycle $\alpha(x, g) = g$ is Zariski dense. This is equivalent to the assertion that there is no measurable G-map $X \to G/H$ where $H \subset G$ is proper algebraic.

A cocycle $\alpha: X \times G \to L$ is called orbital if for (almost) all $x \in X$, $\alpha(x, g) = e$ for all $g \in G_x$, the latter being the stabilizer of x in G. If $\alpha \sim \beta$, then α is orbital if and only if β is orbital. We let $Z^1_{\text{orb}}((X, G); L)$ (resp. $H^1_{\text{orb}}((X, G); L)$) denote the set of (resp., equivalence classes of) orbital cocycles.

Lemma 3.9. *Suppose* G *acts on* (X, μ). *Let* $p: X \times S^1 \to X$ *be projection. Then the map*

$$p^*: Z^1_{\text{orb}}((X, G); L) \to Z^1_{\text{orb}}((X \times S^1, G \times S^1); L)$$

given by $(p^*\alpha)((x,s),(g,z)) = \alpha(x,g)$ induces a bijection on orbital cohomology. The inverse is the map on cohomology induced by $i: X \to X \times S^1$, $i(x) = (x,e)$, i.e. is given by $(i^*\beta)(x,g) = \beta((x,e),(g,e))$. If the G-action on X is ergodic, and L is (the k-points of) an algebraic group, then the algebraic hull of α is equal to the algebraic hull of $p^*\alpha$.

Proof. All but the last assertion follows from [11, 20]. The last assertion is easily verified.

Corollary 3.10. *If (X,G) and (Y,H) are stably orbit equivalent ergodic actions, then there is a bijection $H^1_{orb}((X,G);L) \to H^1_{orb}((Y,H);L)$ preserving algebraic hulls.*

Example 3.11. Consider the situation in Proposition 3.3(c). Suppose further that the G-action on X is ergodic and essentially free (i.e. almost all stabilizers are trivial). We then have a cocycle $\alpha: Y \times H \to G$ characterized by $\theta(y)\alpha(y,h) = \theta(yh)$ (cf. [28, 4.2.8]). Under the isomorphism of Corollary 3.10, this corresponds to the cocycle $\beta: X \times G \to G$ given by $\beta(x,g) = g$. In particular, if G is semisimple with no compact factors, and the measure on X is finite and invariant, then α is Zariski dense in G by Example 3.8.

If G is a connected semisimple Lie group for which all simple factors have **R**-rank at least 2, then we have very explicit information on the cocycles with a simple algebraic hull. If $\pi: G \to L$ is a continuous homomorphism, let $\alpha_\pi(x,g) = \pi(g)$, so that $\alpha_\pi: X \times G \to L$ is a cocycle. The following result is fundamental for the proofs of Theorems A, B, and D.

Theorem 3.12 [26, 27, 28] (Superrigidity for cocycles). *Let G be a connected semisimple Lie group with finite center such that every simple factor of G has **R**-rank at least 2. Let (X,μ) be an ergodic G-space with finite invariant measure. Let $\alpha: X \times G \to L$ be a cocycle.*

(a) *If L is a simple adjoint noncompact Lie group and α is Zariski dense in L, then there is a (smooth) homomorphism $\pi: G \to L$ (necessarily surjective) such that $\alpha \sim \alpha_\pi$.*

(b) *If L is a simple adjoint complex algebraic group, and α is Zariski dense in L, then either: (i) there is a smooth homomorphism $\pi: G \to L$ such that $\alpha \sim \alpha_\pi$; or (ii) $\alpha \sim \beta$ where β takes all values in a compact subgroup of L.*

(c) *If L is a group of k-points of a k-group where k is a totally disconnected local field of characteristic 0, then $\alpha \sim \beta$ where β takes all values in a compact subgroup of L.*

This result is a generalization of Margulis' superrigidity theorem [15, 28]. Theorem 3.12 was first proven under slightly different hypotheses in [26] (see [27, 28] for the present hypotheses), where it was used to prove results about orbit equivalence of actions of semisimple groups. Here we shall need the following variant of the application given in [26].

Corollary 3.13. *Let G and X be as in 3.12 (and assume the G-action on X is effective). Let H be a connected semisimple adjoint Lie group with no compact factors, and Y an ergodic H-space which is essentially free. Suppose further that the action on Y is Zariski dense in H. (We do not assume, however, that the H-action has a finite invariant measure.) Suppose the G-action on X and H-action on Y are stably orbit equivalent. Then there is a surjective (smooth) local isomorphism $\pi: G \to H$ and a measure class preserving map $\psi: X \to Y$ such that $\psi(xg) = \psi(x)\pi(g)$. In particular, Y has a finite H-invariant measure (in the given measure class).*

Proof. This follows from the proof of [28, Theorem 5.2.1]. In this proof, only the Zariski density of the action of H and Y is used, not the fact that there is a finite invariant measure.

Another very useful result about cocycles for such G derives from the fact that G has Kazhdan's property.

Theorem 3.14 [28, Theorem 9.1.1]. *Assume G is a locally compact Kazhdan group (e.g., G is a connected semisimple Lie group with finite center for which every simple factor has **R**-rank at least 2). Let X be an ergodic G-space with a finite invariant measure, and $\alpha: X \times G \to L$ a cocycle, where L is an amenable locally compact group. Then α is equivalent to a cocycle taking all values in a compact subgroup of L.*

We can now indicate the relevance of these considerations for the proof of Theorems A and B. By Example 3.4, the holonomy action of Γ on G is stably orbit equivalent to a finite measure preserving ergodic action of a semisimple Lie group, each of whose simple factors has **R**-rank at least 2. Theorems 3.12 and 3.14 give us some control over the cohomology of the latter action and by Corollary 3.10, we have information on H^1 for the Γ-action on G. Each homomorphism of Γ in turn defines an element of H^1, and thus we can hope to derive information about homomorphisms of Γ. This, of course, has the potential of leading one to an arithmeticity theorem.

4. A HOMOMORPHISM EXTENSION THEOREM

In this section we describe a basic result on the extension of homomorphisms of dense subgroups to homomorphisms of the ambient group. The technique is basically due to Margulis and appears in approximately the following form in our discussion of Margulis' work in [28].

Theorem 4.1. *Let Λ be a dense subgroup of a locally compact group L. Let k be a local field of characteristic 0, M an (algebraically) connected k-group and $\pi: \Lambda \to M_k$ a homomorphism with $\pi(\Lambda)$ Zariski dense. Suppose there is a measurable Λ-map $\phi: L \to M_k/N_k$ where $N \subset M$ is a k-subgroup for which M_k acts effectively on M_k/N_k. Then π extends to a continuous homomorphism $L \to M_k$.*

Proof. This is implicit in the argument of [28, p. 128] but we shall include some details for completeness.

Let $F(L, M_k/N_k)$ denote the space of measurable maps $\phi: L \to M_k/N_k$, two such maps being identified if they agree almost everywhere. This space has the topology of convergence in measure on compact sets. The group M_k acts on $F(L, M_k/N_k)$ pointwise in the image, and L acts on this space by translation on the domain. Since $\Lambda \subset L$ is dense and M_k and N_k are algebraic, we can apply the argument of [28, p. 128]. This will show that there is a point $x_0 \in M_k/N_k$ and a measurable (and hence continuous) homomorphism $\sigma: L \to M_k$ such that for $h \in L$, $\phi(h) = x_0 \cdot \sigma(h)$, provided we can establish that $\{m \in M_k | x \cdot m = x \text{ for all } x \in \phi(L)\}$ is trivial. However, $\phi(L) \cdot \pi(\Lambda) \subset \phi(L)$ since ϕ is a Λ-map and hence the Zariski closure of $\phi(L)$ is $\pi(\Lambda)$-invariant (more precisely, the Zariski closure of the essential range of ϕ). Since $\pi(\Lambda)$ is Zariski dense in M, this implies $\phi(L)$ is Zariski dense. Thus, if $m \in M_k$ pointwise fixes $\phi(L)$, it pointwise fixes M_k/N_k which by our effectiveness assumption implies m is trivial. To complete the proof, it suffices to show that σ extends π. For $\gamma \in \Lambda$ and (a.e.) $h \in L$, we have

$$\phi(h\gamma) = x_0 \cdot \sigma(h\gamma) = x_0 \cdot \sigma(h)\sigma(\gamma) = \phi(h)\sigma(\gamma),$$

and $\phi(h\gamma) = \phi(h)\pi(\gamma)$. Therefore, for almost all h we have $\phi(h)\sigma(\gamma) = \phi(h)\pi(\gamma)$. Therefore $\sigma(\gamma)\pi(\gamma)^{-1}$ pointwise fixes the essential range of ϕ. As we remarked above, this set is Zariski dense, so $\sigma(\gamma)\pi(\gamma)^{-1}$ acts trivially on M_k/N_k, showing $\sigma(\gamma) = \pi(\gamma)$.

Suppose now that G is a connected Lie group. Then there is a unique maximal solvable normal subgroup $R \subset G$ such that G/R is a connected adjoint semisimple Lie group. Namely, let $p: G \to G/\operatorname{rad}(G)$ be the natural projection where $\operatorname{rad}(G)$ is the radical. Then let $R = p^{-1}(Z(G/\operatorname{rad}(G)))$. We then have the following corollary of Theorem 4.1.

Corollary 4.2. *Let G be a connected Lie group, R as above, $p: G \to G/R$. Let $\Gamma \subset G$ be a dense subgroup, and let M be a connected semisimple adjoint real or complex algebraic group. Suppose $N \subset M$ is an algebraic subgroup not containing a nontrivial normal subgroup of M. Let $\pi: \Gamma \to M$ be a homomorphism with $\pi(\Gamma)$ Zariski dense. If there is a measurable Γ-map $\phi: G \to M/N$, then there is a rational homomorphism $\sigma: G/R \to M$ such that $\pi = \sigma \circ (p|\Gamma)$.*

Proof. By Theorem 4.1, π extends to a continuous homomorphism $\sigma: G \to M$, and since $\pi(\Lambda)$ is Zariski dense in M and M is semisimple and adjoint, $\sigma|R$ is trivial.

For totally disconnected groups we obtain:

Corollary 4.3. *Let G be a connected Lie group, and $\Gamma \subset G$ a dense subgroup. Let k be a totally disconnected local field of characteristic 0, and M an (algebraically) connected simple adjoint k-group. Suppose $\pi: \Gamma \to M_k$ is* *a homomorphism with* $\pi(\Gamma)$ *Zariski dense. If* $N \subset M$ *is a proper* k-*subgroup, then there is no measurable* Γ-*map* $G \to M_k/N_k$.

5. FIRST RESULTS ON R

We now begin the proof of Theorems A, B, and D. To fix notation, we assume G is a connected Lie group, and $\Gamma \subset G$ a countable dense subgroup. Let $\text{rad}(G)$ be the radical and $R \supset \text{rad}(G)$ the maximal solvable normal subgroup with $G/R = L$ a connected semisimple adjoint Lie group (see §4). While R is solvable, it is not necessarily connected. We let H be a connected semisimple adjoint Lie group each of whose simple factors has **R**-rank at least 2. Let X be an ergodic, essentially locally free (i.e. almost all stabilizers are discrete) H-space with a finite invariant measure. We assume the actions of H on X and Γ on G are stably orbit equivalent. In this section we obtain some first results on R and $\Gamma \cap R$. We shall of course subsequently obtain much sharper information.

Lemma 5.1. $R \neq G$, *i.e.* L *is nontrivial.*

Proof. Suppose $R = G$. Since R is solvable, the action of Γ on R is amenable. It follows that there is a cocycle $R \times \Gamma \to \mathbf{R}$ whose algebraic hull is $\mathbf{R}$. (This can be deduced from [6], which shows that an amenable action is orbit equivalent to a **Z**-action, and the construction of cocycles on **Z**-actions in [23]. From the latter, one obtains a cocycle α for which the skew product action of Γ on $R \times \mathbf{R}$, given by $(r,t) \cdot \gamma = (r\gamma, t + \alpha(r, \gamma))$ is ergodic. This easily implies that the algebraic hull is $\mathbf{R}$.) By Corollary 3.10, there is a cocycle $\alpha: X \times H \to \mathbf{R}$ with algebraic hull $\mathbf{R}$. However, this contradicts Theorem 3.14.

Remark. The proof shows that the lemma is valid if H is any Kazhdan group, not necessarily a semisimple group with all simple factors of **R**-rank at least 2.

The next result we establish under the additional assumption that the H-action is essentially free.

Lemma 5.2. *Assume the* H-*action on* X *is essentially free. Then* $\Gamma \cap R$ *is discrete.*

Remark. This argument will apply for H any semisimple Lie group with no compact factors.

Proof. Let $R_1 = \overline{(\Gamma \cap R)}$. Since R is normal in G, Γ normalizes R_1, and since Γ is dense in G, $R_1 \subset G$ is normal. Let $E = G/R_1$ and $q: G \to E$ be the natural map. We remark that R_1 is solvable and hence that $\Gamma \cap R$ is an amenable dense subgroup of R_1.

Let $\alpha: G \times \Gamma \to H$ be the cocycle given by Example 3.11. Thus, we choose a Borel isomorphism $\theta: G \to T \subset X$ and define α by $\theta(g\gamma) = \theta(g)\alpha(g, \gamma)$. (The existence of α depends upon the assumption that the H-action is essentially free.) By the remarks in Example 3.11, α is Zariski dense in H. Let

$\alpha_1 = \alpha | G \times (\Gamma \cap R)$. Let $P \subset H$ be a minimal parabolic subgroup, and $M(H/P)$ the space of probability measures on H/P with the weak-$*$-topology. (We recall H/P is compact, and hence $M(H/P)$ is as well.) Since $\Gamma \cap R$ is amenable, the action of $\Gamma \cap R$ on G is amenable [28], and hence there is an α_1-invariant function $\phi : G \to M(H/P)$ (see §3, and for a complete discussion see [28, Chapter 4]). Since $\Gamma \cap R$ acts ergodically on R_1 and H act smoothly on $M(H/P)$ in the sense of [28] (i.e. all orbits are locally closed; see [28, Chapter 3] for a proof) we can apply the cocycle reduction lemma [28, 4.2.11]. This implies that there is an α_1-invariant function $\phi : R_1 \times (\Gamma \cap R) \to H/H_1$ where H_1 is the stabilizer of an element of $M(H/P)$. By a result of Moore [18] (see also [28, Chapter 3]), H_1 is an amenable algebraic group. Choose $A \subset H$ to be an amenable algebraic group of minimal dimension for which there is an α_1-invariant $\phi : R_1 \to H/A$. Since the $\Gamma \cap R$-actions on all the fibers of $q : G \to E$ are mutually isomorphic, we can find a measurable α_1-invariant $\phi : G \times (\Gamma \cap R) \to H/A$.

Now let $N(A^0)$ be the normalizer in H of the identity component of A, so that $N(A^0)$ is also an algebraic subgroup of H. Let $p : H/A \to H/N(A^0)$ be the natural map. The uniqueness argument of [28, 6.2.8], using the minimality of the dimension property of A, shows that if ϕ_1, ϕ_2 are both H/A-valued α_1-invariant functions on R_1, then $p \circ \phi_1 = p \circ \phi_2$. It follows that if $\phi_1, \phi_2 : G \to H/A$ are α_1-invariant, then $p \circ \phi_1 = p \circ \phi_2$. Since Γ normalizes $\Gamma \cap R$, if $\phi : G \to H/A$ is α_1-invariant, then $\phi \cdot \gamma$ will be α_1-invariant as well for any $\gamma \in \Gamma$. Here $\phi \cdot \gamma$ is given by $(\phi \cdot \gamma)(g) = \phi(g\gamma)\alpha(g,\gamma)^{-1}$. Thus, $p \circ \phi = p \circ (\phi \cdot \gamma)$. In other words, $p \circ \phi$ is actually α-invariant, not just α_1-invariant. We thus have an α-invariant map $G \to H/N(A^0)$. Since α is Zariski dense in G, this implies $N(A^0) = H$, and since A is amenable (and H has no compact factors), we have that A^0 is trivial, i.e. A is finite.

Thus, we may suppose $\alpha_1 = \alpha | G \times (\Gamma \cap R)$ is equivalent to a cocycle taking all values in a finite subgroup $A \subset H$. Let $\psi : G \to H$ be measurable with $\beta(g,\gamma) = \psi(g)\alpha(g,\gamma)\psi(g\gamma)^{-1} \in A$ for all $g \in G$, $\gamma \in \Gamma \cap R$. We recall that α satisfies $\theta(g\gamma) = \theta(g)\alpha(g,\gamma)$, where $\theta : G \to X$. Thus, if we define $\lambda : G \to X$ by $\lambda(g) = \theta(g) \cdot \psi(g)^{-1}$, we have $\lambda(g)\beta(g,\gamma) = \lambda(g\gamma)$. Thus, $[\lambda(g)] = [\lambda(g\gamma)]$ in X/A. Since $\Gamma \cap R$ is ergodic on the fibers of q, and X/A is a countably separated Borel space owing to the finiteness of A [28, 2.1.21], $[\lambda]$ is constant on the fibers of q, i.e., all $\lambda(g)$, $g \in R_1$, lie in the same A-orbit. Hence, all $\theta(g)$, $g \in R_1$, lie in the same H-orbit. However, by the choice of θ, this means $\Gamma \cap R$ acts transitively on R_1, i.e. $\Gamma \cap R = R_1$. Hence $\Gamma \cap R$ is closed, and therefore discrete.

Corollary 5.3. *If the H-action is essentially free, $\Gamma \cap R$ is central in G.*

Proof. By Lemma 5.2, it is discrete. Since it is normalized by Γ and Γ is dense in G, it is also normal in G.

Corollary 5.4. *To prove Theorems A and D, it suffices to assume $\Gamma \cap R$ is trivial.*

Proof. Let $G' = G/\Gamma \cap R$ and $\Gamma' = \Gamma/\Gamma \cap R$. Then $\Gamma' \subset G'$ is a dense subgroup, and the action of Γ' on G' is stably orbit equivalent to the action of Γ on G. In the case of Theorem A, if $\mathscr{F}$ has good development over G, and Γ is the holonomy group in G, then $\mathscr{F}$ will have good development over G' and Γ' will be the holonomy group in G'. It follows readily that if we obtain the desired results for G' and Γ', we have them for G and Γ as well.

6. Homomorphisms of Γ into simple Lie groups

We continue with the notation established at the beginning of §5. We let $p: G \to G/R = L$ be projection, and let $\Lambda = p(\Gamma)$. Thus, L is nontrivial (by Lemma 5.1) and $\Lambda \subset L$ is a countable dense subgroup of a connected semisimple adjoint Lie group. If Q is a locally compact group and $h: \Gamma \to Q$ is a homomorphism, we let $\alpha_h: G \times \Gamma \to Q$ be the cocycle $\alpha_h(g, \gamma) = h(\gamma)$. Our main goal in this section is Theorem 6.4, which asserts that if Q is a noncompact simple adjoint Lie group, then the number of homomorphisms of Γ into Q (modulo conjugacy in Q) with Zariski dense image is finite. Basic to proving this is understanding the relation between h and α_h.

We first need the following easy fact.

Lemma 6.1. *Suppose that for $i = 1, 2$, Q_i is a connected simple adjoint (real or complex) Lie group, and $h_i: \Gamma \to Q_i$ is a homomorphism with $h_i(\Gamma)$ Zariski dense in Q_i. Let $h = (h_1, h_2): \Gamma \to Q_1 \times Q_2$ and let Q be the algebraic hull of $h(\Gamma)$. Then either:*

(i) $Q = Q_1 \times Q_2$; *or*

(ii) Q^0 *is the graph of a smooth isomorphism* $\pi: Q_1 \to Q_2$, *and hence* $\pi \circ h_1 = h_2$.

Proof. Since Q_i is connected and $h_i(\Gamma)$ is Zariski dense, Q^0 projects onto both Q_1 and Q_2. Write $Q^0 = R \ltimes U$ where R is reductive and U is unipotent. Then the projection of U into Q_i is normal and unipotent, hence trivial. Similarly, the projection of the center $Z(R)$ onto Q_i is central and hence trivial. Thus, Q^0 is semisimple and adjoint. Hence, there is $\tilde{Q}_i \subset Q^0$ which is normal such that $\tilde{Q}_i$ maps isomorphically onto Q_i. If $\tilde{Q}_1 = \tilde{Q}_2$, we clearly have conclusion (ii). If not, $Q^0 \supset \tilde{Q}_1 \times \tilde{Q}_2$, and we clearly have assertion (i).

Theorem 6.2. *Let G, L, Γ be as above and Q a connected simple adjoint (real or complex) Lie group. For $i = 1, 2$, let $h_i: \Gamma \to Q$ be a homomorphism with $h_i(\Gamma)$ Zariski dense in Q. Suppose $\alpha_{h_1} \sim \alpha_{h_2}$. Then either:*

(a) *there is a continuous automorphism π of Q such that $\pi \circ h_1 = h_2$; or*

(b) *for $i = 1, 2$, h_i extends to a continuous homomorphism $\bar{h}_i: G \to Q$, and hence defines, in the real case, a rational surjection $\bar{h}_i: L \to Q$, and in the complex case a rational surjection $L_{\mathbb{C}} \to Q$.*

Proof. Let $Q_1 \subset Q \times Q$ be given by $Q_1 = \{(x,x)|x \in Q\}$, so that $Q_1 \subset Q \times Q$ is an algebraic subgroup. Then h_1 and h_2 are conjugate by an element of Q if and only if the homomorphism $h = (h_2, h_1): \Gamma \to Q \times Q$ is contained in a $Q \times Q$-conjugate of Q_1. Now assume $\alpha_{h_1} \sim \alpha_{h_2}$. Then there is a measurable map $\phi: G \to Q$ such that $\phi(g)h_1(\gamma)\phi(g\gamma)^{-1} = h_2(\gamma)$ for (almost) all $g \in G$, and all $\gamma \in \Gamma$. The group $Q \times Q$ acts on Q by $z \cdot (x,y) = x^{-1}zy$. This action is transitive and the stabilizers are precisely the conjugates of Q_1 in $Q \times Q$. The cohomology equation above can be rewritten as $\phi(g\gamma) = \phi(g) \cdot (h_2(\gamma), h_1(\gamma))$. In other words, $\phi: G \to Q$ is a Γ-map where Q is considered as the homogeneous space $Q = (Q \times Q)/Q_1$, and Γ acts on Q via h. By Lemma 6.1, if $h(\Gamma)$ is not Zariski dense in $Q \times Q$, we have conclusion (a). Thus, we may assume Zariski density. Since Q_1 contains no nontrivial normal subgroup of $Q \times Q$, we can apply Corollary 4.2 and deduce that h (and hence h_1 and h_2) extends continuously to G.

We shall apply this result to study $\mathrm{Hom}(\Gamma, Q)$. We first observe:

Lemma 6.3. *If Q is a connected simple adjoint (real or complex) Lie group, and $h: \Gamma \to Q$ is a homomorphism with $h(\Gamma)$ Zariski dense, then the following are equivalent:*

(1) *h does not extend to G; and*
(2) *α_h is Zariski dense in Q.*

Proof. If α_h is not Zariski dense, then there is an α_h invariant $\phi: G \to Q/Q_1$, where $Q_1 \subset Q$ is proper algebraic. This simply means that ϕ is a Γ-map. Since $h(\Gamma)$ is Zariski dense we can apply Corollary 4.2 to deduce that h extends to G. Conversely, if h extends, we have $h(g)\alpha_h(g,\gamma)h(g\gamma)^{-1} = e$, so that α_h has algebraic hull $\{e\}$.

Theorem 6.4 (with notation as in the beginning of §5). *For any noncompact, connected, simple, adjoint Lie group Q, the number of conjugacy classes of homomorphisms $\Gamma \to Q$ with Zariski dense image is finite. Furthermore, if there is such a homomorphism which does not extend to G, then Q is a factor of H.*

Proof. The homomorphisms $h: \Gamma \to Q$ with Zariski dense image fall into two groups: (I) those with α_h Zariski dense, and (II) those with α_h not Zariski dense. By Lemma 6.3, those in group (II) extend to continuous surjective homomorphisms $G \to Q$, and these factor to surjections $L \to Q$. Thus, there are only finitely many conjugacy classes in group (II). For those in group (I), we have from Theorem 3.12 and Corollary 3.10 that $\{\alpha_h | h \text{ in group (I)}\}$ forms only finitely many equivalence classes of cocycles, and that if this set is nonempty, then Q is a factor of H. By Theorem 6.2, this implies $\{h | h \text{ in group (I)}\}$ is finite modulo $\mathrm{Aut}(Q)$, and since Q is simple, it is finite modulo conjugacy by elements of Q.

7. HOMOMORPHISMS INTO TOTALLY DISCONNECTED SIMPLE GROUPS

Theorem 7.1. *Let* G, Γ *as above. Let* M *be a* k-*simple connected* k-*group, where* k *is a totally disconnected local field of characteristic* 0. *Then for any homomorphism* $\pi: \Gamma \to M_k$ *with* $\pi(\Gamma)$ *Zariski dense we have* $\overline{\pi(\Gamma)}$ *is compact.*

Proof. We may clearly assume M is an adjoint group. As above, let α_π be the cocycle $\alpha_\pi: G \times \Gamma \to M_k$ given by $\alpha_\pi(g, \gamma) = \pi(\gamma)$.

Lemma 7.2. α_π *is Zariski dense in* M_k.

Proof. If not, there is an α_π-invariant function $\phi: G \to M_k/N_k$ where $N_k \subset M_k$ is (the k-points of) a proper k-subgroup, i.e., ϕ is a Γ-map. However, this contradicts Corollary 4.3.

Continuing the proof of 7.1, we may now apply assertion (c) of Theorem 3.12 and Corollary 3.10. We deduce that α_π is equivalent to a cocycle taking all values in a compact subgroup of M_k. We are thus reduced to proving:

Lemma 7.3. *Let* k *be a local field of characteristic* 0, M *an (algebraically) connected* k-*simple adjoint* k-*group, and* $\pi: \Gamma \to M_k$ *a homomorphism with* $\pi(\Gamma)$ *Zariski dense. Suppose* α_π *is Zariski dense and that* α_π *is equivalent to a cocycle taking all values in a compact subgroup of* M_k. *Then* $\overline{\pi(\Gamma)}$ *is compact.*

Proof. Let $Q \subset M$ be a minimal parabolic k-subgroup. Thus, M_k/Q_k is compact. Let $K \subset M_k$ be a compact subgroup such that there is a measurable $\phi: G \to M_k$ for which $\phi(g)\alpha_\pi(g, \gamma)\phi(g\gamma)^{-1} \in K$. Let $M(M_k/Q_k)$ be the space of probability measures on M_k/Q_k with the weak-* topology. Since K is compact, there is a K-invariant $\mu \in M(M_k/Q_k)$. Let $\lambda: G \to M(M_k/Q_k)$ be defined by $\lambda(g) = \mu \cdot \phi(g)$, where we write the action of M_k on $M(M_k/Q_k)$ on the right. Then K-invariance of μ implies that λ is α_π-invariant. Equivalently, λ is a Γ-invariant element of the function space $F(G, M(M_k/Q_k))$, where, as usual, this is the space of measurable functions, two being identified if they agree almost everywhere. The Γ action is given by right translation on G and the action on $M(M_k/Q_k)$ given by π. The action of G on $F(G, M(M_k/Q_k))$ given by $(g \cdot f)(a) = f(g^{-1}a)$ commutes with the Γ-action. In particular, if we let $A = F(G, M(M_k/Q_k))^\Gamma$ be the space of Γ-invariants, then A is a compact convex set (with the weak-* topology as in [25]; see [25] for a full discussion) on which G acts linearly and continuously. Let $p: G \to L$ be the projection. Let $P \subset L$ be the product of the compact factors of L together with a minimal parabolic subgroup for each noncompact factor of L. (If L is compact, $P = L$. This is degenerate for the remainder of the argument, but the argument still applies.) Since P is amenable and $\ker p = R$ is solvable, $P^* = p^{-1}(P)$ is also amenable. Hence, there is a P^*-invariant element in A. Therefore, if we let $\tilde{\alpha}_\pi$ be the cocycle $\tilde{\alpha}_\pi: P \backslash L \times \Gamma \to M_k$ given by $\tilde{\alpha}_\pi(x, \gamma) = \pi(\gamma)$, such a P^*-invariant element of A will define a $\tilde{\alpha}_\pi$-invariant function $\psi: P \setminus L \to M(M_k/Q_k)$. The M_k action on $M(M_k/Q_k)$ has locally closed orbits and the stabilizers are either compact or contained in the k-points of a proper algebraic k-subgroup of M_k [28, Chapter 3]. Since Γ is dense in G (and hence in L), it acts ergodically on $P \setminus L$. Therefore, as in [28, p. 93], we deduce that either:

(a) there is a measurable Γ-map $\psi: P \setminus L \to M_k/N_k$ where N is a proper k-subgroup; or

(b) there is a measurable Γ-map $\psi: P \setminus L \to M_k/B$ where $B \subset M_k$ is a compact subgroup.

Case (a) is impossible since α_π is Zariski dense in M_k. In case (b), we observe that L has a conull orbit in $P \setminus L \times P \setminus L$, and hence Γ acts ergodically on $P \setminus L \times P \setminus L$. From [28, 5.9.1], we see that (b) implies that $\overline{\pi(\Gamma)}$ is compact.

8. HOMOMORPHISMS OF Γ INTO COMPLEX SIMPLE GROUPS

The arguments of §§6 and 7 now allow us to prove the following result about homomorphisms into complex groups.

Theorem 8.1. *Let Γ, G as above. Let Q be a connected simple adjoint algebraic group (over $\mathbf{C}$). Then the number of conjugacy classes of homomorphisms $h: \Gamma \to Q$ such that* (i) *$h(\Gamma)$ is Zariski dense in Q, and* (ii) *$\overline{h(\Gamma)}$ is not compact, is finite.*

Proof. As in the proof of Theorem 6.4, we consider the two groups of homomorphisms $h: \Gamma \to Q$ with $h(\Gamma)$ Zariski dense: (I), those h with α_h Zariski dense in Q; and (II), those with α_h not Zariski dense. By Lemma 6.3, those in group (II) extend to continuous homomorphisms $G \to Q$, and to continuous surjections $L_{\mathbf{C}} \to Q$. Thus, there are only finitely many conjugacy classes in group (II). For those in group (I) with the additional property that $\overline{h(\Gamma)}$ is not compact, we have from Lemma 7.3 that α_h is not equivalent to a cocycle into a compact subgroup of Q. Therefore, by Theorem 3.12(b), those $\{\alpha_h\}$ in group (I) with $\overline{h(\Gamma)}$ not compact form only finitely many equivalence classes of cocycles. By Theorem 6.2, we deduce that there are only finitely many conjugacy classes of such homomorphisms.

9. ARITHMETICITY IN THE SEMISIMPLE COMPONENT, AND THE PROOF OF THEOREM B

In this section we prove that Λ, the image of Γ in L, is a dense arithmetic subgroup of L, under the additional assumption that the H-action on X is essentially free. Along the way, without this additional assumption, we shall prove Theorem B. The first step in proving arithmeticity is standard, given the homomorphism results of §§6–8. We continue our notation from preceding sections. (We do not yet assume essential freeness of the H-action.)

Lemma 9.1. *There is a number field* k, $Q \subset k \subset \mathbf{R}$, $[k : \mathbf{Q}] < \infty$, *with ring of integers* $\mathcal{O}$, *a semisimple adjoint k-group* W, *and an isomorphism of Lie groups* $L \cong W_{\mathbf{R}}^0$ *such that for some subgroup* $\Lambda' \subset \Lambda$ *of finite index, we have* $\Lambda' \subset W_{\mathcal{O}}$.

Proof. We follow the general outline of [28, Chapter 6]. We first claim $\mathrm{Tr}(\mathrm{Ad}_L(\gamma)) \in \overline{\mathbf{Q}}$ for all $\gamma \in \Lambda$. For this, it clearly suffices to see $\mathrm{Tr}(\mathrm{Ad}_{L_1}(p_1(\gamma))) \in \overline{\mathbf{Q}}$ for all $\gamma \in \Lambda$, and any simple factor $p_1 : L \to L_1$ of L. We can write $L_1 = M_{\mathbf{R}}^0$ where $M \subset \mathrm{GL}(n, \mathbf{C})$ is an **R**-simple adjoint **Q**-group. For any $\sigma \in \mathrm{Gal}(\mathbf{C}/\mathbf{Q})$ we have a (discontinuous in general) automorphism of $\mathrm{GL}(n, \mathbf{C})$, and since M is a **Q**-group, we have an automorphism of M and $\mathrm{Ad}_M(M)$. We denote these automorphisms by σ as well. Let $h_\sigma : \Gamma \to M$ be the composition $\Gamma \to \Lambda \hookrightarrow L \to L_1 \to M \xrightarrow{\sigma} M$. Then $h_\sigma(\Gamma)$ is Zariski dense in M. Suppose $\mathrm{Tr}(\mathrm{Ad}_{L_1}(p_1(\gamma)))$ is transcendental for some $\gamma \in \Lambda$. Then $\{\sigma(\mathrm{Tr}(\mathrm{Ad}_{L_1}(p_1(\gamma)))) | \sigma \in \mathrm{Gal}(\mathbf{C}/\mathbf{Q})\}$ is dense in **C**. This set is of course identical to $\{\mathrm{Tr}(\mathrm{Ad}_M(h_\sigma(\gamma))) | \sigma \in \mathrm{Gal}(\mathbf{C}/\mathbf{Q})\}$ (where we lift γ back to an element of Γ). This implies that for infinitely many such σ, $h_\sigma(\Gamma)$ is not contained in a compact subgroup of M, and that among these, there are infinitely many nonconjugate h_σ. This contradicts Theorem 8.1.

Given that these traces are all algebraic, the lemma of Vinberg (see [28, 6.1.7]) implies that we can realize L as $L = W_{\mathbf{R}}^0$ where W is a $\overline{\mathbf{Q}} \cap \mathbf{R}$-group, and $\Lambda \subset W_{\overline{\mathbf{Q}} \cap \mathbf{R}}$.

Since Λ is finitely generated, we have $\Lambda \subset W_k$ for some $k \subset \mathbf{R}$ with $[k : \mathbf{Q}] < \infty$, and since Λ is Zariski dense, W is defined over k. Once again using the fact that Λ is finitely generated, we can find finitely many (finite) primes $\mathcal{P}_i$ of $\mathcal{O}$ such that, denoting by k_i (resp., $\mathcal{O}_i$) the completion of k (resp., $\mathcal{O}$) at $\mathcal{P}_i$, we have that the map $\Lambda / \Lambda \cap W_{\mathcal{O}} \to \prod_i W_{k_i} / W_{\mathcal{O}_i}$ is injective. Since k_i is totally disconnected, $W_{\mathcal{O}_i}$ is a compact open subgroup, and since the map $\Lambda \to W_{k_i}$ has Zariski dense image, it follows from Theorem 7.1 that for each i the image of Λ in $W_{k_i} / W_{\mathcal{O}_i}$ is finite. Therefore $\Lambda' = \Lambda \cap W_{\mathcal{O}}$ is of finite index in Λ. This proves the lemma.

We now consider the **Q**-group $R_{k/\mathbf{Q}}(W)$ obtained from W by restriction of scalars [28, Chapter 6]. We let $J = R_{k/\mathbf{Q}}(W)_{\mathbf{R}}^0$. Then J is a connected semisimple adjoint Lie group. The group L is a factor of J and we have a projection $r : J \to L$ such that $r(J_{\mathbf{Q}}) = W_k \cap L$, and $r(J_{\mathbf{Z}}) = W_{\mathcal{O}} \cap L$. There is also an isomorphism $\alpha : W_k \to R_{k/\mathbf{Q}}(W)_{\mathbf{Q}}$ such that $\alpha(W_{\mathcal{O}}) = R_{k/\mathbf{Q}}(W)_{\mathbf{Z}}$ (see [28] for details).

Let M^* be the algebraic hull of $\alpha(\Lambda')$ in $R_{k/\mathbf{Q}}(W)$. Since the projection of $\alpha(\Lambda')$ onto each simple factor of J is Zariski dense, the proof of Lemma 6.1 implies that M^* is semisimple and adjoint, and since $\alpha(\Lambda') \subset J_{\mathbf{Q}}$, M^* is a **Q**-group. Let $M = (M_{\mathbf{R}}^*)^0$. By passing to a normal subgroup $\Lambda'' \subset \Lambda$ of finite index, we can assume $\alpha(\Lambda'') \subset M_{\mathbf{Z}}^* \cap M$. Since $r(\alpha(\Lambda''))$ is Zariski dense in

L, we have that M projects onto L, so we can write $M = L \times L' \times K$ where L' is the product of the remaining noncompact simple factors of M, and K is the product of the remaining compact simple factors. Clearly the projection of $\alpha(\Lambda'')$ into $L \times L'$ is discrete. We choose a normal subgroup $L'' \subset L'$ such that the projection of $\alpha(\Lambda'')$ of $L \times L''$ is discrete, but the projection into $L \times L'''$ is not discrete for any proper normal subgroup $L''' \subset L''$. We remark that since Λ (and hence Λ'') is dense in L, L'' is not trivial. Recall the map $p \colon G \to L$, and let $\Gamma'' = p^1(\Lambda'')$, so that $\Gamma'' \subset \Gamma$ is normal and of finite index. We wish to apply our earlier results to the action of Γ'' on G. Therefore, we need:

Lemma 9.2. *The Γ'' action on G is stably orbit equivalent to an ergodic action of H preserving a finite invariant measure.*

Proof. Since G is connected, Γ'' is clearly dense in G. Let c be the cocycle of the Γ-action on G defined by the homomorphism $\lambda \colon \Gamma \to \Gamma/\Gamma''$, i.e. $c(g, \gamma) = \lambda(\gamma)$. By Corollary 3.10 and the discussion preceding it, c corresponds to an orbital cocycle $\tilde{c} \colon X \times H \to \Gamma/\Gamma''$. It is not hard to verify that the action of H on $X \times \Gamma/\Gamma''$ given by $(x, [\gamma]) \cdot h = (xh, [\gamma]\tilde{c}(x, h))$ is then stably orbit equivalent to the action of Γ on $G \times \Gamma/\Gamma''$. This latter action is in turn stably equivalent to the action of Γ'' on G. Finally, we remark that the action of H on $X \times \Gamma/\Gamma''$ is ergodic as ergodicity is an invariant of stable orbit equivalence.

Proof of Theorem B. Let Q be an **R**-simple factor of L''. Then Q must project injectively into some factor W^σ of $R_{k/\mathbf{Q}}(W) = \prod_\sigma W^\sigma$ where σ runs through the distinct embeddings of k in **C**. Thus, $\dim_\mathbf{R} Q \leq \dim_\mathbf{R} W^\sigma = 2 \dim_\mathbf{R} L$. Now consider the homomorphism $\pi_Q \colon \Gamma'' \to Q$ which is the composition of $p \circ \alpha$ with projection onto Q. This has Zariski dense image. Furthermore, π_Q does not extend to a continuous homomorphism of G, for otherwise such a homomorphism would define a rational homomorphism $\overline{\pi}_Q \colon L \to Q$, and projection of $\alpha(\Lambda'')$ onto $L \times Q$ would be contained in the graph of $\overline{\pi}_Q$. This would contract Zariski density of $\alpha(\Lambda'')$ in M. It now follows from Lemma 6.3 that the cocycle $\alpha_{\pi_Q} \colon G \times \Gamma'' \to Q$ is Zariski dense in Q. By Corollary 3.10, Lemma 9.2, and Theorem 3.12, we deduce that there is a rational surjection $H \to Q$. In particular, $\dim Q \geq d(H) = \min\{\dim H' | H'$ is a simple factor of $H\}$. Therefore, $\dim L \geq d(H)/2$, completing the proof of Theorem B.

We now return to the situation in Theorems A and D. We shall henceforth assume, by virtue of Corollary 5.4, that $\Gamma \cap R$ is trivial. Otherwise, we retain all the notation of this section.

Let $q \colon M \to L \times L''$ be projection. We can write the homomorphism $q \circ \alpha \colon \Lambda'' \to L \times L''$ as $(q \circ \alpha)(\gamma) = (\gamma, \alpha'(\gamma))$ where $\alpha' \colon \Lambda'' \to L''$ is a homomorphism. Define $\beta \colon \Gamma'' \to G \times L''$ by $\beta(\gamma) = (\gamma, \alpha'(p(\gamma)))$. Then β is an injective homomorphism and the projection of $\beta(\Gamma'')$ onto $L \times L''$ is simply $q(\alpha(\Lambda''))$.

In particular, this projection is discrete and since $\beta(\Gamma'') \cap \ker(G \times L'' \to L \times L'')$ is $(\Gamma'' \cap R) \times \{e\}$ and hence trivial, we have that $\beta(\Gamma'') \subset G \times L''$ is discrete as well.

Lemma 9.3. *The L''-action on $(G \times L'')/\beta(\Gamma'')$ is stably orbit equivalent to an ergodic action of H preserving a finite invariant measure. Furthermore, this action of L'' is Zariski dense in L''.*

Proof. For the first assertion, we begin by observing that the L'' action on $(G \times L'')/\beta(\Gamma'')$ is stably orbit equivalent to the $\beta(\Gamma'')$ action on $(G \times L'')/L''$. The projection of $G \times L''$ onto G takes $\beta(\Gamma'')$ injectively onto Γ'', so the action in question is stably orbit equivalent to the action of Γ'' on G. Thus, the first assertion of Lemma 9.3 follows from Lemma 9.2.

We now turn to the second assertion of Lemma 9.3. Suppose there is a measurable L''-map $\phi: (G \times L'')/\beta(\Gamma'') \to L''/T$ where $T \subset L''$ is a proper algebraic subgroup. Choose $T \subset L''$ to be of minimal dimension among those algebraic subgroups for which there exists such a map. By the uniqueness argument of [28, 6.2.8] (cf. the proof of Lemma 5.2), if ϕ_1, ϕ_2 are two such maps, then $\omega \circ \phi_1 = \omega \circ \phi_2$ where $\omega: L''/T \to L''/N_{L''}(T^0)$ is projection. Since G commutes with L'', $g \circ \phi$ is also an L''-map, and hence $\omega \circ (g \cdot \phi) = \omega \circ \phi$. Therefore, $\omega \circ \phi$ is a G-map as well as an L''-map. We deduce that $\beta(\Gamma'') \subset G \times N_{L''}(T^0)$. Since the projection of $\beta(\Gamma'')$ into L'' is Zariski dense, it follows that $T^0 \subset L''$ is normal. Write $L'' = T^0 \times T_1$ where $T_1 \subset L''$ is also normal. Then T_1 acts with closed orbits and finite stabilizers on L''/T. (Recall $T^0 \subset T$ is of finite index.) It follows that (perhaps by passing to a conull set) the action of T_1 on $(G \times L'')/\beta(\Gamma'')$ is smooth in the sense of [28, Chapter 2], and hence that the $\beta(\Gamma'')$ action on $(G \times L'')/T_1$ is smooth, i.e., the $\beta(\Gamma'')$ action on $G \times T^0$ is smooth. Since $\beta(\Gamma'')$ is a countable subgroup of $G \times T^0$, this implies that the image of $\beta(\Gamma'')$ in $G \times T^0$ is discrete. Since $R \subset G$ is a solvable normal subgroup, this implies that the image of $\beta(\Gamma'')$ in $G/R \times T^0 = L \times T^0$ is also discrete. (Namely, discreteness of $\beta(\Gamma'')$ implies that the action of $\beta(\Gamma'')$ on $G \times T^0$ is amenable [28]. Since R is solvable normal, the action of $\beta(\Gamma'')$ on $(G/R) \times T_0$ is also amenable; cf. [31]. The main result of [31] implies that the identity component of the closure of the image of $\beta(\Gamma'')$ (i.e. the image of $\alpha(\Lambda'')$) in $L \times T^0$ is solvable. Call this connected solvable group S. It is of course normalized by the image of $\alpha(\Lambda'')$. However, $\alpha(\Lambda'')$ is Zariski dense in M, which implies that S is normal in $L \times T^0$. It must therefore be trivial, so we deduce that the image of $\alpha(\Lambda'')$ in $L \times T^0$ is discrete; cf. Auslander's theorem [19].) However, by the choice of L'', we have $T^0 = L''$, so $T = L''$, verifying Zariski density, and completing the proof of the lemma.

Theorem 9.4. *$\Lambda \subset L$ is a dense arithmetic subgroup. Furthermore, Λ'' is isomorphic to a lattice in $L \times H$.*

Proof. We first claim that the discrete subgroup $q(\alpha(\Lambda'')) \subset L \times L''$ is actually a lattice. Since we have a surjective $G \times L''$-map

$$(G \times L'')/\beta(\Gamma'') \to (L \times L'')/q(\alpha(\Lambda'')),$$

it suffices to see that $\beta(\Gamma'')$ is a lattice in $G \times L''$. We shall need the following fact.

Lemma 9.5. *Let A, B be Lie groups and $D \subset A \times B$ a discrete subgroup. Suppose B acts ergodically on $(A \times B)/D$ with respect to the smooth measure class. If there is a finite B-invariant measure μ of smooth class, then D is a lattice in $A \times B$.*

Proof. Since μ is a finite B-invariant ergodic measure, any other B-invariant measure in the same measure class with the same total mass must be equal to μ. (Namely, for any such measure ν, let $f = d\nu/d\mu$. This is B-invariant, hence constant.) However, for $a \in A$, $a_*\mu$ will be such a measure, and hence μ is A-invariant as well.

Returning to the proof of Theorem 9.4, to see that $\beta(\Gamma'')$ is a lattice, by Lemma 9.5 it suffices to prove that there is a finite L''-invariant measure of smooth class on $(G \times L'')/\beta(\Gamma'')$. However, this follows from Lemma 9.3 and Corollary 3.13. This also shows that L'' is isomorphic to H. To complete the proof of 9.4, it suffices to show that $L' = L''$. For then $\ker(q)$ is compact and hence $\alpha(\Lambda'')$ is a lattice in M. Since we already know $\alpha(\Lambda'') \subset M_{\mathbf{Z}}^* \cap M$, we would then have an inclusion of lattices, and so $\alpha(\Lambda'')$ would be of finite index in $M_{\mathbf{Z}}^* \cap M$, verifying arithmeticity. If $L' \neq L''$, then there is a noncompact simple normal subgroup Y of L' which is not contained in L''. We recall that by construction $q|\alpha(\Lambda'')$ is injective and the projection of $\alpha(\Lambda'')$ into Y is Zariski dense. Since $q(\alpha(\Lambda''))$ is a lattice in $L \times H$ projecting densely to L, and all simple factors of H have **R**-rank at least 2, we can apply Margulis' superrigidity theorem [15, 28] to deduce that there is a rational homomorphism $L \to Y$ extending the map of $q(\alpha(\Lambda''))$ given by projecting $\alpha(\Lambda'')$ into Y. Then the projection of $\alpha(\Lambda'')$ into $L \times Y$ is contained in the graph of this homomorphism, contradicting Zariski density of $\alpha(\Lambda'')$ in M.

10. Completion of the Proofs of Theorems A and D

To prove Theorems A and D, it remains only to show that R is finite and central. (We are still under the assumption that $\Gamma \cap R$ is trivial, given by Corollary 5.4.) By Theorem 9.4, identifying Γ with $\Lambda \subset L$, we can assume that $\Gamma = \rho(D)$ where $D \subset L \times H$ is a lattice, $\rho \colon L \times H \to L$ is projection, and $\rho|D$ is injective. Consider the map $p \colon G \to L$ of ergodic Γ-spaces. This is a principal R-bundle on which Γ acts by principal bundle automorphisms. Hence, if we measurably trivialize the bundle, the Γ-action on G will be given by a cocycle $\alpha \colon L \times \Gamma \to R$, i.e., we have measurably that $G = L \times R$ and Γ

acts by $(l,r)\cdot\gamma = (l\gamma, r\alpha(l,r))$. The Γ action on L is the same as the D-action on $L \cong (L \times H)/H$, which is stably orbit equivalent to the action of H on $(L \times H)/D$. The latter is a finite measure preserving action of a Kazhdan group. By Corollary 3.10 and Theorem 3.14, the cocycle α is equivalent to a cocycle β taking all values in a compact subgroup $R_1 \subset R$. Thus, the Γ-action on G is measurably equivalent to the action on $L \times R$ given by $(l,r)\cdot\gamma = (l\gamma, r\beta(l,\gamma))$. Since this action is ergodic (because Γ is dense in G), it follows that $R_1 = R$, i.e. R is compact. Since R is solvable, we deduce that R^0 is abelian. Then the map $R \to G/[G,G]$ has finite kernel. Since Γ is isomorphic to a lattice in $L \times H$, projecting densely onto all factors of **R**-rank less than 2, the image of Γ in the abelian group $G/[G,G]$ is finite [16], and since Γ is dense in G, $G/[G,G]$ is finite. Thus, R is finite, hence central, and this proves the theorems.

11. Foliations by quaternionic hyperbolic space

In this section we show how one can obtain results similar to Theorem B for foliations by certain rank one locally symmetric spaces, namely those for which the isometry group of the universal cover is a Kazhdan group.

Theorem 11.1. *Let $\mathscr{F}$ be a Riemannian foliation of a compact manifold M. Assume there is a metric on M for which the leaves are locally isomorphic to either a quaternionic hyperbolic space (of dimension at least 8) or to the Cayley hyperbolic plane. If* codim $\mathscr{F} \leq 3$, *then there is a finite covering $\widetilde{M}$ of M such that $\widetilde{\mathscr{F}}$ is the foliation defined by a fiber bundle projection. If $\mathscr{F}$ is a Lie foliation with* codim $\mathscr{F} \leq 6$, *the same conclusion is true.*

The isometry group of the symmetric spaces in question are Kazhdan [14, 28]. Therefore, the discussion above for the higher rank case shows that it is sufficient to prove the following result about stable orbit equivalence.

Theorem 11.2. *Let G be a (nontrivial) connected Lie group with* $\dim G \leq 6$, *and $\Gamma \subset G$ a finitely generated dense subgroup. Suppose H is a Kazhdan group and (X,μ) an ergodic H-space with finite invariant measure. Then the Γ-action on G is not stably orbit equivalent to the H-action on X.*

Example 11.3. Before proceeding with the proof, we observe that we may have a connected Lie group G and a fintiely generated dense subgroup $\Gamma \subset G$ such that neither G nor Γ is Kazhdan, but the Γ-action on G is stably orbit equivalent to a finite measure preserving action of a Kazhdan group. Namely, if $\Gamma \subset G \times G'$ is an irreducible lattice where G' is Kazhdan but G is not, then the Γ-action on G is stably orbit equivalent to the G'-action on $(G \times G')/\Gamma$. One can easily have examples with $G = \mathrm{SO}(1,n)$ and $G' = \mathrm{SO}(2,n-1)$ for $n \geq 4$.

For the proof of Theorem 11.2, we will need the following result about cocycles.

Theorem 11.4 [29]. *Let H be a Kazhdan group and (X, μ) an ergodic H-space with a finite invariant measure. Let k be a local field, $\mathrm{char}(k) = 0$, and $\alpha: X \times H \to \mathrm{PSL}(2, k)$. Then α is equivalent to a cocycle taking all values in a compact subgroup of $\mathrm{PSL}(2, k)$.*

For the proof of 11.4, see [29].

Proof of Theorem 11.2. The proof follows the main lines of the proof of Theorem B using Theorem 11.4 in place of Theorem 3.12. We shall therefore just sketch the argument. Let $R \subset G$ be as in the proofs of Theorems A, B and D. By the remark following Lemma 5.1, we can assume $G \neq R$, so $L' = G/R$ is nontrivial. Since $\dim L' \leq 6$, there is a simple factor L of L' which is a subgroup of $\mathrm{PSL}(2, \mathbf{C})$. For ease of notation, we let $\mathrm{PSL}(2, \mathbf{C}) = M$; Ad will denote Ad_M. Let Λ be the (dense) image of Γ in L. For each $\sigma \in \mathrm{Gal}(\mathbf{C}/\mathbf{Q})$ we obtain an automorphism (usually discontinuous) of M, which we still denote by σ, and hence a homomorphism $h_\sigma: \Gamma \to M$. This image is Zariski dense in M (where M is viewed as an algebraic group over $\mathbf{C}$, not as a real algebraic group). Let $\alpha_\sigma: G \times \Gamma \to M$ be the cocycle $\alpha_\sigma(g, \gamma) = h_\sigma(\gamma)$. Then by Lemma 6.3, either α_σ is Zariski dense in M or h_σ extends to a continuous homomorphism $G \to M$, which must factor to a homomorphism $L' \to M$. In case α_σ is Zariski dense, we have from Theorem 11.4 and Lemma 7.3 that $h_\sigma(\Gamma)$ has compact closure. We deduce, as in the proof of Lemma 9.1, that $\mathrm{Tr}(\mathrm{Ad}(\gamma)) \in \overline{\mathbf{Q}}$ for all $\gamma \in \Lambda$. Then, as in 9.1 (see also [1, p. 135]) we may assume $\mathrm{Ad}(\Lambda) \subset \mathrm{Ad}(M)_k$ where $k \subset \mathbf{R}$ is a finite Galois extension of $\mathbf{Q}$. By Theorem 11.4 and Lemma 7.3, we again deduce, as in the proof of Lemma 9.1, that $\mathrm{Ad}(\Lambda') \subset \mathrm{Ad}(M)_{\mathscr{O}}$ for some subgroup $\Lambda' \subset \Lambda$ of finite index, where $\mathscr{O}$ is the ring of integers in k. Let Γ' be the pull-back of Λ' to Γ, so Γ/Γ' is also finite. Let $\alpha: \mathrm{Ad}(M)_{\mathscr{O}} \to R_{k/\mathbf{Q}}(\mathrm{Ad}(M))_{\mathbf{Z}}$ be the standard map (see [28]). Let N be the algebraic hull of $\alpha(\Lambda')$. Then N is semisimple, projects onto $\mathrm{Ad}(M)$, and we can write $N \cong \mathrm{Ad}(M) \times M'$. Since $\alpha(\Lambda')$ is discrete, and $\mathrm{Ad}(\Lambda')$ is not discrete, the projection of $\alpha(\Lambda')$ onto M' cannot be precompact. Thus, there is some $\mathbf{R}$-simple factor M'' of M' in which the projection of $\alpha(\Lambda')$ is not precompact. There is some $\sigma \in \mathrm{Gal}(k/\mathbf{Q})$ so that the projection of $M'' \subset R_{k/\mathbf{Q}}(\mathrm{Ad}\, M)$ onto $\mathrm{Ad}(M)^\sigma$ is an isomorphism of Lie groups. By Theorem 11.4 and Lemma 7.3, we deduce that for such a σ, the cocycle $\alpha_\sigma: G \times \Gamma' \to M^\sigma$ is not Zariski dense. Therefore, by Lemma 6.3, h_σ extends to a smooth homomorphism $G \to M^\sigma$, hence defines a rational homomorphism $L \to M''$. This contradicts the Zariski density of $\alpha(\Lambda')$ in M, and completes the proof.

REFERENCES

1. H. Bass, *Finitely generated subgroups of* $\mathrm{GL}(2)$, *in the Smith Conjecture* (J. Morgan and H. Bass, eds.), Academic Press, New York, 1984.

2. R. A. Blumenthal, *Transversally homogeneous foliations*, Ann. Inst. Fourier (Grenoble) **29** (1979), 143–158.

3. A. Borel, *Density properties for certain subgroups of semisimple Lie groups without compact factors*, Ann. of Math. (2) **72** (1960), 179–188.

4. Y. Carriere, *Feuilletages Riemanniens a croissance polynomiale*, preprint, Lille, 1986.

5. Y. Carriere and E. Ghys, *Relations d'equivalence moyenables sur les groupes de Lie*, C. R. Acad. Sci. Paris Sér. I Math. **300** (1985), 677–680.

6. A. Connes, J. Feldman, and B. Weiss, *An amenable equivalence relation is generated by a single transformation*, Ergodic Theory Dynamical Systems **1** (1981), 431–450.

7. C. Ehresmann, *Structures feuilletees*, Proc. Fifth Canadian Math. Congress, Univ. Toronto Press, Toronto, 1961, pp. 109–172.

8. D. B. A. Epstein, *Transversely hyperbolic 1-dimensional foliations*, Asterisque **116** (1984), 53–69.

9. E. Fedida, *Sur les feuilletages de Lie*, C. R. Acad. Sci. Paris Ser. I Math. **272** (1971), 999–1002.

10. J. Feldman and C. C. Moore, *Ergodic equivalence relations, cohomology, and von Neumann algebras. I*, Trans. Amer. Math. Soc. **234** (1977), 289–324.

11. J. Feldman, P. Hahn and C. C. Moore, *Orbit structure and countable sections for actions of continuous groups*, Adv. in Math. **28** (1978), 186–230.

12. E. Ghys, *Groupes d'holonomie des feuilletages de Lie*, Nederl. Akad. Wetensch. Proc. Ser. A **88** (1985), 173–182.

13. A. Haefliger, *Structures feuilletees et cohomologie a valeur dans un faisceau de groupoides*, Comment. Math. Helv. **32** (1958), 248–329.

14. B. Kostant, *On existence and irreducibility of certain series of representations*, Bull. Amer. Math. Soc. **75** (1969), 627–642.

15. G. A. Margulis, *Discrete groups of motions of manifolds of non-positive curvature*, Amer. Math. Soc. Transl. **109** (1977), 33–45.

16. ____, *Finiteness of quotient groups of discrete subgroups*, Functional Anal. Appl. **13** (1979), 178–187.

17. P. Molino, *Geometrie globale des feuilletages riemanniens*, Nederl. Akad. Wetensch. Proc. Ser. A **85** (1982), 45–76.

18. C. C. Moore, *Amenable subgroups of semisimple groups and proximal flows*, Israel J. Math. **34** (1979), 121–138.

19. M. S. Raghunathan, *Discrete subgroups of Lie groups*, Springer-Verlag, New York, 1972.

20. A. Ramsay, *Virtual groups and group actions*, Adv. in Math. **6** (1971), 253–322.

21. G. Reeb, *Sur certains proprietes topologiques des varietes feuilletees*, Actualités Sci. Indust, Hermann, Paris, 1952.

22. B. Reinhart, *Foliated manifolds with bundle-like metrics*, Ann. of Math. (2) **69** (1959), 119–132.

23. K. Schmidt, *Cocycles of ergodic transformation groups*, MacMillan, India, 1977.

24. W. Thurston, *The geometry and topology of 3-manifolds*, Princeton Univ. lecture notes.

25. R. J. Zimmer, *Amenable ergodic group actions and an application to Poisson boundaries of random walks*, J. Funct. Anal. **27** (1978), 350–372.

26. ____, *Strong rigidity for ergodic actions of semisimple Lie groups*, Ann. of Math. (2) **112** (1980), 511–529.

27. ____, *Orbit equivalence and rigidity of ergodic actions of Lie groups*, Ergodic Theory Dynamical Systems **1** (1981), 237–253.

28. ____, *Ergodic theory and semisimple groups*, Birkhauser, Boston, 1984.

29. ____, *Kazhdan groups acting on compact manifolds*, Invent. Math. **75** (1984), 425–436.

30. ____, *Ergodic theory and the automorphism group of a G-structure* (Proc. Conf. in honor of G. W. Mackey, Berkeley, 1984), M.S.R.I. Publications, Springer, 1987.

31 ____, *Amenable actions and dense subgroups of Lie groups*, J. Funct. Anal. **27** (1987), 58–64.

DEPARTMENT OF MATHEMATICS, UNIVERSITY OF CHICAGO, CHICAGO, ILLINOIS 60637

JOURNAL OF THE
AMERICAN MATHEMATICAL SOCIETY
Volume 2, Number 2, April 1989

REPRESENTATIONS OF FUNDAMENTAL GROUPS
OF MANIFOLDS WITH A SEMISIMPLE
TRANSFORMATION GROUP

ROBERT J. ZIMMER

1. INTRODUCTION

In this paper, we examine the relationship between the topology of a manifold M, specifically the finite dimensional representation theory of the fundamental group $\pi_1(M)$, and the Lie groups that can act on M. More precisely, let G be a connected semisimple Lie group of higher real rank, and suppose G acts continuously on a (topological) manifold M, preserving a finite measure. The main theme of this paper is that the representation theory of $\pi_1(M)$ in low dimensions is to a large extent controlled by that of G (the latter of course being well understood). In particular, under natural hypotheses (e.g., that the action of G on M is engaging, i.e., there is no loss of ergodicity in passing to finite covers; see Definition 3.1 below), we prove that if G has no nontrivial representations below dimension d, then every representation of $\pi_1(M)$ below dimension d is finite; that is, it factors through a finite quotient group. Under different but related hypotheses (namely that the action is topologically engaging, that is, roughly speaking, that the action is proper on the universal cover; see Definition 3.2), we show that $\pi_1(M)$ admits no faithful representation over $\overline{\mathbf{Q}}$ below dimension d. These results of course impose severe restrictions on the manifolds on which G can act. The hypotheses of engaging or topological engaging are quite mild, and one or the other is satisfied in every known nontrivial example. We also remark that Gromov has shown [5] that every real analytic connection preserving action of G is topologically engaging.

Rather than considering representations, one can consider, more generally, homomorphisms into a general algebraic group, and the above results become special cases of the following theorems. (These results, as well as most of the others in this paper, hold for spaces much more general than the class of topological manifolds. See the beginning of §2.)

Theorem 5.1. *Let G be a connected semisimple Lie group, each of whose simple factors has real rank at least 2. Suppose G acts continuously on a manifold*

Received by the editors May 10, 1988.
1980 *Mathematics Subject Classification* (1985 *Revision*). Primary 57S20.
The author's research was partially supported by an NSF Grant.

M , *preserving a finite measure, and that the action is engaging. Assume* $\pi_1(M)$ *is finitely generated. Let* H *be an algebraic group* (*over* **C**), *and suppose every Lie algebra homomorphism* $\mathfrak{g}_\mathbb{C} \to \mathfrak{h}$ *is trivial. Then every homomorphism* h: $\pi_1(M) \to H$ *is finite.*

Theorem 5.2. *Let* G *be a connected semisimple Lie group, each of whose simple factors has real rank at least* 2 . *Suppose* G *acts continuously on a manifold* *M* , *preserving a finite measure, and that the action is topologically engaging. Assume* $\pi_1(M)$ *is finitely generated and infinite. Let* H *be an algebraic group* (*over* $\overline{\mathbf{Q}}$), *and suppose every Lie algebra homomorphism* $\mathfrak{g}_\mathbb{C} \to \mathfrak{h}$ *is trivial. Then there is no faithful homomorphism* h: $\pi_1(M) \to H_{\overline{\mathbf{Q}}}$.

Via similar techniques, we can prove a variety of related results under varying hypotheses. For example, if H is defined over **R**, it is natural to suppose only that there is no nontrivial Lie algebra homomorphism $\mathfrak{g} \to \mathfrak{h}_\mathbf{R}$. It is also natural to consider the case of homomorphisms of $\pi_1(M)$ into p-adic groups. We obtain satisfactory results in these cases as well, and these are spelled out in §5. We also consider the case of actions of a lattice subgroup $\Gamma \subset G$. The situation here is more complex, and even the definitions of engaging and topologically engaging actions require more care (cf. §8). For many such Γ, we have the following cohomological condition (C): $H^2(\Lambda, \mathbf{R}) = 0$ for every finite index subgroup $\Lambda \subset \Gamma$. Then we show

Theorem 8.5. *Let* G *be as above, and let* $\Gamma \subset G$ *be a lattice subgroup, satisfying cohomological condition* (C). *Let* d *be the minimal dimension of a nontrivial real representation of the Lie algebra* $\mathfrak{g}$. *Then there is no engaging or topologically engaging action of* Γ *on any compact manifold with an infinite abelian fundamental group of rank strictly less than* d .

The action of G on G/Γ is both engaging and topologically engaging. In this case, Theorems 5.1 and 5.2 follow from Margulis' superrigidity theorem. We can thus view Theorems 5.1 and 5.2 as showing that certain features of lattices in higher real rank semisimple groups are still present in the fundamental group of any manifold on which G acts in an engaging, volume preserving way.

In [14], we examined the holonomy group of a Lie foliation with symmetric spaces as leaves. We can describe this roughly in our present context by saying that under suitable geometric hypotheses on a G-action (i.e., the existence of a transverse Lie structure) that the fundamental group of the ambient manifold must have an arithmetic quotient group of a very precise type. Thus, in [14] we assert the existence of a certain type of representation of $\pi_1(M)$, while, of course, in the present paper, we show nonexistence of certain types of representations. Gromov has also considered these issues in [5], obtaining, under geometric hypotheses, "lower bounds" on $\pi_1(M)$. Gromov's paper makes very clear the importance of the role of properness on the universal cover of M, and motivated our definition of topological engaging.

The author would like to thank A. Ash, A. Borel, K. Corlette, G. Prasad, and T. N. Venkataramana for helpful conversations related to this paper, and M. Gromov for communicating his ideas.

2. PRELIMINARIES OF ERGODIC THEORY

Our results will hold for actions on any space that has a good theory of covering spaces and good measure theory. Hence, throughout we shall assume M *is a connected, locally path-connected, semilocally 1-connected, locally compact, separable, metrizable space, with a finitely generated fundamental group. For brevity, we shall simply call such a space "standard."*

We can view the universal covering space of M as a principal $\pi_1(M)$-bundle. Any action of a connected group G on M lifts to $\widetilde{M}$ in such a way as to commute with $\pi_1(M)$. In other words, $\widetilde{G}$ acts by principal bundle automorphisms of $\widetilde{M}$. For any homomorphism of $\pi_1(M)$ into a group H, we obtain an associated principal H-bundle on which $\widetilde{G}$ will also act by principal bundle automorphisms. If G is semisimple and H is algebraic, there is a great deal known about the measure theoretic structure of such actions [12, 13]. Our approach to proving Theorems 5.1 and 5.2 and related results will be to exploit this information. An interesting feature of the argument is the necessity to consider such bundles where the structure group is p-adic, even if one is only interested in real or complex representations. The reader familiar with the general theory of representations of finitely generated groups will not find this surprising. Since we will be dealing with the measure theoretic aspects of such bundles, it is convenient to view them as products, which can always be done measurably. The action of $\widetilde{G}$ is then described via a cocycle. We now review some of the basic notions concerning measurable cocycles for group actions. See [13] for a leisurely account with an eye toward geometric applications.

Suppose G and L are locally compact groups and that M is a measure space on which G acts measurably on the right, leaving the measure quasi-invariant. A Borel function $\alpha: M \times G \to L$ is called a cocycle if for all g, $h \in G$, $\alpha(m, gh) = \alpha(m, g)\alpha(mg, h)$ for (almost) all $m \in M$. Two such cocycles α, β are called equivalent or cohomologous (and we write $\alpha \sim \beta$) if there is a measurable $\varphi: M \to L$ such that for each $g \in G$ we have $\varphi(m)\alpha(m, g)\varphi(mg)^{-1} = \beta(m, g)$ a.e. If $P \to M$ is a principal L-bundle on which G acts by principal bundle automorphisms, then under any measurable trivialization $P \cong M \times L$, the action of G on P will be given by $(m, \lambda)g = (mg, \alpha(m, g)^{-1}\lambda)$ for some cocycle α. Choosing a different measurable trivialization is equivalent to choosing an equivalent cocycle. If $H \subset L$ is a closed subgroup, it is of interest to know when α is cohomologous to a cocycle taking all values in H. This is equivalent to the assertion that there is a measurable G-invariant reduction of P to the group H. If Y is a right L-space, a measurable $\varphi: M \to Y$ is called α-invariant if $\varphi(m)\alpha(m, g) = \varphi(mg)$. We then have the following simple proposition.

Proposition 2.1. *The following are equivalent.*

   (i) α *is equivalent to a cocycle taking all values in* H .

   (ii) *There is an α-invariant* $\varphi\colon M \to H\backslash L$.

    The basic information we shall use regarding cocycles for semisimple group actions is given in the following theorem. It is a part of a general superrigidity theorem for cocycles, proved in [10, 11, 12].

Theorem 2.2. *Suppose G is a connected semisimple Lie group of higher real rank, i.e., the $\mathbf{R}$-rank of every simple factor of G is at least 2 . Suppose the action of G on M preserves a finite measure, and let $\alpha\colon M \times G \to H$ be a cocycle. Then α is equivalent to a cocycle into a compact subgroup of H under any of the following assumptions.*

    (i) H *is a real or complex algebraic group and there is no Lie algebra homomorphism* $\mathfrak{g} \to \mathfrak{h}$.

    (ii) H *is the set of p-adic points of a p-adic algebraic group.*

    (iii) H *is amenable, and the G-action on M is ergodic.*

    A useful device in the study of cocycles is the notion of the Mackey range of a cocycle [9, 12], which we now recall. Given a cocycle $\alpha\colon M \times G \to H$, we let $P = M \times H$ and consider the action of G on P defined by the cocycle, as given above. We let X be the space of ergodic components of the G-action. The action of H on P by right translation on the second factor commutes with G, and hence we have an action of H on X, which will preserve the natural measure class. This H-action on X is called the Mackey range of α, and it will be ergodic if the G-action on M is ergodic. A useful feature of this construction is that α is equivalent to a cocycle taking values in a closed subgroup $H_1 \subset H$ if and only if there is a measurable H-map $X \to H/H_1$. (See [9] for a proof.) Here is a related fact we shall need.

Lemma 2.3. *Suppose $\alpha\colon M \times G \to H$ is a cocycle and that $\pi\colon H \to L$ is a homomorphism. Suppose Y is a right L-space and that there is a $\pi \circ \alpha$-invariant function $\varphi\colon M \to Y$. Let X be the Mackey range of α. Then there is a measurable H-map $X \to Y$.*

Proof. We have the equation $\varphi(m)\pi(\alpha(m,g)) = \varphi(mg)$ for all g,m. Define $\theta\colon M \times H \to Y$ by $\theta(m,h) = \varphi(m)\pi(h)$. This is clearly an H-map, and so it suffices to see it factors to X; i.e., it suffices to see θ is G-invariant. However,

$$\theta((m,h)g) = \theta(mg,\alpha(m,g)^{-1}h)$$
$$= \varphi(mg)\pi(\alpha(m,g)^{-1})\pi(h) = \varphi(m)\pi(h) = \theta(m,h) .$$

    Suppose now that N is a locally compact separable space on which G acts continuously. We recall that the action is called proper if for any pair of compact subsets $A,B \subset N$, we have $\{g \in G | Ag \cap B$ nonempty$\}$ has compact closure. An action on a standard Borel space is called tame (or "smooth in the sense of ergodic theory") if the quotient space has a countably generated Borel structure [12, 13]. For N locally compact and separable, this is equivalent to all orbits being locally closed, or equivalently [12] the quotient space is T_0. Any proper action is tame. If μ is any quasi-invariant measure on N, we call the action μ-tame if there is a G-invariant Borel conull set on which the action is tame, μ-proper if it is μ-tame and almost every stabilizer is compact. Thus, a proper action satisfies all these conditions for any μ. For a μ-tame action, the ergodic components are precisely the orbits [12]. For ease of reference, we record a special case.

Lemma 2.4. *If* $\alpha \colon M \times G \to H$ *is a cocycle, and the* G-*action on* $P = M \times H$ *is tame, then the Mackey range of* α *is the action of* H *on* P/G.

Proposition 2.5. (a) *Suppose* $(N, \nu) \to M$ *is a measurable* H-*map where the action of* H *on* M *is proper. Then the action on* N *is* ν-*proper.*

(b) *Suppose* $G \times H$ *acts on* N, *and that* G *and* H *each act tamely. Let* μ *be a* $G \times H$ *quasi-invariant measure on* N, *and* μ_1, μ_2 *the projections onto* N/G, N/H, *respectively. Then the* G-*action on* N/H *is* μ_2-*tame if and only if the* H-*action on* N/G *is* μ_1-*tame.*

Proof. (a) follows easily from [12, Chapter 2] and (b) from the definitions.

3. ENGAGEMENT

Suppose G is a connected Lie group acting on a standard space (e.g., a topological manifold with finitely generated fundamental group) M. The action lifts to a unique action of $\widetilde{G}$ on $\widetilde{M}$ commuting with $\pi_1(M)$. We also have an action of $\widetilde{G}$ (in fact, of some finite covering of G) on any finite cover M' of M.

Definition 3.1. Suppose μ is a quasi-invariant measure for the G-action on M. The action of G on M is called μ-engaging (or simply engaging if μ is understood) if there is no loss of ergodicity in passing to finite covers of M; more precisely, if for every finite cover M', every measurable $\widetilde{G}$-invariant function on M' is lifted from a function on M.

In particular, if G acts ergodically on M, the condition that the action be engaging is simply that $\widetilde{G}$ is ergodic on every finite cover of M.

Definition 3.2. The action of G on M is called topologically engaging if there is some $\tilde{g} \in \widetilde{G}$ that acts tamely on $\widetilde{M}$ (i.e., has locally closed orbits, e.g., acts properly) and that projects to an element $g \in G$ that does not lie in a compact subgroup.

Example 3.3. Suppose that G is a semisimple Lie group without compact factors and that we have an embedding $G \to H$ where H is another semisimple Lie group without compact factors. Suppose $\Gamma \subset H$ is a cocompact discrete irreducible subgroup. Then G acts on H/Γ and the action is both engaging and topologically engaging. Variants of this construction provide most of the known ergodic volume preserving actions of G on a compact manifold.

Example 3.4. Gromov [5] proves that if G is a semisimple Lie group with no compact factors and finite fundamental group and that if G acts real analytically on M preserving a pseudo-Riemannian metric, then $\widetilde{G}$ acts properly on $\widetilde{M}$. In particular, the action is topologically engaging.

Now let $\Gamma = \pi_1(M)$. Since G acts by automorphisms of the principal Γ-bundle $\widetilde{M} \to M$, we obtain a cocycle $\alpha \colon M \times \widetilde{G} \to \Gamma$. We now consider the relationship of the conditions in Definitions 3.1 and 3.2 with properties of the Mackey range of α.

Proposition 3.5. *Suppose the action of G on M is engaging. Then α is not equivalent to a cocycle taking values in a subgroup $\Lambda \subset \Gamma$ with the following properties:*

  (a) Γ/Λ *is infinite*;
  (b) *there is a subgroup $\Lambda_0 \subset \Lambda$ of finite index that is normal in Γ and with Γ/Λ_0 residually finite.*

Equivalently (by (2.1) the Mackey range X does not admit a measurable Γ-map $X \to \Gamma/\Lambda$, where Λ satisfies (a) and (b).

Proof. Since Γ/Λ_0 is infinite and residually finite, we can choose a finite quotient $\Gamma/\Lambda_1, \Lambda_1 \supset \Lambda_0$, whose cardinality is greater than that of Λ/Λ_0. Let $p \colon \Gamma \to \Gamma/\Lambda_1$ be the quotient map. Then the Galois covering of M corresponding to Λ_1 is measurably isomorphic to $M \times \Gamma/\Lambda_1$ with the $\widetilde{G}$-action given by the cocycle $p \circ \alpha$. If α takes values in Λ, then $M \times (\Lambda\Lambda_1/\Lambda_1) \subset M \times \Gamma/\Lambda_1$ will be a $\widetilde{G}$-invariant measurable set. Since the cardinality of $\Lambda\Lambda_1/\Lambda_1$ is less than that of Γ/Λ_1, the characteristic function of this set does not lift from M.

Proposition 3.6. *Suppose the action of G on M is topologically engaging (and faithful) where G is semisimple without compact factors. Suppose there is a finite G-invariant ergodic measure μ on M. Then the Γ action on the Mackey range of α is not of the form Γ/F where F is a finite subgroup.*

Proof. Let $H \subset \widetilde{G}$ be the group generated by $\breve{g}$. Then H is closed and projects to a subgroup of noncompact closure in G. By Moore's theorem [12], H cannot have almost all orbits be periodic, and hence the action of H on M is not μ-tame. (In fact, if G is simple, H will act properly ergodically.) Let $\widetilde{X}$ be the Mackey range of $\tilde{\alpha} = \alpha | M \times H$. The action of H on $\widetilde{M}$ is tame by the definition of topological engaging, so $\widetilde{X} = \widetilde{M}/H$. The action of Γ on $\widetilde{M}$ is clearly tame with quotient M, and hence by Proposition 2.5(b) the action of Γ on $\widetilde{X}$ cannot be ν-tame, where ν is the natural measure on $\widetilde{X}$. However, by the definition of the Mackey range, we have a measurable Γ-map $\widetilde{X} \to X$. If $X = \Gamma/F$, this would then contradict Proposition 2.5(a).

The first assertion of the following proposition is obvious, and the second follows by routine measure theoretic arguments.

Proposition 3.7. (a) *Suppose* $M' \to M$ *is a finite cover. If the action of* $\widetilde{G}$ *on* M *is engaging, so is the action on* M'. *The action on* M *is topologically engaging if and only if the action on* M' *is as well.*

(b) *Suppose the action on* M *is* μ-*engaging. Let* $\mu = \int^{\oplus} \mu_t$ *be the decomposition of* μ *into ergodic measures on* M. *Then for almost all* t, *the G-action is* μ_t *engaging.*

4. Reduction to homomorphisms over $\overline{\mathbf{Q}}$

We give a criterion for every homomorphism from a finitely generated group into an algebraic group to be finite. Although one can prove sharper statements, we content ourselves here with one that is easily established and which will suffice for our present purposes.

Theorem 4.1. *Let* G *be a connected algebraic group (over* $\mathbf{C}$) *and* Γ *a finitely generated group. Suppose that for every algebraic group* L, *which is a subquotient of* G, *every* $\overline{\mathbf{Q}}$-*structure on* L *and every finite index subgroup* $\Lambda \subset \Gamma$, *we have that every homomorphism* $\Lambda \to L_{\overline{\mathbf{Q}}}$ *is finite. Then every homomorphism* $\Gamma \to G$ *is finite.*

Proof. Fix $\pi: \Gamma \to G$, and let H be the connected component of the algebraic hull of the image. By passing to a subgroup of finite index in Γ, we may assume H is connected and that the image of π is Zariski dense in H. Let R be the radical of H, and $L = H/R$. We first claim that it suffices to see that the image of Γ in L is finite. For if this is so, then on a subgroup Λ of finite index, we have $\pi(\Lambda) \subset R$. Since R is solvable and $\pi(\Lambda)$ is finitely generated, if $\pi(\Lambda)$ is not finite it will have a subgroup of a finite index that maps onto $\mathbf{Z}$. This would obviously contradict the hypotheses. Thus, we shall assume $\pi: \Gamma \to L$ and that the image is Zariski dense. Since L is semisimple, we may fix a realization of $L \subset \mathrm{GL}(n, \mathbf{C})$ as an algebraic group defined over $\mathbf{Q}$. Since the space of homomorphisms $\Gamma \to L$ is then a variety over $\mathbf{Q}$, we can approximate any $\pi: \Gamma \to L$ by a sequence $\pi_j: \Gamma \to L_{\overline{\mathbf{Q}}}$. Since $\pi_j(\Gamma)$ is finite, we have $|\mathrm{tr}(\pi_j(\lambda))| \leq n$ for all $\lambda \in \Gamma$, and hence $|\mathrm{tr}(\pi(\lambda))| \leq n$ as well. It follows that $\mathrm{tr}(\pi(\lambda)) \in \overline{\mathbf{Q}}$ for all λ. If not, we can choose $\sigma \in \mathrm{Gal}(\mathbf{C}/\overline{\mathbf{Q}})$ such that $|\sigma(\mathrm{tr}(\pi(\lambda)))| > n$, i.e., $|\mathrm{tr}((\sigma \circ \pi)(\lambda))| > n$, where, as usual, we let σ act on complex matrices. Since L is defined over $\mathbf{Q}$, $\sigma \circ \alpha: \Gamma \to L$ as well, showing that this is impossible. It follows by a result of Vinberg [12, Lemma 6.1.7] that there is a realization of L as a linear group defined over $\overline{\mathbf{Q}}$ in such a way that $\pi(\lambda) \in L_{\overline{\mathbf{Q}}}$ for all $\lambda \in \Gamma$. By hypothesis, this shows that π is finite.

5. Proof of the main theorems

Theorem 5.1. *Suppose* G *is a connected semisimple Lie group each of whose simple factors has real rank at least* 2. *Let* H *be a complex algebraic group such that every homomorphism of Lie algebras* $\mathfrak{g}_{\mathbf{C}} \to \mathfrak{h}$ *is trivial. Assume there* *is an action of G on M that preserves a finite measure and is engaging. Then every homomorphism $\pi_1(M) \to H$ is finite.*

We remark that the conclusion implies that there are only finitely many conjugacy classes of homomorphisms into a given H, and hence that all such homomorphisms factor through a fixed (for a given H) finite quotient of Γ.

Proof. By Proposition 3.7(b), we can assume the measure on M is ergodic under G. Let $\Gamma = \pi_1(M)$. By Theorem 4.1, it suffices to assume that H is defined over $\overline{\mathbf{Q}}$ and to see that any $\pi: \Gamma \to H_{\overline{\mathbf{Q}}}$ is finite for then this would apply to any finite index subgroup of Γ and any subquotient of H. We may clearly further assume that $\pi(\Gamma)$ is Zariski dense in H. Since $\pi(\Gamma) \subset H_{\overline{\mathbf{Q}}}$ and Γ is finitely generated, there is a number field k ($[k:\mathbf{Q}] < \infty$) such that $\pi(\Gamma) \subset H_k$. In particular, H is defined over k. Let $L = H_{k/\mathbf{Q}}$, the algebraic $\mathbf{Q}$-group obtained from H by restriction of scalars [12]. Every simple factor of L is isomorphic to a simple factor of H (cf. [6]), and hence every local homomorphism $G \to L$ is trivial as well. We have a natural identification of H_k with $L_{\mathbf{Q}}$, and hence we can view π as a homomorphism into $L_{\mathbf{Q}}$. Since Γ is finitely generated, there is a finite set of rational primes such that all denominators of all $\pi(\Gamma)$ have all prime factors in S. By diagonally embedding $L_{\mathbf{Q}}$, we obtain a representation $\sigma: \Gamma \to \widehat{L} = L_{\mathbf{R}} \times \prod_{p \in S} L_{\mathbf{Q}_p}$. By construction, $\ker(\sigma) = \ker(\pi)$, and it is standard that $\sigma(\Gamma)$ is discrete. Let $\alpha: M \times \widetilde{G} \to \Gamma$ be the cocycle given by the action on the universal cover. By Theorem 2.2, $\sigma \circ \alpha$ is equivalent to a cocycle taking all values in a compact subgroup $K \subset \widehat{L}$, and this implies by Lemma 2.3 that the Mackey range X of α admits a measurable Γ-map $\theta: X \to \widehat{L}/K$. Since $\sigma(\Gamma)$ is discrete and K is compact, the action of Γ on $\widehat{L}/K$ is tame. Since the action of Γ on X is ergodic, it follows that the image of θ lies in a single Γ-orbit; i.e., we can view θ as a measurable Γ-map $X \to \sigma(\Gamma)/F$ where F is the intersection of $\sigma(\Gamma)$ with some conjugate of K. In particular, F is finite. Letting $\Lambda = \sigma^{-1}(F)$ and $\Lambda_0 = \ker(\sigma)$, we observe that if $\pi(\Gamma)$ is not finite, then Γ/Λ_0 is infinite, linear (and therefore residually finite [8]), and Λ/Λ_0 is finite. Since the action is engaging, this contradicts Proposition 3.5.

Theorem 5.2. *Suppose G is a connected semisimple Lie group each of whose simple factors has real rank at least 2. Let H be an algebraic $\overline{\mathbf{Q}}$-group such that every homomorphism of Lie algebras $\mathfrak{g}_{\mathbf{C}} \to \mathfrak{h}$ is trivial. Assume there is an action of G on M that preserves a finite measure and is topologically engaging. If $\pi_1(M)$ is infinite, then there is no injective homomorphism $\pi_1(M) \to H_{\overline{\mathbf{Q}}}$.*

Proof. The proof of Theorem 5.1 applies, and we will now have in addition that Λ_0 is trivial since we have an injective homomorphism. This contradicts the assumption of topological engaging by Proposition 3.6.

We now consider the situation in which H is an algebraic group defined over $\mathbf{R}$, and we only assume there is nontrivial homomorphism $\mathfrak{g} \to \mathfrak{h}_{\mathbf{R}}$. To see the meaning of the distinction of this hypothesis from that of nonexistence of homomorphisms $\mathfrak{g} \to \mathfrak{h}$, let $H_{\mathbf{R}}$ be the Lorentz group $O(1,n)$. Then for any simple Lie group G with $\mathbf{R}\text{-rank}(G) \geq 2$, there is no nontrivial homomorphism $\mathfrak{g} \to \mathfrak{h}_{\mathbf{R}}$, but, of course, there may well be a nontrivial homomorphism $\mathfrak{g} \to \mathfrak{h} = o(n+1, \mathbf{C})$. Thus, for example, the following theorems will give us much more information on actions on spaces with a fundamental group that embeds in a Lorentz group.

The following three theorems can be proven by a routine modification of the argument given in the proofs of Theorems 5.1 and 5.2.

Theorem 5.3. *Suppose G is a connected semisimple Lie group each of whose simple factors has real rank at least 2 . Let H be an algebraic $\mathbf{R}$-group such that every homomorphism of Lie algebras $\mathfrak{g} \to \mathfrak{h}_{\mathbf{R}}$ is trivial. Assume there is an action of G on M that preserves a finite measure.*

   (a) *Suppose the action is engaging. Then every homomorphism $\pi_1(M) \to H_{\mathbf{R}}$ with discrete image is finite.*
   (b) *Suppose the action is topologically engaging and $\pi_1(M)$ is infinite. Then there is no injective homomorphism $\pi_1(M) \to H_{\mathbf{R}}$ with discrete image.*

Theorem 5.4. *Suppose G is a connected semisimple Lie group each of whose simple factors has real rank at least 2 . Let H be an algebraic $\mathbf{Q}$-group such that every homomorphism of Lie algebras $\mathfrak{g} \to \mathfrak{h}_{\mathbf{R}}$ is trivial. Assume there is an action of G on M that preserves a finite measure.*

   (a) *Suppose the action is engaging. Then every homomorphism $\pi_1(M) \to H_{\mathbf{Q}}$ is finite.*
   (b) *Suppose the action is topologically engaging and $\pi_1(M)$ is infinite. Then there is no injective homomorphism $\pi_1(M) \to H_{\mathbf{Q}}$.*

Theorem 5.5. *Suppose G is a connected semisimple Lie group each of whose simple factors has real rank at least 2 . Assume there is an action of G on M that preserves a finite measure.*

   (a) *Suppose the action is engaging. Then every homomorphism of $\pi_1(M)$ into a product of finitely many p-adic groups with discrete image has finite image.*
   (b) *Suppose the action is topologically engaging and $\pi_1(M)$ is infinite. Then $\pi_1(M)$ is not isomorphic to a discrete subgroup of a product of finitely many p-adic groups (where p may vary).*

For future reference, we also state a purely ergodic theoretic result, which follows by the same argument.

Theorem 5.6. *Let G be as in Theorem 5.1, and suppose that S is an ergodic G-space with a finite invariant measure. Let $d(G)$ be the minimal dimension of a nontrivial (complex) representation of $\mathfrak{g}$. Suppose $\Lambda \subset \mathrm{GL}(n, \overline{\mathbf{Q}})$ is a finitely* *generated subgroup, where $n < d(G)$. If $\alpha: S \times G \to \Lambda$ is a cocycle, then α is equivalent to a cocycle into a finite subgroup of Λ.*

6. On the unitary dual of $\pi_1(M)$

It is not necessarily true that if there is an engaging or topologically engaging action of G (which we assume to be as above, and, in particular, to satisfy Kazhdan's property (T)) on M then $\pi_1(M)$ is also Kazhdan. Namely, suppose $\Gamma \subset G \times H$ is an irreducible lattice where G is Kazhdan but H is not. Such examples are easily constructed in $\mathrm{SO}(p,q) \times \mathrm{SO}(1, p+q-1)$ for instance. Then the action of G on $M = (G \times H)/\Gamma$ is both engaging and topologically engaging, but as Γ has a dense image homomorphism into a non-Kazhdan group, Γ is not Kazhdan. On the other hand, the following remark shows that the identity is isolated in the space of finite unitary representations, i.e., those that factor through a finite quotient of Γ. These and related properties of a group have been discussed in a number of places, and we refer to [7] as an example and indication of other references.

Proposition 6.1. *Suppose G is a connected semisimple Kazhdan Lie group and that there is an engaging finite measure preserving action of G on M. Then the identity representation of $\pi_1(M)$ is isolated in the space of finite representations of $\pi_1(M)$.*

Proof. If not, we can choose finite quotients $p_n: \Gamma \to F_n$ and nontrivial irreducible representations (π_n, V_n) of F_n such that $\pi_n \circ p_n \to I$. Let $\alpha: M \times \tilde{G} \to \pi_1(M)$ be the cocycle defined by the action on the universal cover. We can suppose the measure on M is ergodic. Form the unitary representation σ_n of $\tilde{G}$ on $L^2(M; V_n)$ by $(\sigma_n(g)f)(m) = \pi_n(p_n(\alpha(m,g)))f(mg)$. Then $\{\sigma_n\}$ is readily seen to weakly contain I (cf. [12, proof of 9.1.1]), and since $\tilde{G}$ is Kazhdan, there is some n for which σ_n has a nonzero invariant vector. Since (π_n, V_n) is contained in the regular representation of F_n, it is easy to see that this implies there is a nonconstant $\tilde{G}$-invariant function in $L^2(M \times F_n)$, where the action of $\tilde{G}$ on $M \times F_n$ is defined by the cocycle α. This means the action on the finite cover of M corresponding to F_n is not ergodic, so that the action is not engaging.

7. Amalgamated products

Using a result of Adams and Spatzier [1], we can obtain further restrictions on fundamental groups of G-spaces. Namely, suppose G is a Kazhdan group acting ergodically with a finite invariant measure on S. Let $H = H_1 *_K H_2$ be an amalgamated product of discrete groups. Then one of the main results of [1] asserts that any cocycle $\alpha: S \times G \to H$ is equivalent to a cocycle into H_1

or H_2. We therefore can deduce

Theorem 7.1. *Suppose G is as in Theorem 5.1. Suppose, for $i = 1,2$, that Γ_i is a discrete group that is either amenable or isomorphic to an infinite finitely generated subgroup of $\mathrm{GL}(n,\overline{\mathbf{Q}})$, where $n < d(G)$ (and the latter is as in Theorem 5.6). Then there is no engaging or topologically engaging action of G on M with a finite invariant measure for any M with $\pi_1(M) = \Gamma_1 *_K \Gamma_2$, where K is some common subgroup.*

We leave the other variants of this theorem to the reader. (Cf. [8] for some useful relevant information on amalgamated products.)

8. ACTIONS OF DISCRETE GROUPS

There are a number of problems that arise when one tries to extend some of the above results to the case in which we assume only that there is an action of a lattice subgroup of G rather than an action of G itself. The first problem is considered in the following.

Definition 8.1. Suppose Γ is a discrete group acting on M. We call the action admissible if there is a finite index subgroup $\Lambda \subset \Gamma$ so that the action of Λ lifts to an action on $\widetilde{M}$, commuting with the action of $\pi_1(M)$.

Example 8.2. (a) The action of $\mathrm{SL}(n,\mathbf{Z})$ on T^n by group automorphisms is not admissible, even though the action lifts.

(b) If the Γ action extends to an action of a connected simply connected group, the action is admissible.

We shall also need some information on $H^2(\Gamma,\mathbf{R})$. Let us say that Γ satisfies cohomology condition (C) if $H^2(\Lambda,\mathbf{R}) = 0$ for every finite index subgroup $\Lambda \subset \Gamma$. If the symmetric space associated to the semisimple group G is Hermitian, then this will not be true, but in most other situations this will be the case. For example, it will be true for any cocompact lattice in $\mathrm{SL}(n,\mathbf{R})$ if $n > 4$, and there are strong positive results for noncocompact lattices as well. In particular, this is true for $\mathrm{SL}(n,\mathbf{Z})$ for $n \geq 6$. See Borel's papers [2, 3], and Borel and Wallach [4] for an extensive discussion.

Proposition 8.3. *Suppose G is as in Theorem 5.1 and $\Gamma \subset G$ is a lattice satisfying condition (C). Let d be the smallest dimension of a nontrivial representation of the Lie algebra of G. If Γ acts on M and $\pi_1(M) = \mathbf{Z}^n$ for $n < d$, then the Γ action is admissible.*

Proof. Letting D be the group of homeomorphisms of $\widetilde{M}$ obtained by taking all possible lifts of all elements of Γ acting on $\widetilde{M}$, we obtain an extension $0 \to \mathbf{Z}^n \to D \to \Gamma \to 0$. In particular, we have a homomorphism $\Gamma \to \mathrm{Aut}(\mathbf{Z}^n) = \mathrm{SL}(n,\mathbf{Z})$. By Margulis superrigidity [12], since $n < d$ this homomorphism is trivial on a subgroup $\Lambda \subset \Gamma$ of finite index, and hence $0 \to \mathbf{Z}^n \to D' \to \Lambda \to 0$ is a central extension, where D' is the inverse image of Λ. This is defined by an element of $H^2(\Lambda,\mathbf{Z}^n)$. Since $H^2(\Lambda,\mathbf{R}) = 0$, we have a surjection

$H^1(\Lambda, T^n) \to H^2(\Lambda, \mathbf{Z}^n)$. Since Λ is Kazhdan, every element of $H^1(\Lambda, T^n)$ is trivial on a subgroup of finite index, and hence this central extension is trivial on a subgroup of finite index.

We remark that we actually need assume only that $\pi_1(M)$ is finitely generated and abelian for this proof to work.

Definition 8.4. (a) Suppose the action of Γ on M is admissible. The action is called engaging if there is a lift of a finite index subgroup Λ to $\widetilde{M}$ (commuting with the fundamental group) so that on every finite cover $M' \to M$, all Λ-invariant measurable functions are lifted from M.

(b) Call the action on M topologically engaging if there is some $\lambda \in \Lambda$ such that λ acts tamely on $\widetilde{M}$ and has almost all ergodic components on M properly ergodic. (That is, the ergodic components are not orbits. This will be the case, for example, if the Γ action on M is mixing; i.e., the matrix coefficients of the unitary representation on $L^2(M)$ vanish at ∞.)

Theorem 8.5. *Let G, Γ, and d be as in Proposition 8.3. Then there is no engaging or topologically engaging finite measure preserving action of Γ on M if $\pi_1(M)$ is an infinitely generated abelian group of rank less than d.*

Proof. Every cocycle $\alpha\colon M \times \Gamma \to \pi_1(M)$ is equivalent to one into a finite subgroup, using Kazhdan's property [12, 9.1.1]. This is incompatible with either engagement assumption.

In particular, Theorem 8.5 applies to actions of lattices in $\mathrm{SL}(n, \mathbf{Z})$ on tori of smaller rank, at least for n sufficiently large. In this context this result has been independently established by D. Witte (private communication).

REFERENCES

1. S. Adams and R. Spatzier, *Kazhdan groups, cocycles, and trees* (to appear).

2. A. Borel, *Stable real cohomology of arithmetic groups*, Ann. Sci. École Norm Sup. (4) 7 (1974), 235–272.

3. ———, *Stable real cohomology of arithmetic groups*, II, Progr. Math., Vol. 15, Birkhauser, Boston, 1981, pp. 21–55.

4. A. Borel and N. Wallach, *Continuous cohomology, discrete subgroups, and representations of reductive groups*, Ann. of Math. Stud., Vol. 94, Princeton Univ. Press, Princeton, NJ, 1980.

5. M. Gromov, *Rigid transformation groups* (to appear).

6. F. E. A. Johnson, *On the existence of irreducible discrete subgroups in isotypic Lie groups of classical type*, Proc. London Math. Soc. 56 (1988), 51–77.

7. A. Lubotzky and R. J. Zimmer, *Variants of Kazhdan's property for subgroups of semisimple groups*, Israel J. Math. (to appear).

8. W. Magnus, *Residually finite groups*, Bull. Amer. Math. Soc. 75 (1969), 306–316.

9. R. J. Zimmer, *Induced and amenable actions of Lie groups*, Ann. Sci. École Norm. Sup. (4) 11 (1978), 407–428.

10. ———, *Strong rigidity for ergodic actions of semisimple Lie groups*, Ann. of Math. (2) 112 (1980), 511–529.

11. ____, *Orbit equivalence and rigidity of ergodic actions of Lie groups*, Ergodic Theory Dynamical Systems **1** (1981), 237–253.

12. ____, *Ergodic theory and semisimple groups*, Birkhauser, Boston, 1984.

13. ____, *Ergodic theory and the automorphism group of a G-structure*, Group representations, ergodic theory, operator algebras, and mathematical physics (C. C. Moore, ed.), Springer, New York, 1987, pp. 247–278.

14. ____, *Arithmeticity of holonomy groups of Lie foliations*, J. Amer. Math. Soc. **1** (1988), 35–58.

DEPARTMENT OF MATHEMATICS, UNIVERSITY OF CHICAGO, CHICAGO, ILLINOIS 60637

ISRAEL JOURNAL OF MATHEMATICS, Vol. 74, Nos. 2-3, 1991

SUPERRIGIDITY, RATNER'S THEOREM, AND FUNDAMENTAL GROUPS

BY

ROBERT J. ZIMMER

Department of Mathematics, University of Chicago,
5734 University Avenue, Chicago, IL 60637, USA

ABSTRACT

We discuss the implications of superrigidity and Ratner's theorem on invariant measures on homogeneous spaces for understanding the fundamental group of manifolds with an action of a semisimple Lie group.

If a non-compact simple Lie group acts on a manifold M, possibly preserving some geometric structure, it is natural to enquire as to the possible relationship between the structure of G and that of $\pi = \pi_1(M)$. Results in this direction appear in the work of Gromov [1], the author [7], and Spatzier and the author [3].

A basic tool in some of these results is the superrigidity theorem for cocycles [4], or, in more invariant terms, the superrigidity theorem for actions of G on principal bundles. In this paper we show how to obtain further results on this question by combining techniques of superrigidity with Ratner's recent solution to the Raghunathan conjecture on invariant measures on homogeneous spaces. This appears to be a new direction of application of Ratner's fundamental work. An interesting feature of our results is that the results in [3],[7],[9] give "lower bounds" on the possible fundamental groups given the presence of a G-action, while in this work we are able to give a type of "upper bound".

Let G be a connected simple Lie group with finite center and with $\mathbf{R}$-rank$(G) \geq 2$. We assume G acts continuously on a compact manifold M preserving a probabil-

Research partially supported by the National Science Foundation and the Israel-U.S. Binational Science Foundation.
Received December 21, 1990

ity measure. Then $G^{\sim}$ acts on $M^{\sim}$. We recall briefly the notions of engaging and topologically engaging from [7]. Namely, the action is called engaging if there is no loss of ergodicity in passing to finite covers of M, and topologically engaging if some element $g \in G^{\sim}$ both projects to an element not contained in a compact subgroup of G and acts tamely on $M^{\sim}$, e.g., acts properly on a conull set. A fundamental result of Gromov is that if the G action is real analytic and preserves a real analytic connection, then the action of $G^{\sim}$ on the universal cover of M is proper on a conull set, and in particular is topologically engaging. All known examples of smooth G actions preserving a volume are both engaging and topologically engaging. The action of $G^{\sim}$ on the principal bundle $M^{\sim} \to M$ yields a measurable cocycle $\alpha \cdot G^{\sim} \times M \to \pi$, with the property that for each $g \in G^{\sim}$, $\alpha(g, \cdot)^{\pm 1}$ is a bounded function on M. Therefore, for any representation $\sigma : \pi \to$ GL$(n,\mathbf{C})$, we clearly have $\sigma(\alpha(g, \cdot)^{\pm 1})$ is bounded. This enables us to freely apply the results of [8], which yields information on the structure of such cocycles under these (in fact weaker) boundedness conditions.

We now recall the statements of superrigidity and of Ratner's theorem in the form in which we shall need them.

THEOREM 1. (Superrigidity [4],[8]) *Let G and M be as above, and let H be a product of finitely many (rational points of) connected algebraic groups defined over (possibly varying) local fields of characteristic 0. Let $\lambda : G^{\sim} \times M \to H$ be a cocycle with the property that $\lambda(g, \cdot)^{\pm 1}$ is bounded for each g. Assume the action of G on M is ergodic.*

(a) *Assume the algebraic hull of the cocycle [4] is (algebraically) connected. Then there is a continuous homomorphism $\theta : G^{\sim} \to H$ and a compact subgroup $K \subset H$ commuting with $\theta(G^{\sim})$ such that λ is equivalent to a cocycle of the form $\beta(g,m) = \theta(g)c(g,m)$ where $c(g,m) \in K$.*

(b) *With no connectivity assumption on the algebraic hull, we can obtain the same conclusion by lifting λ to a cocycle on a finite extension of M.*

THEOREM 2. (Ratner [2]) *Let H be a connected Lie group, $\Gamma \subset H$ a discrete subgroup, and $G \subset H$ a connected simple non-compact subgroup. Let ν be a finite G-invariant ergodic measure on H/Γ, where the action is given by the embedding of G in H. Then there is a closed connected subgroup L with $G \subset L \subset H$ and a point $x \in H/\Gamma$ say with stabilizer in H being $h\Gamma h^{-1}$ such that:*

(i) *$L \cap h\Gamma h^{-1}$ is a lattice in L; and*

(ii) *ν is the measure on H/Γ corresponding to the invariant volume on $L/L \cap h\Gamma h^{-1}$ under the natural bijection $L/L \cap h\Gamma h^{-1} \cong Lx \subset H/\Gamma$.*

The next lemma is the observation that one can combine these results.

LEMMA 3.   *Let G and H be as in Theorem 1, and write $H = H_\infty \times H_f$ where H_∞ is the product of the real and complex terms, and H_f the product of the totally disconnected terms. Let $\Gamma \subset H$ be a discrete subgroup, and assume that $\Gamma \cap H_f = \{e\}$. Let M be an ergodic G-space with a finite invariant measure, and let $\lambda : G^\sim \times M \to \Gamma \subset H$ be a cocycle with the property that $\lambda(g, \cdot)^{\pm 1}$ is bounded for each g. We also assume that this cocycle is not equivalent (as a cocycle into Γ) to a cocycle into a finite subgroup of Γ. Then, either with the assumption that the algebraic hull of λ is Zariski connected, or alternatively, by passing to a finite ergodic extension of M, we have:*

There is a non-trivial homomorphism $\theta : G^\sim \to H_\infty$ (which automatically factors to a homomorphism of a finite cover of G), and a closed subgroup L, $\theta(G^\sim) \subset L \subset H_\infty$, and a compact subgroup $C \subset L$ commuting with $\theta(G^\sim)$ such that:

(i)   *Γ contains a subgroup isomorphic to a lattice $\Gamma' \subset L$; in fact we can take Γ' to be the intersection of L with a conjugate of $p_\infty(\Gamma)$, where p_∞ is the projection of H onto H_∞.*

(ii)   *There is a measure preserving $G^\sim$-map $M \to C \backslash L / \Gamma'$, where the measure on the latter derives from the projection of the L-invariant volume form on L/Γ'; in particular $\theta(\pi_1(G))$ acts trivially on $C \backslash L / \Gamma'$, and hence we have a G-map $M \to C \backslash L / \Gamma'$.*

For the proof we will need the following easy fact.

LEMMA 4.   *Let $\alpha : M \times G \to \Gamma$ be a cocycle for any ergodic group action taking values in a countable group Γ. Suppose $i : \Gamma \to L$ is an embedding of Γ as a discrete subgroup of a locally compact group L. If $i \circ \alpha$ is equivalent to a cocycle into a compact subgroup $K \subset L$, the α is equivalent to a cocycle into a finite subgroup of Γ.*

PROOF OF LEMMA 3.   Choose θ, β, c, and K as in Theorem 1. By Lemma 4, θ is non-trivial. Thus, there is a measurable map $\psi : M \to H$ such that $\psi(gm)\alpha(g,m)\psi(m)^{-1} = c(m,g)\theta(g)$. Rewriting this as $c(m,g)^{-1}\psi(gm)\alpha(g,m) = \theta(g)\psi(m)$, we see that in $K \backslash H / \Gamma$ we have $\omega(gm) = \theta(g)\omega(m)$ where ω is the composition of ψ with the projection $H \to K \backslash H / \Gamma$. If μ is the G-invariant measure on M, let $\omega_*(\mu)$ be the projection of this measure to $K \backslash H / \Gamma$. Thus, we have a $\theta(G^\sim)$-invariant, ergodic, probability measure on $K \backslash H / \Gamma$. Let ν_1 be the lift of this measure to H / Γ by taking the image of Haar measure on each of the fibers of $H / \Gamma \to K \backslash H / \Gamma$. This measure is a $\theta(G^\sim)$-invariant probability measure but it is not *a priori* ergodic. However, the arguments of [5] show that any ergodic component, say ν, still projects onto $\omega_*(\mu)$. Fix such a ν, which is now a $\theta(G^\sim)$-invariant, ergodic, probability measure on $(H_\infty \times H_f)/\Gamma$.

Let $T \subset H_f$ be a compact open subgroup. By enlarging K if necessary, we may assume T to be chosen such that $T \subset K$. Let $\Gamma_\infty = \Gamma \cap (H_\infty \times T)$. Since T is compact, the projection of Γ_∞ to H_∞ (which we recall is injective on Γ by assumption) is still discrete. We can identify $(H_\infty \times T)/\Gamma_\infty \subset (H_\infty \times H_f)/\Gamma$ as an open $G^\sim$-invariant subset. Since T is open in H_f, we can find a countable number of translates of $(H_\infty \times T)/\Gamma_\infty$ by elements of H_f whose union covers all of $(H_\infty \times H_f)/\Gamma$. Since H_f commutes with $\theta(G^\sim)$ these translates are all open and $G^\sim$ invariant, and it follows that ν is supported on one of these translates, say translation by $h \in H_f$. We may thus translate ν itself back via h^{-1} to $(H_\infty \times T)/\Gamma_\infty$, and obtain a $G^\sim$-invariant ergodic measure on this set, and by projection, a $G^\sim$-invariant ergodic measure, say ν_2 on H_∞/Γ_∞, where we have identified Γ_∞ with its image under projection. Assertion (i) now follows immediately from Ratner's theorem. To see (ii), we simply observe that we may view ω as a $G^\sim$-map $M \to K \backslash (H \times hT)/\Gamma_\infty$, and via projection we obtain an equivariant map $M \to p(K) \backslash H_\infty/\Gamma_\infty$, where p is the projection. Since the image of μ under this map is clearly the same as the image of ν_2, (ii) follows.

As discussed in [7], for an action which is engaging or topologically engaging, the $\pi_1(M)$ valued cocycle, say α, defined by the lift of the action to the universal cover of M is not equivalent to a cocycle into a finite subgroup.

COROLLARY 5. *Let G be as in Theorem 1, and suppose G acts on a compact manifold M, preserving a finite measure. Suppose the action is either engaging, topologically engaging, or real analytic connection and volume preserving. If $\pi_1(M)$ is isomorphic to a discrete subgroup of a Lie group, it contains a lattice in a Lie group L which contains a subgroup locally isomorphic to G.*

In particular, this applies if we take $\pi_1(M)$ to be a subgroup of $GL(n, \Theta)$ for some n, where Θ is the ring of algebraic integers.

COROLLARY 6. *Let G and M be as in Corollary 5. Suppose $\pi_1(M)$ is isomorphic to a lattice Γ in G. Then the action is measurably isomorphic to an action induced from an action of a lattice commensurable with Γ.*

PROOF.   Since Γ is Zariski dense in G, we must clearly have $L = G$ and C being trivial in Lemma 3.

From Corollary 6, we also recover the main results of [9].

COROLLARY 7.   *Let G and M be as in Corollary 5. Suppose there is a faithful representation $\pi_1(M) \to \mathrm{GL}(n, \mathbf{Q}^-)$ for some n. Then $\pi_1(M)$ contains a lattice in a Lie group L where L contains a group locally isomorphic to G.*

This follows from the fact that any finitely generated subgroup of $\mathrm{GL}(n, \mathbf{Q}^-)$ is isomorphic to a discrete subgroup in a product of algebraic groups as in Lemma 3. From Corollary 7 we can also recover a number of the results of [7] concerning faithful representations.

One obtains a rather different type of result by combining Lemma 3 with some considerations of entropy. Namely, from Lemma 3 one can sometimes deduce that simply by knowing the fundamental group of M, any (engaging or topologically engaging) G-action on M must have entropy bounded below by specific algebraic data. This in turn can place dimension restrictions on M given the fundamental group. To describe this precisely, we introduce some notation.

DEFINITION 8.   Let $\Gamma \subset L$ be a lattice in a connected Lie group L, and let G be another Lie group. If $H \subset L$ is a closed connected subgroup and $\Lambda \subset H$ is a lattice in H, we call the pair (H, Λ) G-related (or (G, θ)-related if more precision is required) if there is some non-trivial homomorphism $\theta : G^- \to L$ with $\theta(G^-) \subset H \subset L$, and some conjugate $\Gamma' = \lambda \Gamma \lambda^{-1}$ of Γ in L with $H \cap \Gamma' = \Lambda$. We shall also call a subgroup $\Gamma_1 \subset \Gamma$ G-related if there is a G-related (H, Λ) with $\Gamma_1 = \lambda^{-1} \Lambda \lambda$, with λ as above.

We can rephrase Lemma 3 in this context as follows. We recall that a homomorphism of groups is called an isogeny if it is surjective with finite kernel.

COROLLARY 9.   *Let G and M be as in Corollary 5. Assume G acts ergodically on M. Suppose there is an isogeny $\pi_1(M) \to \Gamma$, where Γ is a lattice in a connected Lie group L. Then there is a finite ergodic extension M' of M, a G-related pair (H, Λ) in L, a compact subgroup $C \subset H$ commuting with the image of θ, and a measurable G-equivariant measure preserving map $M' \to C \backslash H / \Lambda$.*

If G acts on a space X with invariant probability measure, we let $h(g, X)$ denote the Komogorov–Sinai entropy of the transformation defined by g. Since entropy is unchanged by a finite ergodic extension, or more generally by an isometric extension (i.e., via a homogeneous space of a compact group [5]), it follows that with the notation of Corollary 9 we have for each $g \in G$, $h(g, M) \geq h(g, H/\Lambda)$. We recall that the entropy on H/Λ can be computed from purely algebraic information.

Namely, if A is a matrix, we define its entropy $h(A)$ to be $\sum \log(|\lambda|)$, where λ runs through the eigenvalues of A of magnitude at least 1. Then for any $a \in H$, $h(a, H/\Lambda) = h(\mathrm{Ad}_H a)$. Further, given any G-invariant probability measure on M, its entropy can be computed by integrating the entropy over the ergodic components. Therefore, we have:

COROLLARY 10. *Let G and M be as in Corollary 5. Suppose there is an isogeny $\pi_1(M) \to \Gamma$, where Γ is a lattice in a connected Lie group L. Then with respect to any G-invariant probability measure on M, we have for any $g \in G^\sim$ that*

$$h(g, M) \geq \min\{h(\mathrm{Ad}_H(\theta(g))) \mid \text{where } H \subset L \text{ is } (G, \theta)\text{-related}\}.$$

Now consider the case in which M is a smooth compact manifold with a G-invariant volume. The entropy of each $g \in G$ can the be computed via Pesin's formula, which in light of Theorem 1 yields a close relationship between $\dim(M)$ and possible values of the entropy. See [4] for a discussion. We can formulate this as follows. For each positive integer m, and $g \in G^\sim$, let

$$c_m(g) = \max\{h(\pi(g)) \mid \pi : G^\sim \to \mathrm{GL}(m, \mathbf{R}) \text{ is a linear representation}\}.$$

This of course is completely calculable in principle, and is certainly easily calculated in low dimensions. Then by [4] we have for any smooth volume preserving action of G on M with $\dim(M) \leq m$ that $h(g, M) \leq c_m(g)$. We therefore deduce:

COROLLARY 11. *Let G be as in Theorem 1. Let M be a compact smooth manifold of dimension m, with a volume preserving action of G. Assume the action is either engaging or topologically engaging (e.g., C^ω connection preserving). Suppose there is an isogeny $\pi_1(M) \to \Gamma$, where Γ is a lattice in a connected Lie group L. Then*

$$c_m(g) \geq \min\{h(\mathrm{Ad}_H(\theta(g))) \mid \text{where } H \subset L \text{ is } (G, \theta)\text{-related}\}.$$

This result can be viewed as giving an upper bound on certain features of Γ if there is an action of G on a compact manifold of dimension at most m and with fundamental group Γ. Corollary 11 is of little use in the extreme case in which there is a representation θ such that $\theta(G)$ intersects Γ in a lattice in $\theta(G)$. (The conclusion in that case is still of interest, but it can be deduced in a much more elementary manner, and in fact is true for any non-compact simple Lie group. See [6], e.g.) On the other hand, for a lattice Γ with very few G-related groups, Corollary 11 can be quite strong. An example at this extreme is given by the following result.

THEOREM 12.   *Let n be a prime. Then there is a cocompact lattice $\Gamma \subset \mathrm{SL}(n, \mathbf{R})$ such that any Lie group L intersecting Γ in a lattice in L must be solvable.*

This result follows immediately from:

THEOREM 13. (Kottwitz)   *Let p be a prime and D be a central division algebra over $\mathbf{Q}$ of dimension p^2. Let G be the algebraic $\mathbf{Q}$-group isomorphic over $\mathbf{R}$ to $\mathrm{SL}(p, \mathbf{R})$ and whose $\mathbf{Q}$-points consist of the elements of reduced norm 1 in $D^\times$. Let H be a connected reductive $\mathbf{Q}$-subgroup of G, with $H \neq G$. Then H is a torus.*

To see that Theorem 12 follows from Theorem 13, we simply take $\Gamma = G(\mathbf{Z})$, where G is as in Theorem 13. Thus Γ is a cocompact lattice in $G(\mathbf{R}) = \mathrm{SL}(p, \mathbf{R})$. If $L \subset \mathrm{SL}(p, \mathbf{R})$ is connected and intersects Γ in a lattice, then $M =$ the algebraic hull of $L \cap \Gamma$ is a $\mathbf{Q}$-group. Applying Theorem 13 to a Levi factor we deduce that the connected component of M is solvable, and hence that L contains a solvable lattice. This implies that L itself is solvable.

PROOF OF THEOREM 13.   For ease of notation we shall use H, G to denote the $\mathbf{Q}$-points of these groups. The maximal $\mathbf{Q}$-tori in $D^\times$ are of the form $E^\times$ for fields E with $\mathbf{Q} \subset E \subset D$ and $[E : \mathbf{Q}] = p$. Let $\mathcal{G} = \mathrm{Gal}(\mathbf{Q}^-/\mathbf{Q})$. The $\mathbf{Q}$-subtori of $E^\times$ correspond bijectively to the $\mathcal{G}$-invariant subspaces of the finite dimensional $\mathbf{Q}$-vector space $V = X^*(E^\times) \otimes_{\mathbf{Z}} \mathbf{Q}$. Note that V is a permutation representation of $\mathcal{G}$ with the $\mathcal{G}$-set $\mathrm{Hom}_{\mathbf{Q}\text{-alg}}(E, \mathbf{Q}^-)$ as basis. Let J denote the image of $\mathcal{G}$ in the group of permutations of the p element set $\mathrm{Hom}_{\mathbf{Q}\text{-alg}}(E, \mathbf{Q}^-)$. Then $p = [E : \mathbf{Q}]$ divides $|J|$, so J contains an element σ of order p, which must act on $\mathrm{Hom}_{\mathbf{Q}\text{-alg}}(E, \mathbf{Q}^-)$ as a p-cycle. Any $\mathcal{G}$ invariant subspace of V is invariant for the cyclic group C of order p generated by σ, and since V is isomorphic as a C-module to the group algebra $\mathbf{Q}[C]$, it follows that the only C-invariant subspaces of V correspond to the four obvious $\mathcal{G}$-invariant subspaces. Therefore, the only $\mathbf{Q}$-subtori of $E^\times$ are: $\{e\}$, $\mathbf{Q}^\times$, $E^\times$, and $T_E = \ker\{N : E^\times \to \mathbf{Q}^\times\}$, where in the latter N is the norm map. The maximal $\mathbf{Q}$-tori in G are the tori T_E; and hence their only $\mathbf{Q}$-subtori are $\{e\}$ and T_E.

Now consider a maximal $\mathbf{Q}$-torus in the connected reductive $\mathbf{Q}$-subgroup H. It is a subtorus of some T_E, and hence is either $\{e\}$ or T_E. If it is $\{e\}$, then H is also trivial. If it is T_E, then H has the same reductive rank as G, which, since H is an inner form of $\mathrm{SL}(p)$, implies that $H = G \cap D'$ for some semisimple $\mathbf{Q}$-subalgebra of D containing E. But since $\dim_E D'$ must divide $\dim_E D = p$, the subalgebra D' must be either D or E. The former case does not arise since $H \neq G$, and the latter case gives $H = T_E$.

In the context of Corollary 11, we consider the following example. Let n be prime and large, and let Γ be as in Theorem 12. Let $G = \mathrm{SL}(3,\mathbf{R})$. Let Ad be the adjoint representation of $\mathrm{SL}(n,\mathbf{R})$. Then for any non-trivial homomorphism $\theta : G \to \mathrm{SL}(n,\mathbf{R})$, we have a real analytic engaging ergodic action of G on the $n^2 - 1$ dimensional manifold $\mathrm{SL}(n,\mathbf{R})/\Gamma$. Let M be a compact manifold of dimension m. Then by Corollary 11 and Theorem 12 we deduce

$$c_m(g) \geq \min\{h(\mathrm{Ad}(\theta(g))) \,|\, \theta : G \to \mathrm{SL}(n,\mathbf{R}) \text{ is a non-trivial representation}\}.$$

For m small enough relative to n it is clear that this is impossible. (One can, in principle, compute both these numbers precisely.) One can of course do the same for any simple Lie group G of higher real rank. Thus:

COROLLARY 14. *Let G be a connected simple Lie group with finite center and* **R**-rank$(G) \geq 2$. *Fix a positive integer m. Then there is a finitely generated group Γ (which we may take to be the group Γ in Theorem 12) with the properties:*

(i) *There exists a compact real analytic manifold X with an isogeny $\pi_1(X) \to \Gamma$ such that there is a real analytic, connection preserving, volume preserving, engaging, ergodic action of G on X.*

(ii) *For any compact smooth manifold M with $\dim(M) \leq m$ and for which there is an isogeny $\pi_1(M) \to \Gamma$, there is no smooth volume preserving action of G on M which is either engaging, or topologically engaging, or real analytic preserving a real analytic connection.*

Corollary 14 represents a new type of phenomenon. In [7],[3] conditions are given under which a group cannot appear as the fundamental group of such an M in any dimension. Here we see that some groups may appear as the fundamental group, but only in a sufficiently large dimension.

ACKNOWLEDGEMENT

We would like to thank R. Kottwitz for the proof of Theorem 13, A. Lubotzky for numerous conversations on this work, and A. Borel and G. Prasad for helpful communication concerning this work.

REFERENCES

1. M. Gromov, *Rigid transformation groups*, in *Geometrie Differentielle* (D. Bernard and Y. Choquet-Bruhat, eds.), Hermann, Paris, 1988.
2. M. Ratner, *On Raghunathan's measure conjecture*, to appear.

3. R. Spatzier and R. J. Zimmer, *Fundamental groups of negatively curved manifolds and actions of semisimple groups*, Topology, to appear.

4. R. J. Zimmer, *Ergodic Theory and Semisimple Groups*, Birkhauser, Boston, 1984.

5. R. J. Zimmer, *Extensions of ergodic group actions*, Ill. J. Math. **20** (1976), 373–409.

6. R. J. Zimmer, *Ergodic theory and the automorphism group of a G-structure*, in *Group Representations, Ergodic Theory, Operator Algebras, and Mathematical Physics* (C.C. Moore, ed.), Springer, New York, 1987.

7. R. J. Zimmer, *Representations of fundamental groups of manifolds with a semisimple transformation group*, J. Am. Math. Soc. **2** (1989), 201–213.

8. R. J. Zimmer, *On the algebraic hull of the automorphism group of a principal bundle*, Comm. Math. Helv. **65** (1990), 375–387.

9. R. J. Zimmer, *Unitary spectrum and the fundamental group for actions of semisimple Lie groups*, Math. Ann. **287** (1990), 697–701.

Topology Vol. 30, No. 4, pp. 591–601, 1991.
Printed in Great Britain

0040-9383/91 $3.00 + .00
© 1991 Pergamon Press plc

FUNDAMENTAL GROUPS OF NEGATIVELY CURVED MANIFOLDS AND ACTIONS OF SEMISIMPLE GROUPS

Ralf J. Spatzier† and Robert J. Zimmer‡

(*Received in revised form* 17 *July* 1990)

1. INTRODUCTION

THIS PAPER is part of the investigation of the fundamental groups of manifolds on which a non-compact semisimple Lie group may act. Earlier work in this direction has dealt largely with the linear representation theory of the fundamental group. In this paper we consider homomorphisms into isometry groups of negatively curved manifolds, and in particular the question as to when a semisimple Lie group may act on a manifold whose fundamental group is isomorphic to that of a negatively curved manifold.

THEOREM A. *Let G be a connected simple Lie group with finite center, finite fundamental group, and $\mathbb{R}$-rank$(G) \geq 2$. Let M be a compact manifold and suppose there is a real analytic action of G on M preserving a (real analytic) connection and a finite measure. Then $\pi_1(M)$ is not isomorphic to the fundamental group of any complete Riemannian manifold N with negative curvature bounded away from 0 and $-\infty$.*

The result extends to semisimple groups with $\mathbb{R}$-rank$(G) \geq 2$ as long as we make the further technical assumption of irreducibility, i.e., the non-compact normal subgroups of G act ergodically on the G-ergodic components.

In the case in which we further assume that N is locally symmetric, the conclusion of Theorem A follows from [15]. With no such locally symmetric assumption on N but in the special case that $M = G/\Gamma$ for $\Gamma \subset G$ a cocompact lattice and the symmetric space associated to G is Hermitian, the result has been recently obtained by Jost and Yau [7] as a corollary of results on the fundamental groups of more general Kähler manifolds. In fact, the present paper was motivated by [7], first to remove the hypothesis that the symmetric space associated to G be Hermitian, and second to establish the result for fundamental groups of manifolds on which G preserves a connection but is not necessarily transitive. (We also remark in this context that the hypothesis in Theorem A that $\pi_1(G)$ is finite can often be dispensed with simply by passing to a suitable lower dimensional subgroup with this property.) Our techniques, however, are completely different from those of Jost–Yau. While in [7] the essential technique was that of exploiting the presence of a certain harmonic map, we shall use in a fundamental way the techniques of algebraic ergodic theory.

One can obtain other generalizations of the results of [7] under some varied hypotheses. We recall from [15] that an action of a connected group G on a manifold M is called engaging if there is no loss of ergodicity in passing to finite covers; i.e., if $M' \to M$ is a finite

†Research partially supported by NSF Grant DMS-88-03053 and an Alfred P. Sloan Foundation Grant.
‡Research partially supported by NSF Grant.

592   Ralf J. Spatzier and Robert J. Zimmer

cover, then any measurable $\tilde{G}$-invariant function on M' is lifted from a function on M. (Here $\tilde{G}$ is the universal cover of G.)

THEOREM B. *Let G be a connected simple Lie group with finite center and $\mathbb{R}$-rank(G) ≥ 2. Suppose there is a continuous, finite measure preserving engaging action of G on a compact manifold M preserving a finite measure. Let N be a complete, simply connected Riemannian manifold with sectional curvature $-b^2 \leq K \leq -a^2$, where $a, b \in \mathbb{R}$, $a \neq 0$. Let $I(N)$ be the isometry group of N, and $h: \pi_1(M) \to I(N)$ be a homomorphism. If $h(\pi_1(M))$ is discrete and residually finite then it is finite. (In fact, $h(\pi_1(M))$ cannot be a discrete subgroup with finite quotients of arbitrarily large cardinality.)*

The information we have about homomorphism into $I(N)$ with non discrete image (say under the hypotheses of Theorems A or B) is not as complete. We recall that a group is called F-simple if every normal subgroup is either finite or of finite index. Margulis has shown that there are large classes of natural F-simple groups [8, 14]. Combining Theorem B with the results of [15], one can show

COROLLARY C. *Under the hypotheses of Theorem B with $\pi_1(M)$ F-simple and residually finite, and $\dim(N)$ sufficiently small with respect to G, then any homomorphism $\pi_1(M) \to I(N)$ has finite image. Without the dimension assumption on N we have that either the image is finite or the Lie algebra of $I(N)$ has a simple subalgebra (necessarily of $\mathbb{R}$-rank ≤ 1) whose complexification contains the complexification of $\mathfrak{g}$.*

There are two central ingredients to the proofs of Theorem A, B. The first are the techniques developed in [13], where certain questions of measurable orbit equivalence are investigated. Although we shall use a number of the arguments of [13], the key point is a generalization of a result proven by Margulis in his investigation of normal subgroups of lattices [8, 14]. Namely, Margulis proved that with G as in Theorem B, $P \subset G$ a parabolic subgroup, and $\Gamma \subset G$ a lattice, then any measurable quotient of the Γ-action on G/P is (measurably) of the form G/P' where $P' \supset P$ is another parabolic subgroup. In [13], this was generalized to the following assertion. If M is any G-space with a finite invariant ergodic measure and X is a measurable G-space for which there are measure class preserving G-maps $M \times G/P \to X \to M$ whose composition is projection, then X is of the form $M \times G/P'$ for some P'. This result, together with other properties of the action of G on $M \times G/P'$ will be basic to the proofs of Theorems A, B.

The second central ingredient is various properties of the action of the isometry group $I(N)$ on ∂N, the boundary of N (i.e., asymptotic classes of geodesics on N). Some of these properties we shall need are known. However the basic new result in this direction we need is the following.

THEOREM 3.1. *Let N be a complete simply connected Riemannian manifold with sectional curvature $-b^2 \leq K \leq -a^2$, $a, b \in \mathbb{R}$, $a \neq 0$. Let $\Gamma \subset I(N)$ be a discrete subgroup and μ any quasi-invariant measure for the Γ action on ∂N. Then the action of Γ on $(\partial N, \mu)$ is amenable.*

(See [14] for a discussion of amenable actions.) If N is symmetric, then ∂N has a natural smooth structure and amenability of Γ on ∂N with the Lebesgue measure class can be established quite directly and algebraically. (See [14], e.g.) If N is not symmetric, then ∂N is in general only a topological sphere. It does however carry a natural measure class that is quasi-invariant under $I(N)$, and in [10] it is established that if $\Gamma \subset I(N)$ is discrete and cocompact, then the Γ action on ∂N is amenable with respect to this measure class. It will be

NEGATIVELY CURVED MANIFOLDS   593

important for our present purposes to have this for any quasi-invariant measure, and the main idea of the proof of [10] will also be important in the proof of Theorem 3.1. The assumption in [10] that Γ be cocompact was used there via an application of Bowen's result on the hyperfiniteness of the unstable foliation of an Anosov flow on a compact manifold. Bowen's proof in turn depends upon symbolic dynamics and does not easily generalize. Here, instead, we use geometry directly to obtain polynomial growth for this foliation, which yields amenability.

In Theorem A, the main use of the assumption that the action preserve an analytic connection is that the action is essentially proper on the universal cover. This result is due to Gromov [5, §6.1]. In fact, the proof of Theorem A shows that one obtains the same conclusions for any action preserving a finite measure such that the action on the universal cover is proper. This condition can actually be weakened further to that of topological engaging. (Cf. [15, Definition 2.2])

2. PRELIMINARIES

We collect here some ergodic theoretic information, mostly known, which we will need for the proofs of Theorems A, B.

Let $P \to M$ be a continuous principal H-bundle, and suppose G acts on P (say on the right) by principal bundle automorphisms. Choosing a measurable trivialization of P, say $P = M \times H$ (measurably), the G action will be given by $(m, h)g = (mg, \alpha(m, g)^{-1}h)$ where $\alpha \colon M \times G \to H$ is a cocycle, i.e., it satisfies the identity $\alpha(m, g_1 g_2) = \alpha(m, g_1)\alpha(mg_1, g_2)$. The commuting action of H is of course simply given by right translation on the second factor. Two such cocycles $\alpha, \beta \colon M \times G \to H$ are called equivalent if there is a measurable $\varphi \colon M \to H$ such that $\varphi(m)\alpha(m, g)\varphi(mg)^{-1} = \beta(m, g)$; this is equivalent to the assertion that α, β correspond to the same action via a different choice of measurable trivialization. If H acts continuously (on the right) on the space X, then one forms the associated bundle E with fiber X, namely $E = (P \times X)/H$. There is a natural action of G on this space by fiber bundle automorphisms, which under a measurable trivialization takes the form $E = M \times X$, $(m, x)g = (mg, x\alpha(m, g))$. A measurable G-invariant section of E corresponds under this trivialization to a measurable map $\varphi \colon M \to X$ such that $\varphi(mg) = \varphi(m)\alpha(m, g)$. We shall call such a map α-invariant.

A basic idea we shall use is that of an amenable action [14]. The following is a direct consequence of the definition (and is almost equivalent to it).

LEMMA 2.1. *Suppose M is an amenable G-space, $\alpha \colon M \times G \to H$ is a cocycle and Y is a compact convex H-space (on which H acts by affine transformations). Then there is a measurable α-invariant map $\varphi \colon M \to Y$.*

A basic construction we shall require is that of the Mackey range of a cocycle. If $\alpha \colon M \times G \to H$ is a cocycle, and G acts on $M \times H$ via the above formula, let X be the space of ergodic components of the G-action. Then H acts on X, and this action is called the Mackey range. If the action of G on $M \times H$ is tame (e.g. if the action of G comes from an action on a principal bundle which is proper or more generally has locally closed orbits), then X is the orbit space $X = (M \times H)/G$. (See [14] for details.) If Y is a continuous H-space, and $\varphi \colon M \to Y$ is α-invariant, then the map $M \times H \to Y$ given by $(m, h) \to \varphi(m)h$ is obviously G-invariant for the action of G on $M \times H$ defined by α, and is an H-map. It thus induces an H-map $\tilde{\varphi} \colon X \to Y$. Thus, we have:

LEMMA 2.2. *If* $\alpha : M \times G \to H$ *is a cocycle with Mackey range* X, *and* Y *is a continuous* H-*space, then* $\varphi \mapsto \tilde{\varphi}$ *defines a bijection between* α-*invariant functions* $M \to Y$ *and* H-*maps* $X \to Y$. (*All functions here are identified if they agree a.e..*)

Suppose now that the action of G on M is ergodic. Then the action of H on the Mackey range will also be ergodic. Suppose now that H is a countable discrete group. An action of H on a measure space M is called recurrent if for every $A \subset M$ of positive measure, and a.e. $m \in A$, there is a sequence $h_n \in H$, $h_n \to \infty$ in H, such that $mh_n \in A$. The following is routine.

LEMMA 2.3. *Suppose the countable discrete group* H *acts ergodically on* M. *The following are equivalent:*

(i) *the action is not recurrent;*
(ii) *the action is equivalent (modulo null sets) to the* H-*action on* H/F *where* $F \subset H$ *is a finite group.*

Condition (ii) is related to the Mackey range via:

LEMMA 2.4. *If* G *acts ergodically on* M *and* $\alpha : M \times G \to H$ *is a cocycle where* H *is a countable discrete group, then the following are equivalent:*

(i) *the Mackey range is* H/F *for some finite subgroup* F.
(ii) α *is equivalent to a cocycle taking all values in a finite subgroup of* H.

Now let N be a complete, simply connected Riemannian manifold with negative curvature bounded away from 0. Let ∂N be the boundary and $M(\partial N)$ the space of probability measures on ∂N. Let $I(N)$ be the isometry group of N (which is a Lie group). Then $I(N)$ acts on ∂N and $M(\partial N)$. The following is a basic fact about negative curvature.

LEMMA 2.5. *Suppose* $h_n \in I(N)$ *and* $h_n \to \infty$. *Let* $\mu \in M(\partial N)$ *and suppose* $h_n(\mu) \to \nu$. *Then* ν *is supported on at most two points.*

For a proof, see [13]. We shall apply this, as in [13], as follows.

LEMMA 2.6. *Suppose* $\alpha : M \times G \to \Lambda$ *is a cocycle where* G *acts ergodically on* M *and* $\Lambda \subset I(N)$ *is a discrete subgroup. Suppose* $\varphi : M \to M(\partial N)$ *is* α-*invariant. Then either:*

(a) *is equivalent to a cocycle into a finite subgroup of* Λ;
or, (b) *for a.e.* $m \in M$, $\varphi(m)$ *is supported on at most two points.*

Proof. This basically follows from the argument in [13, Theorem 3.7]. For completeness, we indicate a quick proof. If (a) does not hold, then by Lemmas 2.2, 2.3, and 2.4 there is a measurable Λ-map $\tilde{\varphi} : X \to M(\partial N)$ corresponding to φ, where X is a recurrent Λ-space. By the construction of $\tilde{\varphi}$ it clearly suffices to see that for a.e. x, $\tilde{\varphi}(x)$ is supported on at most two points. Set $\psi = \tilde{\varphi}$. Let $M_0 \subset M(\partial N)$ be the measures supported on at most two points, and M_1 the complement of $M_0 \subset M(\partial N)$. Suppose there is a set $A \subset X$ of positive measure such that $\psi(x) \in M_1$ for all $x \in A$. Since $M_0 \subset M(\partial N)$ is closed and $M(\partial N)$ is metrizable, one can write $M_0 = \cap U_n$ where $U_n \supset M_0$ is open and $U_{n+1} \subset U_n$. Then for some j we can choose $B \subset A$ of positive measure such that for all $x \in B$ we have $\psi(x) \notin U_j$. By recurrence, for a.e. $x \in B$ we can find a sequence $\lambda_n \in \Lambda$, $\lambda_n \to \infty$, such that $x \cdot \lambda_n \in B$, so in particular, $\psi(x\lambda_n) = \psi(x)\lambda_n \notin U_j$. By passing to a subsequence we can assume $\psi(x)\lambda_n$ converges in

$M(\partial N)$. From Lemma 2.5, this limit is in M_0, which is clearly impossible since $\psi(x)\lambda_n \notin U_j$. This proves the lemma.

A basic difference between negative curvature and symmetric spaces of higher rank is the contrast of Lemma 2.5 to the following simple result.

LEMMA 2.7. [13, Prop. 4.4] *Let G be a semisimple Lie group with $\mathbb{R}$-rank$(G) \geq 2$ and no compact factors. Let $P \subset G$ be a proper parabolic subgroup. Then there is an $\mathbb{R}$-split (non-trivial and hence non-compact) abelian subgroup $A \subset G$ and a non-atomic probability measure $v \in M(G/P)$ which is A-invariant.*

Finally, we recall one of the basic results of [13] which plays a key role in the present paper, and to which we alluded in the introduction.

LEMMA 2.8. [13, Theorem 4.1] *Let G be a connected simple Lie group with finite center and $\mathbb{R}$-rank$(G) \geq 2$. Let M be an ergodic G-space with a finite invariant measure. Let $P \subset G$ be a parabolic subgroup. If Z is an ergodic G-space for which there exists measure class preserving maps $M \times G/P \to Z \to M$ whose composition is projection, then there is a parabolic subgroup $P' \supset P$ such that Z is isomorphic as a measurable G-space to $M \times G/P'$ in such a way that the above maps correspond to the obvious ones.*

(We remark that in [13] it was assumed that G has trivial center. However, the argument works in the finite center case just as well.)

3. AMENABILITY OF BOUNDARY ACTIONS

The point of this section is to prove the following theorem which generalizes the main result of [10] and is of central importance to the proofs of Theorems A, B.

THEOREM 3.1. *Let N be a complete simply connected Riemannian manifold with sectional curvature $-b^2 \leq K \leq -a^2$, $a, b \in \mathbb{R}$, $a \neq 0$. Let $\Gamma \subset I(N)$ be a discrete group of isometries and $\mu \in M(\partial N)$ be quasi-invariant under Γ. Then the Γ-action on $(\partial N, \mu)$ is amenable.*

We remark that amenability for every quasi-invariant measure implies that the stabilizers are all amenable [14], although the converse is far from true. However, we will prove Theorem 3.1 by proving first amenability of all stabilizers and then amenability of the equivalence relation defined by the action. This will suffice by virtue of the following simple remark.

LEMMA 3.2. *Let Γ be a countable discrete group acting on a measure space (X, μ) where μ is quasi-invariant. Let R be the equivalence relation on X defined by the orbits. If R is amenable (in the sense of [12]) and the stabilizer Γ_x is amenable for a.e. $x \in X$, then the Γ-action is amenable.*

Proof. We simply apply the definition. Let $A = \{A_x\}$ be an affine Γ-space over X. Thus, α is a cocycle defined on $X \times \Gamma$ and $\alpha(x, \gamma): A_{x\gamma} \to A_x$ is an affine transformation where A_x is compact and convex. For each $x \in X$, let $\pi_x = \alpha | \{x\} \times \Gamma_x$, so π_x is a representation of Γ_x by affine transformation of A_x. Let $B_x \subset A_x$ be the set of $\pi_x(\Gamma_x)$-fixed points. Then restricting $\alpha(x, \gamma)$ to $B_{x\gamma}$ we obtain an affine Γ-space $B = \{B_x\}$ over X on which the stabilizers act trivially. Thus, the cocycle acting on B can be considered as being defined on R. Since R is amenable, there is an α-invariant section in $\{B_x\} \subset \{A_x\}$, and this establishes amenability.

For further background information on Hadamard manifolds, (i.e. complete simply connected Riemannian manifolds of non-positive curvature) and their isometries we refer to [1]. Now assume the hypotheses of Theorem 3.1. Let $B = \partial N$.

LEMMA 3.3. *Let N and Γ be as above, and let $x \in \partial N$. Then the isotropy subgroup Γ_x of x in Γ is amenable.*

Proof. First suppose that Γ_x does not contain a hyperbolic isometry. Then Γ_x preserves each horosphere based at x [1, Lemmata 6.6 and 6.8]. Fix a stable horosphere H with center at x, and let $p \in H$. Let $\Delta \subset \Gamma_x$ be any subgroup generated by a finite set F. Then the elements in F move p by at most a distance C. Since the curvatures are bounded by $K \leq -a^2$, the geodesic flow g_t contracts H uniformly by at least e^{-ta}. Thus the elements in F move $g_t p$ by arbitrarily little for large enough t. By the Margulis' Lemma, it follows that Δ is almost nilpotent, and in particular amenable. Since an increasing union of amenable groups is amenable, the lemma follows in this case.

It remains to argue the case when Γ_x contains a hyperbolic element γ. Let c be its axis. If $\delta \in \Gamma_x$ then δ fixes the other endpoint of c by [3, Proposition 6.8]. Since N is negatively curved, there is a unique geodesic connecting two distinct points at infinity. Therefore δ preserves c. Now consider the restriction homomorphism from Γ_x to translations of c. As the kernel is discrete and a subgroup of the orthogonal group at $c(0)$, it is finite. Since the image is discrete the claim follows.

Remark. The conclusion of Lemma 3.3 is false if we do not assume that the curvature is bounded below. This follows from the examples of warped products in [3, Proposition 9.1]. The point this assumption is used in the argument is in the application of the Margulis lemma. (See [1, Section 8.3] for a proof of the latter.) These examples also explicitly show the need for the assumption that the curvature be bounded below in Theorems A, B, C, and 3.1.

The next result is well known for compact manifolds of negative curvature or more generally for the stable leaves of an Anosov flow on a compact manifold [4].

PROPOSITION 3.4. *Let N be above. Then every horosphere of N has polynomial growth.*

Proof. Let H be a horosphere in N. We denote the distance in H with respect to the induced metric by $h(\cdot, \cdot)$ while d denotes the distance in N. Heintze and Im Hof proved the following comparison between these distances [6]: for all points $p, q \in H$ we have

$$\frac{2}{a} \sinh\left(\frac{a}{2} d(p, q)\right) \leq h(p, q) \leq \frac{2}{b} \sinh\left(\frac{b}{2} d(p, q)\right).$$

Let $B_R^{\text{horo}}(p)$ denote the ball of radius R about p in H in the horospherical metric and $B_R(p)$ the ball of radius R in N about p. By the above equation there is a constant C such that for all large enough R we have the inclusion

$$B_R^{\text{horo}}(p) \subset B_{C + (2/a)\log R}(p).$$

Let B' be the set of unit tangent vectors inward normal to $B_R^{\text{horo}}(p)$, and let $\pi: SN \to N$ denote the canonical projection where SN is the unit tangent bundle of N. Set

$$A = \pi\left(\bigcup_{0 \leq t \leq 1} g_t B'\right)$$

NEGATIVELY CURVED MANIFOLDS   597

where g_t denotes the geodesic flow on the unit tangent bundle of N. Let b denote the Busemann function of H. Recall that $\|\nabla b\| = 1$. By the coarea formula we find

$$\text{vol } A = \int_0^1 \text{vol}^{\text{horo}} \pi(g_t B') \, dt$$

where vol^{horo} denotes the volume functions on the horospheres with respect to the induced metric. Since $K \geq -b^2$ we get for $0 \leq t \leq 1$ that $\text{vol}^{\text{horo}} \pi(g_t B') \geq e^{-(m-1)b} \text{vol}^{\text{horo}} B_R^{\text{horo}}(p)$, where m is the dimension of H. By Bishop's theorem on volume comparison, the growth of the volume of balls in N is at most $\text{vol } B_r(p) \leq C' e^{(m-1)br}$ for some constant C'. Since $A \subset B_{C+1+(2/a)\log R}(p)$ we find

$$C' e^{(m-1)b(C+1)} R^{(m-1)2b/a} \geq \text{vol } B_{C+1+(2/a)\log R}(p) \geq \text{vol } A \geq e^{-(m-1)b} \text{vol}^{\text{horo}}(B_R^{\text{horo}}(p)).$$

Notice that the growth is in fact independent of the particular horosphere.

COROLLARY 3.5. *Let Γ be a discrete group of isometries of N and set $M = \Gamma \backslash N$.*

(i) *The (strong stable) horospherical foliation of the unit tangent bundle of N has polynmomial growth, and is therefore amenable with respect to any quasi-invariant measure.*

(ii) *The weak stable foliation is also amenable with respect to any quasi-invariant measure.*

Proof. (i) follows from a standard averaging argument. (ii) Since the geodesic flow permutes the leaves of the strong stable foliation and $\mathbb{R}$ is amenable, amenability of the weak stable foliation follows by the same argument that shows a group to be amenable if it has a normal amenable subgroup with amenable quotient [14].

PROPOSITION 3.6. *Let Γ be a discrete group of isometries of N. Then the action of Γ on ∂N defines an amenable equivalence relation with respect to any quasi-invariant measure m.*

Proof. The argument is entirely analogous to that in [10]. Therefore we only give an outline.

Let Δ denote the diagonal in $\partial N \times \partial N$. We identify the unit tangent bundle SN of N with $(\partial N \times \partial N - \Delta) \times \mathbb{R}$ in the usual way. Define a measure $\bar{\mu}$ on SN by

$$d\bar{\mu} = dm \times dm \times dt.$$

Clearly, $\bar{\mu}$ is quasi-invariant under the weak stable foliation and under Γ. Let $M = \Gamma \backslash N$, let SM be its unit tangent bundle and let $\rho : SN \to SM$ denote the obvious projection. Then $\mu \overset{\text{def}}{=} \rho_*(\bar{\mu})$ is quasi-invariant under the weak stable foliation of SM. The weak stable foliation and the Γ-action on ∂N are "dual" in the sense that they define equivalence relations with the same quotient space. Since the weak stable foliation is amenable with respect to μ, one can now deduce the amenability of the equivalence relation defined by the Γ-action using this "duality" (cf. [10, Lemma 1]).

Proof of Theorem 3.1. The result follows immediately from 3.2, 3.3 and 3.6.

COROLLARY 3.7. *Let Γ be as in Theorem 3.1. Let the symmetric group on two letters, S_2, act naturally on $\partial N \times \partial N$, so $Q = (\partial N \times \partial N)/S_2$ can be identified with the subsets of ∂N consisting of one or two points. If μ is any measure on Q quasi-invariant under the Γ-action, then the Γ-action on (Q, μ) is amenable.*

Proof. Let $q: \partial N \times \partial N \to Q$ be the natural map. Then μ lifts to an $S_2 \times \Gamma$-quasi-invariant measure $\tilde{\mu}$ on $\partial N \times \partial N$, and it suffices to see that this action is amenable [11]. Letting p be projection onto the first factor, the action on Γ on $(\partial N, p_*(\tilde{\mu}))$ is amenable by Theorem 3.1. That $(\partial N \times \partial N, \tilde{\mu})$ is also amenable, then follows from [11].

4. PROOF OF THE MAIN RESULTS

We fix an ergodic finite G-invariant measure on M. Let $\Gamma = \pi_1(M)$, and $h: \Gamma \to I(N)$ be a homomorphism. We set $\Lambda = h(\Gamma)$ and assume $\Lambda \subset I(N)$ is discrete. We let $\alpha: M \times G \to \Gamma$ be the cocycle coming from the lift of the G-action to the universal cover $\tilde{M}$. (We remark that since $\pi_1(G)$ is finite, we may actually assume G is simply connected.) For any subgroup $H \subset G$, we let X_H be the Mackey range of the cocycle $(h \circ \alpha)|M \times H$, which is a cocycle $M \times H \to \Lambda$. Let $\tilde{\alpha}$ be the lift of α to the G-space $M \times G/P$, where $P \subset G$ is the minimal parabolic, i.e., $\tilde{\alpha}(m, x, g) = \alpha(m, g)$. We recall that $M \times G/P$ is an amenable ergodic G-space [14]. It follows (from Lemma 2.1) that there is an $(h \circ \tilde{\alpha})$-invariant map $\varphi: M \times G/P \to M(\partial N)$. Let $\psi: M \times G/P \to M \times_{h \circ \alpha} M(\partial N)$ be $\psi(m, x) = (m, \varphi(x))$ where $M \times_{h \circ \alpha} M(\partial N)$ is the G-space with action defined by the cocycle $h \circ \alpha$; i.e. $(m, \mu)g = (mg, [h \circ \alpha(m, g)]^{-1} \mu)$. Projecting the product measure class on $M \times G/P$ to $M \times_{h \circ \alpha} M(\partial N)$ via ψ, we obtain a sequence of measure class preserving G-maps $M \times G/P \xrightarrow{\psi} M \times_{h \circ \alpha} M(\partial N) \xrightarrow{\text{proj}} M$ whose composition is projection. Thus, by Lemma 2.8, $M \times_{h \circ \alpha} M(\partial N)$ is isomorphic as a measurable G-space (with the measure constructed above) to $M \times G/P_0$ where $P_0 \supset P$ is another parabolic.

LEMMA 4.1. *With notation as above, the following are equivalent:*

(i) $P_0 = G$.
(ii) X_G *is a non-recurrent Λ-space.*
(iii) $h \circ \alpha$ *is equivalent to a cocycle into a finite subgroup of Λ.*
(iv) $X_G \cong \Lambda/\Lambda_0$ *as measurable Λ-spaces where $\Lambda_0 \subset \Lambda$ is finite.*

Proof. (ii), (iii) and (iv) are equivalent by Lemma 2.3, 2.4. Suppose $h \circ \alpha \sim \beta$ where $\beta(M \times G) \subset \Lambda_0$, where $\Lambda_0 \subset \Lambda$ is finite. Then $M \times_{h \circ \alpha} M(\partial N) \cong M \times_\beta M(\partial N)$. The ergodic measure ν on $M \times_\beta M(\partial N)$ (given by the isomorphism with $M \times G/P_0$ and the product measure on the latter) can be decomposed as $\nu = \int_{x \in M}^{\oplus} \nu_x$, where $\nu_x \in M(M(\partial N))$. Since Λ_0 is finite and ν is ergodic, it is easy to see that ν_x must be supported on finitely many points for a.e. $x \in M$. Since $M \times_\beta M(\partial N) \cong M \times G/P_0$ via an isomorphism commuting with the projections to M, it follows that the standard measure on G/P_0 is supported on finitely many points, which clearly implies $P_0 = G$.

Conversely, suppose $P_0 = G$. This means that there is an $h \circ \alpha$-invariant map $M \to M(\partial N)$. By Lemma 2.2, there is a corresponding map $\psi: X_G \to M(\partial N)$. We claim X_G is not a recurrent Λ-space. If it is, then by Lemma 2.6 we have $\psi(x)$ supported on one or two points for a.e. $x \in X_G$. Thus, we may identify ψ with a map $\psi: X_G \to Q = (\partial N \times \partial N)/S_2$. By Corollary 3.7, Q is an amenable Λ-space with the Λ-quasi-invariant measure $\psi_*(\nu_G)$ where ν_G represents the measure class on X_G. It then follows from [11, Theorem 2.4, p. 359] that X_G is an amenable Λ-space. If X_G is essentially transitive, then $X_G \cong \Lambda/\Lambda_1$, where Λ_1 is amenable [14, Proposition 4.3.2, p. 72]. Therefore, $h \circ \alpha$ is equivalent to a cocycle taking values in Λ_1. Since G is a Kazhdan group it follows [14, Theorem 9.1.1] that $h \circ \alpha$ is equivalent to a cocycle into a finite subgroup of Λ_1, so that in fact $X_G \cong \Lambda/\Lambda_0$ where Λ_0 is finite, contradicting recurrence of X_G. On the other hand, if X_G is not essentially transitive,

NEGATIVELY CURVED MANIFOLDS   599

but rather is properly ergodic, then there is a cocycle $c: X_G \times \Lambda \to \mathbb{Z}$ whose Mackey range is a properly ergodic $\mathbb{Z}$-space [2, 14]. However, by general properties of the Mackey range [9] this implies that there is a cocycle $M \times G \to \mathbb{Z}$ with the same (properly ergodic) Mackey range. However, by Kazhdan's property again, any $\mathbb{Z}$-valued cocycle on $M \times G$ is co-homologicaly trivial, so the Mackey range must be the action of $\mathbb{Z}$ on itself. This contradiction completes the proof.

We shall need the following variant of the implication (iii) $\Rightarrow$ (i) in the above lemma.

LEMMA 4.2. *With notation as above suppose $(h \circ \alpha)|M \times P$ is equivalent to a cocycle into a finite subgroup of Λ. Then $P_0 = G$.*

Proof. We have $M \times_{h \circ \alpha} M(\partial N) \cong M \times G/P_0$ as G-spaces. The group P acts essentially transitively on G/P_0 with a non-compact stabilizer. It follows from Moore's ergodicity theorem [14] that P acts ergodically on $M \times G/P_0$, and hence on $M \times_{h \circ \alpha} M(\partial N)$. The argument of the first part of the proof of Lemma 4.1 now applies (just letting P act rather than G) to prove Lemma 4.2.

We now apply Lemma 2.7 to give a further condition which ensures $P_0 = G$.

LEMMA 4.3. *Suppose X_A is recurrent for every $\mathbb{R}$-split connected (non-trivial) abelian subgroup $A \subset G$. Then $P_0 = G$.*

Proof. We recall the $h \circ \tilde{\alpha}$-invariant map $\varphi: M \times G/P \to M(\partial N)$. This corresponds to an $h \circ \alpha|(M \times P)$-invariant function $\varphi_0: M \to M(\partial N)$. By Lemmas 4.2 and 2.6, if $P_0 \neq G$, then $\varphi_0(m)$ is supported on one or two points for a.e. $m \in M$. It follows that $\varphi(m, x)$ is supported on one or two points for a.e. $(m, x) \in M \times G/P$. Therefore we may view the sequence of G-spaces,

$$M \times G/P \to M \times_{h \circ \alpha} M(\partial N) \to M$$

as (up to null sets),

$$M \times G/P \to M \times_{h \circ \alpha} Q \to M,$$

where $Q = (\partial N \times \partial N)/S_2$ as above. Thus, we have $M \times G/P_0 \cong M \times_{h \circ \alpha} Q$.

Now choose a measure v on G/P_0 which is non-atomic and invariant under some $\mathbb{R}$-split non-trivial abelian $A \subset G$, as in Lemma 2.7. This corresponds under the above isomorphism to an $(h \circ \alpha)|M \times A$-invariant map $\theta: M \to M(Q)$, and by Lemma 2.2 to a Λ-map $\tilde{\theta}: X_A \to M(Q)$. This lifts naturally to a Λ-map $X_A \to M(\partial N \times \partial N)$. Projecting onto each factor we obtain Λ-maps $X_A \to M(\partial N)$. Since X_A is recurrent, the essential range of this map is supported on atomic measures, by Lemma 2.5. The same is therefore true of $\tilde{\theta}$ and hence θ. However, this contradicts the fact that v is non-atomic, verifying the lemma.

Proof of Theorem A. By [5], the action of G is topologically engaging [15], as is the restriction to any non-compact closed subgroup. It follows from [15] that the cocycle $\alpha|M \times A$ is not equivalent to a cocycle into a finite subgroup of Γ, for any non-compact closed $A \subset G$. Suppose h is injective with discrete image. Then obviously $(h \circ \alpha)|M \times A$ is also not equivalent to a cocycle into a finite subgroup. However, from Lemmas 2.3, 2.4, we have that X_A is recurrent, and by Lemma 4.3 that $P_0 = G$. Lemma 4.1 then provides a contradiction, completing the proof.

600   Ralf J. Spatzier and Robert J. Zimmer

Proof of Theorem B. The argument above shows that for any h with discrete image, $h \circ \alpha$ is equivalent to a cocycle into a finite subgroup of Λ. This implies that α is equivalent to a cocycle into a subgroup $\Gamma_1 \subset \Gamma$ such that

(i) $\Gamma_1 \supset \Gamma_0$ with Γ_0 normal in Γ;
(ii) Γ_1/Γ_0 is finite; and
(iii) Γ/Γ_0 is residually finite.

However, if Γ/Γ_0 is infinite this contradicts the fact that the G-action on M is engaging by [15, 3.5].

Remark. Proposition 3.5 of [15] asserts that for an engaging action α is not equivalent to a cocycle into a subgroup $\Gamma_1 \subset \Gamma = \pi_1(M)$ such that (i), (ii), (iii) (as above) hold with Γ/Γ_0 infinite. An examination of the proof of [15, 3.5] shows that one may replace the condition that Γ/Γ_0 be infinite and residually finite by the weaker condition that Γ/Γ_0 have finite quotients of arbitrarily large cardinality.

Proof of Corollary C. Suppose im(h) is discrete. If the image Λ is infinite, then $\Lambda = \Gamma/F$ where F is finite and normal. Since Γ is residually finite, Λ has finite quotients of arbitrarily large cardinality. By Theorem B this is impossible. Thus, Λ is finite.

Suppose now that Λ is not discrete. Let $H = \bar{\Lambda}$ in $I(N)$. Since $\Lambda \cap H^0$ is normal in Λ, by passing to a subgroup of Γ of finite index we may assume H is connected. If $L \subset H$ is a maximal semisimple subgroup, from the fact that N has strictly negative curvature it follows that there is at most one non-compact simple factor L_1 of L and that L_1 has $\mathbb{R}$-rank 1. (Namely, a non-compact abelian subgroup of $I(N)$ has a fixed point set on ∂N consisting of 1 or 2 points. Therefore, a connected centralizer of this group also leaves these points invariant. Therefore, in $\mathbb{R}$-rank at least 2 we would have a non compact simple group fixing a point on ∂N as it can arise as such a centralizer. However, we have already this is impossible, for example via Lemma 3.3.) By the main results of [15], it follows that $\mathfrak{g}_c$ embeds in the complexified Lie algebra of L. Finally, if $\dim I(N) < \dim G$ (which of course follows if $\dim X$ is sufficiently small), then this is impossible, so Λ must in fact be discrete.

Acknowledgements—We would like to thank Michael Anderson and Kevin Corlette for some helpful conversations.

REFERENCES

1. W. BALLMANN, M. GROMOV and V. SCHROEDER: *Manifolds of non-positive curvature*, Birkhauser-Boston (1985).
2. A. CONNES, J. FELDMAN and B. WEISS: An amenable equivalence relation is generated by a single transformation, *Erg. Th. Dyn. Syst.* **1** (1981), 431–450.
3. P. EBERLEIN and B. O'NEILL: Visibility manifolds, *Pac. J. Math.* **46** (1973), 45–109.
4. M. GROMOV: Groups of polynomial growth and expanding maps, *Publ. Math. I.H.E.S.* **53** (1981), 53–73.
5. M. GROMOV: Rigid transformation groups, in: D. Bernard, Y. Choquet-Bruhat, *Géométrie différentielle*, *Travaux en course*, Herman, Paris (1988).
6. E. HEINTZE and H.-C. IM HOF: Geometry of horospheres, *J. Diff. Geom.* **12** (1977), 481–491.
7. J. JOST and S.-T. YAU: Harmonic maps and group representations, preprint.
8. G. A. MARGULIS: Quotient groups of discrete groups and measure theory, *Funct. Anal. Appl.* **12** (1978), 295–305.
9. A. RAMSAY: Virtual groups and group actions, *Adv. Math.* **6** (1971), 253–322.
10. R. J. SPATZIER: An example of an amenable action from geometry, *Erg. Th. Dyn. Syst.* **7** (1987), 289–293.
11. R. J. ZIMMER: Amenable ergodic groups actions and an application to Poisson boundaries of random walks, *J. Funct. Anal.* **27** (1978), 350–372.

NEGATIVELY CURVED MANIFOLDS   601

12. R. J. ZIMMER: Hyperfinite factors and amenable ergodic actions, *Invent. Math.* **41** (1977), 23–31.
13. R. J. ZIMMER: Ergodic theory, semisimple Lie groups, and foliations by manifolds of negative curvature, *Publ. Math. I.H.E.S.* **5** (1982), 37–62.
14. R. J. ZIMMER: *Ergodic theory and semisimple groups*, Birkhauser-Boston (1984).
15. R. J. ZIMMER: Representations of fundamental groups of manifolds with a semisimple transformation group, *J. Amer. Math. Soc.* **2** (1989), 201–213.

Department of Mathematics
State University of New York
Stony Brook, N Y 11777, U.S.A.

Current address:
Department of Mathematics
University of Michigan
Ann Arbor, MI 48103, U.S.A.

Department of Mathematics
University of Chicago
Chicago, IL 60637, U.S.A.

A canonical arithmetic quotient for simple Lie group actions.*

Alexander Lubotzky**
Institute of Mathematics
Hebrew University
Jerusalem, Israel

Robert J. Zimmer
Department of Mathematics
University of Chicago
Chicago, IL 60637

Preprint No. 5
1997/98

*Research partially supported by grants from NSF and the Binational Science Foundation, Israel - USA.
**Partially supported by the Edmund Landau Center for Research in Mathematical Analysis and Related Areas, sponsored by the Minerva Foundation (Germany).

1

1   Introduction

The aim of this paper is to establish the existence of an essentially unique maximal arithmetic virtual quotient action for a broad class of actions of semisimple Lie groups. This includes all finite measure preserving ergodic actions of such groups with finite entropy. This entropy condition is automatically satisfied for actions on compact manifolds. In particular, our results provide arithmetic invariants of such actions, namely an algebraic $\mathbb{Q}$-group and the associated commensurability class of arithemetic subgroups.

We also discuss examples of results which ensure non-triviality of this quotient, and in particular the relationship to invariant geometric structures and representations of fundamental groups.

2   The canonical arithmetic quotient.

The aim of this section is to prove for an ergodic, finite measure preserving, finite entropy action of a semisimple Lie group with no compact factors on a compact manifold that there is an essentially unique maximal arithmetic virtual quotient. Our main result is Theorem 2.16. We begin the discussion with some general notions and notation.

Suppose a locally compact group G acts ergodically on a space X, preserving a finite measure μ. A finite extension of X is a G-space X' with a measure preserving map $X' \to X$ and finite fibers. By ergodicity of the action on X, (almost) all fibers have the same cardinality, say l. The action on X' can thus be defined by a cocycle $\alpha : G \times X \to S_l$ (the latter being the symmetric group), and the action being given on $X' \cong X \times \{1, \ldots, l\}$ (as a measure space) by $g \cdot (x, y) = (gx, \alpha(g, x)y)$. The ergodic components of X' will thus have a similar form (with a possibly smaller l.) Given two finite ergodic extensions $X_1 \to X, X_2 \to X$, one can form the fibered product $X_1 \times_X X_2 \to X$, which will be a finite, but not necessarily ergodic extension. However, any ergodic component will project surjectively onto both X_1 and X_2. Thus, we can always find a finite ergodic extension X_3 of X such that

2

we have a commuting diagram

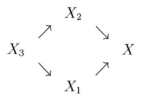

and $X_3 \subset X_1 \times_X X_2$. For a detailed discussion of finite (and more generally compact homogeneous space) extensions, see [Z1].

A basic example (but not the only one of importance) is of course an action on a manifold that lifts to an ergodic action on a finite sheeted covering.

Given an ergodic G-space X, by a quotient of X we mean a G-space Y with invariant measure together with a measure preserving G-map $X \to Y$. By ergodicity of G on X, the map is essentially surjective (i.e. the image is conull) and the action on Y is ergodic. By a virtual quotient of X we mean a quotient of a finite ergodic extension of X.

Example 2.1. Suppose G acts ergodically on X and $G_0 \subset G$ is of finite index and acts ergodically on X. Suppose Y is a G-space and a G_0 quotient for the G-action on X. Then Y is a virtual quotient for the G-action on X. Namely, there is a natural quotient G-map $X \times G/G_0 = X' \to Y$.

We now define a particular class of actions. Let H be an algebraic $\mathbb{Q}$-group, $\Lambda \subset H$ an arithmetic subgroup (i.e. Λ is commensurable with $H_{\mathbb{Z}}$) and $K \subset H_{\mathbb{R}}$ a compact subgroup. Then if H has no $\mathbb{Q}$-characters (e.g., if the reductive Levi component of H is semisimple) then Λ is a lattice, so $K\backslash H_{\mathbb{R}}/\Lambda$ carries a natural finite measure which is preserved by the natural action of $Z_{H_{\mathbb{R}}}(K)$.

Definition 2.2. *By an arithmetic action of a group G we mean an ergodic action of G on a space of the form $K\backslash H_{\mathbb{R}}/\Lambda$ which is defined by a homomorphism $G \to Z_{H_{\mathbb{R}}}(K)$.*

Example 2.3. If G is a non-compact simple Lie group, $G \subset H_{\mathbb{R}}$ where $H_{\mathbb{R}}$ is semisimple with no compact factors, and Λ is an irreducible lattice, then G is ergodic on $H_{\mathbb{R}}/\Lambda$ by Moore's theorem [Z2] and hence defines an

3

arithmetic action of G. (Precise conditions for a subgroup of a general $H_\mathbb{R}$ to act ergodically on $H_\mathbb{R}/\Lambda$ is an extensively studied situation. (c.f. [BM].)

We now collect some information on issues related to arithmetic actions.

For simplicity, for the remainder of this section, for a $\mathbb{Q}$-group H we shall denote $H_\mathbb{R}$ simply by H and refer to $\mathbb{R}$-points of $\mathbb{R}$-groups as real algebraic groups. We also take G to be a semisimple Lie group with no compact factors.

Lemma 2.4. *Let H be a real algebraic $\mathbb{R}$-group and $L \subset H$ a connected Lie subgroup. Suppose $\Gamma \subset L$ is a lattice and that $\Gamma \subset L \cap H_\mathbb{Z}$. Finally, suppose $G \subset L$ and G is ergodic on L/Γ. Then:*

(i) L is a $\mathbb{Q}$-subgroup, and

(ii) $[\mathfrak{l}, \mathfrak{l}] = \mathfrak{l}$, where $\mathfrak{l}$ is the Lie algebra of L.

Proof. We have a G-map $p : L/\Gamma \to \bar{L}/\bar{\Gamma}$ where the closures are in the Zariski topology. Then $p_*(vol_{L/\Gamma})$ is a finite G-invariant ergodic measure on the variety $\bar{L}/\bar{\Gamma}$, and by Borel density [Z2] is supported on a single point. Therefore $L \subset \bar{\Gamma}$, and since $\Gamma \subset L$, we obviously have L is Zariski dense in $\bar{\Gamma}$. Thus, $[L, L]$ is Zariski dense in $[\bar{\Gamma}, \bar{\Gamma}]$. Since $[\mathfrak{l}, \mathfrak{l}]$ is an algebraic Lie algebra, it follows that $[\mathfrak{l}, \mathfrak{l}] = [\text{lie}(\mathfrak{d}), \text{lie}(\mathfrak{d})]$. Thus if $\mathfrak{l} \neq [\mathfrak{l}, \mathfrak{l}]$, we get a map $L/\Gamma \to \bar{\Gamma}/[\bar{\Gamma}, \bar{\Gamma}]\Gamma$ whose image has positive dimension. Since G is ergodic on L/Γ, this implies $\bar{\Gamma}/[\bar{\Gamma}, \bar{\Gamma}]\Gamma$ supports a non-trivial G-ergodic measure, which is impossible since the image of G in $\bar{\Gamma}/[\bar{\Gamma}, \bar{\Gamma}]$ is trivial. Thus, $L = [L, L]$ and is hence algebraic, and since $\Gamma \subset L \subset \bar{\Gamma}$, L is defined over $\mathbb{Q}$.

Definition 2.5. *Let G acting on $K \backslash H/\Lambda$ be an arithmetic action. We say it is of reduced form if G is ergodic on H/Λ.*

Lemma 2.6. *Every arithmetic action of G (which we recall is semisimple with no compact factors) has a finite extension that is an arithmetic action of reduced form.*

Proof. Consider the extension of G-spaces $H/\Lambda \to K \backslash H/\Lambda$. Since the action on the base is ergodic and finite measure preserving, we may choose an ergodic component of the G-action on H/Λ that projects to the standard volume on $K\backslash H/\Lambda$. (c.f. the analysis in [Z1] of ergodic components in compact group extensions.) By Ratner's theorem [R], this measure is supported on some L'-orbit, say $L'/L' \cap h\Lambda h^{-1}$, for some Lie group L' with $G \subset L' \subset H$. Since the projection of this L'-orbit to $K\backslash H/\Lambda$ is of full measure, it follows that there is some $y \in H$ such that $KL'y\Gamma$ is of full dimension in H, and hence KL' is of full dimension. Since K is compact, $KL' = H$. Thus, we have $K \cap L'\backslash L'/L' \cap \Lambda \to K\backslash H/\Lambda$ is a surjective measure preserving map of manifolds of the same dimension and finite volume. This is therefore a finite extension. Replacing L' and its subgroups by $L = h^{-1}L'h$ and the corresponding subgroups, and applying Lemma 2.4, we deduce the desired conclusion.

Lemma 2.7. *If $K\backslash H/\Gamma$ is an arithmetic G-space of reduced form, and $M \subset H$ is a normal $\mathbb{Q}$-subgroup containing G, then $M = H$.*

Proof. H/M is a $\mathbb{Q}$-group and we have a G-map

$$H/\Lambda \to H/M\Lambda \quad (= (H/M)/(H/M)_{\mathbb{Z}}).$$

Since $G \subset M$, the action on the target is trivial, and since G is ergodic on H/Λ (and hence on $H/M\Lambda$), we have $H = M\Lambda$. Since H is connected, $M = H$.

With these preliminaries completed, we now consider arithmetic quotients and virtual quotients of a given ergodic G-space. Supppose H is a $\mathbb{Q}$-group and $\Lambda, \Lambda' \subset H$ are arithmetic (i.e. commensurable with $H_{\mathbb{Z}}$) and hence commensurable. If $G \subset H$ and $K\backslash H/\Lambda$ is a virtual arithmetic quotient of X of reduced form, so is $K\backslash H/\Lambda'$. To see this, it suffices to consider the case $\Lambda' \subset \Lambda$ of finite index. If $X' \to K\backslash H/\Lambda$ is a G-quotient where X' is a finite ergodic extension of X, let X'' be any ergodic component of $X' \times_{K\backslash H/\Lambda} K\backslash H/\Lambda'$. Then X'' will be a finite ergodic extension of X and it has $K\backslash H/\Lambda'$ as a quotient. A similar argument shows that if Λ_1 and Λ_2 are commensurable and $X_i \to K\backslash H/\Lambda_i$ are virtual quotients of X, then there is a finite ergodic extension X' of X with quotients $K\backslash H/\Lambda_1 \cap \Lambda_2$ (and hence both $K\backslash H/\Lambda_i$.) We will speak of this situation as defining commensurable virtual quotients of X. (We remark that if one restricts attention to quotients

5

of a fixed X, one cannot pass in this way between commensurable arithmetic groups. This is only one of the reasons for working with virtual quotients.)

A basic tool in showing the existence of a canonical virtual arithmetic quotient is that of entropy. If G acts ergodically on a space (X, μ) preserving a finite measure μ, for $g \in G$ we denote by $h(g)$ (or $h_X(g)$) the entropy with respect to μ.

Definition 2.8. *We say that G acts with finite entropy if $h(g) < \infty$ for all $g \in G$.*

Example 2.9. If G acts smoothly on a compact manifold, then G acts with finite entropy.

The following is a well-known computation.

Lemma 2.10. *Suppose $G \subset H$ and $M = K \backslash H / \Lambda$ is an arithmetic G-space. Let $A \subset H$ be the maximal $\mathbb{R}$-split torus. Then for $g \in G$,*

$$h_{H/\Lambda}(g) \;=\; h_{K \backslash H / \Lambda}(g) \;=\; \sum \{\log w(g)\}$$

where $w(g)$ is a weight of $Ad_H|G$ with respect to A and $w(g) > 1$.

Corollary 2.11. *If G is semisimple and acts ergodically on X with finite entropy, then the set of possible entropy functions $\{h(g), g \in A\}$ for virtual arithmetic quotients of X is finite.*

Proof. This follows from Lemma 2.10, standard results about the representation theory of G, and the following two facts about extensions. If $X \to Y$ is a quotient of ergodic G-spaces with finite invariant measure, then $h_X(g) \geq h_Y(g)$; and if the extension is finite $h_X(g) = h_Y(g)$.

Definition 2.12. *If $X_i \to Y_i = K_i \backslash H_i / \Lambda_i$ are virtual arithmetic quotients of X, we say that $Y_1 \succ Y_2$ if by passing to commensurable virtual arithmetic quotients $X_i' \to Y_i' = K_i \backslash H_i / \Lambda_i'$, we can find*

  i) a common finite ergodic extension X' of X_i'; and

  ii) a $\mathbb{Q}$-surjection $\theta : H_1 \to H_2$ such that $\theta(K_1) \subset K_2$, $\theta(\Lambda_1') \subset \Lambda_2'$, so that

6

$$
\begin{array}{ccc}
& X_1' & \to & Y_1' \\
\nearrow & & & \\
X' & & & \downarrow \theta \\
\searrow & & & \\
& X_2' & \to & Y_2'
\end{array}
$$

commutes.

We observe that $Y_1 \succ Y_2$ amd $Y_2 \succ Y_1$ if and only if Y_i are commensurable virtual quotients. Furthermore, if Y_i and Z_i are commensurable, then $Y_1 \succ Y_2$ if and only if $Z_1 \succ Z_2$.

For the remainder of this section G is a semisimple Lie group with no compact factors, and X is an ergodic G-space of finite entropy.

Lemma 2.13. *There is a virtual arithmetic quotient action of X, $K \backslash H / \Lambda$, in reduced form such that*

  i) H is semisimple; and

  ii) $K \backslash H / \Lambda$ is maximal (up to commensurability) among all virtual semi-simple arithmetic quotients (i.e. those of the form $C \backslash L / \Gamma$ with L semisimple) of reduced form.

Proof. Choose a semisimple virtual arithmetic quotient of reduced form, $K' \backslash H' / \Lambda'$ which has a maximal entropy function among all such virtual quotients. This exists by Corollary 2.11. Consider the algebraically simply connected algebraic covering group of H' defined over $\mathbb{Q}$, say $q : H \to H'$, such that the inclusion $G \hookrightarrow H'$ lifts to a smooth homomorphism $G \hookrightarrow H$ and such that $K \backslash H / \Lambda$ is a virtual arithmetic quotient for some $K \subset q^{-1}(K')$ and $\Lambda \subset q^{-1}(\Lambda')$. By the descending chain condition on compact subgroups we can further assume K is minimal with this property. We now claim that $K \backslash H / \Lambda$ satisfies assertion (ii).

If not, then we have $X' \to C \backslash L / \Gamma \to K \backslash H / \Lambda$ for some finite extension X' of X, where $C \backslash L / \Gamma \to K \backslash H / \Lambda$ is induced by a $\mathbb{Q}$-surjection $\theta : L \to H$. Let $N = \ker(\theta)$. Then we can write $\mathfrak{l} = \mathfrak{h} \oplus \mathfrak{n}$ to be a direct sum of $\mathbb{Q}$-ideals. The projection of $\mathfrak{g}$ to $\mathfrak{n}$ must be trivial for otherwise the maximality of the entropy function of $K \backslash H / \Lambda$ would be contradicted by Lemma 2.10. However, if we have $\mathfrak{g} \subset \mathfrak{h}$, and $C \backslash L / \Gamma$ is of reduced form, it follows from Lemma 2.7 and the choice of H that $L = H$, i.e. θ is a $\mathbb{Q}$-isomorphism. Finally, by minimality of K, $\theta(C) = K$.

7

Lemma 2.14. *There is a virtual arithmetic quotient in reduced form that is maximal (up to commensurability).*

Proof. Fix $K\backslash H/\Lambda$ to be a virtual semisimple arithmetic quotient satisfying the conclusion of Lemma 2.13. Now consider all larger virtual arithmetic quotients of reduced form, and choose one of maximal entropy function among all such, say $C\backslash M/\Gamma \to K\backslash H/\Lambda$. Finally consider all virtual arithmetic quotients of reduced form $E\backslash P/\Delta \succ C\backslash M/\Gamma$. To prove the Lemma, it suffices to see that the possible values of $\dim(P)$ is bounded.

Let $N = \ker(P \to M)$. Then N is a $\mathbb{Q}$-group. Suppose N is not unipotent. Then letting $\rho : P \to P/\mathrm{Rad}_u(P)$, we see that $\rho(E)\backslash\rho(P)/\rho(\Delta)$ is a virtual arithmetic quotient in reduced form strictly larger than $K\backslash H/\Lambda$. Furthermore, it is of reduced form (since $E\backslash P/\Delta$ is) and since G is semisimple, $P = [P, P]$ by Lemma 2.7. In other words, $\rho(P)$ is also semisimple, which contradicts the choice of H. Thus, we may assume N is unipotent. Since $C\backslash M/\Gamma$ has a maximal entropy function, Lemma 2.10 implies that G centralizes N. Thus, $Z_P(N)$ is a normal $\mathbb{Q}$-subgroup of P, and we have $G \subset Z_P(N)$. By Lemma 2.7 again $Z_P(N) = P$, i.e. N is central in P.

The Lie algebra $\mathfrak{p}$ is defined over $\mathbb{Q}$ and we choose a subspace $\mathfrak{n}' \subset \mathfrak{p}$ such that $\mathfrak{n}'$ is defined over $\mathbb{Q}$ and $\mathfrak{n} \oplus \mathfrak{n}' = \mathfrak{p}$. (We note that $\mathfrak{n} \subset \mathfrak{p}$ is a central ideal, but $\mathfrak{n}'$ is only a linear subspace defined over $\mathbb{Q}$.) Let $B : \mathfrak{n}' \times \mathfrak{n}' \to \mathfrak{n}$ be given by $B(x,y) = \mathrm{proj}_{\mathfrak{n}}([x,y])$. Since $\mathfrak{n} \subset \mathfrak{z}(\mathfrak{p})$, $I = \mathfrak{n}' \oplus \mathfrak{B}(\mathfrak{n}' \times \mathfrak{n}')$ is an ideal in $\mathfrak{p}$ and I is defined over $\mathbb{Q}$. We have $\dim(\mathfrak{n}') \leq \dim(M)$, so if $\dim P > \dim M + (\dim M)^2$, then $I \subset \mathfrak{p}$ is a proper $\mathbb{Q}$-ideal. Thus $J = [I, I]$ is a proper $\mathbb{Q}$-ideal that is also an algebraic Lie subalgebra. Furthermore, since $\mathfrak{n}$ is unipotent we can choose $\mathfrak{n}'$ such that some conjugate of $\mathfrak{g}$ is contained in $\mathfrak{n}'$, hence in I, and since $\mathfrak{g}$ is semisimple, in J. It follows from Lemma 2.7 that some conjugate of G does not act ergodically on P/Δ, and hence G does not either. This contradicts the fact that $E\backslash P/\Delta$ is in reduced form. Therefore, we must have $\dim P \leq \dim M + (\dim M)^2$, and this proves the lemma.

To discuss uniqueness of maximal virtual arithmetic quotients, we first remark that if H is a $\mathbb{Q}$-group, Λ is an arithmetic subgroup, and G acts on H/Λ via $\pi : G \to H$, then G will also act via $g \mapsto h\pi(g)h^{-1}$ for any $h \in H$. If $h \notin H_{\mathbb{Q}}$, the resulting actions will not a priori be commensurable. More

8

generally, we consider the following situation.

Definition 2.15. *Let $K\backslash H/\Lambda$ be an arithmetic G-space defined via a homomorphism $\pi_1 : G \to Z_H(K)$ Let $z \in Z_H(\pi_1(G))$ and $h \in H$. Then G also acts on $hzKz^{-1}h^{-1}\backslash H/\Lambda$ via $\pi_2 : G \to Z_H(hzKz^{-1}h^{-1})$, where $\pi_2(g) = h\pi_1(g)h^{-1}$. We call two such arithmetic actions $\mathbb{R}$-conjugate.*

Clearly, any two $\mathbb{R}$-conjugate arithmetic actions are isomorphic in the category of G-actions, and in particular, one will be a quotient of an ergodic G-space if and only if the other is. We also remark that one is in reduced form if and only if the other is.

We can now state the main result of this section.

Theorem 2.16. *Let G be a semisimple Lie group with no compact factors. Let X be an ergodic G-space with finite invariant measure and finite entropy. Then:*

(1) There is a maximal (up to commensurability) virtual arithmetic quotient of reduced form, say $A(X)$.

(2) $A(X)$ is unique up to $\mathbb{R}$-conjugacy. More precisely, if Y is another such virtual quotient, then $A(X)$ and Y have $\mathbb{R}$-conjugate commensurable virtual arithmetic quotients.

(3) If Y is any virtual arithmetic quotient of X of reduced form, then $Y \prec Z$, where Z is an $\mathbb{R}$-conjugate of $A(X)$.

(4) Any virtual arithmetic quotient of X is a quotient of some finite ergodic extension of $A(X)$.

Proof. Let $K\backslash H/\Lambda$ satisfy the conclusions of Lemma 2.14. Let $C\backslash L/\Gamma$ be any other virtual arithmetic quotient of X in reduced form. Passing to commensurable actions, we can assume there is a finite ergodic extension X' of X and measure preserving G-maps $\psi : X' \to K\backslash H/\Lambda$ and $\Psi : X' \to C\backslash L/\Gamma$. Letting μ be the relevant measure on X', then $\nu = (\psi, \Psi)_*\mu$ is a finite G-invariant ergodic measure for the diagonal action on $K\backslash H/\Lambda \times C\backslash L/\Gamma$, that projects to the standard measure on each factor. Then it is not difficult to see that we can lift ν to an ergodic measure ν' on $H/\Lambda \times L/\Gamma$ that projects to the standard measure on both factors and projects to ν. By Ratner's theorem, this measure is supported on the orbit of a Lie group J, $\mathrm{diag}(G) \subset J \subset H \times L$. It follows that J projects surjectively to both H and L. Thus, we can write the J orbit supporting ν' as J/Δ where $\Delta = (\Lambda \times l\Gamma l^{-1}) \cap J$ for some $l \in L$. Give L the $\mathbb{Q}$-structure obtained by conjugating the given structure by l. Then by Lemma 2.4, J is a $\mathbb{Q}$-group, $\Delta \subset J$ is an arithmetic subgroup, and the projection $J \to H$ is a $\mathbb{Q}$-surjection. Furthermore the projection of J/Δ to $K\backslash H/\Lambda \times C\backslash L/\Gamma = K \times C\backslash H \times L/\Lambda \times \Gamma$ is $(K \times C)\backslash (K \times C)J/\Delta \cong (K \times C) \cap J\backslash J/\Delta$. Thus, letting $D = (K \times C) \cap J$, we have that $D\backslash J/\Delta$ is an arithmetic G-space which is a virtual arithmetic quotient of X in reduced form, and for which we have, via the projection, $D\backslash J/\Delta \succ K\backslash H/\Lambda$. This shows that they must be commensurable. Via a conjugate $\mathbb{Q}$-structure on J, we see also that $D\backslash J/\Delta \succ C\backslash L/\Gamma$. Assertions (1) and (2) then follow directly, and (3), (4) follow from these and earlier results in this section.

3 Examples

If we write $A(X) = K\backslash H/H_{\mathbb{Z}}$, then (the commensurability class of) $H_{\mathbb{Z}}$ is canonically attached to the action of G on X. There are a number of results that relate this discrete group to $\pi_1(X)$.

Theorem 3.1 [Z2] *Let M be a compact real analytic manifold. Supose G is a simple Lie group with $\mathbb{R}$-rank $(G) \geq 2$, and that G acts on M preserving an ergodic volume density and a real analytic connection. Assume further that $\pi_1(M)$ is (abstractly) isomorphic to a subgroup of some arithmetic group. Then there is a $\mathbb{Q}$-group L and a local embedding $G \to L$ such that*

(i) $A(M) \succ C\backslash L/L_{\mathbb{Z}}$.

(ii) $L_{\mathbb{Z}} \subset \pi_1(M)$.

The proof of this result (for which we refer the reader to [Z2]) uses a combinaton of superrigidity for cocycles, Ratner's Theorem, and Gromov's work on rigid transformation groups.

10

In [LZ], we obtianed much sharper results with engaging hypotheses (and without the geometric hypotheses.) We briefly recall some definitions, referring the reader to [LZ] for a more detailed discussion.

Let M be a compact space for which covering space theory holds. Assume we have a continuous action of G on M where G is a connected Lie group. Then $\widetilde{G}$ acts on any covering space M' of M. Suppose there is an ergodic finite invariant measure μ for the G action on M.

Definition 3.2

(i) *The G action on M is called engaging if $\widetilde{G}$ is ergodic on M' for every finite covering $M' \to M$.*

(ii) *The G action on M is called totally engaging if for every non-trivial convering $p : M' \to M$, there is no $\widetilde{G}$-equivariant measurable section of p.*

Remark 3.3

(i) Totally engaging implies engaging.

(ii) Arithmetic actions are totally engaging. (See [LZ].)

To state the main results of [LZ] and their relation to $A(M)$, we need one more definition from [LZ].

Definition 3.4. ([LZ, Definition 4.1]) *Suppose H_i are algebraic k-groups, $i = 1, 2, \ldots$, and $H_i = L_i \ltimes U_i$ are Levi decompositions defined over k. We call H_1 and H_2 k-isotopic if there is a k-isomorphism $L_1 \to L_2$, such that under this isomorphism $\mathfrak{u}_i (= $ Lie algebra of $U_i)$ are k-isomorphic L_i modules.*

The main results of [LZ] exhibit the relationship of $A(M)$ to representations of $\pi_1(M)$. Roughly, they assert that with engaging hypotheses, a representation of $\pi_1(M)$ yields an arithmetic quotient of M and hence of $A(M)$.

11

Theorem 3.5 ([LZ, Theorem 5.1]) *Let G be a connected simple Lie group with $\mathbb{R}$-rank $(G) \geq 2$, and suppose G acts on a compact M, preserving a finite measure and engaging. Let $\sigma : \pi_1(M) \to GL(n, \mathbb{R})$ be any linear representation, with image $\Gamma = \sigma(\pi_1(M))$ an infinite group, and Zariski closure $\overline{\Gamma} \subset GL(n, \mathbb{R})$. Then Γ is $\mathfrak{s}$-arithmetic. More precisely, there is a real algebraic $\mathbb{Q}$-group H, an embedding $\Gamma \hookrightarrow H_{\mathbb{Q}}$ (and hence necessarily in $H_{\mathbb{Z}_S}$ for some finite set of primes S) and subgroups $\Gamma_\infty \subset \Gamma_0 \subset \Gamma$ such that:*

(i) H contains a group $\mathbb{R}$-isotopic to $\overline{\Gamma}$

(ii) $[\mathfrak{h}, \mathfrak{h}] = \mathfrak{h}$.

(iii) $[\Gamma : \Gamma_0] < \infty$

(iv) Γ_∞ is profinitely dense in Γ_0

(v) Γ_∞ is commensurable with $H_{\mathbb{Z}}$ and is a lattice in H.

Furthermore, (perhaps by passing to a finite cover of G), there is a local embedding $G \to H$ such that $C \setminus H / \Gamma_\infty$ is a virtual arithmetic quotient of M. In particular, $C \setminus H / \Gamma_\infty \prec A(M)$, where $A(M)$ is the canonical maximal arithmetic quotient.

Theorem 3.6 ([LZ, Theorem 5.2]) *With the hypotheses of Theorem 3.5, and the additional hypothesis that the action is totally engaging, we may take $\Gamma_\infty = \Gamma_0$. In particular, Γ is arithmetic.*

One of the motivating questions for the developments in the study of actions of simple groups of higher rank can be formulated as follows:

Question 3.7 If $\mathbb{R}$-rank $(G) \geq 2$, when does $A(M) = M$?

Constructions of "non-standard" actions on manifolds (see [B]) have involved constructions along sets of zero measure. Thus, at the level of measure theory, there are still no known examples in higher rank for which equality does not hold. Without the higher rank assumption, however, $A(M)$ may be a very proper quotient of M. This can be seen via the examples of Furman-Weiss [FW] (which they discussed for different purposes.) Namely,

12

let $G = O(1, n)$ and $\Gamma \subset G$ be a cocompact lattice with a surjective homo-morphism $h : \Gamma \to \mathbb{Z}$. Let Y be a compact $\mathbb{Z}$-space with ergodic invariant measure and positive entropy. Let Γ act on Y via h and let $X = (G \times Y)/\Gamma$ be the induced G space. Then X is a compact G-space, with finite entropy. Furthermmore, by continously varying the entropy of the $\mathbb{Z}$-actions on Y, one can continuously (and non-trivially) vary the entropy for elements $g \in G$ acting on X. Since the entropy for arithmetic actions is controlled by Lemma 2.10, this shows that for most such X we will have $A(X) \neq X$, as they cannot have equal entropies. In fact, one can have such examples with $A(X) = G/\Gamma$.

References

[B]   J. Benveniste, "Exotic Geometric Actions of Semisimple Groups and their Deformations," preprint.

[BM] J. Brezin, C.C. Moore, "Flows on homogenous spaces: a new look," *Am. J. Math.*, **103**, pp. 571-613, 1981.

[FW] A. Furman, B. Weiss, "On the ergodic properties of Cartan flows in ergodic actions of $SL(2, \mathbb{R})$ and $SO(n, 1)$," *Erg. Th. Dyn. Sys.*, to appear.

[LZ]   A. Lubotzky, R. J. Zimmer, "Arithmetic structure of fundamental groups and actions of semisimple Lie groups," preprint.

[R]   M. Ratner "On Raghunathan's measure conjecture," *Ann. of Math.*, **134(3)**, pp. 545-607, 1991.

[Z1]   R. J. Zimmer, "Extensions of ergodic actions and generalized discrete spectrum," *Bulletin Amer. Math. Soc.*, **81**, pp. 633-6, 1975.

[Z2]   R. J. Zimmer. *Ergodic Theory and Semisimple Groups*. Birkhauser, Boston, 1984.

[Z3]   R. J. Zimmer. "Superrigidity, Ratner's theorem, and fundamental groups" *Israel J. Math.*, **74**, pp.199-207, 1991.

PERGAMON

Topology 40 (2001) 851–869

TOPOLOGY

www.elsevier.com/locate/top

Arithmetic structure of fundamental groups and actions of semisimple Lie groups[☆]

Alexander Lubotzky[a], Robert J. Zimmer[b,*]

[a]Institute of Mathematics, Hebrew University, Jerusalem, Israel
[b]Department of Mathematics, University of Chicago, Chicago, IL 60637, USA

Received 1 July 1998; accepted 20 August 1999

Abstract

If a simple Lie group acts on a space M with a finite invariant measure, we investigate the arithmetic properties of the fundamental group, and the arithmetic structure of the action. © 2001 Elsevier Science Ltd. All rights reserved.

MSC: 22E40; 22F10; 22F50

Keywords: Arithmetic group; Simple Lie group; Fundamental group; Group action

The aim of this paper is to establish arithmetic properties of the fundamental group of a space on which a non-compact simple Lie group acts. In addition, we establish arithmetic properties of the actions themselves.

More precisely, let G be a connected simple Lie group with $\mathbb{R}$-rank$(G) \geqslant 2$. Let M be a compact space for which covering space theory holds. We assume that we have a continuous action of G on M. Then $\tilde{G}$ acts on any covering space $M' \to M$. We further suppose that there is a finite G-invariant ergodic measure μ on M. The action of G on M is called (μ-)engaging if for every finite covering $M' \to M$, the action of $\tilde{G}$ on M' is ergodic (with respect to the natural lift of μ to M'). The action is called totally engaging if there is no $\tilde{G}$ equivariant measurable section of $M' \to M$ for any non-trivial covering space of M. In general, totally engaging implies engaging. As we shall see, one or both of these conditions holds for the natural actions of G on homogeneous spaces of the form

---

☆Research partially supported by grants from NSF and the Binational Science Foundation, Israel–USA.
* Corresponding author. Tel.: + 1-773-702-1383; fax: + 1-773-702-9595.
E-mail address: r-zimmer@uchicago.edu (R.J. Zimmer).

$M = H/\Gamma$ where G acts via an embedding in H, where H is a Lie group and Γ is a lattice in H. Our main results on fundamental groups are the following.

Let $\pi_1(M) \to \mathrm{GL}(V)$ be any finite-dimensional linear representation over $\mathbb{C}$. Let Γ be the image, and assume Γ is infinite.

Theorem A. *Suppose the action of G on M is totally engaging. Then Γ is an arithmetic group. In fact, Γ is commensurable to $H_{\mathbb{Z}}$, where H is a $\mathbb{Q}$-group with $\mathfrak{g} \hookrightarrow \mathfrak{h}$.*

Theorem B. *Suppose the action of G on M is engaging. Then Γ is $\mathfrak{s}$-arithmetic. In fact, Γ is $\mathfrak{s}$-arithmetic in a $\mathbb{Q}$-group H with $\mathfrak{g} \hookrightarrow \mathfrak{h}$.*

In fact, we show that for engaging actions of G, Γ is arithmetic (not merely $\mathfrak{s}$-arithmetic) if and only if the action is totally engaging. (See Theorem 6.1.)

Here "$\mathfrak{s}$-arithmetic" is a generalization of the standard notion of S-arithmetic group where S is a finite set of primes. In the semi-simple case, such a group is virtually a product of S-arithmetic groups. In general, they will be lattices in a product of real and totally disconnected locally compact groups. These groups are discussed in detail in Section 3.

We remark that Theorem A can be viewed as a generalization of Margulis' arithmeticity theorem. The latter is essentially equivalent to Theorem A when the action of G on M is transitive. In our case, we also need to construct the group H, and an embedding of Γ in H as an arithmetic group. In general, H can be much larger than G, and need not be semi-simple. A more precise and fuller statement of Theorems A and B appear below as Theorems 5.1 and 5.2.

We present in Section 1 below examples showing the necessity of the hypotheses of Theorems A and B.

In addition to these arithmeticity theorems for the fundamental group, we establish arithmetic structure of the action itself. By an arithmetic action of a group G, we mean an action on a space $N = K\backslash H/\Gamma$, where H is a real algebraic $\mathbb{Q}$-group, $\Gamma \subset H$ is an arithmetic subgroup, $K \subset H$ is a compact (perhaps trivial) subgroup, and the G action is defined by a homomorphism $\sigma : G \to H$ so that K centralizes $\sigma(G)$. In [9], we studied arithmetic quotients of a given action. In particular, we showed that a finite entropy action of a non-compact simple Lie group G on a space M has a canonical maximal arithmetic (virtual) quotient action, say $A(M)$. Here, we show that for engaging actions, any linear representation of $\pi_1(M)$ yields an arithmetic quotient of M (and hence of $A(M)$.)

Theorem C. *Let G, M and Γ be as in Theorem B. Let Γ_∞ be the arithmetic subgroup of the $\mathfrak{s}$-arithmetic group Γ. Then M has a virtual arithmetic quotient of the form $K\backslash H/\Gamma_\infty$.*

This theorem also appears in sharper form in Theorem 5.1 below.

Some of the conclusions of Theorems B and C were obtained under stronger assumptions in [21]. In fact, our proof of these results incorporate ideas of Zimmer [21]. One of the basic assumptions in [21] is, in the context of Theorems B and C, that Γ is either discrete or has matrix entries in $\overline{\mathbb{Q}}$ This assumption is eliminated in the present work. This is of particular importance for potential applications where one may have a geometrically constructed representation, e.g. a holonomy

A. Lubotzky, R.J. Zimmer / Topology 40 (2001) 851–869   853

representation, that is a priori neither discrete nor algebraic. We also observe that our conclusions of Theorems B and C in the sharp form of Theorems 5.1 and 5.2, are stronger than those of Zimmer [21], even for representations that satisfy the assumptions in [21].

1. Engaging and totally engaging actions

In this section we discuss the notions of engaging and totally engaging actions. The former was introduced in [19]. It will be useful to consider these notions for actions on general principal bundles with discrete fiber, not only on the coverings of M.

Let $P \to M$ be a principal Γ-bundle where Γ is a discrete group. We assume G is a group that acts by principal Γ-bundle automorphisms such that the action of G on M is ergodic with respect to some G quasi-invariant measure. We remark that if V is a Γ-space with quasi-invariant measure, then the associated bundle $E_V = (P \times V)/\Gamma$ is acted upon naturally by G with a natural measure class left invariant. In particular, for $V = \Gamma/\Gamma_0$ where Γ_0 is a subgroup, one has a natural measure class on $E_{\Gamma/\Gamma_0} \cong P/\Gamma_0$.

Definition 1.1. The G-action on P is engaging if the action on P/Γ_0 is ergodic for every finite index subgroup $\Gamma_0 \subset \Gamma$.

We shall be concerned with G-invariant reductions to subgroups of Γ.

Definition 1.2. If $\Lambda \subset \Gamma$, then the G-action on P is called Λ-reducible if there is a measurable G-invariant reduction of P to Λ; i.e., there is a measurable G- invariant section of $P/\Lambda \to M$. (Here "invariant" is taken to mean invariant modulo null sets.)

This can be reformulated in terms of cocycles. (See [17,18] for general background.) Namely, a measurable trivialization of P, $P \cong M \times \Gamma$, defines a cocycle $\alpha : G \times M \to \Gamma$ by the equation

$$g.(m, \gamma) = (gm, \alpha(g, m)\gamma).$$

It is then easy to verify (see e.g. [18]) that:

Lemma 1.3. *P is Λ-reducible if and only if α is equivalent to a cocycle β such that $\beta(G \times M) \subset \Lambda$. (Here $\alpha \sim \beta$ means $\beta(g, m) = \phi(gm)^{-1}\alpha(g, m)\phi(m)$ for some $\phi : M \to \Gamma$.)*

From Lemma 1.3, we now have the following consequence.

Proposition 1.4. *If the G-action on P is engaging and is also Λ-reducible, then Λ is profinitely dense in Γ. Hence, under any finite dimensional linear representation σ, $\sigma(\Lambda)$ is Zariski dense in $\sigma(\Gamma)$.*

Proof. We recall that profinite density is equivalent to the assertion that for any subgroup $N \subset \Gamma$ of finite index that Λ surjects onto Γ/N. If the action on P is Λ-reducible, choose the cocycle β such that $\beta(G \times M) \subset \Lambda$. Then the action of G on $M \times \Gamma/N \cong P/N$ is given by $g(m, [\gamma]) = (gm, \beta(g, m)[\gamma])$.

Since $M \times \{[e]\}$ is of positive measure, ergodicity of G on P/N clearly implies $\Lambda N = \Gamma$, verifying profinite density. That profinite density implies Zariski density is a result of Margulis and Soifer [11].

In fact, further similar argument shows:

Proposition 1.5. *The action of G on P is engaging if and only if every $\Lambda \subset \Gamma$ for which the G action is Λ-reducible is profinitely dense.*

We shall most often apply Definition 1.1 to the case of $P = \tilde{M}$ and $\Gamma = \pi_1(M)$, or to a quotient of this bundle by a normal subgroup of Γ.

Definition 1.6 (Zimmer [19]). We say the action of G on M is engaging if the action of $\tilde{G}$ on the principal $\pi_1(M)$-bundle $\tilde{M} \to M$ is engaging.

Example 1.7. Let H be a connected Lie group, $\Lambda \subset H$ a lattice, and suppose G is a semisimple Lie group without compact factors that acts ergodically on H/Λ. Then the G action on H/Λ is engaging. This follows as a consequence of the more general Proposition 1.10 below.

Definition 1.8. (i) Suppose G acts on the principal Γ-bundle $P \to M$, acting ergodically on M. We say the action is totally engaging if there is no proper subgroup $\Lambda \subset \Gamma$ for which the action is Λ-reducible.

(ii) If G acts on a manifold M, we say the action is totally engaging if the action of $\tilde{G}$ on $\tilde{M} \to M$ is totally engaging.

Proposition 1.9. *Any totally engaging action is engaging.*

This follows from Proposition 1.5.

Proposition 1.10. *Let H be a connected Lie group, $\Lambda \subset H$ a lattice, and $G \subset H$ a semisimple Lie group with no compact factors. Then the G action on H/Λ is totally engaging.*

Proof. We can write $H/\Lambda = \tilde{H}/\tilde{\Lambda}$ where $\tilde{\Lambda}$ is a lattice in $\tilde{H}$ and is the pull back of Λ to $\tilde{H}$. Thus, we can identify $\widetilde{H/\Lambda}$ with $\tilde{H}$ and $\pi_1(H/\Lambda)$ with $\tilde{\Lambda}$. Suppose the $\tilde{G}$ action is Γ-reducible for some $\Gamma \subset \tilde{\Lambda}$. Then the section $s : H/\Lambda \to \tilde{H}/\Gamma$ defines a finite G-invariant ergodic measure $s_*\mu$ on $\tilde{H}/\Gamma$ that projects to the standard measure μ on H/Λ. By Ratner's theorem, $s_*\mu$ is the measure defined by volume on an L-orbit in $\tilde{H}/\Gamma$ for some Lie group $\tilde{G} \subset L \subset \tilde{H}$. Since the projection of $s_*\mu$ to H/Λ is the volume on the latter, we must have $\dim L = \dim H$; it follows that $L = \tilde{H}$. From the fact that s is a section, it then follows easily that $\Gamma = \tilde{\Lambda}$.

Remark. There are smooth volume preserving actions of non-compact simple Lie groups on compact manifolds that are ergodic but not engaging. These are discussed in detail by Benveniste in [2]. These examples, among a number of illuminating properties, have fundamental groups that are not s-arithmetic. In particular, this demonstrates the need for some hypotheses such as engaging in Theorem B. It is a natural question as to what geometric conditions on an action would imply

A. Lubotzky, R.J. Zimmer / Topology 40 (2001) 851–869   855

engaging. In particular, the results of [2] raise the question as to whether connection-preserving actions must be engaging.

Example 1.11. We present an example which is engaging but not totally engaging. The fundamental group will be S-arithmetic but not arithmetic.

Let G be a connected simply connected semisimple $\mathbb{Q}$-group with $\mathbb{Q}$-rank $= 0$, and each simple factor of G with $\mathbb{R}$-rank $\geqslant 2$. Suppose p is a prime with each simple factor of $G_{\mathbb{Q}_p}$ non-compact. Let K be a maximal compact open subgroup of $G_{\mathbb{Q}_p}$, X the building associated to $G_{\mathbb{Q}_p}$. Thus, we can identify $V = G_{\mathbb{Q}_p}/K \subset X$ with a set of vertices. Let $\Gamma = G_{\mathbb{Z}[1/p]}$. Then $G_{\mathbb{R}}$ acts on the compact spaces

$$(G_{\mathbb{R}} \times V)/\Gamma \subset (G_{\mathbb{R}} \times X)/\Gamma = M.$$

Endow $(G_{\mathbb{R}} \times V)/\Gamma$ with the measure defined by Haar measure on $G_{\mathbb{R}} \times G_{\mathbb{Q}_p}$. We can view this as a finite $G_{\mathbb{R}}$-invariant ergodic measure on M. We remark that the action of $G_{\mathbb{R}}$ on $(G_{\mathbb{R}} \times G_{\mathbb{Q}_p})/\Gamma$ is ergodic if and only if Γ is dense in $G_{\mathbb{Q}_p}$. Since G is simply connected, $G_{\mathbb{Q}_p}$ has no non-trivial subgroups of finite index. Thus, if $\Gamma_0 \subset \Gamma$ is of finite index, the action of $G_{\mathbb{R}}$ on $(G_{\mathbb{R}} \times X)/\Gamma_0$ is also ergodic. These are the finite covers of M, so the action of $G_{\mathbb{R}}$ on M is engaging. (With a little more work, one can easily dispense with the simple connectivity assumption.) On the other hand, let $\Gamma_\infty \subset \Gamma$ be the arithmetic group $G_{\mathbb{Z}}$. Then the embedding

$$G_{\mathbb{R}} \to G_{\mathbb{R}} \times \{[e]\} \subset G_{\mathbb{R}} \times V$$

induces an equivariant bijection

$$G_{\mathbb{R}}/\Gamma_\infty \cong (G_{\mathbb{R}} \times V)/\Gamma \subset M.$$

Consider the covering space of M defined by the subgroup Γ_∞. This is simply

$$(G_{\mathbb{R}} \times X)/\Gamma_\infty \supset G_{\mathbb{R}}/\Gamma_\infty.$$

From this we see that there is a measurable $G_{\mathbb{R}}$-equivariant lift of M to $(G_{\mathbb{R}} \times X)/\Gamma_\infty$, showing that the action is not totally engaging.

We do not know an example of an engaging but not totally engaging action on a manifold.

We shall discuss these conditions further in Section 6, showing the intimate connection between arithmeticity and the totally engaging condition.

We conclude this section with a general result that is very useful when dealing with engaging conditions.

There are numerous general results (some of which we discuss below) on G-invariant reductions of bundles with an algebraic structure group to a subgroup. By considering homomorphisms of Γ into various algebraic subgroups, one would like to translate this into information about reductions of Γ-bundles and hence to the engaging conditions. The basic technique for doing this is the following.

Proposition 1.12. *Suppose L is a locally compact group and $H_1, H_2 \subset L$ are closed subgroups. Let a locally compact G act ergodically on a space (M, μ). Let $\alpha: G \times M \to H_1$ be a cocycle and suppose* *that $i \circ \alpha$ is equivalent to a cocycle into H_2, where $i : H_1 \to L$ is the inclusion. Then*:

(i) *If $H_1 \backslash L / H_2$ is tame [18] (i.e. the H_1 orbits on L/H_2 are locally closed), then α is equivalent to a cocycle into $H_1 \cap lH_2l^{-1}$ for some $l \in L$.*

(ii) *More generally, suppose $\phi : M \to L$ is such that*

$$\phi(gm)^{-1}(i \circ \alpha)(g, m)\phi(m) \in H_2.$$

If $\mathrm{im}(\phi)$ lies (a.e.) in a single $H_1 : H_2$ double coset in L, then α is equivalent to a cocycle into $H_1 \cap lH_2l^{-1}$ for some $l \in L$.

Remark. (i) Proposition 1.12(i) follows from the cocycle reduction lemma [17, Lemma 5.2.11]. Proposition 1.12(ii) follows from its proof, as the first step of the proof of [17, Lemma 5.2.11] is to use tameness to show that $\phi(M)$ lies (a.e.) in a single $H_1 : H_2$ double coset.

(ii) Proposition 1.12 is the basis of the definition of algebraic hull of a cocycle (or action on a principal bundle) [17,18].

(iii) Suppose $H_1 = \Gamma$ is a discrete subgroup of L. If H_2 is compact, and $i \circ \alpha$ is equivalent to a cocycle into H_2, then α is equivalent to a cocycle into a finite subgroup.

(iv) If H_2 is open, then Proposition 1.12(i) always applies.

(v) Proposition 1.12(ii) is the basis of the cohomological application of Ratner's theorem in [15, Proposition 3.6].

2. s-Arithmetic groups

In Section 1, we have seen how S-arithmetic groups give rise to examples of engaging actions of semi-simple real Lie groups. Actually, there are more general examples. In order to present them, let us start with some notations and a definition.

Let k be a number field, S a finite set of primes in k including all the archimedean ones and $\mathcal{O}$ the ring of algebraic integers in k. Denote

$$\mathcal{O}_S = \{x \in k \,|\, v(x) \geq 0 \text{ for every } v \notin S\}.$$

(Here, as usual, we think of the primes as the valuations of k.) An arithmetic group Γ is a group commensurable to $G(\mathcal{O})$ when G is a k-algebraic group. An S-arithmetic group is one commensurable to $G(\mathcal{O}_S)$. Every arithmetic group can be defined by using $\mathbb{Q}$ alone; replace G by $H = \mathrm{Res}_{\mathbb{Q}}^k(G)$ which is a group defined over $\mathbb{Q}$ and for which $H(\mathbb{Z}) \cong G(\mathcal{O})$. This is not the case for S-arithmetic groups: If the set S consists, for some rational prime p, of only a proper subset of the set of primes $\{\pi_i\}_{i \in I}$ of $\mathcal{O}$ lying above p, then $G(\mathcal{O}_S)$ is usually not isomorphic to an S'-arithmetic group for any set S' of rational primes.

To be able to work over $\mathbb{Q}$ and to have the most general notion of S-arithmetic groups we use:

Definition 2.1. A finitely generated group Γ is called an s-arithmetic group if there exists a $\mathbb{Q}$-algebraic group H, with $H(\mathbb{Z})$ infinite, a finite set S of primes of $\mathbb{Q}$ and a subgroup Γ_0 of $H(\mathbb{Z}_S)$,

A. Lubotzky, R.J. Zimmer / Topology 40 (2001) 851–869   857

such that

(i) Γ_0 virtually contains $H(\mathbb{Z})$, i.e., $\Gamma_0 \cap H(\mathbb{Z})$ is of finite index in $H(\mathbb{Z})$.
(ii) Γ_0 is isomorphic to a finite index subgroup of Γ.

Remark 2.2. In Definition 2.1, s is just a name which has nothing to do with the finite set of primes S.

We learned the following result from T.N. Venkataramana. It shows that for H semisimple, s-arithmetic groups are, up to finite index subgroups, finite products of S-arithmetic groups over number fields.

Proposition 2.3. *If H in Definition 2.1 is semi-simple, then there exists finitely many number fields* $k_1, \ldots, k_l$ *and for each* $i = 1, \ldots, l$ *an absolutely almost simple k_i-algebraic group G_i and a finite set S_i of primes of k_i such that, up to a finite index subgroup, Γ is isomorphic to $\prod_{i=1}^{l} G_i(\mathcal{O}_{S_i})$, where S_i is the ring of S_i-integers in k_i.*

We postpone the proof of (2.3) to the end of the section. We remark however, that (2.3) implies that if H is a semi-simple group then the s-arithmetic group Γ is a lattice in a group

$$M = \prod_{i=1}^{l} \prod_{v \in S_i} G_i((k_i)_v)$$

which is a product of a real and p-adic Lie groups.

This corollary holds in a more general context. Before showing this, let us see an example which is not semi-simple.

Example 2.4. Let $U = U_4$ be the unipotent group of 4×4 upper unipotent matrices. So a typical element of U is of the type

$$g = \begin{bmatrix} 1 & a_1 & b_1 & c \\ 0 & 1 & a_2 & b_2 \\ 0 & 0 & 1 & a_3 \\ 0 & 0 & 0 & 1 \end{bmatrix}.$$

Let p and l be two primes and Γ the subgroup of $U(\mathbb{Q})$ defined by the conditions:

$$g \in \Gamma \quad \text{iff} \quad \begin{cases} a_i \in \mathbb{Z} & \text{for } i = 1,2,3, \\ b_i \in \mathbb{Z}[1/p] & \text{for } i = 1,2, \\ c \in \mathbb{Z}[1/p, 1/l] \end{cases}$$

So Γ contains $U(\mathbb{Z})$ and it is contained in $U(\mathbb{Z}[1/p, 1/l])$. Moreover, Γ is a discrete subgroup when it is embedded diagonally in the group $U(\mathbb{R}) \times U(\mathbb{Q}_p) \times U(\mathbb{Q}_l)$. However, it is not a lattice there. The projection of Γ to $U(\mathbb{Q}_p)$ is not dense in $U(\mathbb{Q}_p)$. In fact its closure is equal to $U^+(\mathbb{Q}_p)$ where $U^+(\mathbb{Q}_p)$

858   A. Lubotzky, R.J. Zimmer / Topology 40 (2001) 851–869

is defined by the conditions

$$g \in U^+(\mathbb{Q}_p) \quad \text{iff} \quad \begin{cases} a_i \in \mathbb{Z}_p, \\ b_i, c \in \mathbb{Q}_p. \end{cases}$$

Similarly the closure of Γ in $U(\mathbb{Q}_l)$ is $U^{++}(\mathbb{Q}_l)$ given by

$$g \in U^{++}(\mathbb{Q}_l) \quad \text{iff} \quad \begin{cases} a_i, b_j \in \mathbb{Z}_l, \\ c \in \mathbb{Q}_l \end{cases}.$$

Moreover, it is not difficult to verify that Γ is dense in the product $\Omega = U^+(\mathbb{Q}_p) \times U^{++}(\mathbb{Q}_l)$. One can also easily check that the discrete subgroup Γ is a lattice in

$$U(\mathbb{R}) \times \Omega = U(\mathbb{R}) \times U^+(\mathbb{Q}_p) \times U^{++}(\mathbb{Q}_l).$$

Once can easily now imagine more examples of the kind when the unipotent group U is replaced by a general algebraic group.

Theorem 5.4 below states that if a higher rank real Lie group G acts on a Γ-bundle, then, under suitable assumptions Γ is $\mathfrak{s}$-arithmetic. Moreover, there exists a $\mathbb{Q}$-algebraic group H, with a $\mathbb{R}$-embedding of G into H such that $[H, H] = H$ (and so $H = U \rtimes L$, where U is unipotent, L is semi-simple) and $L(\mathbb{Z})$ is infinite such that Γ virtually contains the group $H(\mathbb{Z})$ and is contained in $H(\mathbb{Z}_s)$ for some set of primes. We will now show that indeed every such $\mathfrak{s}$-arithmetic group Γ gives rise to a Γ-bundle with a G-action, and a finite measure on the base preserved by G.

Lemma 2.5. *Let* $H = U \rtimes L$ *be a connected* $\mathbb{Q}$-algebraic group, such that U is unipotent and L is semi-simple with $L(\mathbb{Z})$ an infinite group. Let S be a finite set of primes and Γ a subgroup of $H(\mathbb{Z}_S)$ which virtually contains $H(\mathbb{Z})$. Let $M = \prod H(\mathbb{Q}_p)$ where p runs over the finite primes in S, and let Ω be the closure of the projection of Γ into M. Then Γ is a lattice in $H(\mathbb{R}) \times \Omega$.

Proof. As $H(\mathbb{Z}_S)$ is a discrete subgroup of $H(\mathbb{R}) \times M$, Γ is clearly a discrete subgroup of $H(\mathbb{R}) \times \Omega$. We need to show that Γ is of finite covolume there. Note that

(i) By strong approximation (see [12, p. 427]), it follows that Ω contains a finite index open subgroup K of $\prod_{p \in S} H(\mathbb{Z}_p)$.
(ii) $H(\mathbb{Z})$ is a lattice in $H(\mathbb{R})$ and so is every finite index subgroup of it. In particular, for $\Gamma_1 = \Gamma \cap H(\mathbb{Z}) \cap K$, there exists a subset V of finite covolume in $H(\mathbb{R})$ such that $\Gamma_1 \cdot V = H(\mathbb{R})$.

We claim now that $\Gamma \cdot (V \times K) = H(\mathbb{R}) \times \Omega$. This shows that Γ is a lattice in $H(\mathbb{R}) \times \Omega$. Indeed, let $(g_1, g_2) \in H(\mathbb{R}) \times \Omega$. By the density of Γ in Ω one can find $\gamma \in \Gamma$ such that

$$\gamma(g_1, g_2) = (\gamma g_1, \gamma g_2) \in H(\mathbb{R}) \times K.$$

Now, we can choose $\gamma_0 \in \Gamma_1 \subseteq \Gamma$ such that $\gamma_0(\gamma g_1) \in V$. Since γ_0 is by the definition of Γ_1 also in K, we have that $(\gamma_0 \cdot \gamma)(g_1, g_2) \in V \times K$ and the proof is complete. $\square$

Corollary 2.6. *With the notation of* (2.5), assume further that G is a semi-simple real Lie group with a $\mathbb{Q}$-embedding into H. Then the embedding of G in $H(\mathbb{R})$ defines an action of G on $(H(\mathbb{R}) \times \Omega)/\Gamma$, which is the base of a Γ-bundle and of finite measure. This action will be engaging if and only if for every finite index subgroup Γ_0 of Γ, the closure of Γ_0 in Ω is Ω. This happens, for example, if H is semi-simple.

A. Lubotzky, R.J. Zimmer / Topology 40 (2001) 851–869   859

We are not sure what is the most general context in which this density property (and hence engaging) holds.

Remark 2.7. As mentioned above, our main theorem is a converse of Corollary 2.6. It says that if G is of higher rank and acts in an engaging way on a Δ-bundle with a compact base, where Δ is a linear group, then Δ is s-arithmetic with H as in (2.5). Note however, that it does not give the complete converse. We assume that the base is compact, but prove only that Δ is a lattice in $H(\mathbb{R}) \times \Omega$ which might be of finite covolume but not necessarily cocompact.

We will return now to the proof of Proposition 2.3 and we start with a lemma:

Lemma 2.8. *Let G be a simply connected absolutely almost simple group defined over a number field k. Let $\Delta = G(\mathcal{O})$ where $\mathcal{O}$ is the set of integers in k, S a finite set of primes of $\mathcal{O}$ and Γ a subgroup of $G(\mathcal{O}_s)$ such that $\Gamma \cap \Delta$ is of finite index in Δ. Assume further that $G(\mathcal{O})$ is infinite. Then there exists a subset S' of S such that Γ is commensurable with $G(\mathcal{O}_{S'})$.*

Proof. For $v \in S$, denote by k_v the completion of k with respect to v. Let S' be the subset of S consisting of all the archimedean ones together with those $v \in S$ for which Γ is dense in $G(k_v)$. If $v \notin S'$ then the closure of Γ in $G(k_v)$ is an open compact subgroup of $G(k_v)$. Indeed, by strong approximation [12, p. 247] and the fact that Γ virtually contains the infinite group $G(\mathcal{O})$, the closure of Γ in $G(k_v)$ is always open. In the simply connected case, every open subgroup is either compact or else is all of $G(k_v)$ [12]. This proves that after replacing Γ by a finite index subgroup, we may assume that Γ is contained in $G(\mathcal{O}_s)$.

We will now prove that Γ is of finite index in $G(\mathcal{O}_{S'})$. Let S'_f be the set of finite primes in S' and Ω the closure of Γ in $\prod_{v \in S'_f} G(k_v)$. Clearly Ω contains the closure of $\Delta' = \Gamma \cap G(\mathcal{O})$, which is of finite index in $G(\mathcal{O})$. By strong approximation, the closure of Δ' contains a product $\prod_{v \in S'_f} M_v$ where each M_v is a compact open subgroup of $G(k_v)$. In particular, for each v the subgroup generated by M_v and its Γ-conjugates lies in Ω. The conjugation action of Γ on $M_v (\subseteq G(k_v))$ factors through the projection of Γ in $G(k_v)$. Therefore, since it is dense in $G(k_v)$ and $G(k_v)$ has no open normal proper subgroups, it follows that Ω contains $G(k_v)$ for each $v \in S'_f$. So Ω is just the product $\prod_{v \in S'_f} G(k_v)$. For the same reason, this product is also the closure of $G(\mathcal{O}_{S'})$.

Let U be the closure of $G(\mathcal{O})$ in $\prod_{v \in S'_f} G(k_v)$ and U' be the closure of $\Delta = \Gamma \cap G(\mathcal{O})$. By our assumptions $[U : U'] < \infty$ and in fact there is $r \in \mathbb{N}$ and $\delta_1, \ldots, \delta_r \in G(\mathcal{O})$ such that

$$G(\mathcal{O}) = \bigcup_{i=1}^{r} \Delta \delta_i \quad \text{and} \quad U = \bigcup_{i=1}^{r} U' \delta_i.$$

Now, U is open in $\prod_{v \in S'_f} G(k_v)$. By virtue of the density of Γ in the above product, for every $g \in G(\mathcal{O}_{S'})$ there exists $\gamma \in \Gamma$ such that $\gamma^{-1} g \in U$. Now, $\gamma^{-1} g \in G(\mathcal{O}_{S'})$ as well and $U \cap G(\mathcal{O}_s) = G(\mathcal{O})$. Thus $\gamma^{-1} g \in G(\mathcal{O})$ and so there exists $\delta \in \Delta$ and $1 \leqslant i \leqslant r$ such that $\gamma^{-1} g = \delta \cdot \delta_i$. Hence $g = \gamma \cdot \delta \cdot \delta_i$. Since $\gamma, \delta \in \Gamma$ we deduce that $G(\mathcal{O}_{S'}) = \bigcup_{i=1}^{r} \Gamma \delta_i$, which shows that $[G(\mathcal{O}_{S'}) : \Gamma] < \infty$. This proves the lemma. □

Now, once Lemma 2.8 is proven for simply connected groups one can deduce a similar result for the non-simply connected case, provided Γ is finitely generated. Indeed, if $\pi : \tilde{G} \to G$ is the simply connected cover of G, then $\pi(\tilde{G}(k))$ is normal in $G(k)$ and $G(k)/\pi(\tilde{G}(k))$ is a torsion abelian group (cf. [8]). If Γ is a finitely generated subgroup of $G(k)$, then a finite index subgroup of it is contained in $\pi(\tilde{G}(k))$. So replacing Γ by a finite index subgroup we can assume that $\Gamma \leqslant \pi(\tilde{G}(k))$. Look now at $\tilde{\Gamma} = \pi^{-1}(\Gamma)$. $\tilde{\Gamma}$, being a subgroup of $\tilde{G}(k)$, is a linear group and hence residually finite. $K = \ker(\pi)$ is a finite subgroup of $\tilde{\Gamma}$, and so $\tilde{\Gamma}$ has a finite index subgroup Γ' intersecting K trivially. Γ' is isomorphic, therefore, to a finite index subgroup of Γ. We can therefore now appeal to (2.8)

Let now H be a $\mathbb{Q}$-algebraic semi-simple group. Then up to a finite kernel (which can be dealt with as in the previous paragraph), we can assume H is a product of $\mathbb{Q}$-simple groups, $H = \prod_{i=1}^{l} H_i$. Each H_i is equal to $\mathrm{Res}_{\mathbb{Q}}^{k_i}(G_i)$ where k_i is a number field and G_i is an absolutely almost simple group defined over k_i. So we can assume $\Gamma \leqslant \prod_{i=1}^{l} G_i(k_i)$ and Γ virtually contains $H(\mathbb{Z}) = \prod_{i=1}^{l} G_i(\mathcal{O}_i)$ where $\mathcal{O}_i$ is the ring of algebraic integers of k_i. For every i, the set S_i' of finite primes of k_i for which Γ is dense in $G_i(k_i)$ (or in a finite index subgroup, when G_i is not simply connected) is finite. Let Ω be the closure of Γ in $\prod_{i=1}^{l} \prod_{v \in S_i'} G_i(k_{i,v})$. Then in a way similar to the proof of (2.8), Ω can be shown to be of finite index in the product. (It is not necessarily the whole product since if the G_i's are not simply connected they have finite index open subgroups.) Then one can continue to argue as in (2.8) to deduce that Γ is commensurable with a finite index subgroup of $\prod_{i=1}^{l} G_i(\mathcal{O}_{S_i})$. This finishes the proof of (2.3).

3. Superrigidity

In this section we summarize superrigidity for actions on principal bundles, i.e. superrigidity for cocycles, and extend this to a formulation we will need. We refer the reader to Zimmer [17] for background on cocycles.

If $P \to M$ is a principal H-bundle on which G acts, then with respect to the trivialization of the bundle defined by a measurable section, the G action will be described by a cocycle $G \times M \to H$. For any cocycle $c: G \times M \to H$, we call c tempered if it is equivalent to a cocycle $\beta: G \times M \to H$ such that for each $g \in G$, $\beta(g, M)$ (up to null sets) is contained in a compact subset of H. This will be the case, for example, for a cocycle coming from a continuous action of G on $P \to M$ in which M is compact. We shall call a cocycle $\beta: G \times M \to H$ superrigid if there is a homomorphism $\sigma: G \to H$, a compact subgroup $C \subset Z_H(\sigma(G))$, and a cocycle $c: G \times M \to C$ such that β is equivalent to the cocycle $(g, m) \mapsto \sigma(g)c(g, m)$. We call β totally superrigid if we can take c to be trivial. If G is a connected simple real algebraic group with $\mathbb{R}$-rank at least 2, and H is algebraic over a local field, then (perhaps by passing to a finite extension of M) any tempered β is superrigid. This is proven in [17,20] when H is defined over a local field of characteristic 0. However, combining these arguments with [10] or [14], one can also prove this in positive characteristic, which we shall need. More precisely:

Theorem 3.1 (Cocycle superrigidity in positive characteristic). *Let G be a connected simple Lie group with finite center and $\mathbb{R}$-rank$(G) \geqslant 2$. Let k be a local field with $\mathrm{char}(k) > 0$. Let H be a group defined over k and $\alpha: G \times M \to H_k$ be a cocycle. Then α is equivalent to cocycle into a compact subgroup of H_k.*

For technical reasons, it will be useful for us to reduce to the case where superrigidity is replaced by total superrigidity.

A. Lubotzky, R.J. Zimmer / Topology 40 (2001) 851–869   861

Given a cocycle $c: G \times M \to C$ where C is compact, one can always pass to an ergodic extension, say $M' \to M$, such that the lift of c to a cocycle $c': G \times M' \to C$ is trivial in cohomology. Namely, c is equivalent to a minimal cocycle $\lambda: G \times M \to D \subset C$ where D is a closed subgroup, which means the skew product action of G on $M \times_\lambda D$, given by $g.(m, d) = (gm, \lambda(g, m)d)$ is ergodic. (See [16] for a full discussion.) The lift of c to $M \times_\lambda D$ is easily seen to be trivial. As discussed in [16], the group D is unique up to conjugacy (and in the case of real Lie groups, coincides with the algebraic hull of c [17,18]). However, if $P \to M$ is a Γ-bundle on which the G action is engaging, it is not immediate that the action on the pullback to M', say $P' \to M'$, is still engaging. We shall need to trivialize the cocycle c arising in superrigidity while at the same time maintaining the engaging property. The following accomplishes this when C is a compact Lie group (which is the only case we shall require).

Lemma 3.2. *Suppose G is a locally compact group with an engaging action on a principal Γ-bundle $P \to M$ (where Γ is discrete). Suppose $c: G \times M \to C$ is a cocycle into a compact Lie group. Then there is:*

(i) *a finite index subgroup $\Gamma' \subset \Gamma$, with associated finite cover $M' \to M$ and principal Γ'-bundle $P \to M'$, and*
(ii) *an ergodic skew product extension X of M' by a compact subgroup of C [16], such that*

    a. *the action of G on the principal Γ'-bundle $P_X \to X$ that is the pullback to X of $P \to M'$ is engaging, and*
    b. *the pullback $c_X: G \times X \to C$ of α is trivial in cohomology.*

Proof. For each finite index subgroup $\Lambda \subset \Gamma$, let $C_\Lambda \subset C$ be the algebraic hull of the cocycle $c_\Lambda: G \times P/\Lambda \to C$ defined by lifting c. The engaging hypothesis ensures that G is ergodic on P/Λ. If $\Delta \subset \Lambda$, then $C_\Delta \subset C_\Lambda$ up to conjugacy. By the descending chain condition on closed subgroups of C, we can choose a finite index $\Gamma' \subset \Gamma$ such that for all $\Lambda \subset \Gamma'$ we have $C_\Lambda = C_{\Gamma'}$ up to conjugacy. Set $D = C_{\Gamma'}$, $M' = P/\Gamma'$, and $\lambda: G \times M' \to D$ a cocycle equivalent to $c_{\Gamma'}$. Let $X = M' \times_\lambda D$. Then c_X is trivial in cohomology. To prove the lemma, it suffices to see that the action of G on the Γ'-bundle P_X is engaging. However, if $\Lambda \subset \Gamma'$ is of finite index then $P_X/\Lambda \cong P/\Lambda \times_\lambda D$ which is ergodic under G since $D = C_\Lambda$ for any such Λ. $\quad\blacksquare$

Corollary 3.3. *Let G be a connected simple Lie group with $\mathbb{R}$-rank$(G) \geqslant 2$. Suppose G acts on a space M with a finite invariant measure, and that $P \to M$ is a principal Γ-bundle with an engaging G-action. Let $\alpha: G \times M \to H$ be a tempered cocycle into a real algebraic group. Then there is a finite index subgroup $\Gamma' \subset \Gamma$ and an ergodic G-space X with finite invariant measure that is an extension $X \to M'$ of $M' = P/\Gamma'$ (and hence M) such that*

(i) *$\alpha_Y: G \times X \to H$ is totally superrigid, and*
(ii) *the G-action on the principal Γ'-bundle $P_X \to X$, the pullback to X of $P \to P/\Gamma'$, is engaging.*

Proof. We apply Lemma 3.2 twice. First, we can replace H by the algebraic hull of α [17]. Let $c_1 = p \circ \alpha$ where $p: H \to H/H^0$ where H^0 is the connected component of the identity. Applying Lemma 3.2 allows us to assume c_1 is trivial, i.e., by passing to a finite ergodic extension of M and a finite index subgroup of Γ, we can assume the algebraic hull is connected. We can then apply superrigidity in characteristic 0 [17,20] to deduce that our cocycle on this extension is superrigid. Second, we apply Lemma 3.2 again to pass to a further extension and a possibly smaller subgroup of finite index to obtain total superrigidity and engaging.

4. Specializations

In this section we develop some specialization theorems for finitely generated groups that we will use to reduce the proofs of our main results to the case of linear groups over $\bar{\mathbb{Q}}$. More precisely, suppose $\Gamma \subset GL(n,\mathbb{C})$ is a finitely generated linear group. Then there is a ring A which is a finitely generated $\mathbb{Q}$-algebra such that $\gamma \in GL(n, A)$ for all $\gamma \in \Gamma$. If $\psi : A \to \bar{\mathbb{Q}}$ is a $\mathbb{Q}$-algebra homomorphism then it induces a homomorphism $\psi : GL(n, A) \to GL(n,\bar{\mathbb{Q}})$ and in particular a homomorphism $\psi : \Gamma \to GL(n,\bar{\mathbb{Q}})$. Then ψ (or $(\psi(\Gamma))$) is called a specialization of Γ.

Definition 4.1. Suppose H_i are algebraic k-groups, $i = 1,2$, and $H_i = L_i \ltimes U_i$ are Levi decompositions defined over k. We call H_1 and H_2 k-isotopic if there is a k-isomorphism $L_1 \to L_2$, such that under this isomorphism $\mathfrak{u}_i(\ =\ $ Lie algebra of (U_i) are k-isomorphic L_i modules.

Our main result about specializations is the following.

Theorem 4.2. Let $\Gamma \subset GL(\mathbb{R}^n)$ be a finitely generated group. Suppose that for each irreducible component π of the semisimplification of this linear realization we have $\mathrm{tr}(\pi(\gamma)) \in \bar{\mathbb{Q}}$ for all $\gamma \in \Gamma$. Let $\bar{\Gamma}$ denote the Zariski closure. Then (after a suitable choice of basis in $\mathbb{R}^n$) there is a specialization ψ of Γ such that:

(i) $\psi(\Gamma) \subset GL(n,\bar{\mathbb{Q}} \cap \mathbb{R})$;
(ii) ψ is faithful on Γ;
(iii) $\overline{\psi(\Gamma)}$ and $\bar{\Gamma}$ are $\mathbb{R}$-isotopic.
Furthermore if $\bar{\Gamma}$ is defined over $\bar{\mathbb{Q}}$, then we have
(iv) $\overline{\psi(\Gamma)} = \bar{\Gamma}$.

We begin the proof of Theorem 4.2 by recalling the following result, which is well known (cf. [1, Section 2] for example).

Lemma 4.3. If $\Lambda \subset GL(n,\mathbb{R})$ is finitely generated and irreducible, and $\mathrm{tr}(\lambda) \in \bar{\mathbb{Q}}$, then Λ is conjugate over $\mathbb{R}$ to a subgroup of $GL(n,\bar{\mathbb{Q}} \cap \mathbb{R})$.

From this lemma and the hypotheses of Theorem 4.2 we deduce the following. Let $\bar{\Gamma} = L \ltimes U$ be a Levi decomposition over $\mathbb{R}$. (More precisely, we are taking L and U to be real algebraic groups.) We can find a flag $0 \subset V_1 \subset \cdots \subset V_r = \mathbb{R}^n$ subspaces W_i such that $V_i \oplus W_{i+1} = V_{i+1}$ and a basis for $\mathbb{R}^n$ which is a union of bases for W_i, $i = 1, \ldots, r$, such that each V_i is $\bar{\Gamma}$ invariant; U acts by the identity on V_{i+1}/V_i; each W_i is L-invariant; the action of Γ on each W_i (via projection to L) is irreducible; and finally, writing $\gamma \in \Gamma$ as $\gamma_s \gamma_u$, where $\gamma_s \in L$ and $\gamma_u \in U$, we have $\gamma_s \in GL(n,\bar{\mathbb{Q}} \cap \mathbb{R})$ with respect to the above basis.

A. Lubotzky, R.J. Zimmer / Topology 40 (2001) 851–869   863

Let $N = \Gamma \cap U$. (This may be trivial.) Then N is a group of unipotent matrices, but is itself not a priori finitely generated. We establish the next lemma to be able to apply results and techniques of Grunewald–Segal [6]. We thank Shahar Mozes for his contribution to the proof of this lemma.

Lemma 4.4. *Let $Z(N)$ be the center of N. Then $Z(N)$ is an abelian group of finite rank, i.e. $Z(N) \otimes_{\mathbb{Z}} \mathbb{Q}$ is finite dimensional over $\mathbb{Q}$.*

Proof. Since Γ is finitely generated, we can find a number field F such that $\gamma_s \in \mathrm{GL}(n, F)$ for all γ. Let $\{\gamma^1, \dots, \gamma^r\}$ be a generating set for Γ. Let $\{x_1, \dots, x_l\} \subset \mathbb{R}$ be the set of real numbers appearing as entries in the matrices for $\{\gamma_u^i\}$. A straightforward induction (and some matrix multiplication) establishes the following. Consider for any matrix with respect to the above basis the set of matrix entries corresponding to $\mathrm{Hom}(W_j, W_i)$, $j \geq i$. Then for any word (of any length) in $\{\gamma^i\}$, such a matrix entry is a polynomial of degree at most $j - i$ in $\{x_1, \dots, x_l\}$ with coefficients in F. (The induction is done on $j - i$.) This implies that each matrix entry for all γ_u is a polynomial in $\{x_1, \dots, x_l, y_1, \dots, y_m\}$ over $\mathbb{Q}$ of degree at most $\max\{n, m\}$ where $\{y_1, \dots, y_m\}$ is a basis for F over $\mathbb{Q}$. This implies that any abelian subgroup of N is of finite rank, and in particular proves the lemma.

Now, let V be the unipotent group of all matrices with respect to the basis of $\mathbb{R}^n$ chosen above so that $T \in V$ if and only if $T|W_i = \mathrm{Id}$, and $T_{ij} = 0$ if $i > j$. Thus, $U \subset V$ and L normalizes V. Let $\mathfrak{u} \subset \mathfrak{v}$ be the corresponding Lie algebras, which are both L-modules. The map $\exp: \mathfrak{v} \to V$ is a bijection which is a $\mathbb{Q}$-regular map, as is the inverse which we denote by $\log$. We have $\exp|\mathfrak{u}: \mathfrak{u} \to U$ is also a bijection, although we recall that $\mathfrak{u}$ itself may or may not be defined over $\mathbb{Q}$ with respect to the standard $\mathbb{Q}$-structure on $\mathfrak{v}$. The matrix entries for the action of Γ_s on $\mathfrak{v}$ lie in $\bar{\mathbb{Q}} \cap \mathbb{R}$. The maps $\exp$ and $\log$ commute with the actions of L on $\mathfrak{v}$ and V.

Recall that for each $\gamma \in \Gamma$, we write $\gamma = \gamma_s \gamma_u$. Choose $\{\gamma^i\}_{i=1,\dots,n} \subset \Gamma$ such that $X_i = \log \gamma_u^i$ is a basis for $\mathfrak{u}$. Extend this to a larger set $\{\gamma^i\}_{i=1\dots n,\dots,m}$ that generates Γ. Let c_{ij}^k be the structural constants for the Lie algebra $\mathfrak{u}$ with respect to the basis $\{X_i; i = 1, \dots, n\}$, i.e. write $[X_i, X_j] = \sum c_{ij}^k X_k$, where $c_{ij}^k \in \mathbb{R}$. For $n < j \leq m$, write $X_j = \sum_{i=1}^n a_{ij} X_i$. Let $X_i = \sum b_i^{jk} E_{jk}$ be the expression for X_i in terms of the standard $\mathbb{Q}$-basis for $\mathfrak{v}$. Finally, let A be the (finitely generated) $\mathbb{Q}$-algebra generated by $\{a_{ij}, c_{ij}^k, b_i^{jk}\}$. Let $\psi: A \to \bar{\mathbb{Q}}$ be a specialization. Let $Y_i = \psi(X_i) = \sum \psi(b_i^{jk}) E_{jk}$. Assume for the moment that $\{Y_i, i = 1, \dots, n\}$ are linearly independent. Let $\mathfrak{u}'$ be the subspace spanned by Y_i. Since $c_{ij}^k \in A$, and $[Y_i, Y_j] = \sum \psi(c_{ij}^k) Y_k$, $\mathfrak{u}'$ is a Lie algebra defined over $\bar{\mathbb{Q}}$. Denote the natural action of $\mathfrak{L}$ on $\mathfrak{b}$ by Ad. Then if $g \in L_{\bar{\mathbb{Q}}}$, we have $\mathrm{Ad}(g)X_i$ has entries in A, and $\psi(\mathrm{Ad}(g)X_i) = \mathrm{Ad}(g)(\psi(X_i))$. Thus, $\mathfrak{u}$ and $\mathfrak{u}'$ are isomorphic L-modules. Letting $U' = \exp(\mathfrak{u}')$, we thus have $L \ltimes U$ and $L \ltimes U'$ are $\mathbb{R}$-isotopic Lie groups. We now claim that the specialization $\psi(\Gamma) \subset L \ltimes U'$. We have $\psi(X_j) = \sum_1^n \psi(a_{ij}) Y_i$, so $\psi(X_j) \in \mathfrak{u}'$ for $j = 1, \dots, m$. Thus, $\exp(\psi(X_j)) = \psi(\exp(X_j)) = \psi(\gamma_u^j) \in U'$. Since $\psi(\gamma) = \gamma \psi(\gamma_u)$, it follows that $\psi(\gamma^j) \in L \ltimes U'$ for γ^j in a generating set, and hence $\psi(\Gamma) \subset L \ltimes U'$. It is then also clear that $\overline{\psi(\Gamma)} = L \ltimes U'$. If $\bar{\Gamma} = L \times U$ is itself defined over $\bar{\mathbb{Q}}$, then so is any specialization of $\mathfrak{u}$ over $\bar{\mathbb{Q}}$, and hence we would have $U' = U$.

Turning to the injectivity of ψ on Γ, we observe that $\psi(\gamma) = \psi(\gamma_s \gamma_u) = \gamma_s \psi(\gamma_u)$. Evidently, this can vanish only if γ_s is trivial, i.e., it suffices to see that $\psi|N$ (where $N = \Gamma \cap U$) is injective. In sum, we have shown that to prove Theorem 4.2, it suffices to prove the following lemma.

Lemma 4.5. *We can choose a specialization* $\psi : A \to \bar{\mathbb{Q}} \cap \mathbb{R}$ *such that*

(i) $\psi(X_i)$ *are linearly independent (over* $\mathbb{R}$*).*
(ii) $\psi|N$ *is injective.*

Proof. Let F be quotient field of A, so $F = \mathbb{Q}(\underline{x})$ where $\underline{x} = (x_1, \ldots, x_l) \in \mathbb{R}^l$. Then $\underline{x}$ generates an absolutely irreducible affine variety $V = \mathrm{spec}(A)$ over $k = F \cap \bar{\mathbb{Q}}$. Noether's normalization theorem supplies $t_1, \ldots, t_r \in A$ which are algebraically independent over k such that A is an integral extension of $B = k[t_1, \ldots, t_r]$. There is a natural map $\eta : \mathrm{spec}(A) \to \mathrm{spec}(B)$ which induces a surjective map from V onto $\mathbb{C}^r$, since A is integral over B. The map $\eta : V(\mathbb{C}) \to \mathbb{C}^r$ is continuous in the Zariski and complex topologies. As $k \subseteq \mathbb{R}$, $\eta(V(\mathbb{R})) \subseteq \mathbb{R}^r$.

The proof of Theorem 3.1 (and Theorem 2.8) in [6] implies that for every number field $K \supseteq k$, there exists an Hilbert set $H \subseteq K^r$ such that if $\psi \in \mathrm{spec}(A)$ and $\eta(\psi) \in H$, then ψ induces an injective map on N. (We apply this with $k = K$.) Moreover, the condition that $\{\psi(X_i)\}$ are linearly independent is an open condition. It is well known that the intersection of a Hilbert set with a Zariski-open set is still a Hilbert set (cf. [3, 6] and the references therein) so we can assume that both conditions (i) and (ii) are satisfied for every ψ such that $\eta(\psi) \in H$. To finish the proof we still need to ensure that the image of ψ is in $\mathbb{R}$. (Note that even if $\eta(\psi) \in k^r \subseteq \mathbb{R}^r$, this does not ensure that the image of ψ is in $\mathbb{R}$, but merely says that $\psi(B) \subseteq \mathbb{R}$.)

To this end we prove part (c) of the following lemma which was provided to us by M. Jarden:

Lemma 4.6. *Suppose* $k \subseteq A \subseteq \mathbb{Q}((\underline{x})) \subseteq \mathbb{R}$ *as before, and* K *a number field with* $k \subseteq K \subseteq \mathbb{R}$.

(a) *There is a real-open neighborhood* u_1 *of* $\underline{x}$ *in* $V(\mathbb{R})$ *and a real-open ball* u_2 *around* $\underline{t} = (t_1, \ldots, t_r) \in \mathbb{R}^r$ *such that* η *maps* u_1 *homeomorphically onto* u_2.
(b) $u_2 \cap H \neq \emptyset$ *for every Hilbert subset* H *of* K^r.
(c) *For each Hilbert subset* H *of* K^r *there exists* $\underline{x}' \in V(\mathbb{R})$ *such that* $\eta(\underline{x}') \in H$.

Proof. (a) From [5, Corollary 9.5] it follows that $\eta : V(\mathbb{R}) \to \mathbb{R}^r$ is a local homeomorphism in the neighborhood of $\underline{x}$. This is just a reformulation of (a). (Actually [5, Corollary 9.5] deals with a Henselian field rather than with $\mathbb{R}$, but one can carry out an analogous proof for $\mathbb{R}$.)

(b) This follows from Lemma 4.1 of [7]. (That lemma deals with valuation, but again an analogous proof works for $\mathbb{R}$. One can also deduce (b) from a more general and more difficult theorem of Geyer [4, Lemma 3.4] in which this density result is proved simultaneously for several valuations and orderings.)

Condition (c) is a consequence of (a) and (b).

5. Statements and proofs of the main results

In this section, we state and prove sharper versions of Theorems A–C of the introduction.

Theorem 5.1. *Let* G *be a connected simple Lie group with* $\mathbb{R}$-rank $(G) \geqslant 2$, *and suppose* G *acts on a compact* M, *preserving a finite measure and engaging. Let* $\sigma : \pi_1(M) \to \mathrm{GL}(n, \mathbb{R})$ *be any linear*

A. Lubotzky, R.J. Zimmer / Topology 40 (2001) 851–869   865

representation, with image $\Gamma = \sigma(\pi_1(M))$ an infinite group, and Zariski closure $\bar{\Gamma} \subset \mathrm{GL}(n,\mathbb{R})$. Then Γ is s-arithmetic. More precisely, there is a real algebraic $\mathbb{Q}$-group H, an embedding $\Gamma \hookrightarrow H_{\mathbb{Q}}$ (and hence necessarily in $H_{\mathbb{Z}_s}$ for some finite set of primes S) and subgroups $\Gamma_\infty \subset \Gamma_0 \subset \Gamma$ such that

(i)   *H contains a group $\mathbb{R}$-isotopic to $\bar{\Gamma}$ (see Definition 4.1).*
(ii) *$[\mathfrak{h},\mathfrak{h}] = \mathfrak{h}$.*
(iii) *$[\Gamma : \Gamma_0] < \infty$.*
(iv) *Γ_∞ is profinitely dense in Γ_0.*
(v)   *Γ_∞ is commensurable with $H_{\mathbb{Z}}$ and is a lattice in H.*

Furthermore, (perhaps by passing to a finite cover of G), there is a local embedding $G \to H$ such that $C\backslash H/\Gamma_\infty$ is a virtual arithmetic quotient of M. In particular, $C\backslash H/\Gamma_\infty \prec A(M)$, where $A(M)$ is the canonical maximal arithmetic quotient [9].

Theorem 5.2. *With the hypotheses of Theorem 5.1, and the additional hypothesis that the action is totally engaging, we may take $\Gamma_\infty = \Gamma_0$. In particular, Γ is arithmetic.*

In particular, we obtain:

Corollary 5.3. *If G is a connected Lie group with $\mathbb{R}$-rank $(G) \geqslant 2$, and G has a totally engaging action on a compact manifold M, then for any representation $\sigma : \pi_1(M) \to \mathrm{GL}(n,\mathbb{R})$ with infinite image, $\sigma(\pi_1(M))$ has a subgroup of finite index that is an arithmetic subgroup of a $\mathbb{Q}$-group H such that $\mathfrak{g} \hookrightarrow \mathfrak{h}$. Thus, if $\pi_1(M)$ is a linear group, it has a subgroup of finite index that is arithmetic in a group H with $\mathfrak{g} \hookrightarrow \mathfrak{h}$.*

In fact, our proof will work for any principal bundle with discrete fiber, not just $\tilde{M}$. More precisely, we have:

Theorem 5.4. *Let π be a discrete group and $P \to M$ a principal π-bundle on which G acts preserving a finite ergodic measure on the compact space M. Let Γ be the image of π under any finite-dimensional linear representation over $\mathbb{C}$. Assume Γ is infinite. If the action of G on $P \to M$ is engaging, the conclusions of Theorem 5.1 hold. If the action is totally engaging, those of Theorem 5.2 hold.*

We now turn to the proof of these theorems.

Our general approach to the proofs of these results will be to reduce to the case in which Γ has algebraic entries, and then to further develop the arguments of [21] using [15, Proposition 3.6] and the results in earlier sections. Rather than reproduce the arguments of [21] in full detail, we shall freely refer to that paper when it is convenient to do so. We now assume all hypotheses of Theorem 5.1.

Lemma 5.5. *Let k be a local field of positive characteristic. Let $\lambda : \Gamma \to \mathrm{GL}(n,k)$ be a representation with discrete image. Then $\lambda(\Gamma)$ is finite.*

Proof. Let $\alpha : \tilde{G} \times M \to \Gamma$ be defined by the action of $\tilde{G}$ on the principal Γ-bundle P, where $P = \tilde{M}/\ker(\sigma)$. By Theorem 3.1 and Definition 1.12(i), $\lambda \circ \alpha$ is equivalent to a cocycle into a finite subgroup of $\lambda(\Gamma)$, which must be profinitely dense in $\lambda(\Gamma)$. Being linear, $\lambda(\Gamma)$ is residually finite, and since it has a profinitely dense finite subgroup, $\lambda(\Gamma)$ is itself finite.

Lemma 5.6. *Any representation of Γ over a local field with positive characteristic has finite image.*

Lemma 5.6 follows from Lemma 5.5 and:

Lemma 5.7. *Let Δ be a finitely generated infinite linear group over a local field F with* $\mathrm{char}(F) > 0$. *Then Δ has a linear representation over F with infinite discrete image.*

Remark. Lemma 5.7 is not true in the case of characteristic 0. For example, $\mathrm{SL}_n(\mathbb{Z}[1/p])$ has no infinite discrete representation over $\mathbb{C}$.

We need a sublemma.

Sublemma 5.8. *If Δ is an infinite finitely generated linear group in* char $p > 0$, *then it has a representation with an infinite image over a global field of* char $p > 0$ *(i.e. of transcendence degree* $= 1$.)

Proof. Let H be the Zariski closure of Δ. If the unipotent radical is of finite index then Δ is virtually nilpotent, hence torsion, hence finite since it is finitely generated. So Δ has an infinite image in a reductive group. If the reductive group has a finite index infinite central torus, then Δ has a finite index central subgroup with an infinite abelian quotient. So Δ certainly would have an infinite representation over a global field. Thus, we may assume Δ has an infinite representation into a semisimple group M. Let ϕ be a faithful irreducible representation of M, so it is also irreducible with respect to Δ. Let D be the ring generated by the traces of $\phi(\Delta)$, and $Q(D)$ be the quotient field. If the transcendence degree of $Q(D)$ is 0, then by [1, Corollary 2.5], $\phi(\Delta)$ is conjugate to a group with entries in a finite extension of $Q(D)$. This is impossible since $\phi(\Delta)$ is infinite and we are in positive characteristic.

So for some $\gamma_0 \in \Delta$, $\mathrm{tr}(\gamma_0)$ is not algebraic. Take now a specialization of D into a global field such that this element of D, (i.e. $\mathrm{tr}(\gamma_0)$) is not algebraic. This ensures a representation with infinite image of Δ, and proves the sublemma.

Proof of Lemma 5.7. We can, by the sublemma, assume that Δ has an infinite representation into a global field. So assume $\Delta \subset \mathrm{GL}_n(K)$, K a global field. By choosing a finite set of primes S_0, Δ is discrete in $\prod_{r \in S_0} \mathrm{GL}_n(K_r)$. But each one of the K_r's is a finite extension of $\mathbb{F}_p((t))$. So altogether we get a faithful discrete representation over $\mathbb{F}_p((t))$, and hence over F which is $\mathbb{F}_q((t))$ for q equal some power of p.

Lemma 5.9. *For any linear representation π of Γ,* $\mathrm{tr}(\pi(\gamma)) \in \bar{\mathbb{Q}}$ *for all* $\gamma \in \Gamma$.

Proof. Since Γ is finitely generated, there is a finitely generated ring A with $\pi(\Gamma) \subset \mathrm{GL}(n, A)$. For any transcendental $a \in A$, there is a ring homomorphism $\psi : A \to F$ where F is a local field of positive characteristic with $\psi(a)$ still transcendental. If some $\gamma \in \Gamma$ had $\mathrm{tr}(\pi(\gamma)) \notin \bar{\mathbb{Q}}$, then ψ would

A. Lubotzky, R.J. Zimmer / Topology 40 (2001) 851–869   867

define a representation π_ψ over F with $\mathrm{tr}(\pi_\psi(\gamma))$ transcendental. This implies that $\pi_\psi(\Gamma)$ is infinite, contradicting Lemma 5.6.

Now apply Theorem 4.2 to Γ. We identify Γ with $\psi(\Gamma)$, but still denote by $\bar{\Gamma}$ the Zariski closure in the original representation. We can use restriction of scalars to find an algebraic $\mathbb{Q}$-group H in which Γ is embedded as a Zariski dense subgroup with $\Gamma \subset H_\mathbb{Q}$. Furthermore, it is easy to check that H must contain a subgroup $\mathbb{R}$-isotopic to $\bar{\Gamma}$.

By applying Corollary 3.3 we can pass to a finite index subgroup Γ_0 and an ergodic G-space X with finite invariant measure such that, letting $\alpha : G \times X \to \Gamma_0$ be the cocycle defined by the engaging action of P_X and $\alpha_\mathbb{R}$ the composition of α with the embedding of $\Gamma_0 \subset H_\mathbb{R}$, we have that $\alpha_\mathbb{R}$ is totally superrigid, defining a homomorphism $\sigma : G \to H_\mathbb{R}$. We also note that we may assume, perhaps by passing to a further subgroup of finite index, that $\Gamma_0 \subset H_\mathbb{R}^0 \subset H_\mathbb{R}$ is Zariski dense in $H_\mathbb{R}$, (replacing the latter by the Zariski closure of Γ_0 if necessary). Now choose a finite set of primes S such that $\Gamma_0 \subset H_{\mathbb{Z}_S}$. Then Γ_0 is discrete in its diagonal embedding $d : \Gamma_0 \to H_\mathbb{R} \times H_f$ where $H_f = \Pi_{p \in S} H_{\mathbb{Q}_p}$. By p-adic superrigidity for cocycles, α_f, the projection of $d \circ \alpha$ onto H_f, is equivalent to a cocycle into a compact subgroup $K \subset H_f$, which we can assume is open. This implies $d \circ \alpha$ is equivalent to a cocycle taking values in $H_\mathbb{R} \times K$ which is open in $H_\mathbb{R} \times H_f$. By Proposition 1.12(i), this implies that $d \circ \alpha$ is equivalent to a cocycle into $d(\Gamma_0) \cap (H_\mathbb{R} \times K')$ for some conjugate K' of K. Let Γ_∞ be the projection of $d(\Gamma_0) \cap (H_\mathbb{R} \times K')$ into $H_\mathbb{R}$. We then have $\Gamma_\infty \subset \Gamma_0 \subset H_\mathbb{R}^0$ with Γ_∞ discrete, and α is equivalent to a cocycle into Γ_∞. By Proposition 1.4, Γ_∞ is profinitely dense in Γ_0, hence Zariski dense in $H_\mathbb{R}$. As in the argument of Zimmer [21], the fact that $\alpha_\mathbb{R}$ is totally superrigid shows that there is a G-equivariant measurable map $\phi : X \to H_\mathbb{R}^0 / \Gamma_\infty$ and, applying Ratner's theorem [13] exactly as in [21], almost all $\phi(X)$ lie in a L-orbit where L is a connected Lie group, say with stabilizer $L \cap h\Gamma_\infty h^{-1}$ that is a lattice in L. The argument of [15, Proposition 3.6] now implies that α is equivalent to a cocycle β into $\Lambda = \Gamma_\infty \cap h^{-1}Lh$. By Proposition 3.4 again, Λ is Zariski dense in $H_\mathbb{R}$.

Let $J = h^{-1}Lh$. Let A be the image of Λ in $J/[J,J]$. The projection of β must be equivalent to a cocycle into a finite subgroup since A is abelian and G has Kazhdan's property [17, Proposition 9.1]. Therefore β is equivalent to a cocycle into a subgroup $\Delta \subset \Lambda$ such that $\Delta_0 = \Delta \cap [J,J]$ is of finite index in Δ. Since the Lie subalgebra $[\mathfrak{j},\mathfrak{j}]$ is algebraic, $[J,J]$ is of finite index in its Zariski closure. However, the Zariski closure of Δ is $H_\mathbb{R}$ and since $\Delta_0 \subset \Delta$ is of finite index and $H_\mathbb{R}$ is algebraically connected, it follows that $[J,J] = H_\mathbb{R}^0$. (Recall J is connected.) Therefore, we deduce $L = H_\mathbb{R}^0$.

We now have $\Lambda = \Gamma_\infty \subset H_\mathbb{R}^0$ and Γ_∞ is a lattice in $H_\mathbb{R}^0$. Since $\Gamma_\infty \subset H_{\mathbb{Z}_S}$ and its projection to H_f has compact closure, it follows that $\Gamma_\infty \cap H_\mathbb{Z}$ is of finite index in Γ_∞. Since it is a lattice, we deduce that Γ_∞ and $H_\mathbb{Z}$ are commensurable. This completes the proof of those parts of Theorem 5.1 that are not explicity stated in [21]. For the remainder of the conclusion, one can see [21] or easily deduce them from the structure described above.

Proof of Theorem 5.2. We need the following two general lemmas concerning totally engaging actions.

Lemma 5.10. *Suppose the G-action on a Γ-bundle $P \to M$ is totally engaging, where G acts on M ergodically with finite invariant measure. (We assume Γ is infinite.) Let Γ_0 be a finite index subgroup. Then the G action on the Γ_0 bundle $P \to P/\Gamma_0$ is totally engaging.*

Lemma 5.11. *Let P, M, Γ, G as in Lemma 5.10. Let $X \to M$ be an ergodic extension with finite invariant measure. Let $P_X \to X$ be the pull-back of P to X. Assume the G action on $P_X \to X$ is engaging. Then it is totally engaging.*

Proof. Suppose there is a subgroup $\Delta \subset \Gamma$ such that there is a G-invariant section $s: X \to P_X/\Delta$. Decompose the G-invariant measure μ on X over M, say $\mu = \int^{\oplus} \mu_m \, dm$, where μ_m is supported on the fiber in X over $m \in M$. Then for each $m, s_*(\mu_m)$ is a finite measure on the discrete set P_m/Δ. For some $\varepsilon > 0$, the set

$$A_m = \{ x \in P_m \mid s_*(\mu_m)(\{x\}) \geqslant \varepsilon \}$$

will be non-empty (and obviously finite) for a set of m of positive measure, and by invariance of μ and ergodicity of G on M, this will be non-empty for a.e. m. Thus, $m \mapsto A_m$ defines a G-invariant section of the bundle $(P/\Delta)^* \to M$ whose fiber consists of finite subsets of P_m/Δ. Since Γ is discrete, the cocycle reduction lemma [17] implies that there is a G-invariant reduction of P to a group $\Delta' \subset \Gamma$ where Δ' stabilizes a finite subset of Γ/Δ. Since the G-action of $P \to M$ is totally engaging, $\Delta' = \Gamma$, and hence $\Delta \subset \Gamma$ is of finite index. Since the action on $P_X \to X$ is engaging, $\Delta = \Gamma$, completing the proof. □

The proof of Lemma 5.10 is similar. A section of $P/\Delta \to P/\Gamma_0$ for some $\Delta \subset \Gamma_0$ in a manner similar to proof of 5.11 yields a reduction of $P \to M$ to a subgroup $\Delta' \subset \Gamma$ that leaves a finite subset in Γ/Δ invariant. As above, this implies that $\Delta \subset \Gamma$ is of finite index, and since totally engaging implies engaging, this is impossible.

To complete the proof of Theorem 5.2, we need now only observe that in the proof of Theorem 5.1, we showed that α is equivalent to a cocycle into $\Gamma_\infty \subset \Gamma_0$. This, with the hypotheses of Theorem 5.2, Lemmas 5.10, 5.11 imply that $\Gamma_\infty = \Gamma_0$, verifying the theorem.

6. On the relationship of arithmeticity and totally engaging actions

In the section we further clarify the relationship of arithmeticity to the engaging conditions.

Theorem 6.1. *Let G be as in Theorem 5.1, and suppose G acts on $P \to M$, a principal Γ-bundle, with M compact and a finite invariant ergodic measure on M. Suppose Γ is arithmetic. If the G action is engaging, it is totally engaging.*

Proof. Let $\alpha: G \times M \to \Gamma$ be the cocycle defined by the action on P. If the action is not totally engaging, there is some $\Lambda \subset \Gamma$ of infinite index such that $\alpha \sim \beta$ with $\beta(G \times M) \subset \Lambda$. Since Λ is a discrete linear group, we can apply the argument of Zimmer [21] and deduce that $\Lambda \supset \Delta$, where Δ is a lattice in some algebraic group and $\beta \sim \delta$ with $\delta(G \times M) \subset \Lambda$. Since the action is engaging, Δ must be profinitely dense in Γ, and hence Zariski dense. Since Δ is a lattice, this implies Δ is of finite index in Γ, which is a contradiction.

A. Lubotzky, R.J. Zimmer / Topology 40 (2001) 851–869   869

Acknowledgements

The authors would like to thank Moshe Jarden, Shahar Mozes, and T.N. Venkataramana for very helpful conversations on this work. This work was carried out while A.L. was a visitor at the University of Chicago and while R.Z. was a visitor at Hebrew University. It was also supported by a grant from the Binational Science Foundation of Israel and the United States. We wish to thank both universities and the BSF for their hospitality and support.

References

[1] H. Bass, Groups of integral representation type, Pacific J. Math. 86 (1980) 15–51.
[2] J. Benveniste, in preparation.
[3] M.D. Fried, M. Jarden, Field Arithmetic, Ergebnisse der Mathematick, Vol. 11, No. 3, Springer, Heidelberg, 1986.
[4] W.D. Geyer, Galois groups of intersections of local fields, Israel J. Math. 30 (1978) 382–396.
[5] B. Green, F. Pop, P. Roquette, On Rumely's local–global principle, Jaresbericht der Deutsche Mathematicker-vereinigung 97 (1995) 43–74.
[6] F. Grunewald, D. Segal, Remarks on injective specializations, J. Algebra 61 (1979) 538–547.
[7] M. Jarden, On ideal theory in high Prüfer domains, Manuscripta Math. 14 (1975) 303–336.
[8] A. Lubotzky, A. Mann, On groups of polynomial subgroup growth, Invent. Math. 104 (1991) 521–544.
[9] A. Lubotzky, R.J. Zimmer, A canonical arithmetic quotent for simple Lie group actions, preprint.
[10] G.A. Margulis, Discrete Subgroups of Semi-Simple Lie Groups, Springer, New York, 1991.
[11] G.A. Margulis, G. Soifer, Maximal subgroups of infinite index in finitely generated linear groups, J. Algebra 69 (1981) 1–23.
[12] V. Platonov, A. Rapinchuk, Algebraic Groups and Number Theory, Academic Press, New York, 1994.
[13] M. Ratner, On Ragunathan's measure conjecture, Ann. Math. 134 (3) (1991) 545–607.
[14] T.N. Venkataramana, Arithmeticity of lattices in the function field case, Invent. Math. 92 (1988) 255–306.
[15] R.J. Zimmer, Discrete subgroups and non-Riemannian homogeneous spaces, J. Amer. Math. Soc. 7 (1994) 159–168.
[16] R.J. Zimmer, Extensions of ergodic actions and generalized discrete spectrum, Bull. Amer. Math. Soc. 81 (1975) 633–636.
[17] R.J. Zimmer, Ergodic Theory and SemiSimple Groups, Birkhauser, Boston, 1984.
[18] R.J. Zimmer, Ergodic theory and the automorphism group of a G-structure, in: C.C. Moore (Ed.), Group Representations, Ergodic Theory, Operator Algebras, and Mathematical Physics, Springer, New York, 1987, pp. 247–278.
[19] R.J. Zimmer, Representations of fundamental groups of manifolds with a semisimple transformation group, J. Amer. Math. Soc. 2 (1989) 201–213.
[20] R.J. Zimmer, On the algebraic hull of an automorphism group of a principal bundle, Comment. Math. Helv. 65 (1990) 375–387.
[21] R.J. Zimmer, Superrigidity, Ratner's theorem, and fundamental groups, Israel J. Math. 74 (1991) 199–207.

Geometriae Dedicata **107**: 47 56, 2004.
© 2004 *Kluwer Academic Publishers. Printed in the Netherlands.*

Entropy and Arithmetic Quotients for Simple Automorphism Groups of Geometric Manifolds*

ROBERT J. ZIMMER
Brown University, Office of the Provost, Box 1862, Providence, RI 02912, U.S.A.
e mail: robert_zimmer@brown.edu

(Received: 25 June 2002)

Abstract. Let G be a connected simple Lie group, and suppose G acts on a compact real ana lytic manifold M. Assume that G preserves a unimodular rigid geometric structure, for exam ple a connection and a volume form. The aim of this paper is to describe the measure theoretic structure of such actions, when $\mathbb{R}$ rank$(G) \geqslant 2$.

Key words. arithmetic quotients, automorphism groups, entropy, rigidity.

1. Introduction

Let G be a connected simple Lie group, and suppose G acts on a compact real analytic manifold M. Assume that G preserves a unimodular rigid geometric structure, for example a connection and a volume form. The aim of this paper is to describe the measure theoretic structure of such actions, when $\mathbb{R}$-rank$(G) \geqslant 2$.

Suppose, more generally, that M is a compact metrizable space for which covering space theory holds. Suppose G acts continuously on M preserving a probability measure μ. The basic examples of such actions are the arithmetic ones. Namely, suppose H is a real algebraic group, $\Gamma \subset H$ is an arithmetic subgroup, $K \subset H$ is a (perhaps trivial) compact subgroup, and $\rho \colon G \to H$ is a continuous homomorphism so that K centralizes of $\rho(G)$. Then G acts on $K \backslash H / \Gamma$ on the left, and such actions are called arithmetic. A basic question is to determine the extent to which every finite measure preserving action of G is close to being arithmetic.

In [Z3], [LZ1], and [LZ2], we studied arithmetic quotients of a given action. In [LZ1] we showed that a finite entropy ergodic action of a noncompact simple Lie group (with no higher rank assumption) on a space M has a canonical maximal arithmetic (virtual) quotient action, say $A(M)$. (We recall that the finite entropy assumption always holds if M is a compact manifold and G acts smoothly.) In [LZ2], we showed (among other results) that under a mild dynamical assumption (namely that the action be 'engaging') every finite-dimensional linear representation of $\pi_1(M)$ yields a nontrivial virtual arithmetic quotient action, when $\mathbb{R}$-rank$(G) \geqslant 2$. (See below for a precise statement.) In other words, the results of [LZ2] assert that

*Work partially supported by the NSF and the Binational Science Foundation (U.S.A. Israel).

$A(M)$ is a large arithmetic quotient, in the sense that every linear representation of $\pi_1(M)$ determines a quotient of $A(M)$.

On the other hand, the results of [LZ2] make no assertions that $A(M)$ is large with respect to the dynamical properties of M. For example, if N is a (virtual) quotient of M, then the entropy always satisfies $h(g, N) \leqslant h(g, M)$. Let us call a quotient action (or virtual quotient, that is quotient of a finite extension) 'fully entropic' if $h(g, N) = h(g, M)$ for all $g \in G$. It is natural to conjecture that under broad conditions, the virtual arithmetic quotient $A(M)$ is a fully entropic quotient of M.

For an action of G on a real analytic compact manifold M preserving a unimodular rigid geometric structure, Gromov [G] (see also [Z2] has constructed a canonical representation of $\pi_1(M)$, which he shows is large in the sense that the Zariski closure of the image contains a group locally isomorphic to G. In [Z3], finer arithmetic information was obtained on $\pi_1(M)$ by combining Gromov's results with superrigidity arguments. In the present paper, we analyze the (virtual) arithmetic quotient that [LZ2] yields when applied to the Gromov representation. Our main result is that this arithmetic quotient is fully entropic. More precisely,

THEOREM A. *Let M be a compact real analytic manifold with a (real analytic) unimodular rigid geometric structure. Let G be a connected simple Lie group with $\mathbb{R}$-rank$(G) \geqslant 2$, and suppose that G acts analytically on M preserving the structure. Assume further the action is engaging. Then there is a (measurable) fully entropic virtual arithmetic quotient $K \backslash H / \Gamma$ for the G-action on M, i.e., for all g, the entropy satisfies $h(g, M) = h(g, k \backslash H / \Gamma)$.*

In fact, $K \backslash H / \Gamma$ is the virtual arithmetic quotient defined by the Gromov representation. In particular, Γ is isomorphic to a subgroup of $\tau(\pi_1(M))$, where τ is the Gromov representation.

2. Preliminaries: Entropy, Superrigidity and Algebraic Hulls

A basic invariant of a measure preserving transformation g of a space M with probability measure is the (Kolmogorov–Sinai) entropy of g. This is a numerical invariant, $h(g) \in \mathbb{R}$, satisfying $0 \leqslant h(g) \leqslant \infty$. When M is compact manifold, g is a smooth diffeomorphism, and the measure is smooth, i.e. given by a smooth volume density, then $h(g)$ is finite and can be computed from the action of g on TM. Namely, fix a Riemannian metric on M, and consider for $x \in M$,

$$\lim_{n \to \infty} \frac{1}{n} \log \|dg_x^n\|.$$

Then this converges a.e. to a function $e(g)(x)$, and if g is ergodic, this function is a constant. Similarly, for each k, $1 \leqslant k \leqslant n = \dim M$, we have

$$e_k(g)(x) = \lim_{n \to \infty} \frac{1}{n} \log \|\Lambda^k dg_x^n\|.$$

The Pesin formula (in the special case we will need it) then asserts that if g is ergodic.

$$h(g) = \max_k e_k(g).$$

If G is a group acting on a measure space M, preserving a probability measure, we denote by $h(G; M)$ the function $h(G; M)$: $G \to \mathbb{R}, h(G, M)(g) = h(g)$.

The limit in the Pesin formula is of course a dynamical version of the formula for the spectral radius. If V is a finite-dimensional real vector space, and $g \in SL(V)$. let

$$e(g) = \lim_{n \to \infty} \frac{1}{n} \log \|g^n\|.$$

Then $e(g) = \log \|g\|_{sp}$. We define the entropy of g to be

$$h(g) = \max_k e(\Lambda^k(g)).$$

If G is a group. V a finite dimensional real vector space, and $\pi \colon G \to SL(V)$ is a representation, we let $h(G, \pi)$ or $h(G; V)$, denote the function $g \to h(\pi(g))$.

EXAMPLE 2.1. Let H be a connected Lie group, $\Gamma \subset H$ a lattice, and $K \subset H$ a compact subgroup. Suppose $\pi \colon G \to H$, with $K \subset Z_H(\pi(G))$. Let $M = K \backslash H / \Gamma$. Then G acts on M preserving a finite measure. Let $\mathfrak{h}$ denote the Lie algebra of H, which is a G-module via the representation $\text{Ad}_H \circ \pi$. Then $h(G; M) = h(G, \text{Ad}_H \circ \pi)$. This follows easily from the Pesin formula when Γ is cocompact (recalling that K centralizes $\pi(G)$) and is well known as well in the noncocompact case.

If G is a simple Lie group with $\mathbb{R}$-rank$(G) \geqslant 2$, the superrigidity theorem for cocycles implies that for any ergodic action on a manifold, the entropy is given by the entropy of a finite-dimensional G-module whose dimension is the same as that of manifold [Z1]. More precisely, suppose $P \to M$ is a principal H-bundle where H is locally compact, and G acts on P by principal bundle automorphisms. If $\pi \colon G \to H$, a section $s \colon M \to P$ is called π-simple if $s(gm) = gs(m)\pi(g)^{-1}$. We call s almost π-simple if there is a compact subgroup $C \subset Z_H(\pi(G))$ such that $s(gm) = gs(m)\pi(g)^{-1}c(g, m)$, where $c(g, m) \in C$. The cocycle superrigidity theorem asserts that if $\mathbb{R}$-rank$(G) \geqslant 2$, G acts ergodically on M preserving a finite measure, M is compact, and H is a product of algebraic groups over local fields, then there is always an almost π-simple section for some $\pi \colon G \to H$. We call such a π the super-rigidity homomorphism for the action of G on P.

Suppose now that H is a Lie group and there is a representation $\rho \colon H \to GL(V)$ such that the associated vector bundle $E_V = (P \times V)/H \to M$ is G-equivariantly isomorphic to $TM \to M$. In this case, we have

COROLLARY 2.2. $h(G, M) = h(G; \rho \circ \pi, V)$.

Suppose now that G is any group acting by automorphisms of a principal H-bundle $P \to M$, where H is a real algebraic and the G action on M is ergodic. A basic invariant attached to such an action is the 'algebraic hull', which is (a conjugacy class of) a real algebraic subgroup L with the following properties:

(i)   There is a measurable G-invariant reduction of P to L;
(ii) There is no measurable G-invariant reduction of P to L' where $L' \subsetneq L$;
(iii) L is unique up to conjugacy satisfying (i) and (ii).

The algebraic hull controls the measurable G-variant sections for a class of associated fiber bundles in the following sense. Let V be a real algebraic variety on which H acts algebraically. Let E_V be the bundle associated to P and the H-action on V. A section of E_V can be identified with an H-map $\varphi \colon P \to V$, and G-invariance of the section is simply G-invariance of this map. Then $\varphi(P)$ lies (a.e.) in a single H-orbit, say $H \cdot v_0$. Furthermore, the stabilizer $H_{v_0} \supset L$ (up to conjugacy.)

When G is a simple Lie group with finite center and $\mathbb{R}$-rank$(G) \geqslant 2$, the superrigidity theorem for cocycles puts significant restriction on the algebraic hull (see [Z4]). Namely, in this case, assuming the G-action on M preserves a finite measure, the algebraic hull L must be reductive with compact center. Furthermore, L must be an almost direct product of normal subgroups, $L = G^* \cdot C$, where G^* is locally isomorphic to G and C is compact. The superrigidity homomorphism is then a homomorphism $\pi \colon G \to G^* \subset L$.

DEFINITION 2.3. In this context, we shall refer to $G^* \subset L$ as the essential hull for the G action on P.

3. Entropy of an Arithmetic Quotient

We now consider the context of [LZ2]. We suppose G acts on M, that the action is engaging, and that $\sigma \colon \pi_1(M) \to GL(V)$ is a finite-dimensional real representation. Let $\Gamma = \sigma(\pi_1(M))$, and $\overline{\Gamma}$ the Zariski closure. The construction and argument of [LZ2] produces a virtual arithmetic quotient M_σ of M. (We note that the quotient as constructed in [LZ2] is not canonical, as it depends upon the choice of a specialization. However, whichever specialization is chosen, the resulting arithmetic quotient is itself a quotient of the canonical arithmetic quotient $A(M)$.) We have the following lower bound on the entropy of M_σ.

PROPOSITION 3.1. $h(G; M_\sigma) \geqslant h(G; \mathrm{Ad}_{\overline{\Gamma}} \circ \pi, \mathrm{lie}(\overline{\Gamma}))$.

Here $\pi \colon G \to \overline{\Gamma}$ is the superrigidity representation of the principal $\overline{\Gamma}$-bundle associated to $\tilde{M}$ and the homomorphism $\sigma \colon \pi_1(M) \to \overline{\Gamma}$; and $\mathrm{lie}(\overline{\Gamma})$ is the Lie algebra of $\overline{\Gamma}$.

Proof. We extract some of the basic structural features of the construction of [LZ2]. And refer the reader to [LZ2] for further details.

Let $\overline{\Gamma} = L \times U$ be the Levi decomposition, and take $\overline{\Gamma} \subset \mathrm{GL}(n, \mathbb{R})$. We may assume $\pi \colon G \to \overline{\Gamma}$ satisfies $\pi(G) \subset L$. Now choose a specialization ψ as in [LZ2], so that $\psi(\Gamma) \subset \mathrm{GL}(n.\dot{\mathbb{Q}})$, and in fact $\psi(\Gamma) \subset \mathrm{GL}(n, k)$, where $k \subset \mathbb{R}$ and $[k : \mathbb{Q}] < \infty$. Let $\overline{\psi(\Gamma)}$ be the Zariski closure and write $\overline{\psi(\Gamma)} = L_\psi \times U_\psi$. Then, following [LZ2], we may assume $L_\psi = L$ (so in particular $\pi(G) \subset L_\psi$), and that $\overline{\psi(\Gamma)}$ and $\overline{\Gamma}$ are $\mathbb{R}$-isotopic, i.e. lie (U) and lie (U_ψ) are isomorphic L-modules. We let H be the Zariski closure of the image of $\psi(\Gamma)$ in the diagonal embedding of $\psi(\Gamma)$ in $R_{k/\mathbb{Q}}\bigl(\overline{\psi(\Gamma)}\bigr)$. Then the arithmetic quotient M_σ is of the form $M_\sigma = K \backslash H / \Lambda$, where $\Lambda \subset \overline{\psi(\Gamma)}$ is an arithmetic subgroup. In particular, $h(G; M_\sigma) = h(G; \mathrm{Ad}_H \circ \pi', \mathfrak{h})$ where π' is the relevant superrigidity representation.

By projecting $R_{k/\mathbb{Q}}\bigl(\overline{\psi(\Gamma)}\bigr)$ onto the first factor, we obtain a surjection $H \to \overline{\psi(\Gamma)}$, and π' projects to π. Thus,

$$h(G; \mathrm{Ad}_H \circ \pi', \mathfrak{h}) \geqslant h\Bigl(G; \mathrm{Ad}_{\overline{\psi(\Gamma)}} \circ \pi, \mathrm{lie}\,\bigl(\overline{\psi(\Gamma)}\bigr)\Bigr).$$

Since $L = L_\psi$ and $\overline{\psi(\Gamma)}$ and $\overline{\Gamma}$ are $\mathbb{R}$-isotopic, it follows that

$$h\Bigl(G; \mathrm{Ad}_{\overline{\psi(\Gamma)}} \circ \pi, \mathrm{lie}\bigl(\overline{\psi(\Gamma)}\bigr)\Bigr) = h(G; \mathrm{Ad}_{\overline{\Gamma}} \circ \pi, \mathrm{lie}(\overline{\Gamma})).$$

This proves the proposition. □

4. Rigid Transformation Groups

We review and elaborate upon those portions of Gromov's theory of rigid transformations groups we shall need. We refer the reader to Gromov's original work [G], or to [Z2] where the development is more closely related to the issues we face here.

We let M be a compact manifold and ω a rigid unimodular geometric structure of algebraic type. (For example, take ω to be a pseudo-Riemannian metric, or more generally a connection and volume density.) Throughout the remainder of the paper, we assume all structures and actions are real analytic. We assume G is a noncompact simple Lie group acting on M preserving ω.

Let $\mathfrak{k}$ be the Lie algebra of vector fields on $\tilde{M}$ that preserve $\tilde{\omega}$ (namely ω lifted to $\tilde{M}$.) By rigidity of ω, $\dim \mathfrak{k} < \infty$. We have a natural embedding $\mathfrak{g} \to \mathfrak{k}$ defined by the G-action on M (and, hence, the $\tilde{G}$ action on $\tilde{M}$). Let $\mathfrak{n} \subset \mathfrak{n} \subset \mathfrak{k}$ be the subalgebras defined by $\mathfrak{n} = \mathrm{normalizer}_\mathfrak{k}(\mathfrak{g})$, and $\mathfrak{h} = \mathrm{centralizer}_\mathfrak{k}(\mathfrak{g})$. Then clearly $\mathfrak{g} \subset \mathfrak{n}$ and by semisimplicity of $\mathfrak{g}$, we have $\mathfrak{n} = \mathfrak{g} \oplus \mathfrak{h}$. The vector fields in $\mathfrak{g}$ are of course complete. However, those in $\mathfrak{h}$ are not, a-priori, necessarily complete. If H is the corresponding simply connected Lie group, we have a local action of H on $\tilde{M}$ defined $\mathfrak{h}$.

Let $\Gamma = \pi_1(M)$. Then Γ acts on $\tilde{M}$, preserves $\tilde{\omega}$, and commutes with $\mathfrak{g}$. Thus, Γ also preserves $\mathfrak{k}$, $\mathfrak{n}$, and $\mathfrak{h}$.

LEMMA 4.1 (Gromov). $\mathfrak{n}$ *is transitive on an open dense conull set in* $\tilde{M}$. *In fact, the same is true for* $\mathfrak{h}$.

52   ROBERT J. ZIMMER

Proof. Transitivity on $\mathfrak{n}$ on an open conull set follows just as in the argument of [Z2, Theorem 2.4]. Namely, let ω' be the structure consisting of the pair whose value at $x \in M$ is $(\omega(x), \mathfrak{g}^{(k)}(x))$, where $\mathfrak{g}^{(k)}(x)$ is the Lie algebra of k-jets of vector fields in $\mathfrak{g}$, where k is sufficiently large. Then G preserves this rigid structure. By ergodicity and the argument of [Z2, Theorem 2.4] the 'infinitesimal automorphisms' of $(\omega, \mathfrak{g}^{(k)})$ are transitive on an open conull set, and hence so are the local automorphisms. The corresponding vector fields are simply elements of $\mathfrak{n}$.

To see $\mathfrak{h}$ is also transitive, we write $\mathfrak{n} = \mathfrak{g} \oplus \mathfrak{h}$, and observe that the $\mathfrak{h}$-orbit actually contains the $\mathfrak{g}$-orbit a.e. by a basic theorem of Gromov (see [Z2, Corollary 4.3]). Since $\mathfrak{n}$ is transitive on an open conull set, $\mathfrak{h}$ must therefore be so as well.   □

We now examine the stabilizers of the $\mathfrak{h}$ and $\mathfrak{n}$ actions. For each $y \in \tilde{M}$, let $\mathfrak{n}_y \subset \mathfrak{n}$ and $\mathfrak{h}_y \subset \mathfrak{h}$ be the stabilizers of y in $\mathfrak{n}$ and $\mathfrak{h}$. The map $\psi \colon \tilde{M} \to \mathrm{Gr}(\mathfrak{h})$, $\psi(y) = \mathfrak{h}_y$ is G-invariant since $\mathfrak{h}$ centralizes G. Furthermore, under the Γ-action on $\mathfrak{h}$, ψ is Γ-equivariant. Let $\overline{\Gamma}$ be the Zariski closure of the image of Γ under the Gromov representation, i.e. $\overline{\Gamma} \subset \mathrm{Aut}(\mathfrak{h})$. The ψ defines measurable map $M \to \mathrm{Gr}(\mathfrak{h})/\overline{\Gamma}$ that is G-invariant, and hence constant (a.e.) by ergodicity of G on M. Thus, ψ is a map who image is essentially contained in a single $\overline{\Gamma}$-orbit, and as such can be considered as a (a.e. defined, measurable) map $\psi \colon \tilde{M} \to \overline{\Gamma}/\Gamma^*$, where Γ^* is the algebraic subgroup of $\overline{\Gamma}$ that is the stabilizer of $\mathfrak{h}_0 \subset \mathfrak{h}$, where $\mathfrak{h}_0 \subset \mathfrak{h}$ is the subalgebra fixing a (suitable) point in $\tilde{M}$. Let $P_{\overline{\Gamma}}$ be the principal $\overline{\Gamma}$-bundle associated to $\tilde{M}$ and the homomorphism $\Gamma \to \overline{\Gamma}$. Summarizing,

LEMMA 4.2. *The map $\psi \colon \tilde{M} \to \overline{\Gamma}/\Gamma^*$ defines a G-invariant measurable reduction of $P_{\overline{\Gamma}} \to M$ to the group Γ^*.*

Denote by $ev \colon \mathfrak{h} \to T\tilde{M}$ the evaluation map. As described in [Z2], this defines a G-map of vector boundless $E_{\overline{\Gamma}, \mathfrak{h}} \to TM$, where $E_{\overline{\Gamma}, \mathfrak{h}}$ is the bundle associated to $P_{\overline{\Gamma}}$ and the $\overline{\Gamma}$-module $\mathfrak{h}$. By Lemma 4.1, this map is actually surjective (on a.e. fiber.) Therefore, $E_{\Gamma^*, \mathfrak{h}} \to TM$ is also surjective. Since Γ^* leaves $\mathfrak{h}_0$ invariant, we also have a quotient map $E_{\Gamma^*, \mathfrak{h}} \to E_{\Gamma^*, \mathfrak{h}/\mathfrak{h}_0}$. The following is straightforward from the construction of $\mathfrak{h}_0$ and Γ^*.

LEMMA 4.3. *$E_{\Gamma^*, \mathfrak{h}} \to TM$ factors to a G-equivariant (measurable) isomorphism of vector bundles. $E_{\Gamma^*, \mathfrak{h}/\mathfrak{h}_0} \to TM$.*

We let $G^* \subset \Gamma^*$ be the essential hull for the G-action on P_{Γ^*} (see 2.3) and $\pi \colon G \to G^*$ the superrigidity homomorphism. By Lemma 4.3 and Corollary 2.2, we then have

COROLLARY 4.4. *Let $\pi \colon G \to G^*$ be the superrigidity representation. Let ρ_0 be the representation of G^* on $\mathfrak{h}/\mathfrak{h}_0$. Then $h(G, M) = h(G; \rho_0 \circ \pi, \mathfrak{h}/\mathfrak{h}_0)$.*

We now turn to the stabilizer $\mathfrak{n}_y \subset \mathfrak{n}$. Let $\theta\colon \tilde{M} \to \mathrm{Gr}(\mathfrak{n})$ be given by $\theta(y) = \mathfrak{n}_y$. As with ψ, θ is Γ-equivariant. The G-action on $\mathfrak{n} = \mathfrak{g} \oplus \mathfrak{h}$ is, of course, nontrivial, and θ is also G-equivariant, where the G-action on $\mathrm{Gr}(\mathfrak{n})$ is now nontrivial. Since Γ commutes with $\mathfrak{g}$, the Γ action on $\mathfrak{n} = \mathfrak{g} \oplus \mathfrak{h}$ is the direct sum of the trivial representation on $\mathfrak{g}$ with the representation on $\mathfrak{h}$. Thus, the algebraic hull of Γ in $\mathrm{Aut}(\mathfrak{n})$ can be naturally identified with $\overline{\Gamma}$, the algebraic hull of Γ in $\mathrm{Aut}(\mathfrak{h})$. We shall denote by $\overline{\Gamma}$ either of these groups, the meaning being clear from the context. The map θ, being $\Gamma \times G$-equivariant, defines a G-map $\overline{\theta}\colon M \to \mathrm{Gr}(\mathfrak{n})/\overline{\Gamma}$. Since G preserves a finite measure μ on M, $\overline{\theta}_*(\mu)$ is finite and invariant under G. The action of G on $\mathrm{Gr}(\mathfrak{n})/\overline{\Gamma}$ is quasi-algebraic, i.e. tame with algebraic stabilizers. It follows from the Borel density theorem that $\overline{\theta}_*(\mu)$ is supported on a point. Thus, $\theta(\tilde{M})$ is essentially contained in a single $\overline{\Gamma}$-orbit. I.e., $\mathfrak{n}_y \in \overline{\Gamma} \cdot \mathfrak{n}_0$ for some $\mathfrak{n}_0$. We may clearly take $\mathfrak{n}_0$ and $\mathfrak{h}_0$ to correspond to the same point in $\tilde{M}$ (say m_0) and in particular $\mathfrak{h}_0 \subset \mathfrak{n}_0$. Since $\mathfrak{h}$ is an ideal in $\mathfrak{n}$, we may also remark that $\mathfrak{n}_0$ acts on $\mathfrak{h}/\mathfrak{h}_0$ via $\mathrm{ad}_\mathfrak{n}$.

We now summarize a fundamental result of Gromov [G] in a form we shall require. It is a slight variant of [Z2, Theorem 4.2] and is easily obtained from that formulation. The essential idea of Gromov is to establish nontriviality of the algebraic hull, and at the same time construct special elements of $\mathfrak{n}$ and $\mathfrak{h}$. (We have in fact already used this result of Gromov in the proof of Lemma 4.1, deducing essential transitivity of $\mathfrak{h}$ from that on $\mathfrak{n}$.)

LEMMA 4.5 (Gromov [G]: see also [Z2, Theorem 4.2 and Corollary 4.3]). *There is a Lie algebra* $\mathfrak{g}_\Delta \subset \mathfrak{n}$ *with the following properties:*

 (i) *elements of* $\mathfrak{g}_\Delta$ *vanish at* $m_0 \in \tilde{M}$; *i.e.* $\mathfrak{g}_\Delta \subset \mathfrak{n}_0$.
 (ii) $\mathfrak{g}_\Delta$ *is isomorphic to* $\mathfrak{g}$, *and in fact is the graph in* $\mathfrak{n} = \mathfrak{g} \oplus \mathfrak{h}$ *of an injective homomorphism* $\sigma\colon \mathfrak{g} \to \mathfrak{h}$. *We let* $\mathfrak{g}'$ *be the image in* $\mathfrak{h}$.
 (iii) $\mathrm{ad}_\mathfrak{n}(\mathfrak{g}_\Delta)$ *acting on* $\mathfrak{h}/\mathfrak{h}_0$ *is contained in the Lie algebra* $\mathrm{lie}\,(L)$ *of the hull* $L \subset \overline{\Gamma}$, *and in particular,* $\mathrm{ad}_\mathfrak{n}(\mathfrak{g}_\Delta) = \mathfrak{g}^* \subset \mathrm{lie}(L) \subset \mathrm{der}(\mathfrak{h}; \mathfrak{h}_0)$, *where the latter is the derivations of* $\mathfrak{h}$ *leaving* $\mathfrak{h}_0$ *invariant.*

Remarks 4.6.

 (i) It follows from 4.5(*i*) and (*ii*) that $\mathfrak{n}_0 = \mathfrak{g}_\Delta + \mathfrak{h}_0$.
 (ii) The group $\overline{\Gamma} \subset \mathrm{Aut}(\mathfrak{h})$ consists a priori of outer automorphisms and, hence, so do the groups L and G^* (the algebraic hull and essential hull of the G-action.) One consequence of 4.5 (*iii*) is the G^* (or at least its connected component) consists of inner automorphisms, namely these corresponding to the Lie algebra $\mathrm{ad}_\mathfrak{h}(\mathfrak{g}')$.
 (iii) We now have two (local) representations of $\mathfrak{g}$ into $\mathrm{gl}(\mathfrak{h}/\mathfrak{h}_0)$, namely,
   (1) the derivative of $\rho_0 \circ \pi$, where π, as in Corollary 4.4, is the superrigidity representation;
   (2) the map $X \mapsto \rho_0 \circ \mathrm{ad}_\mathfrak{h}(X')$, where $X \mapsto X'$ is the homomorphism $\sigma\colon \mathfrak{g} \to \mathfrak{g}'$ defined in 4.5(*ii*), and $\rho_0\colon \mathrm{der}(\mathfrak{h}; \mathfrak{h}_0) \to \mathrm{der}(\mathfrak{h}_0)$.

54   ROBERT J. ZIMMER

We claim that in fact these representations are the same. In other words, the super-rigidity representation, and Gromov's representation $\mathfrak{g} \to \mathfrak{g}'$ act identically on $\mathfrak{h}/\mathfrak{h}_0$.

To see this, we first observe that both have image $\mathfrak{g}^*$, and so they differ at most by an automorphism of $\mathfrak{g}^*$. To see this automorphism is trivial, we need one further piece of information about each representation. The action of G on M preserves a finite mea-sure and, hence, the stabilizers are a.e. discrete. It follows that we may assume $m_0 \in \tilde{M}$ satisfies this property. Thus, we have a natural embedding $\mathfrak{g} \hookrightarrow T\tilde{M}_{m_0} \cong \mathfrak{h}/\mathfrak{h}_0$. Since the group G acts on tangent bundle to the orbits in M via Ad, the restriction of $d(\rho_0 \circ \pi)$ to $\mathfrak{g} \hookrightarrow \mathfrak{h}/\mathfrak{h}_0$ is $\text{ad}_{\mathfrak{g}}$. Similarly, this holds for $\rho_0 \circ \text{ad}_{\mathfrak{h}} \circ \sigma$ via the analysis of the Gromov representation σ as formulated in [Z2, Theorem 4.2]. It then follows easily that the representation in question are identical.

COROLLARY 4.7. $h(G, M) = h(G; (\rho_0 \circ \text{ad}_{\mathfrak{h}} \circ \sigma)^{\tilde{}}, \mathfrak{h}/\mathfrak{h}_0)$. (Here, $(\ )^{\tilde{}}$ denotes integration of a Lie algebra representation to G.)

Proof. This follows from the above discussion and Corollary 4.4. □

Remark 4.8. from 4.6(ii) and (iii), we also deduce that $\pi: G \to G^* \subset \text{Aut}(\mathfrak{h})$ and $\sigma: \mathfrak{g} \to \mathfrak{g}' \subset \mathfrak{h}$ are related by $d\pi = \text{ad}_{\mathfrak{h}} \circ \sigma$.

5. Completion of Proof of Theorem A

The previous sections reduce the proof of Theorem A to a comparison of the entropy of two finite-dimensional G-modules.

LEMMA 5.1. *To prove Theorem A, it suffices to show*

$$h(G : \text{Ad}_{\overline{\Gamma}} \circ \pi, \text{lie}(\overline{\Gamma})) \geqslant h(G; (\rho_0 \circ \text{ad}_{\mathfrak{h}} \circ \sigma)^{\tilde{}}, \mathfrak{h}/\mathfrak{h}_0).$$

Here π is the superrigidity representation, σ the homomorphism $\mathfrak{g} \longrightarrow \approx \mathfrak{g}' \subset \mathfrak{h}$ defind in Section 4, and $(\ )^{\tilde{}}$ denotes integration of the representation of $\mathfrak{g}$ to G.

Proof. This follows from Corollary 4.7, Proposition 3.1, and the fact that $h(G, M_\sigma) \leqslant h(G, M)$. □

The remainder of this section is devoted to proving the required inequality in Lemma 5.1.

We return to an examination of the space $\text{Gr}(\mathfrak{n})$ and the element $\mathfrak{n}_0 \in \text{Gr}(\mathfrak{n})$. We let N and H be Lie groups with Lie algebra $\mathfrak{n}$ and $\mathfrak{h}$. We have local actions of these groups on $\tilde{M}$. Therefore, if $a \in N$ and $y \in \tilde{M}$, we have $\mathfrak{n}_{ay} = \text{Ad}(a)\mathfrak{n}_y$ whenever ay is defined.

LEMMA 5.2. $N \cdot \mathfrak{n}_0 \subset \overline{\Gamma} \cdot \mathfrak{n}_0$ *locally near* $\mathfrak{n}_0$.
Proof. This follows from the fact that a.e. $\mathfrak{n}_y \in \overline{\Gamma} \cdot \mathfrak{n}_0$, as discussed in Section 4.

We now consider the action of N and $G \times \overline{\Gamma}$ on $\text{Gr}(\mathfrak{n})$. We have an embedding $G_\Delta \hookrightarrow N$ corresponding to the inclusion $\mathfrak{g}_\Delta \subset \mathfrak{n}$. We also have $\text{gr}(\pi) \subset G \times \overline{\Gamma}$. G_Δ

and $\mathrm{gr}(\pi)$ are of course (locally) isomorphic, and by Remark 4.8 we may identify the two actions on $\mathfrak{n}$. We will thus view $G_\Delta \subset G \times \overline{\Gamma}$ and $G_\Delta \subset N$, with the actions of $G \times \overline{\Gamma}$ and N on $\mathrm{Gr}(\mathfrak{n})$ restricting to the same action of G_Δ. We note that G_Δ fixes $\mathfrak{n}_0$ since $\mathfrak{g}_\Delta \subset \mathfrak{n}_0$.   □

LEMMA 5.3. *We have an embedding of G_Δ-modules:*

$$\mathfrak{n}/\mathrm{norm}_\mathfrak{n}(\mathfrak{n}_0) \hookrightarrow (\mathfrak{g} \times \mathrm{lie}(\overline{\Gamma}))/\mathrm{stab}_{\mathfrak{g} \times \mathrm{lie}(\overline{\Gamma})}(\mathfrak{n}_0).$$

Proof. As vector spaces, this is simply the inclusion

$$T(N \cdot \mathfrak{n}_0)_{\mathfrak{n}_0} \subset T((G \times \overline{\Gamma}) \cdot \mathfrak{n}_0)_{\mathfrak{n}_0}$$

given via Lemma 5.2. They are both G_Δ-modules since G_Δ fixes $\mathfrak{n}_0$.   □

Now let $G' \subset H$ correspond to $\mathfrak{g}' \subset \mathfrak{h}$.

LEMMA 5.4. *There is an embedding of G'-modules:*

$$\mathfrak{h}/\mathrm{stab}_\mathfrak{h}(\mathfrak{n}_0) \hookrightarrow \mathrm{lie}(\overline{\Gamma})/\mathrm{stab}_{\mathrm{lie}(\overline{\Gamma})}(\mathfrak{n}_0).$$

Proof. We first remark that $\mathrm{stab}_\mathfrak{h}(\mathfrak{n}_0)$ and $\mathrm{stab}_{\mathrm{lie}(\overline{\Gamma})}(\mathfrak{n}_0)$ are both G'-modules since they are intersections of G_Δ-invariant submodules and are contained in spaces on which G_Δ acts by G'.

Now consider the inclusion $\mathfrak{h} \hookrightarrow \mathfrak{n}$. This induces an inclusion $\mathfrak{h}/\mathrm{stab}_\mathfrak{h}(\mathfrak{n}_0) \hookrightarrow \mathfrak{n}/\mathrm{norm}_\mathfrak{n}(\mathfrak{n}_0)$. Since $\mathfrak{g}_\Delta \subset \mathfrak{n}_0$, and $\mathfrak{g}_\Delta + \mathfrak{h} = \mathfrak{n}$, it follows that this is an isomorphism of G_Δ-modules. However, G_Δ acts on $\mathfrak{h}/\mathrm{stab}_\mathfrak{h}(\mathfrak{n}_0)$ via G'.

Similarly, the map $\mathrm{lie}(\overline{\Gamma}) \hookrightarrow \mathfrak{g} \times \mathrm{lie}(\overline{\Gamma})$ induces an isomorphism

$$\mathrm{lie}(\overline{\Gamma})/\mathrm{stab}_{\mathrm{lie}(\overline{\Gamma})}(\mathfrak{n}_0) \cong (\mathfrak{g} \times \mathrm{lie}(\overline{\Gamma}))/\mathrm{stab}_{\mathfrak{g} \times \mathrm{lie}(\overline{\Gamma})}(\mathfrak{n}_0)$$

(since $\mathrm{lie}(\overline{\Gamma}) + \mathfrak{g}_\Delta = \mathfrak{g} \times \mathrm{lie}(\overline{\Gamma})$), and G_Δ acts on $\mathrm{lie}(\overline{\Gamma})$ by G'.

Therefore, Lemma 5.4 follows from Lemma 5.3.   □

COROLLARY 5.5.

$$h(G', \mathfrak{h}/\mathrm{stab}_\mathfrak{h}(\mathfrak{n}_0)) \leqslant h(G', \mathrm{lie}(\overline{\Gamma})).$$

Proof. In fact,

$$h(G'.\mathfrak{h}/\mathrm{stab}_\mathfrak{h}(\mathfrak{n}_0)) \leqslant h\Big(G', \mathrm{lie}(\overline{\Gamma})/\mathrm{stab}_{\mathrm{lie}(\overline{\Gamma})}(\mathfrak{n}_0)\Big).$$

by Lemma 5.4.

We now analyze $\mathfrak{h}/\mathrm{stab}_\mathfrak{h}(\mathfrak{n}_0)$. We have an exact sequence of vector spaces:

$$0 \to \mathrm{stab}_\mathfrak{h}(\mathfrak{n}_0)/\mathfrak{h}_0 \to \mathfrak{h}/\mathfrak{h}_0 \to \mathfrak{h}/\mathrm{stab}_\mathfrak{h}(\mathfrak{n}_0) \to 0, \qquad\qquad (*)$$

which is also an exact sequence of G'-modules.

56   ROBERT J. ZIMMER

LEMMA 5.6. G' *acts trivially on* $\mathrm{stab}_{\mathfrak{h}}(\mathfrak{n}_0/\mathfrak{h}_0)$.

Proof. Let $Y \in \mathfrak{h}$, and suppose Y normalizes $\mathfrak{n}_0$. Given any $X \in \mathfrak{g}$, we have $X + X' \in \mathfrak{g}_\Delta$ and $X + X' \in \mathfrak{n}_0$. Thus, $[Y, X+X'] \in \mathfrak{n}_0 = \mathfrak{g}_\Delta + \mathfrak{h}_0$ (by 4.6(i).) Now X and Y commute, $[Y, X + X'] = [Y, X'] \in \mathfrak{h}$; i.e. $[Y, X'] \in \mathfrak{h} \cap \mathfrak{n}_0 = \mathfrak{h}_0$. In addition, $[Y, \mathfrak{h}_0] \in \mathfrak{h} \cap \mathfrak{n}_0 = \mathfrak{h}_0$, so $\mathfrak{h}_0$ is an ideal in $\mathrm{stab}_{\mathfrak{h}}(\mathfrak{n}_0)$. The assertion $[Y, X'] \in \mathfrak{h}_0$, i.e. $\mathrm{ad}_{\mathfrak{h}}(X')(Y) \in \mathfrak{h}_0$ implies that $\mathrm{ad}_{\mathfrak{h}}(X')$ acts trivially on the Lie algebra $\mathrm{stab}_{\mathfrak{h}}(\mathfrak{n}_0)/\mathfrak{h}_0$. Thus, G' acts trivially.   □

COROLLARY 5.7. $h(G', \mathfrak{h}/\mathrm{stab}_{\mathfrak{h}}(\mathfrak{n}_0)) = h(G', \mathfrak{h}/\mathfrak{h}_0)$.

Proof. This follows from Lemma 5.6 and the exact sequence (∗).   □

Finally, to prove Theorem A, we observe that Corollaries 5.5 and 5.7 imply the inequality required in Lemma 5.1, since the required representations in 5.1 are given by the G' actions.

Acknowledgement

This work was begun and completed while the author was a visitor at the Hebrew University in Jerusalem. The author wishes to express his appreciation for this hospitality.

References

[G]   Gromov, M.: Rigid transformation groups, In: D. Bernard and Y. Choquet Bruhat, (eds), *Geometrie differentialle*, Hermann, Paris, 1988.

[LZ1]   Lubotzky, A. and Zimmer, R. J.: A canonical arithmetic quotient for simple Lie group actions, In: S. Dani, (ed.), *Lie Groups and Ergodic Theory*, 1999.

[LZ2]   Lubotzky, A. and Zimmer, R. J.: Arithmetic structure of fundamental groups and actions of semisimple Lie groups, to appear in *Topology*.

[Z1]   Zimmer, R. J.: *Ergodic Theory and Semisimple Groups*, Birkhäuser, Boston, 1984.

[Z2]   Zimmer, R. J.: Automorphism groups and fundamental groups of geometric mani folds, *Proc. Sympos. Pure Math.* **54** (1993), 693 710.

[LZ3]   Zimmer, R. J.: Superrigidity, Ratner's theorem, and fundamental groups, *Israel J. Math.* **74** (1991), 199 207.

[LZ4]   Zimmer, R. J.: On the algebraic hull of the automorphism group of a principal bundle, *Comment. Math. Helv.* **65** (1990), 375 387.

Comment. Math. Helv. 77 (2002) 326–338
0010-2571/02/020326-13 $ 1.50+0.20/0

Commentarii Mathematici Helvetici

Geometric lattice actions, entropy and fundamental groups

David Fisher[1] and Robert J. Zimmer[2]

Abstract. Let Γ be a lattice in a noncompact simple Lie Group G, where $\mathbb{R} - \text{rank}(G) \geq 2$. Suppose Γ acts analytically and ergodically on a compact manifold M preserving a unimodular rigid geometric structure (e.g. a connection and a volume). We show that either the Γ action is isometric or there exists a "large image" linear representation σ of $\pi_1(M)$. Under an additional assumption on the dynamics of the action, we associate to σ a virtual arithmetic quotient of full entropy.

Mathematics Subject Classification (2000). 57S20, 53C24, 22F50.

Keywords. Lattices in Lie groups, rigidity, rigid geometric structures, entropy.

1. Introduction

Suppose G is a noncompact simple Lie group and G acts analytically on a compact manifold M preserving a unimodular rigid geometric structure, for example either a connection and a volume form or a pseudo-Riemannian metric. Assuming the action is ergodic, Gromov [G] constructed a linear representation $\sigma : \pi_1(M) \to GL_n(\mathbb{R})$ such that the Zariski closure of $\sigma(\pi_1(M))$ contains a group locally isomorphic to G. One cannot hope for an analogous result for lattices in all semi-simple groups, since lattices in rank 1 groups often admit homomorphisms to $\mathbb{Z}$, and many counterexamples can be constructed for $\mathbb{Z}$ actions. In addition, even for lattices in higher rank simple groups, there exist isometric actions on manifolds with finite fundamental group. Here we prove a form of Gromov's theorem for lattices, which shows that for actions of higher rank lattices either the action is isometric or there exists a representation like Gromov's. More precisely:

Theorem 1.1. *Let $\Gamma < G$ be a lattice, where G is a simple group and $\mathbb{R} - \text{rank}(G) \geq 2$. Suppose Γ acts analytically and ergodically on a compact manifold M preserving a unimodular rigid geometric structure. Then either*

---
[1]Partially supported by NSF grant DMS-9902411.

[2]Partially supported by NSF grant.

i) *the action is isometric and $M = K/C$ where K is a compact Lie group and the action is by right translation via $\rho : \Gamma \hookrightarrow K$, a dense image homomorphism, or*

ii) *there exists an infinite image linear representation $\sigma : \pi_1(M) \to GL_n\mathbb{R}$, such that the algebraic automorphism group of the Zariski closure of $\sigma(\pi_1(M))$ contains a group locally isomorphic to G.*

To prove the theorem, one first applies Gromov's result to the induced G action on $(G \times M)/\Gamma$ to obtain a linear representation σ of $\Lambda = \pi_1((G \times M)/\Gamma)$. However, $\pi_1(M)$ is a normal subgroup of Λ, and in fact, if $\pi_1(G)$ is trivial, we have a short exact sequence:

$$1 \longrightarrow \pi_1(M) \longrightarrow \Lambda \longrightarrow \Gamma \longrightarrow 1.$$

To show that the restriction of σ to $\pi_1(M)$ is trivial only when the Γ action is isometric, we use the techniques of [Z5] to compare the entropy of the G action on $(G \times M)/\Gamma$ with the image $\sigma(\Lambda)$. If G is not simply connected, let $\pi : \tilde{G} \to G$ be the universal cover and $\tilde{\Gamma} \to \Gamma$ the pre-image of Γ under the covering map π. The exact sequence above now holds with $\tilde{\Gamma}$ in place of Γ. By viewing the Γ action as a $\tilde{\Gamma}$ action via the homomorphism $\tilde{\Gamma} \to \Gamma$, we can always assume that our lattice Γ is a lattice in a simply connected group $G = \tilde{G}$. We will make this assumption throughout this paper.

There is one earlier result along these lines for lattice actions. In [Z3], the second author shows that, under the assumptions of Theorem 1.1, if we assume $\pi_1(M)$ is trivial, then the action is isometric. As a corollary, we obtain the following generalization of this fact:

Corollary 1.2. *Let Γ act on M as in Theorem 1.1. Further assume there are no infinite image linear representation of $\pi_1(M)$. Then the Γ action on M is isometric and $\pi_1(M)$ is finite.*

An ergodic action of G on M is said to be engaging if any lift of the action to a finite cover of M is still ergodic. In [LZ], given an engaging action on M of a simple group G, $\mathbb{R} - \text{rank}(G) \geq 2$, the authors associate a virtual arithmetic quotient to any linear representation of $\pi_1(M)$. Using the explicit construction of Gromov's representation σ, the second author has shown that for G actions which are analytic and preserve a rigid geometric structure, the associated virtual arithmetic quotient has the same entropy as the original G action [Z5]. Using the results of [F], we prove the following analogous result here:

Theorem 1.3. *Assume Γ acts on M as in Theorem 1.1. Further assume the action is engaging. Then either*

i) *The Γ action is isometric and is described as in Theorem 1.1, or*

ii) *For some finite index subgroup $\Gamma' < \Gamma$ and some finite cover, M' of M, there is a measurable Γ' equivariant map $\varphi : M' \to K\backslash L/L_{\mathbb{Z}}$. If h_- is the entropy function of the relevant Γ action, we have $h_M(\gamma) = h_{K\backslash L/L_{\mathbb{Z}}}(\gamma)$ for all $\gamma \in \Gamma'$.*

Here the Γ' action on $K \backslash L / L_\mathbb{Z}$ is a generalized affine action.

We recall that an affine diffeomorphism f of a homogeneous space A/B is simply one that is covered by a diffeomorphism $\tilde{f}$ of A, where $\tilde{f}$ is the composition of a group automorphism L and left translation by an element of A. For this to make sense it is clear that $L(B) = B$. A group action on a homogeneous space is said to be affine if every element in the group acts by an affine diffeomorphism. The generalized affine action of Theorem 1.3. refers to the fact that we are not acting on a homogeneous space. Instead the action is on a quotient of the homogeneous space by a compact group of affine diffeomorphisms via diffeomorphisms covered by affine diffeomorphisms of the homogeneous space. More precisely:

Definition 1.4. Let A/B be a homogeneous space and D and F two commuting groups of affine diffeomorphisms of A/B, with D compact. The we call the F action on $D \backslash A / B$ a *generalized affine action*.

The proof of Theorem 1.3. gives a more detailed description of the action. The action is shown to be arithmetic in the sense of Definition 3.2. of [F]. That any generalized affine action of a lattice in a higher rank simple group is arithmetic is also a straightforward exercise from Margulis' superrigidity theorem and the structure theory of algebraic groups.

Note that two frequently studied examples of generalized affine actions of higher rank lattices are actions by left translations on homogeneous spaces and affine actions on nilmanifolds. In the latter case the affine diffeomorphisms considered generally have no translational part (at least on a subgroup of finite index). In general, one can construct actions that fall into neither of these two categories, i.e. affine actions where group elements act by compositions of non-trivial translations and non-trivial automorphisms.

We now sketch the proof of Theorem 1.1. For ease of exposition, we assume that $\Lambda = \pi_1\big((G \times M)/\Gamma\big) = \Gamma \ltimes \pi_1(M)$, i.e. that the extension above is split. In order to apply Gromov's result to the G action on $(G \times M)/\Gamma$ we need to produce a rigid geometric structure ω invariant under this action. This will be locally a product of the given structure on M with the natural pseudo-Riemannian structure along the G orbits given by the Killing form on $\mathfrak{g}$. In fact, there does not seem to be a definition in the literature of a product of geometric structures in the sense of Gromov or a proof that a product of rigid structures is rigid. In the case of Cartan's geometric structures of finite type this is classical, and we will give the analogous definitions and proofs for the more general setting in Section 3.

An important property of this particular geometric structure on $(G \times M)/\Gamma$ is that we can explicitly identify a Lie algebra $\mathfrak{g}$ of local Killing fields of the structure commuting with the G action. These come from lifting the structure ω to a structure $\tilde{\omega}$ on the universal cover $G \times \tilde{M}$ and then differentiating the right G action on G, which preserves the structure by construction.

Gromov's representation σ is constructed on all Killing fields V of $\tilde{\omega}$ commuting with the lift of the G action. Ideally, one would like to be able to say that V splits as the sum of $\mathfrak{g}$ above and a collection of Killing fields W tangent to the M fibers in $(G \times M)/\Gamma$. This would immediately give a representation of $\pi_1(M)$ on W. Regrettably, it is not clear how to construct such Killing fields in a natural way that would allow one to show that this representation is non-trivial, and one must approach the problem indirectly. It follows from the structure of the $\pi_1\big((G \times M)/\Gamma\big)$ action on $G \times \tilde{M}$ (as described in [F]) that $\sigma(\pi_1(M))$ acts trivially on the Lie algebra $\mathfrak{g}$ of Killing fields constructed above and that $\sigma(\Gamma)$ acts via the adjoint representation of G. What remains to be shown is that if $\sigma(\pi_1(M))$ is not infinite, then the action is isometric.

To do this, we compare the entropies of various actions and representations. First, in Section 2, we observe that the entropy of the induced action is the sum of the entropy of the Γ action on M and the entropy of the G action on G/Γ. (This is precisely true if we induce and then restrict to the Γ action. A stronger statement will be made precise in Section 2.)

From this it follows that if the entropy of the induced action equals the entropy of the G action on G/Γ, the Γ action must have zero entropy. An adaptation of standard arguments then shows that actions preserving a rigid geometric structure and having zero entropy are isometric.

To complete the proof, we show that unless the $\sigma(\pi_1(M))$ has large Zariski closure, the entropy of the induced action is indeed equal to the entropy of the G action on G/Γ. This is done using an estimate from [Z5]. Gromov has shown that the Lie algebra, $\mathfrak{l}$, of the Zariski closure of $\sigma(\pi_1((G \times M)/\Gamma))$ has a natural structure as a G module. In [Z5], the second author shows that the entropy of this module is an upper bound on the entropy of the G action. (The notion of entropy of a module is made precise in Section 4.) The Lie algebra of the Zariski closure of $\sigma(\pi_1(M))$ is a G submodule of $\mathfrak{l}$. If the entropy of this submodule is zero, then the entire entropy of $\mathfrak{l}$ comes from the Lie algebra of the Zariski closure of $\sigma(\Gamma)$. An explicit analysis of the construction of Gromov's representation shows that the entropy of this module is equal to the entropy of the G action on G/Γ. This analysis uses the existence of our explicitly constructed Lie algebra $\mathfrak{g}$ of Killing fields along the G orbits in $(G \times M)/\Gamma$. This shows that the Lie algebra of the Zariski closure of $\sigma(\pi_1(M))$ is a non-trivial G module unless the Γ action is isometric, and we are done.

To prove Theorem 1.3 we use a similar analysis of Gromov's representation for induced actions to adapt the arguments of [F] and [Z5] to the present setting. We first show that the induced action has a full entropy quotient and then use techniques of [F] to show that this quotient has the structure of an induced action. The techniques of [F] produce an arithmetic quotient Y of the Γ action, and show that the quotient for the induced action on $(G \times M)/\Gamma$ is actually the induced action on $(G \times Y)/\Gamma$. Theorem 2.3 then tells us that the Γ entropy on Y must equal the Γ entropy on M.

The first author would like to thank E. Benveniste, R. Feres and G. Margulis for enlightening conversations regarding rigid geometric structures. Both authors thank the Newton Institute for Mathematical Sciences for it's hospitality during the time when this work was completed and the first author also thanks ETH Zurich.

2. Computing entropy

In this section, we discuss the computation of entropy for actions of lattices and the relation to the entropy of the induced action. The basic tool is the use of superrigidity to linearize the derivative cocycle. Let $P \to M$ be a principal H bundle where H is a connected real algebraic group. Assume a group G acts on P by automorphisms preserving a finite measure on M. Let $\sigma : G \to H$ be a homomorphism. We say that a measurable section $s : M \to P$ is σ-simple if there exists a cocycle $c : M \times G \to K$ where $K < Z_H(\sigma(G))$ is compact and

$$s(gm) = gs(m)\sigma(g)^{-1}c(m, g)$$

for all $g \in G$ and almost all $m \in M$. Let G be a simple Lie group, $\mathbb{R} - \text{rank}(G) \geq 2$, let M be a compact space and assume the G action on M is ergodic and continuous. Superrigidity for cocycles implies that there exists a finite ergodic extension $X \to M$, so that for the action on P', the pullback of P to X, there exists a representation $\sigma : G \to H$ and a σ-simple section s. In fact, we need not assume that M is compact but only that the cocycle defined by the G action on P is quasi-integrable, a boundedness condition that will hold in all situations of interest here (see [Z6] for details). We will refer to σ as the superrigidity representation. If $\Gamma < G$ is a lattice and we have an action of Γ on P which is ergodic on M (with no G action) then we have a very similar conclusion. Either, for some finite ergodic extension there is a linear representation $\sigma : G \to H$ and a σ-simple section or the algebraic hull of the Γ action on P is compact. Here again we require either M compact or the cocycle defined by the Γ action quasi-integrable.

Applying superrigidity to the derivative action on the frame bundle, we can compute the entropy directly via Pesin's formula as in the following two theorems from [Z1]:

Theorem 2.1. *Suppose G is a simple Lie group with $\mathbb{R} - \text{rank}(G) \geq 2$. Suppose G acts ergodically on a manifold M by diffeomorphisms preserving a C^2 volume form such that the volume of M is finite. Further assume that the derivative cocycle is quasi-integrable. For each $g \in G$, let $h(g)$ be the entropy. Let $\sigma : G \to SL_n\mathbb{R}$ be the superrigidity representation for the action on the frame bundle of M, $n = \dim M$. Then $h(g) = \Sigma \log^+ \|\lambda\|$, where λ runs over all eigenvalues of $\sigma(g)$.*

Theorem 2.2. *Suppose $\Gamma < G$ is a lattice, G as above. Suppose Γ acts ergodically by diffeomorphisms on a compact manifold M preserving a C^2 volume form. For* *each $\gamma \in \Gamma$ let $h(\gamma)$ be the entropy. Let $\sigma : G \to SL_n\mathbb{R}$ be the superrigidity representation for the action on the frame bundle of M, $n = \dim M$. Then $h(\gamma) = \Sigma \log^+ \|\lambda\|$, where λ runs over all eigenvalues of $\sigma(\gamma)$, or the algebraic hull of the derivative cocycle is compact and $h(\gamma) = 0$ for all $\gamma \in \Gamma$.*

The proof of this is essentially in [Z1], though there M is always assumed to be compact in Theorem 2.1. Compactness is used to be able to apply Pesin's formula which is often only stated for actions on compact manifolds, but which holds whenever the action preserves a finite volume and the derivative cocycle is quasi-integrable [P1, P2]. We apply Theorem 2.1 to the derivative cocycle of the G action on $(G \times M)/\Gamma$. Though $(G \times M)/\Gamma$ is not compact when Γ is not cocompact, the derivative cocycle is still quasi-integrable by the results in [L] if G is algebraic and by results in Section 7 of [F] if G is not algebraic.

Suppose Γ acts on M by C^2 diffeomorphisms preserving volume. The induced G action on $(G \times M)/\Gamma$ is also C^2 and volume preserving, and the G action clearly preserves the decomposition $T\big((G \times M)/\Gamma\big) = T(\mathfrak{G}) \times TM$ where $T(\mathfrak{G})$ is the tangent space to the G orbits and the direct product is given by the fact that $(G \times M)/\Gamma$ is a flat fiber bundle over G/Γ with fiber M, i.e. locally a product of G/Γ with M. Let h_M, $h_{G/\Gamma}$ and $h_{(G\times M)/\Gamma}$ be the entropy functions for the actions of Γ on M and G on G/Γ and $(G \times M)/\Gamma$ respectively. Note that since h_M is given as in Theorem 2.2, h_M is defined for all $g \in G$, even though we only have an action of Γ. By showing that the superrigidity representation preserves the direct sum decomposition of $T\big((G \times M)/\Gamma\big)$, we prove the following formula:

Theorem 2.3. $h_{(G\times M)/\Gamma}(\gamma) = h_{G/\Gamma}(\gamma) + h_M(\gamma)$

Proof. Since the G action preserves the splitting $T\big((G \times M)/\Gamma\big) = T(\mathfrak{G}) \times TM$, we can view the derivative cocycle as taking values in the bundle $P(\mathfrak{G}) \times P(M)$ of frames respecting this decomposition. The cocycle into $P(\mathfrak{G})$ is given by the adjoint action of G, exactly as in the case of G acting on G/Γ. We are therefore reduced to studying the G action on the bundle $\big(G \times P(M)\big)/\Gamma \to (G \times M)/\Gamma$.

Applying superrigidity for cocycles as described above, we see that possibly after passing to a finite ergodic extension of $(G \times M)/\Gamma$, there is a measurable section $s : (G\times M)/\Gamma \to \big(G\times P(M)\big)/\Gamma$ and a linear representation $\pi : G \to SL_n\mathbb{R}$, where $n = \dim M$ such that

$$s(gx) = gs(x)\pi(g)^{-1}c(x, g)$$

for all $g \in G$ and almost every $x \in (G \times M)/\Gamma$, where c is a cocycle taking values in a compact group which commutes with $\pi(G)$. Note that π being trivial is equivalent to the algebraic hull of the cocycle being compact. This shows us that to compute the entropy of the G action on $(G \times M)/\Gamma$ we can take $\sigma = Ad \oplus \pi$ and compute as in Theorem 2.1. To complete the proof, we need only see that π can also be used via Theorem 2.2, to compute the entropy of the Γ action on M.

Let $[g] \in G/\Gamma$, then $g\Gamma g^{-1}$ acts on the M fiber over $[g]$ in $(G \times M)/\Gamma$ via translation to the fiber over the identity, the given Γ action on M, and then translation back. Since s is defined for almost all $x \in (G \times M)/\Gamma$, writing $x = [g, m]$ and applying Fubini's theorem, we see that for almost all $[g_0] \in G/\Gamma$, s is defined for almost every $[g_0, m]$ in the M fiber over $[g_0]$. Possibly after conjugating the action by g_0^{-1}, we can therefore assume that s is defined for almost every $m \in M$ in the fiber over $[e]$. Restricting s to $[e, m]$ and looking only at $\gamma \in \Gamma$, the formula above implies that

$$s(\gamma[e, m]) = \gamma s([e, m]) \pi^{-1}(\gamma) c([e, m]), \gamma)$$

for all $\gamma \in \Gamma$ and almost every $m \in M$. This shows that if we use the restriction of s to trivialize $P(M) \to M$ we can compute the Lyapunov exponents of the derivative cocycle via those of $\pi^{-1}(\gamma) c(m, \gamma)$ and since the image of c is compact, we can compute the entropy of the Γ action on M via the eigenvalues of the representation π. $\square$

Remark. We have actually proved more than just the formula. We have shown that, for the derivative cocycle, the superrigidity representation for the induced action is cohomologous to the direct sum of the superrigidity representation for the Γ action and the representation Ad_G.

Corollary 2.4. *Suppose Γ acts on a compact manifold M preserving a unimodular rigid geometric structure. Assume further that the action is ergodic and that $h_{(G \times M)/\Gamma}(g) = h_{G/\Gamma}(g)$, for all $g \in G$. Then the action is Riemannian isometric.*

Proof. By Theorem 2.3, this implies that $h_M(\gamma) = 0$ for all $\gamma \in \Gamma$. Either by looking at the proof of that theorem or via Theorem 2.2, this implies the algebraic hull of the derivative cocycle is compact. By the proof of Theorem 4.5 from [Z4] this implies that there is a Γ invariant finite measure on $P^k(M)$, the kth order frame bundle, for any k. In fact this measure is the volume on a (measurable) K bundle over M, where K is a compact subgroup of the higher order frame group. However, Gromov has observed that any closed subgroup of the isometry group of a rigid geometric structure is proper and free on $P^k(M)$ for k large enough [G, Section 0.4]. Since the closure of Γ acts properly on a finite measure space, the closure of Γ in the automorphism group of the structure is compact (see [Z4] proof of Theorem 5.2 for more details) and therefore the action is isometric. $\square$

3. Products of geometric structures

In this section we will develop the ideas we need concerning geometric structures, particularly products of structures on (local) products of manifolds. Throughout D_n^s will denote the group of s-jets of diffeomorphisms of $\mathbb{R}^n$ fixing 0 and $P^s(M)$

will be the sth order frame bundle of the manifold M. This is a D_n^s principal bundle where $n = \dim M$.

Definition 3.1. A geometric structure, ω, on a manifold M, with $\dim M = n$, consists of:

    a) an algebraic variety V on which D_n^s acts for some s

    b) a map $\omega : P^s(M) \to V$ which is D_n^s equivariant.

The structure is said to be of order s and is $C^r (C^\omega)$ when the map w is $C^r (C^\omega)$. The structure is called unimodular if it defines a volume form on M.

If $V = D_n^s / H$ where $H < D_n^s$ is an algebraic subgroup, this reduces to the notion of a H structure. Note that Gromov calls ω as above a rigid A-structure and defines a more general notion of rigid geometric structure.

Given ω, let $\mathrm{Iso}_\omega^r(x)$ be the set of r-jets of diffeomorphisms of M fixing the point x, and fixing ω up to order r at x.

Definition 3.2. A geometric structure is called rigid of order $r + 1$ if the natural map $\mathrm{Iso}^{r+1}(x) \to \mathrm{Iso}^r(x)$ is injective, for all $x \in M$. If we are not concerned with the order, we simply call the structure rigid.

Example 3.3. a) A pseudo-Riemannian metric is rigid of order 1, since it determines a total framing of $P^1(M)$.

    b) Let G be a simple Lie group, $\Gamma < G$ a lattice. Since the G action on G/Γ preserves the Killing form on $T(G/\Gamma)$, it preserves a rigid geometric structure: the pseudo-Riemannian metric defined by the Killing form.

Let V be a smooth manifold of dimension n and let $P_n^s V$ be the space of s-jets of diffeomorphisms of $(\mathbb{R}^n, 0)$ into V. If D_n^r acts on V then D_n^{r+s} acts on $P_n^s V$. Furthermore if V is an algebraic variety, then so is $P_n^s V$ and the D_n^{r+s} action is algebraic if the D_n^r action is. Given a geometric structure $\omega : P^r(M) \to V$, we have a naturally defined map $\omega^s : P^{r+s}(M) \to P_n^s(V)$ which we call the sth prolongation of the geometric structure. It is easy to verify that ω^s is a geometric structure and if ω is rigid so is ω^s. (See [Fe] or [G] for details).

Let M and N be differentiable manifolds of dimension m and n respectively. Suppose $a : P^{r_1}(M) \to V_1$ and $b : P^{r_2}(N) \to V_2$ are geometric structures. By passing to a prolongation of the structure of lower order, we may assume $r_1 = r_2$. To produce a geometric structure on $M \times N$, we begin with the map

$$a \times b : P^r(M) \times P^r(N) \to V_1 \times V_2$$

and, using the inclusion $D_m^r \times D_n^r < D_{m+n}^r$, induce this to a map

$$(a \times b)' : (D_{n+m}^r \times P^r(M) \times P^r(N))/(D_n^r \times D_m^r) \to (D_{n+m}^r \times V_1 \times V_2)/(D_n^r \times D_m^r).$$

Given an action of a closed subgroup $H < L$ on a space X, we can define the induced action of L. The space acted upon is $(L \times X)/H$ where the H action we divide by is given by $(l, x)h = (lh, h^{-1}x)$. The L action on the space is defined by the left L action on the first factor, which is well-defined on the quotient since it commutes with the H action defined above. Note that this definition only works for left actions of H on X, analogous definitions allow us to induce right actions to right actions. If the action of H on X is algebraic and H is an algebraic subgroup of an algebraic group L, then the induced action is an algebraic action on an algebraic variety. Also note that $(D^r_{n+m} \times P^r(M) \times P^r(N))/(D^r_n \times D^r_m)$ with the induced action is canonically isomorphic to $P^r(M \times N)$. Therefore $(a \times b)'$ is in fact a map

$$(a \times b)' : P^r(M \times N) \to V$$

where $V = (D^r_{n+m} \times V_1 \times V_2)/(D^r_n \times D^r_m)$ is an algebraic variety with an algebraic action of D^r_{n+m} for which $(a \times b)'$ is equivariant.

Proposition 3.4. *If $a : P^r(M) \to V_1$ and $b : P^s(M) \to V_2$ are rigid geometric structures, then the product structure $(a \times b)' : P^{\max(r,s)}(M \times N) \to V$ is also rigid.*

Proof. By passing to a prolongation of one structure, it suffices to consider $r = s$. A k-jet of a diffeomorphism fixing a point x leaves $(a \times b)'$ invariant if and only if it leaves $(a \times b)' \mid P^r(M) \times P^r(N)$ invariant. This restriction is exactly $a \times b : P^r(M) \times P^r(N) \to V_1 \times V_2$. Direct computation of derivatives shows that $\mathrm{Iso}^r_{a \times b} = \mathrm{Iso}^r_a \times \mathrm{Iso}^r_b$ where we view an r-jet, $j^r(f)$ (resp. $j^r(g)$) of a diffeomorphism of N at x_2 (resp. M at x_1) as an r-jet of $M \times N$ at $(x_1, x_2) = x$ as $j^r(\mathrm{Id} \times f)$ (resp. $j^r(g \times \mathrm{Id})$).

If a structure ω is rigid of order r, it follows that $\mathrm{Iso}^{r+j}_\omega \to \mathrm{Iso}^r_\omega$ is injective for all $j \geq 1$ (see [G, remark on page 68] or [Fe, Proposition 3.1]). Therefore, it follows that $(a \times b)'$ is rigid (and that it is rigid of the same order as whichever of a and b has the higher order of rigidity). $\square$

Corollary 3.5. *Let G be a semisimple Lie group, $\Gamma < G$ a lattice. Suppose Γ acts on M preserving a rigid geometric structure ψ. Then the G action on $(G \times M)/\Gamma$ also preserves a rigid geometric structure.*

Proof. The pseudo-Riemannian metric defined by the Killing form on G is bi-invariant and so defines a $G \times G$ invariant structure $a : P^1(G) \to V$. By the proposition above, $(a \times \psi)'$ is a rigid geometric structure on $G \times M$ which is invariant under the right and left G actions and the Γ action on M. Therefore it descends to a G invariant geometric structure φ on $(G \times M)/\Gamma$. Since $(a \times \psi)'$ is rigid so is φ, since rigidity is a local property. $\square$

Vol. 77 (2002)   Geometric lattice actions, entropy and fundamental groups   335

4. Proofs

The proof of Theorem 1.1 now follows from Corollary 2.4 and an estimate of the "entropy" of Gromov's representation computed in [Z5]. Given any finite dimensional G module D, we can define the entropy of the G action by $h_D(g) = \Sigma \log^+ \|\lambda\|$ where the sum is taken over eigenvalues for the action of g on D.

Assume G is a simple non-compact Lie group and G acts on a compact manifold X preserving a volume form and a rigid geometric structure ω. Let $\tilde{G}$ be the universal cover of G. The action on X lifts to a $\tilde{G}$ action on $\tilde{X}$, the universal cover of X. Let $\tilde{\omega}$ be the lift to $\tilde{X}$ of the geometric structure. Then $\tilde{G}$ clearly preserves $\tilde{\omega}$. In [G] Gromov constructs a linear representation σ of $\pi_1(X)$ on the set of Killing fields V of $\tilde{\omega}$ centralizing the $\tilde{G}$ action i.e. a representation $\sigma : \pi_1(X) \to \mathrm{Aut}(V)$. Let H be the Zariski closure of $\sigma(\pi_1(X))$. Gromov proves that H contains a group G' locally isomorphic to G, making $\mathfrak{h} = \mathrm{Lie}(H)$ into a G' module via the restriction of Ad_H to G'. For an accessible presentation of the proof see [Z3]. In this context, Lemma 5.1 of [Z5] shows that:

$$h_{\mathfrak{h}}(g) \geq h_X(g) \qquad \forall g \in G. \tag{$*$}$$

Proof of Theorem 1.1. As discussed in the introduction, we apply Gromov's result discussed above to the G action on $(G \times M)/\Gamma$, assuming G is simply connected. The rigid structure, ω, on $(G \times M)/\Gamma$ is as described in Corollary 3.5. Let $\Lambda = \pi_1((G \times M)/\Gamma)$ and recall that

$$1 \to \pi_1(M) \to \Lambda \xrightarrow{p} \Gamma \to 1.$$

By Proposition 6.1 of [F], Λ is isomorphic to the group of lifts of the Γ action on M to $\tilde{M}$ and the action on $G \times \tilde{M}$ is given by $(g, m)\lambda = (gp(\lambda)^{-1}, \lambda m)$. We will also need one obvious fact about the lift of our rigid structure ω to $G \times \tilde{M}$. Since here it is G bi-invariant by construction, the Lie algebra of vector fields, $\mathfrak{g}$, generating the right G action on G (i.e. the left invariant vector fields) are Killing fields of $\tilde{\omega}$ that centralize the lift of our G action from $(G \times M)/\Gamma$. These vector fields are clearly invariant under the $\pi_1(M)$ action on $G \times \tilde{M}$. In Gromov's construction of σ, the group $G' < H$ is described quite explicitly. It's Lie algebra $\mathfrak{g}'$ is a Lie algebra of Killing fields of $\tilde{\omega}$ commuting with the action of G. Even more is true. Let $\mathfrak{n}$ be the algebra of Killing fields of $\tilde{\omega}$ normalizing the G action. Then $\mathfrak{n} = \mathfrak{g} \oplus V$ where V is the algebra of Killing fields commuting with the G action, and $\mathfrak{g}$ is the algebra of Killing fields which are derivatives of the G action. For a point $x_0 \in X$, Gromov proves that there exists a Lie algebra $\mathfrak{g}_\Delta < \mathfrak{n}$ such that 1) elements of $\mathfrak{g}_\Delta$ fix x_0 and 2) $\mathfrak{g}_\Delta$ is isomorphic to $\mathfrak{g}$ and is, in fact, the graph of a isomorphism $\tau : \mathfrak{g} \to V$. Then $\mathfrak{g}'$ is the image $\tau(\mathfrak{g}) < V$. Conditions 1) and 2) above canonically define $\mathfrak{g}_\Delta$ and $\mathfrak{g}'$. In our situation, we can realize $\mathfrak{g}_\Delta$ as those Killing fields at $x_0 = ([g_0], m)$

generated by the action of $g_0 \circ \mathrm{Ad}G \circ g_0^{-1}$ and $\mathfrak{g}'$ as the vector fields generating the right G action on $G \times M$.

The representation σ of Λ is defined on all Killing fields of $\tilde{\omega}$ commuting with the left G action, and so contains as a subrepresentation the representation σ' on the Lie algebra $\mathfrak{g}'$ of left invariant vector fields on G. Since $\pi_1(M)$ acts trivially on these vector fields, it is easy to see that this representation is just $(\mathrm{Ad}_G)\big|_\Gamma \circ p$. In particular, the Zariski closure of $\sigma'(\Lambda)$ is a group locally isomorphic to G.

Let H be the Zariski closure of $\sigma(\Lambda)$ and $L \lhd H$ be the Zariski closure of $\sigma(\pi_1(M))$. Recall that H and L are subgroups of $\mathrm{Aut}(V)$ where V is the Lie algebra of Killing fields of ω commuting with the G action on $G \times \tilde{M}$.

Let $\mathfrak{h} = \mathrm{Lie}(H), \ell = \mathrm{Lie}(L)$. Then $\ell < \mathfrak{h}$ is an ideal and is also a G' submodule for the G' action on $\mathfrak{h}$ given by $(\mathrm{Ad}_H)\big|_{G'}$. To prove Theorem 1.1, it suffices to see that if ℓ is a trivial G module, then the Γ action is isometric.

If ℓ is a trivial G' module, then $h_{\mathfrak{h}}(g) = h_{\mathfrak{h}/\ell}(g) \ \forall g \in G'$. Let $\pi : H \to H/L$ be the natural projection. Then $\pi \circ \sigma$ is a homomorphism of Γ into an algebraic group, which has Zariski dense image, since H is the Zariski closure of $\sigma(\Lambda)$. Now $\pi \circ \sigma$ contains $\pi \circ \sigma'$ as a subhomomorphism. Clearly $\pi_1(M)$ acts trivially on the subspace $\mathfrak{g}'$ of Killing fields on which σ' is defined, so L acts trivially there as well. (Acting trivially is an algebraic condition.) This means that $\sigma'(\Lambda) \cap L$ is trivial, so the Zariski closure of $\pi \circ \sigma'(\Lambda)$ is also a group locally isomorphic to G.

Now $\pi \circ \sigma$ factors through a homomorphism of Γ, a lattice in simple Lie group with $\mathbb{R} - \mathrm{rank}(G) \geq 2$, so the Zariski closure of $\pi \circ \sigma(\Lambda)$ is, by Margulis' Superrigidity Theorem, locally isomorphic to a product of G with a compact semi-simple group. Therefore it is a compact extension of the Zariski closure of $\pi \circ \sigma'(\Lambda)$. So $\mathfrak{h}/\ell$ as a G'-module is just $\mathfrak{g} \oplus \mathfrak{k}$, where $\mathfrak{k}$ is a compact semi-simple Lie algebra that is necessarily trivial as a G' module. Therefore $h_{\mathfrak{h}/\ell}(g) = h_{\mathfrak{g} \oplus \mathfrak{k}}(g) = h_{\mathfrak{g}}(g)$, and furthermore $h_{\mathfrak{h}}(g) = h_{\mathfrak{g}}(g), \forall g \in G$. But by $(*)$ above and Theorem 2.3

$$h_{\mathfrak{h}}(g) \geq h_{(G \times M)/\Gamma}(g) = h_{G/\Gamma}(g) + h_M(g) = h_{\mathfrak{g}}(g) + h_M(g)$$

for all $g \in G$. Since $h_{\mathfrak{h}}(g) = h_{\mathfrak{g}}(g), \forall g \in G$, we see that $h_{(G \times M)/\Gamma}(g) = h_{G/\Gamma}(g)$, $\forall g \in G$, so by Corollary 2.4 the Γ action on M is isometric.

In the equation above, we refer to $h_M(g)$ where only Γ, and not G, acts on M. However, the equation still makes sense, since the entropy of the Γ action on M is really computed, as described in Theorem 2.2 via the eigenvalues of a linear representation of G. So by $h_M(g)$ here, we really mean the entropy of this linear representation, or 0, if the representation is trivial. The equation is formally justified by the remark following Theorem 2.3. $\square$

In [Z5], the second author proves:

Theorem 4.1. *Let X be a compact real analytic manifold with a real analytic unimodular rigid geometric structure. Let G be a connected simple Lie group with*

$\mathbb{R} - rank(G) \geq 2$ *and suppose G acts analytically and ergodically on X preserving the structure. Further assume the action is engaging. Then there is a finite ergodic extension X' of X and a measurable G equivariant map $\varphi : X' \to K\backslash H/\Lambda$ where $G < Z_K(H)$ and K is compact. Furthermore $h_X(g) = h_{K\backslash H/\Lambda}(g)$ for all $g \in G$.*

The quotient $\varphi : X' \to K\backslash H/\Lambda$ is constructed using the techniques of [LZ] from Gromov's representation of $\pi_1(X)$. Our Theorem 1.3 follows by combining the theorem above, the entropy results for induced actions of Section 2, and results of [F] on constructing quotients of lattice actions and induced actions.

First, we recall some of the ideas used to prove Theorem 4.1. Let $\sigma : \pi_1(X) \to GL_n(\mathbb{R})$ be Gromov's representation discussed above. Let A be the finitely generated $\mathbb{Q}$ algebra such that $\sigma(\gamma) \in GL(n, A)$ for all $\gamma \in \pi_1(X)$. Let $\bar{\mathbb{Q}}$ denote the algebraic integers. In [LZ], a specialization $\psi : A \to \bar{\mathbb{Q}}$ is constructed such that the map induced by ψ on $GL_n(A)$ is an isomorphism when restricted to $\sigma(\pi_1(X))$ and $\psi \circ \sigma(\pi_1(X))$ is an $\mathfrak{s}$-arithmetic subgroup of a perfect $\mathbb{Q}$-group H. Furthermore $\psi \circ \sigma(\pi_1(X)) \cap H_\mathbb{Z} = \Lambda$ is of finite index in $H_\mathbb{Z}$ and is a lattice in H. Lubotzky and Zimmer also show that there is a finite ergodic extension X' of X such that there is a measurable map $\varphi : X' \to K\backslash H/\Lambda$. The key idea of [Z5] is to use the way in which σ and ψ are constructed to see that the entropies are equal.

Proof of Theorem 1.3. We apply Theorem 4.1 to the action of G on $(G \times M)/\Gamma$. We can do this despite the fact that $(G \times M)/\Gamma$ is not compact, since this is only used to show that certain cocycles are quasi-integrable. (See the discussion following Theorem 2.2 for entropy considerations and the derivative cocycle and Section 7 of [F] for cocycles coming from representations of the fundamental group.) Letting $\hat{\sigma} = \psi \circ \sigma$, as discussed above, the proof of the main theorem of [LZ] implies that $\hat{\sigma}(\pi_1((G \times M)/\Gamma))$ is $\mathfrak{s}$-arithmetic in an algebraic $\mathbb{Q}$-group H, and if $\Lambda = \hat{\sigma}(\pi_1((G \times M)/\Gamma)) \cap H_\mathbb{Z}$, that we have a map from a finite extension X of $(G \times M)/\Gamma$ to $K\backslash H/H_\mathbb{Z}$ that is G equivariant.

As in the proof of Theorem 1.1, let L' be the Zariski closure of $\sigma(\pi_1(M))$ and let H' be the Zariski closure of $\sigma(\pi_1((G \times M)/\Gamma))$. As above let σ' be the subrepresentation of σ defined on the vector fields $\mathfrak{g}$ given by differentiating the right action of G on $G \times \tilde{M}$. Recall that $\sigma'(\pi_1((G \times M)/\Gamma)) \cap L' = \emptyset$, and since $\sigma'(\pi_1((G \times M)/\Gamma))$ is virtually isomorphic to Γ, it follows that $[\sigma(\pi_1(M)) : \sigma(\pi_1((G \times M)/\Gamma))] = \infty$, unless $\sigma(\pi_1(M))$ is finite and the Γ action on M is isometric. Since the specialization ψ is an isomorphism on $\sigma(\pi_1((G \times M)/\Gamma))$, it is immediate that $[\hat{\sigma}(\pi_1(M)) : \hat{\sigma}(\pi_1((G \times M)/\Gamma))] = \infty$ as well. It is clear from the discussion above of the proof of Theorem 4.1 that H is the Zariski closure of $\hat{\sigma}(\pi_1((G \times M)/\Gamma))$. Let $L \triangleleft H$ be the Zariski closure of $\hat{\sigma}(\pi_1(M))$.

By Theorem 8.1 of [F], we see that $H = (G \times C) \ltimes L$ and $\sigma(\pi_1((G \times M)/\Gamma)) = \Gamma \ltimes \Delta$ (up to finite index) where $\Delta = L \cap \sigma(\pi_1((G \times M)/\Gamma)$ and C is compact. We have a measurable map $\varphi : X \to K\backslash H/\Lambda$ and X, as a finite extension of $(G \times M)/\Gamma$, is $(G \times M')/\Gamma'$ where M' is finite (possibly disconnected) cover of M and $\Gamma' < \Gamma$ is of finite index. (See [F], proof of Lemma 4.3, for details.) Let $\Delta_\infty = \Delta \cap L_{\mathbb{Z}}$. Then by Theorem 8.5 of [F], we see that $K \backslash H / \Lambda = K \backslash (G \times C) \ltimes L / \Gamma' \ltimes \Delta_\infty$ and that φ takes almost every M' fiber of X to a $K \backslash C \times L / \Delta_\infty$ fiber in $K \backslash H / \Lambda$.

The G action on $K \backslash (G \times C) \ltimes L / \Gamma' \ltimes \Delta_\infty$ is isomorphic to the G action induced from the Γ' action on $K \backslash C \times L / \Delta_\infty$. So by Theorem 2.3 above, the entropy function for the Γ' action on M' and for the Γ' action on $K \backslash C \times L / \Delta_\infty$ must be equal, since by Theorem 4.1, the entropy functions for the induced actions are equal. $\qquad \square$

References

[F]   D. Fisher, On the arithmetic structure of lattice actions on compact spaces, to appear *Ergodic Th. and Dyn. Sys.*

[Fe] R. Feres, Rigid geometric structures and actions of semi-simple Lie groups, preprint, to appear in *Proceedings of Strasbourg Conference,* (ed. P. Foulon).

[G]   M. Gromov, Rigid transformation groups, *Géometrie Différentiale* (D. Bernard and Y. Choquet-Bruhat, eds.), Hermann, Paris 1988.

[L]   J. Lewis, The algebraic hull of a measurable cocycle, to appear *Geom. Ded.*

[LZ] A. Lubotzky and R. J. Zimmer, Arithmetic structure of fundamental groups and actions of semisimple groups, *Topology* **40** (2001), 851–869.

[P1] Y. Pesin, Lyapunov characteristic exponents and smooth ergodic theory, *Russian Math. Surveys* **32** (1977), 55–114.

[P2] Y. Pesin, personal communication.

[Z1] R. J. Zimmer, *Ergodic Theory and Semisimple Groups*, Birkhäuser, Boston, 1984.

[Z2] R. J. Zimmer, Superrigidity, Ratner's theorem, and fundamental groups, *Israel J. Math.* **74** (1991), 199–207.

[Z3] R. J. Zimmer, Automorphism groups and fundamental groups of geometric manifolds, *Proc. Symp. in Pure Math.* **54** (1993), vol. 3.

[Z4] R. J. Zimmer, Lattices in semisimple groups and invariant geometric structures of compact manifolds, *Discrete Groups in Geometry and Analysis* (ed. Roger Howe), Birkhäuser, Boston, 1987.

[Z5] R. J. Zimmer, Entropy and arithmetic quotients for simple automorphism groups of geometric manifolds, to appear *Geom. Ded.*

[Z6] R. J. Zimmer, On the algebraic hull of an automorphism group of a principal bundle, *Comm. Math. Helv.* **65** (1990), 375–387.

David Fisher
Department of Mathematics
Yale University
PO Box 208283
New Haven, CT 06520-8283
USA
e-mail: david.fisher@yale.edu

Robert J. Zimmer
Department of Mathematics
University of Chicago
Chicago
IL 60637
USA

(Received: December 14, 2000)

7

Geometric Structures: Automorphisms of Geometric Manifolds and Rigid Structures; Locally Homogeneous Manifolds

Invent. math. 83, 411–424 (1986)

Inventiones
mathematicae
© Springer-Verlag 1986

On the automorphism group
of a compact Lorentz manifold
and other geometric manifolds*

Robert J. Zimmer

Department of Mathematics, University of Chicago, Chicago, IL 60637, USA

1. Introduction

The aim of this paper is to establish some basic results concerning the automorphism group of a general G-structure on a compact manifold, with particular emphasis on the automorphism group of a compact Lorentz manifold.

Let M be an n-manifold and $P \to M$ a G-structure on M where G is a real algebraic group. The G-structure may be of higher order, but we assume that G also defines a volume density on M, which of course is simply the assertion that the first order part of G is contained in $SL(n, \mathbf{R})$, the real matrices with $|\det| = 1$. We let $\mathrm{Aut}(P) \subset \mathrm{Diff}(M)$ denote the automorphism group of the G-structure.

Theorem A. *Let M be a compact manifold and $P \to M$ a G-structure where G is a real algebraic group defining a volume density on M. Let H be a connected non-compact simple Lie group and suppose that H acts non-trivially on M so as to preserve the G-structure. (Thus, we have a non-trivial homomorphism $H \to \mathrm{Aut}(P)$.) Then there is an embedding of Lie algebras $\mathfrak{h} \to \mathfrak{g}$. Moreover, if $\mathfrak{g} \subset \mathfrak{g}l(V)$ is the given linear realization of $\mathfrak{g}$ (so, for example, $V = \mathbf{R}^{\dim M}$ for a first order structure), then the corresponding (faithful) representation $\mathfrak{h} \to \mathfrak{g}l(V)$ contains $\mathrm{ad}_{\mathfrak{h}}$ as a direct summand.*

We remark that Theorem A is easily seen to be false without the assumption that the G-structure defines a volume density, or without the compactness assumption. The first assertion of Theorem A was first proven in [15] under the assumption that $\mathbf{R}$-rank$(H) \geq 2$. This latter assumption was needed in [15] because the argument there involved use of the superrigidity theorem for cocycles [14]. Here we present a much more elementary argument that applies to any non-compact simple H.

By a Lorentz metric ω on M^n we mean, as usual, a smooth assignment of a symmetric non-degenerate bilinear form of signature $(1, n-1)$ to each tangent

---

* Research partially supported by NSF Grant DMS-8301882

space of M. Of course, this is equivalent to a $O(1, n-1)$ structure on M. We denote by $\text{Aut}(M, \omega)$ the diffeomorphisms of M preserving ω, so that $\text{Aut}(M, \omega) = \text{Aut}(P)$ where P is the corresponding $O(1, n-1)$ structure. It is known that in this case $\text{Aut}(M, \omega)$ is a (finite dimensional) Lie group [4, I.5.1]. However, there has not been a great deal known about exactly what Lie group the connected component $\text{Aut}(M, \omega)_0$ might be. Here we give a description of this group depending on whether or not it is amenable, i.e. has a co-compact radical, or equivalently, every subgroup which is a simple Lie group is actually compact.

Theorem B. *Let (M, ω) be a compact Lorentz manifold. Then:*

1) *If $\text{Aut}(M, \omega)_0$ is not amenable, then $\text{Aut}(M, \omega)_0$ is locally isomorphic to $SL(2, \mathbf{R}) \times K$ where K is a compact Lie group. In fact, for the universal cover we have $\widetilde{\text{Aut}(M, \omega)_0} \cong \widetilde{SL(2, \mathbf{R})} \times C \times \mathbf{R}^r$ where C is a compact simple Lie group and $\mathbf{R}^r$ acts on M via a quotient torus.*

2) *If $\text{Aut}(M, \omega)_0$ is amenable, then the nilradical of $\text{Aut}(M, \omega)_0$ is at most a 2-step nilpotent group.*

The assertion in (1) is very sharp at least up to local isomorphism. Every group of the form $PSL(2, \mathbf{R}) \times K$ where K is a compact Lie group, can appear in the connected component of the identity in the automorphism group of a compact Lorentz manifold. However, while assertion (2) is quite restrictive, it is not clear that the conditions in (2) are the best possible. We do however, have the following.

Proposition C. *Let N be a Heisenberg group (and hence a 2-step nilpotent group). Then N admits a free action by automorphisms of a compact Lorentz manifold.*

We remark that N cannot be embedded in any of the groups described in assertion (1) of Theorem B.

The proof of Theorem B depends upon Theorem A and a number of other considerations, in particular, the behavior of measures on the $(n-2)$-sphere under the natural conformal action of $O(1, n-1)$ and its relation to the recurrence properties of $\text{Aut}(M, \omega)_0$ deriving from the invariant volume density. Some of the techniques involved in the proof of Theorem B enable us to sharpen results in [16–18] concerning actions of discrete groups. In particular, we have the following result for actions preserving a Lorentz metric.

Theorem D. *Let Γ be a discrete Kazhdan group (i.e. group with Kazhdan's property (T)) [14]. Let M be a compact Lorentz manifold and suppose that Γ acts on M so as to preserve the Lorentz metric. Then Γ leaves a Riemannian metric invariant and hence the action factors to $\Gamma \to K \to \text{Diff}(M)$ where K is a compact Lie group.*

This result was proved in [18] under the additional hypothesis that Γ acts ergodically. Combining Theorem D with algebraic results in [18], we obtain:

Theorem E. *Let Γ be a discrete Kazhdan group. Let (M, ω) be a compact Lorentz 3-manifold. Then any action $\Gamma \to \text{Aut}(M, \omega)$ is trivial on a subgroup of finite index.*

We can also sharpen results of [17] concerning actions of lattices in higher split rank semisimple groups. If g is a semisimple real Lie algebra let $d(g)$ be the minimal dimension of a non-trivial, real representation of g and $n(g)$ the minimal dimension of a simple factor of g.

Theorem F. *Let H be a connected semisimple Lie group with finite center such that* **R**-rank$(H') \geq 2$ *for every simple factor H' of H. Let $\mathfrak{h}$ be its Lie algebra. Let $\Gamma \subset H$ be a lattice subgroup. Let M be a compact manifold with a volume density. Suppose Γ acts on M so as to preserve the volume and a G-structure of finite type, where G is a real algebraic group. Then either:*

a) *There is a non-trivial homomorphism $\mathfrak{h} \to g$ (hence on embedding if $\mathfrak{h}$ is simple); or*

b) *Γ leaves a Riemannian metric invariant.*

Theorem G. *Let H, Γ, M be as in Theorem F. Suppose M^n is a compact n-manifold with*

i) $n < d(\mathfrak{h})$ *and*

ii) $n(n+1) < 2n(\mathfrak{h})$.

Then any action of Γ on M preserving a volume density and any G-structure of finite type (G real algebraic) is actually an action of a finite quotient of Γ.

Theorem H. *Let k be a totally disconnected local field of characteristic 0 and H an almost k-simple k-group with k-rank$(H) \geq 2$. Let $\Gamma \subset H_k$ be a lattice. Then any action of Γ on a compact manifold preserving a volume density and a G-structure of finite type (G real algebraic) must leave a smooth Riemannian metric invariant.*

Theorems F, G, H are implicit in [17] (Cf. [16]) under the further assumption that all Γ actions under consideration are ergodic.

The outline of this paper is as follows. In Sect. 2 we discuss the notion of the algebraic hull of a linear cocycle, a notion first introduced in [12], and which plays a central role in the proofs of Theorems A, B. Section 3 contains the proof of Theorem A, and Sect. 4 discusses simple and nilpotent automorphism groups of Lorentz manifolds in preparation for the proof of Theorem B. Section 5 shows the existence of a (smooth) invariant Riemannian metric for actions preserving both a G-structure of finite type and measurable Riemannian metrics on suitably many naturally defined bundles. This is used in the completion of the proof of Theorem B which is carried out in Sect. 6, and is the main technical fact that allows the present sharpening of our earlier work on actions of discrete groups in the form of Theorems D, E, F, G, H. This last point is discussed in Sect. 8. Section 7 contains some examples, and in particular, verifies Proposition C.

2. The algebraic hull of a linear cocycle

Suppose $P \to M$ is a principal G-bundle and that H acts on P by principal bundle automorphisms. We shall write the G-action on P on the right and the H-action on the left. The bundle has a measurable section and so measurably we can write $P \cong M \times G$. The action of H on P is then given by $h \cdot (m, g) = (h\,m, \alpha(m, h)\,g)$ where $\alpha \colon M \times H \to G$ is a measurable cocycle. (See [14] for a detailed discussion of measurable cocycles.) The G action on P is of course simply given by right translation on the second factor. We shall be concerned with properties of such cocycles α, or in other words, with the measurable properties of the H-action on P.

If S is a (standard) measurable H-space with a quasi-invariant measure, we recall that two cocycles α, $\beta \colon S \times H \to G$ are called equivalent if there is a measurable $\phi \colon S \to G$ such that for each h, $\phi(h\,s)\,\alpha(s, h)\,\phi(s)^{-1} = \beta(s, h)$ a.e. Different measurable trivializations of the bundle P yield equivalent cocycles. If X is a left G-space, a measurable function $\phi \colon S \to X$ will be called α-invariant if for each h, $\alpha(s, h)\,\phi(s) = \phi(h\,s)$ a.e. If $X = G/G_0$ where $G_0 \subset G$ is a closed subgroup, then there is an α-invariant $\phi \colon S \to X$ if and only if $\alpha \sim \beta$ with $\beta(S \times H) \subset G_0$. (See [14] for discussion.) For a cocycle coming from a measurable trivialization of P, this simply asserts that there is a measurable H-invariant reduction of the bundle P to G_0. For example, if $G = GL(n, \mathbf{R})$, $G_0 = O(n, \mathbf{R})$, we record the following fact.

Lemma 2.1. *Let $E \to M$ be a vector bundle on which H acts by vector bundle automorphisms. If there is a measurable H-invariant Riemannian metric on E, then the cocycle α corresponding to a measurable trivialization of the associated principal frame bundle is equivalent to a cocycle taking all values in $O(n, \mathbf{R})$. Conversely if $P \to M$ is a principal G-bundle, $E \to M$ an associated (real) vector bundle corresponding to a linear representation of G, and α is equivalent to a cocycle into a compact subgroup of G, then there is an H-invariant measurable Riemannian metric on E.*

If $\alpha \colon S \times H \to G$ is a cocycle where G is a Lie group, there is in general no essentially unique "smallest" closed subgroup $G_1 \subset G$ for which $\alpha \sim \beta$ with $\beta(M \times H) \subset G_1$. However, if we restrict our attention only to real algebraic groups and ergodic actions, then such a group does exist. (We recall that an action on a measure space is ergodic if every invariant measurable set is either null or conull.)

Proposition 2.2. [12] *Let $\alpha \colon S \times H \to GL(n, \mathbf{R})$ be a cocycle where H acts ergodically on S. Then there is a real algebraic subgroup $G_1 \subset GL(n, \mathbf{R})$ such that $\alpha \sim \beta$ where $\beta(S \times H) \subset G_1$ and G_1 is minimal among algebraic subgroups with this property. Furthermore, G_1 is unique up to conjugacy in $GL(n, \mathbf{R})$.*

This result is in fact true if $\mathbf{R}$ is replaced by any local field of characteristic 0. We call G_1 (or more precisely its conjugacy class in $GL(n, \mathbf{R})$) the algebraic hull of α. The proof of Proposition 2.2 follows easily from the descending chain condition on algebraic subgroups and the following result.

Lemma 2.3. *Let $\alpha \colon S \times H \to GL(n, \mathbf{R})$ be a cocycle where H acts ergodically on S. Suppose $G_i \subset GL(n, \mathbf{R})$ are algebraic subgroups, $i = 1, 2$, and $\alpha \sim \alpha_i$ where $\alpha_i(S \times H) \subset G_i$. Then there is $g \in GL(n, \mathbf{R})$ such that $\alpha \sim \beta$ where $\beta(S \times H) \subset G_1 \cap g\,G_2\,g^{-1}$.*

Proof. Let $G = GL(n, \mathbf{R})$, and $\phi_i \colon S \to G/G_i$ an α-invariant map. Then $\phi = (\phi_1, \phi_2) \colon S \to G/G_1 \times G/G_2$ is α-invariant, for the product action of G. Let $p \colon$

$G/G_1 \times G/G_2 \to (G/G_1 \times G/G_2)/G$ be the natural quotient map. Then $p \circ \phi$ is an H-invariant map, and since G, G_1, G_2 are algebraic, the quotient space $(G/G_1 \times G/G_2)/G$ is a countably separated Borel space. (Cf. [14, Chapters 2, 3].) By ergodicity of H on S, $p \circ \phi$ is essentially constant, i.e. we can assume $\phi(S)$ lies in a single G-orbit in $G/G_1 \times G/G_2$. This orbit can be expressed as $G/G_1 \cap g G_2 g^{-1}$ for some g, and ϕ can be considered as an α-invariant map $S \to G/G_1 \cap g G_2 g^{-1}$. This proves the lemma.

We also record the following alternative formulation.

Lemma 2.4. *If* $\alpha: S \times H \to GL(n, \mathbf{R})$ *is a measurable cocycle with algebraic hull* $G \subset GL(n, \mathbf{R})$, *and* $\alpha \sim \beta$ *with* $\beta(S \times H) \subset L$ *and* L *algebraic, then there is a conjugate of* G *contained in* L, *i.e. we can assume* $G \subset L$.

3. Simple automorphism groups of G-structures: Proof of Theorem A

In this section we apply the result of the preceding section to the derivative cocycle of an action of a simple Lie group to deduce Theorem A. We begin with the following basic observation from [15].

Theorem 3.1. [15] *Let* H *be a connected non-compact simple Lie group and suppose* H *acts smoothly on a compact manifold* M *so as to preserve a volume density. Then for almost all* $m \in M$, *either* H_m *(the stabilizer of* m *in* H*) is discrete or* $H_m = H$.

Theorem 3.1 is in fact valid for any finite measure preserving action of H on a (standard) measure space. If H acts transitively on M, Theorem 3.1 reduces to the Borel density theorem [1, 14]. Thus, 3.1 can be considered as a generalization of Borel's theorem, and the proof of 3.1 in [15] in fact makes use of Borel's result.

We shall also need the following consequence of the Borel density theorem. If G is a connected semisimple Lie group with Lie algebra $\mathfrak{g}$, we let $G^* \subset GL(\mathfrak{g})$ be the algebraically connected real algebraic group such that $\mathrm{Ad}_G(G) = G_0^*$ (the identity component in the Hausdorff topology).

Lemma 3.2. *Let* G *be a connected simple Lie group,* S *an ergodic* G-*space with an invariant probability measure. Let* $\alpha: S \times G \to GL(\mathfrak{g})$ *be given by* $\alpha(s, g) = \mathrm{Ad}_G(G)$. *Then the algebraic hull of* α *is* G^*.

Proof. Let H be the algebraic hull of α, so $H \subset G^*$. Since $\alpha \sim \beta$ where $\beta(S \times G) \subset H$, there is an α-invariant function $\phi: S \to GL(\mathfrak{g})/H$. By the definition of α, this simply means that ϕ is a G-map, where G acts on $GL(\mathfrak{g})/H$ via $\mathrm{Ad}(G)$. Let $p: GL(\mathfrak{g})/H \to \mathrm{Ad}(G) \backslash GL(\mathfrak{g})/H$, so that $p \circ \phi$ is G-invariant. Since the action on S is ergodic and $\mathrm{Ad}(G) \backslash GL(\mathfrak{g})/H$ is a countably separated Borel space (cf. proof of Lemma 2.3 and [14, Chapters 2, 3]), $p \circ \phi$ is essentially constant. Hence there is a single $\mathrm{Ad}(G)$-orbit in $GL(\mathfrak{g})/H$ containing almost all $\phi(s)$, $s \in S$. Thus, we can view ϕ as a G-map $S \to \mathrm{Ad}(G)/g H g^{-1} \cap \mathrm{Ad}(G)$ for some $g \in GL(\mathfrak{g})$. Since S has a G-invariant probability measure, so does $\mathrm{Ad}(G)/g H g^{-1} \cap \mathrm{Ad}(G)$, and since $\mathrm{Ad}(G)$ is simple and H is algebraic, the Borel density theorem [1, 14] implies $gHg^{-1} \supset G^*$. Since $H \subset G^*$, we have $H = G^*$.

With these preliminaries, we can now prove Theorem A.

Proof of Theorem A. By Theorem 3.1, we can choose an ergodic component $E \subset M$ for the H-action such that the H action is essentially (i.e., almost everywhere) locally free. For $m \in E$, let $V_m \in TM_m$ be the tangent space to the H-orbit. Since the action is essentially locally free, for (almost) every $m \in E$ we have an identification of $\mathfrak{h}$ with V_m given by the derivative of the map $H \to E$, $h \to hm$. Thus, if we let α be the derivative cocycle on M restricted to $E \times H$, we can choose a measurable trivialization of TM such that for $m \in E$, $h \in H$, $\alpha(m, h)$ will have the form $\left(\begin{array}{c|c} \mathrm{Ad}_H(h) & * \\ \hline 0 & * \end{array} \right)$. Let $L \subset GL(n, \mathbf{R})$ be the subgroup $\left(\begin{array}{c|c} H^* & * \\ \hline 0 & * \end{array} \right)$, and let H_1 be the algebraic hull of α. Thus, $H_1 \subset L$. Let $p: L \to GL(\mathfrak{h})$ be the restriction of an element of L to $\mathfrak{h}$. Then $p \circ \alpha: E \times H \to H^*$ is a cocycle whose algebraic hull has connected component $p(H_1)_0$, and hence by Lemma 3.2, $p(H_1)_0 = \mathrm{Ad}_H(H)$. Since H is simple, there is a closed subgroup $\tilde{H} \subset H_1$ such that $p|\tilde{H}: \tilde{H} \to \mathrm{Ad}_H(H)$ is a local isomorphism. By Lemma 2.4, there is $h \in GL(n, \mathbf{R})$ such that $\mathrm{Int}(h)(H_1) \subset G$ where $\mathrm{Int}(h)$ is conjugation by h. It follows that $\mathrm{Int}(h) \circ d((p|\tilde{H})^{-1})_e$ is the required Lie algebra embedding.

4. Simple and nilpotent automorphism groups of Lorentz manifolds

In preparation for the proof of Theorem B, in this section we shall establish some results on automorphism groups of a Lorentz manifold which are either simple or nilpotent.

Theorem 4.1. *Let H be a connected simple non-compact Lie group and suppose there is a non-trivial action of H on a compact manifold preserving a Lorentz structure. Then H is locally isomorphic to $SL(2, \mathbf{R})$.*

Proof. By Theorem A, we have a Lie algebra embedding $\mathfrak{h} \to so(1, n)$ (where $\dim M = n + 1$), such that the resulting representation of $\mathfrak{h}$ on $\mathbf{R}^{n+1}$ contains $\mathrm{ad}_\mathfrak{h}$ as a subrepresentation. Let ω_0 be the standard non-degenerate symmetric bilinear form on $\mathbf{R}^{n+1}$ with signature $(1, n)$. We have a subspace $V \subset \mathbf{R}^{n+1}$ which is $\mathfrak{h}$ invariant and on which $\mathfrak{h}$ acts equivalently to $\mathrm{ad}_\mathfrak{h}$. We shall thus identify V with $\mathfrak{h}$. Then $\omega_0|\mathfrak{h}$ is an invariant symmetric bilinear form on $\mathfrak{h}$. Since any totally isotropic subspace of $\mathbf{R}^{n+1}$ for ω_0 is at most one dimensional, $\omega_0|\mathfrak{h}$ is non-trivial. From the fact that $\mathfrak{h}$ embeds in $so(1, n)$, we deduce that H does not have Kazhdan's property (T) [18, p. 432]. Thus we can assume $\mathfrak{h} \cong so(1, m)$ or $su(1, m)$ for some m. If $\mathfrak{h} \not\cong so(1, 3)$, then the complexification of $\mathfrak{h}$ is simple. It follows that for $\mathfrak{h} \neq so(1, 3)$, $\omega_0|\mathfrak{h}$ is a multiple of the Killing form [3, p. 104, Ex. 9]. Under the Cartan decomposition $\mathfrak{h} = \mathfrak{k} \oplus \mathfrak{p}$, the Killing form is positive definite on $\mathfrak{p}$ and negative definite on $\mathfrak{k}$. Since $\omega_0|\mathfrak{h}$ is a multiple of the Killing form, either $\dim \mathfrak{k} = 1$ or $\dim \mathfrak{p} = 1$. It is clear that this occurs if and only if $\mathfrak{h} = so(1, 2) \cong sl(2, \mathbf{R})$. Therefore, it remains to show that $\mathfrak{h} = so(1, 3)$ is impossible.

Let $B = \omega_0 | so(1, 3)$. If B is degenerate, the kernel of B is a one dimensional $\mathrm{Ad}(SO(1,3))$-invariant subspace. Since $SO(1,3)$ is simple, this is impossible. We may thus assume that B is non-degenerate. Thus B is either positive definite or has signature $(1,5)$. (Recall $\dim so(1,3) = 6$.) Since B is $\mathrm{Ad}(SO(1,3))$-invariant, the former case is impossible. If the signature is $(1,5)$, we have an embedding π: $\mathrm{Ad}_{SO(1,3)}(SO(1,3)) \to SO(1,5)$, and hence, possibly after a conjugation in $SO(1,5)$, we can assume that $\pi(\mathrm{Ad}_{SO(1,3)}(SO(3,\mathbf{R})) \subset SO(5,\mathbf{R})$ where $SO(5,\mathbf{R})$ embeds in $SO(1,5)$ in the standard way. Thus, we can decompose $so(1,3) = so(3,\mathbf{R}) \oplus V \oplus W$ where all of the summands are $\mathrm{Ad}_{SO(1,3)}(SO(3,\mathbf{R}))$-invariant, and $\dim V = 1$. Hence, $V \oplus W$ is a 3-dimensional space of $SO(3,\mathbf{R})$-invariant vectors in $so(1,3)$, which is obviously impossible. This completes the proof of Theorem 4.1.

Theorem 4.2. *Let N be a connected nilpotent Lie group and suppose there is a locally faithful action (i.e., action with discrete kernel) of N on a compact manifold M, preserving a Lorentz structure ω. Then N is at most a 2-step nilpotent group.*

Proof. We can clearly assume that N is simply connected, and hence has the structure of a unipotent real algebraic group. For $m \in M$, let N_m be the stabilizer of m in N, N_m^0 the connected component and $L(N_m)$ $(= L(N_m^0))$ the Lie algebra. The map $m \to L(N_m)$ is a measurable N-map Φ: $M \to Gr(L(N))$ where $Gr(L(N))$ is the disjoint union of the Grassmann varieties of $L(N)$ and N acts on $Gr(L(N))$ via the adjoint representation. Write the ergodic decomposition of the N-action on (M, μ) (where μ is the measure defined by the volume density defined by ω) as $(M, \mu) = \int_{t \in I}^{\oplus} E_t \, d\mu_t$. Thus for each $t \in I$, μ_t is an N-invariant probability measure. It follows that for each t, Φ: $E_t \to Gr(L(N))$ is an N-map, and since the N-action on E_t is ergodic and the N-action on $Gr(L(N))$ is smooth (in the sense of ergodic theory: the quotient space is a countably separated Borel space) owing to the rationality of Ad [14, Chapters 2, 3], almost all of $\Phi(E_t)$ is contained in a single N-orbit in $Gr(L(N))$. This orbit (for a fixed t) is of the form N/N_1 where N_1 is the normalizer of N_m^0 for some $m \in M$. In particular, N_1 is itself an algebraic subgroup. We have an N-map Φ: $E_t \to N/N_1$, and it follows that there is an N-invariant probability measure on N/N_1. Since N is simply connected and $N_1 \subset N$ is algebraic, it follows that $N_1 = N$. In other words, for each t, N_m^0 is a normal connected (hence algebraic) subgroup of N for μ_t-almost all $m \in E_t \subset M$. Hence N_m^0 depends (a.e.) only on t. We shall write $N(t) \subset N$ to be the normal algebraic subgroup such that $N(t) = N_m^0$ for μ_t-almost all $m \in M$.

Since the action is locally faithful, we have $\bigcap_{m \in M_0} N_m^0 = \{e\}$ for any dense set $M_0 \subset M$, and hence we have $\bigcap_{t \in I} N(t) = \{e\}$. By the descending chain condition on algebraic subgroups, there is a finite set $F \subset I$ such that $\bigcap_{t \in F} N(t) = \{e\}$. Thus if we take the product (i.e. diagonal) action of N on $X = \prod_{t \in F} E_t$, we have that N acts essentially locally freely on X with respect to the product measure $\prod_{t \in F} \mu_t$

(i.e. almost all points have discrete stabilizers.) We let $E \subset X$ be an ergodic component such that N acts essentially locally freely on E. We can view E as an N-invariant measurable subset of $M' = \prod_{t \in F} M$, the product of card(F) copies of M.

Let α be the restriction to $E \times N$ of the derivative cocycle for the action of N on M'. Thus, $\alpha(E \times N) \subset \prod_{t \in F} GL(n, \mathbf{R})$, and since N preserves a Lorentz metric on M, we can clearly assume that $\alpha(E \times N) \subset \prod_{t \in F} SO(1, n-1)$. Since N is amenable, the algebraic hull, L, of α is amenable [14, Theorem 9.2.3], [13]. By Moore's classification of amenable algebraic subgroups of semisimple groups [5], and the fact that the nilradical of the proper parabolic subgroup of $SO(1, n-1)$ is abelian, it follows that L has an abelian, normal, unipotent algebraic subgroup B such that L_0/B consists of semisimple elements, where L_0 is the identity component. On the other hand, N acts essentially locally freely on E. Arguing as in the proof of Theorem 4.1, we can realize the Lie algebra $L(N)$ as a linear subspace $L(N) \subset \sum_{t \in F}^{\oplus} \mathbf{R}^n$ such that α is equivalent to a cocycle β taking values in the subgroup $L' = \left(\begin{array}{c|c} \mathrm{Ad}_N(N) & * \\ \hline 0 & * \end{array} \right)$ and such that $\beta(m, g)|L(N) = \mathrm{Ad}_N(g)$ for almost all $m \in E$ and all $g \in N$. Since the algebraic hull of α is L, we may assume (replacing L by a conjugate if necessary) that $L \subset L'$. It follows that the group $(L|L(N)) \subset GL(L(N))$ is the algebraic hull of the cocycle γ on $E \times N$ given by $\gamma(m, g) = \mathrm{Ad}_N(g)$. Since $L|L(N)$ is unipotent, B is abelian and L_0/B consists of semisimple elements, it follows that $L|L(N)$ is abelian. To prove that N is at most a 2-step nilpotent group (or equivalently that $\mathrm{Ad}_N(N)$ is abelian), it suffices to see that the algebraic hull of γ is actually $\mathrm{Ad}_N(N)$.

Lemma 4.3. (Cf. Lemma 3.2). *Let N be a connected, simply connected, nilpotent Lie group with Lie algebra $\mathfrak{h}$, and suppose N acts ergodically by measure preserving transformations of a probability space (S, μ). Let $\gamma: S \times N \to GL(\mathfrak{n})$ be given by $\gamma(s, g) = \mathrm{Ad}(g)$. Then the algebraic hull of γ is $\mathrm{Ad}(N)$.*

Proof. Let $H \subset \mathrm{Ad}(N)$ be the algebraic hull. Since $\mathrm{Ad}(N)$ is unipotent, H is connected. Arguing exactly as in the proof of Lemma 3.2, we deduce that there is an N-map $\phi: S \to \mathrm{Ad}(N)/\mathrm{Ad}(N) \cap gHg^{-1}$ for some $g \in GL(\mathfrak{n})$. Since $\mathrm{Ad}(N)$ and gHg^{-1} are unipotent algebraic groups, $\mathrm{Ad}(N) \cap gHg^{-1}$ is a connected (unipotent) algebraic group. The existence of an N-invariant probability measure on S implies the existence of one on $\mathrm{Ad}(N)/\mathrm{Ad}(N) \cap gHg^{-1}$ and hence $\mathrm{Ad}(N) \cap gHg^{-1} = \mathrm{Ad}(N)$. Thus we have both $H \subset \mathrm{Ad}(N)$ and $\mathrm{Ad}(N) \subset gHg^{-1}$ which implies $\mathrm{Ad}(N) = H$. This completes the proof.

5. Measurable invariant metrics and G-structures of finite type

If a group acts on a manifold so as to preserve a measurable Riemannian metric, it does not follow in general that the action preserves a smooth metric. There are a number of situations, however, in which the existence of a measurable invariant metric together with some further properties of the action does imply the existence of a smooth metric. (See [16–20], e.g.) In particular, in [17], we showed that if the action is ergodic and preserves a G-structure of finite type (in the sense of Cartan; see [4, 8]), then the existence of a measurable invariant metric not just on the tangent bundle but on a number of other natural vector bundles as well implies the existence of an invariant C^0-metric (on the tangent bundle). The argument of [20] showed how to strengthen this to obtain a smooth metric. Here we present a more general result with a simpler proof.

If $P \to M$ is a G structure on M, let $P^{(k)} \to P^{(k-1)}$ be the k-th prolongation of this structure [4] (so that $P^{(1)} = P$). Then $P^{(k)}$ is a principal $G^{(k)}$-bundle where $G^{(k)}$ is the corresponding prolongation of G. We recall that G is of finite type if there exists k such that there is an Aut(P)-invariant framing of the manifold $P^{(k)}$. Thus if G defines a connection in a natural way (e.g. $G = O(p,q)$ defines the Levi-Civita connection), then we have this condition for $k = 1$. If P is a G-structure of finite type and Aut$(P) \subset \mathrm{Diff}(M)$ is the automorphism group of this structure, then Aut(P) is a Lie group in such a way that (Aut$(P), M)$ is a Lie transformation group. This structure on Aut(P) is given as follows [4, Theorem I.3.2ff., Theorem I.5.1]: Let k be as above. Then Aut(P) acts freely on $P^{(k)}$ and every orbit is a closed submanifold. The manifold structure on Aut(P) is that defined via the bijection $g \to gx$ of Aut(P) with Aut$(P) \cdot x$, for any $x \in P^{(k)}$.

Theorem 5.1. *Let $P \to M$ be a G-structure of finite type on a compact manifold and let k be as above. Suppose $H \subset \mathrm{Aut}(P)$ is a closed subgroup and that H preserves a volume density on M. If H preserves a measurable Riemannian metric on M and on the manifolds $P^{(r)}$, $1 \leq r \leq k$, then H is compact.*

Proof. Since every orbit of H on $P^{(k)}$ is closed, every H-ergodic measure on $P^{(k)}$ is supported on an orbit [14, Chapter 2]. The existence of a measurable invariant metric on a manifold implies that the algebraic hull of the derivative cocycle on each ergodic component is compact (Lemma 2.1). By the results of [9] (Cf. [16, p. 23]), we can then see inductively that there is an H-invariant ergodic probability measure on each $P^{(r)}$. By our remarks above, there is an H-orbit in $P^{(k)}$ supporting an invariant probability measure. Since H acts freely on $P^{(k)}$, H must have an H-invariant probability measure which implies that H is compact.

Corollary 5.2. *Let M be a compact manifold with an $O(p,q)$-structure ω, and $H \subset \mathrm{Aut}(M, \omega)$ a closed subgroup. If H preserves a measurable Riemannian metric on M (i.e. on $TM \to M$), then H is compact.*

(As the proof shows, this result is true for any connection on a compact manifold, not just the Levi-Civita connection of an $O(p,q)$-structure, and any volume preserving closed subgroup $H \subset \mathrm{Aff}(M)$, the affine transformations of M with respect to the connection.)

Corollary 5.3. i) *Let Γ act smoothly on a compact manifold M such that Γ preserves a G-structure of finite type and a volume density. Let k be as above. If there is a measurable Γ-invariant metric on M and on $P^{(r)}$ for all r, then there is a smooth Γ-invariant Riemannian metric on M.*

ii) *If a group Γ preserves an $O(p,q)$-structure and a measurable Riemannian metric on M (M compact), then there is a smooth Γ-invariant metric on M.*

Proof. Let $\bar{\Gamma}$ be the closure of Γ in $\mathrm{Aut}(P)$, so that $\bar{\Gamma}$ is a Lie group. By the theorem, it suffices to see that $\bar{\Gamma}$ be leaves the relevant measurable metrics invariant, and in (i), the volume density invariant. The action of $\bar{\Gamma}$ on the space of measurable metrics and on the space of volume densities is Borel (where spaces of sections have the topology of convergence in measure, for example) and hence stabilizers are closed [14, Corollary 2.1.20].

6. Completion of the proof of Theorem B

By virtue of the results of Sect. 4 and the structure of Lie groups, to prove Theorem B it suffices to establish the following result.

Lemma 6.1. *Let G be the universal cover of $SL(2, \mathbf{R})$, N a connected solvable Lie group on which G acts by automorphisms (perhaps trivially) and $H = G \ltimes N$. Let (M, ω) be a compact Lorentz manifold and $\pi\colon H \to \mathrm{Aut}(M, \omega)_0$ be defined by a smooth locally faithful action (i.e. with discrete kernel) with $\pi(H)$ closed. Then $\pi(N)$ is compact.*

We begin the proof of Lemma 6.1 by recalling some facts about the action of $O(1, n)$ on S^{n-1}. We recall that if P is the minimal parabolic subgroup of $O(1, n)$, then $O(1, n)/P \cong S^{n-1}$, and that this gives a realization of $O(1, n)$ as the group of conformal diffeomorphisms of the standard $(n-1)$-sphere. (See [6], e.g.) Let $M(S^{n-1})$ be the space of probability measures on S^{n-1}, so that $M(S^{n-1})$ is a compact convex set with the weak-$*$-topology. Then $O(1, n)$ acts continuously on $M(S^{n-1})$. A basic fact about this action is that if $g_n \in O(1, n)$, $g_n \to \infty$, then there is a subsequence g_{n_j} and a uniquely determined pair of points $x, y \in S^{n-1}$ such that for any $\mu \in M(S^{n-1})$, we have $g_{n_j} \mu \to \nu \in M(S^{n-1})$ where ν is supported on $\{x, y\}$. (Cf. [14, Chapter 3], [21, 6, 2].)

Suppose now that Λ is a group acting ergodically on a probability space (X, μ), preserving μ, and that $\alpha\colon X \times \Lambda \to L$ is a cocycle where L is a locally compact group. We call α strongly unbounded if for any measurable $A \subset X$ with $\mu(A) > 0$, we have that for almost all $x \in A$ there is a sequence $\lambda_n \in \Lambda$ such that $x\lambda_n \in A$ and $\alpha(x, \lambda_n) \to \infty$ in L. The following result appears in [7]. We indicate a brief proof.

Lemma 6.2 (Schmidt). *Under the above assumptions, α is strongly unbounded if and only if α is not equivalent to any cocycle β such that $\beta(X \times \Lambda)$ is contained in a compact subgroup of L.*

Proof. For any measurable $\phi\colon X \to L$, we can write $X = \bigcup X_i$ such that $\phi|X_i$ is bounded. From this it is easy to see that the property of strong unboundedness is a cohomology invariant. Thus, if α is equivalent to a cocycle taking values in a compact subgroup, α is clearly not strongly unbounded. Conversely, if α is not strongly unbounded, there is a measurable $A \subset X$, $\mu(A) > 0$, such that for a set of $x \in A$ of positive measure the set $C_x = \{\alpha(x, \lambda) | x \lambda \in A\}$ is precompact in L. Further, if $\gamma \in \Lambda$ such that y, $y\gamma \in A$, then $\alpha(y, \gamma) C_{y\gamma} = C_y$. Ergodicity then implies that C_x is precompact for almost all $x \in A$. The argument of [10, Theorem 3.1] then shows that there exists a measurable $\phi \colon A \to L$ and a compact subgroup $K \subset L$ such that on a conull subset of $\{(y, \lambda) \in A \times \Lambda | y \lambda \in A\}$ we have $\phi(y) \alpha(y, \lambda) \phi(y\lambda)^{-1} \in K$. Since the action is ergodic, a standard argument (see, e.g. [22]) then shows that ϕ can be defined on all of X such that the same assertion is true.

Returning to the proof of Theorem B, by Theorem 5.1 it suffices to show that if $\alpha \colon M \times H \to O(1, n)$ is the derivative cocycle for the action of H on M, then $\alpha | M \times N$ is (measurably) equivalent to a cocycle taking values in the maximal compact subgroup $O(n, \mathbf{R}) \subset O(1, n)$. We can write $M = M_1 \cup M_2$ where M_i are measurable N-invariant sets such that $\alpha | M_1 \times N$ is equivalent to a cocycle into $O(n, \mathbf{R})$, and for almost all N-ergodic components $E \subset M_2$, $\alpha | E \times N$ is not equivalent to a cocycle into $O(n, \mathbf{R})$. Since G normalizes N, M_i are G-invariant as well. It clearly suffices to show that M_2 is null.

Since N is a solvable group, the N-action on M_2 is amenable [11, 14], and hence there is a measurable $(\alpha | M_2 \times N)$-invariant $\phi \colon M_2 \to M(S^{n-1})$. By the choice of M_2 and Lemma 6.2, it follows for (almost) any N-ergodic component $E \subset M_2$, $\alpha | E \times N$ is strongly unbounded. Hence, by the remarks following the statement of Lemma 6.1, it follows that for almost all $m \in M_2$, $\phi(m)$ is supported on at most two points. Furthermore, we can write $M_2 = M_3 \cup M_4$ where M_3, M_4 are measurable N-invariant sets such that

i) there is an $(\alpha | M_3 \times N)$-invariant $\psi \colon M_3 \to M(S^{n-1})$ such that for a.e. $m \in M_3$, $\operatorname{supp}(\psi(m))$ is a 2 point set;
and

ii) for $m \in M_4$, $\phi(m)$ is supported on a point and $\phi | M_4$ is the unique $(\alpha | M_4 \times N)$-invariant measurable function $M_4 \to S^{n-1}$.

In addition, in case (i) we can assert that although ψ may not be unique, $m \to \operatorname{supp} \psi(m) \subset (S^{n-1} \times S^{n-1})/S(2)$ is the unique $\alpha | M_3 \times N$-invariant function from M_3 into 2 point sets. (These assertions follow easily from strong unboundedness and the paragraph following the statement of Lemma 6.1.) Once again, since G normalizes N, M_3 and M_4 are (essentially) G-invariant as well.

Suppose $\mu(M_4) > 0$. Since G normalizes N, $\phi \colon M_4 \to S^{n-1}$ is also G-invariant by the uniqueness property in (ii) above. I.e., ϕ is actually $(\alpha | M_4 \times G)$-invariant. Since $S^{n-1} \cong O(1, n)/P$, this implies that $\alpha | M_4 \times G$ is equivalent to a cocycle into P. However, if G acts non-trivially on M_4, then G acts non-trivially on a set of G-ergodic components in M_4 of positive measure, and the argument in the proof of Theorem A (Sect. 3) shows that for every such ergodic component E that the algebraic hull of $\alpha | E \times G$ contains a group locally isomorphic to G. Since G is not amenable and P is amenable, this is impossible. Thus, $\mu(M_4) > 0$ implies that G acts trivially on M_4. This in turn implies that for almost every $m \in M_4$, we have that $g \to \alpha(m, g)$ defines a representation $G \to O(1, n)$. Since α is equivalent to a cocycle into P, this implies that for almost all $m \in M_4$, $\alpha(m, g)$

$=I$ for all $g \in G$. But for an automorphism g of a Lorentz manifold, the existence of a point m for which $gm=m$ and $dg_m=I$ implies that g is the identity on M. (See the discussion preceding Theorem 5.1 and the references there to [4].) Since G acts non-trivially on M, this is impossible. Therefore, we deduce that $\mu(M_4)=0$. A similar argument shows that $\mu(M_3)=0$. (Observe that the stabilizer of any 2 point set in S^{n-1} is also amenable.) Thus, M_2 is a null set, and this completes the proof of Lemma 6.1.

7. Some examples

Example 7.1. We first remark that if $G=PSL(2,\mathbf{R})$ and ω is the Killing form on $\mathfrak{g}$, then $G \cong \mathrm{Ad}(G)=O(\mathfrak{g},\omega)_0$. The form ω is non-degenerate, symmetric, of signature $(1, 2)$, i.e., is a Lorentz metric on $\mathfrak{g}$. If $\Gamma \subset G$ is a cocompact lattice, $\mathrm{Ad}(\Gamma)$ obviously leaves ω invariant, and hence there is a G-invariant Lorentz metric on G/Γ. If K is any compact Lie group and X is any homogeneous space of K, there is a K-invariant Riemannian metric on X. The product of a Riemannian manifold and a Lorentz manifold is Lorentz, and hence if K acts effectively on X, $G \times K$ acts effectively on $G/\Gamma \times K$, preserving a Lorentz metric.

Example 7.2. Let N be the 3 dimensional Heisenberg group with Lie algebra $\mathfrak{n} = \mathbf{R}x \oplus \mathbf{R}y \oplus \mathbf{R}z$, where $\mathbf{R}z$ is the center and $[x,y]=z$. Let $\Gamma=N_{\mathbf{Z}}$ be the integral Heisenberg group, so that Γ is discrete and N/Γ is compact. Then $\Gamma/Z(\Gamma) \cong \mathbf{Z}^2$ (where $Z(\Gamma)$ is the center of Γ) and hence $\Gamma/Z(\Gamma)$ acts freely by rotations on the circle S^1. Form the induced action of N on the bundle $X = N/\Gamma \times_\Gamma S^1 \xrightarrow{P} N/\Gamma$. The N action on X is trivial on $Z(\Gamma)$ but the corresponding action of $N/Z(\Gamma)$ is free. For each $x \in X$ we have a splitting of $TX_x = \mathbf{R}w \oplus Y$ where Y is the tangent space to the N-orbit through x and w is the unit tangent vector to the fiber of p through x with positive orientation. Since the N action is locally free, each Y can be identified with $\mathfrak{n}$. We thus have an identification of $TX_x \cong \mathbf{R}w \oplus \mathbf{R}x \oplus \mathbf{R}y \oplus \mathbf{R}z$. If $p(x)=[e]$, let B_x be the linear form on TX_x whose matrix with respect to the basis $\{x,y,z,w\}$ is
$$\begin{pmatrix} 1 & 0 & 0 & 0 \\ 0 & 1 & 0 & 0 \\ 0 & 0 & 0 & 1 \\ 0 & 0 & 1 & 0 \end{pmatrix}.$$ The fiber $p^{-1}([e])$ is obviously Γ-invariant and $x \to B_x$ is Γ-invariant on $p^{-1}(x)$. It follows by a standard argument that B extends to an N-invariant Lorentz metric on X. A similar construction applies to any higher dimensional Heisenberg group as well.

Example 7.3. Let N be as in 7.2 and let $\sigma \in \mathrm{Gal}(\mathbf{Q}(\sqrt{2})/\mathbf{Q})$ be the non-trivial element. Then σ extends naturally to an automorphism of $GL(n,\mathbf{Q}(\sqrt{2}))$ which we continue to denote by σ. Let $\Lambda \subset N \times N$ be given by $\Lambda=\{(\gamma,\sigma(\gamma)) \mid \gamma \in N_{\mathbf{Z}[\sqrt{2}]}\}$.

On the automorphism group of a compact Lorentz manifold   423

Then $\Lambda \subset N \times N$ is discrete and $(N \times N)/\Lambda$ is compact. Let $\{x, y, z\}$ be the above basis for $\mathfrak{n} \times \{0\}$ and x', y', z' the corresponding basis for $\{0\} \times \mathfrak{n}$. Then with respect to the basis $\{x, y, z, z', y', x'\}$ of $\mathfrak{n} \times \mathfrak{n}$, the matrix

$$\begin{pmatrix} 1 & 0 & 0 & 0 & 0 & 0 \\ 0 & 1 & 0 & 0 & 0 & 0 \\ 0 & 0 & 0 & 1 & 0 & 0 \\ 0 & 0 & 1 & 0 & 0 & 0 \\ 0 & 0 & 0 & 0 & 1 & 0 \\ 0 & 0 & 0 & 0 & 0 & 1 \end{pmatrix}$$

defines a non-degenerate symmetric bilinear form of signature $(1, 5)$ on $\mathfrak{n} \times \mathfrak{n}$ and it is easy to check that this form is $\mathrm{Ad}(N \times N)$-invariant. Thus, there is an $N \times N$-invariant Lorentz metric on $(N \times N)/\Lambda$. This action is of course not free, but since the intersection of $N \times \{e\}$ with every conjugate of Λ is trivial, N acts freely on $(N \times N)/\Lambda$ preserving a Lorentz metric. Once again, a similar argument applies to all higher dimensional Heisenberg groups as well, verifying Proposition C in the introduction.

8. Applications to arithmetic groups and Kazhdan groups

As indicated in the introduction, the results of Sect. 5 enable us to sharpen the results of [16, 18] to obtain Theorems D, E, F, G, H. We first observe:

Corollary 8.1. *Let Γ be a Kazhdan group acting on a compact manifold M, preserving a volume density and a G-structure of finite type (possibly of higher order.) If there is a measurable invariant metric on M, then there is a smooth invariant metric. If $\dim M \leq 3$, then the action is by a finite quotient of Γ.*

Proof. Since $P^{(r)}$ is a reduction of the frame bundle of $P^{(r-1)}$, there is a Γ-invariant unipotent structure on each $P^{(r)}$. Any Kazhdan group acting so as to preserve a finite measure and a unipotent structure must preserve a measurable metric [14, Theorem 9.1.1]. We can then see inductively that Γ preserves a measurable metric on each $P^{(r)}$. The first assertion of the corollary then follows from Corollary 5.3. The second assertion follows from [18, Theorem 7] and the fact that any compact Lie group K acting effectively on a compact manifold M with $\dim M \leq 3$ satisfies $\dim K \leq 6$, and hence K is locally isomorphic to a product of a torus and at most two copies of $SO(3, \mathbf{R})$.

Theorems D, E then follow from the existence of a measurable invariant metric for a Kazhdan group acting on a compact manifold preserving a Lorentz metric [18].

Theorems F, G, H all follow in a similar manner, once one can establish the existence of measurable invariant metrics, which is carried out in [16]. For the dimension estimates in Theorem G, see [16, Theorems 2.8, 6.3).

424   R.J. Zimmer

References

1. Borel, A.: Density properties for certain subgroups of semisimple Lie groups without compact factors. Ann. Math. **72**, 179–188 (1960)
2. Eberlein, P., O'Neill, B.: Visibility manifolds. Pac. J. Math. **46**, 45–109 (1973)
3. Jacobson, N.: Lie algebras. New York: Interscience 1962
4. Kobayashi, S.: Transformation groups in differential geometry. New York: Springer 1972
5. Moore, C.C.: Amenable subgroups of semisimple groups and proximal flows. Isr. J. Math. **34**, 121–138 (1979)
6. Mostow, G.D.: Quasi-conformal mappings in n-space and the rigidity of hyperbolic space forms. Publ. Math. I.H.E.S. **34**, 53–104 (1967)
7. Schmidt, K.: Cocycles of ergodic transformation groups, India: MacMillan Co. 1977
8. Sternberg, S.: Lectures on differential geometry. Englewood Cliffs, N.J.: Prentice-Hall 1964
9. Zimmer, R.J.: Extensions of ergodic group actions. Ill. J. Math. **20**, 373–409 (1976)
10. Zimmer, R.J.: On the cohomology of ergodic group actions. Isr. J. Math. **35**, 289–300 (1980)
11. Zimmer, R.J.: Amenable ergodic group actions and an application to Poisson boundaries of random walks. J. Funct. Anal. **27**, 350–372 (1978)
12. Zimmer, R.J.: An algebraic group associated to an ergodic diffeomorphism. Compos. Math. **43**, 59–69 (1981)
13. Zimmer, R.J.: Induced and amenable actions of Lie groups. Ann. Sci. Ec. Norm. Super. **11**, 407–428 (1978)
14. Zimmer, R.J.: Ergodic theory and semisimple groups. Boston: Birkhäuser 1984
15. Zimmer, R.J.: Semisimple automorphism groups of G-structures. J. Differ. Geom. **19**, 117–123 (1984)
16. Zimmer, R.J.: Volume preserving actions of lattices in semisimple groups on compact manifolds. Publ. Math. I.H.E.S. **59**, 5–33 (1984)
17. Zimmer, R.J.: Actions of lattices in semisimple groups preserving a G-structure of finite type. Ergodic Theory Dyn. Syst. (To appear)
18. Zimmer, R.J.: Kazhdan groups acting on compact manifolds. Invent. Math. **75**, 425–436 (1984)
19. Zimmer, R.J.: On discrete subgroups of Lie groups and elliptic geometric structures. Rev. Math. Iber. Am., vol. 1 (To appear)
20. Zimmer, R.J.: Lattices in semisimple groups and distal geometric structures. Invent. Math. **80**, 123–137 (1985)
21. Zimmer, R.J.: Ergodic theory, semisimple Lie groups, and foliations by manifolds of negative curvature. Publ. Math. I.H.E.S. **55**, 37–62 (1982)
22. Feldman, J., Moore, C.C.: Ergodic equivalence relations, cohomology, and von Neumann algebras, I. Trans. Am. Math. Soc. **234**, 289–324 (1977)

Oblatum 26-XI-1985

J. DIFFERENTIAL GEOMETRY
19 (1984) 117–123

SEMISIMPLE AUTOMORPHISM GROUPS
OF G-STRUCTURES

ROBERT J. ZIMMER

1. Statement of the main result

Let M be a compact manifold of dimension n and suppose we are given a G-structure (possibly of higher order) on M. Thus $G \subset \mathrm{GL}(n, \mathbf{R})$ if the G-structure is of first order, and in general $G \subset \mathrm{GL}(n, \mathbf{R})^{(k)}$ for some k, where $\mathrm{GL}(n, \mathbf{R})^{(k)}$ is the group of k-jets at 0 of diffeomorphisms of $\mathbf{R}^n$ fixing 0. We shall suppose G is a real algebraic group (i.e. the $\mathbf{R}$-points of an algebraic $\mathbf{R}$-group). The G-structure is given by a principal G-bundle $P \to M$ which is a reduction of the bundle of $(k-1)$-jets of frames $P^{(k)} \to M$ to G. We let $\mathrm{Aut}(P)$ be the automorphism group of the G-structure, so $\mathrm{Aut}(P) \subset \mathrm{Diffeo}(M)$. In general of course $\mathrm{Aut}(P)$ is infinite dimensional, although it will be a finite dimensional Lie group in a number of important situations. The point of this paper is to examine (irrespective of whether or not $\mathrm{Aut}(P)$ is a Lie group) which semisimple Lie groups can be subgroups of $\mathrm{Aut}(P)$. In other words, what are the obstructions for a smooth action of a semisimple Lie group to preserve a G-structure? If $G \subset O(n, \mathbf{R})$, then the G-structure is essentially a Riemannian metric and, as is well known, $\mathrm{Aut}(P)$ is then compact. Thus any semisimple Lie group admitting an embedding into $\mathrm{Aut}(P)$ must also be compact. If G is of finite type in the sense of E. Cartan (see [8]), then $\mathrm{Aut}(P)$ is a Lie group and in explicit circumstances one can derive bounds on $\dim(\mathrm{Aut}(P))$. Our main result is the following. For a Lie group Q, $L(Q)$ will denote its Lie algebra.

Theorem 1. *Suppose H is a connected simple Lie group with finite center and with $\mathbf{R}$-rank$(H) \geqslant 2$. Let M be a compact n-manifold and $P \to M$ a G-structure where $G \subset \mathrm{GL}(n, \mathbf{R})^{(k)}$ (for some k) is a real algebraic subgroup. Suppose that*

Received May 31, 1983. This work was partially supported by National Science Foundation grant MCS 8004026 and the Alfred P. Sloan Foundation.

H acts smoothly on M with H ⊂ Aut(P) and that H preserves a volume density on M. Then there is an embedding of Lie algebras $L(H) \to L(G)$ (and hence $L(H) \to L(G/\mathrm{rad}(G))$ where $\mathrm{rad}(G)$ is the radical of G).

Thus, roughly speaking, a higher **R**-rank simple Lie automorphism group of P can only be as large as a simple component of G. We remark that the condition that H preserves a volume density will of course follow from the condition $H \subset \mathrm{Aut}(P)$ if the first order part of G is contained in $\mathrm{SL}'(n, \mathbf{R})$ (matrices with $\det^2 = 1$), which is true in many important situations. We also recall that $\mathrm{GL}(n, \mathbf{R})^{(k)} = \mathrm{GL}(n, \mathbf{R}) \ltimes N_{n,k}$ where $N_{n,k}$ is a connected unipotent group. Hence any semisimple subgroup of $G \subset \mathrm{GL}(n, \mathbf{R})^{(k)}$ must actually be a subgroup of $G \cap \mathrm{GL}(n, \mathbf{R})$, i.e. of the first order part of G. In fact, Theorem 1 remains true if we replace $\mathrm{Aut}(P)$ by the group of "automorphisms up to order 1" of P [6].

We do not know whether or not Theorem 1 is true if **R**-rank$(H) = 1$. It seems unlikely that our argument can be made to apply to this case. If one makes the further (strong) assumption that H acts transitively on M then for first order G-structures Theorem 1 can be readily deduced from the Borel density theorem [2], and this argument applies in the **R**-rank 1 case as well. (The proof of Theorem 1 in general in fact makes use of the Borel density theorem, both in the proof of Lemma 6 below and in the proof of Theorem 2, which is quoted from earlier work [11], [13].)

2. Proof of Theorem 1

The proof of Theorem 1 will rely heavily on results from ergodic theory, i.e. on the measure theoretic behavior of the actions in question. We thus begin by recalling two basic results in this direction. The first is a consequence of super-rigidity for measurable cocycles of semisimple Lie group actions which we proved in [11], a generalization of Margulis' super-rigidity theorem [9] (see also [12]). The second is a consequence of the subadditive ergodic theorem, namely, the existence of an "exponent" for a suitable cocycle of a measure preserving transformation of a measure space.

Suppose $P \to M$ is a principal G-bundle, so that G acts on the right of P and $M \cong P/G$. By choosing a measurable section ϕ of P, we obtain a measurable trivialization of P, namely $F: M \times G \to P$ given by $F(m, g) = \phi(m)g$. The action of $\mathrm{Aut}(P)$ on P can then be described as the action on $M \times G$ given by $h(m, g) = (hm, \alpha(h, m)g)$ for $(m, g) \in M \times G$, $h \in \mathrm{Aut}(P)$, and $\alpha(h, m) \in G$. Further, α must satisfy the cocycle identity

$$\alpha(h_1 h_2, m) = \alpha(h_1, h_2 m)\alpha(h_2, m).$$

Two cocycles α, β: $H \times M \to G$ are called equivalent if there is a measurable function f: $M \to G$ such that for all $h \in H$,

$$\alpha(h, m) = f(m)\beta(h, m)f(hm)^{-1}$$

for almost all $m \in M$. The notions of cocycle and equivalence of cocycles clearly make sense for any measurable action of a group on a measure space. The super-rigidity theorem for cocycles [11], [12] implies via the arguments of [12, Chapter 9] or [13, Theorem 2.8] the following.

Theorem 2. *Suppose H is a connected simple Lie group with finite center and* **R**-*rank*$(H) \geq 2$, *and S is a* (*standard*) *Borel H-space with a finite H-invariant measure. Let $G \subset \mathrm{GL}(n, \mathbf{R})$ be a real algebraic group and suppose α: $H \times S \to G$ is a* (*measurable*) *cocycle. Then either*:

(a) *considered as a cocycle α: $H \times S \to \mathrm{GL}(n, \mathbf{R})$, α is equivalent to a cocycle that takes values in $O(n, \mathbf{R})$; or*

(b) *there is an embedding of Lie algebras $L(H) \to L(G)$.*

The other main ergodic theoretic result we shall need is the existence of the Lyapunov exponent of a tempered cocycle of an action of $\mathbf{Z}$ (see [3], [5], [7] for proofs). Thus, let $\mathbf{Z}$ act on a (standard) probability space (S, μ) and suppose α: $\mathbf{Z} \times S \to \mathrm{GL}(n, \mathbf{R})$ is a cocycle. We call α tempered if for each $n \in \mathbf{Z}$ (equivalently for $n = 1$), $s \to \|\alpha(n, s)\|$ and $s \to \|\alpha(n, s)^{-1}\|$ are in $L^\infty(S)$. We then have the following result.

Theorem 3 [5], [7]. *If α: $\mathbf{Z} \times S \to \mathrm{GL}(n, \mathbf{R})$ is a tempered cocycle, then*

$$e_\alpha(s) = \lim_{n \to \infty} \frac{1}{n} \log^+ \|\alpha(n, s)\|$$

exists for almost all s. (*Here $\log^+ = \max\{0, \log\}$.*) *We call e_α the exponent of α.*

We now make some comments on this result that we will need.

Lemma 4. *Suppose $\alpha \sim \beta$ where α and β are tempered. Then $e_\alpha = e_\beta$ (a.e.).*

Proof. Let $\alpha(n, s) = f(s)\beta(n, s)f(ns)^{-1}$. Choose $A \subset S$ to be of positive measure so that for some $B \in \mathbf{R}$, $\|f(s)\|, \|f(s)^{-1}\| \leq B$ for $s \in A$. Let $N(A, s) = \{n \in \mathbf{Z}^+ \mid ns \in A\}$. Since there is an invariant probability measure, Poincaré recurrence implies $N(A, s)$ is infinite for almost all $s \in A$. Thus, for almost all $s \in A$ we have

$$e_\alpha(s) = \lim_{\substack{n \to \infty \\ n \in N(A, s)}} \frac{1}{n} \log^+ \|f(s)\beta(n, s)f(ns)^{-1}\|$$

$$\leq \varlimsup_{\substack{n \to \infty \\ n \in N(A, s)}} \frac{1}{n} \left(2 \log^+ B + \log^+ \|\beta(n, s)\|\right)$$

$$\leq \varlimsup_{\substack{n \to \infty \\ n \in N(A, s)}} \frac{1}{n} \log^+ \|\beta(n, s)\| = e_\beta(s).$$

The reverse inequality follows in a similar fashion and hence $e_\alpha(s) = e_\beta(s)$ for a.e. $s \in A$. Letting $B \to \infty$, we can choose $\mu(S)$ arbitrarily close to 1, and hence $e_\alpha = e_\beta$ a.e.

Suppose now that M is a compact n-manifold, and $\phi\colon M \to M$ is a diffeomorphism. Let ξ be a measurable Riemannian metric on M, i.e. for each $m \in M$ an inner product ξ_m on the tangent space TM_m such that ξ_m varies measurably in $m \in M$. If we measurably trivialize TM by measurably choosing an orthonormal basis of TM_m with respect to ξ_m, the action of $\mathbf{Z}$ on TM defined by the powers of ϕ and $d\phi$ yields a cocycle. $\alpha\colon \mathbf{Z} \times M \to GL(n, \mathbf{R})$. We will say that α is associated to the measurable metric ξ. The equivalence class of this cocycle is independent of the choice of ξ and the choice of orthonormal basis. If ξ is a continuous metric it is clear that α will be tempered owing to compactness of M. For inner products η, σ on a finite dimensional real vector space we let $M(\eta/\sigma) = \max_{\|x\|_\sigma = 1}\{\|x\|_\eta\}$. Thus if η, σ are measurable Riemannian metrics on M, $M(\eta/\sigma)(s) = M(\eta_s/\sigma_s)$ is a measurable function on M. If η is a measurable Riemnannian metric, we will call η bounded if $M(\eta/\xi)$ and $M(\xi/\eta)$ are in $L^\infty(M)$ for some smooth metric ξ. It is clear that this is independent of the smooth metric involved. The following is straightforward.

Lemma 5. *Suppose η is a bounded measurable Riemannian metric on a compact manifold M, and α is an associated cocycle for some diffeomorphism of M. Then α is tempered.*

To prove Theorem 1 it suffices to consider the case of a first order G-structure (cf. the remarks in the next to last paragraph of §1). Thus, we need only eliminate the possibility of conclusion (a) in Theorem 2 where $\alpha\colon H \times M \to GL(n, \mathbf{R})$ is the cocycle defined by the derivative via a measurable trivialization of TM. For $h \in H$, set $\alpha_h = \alpha \,|\, \{h^n\} \times M$ so we may view α_h as a cocycle $\mathbf{Z} \times M \to GL(n, \mathbf{R})$. Since $O(n, \mathbf{R})$ is compact, clearly any cocycle taking values in $O(n, \mathbf{R})$ is tempered and has 0 exponent, and hence condition (a) of Theorem 2 would imply via Lemmas 4 and 5 that for all $h \in H$, $e_{\alpha_h} = 0$ a.e. for any cocycle α associated to any bounded measurable Riemannian metric on M. Thus, the proof will follow once we establish the existence of a bounded measurable Riemannian metric with associated cocycle α such that for some $h \in H$, $e_{\alpha_h} \neq 0$.

We will need the following result which has been observed independently by C. C. Moore (private communciation).

Lemma 6. *Let H be a connected simple noncompact Lie group and S a (standard) H-space with finite invariant measure. For $s \in S$, let H_s be the stabilizer of s in H. Then there is a conull set $S_0 \subset S$ such that for $s \in S_0$ either $H_s = H$ or H_s is discrete.*

Proof. Via the ergodic decomposition of the action it suffices to see that the assertion is true in each ergodic component. Thus, we may as well assume that H acts ergodically on S. Let $\mathbb{S}$ be the set of closed subgroups of H. Then $\mathbb{S}$ has the structure of a standard Borel space and the map $s \to H_s$ is a Borel map $S \to \mathbb{S}$ [10]. Let $V = \bigcup_{k=0}^{n} \mathrm{Gr}_k(L(H))$ be the union of the Grassmann varieties of k-planes in the Lie algebra $L(H)$. Then the map $\mathbb{S} \to V$ given by $Q \to L(Q)$ is also Borel. Thus $\phi(s) = L(H_s)$ is a Borel map, and this is clearly an H-map where H acts on $L(H)$ (and hence on V) by the adjoint representation. As observed in [1], the action of H on V is smooth in the sense of ergodic theory, i.e., the orbit space V/H is a countably separated Borel space. (See [4], [12] for a discussion of this notion). (This follows immediately from the fact that $\mathrm{Ad}(H)$ is a subgroup of finite index in a real algebraic group and that the action of $\mathrm{Ad}(H)$ on V is algebraic; thus the orbits are locally closed and this implies that the action is smooth.) We let $\bar{\phi}$ be the composition of ϕ with the natural projection $V \to V/H$. Thus $\bar{\phi} \colon S \to V/H$ is a measurable H-invariant map. Since H acts ergodically on S and V/H is a countably separated Borel space, $\bar{\phi}$ is essentially constant. In other words, by passing to a conull H-invariant subset of S, we can assume $\phi(S)$ lies in a single H-orbit in V. This orbit can be identified with H/H_1, where H_1 is the stabilizer of a point in this orbit, and we thus have a measurable H-map $\phi \colon S \to H/H_1$. If the orbit $H/H_1 \subset V$ actually lies in $\mathrm{Gr}_k(L(H))$ for $k = 0$ or $k = \dim L(H)$, then $L(H_s) = 0$ or $L(H_s) = L(H)$ for almost every s, as asserted. Suppose on the other hand that $H/H_1 \subset \mathrm{Gr}_k(L(H))$ for $0 < k < \dim L(H)$. Since H is simple, we must have $H_1 \neq H$. Furthermore $\mathrm{Ad}_H(H_1)$, being the stabilizer in $\mathrm{Ad}_H(H)$ of a k-plane in $L(H)$, is of finite index in a real algebraic group. Thus $\mathrm{Ad}_H(H_1)$, and hence H_1 (since H has a finite center), has only finitely many connected components. However, if μ is a finite H-invariant measure on S, then $\phi_* M$ is a finite H-invariant measure on H/H_1. By the Borel density theorem [2], [12], either H_1 is discrete or $H_1 = H$. Since $H_1 \neq H$, H_1 is discrete, and having only finitely many components, H_1 is finite. This obviously contradicts the existence of a finite invariant measure on H/H_1. Thus $k - 0$ or $\dim(L(H))$ and the proof of the lemma is complete.

Returning to the proof of Theorem 1, by the preceding lemma we can choose an H-invariant set $M_0 \subset M$ of positive measure such that the stabilizers are discrete. For any $m \in M_0$, let $V_m \subset TM_m$ be the tangent space to the H-orbit through m. Fix an inner product on the Lie algebra $L(H)$. For each $m \in M_0$, the map $H \to M, h \to hm$ induces an isomorphism of $L(H)$ with V_m, and hence an inner product ξ_m on V_m. The assignment $m \to \xi_m$ is measurable in the obvious sense. Since any measurable function is bounded on a set of large measure, it is clear that there is a bounded measurable Riemannian metric η on

122   ROBERT J. ZIMMER

M and a subset $M_1 \subset M_0$ of positive measure such that for $m \in M_1$, η and ξ are equal when restricted to V_m. Suppose now that $h \in H$ and $m \in M_0$ with $hm \in M_0$ as well. Then $dh: TM_m \to TM_{hm}$, $dh(V_m) = V_{hm}$, and it is easy to see that under the above identifications of V_m and V_{hm} with $L(H)$, the map $dh \mid V_m$ corresponds to $\mathrm{Ad}(h)$. Fix some $h \in H$ such that, letting λ be the maximum absolute value of the eigenvalues of $\mathrm{Ad}(h)$, we have $\lambda > 1$. (Since H is a noncompact simple Lie group, such an element always exists.) Since H acts in a finite volume preserving manner on M and M_1 is of positive measure, Poincaré recurrence implies there is a subset $M_2 \subset M_1$, M_2 conull in M_1, so that for every $x \in M_2$ there is a sequence of distinct positive integers n_i (depending on x) with $h^{n_i}x \in M_2$ for all i. Let α be a (tempered) cocycle corresponding to the bounded Riemannian metric η. Then for $s \in M_2$ we have

$$e_{\alpha_h}(s) = \lim_n \frac{1}{n} \log^+ \|\alpha(h^n, s)\|$$

$$= \lim_i \frac{1}{n_i} \log^+ \|\alpha(h^{n_i}, s)\|.$$

By the choice of M_2 and the construction of η,

$$\|\alpha(h^{n_i}, s)\|_\eta \geqslant \|\mathrm{Ad}(h^{n_i})\|.$$

By the choice of h,

$$\lim_n \|\mathrm{Ad}(h^n)\|^{1/n} = \lambda > 1.$$

It follows that for $s \in M_2$, $e_{\alpha_h}(s) > 0$, and this contradiction completes the proof.

The author would like to thank M. Markowitz for a number of useful conversations related to this work.

References

[1] L. Auslander & C. C. Moore, *Unitary representations of solvable Lie groups*, Mem. Amer. Math. Soc. No. 62, 1966.
[2] A. Borel, *Density properties for certain subgroups of semisimple groups without compact factors*, Ann. of Math. **72** (1960) 179–188.
[3] Y. Derriennic, *Sur le théorème ergodique sous-additif*, C.R. Acad. Sci. Paris Sér. A, **281** (1975) 985–988.
[4] E. G. Effros, *Transformation groups and C*-algebras*, Ann. of Math. **81** (1965) 38–55.
[5] H. Furstenberg & H. Kesten, *Products of random matrices*, Ann. of Math. Statist. **31** (1960) 457–469.
[6] V. Guillemin, *Integrability problem for G-structures*, Trans. Amer. Math. Soc. **116** (1965) 554–560.
[7] J. F. C. Kingman, *The ergodic theory of subadditive stochastic processes*, J. Royal Statist. Soc. Ser. B, **30** (1968) 499–510.

AUTOMORPHISM GROUPS OF G-STRUCTURES   123

[8] S. Kobayashi, *Transformation groups in differential geometry*, Springer, New York, 1972.

[9] G. A. Margulis, *Discrete groups of motions of manifolds of non-positive curvature*, Amer. Math. Soc. Transl., No. 109, 1977, 33–45.

[10] A. Ramsay, *Virtual groups and group actions*, Advances in Math. **6** (1971) 253–322.

[11] R. J. Zimmer, *Strong rigidity for ergodic actions of semisimple Lie groups*, Ann. of Math. **112** (1980) 511–529.

[12] _____, *Ergodic theory and semisimple groups*, to appear.

[13] _____, *Volume preserving actions of lattices in semisimple groups on compact manifolds*, Inst. Hautes Études Sci. Publ. Math., to appear.

UNIVERSITY OF CHICAGO

Proceedings of Symposia in Pure Mathematics
Volume **54** (1993), Part 3

Automorphism Groups and Fundamental Groups of Geometric Manifolds

ROBERT J. ZIMMER

1. Statement of main results

The aim of this paper is to survey some of the recent results concerning the relationship of the structure of the fundamental group of a manifold and that of the automorphism group of a geometric structure on the manifold. We shall focus our attention here on rigid geometric structures in the sense of Gromov [**G2**] (defined below), which is a generalization of the classical notion of structure of finite type in the sense of Cartan [**C, GS**]. The most natural relevant case of such a structure is that of an affine connection, and if the reader is so inclined he may restrict attention to that case without losing, for the most part, the essence of the nature of the results or of the techniques of the proofs. There is also considerable information available in the case of general nonrigid structures, but for this we shall refer the reader to [**Z6, Z4**]. We shall also not discuss the special (nonrigid) situation of complex structures, on which of course there is a large literature.

Let M be a smooth n-manifold. We shall often need M to be real analytic, but shall be explicit about when this assumption is being made. We let $G^{(r)}$ be the group of r-jets of diffeomorphisms of $\mathbb{R}^n$ fixing the origin, and $P^{(r)}(M)$ the principal $G^{(r)}$-bundle of r-frames on M. We recall that an r-frame at $m \in M$ is simply an r-jet of a local diffeomorphism $(\mathbb{R}^n, 0) \to (M, m)$. Thus for $r = 1$, $G^{(r)}$ is naturally isomorphic to $\mathrm{GL}(n, \mathbb{R})$ and $P^{(1)}(M)$ is the usual frame bundle. Suppose V is a manifold on which $G^{(r)}$ acts smoothly. We let $E_V \to M$ be the associated fiber bundle with fiber V. By a structure of order r and type V on M we mean a section of the bundle E_V. Such a structure will be called of algebraic type if V is a real algebraic variety on which $G^{(r)}$ (which we recall is a real algebraic group)

1991 *Mathematics Subject Classification.* Primary 58D19.
Research partially supported by the National Science Foundation.
This paper is in final form and no version will be submitted for publication elsewhere.

acts regularly. By a G-structure of order r we mean a structure of type $G^{(r)}/G$, where $G \subset G^{(r)}$ is a closed subgroup. This is of course equivalent to the assertion that one has a reduction of the bundle $P^{(r)}(M)$ to G. A G-structure will be of algebraic type if G is an algebraic subgroup.

We call a G-structure unimodular if the image of G in $\mathrm{GL}(n, \mathbb{R}) = G^{(1)}$ under the natural projection of jets $G^{(r)} \to G^{(1)}$ is contained in the group of matrices whose determinant is ± 1. (In other words the G-structure defines a volume density.) More generally, a structure ω of type V will be called unimodular if for each $m \in M$, the $G^{(r)}$-orbit in V determined by $\omega(m)$ has stabilizers whose image in $\mathrm{GL}(n, \mathbb{R}) = G^{(1)}$ under the natural projection of jets $G^{(r)} \to G^{(1)}$ is contained in the group of matrices whose determinant is ± 1. Thus, any unimodular structure of type V determines a volume density on M.

The group $\mathrm{Diff}(M)$ acts naturally on $P^{(r)}(M)$, hence on E_V for any $G^{(r)}$-space V, and hence on the space of structures of type V. If ω is such a structure, we let $\mathrm{Aut}(M, \omega)$ be the stabilizer of ω in the group $\mathrm{Diff}(M)$. In the case of rigid structures, $\mathrm{Aut}(M, \omega)$ will be a Lie group (but possibly infinite and of dimension 0), and it is of course a basic classical question to understand the relationship between ω, M, and $\mathrm{Aut}(M, \omega)$. While our main interest here will be specifically in the relationship of $\mathrm{Aut}(M, \omega)$ and $\pi_1(M)$, we mention one other basic result, first proved in [Z3], that we will need to establish along the way. We emphasize that rigidity of the structure is not required in this theorem.

THEOREM 1.1 [Z3]. *Suppose H is a noncompact connected simple Lie group, M is compact, and ω is a unimodular G-structure of algebraic type. If there is a nontrivial action $H \to \mathrm{Aut}(M, \omega)$, then there is an embedding of Lie algebras $\mathfrak{h} \to \mathfrak{g}$. Furthermore, under this inclusion the corresponding representation $\mathfrak{h} \to \mathfrak{gl}(n, \mathbb{R})$ (obtained by composition with the derivative of projection of jets $G \to \mathrm{GL}(n, \mathbb{R})$) contains $\mathrm{ad}_{\mathfrak{h}}$ as a subrepresentation.*

By taking $G = \mathrm{SO}(p, q)$, for example, we obtain considerable information on the possible automorphism groups of pseudo-Riemannian manifolds. This becomes particularly striking in the case of Lorentz manifolds, where one deduces

COROLLARY 1.2 [Z3]. *Let M be a compact Lorentz manifold. Then the connected component of the identity of $\mathrm{Aut}(M)$ is either:*
  (a) *locally isomorphic to $\mathrm{SL}(2, \mathbb{R}) \times K$ where K is compact; or*
  (b) *a compact extension of a 2-step nilpotent group.*

We shall indicate the proof of Theorem 1.1 in §3 below. For the proof of the corollary, see [Z3].

Let ω be a structure of order r and type V. For $k \geq r$ and $x, y \in M$, we let $\mathrm{Aut}^{(k)}(\omega, x, y)$ be the k-jets of local diffeomorphisms of M taking x to y and $\omega(x)$ to $\omega(y)$ up to order k. We set $\mathrm{Aut}^{(k)}(\omega, m, m) =$

$\mathrm{Aut}^{(k)}(\omega, m)$, the k-jets at m of local diffeomorphisms of M fixing m and ω at m up to order k.

DEFINITION 1.3 [G2]. The structure ω is k-rigid if for each $m \in M$, and $j \geq k$, the natural projection map $\mathrm{Aut}^{(j)}(\omega, m) \to \mathrm{Aut}^{(k)}(\omega, m)$ is injective. A structure is called rigid if it is k-rigid for some k.

Thus, k-rigidity entails the assertion that the k-jet of an infinitesimal automorphism at a point determines the infinitesimal automorphism at that point.

EXAMPLE 1.4. Pseudo-Riemannian structures are rigid structures of order 1. Affine connections are rigid structures of order 2. More generally, any G-structure of finite type in the sense of Cartan is rigid.

We can now state some of the main results concerning the relation with fundamental groups. The remaining sections of the paper will essentially be devoted to developing the ideas necessary to explain the proofs of these results.

For Theorems 1.5–1.10, we let M be a compact real analytic manifold, and ω a real analytic, rigid, unimodular structure of algebraic type on M. We let $\mathrm{Aut}^{\mathrm{an}}(M, \omega)$ be the real analytic automorphisms. (For example, we can take ω to be an analytic affine connection and volume density, in which case $\mathrm{Aut}^{\mathrm{an}}(M, \omega)$ is simply the group of affine, volume-preserving analytic diffeomorphisms.) Let H be a connected noncompact simple Lie group with finite center. We suppose there is a nontrivial action $H \to \mathrm{Aut}^{\mathrm{an}}(M, \omega)$, which without loss of generality we may assume is faithful.

THEOREM 1.5 (Gromov [G2]). *There exists a representation $\sigma: \pi_1(M) \to \mathrm{GL}(q, \mathbb{R})$ for some q such that the algebraic hull (i.e., Zariski closure) of $\sigma(\pi_1(M))$ contains a group locally isomorphic to H.*

For example, by a theorem of C. C. Moore [M], the algebraic hull of an amenable group is amenable. Thus, we deduce from Theorem 1.5 that no compact analytic manifold with amenable fundamental group admits an H-action preserving an analytic connection and volume. In particular, of course, M cannot be simply connected. The proof of Theorem 1.5 will be discussed in §4. In the simply connected case, Corollary 1.2 can be strengthened as follows.

THEOREM 1.6 (d'Ambra [d'A]). *Let M be a simply connected real analytic Lorentz manifold. Then the group of automorphisms is compact.*

Returning now to the hypotheses described before Theorem 1.5, we have

THEOREM 1.7 [Z7]. *Assume $\mathbb{R}$-rank$(H) \geq 2$ and that $\pi_1(H)$ is finite. Suppose that $\pi_1(M)$ admits a faithful linear representation $\sigma: \pi_1(M) \to \mathrm{GL}(q, \mathbb{C})$ for some q such that either:*
(i) *$\sigma(\pi_1(M))$ is discrete; or*
(ii) *$\sigma(\pi_1(M)) \subset \mathrm{GL}(q, \overline{\mathbb{Q}})$.*

Then $\pi_1(M)$ contains a lattice in a linear Lie group L, where L contains a group locally isomorphic to H.

The proof of Theorem 1.7 will be discussed §5. If $\pi_1(H)$ is not finite, one can often obtain strong information directly from 1.7 by passing to a subgroup $H_1 \subset H$ for which $\pi_1(H_1)$ is finite and which still satisfies the $\mathbb{R}$-rank condition.

COROLLARY 1.8. *With the hypotheses of Theorem 1.7, $\pi_1(M)$ is not isomorphic to a discrete subgroup of a Lie group of real rank strictly smaller than that of H. In particular, it is not isomorphic to a discrete subgroup of a Lie group of real rank one.*

THEOREM 1.9 [SZ]. *Assume $\mathbb{R}$-rank$(H) \geq 2$, and $\pi_1(H)$ is finite. Then $\pi_1(M)$ is not isomorphic to the fundamental group of a complete Riemannian manifold of negative sectional curvature bounded away from 0 and $-\infty$.*

While the above results give obstructions to the appearance of a group Λ as the fundamental group of a manifold admitting a suitable H-action, there are groups which may appear as such a fundamental group, but only in manifolds of large dimension. Namely:

THEOREM 1.10 [Z7]. *Fix H as above with $\mathbb{R}$-rank$(H) \geq 2$, and fix a positive integer m. Then there is a finitely generated group Λ with the properties:*

(1) If $\dim M \leq m$ (and M is as above), there is no isogeny (i.e., surjection with finite kernel) $\pi_1(M) \to \Lambda$; and

(2) There exists a compact analytic manifold X (necessarily of dimension greater than m) with an isogeny $\pi_1(X) \to \Lambda$ such that there is a real analytic action of H on X preserving an analytic pseudo-Riemannian metric.

The group Λ may in fact be taken to be a suitable cocompact lattice in $\mathrm{SL}(p, \mathbb{R})$ for large enough p, where p is a prime. We sketch the proof in §5.

We can obtain additional information on $\pi_1(M)$ if we make further dynamical assumptions on the H-action. See for example Theorem 5.3 below.

These results all concern connected subgroups of $\mathrm{Aut}^{\mathrm{an}}(M, \omega)$. For actions of discrete groups, one sees immediately that the situation is more complex. For example, the natural action of $\mathrm{SL}(n, \mathbb{Z})$ on T^n shows that Theorem 1.5 is no longer true, since one has actions on manifolds with abelian fundamental group. In this case, the action of $\mathrm{SL}(n, \mathbb{Z})$ on the fundamental group of T^n is of course nontrivial, which one can never have for actions of connected groups. The relevance of this remark is highlighted by the following (new) result whose proof we give in §6.

THEOREM 1.11. *Let (M, ω) be a compact analytic manifold with a real analytic, rigid, unimodular structure of algebraic type, and suppose $\pi_1(M)$ is isomorphic to an infinite finitely generated nilpotent group. Let $n \geq 6$. Then* *there is no analytic ergodic action of* $\Gamma = \mathrm{SL}(n, \mathbb{Z})$ *on* M *preserving* ω *such that the natural homomorphism* $\Gamma \to \mathrm{Out}(\pi_1(M))$ *is trivial.*

We remark that the same is true if $\mathrm{SL}(n, \mathbb{Z})$ is replaced by any finite index subgroup of $\mathrm{SL}(n, \mathbb{Z})$, or more generally by a finite index subgroup Γ of the group of integer points of a simple algebraic group over $\mathbb{Q}$ with $\mathbb{Q}$-rank at least 2, and which satisfies $H^2(\Gamma, \mathbb{R}) = 0$. This latter condition holds for most groups for which the associated symmetric space is non-Hermitian [**B**].

In the simply connected case, one can obtain stronger results. See Theorem 6.1 below.

2. Infinitesimal, local, and global automorphisms

In this section we summarize (mostly without proof) some basic results relating general properties of infinitesimal, local, and global automorphisms.

For any structure ω, and $x, y \in M$, by $\mathrm{Aut}^{\mathrm{loc}}(\omega, x, y)$ we mean the group of germs of automorphisms of ω which are defined in a neighborhood of x and take x to y. We set $\mathrm{Aut}^{\mathrm{loc}}(\omega, m, m) = \mathrm{Aut}^{\mathrm{loc}}(\omega, m)$.

Throughout this section we take M *to be a manifold with a rigid structure* ω.

THEOREM 2.1 (Gromov). (a) *There is a positive integer* k, *and an open dense set* $U \subset M$ *such that for* $x, y \in U$, *every element of* $\mathrm{Aut}^{(k)}(\omega, x, y)$ *extends uniquely to an element of* $\mathrm{Aut}^{\mathrm{loc}}(\omega, x, y)$. *In particular, for* $m \in U$, *every element of* $\mathrm{Aut}^{(k)}(\omega, m)$ *extends uniquely to an element of* $\mathrm{Aut}^{\mathrm{loc}}(\omega, m)$.

(b) *If in addition* (M, ω) *is compact and analytic, we may take* $U = M$.

A proof of this result is given in [**G1**] or [**G2**]. Theorem 2.1 is similar in spirit and closely related to earlier basic work of Singer [**S**], in which he proves that a Riemannian manifold which is homogeneous up to some large (but finite) order is actually locally homogeneous.

If X is a vector field defined on an open subset of M, then X is called a Killing field or a symmetry field of ω if $L_X \omega = 0$. We let $\mathrm{Kill}(M, \omega)$ be the space of globally defined Killing fields, and for $m \in M$, we let $\mathrm{Kill}^{\mathrm{loc}}(\omega, m)$ be the space of germs at m of Killing fields defined in a neighborhood of m.

THEOREM 2.2. *Suppose further that* M *is analytic and simply connected, and that* ω *is analytic. Then for every* $m \in M$, *every element of* $\mathrm{Kill}^{\mathrm{loc}}(\omega, m)$ *extends uniquely to an element of* $\mathrm{Kill}(M, \omega)$.

This theorem is due to Nomizu [**N**] for the case in which ω is a complete parallelism. It was observed that this is still true by basically the same argument for structures of finite type by Amores [**A**], and for rigid structures by Gromov [**G2**]. The next result is a straightforward generalization of classical assertions about structures of finite type.

THEOREM 2.3. *If ω is a rigid structure, then* Aut(M, ω) *is a Lie group in such a way that the action on M is smooth.*

One difficulty in applying Theorem 2.3 is the fact that in general Aut(M, ω) may be an infinite discrete group. It is obviously of interest to have natural conditions under which this group is connected, or at least has finitely many components.

THEOREM 2.4 (Gromov [**G2**]). *If ω is a rigid analytic structure of algebraic type, and M is simply connected and compact, then* Aut(M, ω) *has finitely many components, as does the stabilizer of each point in m.*

PROOF (sketch). Suppose ω is of order r and type V. Consider the kth jet bundle $E_V^{(k)} \to M$ of the basic associated bundle $E_V \to M$. We can naturally view this as a bundle having as fiber a naturally associated algebraic variety $V^{(k)}$ and with structure group $G^{(r+k)}$ which acts on $V^{(k)}$ regularly. The k-jet extension of the structure ω, say $j^k \omega$, defines continuous map $\varphi: M \to V^{(k)}/G^{(r+k)}$ which will be Aut(M, ω)-invariant. The assertion $\varphi(x) = \varphi(y)$ is equivalent to the assertion that Aut$^{(k)}(\omega, x, y)$ is nonempty. It follows from Theorem 2.1 that for k sufficiently large the fibers of φ are precisely the orbits of the pseudogroup Aut$^{\text{loc}}(M, \omega)$. One can also show that these are submanifolds. The points in $V^{(k)}/G^{(r+k)}$ are not necessarily closed, but by basic results about algebraic group actions each point is open in its closure, and every closed set contains a closed point. It follows that there is a closed fiber of φ, i.e., a closed submanifold $N \subset M$ which is an orbit of the pseudogroup Aut$^{\text{loc}}(M, \omega)$. Therefore each $m \in N$ has an open neighborhood in N contained in an orbit of the local flow generated by a local Killing field near m. By Theorem 2.2 and the completeness of the Killing fields (since M is compact), we deduce that each $m \in N$ has an open neighborhood in N contained in the Aut(M, ω)-orbit of m. Therefore, the Aut(M, ω) orbit of m has only finitely many connected components, and so to prove the theorem it suffices to see that all stabilizers in Aut(M, ω) of points in M have finitely many components. However by completeness of Killing fields and Theorem 2.2, every element of Aut$^{\text{loc}}(\omega, m)^0$, the connected component of the identity in Aut$^{\text{loc}}(\omega, m)$, extends to a unique element of the global stabilizer Aut$(M, \omega)_m$. Thus, we have inclusions Aut$^{\text{loc}}(\omega, m)^0 \subset$ Aut$(M, \omega)_m \subset$ Aut$^{\text{loc}}(\omega, m)$. However by Theorem 2.1, we can identify Aut$^{\text{loc}}(\omega, m)$ with Aut$^{(k)}(\omega, m)$ for k sufficiently large. This is an algebraic group, and hence has only finitely many components.

3. Actions on principal bundles, the algebraic hull, and the generalized Borel density theorem

In this section we present some basic information on actions of groups on principal bundles, and in particular discuss the notion of the algebraic hull of such a group action. We use the Borel density theorem to show how to obtain information about this invariant.

Let $P \to M$ be a principal G-bundle and suppose a group H acts on P by principal bundle automorphisms, i.e., it acts on P commuting with the action of G. If G is an algebraic group, the algebraic hull of the action of H on P is (the conjugacy class of) an algebraic subgroup that describes the smallest algebraic subgroup for which P has a measurable H-invariant reduction. This in turn controls the nature of the invariant sections of any associated bundle with an algebraic variety as the fiber. To define the algebraic hull, it is easier to first consider the case in which we have an ergodic measure μ on M for the action of H. The general case can easily be reduced to this case via the ergodic decomposition. (See [Z5] for a discussion in this context.) We also remark that the algebraic hull can be defined for actions on principal bundles where the structure group is (the group of k-points of) an algebraic group over any local field of characteristic 0. While this is useful even for considerations of a differential geometric nature, we shall for simplicity restrict our attention here to real groups.

THEOREM 3.1. *If G is real algebraic and H acts by principal bundle automorphisms of $P \to M$ with a quasi-invariant ergodic measure μ on M, then there is an algebraic subgroup $L \subset G$, unique up to conjugacy, with the following property*:

(i) *There is a measurable H-invariant section of the associated bundle $P/L \to M$, but there is no such invariant section for any proper algebraic subgroup of L.*
Furthermore, L satisfies:

(ii) *If V is any real algebraic variety on which G acts regularly, and there is a measurable H-invariant section φ of the associated bundle $E_V \to M$, then for μ-almost every $m \in M$, the G-orbit in V corresponding to $\varphi(m)$ has each of its stabilizers containing a conjugate of L.*
In particular, if there is an (even measurable) H-invariant reduction of P to an algebraic subgroup G_1, then G_1 contains a conjugate of L.

DEFINITION 3.2. (i) Under the hypotheses of Theorem 3.1, the group L (or more precisely its conjugacy class in G) is called the algebraic hull of the action of H on P.

(ii) If we remove the ergodicity hypothesis, but assume we still have a given quasi-invariant measure (e.g., any smooth measure for a smooth action on a manifold), then by the algebraic hull we will mean the measurable map $M \to \{$Conjugacy classes of algebraic subgroups of $G\}$ obtained from (i) and the ergodic decomposition. (Thus, this map is constant on the ergodic components on M.)

We remark that in (ii), the algebraic hull is well defined as such a measurable map up to equality on conull sets.

For a proof and discussion of Theorem 3.1, see [Z1]. The discussion there also shows that the notion of algebraic hull is more generally defined for actions of groups on measurable principal bundles, and that the algebraic hull is invariant under measurable isomorphism of principal G-bundles on which H acts by principal bundle automorphisms. The following two propositions are clear, but we record them for ease of future use.

PROPOSITION 3.3. *If $P \to M$ is a measurable principal G_1-bundle, $Q \to M$ is a measurable principal G_2-bundle, $\rho: G_1 \to G_2$ is a surjection of algebraic groups, $f: P \to Q$ is a measurable ρ-compatible bundle map (covering the identity), H acts on P, Q by bundle automorphisms commuting with f, and $L_i \subset G_i$ are the respective algebraic hulls, then we can choose L_i (i.e., fix an element in the conjugacy class) such that $\rho(L_1) \subset L_2$ and is a Zariski dense subgroup of finite index. (Over $\mathbb{C}$, it would be a surjection, but here one needs to account for the fact that images of real algebraic groups need only be of finite index in their Zariski closure.)*

PROPOSITION 3.4. *Let $P \to M$ be a principal Γ-bundle where Γ is not necessarily algebraic. (E.g., we shall later need the case of Γ discrete.) Assume H acts by principal bundle automorphisms. Let $\sigma: \Gamma \to G$ be a homomorphism into an algebraic group, and let L be the Zariski closure of the image. Let $Q \to M$ be the associated principal G-bundle, on which we then have a natural induced action of H. Then the algebraic hull of the action of H on Q is contained in L (for almost every point m).*

We now recall the Borel density theorem. We restrict our attention to simple Lie groups, although it can be formulated more generally. Suppose H is a connected simple noncompact real algebraic group. If $L \subset H$ is a closed subgroup and H/L has a finite H-invariant measure, then Borel's theorem asserts that L is Zariski dense in H. (See [Z1] for a proof.) Equivalently, if L is an algebraic subgroup and H/L has a finite invariant measure, then $L = H$. Yet another equivalent formulation is the assertion that if H acts regularly on a variety, any H-invariant probability measure is supported on the fixed points of H. This has the following consequence.

PROPOSITION 3.5. *Let H be as above, and $\rho: H \to G$ a nontrivial rational homomorphism of real algebraic groups. Let H act on M with a finite invariant measure, and let $P = M \times G$ be the product principal G-bundle, on which H acts by the product action. Then the algebraic hull is the Zariski closure of $\rho(H)$, and in particular it contains a group locally isomorphic to H.*

PROOF. An H-invariant section of P/L for an algebraic subgroup $L \subset G$ is simply an H-map $M \to G/L$. Since the invariant measure on M pushes forward to one on G/L, Borel density implies that almost every point in the image is contained in the H-fixed points. Thus, L contains a conjugate of $\rho(H)$.

COROLLARY 3.6. *Suppose H is a simple noncompact Lie group with finite center acting (nontrivially) on a manifold M with a finite invariant smooth measure. Then there is an open dense conull set for which the stabilizers are discrete.*

PROOF. It suffices to see this set is conull, since it is clearly open. If $\mathfrak{h}_m$ is the Lie algebra of the stabilizer of m, then $m \to \mathfrak{h}_m$ is an H-map where H acts on the latter via Ad. Thus, for almost all m, $\mathfrak{h}_m$ is invariant under $\mathrm{Ad}(H)$, and hence by simplicity is either 0 or $\mathfrak{h}$. However, if H fixes a set of positive measure, so does the maximal compact subgroup, and hence by local linearizability for compact groups, the maximal compact subgroup, and hence H, acts trivially.

We now prove Theorem 1.1. A generalization of this argument, which we discuss in the next section, will be a basic step in the proofs of the other main results of §1.

PROOF OF THEOREM 1.1. For each $m \in M$, let $V_m \subset TM_m$ be the tangent space to the H-orbit. We have a natural isomorphism (for a.e. m) $\mathfrak{h} \cong V_m$ in such a way that for $h \in H$, $dh_m|V_m \colon V_m \to V_{hm}$ corresponds to $\mathrm{Ad}(h)$. Thus, by Propositions 3.3, 3.5, the algebraic hull of the action on the first-order frame bundle, say L, contains a group isomorphic to $\mathrm{Ad}(H)$. However, since there is an invariant G-structure, we have $L \subset G$, completing the proof.

4. Application to global Killing fields

In this section we use the techniques of the preceding section, combined with the results of §2, to obtain results on the space of globally defined Killing vector fields on the universal cover of M. As we shall see, this is a key point in relating the group H to $\pi_1(M)$. The development in this section is basically due to Gromov [G2].

There will be two classes of actions on principal bundles that will be most relevant for our considerations. If H acts smoothly on a manifold M, we always have such an action on $P^{(r)}(M) \to M$, and if H is simply connected we always have such an action on $M^{\sim} \to M$ viewed as a principal $\pi_1(M)$-bundle. In the latter case the structure group is of course not algebraic in general; however we obtain bundles with an algebraic structure group by forming the associated principal bundle for any representation $\pi_1(M) \to G$ where G is algebraic.

For the first class of bundles, we immediately see the relevance of the algebraic hull for obtaining further information on the automorphism group.

PROPOSITION 4.1. *Suppose H is any group acting on M preserving a structure ω of algebraic type on M. For every $k \geq 1$, let $L_k \subset G^{(k)}$ be the algebraic hull for the action on $P^{(k)}(M)$. Then for almost every $m \in M$, and every $k \geq 1$, $\mathrm{Aut}^{(k)}(\omega, m)$ contains a group isomorphic to L_k. Therefore in* *the rigid, analytic case, for k sufficiently large* $\mathrm{Aut}^{\mathrm{loc}}(\omega, m)$ *contains a group isomorphic to* L_k. *Thus if* $\tilde{m} \in \tilde{M}$ *projects to* $m \in M$, $\mathrm{Kill}(\tilde{M}, \tilde{\omega})_{\tilde{m}}$, *the globally defined Killing fields on* $\tilde{M}$ *for the lifted structure* $\tilde{\omega}$ *which vanish at* $\tilde{m}$, *contains a Lie algebra isomorphic to* $\mathfrak{L}_k$, *the Lie algebra of* L_k.

PROOF. By definition, the algebraic hull fixes the prolongation of ω to order k, and the corresponding subgroup of $\mathrm{Aut}^{(k)}(\omega, m)$ is simply obtained by conjugating from $(\mathbb{R}^n, 0)$ to (M, m). The last assertions then follow from the results of §2.

We now assume H to be a noncompact simple Lie group with finite center, and show how the results of §3 can be used to improve Proposition 4.1 for actions of these groups. There will be two essential improvements. First, we will be able to assert that the algebraic hulls L_k must each contain a group locally isomorphic to H. This of course immediately gives more information in 4.1. However, as in the proof of Theorem 1.1 in §3, we shall also see that the algebraic hull must be well behaved with respect to the tangent spaces (and higher order tangent spaces) to the orbits of the action of H.

THEOREM 4.2. *Suppose H is a noncompact simple Lie group with finite center acting (nontrivially) on M preserving a unimodular structure ω of algebraic type on M, where M is compact. Then:*

(1) *For almost every $m \in M$, and every $k \geq 1$, the algebraic hull L_k as well as $\mathrm{Aut}^{(k)}(\omega, m)$ contains a group locally isomorphic to H.*

(2) *Suppose further that ω is also analytic and rigid. Then:*

(i) *for almost every m and for k sufficiently large $\mathrm{Aut}^{\mathrm{loc}}(\omega, m)$ contains a group locally isomorphic to H.*

(ii) *Identify the Lie algebra $\mathfrak{h}$ of H with a Lie algebra of globally defined Killing fields on $\tilde{M}$ via the H action on M. For k sufficiently large, and almost every $m \in M$, let L be the algebraic hull at m (i.e., of the ergodic component containing m) of the action on $P^{(k)}(M)$. Suppose $\tilde{m} \in \tilde{M}$ projects to m. Then there is a Lie algebra $\mathfrak{L}$ of globally defined Killing fields on $\tilde{M}$ (depending on $\tilde{m}$) with the following properties:*

(a) $\mathfrak{L}$ *is isomorphic with the Lie algebra of L.*

(b) *All elements of $\mathfrak{L}$ vanish at $\tilde{m}$.*

(c) $\mathfrak{L}$ *normalizes $\mathfrak{h}$, i.e., $[\mathfrak{L}, \mathfrak{h}] \subset \mathfrak{h}$. Furthermore, the associated map $\mathfrak{L} \to \mathrm{Der}(\mathfrak{h})$ is in fact a surjection onto $\mathrm{ad}(\mathfrak{h})$.*

PROOF. (1) follows from 4.1 and the proof of Theorem 1.1 in §3. (2.i) then follows from Theorem 2.1. We take $\mathfrak{L}$ to be the local Lie algebra corresponding to $L \subset \mathrm{Aut}^{\mathrm{loc}}(\omega, m)$, each element of which extends uniquely to a globally defined Killing field by Theorem 2.2. To see (2.ii.c), we argue as in the proof of Theorem 1.1 in §3. Namely, let $T^{(k)}(M)$ be the bundle of k-jets of vector fields. This is an associated bundle to the principal bundle $P^{(k)}(M) \to M$. For each $m \in M$, we let $V_m \subset T^{(k)}(M)$ be the space of

k-jets of vector fields corresponding to the action of $\mathfrak{h}$ on M. We have for almost every m a natural identification $\mathfrak{h} \cong V_m$ in such a way that for $h \in H$, $d^{(k)} h_m | V_m \colon V_m \to V_{hm}$ corresponds to $\mathrm{Ad}(h)$ It follows that the algebraic hull L bay be chosen so that it also leaves V_m invariant and (by Proposition 3.4), if L^0 is the connected component of L, then $L^0 | V_m$ will act by $\mathrm{Ad}(H)$. Since local or global Killing fields are uniquely determined by their k-jets at a point for k sufficiently large, it is clear that this implies assertion (2.ii.c).

Theorem 4.2 shows that the existence of a significant collection of global Killing fields on $M^\sim$ that do not come directly from the original action on M. As a result, we have

COROLLARY 4.3. *Suppose H is a noncompact simple Lie group with finite center acting (nontrivially) on a compact manifold M preserving a unimodular, rigid, analytic, structure ω of algebraic type. Identify the Lie algebra $\mathfrak{h}$ of H with a Lie algebra of globally defined Killing fields on $M^\sim$ via the action of H on M. Let $\mathfrak{z}$ be the centralizer of $\mathfrak{h}$ in the Lie algebra of all globally defined Killing fields on $M^\sim$. For $x \in M^\sim$, let $\mathfrak{z}(x)$ and $\mathfrak{h}(x)$ be the image of $\mathfrak{z}$ and $\mathfrak{h}$ respectively under the evaluation map at x. Then for almost all x, $\mathfrak{z}(x) \supset \mathfrak{h}(x)$.*

PROOF. Let $x \in M^\sim$ satisfy the conclusions of Theorem 4.2. Let L be the local Lie group of diffeomorphisms fixing x corresponding to $\mathfrak{L}$. For $h \in H$ sufficiently small, we can choose a small $g \in L$ such that g acts on $\mathfrak{h}$ by $\mathrm{Ad}(h^{-1})$. It follows that $h \circ g$ is a locally defined diffeomorphism near x that acts trivially on $\mathfrak{h}$, and hence commutes with H. Furthermore, $h \circ g(x) = h(x)$. It follows that if we let Z be the local group at x corresponding to $\mathfrak{z}$, then the local Z-orbit at x contains an open set in the local H-orbit of x. Hence, $\mathfrak{z}(x) \supset \mathfrak{h}(x)$.

While both 4.2 and 4.3 are of interest in terms of establishing the existence of new automorphisms and Killing fields, we now show how 4.3 can be used to obtain information on $\pi_1(M)$, and in particular, how it easily implies Theorem 1.5.

We first make the following simple (essentially tautological) remark which allows the connection of the above results to the fundamental group.

PROPOSITION 4.4. *Let M be a smooth manifold. Suppose V is a finite-dimensional vector space consisting of vector fields on $M^\sim$ and suppose that V is invariant under the natural $\pi_1(M)$-action on $M^\sim$. Let $E_V \to M$ be the associated bundle with fiber V. Then the evaluation map $M^\sim \times V \to TM^\sim$ factors to a vector bundle map $E_V = (M^\sim \times V)/\pi_1(M) \to (TM^\sim)/\pi_1(M) = TM$. If H is a connected, simply connected group acting smoothly on M, and each element of V is fixed under the induced action on $M^\sim$, then this map $E_V \to TM$ also commutes with the H-actions.*

We now prove Theorem 1.5.

PROOF OF THEOREM 1.5. Choose a point $x \in M^{\sim}$ satisfying the conclusions of 4.3. Choose V in Proposition 4.4 to be the Lie algebra $\mathfrak{z}$ in 4.3. Since H commutes with $\pi_1(M)$ on $M^{\sim}$, and $\mathfrak{z}$ is the centralizer of $\mathfrak{h}$, it follows that $\mathfrak{z}$ is $\pi_1(M)$-invariant. By Corollary 4.3, the image of $E_{\mathfrak{z}} \to TM$ contains the (measurable) subbundle consisting of the tangent space to the H-orbits. The algebraic hull for the action on this subbundle contains $\mathrm{Ad}(H)$ by Proposition 3.5, and hence so does the algebraic hull of the action on $E_{\mathfrak{z}}$, by Proposition 3.3. However by Proposition 3.4, this algebraic hull in turn is contained in the Zariski closure of the image of $\pi_1(M)$ under the representation $\pi_1(M) \to \mathrm{GL}(\mathfrak{z})$. Thus $\pi_1(M) \to \mathrm{GL}(\mathfrak{z})$ is the sought after representation.

The above argument also establishes the following result we shall need later.

COROLLARY 4.5. *Suppose H is a noncompact simple Lie group with finite center acting on a compact manifold M preserving a unimodular, rigid, analytic structure ω of algebraic type on M. Assume further that $\pi_1(H)$ is finite. Then for almost every $x \in M^{\sim}$, the $H^{\sim}$ orbit of x is proper; i.e., the stabilizer $H_x^{\sim}$ is compact and the map $H^{\sim} \to M^{\sim}$, $h \to hx$, induces a homeomorphism $H^{\sim}/H_x^{\sim} \to H^{\sim}x$.*

PROOF. If we let U be the open conull set given by Theorem 3.6, then it is clear that every orbit in $P(M)$ that projects to a point in U satisfies this properness condition, since H acts on the tangent space to the orbit foliation by Ad, and the latter is a surjection with finite kernel onto a closed subgroup. By 4.3 and 4.4, the same will be true for the frame bundle of the vector bundle $E_{\mathfrak{z}} \to M$. Since the latter is an associated bundle to the universal cover via a representation $\pi_1(M) \to \mathrm{GL}(\mathfrak{z})$, is is easy to check that the required properness condition holds on $M^{\sim}$. ∎

5. Applications of superrigidity

The results of §§3 and 4 apply to any noncompact, connected, simple Lie group H of finite center. Under the additional assumption that $\mathbb{R}\text{-}\mathrm{rank}(H) \geq 2$, we shall obtain further results. The key additional fact that we obtain from this assumption is superrigidity, which yields a nearly complete description of the measure-theoretic structure of H-actions on principal bundles with algebraic structure group. This result, first proved in [Z2], is a generalization of Margulis's superrigidity theorem, which is essentially equivalent to the special case in which the base manifold M is H/Γ, where Γ is a lattice subgroup of H. See [Z1] for a discussion of both of these versions of superrigidity and [Z4] for some other geometric applications.

If $P \to M$ is a principal G-bundle on which H acts by automorphisms, a measurable section of P (which always exists) yields a measurable trivialization $P \cong M \times G$. The action of H on P is then given by $h \cdot (m, g) = (hm, \alpha(h, m)g)$, where $\alpha: H \times M \to G$ satisfies the cocycle identity. The superrigidity theorem asserts that, under suitable hypotheses, one can choose the measurable trivialization such that the cocycle α is of a particularly transparent form.

THEOREM 5.1 (Superrigidity [Z1, Z8]). *Let* $P \to M$ *be a principal G-bundle where G is a real algebraic group and M is compact. Let H be a connected, simple Lie group with* $\mathbb{R}\text{-rank}(H) \geq 2$, *and suppose H acts by principal bundle automorphisms of P, preserving a finite measure on M. Lifting the action to $H^{\sim}$, we may assume H is also simply connected.*

(a) *Assume the H-action on M is ergodic and the algebraic hull of the action on P is (Zariski) connected. Then one may choose the measurable trivialization $P \cong M \times G$ so that the corresponding cocycle α has the following description. There is a continuous homomorphism $\theta : H \to G$ and a compact subgroup $K \subset G$ commuting with $\theta(H)$ such that $\alpha(h, m) = \theta(h)c(h, m)$ where $c(h, m) \in K$.*

(b) *Without the hypothesis of ergodicity, one may apply (a) to each ergodic component. Without the hypothesis that the algebraic hull is connected, one may obtain the same conclusion by pulling back P over a finite (measurable) extension of M (for each ergodic component).*

Assertion (a) is of course the central one, and (b) is basically one of technical adjustments that rarely cause a serious problem. We shall basically ignore these technicalities in the sequel. For a proof in the case the algebraic hull is semisimple, see [Z1]. For the extension to the form with the above hypotheses, see [Z8].

We shall also need a basic theorem of Ratner, which verified a conjecture of Raghunathan.

THEOREM 5.2 (Ratner [R]). *Let G be a connected Lie group, $\Gamma \subset G$ a discrete subgroup, and $H \subset G$ a connected simple noncompact Lie group. Let ν be a finite H-invariant ergodic measure on G/Γ, where the action of H is given by the embedding in G. Then there is a closed connected subgroup L with $H \subset L \subset G$ and a point $x \in G/\Gamma$, say with stabilizer in G being $g\Gamma g^{-1}$, such that:*

(i) $L \cap g\Gamma g^{-1}$ *is a lattice subgroup in L; and*

(ii) ν *is the measure on G/Γ corresponding to the L-invariant volume on $L/L \cap g\Gamma g^{-1}$ under the natural bijection $L/L \cap g\Gamma g^{-1} \cong Lx \subset G/\Gamma$.*

We now indicate the proof of Theorem 1.7.

PROOF OF THEOREM 1.7 (sketch). We consider first the easier case in which there is some faithful representation $\sigma : \pi_1(M) \to \mathrm{GL}(q, \mathbb{C})$ with discrete image. We let $\mathrm{GL}(q, \mathbb{C}) = G$, and $\Gamma = \pi_1(M)$ which we also identify with its image in G. Fixing a measurable trivialization of the Γ-bundle $M^{\sim} \to M$ yields a measurable trivialization of the associated principal G-bundle for which the associated cocycle $\beta : H \times M \to G$ actually takes values in Γ. Applying Theorem 5.1 (and ignoring the technicalities of 5.1.(b)), we can choose another measurable trivialization so that the corresponding cocycle α

satisfies the conclusions of 5.1. Since any two measurable trivializations of a principal bundle differ by a measurable map from the base to the structure group, α and β will differ by such a function. I.e., there is a measurable $\varphi: M \to G$ such that $\varphi(hm)\beta(h, m)\varphi(m)^{-1} = \alpha(h, m) = c(h, m)\theta(h)$, where c and θ are as in 5.1. There are now two cases to consider.

If θ is trivial, then we can write $\beta(h, m) = \varphi(hm)^{-1}c(h, m)\varphi(m)$. This implies that there is some set $X \subset M$ of positive measure and a compact set $A \subset G$ such that whenever $m, hm \in X$ we have $\beta(h, m) \in A$. Since there is a finite H-invariant measure, Poincaré recurrence implies that for almost every $m \in X$, we can find a sequence $h_n \in H$, $h_n \to \infty$, so that $h_n m \in X$. However, since the image of Γ in G is discrete, this contradicts the properness assertion of Corollary 4.5.

Suppose now that θ is not trivial. Then we have $c(h, m)^{-1}\varphi(hm)\beta(h, m)$ $= \theta(h)\varphi(m)$. Letting $\lambda: M \to K\backslash G/\Gamma$ be the composition of φ with the natural projection onto the double coset space, we see that λ is an H-map. Thus, pushing the measure on M forward by λ, there is a $\theta(H)$-invariant finite measure on $K\backslash G/\Gamma$. Lifting by Haar measure of K, there is a $\theta(H)$-invariant finite measure on G/Γ, and the conclusion of theorem now follows by applying Theorem 5.2.

The proof in case (ii) in the statement of Theorem 1.7, i.e., when we assume there is a faithful (but not necessarily discrete) homomorphism into $GL(q, \mathbb{Q}^-)$, is similar but a bit more involved. The main point is that any finitely generated subgroup of $GL(q, \mathbb{Q}^-)$ has a natural realization as a discrete subgroup of a product of finitely many Lie groups with finitely many algebraic groups over p-adic fields. One can prove Theorem 5.1 for bundles with p-adic structure group and then use the same sort of argument. See [Z7] for details.

We now consider the proof of Theorem 1.10. Once again, we provide only an indication of the proof.

PROOF OF THEOREM 1.10 (sketch). Consider the proof of Theorem 1.7 above where we consider the special case in which $\Gamma = \pi_1(M)$ is a lattice in a Lie group G (which we now take only to be a subgroup of some $GL(q)$). Suppose Γ has the following property, to which we will return in moment:

(*)   If $L \subset G$ is a proper connected subgroup which intersects Γ in a lattice in L, then L is solvable.

If this is the case, the measure ν in the proof of 1.7 must be the standard G-invariant measure on G/Γ. This implies that we have a measure-preserving H-map $M \to K\backslash G/\Gamma$. Therefore, for each $h \in H$, the (measure-theoretic) entropy of h acting on M is at least that of h acting on $K\backslash G/\Gamma$. The latter is explicitly computable in terms of eigenvalues of $\mathrm{Ad}_G(h)$. In other words, if Γ satisfies property (*) above, we obtain a lower bound on possible values of entropy for H acting on M with this given fundamental group. On the other hand, the values of entropy on M can be read off from the exponents of the action, which by superrigidity can be described in terms of the eigenvalues of $\theta(h)$, where θ is as in 5.1. If the dimension of M is bounded above, this explicitly bounds the entropy from above. Comparing these bounds we see that if the dimension of M is too small, the bounds are incompatible, and hence there is no such action. Finally, that there exist cocompact lattices satisfying $(*)$ in $SL(p, \mathbb{R})$ for all primes p is an arithmetic theorem of Kottwitz. See [Z7] for a proof of Kottwitz's theorem and further details of the above proof.

With further dynamical assumptions on the H-action, we obtain further information on $\pi_1(M)$. Namely, let us call an H-action an Anosov action if there is some $\mathbb{R}$-split semisimple element $h \in H$ (i.e., $\mathrm{Ad}(h)$ is diagonalizable over $\mathbb{R}$) whose action on M is hyperbolic normal to the H-orbits. With the assumption of a smooth invariant measure, any locally free Anosov action will be ergodic, as will its lift to any finite cover of M. This condition, which is called "engaging" in [Z6], enables one to prove

THEOREM 5.3 [Z6]. *Let H be a connected simple Lie group with $\mathbb{R}$-rank(H) ≥ 2. Suppose there is an engaging (e.g. Anosov) action of H on a compact M preserving a unimodular structure. Let d be the smallest dimension of a nontrivial representation of the Lie algebra $\mathfrak{h} \to \mathfrak{gl}(d, \mathbb{C})$. Then any representation $\pi_1(M) \to GL(r, \mathbb{C})$ where $r < d$ has finite image.*

The proof uses superrigidity in a spirit similar to the proof of Theorem 1.7 above. See [Z6] for details.

6. Actions of discrete groups

If M is a simply connected compact manifold with a rigid, unimodular, analytic structure ω of algebraic type, and we have an action $\Gamma \to \mathrm{Aut}^{\mathrm{an}}(M, \omega)$ where Γ is a discrete group, we may sometimes apply Theorem 2.4 and results about the finite-dimensional representation theory of Γ to obtain information about the action.

THEOREM 6.1. *With the assumptions as above, assume $\Gamma \subset H$ is a lattice subgroup where H is a connected, simple Lie group of finite center, finite fundamental group, and $\mathbb{R}$-rank$(H) \geq 2$. Then Γ preserves a Riemannian metric on M, or equivalently, the image of Γ in $\mathrm{Aut}^{\mathrm{an}}(M, \omega)$ is contained in a compact subgroup. In particular, if $\Gamma = SL(n, \mathbb{Z})$, $n \geq 3$, then the action factors through a finite quotient of Γ. (This latter assertion is more generally true if H is the real points of a simple $\mathbb{Q}$-group with $\mathbb{Q}$-rank$(H) \geq 2$, and Γ is a corresponding arithmetic subgroup of integral points.)*

PROOF. By Theorem 2.4 $\mathrm{Aut}^{\mathrm{an}}(M, \omega)$ is almost connected. By Margulis's superrigidity theorem [M, Z1], if the image of Γ is not compact, $\mathrm{Aut}^{\mathrm{an}}(M, \omega)$ contains a group locally isomorphic to H. But this is impossible by Theorem 1.5 since M is simply connected.

We now consider the nonsimply connected case, where one can no longer appeal to Theorem 2.4. If Γ acts on M, we have a natural induced homomorphism $\Gamma \to \text{Out}(\pi_1(M))$. In certain cases, we can generalize the results for connected group actions to the discrete case if we assume triviality of this homomorphism. An example is Theorem 1.11 and we now turn to the proof of this result. We need the following general fact.

PROPOSITION 6.2. *Suppose a group Γ acts on a compact manifold M preserving a rigid structure ω and a finite ergodic measure. Let L_k be the algebraic hull for the action on the bundle $P^{(k)}(M) \to M$. If there is a sufficiently large k for which L_k is compact, then the image of Γ in $\text{Aut}(M, \omega)$ has compact closure.*

PROOF. Since there is a measurable Γ-invariant reduction of $P^{(k)}(M)$ to L_k, there is a finite Γ-invariant measure on $P^{(k)}(M)$. However for k sufficiently large, since ω is rigid, $\text{Aut}(M, \omega)$ (and hence any closed subgroup) acts properly on $P^{(k)}(M)$. However, it is easy to see that a group that acts properly on a locally compact space preserving a finite measure must be compact.

PROOF OF THEOREM 1.11. Since $H^2(\Gamma, \mathbb{R}) = 0$ for all lattices in $\text{SL}(n, \mathbb{R})$ for $n \geq 6$, [B], and every homomorphism from Γ into an abelian group has finite image [Z1], every element of $H^2(\Gamma, \mathbb{Z})$ is trivial on a subgroup of finite index. If we let Λ be the group of diffeomorphisms of $M^{\sim}$ covering elements of Γ, we have an exact sequence $0 \to \pi_1(M) \to \Lambda \to \Gamma \to 0$. A straightforward inductive argument then shows that by passing to a subgroup of Γ of finite index, this is a split extension. Therefore, replacing Γ by a subgroup of finite index, we can assume the action lifts to $M^{\sim}$ in such a way as to commute with $\pi_1(M)$. This in turn defines a cocycle $\Gamma \times M \to \pi_1(M)$, which since Γ has Kazhdan's property T and $\pi_1(M)$ is nilpotent must be equivalent to a cocycle taking values in a finite subgroup of $\pi_1(M)$. (See [Z1, Theorem 9.1.1] for a proof.) This implies that the ergodic components of the Γ action on $M^{\sim}$ are actually finite Γ-invariant measures, and in fact each one is simply the smooth measure on $M^{\sim}$ restricted to a Γ-invariant set of positive smooth measure.

Since every homomorphism of Γ into a compact Lie group has finite image (and finite groups cannot act ergodically on a manifold of positive dimension), we deduce from 6.2 that for large k we must have the algebraic hull L_k being nonfinite for every ergodic component. It follows from Proposition 4.1 that the space of global Killing fields on $M^{\sim}$, say W, is nonzero. In fact, if for $x \in M^{\sim}$ we let $W_x = \{X \in W | X(x) = 0\}$, then 4.1 shows that for almost all x we have $W_x \neq 0$. We have a natural representation $\Gamma \to \text{GL}(W)$ and by Margulis's superrigidity theorem the algebraic hull of the image will be locally isomorphic to $\text{SL}(n, \mathbb{R})$. The map $x \to W_x$ is a Γ-map $\varphi: M^{\sim} \to \text{Grass}(W)$. Since each ergodic component of Γ on $M^{\sim}$ has finite measure, by the Borel density theorem the image of φ must be supported on $\text{SL}(n, \mathbb{R})$ fixed points in $\text{Grass}(W)$. Hence, φ is essentially constant on ergodic components. I.e., for almost all x, any $X \in W_x$ must vanish on the ergodic component through x. However, each ergodic component has positive measure with respect to the smooth measure, and this implies that $X = 0$, providing a contradiction.

Proposition 6.2 can also be used to prove the following result. See [Z4] for a complete proof and discussion of related results.

THEOREM 6.3 [Z4]. *Let H be a connected simple Lie group with $\mathbb{R}$-rank$(H) \geq 2$, and let $\Gamma \subset H$ be a lattice subgroup. Let M be a compact manifold and ω a rigid unimodular G-structure of algebraic type on M. Suppose there is an action $\Gamma \to \operatorname{Aut}(M, \omega)$. Then either:*

(i) *There is an embedding of Lie algebras $\mathfrak{h} \to \mathfrak{g}$; or*

(ii) *The image of Γ is contained in a compact subgroup of $\operatorname{Aut}(M, \omega)$.*

COROLLARY 6.4 [Z4]. *Let $\Gamma \subset \operatorname{SL}(n, \mathbb{R})$ be a lattice, $n \geq 3$. Let M be a compact manifold with $\dim(M) < n$, and ω a rigid unimodular G-structure of algebraic type on M. Then any action $\Gamma \to \operatorname{Aut}(M, \omega)$ factors through a finite quotient of Γ.*

REFERENCES

[A] A. M. Amores, *Vector fields of a finite type G-structure*, J. Differential Geom. **14** (1979), 1–6.

[B] A. Borel, *Stable real cohomology of arithmetic groups*, Ann. Sci. École Norm. Sup. **7** (1974), 235–272.

[C] E. Cartan, *Oevres completes*, Gauthier-Villars, Paris, 1952–1955.

[D'A] G. D'Ambra, *Isometry groups of Lorentz manifolds*, Invent. Math. **92** (1988), 555–565.

[G1] M. Gromov, *Partial differential relations*, Springer, New York, 1986.

[G2] ____, *Rigid transformation groups*, Geometrie Differentielle (D. Bernard and Y. Choquet-Bruhat, editors), Hermann, Paris, 1988.

[GS] V. Guillemin and S. Sternberg, *Deformation theory of pseudogroup structures*, Mem. Amer. Math. Soc. No. 64, 1966.

[M] G. A. Margulis, *Discrete subgroups of Lie groups*, Springer, New York, 1991.

[Mo] C. C. Moore, *Amenable subgroups of semisimple groups an proximal flows*, Israel J. Math. **34** (1979), 121–138.

[N] K. Nomizu, *On local and global existence of Killing fields*, Ann. of Math. (2) **72** (1970), 105–112.

[R] M. Ratner, *On Raghunathan's measure conjecture*, Ann. of Math. **134** (1991), 545–607.

[S] I. M. Singer, *Infinitesimally homogeneous spaces*, Comm. Pure Appl. Math. **13** (1960), 685–697.

[SZ] R. Spatzier and R. J. Zimmer, *Fundamental groups of negatively curved manifolds and actions of semisimple groups*, Topology **30** (1991), 591–601.

[Z1] R. J. Zimmer, *Ergodic theory and semisimple groups*, Birkhäuser, Boston, 1984.

[Z2] ____, *Strong rigidity for ergodic actions of semisimple Lie groups*. Ann. of Math. (2) **112** (1980), 511–529.

[Z3] ____, *On the automorphism group of a compact Lorentz manifold and other geometric manifolds*, Invent. Math **83** (1986), 411–426.

[Z4] ____, *Lattices in semisimple groups and invariant geometric structures on compact manifolds*, Discrete Groups in Geometry and Analysis (R. Howe, editor), Birkhäuser, Boston, 1987, pp. 152–210.

[Z5] ____, *Ergodic theory and the automorphism group of a G-structure*, Group Representations, Ergodic Theory, Operator Algebras, and Mathematical Physics (C. C. Moore, editor), Springer, New York, 1987, 247–278.

710   R. J. ZIMMER

[Z6] ____, *Representations of fundamental groups of manifolds with a semisimple transformation group*, J. Amer. Math. Soc. **2** (1989), 201–213.

[Z7] ____, *Superrigidity, Ratner's theorem, and fundamental groups*, Israel J. Math. **74** (1991), 199–207.

[Z8] ____, *On the algebraic hull of the automorphism group of a principal bundle*, Comment. Math. Helv. **65** (1990), 375–387.

UNIVERSITY OF CHICAGO

JOURNAL OF THE
AMERICAN MATHEMATICAL SOCIETY
Volume 7, Number 1, January 1994

DISCRETE GROUPS
AND NON-RIEMANNIAN HOMOGENEOUS SPACES

ROBERT J. ZIMMER

I. Introduction

A basic question in geometry is to understand compact locally homogeneous manifolds, i.e., those compact manifolds that can be locally modelled on a homogeneous space $J\backslash H$ of a finite-dimensional Lie group H. This means that there is an atlas on a manifold M consisting of local diffeomorphisms with open sets in $J\backslash H$ where the transition functions between these open sets are given by translations by elements of H. A basic example is the case of $M = J\backslash H/\Gamma$ where $\Gamma \subset H$ is a discrete subgroup acting freely and properly discontinuously on J/H, with a compact quotient. (These examples can be abstractly characterized as the "complete" case. See [G].) We then call Γ a cocompact lattice on $J\backslash H$. When J is compact, $J\backslash H$ has an H-invariant Riemannian metric, so that M itself becomes a Riemannian manifold in a natural way. Thus, the class of locally homogeneous Riemannian manifolds includes (but is not exhausted by) the locally symmetric spaces. For $H = \mathrm{Aff}(\mathbb{R}^n)$ and $J = \mathrm{GL}(n, \mathbb{R})$, the spaces $J\backslash H/\Gamma$ are the complete affinely flat manifolds, a non-Riemannian example which has been extensively studied.

When H is simple and J is not compact, the situation is far from being well understood. Some special series of such homogeneous spaces admitting a cocompact lattice have been constructed by Kulkarni [Ku] and by T. Kobayashi [K]. For example, Kulkarni shows the existence of such a discrete group Γ for the homogeneous spaces $\mathrm{SO}(1, 2n)\backslash \mathrm{SO}(2, 2n)$ and Kobayshi does the same for $\mathrm{U}(1, n)\backslash \mathrm{SO}(2, 2n)$. On the other hand, the constructions they give for these (and other similar) series are quite special, and there has been the general suspicion that most homogeneous spaces $J\backslash H$ with H simple and J not compact do not admit a cocompact lattice. (We recall that, when J is compact, i.e. the Riemannian case, such Γ always exist by a result of Borel [B].) One general situation in which it is known that no such Γ exists is the case in which $J\backslash H$ is noncompact and of reductive type and $\mathbb{R}\text{-}\mathrm{rank}(H) = \mathbb{R}\text{-}\mathrm{rank}(J)$. This situation was studied by Calabi-Markus [CM], Wolf [W], and Kobayashi [K], and the nonexistence of Γ in this situation is a special case of a more general result known as the Calabi-Markus phenomenon. Kobayashi and Ono [KO] use the Hirzebruch proportionality principle to show that some other special series do

Received by the editors September 28, 1992.
1991 *Mathematics Subject Classification.* Primary 20H15, 22E40, 53C15, 53C30.
Research partially supported by NSF grant DMS 9107285.

not admit cocompact lattices. Recently Benoist and Labourie [BL] have shown that if H is simple and the center $Z(J)$ contains a nontrivial $\mathbb{R}$-split one-parameter subgroup then no such Γ exists. Their result in fact applies more generally to show that there is no compact M locally modelled on $J \backslash H$.

These results are quite special of course and, in particular, give no insight into the question whether such Γ exist for the basic test case of $\mathrm{SL}(p, \mathbb{R}) \backslash \mathrm{SL}(n, \mathbb{R})$ where $p < n$ and the embedding is the standard one. The goal of this paper is to prove a (still special) result that clarifies the situation for the above examples in case $p < n/2$ and for many other natual examples as well. Here is the main result.

Theorem 1.1. *Let* H *be a real algebraic group and* $J \subset H$ *an algebraic subgroup. Suppose:*

(i) H *and* J *are both unimodular;*

(ii) *There is a discrete group* $\Gamma \subset H$ *that is a cocompact lattice for* $J \backslash H$; *and*

(iii) *The centralizer* $Z_H(J)$ *of* J *in* H *contains a subgroup* G *with the following properties:*

(a) G *is not contained in a proper normal subgroup of* H,

(b) G *is a semisimple Lie group each of whose simple factors has* $\mathbb{R}$*-rank at least* 2,

(c) *every local homomorphism of* G *to* J *is trivial.*

Then J *is compact.*

Corollary 1.2. *Let* H *be an almost simple real algebraic group. Suppose* $J \subset H$ *is a unimodular subgroup such that* $Z_H(J)$ *contains a simple Lie group of* $\mathbb{R}$*-rank at least* 2 *that does not have a local embedding in* J. *If* $J \backslash H$ *admits a cocompact lattice, then* J *is compact.*

Corollary 1.3. *If* $2 \leq p < n/2$, $n > 4$, *then* $\mathrm{SL}(p, \mathbb{R}) \backslash \mathrm{SL}(n, \mathbb{R})$ *does not admit a cocompact lattice. The same is true for* $J \backslash \mathrm{SL}(n, \mathbb{R})$ *where* J *is any unimodular noncompact subgroup of* $\mathrm{SL}(p, \mathbb{R})$.

The remainder of the paper is devoted largely to the proof of Theorem 1.1. The proof involves in a crucial way results and arguments of algebraic ergodic theory, using in particular superrigidity for actions on principal bundles [Z1] and Ratner's theorem showing the validity of Raghunathan's measure conjecture [R]. Some parts of the argument are closely related to the arguments of [Z5].

We conclude this introductory section with some remarks and questions.

Questions. (1) With $J \backslash H$ as in Theorem 1.1 and J noncompact, do there exist compact manifolds locally modelled on $J \backslash H$? While some of the ideas in the proof of Theorem 1.1 can be applied in this more general situation, more work is required to resolve this general case.

(2) Consider for simplicity the case of H simple and $J \subset H$ unimodular. It is natural to ask if Corollary 1.2 remains true under weaker hypotheses on $Z_H(J)$. The proof of Theorem 1.1 and the results of [CZ] show that the conclusion of Corollary 1.2 still remains true if we only assume $Z_H(J)$ contains a noncompact simple group with Kazhdan's property, e.g., $\mathrm{Sp}(1, n)$. On the other hand, there are no known examples of $J \backslash H$ admitting a compact lattice where H is simple, J is noncompact and unimodular, and $Z_H(J)$ contains

$GL(2, \mathbb{R})$. It would, of course, be of interest to know if the conclusion of Corollary 1.2 remains true in this case as well.

(3) Are there cocompact lattices for $SL(n - 2, \mathbb{R}) \backslash SL(n, \mathbb{R})$ or $SL(n - 1, \mathbb{R}) \backslash SL(n, \mathbb{R})$? Even the case $SL(2, \mathbb{R}) \backslash SL(3, \mathbb{R})$ remains open.

The author's interest in the general issues this paper deals with was stimulated by the paper of Benoist-Labourie and many conversations with Labourie. We would like to thank Y. Benoist, P. Foulon, and F. Labourie for some very useful conversations that helped clarify some important issues involved in the main result. Special thanks are also due to S. G. Dani with whom I spent many productive hours discussing the arguments of the proof, and to F. Labourie for pointing out a gap in the original version of this paper. Much of the work on this paper was completed at the Institut des Hautes Études Scientifiques which we thank for its hospitality.

2. Preliminaries

In this section we provide one simple background result we shall need and briefly recall some standard notions we shall use freely.

Proposition 2.1. *Let H be a unimodular Lie group, $J \subset H$ a closed unimodular subroup, and $\Gamma \subset H$ a discrete subgroup that is a cocompact lattice for $J \backslash H$. Then there is a volume form on the compact manifold $M = J \backslash H / \Gamma$ that is invariant under the natural left action of $Z_H(J)$ on M.*

Proof. Since both H and J are unimodular, there is an H-invariant volume form ω on $J \backslash H$ that is unique up to a constant multiple. If $g \in Z_H(J)$, then g acts on $J \backslash H$ on the left commuting with the right action of H. Since $g^* \omega$ is then also H-invariant, it must be a scalar multiple of ω, say $g^* \omega = c(g) \omega$. Since ω is Γ-invariant, it factors to a volume form ω_0, and on M we then clearly also have $g^* \omega_0 = c(g) \omega_0$. Since M is compact, ω_0 defines a finite measure μ on M and $g_* \mu = c(g) \mu$. Since the total measure is preserved, $c(g) = 1$, and ω_0 is $Z_H(J)$-invariant.

We shall deal extensively with actions of a Lie group by automorphisms of a principal bundle. We recall some standard notation regarding this situation. Let M be a manifold and $P \to M$ a principal H-bundle where H is a Lie group, and suppose G acts on P by principal bundle automorphisms. If V is any left H-space, we can form the associated bundle $E_V = (P \times V)/H \to P/H = M$ with V as fiber, and G acts by bundle automorphisms of $E_V \to M$. Any measurable section s of $P \to M$ (which always exists) defines a measurable trivialization $P \cong M \times H \to M$. The G action on $M \times H$ corresponding to the action on P under this trivialization will be given by $g \cdot (m, h) = (gm, \alpha(g, m)h)$ where $\alpha : G \times M \to H$ is a measurable cocycle. We recall that two cocycles $\alpha, \beta : G \times M \to H$ are equivalent if there is a measurable $\varphi : M \to H$ such that $\varphi(gm)^{-1} \alpha(g, m) \varphi(m) = \beta(g, m)$. Different trivializations of P yield cocycles equivalent to α, and, conversely, any cocycle equivalent to α comes from a trvialization. There are a large number of results concerning the structure of measurable cocycles or, equivalently, the measurable structure of actions on principal bundles. In particular, the superrigidity theorem for cocycles [Z1–Z3] is such a result, and there are several such results related to amenability and Kazhdan's property [Z7]. The notion of algebraic hull of a cocycle (or of an action on a principal bundle) is a basic invariant. We refer the reader to [Z1, Z4, Z9] for background information on all these and related issues.

3. PROOF OF MAIN THEOREM

Let H be a connected simple Lie group with finite center and $J \subset H$ an algebraic unimodular subgroup. We can assume J is connected. Let $\Gamma \subset H$ be a discrete group, which we assume acts freely and properly discontinuously on $J \backslash H$ such that $M = J \backslash H / \Gamma$ is compact. We assume $Z_H(J)$ contains a simple Lie group of real rank at least 2. We write the Lie algebra $\mathfrak{z}_\mathfrak{h}(\mathfrak{j}) = \mathfrak{l} \ltimes \mathfrak{r}$ where $\mathfrak{r}$ is the radical and $\mathfrak{l}$ is a semisimple Levi factor. We let $G \subset Z_H(J)$ be the connected semisimple group corresponding to the sum of all simple factors of $\mathfrak{l}$ of real rank at least 2. Then G is a semisimple Lie group each of whose simple factors has $\mathbb{R}$ rank of at least 2, and if $\sigma : G \to Z_H(J)$ is any faithful representation then $\sigma(G)$ and G are conjugate in $Z_H(J)$.

We view $J \backslash H \to M$ as a principal Γ-bundle on which G acts by Γ-bundle automorphisms. We have a natural embedding $\Gamma \to H$, and we let $Q \to M$ be the associated H-bundle, i.e., $Q = (J \backslash H \times H) / \Gamma$. The group G then acts on Q by H-bundle automorphisms, and we wish to apply superrigididy for actions on principal bundles. To this end, we let (E, ν) be the space of ergodic components of the G-action on M. Thus, (E, ν) is a measure space and there is a G-invariant measurable map $p : M \to E$ such that, letting μ be the measure on M, we have $\mu = \int_{e \in E}^{\oplus} \mu_e d\nu(e)$, where μ_e is a measure on $M_e = p^{-1}(e)$. Furthermore, a.e. μ_e is a G-invariant ergodic probability measure. For each e, we let $Q_e \to M_e$ the (measurable) principal H-bundle obtained by restricting Q to the fibers over M_e. Since the action of G on M_e is ergodic, there is a well-defined algebraic hull for the action of G. We recall that there this is the unique conjugacy class of algebraic subgroups of H such that there is a measurable G-invariant reduction of the bundle to an element of this conjugacy class but not to any smaller algebraic subgroup. We shall use basic properties of algebraic hulls freely and refer the reader to [Z4] for discussion of this notion.

We shall also find the following notation (from [Z9]) convenient.

Definition 3.1. If $P \to M$ is a principal Γ-bundle with Γ discrete, $\Lambda \subset \Gamma$ is a subgroup, and G acts by automorphisms, we say that the G-action is Λ-reducible if there is a measurable G-invariant reduction of P to Λ.

We then have the following easy fact. (Cf. [Z9].)

Lemma 3.2. Suppose $P \to M$ is a principal Γ-bundle and that G acts ergodically on M. Suppose $\Gamma \to H$ is a discrete faithful representation into an algebraic group H. Let Q be the associated H-bundle. If the algebraic hull for the G action on P is compact, then $P \to M$ is Λ-reducible for some finite $\Lambda \subset \Gamma$.

(The proof of Lemma 3.2 also follows from the proof of Proposition 3.6 below.)

Proposition 3.3. *For almost all* $e \in E$*, the algebraic hull of the action of* G *on* $Q_e \to M_e$ *is noncompact.*

Proof. If this were not the case, then there would be a set of $e \in E$ positive measure for which the algebraic hull is compact. By Lemma 3.2, on a set $e \in E$ of positive measure, the G action on $(J\backslash H)_e \to M_e$ is Λ_e-reducible for some finite $\Lambda_e \subset \Gamma$. It follows that each such $(J\backslash H)_e$ supports a finite G-invariant measure projecting to M_e, and hence (by integrating) $J\backslash H$ supports a finite G-invariant measure projecting to the restriction of μ to a subset of positive μ-measure in M. Since the fibers of $J\backslash H \to M$ are discrete, it follows that $J\backslash H$ admits a finite G-invariant measure that is the restriction of a smooth measure to a G-invariant set. Since the action of G on $J\backslash H$ is an action on a variety, the Borel density theorem implies that finite invariant measures are supported on fixed points. Therefore, we would have a set of fixed points of positive Lebesgue measure. Since the action is algebraic, this implies that G acts trivially on $J\backslash H$, which is clearly impossible because G is not contained in a proper normal subgroup of H.

Remarks. (a) Exactly the same argument can be applied to the action of any simple normal subgroup of G.

(b) In general, by passing to a finite ergodic extension, the algebraic hull will become a subgroup of finite index. Thus, Proposition 3.3 remains true if $Q_e \to M_e$ is replaced by the pullback of Q_e to a finite ergodic extension of M_e.

We now fix an ergodic component e that satisfies the conclusions of Proposition 3.3 and apply superrigidity for cocycles (i.e., for actions on principal bundles) in the form of [Z1, Z3]. We obtain:

Theorem 3.4. *Fix* e *as above. Then there is a finite ergodic extension* $M'_e \to M_e$ *of* M_e*, a locally faithful representation* $\sigma : \tilde{G} \to H$*, a compact subgroup* $K \subset H$ *with* $K \subset Z_H(\sigma(\tilde{G}))$ *and a measurable function* $\psi : M'_e \to H$ *such that*

$$\psi(gm)^{-1}\alpha'(g,m)\psi(m) = \sigma(g)c(g,m)$$

for $g \in \tilde{G}$*,* $m \in M'_e$ *where* $\alpha : G \times M_e \to \Gamma$ *is the cocycle corresponding to* $(J\backslash H)_e \to M_e$*,* α' *is its lift to* $\tilde{G} \times M'_e$*, and* $c(g,m) \in K$*.*

We remark that σ is nontrivial by Proposition 3.3 (and Remark (b)) and, in fact, is locally faithful by Remark (a).

By passing to a further ergodic extension with finite invariant measure, we can obtain the same conclusion but with c trivial. Namely, let $K_0 \subset K$ be the algebraic hull of c. Then c is equivalent to a cocycle c_0 taking values in K_0, and we can form the G-space $X_e = M'_e \times_{c_0} K_0$. Since K is compact, this will be ergodic [Z6] (with respct to the finite G-invariant product measure). The pullback of c_0 to a cocycle $\tilde{G} \times X_e \to K_0$ will be trivial. Therefore, we deduce

Corollary 3.5. *There is a measurable function* $\varphi : X_e \to H$ *such that*

$$\varphi(gx)^{-1}\alpha_X(g,m)\varphi(gx) = \sigma(g)$$

where α_X is the lift of α to a cocycle $\widetilde{G} \times X \to \Gamma$.

Now let $\lambda : X_e \to \Gamma\backslash H$ be the composition of φ with the projection of H to $\Gamma\backslash H$. Since α_X takes values in Γ, we deduce $\lambda(gx) = \lambda(x) \cdot \sigma(g)^{-1}$, i.e., that λ is a $\widetilde{G}$ map. The measure $\lambda_*(\mu_e)$ is then a finite ergodic $\sigma(\widetilde{G})$-invariant measure on $\Gamma\backslash H$. As in [Z5] we can apply Ratner's theorem [R] to this situation.

Proposition 3.6. *There is a connected Lie group $L \subset H$ such that:*

(i) $\Lambda = L \cap \Gamma$ *is a lattice in L, and*

(ii) *the G action on the principal Γ-bundle $P_X \to X$, the pullback of $(J\backslash H)_e \to M_e$ to X, is Λ-reducible.*

Proof. By Ratner's Theorem [R], there is a connected Lie group L, $\sigma(G) \subset L \subset H$, such that $\lambda_*(\mu_e)$ lies on an L-orbit in $\Gamma\backslash H$ of the form $h\Gamma h^{-1} \cap L\backslash L$, where $h \in H$ and $h\Gamma h^{-1} \cap L$ is a lattice in L. In other words, there is some $h \in H$ such that $\varphi(X_e) \subset \Gamma h L$ (a.e.). Thus, we can write $\varphi(x) = \omega(x) h \theta(x)$ where $\omega : X_e \to \Gamma$ and $\theta : X_e \to L$ are measurable. We then have

$$\theta(gx)^{-1} h^{-1} \omega(gx)^{-1} \alpha_X(g, x) \omega(x) h \theta(x) = \sigma(g).$$

The cocycle $\beta(g, x) = \omega(gx)^{-1} \alpha_X(g, x) \omega(x)$ is equivalent (as a cocycle into Γ) to α_X and satisfies

$$\beta(g, x) = h\theta(gx)\sigma(g)\theta(x)^{-1} h^{-1} \in hLh^{-1}.$$

Therefore, $\beta(G \times X) \subset \Gamma \cap hLh^{-1}$. Hence, the G-action on P_X is Λ-reducible for $\Lambda = \Gamma \cap hLh^{-1}$ and Λ is a lattice in hLh^{-1}.

Lemma 3.7. *Let L be as in Proposition 3.6. Then L acts properly on $J\backslash H$.*

Proof. Γ acts properly by assumption, and hence so does Λ. Since Λ is a lattice in L, L will also act properly. For convenience, we recall the proof of this latter assumption.

We first recall there is a compact symmetric set $B \subset L$ such that $L = B\Lambda B$. Namely, choose a compact symmetric $B \subset L$ whose projection [B] to L/Λ has measure at least $3/4$. Then, for any $h \in L$, $h([B]) \cap [B]$ has positive measure; i.e., $hB \cap B\Lambda$ is nontrivial and so $h \in B\Lambda B$. Now take any $A \subset J\backslash H$ that is compact. If $h \in L$ and $Ah \cap A \neq \varnothing$, write $h = b\lambda c$ where $\lambda \in \Lambda$, and $b, c \in B$. Then $(AB)\lambda \cap Ac^{-1} \neq \varnothing$, i.e., $(AB)\lambda \cap AB \neq \varnothing$.

Letting $\Lambda_0 \subset \Lambda$ be $\{\lambda \in \Lambda | AB\lambda \cap AB \neq \varnothing\}$, we have Λ_0 is finite since AB is compact, and $h \in B\Lambda_0 B$ which is compact.

Proposition 3.8. *There is a discrete subgroup $F \subset \Gamma$ with the following properties:*

(i) F *is finitely generated.*

(ii) *The action of G on $P_X \to X$ is F-reducible.*

(iii) *If we let N be the Zariski closure of F, then N acts properly on $J\backslash H$.*

Proof. The group $[L, L] \subset H$ is algebraic. Let $\beta : G \times X \to \Lambda$ be a cocycle corresponding to the G-invariant Λ-reduction of P_X, and $\pi : L \to L/[L, L]$.

Then $\pi \circ \beta$ is a cocycle taking values in the countable abelian group $\pi(\Lambda)$. It follows from [Z1, Theorem 9.1.1] (using Kazhdan's property) that $\pi \circ \beta$ is equivalent to a cocycle taking values in a finite subgroup, and hence β is equivalent to a cocycle taking values in a subgroup $\Lambda_1 \subset \Lambda$ such that $\Lambda_1 \supset \Lambda \cap [L, L]$ as a subgroup of finite index. It follows that Λ_1 is contained in an algebraic group L_1 such that $L_1 \cap [L, L]$ is of finite index in L_1. Since G is a Kazhdan group, any cocycle into a countable group is equivalent to one taking values in a finitely generated subgroup [Z8, Lemma 2.2]. Therefore, there is a finitely generated $F \subset \Lambda_1$ such that the G-action on $P_X \to X$ is F-reducible. Letting N be the Zariski closure of F, we have $N \subset L_1$, and hence $N/N \cap L$ is finite. Properness of the N action on $J \backslash H$ then follows from Lemma 3.7.

Now let $Q_X \to X$ be the principal H-bundle obtained by pulling back $Q_e \to (J \backslash H)_e$ to X. Alternatively, Q_X is the principal H-bundle associated to the Γ-bundle $P_X \to X$ and the embedding $\Gamma \to H$. Since the G-action on P_X is F-reducible, it is clear that the algebraic hull of the G-action on Q_X is contained in N. On the other hand, this algebraic hull is exactly $\overline{\sigma(\widetilde{G})}$ (the Zariski closure) in which $\sigma(\widetilde{G})$ is of finite index. (This assertion follows from the Borel density theorem. See, e.g., [Z4].) Thus a conjugate of $\sigma(\widetilde{G})$ is contained in N. We now wish to show that, in fact, some conjugate of G itself is contained in N. (This would follow of course if we knew σ was conjugate to the original realization $\widetilde{G} \to G \subset H$, but this is what we have not yet established.)

Lemma 3.9. *G is contained in a conjugate of N.*

Proof. Since there is a G-invariant reduction of $P_X \to X$ to F and $F \subset N$, the algebraic hull of the G-action on Q_X is clearly contained in N.

Consider now the right H-action on $J \backslash H$. For each $y \in J \backslash H$, let $\mathfrak{h}_y \subset \mathfrak{h}$ be the Lie algebra of the stabilizer of y. Then $y \mapsto \mathfrak{h}_y$ defines a Γ-map $\eta : J \backslash H \to \mathrm{Gr}(\mathfrak{h})$ (in fact, an H-map) where $\mathrm{Gr}(\mathfrak{h})$ is the Grassmann variety of $\mathfrak{h}$. Since G acts on the left of $J \backslash H$, it is straightforward to check that $\eta(gy) = \eta(y)$ for all $y \in J \backslash H$ and $g \in G$.

Write $J = R \ltimes U$ a Levi decomposition, so U is the unipotent radical. Let $\mathfrak{u}$ be the Lie algebra of U, on which J acts. We can write $o \subset \mathfrak{u}_1 \subset \mathfrak{u}_2 \subset \cdots \subset \mathfrak{u}_r = \mathfrak{u}$, where $\mathfrak{u}_i \subset \mathfrak{u}$ is a J-invariant ideal and the J-action on $\mathfrak{u}_{i+1}/\mathfrak{u}_i$ is irreducible (and, in particular, is trivial on U). Since each $\mathfrak{u}_i$ is J-invariant, we can define an H-equivariant map $w_i : J \backslash H \to \mathrm{Gr}(\mathfrak{u})$ (the Grassmann variety of $\mathfrak{u}$) by $w_i(y) = \mathrm{Ad}(h)^{-1} \mathfrak{u}_i$ where $y = Jh$. For any $g \in G$, let $l(g)$ be left translation on $J \backslash H$. Then $w_j \circ l(g)$ will also be H-equivariant and $[w_j \circ l(g)]([e]) = w_j(g) = \mathrm{Ad}(g)^{-1} \mathfrak{u}_j = \mathfrak{u}_j$ (since $\mathrm{Ad}(G)$ is trivial on $\mathfrak{u}$) $= w_j([e])$. Since $w_j \circ l(g)$ and w_j agree at a point and are H-invariant, they are equal.

Define $w : J \backslash H \to \mathrm{Gr}(\mathfrak{h}) \times \prod_i \mathrm{Gr}(\mathfrak{u}) = V$ by $w = (\eta, w_1, \ldots, w_r)$. This is then an H-equivariant, G-invariant map. The map $y \mapsto (y, w(y))$ therefore factors to a map $s : J \backslash H / \Gamma \to (J \backslash H \times V) / \Gamma$. Since the image of $y \mapsto w(y)$ clearly lies in the orbit of $(j, \mathfrak{u}_1, \ldots, \mathfrak{u}_r)$ under H, the map s can be viewed as a section of the bundle $(J \backslash H \times H/H_0) / \Gamma \to J \backslash H / \Gamma$ where H_0 is the stabilizer of $(j, \mathfrak{u}_1, \ldots, \mathfrak{u}_r)$ in H. It follows from [Z4] that the algebraic hull of the

G-action on the principal H-bundle $(J\backslash H \times H)/\Gamma \to J\backslash H/\Gamma$ is contained in H_0.

It follows that the same is true for the algebraic hull of the G-action on the pullback of this bundle to X, which is clearly isomorphic to Q_X. As we have already observed, this algebraic hull is contained in N. We clearly have that $H_0 = N_H(\mathfrak{j}) \cap \bigcap_i N_H(\mathfrak{u}_i)$. It follows that for some $h \in H$ and some conjugate G_1 of $\sigma(\tilde{G})$ that

$$G_1 \subset hNh^{-1} \cap N_H(\mathfrak{j}) \cap \bigcap_i N_H(\mathfrak{u}_i).$$

By Proposition 3.8, N, and hence any conjugate of N, acts properly on $J\backslash H$. Therefore, G_1 acts properly on the right of $J\backslash H$, so the intersection of any conjugate of G_1 with J is compact. Since G_1 is semisimple with no compact factors and G_1 normalizes $\mathfrak{j}$, we have $\mathfrak{g}_1 \cap \mathfrak{j} = (0)$.

Now consider the subgroup $G_1 J = G_1 RU$. Note that each $\mathfrak{u}_j$ is an ideal, i.e. is invariant under all $G_1 J$. Since G_1 and R are reductive, G_1 normalizes RU, and $\mathfrak{g}_1 \cap \mathfrak{j} = (0)$, in the quotient group $G_1 RU/U$ we can conjugate $G_1 U/U$ so that it centralizes RU/U. (We are using here the fact that there are no nontrivial local homomorphisms from G to J.) Choosing a lift of this semisimple group in $G_1 RU$, we deduce the following: there is a conjugate G_2 of G_1 in $G_1 RU$ such that G_2 centralizes R. Furthermore, $G_2 R$ acts on each $\mathfrak{u}_{i+1}/\mathfrak{u}_i$, and the representation of R on this space is irreducible. It follows that G_2 acts by scalars on $\mathfrak{u}_{i+1}/\mathfrak{u}_i$ and, hence, that G_2 acts on $\mathfrak{u}$ by a solvable group. Since G_2 is simple, G_2 acts trivially on $\mathfrak{u}$. Thus G_2 commutes with both R and U, and hence $G_2 \subset Z_H(\mathfrak{j})$. This shows that some conjugate of $\sigma(\tilde{G})$ is contained in $hNh^{-1} \cap Z_H(J)$. But by remarks at the beginning of this section, since any representation $\tilde{G} \to Z_H(J)$ has image equal to a conjugate of G in $Z_H(J)$, it follows that for some (possibly different) h, $hNh^{-1} \cap Z_H(J) \supset G$. This completes the proof.

Lemma 3.10. *There is a finitely generated discrete group $F \subset \Gamma$ such that*:

    (a) *There is a G-invariant subset of finite (smooth) measure in $J\backslash H/F$*;
    (b) *If N is the Zariski closure F, then N acts properly on $J\backslash H$; and*
    (c) *G is contained in N.*

Proof. Consider the diagram

$$
\begin{array}{ccc}
P_X & \longrightarrow & P_e = (J\backslash H)_e \\
\downarrow & & \downarrow \\
X & \longrightarrow & M_e
\end{array}
$$

This yields a diagram

$$
\begin{array}{ccc}
P_X/F & \longrightarrow & (J\backslash H)_e/F \\
\downarrow & & \downarrow \\
X & \longrightarrow & M_e
\end{array}
$$

By Proposition 3.8, there is a G-invariant section of $P_X/F \to X$ and, hence, a finite G-invariant measure θ_e in P_X/F that projects to the given measure on X. Pushing this measure forward to $(J\backslash H)_e/F$, we obtain a finite G-invariant measure on this space that projects to μ_e. This is all still within one ergodic component. The space of finitely generated subgroups of Γ is countable. Therefore, there is a set of ergodic components of positive ν-measure, say E_0, on which the assignment $e \mapsto F$ will be constant. The measure $\theta = \int_{e \in E_0}^{\oplus} \theta_e$ is a finite G-invariant measure on $J\backslash H/F$ that projects to a set of positive measure in M (namely, the union of the ergodic components $e \in E_0$). Since the bundle $J\backslash H/F \to M = J\backslash H/\Gamma$ has discrete fibers, θ is the restriction of the smooth measure on $J\backslash H/F$ to a set of positive measure. The remaining assertions of Lemma 3.10 now follow from Proposition 3.8, Lemma 3.9, and a conjugation.

Lemma 3.11. *G fixes every point of $J\backslash H/N$.*

Proof. By Rosenlicht's theorem [Ro], there is a Zariski open ($\mathbb{C}$-connected) $J \times G$-invariant subset U of H/N such that $J\backslash U$ is a variety on which G acts algebraically. Pushing the measure of Lemma 3.10(a) forward under the map $J\backslash H/F \to J\backslash H/N$, we deduce that there is a finite G-invariant measure on $J\backslash U$ that is the restriction of a smooth measure to a subset of positive measure. Since the G-action on $J\backslash U$ is algebraic, the Borel density theorem implies any finite invariant measure is supported on G-fixed points. Therefore, we deduce that there is a set of G-fixed points in $J\backslash U$ of positive smooth measure, and, since the action is algebraic, G pointwise fixes $J\backslash U$. This means there is a dense set of G-fixed points in $J\backslash H/N$. However, since the N action on $J\backslash H$ is proper, $J\backslash H/N$ is Hausdorff, and hence G fixes all points of $J\backslash H/N$.

Completion of Proof of Theorem. Choose a 1-parameter $\mathbb{R}$-split subgroup $a(t)$ in G. Since the root spaces for $a(t)$ in $\mathfrak{h}$ corresponding to nonzero roots generate a normal subgroup of H, it suffices to see that all these root spaces are in $\mathfrak{n}$. If $\mathfrak{u} \subset \mathfrak{h}$ is such a 1-dimensional eigenspace and $\mathfrak{u} \subset \mathfrak{j} + \mathfrak{n}$, then we must have $\mathfrak{u} \subset \mathfrak{n}$ since G (and in particular $a(t)$) centralizes $\mathfrak{j}$. Therefore, we need only consider $\mathfrak{u} \not\subset \mathfrak{j} + \mathfrak{n}$. Let $U(s) \subset H$ be the corresponding 1-parameter subgroup. Then

$$a(t)U(s)N = a(t)U(s)a(t)^{-1}N$$
$$= U(e^{\lambda t}s)N \quad (\text{since } a(t) \in G \subset N)$$

for some $\lambda \in \mathbb{R}$, $\lambda \neq 0$ This means that $U^+ = \{U(s)N | s > 0\}$ is contained in a single G-orbit in H/N. However, since G acts trivially on $J\backslash H/N$, it follows that U^+ is contained in a single J-orbit. By Lemma 3.10, N acts properly on $J\backslash H$, so $J\backslash H/N$ is Hausdorff and the J-orbits in H/N are closed. Therefore, U^+ is contained in the J-orbit of eN, i.e. in $JN \subset H$. However, JN is a closed submanifold whose tangent space at $e \in H$ is $\mathfrak{j} + \mathfrak{n}$, and, since $\mathfrak{u} \not\subset \mathfrak{j} + \mathfrak{n}$, we have $U(s) \notin JN$ for $s > 0$ but sufficiently small. This contradiction shows all such $\mathfrak{u}$ are in $\mathfrak{n}$ and, hence, that $\mathfrak{n} = \mathfrak{h}$, $N = H$, and J is compact (by Lemma 3.10).

168   R. J. ZIMMER

REFERENCES

[BL]   Y. Benoist and F. Labourie, *Sur les espaces homogènes modelés de variétés compactes*, preprint.

[B]   A. Borel, *Compact Clifford-Klein forms of symmetric spaces*, Topology **2** (1963), 111–122.

[CM]   E. Calabi and L. Markus, *Relativistic space forms*, Ann. of Math. (2) **75** (1962), 63–76.

[CZ]   K. Corlette and R. J. Zimmer, *Superrigidity for cocycles and hyperbolic geometry*, preprint.

[G]   W. Goldman, *Projective geometry on manifolds*, Univ. of Maryland Lecture Notes, 1988.

[K]   T. Kobayashi, *Proper action on a homogeneous space of reductive type*, Math. Ann. **285** (1989), 249–263.

[KO]   T. Kobayashi and K. Ono, *Note on Hirzebruch's proportionality principle*, J. Fac. Sci. Univ. Tokyo Sect. IA Math. **37** (1990), 71–87.

[Ku]   R. S. Kulkarni, *Proper actions and pseudo-Riemannian space forms*, Adv. Math **40** (1981), 10–51.

[R]   M. Ratner, *On Raghunathan's measure conjecture*, Ann. of Math. (2) **134** (1991), 545–607.

[Ro]   M. Rosenlicht, *A remark on quotient spaces*, An. Acad. Brasil Ciênc. **35** (1963), 487–489.

[W]   J. A. Wolf, *The Clifford-Klein space forms of indefinite metric*, Ann. of Math. (2) **75** (1962), 77–80.

[Z1]   R. J. Zimmer, *Ergodic theory and semisimple groups*, Birkhäuser, Boston, MA, 1984.

[Z2]   ——, *Strong rigidity for ergodic actions of semisimple Lie groups*, Ann. of Math (2) **112** (1980), 511–529.

[Z3]   ——, *On the algebraic hull of the automorphism group of a principal bundle*, Comment. Math. Helv. **65** (1990), 375–387.

[Z4]   ——, *Ergodic theory and the automorphism group of a G-structure*, Group Representations, Ergodic Theory, Operator Algebras, and Mathematical Physics (C. C. Moore, ed.), Springer Verlag, New York, 1987.

[Z5]   ——, *Superrigidity, Ratner's theorem, and fundamental groups*, Israel J. Math. **74** (1991), 199–207.

[Z6]   ——, *Extensions of ergodic groups actions*, Illinois J. Math. **20** (1976), 373–409.

[Z7]   ——, *Kazhdan groups acting on compact manifolds*, Invent. Math. **75** (1984), 425–436.

[Z8]   ——, *Groups generating transversals to semisimple Lie group actions*, Israel J. Math. **73** (1991), 151–159.

[Z9]   ——, *Topological superrigidity*, preprint.

DEPARTMENT OF MATHEMATICS, UNIVERSITY OF CHICAGO, 5734 UNIVERSITY AVENUE, CHICAGO, ILLINOIS 60637

E-mail address: zimmer@math.uchicago.edu

Geometric And Functional Analysis   1016-443X/95/0600955-11$1.50+0.20/0

Vol. 5, No. 6 (1995)   © 1995 Birkhäuser Verlag, Basel

ON MANIFOLDS LOCALLY MODELLED
ON NON-RIEMANNIAN HOMOGENEOUS SPACES

F. Labourie, S. Mozes and R.J. Zimmer

1. Introduction and Statement of Main Results

In this paper we continue the investigation of compact manifolds locally modelled on a homogeneous space of a finite dimensional Lie group. We recall that a manifold M is locally modelled on a homogeneous space $J\backslash H$ if there is an atlas on M consisting of local diffeomorphisms with open sets in $J\backslash H$ and where the transition functions are given by restrictions to open sets of translations by elements of H acting on $J\backslash H$. A basic example is given by a cocompact lattice in $J\backslash H$, namely a discrete subgroup $\Gamma \subset H$ such that Γ acts freely and properly discontinuously on $J\backslash H$ and such that $M = J\backslash H/\Gamma$ is compact. In this case, M is then locally modelled on $J\backslash H$. It is a basic open question to understand those homogeneous spaces $J\backslash H$ for which there is a compact form (i.e. space locally modelled on $J\backslash H$) or a cocompact lattice.

While there have been numerous approaches to this question (see, e.g. the references and brief discussion in [Z1]), one of these has involved the presence of a non-trivial centralizer $G \subset Z_H(J)$ for J. In the case in which H is simple and J actually has a non-trivial $\mathbb{R}$-diagonalizable 1-parameter subgroup in its center, then the main result of [BL] is that there are no compact forms for $J\backslash H$. The technique involves using the center to construct a suitable symplectic form. In [Z1], the case in which J may have trivial $\mathbb{R}$-split center but yet have a significant centralizer in H was studied. The main result of [Z1] is the assertion that if there is a cocompact lattice for $J\backslash H$ then under suitable hypotheses J must be compact. (In the event J is compact it is known that there are always cocompact lattices, at least if H is semisimple.) These hypotheses are

i. H and J are unimodular real algebraic groups;
ii. The centralizer $Z_H(J)$ contains a group G such that:
    (a) G is not contained in a proper normal subgroup of H;
    (b) G is a semisimple Lie group each of whose simple factors has $\mathbb{R}$-rank at least 2; and,

The research of S.M. was partially supported by an NSF grant. The research of R.J.Z. was partially supported by an NSF grant.

(c) Every homomorphism $\tilde{G} \to J$ is trivial.

As a consequence, one obtained in [Z1] the conclusion that $\mathrm{SL}(m,\mathbf{R})\backslash\mathrm{SL}(n,\mathbf{R})$ (for the standard embedding) does not admit a cocompact lattice if $n > 5$ and $m < n/2$. However, the arguments in [Z1] do not give the conclusion that there is no compact form in these cases. It also leaves open the question of existence of cocompact lattices and compact forms if $m \geq n/2$. The techniques of [Z1] involve superrigidity for cocycles and Ratner's theorem on invariant measures on homogeneous spaces.

The aim of this paper is three-fold.

First, we dramatically simplify the proof of the main result of [Z1]. Our approach here enables us to use Moore's ergodicity theorem ([Z3]) in place of Ratner's theorem which is required in the proof given in [Z1]. (However, we remark that the proof in [Z1] using Ratner's theorem, although it is not explicitly stated there, reduces the general question of the existence of cocompact lattices, without assumption (c) above, to a purely algebraic question about Lie groups, i.e. one with nothing to do a priori with discrete subgroups. This seems to be a potentially useful approach to this significantly more general situation.)

Second, using these arguments, we extend the results of [Z1] to the case in which assumption (c) is replaced by:

(c') : (i) Every non-trivial homomorphism $\tilde{G} \to J$ has compact centralizer in J;

     (ii) There is a non-trivial $\mathbf{R}$-split 1-parameter group in $Z_H(JG)$ that is not contained in a normal subgroup of H.

In particular, we show that $\mathrm{SL}(m)\backslash\mathrm{SL}(2m)$ does not admit a cocompact lattice, nor does $J\backslash\mathrm{SL}(2m)$ where $J \subset \mathrm{SL}(m)$ is either $\mathrm{SO}(p,q)$, $p+q = m$, or $\mathrm{Sp}(2r,\mathbf{R})$ with $2r = m$.

Third, we generalize these results to the case of compact forms. Thus our main result can be stated as follows:

THEOREM 1.1. *Let H be a real algebraic group and $J \subset H$ an algebraic subgroup. Suppose:*

i. *H and J are unimodular;*

ii. *There is a compact manifold M locally modelled on $J\backslash H$;*

iii. *The centralizer $Z_H(J)$ of J in H contains a group G with the following properties:*

    (a) *G is a semisimple Lie group each of whose simple factors has $\mathbf{R}$-rank at least 2;*

    (b) *Every non-trivial homomorphism $\tilde{G} \to J$ has a compact centralizer in J;*

(c) If there is a non-trivial homomorphism $\tilde{G} \to J$, then there is a non-trivial 1-parameter $\mathbf{R}$-split subgroup $B \subset Z_H(JG)$ such that neither B nor G is contained in a proper normal subgroup of H.

Then J is compact.

In particular, we have the following generalization of [Z1, Corollary 1.3].

COROLLARY 1.2. If $n \geq 5$, $p \leq n/2$, then there is no compact manifold locally modelled on $SL(p,\mathbf{R})\backslash SL(n,\mathbf{R})$. The same is true for $J\backslash SL(n,\mathbf{R})$ where J is any unimodular non-compact subgroup of $SL(p,\mathbf{R})$ $(p \leq n/2)$.

2. Cocompact Lattices

In this section we give an alternate proof for the results of [Z1] utilizing the conclusions of superrigidity for cocycles in a simpler way. Further considerations regarding uniqueness of the section defined by superrigidity allow us to generalize the homogeneous space under consideration and to deduce Theorem 1.1 for cocompact lattices.

We will use the following consequence of Moore's theorem ([Z3]).

LEMMA 2.1. Let H be a Lie group and $\Gamma \subset H$ a discrete subgroup. Let $B \subset H$ be a non-trivial 1-parameter subgroup such that $Ad_H(B)$ is $\mathbf{R}$-split (i.e. $\mathbf{R}$-diagonalizable) and B is not contained in a proper normal subgroup. Suppose there is a set $Y \subset H/\Gamma$ of finite positive measure (with respect to the natural H-invariant volume on H/Γ) that is B-invariant. Then Γ is a lattice in H and $Y = H/\Gamma$.

Proof: $\chi_Y \in L^2(H/\Gamma)$ is B-invariant. If H is semisimple, then Moore's theorem implies χ_Y (and hence Y) is H-invariant, from which the lemma follows. In general, the proof of Moore's theorem shows that χ_Y is invariant under all 1-parameter subgroups corresponding to non-0 root spaces of B. However. together with B, these generate a normal subgroup of H. By assumption, this must be equal to H, so again Y is H-invariant, proving the lemma.   □

We now give a proof of the main result of [Z1].

THEOREM 2.2. Assume the hypotheses of Theorem 1.1 with additional assumptions that $M = J\backslash H/\Gamma$ for a cocompact lattice Γ in $J\backslash H$, and that every local homomorphism $G \to J$ is trivial. Then J is compact and Γ is a lattice in H.

Proof: If Γ is a cocompact lattice in $J\backslash H$, then $H/\Gamma \to J\backslash H/\Gamma$ is a principal J bundle on which the centralizer $Z_H(J)$ acts by principal bundle automorphisms preserving a volume density on the compact manifold $J\backslash H/\Gamma$

([Z1]). Let G be as in Theorem 1.1, with the additional assumption that every local homomorphism $G \to J$ is trivial (i.e. we have the situation in [Z1]). By superrigidity for cocycles, the cocycle $\alpha : J \backslash H / \Gamma \times G \to J$ defined by the action of G on H/Γ is then equivalent to a cocycle into a compact subgroup $K \subset J$. This means there is a measurable G-invariant section s of $K \backslash H / \Gamma \to J \backslash H / \Gamma$. Letting μ be the finite G-invariant measure on $J \backslash H / \Gamma$ defined by the G-invariant volume density, $s_*(\mu)$ will be a finite G-invariant measure on $K \backslash H / \Gamma$. Since K is compact, we can lift this via the bundle map $H/\Gamma \to K \backslash H / \Gamma$ to a finite G-invariant measure ν on H/Γ that projects to a smooth measure on $K \backslash H / \Gamma$. For any $j \in J$, $j_*\nu$ will also be G-invariant. We can write

$$\nu = \int^{\oplus} \nu_t d\mu(t)$$

where $t \in J \backslash H / \Gamma$ and ν_t is supported on a compact set (in fact a K-orbit) in the fiber over t in H/Γ. Thus, if $X \subset J$ is a compact set of positive Haar measure,

$$\int_{j \in X} j_*(\nu) dj$$

will be a finite G-invariant measure that is simply the restriction of the natural smooth measure on H/Γ to a set of finite positive measure. We now apply Lemma 2.1 to deduce the theorem.   □

We now discuss some uniqueness results concerning superrigidity that will allow us to deduce Theorem 1.1 for cocompact lattices in general (i.e. without the assumption that all local homomorphisms $G \to J$ are trivial). We shall work first in the general framework of principal bundles, as we will need these results for the proof in the case of general compact forms as well.

Let $P \to M$ be a principal J bundle where M is a compact manifold and J is a connected real algebraic group. Let G be a connected semisimple Lie group such that every simple factor is of $\mathbb{R}$-rank at least 2. By passing to the universal cover when necessary, we may assume G is simply connected. Let L be an algebraic group which can be written as $L = BG$ where B is abelian and B centralizes G. We assume L acts on P by principal J-bundle automorphisms preserving a finite measure μ on M. Let $\pi : G \to J$ be a homomorphism (which necessarily factors via the algebraic universal cover of $Ad(G)$), and K a fixed maximal compact subgroup of $Z_J(\pi(G))$. We say (following the terminology of [4]) that a section s of $P \to M$ is π-simple if

$$s(gm) = gs(m)\pi(g)^{-1}c(g,m)$$

where $c(g,m) \in K$. The conclusion of superrigidity for cocycles, see [Z3,4], is that for G as above there always exists a π-simple section for some π

and some K, as long as the G-action is ergodic and the algebraic hull for the G-action is connected. For clarity of exposition we shall assume throughout sections 2 and 3 that the algebraic hull is indeed connected. In section 4 we indicate the necessary modifications for treating the general non-connected case. In the case when the G-action on M is not ergodic, we can decompose the action into ergodic components and apply superrigidity for cocycles to each ergodic component. If there is a $\bar{\pi}$-simple section where $\bar{\pi}$ is conjugate to π, then there is also a π-simple section. Since there are only finitely many conjugacy classes of homomorphisms $G \to J$ we deduce, without the assumption of ergodicity, the following.

LEMMA 2.3. *There is a G-invariant subset $X \subset M$ of positive measure, a homomorphism $G \to J$, and a π-simple section of $P|_X \to X$.*

We now observe:

LEMMA 2.4. *We can take the set $X \subset M$ in Lemma 2.3 to be L-invariant.*

Proof: Let s be the section of $P|_X \to X$ given by Lemma 2.3. For any $b \in B$, the section $s^b(m) = b^{-1}s(bm)$ is also π-simple for the bundle $P|_{b^{-1}X} \to b^{-1}X$. In this way, we can extend s to a π-simple section on $BX = LX$. □

LEMMA 2.5. *Suppose s and t are both π-simple sections for an ergodic G-space with finite invariant measure. Then there is an element $h \in Z_J(\pi(G))$ and functions $k_1, k_2 : M \to K$ such that $s(m) = t(m)k_1(m)hk_2(m)$ a.e.*

Proof: Let $s(m) = t(m)j(m)$ where $j : M \to J$. Write

$$s(gm) = gs(m)\pi^{-1}(g)c(g,m)$$

and

$$t(gm) = gt(m)\pi(g)^{-1}d(g,m)$$

where $c(g,m), d(g,m) \in K \subset Z_J(\pi(G))$ and K is compact. Then it follows that

$$j(gm) = d(g,m)^{-1}\pi(g)j(m)\pi(g)^{-1}c(g,m) . \qquad (*)$$

Since K centralizes $\pi(G)$, G acts by conjugation on $K\backslash J/K$, and if we let $p : J \to K\backslash J/K$ be the natural map, then $(*)$ implies that $p \circ j : M \to K\backslash J/K$ is a G-map. Since J and G are algebraic and K is compact, G acts tamely (i.e. with locally closed orbits) on $K\backslash J/K$ and has algebraic stabilizers. Since the G action on M is ergodic with finite invariant measure, say μ, so is the action with the push forward measure $(p \circ j)_*\mu$. It follows that $p \circ j$ is essentially constant and that the essential image point is a G-fixed point. (This follows from the formulation of the Borel density theorem which asserts for an algebraic action of a semisimple group with no compact factors on a variety that finite invariant measures are supported on fixed points.) Let KhK be such a point. Then $\pi(g)h\pi(g)^{-1} \in KhK$ for all $g \in G$. Thus G acts on KhK and it is easy to see that for this action G preserves the product Haar measure. Thus, by Borel density again, G fixes h, i.e. $h \in Z_J(\pi(G))$. Since $j(m) \in KhK$ for a.e. m, the lemma follows.   □

LEMMA 2.6. *Even without the assumption of ergodicity, for any two π-simple sections s, t we have*

$$s(m) = t(m)j(m)$$

where $j(m) \in Z_J(\pi(G))$.

Proof: Apply Lemma 2.5 to each ergodic component.   □

LEMMA 2.7. *Consider the situation of Lemmas 2.3, 2.4. Assume $Z_J(\pi(G))$ is compact. Let $\beta : B \times X \to J$ be the cocycle corresponding to the action of B on $P|_X$. Then β is equivalent to a cocycle into $Z_J(\pi(G))$.*

Proof: Let s be a π-simple section for the action of G. Since B commutes with G, s^b is also π-simple for the G-action, where $s^b(x) = b^{-1}s(bx)$. By Lemma 2.6, $s^b(x) = s(x)j(b, m)$, i.e. $s(bx) = bs(x)j(b, m)$. Since $j(b, m) \in Z_J(\pi(G))$, the lemma follows.   □

We can now prove Theorem 1.1 for cocompact lattices.

THEOREM 2.8. *Theorem 1.1 holds if $M = J\backslash H/\Gamma$ where Γ is a cocompact lattice for $J\backslash H$.*

Proof: We apply the above considerations to the action of BG on the principal J-bundle $H/\Gamma \to J\backslash H/\Gamma$. (We note that B also centralizes J and hence acts by bundle automorphisms.) Choose π and X as in Lemmas 2.3, 2.4. If π is trivial, we apply the argument of Theorem 2.2. If not, then by Lemma 2.7, $\beta|_{B \times X}$ is equivalent to a cocycle into a compact subgroup of J. Exactly as in the proof of Theorem 2.2 (but with the action of B rather than G), we deduce that there is a set of finite positive measure in H/Γ that is B-invariant. Applying Moore's theorem as in the proof of Theorem 2.2 completes the argument.

3. Proof of Theorem 1.1

We now give the additional arguments necessary to prove Theorem 1.1 in the general case of a compact manifold M locally modelled on $J\backslash H$. We begin with some general facts about locally homogeneous manifolds.

Let M be as in Theorem 1.1. Let $\Gamma = \pi_1(M)$. Then a classical construction of the developing map (see [G] or [R], e.g.) shows that there is a homomorphism $\Gamma \to H$ (the holonomy homomorphism) and a Γ-equivariant local diffeomorphism (not necessarily either injective or surjective) $\tilde{M} \to J\backslash H$ (with Γ acting on the right). We view $H \to J\backslash H$ as a principal J-bundle (with left J-action) and pull this back to $\tilde{M}$ to obtain a principal J-bundle $P^* \to \tilde{M}$ on which Γ acts by principal J-bundle automorphisms. Let $P = P^*/\Gamma$, so that $P \to M$ is a principal J-bundle (with J acting on the left). The map $P^* \to H$ is a $J \times \Gamma$-equivariant local diffeomorphism.

If N is a manifold, we let $\mathrm{Vect}(N)$ denote the Lie algebra of smooth vector fields on N. We identify the Lie algebra $\mathfrak{h} \hookrightarrow \mathrm{Vect}(H)$ as the right-invariant vector fields. In particular, Γ leaves $\mathfrak{h}$ pointwise fixed. Since $P^* \to H$ is a local diffeomorphism, the Lie algebra $\mathfrak{h}$ lifts canonically to a Lie algebra embedding $\mathfrak{h} \to \mathrm{Vect}(P^*)$. We denote the image by $\mathfrak{h}_{P^*}$ and corresponding elements by X_{P^*}. The Lie algebra $\mathfrak{j}_{P^*} \subset \mathfrak{h}_{P^*}$ is simply the Lie algebra defined by the left action of J on P^*. In particular, for $X \in \mathfrak{j}$, X_{P^*} is complete (which is not a priori necessarily true for a general $X \in \mathfrak{h}$). Because all $X \in \mathfrak{h}_{P^*}$ are Γ-invariant, there is a natural induced Lie algebra of vector fields on P which we denote by $\mathfrak{h}_P$. Clearly all $X_P \in \mathfrak{h}_P$ are non-vanishing. For $X \in \mathfrak{j}$, X_P is complete, the Lie algebra $\mathfrak{j}_P$ corresponding to the left action of J on P. If $X \in \mathfrak{z}_\mathfrak{h}(\mathfrak{j})$, then X_P is J-invariant and thus defines a vector field X_M on M. I.e., $X \mapsto X_M$ is an embedding of Lie algebras $\mathfrak{z}_\mathfrak{h}(\mathfrak{j}) \hookrightarrow \mathrm{Vect}(M)$. Since M is compact, we obtain a corresponding action of $\widetilde{Z_H(J)^\circ}$ on M, and hence on P as well.

We can now use exactly the arguments applying superrigidity for cocycles in section 2 to deduce that there is a 1-parameter $\mathbf{R}$-split subgroup, say $\alpha(t)$ (which, as in section 2, is either B or a subgroup of G), for which there is a subset $Y \subset P$ of finite positive measure (with respect to the natural smooth measure on P) that is $\alpha(t)$-invariant. To show that J is compact it suffices to show that Y is essentially (i.e. modulo null sets) J-invariant. Namely, J acts properly on P, and if there is a subset of finite positive measure that is invariant for a measure preserving proper action, then J is compact. (Alternatively there is a J-invariant set of finite positive measure on each fiber of P, and hence a J-invariant subset of J of finite positive Haar measure. This implies J is compact.)

For the remainder of the proof, we shall identify $X \in \mathfrak{h}$ with $X_P \in \mathrm{Vect}(P)$. Let $f = \chi_Y$ be the characteristic function of Y. Then for $X \in \mathfrak{h}$, we can consider the distributional derivative $D_X(f)$. To see that Y is J-invariant, it suffices to see that for all $X \in \mathfrak{j}$, $D_X(f) = 0$. We claim in fact that $D_X(f) = 0$ for all $X \in \mathfrak{h}$. It suffices to show this for X in a set that generates $\mathfrak{h}$ as a Lie algebra, and since the root spaces of α with non-trivial root have this generating property (as in section 2 by virtue of iii(c) in the hypothesis of Theorem 1.1), it suffices to show $D_X(f) = 0$ for any such root vector X of α.

LEMMA 3.1. *If X is a root vector for α with non-trivial root, then for a.e. $y \in Y$, the integral curve $u_y(t)$, for X through y is defined for all t.*

Proof: For $y \in Y$, let $r(y) = \sup\{t \in \mathbf{R} \mid u_y(t)$ is defined on $[0, t)\}$. We claim that for a.e. y, $r(y) = +\infty$. If not, then there is a set $Z \subset Y$ of positive measure and $R > 0$ such that $r(y) < R$ for all $y \in Z$. We can then choose $Z_1 \subset Z$ of positive measure and $r > 0$ such that $r(y) > r$ for $y \in Z_1$. Since $\alpha(t)$ preserves a finite measure on Y, we can apply Poincaré recurrence and deduce that for a.e. $y \in Z_1$, there is a sequence $t_n \to +\infty$ such that $\alpha(t_n)y \in Z_1$. If $\mathrm{Ad}(\alpha(t))X = e^{\lambda t}X$ where $\lambda > 0$, then we have $r(\alpha(t_n)y) > re^{\lambda t_n} \to +\infty$. This contradicts $r(\alpha(t_n)y) < R$. Thus $r(y) = +\infty$ for a.e. $y \in Y$. If $\lambda < 0$, we use the same argument with $t_n \to -\infty$. A similar argument shows $u_y(t)$ is defined for all $t < 0$, completing the proof of the lemma.   □

We can summarize the situation as follows.

LEMMA 3.2. *Let $Y^* = \{u_y(t)y \mid y \in Y, t \in \mathbf{R}, u_y$ the integral curve through y for X as above$\}$. Then $Y^* \subset P$ is a subset of (possibly infinite) positive measure. There is an action of the group $\{\alpha(t)u(s) \mid t, s \in \mathbf{R}\}$ on this set which is measure preserving. The group $\{\alpha(t)u(s)\}$ is isomorphic to the $ax + b$ group, $\mathbf{R}_\times^+ \ltimes \mathbf{R}$. There is a set of finite positive measure $Y \subset Y^*$ that is $\{\alpha(t)\}$-invariant.*

We now complete the proof of Theorem 1.1.

To see that $D_X(f) = 0$, we want to show that Y is $\{u(s)\}$-invariant. Consider the unitary representation of $\{\alpha(t)u(s)\}$ on $L^2(Y^*)$. The vector $f \in L^2(Y^*)$ (since Y has finite measure) and is $\{\alpha(t)\}$-invariant. By the unitary representation theory of the $ax + b$ group, this implies that f is invariant under all $\{\alpha(t)u(s)\}$, completing the proof. (See [Z3, Corollary 2.3.7] for precisely this result on the $ax + b$ group.)

Remark 3.3: Rather than call upon the unitary representation theory of the $ax + b$ group, one could proceed by adapting the standard uniform contraction-expansion argument to deduce invariance of f along the foliation given by the non-vanishing vector field X. However, the result on unitary representations we used is a convenient and easily usable way of incorporating this argument when one is in a group theoretic context. An examination of the proof of this representation theoretic result ([Z3, Corollary 2.3.7]) shows that it is in fact just a Fourier transform version of the "more direct" contraction-expansion argument.

4. Non-connected Algebraic Hulls

We have two commuting groups G and B acting on a principal J-bundle $\pi : P \to M$. Let E be the space of ergodic components for the G-action on M. For each $e \in E$, let M_e be the corresponding component, let $R_e \subset J$ be the (real) algebraic hull of the action of G on the principal J-bundle $\pi^{-1}(M_e) \to M_e$. For each $e \in E$, R_e is a reductive group with compact center. (The algebraic hulls are "real algebraic", i.e. $\mathbf{R}$-point of $\mathbf{R}$-groups.) If R_e is connected for a set of $e \in E$ of positive measure then the argument in section 2 applies. If not then we proceed as follows.

LEMMA 4.1. *Suppose B and G commute and we have an ergodic BG action on a space X and let $X \to E$ be the ergodic decomposition into G-ergodic components. Suppose $\alpha : BG \times J \to J$ is a cocycle. Let $R_e \subset J$ be the algebraic hull of $\alpha|_{G \times X_e}$, for each $e \in E$. Then*
1) Almost all R_e, $e \in E$, are equal. Denote it by R.
2) The algebraic hull of α is contained in $N_J(R)$.

Proof: Let j, r_e be the Lie algebras of J, R_e, respectively. The map $\varphi : E \to Gr(j)/Ad(J)$ given by $\varphi(e) = r_e$ is measurable. It is B-invariant since B centralizes G. (I.e. the algebraic hulls, R_e and R_{be} are the "same", namely are the same conjugacy class.) Since B is ergodic on E (since GB is ergodic on X), and $Ad(J)$ acts tamely on $Gr(j)$, φ is constant a.e. Thus, we can, by changing α only up to coboundary, assume $\alpha|_{G \times X_e}$ takes values in a group R_e so that $R_e^0 = R^0$ are all identical.

Now consider the set $\mathcal{S}$ of algebraic subgroups $S \subset J$, s.t. $S^0 = R^0$. Then any element of J that conjugates two such subgroups is in $N(R^0)$. Further, $N(R^0)$ acts tamely on $\mathcal{S}$ (since this action can be embedded in the action of $N(R^0)$ on finite subsets of J/R^0 which is algebraic). It follows in a manner similar to the previous paragraph, that all R_e are conjugate, and hence by a cohomology change, we can assume all R_e are equal, say $R_e = R$. This proves (1).

To show (2): Fix $b \in B$ and $e \in E$. For $x \in X_e$, let $\varphi(x) = \alpha(b, x)$. Then since b centralizes G, we have from the cocycle identity:

$$\varphi(gx)^{-1}\alpha(g, bx)\varphi(x) = \alpha(g, x) \, . \tag{$*$}$$

This implies that the map $\overline{\varphi} : X_e \to R\backslash J/R$ given by projection of $\varphi : X \to J$ satisfies $\overline{\varphi}(gx) = \overline{\varphi}(x)$ for $x \in X_e$. Thus $\overline{\varphi}$ is constant a.e. Hence, we can write

(i) $\varphi(x) = \theta_1(x)j_0\theta_2(x)$ where $\theta_i : X_e \to R$ and $j_0 \in J$.

We also then have

(ii) $\varphi(gx) = \theta_1(gx)j_0\theta_2(gx)$

and substituting (i) and (ii) into $(*)$ and simplifying, we have

$$\theta_1(gx)^{-1}\alpha(g, bx)\theta_1(x) = j_0\theta_2(gx)\alpha(g, x)\theta_2(x)^{-1}j_0^{-1} \ .$$

In particular $j_0\beta(g, x)j_0^{-1} \in R$, where β is the cocycle (equivalent to α) given by $\beta(g, x) = \theta_2(gx)\alpha(t, x)\theta_2(x)^{-1}$. Since the algebraic hull of β equals R and β takes values in R, we deduce in particular that the Zariski closure of the group generated by $\{\beta(g, x)\}$ is in fact R. Thus $j_0 \in N_J(R)$ and from equation (1), we deduce $\alpha(b, x) \in N_J(R)$. This holds for all $(b, x) \in B \times X$, and since $\alpha(G \times X) \subset R$, assertion (2) is proved.

Now return to our geometric situation in which G is simple of higher **R**-rank. Fix an ergodic component $X \subset M$ for the $B \cdot G$ action. Let R be as in the lemma, i.e. the algebraic hull of the G action on a.e. G-ergodic component of X. Then, by the lemma, the algebraic hull for the action of $B \cdot G$ on $\pi^{-1}(X) \to X$ is contained in $N_J(R)$. By superrigidity, we know that R^0 is isomorphic to $\overline{\pi}(G)$ (in which $\pi(G)$ is of finite index) times a compact subgroup for some homomorphism π.

Notice that:

LEMMA 4.2. *If $Z_J(\pi(G))$ is compact, then $N_J(R)/R^0$ is a compact group.*

We let $\alpha : B \cdot G \times X \to J$ be the cocycle corresponding to the action on $\pi^{-1}(X)$, with $\text{Im}(\alpha) \subset N_J(R)$. Let $\overline{X} = X \times_\alpha N_J(R)/R^0$. Thus, $B \cdot G$ acts preserving the product measure. (However the action of BG on $\overline{X}$ is not necessarily ergodic.) Let $\overline{\alpha}$ be the lift of α to $\overline{X}$. Thus $\overline{\alpha}$ is the cocycle of the J-bundle

$$\overline{P}$$
$$\downarrow$$
$$\overline{X}$$

where $\overline{P}$ is the pull-back to $\overline{X}$ of $\pi^{-1}(X) \to X$. Standard arguments (see [Z3, Proposition 9.2.6]) show that for each ergodic component of the action of G on $\overline{X}$, the algebraic hull is R^0. We can thus obtain a π-simple section on each ergodic component via superrigidity. Arguing as in earlier sections, this implies that there is either a B or G finite invariant measure on $\overline{P}$ that projects to a set of positive measure on $\overline{X}$ (where $\overline{X}$ has the product measure). It follows that either B or G has a finite invariant measure on

$\pi^{-1}(X) \to X$ that projects to a set of positive measure on X. Thus, the same is true for $P \to M$.

Let ν be this finite measure on P. This can be described as $\int_{m \in M_0}^{\oplus} \nu_m$ where $M_0 \subset M$ is of positive measure and is B or G-invariant. Fixing a positive definite inner product on j (=Lie algebra of J) defines a $B \cdot G$ invariant Riemannian metric along the fibers of P. Fix $\varepsilon, r > 0$. Let $A_m^{\varepsilon, r} \subset \pi^{-1}(m)$ be the union of all balls of radius ε in $\pi^{-1}(m)$ whose ν_m-measure is at least r. Then for suitable ε, r, $\bigcup_m A_m^{\varepsilon, r}$ will be a B or G-invariant set of positive finite Lebesgue measure. The argument now proceeds as before.

References

[BL]   Y. BENOIST, F. LABOURIE, Sur les espaces homogènès modèles de variétés compactes, Pub. Math. I.H.E.S. 76 (1992), 99-109.

[G]   W. GOLDMAN, Projective geometry on manifolds, Univ. of Maryland Lecture Notes, 1988.

[R]   J. RATCLIFFE, Foundations of Hyperbolic Manifolds, Springer-Verlag, New York, 1994.

[Z1]   R.J. ZIMMER, Discrete groups and non-Riemannian homogeneous spaces, Jour. Amer. Math. Soc. 7 (1994), 159-168.

[Z2]   R.J. ZIMMER, Topological superrigidity, preprint.

[Z3]   R.J. ZIMMER, Ergodic Theory and Semisimple Groups, Birkhauser, Boston, 1984.

[Z4]   R.J. ZIMMER, Orbit equivalence and rigidity of ergodic actions of Lie groups, Ergodic Theory and Dynamical Systems 1 (1981), 237-253.

François Labourie   Shahar Mozes   Robert J. Zimmer
C.N.R.S.   Institute of Mathematics   Department of Mathematics
Ecole Polytechnique   Hebrew University   University of Chicago
Palaiseau   Jerusalem   Chicago, IL 60637
France   Israel   USA

Submitted: October 1994
Revised Version: September 1995

Mathematical Research Letters **2**, 75–77 (1995)

ON THE NON-EXISTENCE OF COCOMPACT LATTICES
FOR $SL(n)/SL(m)$

François Labourie and Robert J. Zimmer

In this note we continue the investigation of manifolds locally modelled on non-Riemannian homogeneous spaces, along the lines developed in [3], [4]. If H/J is a homogeneous space of a Lie group H, a natural but special class of compact forms of H/J (i.e., compact manifolds locally modelled on H/J) are those compact manifolds of the form $\Gamma\backslash H/J$ where $\Gamma \subset H$ is a cocompact lattice for H/J, i.e., a discrete subgroup acting freely and properly discontinuously on H/J. It remains an open problem to classify those H/J that admit such a discrete subgroup or those admitting a compact form. This problem is not completely solved even for the basic test case of $H = SL(n, \mathbb{R})$, $J = SL(m, \mathbb{R})$, $2 \leq m < n$, and where the embedding $J \subset H$ is the standard one. (See [2] for a survey on the general problem.) In this note we prove:

Theorem 1. *If $m \geq 2$ and $n \geq m+3$, then $SL(n, \mathbb{R})/SL(m, \mathbb{R})$ does not admit a cocompact lattice.*

Remarks.

(1) It is natural to believe that there are no compact forms not only no cocompact lattices.
(2) The results of [3] establish that for $n \geq 2m$, $m \geq 2$, $n \geq 5$, there are no compact forms.
(3) For $2n/3 < m$, the assertion in Theorem 1 has been proven by T. Kobayashi [1] by completely different methods.

Proof. Suppose such a group Γ existed. Set $H = SL(n, \mathbb{R})$, $J = SL(m, \mathbb{R})$, and $G = SL(n-m, \mathbb{R})$, with the natural embedding in the centralizer of J in H. Since G centralizes J we have a natural action of G on the compact manifold $M = \Gamma\backslash H/J$ given by $g \cdot (\Gamma h J) = \Gamma h g^{-1} J$, and this action preserves a smooth volume form on M [4]. The projection $\Gamma\backslash H \to \Gamma\backslash H/J$ is a principal bundle with structure group J on which G acts by principal bundle automorphisms. Following the arguments of [3], [4], since

Received December 21, 1994.

Research of second author partially supported by NSF grant.

$\mathbb{R}$-rank$(G) \geq 2$ from the assumption $n \geq m+3$, we can apply superrigidity for actions on principal bundles. We deduce that there are

(1) a homomorphism $\pi : G \to J$,
(2) a compact subgroup $K \subset Z_J(\pi(G))$, and
(3) a measurable section $s : A \to \Gamma \backslash H$, where $A \subset \Gamma \backslash H / J$ is a G-invariant set of positive measure, such that

$$s(gm) = gs(m)\pi(g)^{-1}c(g,m),$$

where $c(g,m) \in K$.

We can, of course, view π as a representation $\pi : G \to SL(m,\mathbb{R})$ and consider separately the following cases:

(1) π is irreducible;
(2) π is reducible.

In case (i), it follows that the centralizer $Z_J(\pi(G))$ is itself compact. One can then apply the main result of [3] to deduce that this is impossible using the existence of an $\mathbb{R}$-split one parameter subgroup in $Z_H(JG)$. (See [3] for details.) It thus suffices to consider the case in which π is reducible. We shall need the following simple algebraic fact.

Lemma 2. *Let $SL(2,\mathbb{R}) \subset SL(3,\mathbb{R})$ be the standard embedding and $A \subset SL(2,\mathbb{R})$ the diagonal matrices. Then for any finite dimensional representation of $SL(3,\mathbb{R})$, there is a non-zero vector fixed by A.*

We postpone the proof of Lemma 2 and proceed with the proof of Theorem 1.

If we let μ be the finite G-invariant volume on M restricted to A, then $\int_{c \in K} c_*(s_*\mu)d\mu_K(c)$ is a finite $gr(\pi)$-invariant measure on $\Gamma \backslash H$ where μ_K is the Haar measure of K and $gr(\pi)$ is the graph of π. (Cf. [3], proof of Theorem 2.2.) It follows that for any subgroup $L \subset gr(\pi)$ and any conjugate hLh^{-1} in H, there is a hLh^{-1}-invariant probability measure on $\Gamma \backslash H$. Since Γ acts properly on H/J, J acts properly on $\Gamma \backslash H$, and hence, any closed subgroup of J that preserves a probability measure on $\Gamma \backslash H$ must be compact. (Cf. [5], proof of Lemma 2.7.) Therefore, to show the existence of Γ is impossible, it suffices to find a non-compact closed $L \subset gr(\pi)$ and an element $h \in H$ such that $hLh^{-1} \subset J$.

Consider $SL(2,\mathbb{R}) \subset G = SL(n - m,\mathbb{R}) \subset H$ via the standard embedding. Thus, this copy of $SL(2,\mathbb{R})$ acts on the standard basis $e_1, \ldots, e_n$ of $\mathbb{R}^n$ by fixing all e_i except e_{m+1} and e_{m+2}. Since π is reducible, by lemma 2 there are two linearly independent vectors in $\mathbb{R}^m$ that are invariant under $\pi \mid A$, where $A \subset SL(2,\mathbb{R})$ is the split Cartan subgroup. Conjugating π by an element $j \in J = SL(m,\mathbb{R})$, we can assume these vectors are the first two standard basis vectors. Let $L = gr(\pi \mid A)$. Then it is clear that if w is the element of the Weyl group of $SL(n, \mathbb{R})$ with respect to the standard basis that interchanges e_k and e_{k+m} for $k = 1, 2$ and fixes all the others, then $wjLj^{-1}w^{-1} \subset J$. As we have already observed, this suffices to complete the proof of Theorem 1. $\square$

Proof of Lemma 2. We work with the Lie algebras $\mathfrak{a} \subset \mathfrak{sl}(2, \mathbb{R}) \subset \mathfrak{sl}(3, \mathbb{R})$. Let (π, V) be a representation of $\mathfrak{sl}(3, \mathbb{R})$, which we may assume is non-trivial. Write $\pi \mid \mathfrak{sl}(2, \mathbb{R}) = (\pi_{\text{even}}, V_{\text{even}}) \oplus (\pi_{\text{odd}}, V_{\text{odd}})$ where π_{even} is the sum of those irreducible components of $\pi \mid \mathfrak{sl}(2, \mathbb{R})$ with even highest weight and π_{odd} the corresponding sum for odd highest weight. By the standard theory of representations for $\mathfrak{sl}(2, \mathbb{R})$, it suffices to see that π_{even} is non-zero. However, there is $X \in \mathfrak{sl}(3, \mathbb{R})$ (namely the matrix E_{13}) that is an eigenvector of weight 1 under representation $ad_{\mathfrak{sl}(3,\mathbb{R})} \mid \mathfrak{a}$. Thus, $\pi(X)(V_{\text{odd}}) \subset V_{\text{even}}$. If $V_{\text{even}} = (0)$, then $\pi(X) = 0$ which is impossible by the simplicity of $\mathfrak{sl}(3, \mathbb{R})$. This completes the proof. $\square$

Acknowledgement

This work was completed while the second author was a visitor at l'Université de Paris-sud, Orsay, and he wishes to express his appreciation for the hospitality shown him.

References

1. T. Kobayashi, *A necessary condition for the existence of compact Clifford-Klein forms of homogeneous spaces of reductive type*, Duke Math. J. **67** (1992), 653–664.
2. F. Labourie, *Quelques résultats récents sur les espaces localement homogènes compacts,* preprint, 1993.
3. F. Labourie, S. Mozes and R. J. Zimmer, *On manifolds locally modelled on non-Riemannian homogeneous spaces*, preprint.
4. R. J. Zimmer, *Discrete groups and non-Riemannian homogeneous spaces*, Jour. Amer. Math. Soc. **7** (1994), 159–168.
5. ——— *Ergodic theory and the automorphism group of a G-structure,* in Group representations, ergodic theory, operator algebras, and mathematical physics, C. C. Moore, ed., Springer, New York, 1987.

DÉPARTMENT DE MATHÉMATIQUES, UNIVERSITÉ PARIS-SUD, 91405 ORSAY, FRANCE
E-mail address: labourieorphee.polytechnique.fr

DEPARTMENT OF MATHEMATICS, UNIVERSITY OF CHICAGO, CHICAGO, IL 60637
E-mail address: zimmer@math.uchicago.edu

8

Stationary Measures and Structure Theorems for Lie Group Actions

Annals of Mathematics, **156** (2002), 565–594

A structure theorem for actions of semisimple Lie groups

By Amos Nevo and Robert J. Zimmer[*]

Abstract

We consider a connected semisimple Lie group G with finite center, an admissible probability measure μ on G, and an ergodic (G,μ)-space (X,ν). We first note (Lemma 0.1) that (X,ν) has a unique maximal projective factor of the form $(G/Q,\nu_0)$, where Q is a parabolic subgroup of G, and then prove:

1. *Theorem* 1. If every noncompact simple factor of G has real rank at least two, then the maximal projective factor is nontrivial, unless ν is a G-invariant measure.

2. *Theorem* 2. For any G of real rank at least two, if the action has positive entropy and fails to have nontrivial projective factor, then (X,ν) has an equivariant factor space with the same properties, on which G acts via a real-rank-one factor group.

3. *Theorem* 3. Write $\nu = \nu_0 * \lambda$, where λ is a P-invariant measure, $P = MSV$ a minimal parabolic subgroup [F2], [NZ1]. If the entropy $h_\mu(G/P,\nu_0)$ is finite, and every nontrivial element of S is ergodic on (X,λ) (or just a well chosen finite set, Theorem 9.1), then (X,ν) is a measure-preserving extension of its maximal projective factor.

4. The foregoing results are best possible (see §11, in particular Theorem 11.4).

We also give some corollaries and applications of the main results. These include an entropy characterization of amenable actions, an explicit entropy criterion for the invariance of ν, and construction of a projective factor for an action of a lattice in G on a compact metric space.

[*]A. Nevo was supported, in part, by BSF. R. J. Zimmer was supported, in part, by NSF and BSF.

1991 *Mathematics subject classification*: 22D40, 28D15, 47A35, 57S20, 58E40, 60J50.

Introduction and statements of results

Let G be a connected noncompact semisimple Lie group, with finite center. Let μ denote an admissible probability measure on G, namely the support of μ generates G as a semigroup, and μ^{*k} is absolutely continuous with respect to Haar measure for some $k \geq 1$. We let $(X, \mathcal{B}, \nu)$ denote a standard Borel-measurable space, with probability measure ν, and assume that G has a Borel-measurable action on $(X, \mathcal{B})$. We assume that ν is a μ-stationary measure, namely $\nu = \int_G g\nu d\mu(g)$, or equivalently $\int_X f(x)d\nu = \int_G \int_X f(gx)d\nu(x)d\mu(g)$ for every bounded Borel function f. We will refer to (X, ν) as a (G, μ)-space, and recall that every measure ν stationary under an admissible measure μ is G-quasi-invariant (see e.g. [NZ1, Lemma 1.1]). The measure ν will be called ergodic if every set $A \in \mathcal{B}$ satisfying $\nu(A \triangle gA) = 0$ for all $g \in G$ has measure zero or one.

Let P denote a minimal parabolic subgroup of G. There are 2^r closed subgroups Q of G containing P, where $r = \mathbb{R}$-rank (G) (see e.g. [M, Ch. I, 1.2.3]). The G-spaces of the form G/Q where Q contains P, are therefore all the G-factors of the maximal (Poisson) boundary G/P, and will be referred to below as projective homogeneous G-spaces. For a (G, μ)-space (X, ν), any G-equivariant factor of the form $(G/Q, \nu_0)$ will be called below a projective factor. We now note the following general fact, which will be proved in Section 8 below:

LEMMA 0.1. *Let G be a connected noncompact semisimple Lie group with finite center, μ an admissible probability measure on G, and (X, ν) a (G, μ)-space. Then there exists a parabolic subgroup $Q \subset G$, satisfying the following conditions:*

1. *There exists a G-equivariant measurable factor map*

$$\varphi : (X, \nu) \to (G/Q, \nu_0),$$

   where ν_0 is the unique μ-stationary measure on G/Q. (Here φ is defined on a G-invariant ν-conull set contained in X.)

2. *G/Q is the unique maximal projective factor of (X, ν); i.e. any other projective factor of (X, ν) is a factor of G/Q. In particular Q is uniquely determined (up to conjugation).*

3. *The map φ is uniquely determined up to ν-null sets.*

We now formulate the main results of the present paper. We begin with the following existence theorem for nontrivial projective factors.

THEOREM 1. *Let G be a connected noncompact semisimple Lie group with finite center, all of whose simple noncompact factors are of real rank at least two. Let μ be an admissible probability measure on G, and let (X, ν) be an ergodic (G, μ)-space. Then the maximal projective factor G/Q is trivial if and only if the stationary measure ν is G-invariant.*

We note that the real rank assumption stated in Theorem 1 is essential. Theorem 1 does not hold for semisimple groups of real rank at least two which have factors of $\mathbb{R}$-rank one. Indeed, in [NZ2, Thm. B] it was shown that for every simple group H of real rank one and every admissible measure μ, there exists a smooth compact manifold (Y, ν) which is an ergodic (H, μ)-space, such that ν is not H-invariant, but (Y, ν) does not have a nontrivial projective factor. To construct faithful actions of higher rank groups, let $\Gamma \subset H$ be a lattice subgroup, and consider the product action of $(H \times H, \mu \times \mu)$ on $(H/\Gamma \times Y, m_{H/\Gamma} \times \nu)$. This action is an ergodic and faithful action of the group $G = H \times H$ which has real rank two. It is easily seen this action does not admit a nontrivial projective factor, so the conclusion of Theorem 1 fails for G.

Let us note however that by construction, in the previous example the space under consideration, namely $X = (H/\Gamma \times Y, m_{H/\Gamma} \times \nu)$ admits a G-equivariant factor space (namely (Y, ν)), on which a real-rank-one factor group H of G acts, without an invariant measure but with trivial maximal projective factor. It is a remarkable phenomenon that this state of affairs holds in complete generality. Indeed, as will be noted in Theorem 2 below, all actions that fail to have a nontrivial projective factor are of this form (or have an invariant measure).

Before stating Theorem 2 let us recall the fundamental invariant associated with the (G, μ)-space (X, ν) namely the entropy, given by (see [F2], [NZ2])

$$ h_\mu(X, \nu) = \int_G \int_X - \log \frac{dg^{-1}\nu}{d\nu}(x) d\nu(x) d\mu(g) \ . $$

We refer to [NZ2] for a discussion of the basic properties of entropy for semisimple Lie group actions, and some examples and applications. In particular, we recall that the stationary measure ν on X is G-invariant if and only if $h_\mu(X, \nu) = 0$.

We can now state the following generalization of Theorem 1, which shows that for an action with positive entropy, the only obstruction to the existence of a nontrivial projective factor is the appearance of certain actions of a real-rank-one group, as a factor of the original action.

THEOREM 2. *Let G be a connected semisimple Lie group with finite center, μ admissible, and (X, ν) an ergodic (G, μ)-space with positive entropy. Suppose (X, ν) fails to have a nontrivial projective factor. Then there exists a factor group G_1 of G with $\mathbb{R}$-rank $(G_1) = 1$, a nontrivial (G_1, μ_1)-space*

(X_1, ν_1), and a factor map $\rho : (X, \nu) \to (X_1, \nu_1)$, equivariant with respect to G and G_1. Here $\mu_1 = p_*(\mu)$, $p : G \to G_1$ the natural map, and (X_1, ν_1) has positive entropy and does not admit a nontrivial projective factor.

We note that given the real-rank-one counterexamples constructed in [NZ1, Thm. B] and mentioned briefly above, Theorem 1 and Theorem 2 together constitute the best possible result regarding the existence of nontrivial projective factors for actions of arbitrary semisimple Lie groups with stationary measure.

We now turn to some applications of Theorems 1 and 2. First recall that, as noted in Lemma 0.1, if $Q \subset G$ is a parabolic subgroup, then the μ-stationary measure ν_0 on G/Q is unique (see [F2, Thm. 2.3]). We denote the entropy $h_\mu(G/P, \nu_0)$ of the maximal boundary G/P, P a minimal parabolic subgroup, by $H(\mu)$. An admissible measure μ on G will be called a measure of finite entropy if $H(\mu)$ is finite. We note that every admissible measure of compact support has finite entropy (see [N1]). Furthermore, we recall that $h_\mu(X, \nu) \leq h_\mu(G/P, \nu_0)$ for any (G, μ)-space (X, ν) (see [NZ2, Cor. 2.5]).

We now define the following two numerical invariants associated with an admissible measure of finite entropy.

1. $H_{\min}(\mu) = \min\{h_\mu(G/Q, \nu_0) \mid Q$ a proper parabolic subgroup of $G\}$. Note that $H_{\min}(\mu) = \min\{h_\mu(G/Q, \nu_0) \mid Q$ a maximal proper parabolic subgroup of $G\}$, by monotonicity of entropy under passage to factors (see [NZ2]).

2. $H_{na}(\mu) = \max\{h_\mu(G/Q, \nu_0) \mid Q$ a nonamenable parabolic subgroup of $G\}$. Note that $0 < H_{na}(\mu) < H(\mu)$, provided $\mathbb{R}$-rank $(G) \geq 2$, by strict monotonicity of entropy under passage to factors (see [NZ2]).

We can now formulate the following entropy criterion for the existence of a G-invariant probability measure, which follows from Theorem 1.

COROLLARY 1.   Let (G, μ) and (X, ν) be as in Theorem 1. If $h_\mu(X, \nu) < H_{\min}(\mu)$ then ν is G-invariant, and $h_\mu(X, \nu) = 0$.

Note that by Theorem 2 this corollary holds for ergodic actions with a stationary measure of any connected semisimple group with finite center and real rank at least two, if the condition that the action does not factor over an action of a real-rank-one factor group is satisfied.

For locally compact second countable groups satisfying property T of Kazhdan (a class which includes the simple Lie groups $\mathrm{Sp}(n, 1)$, $n \geq 2$ of $\mathbb{R}$-rank one) a more general but less explicit result holds. Namely, every ergodic quasi-invariant measure (not necessarily equivalent to a stationary measure) with sufficiently small entropy is necessarily invariant (see [N2]).

Another corollary of Theorem 1 pertains to actions of lattice subgroups of G on compact metric spaces.

COROLLARY 2. *Let G be as in Theorem 1, and let Γ be a lattice subgroup of G. Let Y be any compact metric Γ-space, and assume Y does not have a Γ-invariant probability measure. Then there exists a measurable Γ-equivariant factor map $\varphi : (Y, \eta) \to (G/Q, \eta_0)$, where Q is a proper parabolic subgroup of G, and η, η_0 are Γ-quasi-invariant measures.*

Proof. Consider the G space $X = G/\Gamma \times Y$ induced to G (see §5). Since Y does not have a Γ-invariant probability measure, X does not have a G-invariant probability measure (see [NZ1, §6], for more details). For any admissible measure μ on G, there exists an ergodic μ-stationary measure ν on X, projecting to the G-invariant measure on G/Γ under the natural map $p : X \to G/\Gamma$, namely $\nu = \int_{G/\Gamma} \eta_{x\Gamma} dm_{G/\Gamma}(x\Gamma)$. By Theorem 1, (X, ν) has a nontrivial projective factor G/Q. By standard arguments (see [NZ1, §6]), restricting the map $\varphi : X \to G/Q$ to an appropriate fiber of p (and conjugating the actions if necessary) we obtain a Γ-equivariant map $\varphi : (Y, \eta) \to (G/Q, \eta_0)$, where $\eta_0 = \varphi_*(\eta)$. $\square$

In general, the nontrivial maximal projective factor $(G/Q, \nu_0)$ of (X, ν) constructed in Theorems 1 and 2 may have strictly smaller entropy, namely $h_\mu(X, \nu) > h_\mu(G/Q, \nu_0) > 0$. We refer to [NZ2] for some examples of this phenomenon.

Under suitable ergodicity assumptions the conclusions of Theorems 1 and 2 can be significantly strengthened, and the existence of a projective factor of full entropy can be established.

THEOREM 3. *Let G be a connected semisimple Lie group of real rank at least two with finite center. Let μ be an admissible measure on G of finite entropy, and (X, ν) a (G, μ)-space. Let $\nu = \nu_0 * \lambda$, where λ is the P-invariant measure on $X_0 \subset X$ corresponding to ν (see §5). Let S denote a maximal $\mathbb{R}$-split torus of G contained in P. If every nontrivial $s \in S$ is ergodic on (X_0, λ), then the unique maximal projective factor $\varphi : (X, \nu) \to (G/Q, \nu_0)$ is an extension with relatively invariant measure. In particular, the maximal projective factor has full entropy, namely $h_\mu(X, \nu) = h_\mu(G/Q, \nu_0)$, λ is Q-invariant, and (X, ν) is induced from the probability measure-preserving action of Q on (X_0, λ).*

Actually, it suffices that a certain *finite* set of elements in S act by ergodic transformations of (X_0, λ). The statement of this result is somewhat more technical and is deferred to Section 9 (Theorem 9.1).

We can now formulate the following entropy criterion for amenability of the (G, μ)-space (X, ν) (see [Z1] for the basic facts regarding amenability). This criterion implies that the condition of sufficiently high entropy (more precisely, $h_\mu(X, \nu) > H_{na}(\mu)$), ensures that the maximal projective factor has full, and maximal, entropy, and thus the action is amenable.

COROLLARY 3.   *Let (G, μ) and (X, ν) be as in Theorem 3, and G/Q the maximal projective factor of (X, ν). If $h_\mu(X, \nu) > H_{na}(\mu)$, then*

1. *The parabolic subgroup Q is a conjugate of the minimal parabolic subgroup, P.*

2. *(X, ν) is an extension of $(G/P, \nu_0)$ with relatively G-invariant measure, or equivalently, (X, ν) is induced from a probability measure-preserving action of P.*

3. *(X, ν) is an amenable action of G.*

4. *$h_\mu(X, \nu) = h_\mu(G/P, \nu_0) = H(\mu)$.*

Proof. The corollary follows immediately from Theorem 3 above and Theorem 4.1 of [NZ2].   □

We note that countably many nonisomorphic actions of certain higher-rank groups for which the maximal projective factor is nontrivial but has less than full entropy were constructed in [NZ2]. The method is by induction from an action of a proper parabolic subgroup Q of G on a space Y, where Q acts on Y via a real rank one factor group. In fact, the construction of such counterexamples by induction from parabolic subgroups turns out to be the most general one possible, at least when the simple factors of G have real rank at least two. This partial converse to Theorem 3 is formulated and proved in Section 11 (Theorem 11.4).

1. Some applications of the structure theorems

Let G be a connected semisimple Lie group with finite center, no compact factors, and real rank at least two. Let (X, m) be an irreducible measure-preserving action of G on a probability space. Let $\alpha : (X \times G/P, m \times \nu_0) \to (Y, \nu)$ be a G-equivariant factor map, where ν_0 is the unique K-invariant probability measure on G/P (K a fixed maximal compact subgroup).

In [NZ3] it is shown, using Theorem 3 of the present paper (or [NZ1, Thm. A]), that (Y, ν) is a measure-preserving extension of its maximal projective factor. This result is used to give an alternative proof of the intermediate factors theorem [Z2, Thm. 4.1], which asserts the following. If there exists a G-equivariant factor map $\beta : (Y, \nu) \to (X, m)$ such that $\beta \circ \alpha = \text{proj}_X$, then (Y, ν) is equivariantly isomorphic to $(X \times G/Q, m \times \nu_0)$ for some parabolic subgroup Q.

SEMISIMPLE LIE GROUPS   571

As is well known [Z2], the intermediate factors theorem implies Margulis' theorem on the classification of Γ-factors of the maximal boundary G/P (see [M, Ch. IV, Thm. 2.11]), which states the following. Let Γ be an irreducible lattice in a connected semisimple Lie group with finite center of real rank at least two without compact factors. Then every Γ-equivariant factor $\varphi : (G/P, m) \to (Y, \eta)$, where m is G-quasi-invariant, satisfies: (Y, η) is isomorphic as a Γ-space to $(G/Q, m)$, for some parabolic subgroup $Q \subset G$.

We note that Theorem 3 also generalizes Theorem A of [NZ1], by weakening the P-mixing condition that appears in its assumptions to ergodicity of (finitely many, according to Theorem 9.1) elements of $S \subset P$ on (X_0, λ). As a consequence, a number of the results in [NZ2] that depend on Theorem A can be improved. This applies, in particular, to the rigidity of entropy phenomenon ([NZ2, Thm. 2.7]) which holds under the ergodicity assumptions of Theorem 3 (or Theorem 9.1), and does not require P-mixing. The proofs of Theorem 3 and [NZ1, Thm. A] are quite different, and we refer to [NZ4] for more on this point, and for a general survey on entropy and structure theory.

We remark that Theorems 1, 2 and 3 (as well as their consequences, in particular, the generalization of the intermediate factors theorem) can be generalized appropriately to apply to arbitrary higher rank semisimple algebraic groups over local fields of characteristic zero. The same applies to certain groups of automorphisms of nontrivial products of semihomogeneous trees. Consequences of such a generalization include triviality of stabilizers in measure-preserving actions of S-arithmetic groups, and applications to orbit equivalence. These results will be considered elsewhere.

Remark about ergodicity. Given a compact metric G-space, there exists a one-to-one affine isomorphism between the set $P_P(X)$ of probability measures invariant under a given minimal parabolic group, and the set $P_\mu(X)$ of measures stationary under a given admissible measure μ on G (see [F2, §2], and [NZ1, Thm. 1.4, Cor. 1.5]). Therefore the extreme points of both compact convex sets correspond to one another. In both sets, extremality easily implies ergodicity, and the converse of course holds for invariant measures. It is easily seen that in any nontrivial convex decomposition $\nu = a\nu_1 + (1 - a)\nu_2$ in $P_\mu(X)$, both ν_1 and ν_2 must be equivalent to ν if ν is ergodic. The space of stationary measures equivalent with a given ergodic ν can thus be identified with the space of positive eigenfunctions in $L^1(\nu)$ (with norm one) of the operator $\pi_1(\mu)f(x) = \int_G f(g^{-1}x)\frac{dg\nu}{d\nu}(x)d\mu(g)$. This space consists of the constant function one if and only if the ergodic μ-stationary measure ν is extremal in $P_\mu(X)$. It is therefore important to note that throughout the paper, we always assume that the stationary measure satisfies only the condition of ergodicity, rather than the (perhaps stronger) condition of extremality.

2. Structure theory of G/P

Let G be a connected noncompact semisimple Lie group with finite center Z. Then G/Z can be identified with the connected component of the identity of the group of real points of a connected semisimple algebraic group $\mathbb{G}$ defined over $\mathbb{R}$. Given a (G, μ)-space (X, ν) we can pass to the G-factor space of Z-orbits, without affecting the assumptions or conclusions of any of the results stated. Therefore we can and will assume without loss of generality that G has trivial center, and use the associated structure of an $\mathbb{R}$-algebraic group. We let:

$$
\begin{aligned}
S &= \text{maximal } \mathbb{R}\text{-split torus in } G, \text{ with Lie algebra } \mathfrak{s}. \\
\Phi^+ &= \text{the set of positive roots of } \mathfrak{s} \text{ in } \mathfrak{g}, \text{with a given fixed ordering.} \\
P_\theta &= \text{the standard parabolic subgroup corresponding to } \theta \subset \Phi^+.
\end{aligned}
$$

We recall the construction of a standard parabolic subgroup P_θ associated with a subset $\theta \subset \Delta$, where $\Delta \subset \Phi^+$ is the set of simple positive roots. We follow the notation of [M, Ch.I §§1.1, 1.2 and Ch. II §3]. If $\mathbb{T}$ is a maximal torus in $\mathbb{G}$ containing S, then we let $\eta(\theta)$ be the set of roots of $\mathbb{T}$ in $\mathbb{G}$, whose restrictions to S belong to θ. Let $[\theta]$ denote the set of roots in Φ expressible as integral linear combination of elements of θ. V_θ denotes the set of real points of the unipotent group generated by the unipotent subgroups $\mathbb{U}_\alpha$ for $\alpha \in \eta(\Phi^+ \setminus [\theta])$. P_θ denotes the set of real points of the group generated by $\mathbb{T}$ and the groups $\mathbb{U}_\alpha$ for $\alpha \in \eta([\theta] \cup \Phi^+ \cup \{0\})$. S_θ denotes the connected subgroup of S whose Lie algebra is intersection of the kernels of the roots $\alpha \in \theta$. Then $P_\theta = Z_G(S_\theta) \ltimes V_\theta$, where V_θ, P_θ and $Z_G(S_\theta) = L_\theta$ are all real points of connected algebraic groups defined over $\mathbb{R}$. L_θ is a reductive group, and $L_\theta = M_\theta S_\theta$, where M_θ is semisimple.

The unipotent radical $\overline{V}_\theta$ of the opposite parabolic subgroup $\overline{P}_\theta$ is defined as the real points of the group generated by the unipotent subgroups $\mathbb{U}_\alpha$ for $\alpha \in \eta(\Phi^- \setminus [\theta])$, and $\overline{P}_\theta = Z_G(S_\theta) \ltimes \overline{V}_\theta$. Here $\overline{V}_\theta \cap P_\theta = \{e\}$, and $P_\theta \cap \overline{P}_\theta = Z_G(S_\theta) = L_\theta$.

Clearly, if $\theta = \emptyset$, then $S_\emptyset = S$, $Z_G(S_\emptyset) = M_\emptyset S$, $V_\emptyset = V$ is the set of real points of the unipotent group generated by all the unipotent subgroups $\mathbb{U}_\alpha$ for all $\alpha \in \eta(\Phi^+)$, and $P_\emptyset = P$ is the minimal parabolic subgroup. Also, for $\theta = \Delta$, $S_\Delta = \{e\}$, $V_\Delta = \{e\}$ and $P_\Delta = G$, by definition.

Let $p_\theta : \overline{V}_\theta \to G/P_\theta$ be the restriction to $\overline{V}_\theta$ of the natural map $G \to G/P_\theta$, and denote $p = p_\emptyset$. Then p_θ takes $\overline{V}_\theta$ diffeomorphically onto an open dense conull set in G/P_θ, such that $p_\theta(m_{\overline{V}_\theta})$ is a measure in the G-quasi-invariant measure class on G/P_θ. (Here $m_{\overline{V}_\theta}$ is Haar measure on $\overline{V}_\theta$.)

Define $S'_\theta = \{s \in Z(L_\theta) \mid \text{Int}(s) \text{ is contracting on } V_\theta, \text{Int}(s)^{-1} \text{ is contracting on } \overline{V}_\theta\}$, where $\text{Int}(s)$ denotes conjugation by s. Note that $S'_\theta \neq \emptyset$ if and only if $\Phi^+ \setminus [\theta]$ is nonempty, if and only if $\theta \subsetneq \Delta$, if and only if $P_\theta \subsetneq G$.

Define also $\overline{U}_\theta = L_\theta \cap \overline{V} = P_\theta \cap \overline{V}$ and recall that $\overline{V} \cong \overline{U}_\theta \ltimes \overline{V}_\theta$ is a semidirect decomposition. (In particular $\overline{V} = \overline{V}_\theta \cdot \overline{U}_\theta \cong \overline{U}_\theta \ltimes \overline{V}_\theta$ has a product structure as a measure space given by this decomposition.) Note that the decomposition is nontrivial if and only if $P \subsetneqq P_\theta \subsetneqq G$, and in particular, only for groups of real rank at least 2. Indeed, $\overline{U}_\Delta = \overline{V}_\emptyset = \overline{V}$.

Recall also that the natural projection $\pi_\theta : \overline{V} \cong \overline{U}_\theta \ltimes \overline{V}_\theta \to \overline{V}_\theta$ given by $\pi_\theta(\overline{u}_\theta, \overline{v}_\theta) = \overline{v}_\theta$, satisfies : $p_\theta \circ \pi_\theta(\overline{v}) = \overline{v}P_\theta = \overline{v}_\theta P_\theta$, for all $\overline{v} = (\overline{u}_\theta, \overline{v}_\theta) \in \overline{V}$.

Preliminary reduction. Before starting the proofs of the results stated, we make one preliminary reduction. We will assume throughout the proofs that G does not have nontrivial compact factors. This assumption will be useful, for example, when applying 3.1 which was formulated for such groups. There is no loss of generality in this assumption, as we will show at the end of Section 11.

3. Stability groups, tangent subalgebras and the Gauss map

We now recall some generalities regarding measurable actions and stability groups. Let G be any locally compact second countable group, and let (X, η) be a standard measurable G-space with quasi-invariant ergodic measure η. As is well known (see [V]), the stability group G_x of almost every point $x \in X$ is closed in G. By [AM, Ch. II, §2], the space $\mathcal{S}(G)$ of closed subgroups of G carries a natural compact metrizable topology, and G acts continuously on $\mathcal{S}(G)$ ([AM, Ch. II. Prop. 2.2]).

Consider the map $\psi : x \mapsto G_x$. By [AM, Ch. II, Prop. 2.3], ψ is measurable as a map to the Borel G-space $\mathcal{S}(G)$. If G is a Lie group, consider the map $d\psi : X \to \mathrm{Gr}\,(\mathrm{Lie}(G)) = \cup_{0 \le n \le \dim G}\mathrm{Gr}_n(\mathrm{Lie}(G))$, given by $d\psi(x) = \mathrm{Lie}(G_x)$. Here $\mathrm{Gr}_n(\mathrm{Lie}(G))$ is the Grassmann variety of n-planes in the Lie algebra $\mathrm{Lie}(G)$ of G, on which G acts via the adjoint representation. $d\psi$ is also a Borel map, and to see that it suffices to show that the natural map $\mathcal{S}(G) \to \mathrm{Gr}(\mathrm{Lie}(G))$, $H \mapsto \mathrm{Lie}(H)$ is Borel. But G acts continuously on the compact metrizable space $M = \mathcal{S}(G)$, and the map $\delta : m \mapsto \dim G_m$ is upper semi-continuous, hence Borel. On each of the sets $M_d = \{m \in M \mid \delta(m) = d\}$ $d\psi$ is continuous (see [St1, Lemma 2.1]), and hence $d\psi$ is a Borel map on M. It is evident that the map is also G-equivariant.

If G is in addition an algebraic group, then its action on the Grassmann variety is algebraic, and hence every orbit of G is locally closed (see e.g. [M, Ch. I, Prop. 2.1.4]). It follows that the image η' of the measure η (under $d\psi$) is supported on a single G-orbit in $\mathrm{Gr}(\mathrm{Lie}(G))$, since η is an ergodic measure and the action of G on $\mathrm{Gr}(\mathrm{Lie}(G))$ is tame; i.e. the orbit space is countably separated (see [Z3]).

Assume now that G is a connected real algebraic group, and that there exists a set of positive measure $Z \subset X$, all of whose points have a stability group of a given dimension $k > 0$. Then there exists a fixed closed connected

Lie subgroup H of G of dimension k, such that the identity component of the stability group of almost every $x \in X$ is a conjugate of H. Indeed, the ergodicity of η implies that Z has measure one since it is clearly G-invariant. Furthermore, the fact that η' is supported on a single orbit implies that the connected component of almost all the stability groups are conjugate, since their Lie algebras form a transitive G-space. In this case then, there exists a G-equivariant measurable factor map $\phi : X \to G/N_G(H)$. Here $N_G(H)$ is the normalizer of H in G, namely the stabilizer of the subspace Lie (H) under the adjoint representation, and it is clearly a real algebraic group. We now consider:

PROPOSITION 3.1.   *Let G be a connected semisimple Lie group with finite center and no compact factors, μ an admissible measure on G, and (X, ν) an ergodic (G, μ)-space. If there exists a set of positive measure $Z \subset X$ where $\dim G_x > 0$ for $x \in Z$, then one of the following alternatives hold:*

1. *Either the identity component G_x^o of the stability group G_x is ν-almost surely a fixed normal subgroup H, and then the action of G on (X, ν) has kernel of positive dimension, containing H,*

2. *or there exists a measurable G-equivariant factor map $\varphi : (X, \nu) \to (G/Q, \nu_0)$, where G/Q is a nontrivial projective factor.*

Proof. By the discussion preceding the proposition, the identity components (in the Hausdorff topology) of the stability groups of almost all points in X are mutually conjugate. Let H denote a group in this conjugacy class, and then there exists a measurable G-equivariant map $d\psi : X \to G/N_G(H)$.

Now note that $G/N_G(H)$ is trivial if and only if H is a normal subgroup of G, if and only if the identity component of the stability group of almost every point coincides with H. In that case it follows that the action of G on (X, ν) has a kernel containing H, and in particular of positive dimension. (This possibility can certainly occur, for example taking an action of $SL_2(\mathbb{R}) \times SL_2(\mathbb{R})$ where the second factor acts trivially.) Therefore, if every normal subgroup of positive dimension acts nontrivially, then H is not normal, and $N_G(H)$ is a proper subgroup of G. The assumption that (X, ν) is a (G, μ)-space implies that the algebraic variety $G/N_G(H)$ carries a μ-stationary measure. Therefore, by Proposition 3.2 below, $N_G(H)$ is contained in a proper parabolic subgroup Q, so (X, ν) has the nontrivial projective factor G/Q.   □

To complete the proof of Proposition 3.1, we now consider the following:

PROPOSITION 3.2.   *Let G be a connected semisimple Lie group with finite center and μ an admissible measure on G. Suppose F is a real algebraic proper subgroup of G, and G/F carries a μ-stationary measure ν. If $P = MSV$ is a* *minimal parabolic subgroup, then F contains a conjugate of SV and is hence co-compact in G. Furthermore, if G has no nontrivial compact factors, then F is contained in a proper parabolic subgroup.*

Proof. Let $P = MSV$ denote a minimal parabolic subgroup. By [F2, Thm. 2.1] (see also [NZ1, Thm. 1.4]), there exists a P-invariant probability measure λ on G/F. G/F being an algebraic variety, and the P-action being algebraic, λ is supported on SV-fixed points in G/F. Hence F contains (a conjugate of) SV, and so F is a co-compact algebraic subgroup of G.

To complete the proof we use the argument of [BN, Lemma 3.2]. Clearly, if F is reductive and contains SV, it also contains $\overline{V}$, and hence coincides with G, if G has no compact factors. Therefore, since F is assumed proper, F is not reductive, and hence is contained in the normalizer of its nontrivial unipotent radical. The latter group is a proper parabolic subgroup defined over $\mathbb{R}$, by [BT, 5 , §3.1]. We remark that an alternative proof can be obtained by appealing to the classification of co-compact subgroups in [W].   □

4. The spectral theory of P-actions

We now consider some facts from the spectral theory of measure-preserving P-actions, which will be used in the proof of Theorems 1 and 2. We keep the notation of Section 2.

PROPOSITION 4.1.   *Let G be a connected semisimple Lie group with finite center. Let $P \subset G$ be a minimal parabolic subgroup, and (X_0, λ) a probability-measure-preserving P-action. Suppose that for some $s \in S'_\theta$, the action of s on (X_0, λ) is not ergodic. Then there exists a nontrivial P-factor-space (X'_0, λ') of (X_0, λ), on which the group V_θ acts trivially. In particular, in (X'_0, λ'), almost every point has a stability group of positive dimension in P.*

Proof. First note that V_θ is a normal subgroup of P, since (see §2) $P \subset P_\theta = Z_C(S_\theta) \ltimes V_0$

Recall that the generalized Mautner phenomenon (see e.g. [M, p. 88]), asserts that in any strongly continuous unitary representation $(\pi, \mathcal{H})$ of P, any vector invariant under $\pi(s)$ (and hence of course also under $\pi(s^{-1})$) is also invariant under $\pi(V_\theta)$. Since V_θ is normal in P, the space $\mathcal{H}_\theta$ of $\pi(V_\theta)$-invariant vectors is a P-invariant space. Hence we can consider the representation of P on $\mathcal{H}_\theta$, and of course, the action of $\pi(V_\theta)$ in this space is trivial. In the particular case where $\mathcal{H} = L^2(X_0, \lambda)$ for some ergodic measure-preserving action of P on (X_0, λ), the σ-algebra of V_θ-invariant sets is a P-invariant σ-algebra. By the Mackey point realization theorem [Ma], there exists a factor P-space (X'_0, λ') of (X_0, λ) such that the subrepresentation of P on $\mathcal{H}_\theta$ is realized on $L^2(X'_0, \lambda')$, and the action of V_θ on (X'_0, λ') is trivial. Finally, the space (X'_0, λ') is trivial if and only if every function in $L^2(X_0, \lambda)$ which is invariant under V_θ is in fact constant. In that case, V_θ is ergodic on (X_0, λ), and hence s is ergodic also by the Mautner phenomenon, contrary to our assumption. $\square$

5. Induced actions, stationary measures and the existence of factors

Let us recall the following facts regarding induced actions. If $H \subset G$ is any closed subgroup, and Y is an H-space with quasi-invariant measure λ, then there is a naturally defined action of G induced by the H-action of Y. Namely, H acts on $G \times Y$ by

$$h \cdot (g, y) = (gh^{-1}, hy),$$

and commutes with the G action given by

$$g_0 \cdot (g, y) = (g_0 g, y).$$

Thus, we obtain a G-action on the space of H-orbits $(G \times Y)/H$, which is the induced space. We denote the induced action by $G \times_H Y$, and it is naturally a bundle over G/H with fiber Y. We denote the H-orbit containing (g, y) by $[g, y]$.

There is a uniquely determined measure class on each fiber, namely that of λ, and hence a natural quasi-invariant measure class on $G \times_H Y$. If λ is H-invariant (not just quasi-invariant) and we fix a measure on G/H, then there is a uniquely determined measure on $G \times_H Y$. We note that if Y is a compact metric H-space, and H is co-compact in G, then $G \times_H Y$ is a compact metric space with a jointly continuous G-action.

Let now X be a compact metric G-space, and X_0 a closed H-invariant subset. Consider the induced action $G \times_H X_0$. Here there exists a natural multiplication map $\xi : G \times_H X_0 \to X$, given by $\xi[g, x_0] = gx_0$, which is a well-defined continuous G-equivariant map.

Induced actions can also be described in terms of cocycles (see [Z3] for further details). To that end, let $\tau : G/H \to G$ be a section of the natural projection, which is bounded on compact sets, and satisfies $\tau([H]) = e$. Let $\beta : G \times G/H \to H$ be the associated cocycle; namely,

$$\beta(g, z) = \tau(gz)^{-1} g \tau(z).$$

If Y is an H-space, define an action of G on $G/H \times Y$ by

$$g \cdot (z, y) = (gz, \beta(g, z)y).$$

When endowed with this action, we denote $G/H \times Y$ by $G/H \times_\beta Y$. It is easily verified (see [Z3]) that $G \times_H Y$ is measurably G-isomorphic to $G/H \times_\beta Y$.

The map $j : G/H \times_\beta Y \to G \times_H Y$, given by $j(gH, y) = [\tau(gH), y]$, is a G-equivariant measurable isomorphism. Below we will often identify the two spaces using this isomorphism.

Assume now that X is a compact metric space with a continuous G-action, that G is a connected semisimple Lie group with finite center, and ν is μ-stationary. Then there exists a convex isomorphism between the compact convex sets $P_\mu(X)$ of μ-stationary measures and $P_P(X)$ of P-invariant measures. Consequently, there exists a probability measure λ on X, which is the unique P-invariant probability measure corresponding to ν under the isomorphism. For a detailed discussion of this fact we refer to [NZ1, Thm. 1.4] or [F2], and we refer also to [NZ3] for the corresponding statements in the measurable setting. The correspondence between ν and λ is given explicitly by convolution, and is described as follows. Let X_0 denote the (closed) support of λ, which is a P-invariant set, so that (X_0, λ) is a measure-preserving P-action. Then

PROPOSITION 5.1 ([NZ1, Prop. 2.5]).   *The G-space (X, ν) is a continuous factor of the G-space $(G \times_P X_0, \nu_0 \times \lambda)$; i.e. there exists a continuous G-equivariant map $\varphi : G \times_P X_0 \to X$, satisfying $\varphi_*(\nu_0 \times \lambda) = \nu$. Here ν_0 is the unique μ-stationary measure on G/P, and we view $G \times_P X_0$ as a bundle over G/P, with $\nu_0 \times \lambda$ denoting the unique measure projecting onto ν_0 whose conditional fiber measures coincide with λ.*

We can write unambiguously $\nu = \nu_0 * \lambda$, and this notation will be used below.

We note that Proposition 5.1 will serve as the starting point of our analysis. Having represented an arbitrary (G, μ)-space (X, ν) as a factor of the induced action $(G \times_P X_0, \nu_0 \times \lambda)$, we will identify G/P with the unipotent radical $\overline{V}$ of the opposite parabolic $\overline{P}$ (up to null sets). We will then study the dynamics of elements in the split torus $S \subset P$ acting on $\overline{V} \times X_0$, making use of their contracting properties, and the special form that the action takes in these coordinates.

6. The basic construction: Inducing factors from P-actions to G-actions.

Let $\psi_0 : (X_0, \lambda) \to (X_0', \lambda')$ be a P-equivariant measurable factor map, where λ (and λ') are P-invariant probability measures. Consider the induced G-spaces, namely $(G/P \times_\beta X_0, \nu_0 \times \lambda)$ and $(G/P \times_\beta X_0', \nu_0 \times \lambda')$. The latter is obviously a G-equivariant factor of the former, with the factor map ψ given by $\psi(yP, x_0) = (yP, \psi_0(x_0))$. We let $\mathcal{L} = \mathcal{L}(Y, \eta)$ denote the σ-algebra of equivalence classes of measurable sets modulo η-null sets on a given measure space (Y, η). Let $\widetilde{\mathcal{L}}(G/P \times_\beta X_0') = \psi^{-1}(\mathcal{L}(G/P \times_\beta X_0'))$ denote the sub-σ-algebra of $\mathcal{L}(G/P \times_\beta X_0)$ consisting of (classes modulo null sets of) lifts by ψ of subsets $A \subset G/P \times_\beta X_0'$, namely $\tilde{A} = \psi^{-1}(A)$. Consider also the sub-σ-algebra $\widetilde{\mathcal{L}}(X) = \varphi^{-1}(\mathcal{L}(X))$ consisting of the lifts by φ of subsets $B \subset X$, namely $\tilde{B} = \varphi^{-1}(B)$, where $\varphi : G/P \times_\beta X_0 \to X$ is given by Proposition 5.1. By the Mackey point realization theorem [Ma], there exists a measurable (G, μ)-space (X', ν'), which a common factor of the G-spaces (X, ν) and $(G/P \times_\beta X_0', \nu_0 \times \lambda')$, such that $\mathcal{L}(X', \nu')$ is G-isomorphic to the intersection $\widetilde{\mathcal{L}}(X) \cap \widetilde{\mathcal{L}}(G/P \times_\beta X_0')$. The common factor corresponding to the intersection σ-algebra is the maximal common factor, and so the intersection σ-algebra is trivial if and only if (X, ν) and $(G/P \times_\beta X_0, \nu_0 \times \lambda)$ have only the trivial factor in common.

Keeping the notation just introduced, we can now formulate the following alternative:

PROPOSITION 6.1.   *Let G be a connected semisimple Lie group with finite center and no compact factors. Let μ be admissible, and (X, ν) an ergodic (G, μ)-space. Write $\nu = \nu_0 * \lambda$, where λ is a P-invariant measure. Let $\psi_0 : (X_0, \lambda) \to (X_0', \lambda')$ be a P-factor map, and assume that in the P-action on (X_0', λ'), λ'-almost all points have stability groups of positive dimension. Suppose the intersection sub-σ-algebra given by*

$$\psi^{-1}(\mathcal{L}(G/P \times_\beta X_0')) \cap \varphi^{-1}(\mathcal{L}(X)) \subset \mathcal{L}(G/P \times X_0)$$

is nontrivial (so that the maximal common factor (X', ν') constructed above is nontrivial). Then either the factor (X', ν') has a nontrivial projective factor, or G acts on (X', ν') with kernel of positive dimension.

Proof. Consider the factor (X', ν') of $(G/P \times_\beta X_0', \nu_0 \times \lambda')$ constructed above. Represent ν' in the form $\nu' = \nu_0 * \lambda''$, λ'' a P-invariant measure supported on a set X_0'', using [NZ3, Lemma 2.3]. By [NZ3, Prop. 2.4], the P-action on (X_0'', λ'') is a factor of the P-action on (X_0', λ'). Hence in the P-action on (X_0'', λ''), λ''-almost all points have stability group of positive dimension. The fact that $\nu' = \nu_0 * \lambda''$ implies that ν'-almost every point of X' is of the form $x' = \tau(yP)x_0''$ for some $yP \in G/P$ and $x_0'' \in X_0''$. It therefore follows that ν'-almost all points $x' \in X'$ have a stability groups of positive dimension. As noted in Section 3, the identity components of the stability groups are conjugate, for ν'-almost every point, to a fixed subgroup H. By our assumption, (X', ν') is nontrivial, and if the group H is not normal in G, we conclude by Proposition 3.1 that (X', ν') has a nontrivial projective factor G/Q. Otherwise H is normal, so H fixes ν'-almost all points in X', and so the kernel of the G-action on X' has positive dimension. $\square$

For a given $\theta \subset \Delta \subset \Phi^+$ with $[\theta] \neq \Phi$, and $s \in S'_\theta$, we can consider the σ-algebra of V_θ-invariant sets in (X_0, λ), and the corresponding factor space (X'_0, λ') of Prop. 4.1, on which the group V_θ acts trivially. Note that by the generalized Mautner phenomenon (see §4), this space is nontrivial if the s-action on (X_0, λ) is nonergodic. Below we will apply the construction preceding Proposition 6.1 to this choice of (X'_0, λ'), and thus obtain a (G, μ)-space (X', ν'), which is a G-equivariant factor of (X, ν). Here $\nu' = \nu_0 * \lambda''$, λ'' a P-invariant measure supported on $X''_0 \subset X'$, and (X''_0, λ'') a P-factor action of (X'_0, λ'). For notational simplicity, the dependence of the factor $\zeta : (X, \nu) \to (X', \nu')$ on θ and $s \in S'_\theta$ will not be mentioned explicitly in the notation.

7. The S-action on $G/P \times_\beta X_0 = \overline{V} \times_\alpha X_0$

We will now use the fact that the group $\overline{V}$, the unipotent radical of the parabolic subgroup $\overline{P}$ opposite to P, has an open dense ν_0-conull orbit on G/P, with trivial stability group. Restricting attention to this orbit, we can consider the natural section $\iota : \overline{V}P \to \overline{V}$, given by $\iota(\overline{v}P) = \overline{v}$. The orbit $\overline{V}P$ is obviously a $\overline{P}$-invariant set, since $\overline{P} = MS\overline{V}$, $\overline{V}$ is normal in $\overline{P}$, and $MS = P \cap \overline{P}$. As a measurable $\overline{P}$-space with quasi-invariant measure, we can identify $(G/P \times_\beta X_0, \nu_0 \times \lambda)$ with the space $(\overline{V} \times_\alpha X_0, m_{\overline{V}} \times \lambda)$. Here the cocycle $\alpha : \overline{P} \times \overline{V} \to P \cap \overline{P} = MS = \mathrm{St}_{\overline{P}}([P])$ is the cocycle corresponding to the section ι (see §5), and $m_{\overline{V}}$ is Haar measure on $\overline{V}$. Under this identification, the action of $\overline{V}$ on $\overline{V} \times_\alpha X_0$ is given by translation in the first variable: $\overline{v}(\overline{v}_1, x_0) = (\overline{v} \cdot \overline{v}_1, x_0)$. The action of $S \subset P \cap \overline{P}$ on $\overline{V} \times_\alpha X_0$ is given by conjugation in the first variable, together with the S-action on X_0, namely: $s(\overline{v}, x_0) = (\mathrm{Int}(s)(\overline{v}), sx_0)$.

The multiplication map $\xi : \overline{V} \times_\alpha X_0 \to X$, given by $\xi(\overline{v}, x_0) = \overline{v}x_0$ is a continuous map. Given any continuous function $f \subset C(X)$, we can consider the lift $\tilde{f} = f \circ \xi$ of f to the space $\overline{V} \times_\alpha X_0$, which is also continuous.

Fix $\theta \subset \Delta$ with $[\theta] \neq \Phi$, and $s \in S'_\theta$. Then $\overline{V}$ has a product decomposition of the form $\overline{V} = \overline{U}_\theta \ltimes \overline{V}_\theta$. Here $\mathrm{Int}(s^{-1})$ acts on $\overline{U}_\theta$ as the identity, and $\mathrm{Int}(s^{-1})$ acts on $\overline{V}_\theta$ via a contracting automorphism (see §2). In particular, for every compact subset $C \subset \overline{V}_\theta$, and any open set $E \subset \overline{V}_\theta$ containing the identity, $\mathrm{Int}(s^{-k})(C) \subset E$ for all $k \geq k_0(C, E)$. We remark that when $\theta = \emptyset$, and in this case only, $\overline{U}_\theta = \{e\}$. In particular, $\overline{U}_\theta = \{e\}$ in the case G has real rank one.

Consider now the action of $s \in S$ on the function $\tilde{f}$, given by $s\tilde{f}(\overline{v}, x_0) = \tilde{f}(\mathrm{Int}(s^{-1})(\overline{v}), s^{-1}x_0)$. In particular, if $s \in S'_\theta$, and we write $\overline{v} = (\overline{u}_\theta, \overline{v}_\theta)$ according to the decomposition $\overline{V} = \overline{U}_\theta \ltimes \overline{V}_\theta$ above, then $s\tilde{f}(\overline{u}_\theta, \overline{v}_\theta, x_0) = \tilde{f}(\overline{u}_\theta, \mathrm{Int}(s^{-1})(\overline{v}_\theta), s^{-1}x_0)$.

The dynamics of the action of an element $s \in S$ on $\overline{V} \times X_0$ is described in these coordinates by the following

PROPOSITION 7.1.   *Fix $\theta \subset \Delta$ with $[\theta] \neq \Phi$, and $s \in S'_\theta$. Then for every $f \in C(X)$, and all $(\overline{u}_\theta, \overline{v}_\theta) \in \overline{U}_\theta \ltimes \overline{V}_\theta$,*

$$\lim_{N \to \infty} \frac{1}{N+1} \sum_{n=0}^{N} s^n \tilde{f}(\overline{u}_\theta, \overline{v}_\theta, x_0) = \mathcal{E}_s \tilde{f}(\overline{u}_\theta, e, x_0) \ .$$

Here $\mathcal{E}_s : L^2(X_0, \lambda) \to L^2(X_0, \lambda)$ is the conditional expectation on the space of s-invariant functions, and the convergence is pointwise λ-almost everywhere, and in the $L^2(X_0, \lambda)$-norm. Furthermore, the limit function on $\overline{V} \times_\alpha X_0$ given by $(\overline{u}_\theta, \overline{v}_\theta, x_0) \mapsto \mathcal{E}_s \tilde{f}(\overline{u}_\theta, e, x_0)$ is still in $\tilde{L}^\infty(X)$.

Proof. Let $f \in C(X)$ be given. Fix a compact neighbourhood $E \subset \overline{V}_\theta$ containing the identity, and fix also an element $\overline{u}_\theta \in \overline{U}_\theta$. Consider the collection of continuous functions on X_0 given by $\tilde{f}(\overline{u}_\theta, \overline{v}_\theta, x_0) = f(\overline{u}_\theta \cdot \overline{v}_\theta \cdot x_0)$, as $\overline{v}_\theta$ ranges over E. Since f in continuous on X, the action of G on X is continuous, and E is compact, it is evident that this is an equicontinuous collection of functions, and in particular, given $\delta > 0$, there exists a neighbourhood $E_\delta \subset E$ of the identity in $\overline{V}_\theta$, such that if $\overline{v}_\theta \in E_\delta$, then $\left| \tilde{f}(\overline{u}_\theta, \overline{v}_\theta, x_0) - \tilde{f}(\overline{u}_\theta, e, x_0) \right| \leq \delta$, for all $x_0 \in X_0$.

Let $\overline{v}_\theta \in \overline{V}_\theta$ be given, and consider the expression

$$\left| \tilde{f}(\overline{u}_\theta, \mathrm{Int}(s^{-n})(\overline{v}_\theta), s^{-n}x_0) - \tilde{f}(\overline{u}_\theta, e, s^{-n}x_0) \right| \ .$$

Since $\mathrm{Int}(s^{-n})(\overline{v}_\theta) \longrightarrow e$ as $n \to \infty$, for all large enough n, $\mathrm{Int}(s^{-n})(\overline{v}_\theta) \in E_\delta$, and then the foregoing expression is bounded by δ, for all $x_0 \in X_0$. Therefore the average over $0 \leq n \leq N$ of the foregoing expressions is bounded by 2δ (say) for all large enough n. Now by the pointwise ergodic theorem for the action of s on (X_0, λ), we have

$$\lim_{N \to \infty} \frac{1}{N+1} \sum_{n=0}^{N} \tilde{f}(\overline{u}_\theta, e, s^{-n}x_0) = \mathcal{E}_s \tilde{f}(\overline{u}_\theta, e, x_0)$$

where $\mathcal{E}_s$ is the conditional expectation on the subspace of s-invariant sets, and the convergence is λ-almost everywhere and in the $L^2(X_0, \lambda)$-norm. Consequently, we conclude that the averages given in the statement of the theorem converge to the limit stated.

Now note that all the functions $s^n \tilde{f}$ are in $\tilde{L}^\infty(X)$, and the same is true of their averages $1/(N+1) \sum_{n=1}^{N} s^n \tilde{f}$. The averages are all bounded by $\left\| \tilde{f} \right\|_\infty$ and converge pointwise almost everywhere, with respect to $m_{\overline{V}} \times \lambda$, as was just shown. The measure $m_{\overline{V}} \times \lambda$ on $\overline{V} \times X_0$ is mapped to a measure equivalent to $\nu_0 \times \lambda$ on $G/P \times X_0$ under the natural identification of the two spaces. Since there exists a factor map $\varphi : (G/P \times X_0, \nu_0 \times \lambda) \to (X, \nu)$, $\tilde{L}^\infty(X, \nu)$ is characterized in $L^\infty(G/P \times X_0, \nu_0 \times \lambda)$ as the set of functions whose value at (yP, x_0) depends only on $\varphi(yP, x_0)$. In other words, $\tilde{L}^\infty(X, \nu)$ is the range of a conditional expectation. Therefore $\tilde{L}^\infty(X)$ is closed under bounded pointwise convergence. Hence the limit of the averages of $s^n \tilde{f}$ is also in $\tilde{L}^\infty(X)$, and the proposition is proved.   □

8. Proof of Theorem 1

Recall that we assume that G is connected, semisimple, with trivial center and no nontrivial compact factors. Further, μ is admissible, and (X, ν) is an ergodic (G, μ)-space. Write $\nu = \nu_0 * \lambda$, and consider the action of P on (X_0, λ), and of $\overline{P}$ on $\overline{V} \times_\alpha X_0 \cong G/P \times_\beta X_0$. We divide the proof to two cases.

Case I. Suppose first that for some $\theta \subset \Delta$ with $[\theta] \neq \Phi$, some element $s \in S'_\theta$ and some continuous function $f \in C(X)$, the lift $\tilde{f}$ to $\overline{V} \times_\alpha X_0$ satisfies the following condition.

> There exists a set of positive Haar measure $F \subset \overline{U}_\theta$ such that each of the functions $x_0 \mapsto \mathcal{E}_s \tilde{f}(\overline{u}_\theta, e, x_0)$ is a λ-nonconstant function of $x_0 \in X_0$, provided $\overline{u}_\theta \in F$.

Since each of these functions (on X_0) is in any case s-invariant (by definition of $\mathcal{E}_s$), it is also invariant under the group $V_\theta \subset P$, by the generalized Mautner phenomenon (see §4). In particular, s and V_θ are not ergodic on (X_0, λ). Therefore by Proposition 4.1, (X_0, λ) has a nontrivial P-factor space (X'_0, λ'), whose σ-algebra is (isomorphic to) the σ-algebra of V_θ-invariant sets. The group V_θ acts trivially on (X'_0, λ'). As in Section 6, we construct the space (X', ν'), which is the largest common G-factor of $(G/P \times_\beta X'_0, \nu_0 \times \lambda')$ and (X, ν).

We now show that under the assumption in Case I, (X', ν') is nontrivial. Indeed, triviality of (X', ν') is equivalent to the fact that the intersection of the sub-σ-algebras $\widetilde{\mathcal{L}}(G/P \times_\beta X'_0)$ and $\widetilde{\mathcal{L}}(X)$ is the trivial σ-algebra. By Proposition 7.1, the function

$$(\overline{u}_\theta, \overline{v}_\theta, x_0) \mapsto \mathcal{E}_s \tilde{f}(\overline{u}_\theta, e, x_0)$$

is in $\tilde{L}^\infty(X)$. Furthermore, for each $(\overline{u}_\theta, \overline{v}_\theta)$, the function $x_0 \mapsto \mathcal{E}_s \tilde{f}(\overline{u}_\theta, e, x_0)$ is an s-invariant function on (X_0, λ), and hence also V_θ-invariant. In addition, for $\overline{u}_\theta$ in a set of positive Haar measure F in $\overline{U}_\theta$, the function $x_0 \mapsto \mathcal{E}_s \tilde{f}(\overline{u}_\theta, e, x_0)$ is a λ-nonconstant function of x_0, by assumption. Hence the function

$$(\overline{u}_\theta, \overline{v}_\theta, x_0) \mapsto \mathcal{E}_s \tilde{f}(\overline{u}_\theta, e, x_0)$$

is a nonconstant function in the intersection of $\tilde{L}^\infty(X)$ and $\tilde{L}^\infty(\overline{V} \times_\alpha X'_0)$. Therefore, the intersection $\tilde{\mathcal{L}}(G/P \times_\beta X'_0) \cap \tilde{\mathcal{L}}(X)$ is a nontrivial σ-algebra, and so (X', ν') is nontrivial.

We will now use Proposition 6.1, which provides a nontrivial projective factor for (X', ν') under the assumption that the G-action on it is faithful, and otherwise asserts that the G-action on (X', ν') has kernel of positive dimension.

Under the assumptions of Theorem 1, the action of G on (X', ν') is ergodic, but perhaps has nontrivial kernel. We therefore divide G by the kernel N of the action of G on (X', ν'), obtaining a (faithful) action of G/N.

Now furnish (X', ν') with a compact metric structure of its own [V], giving the already existing standard Borel structure. Repeat the argument above for G/N acting on (X', ν') (note that ν' is μ'-stationary, μ' being the image of μ under the canonical map $G \to G/N$). Since G has finitely many factor groups, we can continue a finite number of steps, until one of the following two alternatives is realized. Either the action of the factor group falls in Case II, or we arrive at a faithful action of a factor group of G on a nontrivial space (X'_k, ν'_k) with stability groups of positive dimension (which are not normal). We then apply Proposition 6.1 and conclude the existence of a nontrivial projective factor.

Therefore, in Case I we have indeed shown that a nontrivial projective factor exists (or a factor action of a factor group falls in Case II, which will be discussed presently). Note however that the factor obtained this way may have less than full entropy.

Case II. We can now assume the following.

G is of real rank at least two, and for every continuous function $f \in C(X)$, for every $\theta \subset \Delta$ with $[\theta] \neq \Phi$, for every $s \in S'_\theta$ and for $m_{\overline{U}_\theta}$-almost all $\overline{u}_\theta \in \overline{U}_\theta$, the function $x_0 \mapsto \mathcal{E}_s \tilde{f}(\overline{u}_\theta, e, x_0)$ is a λ-constant function on X_0.

We begin by noting the fact that when the function is constant, then this constant is easily found. It must be $\int_{X_0} \tilde{f}(\overline{u}_\theta, e, x_0) d\lambda(x_0)$, as we see by integrating with respect to λ.

Furthermore, we note for future reference that in the argument below we will only use the existence, for each $\theta \subset \Delta$ with $[\theta] \neq \Phi$, of *at least one* $s \in S'_\theta$ for which $x_0 \mapsto \mathcal{E}_s \tilde{f}(\overline{u}_\theta, e, x_0)$ is λ-constant for a.e. $\overline{u}_\theta$, and every $f \in C(X)$. This assumption is certainly fulfilled if for every $\theta \subset \Delta$ with $[\theta] \neq \Phi$, there exists $s \in S'_\theta$ which is ergodic on (X_0, λ). Indeed, in that case every s-invariant function on (X_0, λ) is necessarily constant.

Consider again the behaviour of the function

$$(\overline{u}_\theta, \overline{v}_\theta, x_0) \mapsto \mathcal{E}_s \tilde{f}(\overline{u}_\theta, e, x_0) = \int_{X_0} \tilde{f}(\overline{u}_\theta, e, x_0) d\lambda(x_0)$$

as a function on $(\overline{U}_\theta \ltimes \overline{V}_\theta) \times_\alpha X_0 = \overline{V} \times_\alpha X_0$. Clearly, the function is independent of the $\overline{v}_\theta$ and x_0 variables, and depends only on the $\overline{u}_\theta$ variable. Furthermore, it is a continuous function on $\overline{V} \times X_0$, and belongs to $\tilde{L}^\infty(X)$.

If the function is an $m_{\overline{U}_\theta}$-nonconstant function on $\overline{U}_\theta$, then we have clearly obtained a nonconstant function in $\tilde{L}^\infty(X)$ which is also in the space $\tilde{L}^\infty(\overline{V})$, lifted via the natural projection $\overline{V} \times_\alpha X_0 \to \overline{V}$. This implies of course that $\tilde{f}$ belongs to the intersection $\tilde{L}^\infty(G/P) \cap \tilde{L}^\infty(X)$, where $L^\infty(G/P)$ is lifted via the natural projection $G/P \times_\beta X_0 \to G/P$. Hence the intersection of the σ-algebras $\tilde{\mathcal{L}}(G/P)$ and $\tilde{\mathcal{L}}(X)$ is a nontrivial G-invariant sub-σ-algebra of both. By the Mackey point realization theorem [Ma], (X, ν) and $(G/P, \nu_0)$ have a nontrivial common factor, which is of course of the form $(G/Q, \nu_0)$. Hence there exists a nontrivial projective factor for (X, ν), in this case.

Otherwise, the function above is constant in the variable $\overline{U}_\theta$ as well, and so is a constant function, for every $\theta \subset \Delta$ with $[\theta] \neq \Phi$, and every $f \in C(X)$. We now claim that when $\mathbb{R}$-rank $(G) \geq 2$, it follows that the measure λ is invariant under G. Indeed, for every $f \in C(X)$, we have

$$\int_{X_0} f(\overline{u}_\theta x_0) d\lambda = \int_{X_0} \tilde{f}(\overline{u}_\theta, e, x_0) d\lambda = \int_{X_0} \tilde{f}(e, e, x_0) d\lambda = \int_{X_0} f(x_0) d\lambda.$$

The first equality follows from the definition of the map $f \mapsto \tilde{f}$, since $\overline{V} \times X_0 \to X$ is given by the multiplication map $\xi : (\overline{v}, x_0) \mapsto \overline{v} x_0$. For the second equality we use the fact that since $(\overline{u}_\theta, e, x_0) \mapsto \int_{X_0} \tilde{f}(\overline{u}_\theta, e, x_0) d\lambda(x_0)$ is a continuous function which is $m_{\overline{U}_\theta}$-almost surely constant, it is in fact constant and equal to its value at e. Clearly the foregoing equation implies that the measure $\lambda \in P(X)$ is invariant under $\overline{U}_\theta$. Now if the real rank is at least two, the groups $\overline{U}_\theta$, for $[\theta] \subsetneq \Phi$, generate the group $\overline{V}$ (see [M, Ch. I, proof of 1.2.2]). Hence the measure λ is invariant under $\overline{V}$, and also under P, and these two groups generate G (see [M, Ch. I, 1.2.1]). But since $\nu = \nu_0 * \lambda$, the G-invariance of λ implies that $\nu = \lambda$ and ν is G-invariant. If ν is not G-invariant, we have therefore established the existence of a nontrivial projective factor in Case II as well. $\qquad\square$

Remarks regarding the proof.

1. Note that Theorem 1 implies, in particular, the existence of a nontrivial projective factor for an ergodic action of a *simple* group of $\mathbb{R}$-rank at least two. However, this fails for simple groups of $\mathbb{R}$-rank one by [NZ2, Thm B]. As stated in Theorem 2, the possible appearance of certain factor actions of real-rank-one factor groups is the only obstruction to the existence of a nontrivial projective factor for an action with positive entropy.

2. Note that Case I above does not use the higher real rank assumption. In particular, if G is simple of real rank one, and the condition stated in Case I is satisfied, then there exists a nontrivial projective factor. This factor must be G/P, so the action is in fact amenable and the extension is measure preserving (see [NZ2, §4]). Since (X, ν) is an ergodic induced action, P must be ergodic on (X_0, λ).

3. If $G = \mathrm{PSL}_2(\mathbb{R})$, we can conclude that Case I above does not arise. Indeed, the condition in Case I implies that s is not ergodic on (X_0, λ). By the Mautner phenomenon it follows that the group SV is not ergodic on (X_0, λ). The argument in Case I then implies that (X, ν) has a nontrivial projective factor, which is necessarily G/P. But then (X, ν) is induced from a measure-preserving action of $P = SV$, so λ is P-ergodic, which is a contradiction.

To complete the proof of Theorem 1, it suffices to establish the existence of the unique maximal projective factor of any (G, μ)-space (X, ν).

LEMMA 0.1. *Let G be a connected noncompact semisimple Lie group with finite center, μ an admissible probability measure on G, and (X, ν) a (G, μ)-space. Then there exists a unique maximal projective factor $\varphi : (X, \nu) \to (G/Q, \nu_0)$, with φ uniquely determined up to ν-null sets.*

Proof. Let G/P_{θ_1} and G/P_{θ_2} be two projective factors of (X, ν). Then there exists a measurable G-equivariant map $(\varphi_1, \varphi_2) : (X, \nu) \to G/P_{\theta_1} \times G/P_{\theta_2}$. The measure ν on X is mapped to a stationary measure ν' on $G/P_{\theta_1} \times G/P_{\theta_2}$. Consider now the natural map $\pi : G/P \times G/P \to G/P_{\theta_1} \times G/P_{\theta_2}$. By [F2, Cor. to Lemma 2.2] the induced map on stationary measures is onto. Furthermore, by [F2, Thm. 2.2], there is a unique stationary measure on $G/P \times G/P$ and it is supported on the diagonal. Hence the measure ν' is supported on the G-orbit $\{(gP_{\theta_1}, gP_{\theta_2}) \mid g \in G\}$, which is the image of the diagonal under the map π. Hence ν' is supported on a single G-orbit in the product. Therefore there exists a measurable factor map from (X, ν) to G/H, where $H = gP_{\theta_1}g^{-1} \cap gP_{\theta_2}g^{-1} = Q$ is the stability group of a point in the G-orbit carrying ν'. But H is a conjugate of the standard parabolic subgroup $P_{\theta_1} \cap P_{\theta_2} = P_{\theta_1 \cap \theta_2}$. Hence both factors G/P_{θ_1} and G/P_{θ_2} are factors of G/Q, and it follows that there exists a unique maximal projective factor $(G/Q, \nu_0)$ of (X, ν), and every projective factor of (X, ν) is also a factor of G/Q.

The map $\varphi : (X, \nu) \to (G/Q, \nu_0)$ is unique up to ν-null sets, since given another G-factor map $\varphi' : (X, \nu) \to (G/Q, \nu_0)$, we obtain a G-equivariant map: $(\varphi, \varphi') : (X, \nu) \to G/Q \times G/Q$. Here the image ν' of ν must be supported on the diagonal, by the argument of the preceding paragraph. Hence $\varphi(x) = \varphi'(x)$ for ν-almost all $x \in X$, and the proof of the lemma is complete. $\square$

9. Proof of Theorem 3

Theorem 3 is an immediate consequence of the following stronger version:

THEOREM 9.1.   *Let* (G, μ), (X, ν), (X_0, λ) *be as in Theorem 3. Let* S *denote a maximal* $\mathbb{R}$-*split torus of* G *contained in* P. *If for every* $\theta \subset \Delta$ *with* $[\theta] \neq \Phi$ *there exists an element* $s \in S'_\theta$ *which is ergodic on* (X_0, λ), *then the unique maximal projective factor* $\varphi : (X, \nu) \to (G/Q, \nu_0)$ *is an extension with relatively invariant measure.*

Before starting the proof, note that under the assumption that for every $\theta \subset \Delta$ with $[\theta] \subsetneq \Phi$ there exists some $s \in S'_\theta$ acting ergodically on (X_0, λ), we are back in case II of the preceding section (see also the remark there). In particular, for each $\tilde{f} \in \tilde{C}(X)$, the function

$$(\overline{u}_\theta, \overline{v}_\theta, x_0) \mapsto \mathcal{E}_s \tilde{f}(\overline{u}_\theta, e, x_0) = \int_{X_0} \tilde{f}(\overline{u}_\theta, e, x_0) d\lambda(x_0)$$

is a function in $\tilde{L}^\infty(X)$, depending only on the variable $\overline{u}_\theta$. We now formulate the following:

PROPOSITION 9.2.   *Let* (G, μ), (X, ν) *and* (X_0, λ) *be as in Theorem 3. Given a fixed* $\theta \subset \Delta$ *with* $[\theta] \neq \Phi$, *the following alternative holds:*

1. *Either for every* $f \in C(X)$, *the function above is a constant function of* $\overline{u}_\theta$, *and then* λ *is invariant under the group* $\overline{U}_\theta$,

2. *or there exists some* $f \in C(X)$ *such that the function above is an* $m_{\overline{U}_\theta}$-*nonconstant function on* $\overline{U}_\theta$, *and then there exists a nonconstant function in* $\tilde{L}^\infty(X) \cap \tilde{L}^\infty(G/P)$, *which is not in* $\tilde{L}^\infty(G/P_\theta)$. *(In particular, there exists a nontrivial factor of* $(G/P, \nu_0)$ *which is also a factor of* (X, ν).)

Proof. As was seen in the previous section (see the proof of Case II), if the function in question is constant for every $f \in C(X)$, then the constant is $\int_{X_0} f(x_0) d\lambda$, and λ is invariant under the group $\overline{U}_\theta$.

The function in question is independent of the $\overline{V}_\theta \times X_0$ coordinates; hence in particular it is in $\tilde{L}^\infty(G/P)$. It is also in $\tilde{L}^\infty(X)$ by Proposition 7.1. Furthermore, in the coordinates $\overline{V} = \overline{U}_\theta \ltimes \overline{V}_\theta$, a function is in $\tilde{L}^\infty(G/P_\theta)$ if and only if it is independent of the variable $\overline{u}_\theta \in \overline{U}_\theta$. Hence, if for some $f \in C(X)$ the function in question is nonconstant, then $\tilde{\mathcal{L}}(X) \cap \tilde{\mathcal{L}}(G/P)$ is a nontrivial G-invariant sub-σ-algebra of $\tilde{\mathcal{L}}(G/P)$, which is not contained in $\tilde{L}^\infty(G/P_\theta)$. The last claim follows from the Mackey point realization theorem [Ma].   □

Proof of Theorem 9.1. Consider the maximal projective factor $(G/Q, \nu_0)$ of (X, ν), given by Lemma 0.1. We claim that under the hypotheses of Theorem 3, (X, ν) is an extension of $(G/Q, \nu_0)$ with relatively G-invariant measure. Clearly, if $Q = P$ there is nothing to prove (provided the entropy $H(\mu)$ is finite — see [NZ2, Prop. 1.9 and Cor. 2.5]).

We therefore assume that $P \subsetneq Q \subset G$, and begin by showing that the measure λ is invariant under Q. Using the assumption that G has real rank at least two, we fix $\emptyset \subsetneq \theta \subsetneq \Delta$ with $P_\theta \subset Q$.

If λ is not invariant under $\overline{U}_\theta$, then there exists by Proposition 9.2 some function in $\tilde{L}^\infty(G/P) \cap \tilde{L}^\infty(X)$, which is not in $\tilde{L}^\infty(G/P_\theta)$. (Recall that, as noted above, in the coordinates $\overline{V} = \overline{U}_\theta \ltimes \overline{V}_\theta$, a function is in $\tilde{L}^\infty(G/P_\theta)$ if and only if it is independent of the variable $\overline{u}_\theta \in \overline{U}_\theta$.) Now $P_\theta \subset Q$, so $G/P_\theta \to G/Q$, and $\mathcal{L}(G/Q) \subset \mathcal{L}(G/P_\theta)$. Hence the intersection $\tilde{\mathcal{L}}(G/P) \cap \tilde{\mathcal{L}}(X)$ contains a function not in $\mathcal{L}(G/Q)$. But then the intersection is larger than $\tilde{\mathcal{L}}(G/Q)$ which is a contradiction to the maximality of the projective factor G/Q. Therefore λ is invariant under $\overline{U}_\theta$, for all $\theta \subset \Delta$ with the property that the parabolic subgroup P_θ satisfies $P \subsetneq P_\theta \subsetneq G$, and $P_\theta \subset Q$.

Now $P_\theta \cap \overline{V} = L_\theta \cap \overline{V} = \overline{U}_\theta$, and P_θ is generated by P and $\overline{U}_\theta$, as follows from the Langlands decomposition of parabolic subgroups. Indeed, $P_\theta = M_\theta \cdot S_\theta \cdot V_\theta$, and the Lie algebra of M_θ is the direct sum of the Lie algebras of $S_\theta^\perp$, $M_\emptyset$, U_θ and $\overline{U}_\theta$ (see e.g. [Kn, pp. 415, 417]). All groups listed in this paragraph are subgroups of $P = P_\emptyset$, except $\overline{U}_\theta$, so that P_θ is indeed generated by P and $\overline{U}_\theta$.

It follows that λ is invariant under P_θ, for all θ satisfying conditions $P \subsetneq P_\theta \subsetneq G$, and $P_\theta \subset Q$. By [M, Ch. I, 1.2.4], the subgroups P_θ generate Q, and hence λ is Q-invariant.

By Proposition 2.2 of [NZ2] (see [NZ3] for details), the factor map $\varphi : (X, \nu) \to (G/Q, \nu_0)$ takes λ to a P-invariant measure λ' on G/Q. Since φ is G-equivariant, it follows that λ' is Q-invariant, and hence coincides with the unique Q-invariant probability measure on G/Q, namely the δ-measure at the coset $Q \in G/Q$. It follows that X_0 is contained in the fiber of the map φ lying over the point Q (up to a λ-null set). Hence $\nu = \int_{G/Q} g\lambda d\nu_0(gQ)$ is the disintegration of ν with respect to ν_0. It follows that (X, ν) is G-isomorphic to the action $(G/Q \times_\beta X_0, \nu_0 \times \lambda)$, namely to the action induced to G by the measure-preserving action of Q on (X_0, λ), and the proof of the theorem is complete. $\qquad\square$

10. Proof of Theorem 2

The crucial property used in the proof of Theorem 9.1 (and Theorem 3) is the dichotomy stated in Proposition 9.2. We now formulate a more general form of this dichotomy, without assuming explicitly that certain elements of P are ergodic on (X_0, λ). This will be essential to the proof of Theorem 2.

SEMISIMPLE LIE GROUPS   587

Let (X, ν) be an ergodic (G, μ)-space, X compact metric, μ admissible and G any connected semisimple Lie group with finite center and no compact factors. Fix $\theta \subset \Delta$ with $[\theta] \neq \Phi$, $s \in S'_\theta$ and consider, for each $\tilde{f} \in \tilde{C}(X)$, the functions:

$$(\overline{u}_\theta, \overline{v}_\theta, x_0) \mapsto \mathcal{E}_s \tilde{f}(\overline{u}_\theta, e, x_0)$$

and

$$(\overline{u}_\theta, \overline{v}_\theta, x_0) \mapsto \int_{X_0} \tilde{f}(\overline{u}_\theta, e, x_0) d\lambda(x_0).$$

The first is a function in $\tilde{L}^\infty(X)$, depending on the variables $\overline{u}_\theta \in \overline{U}_\theta$ and $x_0 \in X_0$. The second function obviously does not depend on $x_0 \in X_0$. However, unlike the case where $s \in S'_\theta$ is ergodic on (X_0, λ) discussed above, the second function may fail to belong to $\tilde{L}^\infty(X)$. Recall also the factor map $\zeta : (X, \nu) \rightarrow (X', \nu')$ associated with θ and $s \in S'_\theta$, which was constructed in Section 6.

We now formulate the following:

PROPOSITION 10.1.   *Keep the assumptions of the previous paragraph on* (G, μ), (X, ν) *and* (X_0, λ). *Assume* ν *is not G-invariant, and fix* $\theta \subset \Delta$ *with* $[\theta] \neq \Phi$, $s \in S'_\theta$, *and the associated space* (X', ν') *constructed in Section 6. Then the following alternative holds*:

1. *Either for every* $f \in C(X)$, *the function*

$$(\overline{u}_\theta, \overline{v}_\theta, x_0) \mapsto \int_{X_0} \tilde{f}(\overline{u}_\theta, e, x_0) d\lambda(x_0)$$

   is a constant function, and then λ *is invariant under the group* $\overline{U}_\theta$,

2. *or there exists some* $f \in C(X)$ *such that the function in 1) above is an* $m_{\overline{U}_\theta}$-*nonconstant function on* $\overline{U}_\theta$, *and then the measure* ν' *on the factor space* (X', ν') *is not G-invariant*.

Proof. Under the assumption in Case 1, for every $f \in C(X)$, the continuous function $\int_{X_0} \tilde{f}(\overline{u}_\theta, e, x_0) d\lambda(x_0)$ is essentially constant, and so by definition of $\tilde{f}$ we have (for more details on this argument see Section 8, proof of Case II):

$$\int_{X_0} f(\overline{u}_\theta x_0) d\lambda = \int_{X_0} \tilde{f}(\overline{u}_\theta, e, x_0) d\lambda = \int_{X_0} \tilde{f}(e, e, x_0) d\lambda = \int_{X_0} f(x_0) d\lambda .$$

It follows that λ is invariant under $\overline{U}_\theta$.

In Case 2, we now assume for contradiction that (X', ν') is a measure preserving action. Writing, as in Section 6, $\nu' = \nu_0 * \lambda''$ (λ'' supported on $X_0'' \subset X'$), we note that the factor map $\zeta : (X, \nu) \rightarrow (X', \nu')$ satisfies $\zeta_*(\nu) = \zeta_*(\lambda) = \lambda'' = \nu'$. This fact follows immediately from [NZ3, Prop. 2.3]. Indeed,

$\nu' = \nu_0 * \lambda''$ but here since ν' is G-invariant we have also $\nu' = \nu_0 * \nu'$, so that by uniqueness of the representation, $\nu' = \lambda''$. But $\lambda'' = \zeta_*(\lambda)$, so that $\zeta_*(\lambda) = \lambda'' = \nu' = \zeta_*(\nu)$.

More concretely, in terms of the factor map $\varphi : G/P \times X_0'' \to X'$ of Proposition 5.1, invariance of a stationary measure ν' on a space X' is equivalent with the following property. All functions $\tilde{f}(yP, x_0'')$ lifted from (X', ν') satisfy that the function $u_f(yP) = \int_{X_0''} \tilde{f}((yP, x_0'')d\lambda''(x_0'')$ is independent of $yP \in G/P$, for ν_0-almost all yP. Indeed, invariance of ν' is equivalent with $\int_{G/P} u_f(gyP)d\nu_0(yP)$ being independent of the translation $g \in G$, for all $f \in \tilde{L}^\infty(X')$. The latter property is in turn equivalent (G/P being a boundary space) to the function u_f being a constant, as asserted. Clearly, the constant is then necessarily $\int_{X'} f d\nu' = \int_{G/P} u_f(yP)d\nu_0(yP)$.

For every $f \in C(X)$, the function $(\overline{u}_\theta, \overline{v}_\theta, x_0) \mapsto \mathcal{E}_s \tilde{f}(\overline{u}_\theta, e, x_0)$ is in $\tilde{L}^\infty(X', \nu')$, as noted in the proof of Case I in Section 8. Furthermore, since $\mathcal{E}_s$ is a conditional expectation we clearly have:

$$\int_{X_0} \mathcal{E}_s \tilde{f}(\overline{u}_\theta, e, x_0)d\lambda(x_0) = \int_{X_0} \tilde{f}(\overline{u}_\theta, e, x_0)d\lambda(x_0).$$

The assumption in the present case provides us with a function $f \in C(X)$ where $\int_{X_0} \tilde{f}(\overline{u}_\theta, e, x_0)d\lambda(x_0)$ is not essentially constant on $\overline{U}_\theta$. By the previous equality, $\mathcal{E}_s \tilde{f}(\overline{u}_\theta, e, x_0)$ which is in $\tilde{L}^\infty(X', \nu')$, satisfies the same property.

We now use the fact that $\zeta : (X_0, \lambda) \to (X_0'', \lambda'')$ is a P-factor map (in particular, $\zeta_*(\lambda) = \lambda'$). Furthermore, we have the maps $\varphi : (G/P \times_\beta X_0, \nu_0 \times \lambda) \to (X, \nu)$ and $\varphi' : (G/P \times_\beta X_0'', \nu_0 \times \lambda'') \to (X', \nu')$ given by Proposition 5.1, as well as the maps $\psi : (G/P \times_\beta X_0, \nu_0 \times \lambda) \to (G/P \times_\beta X_0'', \nu_0 \times \lambda'')$ (see §6), and $\zeta : (X, \nu) \to (X', \nu')$. By definition of the maps, $\zeta \circ \varphi = \varphi' \circ \psi$. Since the function $\mathcal{E}_s \tilde{f}(\overline{u}_\theta, e, x_0)$ is in $\tilde{L}^\infty(X', \nu')$, we use the commutativity of the maps and conclude that as a function of x_0, it depends only on $x_0'' = \zeta(x_0)$, namely the function is in $\tilde{L}^\infty(X_0'', \lambda'')$. It follows that

$$\int_{X_0} \mathcal{E}_s \tilde{f}(\overline{u}_\theta, e, x_0)d\lambda(x_0) = \int_{X_0''} \mathcal{E}_s \tilde{f}(\overline{u}_\theta, e, x_0'')d\lambda''(x_0'').$$

Finally, since $\lambda'' = \nu'$, and $\mathcal{E}_s \tilde{f} \in \tilde{L}^\infty(X', \nu')$, we have as noted above, that (for almost every $\overline{u}_\theta \in \overline{U}_\theta$)

$$\int_{X_0''} \mathcal{E}_s \tilde{f}(\overline{u}_\theta, e, x_0'')d\lambda''(x_0'') = \int_{X'} \mathcal{E}_s \tilde{f} d\nu'.$$

Clearly, the last integral does not depend on $\overline{u}_\theta \in \overline{U}_\theta$, and we have obtained a contradiction to our assumption. Therefore ν' is not G-invariant and the proposition is proved. $\qquad \square$

We now come to the proof of Theorem 2, and repeat its formulation for convenience.

THEOREM 2. *Let G be a connected semisimple Lie group with finite center, μ admissible, and (X, ν) an ergodic (G, μ)-space with positive entropy. Suppose (X, ν) fails to have a nontrivial projective factor. Then there exists a factor group G_1 of G with $\mathbb{R}$-rank $(G_1) = 1$, a nontrivial (G_1, μ_1)-space (X_1, ν_1), and a factor map $\rho : (X, \nu) \to (X_1, \nu_1)$, equivariant with respect to G and G_1. Here $\mu_1 = p_*(\mu)$, $p : G \to G_1$ the natural map, and (X_1, ν_1) has positive entropy and does not admit a nontrivial projective factor.*

Proof. If the real rank of G is one, then of course there is nothing to prove, so we assume that the real rank of G is at least two. Consider again the factor (X', ν') constructed for a given $\theta \subset \Delta$ (and $s \in S'_\theta$) in Section 6, and the dichotomy of Proposition 10.1. If for every θ with $[\theta] \subsetneq \Phi$ (and all $s \in S'_\theta$) the first case of this alternative holds, then λ is invariant under every such $\overline{U}_\theta$. Since the real rank of G is at least two, it follows that the groups $\overline{U}_\theta$ generate $\overline{V}$ (see [M, Ch. I, 1.2.2]). It follows that λ is invariant under P as well as $\overline{V}$, hence G-invariant, so that ν is G-invariant in this case, contrary to our assumption.

Therefore, for at least one θ with $[\theta] \neq \Phi$ (and some $s \in S'_\theta$), the second alternative of Proposition 10.1 holds. Consequently, we can assume that for some $\theta \subset \Delta$ with $[\theta] \neq \Phi$ (and some $s \in S'_\theta$), the space (X', ν') has positive entropy, and in particular is nontrivial. In this space, ν'-almost all x' has stability group $G_{x'}$ of positive dimension, as noted in the proof of Proposition 6.1. Almost all groups $G_{x'}$ are conjugate, say to the group L, by ergodicity of ν. If L is not normal in G, we have a nontrivial projective factor, by Proposition 3.1. Therefore L must be normal in G, so the action of G on (X', ν') has nontrivial kernel N containing L. Let μ_1 be the measure on G/N obtained from μ via the natural map $G \to G/N$. Then $h_{\mu_1}(X', \nu') > 0$, and (X', ν') does not have a nontrivial G/N-equivariant projective factor, by assumption. If the factor group G/N has real rank one, we are done. Otherwise, we repeat this argument again for the action of the factor group G/N on (X', ν'). We arrive after finitely many steps at a factor group G_1 acting ergodically on a nontrivial factor space (X_1, ν_1) with positive entropy and without a nontrivial projective factor, for which the first alternative of Proposition 10.1 must hold. Since on (X_1, ν_1) the stationary measure is not G_1-invariant, we can repeat the argument used in the first paragraph of the present proof, and conclude that G_1 is necessarily simple of real rank one, and the theorem is proved. $\square$

11. Optimality, and some further remarks

Theorems 1 and 2, taken together with the counterexamples of [NZ1, Thm. B], are clearly the best possible result on the existence of nontrivial projective factors for general actions of arbitrary connected semisimple Lie group with finite center.

As a complement to Theorem 2, we note the following.

PROPOSITION 11.1.   *Let the assumptions be as in Theorem 2. Let $G = G_1 \cdot G_2$, where G_1 is the product of all simple real rank one factors, and G_2 the product of all other simple factors. If (X, ν) fails to have a nontrivial projective factor, then G_2 preserves the measure ν on X.*

The proof of Proposition 11.1 depends on the following two observations.

LEMMA 11.2.   *If $G = G_1 \cdot G_2$ is a decomposition of a connected semisimple Lie group G to a product of two commuting subgroups, and (X, ν) is a (G, μ)-space, then the measure ν is equivalent to a probablity measure stationary under an admissible measure μ_2 supported on G_2.*

Proof. First recall that the ν is equivalent to a measure η stationary under an admissible left-K-invariant measure on G, by [NZ1, Cor. 1.5]. Then by [NZ1, Thm. 1.4] $\eta = m_K * \lambda = m_{K_1} * m_{K_2} * \lambda$, where m_{K_i} are normalized Haar measures on K_i, and K_i are maximal compact subgroups of G_i satisfying $K = K_1 \cdot K_2$. The measure $\lambda_1 = m_{K_1} * \lambda$ is a P_2-invariant measure, where P_2 is a minimal parabolic subgroup of G_2, since P_2 of course commutes with G_1. Therefore $\eta = m_{K_2} * (m_{K_1} * \lambda) = m_{K_2} * \lambda_1$ is stationary under any left-K_2-invariant measure μ_2 on G_2, again by [NZ1, Thm. 1.4]. $\square$

LEMMA 11.3.   *Let G be a connected semisimple Lie group with finite center, μ admissible and (X, ν) an ergodic (G, μ)-space. If H is a connected normal subgroup of G, and almost every ergodic component of the H-action has a nontrivial projective H-factor, then the G-action has a nontrivial projective G-factor.*

Proof. Let us change the measure ν to an equivalent probability measure $\tilde{\nu}$, stationary under an admissible left-K-invariant measure $\tilde{\mu}$ on G, using [NZ1, Thm. 1.4]. This does not affect the assumptions or the conclusion of the proposition, but we can now assume, using Lemma 11.2, that $\tilde{\nu}$ is stationary under an admissible probability measure $\tilde{\mu}_2$ on $G_2 = H$.

Let E be the space of ergodic components for the H action. Since $\tilde{\mu}_2 * \tilde{\nu} = \tilde{\nu}$, the measure obtained on almost every ergodic component e is a $\tilde{\mu}_2$-stationary measure. Therefore, for almost every $e \in E$, we can consider the maximal projective factor of the H-action on e, denoted $\varphi_e : e \to H(e)/J(e)$. Since there are only finitely many conjugacy classes of parabolic subgroups, there is a subset $F \subset E$ of positive measure for which we can assume the maximal projective factor is a fixed H/J. Let $H' = C_G(H)$ be the complement of H in G. For each $g \in H'$, g permutes the elements of E and commutes with H, and so the map $\varphi_{ge} \circ g : e \to H(ge)/J(ge)$ is an H-equivariant map from e to a projective variety. Therefore $H(ge)/J(ge)$ is an H-equivariant factor of

$H(e)/J(e)$, by maximality of $H(e)/J(e)$. Using the same argument for g^{-1}, it follows that F is an H'-invariant set. By ergodicity of H' on E (which follows from ergodicity of G on X), it follows that $F = E$. In other words, there is a projective factor H/J for the H-action on X that is maximal on almost every ergodic component of the H-action. Furthermore, the map from $\varphi : X \to H/J$ is actually H'-invariant. Indeed, for $g \in H'$, $\varphi \circ g$ is an H-equivariant map to H/J (since H' centralizes H), which must coincide with φ by uniqueness of the maximal quotient map (see Lemma 0.1). Thus, if we write $H/J = G/Q$ where Q contains H', it follows that G/Q is actually a projective G-quotient of X. □

Proof of Proposition 11.1. By Lemma 11.3, we see that G_2 necessarily preserves ν. Indeed, otherwise there will be a nontrivial projective G_2-factor because of Theorem 1 and the rank assumption on G_2, and therefore also a nontrivial G-factor to (X, ν). The proposition is therefore proved. □

Note, finally, that the factor constructed in Theorem 2 is a factor of the space of ergodic components of the G_2-action on (X, ν). Indeed, G_2 acts trivially on the factor constructed, and the space of ergodic components is the unique maximal factor of (X, ν) on which G_2 act trivially.

We now formulate the following complement to Theorem 1 (and Theorem 3), asserting that actions failing to have a projective factor of full entropy must all have a definite form. Namely, such an action is induced from an action of a proper parabolic subgroup Q with the same property. Here, by a projective factor of a parabolic subgroup Q we mean a space of the form Q/Q_1, where Q_1 is a parabolic subgroup of G contained in Q.

THEOREM 11.4. *Let (G, μ) and (X, ν) be as in Theorem 1. Suppose (X, ν) has positive entropy and fails to have a projective factor of full entropy. Then (X, ν) has a factor of the form $(G/Q \times_\beta Y, \nu_0 \times \eta)$, with the following properties. The subgroup Q is a proper and nonminimal parabolic subgroup, Y is a compact metric Q-space, η is not equivalent to a probability measure invariant under Q, and (Y, η) does not admit a nontrivial projective factor of the form Q/Q_1.*

Proof. Let $(G/Q, \nu_0)$ be the maximal projective factor of (X, ν). G/Q is a nontrivial space by Theorem 1, so $Q \subsetneq G$. By standard arguments (see [Z1]) (X, ν) is isomorphic to an action induced from an action of Q on a space (Y, η), namely $(X, \nu) \cong (G/Q \times Y, \nu_0 \times \eta)$. We can take Y to be compact metric Q-space with a continuous Q-action by [V]. Clearly, $P \subsetneq Q$, by Theorem 4.1 of [NZ2]. Furthermore, η is not equivalent to a probability measure invariant under Q, since in that case the entropies of $(G/Q, \nu_0)$ and (X, ν) would be equal (see [NZ2, Prop. 1.9], and also §1.3). Finally, (Y, η) does not admit a nontrivial projective factor of the form Q/Q_1, where Q_1 is a parabolic subgroup of G contained in Q. Indeed, in that case we obtain a factor map $(Y, \eta) \to (Q/Q_1, \eta_0)$, so that (Y, η) is induced from an action of Q_1 on a space (Y_1, η_1). Consequently, we obtain:

$$(G/Q \times Y, \nu_0 \times \eta) \cong (G/Q \times Q/Q_1 \times Y_1, \nu_0 \times \eta_0 \times \eta_1) \to (G/Q \times Q/Q_1, \nu_0 \times \eta_0).$$

Here the second map indicated is projection on the first two components. We conclude that the space $(G/Q \times Q/Q_1)$ with the G-quasi-invariant measure class $\nu_0 \times \eta_0$ is a factor of (X, ν). At the same time, it is G-isomorphic to G/Q_1. But Q_1 is a proper subgroup of Q, so the maximal projective factor of (X, ν) is strictly larger than G/Q, contrary to our assumption. This concludes the proof of the theorem.   □

Remarks.

1. We note that actions with nontrivial projective factor with less than full entropy do indeed occur. For a construction of countably many such actions with distinct entropies for any group G admitting a parabolic subgroup Q which maps onto a group locally isomorphic to $SO(n, 1)$, (e.g. when Q has $PSL_2(\mathbb{R})$ as a factor), we refer to [NZ2, §3, Thm. 3.4].

2. It seems plausible that a structure theorem more refined than Theorem 11.4 can be established iteratively along the same lines, by considering the action of Q on (Y, η). For that one needs to develop the theory put forward in this paper in the generality of arbitrary connected Lie groups, to account for the parabolic subgroups that occur. This problem will be taken up elsewhere.

The case of compact factors. We now show how to prove Theorems 1, 2 and 3 when G has compact factors. Let G_0 denote the maximal compact normal subgroup of G, and consider the action of $G_1 = G/G_0$ on the factor space (X_1, ν_1) of (X, ν) consisting of the space of ergodic components of G_0. The action of G_0 on this space is trivial, and the action of G_1, (which has no nontrivial compact factors) is ergodic. If the space (X_1, ν_1) is trivial, then G_0 is ergodic on (X, ν) and hence (see [Z3]) essentially transitive. In this case (X, ν) is (isomorphic to) a compact homogeneous space $M = G_0/H_0 = G/H$ of G_0, and of G. Let the product of the noncompact components of G be denoted by L. Then L centralizes G_0, and hence each $l \in L$ commutes with the action of G_0 on the homogeneous space M. It follows that $l \in L$ acts by $l[g_0 H_0] = lg_0[H_0] = g_0 l([H_0]) = g_0[y(l)H_0] = g_0 y(l)[H_0]$, where $y(l) \in l[H_0] = [y(l)H_0] \subset G_0$ is a choice of representative. Taking $g_0 = h_0 \in H_0$ we conclude that $h_0 y(l)[H_0] = y(l)[H_0]$ for all $h_0 \in H_0$, so that $y(l)$ belongs to the normalizer of H_0 in G_0. Using again the commutativity of L and G_0, a similar argument shows that the map $\tau : l \mapsto y(l^{-1})$ satisfies $\tau(l_1)H_0 \tau(l_2)H_0 = \tau(l_1 l_2)H_0$ for any choice of representatives $y(l) \in l[H_0]$. Choosing the representatives measurably, $l \mapsto \tau(l)$ is a measurable homomorphism from L to the compact group $N_{G_0}(H_0)/H_0$. Therefore τ is a continuous homomorphism of L to a compact group and is hence trivial. Therefore L acts trivially on (X, ν), and in that case, the measure ν is G-invariant and the conclusions of Theorems 1 and 3 are satisfied. We can therefore assume that that (X_1, ν_1) is a nontrivial L-space, and clearly ν is an invariant measure if and only if ν_1 is. To prove Theorems 1 and 2, it is sufficient to construct a nontrivial projective G_1-factor of (X_1, ν_1), provided ν_1 is not invariant (or an action of a real rank-one factor groups in the case of Theorem 2). Since G_1 has no compact factors, the proof given above to Theorems 1 and 2 applies. As to Theorem 3, we can conclude that $(X_1, \nu_1) \to (G_1/Q_1, \nu_0)$ is a measure-preserving extension, by the arguments above. It follows of course that $(X, \nu) \to (G_1/Q_1, \nu_0)$ is also a measure-preserving extension, since $G_1 = G/G_0$ and G_0 is a compact normal subgroup.

Irreducible actions. The usual definition of irreducibility for a probability-measure-preserving action of a connected semisimple Lie group with finite center is that every normal subgroup of positive dimension acts ergodically. It is natural to inquire what are the actions with stationary measure which are irreducible according to this definition. Theorems 1 and 2 provide the following simple answer. If G is semisimple but not simple, then in an irreducible action the measure is necessarily G-invariant. Indeed, any factor of an irreducible action is irreducible, but a nontrivial projective factor can not be an irreducible action of such a group, and hence must be trivial. But a factor action with nontrivial kernel can not occur either under the assumption of irreducibility, hence the stationary measure must be invariant.

Technion - Israel Institute of Technology, Haifa, Israel
E-mail address: anevo@techunix.technion.ac.il

The University of Chicago, Chicago, IL
Current address: Brown University, Office of the Provost, Providence, RI
E-mail address: robert_zimmer@brown.edu

REFERENCES

[AM]   L. Auslander and C. C. Moore, *Unitary Representations of Solvable Lie Groups,* Memoirs of the A. M. S. **62** (1966).

[BN]   U. Bader and A. Nevo, *Conformal actions of simple Lie groups on compact pseudo-Riemannian manifolds,* preprint.

[BT]   A. Borel and J. Tits, Éléments unipotents et sous-groupes paraboliques de groupes réductifs, I, *Invent. Math.* **12** (1971), 95–104.

[F1]   H. Furstenberg, A Poisson formula for semisimple Lie groups, *Ann. of Math.* **77** (1963), 335–386.

[F2]   ———, Noncommuting random products, *Trans. Amer. Math. Soc.* **108** (1963), 377–428.

594   AMOS NEVO AND ROBERT J. ZIMMER

[GJT] Y. Guivarc'h, L. Ji, and J. C. Taylor, *Compactifications of Symmetric Spaces*, *Progr. in Math.* **156**, Birkhäuser Boston, Inc., Boston, MA, 1998.

[K]   V. A. Kaimanovich, The Poisson formula for groups with hyperbolic properties, *Ann. of Math.* **152** (2000), 659–692.

[Kn]   A. W. Knapp, *Lie Groups Beyond an Introduction*, *Progr. in Math.* **140**, Birkhäuser Boston, Inc., Boston, MA, 1996.

[Ma]   G. W. Mackey, Point realizations of transformation groups, *Illinois J. Math.* **6** (1962), 327–335.

[M]   G. A. Margulis, *Discrete Subgroups of Semisimple Lie Groups*, in *A Series of Modern Surveys in Mathematics* **17**, Springer-Verlag, New York, 1991.

[N1]   A. Nevo, Group actions with positive μ-entropy, preprint.

[N2]   ———, The spectral theory of amenable actions, and invariants of discrete groups, *Geom. Dedicata*, to appear.

[NZ1] A. Nevo and R. J. Zimmer, Homogenous projective factors for actions of semisimple Lie groups, *Invent. Math.* **138** (1999), 229–252.

[NZ2] ———, Rigidity of Furstenberg entropy for semisimple Lie group actions, *Ann. Sci. École Norm. Sup.* **33** (2000), 321–343.

[NZ3] ———, A generalization of the intermediate factor theorem, *J. d'Analyse Math.* **86** (2002), 93–104.

[NZ4] ———, Actions of semisimple Lie groups with stationary measure, *Proc. of the Conference on Rigidity in Geometry and Dynamics* (Isaac Newton Institute, June, 1999), to appear.

[St1] G. Stuck, Minimal actions of semisimple groups, *Ergodic Theory Dynam. Systems* **16** (1996), 821–831.

[V]   V. S. Varadarajan, Groups of automorphisms of Borel spaces, *Trans. Amer. Math. Soc.* **109** (1963), 191–220.

[W]   D. Witte, *Cocompact Subgroups of Semisimple Lie Groups*, in *Lie Algebra and Related Topics* (Madison, WI, 1988) (G. M. Benkart and J. M. Osborn, eds.), *Contemp. Math.* **110**, 309–313, Springer-Verlag, New York, 1990.

[Z1]   R. J. Zimmer, Induced and amenable ergodic actions of Lie groups, *Ann. Sci. École Norm. Sup.* **11** (1978), 407–428.

[Z2]   ———, Ergodic theory, semisimple Lie groups, and foliations by manifolds of negative curvature, *Publ. Math. I. H. E. S.* **55** (1982), 37–62.

[Z3]   ———, *Ergodic Theory and Semisimple Groups*, *Monographs in Math.* **81**, Birkhäuser Verlag, Basel, 1984.

(Received April 10, 2000)

Geometriae Dedicata (2005) 115: 181–199
DOI 10.1007/s10711-005-8194-1

© Springer 2005

Entropy of Stationary Measures and Bounded Tangential de-Rham Cohomology of Semisimple Lie Group Actions

AMOS NEVO[1,*] and ROBERT J. ZIMMER[2,**]
[1]*Department of Mathematics, Israel Institue of Technology, Technion, 32000 Haifa, Israel.*
e-mail:anevo@tx.technion.ac.il
[2] *Brown University, Providence, RI, U.S.A. e-mail:zimmer@math.brown.edu*

(Received: 19 November 2004; accepted in final form: 31 May 2005)

Abstract. Let G be a connected noncompact semisimple Lie group with finite center, K a maximal compact subgroup, and X a compact manifold (or more generally, a Borel space) on which G acts. Assume that ν is a μ-stationary measure on X, where μ is an admissible measure on G, and that the G-action is essentially free. We consider the foliation of $K \backslash X$ with Riemmanian leaves isometric to the symmetric space $K \backslash G$, and the associated tangential bounded de-Rham cohomology, which we show is an invariant of the action. We prove both vanishing and nonvanishing results for bounded tangential cohomology, whose range is dictated by the size of the maximal projective factor G/Q of (X, ν). We give examples showing that the results are often best possible. For the proofs we formulate a bounded tangential version of Stokes' theorem, and establish a bounded tangential version of Poincaré's Lemma. These results are made possible by the structure theory of semisimple Lie groups actions with stationary measure developed in Nevo and Zimmer [*Ann of Math.* **156**, 565–594]. The structure theory assert, in particular, that the G-action is orbit equivalent to an action of a uniquely determined parabolic subgroup Q. The existence of Q allows us to establish Stokes' and Poincaré's Lemmas, and we show that it is the size of Q (determined by the entropy) which controls the bounded tangential cohomology.

Mathematics Subject Classification (2000). 22D40, 28D15, 47A35, 57S20, 58E40, 60J50.

Key words. semisimple Lie groups, parabolic subgroups, stationary measures, entropy, Riemannian foliations, de-Rham cohomology, tangential bounded cohomology.

1. Introduction

Let G be a connected noncompact semisimple Lie group with finite center, let μ be an admissible probability measure on G. Let G act on a compact metrizable space X (or more generally, a standard Borel space) and let ν be a Borel probability measure on X which is μ-stationary, namely satisfies $\mu * \nu = \nu$ or equivalently

---

*Supported by BSF and ISF.
**Supported by BSF and NSF.

$$\int_G \int_X f(gx)\mathrm{d}\nu(x)\mathrm{d}\mu(g) = \int_X f(x)\mathrm{d}\nu(x), \quad \forall f \in L^1(X, \nu).$$

There are two fundamental invariants associated with the (G, μ)-space (X, ν). We begin by discussing the first invariant, which is given by the μ-entropy of the G-quasi-invariant measure ν, namely by the nonnegative quantity (see [8] for a discussion)

$$h_\mu(X, \nu) = \int_G \int_X -\log \frac{\mathrm{d}g^{-1}\nu}{\mathrm{d}\nu}(x)\mathrm{d}\nu(x)\mathrm{d}\mu(g).$$

Let us note the following two facts characterizing the two extreme values that the entropy may assume. First, the entropy $h_\mu(X, \nu)$ vanishes if and only if the measure ν is G-invariant (see, e.g., [5]). Second, the maximal possible value of the entropy is $h_\mu(G/P, \nu_0)$, where $P \subset G$ is a minimal parabolic subgroup, and ν_0 the unique μ-stationary measure on G/P. Furthermore, if this maximum value is finite, then it is achieved if and only if the action of G on (X, ν) is an amenable action [6] (Cor. 2.5, Th. 4.1) [7] (Thm 3).

We now recall a second canonical invariant associated with (X, ν). It was shown in ([7], Lemma 0.1) that there is a uniquely determined standard parabolic subgroup, which we denote by $Q = Q(X, \nu)$, such that (X, ν) admits $(G/Q(X, \nu), \nu_0)$ as a measurable G-equivariant factor, and every other factor of the form $(G/Q', \nu_0)$ where Q' is a standard parabolic subgroup, satisfies $Q(X, \nu) \subset Q'$. The factor space $(G/Q(X, \nu), \nu_0)$ is called the *maximal projective factor* of (X, ν) and constitutes a canonical G-equivariant-isomorphism invariant. Its entropy $h_\mu(G/Q(X, \nu), \nu_0)$, which may be strictly smalled than $h_\mu(X, \nu)$ is of course also an invariant of (X, ν). We remark that $h_\mu(G/Q(X, \nu), \nu_0)$ usually determines the maximal projective factor (namely that's the case for most choices of μ). When the entropies of (X, ν) and its maximal projective factor coincide, we will call (X, ν) a *measure-preserving extension* of its maximal projective factor (see [6] for a full discussion).

Our purpose in the present paper is to show that the structure theorems of [7] can be applied to the study of the bounded tangential de-Rham cohomology of an essentially free action of G on a manifold (or more generally, a standard Borel space) possesing a stationary measure. We will establish both vanishing theorems and nonvanishing theorems for the cohomology spaces, and show that their range is dictated by the size of the maximal projective factor of (X, ν). Remarkably, under the assumption that the action is irreducible, the vanishing and non-vanishing results complement each other. We also present examples showing that the vanishing results are best possible in many cases.

We will employ two distinct main arguments, both taking as their starting point the structure theorems of [7]. First, when the action is a measure preserving extension of its maximal projective factor, the foliation determined by G is equivalent to the foliation determined by a measure-preserving action of parabolic subgroup.

The reductive component of the parabolic group is unimodular and measure-preserving. Therefore, we can apply Stokes's theorem for foliations with transverse invariant measure and finite mass, and produce a nonzero (bounded tangential) cohomology class associated with the volume form on the acting group, since it is unimodular. This argument originated in Ruelle and Sullivan [10] and is discussed more generally in Moore and Schochet [3] (see also [11]).

The second argument is motivated by the technique developed in [1], and establishes vanishing of tangential bounded cohomology by a direct argument involving estimates showing that a (bounded tangential) Poincaré's Lemma holds. Here use is made of the fact that $Q = L \times U$ where L is reductive and U unipotent, and there exists a 1-parameter group in the center of L which acts as contractive automorphisms of U. This flow substitutes for the contractive flow whose vector field is the gradient of the Busemann function on the ideal boundary, used in [1].

We note that when considering foliations associated with semisimple Lie groups actions with stationary measures, our results generalize the recent results of [1] (who have treated the case of arbitrary foliations, but assumed amenability). The main result of [1] translates in our context to the vanishing of the k-dimensional tangential cohomology space for $k > r$, when G has real rank r and acts amenably and essentially freely. In Proposition 5.3 we will give an alternative proof of this fact and in addition show that the r-dimensional tangential bounded cohomology space does not vanish. In fact, the cohomology contains the exterior algebra $A^*(\mathbb{R}^r)$.

Our main results consider general, not necessarily amenable, actions, and are as follows.

THEOREM 1. *Let G be a connected semisimple Lie group with finite center. Assume the action of G on (X, v) is essentially free and (X, v) is a measure-preserving extension of its maximal projective factor. Put $\ell(Q) = \dim(L/K_L)$, where L is the reductive Levi component of $Q(X, v)$. Then:*

(1) $H^{\ell}_{b, \mathrm{d}R}(G, X) \neq 0$.
(2) $H^{k}_{b, \mathrm{d}R}(G, X) = 0$ if $k > \ell$.

Thus both the vanishing and nonvanishing results for the cohomology depend on $Q(X, v)$ in an explicit manner, We note that as the group Q increases (and the entropy $h_\mu(G/Q, v_0)$ decreases), the range where the cohomology vanishes decreases in the foregoing result, until finally for zero entropy (i.e. measure-preserving) actions the conclusion amounts to the nonvanishing of the tangential bounded cohomology space in dimension $\dim G/K$, namely the existence of tangential bounded volume forms which are nonzero in bounded cohomology,

We also establish nonvanishing results in low dimensions using the structure theorems, and some additional arguments. The statement is as follows (for the definition of nondegeneracy of the action, see Section 5.1):

THEOREM 2. *Let G be a connected semisimple Lie group with finite center, Let (X, v) be an essentially free nondegenerate (G, μ)-space of positive entropy, and $(G/Q(X, v), v_0)$ the (non-trivial) maximal projective factor, Write $Q = LU$, where L is the reductive Levy component, and U is unipotent. Then $A^*(\mathbb{R}^z) \subset H^*_{b,dk}(G; X)$, where $z = z(Q(X, v)) > 0$ is the dimension of the vector subgroup $Z_0(L)$ of the Abelian group $Z(L)$. In particular, $H^1_{b,dR}(G; X) \neq 0$.*

In Proposition 5.5 below we will prove for certain G-spaces nonvanishing of tangential bounded cohomology in all dimensions up to the maximal possible value $\ell(Q)$, thus showing that Theorem 1 is best possible.

We also give a cohomological characterization of the vanishing of the entropy, namely

THEOREM 3. *Let G be a connected semisimple Lie group with finite center. Let (X, v) be an ergodic nondegenerate essentially free (G, μ)-space. The following are equivalent:*

(1) *The stationary measure is invariant.*

(2) $H^n_{b,dR}(G, X) \neq 0$, *where $n = \dim G/K$.*

(3) $H^k_{b,dR}(G, X) \neq 0$ *for some $k > \ell_0$, where $\ell_0 = \max\{\dim L/K_L$: L is the reductive Levi component of a proper parabolic subgroup$\}$.*

2. Tangential de-Rham Cohomology of a Foliation

Let M be a standard Borel measure space with a foliation $\mathcal{F}$ into smooth manifolds. Let $\Lambda^*_{\mathcal{F}}$ be the complex of measurable differential forms which are smooth along almost all the leaves of $\mathcal{F}$, with two forms identified if they are equal on almost every leaf. Let d be the exterior derivative (along the leaves), and let $H^*_{dR}(M, \mathcal{F})$ be the cohomology of this complex (we refer to [3] for a detailed description, and to [1] for a useful summary).

We note that the definition of the cohomology spaces depends on a choice of an equivalence class of transverse measures (or at least on a choice of a σ-ideal of 'null sets' on transversals) but we suppress the transverse measure in the notation.

If almost every leaf of $\mathcal{F}$ has a Riemannian metric, then we call $\mathcal{F}$ a 'leafwise Riemannian foliation'. In this case, we call a form $\omega \in \Lambda^*_{\mathcal{F}}$ 'bounded' if $\|\omega\|$ and $\|d\omega\|$ are essentially bounded functions on M. Equivalently, the functions are uniformly bounded off a set of leaves which is null with respect to the given transverse measure class. Here $\|\cdot\|$ is taken with respect to the Riemannian metric. Let $\Lambda^*_{b,\mathcal{F}}$ be the space of bounded forms, which is a subcomplex of $\Lambda^*_{\mathcal{F}}$.

DEFINITION 2.1. $H^*_b(M, \mathcal{F})$ is the cohomology of the complex $(\Lambda^*_{b,\mathcal{F}}, d)$. The space $H^k_b(M, \mathcal{F})$ will be refered to as the cohomology space of bounded tangential k-forms.

We note that, in general, $H_b^*(M, \mathcal{F})$ is an algebra under exterior product.

When M is a compact space and $\mathcal{F}$ is a topological foliation, it is possible to give the following alternative description of bounded tangential cohomology (see e.g. [1] for more details) Let $\mathcal{S}$ be the sheaf of measurable functions on open sets $U \subseteq M$ which are locally constant on the leaves of $\mathcal{F}|_U$. Then $H_{dR}^*(M, \mathcal{F}) = H^*(M, \mathcal{S})$. Similarly, let $\mathcal{S}_b$ be the sheaf of functions on $U \subseteq M$ which are locally bounded on U (namely in $L^\infty_{loc}(U)$), and locally constant on $\mathcal{F}|_U$. Then $H_b^*(M, \mathcal{F}) = H^*(M, \mathcal{S}_b)$.

The natural notion of equivalence associated with measurable bounded tangential cohomology of foliations is as follows.

DEFINITION 2.2. Two measurable foliations $(M_1, \mathcal{F}_1)$ and $(M_2, \mathcal{F}_2)$ are called *pseudo-isometric* if there exists a measurable bijection $M_1 \overset{f}{\leftrightarrow} M_2$ taking the leaves of $\mathcal{F}_1$ diffeomorphically to the leaves of $\mathcal{F}_2$, so that $\|df\|$ and $\|df^{-1}\|$ are in L^∞.

Clearly any pseudo-isometry induces an isomorphism on $H_b^*(M, \mathcal{F})$.

We note that if Q is a connected Lie group and $K \subseteq Q$ a compact subgroup, then any two Q-invariant Riemannian metrics on $K \backslash Q$ are pseudo-isometric. K being a compact subgroup, the space of K-orbits $M = K \backslash S$ where (S, v) is any essentially free Q-space with quasi-invariant measure, carries a natural foliation $\mathcal{F}$ where the leaves can be identified with $K \backslash Q$. To define the foliation, first fix a Riemannian metric τ on the group Q which is invariant under the right action of Q, as well as under the left action of the compact subgroup K. Then τ descends naturally to a Riemannian metric on the homogeneous space $K \backslash Q$, invariant under the (right) Q-action. Any compact subgroup K acts tamely on (S, v) (in the measure-theoretic sense, see [15]) and we consider the map to the orbit space $S \mapsto M = K \backslash S$. For almost every s, the image of the orbit Qs under the map can be identified with $K \backslash Q$, and is defined to be the leaf through $m = Ks$. This identification of $K \backslash Q$ with the leaf allows us to transfer the Riemannian structure τ to the leaf, and this defines unambiguously a Riemannian structure on the leaf. Indeed, choosing any other point $m' = Kq's$ in the same leaf (for some $q' \in Q$), the identification of $K \backslash Q$ with the leaf is now $Kq \mapsto Kqq's$, but since the Riemannian metric τ is invariant under right translations by q', the Riemannian structure on the leaf is well defined.

Let us now note the following simple facts that will be used below.

(1) If an element $\varphi \in Q$ normalizes K, then φ acts naturally on the homogeneous space $K \backslash Q$ via $\varphi(Kq) = K\varphi q$. The Riemmanian structure τ on $K \backslash Q$ identifies each tangent space $T_{Kq}(K \backslash Q)$ with the inner product space $\mathrm{Lie}(K) \backslash \mathrm{Lie}(Q)$ (with the form τ projected from Q). Under this identification, the linear map $(d\varphi)_{Kq} : T_{Kq}(K \backslash Q) \to T_{K\varphi q}(K \backslash Q)$ is given by $\mathrm{ad}_{\mathfrak{k} \backslash \mathfrak{q}}(\varphi) : \mathfrak{k} \backslash \mathfrak{q} \to \mathfrak{k} \backslash \mathfrak{q}$.

(2) Any right-Q-invariant left-K-invariant Riemannian metric on Q defines a leafwise Riemannian structure on the foliation $\mathcal{F}$ of $M = K \backslash S$, and any two

such choices define pseudo-isometric foliations. It follows that $H_b^*(M, \mathcal{F})$ is an invariant of the group action.

DEFINITION 2.3. We let $H_{b,\mathrm{d}R}^*(G; S) = H_b^*(M, \mathcal{F})$ denote the bounded tangential cohomology of the Riemannian foliation with leaves $K\backslash G$, associated with the G-action.

Remark. We note that the foliation depends on the choice of the compact subgroup K, which we choose to be a maximal compact subgroup of G unless stated explicitly otherwise.

Let us first consider the case where $G = \mathbb{R}^n$ and $K = \{0\}$, and note the following simple fact, which will be used below.

PROPOSITION 2.4. *For any essentially free action of $\mathbb{R}^n$ on (S, ν) where ν is quasi-invariant, we have a natural embedding $A^*(\mathbb{R}^n) \hookrightarrow H_{b,\mathrm{d}R}^*(\mathbb{R}^n; S)$, where $A^*(\mathbb{R}^n)$ denotes the exterior algebra of $\mathbb{R}^n$.*

Proof. When $G = \mathbb{R}^n$ and $K = \{0\}$, $(M, \mathcal{F})$ is a foliation by Euclidean spaces. If ω is the usual translation invariant volume form on $\mathbb{R}^n$, then there is no *bounded* $(n-1)$-form α on $\mathbb{R}^n$ with $\mathrm{d}\alpha = \omega$. This fact is evident upon applying Stokes' theorem to a ball of radius r and its boundary sphere, and letting $r \to \infty$.

It follows that the same is true for any nonzero constant (i.e., translation invariant) k-form ω, and the space of such forms is naturally identified with $A^*(\mathbb{R}^n)$. Indeed, given a nonzero k-form ω, there exists a k-dimensional subspace $V \subset \mathbb{R}^n$ such that the restriction of ω to V is (a nonzero multiple of) a volume form. If $\omega = \mathrm{d}\alpha$ for some bounded from α on $\mathbb{R}^n$, then the restrictions of the forms to V have the same property, contradicting the argument of the previous paragraph.

We can view each translation-invariant form α on $\mathbb{R}^n$ as naturally defining an element $\tilde{\alpha}$ of the bounded tangential cohomology $H_{b,\mathrm{d}R}^*(\mathbb{R}^n; S)$. Finally, by definition of the exterior derivative in bounded tangential cohomology, exactness of $\tilde{\alpha}$ would imply that there exists (on almost every leaf) a *bounded* primitive for α and the result follows. □

Remark 2.5.

(1) It would be of interest to understand the cokernel of the map $A^*(\mathbb{R}^n) \hookrightarrow H_{b,\mathrm{d}R}^*(\mathbb{R}^n; S)$ as an invariant of the action.

(2) Proposition 2.4 also applies to ergodic actions with nontrivial stabilizers. The dimension of the stabilizers is almost surely constant under ergodicity, and denoting it by k, the same argument shows that $A^*(\mathbb{R}^{n-k})$ injects into the bounded tangential cohomology of the orbit foliation.

(3) We will assume throughout the paper that the group action is essentially free. It would be interesting to consider also the case of essentially locally free actions, namely when the stability groups are almost surely discrete, and the

associated foliations have locally symmetric leaves. However, when G is semi-simple even for the case of one leaf, namely $K \backslash G / \Gamma$, we are not aware of a characterization of discrete groups $\Gamma \subset G$ for which the volume form is exact in bounded cohomology.

(4) Recall that in the higher rank case, when the action is irreducible and probability-measure-preserving, essential local freeness implies essential freeness by [12], but this is not the case for stationary measures of positive entropy, as we see by inducing a measure preserving essentially locally free (but not essentially free) action of a parabolic subgroup Q to G.

It is instructive to see why the argument of Proposition 2.4 fails for symmetric spaces G/K associated with semisimple groups, which we view as a Riemannian foliation with one leaf. Here translation invariant forms still exist and have constant norm under the Riemannian structure given by the Killing form. These forms define tangential bounded forms since their Riemannian norm is constant, but in contrast to the $\mathbb{R}^n$-case, these form are trivial in bounded cohomology. Indeed, taking hyperbolic space of dimension n for example, one sees that the triviality of the volume form in bounded cohomology is due to the exponential growth of the area of the surface of the sphere, which is comparable to the volume of the ball, so that the volume form is the exterior derivative of a *bounded* $(n{-}1)$-form.

3. Foliated Stokes' Theorem

We now consider the case of a connected semisimple Lie group G with finite center and $K \subseteq G$ a maximal compact subgroup, so that G/K is a Riemannian symmetric space of non-positive curvature. We wish to understand $H^*_{b, \mathrm{dR}}(G; S)$ as an invariant of the action, and, in particular, when the measure on S is ergodic and stationary, how this relates to the entropy and the structure theorem asserting the existence of a nontrivial projective factor.

Given a foliation $\mathcal{F}$, a transverse invariant measure ζ is an assignment of a σ-finite measure to each transversal to $\mathcal{F}$ that is invariant under the holonomy pseudo-group. If $\mathcal{F}$ has a transverse invariant measure and is leafwise Riemannian, then the local measures defined by the volume form along the leaves and the transverse invariant measure ζ combine to give a global σ-finite measure $\tilde{\zeta}$ on M. If $\tilde{\zeta}$- is a finite measure, then we say that the foliation $\mathcal{F}$ together with the transverse invariant measure have finite mass. This will be the case, for example, if M is a compact space, the metric along the leaves is continuous on M and the invariant transverse measure is Radon. In the finite mass case, denoting as above the transverse invariant measure by ζ, every tangential bounded n-form α, where n is the dimension of the leaves, has a well-defined finite $\int_M \alpha \mathrm{d}\zeta$, given as follows. Since α is a top-dimensional form, $\alpha_m = f(m)\mathrm{dvol}$, where vol denotes the canonical Riemannian volume form associated with the Riemannian metric on the leaf containing m. Here $m \mapsto f(m)$ is an essentially bounded function w.r.t. the measure $\tilde{\zeta}$, so we can set $\int_M \alpha \, \mathrm{d}\zeta = \int_M f(m)\mathrm{d}\zeta(m)$.

As noted in the introduction, our nonvanishing result will be based on the following version of Stokes's theorem for foliations.

PROPOSITION 3.1. (Foliated Stokes's theorem [3] (Cor. 4.24), see also [10]). *If $\mathcal{F}$ is a leafwise Riemannian foliation with transverse invariant measure ζ of finite mass, β a bounded tangential n-form and $\sigma \in \Lambda_{b,\mathcal{F}}^{n-1}$ satisfies $\beta = \mathrm{d}\sigma$ (where n is the dimension of the leaves of $\mathcal{F}$), then $\int_M \beta \, \mathrm{d}\zeta = 0$.*

We now note the following consequence of Propositions 3.1 and 2.4, which we will apply repeatedly below.

PROPOSITION 3.2. *Suppose L is a connected Lie group, $K \subseteq L$ a compact subgroup, and that L acts on (S, λ) essentially freely. Assume λ is a finite L-invariant measure and L is unimodular. Then*

(1) *The associated leafwise Riemannian foliation has a transverse invariant measure of finite mass, and $H_b^\ell(L, S) \neq 0$, where $\ell = \dim L/K$.*

(2) *If $L = Z_0 J$ where $Z_0 \cong \mathbb{R}^z$ is the vector subgroup of the center of L, then $A^*(\mathbb{R}^z) \subset H_{b,dR}^*(L, S)$.*

Proof. First, if L is unimodular, preserves a probability measure on S and acts (locally) freely, then the existence of a transverse invariant measure of finite mass for the orbit foliation asociated with L is well-known (see [11], Thm. 3.12, or [9], Lemma 9.1). The fact that the same conclusion holds for the foliation into the Riemannian leaves $K\backslash L$ (assuming that $K\backslash L$ is oriented) is a straightforward variation. Hence every bounded tangential ℓ-form α, where $\ell = \dim L/K$, has a well-defined finite $\int_M \alpha \mathrm{d}\zeta$. Note however the important fact that the foregoing argument fails for nonunimodular groups. Indeed, in this case there is no transverse invariant measure since the Radon–Nikodym cocycle of the foliation is nontrivial (in fact it is given by the modular function of G, see [11], Thm. 3.12).

Second, the Riemannian volume form β along a leaf $K\backslash L$ gives rise to a bounded tangential ℓ-form which we denote by $\tilde{\beta}$. Clearly the total mass of the foliation is given by $\int_M \tilde{\beta}\mathrm{d}\zeta \neq 0$ and, hence, $H_{b,dR}^\ell(L, S) \neq 0$, by Proposition 3.1.

Finally, for the group $Z_0 = Z_0(L) \cong \mathbb{R}^z$, Proposition 2.4 immediately implies that $A^*(\mathbb{R}^z) \subset H_b^*(Z_0, S)$. We can assume without loss of generality that $K \subset J$ and use the natural factor maps $K\backslash L \mapsto J\backslash L = Z_0$ to pull back any translation-invariant closed k-form γ_0 on Z_0 to a k-form γ on $K\backslash L$. The form γ so obtained is closed, since it is a pull-back of a closed form. In addition γ is also a bounded k-form, and in fact it is invariant under right translations on $K\backslash L$. Indeed, since the map used to define γ is a group homomorphism with kernel J, it follows that γ is invariant under J, and is clearly also invariant under Z_0.

Therefore γ defines a bounded tangential k-form $\tilde{\gamma}$ on the foliation $\mathcal{F}_L$. Each leaf of $\mathcal{F}_L$ is foliated into leaves of the orbit foliation of Z_0, and restricting $\tilde{\gamma}$ to these orbits we retrieve the form γ_0 on Z_0. By definition of the exterior derivative in tangential bounded cohomology, it follows that $\tilde{\gamma}$ cannot be exact unless γ_0 is exact, so that $A^*(\mathbb{R}^z) \subset H^*_{b,\mathrm{dR}}(L, S)$ and the proof is complete. $\qquad\square$

We now consider the case where $Q = L \ltimes U$ is a semidirect product with L the reductive Levy component and U unipotent. Then there is a maximal compact subgroup $K \subseteq Q$ which satisfies $K \subseteq L$. We can deduce the following

PROPOSITION 3.3. *Let* $Q = L \ltimes U$, *with* L *reductive and* U *unipotent, Suppose that* Q *preserves a finite measure* λ *on* S *and acts essentially freely. Then*

(1) *Let* $\ell = \ell(Q) = \dim K \backslash L$, *where* $K \subseteq L$ *is a maximal compact subgroup of* Q. *Then* $H^\ell_{b,\mathrm{dR}}(Q, S) \neq 0$.

(2) *Write* $L = Z_0 J$ *where* $Z_0 \cong \mathbb{R}^z$ *is the vector subgroup of* $Z(L)$. *Then* $A^*(\mathbb{R}^z) \subset H^*_{b,\mathrm{dR}}(Q, S)$.

Proof. (1) Consider the measure-preserving action of L on (S, λ), and let $M = K \backslash S$. We then have two foliations of M, $\mathcal{F}_Q$ and $\mathcal{F}_L$, corresponding to the Q- and L-actions, respectively. Each leaf of $\mathcal{F}_Q$ can be identified with $K \backslash Q$, and we have a natural map $K \backslash Q \to K \backslash L$. Let $\ell = \dim K \backslash L$. Then the pull-back of the invariant volume form β_0 on $K \backslash L$ to $K \backslash Q$ defines an ℓ-form β on $K \backslash Q$. The form β is both closed (since it is a pull-back of a closed form), and bounded. In fact, β is invariant under the transitive isometric action of Q on $K \backslash Q$ and hence constant in norm. This assertion follows since the form in question is the pull-back of β_0 which is translation-invariant under the transitive isometric action of L on $K \backslash L$, and in addition it is U-invariant. Indeed, the map used for the pullback arises from the group homomorphism $Q = L \ltimes U \to L$ with kernel U, so that U preserves the forms lifted via this map. We conclude that β defines a bounded tangential ℓ-form $\tilde{\beta} \in \Lambda^\ell_{b,\mathcal{F}_Q}$. Since each leaf of $\mathcal{F}_Q$ is in fact foliated into leaves of $\mathcal{F}_L$, the form $\tilde{\beta}$ can be restricted to the leaves of $\mathcal{F}_L$, and this restriction is just the volume form along the leaves of $\mathcal{F}_L$, namely $\tilde{\beta}_0$. Since Q and therefore L preserve λ and L is unimodular, by the foliated Stokes theorem the form $\tilde{\beta}_0$ is not exact in $\Lambda^*_{b,\mathcal{F}_L}$, so that, by restriction, we deduce that $0 \neq [\tilde{\beta}] \in H^\ell_{b,\mathrm{dR}}(Q; S)$.

(2) The fact that $A^*(\mathbb{R}^z) \subset H^*_b(Q, S)$ follows using Proposition 3.2 to construct nontrivial bounded tangential forms on the foliation $\mathcal{F}_L$ (by lifting from the orbits of Z_0), and then lifting those to nonzero bounded tangential forms on the foliation $\mathcal{F}_Q$ as in the first part of the proof. $\qquad\square$

Remark 3.4.

(1) Note that since Q will not itself be reductive (or unimodular) in general, we cannot deduce using the foregoing argument that $H^n_{b,\mathrm{dR}}(Q, S) \neq 0$, where $n = \dim K \backslash Q$.

(2) Note also that the same argument *cannot* be applied to the restriction of the action to U, even though U is unimodular. This is because the volume form along the leaves of $\mathcal{F}_U$ is (in general) not bounded with respect to the metric on the leaves of $\mathcal{F}_Q$.

4. Bounded Tangential Poincaré Lemma

In the present section we aim to establish a vanishing theorem using a bounded tangential version of the Poincaré Lemma. Our argument is motivated by [1](Section 3), but is considerably simpler, since in our context we are able to introduce the following simplifying assumption. Our interest is in the case $Q = L \ltimes U$ discussed in the previous section, and we can in addition assume that there exists a one-parameter subgroup $\varphi_t \in Z(L)$ which (via conjugation) acts as contracting group automorphisms of U as $t \to \infty$. This assumption will of course be satisfied by any proper parabolic subgroup Q of a semisimple Lie group G, to which we plan to apply the following proposition.

PROPOSITION 4.1. *Let* $Q = L \ltimes U$ *with* L *reductive and* U *unipotent. Let* $\ell = \ell(Q) = \dim K \backslash L$, *where* $K \subseteq L$ *is a maximal compact subgroup of* Q. *Assume that there exists a one-parameter subgroup* $\varphi_t \in Z(L)$ *which (via conjugation) acts as contracting group automorphisms of* U *as* $t \to \infty$.

(1) Considering $K \backslash Q$ *as a Riemannian foliation with one leaf,* $H^r_{b,\mathrm{dR}}(Q; Q) = 0$ *if* $r > \ell$, *namely the bounded de-Rham cohomology of the manifold* $K \backslash Q$ *vanishes for* $r > \ell$.

(2) Suppose that Q *acts essentially freely on* (S, η) *where* η *is quasi-invariant. Then* $H^r_{b,\mathrm{dR}}(Q; S) = 0$ *if* $r > \ell = \dim K \backslash L$.

Proof. (1) We denote the group multiplication in Q by $(l', u')(l, u) = (l'l, u'l'(u))$, where $u \mapsto l(u)$ is given by the implicit homomorphism $L \mapsto Aut(U)$. Now consider the Riemannian manifold $M = K \backslash Q$ as a foliation with one leaf, where the Riemannian structure is given by a left-K-invariant right-Q-invariant Riemannian metric τ on Q. This foliation subdivides into two complementary foliations, one defined by the projection to M of the L-orbits in Q, and the other defined by the projection of the U-orbits (both w.r.t. the left action of Q on itself). More explicitly, given a point $m = Kq = K(l, u) \in M$, the leaf of the Riemannian foliation $\mathcal{F}_L$ through m namely the projection to M of the L-orbit of (l, u), is the set $\{K(l', e)(l, u)); l' \in L\} = K \backslash (L(l, u))$. The leaf through m of the foliation of M determined by U is the projection to M of the U-orbit of (l, u) namely the set $\{K(l, u'); u' \in U\}$. Clearly the tangent spaces to these two foliations at each point constitute a direct sum decomposition of the tangent space $T(K \backslash Q)$.

Let now φ be any element of the group $Z(L)$, the center of L. The action of φ on M is given by $m = K(l, u) \mapsto K(\varphi, e)(l, u) = K(\varphi l, \varphi(u)) = \varphi(m)$. Since $\varphi \in L$, clearly φ maps the leaf $K\backslash(L(l,u))$ of $\mathcal{F}_L$ containing $m = K(l,u)$ to itself. Each leaf of $\mathcal{F}_L$ has the structure of a symmetric space $K\backslash L$, and the map induced by φ on the symmetric space is simply the map $Kl \mapsto K\varphi l$, which is well defined since φ commutes with K. Again since $\varphi \in Z(L)$, the map obviously also coincides with right translation $Kl \mapsto Kl\varphi$.

As to the φ-action on the leaves of the orbit foliation determined by U, since L normalizes U, we see that the leaf $K(l,U)$ (of $\mathcal{F}_U$) containing $m = K(l,u)$ is mapped to the leaf $K(\varphi l, U)$ containing $K(\varphi l, \varphi(u)) = \varphi(m)$, and the map φ induces between the two leaves (when they are identified with U) is of course $u \mapsto \varphi(u)$.

We now choose $\varphi_t \in Z(L)$ as a 1-parameter family of contracting automorphisms of U (under $u \mapsto \varphi_t(u)$). φ_t induces a linear operator on the spaces $D_b^k(M)$ of smooth bounded k-forms on M. We consider the norm on $D_b^r(M)$ defined by taking the Riemannian norm in each tangent space together with the supremum norm over M. For each r-form α on M, and any choice of r vectors $v_i \in T_m(M)$, we claim that as soon as $r > \ell(Q) = \dim K\backslash L$, for all $t > 0$:

$$|(\varphi_t^*\alpha)(v_1,\dots,v_r)| = |\alpha_{\varphi_t(m)}(\mathrm{d}(\varphi_t)_m(v_1),\dots,\mathrm{d}(\varphi_t)_m(v_r)|$$
$$\leqslant C e^{-\delta_r t}\|\alpha\| \cdot \|v_1 \wedge \cdots \wedge v_r\|,$$

where δ_r and C are positive constants depending only the contracting group φ_t.

The condition $r > \ell(Q)$ arises of course since φ_t acts isometrically on each leaf of $\mathcal{F}_L$, and has exponential contraction between the leaves of $\mathcal{F}_U$ (w.r.t. the Riemannian structure on these leaves). Hence as long as the linear span of $v_1,\dots,v_r \in T_m(M)$ intersects the tangent space to the U-orbit of m nontrivially, the norm estimate is evident by a straightforward computation, bearing in mind that $\mathrm{Ad}(\varphi_t)$ acts on $\mathrm{Lie}(U)$ as a diagonalizable linear contraction for $t > 0$. In particular, we can deduce that the linear operator φ_t^* induced by $\varphi_t, t \geqslant 0$ on $D_b^r(M)$ has norm which decays exponentially in $t > 0$.

We now define the map $\Phi: M \times [0,\infty) \to M$ by $\Phi(m,t) = \varphi_t(m)$. Φ induces a linear operator $\Phi^*: D_b^r(M) \to D_b^r(M \times [0,\infty))$, defined as follows. Given a form $\alpha \in D_b^r(M)$, and r tangent vectors $w_1,\dots,w_r \in T_{(m,t)}(M \times [0,\infty))$, by definition

$$(\Phi^*\alpha)_{(m,t)}(w_1,\dots,w_r) = \alpha_{\varphi_t(m)}(\mathrm{d}\Phi_{(m,t)}(w_1),\dots,\mathrm{d}\Phi_{(m,t)}(w_r)).$$

Now $T_{(m,t)}(M \times [0,\infty)) = T_m(M) \oplus T_t([0,\infty))$, and if $w \in T_{(m,t)}(M \times [0,\infty))$ has zero second component, it can be identified with a vector $v_m \in T_m(M)$, and in that case for $f \in C^\infty(M)$ we have

$$\mathrm{d}\Phi_{(m,t)}(w)(f) = v_m(m \mapsto f(\varphi_t(m)) = \mathrm{d}(\varphi_t)_m(v_m)(f).$$

If w has zero first component, then it can be identified with $c(\mathrm{d}/\mathrm{d}s)|_{s=t}$ and then

$$\mathrm{d}\Phi_{(m,t)}(w)(f) = c\frac{\mathrm{d}}{\mathrm{d}s}(s \mapsto f(\varphi_s(m))|_{s=t} = cV_{\varphi_t(m)}(f) = (\mathrm{d}\varphi_t)_m(cV_m)(f),$$

where V is the vector field tangent to the flow φ_t. In order to estimate $|(\Phi^*\alpha)_{(m,t)}(w_1, \ldots, w_r)|$, we now consider the following two possibilities

(1) If all the vectors $w_1, \ldots, w_r \in T_{(m,t)}(M \times [0, \infty))$ have zero second component, then, identifying w_i with $v^i \in T_m(M)$ and using the previous estimate above:

$$|(\Phi^*\alpha)_{(m,t)}(w_1, \ldots, w_r)| = |\alpha_{\varphi_t(m)}((d\varphi_t)_m(v_m^1), \ldots, (d\varphi_t)_m(v_m^r))|$$
$$\leqslant Ce^{-\delta_r t}\|\alpha\| \cdot \|v_1 \wedge \cdots \wedge v_r\| = Ce^{-\delta_r t}\|\alpha\| \cdot \|w_1 \wedge \cdots \wedge w_r\|.$$

(2) Assume $w_1 = v^1 + c\frac{d}{ds}|_{s=t}$ (say) has a nonzero second component. Then $d\Phi_{(m,t)}(w_1) = (d\varphi_t)_m(v_m^1) + c(d\varphi_t)_m(V_m), c \neq 0$. The first summand is discussed in the previous argument, so that using multilinearity we can restrict attention to the second summand. Now V_m is the flow direction, which clearly has nonzero components in both the tangent space to the leaf $K \backslash (L(l, u))$ and the tangent space to the leaf $K(l, U)$. As a result, the subspace spanned by $d\Phi_{(m,t)}(w_1), \ldots, d\Phi_{(m,t)}(w_r)$ cannot be contained in the tangent space $T_{\varphi_t(m)}(K \backslash (L(l, u)))$ to the $\mathcal{F}_L$-leaf containing $\varphi_t(m)$. Therefore if the subspace in question is r-dimensional, it must intersect the tangent space to the $\mathcal{F}_U$-leaf containing $\varphi_t(m)$. Using the exponential contraction of φ_t in the directions tangent to the $\mathcal{F}_U$-leaves as before, the same estimate as in the previous argument follows as before.

Given a form $\alpha \in D_b^r(M)$, there is a unique representation $\Phi^*(\alpha) = \omega^0(t) + \omega^1(t) \wedge dt$, where $\omega^0 \in D_b^r(M \times [0, \infty))$ has interior product zero with $\partial/\partial t$, and $\omega^1 \in D_b^{r-1}(M \times [0, \infty))$ (since Φ^* is bounded). We now define an $(r-1)$-form $\beta = \beta_\alpha$ on M, given as follows. Fix $v_1, \ldots, v_{r-1} \in T_m(M)$ and let

$$\beta_\alpha(v_1, \ldots, v_{r-1}) = \int_0^\infty \Phi^*(\alpha)\left(v_1, \ldots, v_{r-1}, \frac{d}{ds}|_{s=t}\right)dt$$
$$= \int_0^\infty \omega^1(t)(v_1, \ldots, v_{r-1})dt.$$

Observe that the foregoing estimates on $\Phi^*(\alpha)$ show that β_α is a well-defined bounded $(r-1)$-form on M. Indeed, as we saw above:

$$|(\Phi^*\alpha)_{(m,t)}(v_1, \ldots, v_{r-1}, \frac{d}{ds}|_{s=t})| \leq Ce^{-\delta_r t}\|\alpha\| \cdot \|v_1 \wedge \cdots \wedge v_{r-1}\|.$$

We now claim that if α is a closed r-form, then β is a bounded primitive of α. Indeed, $d\beta = \int_0^\infty d_M\omega^1(t)dt$, where d_M is the exterior derivative along $M \times \{0\} \subset M \times [0, \infty)$, and on the other hand:

$$d\int_0^\infty \Phi^*(\alpha)dt = \int_0^\infty d\Phi^*(\alpha)dt = \int_0^\infty \Phi^*(d\alpha)dt.$$

If α is closed, then the last integral is zero. Therefore

$$
0 = \int_0^\infty d\Phi^*(\alpha) dt = \int_0^\infty \left(\frac{\partial \omega^0(t)}{\partial t} + d\omega^1(t) \right) \wedge dt
$$

$$
= \int_0^\infty \frac{\partial \omega^0(t)}{\partial t} \wedge dt + d\beta.
$$

We conclude that

$$
d\beta = -\int_0^\infty \frac{\partial \omega^0(t)}{\partial t} \wedge dt = \lim_{t \to \infty} [(\varphi_0)^*(\alpha) - (\varphi_t)^*(\alpha)] = \lim_{t \to \infty} (\alpha - (\varphi_t)^*(\alpha)) = \alpha,
$$

where the last assertion, namely $\lim_{t \to \infty} (\varphi_t)^*(\alpha) = 0$ follows of course from the norm estimate for the derivative $(\varphi_t)^*$ in its action on r-forms. This concludes the proof in the case of foliations with one leaf.

(2) Consider now the general case where Q acts on (S, η), η quasi-invariant. We have the leafwise Riemannian foliation $\mathcal{F}_Q$ on $M = K \backslash S$ defined by the Q-action (K a maximal compact subgroup of Q, contained in L). As before, since φ_t centralizes L (and hence K), φ_t acts on M, preserving the leaves of $\mathcal{F}_Q$. The action of φ_t on the leaves of $\mathcal{F}_L$ and $\mathcal{F}_U$ is described in the first part of the proof. If a closed bounded tangential r-form exists on the foliation $\mathcal{F}_Q$ which is nonzero in cohomology, then by part (1) and the definition of tangential bounded cohomology, there exists such a form on a set of leaves of positive transverse measure, which is a contradiction. It follows that $H^r_{b,dR}(Q; S) = 0$ if $r > \ell = \dim L/K$. $\qquad \square$

Remark 4.2. Comparing the proof of Proposition 4.1 to that of the 'Poincaré lemma with estimates' given in [1], we see that in our context, the presence of the 1-parameter group of contracting auomorphism φ_t eliminates the need to choose a point in the asymptotic boundary, construct the contracting flow corresponding to the gradient of the Busemann function at the chosen point to obtain bounded primitives, and then use the assumption of amenability of the foliation to measurably select one such bounded primitive for almost every point, as in [1].

5. Actions of Semisimple Lie Groups with Stationary Measure

We now turn to the case in which G is a connected semisimple Lie group with finite center. Suppose Q is a parabolic subgroup of G. We may choose K to be a maximal compact subgroup of G such that $K_Q = K \cap Q$ is a maximal compact subgroup of Q. Further, if $Q = L \ltimes U$ is the Levi decomposition, we may choose K so that $K_Q \subseteq L$.

Suppose Y is an essentially free Q-space with quasi-invariant measure η and $S = (G \times Y)/Q$ is the G-space induced from the Q-action on Y. Then the map $Y \to S$ defined by $y \mapsto Q(e, y)$ induces a pseudo-isometry of the foliation $(K_Q \backslash Y, \mathcal{F}_Q)$ with $(K \backslash S, \mathcal{F}_G)$. Here note that since Q is a parabolic subgroup, $KQ = G$ and hence

$K_Q \backslash Q \cong K \backslash G$, and so $K_Q \backslash Q$, with a suitable Q-invariant metric, is isometric to $K \backslash G$ (see the remark preceding Definition 2.3). Thus in both foliations the leaves are isomorphic to the symmetric space $K \backslash G \cong (K \cap Q) \backslash Q$.

We may therefore apply Proposition 4.1 to $(K \backslash S, \mathcal{F}_G)$. Furthermore, if there is a Q-invariant measure on Y, we may apply Proposition 3.3 as well. The main point of our discussion is that in [7] it was shown that the assumption that the G-action is induced from a parabolic subgroup is not restrictive at all, provided only that v is stationary. In order to proceed, we now briefly recall the relevant facts and definitions regarding stationary measures and the main structure theorems from [5, 7].

Standing assumptions. we will assume throughout this section that G is a connected semisimple Lie group with finite center (and arbitrary positive real rank), and K a maximal compact subgroup of G. $\mu \in P(G)$ will denote an admissible probability measure namely such that some convolution power of μ is absolutely continuous w.r.t Haar measure, and the support of μ generates G as a semi group. We will sometimes assume μ to be of finite entropy, namely $h_\mu(G/P, v_0) < \infty$, where P is a minimal parabolic subgroup, and v_0 the unique μ-stationary measure on G/P. (Recall that $h_\mu(G/P, v_0) < \infty$ whenever μ is admissible and has bounded density of compact support, for example.)

Suppose that G acts on (X, v), where v is μ-stationary and μ admissible. Then there exists a unique P-invariant probability measure λ supported on $S \subset X$ with the property that $v = \int_{G/P} g\lambda \mathrm{d}v_0(gP)$, where P is a minimal parabolic subgroup, and v_0 the unique μ-stationary measure on G/P. The action is called P-mixing if the measure-preserving P-action is mixing on (S, λ) [5].

DEFINITION 5.1. Let (X, v) be a (G, μ)-space, and write $v = \int_{G/P} g\lambda \mathrm{d}v_0(gP)$, where λ is P-invariant. The action is called irreducible if each nontrivial element of the maximal $\mathbb{R}$-split Cartan subgroup $A \subset P$ acts ergodically on (S, λ).

We note that when the measure v is actually G-invariant (so that $\lambda = v$), then the definition of irreducibility given above is equivalent to the standard one familiar in the theory of measure-preserving action of semisimple groups, namely that every non-compact simple factor acts ergodically. Thus the definition above is a natural generalization of the usual notion of irreducibility.

Finally, we recall ([7], Lemma 0.1) that associated to (X, v) there is a uniquely determined standard parabolic subgroup, which we denote by $Q = Q(X, v)$, such that the maximal projective factor of (X, v) coincides with $(G/Q(X, v), v_0)$ (see Section 1). We can now state the following

Structured theorem for actions with stationary measure [7].

(1) *If every noncompact simple factor of G has real rank at least two, then $Q(X, v)$ is a proper parabolic subgroup (and so (X, v) has a nontrivial projective factor), unless v is an invariant measure.*

(2) *For any G of real rank at least two, if the action is irreducible (in particular, if it is P-mixing) then the projective factor $(G/Q(X, v), v_0)$ has full entropy: $h_\mu(X, v) = h_\mu(G/Q(X, v), v_0)$, and Q is the stabilizer of the probability measure λ.*

(3) *For any G of real rank at least two, every action with positive entropy has a nontrivial projective factor (not necessarily with full entropy) unless (X, v) has a factor with positive entropy (and no projective factors) on which G acts via a real-rank-one factor group.*

We now introduce the following

DEFINITION 5.2. Given a group G of any real rank, a (G, μ) space (X, v) of positive entropy which has trivial maximal projective factor will be called degenerate.

Note that Part (3) of the structure theorem asserts that a degenerate space necessarily factors over a positive entropy degenerate action of a real-rank one factor group of G. Hence, if G has all its simple noncompact factors of real rank at least two, then *every* (G, μ)-space in nondegenerate. In addition, by part (2), every irreducible action of any group of real rank at least two is nondegenerate, and is a measure-preserving extension of its maximal projective factor. Finally, a nondegenerate action of a real-rank-one group is necessarily amenable and is a measure-preserving extension of the maximal boundary G/P (see [6], Thm. 4.1).

We will start by proving Theorem 3, whose formulation we recall for the reader's convenience.

THEOREM 3. *Let G be a connected semisimple Lie group with finite center. Let (X, v) be an ergodic nondegenerate essentially free (G, μ)-space. The following are equivalent:*

(1) *The stationary measure is invariant.*

(2) $H^n_{b,dR}(G, X) \neq 0$, *where $n = \dim G/K$.*

(3) $H^k_{b,dR}(G, X) \neq 0$ *for some $k > \ell_0$, where $\ell_0 = \max\{\dim L/K_L: L$ is the reductive Levi component of a proper parabolic subgroup$\}$.*

Proof. (1) implies (2). If the stationary measure is invariant, the foliation of $K \backslash X$ by symmetric space leaves $K \backslash G$ has transverse invariant measure and finite mass (take $G = L$ in Proposition 3.2). By Proposition 3.2, $H^n_{b,dR}(G, X) \neq 0$, where $n = \dim G/K$.

(2) implies (1). If $H^n_{b,dR}(G; S) \neq 0$, then by the bounded tangential Poincaré Lemma of Proposition 4.1, the maximal projective factor of (X, v) must be trivial. Otherwise the cohomology in dimensions greater than $\dim L/K_L$ must vanish, where L is the reductive component of the parabolic subgroup. But if the maximal projective factor is trivial and the action nondegenerate, by the structure theorem the measure v must be invariant.

Finally, clearly (2) implies (3), and the fact that (3) implies (1) follows again by Proposition 4.1.   □

Remark. It is interesting to compare Theorem 3 to [3](Cor. 4.25).

We now assume in addition that (X, ν) is a measure-preserving extension of its maximal projective factor, and prove Theorem 1, whose formulation we recall.

THEOREM 1. *Let G be a connected semisimple Lie group with finite center. Assume the action of G on (X, ν) is essentially free and (X, ν) is a measure-preserving extension of its maximal projective factor. Put $\ell = \dim(L/K_L)$, where L is the reductive Levi component of $Q(X, \nu)$. Then:*

(1) $H_{b,\mathrm{dR}}^{\ell}(G; X) \neq 0$

(2) $H_{b,\mathrm{dR}}^{k}(G; X) = 0 \quad$ *if $k > \ell$.*

Proof. By assumption, $Q = Q(X, \nu)$ preserves the probability measure λ on X. Now $X = (G \times S)/Q$ is induced from the Q-action on (S, λ), and hence the foliation of $K \backslash X$ with leaves $K \backslash G$ is pseudo-isometric to the foliation of $K_Q \backslash S$ with leaves $K_Q \backslash Q$. Since $Q = L \ltimes U$ with L reductive and U unipotent, Proposition 3.3 therefore applies and yields $H_{b,\mathrm{dR}}^{\ell}(G; X) \neq 0$. In addition, Proposition 4.1 also applies and yields $H_{b,\mathrm{dR}}^{k}(G; X) = 0$ if $k > \ell$. $\square$

In the special case where the action of G is amenable, we obtain the following corollary, the second part of which also follows from [1].

PROPOSITION 5.3. *Let G be a connected semisimple Lie group with finite center. Assume the action of G on (X, ν) is essentially free and amenable with finite entropy, and the real rank of G is r. Then*

(1) $A^*(\mathbb{R}^r) \subset H_{b,\mathrm{dR}}^*(G; X)$.

(2) $H_{b,\mathrm{dR}}^r(G; X) = 0$ *if $k > r$.*

Proof. By [6] (Thm. 4.1), (X, ν) factors over $(G/P, \nu_0)$, and the entropies of the two spaces coincide. The Levi decomposition of the minimal parabolic subgroup is given by $P = MA \ltimes N$, where M is compact and A is a maximal $\mathbb{R}$-split torus, isomorphic to $\mathbb{R}^r$. The result therefore follows from Proposition 3.3 and Proposition 4.1. $\square$

Remark 5.4. For general (not necessarily essentially free) amenable ergodic actions, the arguments in the proof of 3.3 imply that $A^*(\mathbb{R}^{r-k}) \subset H_{b,\mathrm{dR}}^*(G; X)$, where k is the dimension of the stability group in A of almost every point in the support $S \subset X$ of the P-invariant measure λ. Examples of such actions are easily obtained by induction from actions of A with positive dimensional stabilizers (e.g. with nontrivial kernel). By [6] (Thm. 4.1) every amenable ergodic (G, μ)-space is induced from some probability-measure-preserving action of a minimal parabolic subgroup P.

When the action on X is induced from the Q-action on a space (Y, η), whether or not η is a Q-invariant measure, we can also obtain a nonvanishing theorem in other dimensions. Namely, let $Q = L \ltimes U$ as above and write $L = Z_0(L)J$, where $Z_0(L)$ is the vector subgroup of the Abelian group $Z(L)$, and J is reductive with compact center. $Z_0 = Z_0(L)$ is then isomorphic to $\mathbb{R}^z$, where $z = z(Q)$ is positive for every proper parabolic subgroup Q.

THEOREM 2. *Let G be a connected semisimple Lie group with finite center. Let (X, v) be an essentially free nondegenerate (G, μ)-space of positive entropy, and $(G/Q(X, v), v_0)$ the (nontrivial) maximal projective factor. Then $A^*(\mathbb{R}^z) \subset H^*_{b,dR}(G; X)$, where $z = z(Q(X, v)) > 0$ is the dimension of the vector subgroup $Z_0(L)$ of $Z(L)$. In particular, $H^1_{b,dR}(G; X) \neq 0$.*

Proof. Consider the foliation associated to the Q-action on Y, with leaves $K_Q \backslash Q \cong Z_0(L) \times (K_Q \backslash J)$, where the product is Riemannian. As in the proof of Proposition 3.3, we see that $Z_0(L)$ (as well as any of its subgroups) will define a nonvanishing bounded cohomology class by taking (for example) the volume form on $Z_0(L)$ and pulling it back to the product $Z_0(L) \times K_Q \backslash J$ via the natural projection. The same of course holds for any other translation-invariant form on $Z_0(L)$. Again as in the proof of Proposition 3.3, the form gives rise naturally to a bounded tangential form on the foliation with leaves $K_Q \backslash Q$, which restrict to the volume form along the orbits of $Z_0(L)$ on $K_Q \backslash Q$. To see that the latter form is nonzero in the tangential bounded cohomology $H^*_{b,dR}(Q, S)$, we use the fact that $Z_0(L)$ is a vector group leaving the probability measure λ quasi-invariant, so that by Proposition 2.4 we can conclude that $A^*(\mathbb{R}^z)$ embeds in $H^*_{b,dR}(Z_0(L), Y)$, and the desired result follows.   □

A similar argument, using the foliated Stokes Theorem of Proposition 3.3 in the presence of an invariant probability measure yields the following (keeping the notation introduced above):

PROPOSITION 5.5. *Let G be a connected semisimple Lie group with finite center. Let (X, v) be an essentially free nondegenerate (G, μ)-space, which is a measure-preserving extension of its the maximal projective factor. We assume that $h_\mu(X, v) > 0$ (i.e. that the stationary measure is not invariant, and so $Q(X, v)$ is a proper subgroup).*

(1) $H^k_{b,dR}(G; X) \neq 0$ for $\dim(K_Q \backslash J) \leqslant k \leqslant \dim(K_Q \backslash L) = \ell(Q)$.

(2) *If $K_Q \backslash J$ has the structure of a Kähler manifold, then $H^k_{b,dR}(G; X) \neq 0$ for $1 \leqslant k \leqslant \dim K_Q \backslash L = \ell(Q)$.*

Proof. (1) There exist z linearly independent translation-invariant 1-forms $\alpha^0_1, \ldots, \alpha^0_z$, on $Z_0(L)$. If β^0 denoted the Riemannian volume form on $K \backslash J$, then $\alpha^0_1 \wedge \cdots \wedge \alpha^0_z \wedge \beta^0$ is the Riemannian volume form on $K \backslash L$. The forms just mentioned can be pulled back to $K\backslash Q$ and give rise to translation invariant forms $\alpha_1, \ldots, \alpha_z, \beta$. These in turn give rise to bounded tangential forms $\tilde{\alpha}_1, \ldots, \tilde{\alpha}_z, \tilde{\beta}$ on the foliation $\mathcal{F}_Q$. By Proposition 3.2, since $\tilde{\alpha}_1 \wedge \cdots \wedge \tilde{\alpha}_z \wedge \tilde{\beta}$ is the form pulled back from the Riemannian volume on $K\backslash L$, it is nonzero in cohomology and, hence, the same holds for the form $\tilde{\beta}$. It follows that all z of the following forms

$$\tilde{\beta}, \ \tilde{\beta} \wedge \tilde{\alpha}_1, \ldots, \tilde{\beta} \wedge \tilde{\alpha}_1 \wedge \cdots \wedge \tilde{\alpha}_z$$

are non-zero in cohomology and the result follows.

(2) In the case when $K_Q\backslash J$ is Kähler, denoting the Kähler form by Ω_0, the jth wedge power $\Omega_0^j, 1 \leqslant j \leqslant \frac{1}{2}\dim(K_Q\backslash J)$ is an invariant closed $2j$-form. Furthermore, $\Omega_0^{\frac{1}{2}\dim(K_Q\backslash J)}$ is an invariant volume form. Arguing as in the first part of the Proposition, we can pull back the forms Ω_0^j on $K_Q\backslash J$ to forms Ω^j on $K_Q\backslash Q$ and then to bounded tangential forms $\tilde{\Omega}^j$. Now note that $\tilde{\Omega}^{\frac{1}{2}\dim(K_Q\backslash L)}$ is nonzero in bounded tangential cohomology, by Proposition 3.3. Therefore we can conclude that $\tilde{\Omega}^j \in H_{b,dR}^{2j}(Q, S) = H_{b,dR}^{2j}(G, X) \neq 0$ in all positive even dimensions less than or equal to $\dim K_Q\backslash L$. Now the dimension of the vector subgroup of $Z(L)$ satisfies $z = z(Q) \geqslant 1$ since Q is a proper parabolic subgroup by assumption, so that there exists at least one tangential bounded 1-form $\tilde{\alpha}_1$ as in part (1). Then $\tilde{\alpha}_1 \wedge \tilde{\Omega}^j \neq 0$, since

$$\tilde{\alpha}_1 \wedge \cdots \wedge \tilde{\alpha}_z \wedge \tilde{\Omega}^{\frac{1}{2}\dim K_Q\backslash J}$$

is nonzero, by Proposition 3.3. We therefore obtain nonvanishing of the bounded tangential cohomology in all odd dimensions (up to the maximum possible dimension, which is $\dim K_Q\backslash L$) as well and the proof is complete. $\square$

Acknowledgements

The authors would like to thank Kevin Corlette for an advance copy of [1], which motivated the present paper and was helpful in its preparation. The first named author would like to thank the department of mathematics in the University of Chicago for its hospitality and financial support during his visits. The second named author would like to thank the Department of Mathematics at the Technion for its hospitality and financial support during his visits.

References

1. Corlette, K., Hernandez-Lamoneda, L. and Iozzi, A.: A vanishing theorem for the tangential de-Rham cohomology of a foliation with amenable fundamental groupoid, *Geom. Dedicata* **103** (2004), 205–223.
2. Gromov, M.: Volume and bounded cohomology, *Pub. Sci. IHES* **56** (1983), 5–99.
3. Moore, C. C. and Schochet C.: *Global Analysis on Foliated Spaces*, Springer-Verlag, New York, 1988.

ENTROPY OF STATIONARY MEASURES 199

4. Nevo, A.: On the spectral theory of amenable actions, and invariants of discrete groups, *Geom. Dedicata* **100** (2003), 187–218.

5. Nevo, A. and Zimmer, R. J.: Homogeneous projective factors for actions of semisimple Lie groups, *Invent. Math.* **138** (1999), 229–252.

6. Nevo, A. and Zimmer, R. J.: Rigidity of Furstenberg entropy for semisimple Lie group actions, *Ann. Sci. Ecole Norm. Sup.* **33** (2000), 321–343.

7. Nevo, A. and Zimmer, R. J.: A structure theorem for actions of semisimple Lie groups, *Ann. Math.* **156**, 565–594.

8. Nevo, A. and Zimmer, R. J.: *Actions of Semisimple Lie Groups with Stationary Measure, Rigidy in Dynamics and Geometry*, Springer, New York, 2002, pp. 321–343.

9. Plante, J. F.: Foliations with measure-preserving holonomy, *Ann. of Math.* **102** (1975), 327–361.

10. Ruelle, D. and Sullivan, D.: Currents, flows, and diffeomorphisms, *Topology* **14** (1975), 319–327.

11. Series, C.: The Poincaré flow of a foliation, *Amer. J. Math.* **102** (1980), 93–128.

12. Stuck, G. and Zimmer, R. J.: Stabilizers for ergodic actions of higher rank semisimple groups, *Ann. of Math.* **139** (1994), 723–747.

13. Zimmer, R. J.: On the cohomology of ergodic group actions, *Israel J. Math.* **35**(1989), 289–300.

14. Zimmer, R. J.: Ergodic theory, semi-simple Lie groups, and foliations by manifolds of negative curvature, *Publ. Math. IHES* **55** (1982), 37–62.

15. Zimmer, R. J.: *Ergodic Theory and Semisimple Groups*, Birkhauser, Boston, 1984.

INVARIANT RIGID GEOMETRIC STRUCTURES AND SMOOTH PROJECTIVE FACTORS

Amos Nevo*
Technion IIT
anevo@tx.technion.ac.il

Robert J. Zimmer
University of Chicago
president@uchicago.edu

December 16, 2013

Abstract

We consider actions of non-compact simple Lie groups preserving an analytic rigid geometric structure of algebraic type on a compact manifold. The structure is not assumed to be unimodular, so an invariant measure may not exist. Ergodic stationary measures always exist, and when such a measure has full support, we show the following.

1) Either the manifold admits a smooth equivariant map onto a homogeneous projective variety, defined on an open dense conull invariant set, or the Lie algebra of the Zariski closure of the Gromov representation of the fundamental group contains a Lie subalgebra isomorphic to the Lie algebra of the acting group. As a corollary, a smooth non-trivial homogeneous projective factor does exist whenever the fundamental group of M admits only virtually solvable linear representations, and thus in particular when M is simply connected, regardless of the real rank.

2) There exist explicit examples showing that analytic rigid actions of certain simple groups (of real rank one) may indeed fail to have a smooth projective factor.

3) It is possible to generalize Gromov's theorem on the algebraic hull of the representation of the fundamental group of the manifold to the case of analytic rigid non-unimodular structures, for actions of simple groups of any real rank.

An important ingredient in the proofs is a generalization of Gromov's centralizer theorem beyond the case of invariant measures.

1   Introduction and statements of main results

Our purpose here is to study analytic actions of simple Lie groups on manifolds which preserve a rigid analytic geometric structure of algebraic type. These structures have been introduced in the foundational work [G1] as a generalization of Cartan's notion of structures of finite type. The study of rigid structures, both for their own sake and also in connection with rigidity theory for group actions [Z3] has attracted quite a bit of attention (see e.g. [CQB] [F][FZ][QB] and references therein). We note however that most of the work carried out on group actions preserving rigid structures concentrated on the case of unimodular structures. This assumption is used extensively in a number of considerations, particularly in providing an invariant volume form and thus an invariant measure on the manifold, to which considerations related to Borel density theorem and its generalizations can be applied. However, such an assumption leaves out some very interesting and natural examples, such as conformal pseudo-Riemannian and other conformal structures (see [BN] and [FrZe] for some recent work in this direction).

---

*Supported by the Institute for Advanced Study, Princeton and ISF grant # 975-05.
[0] *1991 Mathematics subject classification:* 22D40, 28D15, 47A35, 57S20, 58E40, 60J50.
Key words and phrases: semi-simple Lie groups, parabolic subgroups, stationary measure, rigid analytic structure.

1

Here we consider general, not neceessarily unimodular structures, dispensing with the assumption of an invariant volume form, and assuming only the existence of an ergodic stationary measure of full support. We thus introduce the following :

Definition. Rigid analytic stationary systems. *A rigid analytic stationary system (M, ω, G, ν) is a compact connected analytic manifold, with rigid analytic geometric structure of algebraic type ω, a connected non-compact almost simple Lie subgroup $G \subset Aut(M, \omega)$ with finite fundamental group, and an ergodic probability measure ν of full support on M, which is stationary under an admissible measure μ on G (we suppress μ in the notation).*

Our first main result is then as follows :

Theorem 1.1. Fundamental group and the existence of a smooth projective factor.
Let (M, ω, G, ν) be a rigid analytic stationary system. Then one of the following holds :

1. *There exists a generic (i.e., open dense co-null) G-invariant subset $M_0 \subseteq M$ and a smooth G-equivariant map $M_0 \to G/Q$, where Q is a proper parabolic subgroup.*

2. *The Lie algebra of the Zariski closure of the image of the fundamental group under the Gromov representation contains a Lie subalgebra isomorphic to $\mathfrak{g}$.*

In particular, a rigid analytic stationary system admits a non-trivial generic smooth homogeneous projective factor if the fundamental group of M admits only virtually solvable linear representations.

As we shall see below, every smooth G-manifold with ergodic stationary measure of full support admits a unique maximal projective factor of the form G/Q with the factor map defined and smooth on a generic set. Thus Theorem 1.1 shows that this factor is *non-trivial* whenever the fundamental group is amenable, or more generally does not contain a free group on two generators.

As to the necessity of the assumptions in Theorem 1.1, we note that rigid analytic stationary systems do not always admit smooth projective factors, even when the stationary measure is not invariant. This is demonstrated by the following :

Theorem 1.2. Examples for real rank one groups. *Let $G = SO^0(n, 1)$ for some $n \geq 2$, and μ an admissible measure on G. Then there exists an analytic action of G on a compact analytic manifold M, with a rigid analytic geometric structure of algebraic type ω preserved by G, such that*

1. *M has a unique G-ergodic μ-stationary measure ν, and ν is of full support on M and not G-invariant (ν is in addition P-mixing, see [NZ1]).*

2. *M does not admit a non-trivial generic smooth projective factor. In fact it does not even admit a non-trivial measurable projective factor, for any G-quasi-invariant measure. In addition, the G-action on M is everywhere locally free.*

We now turn to a generalization of Gromov's results regarding rigid analytic structures from the measure-preserving case to the stationary case. We will thus be able to obtain significant information regarding the algebraic structure of the fundamental group of the manifold, as well as on the action of G on M. This information is valid for simple groups of any real rank, including one. We formulate below a generalization of Gromov's representation theorem [G1][Z3, Thm. 1.5], formulated originally for unimodular structures (where the maximal measurable projective factor is necessarily trivial). In our present context the structure may not be unimodular, and the maximal projective factor gives rise to a certain parabolic subgroup Q which replaces G in the conclusion of the representation theorem, so that when the stationary measure is invariant the results coincide.

Before stating this generalization we introduce the following notation. Let Q be any proper parabolic subgroup of G, and write the Langlands decomposition as $Q = M_Q A_Q N_Q$. Denote by $Q_0 = M'_Q A_Q N_Q$ the co-compact subgroup of Q obtained by requiring that M'_Q contains exactly the simple factors of M_Q which are non-compact. We recall that any stationary measure ν has a representation as a convolution $\nu = \tilde{\nu}_0 * \lambda$, where λ is a uniquely determined P-invariant probability

2

measure, P a minimal parabolic subgroup, and $\tilde{\nu}_0$ is a probability measure on G. Furthermore, we recall that (M, ν) has a maximal (measurable) projective factor $G/Q(X.\nu)$, where $Q(X, \nu)$ is a uniquely determined standard parabolic subgroup. In general, if we denote by $Q = Q_\lambda$ the stability group of the measure λ, we have $Q_\lambda \subset Q(X, \nu)$ and Q_λ may be a proper subgroup. The extension $M \to G/Q(X, \nu)$ is called a measure-preserving extension if the measure λ is indeed invariant under $Q(X, \nu)$ (see [NZ1] and [NZ2] for a full discussion). We refer to §3 for further details on the notation and terminology used in the statement below.

Our second main result is as follows :

Theorem 1.3. Generalization of Gromov's representation theorem. *Let (M, ω, G, ν) be a rigid analytic stationary system, and assume that it is a measure-preserving extension of its maximal projective factor. Consider the image Γ of the Gromov representation of the fundamental group $\pi_1(M)$ in $GL(\mathfrak{z})$, where $\mathfrak{z}$ is the Lie algebra of Killing fields on $\tilde{M}$ commuting with $\mathfrak{g}$. Let $\overline{\Gamma}$ denote the Zariski closure of Γ. Then precisely one of the following holds :*

1. *The G-action is generically locally free, and then the Lie algebra of $\overline{\Gamma}$ has $\mathfrak{q}_0$ as a sub-quotient. Namely $Lie(\overline{\Gamma})$ contains a sub-algebra which has a factor algebra containing an isomorphic copy of the Lie algebra $\mathfrak{q}_0$ of the group Q_0 associated with Q.*

2. *On a generic set M_0, the connected components of all stability groups are mutually conjugate and of positive dimension, the generic smooth maximal projective factor G/Q_s is non-trivial, and M_0 is diffeomorphic to the induced action $G \times_{Q_s} Y$, where $Y \subset M_0$ is a Q_s-invariant submanifold.*

Note that when the stationary measure is G-invariant, the first case of Theorem 1.3 applies and we retrieve a version of Gromov's representation theorem.

It may happen that the stability group Q_λ reduces to the minimal parabolic subgroup P. But it is often the case that Q_λ is as large as it can possibly be, namely $Q_\lambda = Q(X, \nu)$. This condition is equivalent to (M, ν) being a measure-preserving extension of its maximal (measurable) projective factor. We recall that this holds whenever G is of real rank at least two and the action of G on (M, ν) is irreducible in the sense defined by [NZ2]. In particular, this holds when the P-action on (M, λ) is mixing [NZ1].

We refer to §6 for some further results and corollaries of Theorem 1.1 and Theorem 1.3.

As to the methods of proof, let us note the following essential ingredients. First, we will show that it is possible to generalize Gromov's centralizer theorem beyond the context of simple groups preserving a unimodular structure. We will establish such a generalization below, relying on and generalizing some of the arguments presented in [Z3]. Second, we note that in the presence of an invariant rigid structure of algebraic type, a number of equivariant maps to projective varieties present themselves naturally. Such a map immediately produces an equivariant projective factor, unless it is generically constant. We will formulate this below as a general dichotomy, and apply it to the generic algebraic hull, to the Lie algebra of a generic stability group, and to several maps defined by evaluation of Killing vector fields. We will thus obtain certain useful reductions which will be utilized in the course of the proof of Theorems 1.1 and 1.3.

Acknowledgements. The authors would like to thank David Fisher, Charles Frances and Karin Melnick for their very useful comments.

2   The generic smooth algebraic hull and maximal projective factor

2.1 The generic smooth algebraic hull

Let M be a manifold, G a Lie group acting smoothly on M, and let η be a G-quasi-invariant finite Borel measure on M of full support.

A subset $M_0 \subset M$ is called generic if it is open, dense and co-null. A map $f : M \to N$ is called generic if it defined on some generic subset $M_0 \subset M$ and is smooth on M_0. If M and N are G-spaces a G-equivariant generic map is a generic map defined on some G-invariant generic subset satisfying $f(gm_0) = gf(m_0)$, for $g \in G$ and $m_0 \in M_0$.

If $P \to M$ is a principal H-bundle, then by a *generic reduction* of P to a subgroup $L \subseteq H$ we mean a smooth reduction defined on a generic set, i.e., a smooth H-equivariant map $P_1 \to H/L$, where P_1 is an H-invariant generic set. If G acts on P by principal bundle automorphisms, a G-invariant generic reduction is a G-invariant reduction defined on a G-invariant generic set P_1.

Proposition 2.1. Generic smooth algebraic hull. *Assume G acts topologically transitively on M (for example, ergodically on (M, η), where η has full support), and $P \to M$ is a principal H-bundle on which G acts by bundle automorphisms. Suppose H is real algebraic. Then there is an algebraic subgroup $L \subseteq H$ such that*

1. *There is a G-invariant generic reduction of P to L.*

2. *If $L' \subsetneq L$ and L' is algebraic, then there is no such reduction.*

3. *L is unique up to conjugacy in H with respect to properties (1) and (2).*

4. *If there is a G-invariant generic reduction of P to J, then some conjugate of J contains L.*

Definition 2.2. The group L is called the generic smooth (or C^∞) algebraic hull of the G-action on P.

Remark 2.3.   1. Proposition 2.1, its proof and Definition 2.2 are analogues of the measurable algebraic hull (see e.g. [Z1]). The generic smooth algebraic hull was first introduced in [Z4].

2. Similar arguments establish the existence of the C^r generic algebraic hull, of an action on a principal bundle, for any $r \geq 0$. A full discussion can be found in [F].

Let us illustrate this notion by the following example, which will appear repeatedly below. Let G be a Lie group, and consider its action on $M \times \mathfrak{g}$, the product vector bundle, where G acts on $\mathfrak{g}$ via Ad. Consider the associated principal bundle $M \times GL(\mathfrak{g})$. Then the generic smooth algebraic hull $L \subseteq GL(\mathfrak{g})$ obviously satisfies $L \subseteq \mathrm{Zcl}(\mathrm{Ad}(G))$, Zcl denoting the closure in the Zariski topology.

Recall also that the image of a real algebraic group under a rational representation is of finite index in its Zariski closure. We will apply this fact for G semisimple or for the group $A_Q N_Q$ (where $Q = M_Q A_Q N_Q$ is a parabolic subgroup), and the representations $\mathrm{Ad}(G)$ and $\mathrm{Ad}_{\mathfrak{g}}(A_Q N_Q)$.

2.2 Generic smooth maximal projective factor

Recall that a compactly supported probability measure μ on an lcsc group G is called admissible if some convolution power is absolutely continuous with respect to Haar measure, and the support of μ generates G. If G acts measurably on a standard Borel space X, a Borel probability measure ν is called μ-stationary if $\mu * \nu = \nu$, namely for every $f \in L^1(X, \nu)$

$$\int_G \int_X f(gx) d\nu(x) d\mu(g) = \int_X f(x) d\nu(x) .$$

The pair (X, ν) is called a (G, μ)-space. We can now state

Proposition 2.4. Generic smooth maximal projective factor *Let G be a connected noncompact semisimple Lie group with finite center, and M a manifold on which G acts smoothly. Let μ be an admissible probability measure on G, and ν an ergodic μ-stationary measure of full support. Then there exists a parabolic subgroup $Q \subset G$, with the following properties :*

1. *There exists a generic smooth G-equivariant factor map $\phi : M_0 \to G/Q$, satisfying also $\phi_*(\nu) = \nu_0$, where $M_0 \subset M$ is generic and G-invariant, ν_0 the unique μ-stationary measure on G/Q.*

4

2. G/Q is the unique generic smooth maximal homogeneous projective factor of M, namely any other such factor of M is also a factor of G/Q. In particular the parabolic subgroup Q is uniquely determined up to conjugation.

3. The map ϕ is generically unique, namely any other G-equivariant generic smooth map $\psi : M_1 \to G/Q$ agrees with ϕ on a G-invariant generic set.

Proof. The existence of a unique maximal homogeneous projective factor was proved for the measurable category in [NZ2, Lemma 0.1], and the same proof applies without change in the present set-up as well. $\square$

Definition 2.5. Under the assumptions of Proposition 2.4, G/Q is called the generic smooth maximal projective factor of the G-action on (M, ν).

2.3 The smooth algebraic hull along orbits

In our analysis below, G will act smoothly on a manifold M with tangent bundle TM, and we will have cause to be interested in the sub-bundle TO consisting of the tangents to the G-orbits, particularly when the G-action is locally free. An important role will be played by the generic smooth algebraic hull along the orbits, particularly when the action is generically locally free. Recall first that for a Lie group G and the associated principal bundle $M \times GL(\mathfrak{g})$, (a conjugate of) the generic smooth algebraic hull L must be contained in $\mathrm{Zcl}(\mathrm{Ad}(G)) \subseteq GL(\mathfrak{g})$.

We can now state the following result, which will play an important role below.

Proposition 2.6. Generic smooth algebraic hull along orbits.

Let G be a connected semisimple Lie group with finite center and no compact factors, acting smoothly on a manifold M. Let μ be an admissible probability measure on G.

1. If ν is an ergodic μ-stationary measure on M of full support, then the generic smooth algebraic hull L for the action of G on the bundle $M \times GL(\mathfrak{g})$ is of finite index in $\mathrm{Zcl}(\mathrm{Ad}(G))$ if and only if the generic smooth maximal projective factor of (M, ν) is trivial.

2. Let $Q \subset G$ be a proper parabolic subgroup with Langlands decomposition $Q = M_Q A_Q N_Q$. The measurable algebraic hull L_0 of the action of the group $A_Q N_Q \subset Q$ on $M \times GL(\mathfrak{g})$ w.r.t. an ergodic invariant probability measure λ on M is of finite index in $\mathrm{Zcl}(\mathrm{Ad}_{\mathfrak{g}}(A_Q N_Q))$.

Proof.   1. If L is of finite index in $\mathrm{Zcl}(\mathrm{Ad}(G))$, then M does not admit a generic smooth equivariant map to any variety of the form G/H if H is algebraic of positive codimension. In particular, there exists such a map to the variety G/Q only if $Q = G$. Conversely, if L has positive codimension in $\mathrm{Zcl}(\mathrm{Ad}(G))$ then M does admit a non-trivial projective factor. Indeed, it follows from Proposition 3.2 of [NZ2] that any positive dimensional algebraic variety of the form G/L which supports a μ-stationary measure must factor over G/Q for some $Q \subsetneq G$.

2. $L_0 \subset \mathrm{Zcl}(\mathrm{Ad}_{\mathfrak{g}}(A_Q N_Q))$ and there exists an equivariant map $M \to \mathrm{Zcl}(\mathrm{Ad}_{\mathfrak{g}}(A_Q N_Q))/L_0$, so the algebraic variety $\mathrm{Zcl}(\mathrm{Ad}_{\mathfrak{g}}(A_Q N_Q))/L_0$ has an $A_Q N_Q$-invariant probability measure, the image of λ. As is well-known, $A_Q N_Q$ does not have algebraic subgroups of finite covolume, and any $A_Q N_Q$-invariant probability measure on any algebraic variety must be supported on fixed points of $A_Q N_Q$ in its action on the variety. On the other hand $A_Q N_Q$ has finitely many orbits on the variety, since it is of finite index in its Zariski closure. It follows that L_0 contains $\mathrm{Ad}_{\mathfrak{g}}(A_Q N_Q)$ and is of finite index. $\square$

5

2.4 Projective factors and local freeness

Let G be a connected lie group acting smoothly on a manifold M. For any $m \in M$, we can consider the stability group $st_G(m)$, and its Lie algebra $\mathfrak{st}_G(m)$. Consider the map $m \mapsto \mathfrak{st}_G(m)$, from M to $Gr(\mathfrak{g}) = \coprod_{k=0}^{\dim G} Gr_k(\mathfrak{g})$, the disjoint union of the Grassmann varieties of k-dimensional subspaces of $\mathfrak{g}$. This map is obviously G-equivariant (where G acts on $\mathfrak{g}$ via Ad), and measurable (see e.g. [NZ2, §3] or [Da1]). Given an ergodic G-quasi-invariant measure η on M, the values of this map will be almost surely concentrated in a single G-orbit in $Gr(\mathfrak{g})$, since the action of G on $Gr(\mathfrak{g})$ which is a disjoint union of algebraic varieties is (measure-theoretically) tame.

Now note that if the measure has full support, we can deduce that on a G-invariant *generic* set M_0 the Lie algebras of stability groups of points in M_0 are all in the same $\text{Ad}(G)$-orbit, and in particular have the same dimension, say d_0. Indeed, the G-invariant sets $U_d = \{m \in M \, ; \, \dim \mathfrak{st}_G(m) \le d\}$ are open sets, since the dimension function is upper semi continuous (see [St, §2]). Now U_{d_0-1} is an open and hence must be empty, since the measure assigns positive measure to every non-empty open set, but the dimension of the stability group is almost surely d_0. Thus $U_{d_0} = \{m \, ; \, \dim \mathfrak{st}_G(m) = d_0\}$ is a conull, G-invariant open and dense set.

Denoting the latter set by M_0, the map $m \mapsto \mathfrak{st}_G(m)$ is smooth on M_0 (see e.g. the argument of [St, Lemma 2.10]). Thus we obtain a generic smooth map $M_0 \to G \cdot \mathfrak{s}$, where $\mathfrak{s}$ is the Lie algebra of a closed subgroup S. We can write $G \cdot \mathfrak{s} = G/N_G(S^0)$, where S^0 is the connected component and $N_G(S^0)$ is its normalizer in G. Thus for a connected semisimple Lie group we can conclude that (M, η) has a non-trivial homogeneous smooth projective factor whenever $N_G(S^0)$ is contained in a non-trivial parabolic subgroup Q. We can thus conclude the following, which is a generic smooth version of [NZ2, Prop. 3.1].

Proposition 2.7. Smooth projective factors and local freeness.

Let G be a connected semisimple Lie group with finite center and no compact factors, acting smoothly on a manifold M with ergodic quasi-invariant measure of full support. Then exactly one of the following three alternatives occur :

1. *On a G-invariant generic set $M_0 \subset M$, the identity components of the stability groups $st_G(m)$ are all conjugate (say to S^0) and have positive dimension. The generic smooth maximal projective factor G/Q is non-trivial, and $N_G(S^0) \subset Q \subsetneq G$.*

2. *On a G-invariant generic set the identity components of the stability groups are all equal to a fixed normal subgroup G_1 of G of positive dimension, so that the action of G on the generic set factors to an action of the group G/G_1.*

3. *On a G-invariant generic set, the G-action is locally free, namely the stability groups are discrete.*

Remark 2.8. We note that we will have below several other (a priori, measurable) G-equivariant maps which assign to points in M linear subspaces of a finite dimensional space, and that the argument preceding Proposition 2.7 is completely general, and shows the following. Given such a map ϕ, suppose that the dimension function $\dim \phi(m)$ is upper semi continuous, and the function ϕ is smooth on open sets where $\dim \phi(m)$ is constant. If the linear spaces $\phi(m)$ almost surely lie in one G-orbit (where the measure on M has full support), then there is a *generic* set where $\dim \phi(m)$ is constant and ϕ is smooth. This will be applied to the map ψ below and others, and the verifications of the required properties is routine.

3 Rigid geometric structures : orbits of the centralizer

Let G be a connected Lie group acting on a connected manifold M, preserving a rigid analytic structure of algebraic type [G1], which we denote by ω. Of course, we do not assume ω is unimodular. We let η denote a G-quasi-invariant finite Borel measure on M, usually assumed to be of full support.

We will follow the notation and terminology of [Z3], to which we refer for further discussion and the results cited below. Let $\tilde{\omega}$ be the structure on $\tilde{M}$ lifted from ω, and $\tilde{\mathfrak{k}}$ the space of global vector fields on $\tilde{M}$ preserving $\tilde{\omega}$, namely global Killing fields. Let $\mathfrak{n}$ be the normalizer of $\mathfrak{g}$ in $\tilde{\mathfrak{k}}$, where we have $\mathfrak{g} \subseteq \tilde{\mathfrak{k}}$ by lifting the action of G to its universal covering group $\tilde{G}$. Let $\mathfrak{z}$ be the centralizer of $\mathfrak{g}$ in $\tilde{\mathfrak{k}}$. Thus $\mathfrak{z} \subseteq \mathfrak{n} \subseteq \tilde{\mathfrak{k}}$. Since $\mathfrak{g}$ is semi-simple, $\mathfrak{n} = \mathfrak{g} \oplus \mathfrak{z}$.

$\pi_1(M)$ acts on $\tilde{\mathfrak{k}}$, and $\mathfrak{n}$ and $\mathfrak{z}$ are $\pi_1(M)$-submodules. Elements of $\tilde{\mathfrak{k}}$ are lifted from a vector field on M if and only if they are $\pi_1(M)$-invariant. $\tilde{G}$ also acts on $\tilde{\mathfrak{k}}$, the derived $\mathfrak{g}$-module structure being simply $\mathrm{ad}_{\mathfrak{k}}\big|_{\mathfrak{g}}$. By definition, $\tilde{G}$ commutes with $\mathfrak{z}$. The $\pi_1(M)$- and $\tilde{G}$-actions commute on $\tilde{M}$ and, in particular, on $\tilde{\mathfrak{k}}$ and its submodules. If G acts analytically on the analytic manifold M, rigidity of the structure ω implies that the space of analytic Killing vector fields $\tilde{\mathfrak{k}}$ is finite dimensional and in fact $\mathrm{Aut}(M, \omega)$ is a Lie group acting analytically on M, see [G1], [Z3, Thm. 2.3].

For each $y \in \tilde{M}$, let $\mathrm{Ev}_y : \tilde{\mathfrak{k}} \to T\tilde{M}_y$ be the evaluation map. Write

$$\psi(y) = \{X \in \mathfrak{g} : X(y) \in \mathrm{Ev}_y(\mathfrak{z})\} \subseteq \mathfrak{g}.$$

Then ψ is $\pi_1(M)$-invariant and G-equivariant (where G acts on $\mathfrak{g}$ by Ad), and hence factors through a generic G-map $M \to \mathrm{Gr}(\mathfrak{g})$. If the structure ω is unimodular and M compact, so that there exists a G-invariant finite volume form, then Gromov's centralizer theorem is the assertion that, on a generic set, $\psi(y) = \mathfrak{g}$, provided G is simple. This is a key property in the analysis of [G1] and [Z3].

We will generalize this in three respects, allowing non-unimodular structures, stationary measures rather than invariant ones, and groups more general than simple ones when a probability measure is assumed to be preserved. We begin with formulating the following generalization of Gromov's centralizer theorem [G1][Z3](Cor. 4.3). Consider the rigid structure of algebraic type ω' on M consisting of the pair $(\omega, \mathfrak{g})$. A local automorphism of this structure is a local diffeomorphism which preserves ω and normalizes $\mathfrak{g}$. We define global and infinitesimal (to order k) such automorphisms similarly. Let $M_0 \subseteq M$ be the set of points for which infinitesimal automorphisms of ω' of sufficiently high fixed order at $m \in M_0$ extend to local automorphisms of ω'. [cf. the discussion in [G1] and [Z3](Theorem 2.1).] Then M_0 contains a generic set, and in addition is clearly G-invariant, so we can in fact assume M_0 is generic and G-invariant.

Theorem 3.1. Infinitesimal orbits of the centralizer for general groups. *Let (M, ω) be a connected analytic compact manifold with a rigid analytic structure of algebraic type. Let $H \subset \mathrm{Aut}(M, \omega)$ be a connected real algebraic group with finite center, and no proper real algebraic subgroups of finite co-volume. Assume H preserves an ergodic probability measure λ and is essentially locally free. Let $\mathfrak{z}$ be the centralizer of $\mathfrak{h}$ in the normalizer subalgebra $\mathfrak{n}$. Then, for λ-a.e. $y \in M$, $\mathrm{Ev}_y(\mathfrak{z}) \supseteq \mathrm{Ev}_y(\mathfrak{h})$.*

Proof. Gromov's argument to prove the centralizer theorem in the case of invariant measure does not depend on semi-simplicity of G. Rather, it uses two crucial properties of the group, namely that the adjoint representation has finite kernel and that the Borel density theorem hold (i.e., that the group should have no algebraic subgroups of co-finite volume.) In addition, the argument requires essential local freeness of the action, which holds for any *measure-preserving* action of a simple group. Our assumptions about the group H and its action on M allow the same argument to proceed without change. We refer to [Z3, Thm 4.2, Cor. 4.3] or [G1] for details. $\square$

Theorem 3.1 will be applied to the action of a parabolic subgroup Q of a simple group G, preserving a probability measure on the manifold. In this case, we obtain the following simple relation between global Killing fields centralizing the Q-action and those centralizing the G-action.

Lemma 3.2. *Let $Q = M_Q A_Q N_Q$ be a proper parabolic subgroup of a connected semisimple Lie group G with finite center and no compact factors. Let G act on a compact manifold preserving a rigid analytic structure of algebraic type. If a Killing field on $\tilde{M}$ normalizes $\mathfrak{g}$ and centralizes $\mathfrak{a}_Q \oplus \mathfrak{n}_Q$, then it also centralizes $\mathfrak{g}$, namely $\mathfrak{l} = \mathfrak{z}$.*

Proof. The space $\tilde{\mathfrak{k}}$ of global Killing fields is a finite-dimensional $\mathfrak{g}$-module, so any vector invariant under $\mathfrak{a}_Q \oplus \mathfrak{n}_Q$ is also invariant under $\mathfrak{g}$. This follows from the fact that for a proper parabolic

7

subgroup Q the subgroup $A_Q N_Q$ is epimorphic in G, namely has the same set of invariant vectors as G does, in any finite dimensional representation. $\square$

Consider now a semisimple Lie group $G \subset \mathrm{Aut}(M, \omega)$, where now the underlying measure ν on M is assumed to be stationary, for some admissible measure μ on G, namely consider rigid analytic stationary systems. Note that the G-equivariant map $\psi : M \to Gr(\mathfrak{g})$ defined above (via evaluation) always exists, and allows us to modify Gromov's techniques from [G1] to prove the following.

Theorem 3.3. Projective factors and the evaluation map.
Let (M, ω, G, ν) be a rigid analytic stationary system. Then at least one of the following possibilities holds :

1. *for y in a G-invariant generic set, $\psi(y) = \mathfrak{g}$.*

2. *M has a generic smooth maximal projective factor which is non-trivial.*

Proof. We apply the trichotomy provided by Proposition 2.7 and the results above, as follows.

First, since ν is stationary, we can let λ be the corresponding P-invariant measure, where P is a minimal parabolic subgroup (see [NZ1] or [NZ3] for a full discussion). Then $\nu = \tilde{\nu}_0 * \lambda$, where $\tilde{\nu}_0$ is a (compactly supported, absolutely continuous) probability measure on G.

Consider first the case where G is essentially locally free on (M, ν) (and thus also on $\tilde{M}$). It then follows that, for $\tilde{\nu}_0$-a.e. $g \in G$, gPg^{-1} is essentially locally free on $(M, g_*\lambda)$. Furthermore, for $\tilde{\nu}_0$-almost every g, $g_*\lambda(M_0) = 1$, if $\nu(M_0) = 1$ for the generic set M_0. Thus, replacing P and λ by conjugates under some $g \in G$ if necessary, we may assume that P is essentially locally free and $\lambda(M_0) = 1$. Applying Theorem 3.1 to the action of $A_P N_P$, together with Lemma 3.2 we conclude that $\psi(y) \neq 0$ for λ-almost all points $y \in M_0$. Then since $\tilde{\nu}_0 * \lambda = \nu$, it follows easily that $\psi(y) \neq 0$ for ν-almost every $y \in M_0$. Now clearly $\dim \psi(m)$ is ν-almost surely constant and if this constant is smaller than $\dim \mathfrak{g}$ then we can apply the arguments noted in Remark 2.8 to ψ and conclude that we have a non-trivial generic smooth maximal projective factor.

Otherwise $\psi(y) = \mathfrak{g}$ ν-almost surely, and we can again apply the argument of Remark 2.8 to conclude that on a generic set $\psi(y) = \mathfrak{g}$, so that case (1) obtains.

We can now therefore consider the case where G is not essentially locally free. Since ν is stationary, any generic smooth quotient G/L with L a proper algebraic subgroup has a non-trivial homogeneous projective factor, i.e. a factor of the form G/Q with $Q \subsetneq G$ a parabolic subgroup, as we saw in Proposition 2.7. We therefore obtain a generic smooth non-trivial maximal projective factor, unless the stability groups are normal on a generic set. This is ruled out by our assumption that the G-action is faithful and the proof is complete.

$\square$

Thus, Theorem 3.3 allows us to reduce to the case $\psi(y) = \mathfrak{g}$ generically, a fact we will use below.

4   Stability subalgebras of the normalizer

We now add some further preparations before taking up the proof of Theorem 1.1. We return to considering TM, and from now on assume that the G-action is generically locally free, otherwise there already exists a non-trivial generic smooth maximal projective factor, according to Proposition 2.7 and our assumptions in Theorem 1.1. Then on a generic set we can identify the tangent bundle to the orbits of G, say $T\mathcal{O} \subseteq TM$ with $M \times \mathfrak{g}$, where G acts on $\mathfrak{g}$ via Ad. By Proposition 2.6, either M has a non-trivial generic smooth projective factor or $\mathrm{Zcl}(\mathrm{Ad}(G))$ coincides (up to finite index) with the generic smooth algebraic hull of the action on $M \times \mathfrak{g}$, and hence on $T\mathcal{O}$. Similarly, consider the extension of the G-action to an action on the principal bundle $P^{(k)}(M)$ whose fiber at each point is the space of k-frames. Again, the algebraic hull of this action contains a group locally isomorphic to G. This therefore enables us to obtain the conclusions of [Z3, Thm. 4.2] exactly as in the finite measure-preserving case. Namely, we can assume that the smooth generic algebraic hull

8

on $M \times \mathfrak{g}$ (and $T\mathcal{O}$) is in fact the Zariski closure of $\mathrm{Ad}(G)$ in $GL(\mathfrak{g})$ (otherwise the generic smooth maximal projective factor is already non-trivial).

Remark 4.1. As noted already in the introduction, we are repeatedly using the technique of asserting that either there is a non trivial projective factor, or we have a conclusion similar to the measure-preserving case. This was already applied to three maps from M to projective varieties, namely to the map associating to a point the Lie algebra of its stability group, to the map $M \to G/L$ where L is the algebraic hull on the tangent bundle to the G-orbits, and to the map ψ to the Grassmann variety $Gr(\mathfrak{g})$. We will also use this technique once more in the proof of Theorem 1.1 below, applying it to the map $y \mapsto \mathfrak{n}_y$ defined by the kernel of the evaluation map, where $y \in \tilde{M}$. Of course, in the measure-preserving case, Borel density is used to eliminate the possibility of a projective quotient from the outset.

This technique yield also the following results, which we will use below.

Proposition 4.2. Stability Lie algebras.

Let (M, ω, G, ν) be a rigid analytic stationary system. Then either the generic smooth maximal projective factor is non-trivial, or for y_0 in a generic set, there exists a Lie algebra $\mathfrak{g}_\Delta \subseteq \mathfrak{n}$ with the following properties:

1. *All elements of $\mathfrak{g}_\Delta$ vanish at y_0; i.e., $\mathfrak{g}_\Delta \subseteq \mathfrak{n}_0$ ($\mathfrak{n}_0 = \mathfrak{n}_{y_0}$ is the kernel of the evaluation map $X \mapsto X_{y_0}$).*

2. *$\mathfrak{g}_\Delta$ is isomorphic to $\mathfrak{g}$, and in fact is the graph in $\mathfrak{n} = \mathfrak{g} \oplus \mathfrak{z}$ of an injective homomorphism $\sigma : \mathfrak{g} \to \mathfrak{z}$. We let $\mathfrak{g}' = \sigma(\mathfrak{g})$ be the image in $\mathfrak{z}$.*

3. *Let $\mathfrak{z}_0 = \mathfrak{z}_{y_0}$ denote the kernel of the evaluation map. Let L be the generic smooth algebraic hull of the G-action on the principal $GL(\mathfrak{n})$-bundle over M associated with the representation of $\pi_1(M) \to \Gamma \subset GL(\mathfrak{n})$. Then $\mathrm{ad}_\mathfrak{n}(\mathfrak{g}_\Delta)$ acting on $\mathfrak{z}/\mathfrak{z}_0$ is contained in the Lie algebra $\mathrm{Lie}(L)$ of the algebraic hull, and L is contained in $\overline{\Gamma} = Zcl(\pi_1(M)) \subset GL(\mathfrak{n})$. In particular, $\mathrm{ad}_\mathfrak{z}(\mathfrak{g}_\Delta) = \mathfrak{g}' \subseteq \mathrm{Lie}(L) \subseteq \mathrm{Der}(\mathfrak{z}; \mathfrak{z}_0)$, where the latter is the set of derivations of $\mathfrak{z}$ leaving $\mathfrak{z}_0$ invariant.*

Proof. Starting with a simple G, by Proposition 2.6 if the generic smooth maximal projective factor is trivial, then the action is generically locally free, and the generic smooth algebraic hull of the tangent bundle to the orbits coincides with $Zcl(\mathrm{Ad}(G))$. Given these facts, the method of proof of [Z3, Thm. 4.2, Cor. 4.3, Thm. 1.5] applies without change. □

We let $G' \subseteq L$ and $G_\Delta \subset G \times L$ be the corresponding groups, where G_Δ is the graph of $\sigma : G \to G' \subseteq L$.

Remark 4.3. Let us note that a conclusion similar to that of Proposition 4.2 holds for any connected Lie group H with finite center, provided its action is ergodic w.r.t. a quasi-invariant measure of full support, generically locally free, and the generic smooth algebraic hull along the orbits coincides with with $Zcl(\mathrm{Ad}(H))$ (up to finite index). Under these conditions either the rigid analytic system admits an H-equivariant non-constant map to a projective variety, or the conclusions 1, 2 and 3 holds. When H is simple, such a map necessarily gives rise to a *compact homogeneous* projective factor, as stated.

We now note that the following result on properness of the action on the universal cover, in analogy with [Z3, Cor. 4.5], which proves properness in the unimodular, measure-preserving case.

Proposition 4.4. Generic properness on the universal cover.

Let (M, ω, G, ν) be a rigid analytic stationary system. Then either the generic smooth maximal projective factor is non-trivial, or the $\tilde{G}$-action on $\tilde{M}$ is (locally free and) proper on a generic set M_0. Namely, generically the stability groups in $\tilde{G}$ of points in $\tilde{M}$ are compact, and the orbit map $\tilde{G}/\tilde{G}_y \to \tilde{G} \cdot y$ is a diffeomorphism. The same conclusion obtains also under the assumptions in the last part of Proposition 4.2.

9

Proof. Assume the generic smooth maximal projective factor is trivial. We first claim that the stability groups are generically compact (so in our situation, finite), and the orbit maps $\tilde{G}/\tilde{G}_y \mapsto \tilde{G} \cdot y \subset \tilde{M}$, are homeomorphisms (so that orbits are locally closed). Indeed, this follows from the argument in [Z3, Cor. 4.5], which uses only generic local freeness of the G-action, together with the fact that generically $\psi(y) = \mathfrak{g}$, and its consequence that the local orbits of the centralizer covers the local orbits of $\tilde{G}$. Now if the orbit $\tilde{G}y$ is locally closed, then it is a regular submanifold of $\tilde{M}$ (see [Va, Thm. 2.9.7]), of dimension $\dim G$ and the orbit map is a diffeomorphism. $\qquad\square$

5   Proof of Theorem 1.1, Theorem 1.2 and Theorem 1.3

Proof of Theorem 1.1.

Theorem 1.1 follows immediately from Proposition 4.2. Indeed, assume that the generic smooth maximal projective factor of the rigid analytic stationary system is trivial. Then the group L defined as the generic smooth algebraic hull of the G-action on the principal $GL(\mathfrak{n})$-bundle associated with the representation $\pi_1(M) \to \Gamma \subset GL(\mathfrak{n})$ has the property that it is contained in the Zariski closure $\overline{\Gamma}$ in $GL(\mathfrak{n})$. On the other hand, the Lie algebra of L contains the Lie algebra $\mathfrak{g}'$ which is an image of $\mathfrak{g}$ under an injective homomorphism, and hence is isomorphic to $\mathfrak{g}$.

The last statement follows from the fact that the Zariski closure of a virtually solvable group has a solvable Lie algebra, which therefore cannot contain a copy of the simple Lie algebra of $\mathfrak{g}$ as a subalgebra. $\qquad\square$

Proof of Theorem 1.2.

Fix $G = SO^0(n,1)$, $n \geq 2$, and let Γ be a uniform lattice in G which maps onto a free group on $k \geq 2$ generators. Note that any uniform lattice in G admits a finite index subgroup which maps onto a non-Abelian free group. Now fix a uniform lattice $\Delta \subset SO^0(1,m)$, $m \geq 4$ which has a generating set with k elements. Consider the surjective homomorphism from Γ to Δ obtained by composition, and let Γ act on the $(m-1)$-dimensional sphere via the action of $\Delta \subset SO^0(1,m)$ as a group of conformal transformations. As is well-known, the Δ-action, and hence also the Γ action, is minimal and has no invariant probability measure. Let M denote the action of G induced by the action of Γ on the $(m-1)$-dimensional sphere. It has been shown in [NZ1, Thm. B] that G has a unique stationary measure on M, which is thus ergodic. This measure is not invariant, since the sphere does not admit a Γ-invariant measure. Since the Γ-action is minimal on the sphere, the G-action is minimal on M, and hence any G-quasi-invariant measure has full support. M has the structure of a smooth bundle over G/Γ with typical fiber an $(m-1)$-dimensional sphere, and M also has the structure of an analytic manifold in the obvious way. To produce an invariant rigid analytic geometric structure, take the product of the pseudo-Riemannain structure associated with the Killing form on G/Γ, and the conformal structure on the sphere. When $m \geq 4$ the conformal structure is of finite type and hence rigid, and we refer to [FZ, §3] for a full discussion of the rigidity of the product structure. This concludes the proof of Theorem 1.2. $\qquad\square$

Proof of Theorem 1.3.

We divide the proof into the following steps.

1. Let the Lie group G act by bundle automorphisms on the principal bundle $\tilde{M} \to M$ with fiber $\pi_1(M)$, with an ergodic stationary measure on the base M of full support. Let Γ be the image of the fundamental group $\pi_1(M)$ under the representation into $GL(\mathfrak{z})$. Consider the associated principal $GL(\mathfrak{z})$-bundle over M with fiber $GL(\mathfrak{z})$, on which G acts by bundle automorphisms. Then the (measurable) algebraic hull L_ν of the G-action on the $GL(\mathfrak{z})$-bundle is contained in the Zariski closure $\overline{\Gamma}$ of Γ in $GL(\mathfrak{z})$ (see [Z3, Prop. 3.4]). To prove Theorem 1.3, we first assume that the action is generically locally free, and then it suffices to prove that the Lie algebra $\mathfrak{l}_\nu$ of the algebraic hull L_ν has a factor algebra which contains (an isomorphic copy of) the Lie algebra $\mathfrak{q}_0$ of the group $Q_0 \subset Q$. We are assuming that $Q = Q_\lambda = Q(X, \nu)$ is the parabolic subgroup stabilizing the measure λ canonically associated to ν. Note that when ν is invariant

10

we have $\nu = \lambda$ and so $Q = G$, and we obtain Gromov's representation theorem (see [G1] or [Z3, Thm. 1.5]).

2. Let us denote by Ψ the sub-bundle of the tangent bundle to M whose fiber at a given point $y \in M$ is $\psi(y)$. Recall that $\psi(y)$ consists of the vectors in the tangent space to the G-orbit at y which are obtained via evaluation of the global Killing vector fields in $\mathfrak{z}$ at y. Thus by definition of ψ, Ψ is actually a sub-bundle of the tangent bundle to the G-orbits in M. The (measurable) algebraic hull L'_ν of the G-action on the bundle Ψ is a factor group of L_ν, and we claim that $\mathfrak{l}'_\nu$ contains $\mathfrak{q}_0$.

   To establish that, we consider the ergodic measure-preserving action of Q_0 on (M, λ). Q_0 satisfies the condition of Borel density theorem, and hence by Theorem 3.1, for λ-almost all points y, the evaluation of the global Killing vectors fields centralizing the action of Q_0 covers the local orbits of Q_0. But by Theorem 3.2, since $A_Q N_Q$ is epimorphic for every proper parabolic subgroup Q, it follows that a global Killing field normalizing $\mathfrak{g}$ and centralizing $\mathfrak{q}_0$ necessarily centralizes $\mathfrak{g}$. We conclude that for λ-almost all points y, $\psi(y)$ covers the local orbits of Q_0. The Q_0-action on (M, λ) is essentially locally free, so we deduce by the Borel density property that the (measurable) algebraic hull of the Q_0-action on Ψ contains the Zariski closure of $\mathrm{Ad}(Q_0)$, a group locally isomorphic to Q_0.

3. By assumption the maximal projective factor of (M, ν) in the measurable category is $G/Q(X, \nu)$ and $M \to G/Q(X, \nu)$ is a measure-preserving extension. Thus the G-action on (M, ν) is induced from the Q-action on (M, λ), and it follows that the (measurable) algebraic hull of the G-action on the bundle Ψ over (M, ν) contains the algebraic hull of the Q-action on the bundle Ψ over (M, λ). Thus it also contains the algebraic hull of the Q_0-action on the bundle Ψ over (M, λ). Taking parts 1) and 2) into account, we have established that $\mathfrak{q}_0$ is contained in the Lie algebra L'_ν, which is a factor of the subalgebra $\mathfrak{l}_\nu$ of the Lie algebra of $\overline{\Gamma}$.

4. Finally, if the action is not generically free, then the first case of Proposition 2.7 applies, and generically, the connected component of every stability group is conjugate to a fixed connected group S, with $N_G(S) \subset Q_s$, Q_s defining the generic smooth maximal projective factor, which is of course non-trivial in this case. Let $\phi : M_0 \to G/Q_s$ be the equivariant smooth factor map, and let $Y = \phi^{-1}([Q_s])$. Then Y is a smooth Q_s-invariant regular submanifold, and M_0 is equivariantly diffeomorphic to the G-action on the manifold $(G \times Y)/Q_s$, namely the G-action induced from the action of Q_s on Y. This fact is general and follows simply from the existence of a smooth equivariant map onto a homogeneous space.

This concludes the proof of Theorem 1.3.   □

   Let us formulate some corollaries of Theorem 1.1 and Theorem 1.3, explicating the consequences of restrictions imposed on the fundamental group of M. We begin with the following result, establishing that certain rigid analytic stationary actions must have stability groups of positive dimension.

Theorem 5.1. Manifolds with nilpotent fundamental group. *Let (M, ω, G, ν) be a rigid analytic stationary system, and keep the assumptions of Theorem 1.3. Suppose further that $\pi_1(M)$ is virtually nilpotent. Then:*

1. *The G-action on (M, ν) is not generically locally free.*

2. *The generic smooth maximal projective factor is non-trivial (and in particular, the stationary measure is not invariant).*

Proof. Note that if the action is generically locally free, then by Theorem 1.3 the Zariski closure of the image of $\pi_1(M)$ in $GL(\mathfrak{z})$ must have AN as a sub-quotient, where $P = MAN$ is a minimal parabolic subgroup. But AN is a solvable connected Lie group which is not nilpotent, and the Zariski closure of a virtually nilpotent group is virtually nilpotent.   □

11

We recall that one case where the stationary system (M, ν) is a measure-preserving extension of its maximal measurable projective factor (so that Theorem 5.1 applies) is when the action of G is amenable. Such an action is necessarily induced by an action of a minimal parabolic group (see [NZ2] for further discussion).

Conversely, any action of a minimal parabolic subgroup P with invariant probability measure induces an amenable action of G. However the question of whether the induced action carries a G-invariant rigid analytic structure if the original P-action has an invariant rigid analytic structure has not been resolved. The following consequence of Theorem 5.1 is relevant in this direction.

Corollary 5.2. Rigid analytic actions of a minimal parabolic subgroup. *Let (M_0, ω) be a compact manifold with rigid analytic geometric structure of algebraic type and virtually nilpotent fundamental group. Assume $P \subset Aut(M_0, \omega)$, where P is a minimal parabolic subgroup of a simple non-compact Lie group. Let λ be a P-invariant ergodic measure on M_0 of full support. If the induced action carries a G-invariant rigid analytic structure, then generically points in M_0 have a stability group of positive dimension.*

Proof. Assume for contradiction that the action of P on M_0 is generically locally free. Then so is the action of G on the manifold $M = M_0 \times_P G$, namely the action induced to G, with the associated stationary measure (see [NZ1]). The manifold M still has a virtually nilpotent fundamental group and a rigid analytic geometric structure of algebraic type preserved by G, by assumption. The stationary measure is ergodic and of full support, and M is a measure-preserving extension of G/P. Hence Theorem 5.1 applies, but the first alternative is ruled out by the fact that $\pi_1(M)$ is virtually nilpotent, and the second alternative is ruled out also since the G-action is locally free. This contradiction implies that the P action on M_0 has stability groups of positive dimension, generically. $\square$

Remark 5.3. Finally, let us note some consequences that hold whenever the maximal generic smooth projective factor of a rigid analytic stationary system is non-trivial.

1. The rigid geometric structure in question cannot be unimodular. Indeed, otherwise it defines an invariant volume form of finite total mass, and this would imply that an invariant probability measure exists on the projective factor, which is not the case.

2. As a result, under the first alternative in Theorem 1.1, or under Theorem 5.1, the rigid structure in question cannot be unimodular. More generally, M does not carry any invariant probability measure of full support.

3. It follows also that the top exterior power of the derivative cocycle cannot be cohomologous to a trivial cocycle, with integrable transfer function.

References

[BN]   U. Bader and A. Nevo, *Conformal actions of simple Lie groups on compact pseudo-Riemannian manifolds.* J. Differential Geometry, vol. 60, pp. 355-387 (2002).

[DG]   G. D'ambra and M. Gromov, *Lectures on transformation groups : geometry and dynamics.* Surveys in differential geometry (cambridge MA, 1990), pp. 19-111, Lehigh University, Bethlehem PA, 1991.

[CQB] A. Candel and R. Quiroga-Barranco, *Gromov's centralizer theorem*, Geom. Dedicata, vol. 100, pp. 123-155 (2003).

[Da1] S. G. Dani, *On conjugacy classes of closed subgroups and stabilizers of Borel actions of Lie groups.* Erg. Th. Dyn. Sys. vol. 22, pp. 1696-1714 (2002).

[F]   R. Feres, *Rigid geometric structures and actions of semisimple Lie groups*. Rigidité, groupe fondamental et dynamique, pp. 121-167, Panor. Synthèses, vol. 13, Soc. Math. France, Paris, 2002.

[FZ]   D. Fisher and R. Zimmer, *Geometric lattice actions, entropy and fundamental groups*. Comment. Math. Helv. vol. 77, pp. 326-338 (2002).

[FrZe]   C. Frances and A. Zeghib, *Some remarks on conformal pseudo-Riemannian actions of simple Lie groups*. math. Res. Lett. vol. 12, pp. 49-56 (2005).

[G1]   M. Gromov, *Rigid transformation groups*. Géométrie différentielle (Paris 1986), pp. 65-139, Travaux en Cours, 33, Herman, Paris, 1988.

[NZ1]   A. Nevo, R.J. Zimmer, *Homogenous projective factors for actions of semisimple Lie groups*. Inven. Math. vol. 138, pp. 229-252 (1999).

[NZ2]   A. Nevo, R.J. Zimmer, *A structure Theorem for actions of semisimple Lie groups*. Ann. Math. vol. 156, pp. 565-594 (2002).

[NZ3]   A. Nevo, R. J. Zimmer, *A generalization of the intermediate factors theorem*. J. Anal. math. vol. 86, pp. 93-104 (2002).

[NZ4]   A. Nevo, R.J. Zimmer, *Rigidity of Furstenberg entropy for semisimple Lie group actions*. Ann. Sci. de Ec. Norm. Sup., t. 33, pp. 321-343 (2000).

[QB]   R. Quiroga-Barranco, *Isometric actions of simple Lie groups on pseudo-Riemannian manifolds*. Ann. Math. vol. 164, pp. 941-969 (2006).

[St]   G. Stuck, *Minimal actions of semisimple groups*. Erg. Th. Dyn. Sys. vol. 16, pp. 821-831 (1996).

[St1]   G. Stuck, *Growth of homogeneous spaces, density of discrete groups and Kazhdan's property T*. Invent. Math. vol. 109, pp. 505-517 (1992).

[Va]   V. S. Varadarajan, *Lie groups, Lie algebras and their representations*. Graduate Texts in Mathematics, vol. 102, Springer Verlag, 1984.

[Z1]   R. J. Zimmer, Ergodic Theory and Semisimple Groups, Birkhauser, Boston, 1984.

[Z2]   R. J. Zimmer, *Representations of fundamental groups of manifolds with a semisimple transformation group*, JAMS **2**, pp. 201-13, (1989).

[Z3]   R. J. Zimmer, *Automorphism groups and fundamental groups of geometric manifolds*. Proc. Symp. Pure Math. vol. 54, part 3, pp. 693-710 (1993).

[Z4]   R. J. Zimmer, *Topological super-rigidity*. Lecture notes, University of Chicago, 1992.

13

GROUPS ACTING ON MANIFOLDS: AROUND THE ZIMMER PROGRAM

David Fisher

TO ROBERT ZIMMER ON THE OCCASION OF HIS 60TH BIRTHDAY

Abstract. This paper is a survey on the *Zimmer program*. In its broadest form, this program seeks an understanding of actions of large groups on compact manifolds. The goals of this survey are (1) to put into context the original questions and conjectures of Zimmer and Gromov that motivated the program, (2) to indicate the current state of the art on as many of these conjectures and questions as possible, and (3) to indicate a wide variety of open problems and directions of research.

1. Prologue

Traditionally, the study of dynamical systems is concerned with actions of $\mathbb{R}$ or $\mathbb{Z}$ on manifolds, i.e., with flows and diffeomorphisms. It is natural to consider instead dynamical systems defined by actions of larger discrete or continuous groups. For nondiscrete groups, the Hilbert-Smith conjecture is relevant, since the conjecture states that nondiscrete, locally compact, totally disconnected groups do not act continuously by homeomorphisms on manifolds. For smooth actions, known results suffice to rule out actions of totally disconnected groups; the Hilbert-Smith conjecture has been proven for Hölder diffeomorphisms [154, 203]. For infinite discrete groups, the whole universe is open. One might consider the "generalized Zimmer program" to be the study of homomorphisms $\rho : \Gamma \to \mathrm{Diff}(M)$, where Γ is a finitely generated group and M is a compact manifold. In this survey much emphasis will be on a program proposed by Zimmer. Here one considers a very special class of both

Author partially supported by NSF grants DMS-0541917 and 0643546, a fellowship from the Radcliffe Institute for Advanced Studies, and visiting positions at École Polytechnique, Palaiseau, and Université Paris-Nord, Villateneuse.

Lie groups and discrete groups, namely semisimple Lie groups of higher real rank and their lattices.

Zimmer's program is motivated by several of Zimmer's own theorems and observations, many of which we will discuss below. But in broadest strokes, the motivation is simpler. Given a group whose linear and unitary representations are very rigid or constrained, might it also be true that the group's representations into $\text{Diff}^\infty(M)$ are also very rigid or constrained at least when M is a compact manifold? For the higher rank lattices considered by Zimmer, Margulis's superrigidity theorems classified finite-dimensional representations and the groups enjoy property (T) of Kazhdan, which makes unitary representations quite rigid.

This motivation also stems from an analogy between semisimple Lie groups and diffeomorphism groups. When M is a compact manifold, not only is $\text{Diff}^\infty(M)$ an infinite-dimensional Lie group, but its connected component is simple. Simplicity of the connected component of $\text{Diff}^\infty(M)$ was proven by Thurston using results of Epstein and Herman [224, 49, 115]. Herman had used Epstein's work to see that the connected component of $\text{Diff}^\infty(\mathbb{T}^n)$ is simple and Thurston's proof of the general case uses this. See also further work on the topic by Banyaga and Mather [7, 166, 165, 167, 168], as well as Banyaga's book [8].

A major motivation for Zimmer's program came from his cocycle superrigidity theorem, discussed in detail below in section 5. One can think of a homomorphism $\rho : \Gamma \rightarrow \text{Diff}(M)$ as defining a *virtual homomorphism* from Γ to $GL(\dim(M), \mathbb{R})$. The notion of virtual homomorphisms, now more commonly referred to in our context as cocycles over group actions, was introduced by Mackey [153].

In all of Zimmer's early papers, it was always assumed that we had a homomorphism $\rho : \Gamma \rightarrow \text{Diff}^\infty(M,\omega)$, where ω is a volume form on M. A major motivation for this is that the cocycle superrigidity theorem referred to in the last paragraph only applies to cocycles over measure-preserving actions. That this hypothesis is necessary was confirmed by examples of Stuck, who showed that no rigidity could be hoped for unless one had an invariant-volume form or assumed that M was very low dimensional [223]. Recent work of Nevo and Zimmer does explore the non-volume-preserving case and will be discussed below, along with Stuck's examples and some others due to Weinberger in section 10.

Let G be a semisimple real Lie group, all of whose simple factors are of real rank at least 2. Let Γ in G be a lattice. The simplest question asked by Zimmer was, can one classify smooth volume-preserving actions of Γ on compact manifolds? At the time of this writing, the answer to this question is still unclear. There certainly is a wider collection of actions than Zimmer may have initially suspected, and it is also clear that the moduli space of such actions is not discrete; see [12, 77, 127] or section 9 below. But there are still few enough examples known that a classification remains plausible. And if one assumes some additional conditions on the actions, then plausible conjectures and striking results abound.

We now discuss four paradigmatic conjectures. To make this introduction accessible, these conjectures are all special cases of more general conjectures stated later in the text.

In particular, we state all the conjectures for lattices in $SL(n, \mathbb{R})$ when $n > 2$. The reader not familiar with higher rank lattices can consider the examples of finite-index subgroups of $SL(n, \mathbb{Z})$, with $n > 2$, rather than general lattices in $SL(n, \mathbb{R})$. We warn the reader that the algebraic structure of $SL(n, \mathbb{Z})$ and its finite-index subgroups makes many results easier for these groups than for other lattices in $SL(n, \mathbb{R})$. The conjectures we state concern, respectively, (1) classification of low-dimensional actions, (2) classification of geometric actions, (3) classification of uniformly hyperbolic actions, and (4) the topology of manifolds admitting volume-preserving actions.

A motivating conjecture for much recent research is the following:

Conjecture 1.1. (Zimmer's conjecture). *For any $n > 2$, any homomorphism $\rho : SL(n, \mathbb{Z}) \rightarrow$ Diff(M) has finite image if* $\dim(M) < n - 1$. *The same for any lattice Γ in $SL(n\mathbb{R})$.*

The dimension bound in the conjecture is clearly sharp, as $SL(n, \mathbb{R})$ and all of its subgroups act on $\mathbb{P}^{n-1}$. The conjecture is a special case of a conjecture of Zimmer's that concerns actions of higher rank lattices on low-dimensional manifolds, which we state as Conjecture 4.12 below. Recently much attention has focused on these conjectures concerning low-dimensional actions.

In our second, geometric setting, (a special case of) a major motivating conjecture is the following:

Conjecture 1.2. (affine actions). *Let $\Gamma < SL(n, \mathbb{R})$ be a lattice. Then there is a classification of actions of Γ on compact manifolds that preserve both a volume form and an affine connection. All such actions are algebraically defined in a sense to be made precise below. (See Definition 2.4 below.)*

The easiest example of an algebraically defined action is the action of $SL(n, \mathbb{Z})$ (or any of its subgroups) on $\mathbb{T}^n$. We formulate our definition of algebraically defined action to include all trivial actions, all isometric actions, and all (skew) products of other actions with these. The conjecture is a special case of a conjecture stated below as Conjecture 6.15. In addition to considering more general acting groups, we will consider more general invariant geometric structures, not just affine connections. A remark worth making is that for the geometric structures we consider, the automorphism group is always a finite-dimensional Lie group. The question of when the full automorphism group of a geometric structure is large is well studied from other points of view; see particularly [134]. However, this question is generally most approachable when the large subgroup is connected and much less is known about discrete subgroups. In particular, geometric approaches to this problem tend to use information about the connected component of the automorphism group and give much less information about the group of components, particularly if the connected component is trivial.

One is also interested in the possibility of classifying actions under strong dynamical hypotheses. The following conjecture is motivated by work of Feres-Labourie and Goetze-Spatzier [60, 104, 105] and is similar to conjectures stated in [107, 119]:

Conjecture 1.3. *Let $\Gamma < SL(n, \mathbb{R})$ be a lattice. Then there is a classification of actions of Γ on a compact manifold M that preserve a volume form and in which one element $\gamma \in \Gamma$ acts as an Anosov diffeomorphism. All such actions are algebraically defined in a sense to be made precise below. (Again see Definition 2.4.)*

In the setting of this particular conjecture, a proof of an older conjecture of Franks concerning Anosov diffeomorphisms would imply that any manifold M as in the conjecture was homeomorphic to an infranilmanifold on which γ is conjugate by a homeomorphism to a standard affine Anosov map. Even assuming Franks's conjecture, Conjecture 1.3 is open. One can make a more general conjecture by assuming only that γ has some uniformly partially hyperbolic behavior. Various versions of this are discussed in section 7; see particularly Conjecture 7.10.

We end this introduction by stating a topological conjecture about *all* manifolds admitting smooth volume-preserving actions of a simple higher rank algebraic group. Very special cases of this conjecture are known and the conjecture is plausible in light of existing examples. More precise variants and a version for lattice actions will be stated below in section 8.

Conjecture 1.4. *Let G be a simple Lie group of real rank at least 2. Assume G has a faithful action preserving volume on a compact manifold M. Then $\pi_1(M)$ has a finite index subgroup Λ such that Λ surjects onto an arithmetic lattice in a Lie group H, where H locally contains G.*

The conjecture says, more or less, that admitting a G action forces the fundamental group of M to be large. Passage to a finite-index subgroup and a quotient is necessary; see section 9 and [83] for more discussion. The conjecture might be considered the analogue for group actions of Margulis's arithmeticity theorem.

This survey is organized along the following lines. We begin in section 2 by describing in some detail the kinds of groups we consider and examples of their actions on compact manifolds. In section 3 we digress with a prehistory of motivating results from rigidity theory. Then in section 4, we discuss conjectures and theorems concerning actions on "low dimensional" manifolds. In this section, there are a number of related results and conjectures concerning groups not covered by Zimmer's original conjectures. Here "low dimensional" is (1) small compared to the group as in Conjecture 1.1 and (2) absolutely small, as in dimension being between 1 and 4. This discussion is simplified by the fact that many theorems conclude with the nonexistence of actions or at least with all actions factoring through finite quotients. In section 5 we further describe Zimmer's motivations for his conjectures by discussing the cocycle superrigidity theorem and some of its consequences for smooth group actions. In sections 6 and 7, we discuss, respectively, geometric and dynamical conditions under which a simple classification might be possible. In section 8, we discuss another approach to classifying G and Γ actions using topology and representations of fundamental groups in order to produce algebraically defined quotients of actions. Then in section 9, we describe the known "exotic examples" of the acting groups we consider. This construction reveals some necessary complexity of a high-dimensional classification. We then describe known results and examples of actions not preserving volume in section 10 and some rather surprising group actions on manifolds in section 11. Finally, we end the survey with a collection of questions about the algebraic and geometric structure of finitely generated subgroups of $\mathrm{Diff}(M)$.

Some remarks on biases and omissions. Like any survey of this kind, this work is informed by its author's biases and experiences. There is an additional bias in that this paper emphasizes developments that are close to Zimmer's own work and conjectures. In particular, the study of rigidity of group actions often focuses on the low-dimensional setting in which all group actions are conjectured to be finite or trivial. While Zimmer did substantial work in this setting, he also proved many results and made many conjectures in more general settings in which any potential classification of group actions is necessarily, because of the existence of examples, more complicated.

Another omission is that almost nothing will be said here about local rigidity of group actions, since the author has recently written another survey on that topic [76]. While that survey could already use updating, that update will appear elsewhere.

For other surveys of Zimmer's program and rigidity of large group actions the reader is referred to [67, 143, 258]. Zimmer and Witte Morris [258] is a particularly useful introduction to ideas and techniques in Zimmer's own work. Also of interest are (1) a brief survey of rigidity theory by Spatzier with a more geometric focus [220], (2) an older survey also by Spatzier, with a somewhat broader scope [219], (3) a recent problem list by Margulis on rigidity theory with a focus on measure rigidity [163], and (4) a more recent survey by Lindenstrauss focused on recent developments in measure rigidity [146]. Both of the last two mentioned surveys are particularly oriented toward connections between rigidity and number theory that are not mentioned at all in this survey.

Finally, while all mistakes and errors in this survey are the sole responsibility of its author, I would like to thank many people whose comments led to improvements. These include Danny Calegari, Renato Feres, Etienne Ghys, Daniel Groves, Steve Hurder, Jean-François Lafont, Karin Melnick, Nicolas Monod, Dave Witte Morris, Andrés Navas, Leonid Polterovich, Pierre Py, Yehuda Shalom, Ralf Spatzier, and an anonymous referee.

2. A brief digression: Some examples of groups and actions

In this section we briefly describe some of the groups that will play important roles in the results discussed here. The reader already familiar with semisimple Lie groups and their lattices may want to skip to the second subsection where we give descriptions of group actions. The following convention is in force throughout this paper. For definitions of relevant terms the reader is referred to the following subsections.

Convention. In this article we will have occasion to refer to three overlapping classes of lattices in Lie groups that are slightly different. Let G be a semisimple Lie group and $\Gamma < G$ a lattice. We call Γ a *higher rank lattice* if all simple factors of G have real rank at least 2. We call Γ a *lattice with* (T) if all simple factors of G have property (T). Lastly we call Γ an *irreducible higher rank lattice* if G has real rank at least 2 and Γ is irreducible.

2.1. Semisimple groups and their lattices

By a simple Lie group, we mean a connected Lie group all of whose normal subgroups are discrete, though we make the additional convention that $\mathbb{R}$ and S^1 are not simple. By a semisimple Lie group we mean the quotient of a product of simple Lie groups by some subgroup of the product of their centers. Note that with our conventions, the center of a simple Lie group is discrete and is in fact the maximal normal subgroup. There is an elaborate structure theory of semisimple Lie groups and the groups are completely classified; see [114] or [133] for details. Here we merely describe some examples, all of which are matrix groups. All connected semisimple Lie groups are discrete central extensions of matrix groups, so the reader will lose very little by always thinking of matrix groups.

1. The groups $SL(n,\mathbb{R})$, $SL(n,\mathbb{C})$, and $SL(n,\mathbb{H})$ of n by n matrices of determinant 1 over the real numbers, the complex numbers, or the quaternions.

2. The group $SP(2n,\mathbb{R})$ of $2n$ by $2n$ matrices of determinant 1 that preserve a real symplectic form on $\mathbb{R}^{2n}$.

3. The groups $SO(p,q)$, $SU(p,q)$, and $SP(p,q)$ of matrices that preserve inner products of signature (p,q) in which the inner product is real linear on $\mathbb{R}^{p+q}$, hermitian on $\mathbb{C}^{p+q}$, or quaternionic hermitian on $\mathbb{H}^{p+q}$, respectively.

Let G be a semisimple Lie group that is a subgroup of $GL(n,\mathbb{R})$. We say that G has *real rank k* if G has a k-dimensional abelian subgroup that is conjugate to a subgroup of the real diagonal matrices and no $(k+1)$-dimensional abelian subgroups with the same property. The groups in (1) have rank $n-1$, the groups in (2) have rank n, and the groups in (3) have rank $\min(p,q)$.

Since this article focuses primarily on finitely generated groups, we are more interested in discrete subgroups of Lie groups than in the Lie groups themselves. A discrete subgroup Γ in a Lie group G is called a lattice if G/Γ has finite Haar measure. The lattice is called *cocompact* or *uniform* if G/Γ is compact and *nonuniform* or *not cocompact* otherwise. If $G = G_1 \times \cdots \times G_n$ is a product then we say a lattice $\Gamma < G$ is *irreducible* if its projection to each G_i is dense. It is more typical in the literature to insist that projections to all factors are dense, but this definition is more practical for our purposes. More generally we make the same definition for an *almost direct product*, by which we mean a direct product G modulo some subgroup of the center $Z(G)$. Lattices in semisimple Lie groups can always be constructed by arithmetic methods; see [18] and also [238] for more discussion. In fact, one of the most important results in the theory of semisimple Lie groups is that if G is a semisimple Lie group without compact factors, then all irreducible lattices in G are arithmetic unless G is locally isomorphic to $SO(1,n)$ or $SU(1,n)$. For G of real rank at least 2, this is Margulis's arithmeticity theorem, which he deduced from his superrigidity theorems [158, 198, 161]. For nonuniform lattices, Margulis had an earlier proof that does not use the superrigidity theorems; see [157, 159]. This earlier proof depends on the study of dynamics of unipotent elements on the space G/Γ, and particularly on what is now known as the "nondivergence of unipotent flows." Special cases of the superrigidity theorems were then proven for $Sp(1,n)$ and F_4^{-20} by Corlette and Gromov-Schoen, which sufficed to imply the statement on arithmeticity given above [39, 112]. As we will be almost exclusively concerned with arithmetic lattices, we do not give examples of nonarithmetic lattices here, but refer the reader to [161] and [238] for more discussion. A formal definition of arithmeticity, at least when G is algebraic, is as follows:

Definition 2.1. *Let G be a semisimple algebraic Lie group and $\Gamma < G$ a lattice. Then Γ is arithmetic if there exists a semisimple algebraic Lie group H such that*

1. *there is a homomorphism $\pi : H^0 \to G$ with compact kernel, and*
2. *there is a rational structure on H such that the projection of the integer points of H to G are commensurable to Γ; i.e., $\pi(H(\mathbb{Z})) \cap \Gamma$ is of finite index in both $H(\mathbb{Z})$ and Γ.*

We now give some examples of arithmetic lattices. The simplest is to take the integer points in a simple (or semisimple) group G that is a matrix group, e.g., $SL(n,\mathbb{Z})$ or $Sp(n,\mathbb{Z})$. This exact construction always yields lattices, but also always yields nonuniform lattices. In fact the lattices one can construct in this way have very special properties because they will contain many unipotent matrices. If a lattice is cocompact, it will necessarily contain no unipotent matrices. The standard trick for understanding the structure of lattices in G that become integral points after passing to a compact extension is called *change of base*. For much more discussion see [161, 238, 245]. We give one example to illustrate the process. Let $G = SO(m,n)$, which we view as the set of matrices in $SL(n+m,\mathbb{R})$ that preserve the inner product

$$\left\langle v, w \right\rangle = \left(-\sqrt{2} \sum_{i=1}^{m} v_i w_i \right) + \left(\sum_{i=m+1}^{n+m} v_i w_i \right),$$

where v_i and w_i are the ith components of v and w. This form, and therefore G, is defined over the field $\mathbb{Q}(\sqrt{2})$, which has a Galois conjugation σ defined by $\sigma(\sqrt{2}) = -\sqrt{2}$. If we looks at the points $\Gamma = G(\mathbb{Z}[\sqrt{2}])$, we can define an embedding of Γ in $SO(m,n) \times SO(m+n)$ by taking γ to $(\gamma, \sigma(\gamma))$. It is straightforward to check that this embedding is discrete. In fact, this embeds Γ in $H = (m,n) \times SO(m+n)$ as integral points for the rational structure on H in which the rational points are exactly the points $(\mathbf{M}, \sigma(\mathbf{M}))$, where $\mathbf{M} \in G(\mathbb{Q}(\sqrt{2}))$. This makes Γ a lattice in H and it is easy to see that Γ projects to a lattice in G, since G is cocompact in H. What is somewhat harder to verify is that Γ is cocompact in H, for which we refer the reader to the list of references above.

Similar constructions are possible with $SU(m,n)$ or $SP(m,n)$ in place of $SO(m,n)$ and also with more simple factors and fields with more Galois automorphisms. There are also a number of other constructions of arithmetic lattices using division algebras. See [190, 238] for a comprehensive treatment.

We end this section by defining a key property of many semisimple groups and their lattices. This is property (T) of Kazhdan, and was introduced by Kazhdan in [130] in order to prove that nonuniform lattices in higher rank semisimple Lie groups are finitely generated and have finite abelianization. It has played a fundamental role in many subsequent developments. We do not give Kazhdan's original definition, but one that was shown to be equivalent by work of Delorme and Guichardet [43, 113].

Definition 2.2. *A locally compact group Γ has property (T) of Kazhdan if $H^1(\Gamma, \pi) = 0$ for every continuous unitary representation π of Γ on a Hilbert space. This is equivalent to saying that any continuous isometric action of Γ on a Hilbert space has a fixed point.*

Remarks 2.3.

1. *Kazhdan's definition is that the trivial representation is isolated in the Fell topology on the unitary dual of Γ.*

2. *If a continuous group G has property (T) so does any lattice in G. This result was proved in [130].*

3. *Any semisimple Lie group has property (T) if and only if it has no simple factors locally isomorphic to $SO(1,n)$ or to $SU(1,n)$. For a discussion of this fact and attributions, see [42]. For groups with all simple factors of real rank at least 3, this is proven in [130].*

4. *No noncompact amenable group, and in particular no noncompact abelian group, has property (T). An easy averaging argument shows that all compact groups have property (T).*

Groups with property (T) play an important role in many areas of mathematics and computer science.

2.2. Some actions of groups and lattices

Here we define and give examples of a general class of actions. A major impetus in Zimmer's work is determining optimal conditions for actions to lie in this class. The class we describe is slightly more general than the class Zimmer termed "standard actions" in, e.g., [249]. Let H be a Lie group and $L < H$ a closed subgroup. Then a diffeomorphism f of H/L is called *affine* if there is a diffeomorphism $\tilde{f}$ of H such that $f([h]) = f(h)$, where $\tilde{f} = A \circ \tau_h$ with A an automorphism of H with $A(L) = L$ and τ_h is left translation by some h in H. Two obvious classes of affine diffeomorphisms are left translations on any homogeneous space and linear automorphisms of tori, or more generally automorphisms of nilmanifolds. A group action is called *affine* if every element of the group acts by an affine diffeomorphism. It is easy to check that the full group of affine diffeomorphisms $\mathrm{Aff}(H/L)$ is a finite-dimensional Lie group and an affine action of a group D is a homomorphism $\pi : D \to \mathrm{Aff}(H/L)$. The structure of $\mathrm{Aff}(H/L)$ is surprisingly complicated in general; it is a quotient of a subgroup of the group $\mathrm{Aut}(H) \ltimes H$, where $\mathrm{Aut}(H)$ is the group of automorphisms of H. For a more detailed discussion of this relationship, see [80, section 6]. While it is not always the case that any affine action of a group D on H/L can be described by a homomorphism $\pi : D \to \mathrm{Aut}(H) \ltimes H$, this is true for two important special cases:

1. D is a connected semisimple Lie group and L is a cocompact lattice in H.
2. D is a lattice in a semisimple Lie group G where G has no compact factors and no simple factors locally isomorphic to $SO(1, n)$ or $SU(1, n)$, and L is a cocompact lattice in H.

These facts are [80, Theorems 6.4 and 6.5] where affine actions as in (1) and (2) above are classified.

The most obvious examples of affine actions of large groups are of the following forms, which are frequently referred to as *standard actions*:

1. Actions of groups by automorphisms of nilmanifolds. That is, let N be a simply connected nilpotent group, $\Lambda < N$ a lattice (which is necessarily cocompact), and assume a finitely generated group Γ acts by automorphisms of N preserving . The most obvious examples of this are when $N = \mathbb{R}^n$, $\Lambda = \mathbb{Z}^n$, and $\Gamma < SL(n, \mathbb{Z})$, in which case we have a linear action of Γ on $\mathbb{T}^n$.
2. Actions by left translations. That is, let H be a Lie group and $\Lambda < H$ a cocompact lattice and $\Gamma < H$ some subgroup. Then Γ acts on H/Λ by left translations. Note that in this case need not be discrete.

3. Actions by isometries. Here K is a compact group that acts by isometries on some compact manifold M, and $\Gamma < K$ is a subgroup. Note that here Γ is either discrete or a discrete extension of a compact group.

We now briefly define a few more general classes of actions, which we need in order to formulate most of the conjectures in this paper. We first fix some notations. Let A and D be topological groups, and $B < A$ a closed subgroup. Let $\rho : D \times A/B \to A/B$ be a continuous affine action.

Definition 2.4.

1. *Let A,B,D, and ρ be as above. Let C be a compact group of affine diffeomorphisms of A/B that commute with the D action. We call the action of D on $C\backslash A/B$ a* generalized affine action.

2. *Let A, B, D, and ρ be as in (1) above. Let M be a compact Riemannian manifold and $\iota : D \times A/B \to \mathrm{Isom}(M)$ a C^1 cocycle. We call the resulting skew product D action on $A/B \times M$ a* quasi-affine action. *If C and D are as in (1), and we have a smooth cocycle, $\alpha : D \times C\backslash A/B \to \mathrm{Isom}(M)$, then we call the resulting skew product action of D on $C\backslash A/B \times M$ a* generalized quasi-affine action.

Many of the conjectures stated in this paper will end with the conclusion that all actions satisfying certain hypotheses are generalized quasi-affine actions. It is not entirely clear that generalized quasi-affine actions of higher rank groups and lattices are much more general than generalized affine actions. The following discussion is somewhat technical and might be skipped on first reading.

One can always take a product of a generalized affine action with a trivial action on any manifold to obtain a generalized quasi-affine action. One can also do variants on the following. Let H be a semisimple Lie group and $\Lambda < H$ a cocompact lattice. Let $\pi : \Lambda \to K$ be any homomorphism of Λ into a compact Lie group. Let ρ be a generalized affine action of G on $C\backslash H/\Lambda$ and let M be a compact manifold on which K acts. Then there is a generalized quasi-affine action of G on $(C\backslash H \times M)/\Lambda$.

Question 2.5. *Is every generalized quasi-affine action of a higher rank simple Lie group of the type just described?*

The question amounts to asking for an understanding of compact-group-valued cocycles over quasi-affine actions of higher rank simple Lie groups. We leave it to the interested reader to formulate the analogous question for lattice actions. For some work in this direction, see the paper of Witte Morris and Zimmer [173].

2.3. Induced actions

We end this section by describing briefly the standard construction of an *induced or suspended action*. This notion can be seen as a generalization of the construction of a flow under a function or as an analogue of the more algebraic notion of inducing a representation. Given a group H, a (usually closed) subgroup L, and an action ρ of L on a space X, we can form the space $(H \times X)/L$ where L acts on $H \times X$ by $h \cdot (l, x) = (lh^{-1}, \rho(h)x)$. This space now has a natural H action by left multiplication on the first coordinate. Many properties of the L action on X can be studied more easily in terms of properties of the H action on $(H \times X)/L$. This construction is particularly useful when L is a lattice in H.

This notion suggests the following principle:

Principle 2.6. *Let Γ be a cocompact lattice in a Lie group G. To classify Γ actions on compact manifolds it suffices to classify G actions on compact manifolds.*

The principle is a bit subtle to implement in practice, since we clearly need a sufficiently detailed classification of G actions to be able to tell which ones arise as induction of Γ actions. While it is a bit more technical to state and probably more difficult to use, there is an analogous principle for non-cocompact lattices. Here one needs to classify G actions on manifolds that are not compact but where the G action preserves a finite volume. In fact, one needs only to study such actions on manifolds that are fiber bundles over G/Γ with compact fibers.

The lemma raises the question as to whether or not one should simply always study G actions. While in many settings this is useful, it is not always. In particular, many known results about Γ actions require hypotheses on the Γ action in which there is no useful way of rephrasing the property as a property of the induced action, or, perhaps more awkwardly, require assumptions on the induced action that cannot be rephrased in terms of hypotheses on the original Γ action. We will illustrate the difficulties in employing Principle 2.6 at several points in this paper.

A case in which the implications of the principle are particularly clear is a negative result concerning actions of $SO(1, n)$. We will make clear by the proof what we mean by the following:

Theorem 2.7. *Let $G = SO(1, n)$. Then one cannot classify actions of G on compact manifolds.*

Proof. For every n there is at least one lattice $\Gamma < G$ that admits homomorphisms onto non-abelian free groups, see, e.g., [149], and therefore also onto $\mathbb{Z}$. So we can take any action of F_n or $\mathbb{Z}$ and induce to a G action. It is relatively easy to show that if the induced actions are isomorphic, then so are the actions they are induced from; see, e.g., [77]. A classification of $\mathbb{Z}$ actions would amount to a classification of diffeomorphisms and a classification of F_n actions would involve classifying all n-tuples of diffeomorphisms. As there is no reasonable classification of diffeomorphisms there is also no reasonable classification of n-tuples of diffeomorphisms. □

We remark that essentially the same theorem holds for actions of $SU(1, n)$ in which homomorphisms to $\mathbb{Z}$ exist for certain lattices [131]. Much less is known about free quotients of lattices in $SU(1, n)$; see [147] for one example. For a surprising local rigidity result for some lattices in $SU(1, n)$, see [71, Theorem 1.3].

3. Prehistory

3.1. Local and global rigidity of homomorphisms into finite-dimensional groups

The earliest work on rigidity theory is a series of works by Calabi-Vesentini, Selberg, Calabi, and Weil, which resulted in the following:

Theorem 3.1. *Let G be a semisimple Lie group and assume that G is not locally isomorphic to $SL(2, \mathbb{R})$. Let $\Gamma < G$ be an irreducible cocompact lattice. Then the defining embedding of Γ in G is locally rigid; i.e., any embedding ρ close to the defining embedding is conjugate to the defining embedding by a small element of G.*

Remarks 3.2.

1. *If $G = SL(2, \mathbb{R})$, the theorem is false and there is a large, well-studied space of deformation of Γ in G, known as the Teichmueller space.*
2. *There is an analogue of this theorem for lattices that are not cocompact. This result was proven later and has a more complicated history that we omit here. In this case it is also necessary to exclude G locally isomorphic to $SL(2, \mathbb{C})$.*

This theorem was originally proven in special cases by Calabi, Calabi-Vesentini, and Selberg. In particular, Selberg gives a proof for cocompact lattices in $SL(n, \mathbb{R})$ for $n \geq 3$ in [209], Calabi-Vesentini give a proof when the associated symmetric space $X = G/K$ is Kähler in [28] and Calabi gives a proof for $G = SO(1, n)$, where $n \geq 3$, in [27]. Shortly afterward, Weil gave a complete proof of Theorem 3.1 in [229, 230].

In all of the original proofs, the first step was to show that any perturbation of Γ was discrete and therefore a cocompact lattice. This is shown in special cases in [27, 28, 209] and proven in a somewhat broader context than Theorem 3.1 in [230].

The different proofs of cases of Theorem 3.1 are also interesting in that there are two fundamentally different sets of techniques employed, and this dichotomy continues to play a role in the history of rigidity. Selberg's proof essentially combines algebraic facts with a study of the dynamics of iterates of matrices. He makes systematic use of the existence of singular directions, or Weyl chamber walls, in maximal diagonalizable subgroups of $SL(n, \mathbb{R})$. Exploiting these singular directions is essential to much later work on rigidity, both of lattices in higher rank groups and of actions of abelian groups. It seems possible to generalize Selberg's proof to the case of G an $\mathbb{R}$-split semisimple Lie group with rank at least 2. Selberg's proof, which depended on asymptotics at infinity of iterates of matrices, inspired Mostow's explicit use of boundaries in his proof of strong rigidity [176]. Mostow's work in turn provided inspiration for the use of boundaries in later work of Margulis, Zimmer, and others on rigidity properties of higher rank groups.

The proofs of Calabi, Calabi-Vesentini, and Weil involve studying variations of geometric structures on the associated locally symmetric space. The techniques are analytic and use a variational argument to show that all variations of the geometric structure are trivial. This work is a precursor to much work in geometric analysis studying variations of geometric structures and also informs later work on proving rigidity/vanishing of harmonic forms and maps. The dichotomy between approaches based on algebra/dynamics and approaches that are in the spirit of geometric analysis continues through much of the history of rigidity and the history of rigidity of group actions in particular.

Shortly after completing this work, Weil discovered a new criterion for local rigidity [231]. In the context of Theorem 3.1, this allows one to avoid the step of showing that a perturbation of Γ remains discrete. In addition, this result opened the way for understanding local rigidity of more general representations of discrete groups than the defining representation.

Theorem 3.3. *Let Γ be a finitely generated group, G a Lie group, and $\pi : \Gamma {\rightarrow} G$ a homomorphism. Then π is locally rigid if $H^1(\Gamma, \mathfrak{g}) = 0$. Here $\mathfrak{g}$ is the Lie algebra of G and Γ acts on $\mathfrak{g}$ by $Ad_G \circ \pi$.*

Weil's proof of this result uses only the implicit-function theorem and elementary properties of the Lie group exponential map. The same theorem is true if G is an algebraic group over any local field of characteristic zero. In [231], Weil remarks that if $\Gamma < G$ is a cocompact lattice and G satisfies the hypothesis of Theorem 3.1, then the vanishing of $H^1(\Gamma, \mathfrak{g})$ can be deduced from the computations in [229]. The vanishing of $H^1(\Gamma, \mathfrak{g})$ is proven explicitly by Matsushima and Murakami in [169].

Motivated by Weil's work and other work of Matsushima, conditions for vanishing of $H^1(\Gamma, \mathfrak{g})$ were then studied by many authors. See particularly [169] and [199]. The results in these papers imply local rigidity of many linear representations of lattices.

3.2. Strong rigidity and superrigidity

In a major and surprising development, it turns out that in many instances, local rigidity is just the tip of the iceberg and that much stronger rigidity phenomena exist. We discuss now major developments from the 1960s and 1970s.

The first remarkable result in this direction is Mostow's rigidity theorem; see [174, 175] and references therein. Given G as in Theorem 3.1, and two irreducible cocompact lattices Γ_1 and Γ_2 in G, Mostow proves that any isomorphism from Γ_1 to Γ_2 extends to an isomorphism of G with itself. Combined with the principal theorem of [230], which shows that a perturbation of a lattice is again a lattice, this gives a remarkable and different proof of Theorem 3.1, and Mostow was motivated by the desire for a "more geometric understanding" of Theorem 3.1 [175]. Mostow's theorem is in fact a good deal stronger and controls not only homomorphisms $\Gamma \to G$ near the defining homomorphism, but any homomorphism into any other simple Lie group G' in which the image is lattice. As mentioned above, Mostow's approach was partially inspired by Selberg's proof of certain cases of Theorem 3.1 [176]. A key step in Mostow's proof is the construction of a continuous map between the geometric boundaries of the symmetric spaces associated with G and G'. Boundary maps continue to play a key role in many developments in rigidity theory. A new proof of Mostow rigidity, at least for G_i of real rank 1, was provided by Besson, Courtois, and Gallot. Their approach is quite different and has had many other applications concerning rigidity in geometry and dynamics; see, e.g., [13, 14, 38].

The next remarkable result in this direction is Margulis's superrigidity theorem. Margulis proved this theorem as a tool to prove arithmeticity of irreducible uniform lattices in groups of real rank at least 2. For irreducible lattices in semisimple Lie groups of real rank at least 2, the superrigidity theorem classifies all finite-dimensional linear representations. Margulis's theorem holds for irreducible lattices in semisimple Lie groups of real rank at least 2. Given a lattice $\Gamma < G$ in which G is simply connected, one precise statement of some of Margulis's results is to say that any linear representation σ of Γ *almost extends* to a linear representation of G. By this we mean that there is a linear representation $\tilde{\sigma}$ of G and a bounded-image representation $\bar{\sigma}$ of Γ such that $\sigma(\gamma) = \tilde{\sigma}(\gamma)\bar{\sigma}(\gamma)$ for all γ in G. Margulis's theorems also give an essentially complete description of the representations $\bar{\sigma}$, up to some issues concerning finite-image representations. The proof here is partially inspired by Mostow's work: a key step is the construction of a measurable "boundary map." However, the methods for producing the boundary map in this case are very dynamical. Margulis's original proof used the Oseledec multiplicative ergodic theorem. Later proofs were given by both Furstenberg and Margulis using the theory of group boundaries as developed by Furstenberg from his study of random walks on groups [92, 93]. Furstenberg's probabilistic version of boundary theory has had a profound influence on many subsequent developments in rigidity theory. For more discussion of Margulis's superrigidity theorem, see [158, 160, 161, 245].

Margulis's theorem, by classifying all linear representations, and not just ones with constrained images, leads one to believe that one might be able to classify all homomorphisms to other interesting classes of topological groups. Zimmer's program is just one aspect of this theory; in other directions, many authors have studied homomorphisms to isometry groups of nonpositively curved (and more general) metric spaces. See, e.g., [23, 97, 171].

3.3. Harmonic-map approaches to rigidity

In the 1990s, first Corlette and then Gromov-Schoen showed that one could prove major cases of Margulis's superrigidity theorem also for lattices in $Sp(1,n)$ and F_4^{-20} [39, 112]. Corlette considered the case of representations over Archimedean fields and Gromov and Schoen proved results over other local fields. These proofs use harmonic maps and proceed in three steps: first showing a harmonic map exists, second showing it is smooth, and third using certain special Bochner-type formulas to show that the harmonic mapping must be a local isometry. Combined with earlier work of Matsushima, Murakami, and Raghunathan, this leads to a complete classification of linear representations for lattices in these groups [169, 199]. It is worth noting that the use of harmonic-map techniques in rigidity theory had been pioneered by Siu, who used them to prove generalizations of Mostow rigidity for certain classes of Kähler manifolds [217]. There is also much later work on applying harmonic-map techniques to reprove cases of Margulis's superrigidity theorem by Jost-Yau and Mok-Siu-Yeung, among others; see, e.g., [121, 170]. We remark in passing that the general problem of the existence of harmonic maps for noncompact, finite-volume, locally symmetric spaces has not been solved in general. The results of Saper and Jost-Zuo allow one to prove superrigidity for fundamental groups of many such manifolds, but only while assuming arithmeticity [122, 207]. The use of harmonic maps in rigidity was inspired by the use of variational techniques and harmonic forms and functions in work on local rigidity and vanishing of cohomology groups. The original suggestion to use harmonic maps in this setting appears to go back to Calabi [216].

3.4. A remark on cocycle superrigidity

An important impetus for the study of rigidity of groups acting on manifolds was Zimmer's proof of his cocycle superrigidity theorem. We discuss this important result below in section 5, where we also indicate some of its applications to group actions. This theorem is a generalization of Margulis's superrigidity theorem to the class of *virtual homomorphisms* corresponding to cocycles over measure-preserving group actions.

3.5. Margulis's normal-subgroup theorem

We end this section by mentioning another result of Margulis that has had tremendous importance in results concerning group actions on manifolds. This is the normal-subgroup theorem, which says that any normal subgroup in a higher rank lattice is either finite or of finite index; see, e.g., [161, 245] for more on the proof. The proof precedes by a remarkable strategy. Let N be a normal subgroup of Γ; we show Γ/N is finite by showing that it is amenable and has property (T). This strategy has been applied in other contexts and is a major tool in the construction of simple groups with good geometric properties; see [6, 24, 25, 36]. The proof that Γ/N is amenable already involves one step that might rightly be called a theorem about rigidity of group actions. Margulis shows that if G is a semisimple group of higher rank, P is a minimal parabolic, and Γ is a lattice, then any measurable Γ space X that is a measurable quotient of the Γ action on G/P is necessarily of the form G/Q, where Q is a parabolic subgroup containing P. The proof of this result, sometimes called the projective-factor theorem, plays a fundamental role in work of Nevo and Zimmer on non-volume-preserving actions. See section 10.3 below for more discussion. It is also worth noting that Dani has proven a topological analogue of Margulis's result on quotients [41]; i.e., he has proven that continuous quotients of the Γ action on G/P are all Γ actions on G/Q.

The usual use of Margulis's normal-subgroup theorem in studying rigidity of group actions is usually quite straightforward. If one wants to prove that a group satisfying the normal-subgroup theorem acts finitely, it suffices to find one infinite-order element that acts trivially.

4. Low-dimensional actions: Conjectures and results

We begin this section by discussing results in particular, very low dimensions, namely dimensions 1, 2, and 3.

Before we begin this discussion, we recall a result of Thurston that is often used to show that low-dimensional actions are trivial [225].

Theorem 4.1. (Thurston stability). *Assume Γ is a finitely generated group with finite abelianization. Let Γ act on a manifold M by C^1 diffeomorphisms, fixing a point p and with trivial derivative at p. Then the Γ action is trivial in a neighborhood of p. In particular, if M is connected, the Γ action is trivial.*

The main point of Theorem 4.1 is that to show an action is trivial, it often suffices to find a fixed point. This is because, for the groups we consider, there are essentially no nontrivial low-dimensional linear representations and therefore the derivative at a fixed point is trivial. More precisely, the groups usually only have finite-image low-dimensional linear representations, which allows one to see that the action is trivial on a subgroup of finite index.

4.1. Dimension 1

The most dramatic results obtained in the Zimmer program concern a question first brought into focus by Dave Witte Morris in [234]: can higher rank lattices act on the circle? In fact, the paper [234] is more directly concerned with actions on the line $\mathbb{R}$. A detailed survey of results in this direction is contained in Witte Morris [237], so we do not repeat that discussion here. We merely state a conjecture and a question.

Conjecture 4.2. *Let Γ be a higher rank lattice. Then any continuous Γ action on S^1 is finite.*

This conjecture is well known and first appeared in print in [100]. By results in [234] and [145], the case of non-cocompact lattices is almost known. The paper [234] does the case of higher $\mathbb{Q}$ rank. The latter work of Lifchitz and Witte Morris reduces the general case to the case of quasi-split lattices in $SL(3,\mathbb{R})$ and $SL(3,\mathbb{C})$. It follows from work of Ghys or Burger-Monod that one can assume the action fixes a point, see, e.g., [21, 22, 98, 100, 101], so the question is equivalent to asking if the groups act on the line. We remark that actions of these groups on the circle that fix a point and are C^1 are easily seen to be finite by using Theorem 4.1 and the fact that all the groups in question have finite first homology. An interesting approach might be to study the induced action of Γ on the space of left orders on Γ; for ideas about this approach, we refer the reader to work of Navas and Witte Morris [172, 178].

Perhaps the following should also be a conjecture, but here we ask it only as a question.

Question 4.3. *Let Γ be a discrete group with property (T). Does Γ admit an infinite action by C^1 diffeomorphisms on S^1? Does Γ admit an infinite action by homeomorphisms on S^1?*

By a result of Navas, the answer to the above question is no if C^1 diffeomorphisms are replaced by C^k diffeomorphisms for any $k > \frac{3}{2}$ [179]. A result noticed by the author and Margulis and contained in a paper of Bader, Furman, Gelander, and Monod allows one to adapt this proof for values of k slightly less than $\frac{3}{2}$ [5, 81]. By Thurston's Theorem 4.1, in the C^1 case it suffices to find a fixed point for the action.

We add a remark here pointed out to the author by Navas that perhaps justifies calling Question 4.3 only a question. In [180], Navas extends his results from [179] to groups with relative property (T). This means that no such group acts on the circle by C^k diffeomorphisms where $k > \frac{3}{2}$. However, if we let Γ be the semidirect product of a finite-index free subgroup of $SL(2,\mathbb{Z})$ and $\mathbb{Z}^2$, this group is left orderable, and so acts on S^1; see, e.g., [172]. This is the prototypical example of a group with relative property (T). This leaves open the possibility that groups with property (T) would behave in a similar manner and admit continuous actions on the circle and the line.

Other possible candidates for a group with property (T) acting on the circle by homeomorphisms are the more general variants of Thompson's group F constructed by Stein in [222]. The proofs that F does not have (T) do not apply to these groups [57].

A closely related question is the following, suggested to the author by Andrés Navas along with examples in the last paragraph.

Question 4.4. *Is there a group Γ that is bi-orderable and does not admit a proper isometric action on a Hilbert space, i.e., that does not have the Haagerup property?*

For much more information on the fascinating topic of groups of diffeomorphisms of the circle, see the survey by Ghys [101], the more recent book by Navas [177], and the article by Witte Morris [237].

4.2. Dimension 2

Already in dimension 2 much less is known. There are some results in the volume-preserving setting. In particular, we have the following:

> **Theorem 4.5.** *Let Γ be a nonuniform irreducible higher rank lattice. Then any volume-preserving Γ action on a closed orientable surface other than S^2 is finite. If Γ has $\mathbb{Q}$ rank at least 1, then the same holds for actions on S^2.*

This theorem was proven by Polterovich for all surfaces but the sphere and shortly afterward proven by Franks and Handel for all surfaces [86, 85, 192]. It is worth noting that the proofs use entirely different ideas. Polterovich's proof belongs very clearly to symplectic geometry and also implies some results for actions in higher dimensions. Franks and Handel use a theory of normal forms for C^1 surface diffeomorphisms that they develop in analogy with the Thurston theory of normal forms for surface homeomorphisms with finite fixed sets. The proof of Franks and Handel can be adapted to a setting where one assumes much less regularity of the invariant measure [87].

There are some major reductions in the proofs that are similar, which we now describe. The first of these should be useful for studying actions of cocompact lattices as well. Namely, one can assume that the homomorphism $\rho : \Gamma \to \mathrm{Diff}(S)$ defining the action takes values in the connected component. This follows from the fact that any $\rho : \Gamma \to MCG(S)$ is finite, which can now be deduced from a variety of results [16, 21, 52, 123]. This result holds not only for the groups considered in Theorem 4.5 but also for cocompact lattices and even for lattices in $SP(1,n)$ by results of Sai-Kee Yeung [239]. It may be possible to show something similar for all groups with property (T) using recent results of Andersen showing that the mapping class group does not have property (T) [2]. It is unrealistic to expect a simple analogue of this result for homomorphisms to $\mathrm{Diff}(M)/\mathrm{Diff}(M)^0$ for general manifolds; see subsection 4.4 below for a discussion of dimension 3.

Also, in the setting of Theorem 4.5, Margulis's normal-subgroup theorem implies that it suffices to show that a single infinite-order element of Γ acts trivially. The proofs of Franks-Handel and Polterovich then use the existence of distortion elements in Γ, a fact established by Lubotzky, Mozes, and Raghunathan in [150]. The main result in both cases shows that $\mathrm{Diff}(S,\omega)$ does not contain exponentially distorted elements, and the proofs of this fact are completely different. An interesting result of Calegari and Freedman shows that this is not true of $\mathrm{Diff}(S^2)$ and that $\mathrm{Diff}(S^2)$ contains subgroups with elements of arbitrarily large distortion [29]. These examples are discussed in more detail in section 9.

Polterovich's methods also allow him to see that there are no exponentially distorted elements in $\text{Diff}(M,\omega)^0$ for certain symplectic manifolds (M, ω). Again, this yields some partial results toward Zimmer's conjecture. On the other hand, Franks and Handel are able to work with a Borel measure μ with some properties and show that any map $\rho : \Gamma \rightarrow \text{Diff}(S, \mu)$ is finite.

We remark here that a variant on this is due to Zimmer, when there is an invariant measure supported on a finite set.

Theorem 4.6. *Let Γ be a group with property (T) acting on a compact surface S by C^1 diffeomorphisms. Assume Γ has a periodic orbit on S; then the Γ action is finite.*

The proof is quite simple. First, pass to a finite-index subgroup that fixes a point x. Look at the derivative representation $d\rho_x$ at the fixed point. Since Γ has property (T), and $SL(2, \mathbb{R})$ has the Haagerup property, it is easy to prove that the image of $d\rho_x$ is bounded. Bounded subgroups of $SL(2, \mathbb{R})$ are all virtually abelian and this implies that the image of $d\rho_x$ is finite. Passing to a subgroup of finite index, one has a fixed point where the derivative action is trivial of a group that has no cohomology in the trivial representation. One now applies Thurston's Theorem 4.1.

The difficulty in combining Theorem 4.6 with ideas from the work of Franks and Handel is that while Franks and Handel can show that individual surface diffeomorphisms have large sets of periodic orbits, their techniques do not easily yield periodic orbits for the entire large group action.

We remark here that there is another approach to showing that lattices have no volume-preserving actions on surfaces. This approach is similar to the proof that lattices have no C^1 actions on S^1 via bounded cohomology [21, 22, 98]. That $\text{Diff}(S,\omega)^0$ admits many interesting quasi-morphisms follows from work of Entov-Polterovich, Gambaudo-Ghys, and Py [48, 95, 195, 196, 195]. By results of Burger and Monod, for any higher rank lattice and any homomorphism $\rho : \Gamma \rightarrow \text{Diff}(S, \omega)^0$, the image is in the kernel of all these quasi-morphisms. (To make this statement meaningful and the kernel well defined, one needs to take the homogeneous versions of the quasi-morphisms.) What remains to be done is to extract useful dynamical information from this fact; see Py [194] for more discussion.

In our context, the work of Entov-Polterovich mentioned above really only constructs a single quasi-morphism on $\text{Diff}(S^2,\omega)^0$. However, this particular quasi-morphism is very nice in that it is Lipschitz in the metric known as Hofer's metric on $\text{Diff}(S^2,\omega)^0$. This construction does apply more generally in higher dimensions and indicates connections between quasi-morphisms and the geometry of $\text{Diff}(M,\omega)^0$, for ω a symplectic form, that are beyond the scope of this survey. For an introduction to this fascinating topic, we refer the reader to [191].

Motivated by the above discussion, we recall the following conjecture of Zimmer.

Conjecture 4.7. *Let Γ be a group with property (T); then any volume-preserving smooth Γ action on a surface is finite.*

Here the words "volume preserving" or at least "measure preserving" are quite necessary. As $SL(3, \mathbb{R})$ acts on a S^2, so does any lattice in $SL(3, \mathbb{R})$ or any irreducible lattice in G in which G has $SL(3, \mathbb{R})$ as a factor. The following question seems reasonable; I believe I first learned it from Leonid Polterovich.

Question 4.8. *Let Γ be a higher rank lattice (or even just a group with property (T)) and assume Γ acts by diffeomorphisms on a surface S. Is it true that either (1) the action is finite or (2) the surface is S^2 and the action is smoothly conjugate to an action defined by some embedding $i : \Gamma \rightarrow SL(3, \mathbb{R})$ and the projective action of $SL(3, \mathbb{R})$ on S^2?*

This question seems quite far beyond existing technology.

4.3. Dimension 3

We now discuss briefly some work of Farb and Shalen that constrains actions by homeomorphisms in dimension 3 [54]. This work makes strong use of the geometry of 3-manifolds, but uses very little about higher rank lattices and is quite soft. A special case of their results is the following:

Theorem 4.9. *Let M be an irreducible 3-manifold and Γ be a higher rank lattice. Assume Γ acts on M by homeomorphisms so that the action on homology is nontrivial. Then M is homeomorphic to $\mathbb{T}^3$, $\Gamma < SL(3, \mathbb{Z})$ with finite index, and the Γ action on $H^1(M)$ is the standard Γ action on Z^3.*

Farb and Shalen actually prove a variant of Theorem 4.9 for an arbitrary 3-manifold admitting a homologically infinite action of a higher rank lattice. This can be considered as a significant step toward understanding when, for Γ a higher rank lattice and M^3 a closed 3-manifold, $\rho : \Gamma \rightarrow \text{Diff}(M^3)$ must have image in $\text{Diff}(M^3)^0$. Unlike the results discussed above for dimension 2, the answer is not simply "always." This three-dimensional result uses a great deal of the known structure of 3-manifolds, though it does not use the full geometrization conjecture proven by Perelman, but only the Haken case due to Thurston. The same sort of result in higher dimensions seems quite out of reach. A sample question is the following. Here we let Γ be a higher rank lattice and M a compact manifold.

Question 4.10. *Under what conditions on the topology of M do we know that a homomorphism $\rho : \Gamma \rightarrow \text{Diff}(M)$ has image in $\text{Diff}(M)^0$?*

4.4. Analytic actions in low dimensions

We first mention a direction pursued by Ghys that is in a similar spirit to the Zimmer program and that has interesting consequences for that program. Recall that the Zassenhaus lemma shows that any discrete linear group generated by small enough elements is nilpotent. The main point of Ghys's article [99] is to attempt to generalize this result for subgroups of Diff$^\omega(M)$. While the result is not actually true in that context, Ghys does prove some intriguing variants that yield some corollaries for analytic actions of large groups. For instance, he proves that $SL(n, \mathbb{Z})$ for $n > 3$ admits no analytic action on the 2-sphere. We remark that the attempt to prove the Zassenhaus lemma for diffeomorphism groups suggests that one can attempt to generalize other facts about linear groups to the category of diffeomorphism groups. We discuss several questions in this direction, mainly due to Ghys, in section 13.

For the rest of this subsection we discuss a different approach of Farb and Shalen for showing that real analytic actions of large groups are finite. This method is pursued in [53, 55, 56].

We begin by giving a cartoon of the main idea. Given any action of a group Γ, an element $\gamma \in \Gamma$, and the centralizer $Z(\gamma)$, it is immediate that $Z(\gamma)$ acts on the set of γ fixed points. If the action is analytic, then the fixed sets are analytic and so have good structure and are "reasonably close" to being submanifolds. If one further assumes that all normal subgroups of Γ have finite index, then this essentially allows one to bootstrap results about $Z(\gamma)$ not acting on manifolds of dimension at most $n-1$ to facts about Γ not having actions on manifolds of dimension n provided one can show that the fixed set for γ is not empty. This is not true, as analytic sets are not actually manifolds, but the idea can be implemented using the actual structure of analytic sets in a way that yields many results.

For example, we have the following:

Theorem 4.11. *Let M be a real analytic 4-manifold with zero Euler characteristic; then any real analytic, volume-preserving action of any finite-index subgroup in $SL(n, \mathbb{Z})$ for $n \geq 7$ is trivial.*

This particular theorem also requires a result of Rebelo [201] that concerns fixed sets for actions of nilpotent groups on $\mathbb{T}^2$ and uses the ideas of [99].

The techniques of Farb and Shalen can also be used to prove that certain cocompact higher rank lattices have no real analytic actions on surfaces of genus at least 1. For this result one needs only that the lattice contains an element γ whose centralizer already contains a higher rank lattice. In the paper [53] a more technical condition is required, but this can be removed using the results of Ghys and Burger-Monod on actions of lattices on the circle. (This simplification was first pointed out to the author by Farb.) The point is that γ has fixed points for topological reasons and the set of these fixed points contains either (1) a $Z(\gamma)$-invariant circle or (2) a $Z(\gamma)$-invariant point. Case (1) reduces to case (2) after passing to a finite-index subgroup via the results on circle actions. Case (2) is dealt with by the proof of Theorem 4.6.

4.5. Zimmer's full conjecture, partial results

We now state the full form of Zimmer's conjecture. In fact we generalize it slightly to include all lattices with property (T). Throughout this subsection G will be a semisimple Lie group with property (T) and Γ will be a lattice in G. We define two numerical invariants of these groups. First, for any group F, let $d(F)$ be the lowest dimension in which F admits an infinite-image linear representation. We note that the superrigidity theorems imply that $d(\Gamma) = d(G)$ when Γ is a lattice in G and either G is a semisimple group property (T) or Γ is irreducible higher rank. The second number, $n(G)$, is the lowest dimension of a homogeneous space K/C for a compact group K on which a lattice Γ in G can act via a homomorphism $\rho : \Gamma{\rightarrow}K$. In Zimmer's statement of the conjecture, he conjectures a slightly weaker bound in place of $n(G)$. Namely, he uses the smallest n' such that

$$\frac{n'\left(n'+1\right)}{2} \geq \min\left\{\dim{}_{\mathbb{C}}G' \text{ where } G' \text{ is a simple factor of } G_{\mathbb{C}}\right\}.$$

Using Margulis's superrigidity theorem and elementary facts about isometry groups of compact manifolds, it is easy to see that $n(G) \leq n'(G)$.

To state the more fashionable variant of Zimmer's conjecture, first made explicitly by Farb and Shalen for higher rank lattices in [53], we need an additional number. We let $c(G)$ be the codimension of a maximal parabolic Q in G. The relevance of this number is clear since any lattice Γ in G acts on the manifold G/Q by left translation.

> **Conjecture 4.12.** *Let Γ be a lattice as above. Let $b = \min\{n, d, c\}$ and let M be a manifold of dimension less than b; then any Γ action on M is trivial.*

Zimmer made conjectures only for smooth actions and his motivation for the conjecture depended heavily on the action being at least C^1. However, there is no known reason to avoid actions by homeomorphisms.

It is particularly bold to state this conjecture including lattices in $Sp(1,n)$ and F_4^{-20}. This is usually avoided because such lattices have abundant non-volume-preserving actions on certain types of highly regular fractals. I digress briefly to explain why I believe the conjecture is plausible in that case by discussing how the actions on fractals arise. Namely, such a lattice Γ, being a hyperbolic group, has many infinite proper quotients that are also hyperbolic groups [110]. If Γ' is a quotient of Γ by an infinite-index, infinite normal subgroup N and Γ' is hyperbolic, then the boundary $\partial\Gamma'$ is a Γ space with interesting dynamics and a good (Ahlfors regular) quasi-invariant measure class. However, $\partial\Gamma'$ is only a manifold when it is a sphere. It seems highly unlikely that this is ever the case for these groups and even more unlikely that this is ever the case with a smooth boundary action. The only known way to build a hyperbolic group that acts smoothly on its boundary is to have the group be the fundamental group of a compact negatively curved manifold M with smoothly varying horospheres. If smooth is taken to mean C^∞, this then implies the manifold is locally symmetric [13] and then superrigidity results make it impossible for this to occur with $\pi_1(M)$ a quotient of a lattice by an infinite-index infinite normal subgroup. If smooth only means C^1, then even in this context, no result known rules out M having fundamental group a quotient of a lattice in $Sp(1,n)$ or F_4^{-20}. All results one can prove using harmonic-map techniques in this context only rule out M with nonpositive complexified sectional curvature. We make the following conjecture, which is stronger than what is needed for Conjecture 4.12.

Conjecture 4.13. *Let Γ be a cocompact lattice in $Sp(1,n)$ or F_4^{-20} and let Γ' be a quotient of Γ that is Gromov hyperbolic. If $\partial\Gamma'$ is a sphere, then the kernel of the quotient map is finite and $\partial\Gamma' = \partial\Gamma$.*

For background on hyperbolic groups and their boundaries from a point of view relevant to this conjecture, see [132]. As a cautionary note, we point the reader to subsection 11.3, where we recall a construction of Farrell and Lafont that shows that any Gromov hyperbolic group has an action by homeomorphisms on a sphere.

The version of Zimmer's conjecture that Zimmer made in [249] and [250] was only for volume-preserving actions. Here we break it down somewhat explicitly to clarify the role of d and n.

Conjecture 4.14. (Zimmer). *Let Γ be a lattice as above and assume Γ acts smoothly on a compact manifold M preserving a volume form. Then if $\dim(M) < d$, the Γ action is isometric. If, in addition, $\dim(M) < n$ or if Γ is nonuniform, then the Γ action is finite.*

Some first remarks are in order. The cocycle superrigidity theorems (discussed below) imply that, when the conditions of the conjecture hold, there is always a measurable invariant Riemannian metric. Also, the finiteness under the conditions in the second half of the conjecture follows from cases of Margulis's superrigidity theorem as soon as one knows that the action preserves a smooth Riemannian metric. So from one point of view, the conjecture is really about the regularity of the invariant metric.

We should also mention that the conjecture is proven under several additional hypotheses by Zimmer around the time he made the conjecture. The first example is the following.

Theorem 4.15. *Conjecture 4.14 holds provided the action also preserves a rigid geometric structure, e.g., a torsion-free affine connection or a pseudo-Riemannian metric.*

This is proven in [246] for structures of finite type in the sense of Elie Cartan; see also [250]. The fact that roughly the same proof applies for rigid structures in the sense of Gromov was remarked in [84]. The point is simply that the isometry group of a rigid structure acts properly on some higher order frame bundle and that the existence of the measurable metric implies that Γ has bounded orbits on all frame bundles as soon as Γ has property (T). This immediately implies that Γ must be contained in a compact subgroup of the isometry group of the structure.

Another easy version of the conjecture follows from the proof of Theorem 4.6:

Theorem 4.16. *Let G be a semisimple Lie group with property (T) and $\Gamma < G$ a lattice. Let Γ act on a compact manifold M by C^1 diffeomorphisms where $\dim(M) < d(G)$. If Γ has a periodic orbit on S, then the Γ action is finite.*

A more difficult theorem of Zimmer shows that Conjecture 4.14 holds when the action is distal in a sense defined in [247].

4.6. Some approaches to the conjectures

4.6.1. *Discrete spectrum of actions.* A measure preserving action of a group D on a finite-measure space (X, μ) is said to have *discrete spectrum* if $L^2(X, \mu)$ splits as the sum of finite-dimensional D-invariant subspaces. This is a strong condition that is (quite formally) the opposite of weak mixing; for a detailed discussion see [91]. It is a theorem of Mackey (generalizing earlier results of Halmos and von Neumann for D abelian) that an ergodic discrete spectrum D action is measurably isomorphic to one described by a dense embedding of D into a compact group K and considering a K action on a homogeneous K space. The following remarkable result of Zimmer from [253] is perhaps the strongest evidence for Conjecture 4.14. This result is little known and has only recently been applied by other authors; see [82, 90].

Theorem 4.17. *Let Γ be a group with property (T) acting by smooth, volume-preserving diffeomorphisms on a compact manifold M. Assume in addition that Γ preserves a measurable invariant metric. Then the Γ action has discrete spectrum.*

This immediately implies that no counterexample to Conjecture 4.14 can be weak mixing or even admit a weak-mixing measurable factor.

The proof of the theorem involves constructing finite-dimensional subspaces of $L^2(M, \omega)$ that are Γ invariant. If enough of these subspaces could be shown to be spanned by smooth functions, one would have a proof of Conjecture 4.14. Here by "enough" we simply mean that it suffices to have a collection of finite-dimensional D-invariant subspaces that separate points in M. These functions would then specify a D-equivariant smooth embedding of M into $\mathbb{R}^N$ for some large value of N.

To construct finite-dimensional invariant subspaces of $L^2(M)$, Zimmer uses an approach similar to the proof of the Peter-Weyl theorem. Namely, he constructs Γ-invariant kernels on $L^2(M \times M)$ that are used to define self-adjoint, compact operators on $L^2(M)$. The eigenspaces of these operators are then finite-dimensional, Γ-invariant subspaces of $L^2(M)$. The kernels should be thought of as functions of the distance to the diagonal in $M \times M$. The main difficulty here is that for these to be invariant by "distance," we need to mean something defined in terms of the measurable metric instead of a smooth one. The construction of the kernels in this setting is quite technical and we refer readers to the original paper.

It would be interesting to try to combine the information garnered from this theorem with other approaches to Zimmer's conjectures.

4.6.2. Effective invariant metrics. We discuss here briefly another approach to Zimmer's conjecture, due to the author, which seems promising.

We begin by briefly recalling the construction of the space of "L^2 metrics" on a manifold M. Given a volume form ω on M, we can consider the space of all (smooth) Riemannian metrics on M whose associated volume form is ω. This is the space of smooth sections of a bundle $P \to M$. The fiber of P is $X = SL(n, \mathbb{R})/SO(n)$. The bundle P is an associated bundle to the $SL(n, \mathbb{R})$ subbundle of the frame bundle of M defined by ω. The space X carries a natural $SL(n, \mathbb{R})$-invariant Riemannian metric of nonpositive curvature; we denote its associated distance function by d_x. This induces a natural notion of distance on the space of metrics, given by $d(g_1, g_2)^2 = \int_M d_X(g_1(m), g_2(m))^2 d\omega$. The completion of the sections with respect to the metric d will be denoted $L^2(M, \omega, X)$; it is commonly referred to as the *space of L^2 metrics on M* and its elements will be called *L^2 metrics* on M. That this space is CAT(0) follows easily from the fact that X is CAT(0). For more discussion of X and its structure as a Hilbert manifold, see, e.g., [78]. It is easy to check that a volume-preserving Γ action on M defines an isometric Γ action on $L^2(M, \omega, X)$. Given a generating set S for Γ and a metric g in $L^2(M, \omega, X)$, we write $\mathrm{disp}(g) = max_{\gamma \in S} d(\gamma g, g)$.

Given a group Γ acting smoothly on M preserving ω, this gives an isometric Γ action on $L^2(M, \omega, X)$ that preserves the subset of smooth metrics. Let S be a generating set for Γ. We define an operator $P : L^2(M, \omega, X) \to L^2(M, \omega, X)$ by taking a metric g to the barycenter of the measure $\Sigma_S \delta_{(\gamma g)}$. The first observation is a consequence of the (standard, finite-dimensional) implicit-function theorem.

Lemma 4.18. *If g is a smooth metric, then Pg is also smooth.*

Moreover, we have the following two results.

Theorem 4.19. *Let M be a surface and* Γ *a group with property* (T) *and finite generating set S. Then there exists* $0 < C < 1$ *such that the operator P satisfies the following:*

1. $\mathrm{disp}(Pg) < C\mathrm{disp}(g)$;
2. *for any g, the* $\lim_n(P^n g)$ *exists and is* Γ *invariant.*

A proof of this theorem can be given by using the standard construction of a negative-definite kernel on $\mathbb{H}^2$ to produce a negative-definite kernel on $L^2(S, \mu, \mathbb{H}^2)$. The theorem is then proved by transferring the first property from the resulting Γ action on a Hilbert space. The second property is an obvious consequence of the first and completeness of the space $L^2(M, \omega, X)$.

Theorem 4.20. *Let G be a semisimple Lie group all of whose simple factors have property* (T) *and* $\Gamma < G$ *a lattice. Let M be a compact manifold such that* $\dim(M) < d(G)$. *Then the operator P on* $L^2(M, \omega, X)$ *satisfies the conclusions of Theorem 4.19.*

This theorem is proven from results in [79] using convexity of the distance function on $L^2(M, \omega, X)$. For cocompact lattices a proof can be given using results in [137] instead.

The problem now reduces to estimating the behavior of the derivatives of $P^n g$ for some initial smooth g. This expression clearly involves derivatives of random products of elements of Γ, i.e., derivatives of elements of Γ weighted by measures that are convolution powers of equidistributed measure on S. The main cause for optimism is that the fact that $\mathrm{disp}(P^n g)$ is small immediately implies that the first derivative of any $\gamma \in S$ must be small when measured at that point in $L^2(M, \omega, X)$. One can then try to use estimates on compositions of diffeomorphisms and convexity of derivatives to control derivatives of $P^n g$. The key difficulty is that the initial estimate on the first derivative of γ applied to $P^n g$ is only small in an L^2 sense.

4.7. Some related questions and conjectures on countable subgroups of Diff(M)

In this subsection, we discuss related conjectures and results on countable subgroups of Diff(M). All of these are motivated by the belief that countable subgroups of Diff(M) are quite special, though considerably less special than, say, linear groups. We defer positive constructions of nonlinear subgroups of Diff(M) to section 11 and a discussion of possible algebraic and geometric properties shared by all finitely generated subgroups of Diff(M) to section 13. Here we concentrate on groups that do not act on manifolds, either by theorems or conjecturally.

4.7.1. *Groups with property (T) and generic groups.* We begin by focusing on actions of groups with property (T). For a finitely generated group Γ, we recall that $d(\Gamma)$ is the smallest

dimension in which Γ admits an infinite-image linear representation. We then make the following conjecture:

Conjecture 4.21. *Let Γ be a group with property (T) acting smoothly on a compact manifold M, preserving volume. Then if $\dim(M) < d(\Gamma)$, the action preserves a smooth Riemannian metric.*

For many groups Γ with property (T), one can produce a measurable invariant metric; see [82]. In fact, in [82], Silberman and the author prove that there are many groups with property (T) with no volume-preserving actions on compact manifolds. Key steps include finding the invariant measurable metric, applying Zimmer's theorem on discrete spectrum from section 4.6, and producing groups with no finite quotients and so no linear representations at all.

A result of Furman announced in [89] provides some further evidence for the conjecture. This result is analogous to Proposition 5.1 and shows that any action of a group with property (T) either leaves invariant a measurable metric or has positive *random entropy*. We refer the reader to [89] for more discussion. While the proof of this result is not contained in [89], it is possible to reconstruct it from results there and others in [90].

Conjecture 4.21 and the work in [82] are motivated in part by the following conjecture of Gromov:

Conjecture 4.22. *There exists a model for random groups in which a "generic" random group admits no smooth actions on compact manifolds.*

It seems quite likely that the conjecture could be true for random groups in the density model with density more than $\frac{1}{3}$. These groups have property (T); see Ollivier's book [189] for discussion on random groups. For the conjecture to be literally true would require that a random hyperbolic group have no finite quotients, and it is a well-known question to determine whether there are any hyperbolic groups that are not residually finite, let alone generic ones. If one is satisfied by saying the generic random group has only finite smooth actions on compact manifolds, one can avoid this well-known open question.

4.7.2. *Universal lattices.* Recently, Shalom has proven that the groups $SL(n, \mathbb{Z}[X])$ have property (T) when $n > 3$ (as well as some more general results). Shalom refers to these groups as *universal lattices*. An interesting and approachable question is as follows:

Question 4.23. *Let Γ be a finite-index subgroup in $SL(n, \mathbb{Z}[X])$ and let Γ be acting by diffeomorphisms on a compact manifold M preserving volume. If $\dim(M) < n$ is there a measurable Γ-invariant metric?*

If one gives a positive answer to this question, one is then clearly interested in whether or not the metric can be chosen to be smooth. It is possible that this is easier for these "larger" groups than for the lattices originally considered by Zimmer. One can ask a number of variants of this question, including trying to prove a full cocycle superrigidity theorem for these groups; see below. The question just asked is particularly appealing as it can be viewed as a fixed-point problem for the Γ action on the space of metrics (see section 4.6 for a definition). In fact, one knows one has the invariant metric for the action of any conjugate of $SL(n, \mathbb{Z})$ and only needs to show that there is a consistent choice of invariant metrics over all conjugates. One might try to mimic the approach from [213], though a difficulty clearly arises at the point where Shalom applies a scaling-limit construction. Scaling limits of $L^2(M, \omega, X)$ can be described using nonstandard analysis, but are quite complicated objects and not usually isomorphic to the original space.

4.7.3. *Irreducible actions of products*. Another interesting variant on Zimmer's conjecture is introduced in [90]. In that paper Furman and Monod study obstructions to irreducible actions of product groups. A measure-preserving action of a product $\Gamma_1 \times \Gamma_2$ on a measure space (X, μ) is said to be *irreducible* if both Γ_1 and Γ_2 act ergodically on X. Furman and Monod produce many obstructions to irreducible actions; e.g., one can prove from their results the following:

> **Theorem 4.24.** *Let $\Gamma = \Gamma_1 \times \Gamma_2 \times \Gamma_3$. Assume that Γ_1 has property (T) and no unbounded linear representations. Then there are no irreducible, volume-preserving Γ actions on compact manifolds. The same is true for $\Gamma = \Gamma_1 \times \Gamma_2$ if Γ_1 is as above and Γ_2 is solvable.*

A key step in the proof of this result is to use Zimmer's result on actions with discrete spectrum to show that the Γ action is measurably conjugate to an irreducible isometric action. In particular, the second conclusion follows since no compact group can contain simultaneously a dense (T) group and a dense solvable group. This motivates the following conjecture.

> **Conjecture 4.25.** *Let $\Gamma = \Gamma_1 \times \Gamma_2$. Assume that Γ_1 has property (T) and no unbounded linear representations and that Γ_2 is amenable. Then there are no irreducible, volume-preserving Γ actions on compact manifolds.*

One might be tempted to prove this by showing that no compact group contains both a dense finitely generated amenable group and a dense Kazhdan group. This is however not true. The question was raised by Lubotzky in [148] but has recently been resolved in the negative by Kassabov [126].

In the context of irreducible actions, asking that groups have property (T) is perhaps too strong. In fact, there are already relatively few known irreducible actions of $\mathbb{Z}^2 = \mathbb{Z} \times \mathbb{Z}$! If one element of $\mathbb{Z}^2$ acts as an Anosov diffeomorphism, then it is conjectured by Katok and Spatzier that all actions are algebraic [129]. Even in more general settings where one assumes nonuniform hyperbolicity, there are now hints that a classification might be possible, though clearly more complicated than in the Anosov case; see [124] and references therein. In a sense that paper indicates that the "exotic examples" that might arise in this context may be no worse than those that arise for actions of higher rank lattices.

4.7.4. *Torsion groups and* Homeo(M).

A completely different and well-studied aspect of the theory of transformation groups of compact manifolds is the study of finite subgroups of Homeo(M). It has recently been noted by many authors that this study has applications to the study of "large subgroups" of Homeo(M). Namely, one can produce many finitely generated and even finitely presented groups that have no nontrivial homomorphisms to Homeo(M) for any compact M. This is discussed in, e.g., [19, 82, 232]. As far as I know, this observation was first made by Ghys as a remark in the introduction to [99]. In all cases, the main trick is to construct infinite, finitely generated groups that contain infinitely many conjugacy classes of finite subgroups, usually just copies of $(\mathbb{Z}/p\mathbb{Z})^k$ for all $k > 1$. A new method of constructing such groups was recently introduced by Chatterji and Kassabov [37].

As far as the author knows, even if one fixes M in advance, this is the only existing method for producing groups Γ with no nontrivial (or even no infinite image) homomorphisms to Homeo(M) unless $M = S^1$. For $M = S^1$ we refer back to subsection 4.1 and to [237].

5. Cocycle superrigidity and immediate consequences

5.1. Zimmer's cocycle superrigidity theorem and generalizations

A main impetus for studying rigidity of group actions on manifolds came from Zimmer's theorem on superrigidity for cocycles. This theorem and its proof were strongly motivated by Margulis's work. In fact, Margulis's theorem reduces to the special case of Zimmer's theorem for a certain cocycle $\alpha : G \times G/\Gamma \to \Gamma$. In order to avoid technicalities, we describe only a special case of this result, essentially avoiding boundedness and integrability assumptions on cocycles that are automatic fulfilled in any context arising from a continuous action on a compact manifold. Let M be a compact manifold, H a matrix group, and P an H bundle over M. For readers not familiar with bundle theory, the results are interesting even in the case where $P = M \times H$. Now let a group Γ act on P continuously by bundle automorphisms, i.e., such that there is a Γ action on M for which the projection from P to M is equivariant. Further assume that the action on M is measure preserving and ergodic. The cocycle superrigidity theorem says that if Γ is a lattice in a simply connected, semisimple Lie group G all of whose simple factors are noncompact and have property (T), then there is a measurable map $s : M \to H$, a representation $\pi : G \to H$, a compact subgroup $K < H$ that commutes with $\pi(G)$, and a measurable map $\Gamma \times M \to K$ such that

$$(1) \qquad \gamma \cdot s(m) = k(m,\gamma)\pi(\gamma)s(\gamma \cdot m).$$

It is easy to check from this equation that the map K satisfies the equation that makes it into a cocycle over the action of Γ. One should view s as providing coordinates on P in which the Γ action is *almost a product*. For more discussion of this theorem, particularly in the case where all simple factors of G have higher rank, the reader should see any of [64, 65, 80, 94, 245]. (The version stated here is only proven in [80]; previous proofs all yielded somewhat more complicated statements that require passing to finite ergodic extensions of the action.) For the case of G with simple factors of the form $Sp(1, n)$ and F_4^{-20}, the results follow from work of the author and Hitchman [79], building on earlier results of Korevaar-Schoen and Corlette-Zimmer [40, 137, 138, 139].

As a sample application, let $M = \mathbb{T}^n$ and let P be the frame bundle of M, i.e., the space of frames in the tangent bundle of M. Since $\mathbb{T}^n$ is parallelizable, we have $P = \mathbb{T}^n \times GL(n, \mathbb{R})$. The cocycle superrigidity theorem then says that "up to compact noise" the derivative of any measure-preserving Γ action on $\mathbb{T}^n$ looks measurably like a constant linear map. In fact, the cocycle superrigidity theorems apply more generally to continuous actions on any principal bundle P over M with fiber H, an algebraic group, and in this context produce a measurable section $s : M \to P$ satisfying equation (1). So in fact, cocycle superrigidity implies that for any action preserving a finite measure on any manifold the derivative cocycle looks measurably like a constant cocycle, up to compact noise. That cocycle superrigidity provides information about actions of groups on manifolds through the derivative cocycle was first observed by Furstenberg in [94]. Zimmer originally proved cocycle superrigidity in order to study orbit equivalence of group actions. For recent surveys of subsequent developments concerning orbit equivalence rigidity and other forms of superrigidity for cocycles, see [88, 193, 212].

5.2. First applications to group actions and the problem of regularity

The following result, first observed by Furstenberg, is an immediate consequence of cocycle superrigidity.

Proposition 5.1. *Let G be a semisimple Lie group with no compact factors and with property (T) of Kazhdan and let $\Gamma < G$ be a lattice. Assume G or Γ acts by volume-preserving diffeomorphisms on a compact manifold M. Then there is a linear representation $\pi : G \to GL(\dim(M), \mathbb{R})$ such that the Lyapunov exponents of $g \in G$ or Γ are exactly the absolute values of the eigenvalues of $\pi(g)$.*

This proposition is most striking to those familiar with classical dynamics, where the problem of estimating, let alone computing, Lyapunov exponents is quite difficult.

Another immediate consequence of cocycle superrigidity is the following:

Proposition 5.2. *Let G or Γ be as above, acting by smooth diffeomorphisms on a compact manifold M preserving volume. If every element of G or Γ acts with zero entropy,* *then there is a measurable invariant metric on M. In particular, if G admits no non-trivial representations of* dim(M), *then there is a measurable invariant metric on M.*

As mentioned above in section 4, this reduces Conjecture 4.14 to the following technical conjecture.

Conjecture 5.3. *Let G or* Γ *acting on M be as above. If G or* Γ *preserves a measurable Riemannian metric, then they preserve a smooth invariant Riemannian metric.*

We recall that first evidence toward this conjecture is Theorem 4.15. We remark that in the proof of that theorem, it is not the case that the measurable metric from Proposition 5.2 is shown to be smooth, but instead it is shown that the image of Γ in Diff(M) lies in a compact subgroup. In other work, Zimmer explicitly improves regularity of the invariant metric; see particularly [247].

In general the problem that arises immediately from cocycle superrigidity in any context is understanding the regularity of the straightening section σ. This question has been studied from many points of view, but still relatively little is known in general. For certain examples of actions of higher rank lattices on compact manifolds, discussed below in section 9, the σ that straightens the derivative cocycle cannot be made smooth on all of M. It is possible that for volume-preserving actions on manifolds and the derivative cocycle, σ can always be chosen to be smooth on a dense open set of full measure.

We will return to the theme of regularity of the straightening section in section 7. First we turn to more geometric contexts in which the output of cocycle superrigidity is also often used but usually more indirectly.

6. Geometric actions and rigid geometric structures

In this section, we discuss the role of rigid geometric structures in the study of actions of large groups. The notion of rigid geometric structure was introduced by Gromov, partially in reaction to Zimmer's work on large group actions.

The first subsection of this section recalls the definition of rigid geometric structure, gives some examples, and explains the relation of Gromov's rigid geometric structures to other notions introduced by Cartan. Subsection 6.2 recalls Gromov's initial results relating actions of simple groups preserving rigid geometric structures on M to representations of the fundamental group of M, and extensions of these results to lattice actions due to Zimmer and the author. The third section concerns a different topic, namely the rigidity of connection-preserving actions of a lattice Γ, particularly on manifolds of dimension not much larger than $d(G)$ as defined in section 4. The fourth subsection recalls some obstructions to actions preserving geometric structures, particularly a result known as Zimmer's geometric Borel density theorem and some recent related results of Bader, Frances, and Melnick. We discuss actions preserving a complex structure in subsection 6.5 and end with a subsection on questions concerning geometric actions.

6.1. Rigid structures and structures of finite type

In this subsection we recall the formal definition of a rigid geometric structure. Since the definition is somewhat technical, some readers may prefer to skip it, and read on keeping in mind examples rather than the general notion. The basic examples of rigid geometric structures are Riemannian metrics, pseudo-Riemannian metrics, conformal structures defined by either type of metric, and affine or projective connections. Basic examples of geometric structures that are not rigid are a volume form or symplectic structure. An intermediate type of structure that exhibits some rigidity but that is not literally rigid in the sense discussed here is a complex structure.

If N is a manifold, we denote the kth-order frame bundle of N by $F^k(N)$, and by $J^{s,k}(N)$ the bundle of k-jets at 0 of maps from $\mathbb{R}^s$ to N. If N and N' are two manifolds, and $f : N \to N'$ is a map between them, then the k-jet $j^k(f)$ induces a map $J^{s,k}N \to J^{s,k}N'$ for all s. We let $D^k(N)$ be the bundle whose fiber D^k_p at a point p consists of the set of k-jets at p of germs of diffeomorphisms of N fixing p. We abbreviate $D^k_0(\mathbb{R}^n)$ by D^k_n or simply D^k; this is a real algebraic group. For concreteness, one can represent each element uniquely, in terms of standard coordinates $(\xi_1, \ldots, \xi_n)$ on $\mathbb{R}^n$, in the form

$$(P_1(\xi_1, \ldots, \xi_n), \ldots, P_n(\xi_1, \ldots, \xi_n)),$$

where $P_1, P_2, \ldots, P_n$ are polynomials of degree $\leq k$. We denote the vector space of such polynomial maps of degree $\leq k$ by $P_{n,k}$.

The group D^k_n has a natural action on $F^k(N)$, where n is the dimension of N. Suppose we are given an algebraic action of D^k_n on a smooth algebraic variety Z. Then following Gromov [111], we make the following definition:

Definition 6.1.
1. *An A-structure on N (of order k, of type Z) is a smooth map $\phi : F^k(N) \to Z$ equivariant for the D^k_n actions.*
2. *With notation as above, the rth prolongation of f, denoted ϕ^r, is the map ϕ^r: $F^{k+r}(N) \to J^{n,r}(Z)$ defined by $\phi^r = j^r(\phi) \circ \iota^{r+k}_k$, where $\iota^{r+k}_k : F^{k+r}(N) \to J^{n,r}(F^k(N))$ is the natural inclusion and $j^k(h) : J^{n,r}(F^k(N)) \to J^{n,r}(Z)$ is as before; this is an A-structure of type $J^{n,r}(Z)$ and order $k+r$.*

Equivalently, an A-structure of type Z and order k is a smooth section of the associated bundle $F^k(N) \times_{D^k} Z$ over N. Note that an A-structure on N defines by restriction an A-structure $\phi|_U$ on any open set $U \subset N$.

Remark 6.2. *A-structures were introduced in* [111]; *a good introduction to the subject, with many examples, can be found in* [9]. *A comprehensive and accessible discussion of the results of* [111] *concerning actions of simple Lie groups can be found in* [66].

Note that if N and N' are n-manifolds, and $h : N \to N'$ is a diffeomorphism, then h induces a bundle map $j^k(h) : F^k(N) \to F^k(N')$.

Definition 6.3.

1. If $\phi : F^k(N) \to Z$, $\phi' : F^k(N') \to Z$ are A-structures, a diffeomorphism $h : N \to N'$ is *an isometry from* ϕ *to* ϕ' *if* $\phi' \circ j^k(h) = \phi$.
2. *A local isometry of f is a diffeomorphism* $h : U_1 \to U_2$, *for open sets* $U_1, U_2 \subset N$, *which is an isometry from* $\phi|_{U_1}$ *to* $\phi|_{U_2}$.

For $p \in M$ denote by $I s_p^{loc}(\phi)$ the pseudogroup of local isometries of ϕ fixing p, and for $l \geq k$, we denote by $I s_p^l(\phi)$ the set of elements $j_p^l(h) \in D_p^l$ such that $j_p^l(\phi \circ j_p^k(h)) = \phi^{l-k}$, where both sides are considered as maps $F^{k+l}(N) \to J^{l-k}(Z)$. Note that $I s_p^l(\phi)$ is a group, and there is a natural homomorphism $r_p^{l;m} : I s_p^l(\phi) \to I s_p^m(\phi)$ for $m < l$; in general, it is neither injective nor surjective.

Definition 6.4. *The structure f is called k-rigid if for every point p, the map $r_p^{k+1;k}$ is injective.*

A first remark worth making is that an affine connection is a rigid geometric structure and that any generalized quasi-affine action on a manifold of the form $K \backslash H / \Lambda \times M$ with Λ discrete preserves an affine connection and therefore also a torsion-free affine connection. In order to provide some more examples, we recall the following lemma of Gromov.

Lemma 6.5. *Let V be an algebraic variety and G a group acting algebraically on V. For every k, there is a tautological G-invariant geometric structure of order k on V, given by $\omega : P^k(V) \to P^k(V)/G$. This structure is rigid if and only if the action of G on $P^k(V)$ is free and proper.*

The conclusion in the first sentence is obvious. The second sentence is proven in section 0.4, pages 69–70, of [111].

Examples:

1. The action of $G = SL_n(\mathbb{R})$ on $\mathbb{R}^n$ is algebraic. So is the action of G on the manifold N_1 obtained by blowing up the origin. The reader can easily verify that the action of G on $P^2(N_1)$ is free and proper.

2. We can compactify N_1 by N_2 by viewing the complement of the blowup as a subset of the projective space P^n. Another description of the same action, which may make the rigid structure more visible to the naked eye, is as follows. $SL_{n+1}(\mathbb{R})$ acts on P^n. Let G be $SL_n(\mathbb{R}) < SL_{n+1}(\mathbb{R})$ as block-diagonal matrices with blocks of size n and 1 and a 1×1 block equal to 1. Then G acts on P^n fixing a point p. We can obtain N_2 by blowing up the fixed point p. The G actions on both P^n and N_2 are algebraic and again the reader can verify that the action is free and proper on $P^n(N_2)$.

3. In the construction from (2) above, there is an action of a group H in which $H = SL_n(\mathbb{R}) \ltimes \mathbb{R}^n$ and $G = SL_n(\mathbb{R}) < SL_n(\mathbb{R}) \ltimes \mathbb{R}^n < SL_{n+1}(\mathbb{R})$. The H action fixes the point p and so also acts on N_2 algebraically. However, over the exceptional divisor, the action is never free on any frame bundle, since the subgroup $\mathbb{R}^n$ acts trivially to all orders at the exceptional divisor.

The behavior in example (3) above illustrates the fact that the existence of invariant rigid structures is more complicated for algebraic groups that are not semisimple; see the discussion in section 0.4.C. of [111].

Definition 6.6. *A rigid geometric structure is called a finite-type geometric structure if V is a homogeneous D_n^k space.*

This is by no means the original definition that is due to Cartan and predates Gromov's notion of a rigid geometric structure by several decades. It is equivalent to Cartan's definition of a finite-type structure by work of Candel and Quiroga [30, 31, 197]. Candel and Quiroga also give a development of rigid geometric structures that more closely parallels the older notion of a structure of finite type as presented in, e.g., [134].

All standard examples of rigid geometric structures are structures of finite type. However, example (2) above is not a structure of finite type. The notion of structure of finite type was first given in terms of prolongations of Lie algebras and so yields a criterion that is, in principle, computable. The work of Candel and Quiroga extends this computable nature to general rigid geometric structures.

6.2. Rigid structures and representations of fundamental groups

In this subsection, we restrict our attention to actions of simple Lie groups G and lattices $\Gamma < G$. Many of the results of this section extend to semisimple G, though the formulations become more complicated.

A major impetus for Gromov's introduction of rigid geometric structures is the following theorem from [111].

Theorem 6.7. *Let G be a simple noncompact Lie group and M a compact real analytic manifold. Assume G acts on M preserving an analytic rigid geometric structure ω and a volume form ν. Further assume the action is ergodic. Then there is a linear representation $\rho : \pi_1(M) \to GL(n, \mathbb{R})$ such that the Zariski closure of $\rho(\pi_1(M))$ contains a group locally isomorphic to G.*

We remark that by Tits's alternative, the theorem immediately implies that actions of the type described cannot occur on manifolds with amenable fundamental group or even on manifolds whose fundamental groups do not contain a free group on two generators [226]. Expository accounts of the proof can be found in [66, 255, 258]. The representation ρ is actually defined on Killing fields of the lift $\tilde{\omega}$ of ω to $\tilde{M}$.

This is a worthwhile moment to indicate the weakness of Lemma 2.6. If we start with a cocompact lattice $\Gamma < G$ and an action of Γ on a compact manifold M satisfying the assumptions of Theorem 6.7, we can induce the action to a G action on the compact manifold $N = (G \times M)/\Gamma$. However, in this setting, there is an obvious representation of $\pi_1(N)$ satisfying the conclusion of Theorem 6.7. This is because there is a surjection $\pi_1(N) \to \Gamma$. In fact for most cocompact lattices $\Gamma < G$ there are homomorphisms $\sigma : \Gamma \to K$ in which K is compact and simply connected so that we can have Γ act on K by left translation and obtain examples in which $\pi_1(N) = \Gamma$. For G of higher rank, the following theorem of Zimmer and the author shows that this is the only obstruction to a variant of Theorem 6.7 for lattices.

Theorem 6.8. *Let $\Gamma < G$ be a lattice, where G is a simple group and $\mathbb{R}$–rank$(G) \geq 2$. Suppose Γ acts analytically and ergodically on a compact manifold M preserving a unimodular rigid geometric structure. Then either*

1. *the action is isometric and $M = K/C$, where K is a compact Lie group and the action is by right translation via $\rho : \Gamma \to K$, a dense image homomorphism, or*
2. *there exists an infinite image linear representation $\sigma : \pi_1(M) \to GL_n\mathbb{R}$, such that the algebraic automorphism group of the Zariski closure of $\sigma(\pi_1(M))$ contains a group locally isomorphic to G.*

The proof of this theorem makes fundamental use of a notion that does not occur in its statement: the entropy of the action. A key observation is the following formula for the entropy of the restriction of the induced action to Γ:

$$h_{(G \times M)/\Gamma}(\gamma) = h_{G/\Gamma}(\gamma) + h_M(\gamma).$$

We then use the fact that the last term on the right-hand side is zero if and only if the action preserves a measurable Riemannian metric as already explained in Proposition 5.1.

In this setting, it then follows from Theorem 4.15 that the action is isometric and the other conclusions in (1) are simple consequences of ergodicity of the action.

When $h_M(\gamma) > 0$ for some Γ, we use relations between the entropy of the action and the Gromov representation discovered by Zimmer [257]. We discuss some of these ideas below in section 8.

6.3. Affine actions close to the critical dimension

In this section we briefly describe a direction pursued by Feres, Goetze, Zeghib, and Zimmer that classifies connection-preserving actions of lattices in dimensions close to, but not less than, the critical dimension d. A representative result is the following:

Theorem 6.9. *Let $n > 2$, $G = SL(n, \mathbb{R})$, and $\Gamma < G$ a lattice. Let M be a compact manifold with $\dim(M) \leq n + 1 = d(G) + 1$. Assume that Γ acts on M preserving a volume form and an affine connection. Then either*

1. *$\dim(M) < n$ and the action is isometric;*
2. *$\dim(M) = n$ and the action is either isometric or, upon passing to finite covers and finite index subgroups, smoothly conjugate to the standard $SL(n, \mathbb{Z})$ action on $\mathbb{T}^n$; or*
3. *$\dim(M) = n + 1$ and the action is either isometric or, upon passing to finite covers and finite-index subgroups, smoothly conjugate to the action of $SL(n, \mathbb{Z})$ on $\mathbb{T}^{n+1} = \mathbb{T}^n \times \mathbb{T}$, where the action on the second factor is trivial.*

Variants of this theorem under more restrictive hypotheses were obtained by Feres, Goetze, and Zimmer [62, 103, 248]. The theorem as stated is due to Zeghib [243]. The cocycle superrigidity theorem is used in all of the proofs, mainly to force vanishing of curvature and torsion tensors of the associated connection.

One would expect similar results for a more general class of acting groups and also for a wider range of dimensions. Namely, if we let $d_2(G)$ be the dimension of the second nontrivial representation of G, we would expect a similar result to (3) for any affine, volume-preserving action on a manifold M with $d(G) < d_2(G)$. With the further assumption that the connection is Riemannian (and still only for lattices in $SL(n, \mathbb{R})$) this is proven by Zeghib in [242].

Further related results are contained in other papers of Feres [59, 63]. No results of this kind are known for $\dim(M) \geq d_2(G)$. A major difficulty arises as soon as one has $\dim(M) \geq d_2(G)$, namely that more complicated examples can arise, including affine actions on nilmanifolds and left-translation actions on spaces of the form G/Λ. All the results mentioned in this subsection depend on the particular fact that flat Riemannian manifolds have finite covers that are tori.

6.4. Zimmer's Borel density theorem and generalizations

An obvious first question concerning G actions is the structure of the stabilizer subgroups. In this direction we have the following:

Theorem 6.10. *Let G be a simple Lie group acting essentially faithfully on a compact manifold M preserving a volume form; then the stabilizer of almost every point is discrete.*

As an application of this result, one can prove the following:

Theorem 6.11. *Let G be a simple Lie group acting essentially faithfully on a compact manifold, preserving a volume form and a homogeneous geometric structure with structure group H. Then there is an inclusion $\mathfrak{g} \to \mathfrak{h}$.*

In particular, the theorem provides an obstruction to a higher rank simple Lie group having a volume-preserving action on a compact Lorentz manifold. For more discussion of these theorems, see [258]. This observation led to a major series of works studying automorphism groups of Lorentz manifolds; see, e.g., [1, 140, 240, 241]. We remark here that Theorem 6.11 does not require that the homogeneous structure be rigid.

More recently some closely related phenomena have been discovered in a joint work of Bader, Frances, and Melnick [227]. Their work uses yet another notion of a geometric structure, that of a *Cartan geometry*. We will not define this notion rigorously here; it suffices to note that any rigid homogeneous geometric structure defines a Cartan geometry, as discussed in [227, Introduction]. The converse is not completely clear in general, but most classical examples of rigid geometric structures can also be realized as Cartan geometries. A Cartan geometry is essentially a way of saying that a manifold is *infinitesimally* modeled on some homogeneous space G/P. To recapture the notion of a Riemannian connection, $G = O(n) \ltimes \mathbb{R}^n$ and $P = O(n)$; to recapture the notion of an affine connection, $G = Gl(n, \mathbb{R}) \ltimes \mathbb{R}^n$ and $P = Gl(n, \mathbb{R})$. Cartan connections come with naturally defined curvatures that vanish if and only if the manifold is *locally* modeled on G/P. We refer the reader to the book [214] for a general discussion of Cartan geometries and their use in differential geometry.

Given a connected linear group L, we define its real rank $rk(L)$ as before to be the dimension of a maximal $\mathbb{R}$-diagonalizable subgroup of L, and let $n(L)$ denote the maximal nilpotence degree of a connected nilpotent subgroup. One of the main results of [227] is as follows:

Theorem 6.12. *If a group L acts by automorphisms of a compact Cartan geometry modeled on G/P, then*

    1. $rk(AdL^0) \leq rk(Ad_g P)$ *and*

    2. $n(AdL^0) \leq n(Ad_g P)$.

The main point is that L is not assumed either simple or connected. This theorem is deduced from an embedding theorem similar in flavor to Theorem 6.10.

In addition, when G is a simple group and P is a maximal parabolic subgroup, Bader, Frances, and Melnick prove rigidity results classifying all possible actions when the rank bound in Theorem 6.12 is achieved. This classification essentially says that all examples are algebraic; see [227] for detailed discussion.

An earlier paper by Feres and Lampe also explored applications of Cartan geometries to rigidity and dynamical conditions for flatness of Cartan geometries [61].

6.5. Actions preserving a complex structure

We mention here one recent result by Cantat and a few active related directions of research. In [250], Zimmer asked whether one had restrictions on low-dimensional holomorphic actions of lattices. The answer was obtained in [35] and is as follows:

Theorem 6.13. *Let G be a connected simple Lie group of real rank at least 2 and suppose $\Gamma < G$ is a lattice. If Γ admits a faithful action by automorphisms on a compact Kähler manifold M, then the rank of G is at least the complex dimension of M.*

The proof of the theorem depends primarily on results concerning $\mathrm{Aut}(M)$, particularly recent results on holomorphic actions of abelian groups by Dinh and Sibony [46]. It is worth noting that while the theorem does depend on Margulis's superrigidity theorem, it does not depend on Zimmer's cocycle superrigidity theorem. The condition that the action is faithful can be weakened considerably using Margulis's normal-subgroups theorem.

More recently Cantat, Deserti, and others have begun a program of studying large subgroups of automorphism groups of complex manifolds; see, e.g., [34, 45]. In particular, Cantat has proven an analogue of Conjecture 4.7 in the context of birational actions on complex surfaces.

Theorem 6.14. *Let S be a compact Kähler surface and G an infinite, countable group of birational transformations of S. If G has property (T), then there is a birational map $j : S \to \mathbb{P}^2(\mathbb{C})$ that conjugates G to a subgroup of $\mathrm{Aut}(\mathbb{P}^2(\mathbb{C}))$.*

6.6. Conjectures and questions

A major open problem in generalizing Conjecture 1.2 is the following:

Conjecture 6.15. (Gromov-Zimmer). *Let D be a semisimple Lie group with all factors having property (T) or a lattice in such a Lie group. Then any D action on a compact manifold preserving a rigid geometric structure and a volume form is generalized quasi-affine.*

A word is required on the attribution. Both Gromov and Zimmer made various less precise conjectures concerning the classification of actions as in the conjecture [111, 249]. The exact statement of the correct conjecture was muddy for several years while it was not known whether the Katok-Lewis- and Benveniste-type examples admitted invariant rigid geometric structures. See section 9 for more discussion of these examples. In the context of the results of [11], where Benveniste and the author prove that those actions do not preserve rigid geometric structures, this version of the classification seems quite plausible. It is perhaps more plausible if one assumes the action is ergodic or that the geometric structure is homogeneous.

A possibly easier question that is relevant is the following:

Question 6.16. *Let M be a compact manifold equipped with a homogeneous rigid geometric structure ω. Assume* $\mathrm{Aut}(M, \omega)$ *is ergodic or has a dense orbit. Is M locally homogeneous?*

Gromov's theorem on the open dense implies that a dense open set in M is locally homogeneous even if ω is not homogeneous. The question is whether this homogeneous structure extends to all of M if ω is homogeneous. If ω is not homogeneous, the examples following Lemma 5 show that M need not be homogeneous.

It seems possible to approach the case of Conjecture 6.15 in which ω is homogeneous and D acts ergodically by answering Question 6.16 positively. At this point, one is left with the problem of classifying actions of higher rank groups and lattices on locally homogeneous manifolds. Induction easily reduces one to considering G actions. The techniques Gromov uses to produce the locally homogeneous structure gives slightly more precise information in this setting: one is left trying to classify homogeneous manifolds modeled on H/L in which G acts via an inclusion in H centralizing L. In fact this reduces one to problems about locally homogeneous manifolds studied by Zimmer and collaborators; see particularly [141, 142, 144, 256]. For a more general survey of locally homogeneous spaces, see also [135] and [136].

The following question concerns a possible connection between preserving a rigid geometric structure and having uniformly partially hyperbolic dynamics.

Question 6.17. *Let D be any group acting on a compact manifold M preserving a volume form and a rigid geometric structure. Is it true that if some element d of D has positive Lyapunov exponents, then d is uniformly partially hyperbolic?*

This question is of particular interest for us when D is a semisimple Lie group or a lattice, but would be interesting to resolve in general. The main reason to believe that the answer might be yes is that the action of D on the space of frames of M is proper. Having a proper action on the tangent bundle minus the zero section implies that an action is Anosov; see Mañé's article for a proof [155].

7. Topological superrigidity: Regularity results from hyperbolic dynamics and applications

A major area of research in the Zimmer program has been the application of hyperbolic dynamics. This area might be described by the following maxim: in the presence of hyperbolic dynamics, the straightening section from cocycle superrigidity is often more regular. Some major successes in this direction are the work of Goetze-Spatzier and Feres-Labourie. In the context of hyperbolic dynamical approaches, there are two main settings. In the first, one considers actions of higher rank lattices on a particular class of compact manifolds; in the second, one makes no assumption on the topology of the manifold acted upon.

In this section we discuss a few such results, after first recalling some facts about the stability of hyperbolic dynamical systems in subsection 7.1. We then discuss the best-known rigidity results for actions on tori in subsection 7.2 and after this discuss the work of Goetze-Spatzier and Feres-Labourie in a more general context in subsection 7.3.

The term "topological superrigidity" for this area of research was coined by Zimmer, whose early unpublished notes on the topic dramatically influenced research in the area [244].

This aspect of the Zimmer program has also given rise to a study of rigidity properties for other uniformly hyperbolic actions of large groups, most particularly higher rank abelian groups. See, e.g., [125, 129, 206].

The following subsection recalls basic notions from hyperbolic dynamics that are needed in this section.

7.1. Stability in hyperbolic dynamics

A diffeomorphism f of a manifold X is said to be *Anosov* if there exists a continuous f-invariant splitting of the tangent bundle $TX = E^u_f \oplus E^s_f$ and constants $a > 1$ and $C, C' > 0$ such that for every $x \in X$,

(1) $$\|D f^n(v^u)\| \geq Ca^n \|v^u\| \text{ for all } v^u \in E^u_f(x), \text{ and}$$

(2) $$\|D f^n(v^s)\| \leq C'a^{-n} \|v^s\| \text{ for all } v^s \in E^s_f(x).$$

We note that the constants C and C' depend on the choice of metric, and that a metric can always be chosen so that $C = C' = 1$. There is an analogous notion for a flow f_t, where $TX = T\mathcal{O} \oplus E^u_{f_t} \oplus E^s_{f_t}$, where $T\mathcal{O}$ is the tangent space to the flow direction and vectors in $E^u_{f_t}$ (resp. $E^s_{f_t}$) are uniformly expanded (resp. uniformly contracted) by the flow. This notion was introduced by Anosov and named after Anosov by Smale, who popularized the notion in the United States [3, 218]. One of the earliest results in the subject is Anosov's proof that Anosov diffeomorphisms are *structurally stable*, i.e., that any C^1 perturbation of an Anosov diffeomorphism is conjugate back to the original diffeomorphism by a homeomorphism. There is an analogous result for flows, though this requires that one introduce a notion of time change that we will not consider here. Since Anosov also showed that C^2 volume-preserving Anosov flows and diffeomorphisms are ergodic, structural stability implies the existence of an open set of "chaotic" dynamical systems.

The notion of an Anosov diffeomorphism has had many interesting generalizations, for example, axiom A diffeomorphisms, nonuniformly hyperbolic diffeomorphisms, and diffeomorphisms admitting a dominated splitting. A notion that has been particularly useful in the study of rigidity of group actions is the notion of a partially hyperbolic diffeomorphism as introduced by Hirsch, Pugh, and Shub. Under strong enough hypotheses, these diffeomorphisms have a weaker stability property similar to structural stability. More or less, the diffeomorphisms are hyperbolic relative to some foliation, and any nearby action is hyperbolic relative to some nearby foliation. To describe more precisely the class of diffeomorphisms we consider and the stability property they enjoy, we require some definitions.

The use of the word *foliation* varies with context. Here a *foliation by C^k leaves* will be a continuous foliation whose leaves are C^k injectively immersed submanifolds that vary continuously in the C^k topology in the transverse direction. To specify transverse regularity we will say that a foliation is transversely C^r. A foliation by C^k leaves that is transversely C^k is called simply a C^k foliation. (Note that our language does not agree with that in [117].)

Given an automorphism f of a vector bundle $E \to X$ and constants $a > b \geq 1$, we say f is (a, b) *partially hyperbolic* or simply *partially hyperbolic* if there is a metric on E, a constant $C \geq 1$, and a continuous f-invariant, nontrivial splitting $E = E_f^u \oplus E_f^c \oplus E_f^s$ such that for every x in X,

$$(1) \qquad \| f^n(v^u) \| \geq Ca^n \| v^u \| \text{ for all } v^u \in E_f^u(x),$$

$$(2) \qquad \| f^n(v^s) \| \leq C^{-1}a^{-n} \| v^s \| \text{ for all } v^s \in E_f^s(x), \text{ and}$$

$$(3) \qquad C^{-1}b^{-n} \| v^0 \| < \| f^n(v^0) \| \leq Cb^n \| v^0 \| \text{ for all } v^0 \in E_f^c(x), \text{ and all integers } n.$$

A C^1 diffeomorphism f of a manifold X is (a, b) *partially hyperbolic* if the derivative action Df is (a, b) partially hyperbolic on TX. We remark that for any partially hyperbolic diffeomorphism, there always exists an *adapted metric* for which $C = 1$. Note that E_f^c is called the *central distribution* of f, E_f^u is called the *unstable distribution* of f, and E_f^s the *stable distribution* of f.

Integrability of various distributions for partially hyperbolic dynamical systems is the subject of much research. The stable and unstable distributions are always tangent to invariant foliations that we call the stable and unstable foliations and denote by W_f^s and W_f^u. If the central distribution is tangent to an f-invariant foliation, we call that foliation a *central foliation* and denote it by W_f^c. If there is a unique foliation tangent to the central distribution, we call the central distribution *uniquely integrable*. For smooth distributions unique integrability is a consequence of integrability, but the central distribution is usually not smooth. If the central distribution of an (a, b) partially hyperbolic diffeomorphism f is tangent to an invariant foliation W_f^c, then we say f is r *normally hyperbolic to* W_f^c for any r such that $a > b^r$. This is a special case of the definition of r normally hyperbolic given in [117].

Before stating a version of one of the main results of [117], we need one more definition. Given a group G, a manifold X, two foliations $\mathfrak{F}$ and $\mathfrak{F}'$ of X, and two actions ρ and ρ' of G on X, such that ρ preserves $\mathfrak{F}$ and ρ' preserves $\mathfrak{F}'$, following [117] we call ρ and ρ' *leaf conjugate* if there is a homeomorphism h of X such that

(1) $h(\mathfrak{F}) = \mathfrak{F}'$ *and*

(2)   for every leaf $\mathfrak{L}$ of $\mathfrak{F}$ and every $g \in G$, we have $h(\rho(g)\mathfrak{L}) = \rho'(g)h(\mathfrak{L})$.

The map h is then referred to as a *leaf conjugacy* between $(X, \mathfrak{F}, \rho)$ and $(X, \mathfrak{F}', \rho')$. This essentially means that the actions are conjugate modulo the central foliations.

We state a special case of some of the results of Hirsch-Pugh-Shub on perturbations of partially hyperbolic actions of $\mathbb{Z}$; see [117]. There are also analogous definitions and results for flows. As these are less important in the study of rigidity, we do not discuss them here.

> **Theorem 7.1.** *Let f be an (a, b) partially hyperbolic C^k diffeomorphism of a compact manifold M that is k normally hyperbolic to a C^k central foliation W_f^c. Then for any $\delta > 0$, if f' is a C^k diffeomorphism of M that is sufficiently C^1 close to f, we have the following:*
>
> 1. *f' is (a', b') partially hyperbolic, where $|a - a'| < \delta$ and $|b - b'| < \delta$, and the splitting $TM = E_{f'}^u \oplus E_{f'}^c \oplus E_{f'}^s$, for f' is C^0 close to the splitting for f;*
> 2. *there exist f'-invariant foliations by C^k leaves $W_{f'}^c$ tangent to $E_{f'}^c$, which is close in the natural topology on foliations by C^k leaves to W_f^c;*
> 3. *there exists a (nonunique) homeomorphism h of M with $h(W_f^c) = W_{f'}^c$, and h is C^k along leaves of W_f^c; furthermore h can be chosen to be C^0 small and C^k small along leaves of W_f^c;*
> 4. *the homeomorphism h is a leaf conjugacy between the actions (M, W_f^c, f) and $(M, W_{f'}^c, f')$.*

Conclusion (1) is easy and probably older than [117]. One motivation for Theorem 7.1 is to study the stability of dynamical properties of partially hyperbolic diffeomorphisms. See the survey [26] by Burns, Pugh, Shub, and Wilkinson for more discussion of that and related issues.

7.2. Uniformly hyperbolic actions on tori

Many works have been written considering local rigidity of actions with some affine, quasi-affine, and generalized quasi-affine actions with hyperbolic behavior. For a discussion of this, we refer to [76]. Here we only discuss results that prove some sort of global rigidity of groups acting on manifolds. The first such results were contained in papers of Katok-Lewis and Katok-Lewis-Zimmer [127, 128]. As these are now special cases of later more general results, we do not discuss them in detail here.

In this section we discuss results that only provide continuous conjugacies to standard actions. This is primarily because these results are less technical and easier to state. In this context, one can improve regularity of the conjugacy given certain technical dynamical hypotheses on certain dynamical foliations.

Definition 7.2. *An action of a group Γ on a manifold M is weakly hyperbolic if there exist elements $\gamma_1, \ldots, \gamma_k$ each of which is partially hyperbolic such that the sum of the stable subbundles of the γ_i spans the tangent bundle to M at every point; i.e., $\sum_i E_{\gamma_i} = TM$.*

To discuss the relevant results we need a related topological notion that captures hyperbolicity at the level of fundamental group. This was introduced in [83] by Whyte and the author. If a group Γ acts on a manifold with torsion-free nilpotent fundamental group and the action lifts to the universal cover, then the action of Γ on $\pi_1(M)$ gives rise to an action of Γ on the Malcev completion N of $\pi_1(M)$ that is a nilpotent Lie group. This yields a representation of Γ on the Lie algebra $\mathbf{n}$.

Definition 7.3. *We say an action of Γ on a manifold M with nilpotent fundamental group is π_1 hyperbolic if for the resulting Γ representation on $\mathbf{n}$, we have finitely many elements $\gamma_1, \ldots, \gamma_k$ such that the sum of their eigenspaces with eigenvalue of modulus less than 1 is all of $\mathbf{n}$.*

One can make this definition more general by considering M for which $\pi_1(M)$ has a Γ-equivariant nilpotent quotient; see [83]. We now discuss results that follow by combining work of Margulis-Qian with later work of Schmidt and Fisher-Hitchman [78, 164, 208].

Theorem 7.4. *Let $M = N/\Lambda$ be a compact nilmanifold and let Γ be a lattice in a semisimple Lie group with property (T). Assume Γ acts on M such that the action lifts to the universal cover and is $\pi_1(M)$ hyperbolic; then the action is continuously semiconjugate to an affine action.*

This theorem was proven by Margulis and Qian, who noted that if the Γ action contained an Anosov element, then the conjugacy could be taken to be a homeomorphism. In [83], the author and Whyte point out that this theorem extends easily to the case of any compact manifold with torsion-free nilpotent fundamental group. In [83], we also discuss extensions to manifolds with fundamental group with a quotient that is nilpotent.

Margulis and Qian asked whether the assumption of π_1 hyperbolicity could be replaced by the assumption that the action on M was weakly hyperbolic. In the case of actions of

Kazhdan groups on tori, Schmidt proved that weak hyperbolicity implies π_1 hyperbolicity, yielding the following:

Theorem 7.5. *Let $M = \mathbb{T}^n$ be a compact torus and let Γ be a lattice in a semisimple Lie group with property (T). Any weakly hyperbolic Γ action M and that lifts to the universal cover is continuously semiconjugate to an affine action.*

Remarks 7.6.

1. The contribution of Fisher-Hitchman in both theorems is just in extending cocycle superrigidity to a wider class of groups, as discussed above.

2. The assumption that the action lifts to the universal cover of M is often vacuous because of results concerning cohomology of higher rank lattices. In particular, it is vacuous for cocompact lattices in simple Lie groups of real rank at least 3.

It remains an interesting, open question to take this result and prove that the semiconjugacy is always a conjugacy and is also always a smooth diffeomorphism.

7.3. Rigidity results for uniformly hyperbolic actions

We begin by discussing some work of Goetze and Spatzier. To avoid technicalities we only discuss some of their results. We begin with the following definition.

Definition 7.7. *Let $\rho : \mathbb{Z}^k \times M {\rightarrow} M$ be an action and $\gamma_1, \ldots, \gamma_l$ be a collection of elements that generate for $\mathbb{Z}^k$. We call ρ a Cartan action if*

1. *each $\rho(\gamma_i)$ is an Anosov diffeomorphism,*
2. *each $\rho(\gamma_i)$ has one-dimensional strongest stable foliation, and*
3. *the strongest stable foliations of the $\rho(\gamma_i)$ are pairwise transverse and span the tangent space to the manifold.*

It is worth noting that Cartan actions are very special in three ways. First we assume that a large number of elements in the acting group are Anosov diffeomorphisms, second we assume that each of these has one-dimensional strongest stable foliation, and lastly we assume these one-dimensional directions span the tangent space. All aspects of these assumptions are used in the following theorem. Reproving it even assuming two-dimensional strongest stable foliations would require new ideas.

Theorem 7.8. *Let G be a semisimple Lie group with all simple factors of real rank at least 2 and Γ in G a lattice. Then any volume-preserving Cartan action of Γ is smoothly conjugate to an affine action on an infranilmanifold.*

This is slightly different from the statement in Goetze-Spatzier, where they pass to a finite cover and a finite-index subgroup. It is not too hard to prove this statement from theirs. The proof spans the two papers [104] and [105]. The first paper [104] proves, in a somewhat more general context, that the π-simple section arising in the cocycle superrigidity theorem is in fact Hölder continuous. The second paper makes use of the resulting Hölder Riemannian metric in conjunction with ideas arising in other work of Katok and Spatzier to produce a smooth homogeneous structure on the manifold.

The work of Feres-Labourie differs from other work on rigidity of actions with hyperbolic properties in that it does not make any assumptions concerning the existence of invariant measures. Here we state only some consequences of their results, without giving the exact form of cocycle superrigidity that is their main result.

Theorem 7.9. *Let Γ be a lattice in $SL(n, \mathbb{R})$ for $n \geq 3$ and assume Γ acts smoothly on a compact manifold M of dimension n. Further assume that for the induced action $N = (G \times M)/\Gamma$ we have that*

1. *every $\mathbb{R}$-semisimple 1-parameter subgroup of G acts transitively on N and*
2. *some element g in G is uniformly partially hyperbolic with $E_s \oplus E_w$ containing the tangent space to M at any point;*

then M is a torus and the action on M is a standard affine action.

These hypotheses are somewhat technical and essentially ensure that one can apply the topological version of cocycle superrigidity proven in [60]. The proof also uses a deep result of Benoist and Labourie classifying Anosov diffeomorphisms with smooth stable and unstable foliations [10].

The nature of the hypotheses of Theorem 7.9 makes an earlier remark clear. Ideally one would only have hypotheses on the Γ action, but here we require hypotheses on the induced action instead. It is not clear how to reformulate these hypotheses on the induced action as hypotheses on the original Γ action.

Another consequence of the work of Feres and Labourie is a criterion for promoting invariance of rigid geometric structures. More precisely, they give a criterion for a G action to preserve a rigid geometric structure on a dense open subset of a manifold M provided a certain type of subgroup preserves a rigid geometric structure on M.

7.4. Conjectures and questions on uniformly hyperbolic actions

We begin with a very general variant of Conjecture 1.3.

Conjecture 7.10. *Let G be a semisimple Lie group all of whose simple factors have property (T), and let $\Gamma < G$ be a lattice. Assume G or Γ acts smoothly on a compact manifold*

M preserving volume such that some element g in the acting group is nontrivially uni-formly partially hyperbolic. Then the action is generalized quasi-affine.

There are several weaker variants on this conjecture, where, e.g., one assumes the action is volume preserving or weakly hyperbolic. Even the following much weaker variant seems difficult:

Conjecture 7.11. *Let $M = \mathbb{T}^n$ and Γ be as in Conjecture 7.10. Assume Γ acts on $\mathbb{T}^n$ weakly hyperbolicly and preserving a smooth measure. Then the action is affine.*

This conjecture amounts to conjecturing that the semiconjugacy in Theorem 7.5 is a dif-feomorphism. In the special case in which some element of Γ is Anosov, the semiconjugacy is at least a homeomorphism. If this is true and enough dynamical foliations are one and two dimensional and Γ has higher rank, one can then deduce smoothness of the conjugacy from work of Rodriguez-Hertz on rigidity of actions of abelian groups [206]. Work in progress by the author, Kalinin, and Spatzier seems likely to provide a similar result when Γ contains many commuting Anosov diffeomorphisms without any assumptions on dimensions of foliations.

Finally, we recall an intriguing question from [83] that arises in this context.

Question 7.12. *Let M be a compact manifold with $\pi_1(M) = \mathbb{Z}^n$ and assume $\Gamma < SL(n, \mathbb{Z})$ has finite index. Let Γ act on M fixing a point so that the resulting Γ action on $\pi_1(M)$ is given by the standard representation of $SL(n, \mathbb{Z})$ on $\mathbb{Z}^n$. Is it true that $\dim(M) \geq n$?*

The question is open even if the action on M is assumed to be smooth. The results of [83] imply that there is a continuous map from M to $\mathbb{T}^n$ that is equivariant for the standard Γ action on M. Since the image of M is closed and invariant, it is easy to check that it is all of $\mathbb{T}^n$. So the question amounts to one about the existence of equivariant "space-filling curves," where curve is taken in the generalized sense of continuous map from a lower dimensional manifold. In another context there are equivariant space-filling curves, but they seem quite special. They arise as surface group equivariant maps from the circle to S^2 and come from 3-manifolds that fiber over the circle; see [33].

8. Representations of fundamental groups and arithmetic quotients

In this section we discuss some results and questions related to topological approaches to classifying actions, particularly some related to Conjecture 1.4. In the second subsection, we also discuss related results and questions concerning maximal generalized affine quotients of actions.

8.1. Linear images of fundamental groups

This section is fundamentally concerned with the following question:

Question 8.1. *Let G be a semisimple Lie group all of whose factors are not compact and have property (T). Assume G acts by homeomorphisms on a manifold M preserving a measure. Can we classify linear representations of $\pi_1(M)$? Similarly for actions of $\Gamma < G$ a lattice on a manifold M.*

We remark that in this context, it is possible that $\pi_1(M)$ has no infinite-image linear representations; see discussion in [83] and section 9 below. In all known examples in which this occurs there is an "obvious" infinite-image linear representation on some finite-index subgroup. Also as first observed by Zimmer [251] for actions of Lie groups, under mild conditions on the action, representations of the fundamental group become severely restricted. Further work in this direction was done by Zimmer in conjunction with Spatzier and later Lubotzky [151, 152, 221]. Analogous results for lattices are surprisingly difficult in this context and constitute the author's dissertation [72, 73].

We recall a definition from [251]:

Definition 8.2. *Let D be a Lie group and assume that D acts on a compact manifold M preserving a finite measure μ and ergodically. We call the action engaging if the action of $\tilde{D}$ on any finite cover of M is ergodic.*

There is a slightly more technical definition of engaging for nonergodic actions that says there is no loss of ergodicity on passing to finite covers, i.e., that the ergodic decomposition of μ and its lifts to finite covers are canonically identified by the covering map. There are also two variants of this notion, *totally engaging* and *topologically engaging*; see, e.g., [258] for more discussion.

It is worth noting that for ergodic D actions on M, the action on any finite cover has at most finitely many ergodic components, in fact at most the degree of the cover ergodic components. We remark here that any generalized affine action of a Lie group is engaging if it is ergodic. The actions constructed below in section 9 are not in general engaging.

Theorem 8.3. *Let G be a simple Lie group of real rank at least 2. Assume G acts by homeomorphisms on a compact manifold M, preserving a finite measure μ and engaging. Assume $\sigma : \pi_1(M) \to GL(n, \mathbb{R})$ is an infinite-image linear representation. Then $\sigma(\pi_1(M))$ contains an arithmetic group $H_{\mathbb{Z}}$ where $H_{\mathbb{R}}$ contains a group locally isomorphic to G.*

For an expository account of the proof of this theorem and a more detailed discussion of engaging conditions for Lie groups, we refer the reader to [258].

The extension of this theorem to lattice actions is nontrivial and is in fact the author's dissertation. Even the definition of engaging requires modification, since it is not at all clear that a discrete group action lifts to the universal cover.

> **Definition 8.4.** *Let D be a discrete group and assume that D acts on a compact manifold M preserving a finite measure μ and ergodically. We call the action engaging if for every*
> 1. *finite-index subgroup D' in D,*
> 2. *finite cover M' of M, and*
> 3. *lift of the D' action to M'*
> *the action of D' on M' is ergodic.*

The definition does immediately imply that every finite-index subgroup D' of D acts ergodically on M. In [73], a definition is given that does not require the D action to be ergodic, but even in that context the ergodic decomposition for D' is assumed to be the same as that for D. Definition 8.4 is rigged to guarantee the following lemma:

> **Lemma 8.5.** *Let G be a Lie group and Γ < G a lattice. Assume Γ acts on a manifold M preserving a finite measure and engaging; then the induced G action on $(G \times M)/\Gamma$ is engaging.*

We remark that with our definitions here, the lemma only makes sense for Γ cocompact, but this is not an essential difficulty. We can now state a first result for lattice actions. We let $\Lambda = \pi_1((G \times M)/\Gamma)$.

> **Theorem 8.6.** *Let G be a simple Lie group of real rank at least 2 and Γ < G a lattice. Assume Γ acts by homeomorphisms on a compact manifold M, preserving a finite measure μ and engaging. Assume $\sigma : \Lambda \to GL(n, \mathbb{R})$ is a linear representation whose restriction to $\pi_1(M)$ has infinite image. Then $\sigma(\pi_1(M))$ contains an arithmetic group $H_\mathbb{Z}$ where $Aut(H_\mathbb{R})$ contains a group locally isomorphic to G.*

This theorem is proven by inducing actions, applying Theorem 8.3, and analyzing the resulting output carefully. One would like to assume σ a priori only defined on $\pi_1(M)$, but there seems no obvious way to extend such a representation to the linear representation of Λ required by Theorem 8.3 without a priori information on Λ. This is yet another example of the difficulties in using induction to study lattice actions. As a consequence of Theorem 8.6, we discover that at least $\sigma(\Lambda)$ splits as a semidirect product of Γ and $\sigma(\pi_1(M))$. But this is not clear a priori and not clear a posteriori for Λ.

8.2. Arithmetic quotients

In the context of Theorems 8.3 and 8.6, one can obtain much more dynamical information concerning the relation of $H_\mathbb{Z}$ and the dynamics of the action on M. In particular, there is a compact subgroup $C < H_\mathbb{R}$ and a measurable equivariant map $\phi : M \to C \backslash H_\mathbb{R} / H_\mathbb{Z}$, which we refer to as a *measurable arithmetic quotient*. The papers [72] and [151] prove that there is always a canonical maximal quotient of this kind for any action of G or Γ on any compact manifold, essentially by using Ratner's theorem to prove that every pair of arithmetic quotients are dominated by a common, larger arithmetic quotient. Earlier results of Zimmer also obtained arithmetic quotients, but only under the assumption that there was an infinite-image linear representation with discrete image [254]. The results we mention from [73] and [152] then show that there are "lower bounds" on the size of this arithmetic quotient, provided that the action is engaging, in terms of the linear representations of the fundamental group of the manifold. In particular, one obtains arithmetic quotients in which $H_\mathbb{R}$ and $H_\mathbb{Z}$ are essentially determined by $\sigma(\pi_1(M))$. In fact $\sigma(\pi_1(M))$ contains $H_\mathbb{Z}$ and is contained in $H_\mathbb{Q}$.

The book [258] provides a good description of how to produce arithmetic quotients for G actions and the article [74] provides an exposition of the relevant constructions for Γ actions.

Under even stronger hypotheses, the papers [84] and [257] imply that the arithmetic quotient related to the "Gromov representation" discussed above in subsection 6.2 has the same entropy as the original action. This means that, in a sense, the arithmetic quotient captures most of the dynamics of the original action.

8.3. Open questions

As promised in the introduction, we have the following analogue of Conjecture 1.4 for lattice actions.

Conjecture 8.7. *Let Γ be a lattice in a semisimple group G with all simple factors of real rank at least 2. Assume Γ acts faithfully, preserving volume on a compact manifold M. Further assume the action is not isometric. Then $\pi_1(M)$ has a finite-index subgroup Λ such that Λ surjects onto an arithmetic lattice in a Lie group H in which $\mathrm{Aut}(H)$ locally contains G.*

The following questions about arithmetic quotients are natural.

Question 8.8. *Let G be a simple Lie group of real rank at least 2 and $\Gamma < G$ a lattice. Assume that G or Γ act faithfully on a compact manifold M, preserving a smooth volume.*
   1. *Is there a nontrivial measurable arithmetic quotient?*
   2. *Can we take the quotient map ϕ smooth on an open dense set?*

Due to a construction in [83], one cannot expect that every M admitting a volume-preserving G or Γ action has an arithmetic quotient that is even globally continuous or has

$\pi_1(M)$ admitting an infinite-image linear representation. However, the difficulties created in those examples all vanish upon passage to a finite cover.

Question 8.9. *Let G be a simple Lie group of real rank at least 2 and $\Gamma < G$ a lattice. Assume that G or Γ acts smoothly on a compact manifold M, preserving a smooth volume.*

1. *Is there a finite cover of M' of M such that $\pi_1(M')$ admits an infinite-image linear representation?*

2. *Can we find a finite cover M' of M, a lift of the action to M' (on a subgroup of finite index), and an arithmetic quotient in which the quotient map ϕ is continuous and smooth on an open dense set?*

The examples discussed in the next subsection imply that ϕ is at best Hölder continuous globally. It is not clear whether the questions and conjectures we have just discussed are any less reasonable for $SP(1,n)$, F_4^{-20}, and their lattices.

9. Exotic volume-preserving actions

In this section, I discuss what is known about what are typically called "exotic actions." These are the only known smooth volume-preserving actions of higher rank lattices and Lie groups that are not generalized affine algebraic. These actions make it clear that a clean classification of volume-preserving actions is out of reach. In particular, these actions have continuous moduli and provide counterexamples to any naive conjectures of the form "the moduli spaces of actions of some lattice Γ on any compact manifold M are countable." In particular, I explain examples of actions of either Γ or G that have large continuous moduli of deformations as well as manifolds in which these moduli have multiple connected components.

Essentially all of the examples given here derive from the simple construction of "blowing up" a point or a closed orbit, which was introduced to this subject in [127]. The further developments after that result are all further elaborations on one basic construction. The idea is to use the "blowup" construction to introduce distinguished closed invariant sets that can be varied in some manner to produce deformations of the action. The "blowup" construction is a classical tool from algebraic geometry that takes a manifold N and a point p and constructs from them a new manifold N' by replacing p by the space of directions at p. Let $\mathbb{R}P^l$ be the l-dimensional projective space. To blow up a point, we take the product of $N \times \mathbb{R}P^{\dim(N)}$ and then find a submanifold in which the projection to N is a diffeomorphism off of p and the fiber of the projection over p is $\mathbb{R}P^{\dim(N)}$. For detailed discussion of this construction we refer the reader to any reasonable book on algebraic geometry.

The easiest example to consider is to take the action of $SL(n,\mathbb{Z})$ or any subgroup $\Gamma < SL(n,\mathbb{Z})$ on the torus $\mathbb{T}^n$ and blow up the fixed point, in this case the equivalence class of the origin in $\mathbb{R}^n$. Call the resulting manifold M. Provided Γ is large enough, e.g., Zariski dense in $SL(n, \mathbb{R})$, this action of Γ does not preserve the measure defined by any volume form on M. A clever construction introduced in [127] shows that one can alter the standard blowing-up procedure in order to produce a one-parameter family of $SL(n, \mathbb{Z})$ actions on M, only one of which preserves a volume form. This immediately shows that this action on M admits perturbations, since it cannot be conjugate to the nearby, non-volume-preserving actions. Essentially, one constructs different differentiable structures on M that are diffeomorphic but not equivariantly diffeomorphic.

After noticing this construction, one can proceed to build more complicated examples by passing to a subgroup of finite index, and then blowing up several fixed points. One can also glue together the "blown-up" fixed points to obtain an action on a manifold with more complicated topology. In particular, one can achieve a fundamental group that is an essentially arbitrary free product with amalgamation or HNN extension of the fundamental group of the original manifold over the fundamental group of (blown-up) orbits. In the context described in more detail below, of blowing up along closed orbits instead of points, it is not hard to do the blowing up and gluing in way that guarantees that there are no linear representations of the fundamental group of the "exotic example." To prove the nonexistence of linear representations, one chooses examples in which all groups involved are higher rank lattices and have very constrained linear representation theory. See [83, 127] for discussion of the topological complications one can introduce.

While these actions do not preserve a rigid geometric structure, they do preserve a slightly more general object, an *almost*-rigid structure introduced by Benveniste and the author in [11] and described below.

In [12] it is observed that a similar construction can be used for the action of a simple group G by left translations on a homogeneous space H/Λ, where H is a Lie group containing G and $\Lambda < H$ is a cocompact lattice. Here we use a slightly more involved construction from algebraic geometry, and "blow up" the directions normal to a closed submanifold; i.e., we replace some closed submanifold N in H/Λ by the projective normal bundle to N. In all the cases we consider here, this normal bundle is trivial and so is just $N \times \mathbb{R}P^l$, where $l = \dim(H) - \dim(N)$.

Benveniste used his construction to produce more interesting perturbations of actions of higher rank simple Lie group G or a lattice Γ in G. In particular, he produced volume-preserving actions that admit volume-preserving perturbations. He does this by choosing $G < H$ such that not only are there closed G orbits but the centralizer $Z = Z_H(G)$ of G in H has no trivial connected component. If we take a closed G orbit N, then any translate zN for z in Z is also closed and so we have a continuum of closed G orbits. Benveniste shows that if we choose two closed orbits N and zN to blow up and glue, and then vary z in a small open set, the resulting actions can only be conjugate for a countable set of choices of z.

This construction is further elaborated in [77]. Benveniste's construction is not optimal in several senses, nor is his proof of rigidity. In [77], I give a modification of the construction that produces nonconjugate actions for every choice of z in a small enough neighborhood. By blowing up and gluing more pairs of closed orbits, this allows me to produce actions in which the space of deformations contains a submanifold of arbitrarily high, finite dimension. Further, Benveniste's proof that the deformations are nontrivial is quite involved and applies only to higher rank groups. In [77], I give a different proof of nontriviality of the deformations, using consequences of Ratner's theorem due to Shah and Witte Morris [200, 210, 235]. This shows that the construction produces nontrivial perturbations for any semisimple G and any lattice Γ in G.

As mentioned above, in [11] we show that none of these actions preserve any rigid geometric structure in the sense of Gromov but that they do preserve a slightly more complicated object that we call an *almost-rigid structure*. Both rigid and almost-rigid structures are easiest to define in dimension 1. In this context a rigid structure is a nonvanishing vector field and an almost-rigid structure is a vector field vanishing at isolated points and to finite degree.

We continue to use the notation of subsection 6.1 in order to give the precise definition of an almost-rigid geometric structure.

Definition 9.1. *An A-structure ϕ is called (j, k) almost rigid (or just almost rigid) if for every point p, $r_p^{k,k-1}$ is injective on the subgroup $r^{k+j,k}(I\,s^{k+j}) \subset I\,s^k$.*

Thus k-rigid structures are the $(0, k)$-almost-rigid structures.

Basic example: Let V be an n-dimensional manifold. Let $X_1, \ldots, X_n$ be a collection of vector fields on M. This defines an A-structure ψ of type $\mathbb{R}^{n^2}$ on M. If $X_1, \ldots, X_n$ span the tangent space of V at every point, then the structure is rigid in the sense of Gromov. Suppose instead that there exists a point p in V and $X_1 \wedge \ldots \wedge X_n$ vanishes to order $\leq j$ at p in V. Then ψ is a $(j, 1)$-almost-rigid structure. Indeed, let $p \in M$, and let $(x_1, \ldots, x_n)$ be coordinates around p. Suppose that in terms of these coordinates, $X_l = a_l^m \frac{\partial}{\partial x_j}$. Suppose that $f \in I\,s_p^{j+1}$. We must show that $r_p^{j+1,1}(f)$ is trivial. Let $(f^1, \ldots, f^n)$ be the coordinate functions of f. Then $f \in I\,s_p^{j+1}$ implies that

(2) $$a_k^l - a_k^m \frac{\partial f^l}{\partial x^m}$$

vanishes to order $j + 1$ at p for all k and l. Let (b_k^l) be the matrix so that $b_k^m a_m^l = \det(a_r^s)\delta_k^l$. Multiplying expression (2) by (b_k^l), we see that $\det(a_r^s)(\delta_k^l - \frac{\partial f^l}{\partial x^k})$ vanishes to order $j + 1$. But since by assumption $\det(a_r^s)$ vanishes to order $\leq j$, this implies that $(\partial f^l / \partial x^k)(p) = \delta_k^l$, so $r_p^{j+1,1}(f)$ is the identity, as required.

If confused by the notation, the interested reader may find it enlightening to work out the basic example in the trivial case $n = 1$. Similar arguments can be given to show that frames that degenerate to subframes are also almost rigid, provided the order of vanishing of the form defining the frame is always finite.

Question 9.2. *Does any smooth (or analytic) action of a higher rank lattice Γ admit a smooth (analytic) almost-rigid structure in the sense of [11]? More generally does such an action admit a smooth (analytic) rigid geometric structure on an open dense set of full measure?*

This question is, in a sense, related to the discussion above about regularity of the straightening section in cocycle superrigidity. In essence, cocycle superrigidity provides one with a measurable invariant connection and what one wants to know is whether one can improve the measurable connection to a smooth geometric structure with some degeneracy on a small set. The examples described in this section show, among other things, that one cannot expect the straightening section to be smooth in general, though one might hope it is smooth in the complement of a closed submanifold of positive codimension or at least on an open dense set of full measure.

We remark that there are other possible notions of almost-rigid structures. See Dumitrescu [47] for a detailed discussion of a different useful notion in the context of complex analytic manifolds. Dumitrescu's notion is strictly weaker than the one presented here.

10. Non-volume-preserving actions

This section describes what is known for non-volume-preserving actions. The first two subsections describe examples that show that a classification in this setting is not in any sense possible. The last subsection describes some recent work of Nevo and Zimmer that proves surprisingly strong rigidity results in special settings.

10.1. Stuck's examples

The following observations are from Stuck's paper [223]. Let G be any semisimple group. Let P be a minimal parabolic subgroup. Then there is a homomorphism $P \to \mathbb{R}$. As in the proof of Theorem 2.7, one can take any $\mathbb{R}$ action, view it is a P action, and induce to a G action. If we take an $\mathbb{R}$ action on a manifold M, then the induced action takes place on $(G \times M)/P$. We remark that the G action here is not volume preserving, simply because the G action on G/P is proximal. The same is true of the restriction to any Γ action when Γ is a lattice in G.

This construction shows that classifying G actions on all compact manifolds implicitly involves classifying all vector fields on compact manifolds. It is relatively easy to convince oneself that there is no reasonable sense in which the moduli space of vector fields can be understood up to smooth conjugacy.

10.2. Weinberger's examples

This is a variant on the Katok-Lewis examples obtained by blowing up a point and is similar to a blowing-up construction common in foliation theory. The idea is that one takes an action of a subgroup Γ of $SL(n, \mathbb{Z})$ on $M = \mathbb{T}^n$, removes a fixed or periodic point p, retracts onto a manifold with boundary $\bar{M}$, and then glues in a copy of the $\mathbb{R}^n$ compactified at infinity by the projective space of rays. It is relatively easy to check that the resulting space admits a continuous Γ action and even that there are many invariant measures for the action, but no invariant volume. One can also modify this construction by doing the same construction at multiple fixed or periodic points simultaneously and by doing more complicated gluings on the resulting $\bar{M}$.

This construction is discussed in [54] and a variant for abelian group actions is discussed in [124]. In the abelian case it is possible to smooth the action, but this does not seem to be the case for actions of higher rank lattices.

As far as I can tell, there is no obstruction to repeating this construction for closed orbits as in the case of the algebro-geometric blowup, but this does not seem to be written formally anywhere in the literature.

Also, a recent construction of Hurder shows that one can iterate this construction infinitely many times, taking retracts in smaller and smaller neighborhoods of periodic points of higher and higher orders. The resulting object is a kind of fractal admitting an $SL(n, \mathbb{Z})$ action [118].

10.3. Work of Nevo-Zimmer

In this subsection, we describe some work of Nevo and Zimmer from the sequence of papers [183, 184, 187, 182]. For more detailed discussion see the survey by Nevo and Zimmer [185] as well as the following two articles for related results [186, 188].

Given a group G acting on a space X and a measure μ on G, we call a measure ν on X *stationary* if $\mu*\nu = \nu$. We will only consider the case in which the group generated by the support μ is G; such measures are often called *admissible*. This is a natural generalization of the notion of an invariant measure. If G is an amenable group, any action of G on a compact metric space admits an invariant measure. If G is not amenable, invariant measures need not exist, but stationary measures always do. We begin with the following cautionary example:

Example 10.1. *Let $G = SL(n, \mathbb{R})$ acting on $\mathbb{R}^{n+1}$ by the standard linear action on the first n coordinates and the trivial action on the last. Then the corresponding G action on*

$\mathbb{P}(\mathbb{R}^{n+1})$ *has the property that any stationary measure is supported either on the subspace* $\mathbb{P}(\mathbb{R}^n)$ *given by the first n coordinates or on the subspace* $\mathbb{P}(\mathbb{R})$ *given by zeroing the first n coordinates.*

The proof of this assertion is an easy exercise. The set $\mathbb{P}(\mathbb{R})$ is a collection of fixed points, so it clearly admits invariant measures. The orbit of any point in $\mathbb{P}(\mathbb{R}^{n+1})\backslash\mathbb{P}(\mathbb{R}^n)\sqcup\mathbb{P}(\mathbb{R}))$ is $SL(n,\mathbb{R})/(SL(n,\mathbb{R})\ltimes\mathbb{R}^n)$. It is straightforward to check that no stationary measures can be supported on unions of sets of this kind. This fact should not be a surprise as it generalizes the fact that invariant measures for amenable group actions are often supported on minimal sets.

To state the results of Nevo and Zimmer, we need a slightly stronger notion of admissibility. We say a measure μ on a locally compact group G is *strongly admissible* if the support of μ generates G and μ^{*k} is absolutely continuous with respect to Haar measure on G for some positive k. In the papers of Nevo and Zimmer, this stronger notion is called admissible.

Theorem 10.2. *Let X be a compact G space where G is a semisimple Lie group with all factors of real rank at least 2. Then for any admissible measure μ on G and any μ-stationary measure ν on X, we have either*

1. *ν is G invariant or*
2. *there is a nontrivial measurable quotient of the G space (X, ν) that is of the form $(G/Q, Lebesgue)$, where $Q \subset G$ is a parabolic subgroup.*

The quotient space G/Q is called a *projective quotient* in the work of Nevo and Zimmer. Theorem 10.2 is most interesting for us for minimal actions, where the measure ν is necessarily supported on all of X and the quotient therefore reflects the Γ action on X and not some smaller set. Example 10.1 indicates the reason to be concerned, since there the action on the larger projective space is not detected by any stationary measure.

We remark here that Feres and Ronshausen have introduced some interesting ideas for studying group actions on sets not contained in the support of any stationary measure [68]. Similar ideas are developed in a somewhat different context by Deroin, Kleptsyn, and Navas [44].

In [182], Nevo and Zimmer show that for smooth non-measure-preserving actions on compact manifolds that also preserve a rigid geometric structure, one can sometimes prove the existence of a projective factor in which the factor map is smooth on an open dense set.

We also want to mention the main result of Stuck's paper [223], which we have already cited repeatedly for the fact that non-volume-preserving actions of higher rank groups on compact manifolds cannot be classified.

Theorem 10.3. *Let G be a semisimple Lie group with finite center. Assume G acts minimally by homeomorphisms on a compact manifold M. Then either the action is locally* *free or the action is induced from a minimal action by homeomorphisms of a proper parabolic subgroup of G on a manifold N.*

Stuck's theorem is proven by studying the Gauss map from M to the Grassman variety of subspaces of $\mathfrak{g}$ defined by taking a point to the Lie algebra of its stabilizer. This technique also plays an important role in the work of Nevo and Zimmer.

11. Groups that do act on manifolds

Much of the work discussed so far raises an obvious question: Are there many interesting subgroups of Diff(M) for a general M? So far we have only seen "large" subgroups of Diff(M) that arise in a geometric fashion, from the presence of a connected Lie group in either Diff(M) or Diff($\tilde{M}$). In this section, we describe two classes of examples that make it clear that other phenomena exist. The following problem, however, seems open:

Problem 11.1. *For a compact manifold M with a volume form ω, construct a subgroup $\Gamma < \text{Diff}(M, \omega)$ such that Γ has no linear representations.*

An example that is not often considered in this context is that Aut(F_n) acts on the space Hom(F_n, K) for K any compact Lie group. Since Hom(F_n, K) $\simeq K^n$ this defines an action of Aut(F_n) on a manifold. This action clearly preserves the Haar measure on K^n; see [106]. This action is not very well studied; we know of only [75, 96]. Similar constructions are possible, and better known, with mapping class groups, although in that case one obtains a representation variety that is not usually a manifold. Since Aut(F_n) has no faithful linear representations, this yields an example of truly "nonlinear" action of a large group. This action is still very special and one expects many other examples of nonlinear actions. We now describe two constructions that yield many examples if we drop the assumption that the action preserves volume.

11.1. Thompson's groups

Richard Thompson introduced a remarkable family of groups, now referred to as Thompson's groups. These come in various flavors and have been studied from several points of view; see, e.g., [32, 116]. For our purposes, the most important of these groups are the ones typically denoted T. One description of this group is the collection of piecewise linear diffeomorphisms of a the circle in which the break points and slopes are all dyadic rationals. (One can replace the implicit 2 here with other primes and obtain similar, but different, groups.) We record here two important facts about this group T of piecewise linear homeomorphisms of S^1.

Theorem 11.2. (Thompson; see [32]). *The group T is simple.*

Theorem 11.3. (Ghys-Sergiescu [102]). *The defining piecewise linear action of the group T on S^1 is conjugate by a homeomorphism to a smooth action.*

These two facts together provide us with a rather remarkable class of examples of groups that act on manifolds. As finitely generated linear groups are always residually finite, the group T has no linear representations whatsoever. A simpler variant of Problem 11.1 is as follows:

Problem 11.4. *Does T admit a volume-preserving action on a compact manifold?*

It is easy to see that compactness is essential in this question. We can construct a smooth action of T on $S^1 \times \mathbb{R}$ simply by taking the Ghys-Sergiescu action on S^1 and acting on $\mathbb{R}$ by the inverse of the derivative cocycle. Replacing the derivative cocycle with the Jacobian cocycle, this procedure quite generally converts non-volume-preserving actions on compact manifolds to volume-preserving ones on noncompact manifolds, but we know of no real application of this in the present context. Another variant of Problem 11.4 is

Problem 11.5. *Given a compact manifold M and a volume form ω does $\mathrm{Diff}(M, \omega)$ contains a finitely generated, infinite discrete simple group?*

This question is reasonable for any degree of regularity on the diffeomorphisms.

In Ghys's survey on groups acting on the circle, he points out that Thompson's group can be realized as piecewise $SL(2, \mathbb{Z})$ homeomorphisms of the circle [101]. In [69], Whyte and the author point out that the group of piecewise $SL(n, \mathbb{Z})$ maps on either the torus or the real projective space is quite large. The following are natural questions; see [69] for more discussion.

Question 11.6. *Are there interesting finitely generated or finitely presented subgroups of piecewise $SL(n, \mathbb{Z})$ maps on $\mathbb{T}^n$ or $\mathbb{P}^{n-1}$? Can any such group that is not a subgroup of $SL(n, \mathbb{Z})$ be made to act smoothly?*

11.2. Highly distorted subgroups of Diff(S^2)

In [29], Calegari and Freedman construct a very interesting class of subgroups of $\mathrm{Diff}^\infty(S^2)$. Very roughly, they prove the following:

Theorem 11.7. *There is a finitely generated subgroup G of $\mathrm{Diff}^\infty(S^2)$ that contains a rotation r as an arbitrarily distorted element.*

Here by *arbitrarily distorted*, we mean that we can choose the group G so that the function $f(n) = \|r^n\|_G$ grows more slowly than any function we choose. It is well known that for linear groups, the function $f(n)$ is at worst a logarithm, so this theorem immediately implies that we can find G with no faithful linear representations. This also answers a question raised by Franks and Handel in [87].

More recently, Avila has constructed similar examples in $\mathrm{Diff}^\infty(S^1)$ [4]. This answers a question raised in [29], where such subgroups were constructed in $\mathrm{Diff}^1(S^1)$.

We are naturally led to the following questions. We say a diffeomorphism has *full support* if the complement of the fixed set is dense.

Question 11.8. *For which compact manifolds M does $\mathrm{Diff}^\infty(M)$ contain arbitrarily distorted elements of full support? The same question for $\mathrm{Diff}^\omega(M)$. The same question for $\mathrm{Diff}^\infty(M, \nu)$, where ν is a volume form on M. For the second two questions, we can drop the hypothesis of full support.*

The second and third questions here seem quite difficult and the answer could conceivably be "none." The only examples for which anything is known in the volume-preserving setting are compact surfaces of genus at least 1, where no element is more than quadratically distorted by a result of Polterovich [192]. However this result depends heavily on the fact that in dimension 2, preserving a volume form is the same as preserving a symplectic structure.

11.3. Topological construction of actions on high-dimensional spheres

In this subsection, we recall a construction due to Farrell and Lafont that allows one to construct actions of a large class of groups on closed disks, and so by doubling, to construct actions on spheres [58]. The construction yields actions on very high dimensional disks and spheres and is only known to produce actions by homeomorphisms.

The class of groups involved is the set of groups that admit an *EZ*-structure. This notion is a modification of an earlier notion due to Bestvina of a Z-structure on a group [15]. We do not recall the precise definition here, but refer the reader to the introduction to [58] and remark here that both torsion-free Gromov hyperbolic groups and CAT(0) groups admit *EZ*-structures.

The result that concerns us is the following:

Theorem 11.9. *Given a group Γ with an EZ-structure, there is an action of Γ by homeomorphisms on a closed disk.*

In fact, Farrell and Lafont give a fair amount of information concerning their actions, which are quite different from the actions we are usually concerned with here. In particular, the action is properly discontinuous off a closed subset Δ of the boundary of the disk. So from a dynamical viewpoint Δ carries the "interesting part" of the action, e.g., is the support of any Γ stationary measure. Farrell and Lafont point out an analogy between their construction and the action of a Kleinian group on the boundary of hyperbolic space. An interesting general question is as follows:

Question 11.10. *When can the Farrell-Lafont action on a disk or sphere be chosen smooth?*

12. Rigidity for other classes of acting groups

In this section, we collect some results and questions concerning actions of other classes of groups. In almost all cases, little is known.

12.1. Lattices in semidirect products

While it is not reasonable to expect classification results for arbitrary actions of *all* lattices in *all* Lie groups, there are natural broader classes to consider. To pick a reasonable class, a first guess is to try to exclude Lie groups whose lattices have homomorphisms onto $\mathbb{Z}$ or larger free groups. There are many such groups. For example, the groups $Sl(n, \mathbb{Z}) \ltimes \mathbb{Z}^n$ that are lattices in $Sl(n, \mathbb{R}) \ltimes \mathbb{R}^n$ have property (T) as soon as $n > 2$. As it turns out, a reasonable setting is to consider perfect Lie groups with no compact factors or factors locally isomorphic to $SO(1, n)$ or $SU(1, n)$. Any such Lie group will have property (T) and therefore so will its lattices. Many examples of such lattices are described in [228]. Some first rigidity results for these groups are contained in Zimmer's paper [252]. The relevant full generalization of the cocycle superrigidity theorem is contained in [236]. In this context it seems that there are probably many rigidity theorems concerning actions of these groups already implicit in the literature, following from the results in [236] and existing arguments.

12.2. Universal lattices

As mentioned above in subsection 4.7, the groups $SL(n, \mathbb{Z}[X])$ for $n > 2$ have property (T) by a result of Shalom [213]. His proof also works with larger collections of variables and some other arithmetic groups. The following is an interesting problem.

Problem 12.1. *Prove cocycle superrigidity for universal lattices.*

For clarity, we indicate that we expect that all linear representations and cocycles (up to "compact noise") will be described in terms of representations of the ambient groups, e.g., $SL(n, \mathbb{R})$, and will be determined by specifying a numerical value of X.

Some partial results toward superrigidity are known by Farb [51] and Shenfeld [215]. The problem of completely classifying linear representations in this context does seem to be open.

12.3. $SO(1,n)$ and $SU(1,n)$ and their lattices

It is conjectured that all lattices in $SO(1,n)$ and $SU(1,n)$ admit finite-index subgroups with surjections to $\mathbb{Z}$; see [17]. The conjecture is usually attributed to Thurston and sometimes to Borel for the case of $SU(1,n)$. If this is true, it immediately implies that actions of those lattices can never be classified.

There are still some interesting results concerning actions of these lattices. In [211], Shalom places restrictions on the possible actions of $SO(1,n)$ and $SU(1,n)$ in terms of the fundamental group of the manifold acted upon. This work is similar in spirit to work of Lubotzky and Zimmer described in section 8, but requires more restrictive hypotheses.

In [77], the author exhibits large moduli spaces of ergodic affine algebraic actions constructed for certain lattices in $SO(1,n)$. These moduli spaces are, however, all finite dimensional. In [70], I construct infinite-dimensional moduli of deformations of an isometric action of $SO(1,n)$. Both of these constructions rely on a notion of bending introduced by Johnson and Millson in the finite-dimensional setting [120].

I do not formulate any precise questions or conjectures in this direction as I am not sure what phenomenon to expect.

12.4. Lattices in other locally compact groups

Much recent work has focused on developing a theory of lattices in locally compact groups other than Lie groups. This theory is fully developed for algebraic groups over other local fields. Though we did not mention it here, some of Zimmer's own conjectures were made in the context of S-arithmetic groups, i.e., lattices in products of real and p-adic Lie groups. The following conjecture is natural in this context and does not seem to be stated in the literature.

> **Conjecture 12.2.** *Let G be a semisimple algebraic group defined over a field k of positive characteristic and $\Gamma < G$ a lattice. Further assume that all simple factors of G have k-rank at least 2. Then any Γ action on a compact manifold factors through a finite quotient.*

The existence of a measurable invariant metric in this context should be something one can deduce from the cocycle superrigidity theorems, though it is not clear that the correct form of these theorems is known or in the literature.

There is also a growing interest in lattices in locally compact groups that are not algebraic. We remark here that Kac-Moody lattices typically admit no nontrivial homomorphisms even to Homeo(M); see [82] for a discussion.

The only other interesting class of lattices known to the author is the lattices in the isometry group of a product of two trees constructed by Burger and Mozes [25, 24]. These groups are infinite simple finitely presented groups.

Problem 12.3. *Do the Burger-Mozes lattices admit any nontrivial homomorphisms to* Diff(M) *when M is a compact manifold?*

That these groups do act by homeomorphisms on high-dimensional spheres by the construction discussed in subsection 11.3 was pointed out to the author by Lafont.

12.5. Automorphism groups of free and surface groups

We briefly mention a last set of natural questions. There is a long-standing analogy between higher rank lattices and two other classes of groups. These are mapping class groups of surfaces and the outer automorphism group of the free group. See [20] for a detailed discussion of this analogy. In the context of this article, this raises the following question. Here we denote the mapping class group of a surface by $MCG(\Sigma)$ and the outer automorphism group of F_n by Out(F_n).

Question 12.4. *Assume Σ has genus at least 2 or that $n > 2$. Does $MCG(\Sigma)$ or* Out(F_n) *admit a faithful action on a compact manifold? A faithful action by smooth, volume-preserving diffeomorphisms?*

By not assuming that M is connected, we are implicitly asking the same question about all finite-index subgroups of $MCG(\Sigma)$ and Out(F_n). We recall from section 11 that Aut(F_n) does admit a volume-preserving action on a compact manifold. This makes the question above particularly intriguing.

13. Properties of subgroups of Diff(*M*)

As remarked in subsection 4.4, in the paper [99], Ghys attempts to reprove the classical Zassenhaus lemma for linear groups for groups of analytic diffeomorphisms. While the full strength of the Zassenhaus lemma does not hold in this setting, many interesting results do follow. This immediately leaves one wondering to what extent other properties of linear groups might hold for diffeomorphism groups or at least what analogues of many theorems might be true. This direction of research was initiated by Ghys and most of the questions below are due to him.

It is worth noting that some properties of finitely generated linear groups, like residual finiteness, do not appear to have reasonable analogues in the setting of diffeomorphism groups. For residual finiteness, the obvious example is the Thompson group discussed above, which is simple.

13.1. Jordan's theorem

For linear groups, there is a classical (and not too difficult) result known as Jordan's theorem. This says that for any finite subgroup of $GL(n, \mathbb{C})$, there is a subgroup of index at most $c(n)$ that is abelian. One cannot expect better than this, as S^1 has finite subgroups of arbitrarily large order and is a subgroup of $GL(n, \mathbb{C})$. For proofs of Jordan's theorem as well as the theorems on linear groups mentioned in the next subsection, we refer the reader to, e.g., [205].

Question 13.1. *Given a compact manifold M and a finite subgroup F of* Diff(*M*), *is there a constant c(M) such that F has an abelian subgroup of index c(M)?*

As above, one cannot expect more than this, simply because one can construct actions of S^1 on M and therefore finite abelian subgroups of Diff(*M*)of arbitrarily large order.

For this question, it may be more natural to ask about finite groups of homeomorphisms and not assume any differentiability of the maps. Using the results in, e.g., [156], one can show that at most finitely many simple finite groups act on a given compact manifold. To be clear, one can show this using the classification of finite simple groups. It would be most interesting to resolve Question 13.1 without reference to the classification.

A recent preprint of Mundet i Rieri provides some evidence for a positive answer to the question [204].

13.2. Burnside problem

A group is called *periodic* if all of its elements have finite order. We say a periodic group G has *bounded exponent* if every element has order at most m. In 1905 Burnside proved that finitely generated linear groups of bounded exponent are finite and in 1911 Schur proved that finitely generated periodic linear groups are finite. For a general finitely generated group, this is not true; counterexamples to the *Burnside conjecture* were constructed in a sequence of works by many authors, including Novikov, Golod, Shafarevich, and Ol'shanskii. We refer the reader to the website https://www-history.mcs.st-and.ac.uk/HistTopics/Burnside_problem.html for a detailed discussion of the history. This page also discusses the *restricted Burnside conjecture* resolved by Zelmanov.

In our context, the following questions seem natural:

Question 13.2. *Are there infinite, finitely generated periodic groups of diffeomorphisms of a compact manifold M? Are there infinite, finitely generated bounded exponent groups of diffeomorphisms of a compact manifold M?*

Some first results in this direction are contained in the paper of Rebelo and Silva [202].

13.3. Tits's alternative

In this section, we ask a sequence of questions related to a famous theorem of Tits and more recent variants on it [226].

Theorem 13.3. *Let Γ be a finitely generated linear group. Then either Γ contains a free subgroup on two generators or Γ is virtually solvable.*

The following conjecture of Ghys is a reasonable alternative to this for groups of diffeomorphisms.

Conjecture 13.4. *Let M be a compact manifold and Γ a finitely generated group of smooth diffeomorphisms of M. Then either Γ contains a free group on two generators or Γ preserves some measure on M.*

The best evidence for this conjecture to date is a theorem of Margulis that proves the conjecture for $M = S^1$ [162]. Ghys has also asked whether the more exact analogue of Tits's theorem might be true for analytic diffeomorphisms.

A closely related question is whether or not there exist groups of intermediate growth inside diffeomorphism groups. The growth of a group is the rate of growth of the balls in the Cayley graph. For linear groups, this is either polynomial or exponential. In general, there are groups of intermediate growth, i.e., subexponential but not polynomial. The first examples of these were discovered by Grigorchuk [108]. Some of these examples were shown by Grigorchuk and Maki to act by homeomorphisms on the interval [109]. More recently, it was shown by Navas that these examples do have an action by C^1 diffeomorphisms on the interval [181]. In the opposite direction, Navas showed that these phenomena disappear if one considers instead $C^{1+\alpha}$ diffeomorphisms for any positive α. In that setting, he shows that any finitely generated group of diffeomorphisms of the interval with subexponential growth must be nilpotent. It would be interesting to understand what can occur in higher dimensions.

A recent related line of research for linear groups concerns uniform exponential growth. A finitely generated group Γ has uniform exponential growth if the number of elements in a ball of radius r in a Cayley graph for Γ grows at least as λ^r for some $\lambda > 1$ that does not depend on the choice of generators. For linear groups, exponential growth implies exponential growth by a theorem of Eskin, Mozes, and Oh [50]. There are examples of groups having exponential but not uniform exponential growth due to Wilson [233]. This raises the following question:

Question 13.5. *Let M be a compact manifold and* Γ *a finitely generated subgroup of* Diff(*M*). *If* Γ *has exponential growth, does* Γ *have uniform exponential growth?*

The question seems most likely to have a positive answer if one further assumes that Γ is nonamenable.

References

1. Scot Adams, *Dynamics on Lorentz manifolds*, World Scientific Publishing Co. Inc., River Edge, NJ, 2001. MR MR1875067 (2003c:53101)

2. Jurgen Andersen, *Mapping class groups do not have Kazhdan's property (T)*, preprint, 2007.

3. D. V. Anosov, *Geodesic flows on closed Riemann manifolds with negative curvature*, Proceedings of the Steklov Institute of Mathematics, no. 90 (1967). Translated from the Russian by S. Feder, American Mathematical Society, Providence, RI, 1969. MR0242194 (39 #3527)

4. Artur Avila, *Distortion elements in Diff$^\infty$ (R/Z)*, preprint.

5. Uri Bader, Alex Furman, Tsachik Gelander, and Nicolas Monod, *Property (T) and rigidity for actions on Banach spaces*, Acta Math. **198** (2007), no. 1, 57–105. MR2316269 (2008g:22007)

6. Uri Bader and Yehuda Shalom, *Factor and normal subgroup theorems for lattices in products of groups*, Invent. Math. **163** (2006), no. 2, 415–454. MR2207022 (2006m:22017)

7. Augustin Banyaga, *Sur la structure du groupe des difféomorphismes qui préservent une forme symplectique*, Comment. Math. Helv. **53** (1978), no. 2, 174–227. MR490874 (80c:58005)

8. ____, *The structure of classical diffeomorphism groups*, Mathematics and its Applications, vol. 400, Kluwer Academic Publishers Group, Dordrecht, 1997. MR1445290 (98h:22024)

9. Yves Benoist, *Orbites des structures rigides (d'après M. Gromov)*, Integrable systems and foliations/ Feuilletages et systèmes intégrables (Montpellier, 1995), Progr. Math., vol. 145, Birkhäuser Boston, Boston, MA, 1997, pp. 1–17. MR1432904 (98c:58126)

10. Yves Benoist and François Labourie, *Sur les difféomorphismes d'Anosov affines à feuilletages stable et instable différentiables*, Invent. Math. **111** (1993), no. 2, 285–308. MR1198811 (94d:58114)

11. E. Jerome Benveniste and David Fisher, *Nonexistence of invariant rigid structures and invariant almost rigid structures*, Comm. Anal. Geom. **13** (2005), no. 1, 89–111. MR2154667 (2006f:53056)

12. E. J. Benveniste, *Rigidity and deformations of lattice actions preserving geometric structures*, University of Chicago, unpublished PhD thesis, 1996.

13. G. Besson, G. Courtois, and S. Gallot, *Entropies et rigidités des espaces localement symétriques de courbure strictement négative*, Geom. Funct. Anal. **5** (1995), no. 5, 731–799. MR1354289 (96i:58136)

14. Gérard Besson, Gilles Courtois, and Sylvestre Gallot, *Minimal entropy and Mostow's rigidity theorems*, Ergodic Theory Dynam. Systems **16** (1996), no. 4, 623–649. MR1406425 (97e:58177)

15. Mladen Bestvina, *Local homology properties of boundaries of groups*, Michigan Math. J. **43** (1996), no. 1, 123–139. MR1381603 (97a:57022)

16. Mladen Bestvina and Koji Fujiwara, *Bounded cohomology of subgroups of mapping class groups*, Geom. Topol. **6** (2002), 69–89 (electronic). MR1914565 (2003f:57003)

17. A. Borel and N. Wallach, *Continuous cohomology, discrete subgroups, and representations of reductive groups*, second ed., Mathematical Surveys and Monographs, vol. 67, American Mathematical Society, Providence, RI, 2000. MR1721403 (2000j:22015)

18. Armand Borel, *Compact Clifford-Klein forms of symmetric spaces*, Topology **2** (1963), 111–122. MR0146301 (26 #3823)

19. Martin Bridson and Karen Vogtmann, *Actions of automorphism groups of free groups on homology spheres and acyclic manifolds*, Comment. Math. Helv. 86 (2011), no. 1, 73–90. MR2745276 (2011j: 20104)

20. Martin R. Bridson and Karen Vogtmann, *Automorphism groups of free groups, surface groups and free abelian groups*, Problems on mapping class groups and related topics, Proc. Sympos. Pure Math., vol. 74, Amer. Math. Soc., Providence, RI, 2006, pp. 301–316. MR2264548 (2008g: 20091)

21. M. Burger and N. Monod, *Continuous bounded cohomology and applications to rigidity theory*, Geom. Funct. Anal. **12** (2002), no. 2, 219–280. MR1911660 (2003d:53065a)

22. Marc Burger, *An extension criterion for lattice actions on the circle*, Geometry, Rigidity, and Group Actions, Chicago Lectures in Mathematics, University of Chicago Press, Chicago, 2011, pp. 3–31.

23. ____, *Rigidity properties of group actions on* CAT(0)-*spaces*, Proceedings of the International Congress of Mathematicians, vol. 1, 2 (Zürich, 1994) (Basel), Birkhäuser, 1995, pp. 761–769. MR1403976 (97j:20033)

24. Marc Burger and Shahar Mozes, *Groups acting on trees: From local to global structure*, Inst. Hautes Études Sci. Publ. Math. (2000), no. 92, 113–150 (2001). MR1839488 (2002i:20041)

25. ____, *Lattices in product of trees*, Inst. Hautes Études Sci. Publ. Math. (2000), no. 92, 151–194 (2001). MR1839489 (2002i:20042)

26. Keith Burns, Charles Pugh, Michael Shub, and Amie Wilkinson, *Recent results about stable ergodicity*, Smooth ergodic theory and its applications (Seattle, WA, 1999), Proc. Sympos. Pure Math., vol. 69, Amer. Math. Soc., Providence, RI, 2001, pp. 327–366. MR1858538 (2002m:37042)

27. Eugenio Calabi, *On compact, Riemannian manifolds with constant curvature. I*, Proc. Sympos. Pure Math., vol. 3, American Mathematical Society, Providence, RI, 1961, pp. 155–180. MR0133787 (24 #A3612)

28. Eugenio Calabi and Edoardo Vesentini, *On compact, locally symmetric Kähler manifolds*, Ann. of Math. (2) **71** (1960), 472–507. MR0111058 (22 #1922b)

29. Danny Calegari and Michael H. Freedman, *Distortion in transformation groups*, Geom. Topol. **10** (2006), 267–293 (electronic), with an appendix by Yves de Cornulier. MR2207794 (2007b:37048)

30. A. Candel and R. Quiroga-Barranco, *Gromov's centralizer theorem*, Geom. Dedicata **100** (2003), 123–155. MR2011119 (2004m:53075)

31. Alberto Candel and Raul Quiroga-Barranco, *Parallelisms, prolongations of Lie algebras and rigid geometric structures*, Manuscripta Math. **114** (2004), no. 3, 335–350. MR2076451 (2005f:53056)

32. J. W. Cannon, W. J. Floyd, and W. R. Parry, *Introductory notes on Richard Thompson's groups*, Enseign. Math. (2) **42** (1996), no. 3-4, 215–256. MR1426438 (98g:20058)

33. James W. Cannon and William P. Thurston, *Group invariant Peano curves*, Geom. Topol. **11** (2007), 1315–1355. MR2326947 (2008i:57016)

34. Serge Cantat, *Groupes de transformations birationnelles du plan*, preprint.

35. ____, *Version kählérienne d'une conjecture de Robert J. Zimmer*, Ann. Sci. École Norm. Sup. (4) **37** (2004), no. 5, 759–768. MR2103473 (2006b:22010)

36. Pierre-Emmanuel Caprace and Bertrand Rémy, *Simplicité abstraite des groupes de Kac-Moody non affines*, C. R. Math. Acad. Sci. Paris **342** (2006), no. 8, 539–544. MR2217912 (2006k:20102)

37. Indira Chatterji and Martin Kassobov, *New examples of finitely presented groups with strong fixed point properties,* J. Topol. Anal. **1** (2009), no. 1, 1–12. MR2649346 (2011e:20043)

38. Christopher Connell and Benson Farb, *Some recent applications of the barycenter method in geometry*, Topology and geometry of manifolds (Athens, GA, 2001), Proc. Sympos. Pure Math., vol. 71, Amer. Math. Soc., Providence, RI, 2003, pp. 19–50. MR2024628 (2005e:53058)

39. Kevin Corlette, *Archimedean superrigidity and hyperbolic geometry*, Ann. of Math. (2) **135** (1992), no. 1, 165–182. MR1147961 (92m:57048)

40. Kevin Corlette and Robert J. Zimmer, *Superrigidity for cocycles and hyperbolic geometry*, Internat. J. Math. **5** (1994), no. 3, 273–290. MR1274120 (95g:58055)

41. S. G. Dani, *Continuous equivariant images of lattice-actions on boundaries*, Ann. of Math. (2) **119** (1984), no. 1, 111–119. MR736562 (85i:22009)

42. Pierre de la Harpe and Alain Valette, *La propriété (T) de Kazhdan pour les groupes localement compacts (avec un appendice de Marc Burger)*, Astérisque (1989), no. 175, 158, with an appendix by M. Burger. MR1023471 (90m:22001)

43. Patrick Delorme, *1-cohomologie des représentations unitaires des groupes de Lie semi-simples et résolubles. Produits tensoriels continus de représentations*, Bull. Soc. Math. France **105** (1977), no. 3, 281–336. MR0578893 (58 #28272)

44. Bertrand Deroin, Victor Kleptsyn, and Andrés Navas, *Sur la dynamique unidimensionnelle en régularité intermédiaire*, Acta Math. **199** (2007), no. 2, 199–262. MR2358052

45. Julie Déserti, *Groupe de Cremona et dynamique complexe: Une approche de la conjecture de Zimmer*, Int. Math. Res. Not. (2006), Art. ID 71701, 27. MR2233717 (2007d:22013)

46. Tien-Cuong Dinh and Nessim Sibony, *Groupes commutatifs d'automorphismes d'une variété kählérienne compacte*, Duke Math. J. **123** (2004), no. 2, 311–328. MR2066940 (2005g:32020)

47. Sorin Dumitrescu, *Meromorphic almost rigid geometric structures*, Geometry, Rigidity, and Group Actions, Chicago Lectures in Mathematics, University of Chicago Press, Chicago, 2011, pp. 32–58.

48. Michael Entov and Leonid Polterovich, *Calabi quasimorphism and quantum homology*, Int. Math. Res. Not. (2003), no. 30, 1635–1676. MR1979584 (2004e:53131)

49. D. B. A. Epstein, *The simplicity of certain groups of homeomorphisms*, Compositio Math. **22** (1970), 165–173. MR0267589 (42 #2491)

50. Alex Eskin, Shahar Mozes, and Hee Oh, *On uniform exponential growth for linear groups*, Invent. Math. **160** (2005), no. 1, 1–30. MR2129706 (2006a:20081)

51. Benson Farb, *Group actions and Helly's theorem*, preprint, 2008.

52. Benson Farb and Howard Masur, *Superrigidity and mapping class groups*, Topology **37** (1998), no. 6, 1169–1176. MR1632912 (99f:57017)

53. Benson Farb and Peter Shalen, *Real-analytic actions of lattices*, Invent. Math. **135** (1999), no. 2, 273–296. MR1666834 (2000c:22017)

54. ____, *Lattice actions, 3-manifolds and homology*, Topology **39** (2000), no. 3, 573–587. MR1746910 (2001b:57041)

55. ____, *Groups of real-analytic diffeomorphisms of the circle*, Ergodic Theory Dynam. Systems **22** (2002), no. 3, 835–844. MR1908556 (2003e:37030)

56. Benson Farb and Peter B. Shalen, *Real-analytic, volume-preserving actions of lattices on 4-manifolds*, C. R. Math. Acad. Sci. Paris **334** (2002), no. 11, 1011–1014. MR1913726 (2003e:57055)

57. Daniel S. Farley, *Proper isometric actions of Thompson's groups on Hilbert space*, Int. Math. Res. Not. (2003), no. 45, 2409–2414. MR2006480 (2004k:22005)

58. F. T. Farrell and J.-F. Lafont, *EZ-structures and topological applications*, Comment. Math. Helv. **80** (2005), no. 1, 103–121. MR2130569 (2006b:57022)

59. R. Feres, *Affine actions of higher rank lattices*, Geom. Funct. Anal. **3** (1993), no. 4, 370–394. MR1223436 (96d:22013)

60. R. Feres and F. Labourie, *Topological superrigidity and Anosov actions of lattices*, Ann. Sci. École Norm. Sup. (4) **31** (1998), no. 5, 599–629. MR1643954 (99k:58112)

61. R. Feres and P. Lampe, *Cartan geometries and dynamics*, Geom. Dedicata **80** (2000), no. 1-3, 29–41. MR1762497 (2001i:53059)

62. Renato Feres, *Connection preserving actions of lattices in* SL_nR, Israel J. Math. **79** (1992), no. 1, 1–21. MR1195250 (94a:58039)

63. ____, *Actions of discrete linear groups and Zimmer's conjecture*, J. Differential Geom. **42** (1995), no. 3, 554–576. MR1367402 (97a:22016)

64. ____, *Dynamical systems and semisimple groups: An introduction*, Cambridge Tracts in Mathematics, vol. 126, Cambridge Univ. Press, Cambridge, 1998. MR1670703 (2001d:22001)

65. ____, *An introduction to cocycle super-rigidity*, Rigidity in dynamics and geometry (Cambridge, 2000), Springer, Berlin, 2002, pp. 99–134. MR1919397 (2003g:37041)

66. ____, *Rigid geometric structures and actions of semisimple Lie groups*, Rigidité, groupe fondamental et dynamique, Panor. Synthèses, vol. 13, Soc. Math. France, Paris, 2002, pp. 121–167. MR1993149 (2004m:53076)

67. Renato Feres and Anatole Katok, *Ergodic theory and dynamics of G-spaces (with special emphasis on rigidity phenomena)*, Handbook of dynamical systems, vol. 1A, North-Holland, Amsterdam, 2002, pp. 665–763. MR1928526 (2003j:37005)

68. Renato Feres and Emily Ronshausen, *Harmonic functions over group actions*, Geometry, Rigidity, and Group Actions, Chicago Lectures in Mathematics, University of Chicago Press, Chicago, 2011, pp. 59–71.

69. D. Fisher and K. Whyte, *When is a group action determined by its orbit structure?*, Geom. Funct. Anal. **13** (2003), no. 6, 1189–1200. MR2033836 (2004k:37045)

70. David Fisher, *Cohomology of arithmetic groups and bending group actions*, in preparation.

71. ____, *First cohomology and local rigidity of group actions*, in preparation.

72. David Fisher, *A canonical arithmetic quotient for actions of lattices in simple groups*, Israel J. Math. **124** (2001), 143–155. MR1856509 (2002f:22016)

73. ____, *On the arithmetic structure of lattice actions on compact spaces*, Ergodic Theory Dynam. Systems **22** (2002), no. 4, 1141–1168. MR1926279 (2004j:37004)

74. ____, *Rigid geometric structures and representations of fundamental groups*, Rigidity in dynamics and geometry (Cambridge, 2000), Springer, Berlin, 2002, pp. 135–147. MR1919398 (2003g:22007)

75. ____, *Out(F_n) and the spectral gap conjecture*, Int. Math. Res. Not. (2006), Art. ID 26028, 9. MR2250018 (2007f:22006)

76. ____, *Local rigidity of group actions: Past, present, future*, Dynamics, ergodic theory, and geometry, Math. Sci. Res. Inst. Publ., vol. 54, Cambridge Univ. Press, Cambridge, 2007, pp. 45–97. MR2369442

77. ____, *Deformations of group actions*, Trans. Amer. Math. Soc. **360** (2008), no. 1, 491–505 (electronic). MR2342012

78. David Fisher and Theron Hitchman, *Harmonic maps with infinite dimensional targets and cocycle superrigidity*, in preparation.

79. ____, *Cocycle superrigidity and harmonic maps with infinite-dimensional targets*, Int. Math. Res. Not. (2006), 72405, 19. MR2211160

80. David Fisher and G. A. Margulis, *Local rigidity for cocycles*, (Boston, MA, 2002), Surv. Differ. Geom., 8 Int. Press, Somerville, MA, 2003, pp. 191–234. MR2039990 (2004m:22032)

81. David Fisher and Gregory Margulis, *Almost isometric actions, property (T), and local rigidity*, Invent. Math. **162** (2005), no. 1, 19–80. MR2198325

82. David Fisher and Lior Silberman, *Groups not acting on manifolds*, Int. Math. Res. Not. (2008), no. 16, Art. ID Rnn 060, 11.

83. David Fisher and Kevin Whyte, *Continuous quotients for lattice actions on compact spaces*, Geom. Dedicata **87** (2001), no. 1-3, 181–189. MR1866848 (2002j:57070)

84. David Fisher and Robert J. Zimmer, *Geometric lattice actions, entropy and fundamental groups*, Comment. Math. Helv. **77** (2002), no. 2, 326–338. MR1915044 (2003e:57056)

85. John Franks and Michael Handel, *Area preserving group actions on surfaces*, Geom. Topol. **7** (2003), 757–771 (electronic). MR2026546 (2004j:37042)

86. ____, *Periodic points of Hamiltonian surface diffeomorphisms*, Geom. Topol. **7** (2003), 713–756 (electronic). MR2026545 (2004j:37101)

87. ____, *Distortion elements in group actions on surfaces*, Duke Math. J. **131** (2006), no. 3, 441–468. MR2219247 (2007c:37018)

88. Alex Furman, *A survey of measured group theory*, Geometry, Rigidity, and Group Actions, Chicago Lectures in Mathematics, University of Chicago Press, Chicago, 2011, pp. 296–374.

89. ____, *Random walks on groups and random transformations*, Handbook of dynamical systems, vol. 1A, North-Holland, Amsterdam, 2002, pp. 931–1014. MR1928529 (2003j:60065)

90. Alex Furman and Nicolas Monod, *Product groups acting on manifolds*, preprint, 2007.

91. H. Furstenberg, *Recurrence in ergodic theory and combinatorial number theory*, Princeton Univ. Press, Princeton, NJ, 1981, M. B. Porter Lectures. MR603625 (82j:28010)

92. Harry Furstenberg, *A Poisson formula for semi-simple Lie groups*, Ann. of Math. (2) **77** (1963), 335–386. MR0146298 (26 #3820)

93. ____, *Boundaries of Lie groups and discrete subgroups*, Actes du Congrès International des Mathématiciens (Nice, 1970), tome 2, Gauthier-Villars, Paris, 1971, pp. 301–306. MR0430160 (55 #3167)

94. ____, *Rigidity and cocycles for ergodic actions of semisimple Lie groups (after G. A. Margulis and R. Zimmer)*, Bourbaki Seminar, vol. 1979/80, Lecture Notes in Math., vol. 842, Springer, Berlin, 1981, pp. 273–292. MR636529 (83j:22003)

95. Jean-Marc Gambaudo and Étienne Ghys, *Commutators and diffeomorphisms of surfaces*, Ergodic Theory Dynam. Systems **24** (2004), no. 5, 1591–1617. MR2104597 (2006d:37071)

96. Tsachik Gelander, *On deformtions of F_n in compact Lie groups*, Israel J. Math. **167** (2008), 15–26.

97. Tsachik Gelander, Anders Karlsson, and Gregory A. Margulis, *Superrigidity, generalized harmonic maps and uniformly convex spaces*, Geom. Funct. Anal. **17** (2008), no. 5, 1524–1550. MR2377496

98. Étienne Ghys, *Groupes d'homéomorphismes du cercle et cohomologie bornée*, The Lefschetz centennial conference, part 3 (Mexico City, 1984), Contemp. Math., vol. 58, Amer. Math. Soc., Providence, RI, 1987, pp. 81–106. MR893858 (88m:58024)

99. ____, *Sur les groupes engendrés par des difféomorphismes proches de l'identité*, Bol. Soc. Brasil. Mat. (N.S.) **24** (1993), no. 2, 137–178. MR1254981 (95f:58017)

100. ____, *Actions de réseaux sur le cercle*, Invent. Math. **137** (1999), no. 1, 199–231. MR1703323 (2000j:22014)

101. ____, *Groups acting on the circle*, Enseign. Math. (2) **47** (2001), no. 3-4, 329–407. MR1876932 (2003a:37032)

102. Étienne Ghys and Vlad Sergiescu, *Sur un groupe remarquable de difféomorphismes du cercle*, Comment. Math. Helv. **62** (1987), no. 2, 185–239. MR896095 (90c:57035)

103. Edward R. Goetze, *Connection preserving actions of connected and discrete Lie groups*, J. Differential Geom. **40** (1994), no. 3, 595–620. MR1305982 (95m:57052)

104. Edward R. Goetze and Ralf J. Spatzier, *On Livšic's theorem, superrigidity, and Anosov actions of semisimple Lie groups*, Duke Math. J. **88** (1997), no. 1, 1–27. MR1448015 (98d:58134)

105. ____, *Smooth classification of Cartan actions of higher rank semisimple Lie groups and their lattices*, Ann. of Math. (2) **150** (1999), no. 3, 743–773. MR1740993 (2001c:37029)

106. William M. Goldman, *An ergodic action of the outer automorphism group of a free group*, Geom. Funct. Anal. **17** (2007), no. 3, 793–805. MR2346275 (2008g:57001)

107. Alexander Gorodnik, *Open problems in dynamics and related fields*, J. Mod. Dyn. **1** (2007), no. 1, 1–35. MR2261070 (2007f:37001)

108. R. I. Grigorchuk, *Degrees of growth of finitely generated groups and the theory of invariant means*, Izv. Akad. Nauk SSSR Ser. Mat. **48** (1984), no. 5, 939–985. MR764305 (86h:20041)

109. R. I. Grigorchuk and A. Maki, *On a group of intermediate growth that acts on a line by homeomorphisms*, Mat. Zametki **53** (1993), no. 2, 46–63. MR1220809 (94c:20008)

110. M. Gromov, *Hyperbolic groups*, Essays in group theory, Math. Sci. Res. Inst. Publ., vol. 8, Springer, New York, 1987, pp. 75–263. MR919829 (89e:20070)

111. Michael Gromov, *Rigid transformations groups*, Géométrie différentielle (Paris, 1986), Travaux en Cours, vol. 33, Hermann, Paris, 1988, pp. 65–139. MR955852 (90d:58173)

112. Mikhail Gromov and Richard Schoen, *Harmonic maps into singular spaces and p-adic superrigidity for lattices in groups of rank one*, Inst. Hautes Études Sci. Publ. Math. (1992), no. 76, 165–246. MR1215595 (94e:58032)

113. Alain Guichardet, *Sur la cohomologie des groupes topologiques. II*, Bull. Sci. Math. (2) **96** (1972), 305–332. MR0340464 (49 #5219)

114. Sigurdur Helgason, *Differential geometry, Lie groups, and symmetric spaces*, Graduate Studies in Mathematics, vol. 34, American Mathematical Society, Providence, RI, 2001, corrected reprint of the 1978 original. MR1834454 (2002b:53081)

115. Michael-Robert Herman, *Simplicité du groupe des difféomorphismes de classe C^{∞}, isotopes à l'identité, du tore de dimension n*, C. R. Acad. Sci. Paris Sér. A-B **273** (1971), A232–A234. MR0287585 (44 #4788)

116. Graham Higman, *Finitely presented infinite simple groups*, Department of Pure Mathematics, Department of Mathematics, I.A.S. Australian National University, Canberra, 1974, Notes on Pure Mathematics, No. 8 (1974). MR0376874 (51 #13049)

117. M. W. Hirsch, C. C. Pugh, and M. Shub, *Invariant manifolds*, Lecture Notes in Mathematics, Vol. 583, Springer-Verlag, Berlin, 1977. MR0501173 (58 #18595)

118. Steven Hurder, personal communication.

119. Steven Hurder, *A survey of rigidity theory for Anosov actions*, Differential topology, foliations, and group actions (Rio de Janeiro, 1992), Contemp. Math., vol. 161, Amer. Math. Soc., Providence, RI, 1994, pp. 143–173. MR1271833 (95b:58112)

120. Dennis Johnson and John J. Millson, *Deformation spaces associated to compact hyperbolic manifolds*, Discrete groups in geometry and analysis (New Haven, CT, 1984), Progr. Math., vol. 67, Birkhäuser Boston, Boston, MA, 1987, pp. 48–106. MR900823 (88j:22010)

121. Jürgen Jost and Shing-Tung Yau, *Harmonic maps and superrigidity*, Differential geometry: Partial differential equations on manifolds (Los Angeles, CA, 1990), Proc. Sympos. Pure Math., vol. 54, Amer. Math. Soc., Providence, RI, 1993, pp. 245–280. MR1216587 (94m:58060)

122. Jürgen Jost and Kang Zuo, *Harmonic maps of infinite energy and rigidity results for representations of fundamental groups of quasiprojective varieties*, J. Differential Geom. **47** (1997), no. 3, 469–503. MR1617644 (99a:58046)

123. Vadim A. Kaimanovich and Howard Masur, *The Poisson boundary of the mapping class group*, Invent. Math. **125** (1996), no. 2, 221–264. MR1395719 (97m:32033)

124. Boris Kalinin, Anatole Katok, and Federico Rodriguez Hertz. *Non-uniform measure rigidity*, preprint.

125. Boris Kalinin and Ralf Spatzier, *On the classification of Cartan actions*, Geom. Funct. Anal. **17** (2007), no. 2, 468–490. MR2322492 (2008i:37054)

126. Martin Kassabov, personal communication, 2008.

127. A. Katok and J. Lewis, *Global rigidity results for lattice actions on tori and new examples of volume-preserving actions*, Israel J. Math. **93** (1996), 253–280. MR1380646 (96k:22021)

128. A. Katok, J. Lewis, and R. Zimmer, *Cocycle superrigidity and rigidity for lattice actions on tori*, Topology **35** (1996), no. 1, 27–38. MR1367273 (97e:22009)

129. A. Katok and R. J. Spatzier, *Differential rigidity of Anosov actions of higher rank abelian groups and algebraic lattice actions*, Tr. Mat. Inst. Steklova **216** (1997), no. Din. Sist. i Smezhnye Vopr., 292–319. MR1632177 (99i:58118)

130. D. A. Každan, *On the connection of the dual space of a group with the structure of its closed subgroups*, Funkcional. Anal. i Priložen. **1** (1967), 71–74. MR0209390 (35 #288)

131. David Kazhdan, *Some applications of the Weil representation*, J. Analyse Mat. **32** (1977), 235–248. MR0492089 (58 #11243)

132. Bruce Kleiner, *The asymptotic geometry of negatively curved spaces: Uniformization, geometrization and rigidity*, International Congress of Mathematicians, vol. 2, Eur. Math. Soc., Zürich, 2006, pp. 743–768. MR2275621 (2007k:53054)

133. Anthony W. Knapp, *Lie groups beyond an introduction*, Progress in Mathematics, vol. 140, Birkhäuser Boston Inc., Boston, MA, 1996. MR1399083 (98b:22002)

134. Shoshichi Kobayashi, *Transformation groups in differential geometry*, Classics in Mathematics, Springer-Verlag, Berlin, 1995, Reprint of the 1972 edition. MR1336823 (96c:53040)

135. Toshiyuki Kobayashi, *Discontinuous groups for non-Riemannian homogeneous spaces*, Mathematics unlimited—2001 and beyond, Springer, Berlin, 2001, pp. 723–747. MR1852186 (2002f: 53086)

136. Toshiyuki Kobayashi and Taro Yoshino, *Compact Clifford-Klein forms of symmetric spaces—revisited*, Pure Appl. Math. Q. **1** (2005), no. 3, part 2, 591–663. MR2201328 (2007h:22013)

137. Nicholas Korevaar and Richard Schoen, preprint, 1999.

138. Nicholas J. Korevaar and Richard M. Schoen, *Sobolev spaces and harmonic maps for metric space targets*, Comm. Anal. Geom. **1** (1993), no. 3-4, 561–659. MR1266480 (95b:58043)

139. ____, *Global existence theorems for harmonic maps to non-locally compact spaces*, Comm. Anal. Geom. **5** (1997), no. 2, 333–387. MR1483983 (99b:58061)

140. Nadine Kowalsky, *Noncompact simple automorphism groups of Lorentz manifolds and other geometric manifolds*, Ann. of Math. (2) **144** (1996), no. 3, 611–640. MR1426887 (98g:57059)

141. F. Labourie, *Quelques résultats récents sur les espaces localement homogènes compacts*, Manifolds and geometry (Pisa, 1993), Sympos. Math., 36, Cambridge Univ. Press, Cambridge, 1996, pp. 267–283. MR1410076 (97h:53055)

142. F. Labourie, S. Mozes, and R. J. Zimmer, *On manifolds locally modelled on non-Riemannian homogeneous spaces*, Geom. Funct. Anal. **5** (1995), no. 6, 955–965. MR1361517 (97j:53053)

143. François Labourie, *Large groups actions on manifolds*, Proceedings of the International Congress of Mathematicians, vol. 2 (Berlin, 1998), no. extra vol. 2, 1998, pp. 371–380 (electronic). MR1648087 (99k:53069)

144. François Labourie and Robert J. Zimmer, *On the non-existence of cocompact lattices for* SL(n)/SL(m), Math. Res. Lett. **2** (1995), no. 1, 75–77. MR1312978 (96d:22014)

145. Lucy Lifschitz and Dave Witte Morris, *Bounded generation and lattices that cannot act on the line*, Pure Appl. Math. Q. **4** (2008), no. 1, part 2, 99–126. MR2405997

146. Elon Lindenstrauss, *Rigidity of multiparameter actions*, Israel J. Math. **149** (2005), 199–226, Probability in mathematics. MR2191215 (2006j:37007)

147. Ron Livne, *On certain covers of the universal elliptic curve*, Harvard University, unpublished PhD thesis, 1981.

148. Alexander Lubotzky, *Discrete groups, expanding graphs and invariant measures*, Progress in Mathematics, vol. 125, Birkhäuser Verlag, Basel, 1994, with an appendix by Jonathan D. Rogawski. MR1308046 (96g:22018)

149. ____, *Free quotients and the first Betti number of some hyperbolic manifolds*, Transform. Groups **1** (1996), no. 1-2, 71–82. MR1390750 (97d:57016)

150. Alexander Lubotzky, Shahar Mozes, and M. S. Raghunathan, *The word and Riemannian metrics on lattices of semisimple groups*, Inst. Hautes Études Sci. Publ. Math. (2000), no. 91, 5–53 (2001). MR1828742 (2002e:22011)

151. Alexander Lubotzky and Robert J. Zimmer, *A canonical arithmetic quotient for simple Lie group actions*, Lie groups and ergodic theory (Mumbai, 1996), Tata Inst. Fund. Res. Stud. Math., vol. 14, Tata Inst. Fund. Res., Bombay, 1998, pp. 131–142. MR1699362 (2000m:22013)

152. ____, *Arithmetic structure of fundamental groups and actions of semisimple Lie groups*, Topology **40** (2001), no. 4, 851–869. MR1851566 (2002f:22017)

153. George W. Mackey, *Ergodic theory and virtual groups*, Math. Ann. **166** (1966), 187–207. MR0201562 (34 #1444)

154. Ĭozhe Maleshich, *The Hilbert-Smith conjecture for Hölder actions*, Uspekhi Mat. Nauk **52** (1997), no. 2 (314), 173–174. MR1480156 (99d:57026)

155. Ricardo Mañé, *Quasi-Anosov diffeomorphisms and hyperbolic manifolds*, Trans. Amer. Math. Soc. **229** (1977), 351–370. MR0482849 (58 #2894)

156. L. N. Mann and J. C. Su, *Actions of elementary p-groups on manifolds*, Trans. Amer. Math. Soc. **106** (1963), 115–126. MR0143840 (26 #1390)

157. G. A. Margulis, *Arithmetic properties of discrete subgroups*, Uspehi Mat. Nauk **29** (1974), no. 1 (175), 49–98. MR0463353 (57 #3306a)

158. ____, *Discrete groups of motions of manifolds of nonpositive curvature*, Proceedings of the International Congress of Mathematicians (Vancouver, BC, 1974), vol. 2, Canad. Math. Congress, Montreal, QC, 1975, pp. 21–34. MR0492072 (58 #11226)

159. ____, *Non-uniform lattices in semisimple algebraic groups*, Lie groups and their representations (Proc. Summer School on Group Representations of the Bolyai János Math. Soc., Budapest, 1971), Halsted, New York, 1975, pp. 371–553. MR0422499 (54 #10486)

160. ____, *Arithmeticity of the irreducible lattices in the semisimple groups of rank greater than 1*, Invent. Math. **76** (1984), no. 1, 93–120. MR739627 (85j:22021)

161. ____, *Discrete subgroups of semisimple Lie groups*, Ergebnisse der Mathematik und ihrer Grenzgebiete (3) [Results in Mathematics and Related Areas (3)], vol. 17, Springer-Verlag, Berlin, 1991. MR1090825 (92h:22021)

162. Gregory Margulis, *Free subgroups of the homeomorphism group of the circle*, C. R. Acad. Sci. Paris Sér. I Math. **331** (2000), no. 9, 669–674. MR1797749 (2002b:37034)

163. ____, *Problems and conjectures in rigidity theory*, Mathematics: Frontiers and perspectives, Amer. Math. Soc., Providence, RI, 2000, pp. 161–174. MR1754775 (2001d:22008)

164. Gregory A. Margulis and Nantian Qian, *Rigidity of weakly hyperbolic actions of higher real rank semisimple Lie groups and their lattices*, Ergodic Theory Dynam. Systems **21** (2001), no. 1, 121–164. MR1826664 (2003a:22019)

165. John N. Mather, *Commutators of diffeomorphisms*, Comment. Math. Helv. **49** (1974), 512–528. MR0356129 (50 #8600)

166. ____, *Simplicity of certain groups of diffeomorphisms*, Bull. Amer. Math. Soc. **80** (1974), 271–273. MR0339268 (49 #4028)

167. ____, *Commutators of diffeomorphisms. II*, Comment. Math. Helv. **50** (1975), 33–40. MR0375382 (51 #11576)

168. ____, *Foliations and local homology of groups of diffeomorphisms*, Proceedings of the International Congress of Mathematicians (Vancouver, BC, 1974), vol. 2, Canad. Math. Congress, Montreal, QC, 1975, pp. 35–37. MR0431203 (55 #4205)

169. Yozô Matsushima and Shingo Murakami, *On vector bundle valued harmonic forms and automorphic forms on symmetric Riemannian manifolds*, Ann. of Math. (2) **78** (1963), 365–416. MR0153028 (27 #2997)

170. Ngaiming Mok, Yum Tong Siu, and Sai-Kee Yeung, *Geometric superrigidity*, Invent. Math. **113** (1993), no. 1, 57–83. MR1223224 (94h:53079)

171. Nicolas Monod, *Superrigidity for irreducible lattices and geometric splitting*, J. Amer. Math. Soc. **19** (2006), no. 4, 781–814 (electronic). MR2219304 (2007b:22025)

172. Dave Witte Morris, *Amenable groups that act on the line*, Algebr. Geom. Topol. **6** (2006), 2509–2518. MR2286034 (2008c:20078)

173. Dave Witte Morris and Robert J. Zimmer, *Ergodic actions of semisimple Lie groups on compact principal bundles*, Geom. Dedicata **106** (2004), 11–27. MR2079831 (2005e:22016)

174. G. D. Mostow, *Quasi-conformal mappings in n-space and the rigidity of hyperbolic space forms*, Inst. Hautes Études Sci. Publ. Math. (1968), no. 34, 53–104. MR0236383 (38 #4679)

175. ____, *Strong rigidity of locally symmetric spaces*, Princeton Univ. Press, Princeton, NJ, 1973, Annals of Mathematics Studies, no. 78. MR0385004 (52 #5874)

176. G. D. Mostow, personal communication, 2000.

177. Andrés Navas, *Groups of circle diffeomorphisms*, book in preprint form, 2008.

178. ____, *On the dynamics of (left) orderable groups*, preprint.

179. Andrés Navas, *Actions de groupes de Kazhdan sur le cercle*, Ann. Sci. École Norm. Sup. (4) **35** (2002), no. 5, 749–758. MR1951442 (2003j:58013)

180. ____, *Quelques nouveaux phénomènes de rang 1 pour les groupes de difféomorphismes du cercle*, Comment. Math. Helv. **80** (2005), no. 2, 355–375. MR2142246 (2006j:57003)

181. ____, *Growth of groups and diffeomorphisms of the interval*, Geom. Funct. Anal. **18** (2008), no. 3, 988–1028. MR2439001

182. Amos Nevo and Robert J. Zimmer, *Invariant rigid geometric structures and smooth projective factors*, chap. 8c in this volume.

183. ____, *Homogenous projective factors for actions of semi-simple Lie groups*, Invent. Math. **138** (1999), no. 2, 229–252. MR1720183 (2000h:22006)

184. ____, *Rigidity of Furstenberg entropy for semisimple Lie group actions*, Ann. Sci. École Norm. Sup. (4) **33** (2000), no. 3, 321–343. MR1775184 (2001k:22009)

185. ____, *Actions of semisimple Lie groups with stationary measure*, Rigidity in dynamics and geometry (Cambridge, 2000), Springer, Berlin, 2002, pp. 321–343. MR1919409 (2003j:22029)

186. ____, *A generalization of the intermediate factors theorem*, J. Anal. Math. **86** (2002), 93–104. MR1894478 (2003f:22019)

187. ____, *A structure theorem for actions of semisimple Lie groups*, Ann. of Math. (2) **156** (2002), no. 2, 565–594. MR1933077 (2003i:22024)

188. ____, *Entropy of stationary measures and bounded tangential de-Rham cohomology of semisimple Lie group actions*, Geom. Dedicata **115** (2005), 181–199. MR2180047 (2006k:22008)

189. Yann Ollivier, *A January 2005 invitation to random groups*, Ensaios Matemáticos [Mathematical Surveys], vol. 10, Sociedade Brasileira de Matemática, Rio de Janeiro, 2005. MR2205306

190. Vladimir Platonov and Andrei Rapinchuk, *Algebraic groups and number theory*, Pure and Applied Mathematics, vol. 139, Academic Press Inc., Boston, MA, 1994, Translated from the 1991 Russian original by Rachel Rowen. MR1278263 (95b:11039)

191. Leonid Polterovich, *The geometry of the group of symplectic diffeomorphisms*, Lectures in Mathematics ETH Zürich, Birkhäuser Verlag, Basel, 2001. MR1826128 (2002g:53157)

192. ____, *Growth of maps, distortion in groups and symplectic geometry*, Invent. Math. **150** (2002), no. 3, 655–686. MR1946555 (2003i:53126)

193. Sorin Popa, *Deformation and rigidity for group actions and von Neumann algebras*, International Congress of Mathematicians, vol. 1, Eur. Math. Soc., Zürich, 2007, pp. 445–477. MR2334200

194. Pierre Py, *Some remarks on area-preserving actions of lattices*, Geometry, Rigidity, and Group Actions, Chicago Lectures in Mathematics, University of Chicago Press, Chicago, 2011, pp. 208–228.

195. ____, *Quasi-morphismes de Calabi et graphe de Reeb sur le tore*, C. R. Math. Acad. Sci. Paris **343** (2006), no. 5, 323–328. MR2253051 (2007e:53116)

196. ____, *Quasi-morphismes et invariant de Calabi*, Ann. Sci. École Norm. Sup. (4) **39** (2006), no. 1, 177–195. MR2224660 (2007f:53116)

197. R. Quiroga-Barranco and A. Candel, *Rigid and finite type geometric structures*, Geom. Dedicata **106** (2004), 123–143. MR2079838 (2005d:53068)

198. M. Raghunathan, *Diskretnye podgruppy grupp Li*, Izdat. "Mir", Moscow, 1977, translated from the English by O. V. Švarcman, edited by È. B. Vinberg, with a supplement "Arithmeticity of

irreducible lattices in semisimple groups of rank greater than 1" by G. A. Margulis. MR0507236 (58 #22394b)

199. M. S. Raghunathan, *On the first cohomology of discrete subgroups of semisimple Lie groups*, Amer. J. Math. **87** (1965), 103–139. MR0173730 (30 #3940)

200. Marina Ratner, *On Raghunathan's measure conjecture*, Ann. of Math. (2) **134** (1991), no. 3, 545–607. MR1135878 (93a:22009)

201. Julio C. Rebelo, *On nilpotent groups of real analytic diffeomorphisms of the torus*, C. R. Acad. Sci. Paris Sér. I Math. **331** (2000), no. 4, 317–322. MR1787192 (2001m:22041)

202. Julio C. Rebelo and Ana L. Silva, *On the Burnside problem in* Diff(*M*), Discrete Contin. Dyn. Syst. **17** (2007), no. 2, 423–439. MR2257443 (2007j:53107)

203. Dušan Repovš and Evgenij Ščepin, *A proof of the Hilbert-Smith conjecture for actions by Lipschitz maps*, Math. Ann. **308** (1997), no. 2, 361–364. MR1464908 (99c:57066)

204. I. Mundet i Rieri, *Jordan's theorem for the diffeomorphism group of some manifolds*, preprint, 2008.

205. Derek J. S. Robinson, *A course in the theory of groups*, second ed., Graduate Texts in Mathematics, vol. 80, Springer-Verlag, New York, 1996. MR1357169 (96f:20001)

206. Federico Rodriguez Hertz, *Global rigidity of certain abelian actions by toral automorphisms*, J. Mod. Dyn. **1** (2007), no. 3, 425–442. MR2318497 (2008f:37063)

207. Leslie Saper, *Tilings and finite energy retractions of locally symmetric spaces*, Comment. Math. Helv. **72** (1997), no. 2, 167–202. MR1470087 (99a:22019)

208. B. Schmidt, *Weakly hyperbolic actions of Kazhdan groups on tori*, Geom. Funct. Anal. **16** (2006), no. 5, 1139–1156. MR2276535 (2008b:37045)

209. Atle Selberg, *On discontinuous groups in higher-dimensional symmetric spaces*, Contributions to function theory (Internat. Colloq. Function Theory, Bombay, 1960), Tata Institute of Fundamental Research, Bombay, 1960, pp. 147–164. MR0130324 (24 #A188)

210. Nimish A. Shah, *Invariant measures and orbit closures on homogeneous spaces for actions of subgroups generated by unipotent elements*, Lie groups and ergodic theory (Mumbai, 1996), Tata Inst. Fund. Res. Stud. Math., vol. 14, Tata Inst. Fund. Res., Bombay, 1998, pp. 229–271. MR1699367 (2001a:22012)

211. Yehuda Shalom, *Rigidity, unitary representations of semisimple groups, and fundamental groups of manifolds with rank one transformation group*, Ann. of Math. (2) **152** (2000), no. 1, 113–182. MR1792293 (2001m:22022)

212. ____, *Measurable group theory*, European Congress of Mathematics, Eur. Math. Soc., Zürich, 2005, pp. 391–423. MR2185757 (2006k:37007)

213. ____, *The algebraization of Kazhdan's property (T)*, International Congress of Mathematicians, vol. 2, Eur. Math. Soc., Zürich, 2006, pp. 1283–1310. MR2275645 (2008a:22003)

214. R. W. Sharpe, *Differential geometry*, Graduate Texts in Mathematics, vol. 166, Springer-Verlag, New York, 1997, Cartan's generalization of Klein's Erlangen program, with a foreword by S. S. Chern. MR1453120 (98m:53033)

215. Daniel Shenfeld, *Semisimple representations of universal lattices*, Groups Geom. Dyn. **4** (2010), no. 1, 179–193. MR2566305 (2011c:22021)

216. Y. T. Siu, *Remarks in talk at Margulis's 60th birthday conference*, February 2006.

217. Yum Tong Siu, *The complex-analyticity of harmonic maps and the strong rigidity of compact Kähler manifolds*, Ann. of Math. (2) **112** (1980), no. 1, 73–111. MR584075 (81j:53061)

218. S. Smale, *Differentiable dynamical systems*, Bull. Amer. Math. Soc. **73** (1967), 747–817. MR0228014 (37 #3598)

219. R. J. Spatzier, *Harmonic analysis in rigidity theory*, Ergodic theory and its connections with harmonic analysis (Alexandria, 1993), London Math. Soc. Lecture Note Ser., vol. 205, Cambridge Univ. Press, Cambridge, 1995, pp. 153–205. MR1325698 (96c:22019)

220. ____, *An invitation to rigidity theory*, Modern dynamical systems and applications, Cambridge Univ. Press, Cambridge, 2004, pp. 211–231. MR2090772 (2006a:53041)

221. Ralf J. Spatzier and Robert J. Zimmer, *Fundamental groups of negatively curved manifolds and actions of semisimple groups*, Topology 30 (1991), no. 4, 591–601. MR1133874 (92m:57047)

222. Melanie Stein, *Groups of piecewise linear homeomorphisms*, Trans. Amer. Math. Soc. 332 (1992), no. 2, 477–514. MR1094555 (92k:20075)

223. Garrett Stuck, *Minimal actions of semisimple groups*, Ergodic Theory Dynam. Systems 16 (1996), no. 4, 821–831. MR1406436 (98a:57046)

224. William Thurston, *Foliations and groups of diffeomorphisms*, Bull. Amer. Math. Soc. 80 (1974), 304–307. MR0339267 (49 #4027)

225. William P. Thurston, *A generalization of the Reeb stability theorem*, Topology 13 (1974), 347–352. MR0356087 (50 #8558)

226. J. Tits, *Free subgroups in linear groups*, J. Algebra 20 (1972), 250–270. MR0286898 (44 #4105)

227. Uri Bader, Charles Frances and Karin Melnick, *An embedding theorem for automorphism groups of Cartan geometries*, Geom. Funct. Anal. 19 (2009), no. 2, 233–355. MR2545240 (2011c:53104)

228. Alain Valette, *Group pairs with property (T), from arithmetic lattices*, Geom. Dedicata 112 (2005), 183–196. MR2163898 (2006d:22014)

229. André Weil, *On discrete subgroups of Lie groups*, Ann. of Math. (2) 72 (1960), 369–384. MR0137792 (25 #1241)

230. ____, *On discrete subgroups of Lie groups. II*, Ann. of Math. (2) 75 (1962), 578–602. MR0137793 (25 #1242)

231. ____, *Remarks on the cohomology of groups*, Ann. of Math. (2) 80 (1964), 149–157. MR0169956 (30 #199)

232. Shmuel Weinberger, *Some remarks inspired by the C^0 Zimmer program*, Geometry, Rigidity and Group Actions, Chicago Lectures in Mathematics, University of Chicago Press, Chicago, 2011, pp. 262–282.

233. John S. Wilson, *On exponential growth and uniformly exponential growth for groups*, Invent. Math. 155 (2004), no. 2, 287–303. MR2031429 (2004k:20085)

234. Dave Witte, *Arithmetic groups of higher Q-rank cannot act on 1-manifolds*, Proc. Amer. Math. Soc. 122 (1994), no. 2, 333–340. MR1198459 (95a:22014)

235. ____, *Measurable quotients of unipotent translations on homogeneous spaces*, Trans. Amer. Math. Soc. 345 (1994), no. 2, 577–594. MR1181187 (95a:22005)

236. ____, *Cocycle superrigidity for ergodic actions of non-semisimple Lie groups*, Lie groups and ergodic theory (Mumbai, 1996), Tata Inst. Fund. Res. Stud. Math., vol. 14, Tata Inst. Fund. Res., Bombay, 1998, pp. 367–386. MR1699372 (2000i:22008)

237. Dave Witte Morris, *Can lattices in SL(n, ℝ) act on the circle?*, Geometry, Rigidity and Group Actions, Chicago Lectures in Mathematics, University of Chicago Press, Chicago, 2011, pp. 158–207.

238. ____, *An introduction to arithmetic groups*, book in preprint form.

239. Sai-Kee Yeung, *Representations of semisimple lattices in mapping class groups*, Int. Math. Res. Not. (2003), no. 31, 1677–1686. MR1981481 (2004d:53047)

240. A. Zeghib, *Isometry groups and geodesic foliations of Lorentz manifolds. I. Foundations of Lorentz dynamics*, Geom. Funct. Anal. 9 (1999), no. 4, 775–822. MR1719606 (2001g:53059)

241. ____, *Isometry groups and geodesic foliations of Lorentz manifolds. II. Geometry of analytic Lorentz manifolds with large isometry groups*, Geom. Funct. Anal. 9 (1999), no. 4, 823–854. MR1719610 (2001g:53060)

242. Abdelghani Zeghib, *Le groupe affine d'une variété riemannienne compacte*, Comm. Anal. Geom. 5 (1997), no. 1, 199–211. MR1456311 (98g:53065)

243. ____, *Sur les actions affines des groupes discrets*, Ann. Inst. Fourier (Grenoble) 47 (1997), no. 2, 641–685. MR1450429 (98d:57068)

244. Robert J. Zimmer, *Topological superrigidity*, unpublished notes.

245. ____, *Ergodic theory and semisimple groups*, Monographs in Mathematics, vol. 81, Birkhäuser Verlag, Basel, 1984. MR776417 (86j:22014)

246. ____, *Actions of lattices in semisimple groups preserving a G-structure of finite type*, Ergodic Theory Dynam. Systems **5** (1985), no. 2, 301–306. MR796757 (87g:22011)

247. ____, *Lattices in semisimple groups and distal geometric structures*, Invent. Math. **80** (1985), no. 1, 123–137. MR784532 (86i:57056)

248. ____, *On connection-preserving actions of discrete linear groups*, Ergodic Theory Dynam. Systems **6** (1986), no. 4, 639–644. MR873437 (808g:57045)

249. ____, *Actions of semisimple groups and discrete subgroups*, Proceedings of the International Congress of Mathematicians, vol. 1, 2 (Berkeley, CA, 1986) (Providence, RI), Amer. Math. Soc., 1987, pp. 1247–1258. MR934329 (89j:22024)

250. ____, *Lattices in semisimple groups and invariant geometric structures on compact manifolds*, Discrete groups in geometry and analysis (New Haven, CT, 1984), Progr. Math., vol. 67, Birkhäuser Boston, Boston, MA, 1987, pp. 152–210. MR900826 (88i:22025)

251. ____, *Representations of fundamental groups of manifolds with a semisimple transformation group*, J. Amer. Math. Soc. **2** (1989), no. 2, 201–213. MR973308 (90i:22021)

252. ____, *On the algebraic hull of an automorphism group of a principal bundle*, Comment. Math. Helv. **65** (1990), no. 3, 375–387. MR1069815 (92f:57050)

253. ____, *Spectrum, entropy, and geometric structures for smooth actions of Kazhdan groups*, Israel J. Math. **75** (1991), no. 1, 65–80. MR1147291 (93i:22014)

254. ____, *Superrigidity, Ratner's theorem, and fundamental groups*, Israel J. Math. **74** (1991), no. 2-3, 199–207. MR1135234 (93b:22019)

255. ____, *Automorphism groups and fundamental groups of geometric manifolds*, Differential geometry: Riemannian geometry (Los Angeles, CA, 1990), Proc. Sympos. Pure Math., vol. 54, Amer. Math. Soc., Providence, RI, 1993, pp. 693–710. MR1216656 (95a:58017)

256. ____, *Discrete groups and non-Riemannian homogeneous spaces*, J. Amer. Math. Soc. **7** (1994), no. 1, 159–168. MR1207014 (94e:22021)

257. ____, *Entropy and arithmetic quotients for simple automorphism groups of geometric manifolds*, Geom. Dedicata **107** (2004), 47–56. MR2110753 (2005i:22011)

258. Robert J. Zimmer and Dave Witte Morris, *Ergodic theory, groups and geometry*, CBMS Regional Conference Series in Mathematics, No. 109, American Mathematical Society, Providence, RI, 2008. MR2457556 (2009k:37059).

Afterword

RECENT PROGRESS IN THE ZIMMER PROGRAM

David Fisher

1. Introduction

One could attempt to write an afterword to this volume that addressed the full scope and impact of Zimmer's work on mathematics. The result would be either terse and unreadable or another volume longer than the current one. While Zimmer's contributions and ideas have fostered research in many directions, there is an essential unity to his perspective. So rather than attempting to broadly survey all the impacts, I choose here to focus on a "test case" in which Zimmer's contributions are particularly important and in which there has been dramatic recent progress: the *Zimmer program*.

The *Zimmer program* aims to classify actions of higher rank Lie groups and their lattices on compact manifolds. The program was initiated by Zimmer in a series of papers in the early 1980s and framed explicitly in [Zim1, Zim4]. In the last 35 years, numerous authors have made deep and important contributions to the program, using ideas from across the mathematical landscape. In late 2008, I wrote a long and detailed survey of the state of the art in the program [Fis2]. At the time, I was not terribly optimistic about the prognosis for the area. The most interesting questions seemed inaccessible and activity in the area seemed to be slowing. Much to my surprise, the last decade has proven to be remarkably fruitful for the program and led to a series of breakthroughs that make the program now more vibrant than ever. This includes the progress on Zimmer's conjecture by Brown, Hurtado, and me that Bob so graciously highlights in his introduction to this volume, but also several other related works that made that development possible. I will very briefly describe the developments here and point to the subsections in which they are described. In writing this article, a major goal is to point to places in which similar arguments are used in different

David Fisher was partially supported by NSF grant DMS-1607041.

Thanks to Marc Burger, Manfred Einseidler, Alex Eskin, Simion Filip, Katie Mann, Federico Rodriguez Hertz, and Bob Zimmer for useful comments and corrections. Particular thanks to Aaron Brown and Ralf Spatzier for thorough readings and copious remarks and corrections.

contexts, in order to try to isolate techniques and ideas that are likely to play a key role in further progress on Zimmer's program and in related areas. Due to constraints of space and time, detailed arguments will not be included and so this paper may serve more as "reader's guide" to the area than as an introduction or survey.

A recurring theme in this article will be surprising connections and developments. I start the narrative with some developments concerning rigidity of Anosov $\mathbb{Z}^d$ actions that end in a proof of a conjecture of Katok and Spatzier by Rodriguez Hertz and Wang, using some prior results by Kalinin, Spatzier, and myself. While there has long been some interaction between the Zimmer program and the rigidity of hyperbolic actions of abelian groups, in the past the main successes had concerned results on local rigidity of actions that were a priori hyperbolic or global rigidity results that required very strong assumptions. Rapid progress on rigidity of abelian group actions in my work with Kalinin and Spatzier and the work of Rodriguez Hertz and Wang described in section 3 changed the landscape and provided strong enough tools for $\mathbb{Z}^d$ actions that it made sense to see what would happen for lattice actions. Spatzier and I encouraged Rodriguez Hertz, and Wang to do this, in my case mistakenly believing this would be more or less a corollary of their theorems. This led to the breakthrough paper of Brown, Rodriguez Hertz, and Wang on higher rank Anosov actions in which the key philosophy of *nonresonance implies invariance* was introduced; see section 4. The second application of this philosophy led to a better understanding of invariant measures for low-dimensional actions of lattices in another paper by Brown, Rodriguez Hertz, and Wang. I think it is safe to say that these authors did not appreciate the full importance of what they had done. Key ingredients throughout these developments are derived from much closer connections to measure rigidity. This occurs both in the proofs of the result of Brown, Rodriguez Hertz, and Wang and in terms of the use of dynamics of unipotent flows and particularly Ratner's theorems in the work on Zimmer's conjecture. All of these developments are described in section 7.

In the middle of that narrative an important additional piece of the puzzle was provided by Hurtado's remarkable paper on the Burnside problem, which formalized a key notion of *subexponential growth of derivatives*. This notion is really the natural notion of uniform nonhyperbolicity of a group action and is certainly implicit earlier, but making it explicit and exhibiting it as a fulcrum or turning point in proofs was a key conceptual development; see section 5.

Lastly, another important piece of the puzzle predates all these developments: the strong property (T) of Vincent Lafforgue. While this notion vaguely resembles one contained in a paper I wrote with Margulis and also some ideas in a paper of Gromov, the precise notion and its proof are profound and original developments. This is described in section 6. Lafforgue developed this notion primarily as an obstruction to proving the Baum-Connes conjecture and also as a route to constructing superexpanders, but just as property (T) before it, strong property (T) turns out to be robustly applicable.

This paper is a short and very personal history of these mathematical developments. These developments all come in the context of the Zimmer program and it is a pleasure to describe them here in a volume selecting some of Zimmer's papers. Following Bob's example in his introduction to this volume, the style here will be one of personal narrative in a manner that is somewhat unusual in modern mathematical exposition. I want to explain both some mathematical ideas and the history of how they arose, combined, and led to exciting new developments. All these developments owe a tremendous amount to Bob's insights, theorems, and questions. They also involve the work of several people, each of whom would surely tell this story differently. I would like to take the opportunity to thank them all for their roles in these developments, including the pleasure I have had in writing papers with most of them. They are, in alphabetical order, Aaron Brown, Sebastian Hurtado, Boris Kalinin, Federico Rodriguez Hertz, Ralf Spatzier, and Zhiren Wang.

2. Prehistory

I am going to start the story where it started for me personally. This was at an AIM workshop organized by Lindenstrauss, Katok, and Spatzier titled "Emerging directions in measure rigidity" in 2004. From my point point of view, the start of this long conversation with many people is a remark Ralf Spatzier made during that problem session. Either Katok or Spatzier was discussing their conjecture that every *genuinely higher rank* Anosov $\mathbb{Z}^d$ action was smoothly conjugate to an affine Anosov action on an infranilmanifold. As was usual at the time, discussion turned to the related conjecture, usually attributed to Smale or Franks, that an Anosov diffeomorphism (or $\mathbb{Z}$ action) is topologically conjugate to an affine diffeomorphism on a nilmanifold. The conjecture of Katok and Spatzier is motivated by their work on local rigidity and cocycle rigidity for higher rank Anosov actions [KS3, KS2]. It is well known in the case $d = 1$ that one cannot expect smooth conjugacy since perturbations of affine actions can be seen to no longer have a constant derivative on the periodic orbits and so are not smoothly conjugate to the original action. Spatzier reminded the audience of a deeper obstruction from work of Farrell and Jones, the existence of Anosov diffeomorphisms on "exotic tori," manifolds homeomorphic but not diffeomorphic to tori [FJ]. Farrell and Gogolyev give a more modern and general construction in [FG]. Exotic tori are constructed by taking the connected sum of an exotic sphere and a torus. The highly nontrivial part of the papers just mentioned is showing that one can take an Anosov diffeomorphism and first restrict it to the complement of a neighborhood of a fixed point, connect sums in an exotic sphere, and then extend the Anosov diffeomorphism back over the whole manifold. Some time later, I realized that no existing result prevented the construction of similar examples of $\mathbb{Z}^d$ Anosov actions on exotic tori. I even asked both Katok and Spatzier about this, and their responses led me to believe that no one had thought seriously about the possibility. I then spent some time trying to construct examples, talking about the problem with Chris Connell, Tom Farrell, and Shmuel Weinberger at various moments. The obstruction is a simple one: if you follow the Farrell-Jones construction, it is completely unclear whether the resulting diffeomorphism has a nontrivial centralizer. Given an exotic Anosov diffeomorphism f on a torus, there are diffeomorphisms g commuting with the f at the level of the homotopy data, so as automorphisms of $\pi_1(\mathbb{T}^n) = \mathbb{Z}^n$. But it could easily happen that the commutator of the diffeomorphisms f and g, while homotopic to the identity, is nontrivial and the group generated by f and g in Diff($\mathbb{T}^n$) might still be free or at least nonabelian. Repeated conversation with Spatzier about this issue led to our beginning to work together and with Kalinin, and in the not too long run to the resolution of the Katok-Spatzier conjecture for actions on infranilmanifolds by Rodriguez Hertz and Wang, all described in the next section of this article.

3. Rigidity of Anosov $\mathbb{Z}^d$ actions

At the time of the conversations above, the best result concerning Anosov $\mathbb{Z}^d$ actions on tori was by Federico Rodriguez Hertz. This paper first appeared as a preprint in 2001 but was only revised and published in 2007 [RH]. I must admit that I did not understand this paper well for several years after that, but it was clear at the time that the dynamical conditions needed were quite restrictive, including needing low-dimensional dynamical foliations, and the lack of actions on exotic tori in that context was unsurprising. In the interim between that paper and my work with Kalinin and Spatzier, most of the work on the conjecture of Katok and Spatzier focused on the much harder and still open case of actions on general manifolds. The best work on this topic is by Kalinin and Sadovskya, building on earlier work of Kalinin and Spatzier [KlSp, KlS2, KlS1]. Before continuing, I recall the formal definition of Anosov diffeomorphism and Anosov action.

Let a be a diffeomorphism of a compact manifold M. We recall that a is *Anosov* if there exists a continuous a-invariant decomposition of the tangent bundle $TM = E_a^s \oplus E_a^u$ and constants $K > 0, \lambda > 0$ such that for all $n \in \mathbb{N}$,

$$
\begin{aligned}
\|Da^n(v)\| &\le Ke^{-\lambda n}\|v\| \text{ for all } v \in E_a^s, \\
\|Da^{-n}(v)\| &\le Ke^{-\lambda n}\|v\| \text{ for all } v \in E_a^u.
\end{aligned}
\tag{1}
$$

The distributions E_a^s and E_a^u are called the *stable* and *unstable* distributions of a. Given an action of a group Γ on a compact manifold M via diffeomorphisms, we call the action *Anosov* if there is an element γ in Γ such that $\alpha(\gamma)$ is an Anosov diffeomorphism.

Progress on Anosov $\mathbb{Z}^k$ actions on tori and nilmanifolds began again in two papers I wrote with Boris Kalinin and Ralf Spatzier. The starting point for essentially all work on this problem in the context of action on infranilmanifolds is results of Franks and Manning that show that any Anosov diffeomorphism is topologically conjugate to an affine Anosov diffeomorphism [Fra, Man]. Given an Anosov diffeomorphism f of a torus $\mathbb{T}^n$, we write f_* for the induced linear action of f on $\pi_1(\mathbb{T}^n) = \mathbb{Z}^n$ and note that this defines a linear toral automorphism f_0. Franks showed that for any Anosov diffeomorphism on a torus, there is a homeomorphism ϕ, conjugating an Anosov diffeormorphism f to the linearization f_0 [Fra]. This was generalized by Manning to the case of nilmanifolds [Man]. Since the conjugacy ϕ is essentially unique in its homotopy class, it follows that the conjugacy is also a conjugacy for any diffeomorphism commuting with f and so that any Anosov $\mathbb{Z}^d$ action on an infranilmanifold is topologically conjugate to an affine action. The remaining problem is simply to improve the regularity of the conjugacy. For this exposition, we restrict attention to the case of actions on tori, where ideas are easier to explain. The affine model action is always linear on a finite-index subgroup. Call a linear $\mathbb{Z}^d$ action *irreducible* if it does not split rationally into nontrivial factors. We can always split a linear action into a product of irreducible factors. We call a general Anosov action by $\mathbb{Z}^d$ on a $\mathbb{T}^n$ (or more generally a nilmanifold) *irreducible* if it is topologically conjugate to an affine action that is irreducible. A $\mathbb{Z}^k$ action α on $\mathbb{T}^n$ induces a linear action α_0 on homology called the *linearization* of α and the affine conjugate of an Anosov action is equal to α_0 on a subgroup of finite index. The logarithms of the moduli of the eigenvalues define linear maps $\lambda_i : \mathbb{Z}^k \mapsto \mathbb{R}$ that extend to $\mathbb{R}^k$. A *Weyl chamber* of α_0 is a connected component of $\mathbb{R}^k - \cup_i \ker \lambda_i$. The key hypothesis in the work with Kalinin and Spatzier is the presence of an Anosov diffeomorphism in each Weyl chamber, and a special case of our results in [FKS1, FKS2] is as follows:

Theorem 3.1. *Suppose α is an irreducible C^∞ action of $\mathbb{Z}^k$, $k \geq 2$ on a torus $\mathbb{T}^n$. Further assume there is an Anosov element for α in each Weyl chamber of α_0. Then α is C^∞ conjugate to an affine action with linear part α_0.*

The key point of the assumption of having an Anosov in each Weyl chamber is the following. Just using the Oseledec theorem and Pesin theory, one has that the action α is nonuniformly partially hyperbolic in a particularly nice way. The presence of enough Anosov diffeomorphisms forces all invariant measures to have proportional Lyaponuv exponents and therefore forces essentially all elements of the action to be uniformly hyperbolic in the strongest possible sense. The fact that controlling all invariant measures turns nonuniform estimates obtained from the Oseledec theorem into uniform estimates is an old one in dynamical systems, is also used in Rodriguez Hertz's earlier paper, and is also used in a surprising way in the proof of Zimmer's conjecture. In particular, in this context, it shows that if there are elements of $\mathbb{Z}^d$ in Weyl chamber walls, then restricted to the corresponding *central foliation* one has *subexponential growth of derivatives*. This viewpoint is a bit ahistorical as the notion of subexponential growth of derivatives for a group action was only formalized later by Hurtado in [Hur1]. We will discuss the idea in detail in the less technical setting of that article in section 5 below.

In the first of our papers, this observation is combined with ideas concerning normal forms for group actions and an examination of holonomies to produce a result for so-called *totally nonsymplectic actions* [FKS1]. While the techniques used in this paper have not had any further applications that I know of, it was around this time that I started giving talks emphasizing the close connections to homogeneous dynamics that had last been clear in early work of Katok and Spatzier [KS2, KS1], since our holonomy arguments were inspired by the use of unipotents in work of both Ratner and Lindenstrauss on rigidity of invariant measures. This connection to measure rigidity turns out to have been deep and fruitful as we will see below.

In our second paper [FKS2], we took an entirely different approach that turns out to also have some remarkable similarities to the work on Zimmer's conjecture. Namely, we wrote the conjugacy, at least projected onto certain dynamically defined submanifolds for the action, as a solution to a cohomological equation. We then combined the subexponential growth of derivatives with exponential decay of matrix coefficients to obtain greater regularity of the solution to the cohomological equation. This argument is quite complicated as one needs to work along various foliations, and so the regularity one obtains is best described in terms of wavefront sets, which then allow one to patch the regularity together globally with arguments that are fairly standard in PDE. The remarkable thing is that this argument is quite similar to the last step in the proof of Zimmer's conjecture, where we use strong property (T) in conjunction with subexponential growth of derivatives (in all directions) to find an invariant metric. The proof of strong property (T) depends essentially on exponential decay of matrix coefficients, so the similarity in the arguments appears to be quite deep. There is, of course, an important point of contrast. In our setting the solution to the cohomological equation is given. Lafforgue proves additional estimates of decay of certain twisted matrix coefficients in order to prove the solution exists.

The paper of Rodriguez Hertz and Wang completes the proof of the Katok-Spatzier conjecture for higher rank abelian Anosov actions on nilmanifolds and tori [RHW]. They proceed by showing that the presence of a single Anosov element in a $\mathbb{Z}^d$ action on a nilmanifold or torus implies the existence of an Anosov element in every Weyl chamber and therefore verify the hypothesis of Theorem 3.1 above. It is relatively easy to see that the presence of the single Anosov element forces the entire $\mathbb{Z}^d$ action to be nonuniformly hyperbolic in a strong sense, and so the key contribution is showing that this nonuniform hyperbolicity is uniform. This is a theme that recurs in both their work with Brown on rigidity of Anosov actions of higher rank lattices on tori and nilmanifolds and in the proof of Zimmer's conjecture. I expect this then will continue to recur in further progress in the program.

The first step in their argument is to show that the conjugacy is smooth along certain dynamically defined foliations, namely the foliations along which the map is switching from expansion to contraction as one changes Weyl chambers. The proof of this fact follows quite closely arguments in [FKS2]. The main part of the paper is then to show that this smooth map along dynamical foliations, in addition to being smooth, is nonsingular. That this suffices is shown using a theorem of Mañé, which shows that any quasi-Anosov diffeomorphism is in fact Anosov. Mañé's work seems a fertile source of ideas for understanding further connections between nonuniform hyperbolicity and uniform hyperbolicity.

The basic idea for the hardest part of the paper is that along the singular set, the conjugacy collapses volume. Rodriguez Hertz and Wang derive a contradiction to this fact by using a measure-rigidity-type argument to show that this collapse of volume is impossible. This is by far the most delicate part of the paper and uses that strong stable manifolds are Lipschitz inside stable manifolds. The measure rigidity result is particularly difficult since the measure considered is an invariant measure supported on the (closed) set of singular points. This application of measure rigidity ideas in the nonlinear settings and in particular using them to improve nonuniform hyperbolicity to uniform hyperbolicity marks another instance of an important new trend in Zimmer's program. We will continue to see more instances of this below.

The reader far from hyperbolic dynamics may wonder why we focus on converting nonuniform hyperbolicity to uniform hyperbolicity when the real breakthrough comes on Zimmer's conjecture, which concerns actions with no hyperbolicity. One motivation for the conjecture was Zimmer's cocycle superrigidity theorem, which produced a measurable invariant metric and showed all Lyapunov exponents were zero. While the program traditionally was framed as an attempt to improve regularity of this metric, the breakthrough comes from a change of perspective. Rather than improving regularity of the metric directly, one improves estimates on dynamics of the derivative from nonuniform to uniform.

In closing, it is an old conjecture often attributed to either Smale or Franks that all Anosov diffeomorphisms are conjugate to affine maps on nilmanifolds. In the final comment of [Mar], Margulis draws an intriguing parallel between now resolved cases of Zimmer's conjecture, open problems in the Zimmer program, and this conjecture on Anosov diffeomorphisms. Resolving this conjecture, combined with the work just described, would provide a smooth classification of Anosov actions of higher rank abelian groups.

4. Rigidity of Anosov lattice actions

The next major breakthrough concerned actions of higher rank lattices, not just higher rank abelian groups, and so was truly part of Zimmer's program. It is still deeply embedded in hyperbolic dynamics as it concerns Anosov actions. The paper of Brown, Rodriguez Hertz, and Wang contains not one but two major results, each containing ideas that are relevant to future directions [BRHW2]. We recall that for any group Γ an action is *Anosov* if some individual element γ in the group acts as an Anosov diffeomorphism.

In the first part of the paper, they give a proof that any Anosov action of a higher rank lattice Γ on a torus or nilmanifold M is continuously conjugate to an an affine action subject to the condition that the action lifts to the universal cover. For the case of actions preserving a measure of full support, this reproves a result of Margulis and Qian [MQ], but the proof by Brown, Rodriguez Hertz, and Wang is different and does not use Zimmer's cocycle superrigidity theorem. The philosophy behind this argument is an important new development that is also applied in the proof of Zimmer's conjecture. This philosophy develops a link between the finite-dimensional-representation theory of the group G and the possible actions of G and Γ. A link between the two appeared already in Zimmer's cocycle superrigidity theorem. Zimmer showed that for smooth actions of G or Γ, the derivative of measure-preserving action is always essentially described, at least measurably, by a finite-dimensional representation of G. The presence of an invariant measure is a necessary hypothesis for Zimmer's result. The philosophy of "nonresonance implies invariance" allows one to extend the connection between group actions and finite-dimensional representations to the setting in which there are, a priori, no invariant measures.

To employ this philosophy for actions of a lattice Γ one always needs to pass to the induced G action on $(G \times M)/\Gamma$. This allows one to use the structure of G, namely the root data associated with a choice of Cartan subalgebra. To explain this philosophy better, I recall some basic facts. The Cartan subgroup A of G is the largest subgroup diagonalizable over $\mathbb{R}$; the Cartan subalgebra α is its Lie algebra. It has been known since the work of Élie Cartan that a finite-dimensional linear representation ρ of G is completely determined by linear functionals on α that arise as generalized eigenvalues of the restriction of ρ to A. Here we use that there is always a simultaneous eigenspace decomposition for groups of commuting symmetric matrices and that this makes the eigenvalues into linear functionals. These linear functionals are referred to as the *weights* of the representation. For the adjoint representation of G on its own Lie algebra, the weights are given the special name of *roots*. Corresponding to each root β there is a unipotent subgroup $G_\beta < G$ called a *root subgroup* and it is well known that "large enough" collections of root subgroups generate G. Two linear functionals are called *resonant* if one is a positive multiple of the other. Abstractly, given a G action and an A-invariant object O, one may try to associate with O a class of linear functionals Ω. *Nonresonance implies invariance* is the observation that given any root β of G that is not resonant to an element of Ω, the object O will automatically be invariant under the unipotent root group G^β. If one can find enough such nonresonant roots, the object O is automatically G invariant.

The philosophy that "nonresonance implies invariance" is perhaps more transparent in its application to invariant measures in section 7, but I will describe it here first. To describe this philosophy in action here, we need to develop the picture a bit further. As discussed in the last section, Franks and Manning produced a conjugacy ϕ conjugating an Anosov diffeormorphism f to the linearization f_0. Note that the action on fundamental group gives a linear representation of Γ into $\mathrm{Aut}\,(\pi_1(M))$ restricting to f_0 on the Anosov diffeomorphism f. Using this Γ action on homology, the structure of M as a nilmanifold, and Margulis's superrigidity theorem, one defines an affine G action on $(G \times M)/\Gamma$. As discussed above, the map ϕ is sufficiently unique that it also conjugates the centralizer of the Anosov diffeomorphism to linear maps. Letting A be the maximal Cartan subgroup of G, a variant of that argument produces a conjugacy ϕ between the restriction to A of the original G action and the affine action of A on $(G \times M)/\Gamma$. This affine action is defined in terms of a linear representation of A and therefore gives rise to a collection Ω of linear functionals on A. More concretely, since A acts affinely and $(G \times M)/\Gamma$ is parallelizable, the derivative is constant and these linear functionals are exactly generalized eigenvalues of the derivative. So this semiconjugacy to a linear action gives rise to a new collection Ω of linear functionals on A that one can compare to the roots of G. Brown, Rodriguez Hertz, and Wang show that for any root β that is not resonant to any linear functional α in Ω, the map ϕ is G_β invariant. One key point for the proof is that the linear functionals in Ω associated with Anosov actions cannot be resonant to roots for purely algebraic reasons. The other key point is the existence of elements of A that commute with G_β. This shows that ϕ is G equivariant, as desired.

The second half of the paper proves that this continuous conjugacy is smooth. The argument surprisingly goes back to an old argument in a paper of Katok, Lewis, and Zimmer on actions of $\mathrm{SL}(n, \mathbb{Z})$ on $\mathbb{T}^n$ [KLZ]. Both there and here the goal is to show that the Lyapunov exponents coming out of cocycle superrigidity can be improved to give uniform estimates on derivatives. In this paper, the key point is to control the Lyapunov exponents uniformly along all invariant measures, using that Ratner's theorem classifies invariant measures on the linear side and so also on the nonlinear side by conjugacy. This combination of cocycle superrigidity with control on exponents in all invariant measures also foreshadows elements of the proof of Zimmer's conjecture and will certainly appear in future developments as well.

5. Hurtado's work on the Burnside problem for diffeomorphisms

A key impetus for Brown, Hurtado, and me to work together on Zimmer's conjecture was Hurtado's paper proving a Burnside-type theorem for subgroups of $\mathrm{Diff}(S^2, \omega)$, i.e., for volume-preserving diffeomorphisms of the sphere in dimension 2 [Hur1]. Hurtado proves that any finitely generated subgroup Γ of $\mathrm{Diff}(S^2, \omega)$ that consists entirely of torsion elements is finite. Since the conjecture was known for all other surfaces, the focus on S^2 is natural. The requirement of an invariant-volume form is an artifact of the proof. Farb and Ghys have each independently conjectured finiteness of all finitely generated torsion subgroups of the diffeomorphism group of any compact manifold.

Hurtado formalizes a key notion of subexponential growth of derivatives. Let Γ be a finitely generated group. Let $l : \Gamma \to \mathbb{N}$ denote the word-length function relative to some fixed finite set of generators F. Let $\alpha : \Gamma \to \mathrm{Diff}^1(M)$ be an action of Γ on a compact manifold M by C^1 diffeomorphisms. We say the action α has *uniform subexponential growth of derivatives* if for all $\varepsilon > 0$ there is a C_ε such that for all $\gamma \in \Gamma$ we have

$$\|D\alpha(\gamma)\| \le C_\varepsilon e^{\varepsilon l(\gamma)},$$

where $\| D\alpha(\gamma) \| = \sup_{x \in M} \| D_x \alpha(\gamma) \|$.

In Zimmer's own work, several notions of slow growth of derivatives arose, and showing slow enough growth of derivatives was clearly the key to the conjecture from many points of view. I will discuss this more at the end of section 7 below. In the context of hyperbolic dynamics the notion of subexponential growth here is the correct uniform analogue of an action having all zero Lyapunov exponents. Even though this idea has been used before (even implicitly in my work with Kalinin and Spatzier), Hurtado's paper seems to be the first place it is formalized for group actions, and its formulation proved extremely useful for our subsequent work on Zimmer's conjecture. For a single diffeomorphism it is an easy classical fact that having zero Lyapunov exponents for all invariant measures implies subexponential growth of derivatives. I state this result formally here as it and Hurtado's adaptation of it were starting points for the work on Zimmer's conjecture.

Proposition 5.1. *Let M be a compact manifold and let $\mathbb{Z}$ act smoothly on M. Then the $\mathbb{Z}$ action has subexponential growth of derivatives if and only if every $\mathbb{Z}$-invariant measure on M has zero Lyaponuv exponents.*

To prove the nontrivial implication in the proposition, one takes sequences of orbit segments witnessing the failure of subexponential growth of derivatives, views these as defining measures, and shows the weak* limits of these measures have positive Lyapunov exponents.

Hurtado uses an analogue of this result for the $\mathbb{Z}$ action built by taking the shift action of $\mathbb{Z}$ on the space $\Omega = F^{\mathbb{Z}}$ and looking at the standard skew product action of $\mathbb{Z}$ on $\Omega \times S^2$. An extension of the classical argument says that either the Γ action has subexponential growth of derivatives or there is a $\mathbb{Z}$-invariant measure on $\Omega \times S^2$ with a positive exponent for the derivative along S^2. Since Hurtado assumes an invariant-volume form, the presence of one nonzero exponent forces both exponents along S^2 to be nonzero. At this point, Hurtado invokes a classical result of Katok on hyperbolic dynamics of diffeomorphisms with no zero exponents [Kat]. To use this theorem, he needs to embed $\Omega \times S^2$ in a manifold and construct a hyperbolic diffeomorphism f of the manifold extending the shift action on $\Omega \times S^2$. Katok's theorem then provides a hyperbolic fixed point for f that in this construction gives a hyperbolic periodic point for some element of Γ on S^2. Since all elements of Γ are finite order, this is a contradiction.

Hurtado's paper contains many other intriguing ideas once this result is in place. In fact, he uses the idea of averaging a metric over an action with subexponential growth of derivatives to obtain an invariant metric. Here this only works directly under the additional assumption that the group is amenable, but this too foreshadows elements of the proof of Zimmer's conjecture. Experts appear to have known that subexponential growth of derivatives should be relevant to the Burnside problem before Hurtado's work. The approach is at least partially inspired by Kalinin's matrix-valued Livšic theorem [Kal] and the diffeomorphism-valued Livšic theorem of Kocsard and Potrie [KP], but Hurtado's success in implementing this style of proof was a major impetus for our work on Zimmer's conjecture. The connection to Livšic-type theorems is also related to another paper of Hurtado's in which he produces a surprising example [Hur2]. He produces an action of a free group in which every element is conjugate to an isometry but the full action has exponential growth of derivatives.

6. Strong property (T)

Another key ingredient in the proof of Zimmer's conjecture that was developed much earlier is Lafforgue's notion of strong property (T) introduced in 2007 by Lafforgue as an obstruction to certain strategies for proving the Baum-Connes conjecture [Laf]. The original definition is made in terms of the existence of certain kinds of projections in a certain completion of the group algebra. Instead, we recall a variant that does not involve operator algebras. Given a group Γ and a finite or compact generating set S, we let l be the word length on Γ. In what follows X will denote a Banach space and $B(X)$ will denote the bounded operators on X. We always consider the operator norm topology on $B(X)$ and we always mean the operator norm when we write $\|T\|$ for $T \in B(X)$.

Definition 6.1. *Let Γ be a group with a length function l, X a Banach space, and $\pi : \Gamma \to B(X)$. Given $\varepsilon > 0$, we say π has ε-subexponential norm growth if there exists a constant L such that $\|\pi(\gamma)\| \leq Le^{\varepsilon l(\gamma)}$ for all $\gamma \in \Gamma$. We say π has subexponential norm growth if it has ε-subexponential norm growth for all $\varepsilon > 0$.*

Here we will focus only on the case of actions on Hilbert spaces, though the robustness of strong property (T) outside of that class is important for many applications and is discussed briefly below.

Definition 6.2. *A group Γ has strong property (T) if there exists a constant $t > 0$ and a sequence of measures μ_n supported in the balls $B(n) = \{\gamma \in \Gamma \mid l(\gamma) \leq n\}$ in Γ such that for any representation $\pi : \Gamma \to B(X)$ with t-subexponential norm growth and X a Hilbert space, the operators $\pi(\mu_n)$ converge exponentially quickly to a projection onto the space of invariant vectors. That is, there exists $0 < \lambda < 1$ (independently of π), a projection $P \in B(X)$ onto the space of Γ-invariant vectors, and an $n_0 \in \mathbb{N}$ such that $\|\pi(\mu_n) - P\| < \lambda^n$ for all $n \geq n_0$.*

The original definition of Lafforgue is equivalent not to this but to a similar statement with signed measures in place of measures [dlS1]. All known proofs that a group has strong property (T) produce positive measures and this definition seems more useful, if harder to state in the purely operator-theoretic language [dlS1]. If one makes similar definitions but assumes instead that $L = 1$ in definition 6.1, then the resulting notion is equivalent to property (T) by my work with Margulis in [FM1]. One might say that in that work we insist on *immediate* subexponential norm growth while Lafforgue allows the weaker condition of *eventual* subexponential norm growth. The difference is quite profound. Lafforgue proves that no hyperbolic groups can have strong property (T) while many (even most in certain random senses) of them have property (T). The difference is also profound for applications. One should only expect immediate subexponential growth of derivatives when perturbing isometric actions as in [FM1].

By work of Lafforgue, de la Salle, and de la Salle–de Laat, strong property (T) is known for the groups that interest us here [Laf, dLdlS, dlS2].

Theorem 6.3. *Let G be a semisimple Lie group all of whose simple factors have real rank at least 2 and let $\Gamma < G$ be a lattice. Then G and Γ have strong property (T).*

Combined with additional work of Liao, these results also establish strong property (T) for lattices in higher rank simple groups over other local fields [Lia].

While the notion just described suffices to prove Zimmer's conjecture for smooth actions, to obtain Zimmer's conjecture in lower regularity, one needs to consider non-Hilbertian function spaces. In this context, it suffices to consider the θ-Hilbertian Banach spaces. These are subspaces of spaces obtained by complex interpolation with Hilbert spaces. Most reasonable function spaces arising in dynamics and geometry are θ-Hilbertian except those that are defined in terms of supremum norms and so have no good convexity properties. Strong property (T) for higher rank lattices is known to hold for all θ-Hilbertian Banach spaces and even for significantly broader classes of Banach spaces. It is conjectured to hold for any uniformly convex Banach space.

7. Recent work on Zimmer's conjecture

In this section, I will focus on the recent breakthrough on Zimmer's conjecture, discussing only the case of actions of lattices in $\mathrm{SL}(n, \mathbb{R})$ for $n > 2$ since that simplifies terminology and notation. For an excellent account of both the developments discussed here and the numerology associated with actions of other groups, the reader should consult Cantat's recent Bourbaki Seminaire article [Can]. Another account of the recent developments on Zimmer's conjecture that is probably more introductory and accessible than the one here is contained in a survey by Brown [Bro].

The discussion here will focus on three articles: first the article of Brown, Rodriguez Hertz, and Wang that produces invariant measures for low-dimensional actions [BRHW1], then the article by Brown, Hurtado, and me that proves Zimmer's conjecture for actions of a cocompact lattice [BFH1], at least for $\mathbb{R}$-split classical groups, and lastly the very recent preprint in which Brown, Hurtado, and I establish the conjecture for $\mathrm{SL}(n, \mathbb{Z})$ for $n > 2$ [BFH2]. Throughout this section we fix the notation that $G = \mathrm{SL}(n, \mathbb{R})$, $n > 2$, and $\Gamma < G$ a lattice.

The first results we refer to are those of Brown, Rodriguez Hertz, and Wang on rigidity of invariant measures. A simple version of their main result is as follows:

Theorem 7.1. *Assume Γ acts smoothly on a compact manifold with dimension less than $n - 1$. Then there is a Γ-invariant measure on M.*

Their results are much stronger than this and also find an invariant measure in certain higher dimensions unless the action has a *measurable projective factor*, i.e., a measurable quotient given by the Γ action G/Q, where Q is a parabolic. They also have criteria for which parabolic subgroups can arise in terms of the dimension of M. The theorem has some resemblance to work of Nevo and Zimmer that also produces invariant measures or measurable projective factors [NZ], but the important difference is a computable criterion for triviality of the measurable projective factor in terms of Lyapunov exponents associated with a measure invariant by the maximal Cartan subgroup. While the techniques in Nevo and Zimmer are a difficult generalization of the proofs of Margulis's projective-quotient theorem, the proof here is a less direct rethinking of that proof and can be used to give a proof of the projective-factor theorem that is organized somewhat differently than Margulis's.

I want to describe the strategy in some detail as it is another example of the philosophy that nonresonance implies invariance. The reader may want to reread the first paragraph or two describing this philosophy in section 4 before proceeding, at least to recall the basic background and motivation. Recall that two linear functionals are called resonant if one is a positive multiple of the other. As in section 4 one first passes to the induced action on $X = (G \times M)/\Gamma$. Taking the minimal parabolic P and using that it is amenable, one finds a P-invariant measure μ and the goal is to prove that μ is G invariant. The measure μ is also clearly invariant under the Cartan subgroup A that is contained in P and so one can try to apply the philosophy that nonresonance implies invariance by associating some linear functionals with the pair (A, μ). The linear functionals we consider are the Lyapunov exponents for the A action. More precisely, we consider the Lyapunov exponents for the restriction of the derivative of A action to the subbundle F of $T((G \times M)/\Gamma)$ defined by directions tangent to the M fibers in that bundle over G/Γ. We refer to this collection of linear functionals as *fiberwise Lyapunov exponents*. In this context [BRHW1, Proposition 5.3] shows that, given an A-invariant measure on X that projects to Haar measure on G/Γ, if a root β of G is not resonant with any fiberwise Lyapunov exponent, then the measure is invariant by the root subgroup G_β. The rest of the proof is quite simple. The stabilizer of μ contains P, which implies the projection of μ to G/Γ is Haar measure, so the proposition just described applies. The stablizer G_μ of μ in G is a closed subgroup containing P. We also know that G_μ contains the group generated by the G_β for all roots β not resonant with any fiberwise Lyapunov exponent. We also know that the number of distinct fiberwise Lyapunov exponents is bounded by the dimension of M. Since any closed subgroup of G containing P is parabolic, G_μ is parabolic. So either $G_\mu = G$ or the number of resonant roots needs to be at least the dimension of

G/Q for Q a maximal proper parabolic. This is because given any single root β with $G_\beta \not\subset Q$, the group generated by G_β and Q is G. Our assumption on the dimension of M immediately implies there are not enough fiberwise Lyapunov exponents to produce $\dim(G/Q)$ resonant roots, so μ is G invariant.

The hard part of the proof is contained in [BRHW1, Proposition 5.3]. This follows an outline that is common in measure rigidity, in that the key tool is computing entropy and using deep work of Ledrappier and Young relating entropy to dimensions of invariant measures [Led, LY1]. To apply these techniques, they also need to redevelop the Ledrappier-Young theory along coarse Lyapunov foliations for the group action and a more refined form of the Abramov-Rohklin theorem for entropy of skew products [BRHW3]. We remark here that for lattices in $SL(n, \mathbb{R})$ one can get away with a less difficult application of the work of Ledrappier-Young that was discovered by Hurtado and is explained by Cantat in [Can]; see below for more discussion.

In this context, the main result proven in the work with Brown and Hurtado [BFH1] is as follows:

Theorem 7.2. *Assume Γ is a cocompact lattice and assume Γ acts smoothly on a compact manifold M with dimension less than $n-1$. Then the action factors through a finite quotient of Γ. Suppose Γ acts smoothly on a compact manifold M of dimension $n-1$ preserving a smooth volume form. Then again the action factors through a finite quotient.*

The proof of this theorem again begins by inducing the action to a G action on $(G \times M)/\Gamma$. Since Γ is cocompact it is quite easy to verify that subexponential growth of derivatives for the Γ action on M is equivalent to subexponential growth of derivatives for G along the vector bundle F tangent to the M fibers in $(G \times M)/\Gamma$. The derivative along G is computed by the adjoint action of G on its Lie algebra and so has exponential growth. We denote by F the subbundle of $T((G \times M)/\Gamma)$ that is tangent to the M fibers and refer to the derivative restricted to F as the *fiberwise derivative* and Lyapunov exponents for the derivative restricted to F as *fiberwise Lyapunov exponents*. We would like to control growth of fiberwise derivatives by controlling fiberwise Lyapunov exponents. For any G-invariant measure, it follows from Zimmer's cocycle superrigidity theorem that all fiberwise Lyapunov exponents are zero. But G and Γ are nonamenable so a priori one does not have many G-invariant measures with which to work.

The first key observation in our proof is to recall that we have a Cartan decomposition $G = KAK$ in which A is the diagonal matrices and so isomorphic to $\mathbb{R}^{n-1}$ and $K = SO(n)$ is compact. Using compactness of K to average, it is obvious that there is a K-invariant metric on $(G \times M)/\Gamma$ and so subexponential growth of fiberwise derivatives for G is equivalent to subexponential growth of fiberwise derivatives for A. Using essentially the proof of Proposition 5.1, we can then show that if the G action does not have subexponential growth of fiberwise derivatives, there is an A-invariant measure μ with nonzero fiberwise Lyapunov exponents. The heart of our proof is to promote this measure to one that is G invariant and so obtain a contradiction to Zimmer's cocycle superrigidity theorem.

The desired G invariance is obtained in two steps. There is a natural projection $\pi : (G \times M)/\Gamma \to G/\Gamma$ and the obstruction to applying the proof of Theorem 7.1 directly is that we do not know $\pi_*(\mu)$ is Haar measure or that μ is P invariant. We begin by carefully averaging μ over various unipotent subgroups of G. We need to do this in a manner that

1. preserves the A invariance of μ,
2. preserves the positive fiberwise Lyapunov exponent of μ, and
3. increases the size of the stabilizer of $\pi_*(\mu)$.

Doing this requires that we use Ratner's theorems on measures invariant under unipotent flows as well as some additional facts from her proof about stabilizers of measures and a result about uniqueness of averages due to Ratner for one-parameter unipotent subgroups and to Shah for averages over higher dimensional unipotent subgroups [Rat2, Rat1, Sha].

Once we know that $\pi_*(\mu)$ is Haar, we still cannot argue as in the proof of Theorem 7.1, since we do not know that the stabilizer of the invariant measure is parabolic. However, it turns out there are no closed subgroups of G of dimension less than $n-1$, so for manifolds of dimension less than $n-2$ just counting the number of roots of G whose root groups can be resonant with fiberwise Lyapunov exponents completes the proof. To finish the volume-preserving proof, one uses that the only subgroups of codimension $n-1$ are parabolic and that the roots omitted in one of those cannot be the fiberwise Lyaponuv exponents of a volume-preserving action.

Once we have established that derivatives of the Γ action grow subexponentially, the proof hinges on strong property (T) but is otherwise quite close to arguments in my paper with Margulis on local rigidity of isometric actions [FM1]. Using a variant of the chain rule for higher derivatives, one verifies that subexponential growth of derivatives implies subexponential norm growth for actions of Γ on various Sobolev-type spaces of symmetric 2-forms. Picking a smooth metric g one then looks at the averages of these over the μ_n provided by strong property (T) as in definition 6.2. These converge to a Γ-invariant symmetric 2-form g_0. Using that the convergence to the invariant form is exponential while the derivatives are subexponential, one checks that g_0 is positive definite and so a metric. The Sobolev embedding theorems allow one to see g_0 is smooth and the rest of the proof is standard.

I want to make a short remark on a historical quirk. The proof of Theorem 7.2 in dimension less than $n-1$ can be simplified substantially as explained in [Can, section 8.3]. Though it is not remarked explicitly there, this simplification also applies to volume-preserving actions in dimension $n-1$. This also gives a much simpler proof of Theorem 7.1 in dimension less than $n-1$. This simplified version does not really use the philosophy of nonresonance implying invariance, and does not flow as naturally from the history of ideas discussed in this essay. It is, in fact, somewhat surprising that this easier version, particularly as a proof of Theorem 7.1, was not discovered much earlier. To me this justifies my choice of writing this essay as a history of ideas that indicates how we reached this point, rather than writing an article that simply explains the mathematics that exists at the end of the trajectory. Mathematical ideas may be logical but their development is an idiosyncratic human activity, and the failure to proceed along the shortest line to a goal is often fruitful in surprising ways. I should also remark that the easier proofs produce much less sharp results for most simple Lie groups.

I will discuss briefly the issues that arise in the case of nonuniform lattices. This is discussed in much more detail in the introduction to the recent preprint with Brown and Hurtado proving Zimmer's conjecture for actions of $\mathrm{SL}(m, \mathbb{Z})$, where $m > 2$ [BFH2]. In this case $(G \times M)/\Gamma$ is not compact, so the space of probability measures on it is not weak* compact. This presents difficulties for both producing invariant measures with nonzero Lyapunov exponents and then averaging them to improve their projections. Whether we are taking a given measure and translating it or taking a sequence of empirical measures, it is unclear that the sequences of measures we consider have limits. At the step of averaging over unipotents, this issue is easily controlled using results on quantative nondivergence of unipotent orbits from work of Kleinbock-Margulis [KM, Kle]. It is worth remarking that the first results on nondivergence of unipotent orbits were part of Margulis's dynamical proof of arithmeticity of nonuniform lattices [Mar3, Mar1, Mar2], a proof that preceded his superrigidity theorem by a few years. By the late 1970s there were proofs of arithmeticity for nonuniform lattices via superrigidity that did not pass through homogeneous dynamics, but I find it quite striking that to prove Zimmer's conjecture we needed ideas from both proofs of arithmeticity.

The problem of constructing A-invariant measures to begin with is more serious and seems related to issues and conjectures regarding the set of divergent A orbits in G/Γ as discussed in the paper of Kadayev, Kleinbock, Margulis, and Lindenstrauss [KKLM]. It is not clear that it is possible to resolve this issue directly, even modulo several important conjectures in homogeneous dynamics, but this is not the only obstruction in this setting. Several other problems arise due to the properties of the *return cocycle* $\beta : G \times G/\Gamma \to \Gamma$ in this context. The first of these is that subexponential growth of derivatives for the Γ action is not equivalent to subexponential growth of derivatives for the induced G action on $(G \times M)/\Gamma$. Fixing a basepoint x_0 in G/Γ, if x_n are points in G/Γ with $d(x_n, x_0) = n$, then for any fixed g, the size of $\beta(g, x_n)$ will grow linearly in n. Because of this, subexponential growth of fiber derivatives for the induced action is equivalent to having a Γ-invariant metric on M. Despite this, we still manage to construct a proof using the induced action. Another key difficulty for all approaches is that we are not able to control the "images" of the cocycle $\beta : G \times G/\Gamma \to \Gamma$ even for $\mathrm{SL}(n, \mathbb{Z})$. To understand this remark better, consider first the case in which $G = \mathrm{SL}(2, \mathbb{R})$ and $\Gamma = \mathrm{SL}(2, \mathbb{Z})$. If we take a one-parameter subgroup $c(t) < \mathrm{SL}(2, \mathbb{R})$ and take the trajectory $c(t)x$ for t in some interval $[0, T]$ and assume the entire trajectory on G/Γ lies deep enough in the cusp, then $\beta(a(t), x)$ is necessarily unipotent for all t in $[0, T]$. No similar statement is true for $G = SL(m, \mathbb{R})$ and $\Gamma = SL(m, \mathbb{Z})$. In fact analogous statements are true if and only if Γ has $\mathbb{Q}$-rank 1. This is closely related to the fact that higher $\mathbb{Q}$-rank locally symmetric spaces are 1-connected at infinity. This forces us to "factor" the action into actions of rank-1 subgroups in order to control the growth of derivatives. In the case of $SL(m, \mathbb{Z})$ we prove subexponential growth of derivatives for the m^2 canonical copies of $SL(2, \mathbb{Z})$ in $SL(m, \mathbb{Z})$ given by choices of pairs of coordinates and then use the result of Lubotzky, Mozes, and Raghunathan that $SL(m, \mathbb{Z})$ is boundedly generated by these to finish the proof [LMR1]. We are currently working on a proof for general nonuniform lattices jointly with Dave Witte Morris. One step is to find an analogue of this bounded-generation statement in general. While this is how Lubotzky, Mozes, and Raghunathan prove their results on nondistortion of $SL(m, \mathbb{Z})$ in $SL(m, \mathbb{R})$, it is not how their proof proceeds for general higher rank lattices [LMR2]. Our proof in the general case also makes use of some more recent results in homogeneous dynamics.

Before closing this section, I want to point to a few instances in Zimmer's own work in which the conjecture was proven assuming slow enough growth of derivatives. First in [Zim2], the conjecture is proven for ergodic volume-preserving actions with what one might call *immediate subexponential growth of derivatives*, that is, where we have

$$\|D\alpha(\gamma)\| \le e^{\varepsilon l(\gamma)}.$$

This condition holds in particular for perturbations of isometric actions. In the original proof, Zimmer used that Γ was a higher rank lattice, but later he improved the result to cover all groups with property (T) [Zim5]. A stronger result, not requiring either ergodic or volume preserving, can be proven using the techniques I developed with Margulis in [FM1]. In this essentially perturbative setting, my results with Margulis are clearly more robust than Zimmer's, allowing us to prove foliated results in [FM1] with further applications in [FM2]. In a later paper, Zimmer introduced the notion of distal action and proved the conjecture for ergodic distal actions [Zim3] by an argument similar to the one in [Zim2]. From the current point of view, being distal amounts to having a smooth invariant volume and having (eventual) polynomial growth of derivatives. For smooth volume-preserving ergodic actions, Zimmer's proof here actually gives a proof assuming only subexponential growth of derivatives. Another proof under those same hypotheses can be given by following Zimmer's arguments in [Zim6]. The proof above using strong property (T) is both more robust by not requiring a smooth volume and considerably simpler than following Zimmer's arguments. However, essentially all of Zimmer's arguments in this context apply to ergodic volume-preserving actions of groups with property (T) that also preserve a measuable metric. Many hyperbolic groups, including lattices in $Sp(1, n)$ and random groups in many models, have property (T) and fail to have strong property (T), and Zimmer's results may be useful in the study of actions of those groups. In any case it was already clear from Zimmer's earliest work that proving some kind of uniformly slow growth of derivatives was key to proving the conjecture. What was missing until very recently was the ability to prove any estimate of that kind, an ability provided by the connection to hyperbolic dynamics, Lyapunov exponents, and rigidity of invariant measures in both the homogeneous and inhomogeneous setting.

As a last remark in this section, there is an unresolved issue in the current proofs that is obscured by the special case in which I state all results. Namely, in the current versions of the argument by Brown, Rodriguez Hertz, and Wang, the results one can prove only see the number of root spaces and not their multiplicities. For this reason, for nonsplit groups, the current state of the art is often quite far from optimal. My current sense is that while this difficulty is serious and nontrivial, it is one that will be overcome in time.

8. Future directions

In this section, I will briefly discuss some of the most promising future directions and connections and end with a discussion of some related work that I think may help point the way to a more general diffeomorphism-group-valued representation theory of finitely generated groups.

A key ingredient in the discussion above is close connections to homogeneous dynamics and the study of invariant measures in that context. This area has been intensely studied since the 1960s or 1970s with key contributions by Dani, Einseidler, Furstenberg, Katok, Lindenstrauss, Margulis, Raghunathan, Ratner, Shah, Spatzier, and Witte Morris. Progress in this area has been quite dramatic since Ratner proved her theorem on the classification of measures invariant under unipotent flows in the early 1990s, and even our arguments in the case of nonuniform lattices barely scratch the surface. I want to point to a certain thread of ideas particularly relevant to the recent work on Zimmer's conjecture. A key impetus comes from work of Margulis and Tomanov in which they, in the course of proving an extension of Ratner's theorems, introduce entropy techniques based on the work of Ledrappier and Young [MT, LY2, LY1, Led]. The shorter argument in Cantat's paper for special cases of Theorems 7.1 and 7.2 uses ideas that appear in some form in those papers. Another key development that began soon afterward in work of Katok and Spatzier on invariant measures for higher rank abelian groups is to use a more refined form of the entropy argument, one that sees coarse Lyapunov subspaces and not just Lyapunov subspaces [KS1]. This idea has been used constantly since that time in the study of invariant measures for higher rank abelian groups, including the work of Einsiedler, Katok, and Lindenstrauss [Lin, EK, EKL, EL]. For a summary of those ideas and a particularly accessible account of the entropy lemmas in the homogeneous setting, I recommend the survey by Einsiedler and Lindenstrauss [EL]. A key contribution to the progress reported in section 7 is the work of Brown, Rodriguez Hertz, and Wang, which develops the entropy theory along coarse Lyapunov foliations in the general, nonhomogeneous setting [BRHW3, BRHW1]. This can be seen as part of a promising set of results that bring measure rigidity into the nonlinear setting. The earliest of these results are for actions of abelian groups in the work of Katok, Kalinin, and Rodriguez Hertz (see for example [KKRH]), and these will likely prove relevant, but I want to place more emphasis on a different, more recent direction.

A major development in the homogeneous setting occured with work of Benoist and Quint [BQ], in which they study stationary and invariant measure for quite general groups. These ideas then inspired the remarkable breakthrough work by Eskin and Mirzakhani on stationary and invariant measures for the $SL(2, \mathbb{R})$ action on the Teichmuller moduli space [EM], a setting that one might label semihomogeneous. Using some ideas from [EM], Brown and Rodriguez Hertz have written a magesterial paper on stationary and invariant measures for groups acting on surfaces subject to some natural hyperbolicity conditions [BRH]. This is a surprising connection of rigidity of group actions to Teichmuller dynamics. I hope the discussion of the arguments from [BFH1] convinces readers that there is interest in hyperbolic measures even if one is not a priori interested in hyperbolic dynamics. Roughly speaking, their results show that if one has a hyperbolic stationary measure μ for a generating measure of some group Γ acting by C^2 volume-preserving diffeomorphisms on a compact surface M, one has a trichotomy:

1. μ is supported on a finite set of points,
2.   there is a Γ-invariant line field, or
3. μ is, up to normalization, the restriction of volume to a Γ-invariant set.

In particular, in the first and third case, μ is Γ invariant. This result is potentially relevant to another old conjecture of Zimmer's: if Γ is a group with property (T), then any volume-preserving action of Γ on a surface factors through a finite quotient. In fact, if one can eliminate case (2) above for actions of groups with property (T), then it is relatively easy to prove subexponential growth of derivatives. While it is an easy consequence of a result of Zimmer that no measure-preserving action of a group with property (T) can preserve a line field on a surface, this is not known for non-measure-preserving actions, which makes application of the result somewhat difficult.

In joint work with Eskin, Brown and Rodriguez Hertz are developing a high-dimensional analogue of this theorem that is somewhat more difficult to state. This work should have further ramifications for work on rigidity of group actions in higher dimensions.

Brown, Rodriguez Hertz, and Wang believe that the methods of [BRHW1] can be adapted to show that if a lattice in $SL(n, \mathbb{R})$ acts on a manifold of dimension $n-1$ without invariant measure, then the action is smoothly conjugate to the standard one on a sphere S^{n-1} or a projective space $P(\mathbb{R}^n)$. Combining their ideas with the proof of Zimmer's conjectures in this case should completely classify all smooth actions of lattices in $SL(n, \mathbb{R})$ on manifolds of dimension at most $n-1$. The next test case for our techniques is clearly manifolds of dimension n, and while progress seems possible, this would be a major advance. The key difficulty is that there are many more examples here. Not only can one take the $SL(n, \mathbb{Z})$ action on the torus $\mathbb{T}^n$, but one can blow up periodic orbits for this action and also glue along blown-up periodic orbits to produce manifolds of quite complex topology. These examples were first discovered by Katok and Lewis, and their properties were further explored in work I did with Benveniste and Whyte [KL, FW, BF, Fis1]. A reasonable first case to study is to assume that all stationary measures for the Γ action are invariant and see whether one can prove the action is smoothly conjugate to the standard action of SL $(n, \mathbb{Z})$ on $\mathbb{T}^n$. Another reasonable test case is to see whether having a noninvariant stationary measure of full support implies the action is a skew product action on $P(\mathbb{R}^n) \times S^1$ or $S^{n-1} \times S^1$. Many such actions can be constructed by viewing $P(\mathbb{R}^n)$ as G/P, taking a homomorphism $P \to \mathbb{R}$, and inducing any vector field on S^1 to an action of G on $G/P \times S^1$. It is also true that SL$(n, \mathbb{R})$ acts on the $P(\mathbb{R}^{n+1})$ and S^n and therefore so do all lattices in SL$(n, \mathbb{R})$. In this case there are orbits in which the dynamics is dissipative instead of conservative and new ideas are likely needed.

I want to close this paper by discussing other developments that point much further into the future of the Zimmer program and its variants and generalizations. Zimmer's program is essentially the generalization of the study of discrete subgroups of Lie groups and their finite-dimensional-representation theory to the study of those infinite-dimensional representations that arise via actions on compact manifolds. This analogy has proven very robust and Hurtado's Burnside Theorem is certainly a good example of another direction one can pursue. Here I want to mention another, which is trying to study the *representation variety* or *character variety* in settings in which there are many representations. The word "variety" should not be taken too seriously here as a principal difficulty of this setting is that there is no algebraic geometry that applies to groups of diffeomorphisms or homeomorphisms. Despite this, in very recent work Mann and Wolff have given a complete characterization of the *rigid* representations or isolated points for the Homeo(S^1) character variety of the fundamental group of a compact surface [MW]. These are exactly the representations that arise by restricting the action of a finite-dimensional Lie subgroup of Homeo(S^1) to a cocompact lattice, a surprising and deep connection with the finite-dimensional setting. In this setting the finite-dimensional Lie group that arises is necessarily a finite cover of PSL(2, $\mathbb{R}$). This result is promising, but it is clearly a long way from understanding diffeomorphism or homeomorphism valued representation varieties in any detail.

References

[BQ]   Y. Benoist and J.-F. Quint. Stationary measures and invariant subsets of homogeneous spaces (III). *Ann. of Math. (2)*, 178(3):1017–1059, 2013.

[BF]   E. J. Benveniste and D. Fisher. Nonexistence of invariant rigid structures and invariant almost rigid structures. *Comm. Anal. Geom.*, 13(1):89–111, 2005.

[BRO]   A. Brown. Entropy, smooth ergodic theory and rigidity of group actions. *Preprint*, 2017.

[BFH1]   A. Brown, D. Fisher, and S. Hurtado. Zimmer's conjecture: Subexponential growth, measure rigidity, and strong property (T). *Preprint*, 2016. *arXiv:1608.04995*.

[BFH2]   A. Brown, D. Fisher, and S. Hurtado. Zimmer's conjecture for actions of SL$(m, \mathbb{Z})$. *Preprint*, 2017.

[BRH]   A. Brown and F. Rodriguez Hertz. Measure rigidity for random dynamics on surfaces and related skew products. *J. Amer. Math. Soc.*, 30(4):1055–1132, 2017.

[BRHW1] A. Brown, F. Rodriguez Hertz, and Z. Wang. Invariant measures and measurable projective factors for actions of higher-rank lattices on manifolds. *Preprint*, 2016. *arXiv:1609.05565*.

[BRHW2] A. Brown, F. Rodriguez Hertz, and Z. Wang. Global smooth and topological rigidity of hyperbolic lattice actions. *Ann. of Math. (2)*, 186(3):913–972, 2017.

[BRHW3] A. W. Brown, F. Rodriguez Hertz, and Z. Wang. Smooth ergodic theory of $\mathbb{Z}^d$-actions. *Preprint*, 2016.

[CAN]   S. Cantat. Progrès récents concernant le programme de Zimmer [d'après A. Brown, D. Fisher, et S. Hurtado]. *Preprint*, 2017.

[DLS1]   M. de la Salle. A local characterization of Kazhdan projections and applications. *Preprint*, 2016. *arXiv:1512.06720*.

[DLS2]   M. de la Salle. Strong (T) for higher rank lattices. *Preprint*, 2017.

[DLDLS] T. de Laat and M. de la Salle. Strong property (T) for higher-rank simple Lie groups. *Proc. Lond. Math. Soc. (3)*, 111(4):936–966, 2015.

[EL]   M. Einsiedler and E. Lindenstrauss. Diagonal actions on locally homogeneous spaces. In *Homogeneous flows, moduli spaces and arithmetic*, volume 10 of *Clay Math. Proc.*, pages 155–241. Amer. Math. Soc., Providence, RI, 2010.

[EK]   M. Einsiedler and A. Katok. Rigidity of measures—the high entropy case and non-commuting foliations. *Israel J. Math.*, 148:169–238, 2005. *Probability in mathematics*.

[EKL]   M. Einsiedler, A. Katok, and E. Lindenstrauss. Invariant measures and the set of exceptions to Littlewood's conjecture. *Ann. of Math. (2)*, 164(2):513–560, 2006.

[EL]   M. Einsiedler and E. Lindenstrauss. On measures invariant under tori on quotients of semisimple groups. *Ann. of Math. (2)*, 181(3):993–1031, 2015.

[EM]   A. Eskin and M. Mirzakhani. Invariant and stationary measures for the SL(2,R) action on moduli space. *Preprint*, 2013. *arXiv:1302.3320*.

[FJ]   F. T. Farrell and L. E. Jones. Anosov diffeomorphisms constructed from $\pi_1 \mathrm{Diff}(S^n)$. *Topology*, 17(3):273–282, 1978.

[FG]   F. T. Farrell and A. Gogolev. Anosov diffeomorphisms constructed from $\pi_k(\mathrm{Diff}(S^n))$. *J. Topol.*, 5(2):276–292, 2012.

[FIS1]   D. Fisher. Deformations of group actions. *Trans. Amer. Math. Soc.*, 360(1):491–505, 2008.

[FIS2]   D. Fisher. Groups acting on manifolds: Around the Zimmer program. In *Geometry, rigidity, and group actions*, Chicago Lectures in Math., pages 72–157. Univ. Chicago Press, Chicago, IL, 2011.

[FKS1]   D. Fisher, B. Kalinin, and R. Spatzier. Totally nonsymplectic Anosov actions on tori and nilmanifolds. *Geom. Topol.*, 15(1):191–216, 2011.

[FKS2]   D. Fisher, B. Kalinin, and R. Spatzier. Global rigidity of higher rank Anosov actions on tori and nilmanifolds. *J. Amer. Math. Soc.*, 26(1):167–198, 2013. *With an appendix by James F. Davis*.

[FM1]   D. Fisher and G. Margulis. Almost isometric actions, property (T), and local rigidity. *Invent. Math.*, 162(1):19–80, 2005.

[FM2]   D. Fisher and G. Margulis. Local rigidity of affine actions of higher rank groups and lattices. *Ann. of Math. (2)*, 170(1):67–122, 2009.

[FW]   D. Fisher and K. Whyte. Continuous quotients for lattice actions on compact spaces. *Geom. Dedicata*, 87(1-3):181–189, 2001.

[FRA]   J. Franks. Anosov diffeomorphisms on tori. *Trans. Amer. Math. Soc.*, 145:117–124, 1969.

[HUR1]   S. Hurtado. The Burnside problem for Diff(S^2). *Preprint*, 2016. *arXiv:1607.04603*.

[HUR2]   S. Hurtdao. Examples of diffeomorphism group cocycles with no periodic approximation. *Preprint*, 2017.

[KKLM]   S. Kadyrov, D. Y. Kleinbock, E. Lindenstrauss, and G. A. Margulis. Singular systems of linear forms and non-escape of mass in the space of lattices. *Preprint*, 2016. *arXiv:1407.5310*.

[KAL]   B. Kalinin. Livšic theorem for matrix cocycles. *Ann. of Math. (2)*, 173(2):1025–1042, 2011.

[KKRH]   B. Kalinin, A. Katok, and F. Rodriguez Hertz. Nonuniform measure rigidity. *Ann. of Math. (2)*, 174(1):361–400, 2011.

[KLS1]   B. Kalinin and V. Sadovskaya. Global rigidity for totally nonsymplectic Anosov $\mathbb{Z}^k$ actions. *Geom. Topol.*, 10:929–954, 2006.

[KLS2]   B. Kalinin and V. Sadovskaya. On the classification of resonance-free Anosov $\mathbb{Z}^k$ actions. *Michigan Math. J.*, 55(3):651–670, 2007.

[KLSP]   B. Kalinin and R. Spatzier. On the classification of Cartan actions. *Geom. Funct. Anal.*, 17(2):468–490, 2007.

[KAT]   A. Katok. Lyapunov exponents, entropy and periodic orbits for diffeomorphisms. *Inst. Hautes Études Sci. Publ. Math.*, (51):137–173, 1980.

[KL]   A. Katok and J. Lewis. Global rigidity results for lattice actions on tori and new examples of volume-preserving actions. *Israel J. Math.*, 93:253–280, 1996.

[KLZ]   A. Katok, J. Lewis, and R. Zimmer. Cocycle superrigidity and rigidity for lattice actions on tori. *Topology*, 35(1):27–38, 1996.

[KS1]   A. Katok and R. J. Spatzier. Invariant measures for higher-rank hyperbolic abelian actions. *Ergodic Theory Dynam. Systems*, 16(4):751–778, 1996.

[KS2]   A. Katok and R. J. Spatzier. Differential rigidity of Anosov actions of higher rank abelian groups and algebraic lattice actions. *Tr. Mat. Inst. Steklova*, 216(Din. Sist. i Smezhnye Vopr.):292–319, 1997.

[KS3]   A. Katok and R. J. Spatzier. First cohomology of Anosov actions of higher rank abelian groups and applications to rigidity. *Inst. Hautes Études Sci. Publ. Math.*, (79):131–156, 1994.

[KM]   D. Y. Kleinbock and G. A. Margulis. Flows on homogeneous spaces and Diophantine approximation on manifolds. *Ann. of Math. (2)*, 148(1):339–360, 1998.

[KLE]   D. Kleinbock. Quantitative nondivergence and its Diophantine applications. In *Homogeneous flows, moduli spaces and arithmetic*, volume 10 of *Clay Math. Proc.*, pages 131–153. Amer. Math. Soc., Providence, RI, 2010.

[KP]   A. Kocsard and R. Potrie. Livšic theorem for low-dimensional diffeomorphism cocycles. *Comment. Math. Helv.*, 91(1):39–64, 2016.

[LAF]   V. Lafforgue. Un renforcement de la propriété (T). *Duke Math. J.*, 143(3):559–602, 2008.

[LED]   F. Ledrappier. Propriétés ergodiques des mesures de Sinaï. *Inst. Hautes Études Sci. Publ. Math.*, (59):163–188, 1984.

[LY1]   F. Ledrappier and L.-S. Young. The metric entropy of diffeomorphisms. I. Characterization of measures satisfying Pesin's entropy formula. *Ann. of Math. (2)*, 122(3):509–539, 1985.

[LY2]   F. Ledrappier and L.-S. Young. The metric entropy of diffeomorphisms. II. Relations between entropy, exponents and dimension. *Ann. of Math. (2)*, 122(3):540–574, 1985.

[LIA]   B. Liao. Strong Banach property (T) for simple algebraic groups of higher rank. *J. Topol. Anal.*, 6(1):75–105, 2014.

[LIN]   E. Lindenstrauss. Invariant measures and arithmetic quantum unique ergodicity. *Ann. of Math. (2)*, 163(1):165–219, 2006.

[LMR1]   A. Lubotzky, S. Mozes, and M. S. Raghunathan. Cyclic subgroups of exponential growth and metrics on discrete groups. *C. R. Acad. Sci. Paris Sér. I Math.*, 317(8):735–740, 1993.

[LMR2]   A. Lubotzky, S. Mozes, and M. S. Raghunathan. The word and Riemannian metrics on lattices of semisimple groups. *Inst. Hautes Études Sci. Publ. Math.*, (91):5–53 (2001), 2000.

[MW]   K. Mann and M. Wolff. Rigidity and geometricity for surface group actions on the circle. *Preprint*, 2017.

[MAN]   A. Manning. Anosov diffeomorphisms on nilmanifolds. *Proc. Amer. Math. Soc.*, 38:423–426, 1973.

[MAR1]   G. A. Margulis. Arithmetic properties of discrete subgroups. *Uspehi Mat. Nauk*, 29(1 (175)): 49–98, 1974.

[MAR2]   G. A. Margulis. Arithmeticity of nonuniform lattices in weakly noncompact groups. *Funkcional. Anal. i Priložen.*, 9(1):35–44, 1975.

[MAR3]   G. A. Margulis. Non-uniform lattices in semisimple algebraic groups. *Lie groups and their representations* (Proc. Summer School on Group Representations of the Bolyai János Math, Soc., Budapest, 1971), Halsted, New York, 1975, pages 371–553.

[MT]   G. A. Margulis and G. M. Tomanov. Invariant measures for actions of unipotent groups over local fields on homogeneous spaces. *Invent. Math.*, 116(1-3):347–392, 1994.

[MAR]   G. Margulis. Problems and conjectures in rigidity theory. In *Mathematics: Frontiers and perspectives*, pages 161–174. Amer. Math. Soc., Providence, RI, 2000.

[MQ]   G. A. Margulis and N. Qian. Rigidity of weakly hyperbolic actions of higher real rank semisimple Lie groups and their lattices. *Ergodic Theory Dynam. Systems*, 21(1):121–164, 2001.

[NZ]   A. Nevo and R. J. Zimmer. A structure theorem for actions of semisimple Lie groups. *Ann. of Math. (2)*, 156(2):565–594, 2002.

[RAT1]   M. Ratner. Invariant measures and orbit closures for unipotent actions on homogeneous spaces. *Geom. Funct. Anal.*, 4(2):236–257, 1994.

[RAT2]   M. Ratner. On Raghunathan's measure conjecture. *Ann. of Math. (2)*, 134(3):545–607, 1991.

[RH]   F. Rodriguez Hertz. Global rigidity of certain abelian actions by toral automorphisms. *J. Mod. Dyn.*, 1(3):425–442, 2007.

[RHW]   F. Rodriguez Hertz and Z. Wang. Global rigidity of higher rank abelian Anosov algebraic actions. *Invent. Math.*, 198(1):165–209, 2014.

[SHA]   N. A. Shah. Limit distributions of polynomial trajectories on homogeneous spaces. *Duke Math. J.*, 75(3):711–732, 1994.

[ZIM1]   R. J. Zimmer. Arithmetic groups acting on compact manifolds. *Bull. Amer. Math. Soc. (N.S.)*, 8(1):90–92, 1983.

[ZIM2]   R. J. Zimmer. Volume preserving actions of lattices in semisimple groups on compact manifolds. *Inst. Hautes Études Sci. Publ. Math.*, (59):5–33, 1984.

[ZIM3]   R. J. Zimmer. Lattices in semisimple groups and distal geometric structures. *Invent. Math.*, 80(1):123–137, 1985.

[ZIM4]   R. J. Zimmer. Actions of semisimple groups and discrete subgroups. In *Proceedings of the International Congress of Mathematicians, vol. 1, 2 (Berkeley, Calif., 1986)*, pages 1247–1258. Amer. Math. Soc., Providence, RI, 1987.

[ZIM5]   R. J. Zimmer. Lattices in semisimple groups and invariant geometric structures on compact manifolds. In *Discrete groups in geometry and analysis (New Haven, Conn., 1984)*, volume 67 of *Progr. Math.*, pages 152–210. Birkhäuser Boston, Boston, MA, 1987.

[ZIM6]   R. J. Zimmer. Spectrum, entropy, and geometric structures for smooth actions of Kazhdan groups. *Israel J. Math.*, 75(1):65–80, 1991.

Acknowledgments

Material included in this book has previously appeared in the following publications: "Extensions of Ergodic Group Actions," *Illinois Journal of Mathematics* 20, no. 3 (1976): 373–409, reprinted with permission of the Illinois Journal of Mathematics, published by the University of Illinois at Urbana-Champaign; "Ergodic Actions with Generalized Discrete Spectrum," *Illinois Journal of Mathematics* 20, no. 4 (1976): 555–88, reprinted with permission of the Illinois Journal of Mathematics, published by the University of Illinois at Urbana-Champaign; "Orbit Spaces of Unitary Representations, Ergodic Theory, and Simple Lie Groups," *Annals of Mathematics*, 2nd ser., 106, no. 3 (1977): 573–88; "Amenable Ergodic Group Actions and an Application to Poisson Boundaries of Random Walks," *Journal of Functional Analysis* 27, no. 3 (1978): 350–72, copyright 1978, reprinted with permission from Elsevier; "Induced and Amenable Ergodic Actions of Lie Groups," *Annales scientifiques de l'École normale supérieure*, 4th ser., 11, no. 3 (1978): 407–28; "Hyperfinite Factors and Amenable Ergodic Actions," *Inventiones mathematicae* 41, no. 1 (1977): 23–31, © by Springer-Verlag 1977, reprinted with permission from Springer; "Curvature of Leaves in Amenable Foliations," *American Journal of Mathematics* 105, no. 4 (1983): 1011–22, reprinted with permission from Johns Hopkins University Press; "Amenable Actions and Dense Subgroups of Lie Groups," *Journal of Functional Analysis* 72, no. 1 (1987): 58–64, copyright 1987, reprinted with permission from Elsevier; "Strong Rigidity for Ergodic Actions of Semisimple Lie Groups," *Annals of Mathematics*, 2nd ser., 112, no. 3 (1980): 511–29; "Orbit Equivalence and Rigidity of Ergodic Actions of Lie Groups," *Ergodic Theory and Dynamical Systems* 1, no. 2 (1981): 237–53, reprinted with permission from Cambridge University Press; "Ergodic Actions of Semisimple Groups and Product Relations," *Annals of Mathematics*, 2nd ser., 118, no. 1 (1983): 9–19; "Volume Preserving Actions of Lattices in Semisimple Groups on Compact Manifolds," *Institut des Hautes Études Scientifiques Publications Mathématiques*, no. 59 (1984): 5–33; "Kazhdan Groups Acting on Compact Manifolds," *Inventiones mathematicae* 75, no. 3 (1984): 425–36, © by Springer-Verlag 1984, reprinted with permission from Springer; "Actions of Lattices in Semisimple Groups Preserving a G-Structure of Finite Type," *Ergodic Theory and Dynamical Systems* 5, no. 2 (1985): 301–6, reprinted with permission from Cambridge University Press; "Actions of Semisimple Groups and Discrete Subgroups," in *Proceedings of the International Congress of Mathematicians, August 3–11, 1986*, edited by Andrew M. Gleason (Providence, RI: American Mathematical Society, 1987), 2:1247–58; "Split Rank and Semisimple Automorphism Groups of G-Structures," *Journal of Differential Geometry* 26, no. 1 (1987): 169–73;

"Manifolds with Infinitely Many Actions of an Arithmetic Group" (with Richard K. Lashof), *Illinois Journal of Mathematics* 34, no. 4 (1990): 765–68, reprinted with permission of the Illinois Journal of Mathematics, published by the University of Illinois at Urbana-Champaign; "Spectrum, Entropy, and Geometric Structures for Smooth Actions of Kazhdan Groups," *Israel Journal of Mathematics* 75, no. 1 (1991): 65–80, © by Springer-Verlag 1991, reprinted with permission from Springer; "Cocycle Superrigidity and Rigidity for Lattice Actions on Tori" (with Anatole Katok and James Lewis), *Topology* 35, no. 1 (1996): 27–38, copyright 1996, reprinted with permission from Elsevier; "Volume-Preserving Actions of Simple Algebraic Q-Groups on Low-Dimensional Manifolds" (with Dave Witte Morris), *Journal of Topology and Analysis* 4, no. 2 (2012): 115–20, reprinted with permission from World Scientific Publishing Co., Inc.; "Stabilizers for Ergodic Actions of Higher Rank Semisimple Groups" (with Garrett Stuck), *Annals of Mathematics*, 2nd ser., 139, no. 3 (1994): 723–47; "Arithmeticity of Holonomy Groups of Lie Foliations," *Journal of the American Mathematical Society* 1, no. 1 (1988): 35–58, copyright © 1988, American Mathematical Society, reprinted with permission; "Representations of Fundamental Groups of Manifolds with a Semisimple Transformation Group," *Journal of the American Mathematical Society* 2, no. 2 (1989): 201–13, copyright © 1989, American Mathematical Society, reprinted with permission; "Superrigidity, Ratner's Theorem, and Fundamental Groups," *Israel Journal of Mathematics* 74, no. 2–3 (1991): 199–207, © by Springer-Verlag 1991, reprinted with permission from Springer; "Fundamental Groups of Negatively Curved Manifolds and Actions of Semisimple Groups" (with Ralf J. Spatzier), *Topology* 30, no. 4 (1991): 591–601, copyright 1991, reprinted with permission from Elsevier; "A Canonical Arithmetic Quotient for Simple Lie Group Actions" (with Alexander Lubotzky), in *Proceedings of the International Colloquium on Lie Groups and Ergodic Theory: Mumbai, 1996*, edited by S. G. Dani (Bombay: Tata Institute of Fundamental Research / Narosa Publishing House, 1998), 131–42, reprinted with permission of Tata Institute of Fundamental Research; "Arithmetic Structure of Fundamental Groups and Actions of Semisimple Lie Groups" (with Alexander Lubotzky), *Topology* 40, no. 4 (2001): 851–69, copyright 2001, reprinted with permission from Elsevier; "Entropy and Arithmetic Quotients for Simple Automorphism Groups of Geometric Manifolds," *Geometriae Dedicata* 107 (2004): 47–56, © 2004 Kluwer Academic Publishers, reprinted with permission from Springer; "Geometric Lattice Actions, Entropy and Fundamental Groups" (with David Fisher), *Commentarii Mathematici Helvetici* 77, no. 2 (2002): 326–38; "On the Automorphism Group of a Compact Lorentz Manifold and Other Geometric Manifolds," *Inventiones mathematicae* 83, no. 3 (1986): 411–24, © by Springer-Verlag 1986, reprinted with permission from Springer; "Semisimple Automorphism Groups of G-Structures," *Journal of Differential Geometry* 19, no. 1 (1984): 117–23; "Automorphism Groups and Fundamental Groups of Geometric Manifolds," in *Proceedings of Symposia in Pure Mathematics*, vol. 54, pt. 3, *Differential Geometry: Riemannian Geometry*, edited by Robert Greene and S. T. Yau (Providence, RI: American Mathematical Society, 1993), 693–710, copyright © 1993, American Mathematical Society, reprinted with permission; "Discrete Groups and Non-Riemannian Homogeneous

Spaces," *Journal of the American Mathematical Society* 7, no. 1 (1994): 159–68, copyright © 1994, American Mathematical Society, reprinted with permission; "On Manifolds Locally Modelled on Non-Riemannian Homogeneous Spaces" (with François Labourie and Shahar Mozes), *Geometric and Functional Analysis* 5, no. 6 (1995): 955–65, © 1995 Birkhäuser Verlag, Basel, reprinted with permission from Springer; "On the Non-existence of Cocompact Lattices for $SL(n)/SL(m)$" (with François Labourie), *Mathematical Research Letters* 2, no. 1 (1995): 75–77; "A Structure Theorem for Actions of Semisimple Lie Groups" (with Amos Nevo), *Annals of Mathematics*, 2nd ser., 156, no. 2 (2002): 565–94; "Entropy of Stationary Measures and Bounded Tangential de-Rham Cohomology of Semisimple Lie Group Actions" (with Amos Nevo), *Geometriae Dedicata* 115 (2005): 181–99, © Springer 2005, reprinted with permission from Springer; "Invariant Rigid Geometric Structures and Smooth Projective Factors" (with Amos Nevo), *Geometric and Functional Analysis* 19, no. 2 (2009): 520–35, © 2009 Birkhäuser Verlag, Basel, reprinted with permission from Springer.